D0321812

WITHDRAWN

LIVERPOOL JMU LIBRARY
3 1111 01349 3612

Theory of Machines and Mechanisms

Theory of Machines and Mechanisms

Fourth Edition

John J. Uicker, Jr.
Professor Emeritus of Mechanical Engineering
University of Wisconsin–Madison

Gordon R. Pennock
Associate Professor of Mechanical Engineering
Purdue University

Joseph E. Shigley
Late Professor Emeritus of Mechanical Engineering
The University of Michigan

New York Oxford
OXFORD UNIVERSITY PRESS
2011

Oxford University Press, Inc., publishes works that further Oxford University's
objective of excellence in research, scholarship, and education.

Oxford New York
Auckland Cape Town Dar es Salaam Hong Kong Karachi
Kuala Lumpur Madrid Melbourne Mexico City Nairobi
New Delhi Shanghai Taipei Toronto

With offices in
Argentina Austria Brazil Chile Czech Republic France Greece
Guatemala Hungary Italy Japan Poland Portugal Singapore
South Korea Switzerland Thailand Turkey Ukraine Vietnam

Copyright © 2011, 2003 by Oxford University Press, Inc.; 1995, 1980 by McGraw–Hill

Published by Oxford University Press, Inc.
198 Madison Avenue, New York, New York 10016
http://www.oup.com

Oxford is a registered trademark of Oxford University Press

All rights reserved. No part of this publication may be reproduced,
stored in a retrieval system, or transmitted, in any form or by any means,
electronic, mechanical, photocopying, recording, or otherwise,
without the prior permission of Oxford University Press.

Library of Congress Cataloging-in-Publication Data
Uicker, John Joseph.
Theory of machines and mechanisms/John J. Uicker, Jr., Gordon R. Pennock,
Joseph E. Shigley. – 4th ed.
p. cm.
1. Mechanical engineering. I. Pennock, G. R. II. Shigley, Joseph Edward. III. Title.
TJ145.U33 2010
621.8–dc22 2009024985

ISBN 978-0-19-537123-9

9 8 7 6 5 4 3 2

Printed in the United States of America
on acid-free paper

This textbook is dedicated to the memory of my late father, John J. Uicker, Emeritus Dean of Engineering, University of Detroit, to my late mother; Elizabeth F. Uicker, and to my six children, Theresa A. Zenchenko, John J. Uicker, III, Joseph M. Uicker, Dorothy J. Winger, Barbara A. Peterson, and Joan E. Horne.

—John J. Uicker, Jr.

This work is also dedicated first and foremost to my wife, Mollie B., and my son, Callum R. Pennock. The work is also dedicated to my friend and mentor, the late Dr. An (Andy) Tzu Yang, and to my colleagues in the School of Mechanical Engineering, Purdue University, West Lafayette, Indiana.

—Gordon R. Pennock

Finally, this text is dedicated to the memory of the third author, the late **Joseph E. Shigley,** Professor Emeritus, Mechanical Engineering Department, University of Michigan, Ann Arbor, on whose previous writings much of this text is based.

Contents

Preface

This book is intended to cover that field of engineering theory, analysis, design, and practice that is generally described as mechanisms or as kinematics and dynamics of machines. Although this text is written primarily for students of engineering, much of the material can be of value to practicing engineers throughout their professional careers.

The continued tremendous growth of knowledge, including the areas of kinematics and dynamics of machinery, over the past 50 years has resulted in great pressure on the engineering curricula of many schools for the substitution of "modern" subjects for subjects perceived as weaker or outdated. At some schools, depending on the faculty, this has meant that kinematics and dynamics of machines could only be made available as an elective topic for specialized study by a small number of students. At other schools, the subject has remained a requirement for all mechanical engineering students. Still other schools required a greater emphasis on design at the expense of depth in analysis. Overall, the times have produced a need for a textbook that satisfies the requirements of new and changing course structures.

Much of the new knowledge developed over this period exists in a large variety of technical papers, each couched in its own singular language and nomenclature and each requiring additional background for its comprehension. The individual contributions being published might be used to strengthen engineering courses if the necessary foundation were provided and a common notation and nomenclature were established. These new developments could then be integrated into existing courses to provide a logical, modern, and comprehensive whole. To provide the background that will allow such integration is the purpose of this book.

To develop a broad and basic comprehension, all methods of analysis and development common to the literature of the field are employed. We have used graphical methods of analysis and synthesis extensively throughout the book because the authors are firmly of the opinion that graphical computation provides visual feedback that enhances the student's understanding of the basic nature of and interplay between the equations involved. Therefore, in this book, graphic methods are presented as one possible solution technique, but are always accompanied by vector equations defined by the fundamental laws of mechanics, rather than as graphical "tricks" to be learned by rote and applied blindly. In addition, although graphic techniques, performed by hand, may be lacking in accuracy, they can be performed quickly and even inaccurate sketches can often provide reasonable estimates of a solution or can be used to check the results of analytic or numeric solution techniques.

We also use conventional methods of vector analysis throughout the book, both in deriving and presenting the governing equations and in their solution. Raven's methods using complex algebra for the solution of two-dimensional vector equations are presented throughout the book because of their compactness, because they are employed so frequently in the literature, and because they are so easy to program for computer evaluation. In the chapters dealing with three-dimensional kinematics and robotics, we briefly present an introduction to Denavit and Hartenberg's methods using transformation matrices.

Another feature of this text is the presentation of kinematic coefficients, which are derivatives of motion variables with respect to the input motion rather than with respect to time. The authors believe that these provide several new and important advantages, among which are the following. (1) They clarify for the student those parts of a motion problem that are kinematic (geometric) in their nature and clearly separate them from the parts that are dynamic or speed dependent. (2) They help to integrate the analysis of different types of mechanical systems, such as gears, cams, and linkages, which might not otherwise seem similar.

One dilemma that all writers on the subject of this book have faced is how to distinguish between the motions of two different points of the same moving body and the motions of coincident points of two different moving bodies. In other texts it has been customary to describe both of these as "relative motion"; however, because they are two distinctly different situations and are described by different equations, this causes the student confusion in distinguishing between them. We believe that we have greatly relieved this problem by the introduction of the terms *motion difference* and *apparent motion* and by using different terminology and different notation for the two cases. Thus, for example, this book uses the two terms *velocity difference* and *apparent velocity,* instead of the term "relative velocity," which will not be found when speaking rigorously. This approach is introduced beginning with the concepts of position and displacement, used extensively in the chapter on velocity, and brought to fulfillment in the chapter on accelerations where the Coriolis component *always* arises in, and *only* in, the apparent acceleration equation.

Access to personal computers, programmable calculators, and laptop computers is now commonplace and is of considerable importance to the material of this book. Yet engineering educators have told us very forcibly that they do not want computer programs included in the text. They prefer to write their own programs and they expect their students to do so as well. Having programmed almost all the material in the book many times, we also understand that the book should not become obsolete with changes in computers or programming languages.

We have endeavored to use U.S. Customary units and SI units in about equal proportions throughout the book. However, there are certain exceptions. For example, in Chapter 16 (Dynamics of Reciprocating Engines) only SI units are presented because we believe that engines are now designed for an international marketplace. Therefore, they are now always rated in kilowatts rather than horsepower; they have displacements in milliliters or liters rather than cubic inches; and their cylinder pressures are measured in kilopascals rather than in pounds per square inch.

Part 1 of this book deals mostly with theory, nomenclature, notation, and methods of analysis. Serving as an introduction, Chapter 1 tells what a mechanism is, what a mechanism can do, how mechanisms can be classified, and what some of their limitations are. Chapters 2, 3, and 4 are concerned totally with analysis, specifically with kinematic analysis, because they cover position, velocity, and acceleration analyses, respectively, of single-degree-of-freedom planar mechanisms. Chapter 5, new in the fourth edition, expands this background to include multi-degree-of-freedom planar mechanisms.

Part 2 of the book goes on to demonstrate engineering applications involving the selection, the specification, the design, and the sizing of mechanisms to accomplish specific motion objectives. This part includes chapters on cam systems, gears, gear trains, synthesis of linkages, spatial mechanisms, and an introduction to robotics. Chapter 6 is a study of the geometry, kinematics, and proper design of high-speed cam systems. Chapter 7 studies the geometry and kinematics of spur gears, particularly of involute tooth profiles, their manufacture, and proper tooth meshing. Chapter 8 expands this background to include helical gears, bevel gears, worms, and worm gears. Chapter 9 then studies gear trains, with an emphasis on epicyclic and differential gear trains. Chapter 10 is an introduction to the kinematic synthesis of planar linkages. Chapter 11 is a brief introduction to the kinematic analysis of spatial mechanisms. Chapter 12 presents the forward and inverse kinematics problems associated with robotics.

Part 3 of the book adds the dynamics of machines. In a sense, this part is concerned with the consequences of the mechanism design specifications. In other words, having designed a machine by selecting, specifying, and sizing the various components, what happens during the operation of the machine? What forces are produced? Are there any unexpected operating results? Will the proposed design be satisfactory in all respects? Chapter 13 presents the static force analysis of machines. This chapter now includes new sections focusing on the buckling of two-force members subjected to axial loads. Chapter 14 studies the planar and spatial aspects of the dynamic force analysis of machines. Chapter 15 then presents the vibration analysis of mechanical systems. Chapter 16 is a more detailed study of one particular type of mechanical system, namely the dynamics of both single- and multicylinder reciprocating engines. Chapter 17 next addresses the static and dynamic balancing of rotating and reciprocating systems. Chapter 18 is on the dynamics of cam systems. Finally, Chapter 19 is on the study of the dynamics of flywheels, governors, and gyroscopes.

As with all texts, the subject matter of this book also has limitations. Probably the clearest boundary on the coverage in this text is that it is limited to the study of rigid-body mechanical systems. It does study planar multibody systems with movable connections or constraints between them. However, all elastic effects are assumed to come within the connections; the shapes of the individual bodies are assumed constant. This assumption is necessary to allow the separate study of kinematic effects from that of dynamics. Because each individual body is assumed rigid, it can have no strain; therefore, the study of stress is also outside the scope of this text. It is hoped, however, that courses using this text can provide background for the later study of stress, strength, fatigue life, modes of failure, lubrication, and other aspects important to the proper design of mechanical systems.

Despite the limitations on the scope of this book, it is still clear that it is not reasonable to expect that all of the material presented here can be covered in a single-semester course. As stated above, a variety of methods and applications have been included to allow the instructor to choose those topics that best fit the course objectives and to still provide a reference for follow-on courses and help build the student's library. Yet many instructors have asked for our suggestions regarding a choice of topics that might fit a 3-hour per week, 15-week course. Two such outlines follow, as used by the two authors to teach such courses at their own institutions. It is hoped that these might be used as helpful guidelines to assist others in making parallel choices.

Tentative Schedule I

Kinematics and Dynamics of Machine Systems

Week	Topics	Sections
1	Introduction to Mechanisms	1.1–1.10
	Kutzbach and Grashof Criteria	1.6, 1.9
	Advance-to-Return Time Ratio	1.7
	Overlay Method of Synthesis	10.8
2	Vector Loop-Closure Equation	2.6, 2.7
	Velocity Difference Equation	3.1–3.3
	Velocity Polygons; Velocity Images	3.4
3	Apparent Velocity Equation	3.5, 3.6, 3.8
	Direct and Rolling Contact Velocity	3.7
4	Instantaneous Centers of Velocity	3.13
	Aronhold–Kennedy Theorem of Three Centers	3.14, 3.15
	Use of Instant Centers to Find Velocities	3.16, 3.17
5	Exam 1	
	Acceleration Difference Equation	4.1–4.3
	Acceleration Polygons; Acceleration Images	4.4
6	Apparent Acceleration Equation	4.5, 4.6
	Coriolis Component of Acceleration	
7	Direct and Rolling Contact Acceleration	4.7, 4.8
	Review of Velocity and Acceleration Analyses	
8	Raven's Method of Kinematic Analysis	2.10, 3.10, 4.10
	Kinematic Coefficients	3.12, 4.12
	Computer Methods in Kinematics	11.9
9	Exam 2	
	Static Forces	13.1–13.6
	Two-, Three-, and Four-Force Members	13.7, 13.8
	Force Polygons	13.7–13.8
10	Coulomb Friction Forces in Machines	13.9, 13.10
11	D'Alembert's Principle	14.1–14.4
	Dynamic Forces in Machine Members	14.4, 14.5
12	Introduction to Cam Design	6.1–6.4
	Choice of Cam Profiles; Matching Displacement Curves	6.5–6.8

13	First-Order Kinematic Coefficients; Face Width; Pressure Angle	6.9
	Second-Order Kinematic Coefficients; Pointing and Undercutting	6.10
14	Exam 3	
	Introduction to Gearing	7.1–7.6
	Involute Tooth Geometry; Contact Ratio; Undercutting	7.7–7.9, 7.11
15	Epicyclic and Differential Gear Trains	9.1–9.8
	Review	
	Final Exam	

Tentative Schedule II

Machine Design I

Week	Topics	Sections
1	The World of Mechanisms	1.1–1.6
	Measures of Performance (Indices of Merit)	1.10, 3.20
	Quick Return Mechanisms	1.7
2	Position Analysis. Vector Loops	2.1–2.7
	Newton–Raphson Technique	2.8, 2.11
3	Velocity Analysis	3.1–3.9
	First-Order Kinematic Coefficients	3.11–3.16
	Instant Centers of Zero Velocity	3.13–3.18
4	Rolling Contact, Rack and Pinion, Two Gears	3.11
	Acceleration Analysis	4.1–4.4
	Second-Order Kinematic Coefficients	4.5–4.11
5	Geometry of a Point Path	4.16
	Kinematic Coefficients for Point Path	4.16
	Radius and Center of Curvature	4.17
6	Cam Design	6.1–6.4
	Lift Curve	6.1–6.4
	Exam 1	
7	Kinematic Coefficients of the Follower	6.5
	Roller Follower	6.9, 6.10
	Flat-Face Follower	6.9, 6.10

8	Graphic Approach	13.5, 13.6
	Two-, Three-, and Four-Force Members	13.7, 13.8
	Friction-Force Models	13.9, 13.10
9	Dynamic Force Analysis	14.1–14.3
	Force and Moment Equations	14.4–14.6
	Static Force Analysis	13.1–13.4
10	Power Equation	14.9
	Kinetic, Potential, and Dissipative Energy	14.9
	Equivalent Inertia and Equivalent Mass	14.9
11	Equation of Motion	14.9
	Critical Speeds of a Shaft	15.17
	Exam 2	
12	Exact Equation	15.17
	Dunkerley and Rayleigh–Ritz Approximations	15.17
	Shaking Forces and Moments	16.5
13	Rotating Unbalance	17.3
	Discrete Mass System	17.5
	Distributed Mass System	17.5
14	Reciprocating Unbalance	17.9
	Single-Cylinder Engine	17.9
	Multicylinder Engine	17.10
15	Primary Shaking Forces	17.11
	Secondary Shaking Forces	17.11
	Comparison of Forces	17.9
	Final Exam	

The supplement package for this fourth edition has been designed to support both the student and the instructor in the kinematics and dynamics course. For the student, the CD packaged with this text includes over 100 animations of key figures from the text, marked with a ● symbol. These animations, created by Zhong Hu of South Dakota State University, are presented in both Working Model and .avi file formats, and are meant to help students visualize and comprehend the movement of important mechanisms. For the instructor, a complete solutions manual is available, as well as an instructor's resource CD that contains solutions for 100 problems in the text worked out using MatLab software. Due to a call for increased MatLab resources, we have provided these solutions for professors wishing to incorporate MatLab code into their course. Problems marked with a † signify that there is a MatLab-based solution available on the instructor's CD. Thank you to Bob Williams at Ohio University for your help with these solutions.

The authors thank the reviewers for their very helpful criticisms and recommendations for improvement. They are: Efstratios Nikolaidis, University of Toledo; Fred Choy, University of Akron; Bob Williams, Ohio University; Lubambala Kabengela, UNC Charlotte; Carol Rubin, Vanderbilt University; Yeau-Jian Liao, Wayne State University; Chad O'Neal, Louisiana Tech University; Alba Perez-Garcia, Idaho State University; Zhong Hu, South Dakota State University. The many students who have tolerated previous versions of this book and made suggestions for its improvement also deserve our continuing gratitude.

The authors would also like to offer our sincere thanks to Rachael Zimmerman, Associate Editor for Engineering, and Jennifer Bossert, Senior Production Editor, Oxford University Press, USA, for their continuing cooperation and assistance in bringing this fourth edition to completion.

John J. Uicker, Jr.
Gordon R. Pennock
January 15, 2010

About the Authors

John J. Uicker, Jr., is Professor Emeritus of Mechanical Engineering at the University of Wisconsin–Madison. He received his B.M.E. degree from the University of Detroit and his M.S. and Ph.D. degrees in mechanical engineering from Northwestern University. Since joining the University of Wisconsin faculty in 1967, his teaching and research specialties have been in solid geometric modeling and the modeling of mechanical motion and their application to computer-aided design and manufacture; these include the kinematics, dynamics, and simulation of articulated rigid-body mechanical systems. He was the founder of the UW Computer-Aided Engineering Center and served as its director for its initial 10 years of operation. He has served on several national committees of the American Society of Mechanical Engineers (ASME) and the Society of Automotive Engineers (SAE) and he received the SAE Ralph R. Teetor Educational Award in 1969, the ASME Mechanisms Committee Award in 2004, and the ASME Fellow Award in 2007. He is one of the founding members of the U.S. Council for the Theory of Machines and Mechanisms and of IFToMM, the international federation. He served for several years as editor-in-chief of the federation journal *Mechanism and Machine Theory* . He is also a registered mechanical engineer in the state of Wisconsin and has served for many years as an active consultant to industry.

As an ASEE Resident Fellow he spent 1972–1973 at Ford Motor Company. He was also awarded a Fulbright–Hayes Senior Lectureship and became a visiting professor to Cranfield Institute of Technology in Cranfield, England, in 1978–1979. He is the pioneering researcher on matrix methods of linkage analysis and was the first to derive the general dynamic equations of motion for rigid-body articulated mechanical systems. He has been awarded twice for outstanding teaching, three times for outstanding research publications, and twice for historically significant publications.

Gordon R. Pennock is Associate Professor of Mechanical Engineering at Purdue University, West Lafayette, Indiana. His teaching is primarily in the area of machine design. His research specialties are in theoretical kinematics and the dynamics of mechanical systems. He has applied his research to robotics, rotary machinery, and biomechanics, including the kinematics, statics, and dynamics of articulated rigid-body mechanical systems.

He received his B.Sc. degree (Hons.) from Heriot–Watt University, Edinburgh, Scotland, his M.Eng.Sc. from the University of New South Wales, Sydney, Australia, and his Ph.D. degree in mechanical engineering from the University of California, Davis. Since joining the Purdue University faculty in 1983, he has served on several national committees and international program committees. He is the student section advisor of the ASME at Purdue University, and a member of the Student Section Committee. He is a member of the Commission on Standards and Terminology, the International Federation of the Theory of Machines and Mechanisms. He is also an associate of the Internal Combustion Engine Division, ASME, and served as the Technical Committee Chairman of Mechanical Design,

Internal Combustion Engine Division, from 1993 to 1997. He also served as chairman of the Mechanisms and Robotics Committee, ASME, from 2008 to 2009.

He is a fellow of the ASME, a fellow of the SAE, and a fellow and chartered engineer of the Institution of Mechanical Engineers, United Kingdom. He is a senior member of the Institute of Electrical and Electronics Engineers and a senior member of the Society of Manufacturing Engineers. He received the ASME Faculty Advisor of the Year Award in 1998, and was named the Outstanding Student Section Advisor, Region VI, 2001. The Central Indiana Section recognized him in 1999 by the establishment of the Gordon R. Pennock Outstanding Student Award to be presented annually to the senior student in recognition of academic achievement and outstanding service to the ASME student section at Purdue University. He was presented with the Ruth and Joel Spira Award for outstanding contributions to the School of Mechanical Engineering and its students in 2003. He received the SAE Ralph R. Teetor Educational Award in 1986, the Ferdinand Freudenstein Award at the Fourth National Applied Mechanisms and Robotics Conference in 1995, and the A. T. Yang Memorial Award from the Design Engineering Division of ASME in 2005. He has been at the forefront of many new developments in mechanical design, primarily in the areas of kinematics and dynamics. He has published some 100 technical papers and is a regular conference and symposium speaker, workshop presenter, and conference session organizer and chairman.

Joseph E. Shigley (deceased May 1994) was Professor Emeritus of Mechanical Engineering at the University of Michigan and a fellow in the ASME. He received the Mechanisms Committee Award in 1974, the Worcester Reed Warner medal in 1977, and the Machine Design Award in 1985. He was author of eight books, including *Mechanical Engineering Design* (with Charles R. Mischke) and *Applied Mechanics of Materials.* He was coeditor-in-chief of the *Standard Handbook of Machine Design.* He first wrote *Kinematic Analysis of Mechanisms* in 1958 and then wrote *Dynamic Analysis of Machines* in 1961, and these were published in a single volume titled *Theory of Machines* in 1961; they have evolved over the years to become the current text, *Theory of Machines and Mechanisms*, now in its fourth edition.

He was awarded the B.S.M.E. and B.S.E.E. degrees from Purdue University and received his M.S. at the University of Michigan. After several years in industry, he devoted his career to teaching, writing, and service to his profession, first at Clemson University and later at the University of Michigan. His textbooks have been widely used throughout the United States and internationally.

Theory of Machines and Mechanisms

PART 1

Kinematics and Mechanisms

1 The World of Mechanisms

1.1 INTRODUCTION

The theory of machines and mechanisms is an applied science that is used to understand the relationships between the geometry and motions of the parts of a machine, or mechanism, and the forces that produce these motions. The subject, and therefore this book, is naturally divided into three parts. Part 1, which includes Chapters 1 through 5, is concerned with mechanisms and the kinematics of mechanisms, which is the analysis of their motions. Part 1 lays the groundwork for Part 2, comprising Chapters 6 through 12, in which we study the methods of designing mechanisms. Finally, in Part 3, which includes Chapters 13 through 19, we take up the study of kinetics, the time-varying forces in machines, and the resulting dynamic phenomena that must be considered in their design.

The design of a modern machine is often very complex. In the design of a new engine, for example, automotive engineers must deal with many interrelated questions. What is the relationship between the motion of the piston and the motion of the crankshaft? What will be the sliding velocities and the loads at the lubricated surfaces and what lubricants are available for this purpose? How much heat will be generated and how will the engine be cooled? What are the synchronization and control requirements and how will they be met? What will be the cost to the consumer, both for initial purchase and for continued operation and maintenance? What materials and manufacturing methods will be used? What will be the fuel economy, noise, and exhaust emissions; will they meet legal requirements? Although these and many other important questions must be answered before the design can be completed, obviously they cannot all be addressed in a book of this size. Just as people with diverse skills must be brought together to produce an adequate design, so too must many branches of science be brought together. This book assembles material that falls into the science of mechanics as it relates to the design of mechanisms and machines.

1.2 ANALYSIS AND SYNTHESIS

The study of mechanical systems comprises two completely different aspects, *design* and *analysis*. The concept embodied in the word "design" might be more properly termed *synthesis*, the process of contriving a scheme or a method of accomplishing a given purpose. Design is the process of prescribing the sizes, shapes, material compositions, and arrangements of parts so that the resulting machine will perform the prescribed task.

Although many phases in the design process can be approached in a well-ordered, scientific manner, the overall process is by its very nature as much an art as a science. It calls for imagination, intuition, creativity, judgment, and experience. The role of science in the design process is merely to provide tools to be used by designers as they practice their art.

It is in the process of evaluating the various interacting alternatives that designers find a need for a large collection of mathematical and scientific tools. These tools, when applied properly, can provide more accurate and more reliable information for use in judging a design than one can achieve through intuition or estimation. Thus, the tools can be of tremendous help in deciding among alternatives. However, scientific tools cannot make decisions for designers; designers have every right to exert their imagination and creative abilities, even to the extent of overruling the mathematical recommendations.

Probably the largest collection of scientific methods at the designer's disposal falls into the category called *analysis*. These are the techniques that allow the designer to critically examine an already existing or proposed design to judge its suitability for the task. Thus, analysis in itself is not a creative science but one of evaluation and rating of things already conceived.

We should always bear in mind that, although most of our effort may be spent on analysis, the real goal is synthesis, the design of a machine or system. Analysis is simply a tool. It is, however, a vital tool and will inevitably be used as one step in the design process.

1.3 THE SCIENCE OF MECHANICS

The branch of scientific analysis that deals with motions, time, and forces is called *mechanics* and is made up of two parts, statics and dynamics. *Statics* deals with the analysis of stationary systems—that is, those in which time is not a factor—and *dynamics* deals with systems that change with time.

As illustrated in Fig. 1.1, dynamics is also made up of two major disciplines, first recognized as separate entities by Euler in 1765:[1,*]

> The investigation of the motion of a rigid body may be conveniently separated into two parts, the one geometrical, the other mechanical. In the first part, the transference of the body from a given position to any other position must be investigated without respect to the causes of the motion, and must be represented by analytical formulae, which will define the position of each point of the body. This investigation will therefore be referable solely to geometry, or rather to stereotomy.
>
> It is clear that by the separation of this part of the question from the other, which belongs properly to Mechanics, the determination of the motion from dynamical principles will be made much easier than if the two parts were undertaken conjointly.

---

* Numeric superscripts refer to references at the end of each chapter.

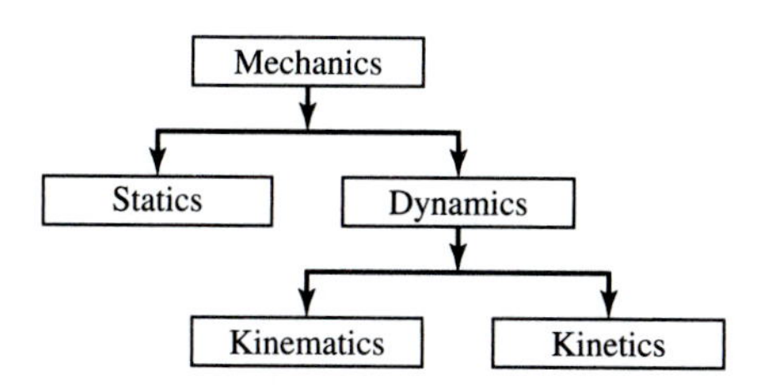

Figure 1.1

These two aspects of dynamics were later recognized as the distinct sciences of *kinematics* (*cinématique* was a term coined by Ampère* and derived from the Greek word *kinema*, meaning motion) and kinetics and deal with motion and the forces producing the motion, respectively.

The initial problem in the design of a mechanical system is therefore understanding the kinematics. *Kinematics* is the study of motion, quite apart from the forces that produce the motion. More particularly, kinematics is the study of position, displacement, rotation, speed, velocity, and acceleration. The study, say, of planetary or orbital motion is also a problem in kinematics, but in this book we shall concentrate our attention on kinematic problems that arise in the design of mechanical systems. Thus, the kinematics of machines and mechanisms is the focus of the next several chapters of this book. Statics and kinetics, however, are also vital parts of a complete design analysis, and they are also covered in later chapters.

It should be carefully noted in the above quotation that Euler based his separation of dynamics into kinematics and kinetics on the assumption that they deal with *rigid* bodies. It is this very important assumption that allows the two to be treated separately. For flexible bodies, the shapes of the bodies themselves, and therefore their motions, depend on the forces exerted on them. In this situation, the study of force and motion must take place simultaneously, thus significantly increasing the complexity of the analysis.

Fortunately, although all real machine parts are flexible to some degree, machines are usually designed from relatively rigid materials, keeping part deflections to a minimum. Therefore, it is common practice to assume that deflections are negligible and parts are rigid when analyzing a machine's kinematic performance and then, after dynamic analysis when loads are known, to design the parts so that the assumption is justified.

1.4 TERMINOLOGY, DEFINITIONS, AND ASSUMPTIONS

Reuleaux† defines a *machine*†† as a "*combination of resistant bodies so arranged that by their means the mechanical forces of nature can be compelled to do work accompanied by certain determinate motions.*" He also defines a mechanism as an "*assemblage of resistant*

* André-Marie Ampère (1775–1836).

† Much of the material in this section is based on definitions originally set down by Franz Reuleaux (1829–1905), a German kinematician whose work marked the beginning of a systematic treatment of kinematics. For additional reading, see A. B. W. Kennedy, *Reuleaux's Kinematics of Machinery*. London: Macmillan, 1876. Reprint, New York: Dover, 1963.

†† There appears to be no agreement at all on the proper definition of a machine. In a footnote, Reuleaux gives 17 definitions; his translator gives 7 more and discusses the whole problem in detail.

bodies, connected by movable joints, to form a closed kinematic chain with one link fixed and having the purpose of transforming motion."

Some light can be shed on these definitions by contrasting them with the term *structure*. A structure is also a combination of resistant (rigid) bodies connected by joints, but the purpose of a structure (such as a truss) is not to do work or to transform motion; it is intended to be rigid. A truss can perhaps be moved from place to place and is movable in this sense of the word; however, it has no *internal* mobility. A structure has no *relative motions* between its various members, whereas both machines and mechanisms do have relative motions. Indeed, the whole purpose of a machine or mechanism is to utilize these relative internal motions in transmitting power or transforming motion.

A machine is an arrangement of parts for doing work, a device for applying power or changing the direction of motion. A machine differs from a mechanism in its purpose. In a machine, terms such as force, torque, work, and power describe the predominant concepts. In a mechanism, although it may transmit power or force, the predominant idea in the mind of the designer is one of achieving a desired motion. There is a direct analogy among the terms structure, mechanism, and machine and the three branches of mechanics illustrated in Fig. 1.1. The term structure is to statics as the term mechanism is to kinematics and as the term machine is to kinetics.

We shall use the word *link* to designate a machine part or a component of a mechanism. As discussed in the previous section, a link is assumed to be completely rigid. Machine components that do not fit this assumption of rigidity, such as springs, usually have no effect on the kinematics of a device but do play a role in supplying forces. Such members are not called links; they are usually ignored during kinematic analysis, and their force effects are introduced during dynamic analysis. Sometimes, as with a belt or chain, a machine member may possess one-way rigidity; such a member would be considered a link when in tension but not under compression.

The links of a mechanism must be connected together in some manner to transmit motion from the *driver*, or input motion, to the *follower*, or output. These connections, joints between the links, are called *kinematic pairs* (or just *pairs*), because each joint consists of a pair of mating surfaces, two elements, with one mating surface or element being a part of each of the joined links. Thus, we can also define a link as *the rigid connection between two or more joint elements*.

Stated explicitly, the assumption of rigidity is that there can be no relative motion (change in distance) between two arbitrarily chosen points on the same link. In particular, the relative positions of pairing elements on any given link do not change. In other words, the purpose of a link is to hold a constant spatial relationship between the elements of its pairs.

As a result of the assumption of rigidity, many of the intricate details of the actual part shapes are unimportant when studying the kinematics of a machine or mechanism. For this reason, it is common practice to draw highly simplified schematic diagrams, which contain important features of the shape of each link, such as the relative locations of pair elements, but that completely subdue the real geometry of the manufactured parts. The slider-crank mechanism of the internal combustion engine, for example, can be simplified for the purposes of analysis to the schematic diagram illustrated later in Fig. 1.3*b*. Such simplified schematics are a great help because they eliminate confusing factors that do not affect the analysis; such diagrams are used extensively throughout this text. However, these schematics also have the drawback of bearing little resemblance to physical hardware. As a result, they may give the impression that they represent only academic constructs rather than real machinery. We should always bear in mind that these simplified diagrams are intended to carry only the minimum necessary information so as not to confuse the issue with all the unimportant detail (for kinematic purposes) or complexity of the true machine parts.

When several links are connected together by joints, they are said to form a *kinematic chain*. Links containing only two pair elements are called *binary* links, those having three are called *ternary* links, and so on. If every link in the chain is connected to at least two other links, the chain forms one or more closed loops and is called a *closed* kinematic chain; if not, the chain is referred to as *open*. When no distinction is made, the chain is assumed to be closed. If the chain consists entirely of binary links then it is *simple-closed*. *Compound-closed* chains, however, include other than binary links and thus form more than a single closed loop.

Recalling Reuleaux's definition of a mechanism, we see that it is necessary to have a closed kinematic chain *with one link fixed*. When we say that one link is fixed, we mean that it is chosen as a frame of reference for all other links; that is, the motions of all points on the linkage will be measured with respect to this link, which is thought of as being fixed. This link in a practical machine usually takes the form of a stationary platform or base (or a housing rigidly attached to such a base) and is called the *frame* or *base link*. The question of whether this reference frame is truly stationary (in the sense of being an inertial reference frame) is immaterial in the study of kinematics but becomes important in the investigation of kinetics, where forces are considered. In either case, once a frame member is designated (and other conditions are met), the kinematic chain becomes a mechanism and, as the driver is moved through various positions, all other links have well-defined motions with respect to the chosen frame of reference. We use the term *kinematic chain* to specify a particular arrangement of links and joints when it is not clear which link is to be treated as the frame. When the frame link is specified, the kinematic chain is called a *mechanism*.

For a mechanism to be useful, the motions between links cannot be completely arbitrary; they must be constrained to produce the proper relative motions, those chosen by the designer for the particular task to be performed. These desired relative motions are obtained by a proper choice of the number of links and the kinds of joints used to connect them. Thus, we are led to the concept that, in addition to the distances between successive joints, the nature of the joints themselves and the relative motions they permit are essential in determining the kinematics of a mechanism. For this reason, it is important to look more closely at the nature of joints in general terms and in particular at several of the more common types.

The controlling factor that determines the relative motions allowed by a given joint is the mating of the shapes of the surfaces or elements. Each type of joint has its own characteristic shapes for the elements and each allows a given type of motion, which is determined by the possible ways in which these elemental surfaces can move with respect to each other. For example, the pin joint in Fig. 1.2*a* has cylindric elements and, assuming that the links cannot slide axially, these surfaces permit only relative rotational motion. Thus, a pin joint allows the two connected links to experience relative rotation about the pin axis. In like manner, other joints each have their own characteristic element shapes and relative motions. These shapes restrict the totally arbitrary motion of two unconnected links to some prescribed type of relative motion and form the constraining conditions or constraints on the mechanism's motion.

It should be pointed out that the element shapes may often be subtly disguised and difficult to recognize. For example, a pin joint might include a needle bearing, so that two mating surfaces, as such, are not distinguishable. Nevertheless, if the motions of the individual rollers are not of interest, the motions allowed by the joints are equivalent and the pairs are of the same generic type. Thus, the criterion for distinguishing different pair types is the relative motions they permit and not necessarily the shapes of the elements, although these may provide vital clues. The diameter of the pin used (or other dimensional data) is also of no more importance than the exact sizes and shapes of the connected links. As stated previously, the kinematic function of a link is to hold a fixed geometric relationship between the pair elements. In a similar way, the only kinematic function of a joint, or pair, is to determine the relative motion between the connected links. All other features are determined for other reasons and are unimportant in the study of kinematics.

When a kinematic problem is formulated, it is necessary to recognize the type of relative motion permitted in each of the pairs and to assign to it some variable parameter(s) for measuring or calculating the motion. There will be as many of these parameters as there are degrees of freedom of the joint in question and they are referred to as the *pair variables*. Thus, the pair variable of a pinned joint will be a single angle measured between reference lines fixed in the adjacent links, whereas a spheric pair will have three pair variables (all angles) to specify its three-dimensional rotation.

Kinematic pairs were divided by Reuleaux into *higher pairs* and *lower pairs*, with the latter category consisting of six prescribed types to be discussed next. He distinguished between the categories by noting that the lower pairs, such as the pin joint, have surface contact between the pair elements, whereas higher pairs, such as the connection between a cam and its follower, have line or point contact between the elemental surfaces. However, as noted in the case of a needle bearing, this criterion may be misleading. We should look instead for distinguishing features in the relative motion(s) that the joint allows.

The six lower pairs are illustrated in Fig. 1.2. Table 1.1 lists the names of the lower pairs and the symbols employed for them by Hartenberg and Denavit[2], together with the number of degrees of freedom and the pair variables for each of the six.

The *revolute* or *turning pair, R* (Fig. 1.2*a*), permits only relative rotation and is often referred to as a pin joint. This pair has one degree of freedom.

The *prismatic pair*, *P*(Fig. 1.2*b*), permits only a relative sliding motion and therefore is often called a sliding joint. This pair also has one degree of freedom.

The *helical* or *screw pair, H* (Fig. 1.2*c*), permits angular rotation and sliding motion. However, it only has one degree of freedom because the rotation and the sliding are related by the helix angle of the thread. Thus, the pair variable may be chosen as either Δs or $\Delta\theta$, but not both. Note that the screw pair reduces to a revolute pair if the helix angle is made zero and to a prismatic pair if the helix angle is made 90°.

The *cylindric pair, C* (Fig. 1.2*d*), permits both angular rotation and an independent sliding motion. Thus, the cylindric pair has two degrees of freedom.

The *spheric* or *globular pair, S* (Fig. 1.2*e*), is a ball-and-socket joint. It has three degrees of freedom, sometimes taken as rotations about each of the coordinate axes.

The *flat* or *planar pair*, sometimes called the *ebene pair*, *F*(Fig. 1.2*f*), is seldom found in mechanisms in its undisguised form, except at a support point. It has three degrees of freedom, that is, two translations and a rotation.

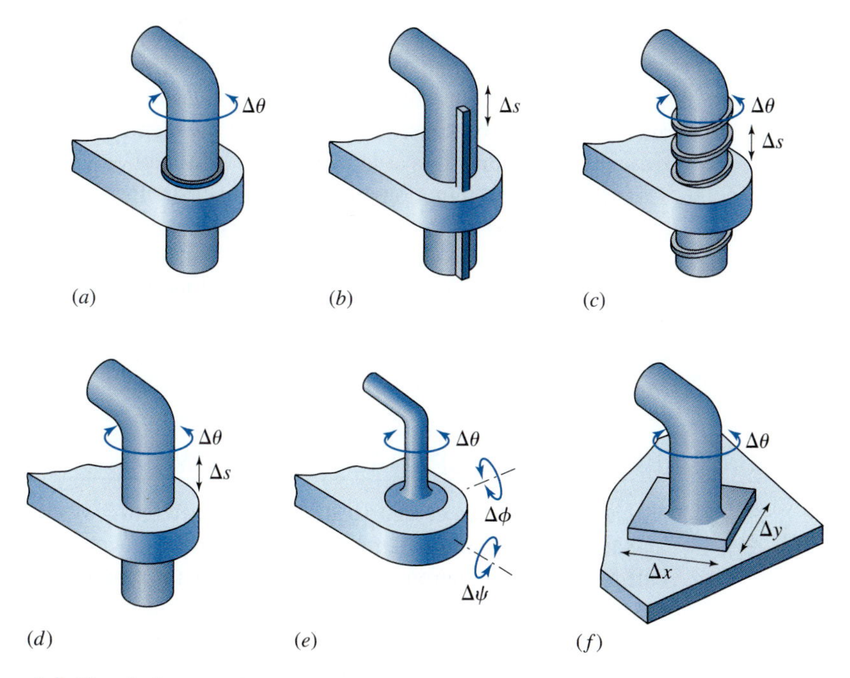

Figure 1.2 The six lower pairs: (*a*) revolute, (*b*) prism, (*c*) helical, (*d*) cylinder, (*e*) sphere, and (*f*) flat.

TABLE 1.1 The lower pairs

Pair	Symbol	Pair Variable	Degrees of Freedom	Relative motion
Revolute	R	$\Delta\theta$	1	Circular
Prism	P	Δs	1	Rectilinear
Helical	H	$\Delta\theta$ or Δs	1	Helical
Cylinder	C	$\Delta\theta$ and Δs	2	Cylindric
Sphere	S	$\Delta\theta,\ \Delta\phi,\ \Delta\psi$	3	Spheric
Flat	F	Δx, Δy, $\Delta\theta$	3	Planar

All other joint types are called higher pairs. Examples are mating gear teeth, a wheel rolling and/or sliding on a rail, a ball rolling on a flat surface, and a cam contacting its follower. Because an infinite variety of higher pairs exist, a systematic accounting of them is not a realistic objective. We shall treat each as a separate situation as it arises.

Among the higher pairs is a subcategory known as *wrapping pairs*. Examples are the connections between a belt and a pulley, a chain and a sprocket, or a rope and a drum. In each case, one of the links has one-way rigidity.

The treatment of various joint types, whether lower or higher pairs, includes another important limiting assumption. Throughout the book, we assume that the actual joint, as manufactured, can be reasonably represented by a mathematical abstraction having perfect geometry. That is, when a real machine joint is assumed to be a spheric pair, for example, it is also assumed that there is no "play" or clearance between the joint elements and that any deviation in the spherical geometry of the elements is negligible. When a pin joint is treated as a revolute, it is assumed that no axial motion can take place; if it is necessary to study the small axial motions resulting from clearances between the real elements, the joint must be treated as cylindric, thus allowing the axial motion.

The term mechanism, as defined earlier, can refer to a wide variety of devices, including both higher and lower pairs. A more limited term, however, refers to those mechanisms having only lower pairs; such a mechanism is called a *linkage*. A linkage, then, is connected only by the lower pairs illustrated in Fig. 1.2.

1.5 PLANAR, SPHERICAL, AND SPATIAL MECHANISMS

Mechanisms may be categorized in several different ways to emphasize their similarities and differences. One such grouping divides mechanisms into planar, spherical, and spatial categories. All three groups have many things in common; the criterion that distinguishes the groups, however, is to be found in the characteristics of the motions of the links.

A *planar mechanism* is one in which all particles describe planar curves in space and all these curves lie in parallel planes; that is, the loci of all points are plane curves parallel to a single common plane. This characteristic makes it possible to represent the locus of any chosen point of a planar mechanism in its true size and shape on a single drawing or figure. The motion transformation of any such mechanism is called *coplanar*. The plane four-bar linkage, the plate cam and follower, and the slider-crank mechanism are familiar examples of planar mechanisms. The majority of mechanisms in use today are planar.

Planar mechanisms utilizing only lower pairs are called *planar linkages*; they include only revolute and prismatic pairs. Although a planar pair (Fig. 1.2*f*) might theoretically be included, this would impose no constraint and thus be equivalent to an opening in the kinematic chain. Planar motion also requires that all revolute axes be normal to the plane of motion and that all prismatic pair axes be parallel to the plane.

As pointed out above, it is possible to observe the motions of all particles of a planar mechanism in true size and shape from a single direction. In other words, all motions can be represented graphically in a single view. Thus, graphical techniques are well suited to their solution. Because spherical or spatial mechanisms do not all have this fortunate geometry, visualization becomes more difficult and more powerful techniques must be used for their analysis.

A *spherical mechanism* is one in which each link has some point that remains stationary as the linkage moves and in which the stationary points of all links lie at a common location; that is, the locus of each point is a curve contained in a spherical surface, and the spherical surfaces defined by several arbitrarily chosen points are all *concentric*. The motions of all particles can therefore be completely described by their radial projections, or "shadows," on the surface of a sphere with a properly chosen center. Hooke's universal joint (also referred to as a Cardan, or cardanic, joint) is perhaps the most familiar example of a spherical mechanism.

Spherical linkages are composed entirely of revolute pairs. A spheric pair (Fig. 1.2*e*), would produce no constraints and would be equivalent to an opening in the chain, whereas all other lower pairs have nonspheric motion. In a spherical linkage, the axes of all revolute pairs must intersect at a point.

Spatial mechanisms include *no* restrictions on the relative motions of the particles. The motion transformation is not necessarily coplanar nor must it be concentric. A spatial mechanism may have particles with loci of double curvature. Any linkage that contains a screw pair (Fig. 1.2*c*), for example, is a spatial mechanism, because the relative motion within a screw pair is helical. Because of the more complex motion characteristics of spatial mechanisms and because these motions cannot be analyzed graphically from a single viewing direction, more powerful mathematical techniques are required for their analysis. Such techniques become necessary and are introduced in Chapters 11 and 12, which cover spatial mechanisms and robotics.

Because the majority of mechanisms in use today are planar, one might question the need for the more complicated mathematical techniques that are used for spatial mechanisms. However, more powerful methods are of value for a number of reasons, even though simpler graphical techniques have been mastered:

1. They provide new, alternative methods that solve problems in a different way. Thus, they provide a means of checking results. Certain problems by their nature may also be more amenable to one method than another.
2. Methods that are analytical in nature are better suited to solution by calculator or digital computer than graphical techniques.
3. Although the majority of mechanisms in use today are planar and well-suited to graphical solution, the remaining mechanisms must also be analyzed, and techniques should be known for analyzing them.
4. One reason why planar linkages are so common is that good methods of analysis for the more general spatial linkages have not been available until relatively recently. Without methods for their analysis, their design and use have not been common, although they may be inherently better suited to certain applications.
5. We will discover that spatial linkages are much more common in practice than their formal description indicates.

Consider a planar four-bar linkage. It has four links connected by four pins whose axes are parallel. This "parallelism" is a mathematical hypothesis; it is not a reality. The axes as produced in a machine shop—in any machine shop, no matter how good—are only approximately parallel. If the axes are far out of parallel, there is binding in no uncertain terms, and the mechanism moves only because the rigid links flex and twist, producing loads in the bearings. If the axes are nearly parallel, the mechanism operates because of the looseness of the running fits of the bearings or flexibility of the links. A common way of compensating for small nonparallelism is to connect the links with self-aligning bearings, actually spherical joints allowing three-dimensional rotation. Such a "planar" linkage is thus a low-grade spatial linkage.

Thus, the overwhelmingly large category of planar mechanisms and the category of spherical mechanisms are special cases, or subsets, of the all-inclusive category of spatial mechanisms. They occur as a consequence of the special orientations of their pair axes.

1.6 MOBILITY

One of the first concerns in either the design or the analysis of a mechanism is the number of degrees of freedom, also called the *mobility* of the device. The mobility* of a mechanism is the number of input parameters (usually pair variables) that must be controlled independently to bring the device into a particular position or orientation. Ignoring for the moment certain exceptions to be mentioned later, it is possible to determine the mobility of a mechanism directly from a count of the number of links and the number and types of joints comprising the system.

To develop this relationship, consider that before they are connected together, each link of a planar mechanism has three degrees of freedom when moving relative to the fixed link. Not counting the fixed link, therefore, an n-link planar mechanism has $3(n-1)$ degrees of freedom before any of the joints are connected. Connecting two links by a joint that has one degree of freedom, such as a revolute pair, has the effect of providing two constraints between the connected links. If the two links are connected by a two-degree-of-freedom pair, it provides one constraint. When the constraints for all pairs are subtracted from the total freedoms of the unconnected links, we find the resulting mobility of the assembled mechanism.

If we denote the number of single-degree-of-freedom pairs as j_1 and the number of two-degree-of-freedom pairs as j_2 then the resulting mobility m of a planar n-link mechanism is given by

$$m = 3(n-1) - 2j_1 - j_2. \tag{1.1}$$

Written in this form, Eq. (1.1) is called the *Kutzbach criterion* for the mobility of a planar mechanism. Its application is illustrated for several simple cases in Fig. 1.3.

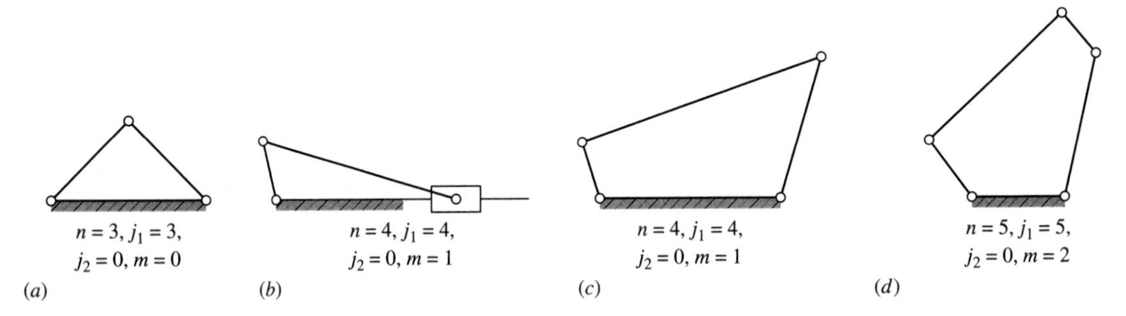

Figure 1.3 Applications of the Kutzbach mobility criterion.

* The German literature distinguishes between *movability* and *mobility*. Movability includes the six degrees of freedom of the device as a whole, as though the ground link were not fixed, and thus applies to a kinematic chain. Mobility neglects these **degrees of freedom** and considers only the internal relative motions, thus applying to a mechanism. The English literature seldom recognizes this distinction, and the terms are used somewhat interchangeably.

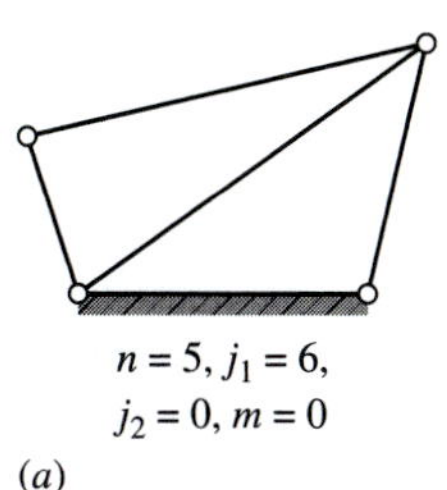

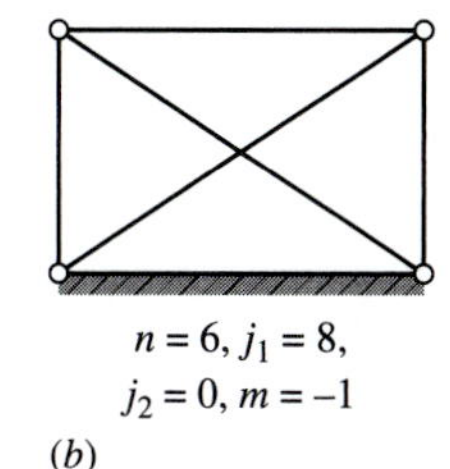

Figure 1.4 Applications of the Kutzbach criterion to structures.

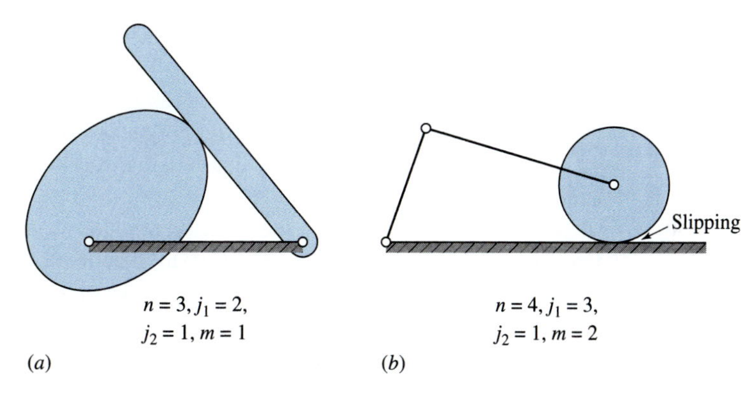

Figure 1.5

If the Kutzbach criterion yields $m > 0$, the mechanism has m degrees of freedom. If $m = 1$, the mechanism can be driven by a single input motion. If $m = 2$, then two separate input motions are necessary to produce constrained motion for the mechanism; such a case is illustrated in Fig. 1.3*d*.

If the Kutzbach criterion yields $m = 0$, as in Figs. 1.3*a* and 1.4*a*, motion is impossible and the mechanism forms a structure.

If the criterion yields $m < 0$, then there are redundant constraints in the chain and it forms a statically indeterminate structure. An Example is illustrated in Fig. 1.4*b*. Note in the examples of Fig. 1.4 that when three links are joined by a single pin, two pairs must be counted; such a connection is treated as two separate but concentric pairs.

Figure 1.5 illustrates examples of the Kutzbach[3] criterion applied to mechanisms with two-degree-of-freedom joints, that is j_2 pairs. Particular attention should be paid to the contact (pair) between the wheel and the fixed link in Fig. 1.5*b*. Here it is assumed that slipping is possible between the two links. If this contact prevents slipping, the joint would be counted as a one-degree-of-freedom pair, that is a j_1 pair, because only one relative motion would be possible between the links.

Sometimes the Kutzbach criterion will give an incorrect result. Note that Fig. 1.6*a* represents a structure and that the criterion properly predicts $m = 0$. However, if link 5 is arranged as in Fig. 1.6*b*, the result is a double-parallelogram linkage with a mobility $m = 1$ although Eq. (1.1) indicates that it is a structure. The actual mobility of $m = 1$ results only if the parallelogram geometry is achieved. Because in the development of

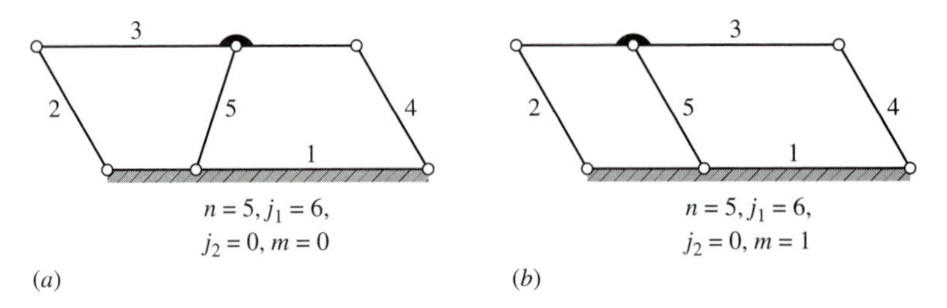

Figure 1.6

the Kutzbach criterion no consideration was given to the lengths of the links or other dimensional properties, perhaps it should not be surprising that exceptions to the criterion exist for particular cases with equal link lengths, parallel links, or other special geometric features.

Although it has exceptions, the criterion remains useful because it is so easily applied. To avoid exceptions it would be necessary to include all the dimensional properties of the mechanism. The resulting criterion would be very complex and would be useless at the early stages of design when dimensions may not be known.

An earlier mobility criterion named after Grübler[4] applies to planar mechanisms with only single-degree-of-freedom joints where the overall mobility of the mechanism is unity. Putting $j_2 = 0$ and $m = 1$ into Eq. (1.1), we find Grübler's criterion for planar mechanisms with constrained motion and only turning or sliding pairs:

$$3n - 2j_1 - 4 = 0. \tag{1.2}$$

From this we can see, for example, that a planar mechanism with a mobility of 1 and only single-degree-of-freedom joints cannot have an odd number of links. Also, we can find the simplest possible mechanism of this type; by assuming all binary links, we find $n = j_1 = 4$. This helps to explain why the four-bar linkage (Fig. 1.3*c*) and the slider-crank mechanism (Fig. 1.3*b*) are so common in application.

Both the Kutzbach criterion, Eq. (1.1), and the Grübler criterion, Eq. (1.2), were derived for the case of planar mechanisms. If similar criteria are developed for spatial mechanisms, which is the subject of Chapter 11, we must recall that each unconnected link has six degrees of freedom and each single-degree-of-freedom joint provides five constraints, each two-degree-of-freedom joint provides four constraints, and so on. Similar arguments then lead to the three-dimensional form of the Kutzbach criterion,

$$m = 6(n - 1) - 5j_1 - 4j_2 - 3j_3 - 2j_4 - j_5 \tag{1.3}$$

and the 3-D Grübler criterion,

$$6n - 5j_1 - 7 = 0. \tag{1.4}$$

The simplest form of a spatial mechanism,* with all single-degree-of-freedom pairs and a mobility of 1, is therefore $n = j_1 = 7$.

EXAMPLE 1.1

Determine the mobility of the planar mechanism illustrated in Fig. 1.7*a*.

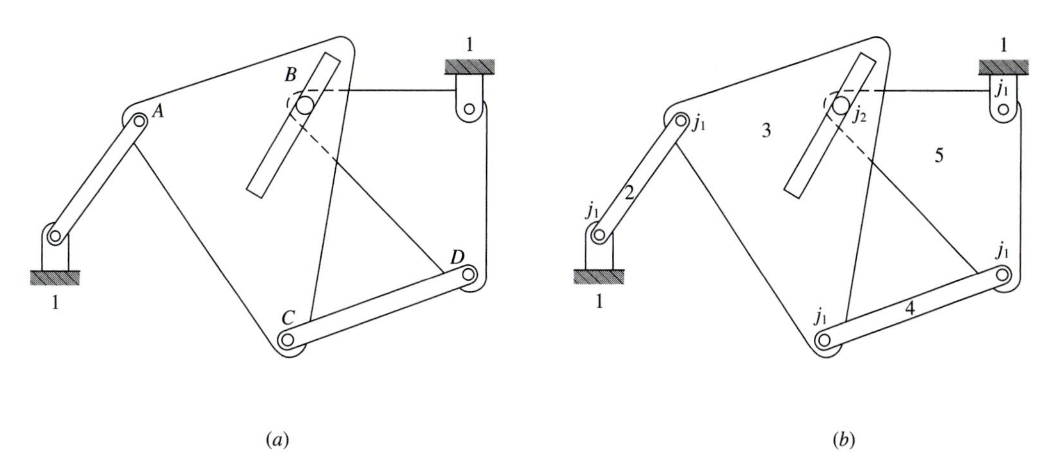

Figure 1.7 Example 1.1.

SOLUTION

The link numbers and the joint types for the mechanism are illustrated in Fig. 1.7*b*. The number of links is $n = 5$, the number of lower pairs is $j_1 = 5$, and the number of higher pairs is $j_2 = 1$. Substituting these values into the Kutzbach criterion, Eq. (1.1), the mobility of the mechanism is

$$m = 3(5 - 1) - 2(5) - 1 = 1. \qquad \textit{Ans.}$$

Note that the Kutzbach criterion gives the correct answer for the mobility of this mechanism; that is, a single input motion is required to give a unique output motion. For example, rotation of link 2 could be used as the input and rotation of link 5 could be used as the output.

EXAMPLE 1.2

For the mechanism illustrated in Fig. 1.8*a*, determine (*i*) the number of lower pairs (j_1 pairs) and the number of higher pairs (j_2 pairs) and (*ii*) the mobility of the mechanism using the

---

* Note that all planar linkages are exceptions to the spatial mobility criterion. They have special geometric characteristics in that all revolute axes are parallel and perpendicular to the plane of motion and all prismatic axes lie in the plane of motion.

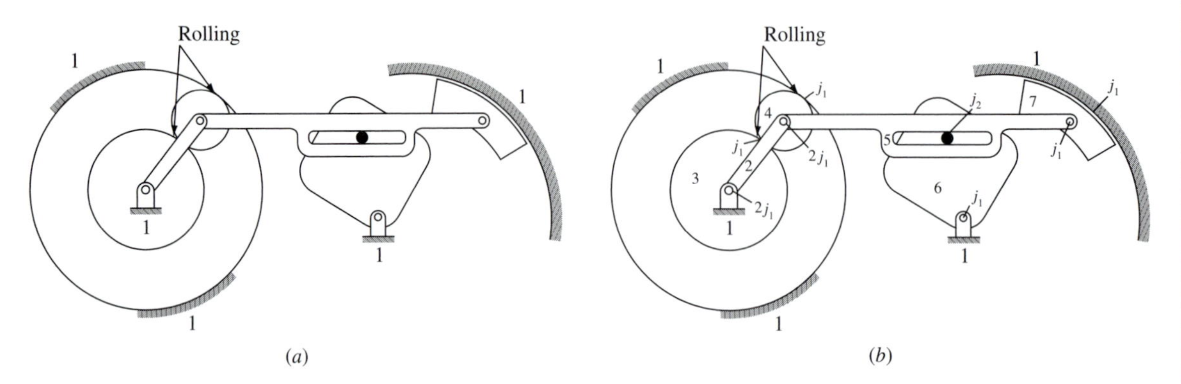

Figure 1.8 Example 1.2.

Kutzbach criterion. Does this criterion provide the correct answer for the mobility of this mechanism? Briefly explain why or why not.

(i) The links and the joint types of the mechanism are labeled in Fig. 1.8*b*. The number of links is $n = 7$, the number of lower pairs is $j_1 = 9$, and the number of higher pairs is $j_2 = 1$. *Ans.*

(ii) Substituting these values into the Kutzbach criterion, Eq. (1.1), the mobility of the mechanism is

$$m = 3(7 - 1) - 2(9) - 1(1) = -1. \qquad \textit{Ans.}$$

However, this answer is not corrrect; that is, the Kutzbach criterion does not give the correct mobility for this mechanism. The mobility of this mechanism is in fact 1; that is, a single input position will give a unique output position.

Reasoning: Links 3 and 4 are superfluous to the constraints of the mechanism. If links 3 and 4 were removed, the motion of the remaining links would be unaffected. With links 3 and 4 removed, the mobility of the mechanism using the Kutzbach criterion is 1. Note that if links 3 and 4 were attached in a general manner, that is, not pinned at their centers, for example, then the mechanism would indeed be locked and the answer $m = -1$ would be correct.

EXAMPLE 1.3

For the mechanism illustrated in Fig. 1.9*a*, determine (*i*) the number of lower pairs and the number of higher pairs and (*ii*) the mobility of the mechanism predicted by the Kutzbach criterion. Does this criteria provide the correct answer for this mechanism? Briefly explain why or why not.

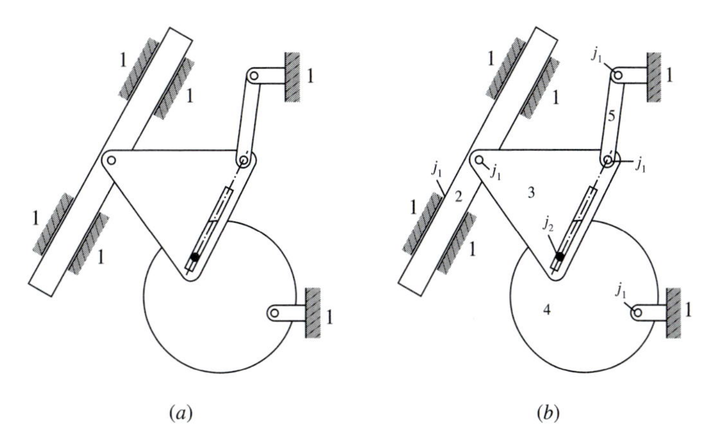

Figure 1.9 Example 1.3.

SOLUTION

(i) The links and the joints of the mechanism are labeled as illustrated in Fig. 1.9*b*. The number of links is 5, the number of lower pairs is 5, and the number of higher pairs is 1. *Ans.*

(ii) Substituting these values into the Kutzbach criterion, Eq. (1.1), the mobility of the mechanism is

$$m = 3(5 - 1) - 2(5) - 1(1) = 1. \qquad \textit{Ans.}$$

For this mechanism, the mobility is indeed 1, which indicates that the Kutzbach criterion gives the correct answer for this mechanism.

1.7 CLASSIFICATION OF MECHANISMS

An ideal system of classification of mechanisms would be one that allows a designer to enter the system with a set of specifications and leave with one or more mechanisms that satisfy those specifications. Although history[5] demonstrates that many attempts have been made, none has been very successful in devising a completely satisfactory method. In view of the fact that the purpose of a mechanism is the transformation of motion, we shall follow Torfason's lead[6] and classify mechanisms according to the type of motion transformation. All together, Torfason displays 262 mechanisms, each of which is capable of variation in dimensions. His categories are as follows.

Snap-Action Mechanisms The mechanisms of Fig. 1.10 are typical of snap-action mechanisms, but Torfason also includes spring clips and circuit breakers.

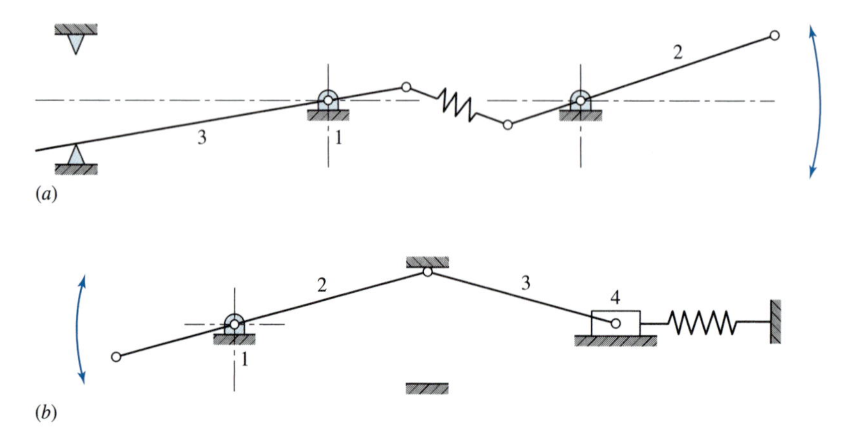

Figure 1.10 Typical snap-action, toggle, or flip-flop mechanisms used for switches, clamps, or fasteners. The elements of a mechanism are always numbered beginning with 1 for the base or frame and 2 for the input or driving element. The mechanism of part (*a*) is bistable; that of (*b*) is a true toggle.

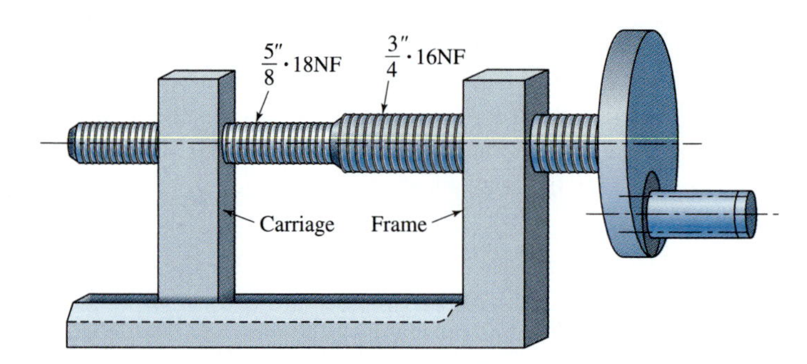

Figure 1.11 A differential screw. You should be able to determine the motion of the carriage resulting from one turn of the handle.

Linear Actuators Linear actuators include the following:

1. Stationary screws with traveling nuts.
2. Stationary nuts with traveling screws.
3. Single- and double-acting hydraulic and pneumatic cylinders.

Fine Adjustments Fine adjustments may be obtained with screws, including the differential screw of Fig. 1.11, worm gearing, wedges, levers, levers in series, and various motion-adjusting mechanisms.

Clamping Mechanisms Typical clamping mechanisms are the C-clamp, the woodworker's screw clamp, cam- and lever-actuated clamps, vises, presses such as the toggle press illustrated in Fig. 1.10*b*, collets, and stamp mills.

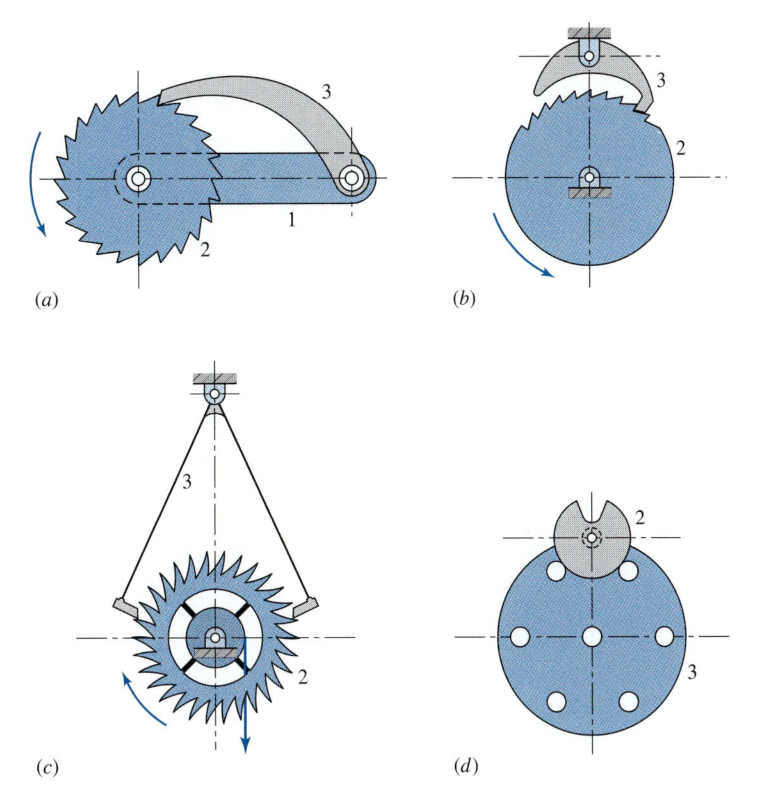

Figure 1.12 Ratchets and escapements.

Locational Devices

Torfason illustrates 15 locational mechanisms. These are usually self-centering and locate either axially or angularly using springs and detents.

Ratchets and Escapements

There are many different forms of ratchets and escapements, some quite clever. They are used in locks, jacks, clockwork, and other applications requiring some form of intermittent motion. Figure 1.12 illustrates four typical applications.

The ratchet in Fig. 1.12*a* allows only one direction of rotation of wheel 2. Pawl 3 is held against the wheel by gravity or a spring. A similar arrangement is used for lifting jacks, which then employ a toothed rack for rectilinear motion.

Figure 1.12*b* is an escapement used for rotary adjustments.

Graham's escapement of Fig. 1.12*c* is used to regulate the movement of clockwork. Anchor 3 drives a pendulum whose oscillating motion is caused by the two clicks engaging escapement wheel 2. One is a push click and the other is a pull click. The lifting and engaging of each click caused by oscillation of the pendulum results in a wheel motion that, at the same time, presses each respective click and adds a gentle force to the motion of the pendulum.

The escapement illustrated in Fig. 1.12*d* has a control wheel 2 that may rotate continuously to allow wheel 3 to be driven (by another source) in either direction.

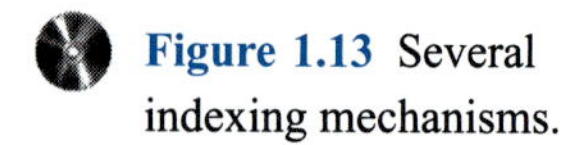
Figure 1.13 Several indexing mechanisms.

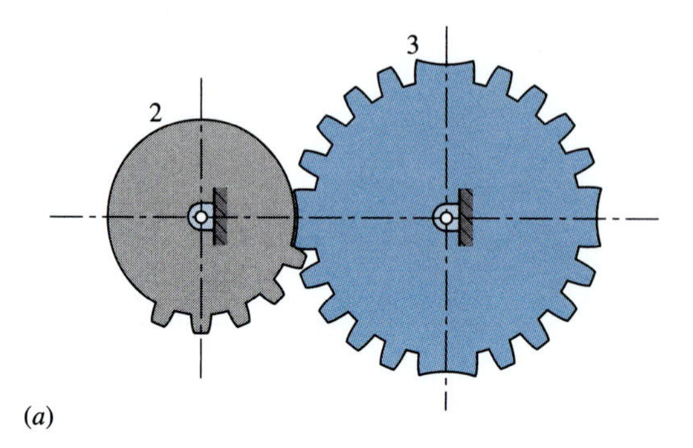

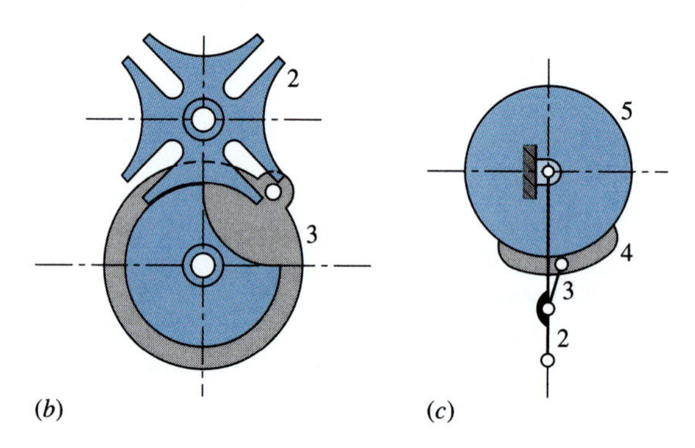

Indexing Mechanisms

The indexer illustrated in Fig. 1.13*a* uses standard gear teeth; for light loads, pins can be used in wheel 2 with corresponding slots in wheel 3, but neither form should be used if the shaft inertias are large.

Figure 1.13*b* illustrates a Geneva-wheel, sometimes called a Maltese-cross, indexer. Three or more slots (up to 16) may be used in driver 2, and wheel 3 can be geared to the output to be indexed. High speeds and large inertias may cause problems with this indexer.

Toothless ratchet 5 in Fig. 1.13*c* is driven by oscillating crank 2 of variable throw. Note the similarity of this indexing mechanism to the ratchet of Fig. 1.12*a*.

Torfason lists nine different indexing mechanisms, and many variations are possible.

Swinging or Rocking Mechanisms

The class of swinging or rocking mechanisms is often termed *oscillators*; in each case, the output member rocks or swings through an angle that is generally less than 360°. However, the output shaft can be geared to a second shaft to produce a larger angle of oscillation.

Figure 1.14*a* is a mechanism consisting of a rotating crank 2 and coupler 3 containing a toothed rack, which meshes with output gear 4 to produce the oscillating motion.

In Fig. 1.14*b*, crank 2 drives member 3, which slides on output link 4, producing a rocking motion. This mechanism is described as a *quick-return linkage* because crank 2 rotates through a larger angle on the forward stroke of link 4 than on the return stroke.

Figure 1.14*c* is a *four-bar linkage* called the *crank-and-rocker mechanism* (see Section 1.9). Crank 2 drives rocker 4 through coupler 3. Of course, link 1 is the frame.

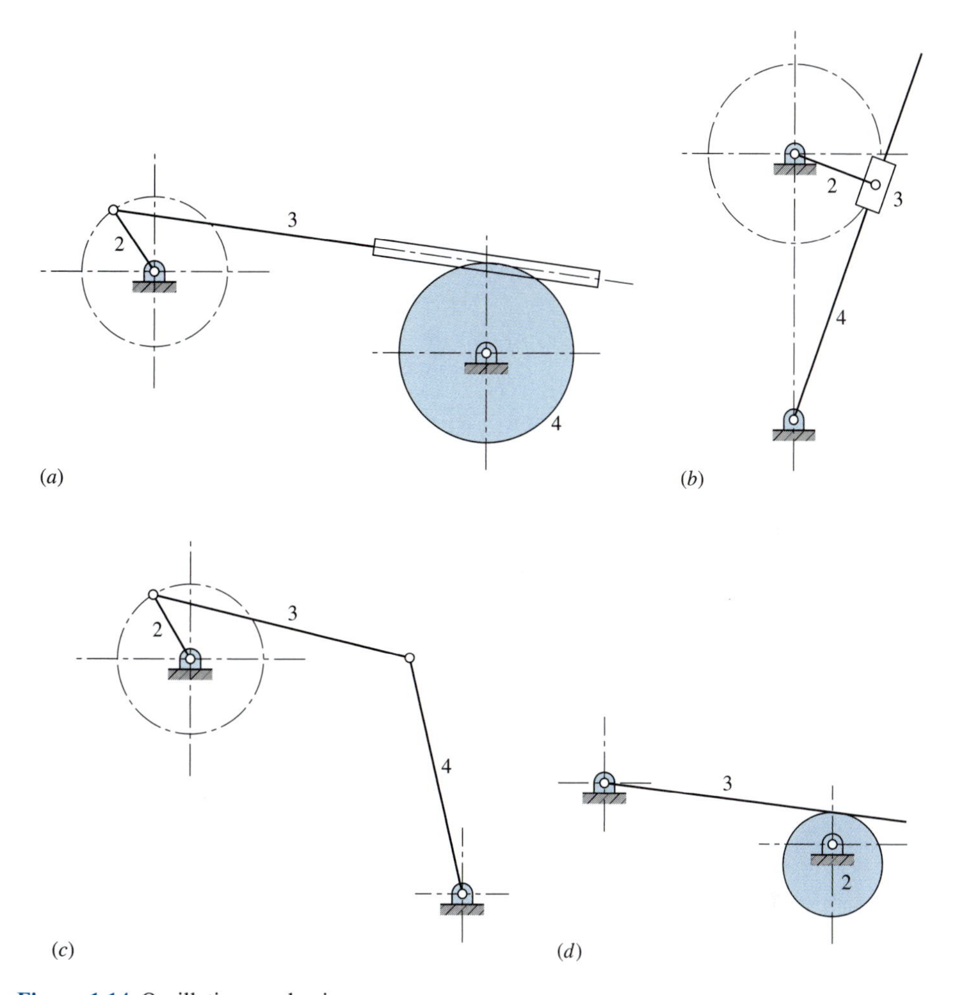

Figure 1.14 Oscillating mechanisms.

The characteristics of the rocking motion depend on the dimensions of the links and the placement of the frame points.

Figure 1.14*d* illustrates a *cam-and-follower mechanism*, in which the rotating *cam* 2 drives link 3, called the *follower*, in a rocking motion. An endless variety of cam-and-follower mechanisms exist, many of which will be discussed in Chapter 6. In each case, the cams can be formed to produce rocking motions with nearly any set of desired characteristics.

Reciprocating Mechanisms Repeating straight-line motion is commonly obtained using pneumatic and hydraulic cylinders, a stationary screw and traveling nut, rectilinear drives using reversible motors or reversing gears, and cam-and-follower mechanisms. A variety of typical linkages for obtaining reciprocating motion are illustrated in Figs. 1.15 and 1.16.

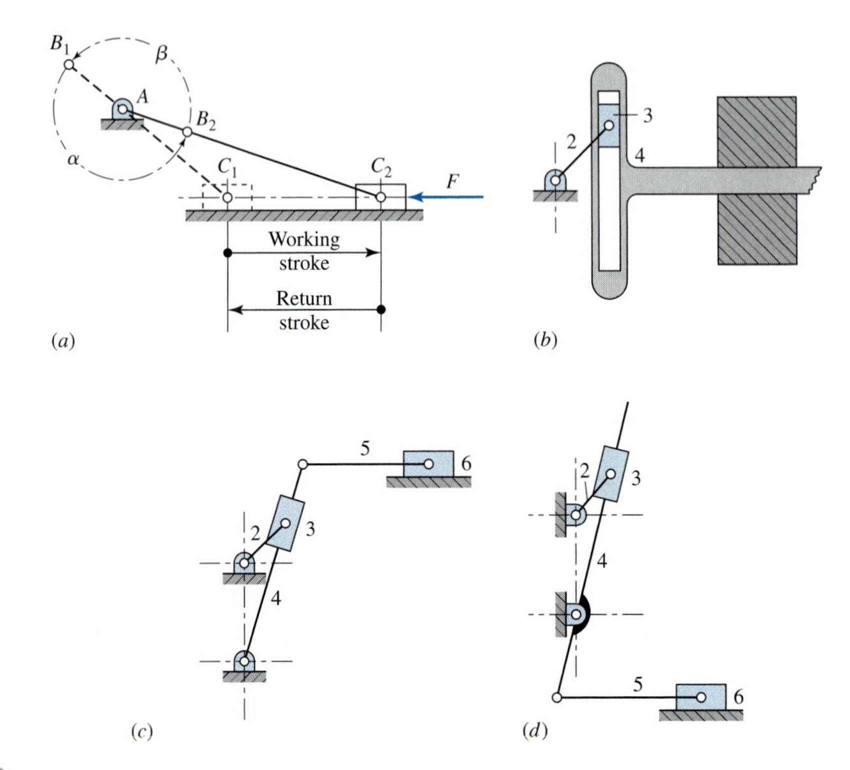

Figure 1.15 Reciprocating mechanisms.

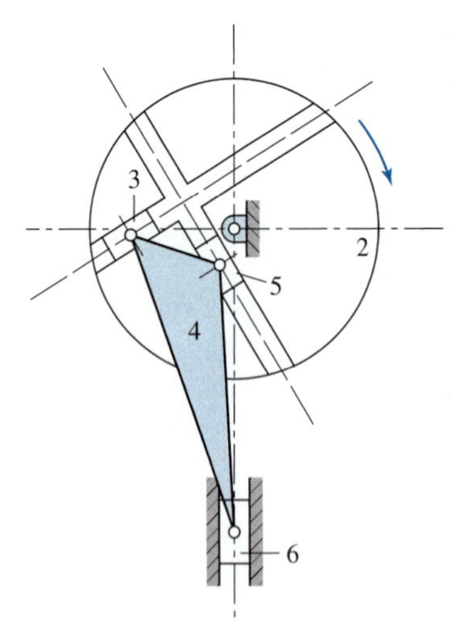

Figure 1.16 The Wanzer needle-bar mechanism (as reported by A. R. Holowenko, *Dynamics of Machinery*. New York: Wiley, 1955, 54).

The *offset slider-crank mechanism* illustrated in Fig. 1.15*a* has velocity characteristics that differ from those of an on-center slider-crank (see Fig 1.3*b*). If connecting rod 3 of an on-center slider-crank mechanism is large relative to the length of crank 2, then the resulting motion is nearly harmonic.

Link 4 of the *Scotch-yoke mechanism* illustrated in Fig. 1.15*b* delivers exact harmonic motion.

The six-bar linkage illustrated in Fig. 1.15*c* is often called the *shaper mechanism*, after the name of the machine tool in which it is used. Note that it is derived from Fig. 1.14*b* by adding coupler 5 and slider 6. The slider stroke has a quick-return characteristic.

Figure 1.15*d* illustrates another version of the shaper mechanism, which is often termed the *Whitworth quick-return mechanism*. The linkage is presented in an upside-down configuration to illustrate its similarity to Fig. 1.15*c*.

In many applications, mechanisms are used to perform repetitive operations such as pushing parts along an assembly line, clamping parts together while they are welded, or folding cardboard boxes in an automated packaging machine. In such applications, it is often desirable to use a constant-speed motor; this will lead us to a discussion of Grashof's law in Section 1.9. In addition, however, we should also give some consideration to the power and timing requirements.

In these repetitive operations there is usually a part of the cycle when the mechanism is under load, called the advance or working stroke, and a part of the cycle, called the return stroke, when the mechanism is not working but simply returning so that it may repeat the operation. For example, consider the offset slider-crank mechanism illustrated in Fig. 1.15*a*. Work may be required to overcome the load F while the piston moves to the right from position C_1 to position C_2 but not during its return to position C_1 because the load may have been removed. In such situations, to keep the power requirements of the motor to a minimum and to avoid wasting valuable time, it is desirable to design the mechanism so that the piston will move much faster through the return stroke than it does during the advance (or working) stroke, that is, to use a higher fraction of the cycle time for doing work than for returning.

A measure of the suitability of a mechanism from this viewpoint, called the *advance-to-return time ratio*, or simply the time ratio, is defined by the formula

$$Q = \frac{\text{time of advance stroke}}{\text{time of return stroke}}. \qquad (a)$$

A mechanism for which the value of Q is high is more desirable for such repetitive operations than one in which the value of Q is lower. Certainly, any such operations would use a mechanism for which Q is greater than unity. Because of this, mechanisms with Q greater than unity are called *quick-return mechanisms*.

Assuming that the driving motor operates at constant speed, it is easy to find the time ratio. As illustrated in Fig. 1.15*a*, the first step is to determine the two crank positions, AB_1 and AB_2, which mark the beginning and the end of the working stroke. Next, noting the direction of rotation of the crank, we can measure the crank angle α traveled through during the advance stroke and the remaining crank angle β of the return stroke. Then, if the period of the motor is τ, the time of the advance stroke is

$$\text{Time of advance stroke} = \frac{\alpha}{2\pi}\tau \qquad (b)$$

and the time of the return stroke is

$$\text{Time of return stroke} = \frac{\beta}{2\pi}\tau. \tag{c}$$

Finally, combining Eqs. (*a*), (*b*), and (*c*), we obtain a simple equation for the time ratio, namely,

$$Q = \frac{\alpha}{\beta}. \tag{1.5}$$

EXAMPLE 1.4

The rocker of a four-bar linkage is required to have a length of 4 in and swing through a total angle of 45°. Also, the time ratio of the linkage is required to be 2.0. Determine a suitable set of link lengths for the remaining three links of the four-bar linkage.

SOLUTION

The time ratio, Eq. (1.5), must be

$$Q = \frac{\alpha}{\beta} = 2.0, \tag{1}$$

where

$$\alpha = 180^\circ + \phi \tag{2}$$

and

$$\beta = 180^\circ - \phi. \tag{3}$$

Substituting Eqs. (2) and (3) into Eq. (1) allows us to solve for

$$\phi = 60^\circ, \quad \alpha = 240^\circ, \quad \text{and} \quad \beta = 120^\circ.$$

Now, referring to Fig. 1.17, we apply the following graphical procedure:

(i) Draw the rocker $r_4 = 4.0$ in to a suitable scale in the two extreme positions; that is, show the swing angle of the rocker of 45°. Label the ground pivot O_4 and label the pin B in the two positions B_1 and B_2.

(ii) Through point B_1 draw an arbitrary line (labeled the X-line). Through B_2 draw a line parallel to the X-line.

(iii) Measure the angle $\phi = 60^\circ$ counterclockwise from the X-line through point B_1. The intersection of this line with the line parallel to the X-line is the input crank pivot O_2.

(iv) The length O_2O_4 of the ground link can be measured from the scaled drawing. That is,

$$r_1 = 1.50 \text{ in.} \qquad \textit{Ans.}$$

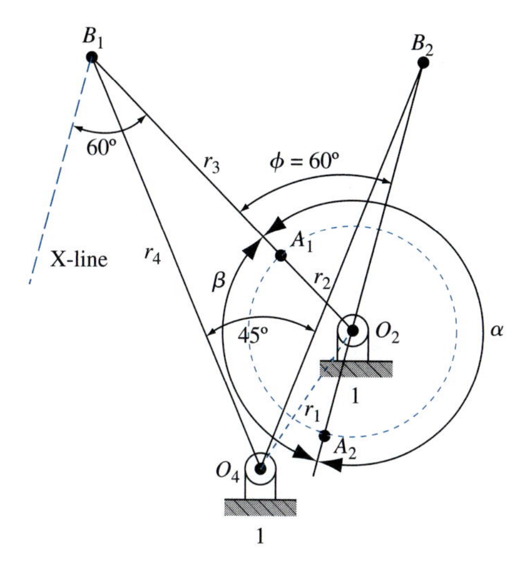

Figure 1.17 Example 1.4. The synthesized four-bar linkage.

(v) The lengths of the crank link and the coupler link can be determined from the measurements

$$O_2B_1 = r_3 + r_2 = 3.50 \text{ in} \quad \text{and} \quad O_2B_2 = r_3 - r_2 = 2.50 \text{ in.}$$

That is,

$$r_2 = 0.5(O_2B_1 - O_2B_2) = 0.50 \text{ in} \quad \text{and} \quad r_3 = 0.5(O_2B_2 + O_2B_1) = 3.00 \text{ in.}$$

Ans.

The solution of the synthesized four-bar linkage is illustrated in Fig. 1.17.

EXAMPLE 1.5

Determine a suitable set of link lengths for a slider-crank linkage such that the stroke is 2.50 in and the time ratio is 1.4.

SOLUTION

The time ratio, Eq. (1.5), must be

$$Q = \frac{\alpha}{\beta} = 1.40, \tag{1}$$

where

$$\alpha = 180° + \phi \tag{2}$$

and

$$\beta = 180° - \phi. \tag{3}$$

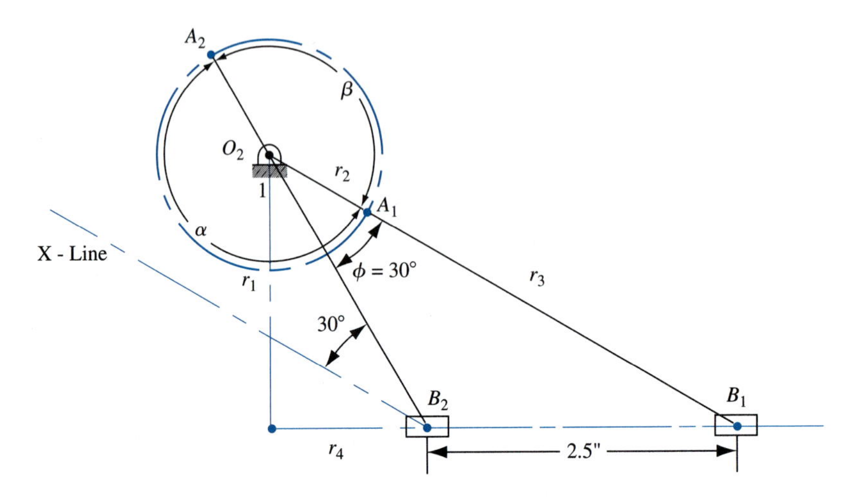

Figure 1.18 Example 1.5. The synthesized slider-crank linkage.

Substituting Eqs. (2) and (3) into Eq. (1) allows us to solve for

$$\phi = 30^\circ, \quad \alpha = 210^\circ, \quad \text{and} \quad \beta = 150^\circ.$$

Now, referring to Fig. 1.18, we apply the following graphical procedure:

(i) Draw the stroke (assumed to be horizontal) of 2.50 in of the slider-crank linkage to a suitable scale. Label the pin B in its two extreme positions B_1 and B_2.

(ii) Through point B_2 draw an arbitrary line (labeled the X-line). Through point B_1 draw a line parallel to the X-line.

(iii) Measure the angle $\phi = 30^\circ$ clockwise from the X-line. The intersection of this line with the line parallel to the X-line is the ground pivot O_2.

(iv) The length of the ground link, that is, the offset (or vertical distance) of the ground pivot from the line of travel of the slider, can be measured from the scaled drawing. That is,

$$r_1 = 2.17 \text{ in} \qquad \textit{Ans.}$$

(v) The lengths of the crank and the coupler links can be determined from the measurements

$$O_2B_1 = r_3 + r_2 = 4.33 \text{ in} \quad \text{and} \quad O_2B_2 = r_3 - r_2 = 2.50 \text{ in.}$$

That is,

$$r_2 = 0.5(O_2B_1 - O_2B_2) = 0.92 \text{ in} \quad \text{and} \quad r_3 = 0.5(O_2B_2 + O_2B_1) = 3.42 \text{ in.}$$

Ans.

The solution of the synthesized slider-crank linkage is illustrated in Fig. 1.18.

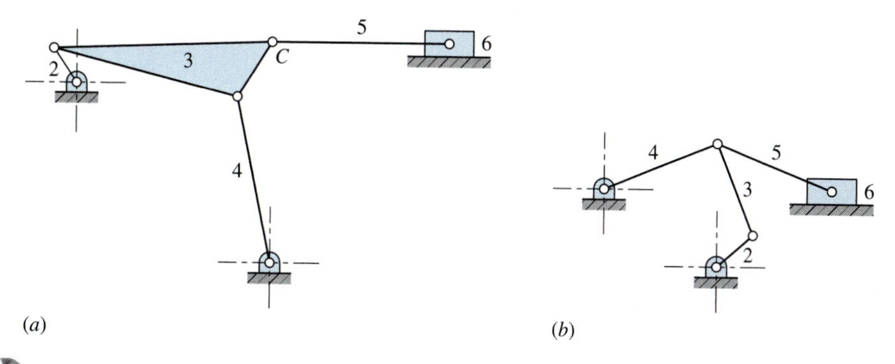

Figure 1.19 Additional six-bar reciprocating mechanisms.

Note that the time ratio of a quick-return mechanism does not depend on the amount of work being done or even on the speed of the driving motor. It is a kinematic property of the mechanism itself and can be found strictly from the geometry of the device.

We also note that there is a proper and an improper direction of rotation for such a device. If the direction of crank rotation were reversed in the example of Fig. 1.18, the roles of α and β would also be reversed and the time ratio would be less than 1.0. Thus, the motor would have to rotate clockwise for this mechanism to have the quick-return property.

Many other mechanisms with quick-return characteristics exist. Another example is the *Whitworth mechanism*, also called the *crank-shaper mechanism*, illustrated in Figs. 1.15*c* and 1.15*d*. Although the determination of the angles α and β is different for each mechanism, Eq. (1.5) would apply to all of them.

Figure 1.19*a* illustrates a six-bar linkage derived from the crank-and-rocker linkage of Fig. 1.14*c* by expanding coupler 3 and adding coupler 5 and slider 6. Coupler point *C* should be located so as to produce the desired motion characteristic of slider 6.

A crank-driven toggle mechanism is illustrated in Fig. 1.19*b*. With this mechanism, a high mechanical advantage is obtained at one end of the stroke of slider 6. The synthesis of a quick-return mechanism, as well as mechanisms with other properties, is discused in some detail in Chapter 10.

Reversing Mechanisms

When a mechanism capable of delivering output rotation in either direction is desired, some form of reversing mechanism is required. Many such devices make use of a two-way clutch that connects the output shaft to either of two drive shafts turning in opposite directions. This method is used in both gear and belt drives and does not require that the drive be stopped to change direction. Gear-shift devices, as in automotive transmissions, are also in quite common use.

Couplings and Connectors

Couplings and connectors are used to transmit motion among coaxial, parallel, intersecting, and skewed shafts. Gears of one kind or another can be used for any of these situations. These will be discussed in Chapters 7 and 8.

Flat belts can be used to transmit motion between parallel shafts. They can also be used between intersecting or skewed shafts if guide pulleys are used, as illustrated in Fig. 1.20*a*.

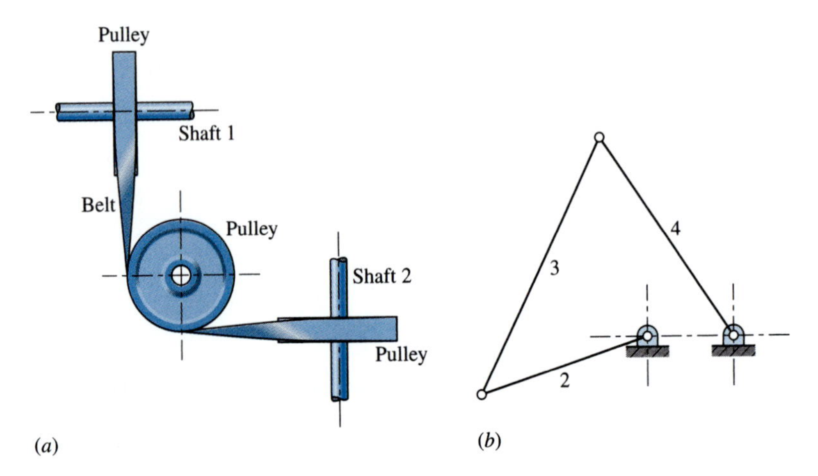

Figure 1.20 Two-shaft coupling mechanisms.

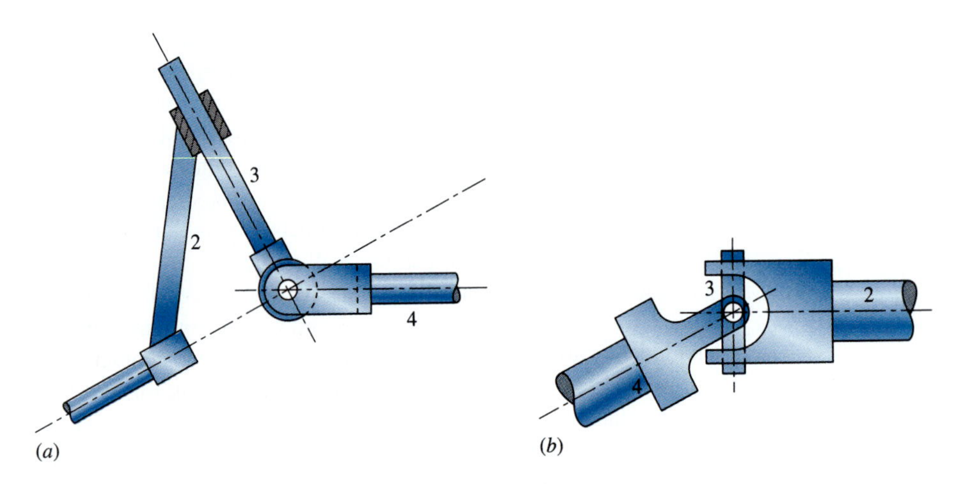

Figure 1.21 Coupling mechanisms for intersecting shafts.

When parallel shafts are involved, the belts can be open or crossed, depending on the direction of rotation desired.

Figure 1.20*b* illustrates the four-bar *drag-link mechanism* used to transmit rotary motion between parallel shafts. Here crank 2 is the driver and link 4 is the output. This is a very interesting linkage; you should try to duplicate it using cardboard strips and thumbtacks for the joints to observe its motion. Can you devise a working model for complete rotation?

The *Reuleaux coupling* illustrated in Fig. 1.21*a* for intersecting shafts is recommended only for light loads.

Figure 1.21*b* illustrates *Hooke's joint* for intersecting shafts. It is customary to use two of these for parallel shafts.

Sliding Connectors Sliding connectors are used when one slider (the input) is to drive another slider (the output). The usual problem is that the two sliders operate in the same plane but in different directions. The possible solutions are as follows:

1. A rigid link pivoted at each end to a slider.
2. A belt or chain connecting the two sliders with the use of a guide pulley or sprocket.
3. Rack gear teeth cut on each slider and the connection completed using one or more gears.
4. A flexible cable connector.

Stop, Pause, and Hesitation Mechanisms In an automotive engine a valve must open, remain open for a short period of time, and then close. A conveyor line may need to halt for a period of time while an operation is being performed and then continue its advance motion. Many similar requirements occur in the design of machines. Torfason classifies these as *stop-and-dwell*, *stop-and-return*, *stop-and-advance*, and so on. Such requirements can often be met using cam-follower mechanisms (see Chapter 6), all indexing mechanisms, including those of Fig. 1.13, ratchets, linkages at the limits of their motion, and gear-and-clutch mechanisms.

The six-bar linkage of Fig. 1.22 is a clever method of obtaining a rocking motion containing a dwell. This mechanism has a four-bar linkage consisting of frame 1, crank 2, coupler 3, and rocker 4 selected such that point C on the coupler generates the coupler curve shown by dashed lines. A portion of this curve fits very closely to a circular arc whose radius is the link length DC. Thus, when point C traverses this portion of the coupler curve, link 6, the output rocker, remains motionless. In the study of four-bar linkages, many similar coupler curves can be found.

Curve Generators The connecting rod or coupler of a planar four-bar linkage may be imagined as an infinite plane extending in all directions but pin connected to the input and output cranks. Then, during motion of the linkage, any point attached to the plane of the coupler generates some path with respect to the fixed link; this path is called a *coupler*

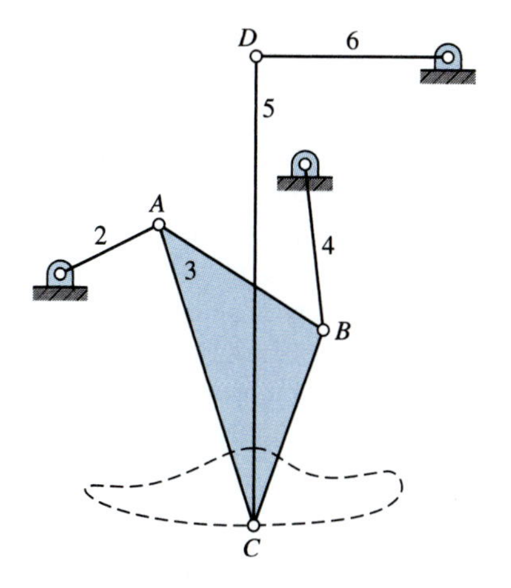

Figure 1.22 Six-bar stop-and-dwell mechanism.

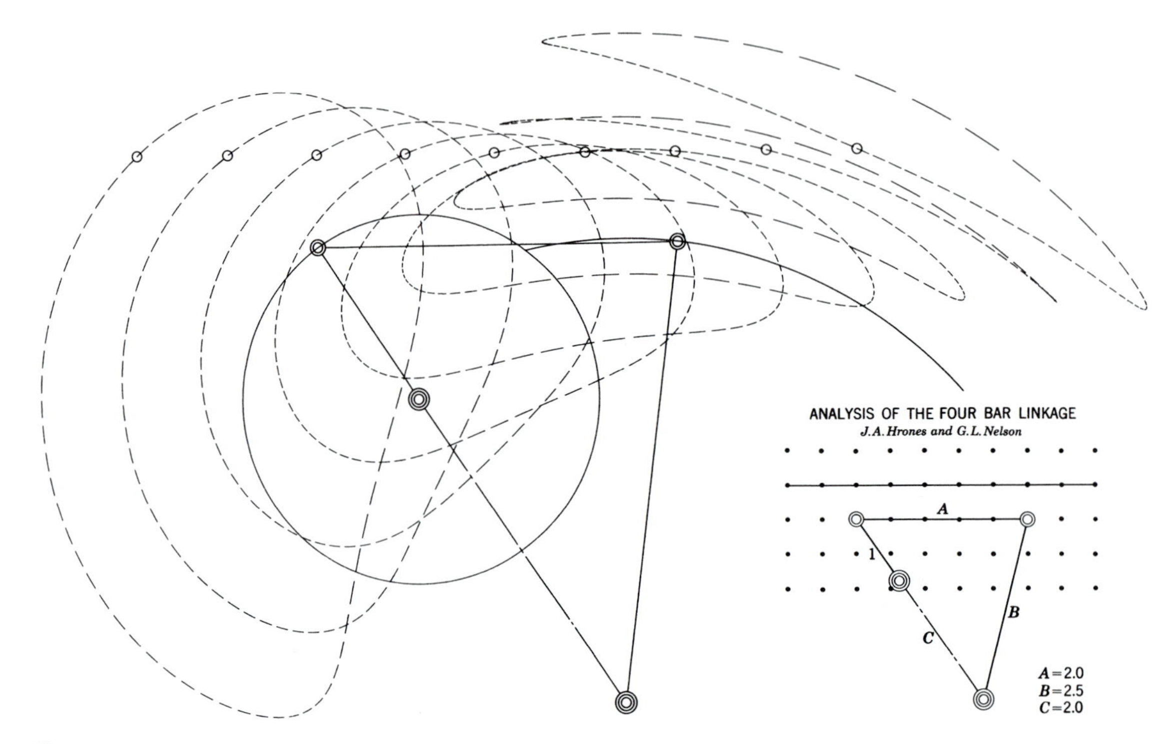

Figure 1.23 A set of coupler curves. (Reproduced by permission of the publishers, The Technology Press, MIT, Cambridge, MA, and Wiley, New York, from J. A. Hrones and G. L. Nelson, *Analysis of the Four-Bar Linkage*, 1951.)

curve. Two of these paths, namely those generated by the pin connectors of the coupler, are true circles with centers at the two fixed pivots. However, other points can be found that trace much more complex curves.

One of the best sources of coupler curves for the four-bar linkage is the Hrones–Nelson atlas.[7] This book consists of a set of 11×17 in drawings containing over 7,000 coupler curves of crank-rocker linkages. Figure 1.23 is a reproduction of a typical page of this atlas. In each case, the length of the crank is a unit length, and the lengths of other links vary from page to page to produce the different combinations. On each page, a number of different coupler points are chosen and their coupler curves are shown. This atlas of coupler curves is invaluable to the designer who needs a linkage to generate a curve with specified characteristics.

The algebraic equation of a coupler curve is, in general, of sixth-order; thus, it is possible to find coupler curves with a wide variety of shapes and many interesting features. Some coupler curves have sections that are nearly straight line segments; others have almost exact circular arc sections; still others have one or more cusps or cross-over themselves like a figure eight. Therefore, it is often not necessary to use a mechanism with a large number of links to obtain a fairly complex motion.

Yet the complexity of the coupler-curve equation is also a hindrance; it means that hand calculation methods can become very cumbersome. Thus, over the years, many mechanisms have been designed by strictly intuitive procedures and proven with cardboard models, without the use of kinematic principles or procedures. Until quite recently, those techniques that did offer a rational approach have been graphical, avoiding the need for tedious computations. Finally, with the availability of digital computers and particularaly with the growth of computer graphics, useful design methods are now emerging that can deal directly with the complicated calculations required without burdening the designer with the computational drudgery (see Section 11.9 for details on some of these methods).

One of the more curious and interesting facts about the coupler curve equation is that the same curve can always be generated by three different four-bar linkages. These are called *cognate linkages*, and the theory is developed in Section 10.10.

Straight-Line Generators In the late 17th century, before the development of the milling machine, it was extremely difficult to machine straight, flat surfaces. For this reason, good prismatic pairs without backlash were not easy to make. During that era, much thought was given to the problem of attaining a straight-line motion as a part of the coupler curve of a linkage having only revolute connections. Probably the best known result of this search is the straight-line mechanism developed by Watt for guiding the piston of early steam engines. Figure 1.24*a* illustrates Watt's linkage to be a four-bar linkage developing an approximate straight line as a part of its coupler curve. Although it does not generate an exact straight line, a good approximation is achieved over a considerable distance of travel.

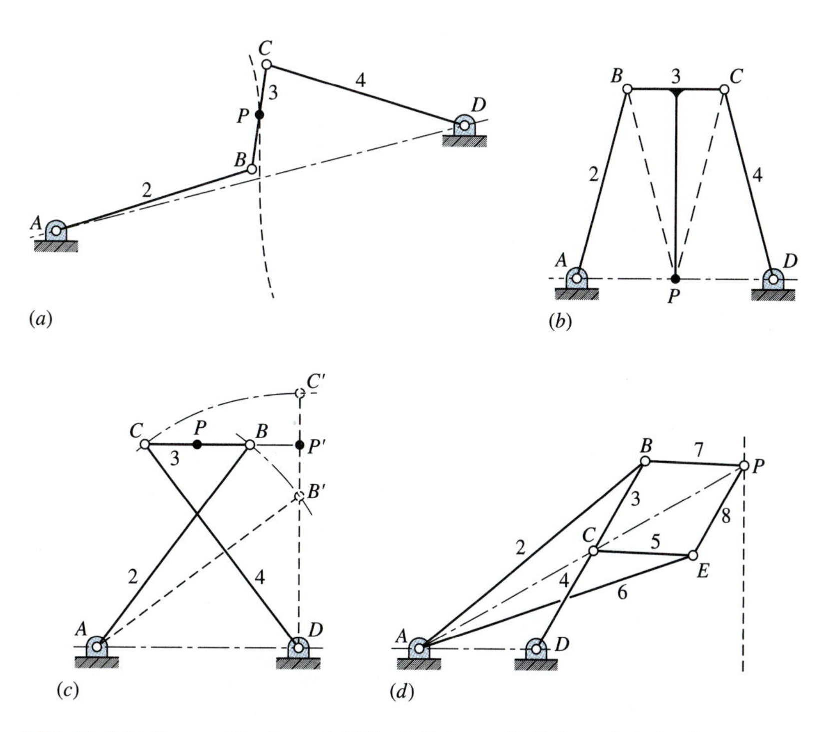

Figure 1.24 Straight-line mechanisms: (*a*) Watt's linkage, (*b*) Roberts' mechanism, (*c*) Chebychev linkage, and (*d*) Peaucillier inversor.

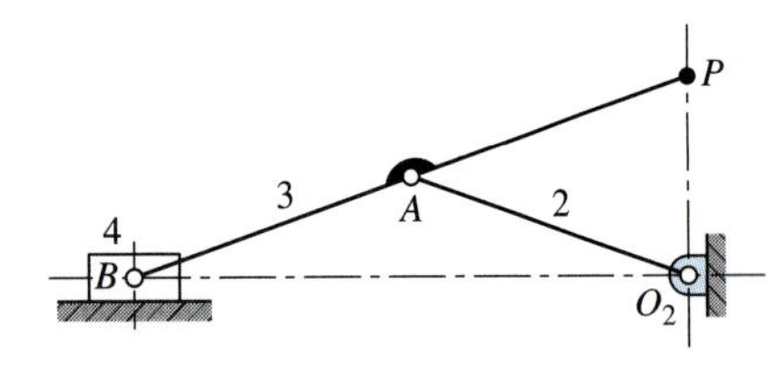

Figure 1.25 Scott–Russell exact straight-line mechanism; $AB = AP = O_2A$.

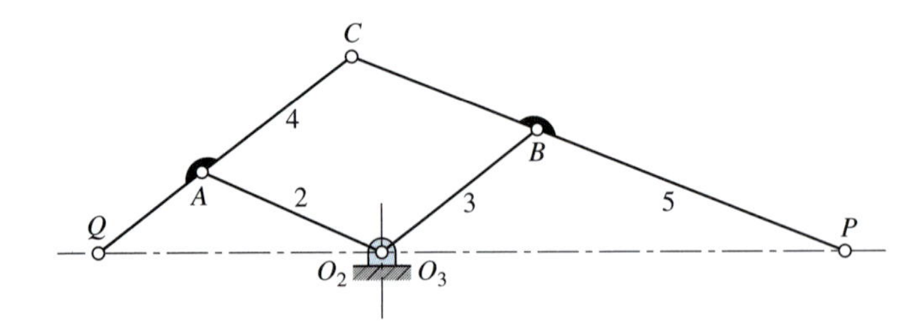

Figure 1.26 The pantograph linkage.

Another four-bar linkage in which the tracing point P generates an approximate straight-line coupler-curve segment is Roberts' mechanism (Fig. 1.24*b*). The dashed lines in Fig. 1.24*b* indicate that the linkage is defined by forming three congruent isosceles triangles; thus, $BC = AD/2$.

The tracing point P of the Chebychev linkage in Fig. 1.24*c* also generates an approximate straight line. The linkage is formed by creating a 3-4-5 triangle with link 4 in the vertical position, as illustrated by the dashed lines; thus, $DB' = 3, AD = 4$, and $AB' = 5$. Because $AB = DC$, $DC' = 5$ and the tracing point P' is the midpoint of link BC. Note that $DP'C$ also forms a 3-4-5 triangle and hence P and P' are two points on a straight line parallel to AD.

Yet another mechanism that generates a straight-line segment is the Peaucillier inversor illustrated in Fig. 1.24*d*. The conditions describing its geometry are that $BC = BP = EC = EP$ and $AB = AE$ such that, by symmetry, points A, C, and P always lie on a straight line passing through A. Under these conditions, $AC \cdot AP = k$, a constant, and the curves generated by C and P are said to be *inverses* of each other. If we place the other fixed pivot D such that $AD = CD$, then point C must trace a circular arc, whereas point P follows an exact straight line. Another interesting property is that if AD is not equal to CD, then point P traces a true circular arc of very large radius.

Figure 1.25 illustrates an exact straight-line mechanism, but note that it employs a slider.

The *pantograph* of Fig. 1.26 is used to trace figures at a larger or smaller size. If, for example, point P traces a map, then a pen at Q will draw the same map at a smaller scale. The dimensions O_2A, AC, CB, and BO_3 must conform to an equal-sided parallelogram.

Torfason also includes robots, speed-changing devices, computing mechanisms, function generators, loading mechanisms, and transportation devices in his classification. Many of these utilize arrangements of mechanisms already presented. Others will appear in some of the chapters to follow.

1.8 KINEMATIC INVERSION

In Section 1.4, we noted that every mechanism has a fixed link called the frame. Until a frame link has been chosen, a connected set of links is called a kinematic chain. When different links are chosen as the frame for a given kinematic chain, the *relative* motions between the various links are not altered, but their *absolute* motions (those measured with

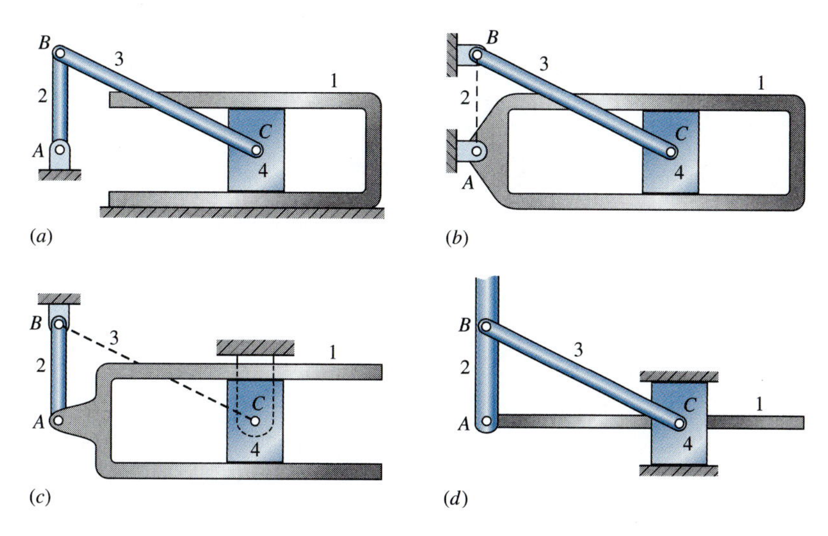

Figure 1.27 Four inversions of the slider-crank mechanism.

respect to the frame link) may be changed dramatically. The process of choosing different links of a chain for the frame is known as *kinematic inversion*.

In an n-link kinematic chain, choosing each link in turn as the frame yields n distinct kinematic inversions of the chain, n different mechanisms. As an example, the four-link slider-crank chain of Fig. 1.27 has four different inversions.

Figure 1.27*a* illustrates the basic slider-crank mechanism found in most internal combustion engines today. Link 4, the piston, is driven by the expanding gases and this movement provides the input; link 2, the crank, rotates as the driven output. The frame is the cylinder block, link 1. By reversing the roles of the input and output, the same mechanism can be used as a compressor.

Figure 1.27*b* illustrates the same kinematic chain; however, it is now inverted and link 2 is stationary. Link 1, formerly the frame, now rotates about the revolute at A. This inversion of the slider-crank mechanism was used as the basis of the rotary engine found in early aircraft.

Another inversion of the same slider-crank chain is illustrated in Fig. 1.27*c*; it has link 3, formerly the connecting rod, as the frame link. This mechanism was used to drive the wheels of early steam locomotives, with link 2 being attached to a wheel.

The fourth and final inversion of the slider-crank chain has the piston, link 4, stationary. Although it is not found in engines, by rotating the figure 90° clockwise this mechanism can be recognized as part of a garden water pump. It will be noted in Fig. 1.27***d*** that the prismatic pair connecting links 1 and 4 is also inverted, that is, the "inside" and "outside" elements of the pair have been reversed.

1.9 GRASHOF'S LAW

A very important consideration when designing a mechanism to be driven by a motor, obviously, is to ensure that the input crank can make a complete revolution. Mechanisms

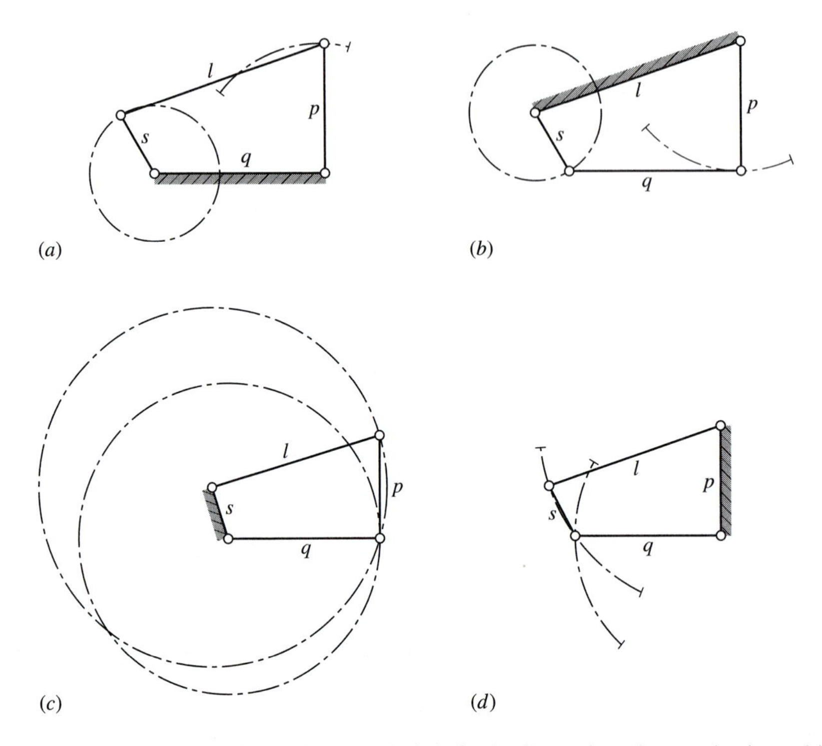

Figure 1.28 Four inversions of the Grashof chain: (*a*, *b*) crank-rocker mechanisms, (*c*) drag-link mechanism, and (*d*) double-rocker mechanism.

in which no link makes a complete revolution would not be useful in such applications. For the four-bar linkage, there is a very simple test of whether this is the case.

Grashof's law states that *for a planar four-bar linkage, the sum of the shortest and longest link lengths cannot be greater than the sum of the remaining two link lengths if there is to be continuous relative rotation between two members*. This is illustrated in Fig. 1.28, where the longest link has length l, the shortest link has length s, and the other two links have lengths p and q. In this notation, Grashof's law states that one of the links, in particular the shortest link, will rotate continuously relative to the other three links if and only if

$$s + l \leq p + q. \tag{1.6}$$

If this inequality is not satisfied, no link will make a complete revolution relative to another.

Attention is called to the fact that nothing in Grashof's law specifies the order in which the links are connected or which link of the four-bar chain is fixed. We are free, therefore, to fix any of the four links. When we do so, we create the four inversions of the four-bar linkage illustrated in Fig. 1.28. All of these fit Grashof's law, and in each the link s makes a complete revolution relative to the other links. The different inversions are distinguished by the location of the link s relative to the fixed link.

If the shortest link s is adjacent to the fixed link, as illustrated in Figs 1.28*a* and 1.28*b*, we obtain what is called a *crank-rocker* linkage. Link s is, of course, the crank because it is able to rotate continuously, and link p, which can only oscillate between limits, is the rocker.

The *drag-link* mechanism, also called the *double-crank* linkage, is obtained by fixing the shortest link s as the frame. In this inversion, illustrated in Fig. 1.28*c*, both links adjacent to s can rotate continuously, and both are properly described as cranks; the shorter of the two is generally used as the input.

Although this is a very common mechanism, you will find it an interesting challenge to devise a practical working model that can operate through the full cycle.

By fixing the link opposite to s, we obtain the fourth inversion, the *double-rocker* mechanism of Fig. 1.28*d*. Note that although link s is able to make a complete revolution, neither link adjacent to the frame can do so; both must oscillate between limits and are therefore rockers.

In each of these inversions, the shortest link s is adjacent to the longest link l. However, exactly the same types of linkage inversions will occur if the longest link l is opposite the shortest link s; you should demonstrate this to your own satisfaction.

Reuleaux approaches the problem somewhat differently but, of course, obtains the same results. In this approach and using Fig. 1.28*a*, the links are named

s, the crank	p, the level
l, the coupler	q, the frame,

where l need not be the longest link. Then the following conditions apply:

$$s + l + p \geq q \tag{1.7}$$

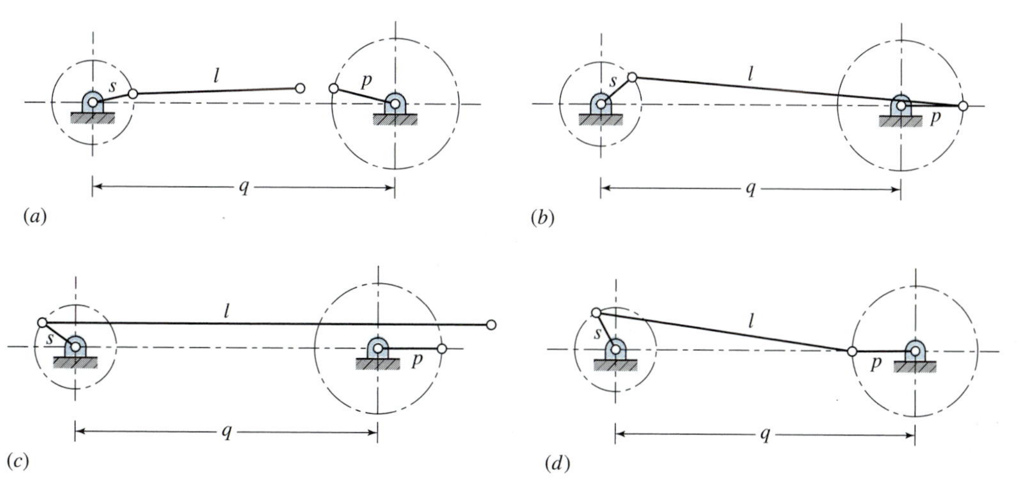

Figure 1.29 (*a*) Eq. (1.7): $s + l + p < q$ and the links cannot be connected; (*b*) Eq. (1.8): $s + l - p > q$ and s is incapable of rotation; (*c*) Eq. (1.9): $s + q + p < l$ and the links cannot be connected; (*d*) Eq. (1.10): $s + q - p < l$ and s is incapable of rotation.

$$s + l - p \leq q \tag{1.8}$$

$$s + q + p \geq l \tag{1.9}$$

$$s + q - p \leq l. \tag{1.10}$$

These four conditions are illustrated in Fig. 1.29 by demonstrating what happens if the conditions are not met.

EXAMPLE 1.6

Determine whether the four-bar linkage illustrated in Fig. 1.30 is a crank–rocker four-bar linkage, a double-rocker four-bar linkage, or a double-crank four-bar linkage.

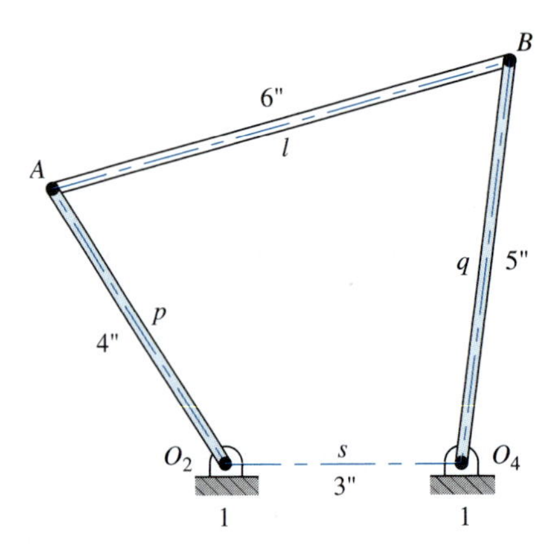

Figure 1.30 Example 1.6. A planar four-bar linkage.

SOLUTION

Substituting the link lengths into Eq. (1.6) gives

$$3 \text{ in} + 6 \text{ in} \leq 4 \text{ in} + 5 \text{ in}$$

or

$$9 \text{ in} \leq 9 \text{ in}.$$

Therefore, the four-bar linkage satisfies Grashof's law; that is, the linkage is a Grashof four-bar linkage. Because the shortest link of the four-bar linkage is grounded, the two links adjacent to the shortest link can both rotate continuously (see Fig. 1.28*c*) and both are properly described as cranks. Therefore, this four-bar linkage is a double-crank or drag–link mechanism. *Ans.*

1.10 MECHANICAL ADVANTAGE

Because of the widespread use of the four-bar linkage, a few remarks are in order here that will help us judge the quality of such a linkage for its intended application. Consider the

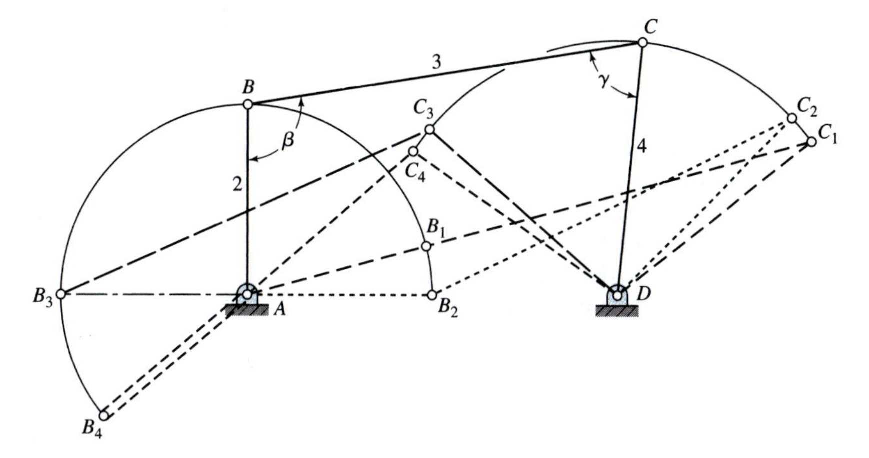

Figure 1.31

four-bar linkage illustrated in Fig. 1.31. Because, according to Grashof's law, this particular linkage is of the crank-rocker variety, it is likely that link 2 is the driver and link 4 is the follower. Link 1 is the frame and link 3 is called the coupler because it couples the motions of the input crank and the output rocker.

The *mechanical advantage* of a linkage is the ratio of the output torque exerted by the driven link to the necessary input torque required at the driver. In Section 3.17, we will prove that the mechanical advantage of the four-bar linkage is directly proportional to the sine of the angle γ between the coupler and the follower and inversely proportional to the sine of the angle β between the coupler and the driver. Of course, both of these angles, and therefore the mechanical advantage, are continuously changing as the linkage moves.

When the sine of the angle β becomes zero, the mechanical advantage becomes infinite; thus, at such a position, only a small input torque is necessary to overcome a very large output torque load. This is the case when the driver AB of Fig. 1.31 is directly in line with the coupler BC; it occurs when the crank is in position AB_1 and again when the crank is in position AB_4. Note that these also define the extreme positions of travel of the rocker DC_1 and DC_4. When the four-bar linkage is in either of these positions, the mechanical advantage is infinite and the linkage is said to be in a *toggle* position.

The angle γ between the coupler and the follower is called the *transmission angle*. As this angle becomes small, the mechanical advantage decreases and even a small amount of friction will cause the mechanism to lock or jam. A common rule of thumb is that a four-bar linkage should not be used to overcome a load in a region where the transmission angle is less than, say, 45° or 50°. The extreme values of the transmission angle occur when the crank AB lies along the line of the frame AD. In Fig. 1.31, the transmission angle is minimal when the crank is in position AB_2 and maximal when the crank has position AB_3. Because of the ease with which it can be visually inspected, the transmission angle has become a commonly accepted measure of the quality of a design of the four-bar linkage.

Note that the definitions of mechanical advantage, toggle, and transmission angle depend on the choice of the driver and driven links. If, in Fig. 1.31, link 4 is used as the driver and link 2 as the follower, the roles of β and γ are reversed. In this case, the linkage has no toggle position, and its mechanical advantage becomes zero when link 2 is in position AB_1 or AB_4 because the transmission angle is then zero. These and other methods of rating the suitability of the four-bar or other linkages are discussed more thoroughly in Section 3.20.

1.11 REFERENCES

[1] Euler, L. Theoria motus corporum solidorum seu rigidorum, *Commentarii Academiae Scientiarum Imperialis Petropolitanae,* 1765. Translated by Willis, R. W. 1841. *Principles of Mechanisms*, John W. Parker, West Strand, London, Cambridge University Press, London.

[2] Hartenberg, R. S., and J. Denavit. *Kinematic Synthesis of Linkages*. New York: McGraw–Hill, 1964, chap. 2.

[3] Kutzbach, K. 1929. Mechanische Leitungsverzweigung, ihre Gesetze und Anwendungen. *Maschinenbau*, 8:710–6.

[4] Grübler, M. F. 1883. Allgemeine Eigenschaften der Zwangläufigen ebenen kinematische Kette: I. *Civilingenieur* 29:167–200.

[5] For an excellent short history of the kinematics of mechanisms, see Hartenberg and Denavit, *Kinematic Synthesis of Linkages*, chap. 1.

[6] See Torfason, L. E. 1990. "A Thesaurus of Mechanisms," in *Mechanical Designer's Notebooks,* Vol. 5, *Mechanisms*, edited by J. E. Shigley and C. R. Mischke. New York: McGraw-Hill, 1990, Chap. 1. Alternatively, see Torfason, L. E. "A Thesaurus of Mechanisms," in *Standard Handbook of Machine Design*, edited by J. E. Shigley and C. R. Mischke. New York: McGraw-Hill, 1986, Chap. 39.

[7] Hrones, J. A., and G. L. Nelson. *Analysis of the Four-Bar Linkage*, Cambridge, MA: Technology Press; New York: Wiley, 1951.

PROBLEMS

1.1 Sketch at least six different examples of the use of a planar four-bar linkage in practice. They can be found in the workshop, in domestic appliances, on vehicles, on agricultural machines, and so on.

†1.2 The link lengths of a planar four-bar linkage are 1, 3, 5, and 5 in. Assemble the links in all possible combinations and sketch the four inversions of each. Do these linkages satisfy Grashof's law? Describe each inversion by name, for example, a crank–rocker mechanism or a drag-link mechanism.

†1.3 A crank-rocker linkage has a 100-mm frame, a 25-mm crank, a 90-mm coupler, and a 75-mm rocker. Draw the linkage and find the maximum and minimum values of the transmission angle. Locate both toggle positions and record the corresponding crank angles and transmission angles.

†1.4 In Fig. P1.4, point C is attached to the coupler; plot its complete path.

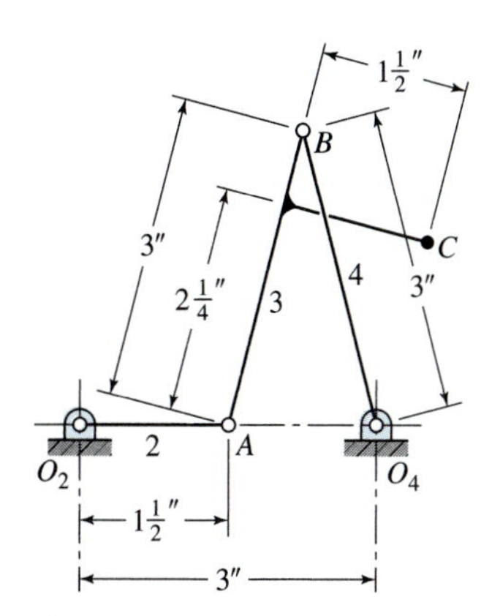

Figure P1.4

†1.5 Find the mobility of each mechanism illustrated in Fig. P1.5.

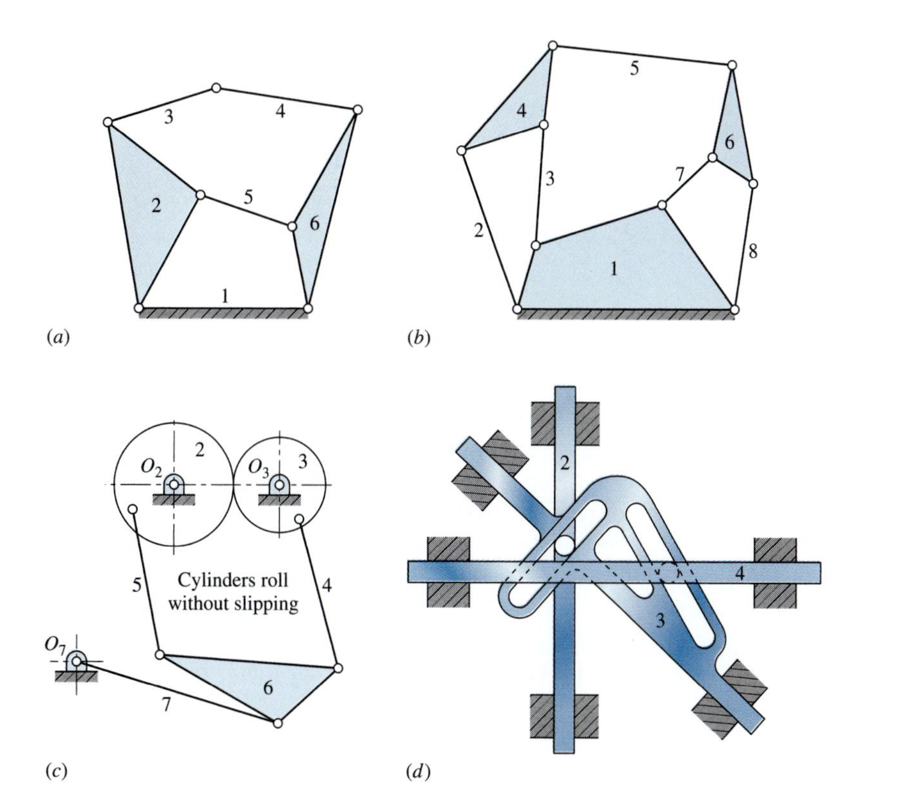

Figure P1.5

1.6 Use the Kutzbach criterion to determine the mobility of the mechanism illustrated in Fig. P1.6.

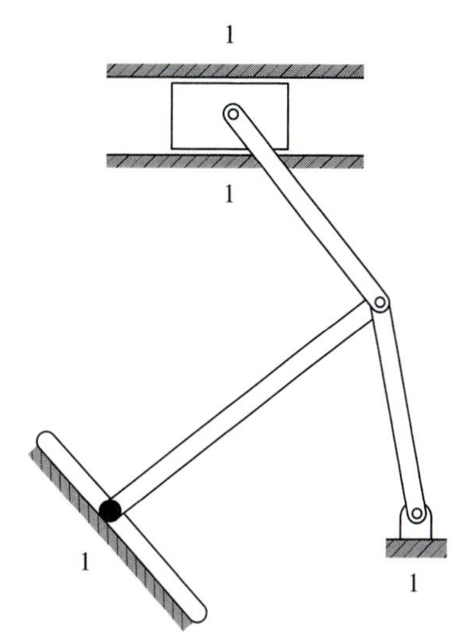

Figure P1.6

1.7 Find a planar mechanism with a mobility of 1 that contains a moving quaternary link. How many distinct variations of this mechanism can you find?

†**1.8** Use the Kutzbach criterion to detemine the mobility of the planar mechanism illustrated in Fig. P1.8. Clearly number each link and label the lower pairs (j_1) and higher pairs (j_2) on Fig. P1.8.

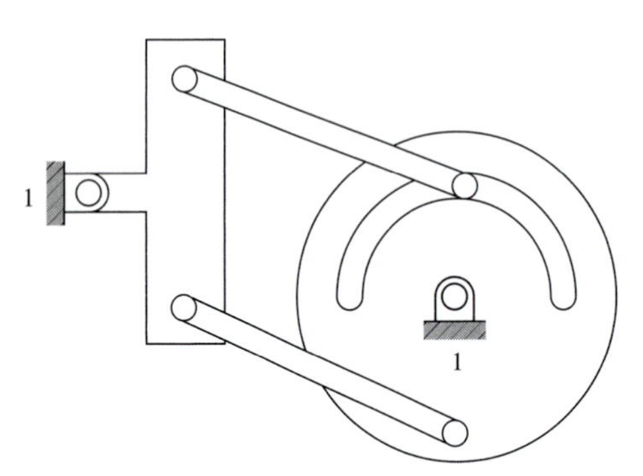

Figure P1.8

†**1.9** For the mechanism illustrated in Fig. P1.9, determine the number of links, the number of lower pairs, and the number of higher pairs. Using the Kutzbach criterion, determine the mobility of the mechanism. Is the answer correct? Briefly explain.

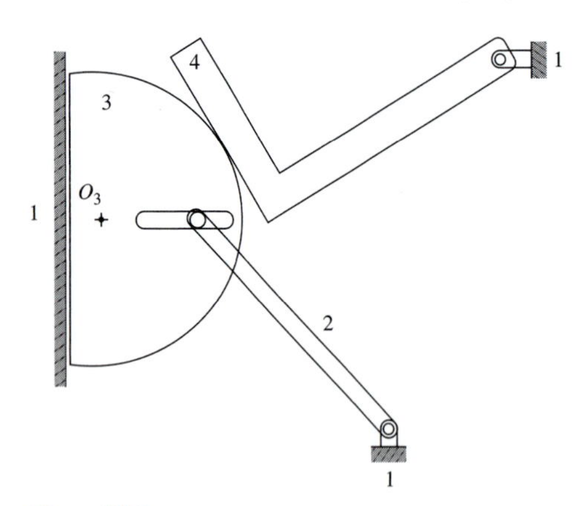

Figure P1.9

1.10 Use the Kutzbach criterion to detemine the mobility of the planar mechanism illustrated in Fig. P1.10. Clearly number each link and label the lower pairs (j_1) and higher pairs (j_2) on Fig. P1.10. Treat rolling contact to mean rolling with no slipping.

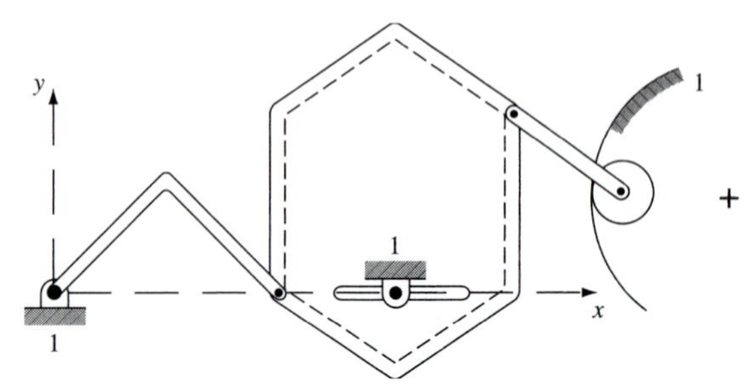

Figure P1.10

1.11 For the mechanism illustrated in Fig. P1.11, determine the number of links, the number of lower pairs, and the number of higher pairs. Using the Kutzbach criterion, determine the mobility. Is the answer correct? Briefly explain.

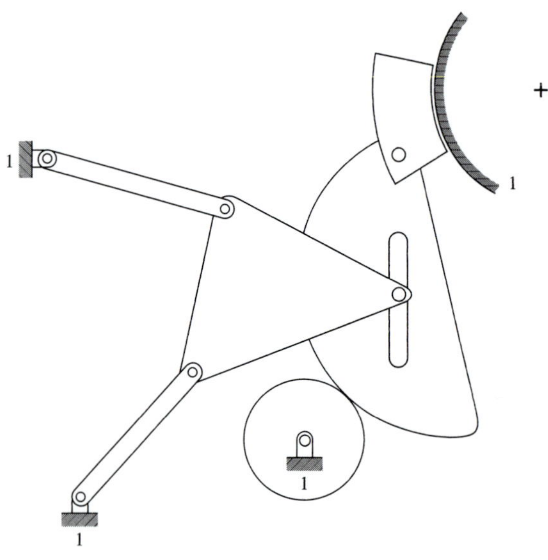

Figure P1.11

1.12 Does the Kutzbach criterion provide the correct result for the planar mechanism illustrated in Fig. P1.12? Briefly explain why or why not.

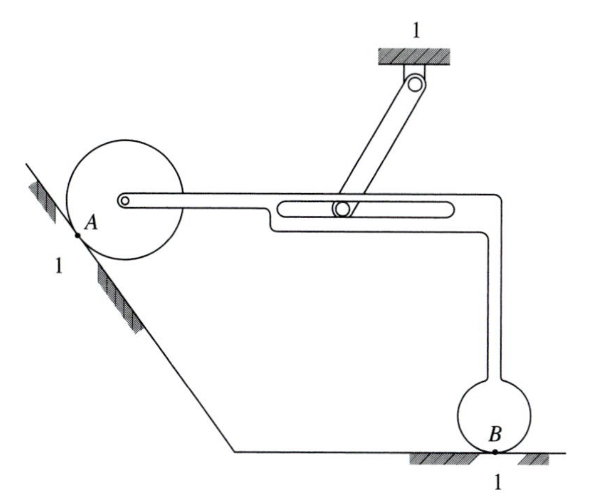

Figure P1.12

1.13 The mobility of the mechanism illustrated in Fig. P1.13 is $m = 1$. Use the Kutzbach criterion to determine the number of lower pairs and the number of higher pairs. Is the wheel rolling without slipping, or rolling and slipping, at point A on the wall?

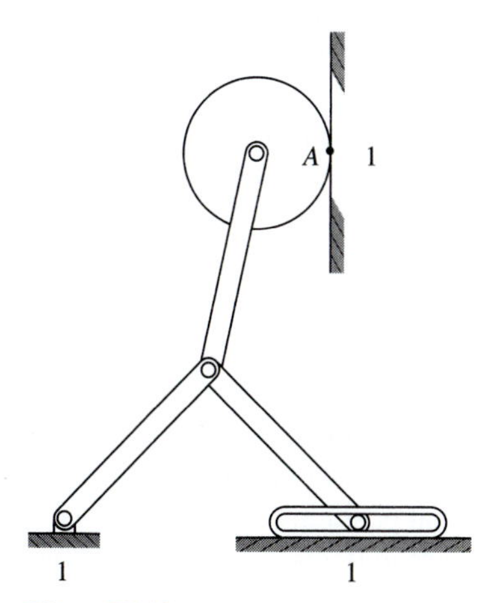

Figure P1.13

1.14 Devise a practical working model of the drag–link mechanism.

†1.15 Find the time ratio of the linkage of Problem 1.3.

†1.16 Plot the complete coupler curve of Roberts' mechanism illustrated in Fig. 1.24*b*. Use $AB = CD = AD = 2.5$ in. and $BC = 1.25$ in.

1.17 If the crank of Fig. 1.11 is turned 10 revolutions clockwise, how far and in what direction does the carriage move?

†1.18 Show how the mechanism of Fig. 1.15*b* can be used to generate a sine wave.

†1.19 Devise a crank-and-rocker mechanism, as in Fig. 1.14*c*, having a rocker angle of 60°. The rocker length is to be 0.50 m.

1.20 A crank–rocker four-bar linkage is required to have a time ratio $Q = 1.2$. The rocker is to have a length of 2.5 in. and oscillate through a total angle of 60°. Determine a suitable set of link lengths for the remaining three links of the four-bar linkage.

2 Position and Displacement

In analyzing motion, the first and most basic problem encountered is that of defining and dealing with the concepts of position and displacement. Because motion can be thought of as a time series of displacements between successive positions, it is important to understand exactly the meaning of the term *position*; rules or conventions must be established to make the definition precise.

Although many of the concepts in Chapter 2 may appear intuitive and almost trivial, many subtleties are explained here that are required for an understanding of the next several chapters.

2.1 LOCUS OF A MOVING POINT

In speaking of the position of a particle or point, we are really answering the question: Where is the point or what is its location? We are speaking of something that exists in nature and we are posing the question of how to express this (in words or symbols or numbers) in such a way that the meaning is clear. We soon discover that position cannot be defined on a truly absolute basis. We must define the position of a point in terms of some agreed-upon frame of reference, some reference coordinate system.

As illustrated in Fig. 2.1, once we have agreed upon the xyz coordinate system as the frame of reference, we can say that point P is located x units along the x axis, y units along the y axis, and z units along the z axis *from the origin* O. In this very statement we see that three vitally important parts of the definition depend on the existence of the reference coordinate system:

1. The *origin* of coordinates O provides an agreed-upon location from which to measure the location of point P.

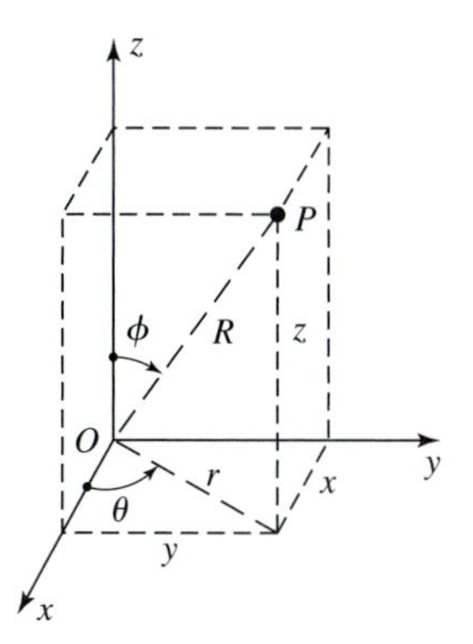

Figure 2.1 A right-hand three-dimensional coordinate system illustrating how point P is located algebraically.

2. The *coordinate axes* provide agreed-upon *directions* along which the measurements are to be made; they also provide known lines and planes for the definition and measurement of angles.
3. The unit distance along any of the axes provides a scale for quantifying distances.

These observations are not restricted to the Cartesian coordinates (x, y, z) of point P. All three properties of the coordinate system are also necessary in defining the cylindrical coordinates (r, θ, z), spherical coordinates (R, θ, ϕ), or any other coordinates of point P. The same properties would also be required if point P were restricted to remain in a single plane and a two-dimensional coordinate system were used. No matter how it is defined, the concept of the position of a point cannot be related without the definition of a reference coordinate system.

We note in Fig. 2.1 that the direction cosines for locating point P are

$$\cos\alpha = \frac{x}{R}, \quad \cos\beta = \frac{y}{R}, \quad \cos\gamma = \frac{z}{R}, \tag{2.1}$$

where the angles α, β, and γ are, respectively, the angles measured from the positive coordinate axes to the directed line OP.

One means of expressing the motion of a point or particle is to define its components along the reference axes as functions of some parameter such as time:

$$x = x(t), \quad y = y(t), \quad z = z(t). \tag{2.2}$$

If these relations are known, the position of P can be found for any time t. This is the general case for the motion of a particle and is illustrated in the following example.

EXAMPLE 2.1

Describe the motion of a particle P whose position changes with time according to the equations $x = a\cos 2\pi t$, $y = a\sin 2\pi t$, and $z = bt$.

SOLUTION

Substitution of values for t from 0 to 2 gives values as indicated in Table 2.1. As illustrated in Fig. 2.2, the point moves with *helical motion* with radius a around the z axis and with a lead of b. Note that if $b = 0$, then $z(t) = 0$, the moving particle is confined to the xy plane, and the motion is a circle with its center at the origin.

TABLE 2.1 Example 2.1

t	x	y	z
0	a	0	0
1/4	0	a	$b/4$
1/2	$-a$	0	$b/2$
3/4	0	$-a$	$3b/4$
1	a	0	b
5/4	0	a	$5b/4$
3/2	$-a$	0	$3b/2$
7/4	0	$-a$	$7b/4$
2	a	0	$2b$

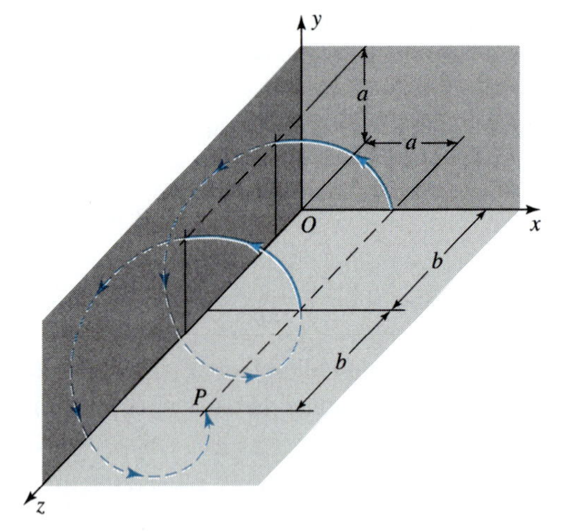

Figure 2.2 Helical motion of a particle.

We have used the words *particle* and *point* interchangeably. When we use the word *point*, we have in mind something that has no dimensions, that is, something with zero length, zero width, and zero thickness. When the word *particle* is used, we have in mind something whose dimensions are small and unimportant, that is, a tiny material body whose dimensions are negligible, a body small enough for its dimensions to have no effect on the analysis to be performed.

The successive positions of a moving point define a line or curve. This curve has no thickness because the point has no dimensions. However, the curve does have length because the point occupies different positions as time changes. This curve, representing the successive positions of the point, is called the *path* or *locus* of the moving point in the reference coordinate system.

If three coordinates are necessary to describe the path of a moving point, the point is said to have *spatial motion*. If the path can be described by only two coordinates—that is, if the coordinate axes can be chosen such that one coordinate is always zero or constant—the path is contained in a single plane and the point is said to have *planar motion*. Sometimes it happens that the path of a point can be described by a single coordinate. This means that two of its spatial position coordinates can be taken as zero or constant. In this case the point moves in a straight line and is said to have *rectilinear motion*.

In each of the three cases described, it is assumed that the coordinate system is chosen so as to obtain the smallest number of coordinates necessary to describe the motion of the point. Thus, the description of rectilinear motion requires one coordinate, a point whose path is a *plane curve* requires two coordinates, and a point whose locus is a *space curve*, sometimes called a *skew curve*, requires three position coordinates.

2.2 POSITION OF A POINT

The physical process involved in observing the position of a point, as illustrated in Fig. 2.3, implies that the observer is actually keeping track of the relative location of two points, P and O, by looking at both, performing a mental comparison, and recognizing that point P has a certain location *with respect to point* O. In this determination two properties are noted, the distance from O to P (based on the unit distance or grid size of the reference coordinate system) and the *relative* angular orientation of the line OP in the coordinate system. These two properties, magnitude and direction, are precisely the properties required for a vector. Therefore, we can also define the *position of a point* as the *vector from the origin of a*

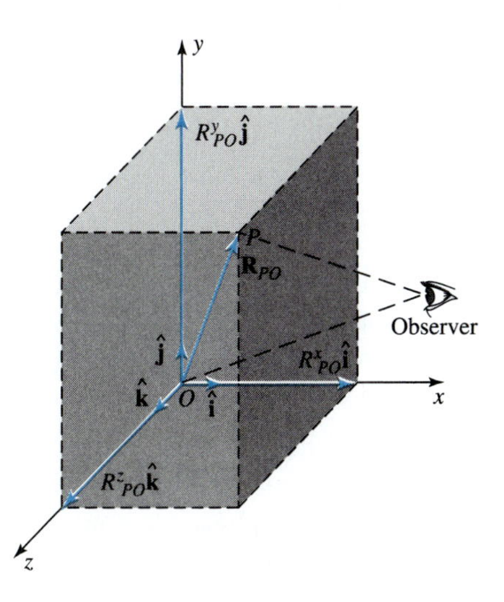

Figure 2.3 The position of a point defined by vectors.

specified reference coordinate system to the point. We choose the symbol $\mathbf{R}_{PO}$ to denote the vector position of point P relative to point O, which is read *the position of P with respect to O*.*

The reference system is, therefore, related in a very special way to what is seen by a specific observer. What is the relationship? What properties must this coordinate system have to ensure that position measurements made with respect to it are actually those of the observer? The key is that the coordinate system *must* be stationary with respect to this particular observer. To phrase it in another way, the observer is always *stationary in* this reference system. This means that if the observer moves, the coordinate system moves too—through a rotation, a distance, or both. If objects or points are fixed in this coordinate system, then these objects always appear stationary to the observer regardless of what movements the observer (and the reference system) may execute. Their positions with respect to the observer do not change, and hence their position vectors remain unchanged.

The actual location of the observer within the frame of reference has no meaning because the positions of observed points are always defined with respect to the origin of the coordinate system.

Often it is convenient to express the position vector in terms of its components along the coordinate axes:

$$\mathbf{R}_{PO} = R^x_{PO}\hat{\mathbf{i}} + R^y_{PO}\hat{\mathbf{j}} + R^z_{PO}\hat{\mathbf{k}}, \tag{2.3}$$

where superscripts are used to denote the direction of each component. As in the remainder of this text, $\hat{\mathbf{i}}, \hat{\mathbf{j}}$, and $\hat{\mathbf{k}}$ are used to designate unit vectors in the x, y, and z axis directions, respectively.

Whereas vectors are denoted throughout the book by boldface symbols, the scalar magnitude of a vector is signified by the same symbol in italics, without boldface.

The magnitude of the position vector, for example, is

$$R_{PO} = |\mathbf{R}_{PO}| = \sqrt{\mathbf{R}_{PO} \cdot \mathbf{R}_{PO}} = \sqrt{(R^x_{PO})^2 + (R^y_{PO})^2 + (R^z_{PO})^2}. \tag{2.4}$$

The unit vector in the direction of $\mathbf{R}_{PO}$ is denoted by the same boldface symbol with a caret, that is

$$\hat{\mathbf{R}}_{PO} = \frac{\mathbf{R}_{PO}}{R_{PO}}. \tag{2.5}$$

A distinction can be made between the direction of a line and the orientation of a directed line, that is, a line which is assigned a positive or a negative sense. In general, a vector has a magnitude, a direction, and a sense. The sense defines the positive or negative attribute of the vector and can be used to distinguish between the direction of a vector and its orientation.

---

* Note that we do not use the slash notation of some other texts, that is, $\mathbf{R}_{PO} \neq \mathbf{R}_{P/O}$. The slash notation is reserved for a different meaning and will be described in Section 2.4.

2.3 POSITION DIFFERENCE BETWEEN TWO POINTS

We now investigate the relationship between the position vectors of two different points. The situation is illustrated in Fig. 2.4. In the preceding section, we reported that an observer fixed in the *xyz* coordinate system would observe the positions of points P and Q by comparing each with the position of the origin. The positions of the two points are defined by the vectors $\mathbf{R}_{PO}$ and $\mathbf{R}_{QO}$.

Inspection of Fig. 2.4 illustrates that these two vectors are related by a third vector, $\mathbf{R}_{PQ}$, the *position difference* between points P and Q. Figure 2.4 illustrates the relationship to be

$$\mathbf{R}_{PQ} = \mathbf{R}_{PO} - \mathbf{R}_{QO}. \tag{2.6}$$

The physical interpretation is now slightly different from that of the position vector itself. The observer is no longer defining the position of P with respect to the origin but is now defining the position of P with respect to the position of Q. Put another way, the position of point P is being defined as if it were in another coordinate system $x'y'z'$ with origin at Q and axes directed parallel to those of the basic reference frame *xyz*, as illustrated in Fig. 2.4. Either point of view can be used for the interpretation, but we should understand both of them because we shall use both in future developments. Finally, it is worth remarking that having the axes of $x'y'z'$ parallel to those of *xyz* is only a matter of convenience and not a necessary condition; it is only necessary that $x'y'z'$ does not rotate with respect to *xyz*. This parallelism is used throughout the book, however, because it causes no loss of generality and it simplifies the visualization when the coordinate systems are in motion.

Having now generalized our concept of relative position to include the position difference between any two points, we reflect again on the above discussion of the position vector itself. We note that it is merely the special case where we agree to measure using the origin of coordinates as the second point. Thus, to be consistent in notation, we have denoted the position vector of a single point P by the dual subscripted symbol $\mathbf{R}_{PO}$. However, in the interest of brevity, we will henceforth agree that when the second subscript is not given explicitly, it is understood to be the origin of the observer's coordinate system,

$$\mathbf{R}_{P} = \mathbf{R}_{PO}. \tag{2.7}$$

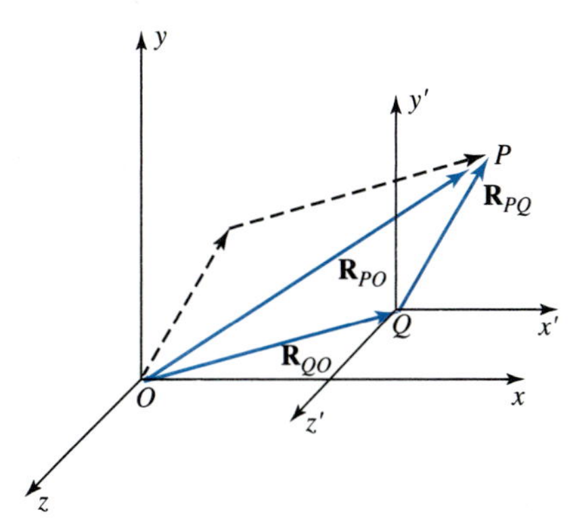

Figure 2.4 Definition of the position-difference vector $\mathbf{R}_{PQ}$.

2.4 APPARENT POSITION OF A POINT

Up to now, in discussing the position vector our point of view has been entirely that of a single observer in a single coordinate system. However, it is sometimes desirable to make observations in a secondary coordinate system, that is, as seen by a second observer in a different coordinate system, and then to convert this information into the basic coordinate system. Such a situation is illustrated in Fig. 2.5.

If two observers (denoted 1 and 2), one using the reference frame $x_1y_1z_1$ and the other using $x_2y_2z_2$, were both asked to give the location of a particle at P, they would report different results. Observer 1 in coordinate system $x_1y_1z_1$ would observe the vector $\mathbf{R}_{PO_1}$, whereas observer 2, using the $x_2y_2z_2$ coordinate system, would report the position vector $\mathbf{R}_{PO_2}$. We note from Fig. 2.5 that these vectors are related by

$$\mathbf{R}_{PO_1} = \mathbf{R}_{O_2O_1} + \mathbf{R}_{PO_2}. \tag{2.8}$$

The difference in the positions of the two origins is not the only incompatibility between the two observations of the position of point P. Because the two coordinate systems are not aligned, the two observers would be using different reference lines for their measurements of direction; observer 1 would report components measured along the $x_1y_1z_1$ axes, whereas observer 2 would measure in the $x_2y_2z_2$ directions.

A third and very important distinction between these two observations becomes clear when we consider that the two coordinate systems may be moving with respect to each other. Whereas point P may appear stationary with respect to one observer, it may be in motion with respect to the other observer; that is, the position vector $\mathbf{R}_{PO_1}$ may appear constant to observer 1, whereas $\mathbf{R}_{PO_2}$ appears to vary as seen by observer 2.

When any of these conditions exists, it will be convenient to add an additional subscript to our notation that will distinguish which observer is being considered. When we are considering the position of P as seen by observer 1 using coordinate system $x_1y_1z_1$, we denote this by the symbol $\mathbf{R}_{PO_1/1}$ or, because O_1 is the origin for this observer,* by $\mathbf{R}_{P/1}$. The observations made by observer 2, done in coordinate system $x_2y_2z_2$, are denoted $\mathbf{R}_{PO_2/2}$

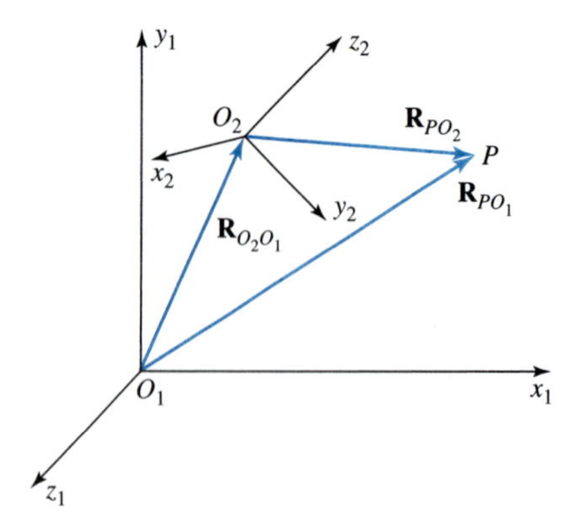

Figure 2.5 Definition of the apparent-position vector $\mathbf{R}_{PO_2}$ of point P.

* Note that $\mathbf{R}_{PO_2/1}$ cannot be abbreviated as $\mathbf{R}_{P/1}$ because O_2 is not the origin used by observer 1.

or $\mathbf{R}_{P/2}$. With this extension of the notation, Eq. (2.8) can be written as

$$\mathbf{R}_{P/1} = \mathbf{R}_{O_2/1} + \mathbf{R}_{P/2}. \tag{2.9}$$

We refer to $\mathbf{R}_{P/2}$ as *the apparent position* of point P to an *observer in coordinate system* 2, and we note that it is by no means equal to the apparent position vector $\mathbf{R}_{P/1}$ seen by observer 1.

We have now made note of certain intrinsic differences between $\mathbf{R}_{P/1}$ and $\mathbf{R}_{P/2}$ and found Eq. (2.9) to relate them. However, there is no reason why components of either vector must be taken along the natural axes of the observer's coordinate system. As with all vectors, components can be found along any desired set of axes.

In applying the apparent-position equation, Eq. (2.9), we must use a single consistent set of axes during the numerical evaluation. Although the observer in coordinate system 2 may find it most natural to measure the components of $\mathbf{R}_{P/2}$ along the $x_2y_2z_2$ axes, these components must be transformed into the equivalent components in the $x_1y_1z_1$ system before the addition is actually performed, that is

$$\begin{aligned}\mathbf{R}_{P/1} &= \mathbf{R}_{O_2/1} + \mathbf{R}_{P/2} \\ &= R^{x_1}_{O_2/1}\hat{\mathbf{i}}_1 + R^{y_1}_{O_2/1}\hat{\mathbf{j}}_1 + R^{z_1}_{O_2/1}\hat{\mathbf{k}}_1 + R^{x_1}_{P/2}\hat{\mathbf{i}}_1 + R^{y_1}_{P/2}\hat{\mathbf{j}}_1 + R^{z_1}_{P/2}\hat{\mathbf{k}}_1 \\ &= (R^{x_1}_{O_2/1} + R^{x_1}_{P/2})\hat{\mathbf{i}}_1 + (R^{y_1}_{O_2/1} + R^{y_1}_{P/2})\hat{\mathbf{j}}_1 + (R^{z_1}_{O_2/1} + R^{z_1}_{P/2})\hat{\mathbf{k}}_1 \\ &= R^{x_1}_{P/1}\hat{\mathbf{i}}_1 + R^{y_1}_{P/1}\hat{\mathbf{j}}_1 + R^{z_1}_{P/1}\hat{\mathbf{k}}_1.\end{aligned}$$

The addition can be performed equally well if all vector components are transformed into the $x_2y_2z_2$ system or, for that matter, into any other consistent set of directions. However, they cannot be added algebraically when they have been evaluated along inconsistent axes. The additional subscript in the apparent position vector, therefore, does not specify a set of directions to be used in the evaluation of components; it merely states the coordinate system in which the vector is defined, the coordinate system in which the observer is stationary.

2.5 ABSOLUTE POSITION OF A POINT

We now turn our attention to the meaning of the term absolute position. We saw in Section 2.2 that every position vector is defined relative to a second point, the origin of the observer's coordinate reference frame. It is one particular case of the position-difference vector studied in Section 2.3, where the reference point is the origin of coordinates.

In Section 2.4, we noted that it may be convenient in certain problems to consider the apparent positions of a single point as viewed by more than one observer using different coordinate systems. When a particular problem leads us to consider multiple coordinate systems, however, the application will lead us to single out one of the coordinate systems as primary or most basic. Most often this is the coordinate system in which the final result is to be expressed, and this coordinate system is usually considered stationary. It is referred to as the *absolute coordinate system*. The absolute position of a point is defined as its apparent position as seen by an observer in the *absolute coordinate system*.

Which coordinate system is designated absolute (most basic) is an arbitrary decision and is unimportant in the study of kinematics. Whether the absolute coordinate system is truly stationary is also a moot point because, as we have seen, all position (and motion) information is measured relative to something else; nothing is truly absolute in the strict sense. When analyzing the kinematics of an automobile suspension, for example, it may be convenient to choose an "absolute" coordinate system attached to the frame of the car and to study the motion of the suspension relative to this. It is then unimportant whether the car is moving; motions of the suspension relative to the frame would be defined as absolute.

It is common convention to number the absolute coordinate system 1 and to use other numbers for other moving coordinate systems. Because we adopt this convention throughout this text, absolute-position vectors are those apparent-position vectors viewed by an observer in coordinate system 1 and carry symbols of the form $\mathbf{R}_{P/1}$. In the interest of brevity and to reduce complexity, we will agree that when the coordinate system number is not given explicitly it is assumed to be 1; thus, $\mathbf{R}_{P/1}$ can be abbreviated $\mathbf{R}_P$. Similarly, the apparent-position equation (2.9) can be written* as

$$\mathbf{R}_P = \mathbf{R}_{O_2} + \mathbf{R}_{P/2}. \tag{2.10}$$

EXAMPLE 2.2

The path of a moving point is defined by the equation $y = 2x^2 - 28$. Find the position difference from point P to point Q when $R_P^x = 4$ and $R_Q^x = -3$.

SOLUTION

The y components of the two vectors can be written as

$$R_P^y = 2(4)^2 - 28 = 4 \quad \text{and} \quad R_Q^y = 2(-3)^2 - 28 = -10.$$

Therefore, the two vectors can be written as

$$\mathbf{R}_P = 4\hat{\mathbf{i}} + 4\hat{\mathbf{j}} \quad \text{and} \quad \mathbf{R}_Q = -3\hat{\mathbf{i}} - 10\hat{\mathbf{j}}.$$

The position difference from point P to point Q is

$$\mathbf{R}_{QP} = \mathbf{R}_Q - \mathbf{R}_P = -7\hat{\mathbf{i}} - 14\hat{\mathbf{j}} = 15.65\angle 243.4^\circ. \qquad \textit{Ans.}$$

2.6 THE LOOP-CLOSURE EQUATION

Our discussion of the position-difference and apparent-position vectors has been quite abstract so far, with the intent being to develop a rigorous foundation for the analysis of

* Reviewing Sections 2.1 through 2.3 will verify that the position difference vector $\mathbf{R}_{PQ}$ was treated entirely from the absolute coordinate system and is an abbreviation of the notation $\mathbf{R}_{PQ/1}$. We have no need to treat the completely general case $\mathbf{R}_{PQ/2}$, the apparent-position difference vector.

motion in mechanical systems. Certainly precision is not without merit, because it is this rigor that permits science to predict a correct result despite the personal prejudices and emotions of the analyst. However, tedious developments are not interesting unless they lead to applications in real-life problems. Although many fundamental principles are yet to be discovered, it may be worthwhile at this point to illustrate the relationship between the relative-position vectors discussed above and some of the typical linkages met in real machines.

As indicated in Chapter 1, one of the most common and most useful of all mechanisms is the planar four-bar linkage. A practical example of this linkage is the clamping device illustrated in Fig. 2.6. A brief study of the assembly drawing indicates that, as the handle of the clamp is lifted, the clamping bar swings away from the clamping surface, thereby opening the clamp. As the handle is pressed, the clamping bar swings down and the clamp closes again. If we wish to design such a clamp accurately, however, things are not quite so simple. It may be desirable, for example, for the clamp to open at a given rate for a certain rate of lift of the handle. Such relationships are not obvious; they depend on the exact dimensions of the various parts and the relationships or interactions between the parts. To

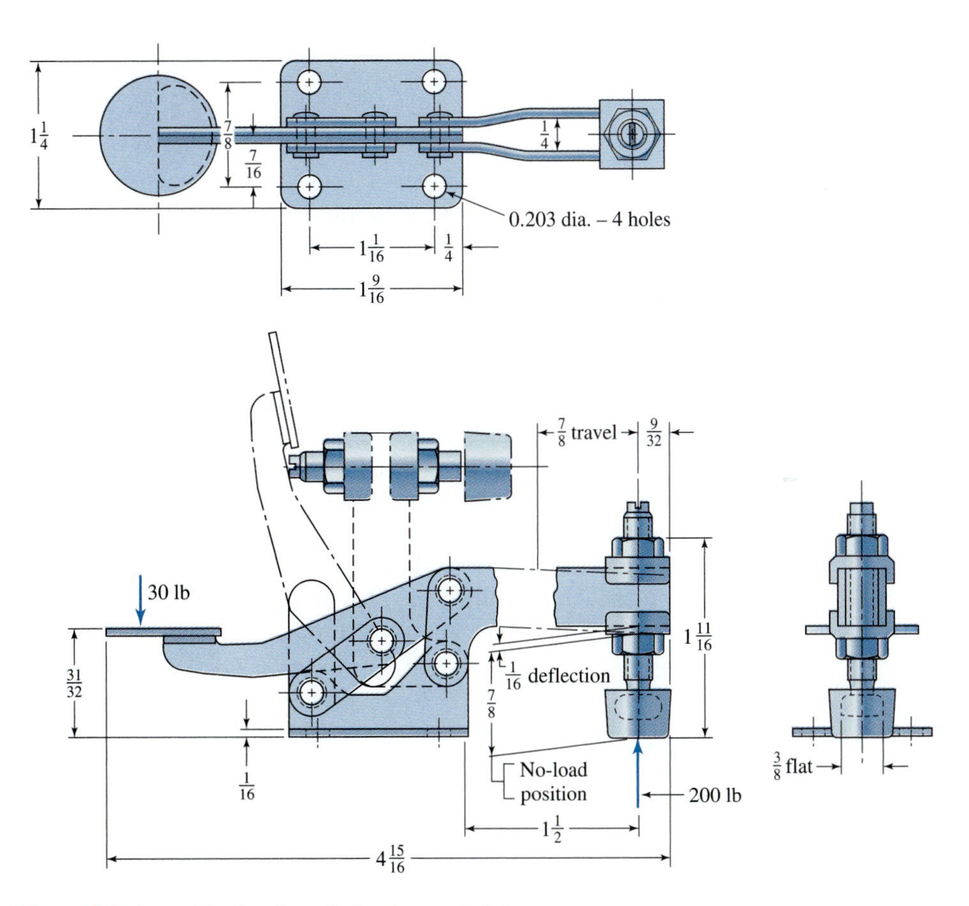

Figure 2.6 Assembly drawing of a hand-operated clamp.

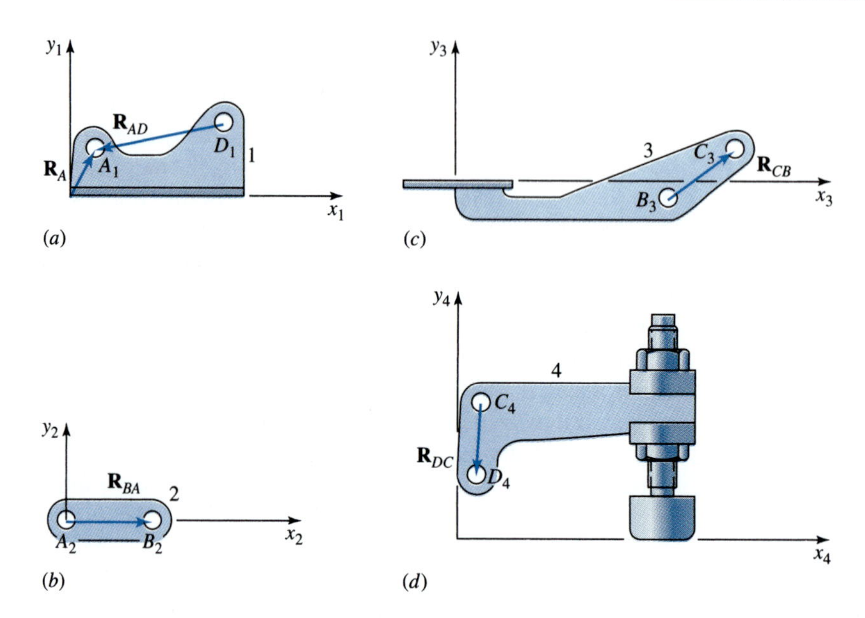

Figure 2.7 Detail drawings of the clamp of Fig. 2.6: (*a*) frame link, (*b*) connecting link, (*c*) handle, and (*d*) clamping bar.

discover these relationships, a rigorous description of the essential features of the device is required. The position-difference and apparent-position vectors can be used to provide such a description.

Figure 2.7 illustrates the detail drawings of the individual links of the disassembled clamp. Although not shown, the detail drawings would be completely dimensioned, thus fixing once and for all the complete geometry of each link. The assumption that each is a rigid link ensures that the position of any point on any one of the links can be determined precisely relative to any other point on the same link by simply identifying the proper points and scaling the appropriate detail drawing.

The features that are lost in the detail drawings, however, are the interrelationships between the individual parts, that is, the constraints that ensure each link will move relative to its neighbors in the prescribed fashion. These constraints are, of course, provided by the four pinned joints. Anticipating that they will be of importance in any description of the linkage, we label these pin centers A, B, C, and D and we identify the appropriate points on link 1 as A_1 and D_1, those on link 2 as A_2 and B_2, and so on. As illustrated in Fig. 2.7, we also pick a different coordinate system rigidly attached to each link.

Because it is necessary to relate the relative positions of the successive joint centers, we define the position difference vectors $\mathbf{R}_{AD}$ on link 1, $\mathbf{R}_{BA}$ on link 2, $\mathbf{R}_{CB}$ on link 3, and $\mathbf{R}_{DC}$ on link 4. We note again that each of these vectors appears constant to an observer fixed in the coordinate system of that particular link; the magnitudes of these vectors can be obtained from the constant dimensions of the links.

A vector equation can also be written to describe the constraints provided by each of the revolute (pinned) joints. Note that no matter which position or which observer is chosen, the two points describing each pin center—for example, A_1 and A_2—remain coincident. Thus,

$$\mathbf{R}_{A_2A_1} = \mathbf{R}_{B_3B_2} = \mathbf{R}_{C_4C_3} = \mathbf{R}_{D_1D_4} = \mathbf{0}. \tag{2.11}$$

Let us now develop vector equations for the absolute position of each of the pin centers. Because link 1 is the frame, absolute positions are those defined relative to an observer in coordinate system 1. Point A_1 is, of course, at the position described by $\mathbf{R}_A$. Next, we mathematically connect link 2 to link 1 by writing

$$\mathbf{R}_{A_2} = \mathbf{R}_{A_1} + \cancelto{0}{\mathbf{R}_{A_2A_1}} = \mathbf{R}_A. \tag{a}$$

Transferring to the other end of link 2, we attach link 3. Therefore

$$\mathbf{R}_B = \mathbf{R}_A + \mathbf{R}_{BA}. \tag{b}$$

Connecting joints C and D in the same manner, we obtain

$$\mathbf{R}_C = \mathbf{R}_B + \mathbf{R}_{CB} = \mathbf{R}_A + \mathbf{R}_{BA} + \mathbf{R}_{CB} \tag{c}$$

$$\mathbf{R}_D = \mathbf{R}_C + \mathbf{R}_{DC} = \mathbf{R}_A + \mathbf{R}_{BA} + \mathbf{R}_{CB} + \mathbf{R}_{DC}. \tag{d}$$

Finally, we transfer back across link 1 to point A,

$$\mathbf{R}_A = \mathbf{R}_D + \mathbf{R}_{AD} = \mathbf{R}_A + \mathbf{R}_{BA} + \mathbf{R}_{CB} + \mathbf{R}_{DC} + \mathbf{R}_{AD}, \tag{e}$$

and from this we obtain

$$\mathbf{R}_{BA} + \mathbf{R}_{CB} + \mathbf{R}_{DC} + \mathbf{R}_{AD} = \mathbf{0}. \tag{2.12}$$

This important equation is called the *loop-closure* equation for the clamp. As illustrated in Fig. 2.8, it expresses the fact that the mechanism forms a closed loop. Therefore, the polygon formed by the position-difference vectors through successive links and joints must remain closed as the mechanism moves. The constant lengths of these vectors ensure that the joint centers remain separated by constant distances, the requirement for rigid links. The rotations between successive vectors indicate the motions within the pinned joints, whereas the rotation of each individual position-difference vector illustrates the rotational

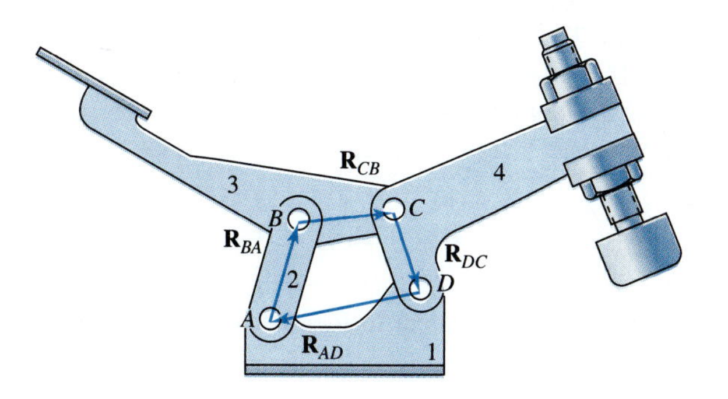

Figure 2.8 The loop-closure equation.

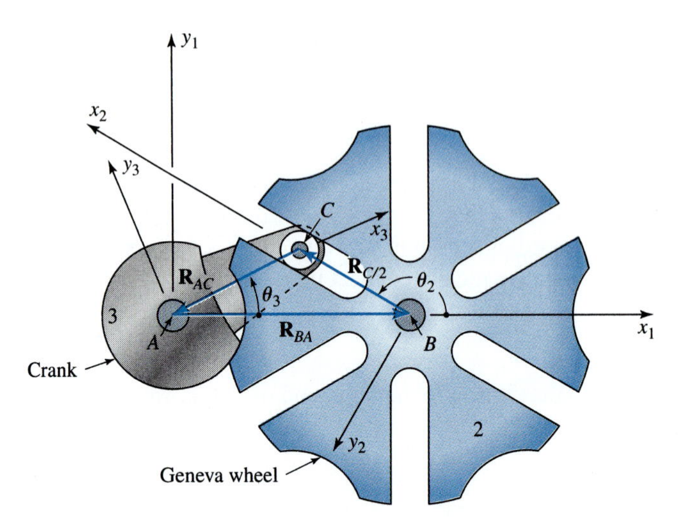

Figure 2.9 The Geneva mechanism or Maltese cross.

motion of a particular link. Thus, the loop-closure equation holds within it all the important constraints that determine how this particular clamp operates. This forms a mathematical description, or *model*, of the linkage, and many of the later developments in this book are based on this model as a starting point.

Of course, the form of the loop-closure equation depends on the type of linkage. This is demonstrated by another example, the *Geneva* mechanism or *Maltese* cross, illustrated in Fig. 2.9. One early application of this mechanism was to prevent overwinding a watch. Today the mechanism finds wide use as an indexing device, for example, in a milling machine with an automatic tool changer.

Although the frame of the mechanism, link 1, is not illustrated in Fig. 2.9, it is an important part of the mechanism because it holds the two shafts with centers A and B a constant distance apart. Thus, we define the vector $\mathbf{R}_{BA}$ to show this dimension. The left crank, link 3, is attached to a shaft, usually rotated at a constant speed, and carries a roller at C, running in the slot of the Geneva wheel. The vector $\mathbf{R}_{AC}$ has a constant magnitude equal to the crank length, the distance from the center of the roller C to the shaft center A. The rotation of this vector relative to link 1 will be used later to describe the angular speed of the crank. The x_2 axis is aligned along one slot of the wheel; thus, the roller is constrained to ride along this slot. The vector $\mathbf{R}_{C/2}$ has the same rotation as the wheel, link 2. Also, its changing length $\Delta\mathbf{R}_{C/2}$ demonstrates the relative sliding motion taking place between the roller on link 3 and the slot on link 2.

From Fig. 2.9 we see that the loop-closure equation for this mechanism is

$$\mathbf{R}_{BA} + \mathbf{R}_{C/2} + \mathbf{R}_{AC} = \mathbf{0}. \tag{2.13}$$

Note that the term $\mathbf{R}_{C/2}$ is equivalent to $\mathbf{R}_{CB}$ because point B is the origin of coordinate system 2.

This form of the loop-closure equation is a valid mathematical model only as long as roller C remains in the slot along x_2. However, this condition does not hold throughout the entire cycle of motion. Once the roller leaves the slot, the motion is controlled by the two mating circular arcs on links 2 and 3. A new form of the loop-closure equation rules that part of the cycle.

Mechanisms can, of course, be connected together, forming a multiple-loop kinematic chain. In such a case more than one loop-closure equation is required to model the system completely. The procedures for obtaining the equations, however, are identical to those illustrated in the above examples.

2.7 GRAPHIC POSITION ANALYSIS

When the paths of the moving points in a mechanism lie in a single plane or in parallel planes, the mechanism is called a *planar* mechanism. Because a substantial portion of the investigations in this book deal with planar mechanisms, the development of special methods suited to such problems is justified. As we will see in the following section, the nature of the loop-closure equation often leads to the solution of simultaneous nonlinear equations when approached analytically and can become quite cumbersome. Yet, particularly for planar mechanisms, the solution is usually straightforward when approached graphically.

First, let us briefly review the process of vector addition. Any two known vectors **A** and **B** can be added graphically, as illustrated in Fig. 2.10*a*. After a scale is chosen, the vectors are drawn tip to tail in either order and their sum **C** is identified:

$$\mathbf{C} = \mathbf{A} + \mathbf{B} = \mathbf{B} + \mathbf{A}. \tag{2.14}$$

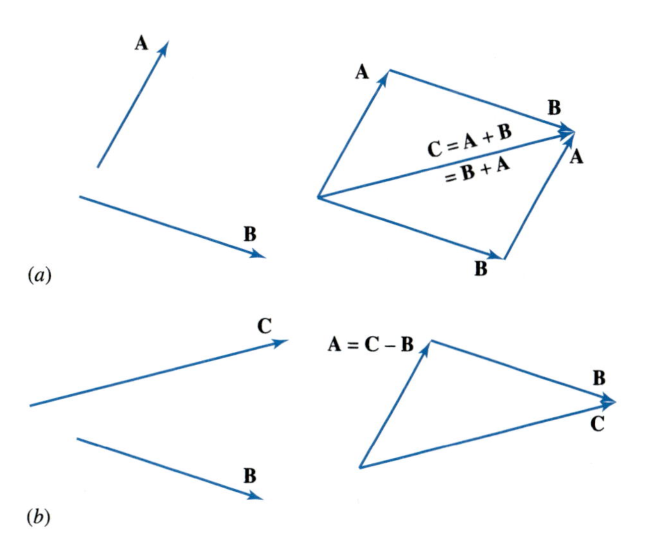

Figure 2.10 (*a*) Vector addition and (*b*) vector subtraction.

Note that the magnitudes and the orientations of both vectors **A** and **B** are used in performing the addition and that both the magnitude and the orientation of the sum **C** are found as a result.

The operation of graphical vector subtraction is illustrated in Fig. 2.10*b*, where the vectors are drawn tip to tip in solving the equation

$$\mathbf{A} = \mathbf{C} - \mathbf{B}. \tag{2.15}$$

These graphical vector operations should be studied carefully and understood because they are used extensively throughout the book.

A three-dimensional vector equation,

$$\mathbf{C} = \mathbf{D} + \mathbf{E} + \mathbf{B}, \tag{a}$$

can be divided into components along any convenient axes, leading to the three scalar equations

$$C^x = D^x + E^x + B^x, \quad C^y = D^y + E^y + B^y, \quad C^z = D^z + E^z + B^z. \tag{b}$$

Because they are components of the same vector equation, these three scalar equations must be consistent. If the three are also linearly independent, they can be solved simultaneously for three unknowns, which may be three magnitudes, three directions, or any combination of three magnitudes and directions. For some combinations, however, the problem is highly nonlinear and quite difficult to solve. Therefore, we shall delay consideration of the three-dimensional problem until it is needed in Chapter 11.

A two-dimensional vector equation can be solved for two unknowns: two magnitudes, two directions, or one magnitude and one direction. Sometimes it is desirable to indicate the known ($\surd$) and unknown (?) quantities above each vector in an equation like this,

$$\overset{?\surd}{\mathbf{C}} = \overset{\surd\surd}{\mathbf{D}} + \overset{\surd\surd}{\mathbf{E}} + \overset{?\surd}{\mathbf{B}}, \tag{c}$$

where the first symbol ($\surd$ or ?) above each vector indicates its magnitude and the second symbol indicates its direction. Another equivalent form is

$$\overset{?}{C}\overset{\surd}{\hat{\mathbf{C}}} = \overset{\surd}{D}\overset{\surd}{\hat{\mathbf{D}}} + \overset{\surd}{E}\overset{\surd}{\hat{\mathbf{E}}} + \overset{?}{B}\overset{\surd}{\hat{\mathbf{B}}}. \tag{d}$$

Either of these equations clearly identifies the unknowns and indicates whether a solution can be achieved. In Eq. (*c*), the vectors **D** and **E** are completely defined and can be replaced by their sum,

$$\mathbf{A} = \mathbf{D} + \mathbf{E}, \tag{e}$$

giving

$$\mathbf{C} = \mathbf{A} + \mathbf{B}. \tag{2.16}$$

In like manner, any plane vector equation, if it is solvable, can be reduced to a three-term equation with two unknowns.

TABLE 2.2 Planar vector equation unknowns

Case	Unknowns
1	Magnitude and direction of the same vector – for example, C, $\hat{\mathbf{C}}$
2	Magnitude of one vector and direction of another vector – for example, A, $\hat{\mathbf{B}}$
3	Magnitudes of two different vectors–for example, A, B
4	Directions of two different vectors–for example, $\hat{\mathbf{A}}$, $\hat{\mathbf{B}}$

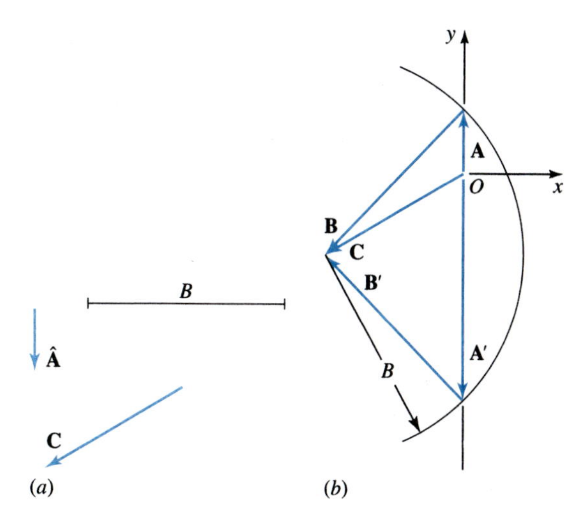

Figure 2.11 Graphical solution of case 2: (*a*) given **C**, $\hat{\mathbf{A}}$, and B; (*b*) solution for A, $\hat{\mathbf{B}}$ and A', $\hat{\mathbf{B}}'$.

Depending on the forms of the two unknowns, four distinct cases occur. These can be classified according to the unknowns; the four cases and the corresponding unknowns are presented in Table 2.2.

We will illustrate the solutions of these four cases graphically.

In case 1, the two unknowns are the magnitude and the direction of the same vector. This case can be solved by straightforward graphical addition or subtraction of the remaining vectors, which are completely defined. This case was illustrated in Fig. 2.10.

For case 2, a magnitude and an orientation from different vectors—say, A and $\hat{\mathbf{B}}$—are to be found:

$$\overset{\surd\surd}{\mathbf{C}} = \overset{?\surd}{\mathbf{A}} + \overset{\surd?}{\mathbf{B}}. \tag{2.17}$$

The solution, illustrated in Fig. 2.11, is obtained as follows:

1. Choose a coordinate system and scale factor and draw vector **C**.
2. Construct a line through the origin of **C** parallel to $\hat{\mathbf{A}}$.
3. Adjust a compass to the scaled magnitude B and construct a circular arc with center at the terminus of **C**.

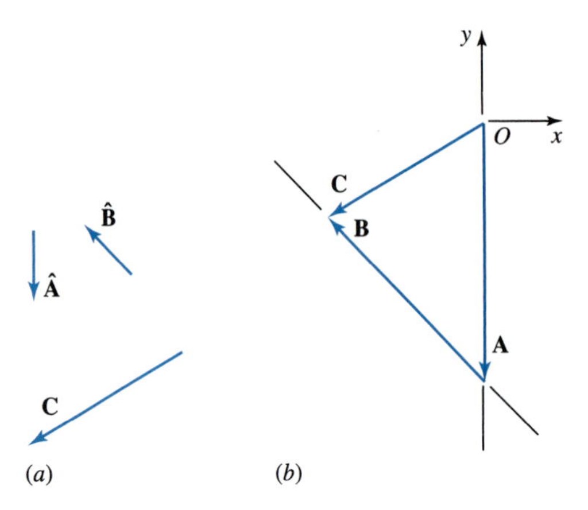

Figure 2.12 Graphical solution of case 3: (*a*) given **C**, $\hat{\mathbf{A}}$, and $\hat{\mathbf{B}}$; (*b*) solution for A and B.

4. The two intersections of the line and the arc define the two sets of solutions A, $\hat{\mathbf{B}}$ and A', $\hat{\mathbf{B}}'$.

For case 3, two magnitudes—say, A and B—are to be found:

$$\overset{\surd\surd}{\mathbf{C}} = \overset{?\surd}{\mathbf{A}} + \overset{?\surd}{\mathbf{B}}\,. \tag{2.18}$$

The solution, illustrated in Fig. 2.12, is obtained as follows:

1. Choose a coordinate system and scale factor and draw vector **C**.
2. Construct a line through the origin of **C** parallel to $\hat{\mathbf{A}}$.
3. Construct another line through the terminus of **C** parallel to $\hat{\mathbf{B}}$.
4. The intersection of these two lines defines both magnitudes, A and B, which may be either positive or negative.

Note that case 3 has a unique solution unless the lines are collinear; if the lines are parallel but distinct, the magnitudes A and B are both infinite.

Finally, for case 4 the orientations of two vectors, $\hat{\mathbf{A}}$ and $\hat{\mathbf{B}}$, are to be found:

$$\overset{\surd\surd}{\mathbf{C}} = \overset{\surd?}{\mathbf{A}} + \overset{\surd?}{\mathbf{B}}\,. \tag{2.19}$$

The steps in the solution are illustrated in Fig. 2.13:

1. Choose a coordinate system and scale factor and draw vector **C**.
2. Construct a circular arc of radius A centered at the origin of **C**.
3. Construct a circular arc of radius B centered at the terminus of **C**.
4. The two intersections of these arcs define the two sets of solutions $\hat{\mathbf{A}}$, $\hat{\mathbf{B}}$ and $\hat{\mathbf{A}}'$, $\hat{\mathbf{B}}'$.

Note that a real solution is possible only if $A + B \geq C$.

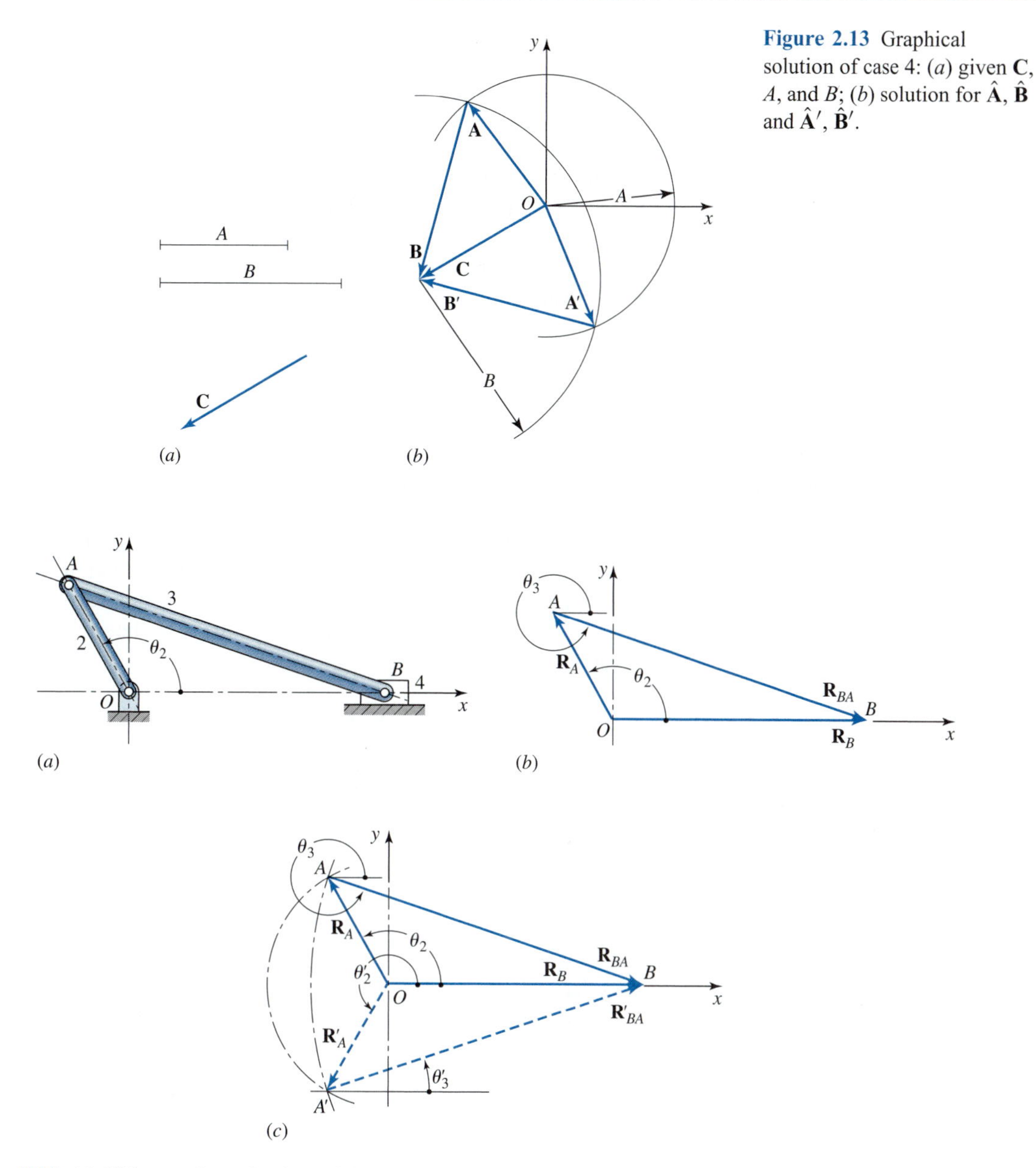

Figure 2.13 Graphical solution of case 4: (*a*) given **C**, *A*, and *B*; (*b*) solution for $\hat{\mathbf{A}}$, $\hat{\mathbf{B}}$ and $\hat{\mathbf{A}}'$, $\hat{\mathbf{B}}'$.

Figure 2.14 (*a*) Slider-crank mechanism with the distance R_B given and angles θ_2 and θ_3 unknown. (*b*) Vectors replace the link centerlines. (*c*) Graphic position analysis.

The Slider-Crank Linkage Let us now apply the procedures of this section to the solution of the loop-closure equation for the slider-crank linkage of Fig. 2.14*a*. Here link 2 is the crank, constrained to rotate about the fixed pivot O. Link 3 is the connecting rod and link 4 is the slider. In Fig. 2.14*b* we have replaced the link centerlines with vectors $\mathbf{R}_A$ for the crank, $\mathbf{R}_B$ for the frame, and $\mathbf{R}_{BA}$ for the connecting rod. Then the loop-closure equation is

$$\mathbf{R}_B = \mathbf{R}_A + \mathbf{R}_{BA}. \tag{f}$$

The problem of position analysis is to determine the values of all position variables (the positions of all points and joints) given the dimensions of each link and the value(s) of the independent variable(s), those chosen to represent the degree(s) of freedom of the mechanism.

In this example we choose to specify the position of slider 4. Then the unknowns are the angles θ_2 and θ_3 corresponding to the orientations of the vectors $\hat{\mathbf{R}}_A$ and $\hat{\mathbf{R}}_{BA}$. After identifying the known dimensions of the links, we can write Eq. (*f*) as

$$\overset{\surd\surd}{\mathbf{R}_B} = \overset{\surd ?}{\mathbf{R}_A} + \overset{\surd ?}{\mathbf{R}_{BA}}. \tag{g}$$

We recognize this as case 4 of the loop-closure equation (see Table 2.2). The graphical solution procedure explained in Fig. 2.13 is now carried out in Fig. 2.14*c*. We note that two possible solutions are found, θ_2 and θ_3 and θ_2' and θ_3', which correspond to two different configurations of the linkage, that is, two positions of the links, both of which are consistent with the given position of the slider. Of course in a graphical solution we know in advance which of the two solutions is desired. In an analytic solution, both sets of results are equally valid roots to the loop-closure equation, and it is necessary to choose between them according to the application.

The Four-Bar Linkage The loop-closure equation for the four-bar linkage of Fig. 2.15*a* is

$$\overset{\surd\surd}{\mathbf{R}_A} + \overset{\surd ?}{\mathbf{R}_{BA}} = \overset{\surd\surd}{\mathbf{R}_C} + \overset{\surd ?}{\mathbf{R}_{BC}} \tag{h}$$

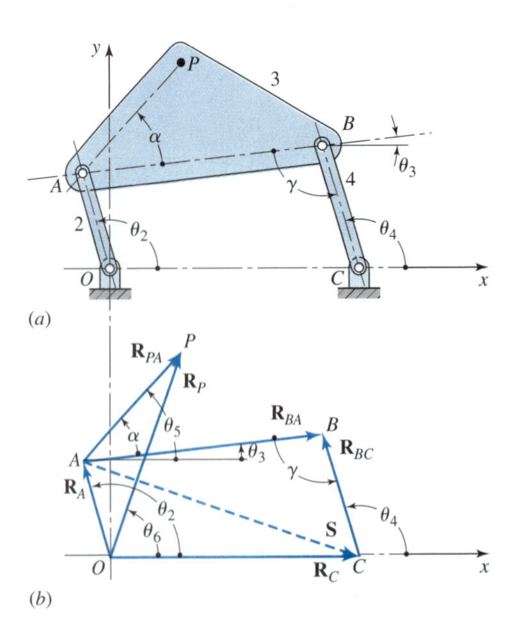

Figure 2.15 Four-bar linkage illustrating coupler point *P*. (*a*) The dimensions of the links together with the orientation of link 2 are given. The problem is to define the positions of the remaining links and of point *P*. (*b*) Vector diagram illustrating the graphical solution for the open configuration of the linkage.

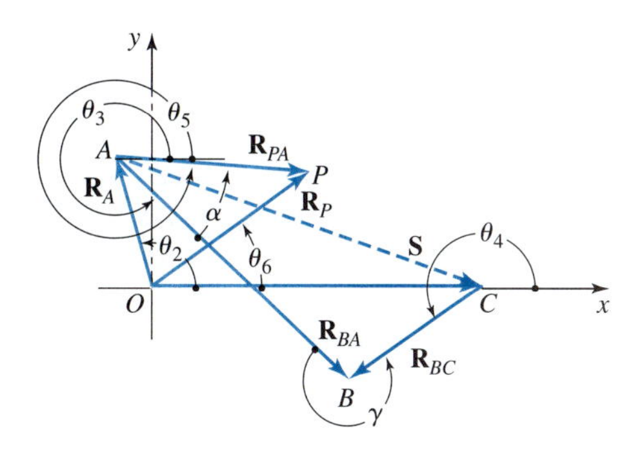

Figure 2.16 Vector diagram illustrating the graphical solution for the crossed configuration of the linkage of Fig. 2.15*a*.

and the position of point P is given by the position-difference equation

$$\overset{??}{\mathbf{R}_P} = \overset{\surd\surd}{\mathbf{R}_A} + \overset{\surd ?}{\mathbf{R}_{PA}}. \tag{i}$$

Note that Eq. (i) contains three unknowns, but this can be reduced to two after Eq. (h) is solved by noting the constant angular relationship between $\hat{\mathbf{R}}_{PA}$ and $\hat{\mathbf{R}}_{BA}$, namely

$$\theta_5 = \theta_3 + \alpha. \tag{j}$$

Begin the graphical solution for this problem by combining the two unknown terms in Eq. (h), thus locating the positions of points A and C as illustrated in Fig. 2.15b and Fig. 2.16.

$$\mathbf{S} = \overset{\surd\surd}{\mathbf{R}_C} - \overset{\surd\surd}{\mathbf{R}_A} = \overset{\surd ?}{\mathbf{R}_{BA}} - \overset{\surd ?}{\mathbf{R}_{BC}}. \tag{k}$$

The solution procedure for case 4 (that is, two unknown orientations) is then used to locate point B. Two possible solutions are found: the open configuration in Fig. 2.15b and the crossed configuration in Fig. 2.16.

Next we apply Eq. (j) to determine the two possible orientations of $\hat{\mathbf{R}}_{PA}$. Equation (i) can then be solved by the procedures for case 1. Two solutions are finally achieved, as shown in Fig. 2.15b and Fig. 2.16, for the position of point P. Both are valid solutions to Eqs. (h) through (j), although the crossed configuration could not be obtained, in this instance, from the open configuration illustrated in Fig. 2.15a without first disassembling the mechanism.

From these two examples (the slider-crank linkage and the four-bar linkage) it is clear that graphical position analysis requires precisely the same constructions that would be chosen naturally in drafting a scale drawing of the mechanism at the posture under consideration. For this reason, the procedure seems trivial and not truly worthy of the title "analysis." Yet this is highly misleading. As we shall see in the upcoming sections of Chapter 2, the position analysis of a mechanism is a nonlinear algebraic problem when approached by analytical or computer methods. It is, in fact, the most difficult problem in kinematic analysis, and this is one primary reason that graphical solution techniques have retained their attraction for planar mechanism analysis.

2.8 ALGEBRAIC POSITION ANALYSIS

In this section we present the classical approach used in the position analysis of the slider-crank mechanism. Figure 2.17 illustrates the offset version that has been chosen for analysis. By making the offset equal to zero, the same equations can be used for the centered or symmetrical version. The notation in Fig. 2.17 illustrates that the connecting rod angle θ_3 has been selected in a manner that avoids the use of negative angles in calculator and computer calculations.

The two problems that occur in the position analysis of the slider-crank mechanism are as follows:

Problem 1: Given the input angle θ_2, find the connecting rod angle θ_3 and the position x_B.
Problem 2: Given the position x_B, find the input angle θ_2 and the connecting rod angle θ_3.

Starting with Problem 1, we define the position of point A with the equation set

$$x_A = r_2 \cos\theta_2 \quad \text{and} \quad y_A = r_2 \sin\theta_2. \tag{2.20}$$

Next, we note that

$$r_2 \sin\theta_2 = r_3 \sin\theta_3 - e \tag{a}$$

so that

$$\sin\theta_3 = \frac{1}{r_3}(e + r_2 \sin\theta_2). \tag{2.21}$$

From the geometry of Fig. 2.17 we see that

$$x_B = r_2 \cos\theta_2 + r_3 \cos\theta_3. \tag{b}$$

Then, using the trigonometric identity

$$\cos\theta_3 = \pm\sqrt{1 - \sin^2\theta_3},$$

we have, from Eq. (2.21),

$$\cos\theta_3 = \pm\frac{1}{r_3}\sqrt{r_3^2 - (e + r_2 \sin\theta_2)^2}. \tag{c}$$

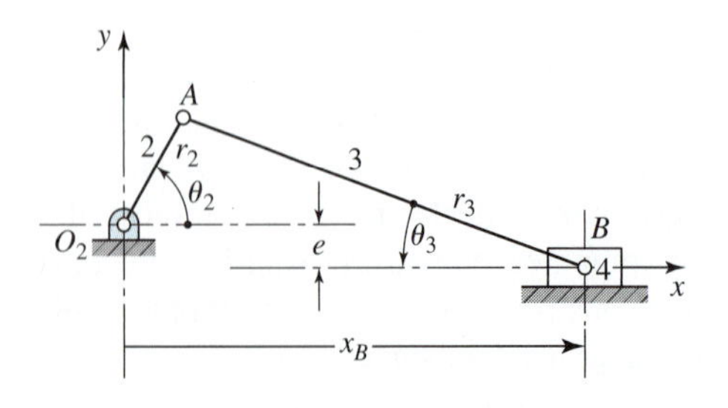

Figure 2.17 Notation used for the offset slider-crank mechanism.

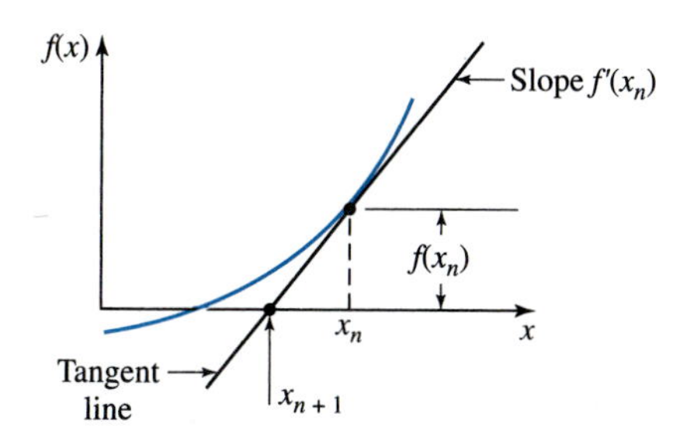

Figure 2.18 The Newton–Raphson method.

We arbitrarily select the positive sign, which we see from Fig. 2.17 corresponds to the solution with the piston to the right of the crank pin. Then, substituting Eq. (*c*) into Eq. (*b*) gives

$$x_B = r_2 \cos\theta_2 + \sqrt{r_3^2 - (e + r_2 \sin\theta_2)^2}. \tag{2.22}$$

Thus, with the angle θ_2 given, the unknowns θ_3 and x_B can be obtained by solving Eqs. (2.21) and (2.22).

Problem 2 requires that, given x_B, we solve Eq. (2.22) for the angle θ_2. This requires the use of a calculator or a computer together with a root-finding technique. Here we select the well-known Newton–Raphson method.[1] This method can be explained by reference to Fig. 2.18.

Figure 2.18 is a graph of some function $f(x)$ versus x. Let x_n be a first approximation (or a rough estimate) of the root where $f(x) = 0$, which we wish to find. A tangent line drawn to the curve at $x = x_n$ intersects the x axis at x_{n+1}, which is a better approximation to the root. The slope of the tangent line is the derivative of the function at $x = x_n$ and is

$$f'(x_n) = \frac{f(x_n)}{x_n - x_{n+1}}. \tag{d}$$

Solving for x_{n+1} gives

$$x_{n+1} = x_n - \frac{f(x_n)}{f'(x_n)}. \tag{2.23}$$

Using a computer, for example, we can start a solution by entering an estimate of x_n, solve for x_{n+1}, use this as the next estimate, and repeat this process as many times as needed to obtain the result with satisfactory accuracy. This accuracy is evaluated by comparing $x_{n+1} - x_n$ with a small number ε after each repetition and stopping when $(x_{n+1} - x_n) < \varepsilon$.

We note that the root-finding programs built into most calculators utilize an approximation to obtain $f'(x_n)$. These will sometimes be of little value in solving Eq. (2.22), so we proceed as follows. In Eq. (2.22) we replace the angle θ_2 with x and let r_2, r_3, e, and x_B be given constants. Then,

$$f(x) = r_2 \cos x + \sqrt{r_3^2 - (e + r_2 \sin x)^2} - x_B \tag{e}$$

and

$$f'(x) = -r_2 \sin x - \frac{(e + r_2 \sin x) r_2 \cos x}{\sqrt{r_3^2 - (e + r_2 \sin x)^2}}. \tag{f}$$

These two equations can now be programmed with Eq. (2.23) to solve for the unknown value of the angle θ_2 when x_B is given. We note that the angle θ_2 will have two possible values. These may be found separately using appropriate initial estimates.

An algebraic solution is possible if the eccentricity e is zero, that is, if the linkage is centered. For this case we can take Eqs. (a) and (b), square them, and add them together. Noting a trigonometric identity, the result is

$$x_B^2 - 2x_B r_2 \cos \theta_2 + r_2^2 = r_3^2. \tag{g}$$

Solving for θ_2 gives

$$\theta_2 = \cos^{-1} \frac{x_B^2 + r_2^2 - r_3^2}{2x_B r_2}. \tag{2.24}$$

The solution of Problem 1 for the centered version is, of course, obtained directly from Eq. (2.22) by making $e = 0$.

The Crank-and-Rocker Mechanism The four-bar linkage illustrated in Fig. 2.19 is called the *crank-and-rocker* (or crank-rocker) mechanism. Thus, link 2, which is the crank, can rotate through a full circle, but the rocker, link 4, can only oscillate. In general, we shall follow the commonly accepted practice of designating the frame or fixed link as link 1. Link 3 in Fig. 2.19 is called the *coupler* or the *connecting rod*. With the four-bar linkage, the position problem generally consists of finding the directions of the coupler link and the output link, or rocker, when the link dimensions and the input crank position are known.

To obtain the analytical solution, we designate s the distance R_{AO_4} in Fig. 2.19. The cosine law can then be written for each of the two triangles O_4O_2A and ABO_4. In terms of

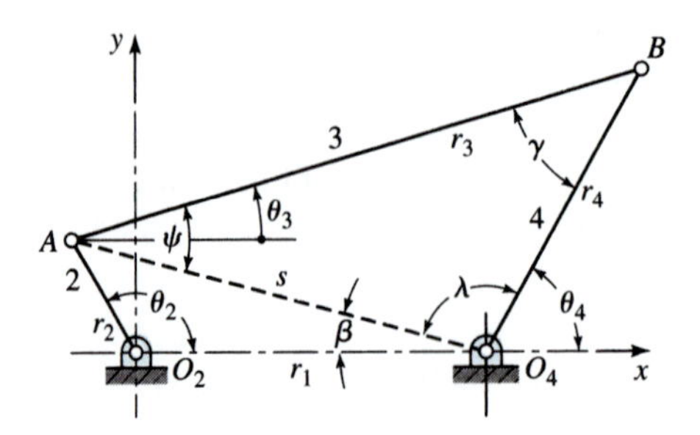

Figure 2.19 The crank-and-rocker four-bar linkage.

the angles and link lengths illustrated in Fig. 2.19, we then have

$$s = \sqrt{r_1^2 + r_2^2 - 2r_1r_2\cos\theta_2} \tag{2.25}$$

$$\beta = \cos^{-1}\frac{r_1^2 + s^2 - r_2^2}{2r_1s} \tag{2.26}$$

$$\psi = \cos^{-1}\frac{r_3^2 + s^2 - r_4^2}{2r_3s} \tag{2.27}$$

$$\lambda = \cos^{-1}\frac{r_4^2 + s^2 - r_3^2}{2r_4s}. \tag{2.28}$$

There will generally be two values of λ corresponding to each value of θ_2. If θ_2 is in the range $0 \leq \theta_2 \leq \pi$, the unknown directions are taken as

$$\theta_3 = \psi - \beta \tag{2.29}$$

$$\theta_4 = \pi - \lambda - \beta. \tag{2.30}$$

However, if θ_2 is in the range $\pi \leq \theta_2 \leq 2\pi$, then

$$\theta_3 = \psi + \beta \tag{2.31}$$

$$\theta_4 = \pi - \lambda + \beta. \tag{2.32}$$

Finally, it is worth noting that Eqs. (2.27) and (2.28) also yield double values because they involve arc cosines. These should always be positive or negative pairs of values; the positive values correspond to the open configuration shown, whereas the negative values correspond to the crossed configuration.

In Section 1.10 the concept of transmission angle is discussed in connection with the subject of mechanical advantage. With Fig. 2.19 we see that this angle is given by the equation

$$\gamma = \pm\cos^{-1}\frac{r_3^2 + r_4^2 - s^2}{2r_3r_4}. \tag{2.33}$$

2.9 COMPLEX-ALGEBRA SOLUTIONS OF PLANAR VECTOR EQUATIONS

In planar problems it is often desirable to express a vector by specifying its magnitude and orientation in *polar notation*:

$$\mathbf{R} = R\angle\theta. \tag{a}$$

In Fig. 2.20*a*, the two-dimensional vector

$$\mathbf{R} = R^x\hat{\mathbf{i}} + R^y\hat{\mathbf{j}} \tag{2.34}$$

LIVERPOOL JOHN MOORES UNIVERSITY
LEARNING SERVICES

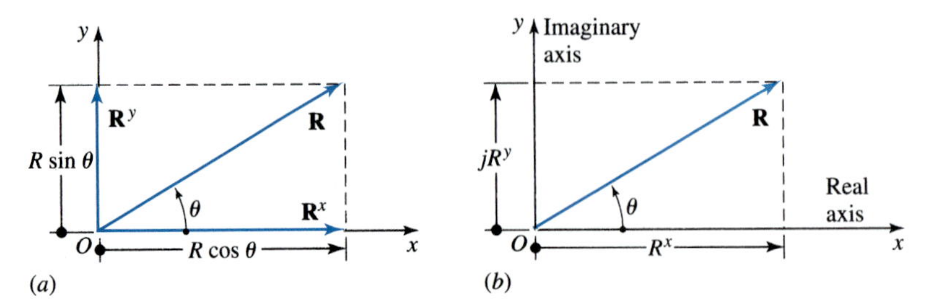

Figure 2.20 Correlation of planar vectors and complex numbers.

has two rectangular components of magnitudes,

$$R^x = R\cos\theta \quad \text{and} \quad R^y = R\sin\theta, \tag{2.35}$$

with

$$R = \sqrt{(R^x)^2 + (R^y)^2} \quad \text{and} \quad \theta = \tan^{-1}\frac{R^y}{R^x}. \tag{2.36}$$

Note that we have made the arbitrary choice here of accepting the positive square root for the magnitude R when calculating from the components of **R**. Therefore, we must be careful to interpret the signs of R^x and R^y individually when deciding upon the quadrant of the angle θ. Note that θ is defined as the angle from the positive x axis to the positive end of vector **R**, measured about the origin of the vector, and this angle is positive when measured counterclockwise.

EXAMPLE 2.3

Express the vectors $A = 10\angle 30°$ and $B = 8\angle - 15°$ in rectangular notation* and find their sum.

SOLUTION

The vectors are illustrated in Fig. 2.21 and can be written as

$$\mathbf{A} = 10\cos 30°\hat{\mathbf{i}} + 10\sin 30°\hat{\mathbf{j}} = 8.66\hat{\mathbf{i}} + 5.00\hat{\mathbf{j}} \qquad \textit{Ans.}$$

$$\mathbf{B} = 8\cos(-15°)\hat{\mathbf{i}} + 8\sin(-15°)\hat{\mathbf{j}} = 7.73\hat{\mathbf{i}} - 2.07\hat{\mathbf{j}} \qquad \textit{Ans.}$$

$$\mathbf{C} = \mathbf{A} + \mathbf{B} = (8.66 + 7.73)\hat{\mathbf{i}} + (5.00 - 2.07)\hat{\mathbf{j}}$$
$$= 16.39\hat{\mathbf{i}} + 2.93\hat{\mathbf{j}}.$$

* Many calculators are equipped to perform polar–rectangular and rectangular–polar conversions directly.

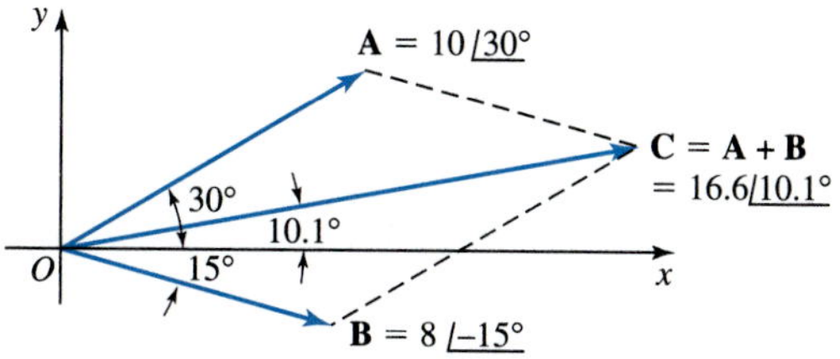

Figure 2.21 Example 2.3.

The magnitude of the resultant, determined from Eq. (2.36), is

$$C = \sqrt{16.39^2 + 2.93^2} = 16.65,$$

and the angle is

$$\theta = \tan^{-1} \frac{2.93}{16.39} = 10.1°.$$

The final result in polar notation is

$$\mathbf{C} = 16.6\angle 10.1°. \qquad \textit{Ans.}$$

2.10 COMPLEX POLAR ALGEBRA

Another way of treating two-dimensional vector problems analytically makes use of complex algebra. Although complex numbers are not vectors, they can be used to represent vectors in a plane by choosing an origin and real and imaginary axes. In two-dimensional kinematic problems, these axes can conveniently be chosen coincident with the x_1y_1 axes of the absolute coordinate system.

As illustrated in Fig. 2.20*b*, the location of any point in the plane can be specified either by its absolute-position vector or by its corresponding real and imaginary coordinates,

$$\mathbf{R} = R^x + jR^y, \tag{2.37}$$

where the operator j is defined as the unit imaginary number,

$$j = \sqrt{-1}. \tag{2.38}$$

The real usefulness of complex numbers in planar analysis stems from the ease with which they can be switched to polar form. Employing complex rectangular notation for the vector **R**, we can write

$$\mathbf{R} = R\angle\theta = R\cos\theta + jR\sin\theta. \tag{2.39}$$

But using the well-known Euler equation from trigonometry,

$$e^{j\theta} = \cos\theta + j\sin\theta, \tag{2.40}$$

we can also write **R** in complex polar form as

$$\mathbf{R} = Re^{j\theta}, \tag{2.41}$$

where the magnitude and orientation of the vector appear explicitly. As we will see in Chapters 3 and 4, expression of a vector in this form is especially useful when differentiation is required.

Some familiarity with useful manipulation techniques for vectors written in complex polar form can be gained by again solving the four cases of the loop-closure equation of Table 2.2. Writing Eq. (2.16) in complex polar form, we have

$$Ce^{j\theta_C} = Ae^{j\theta_A} + Be^{j\theta_B}. \tag{2.42}$$

In case 1 the two unknowns are C and θ_C. We begin the solution by separating the real and imaginary parts of the equation. Substituting Euler's equation, Eq. (2.40), into this equation gives

$$C(\cos\theta_C + j\sin\theta_C) = A(\cos\theta_A + j\sin\theta_A) + B(\cos\theta_B + j\sin\theta_B). \tag{a}$$

Upon equating the real terms and the imaginary terms separately, we obtain two real equations corresponding to the horizontal and vertical components of the two-dimensional vector equation.

$$C\cos\theta_C = A\cos\theta_A + B\cos\theta_B \tag{b}$$

$$C\sin\theta_C = A\sin\theta_A + B\sin\theta_B. \tag{c}$$

The unknown angle θ_C can be eliminated by squaring and adding these two equations. The result is

$$C = \sqrt{A^2 + B^2 + 2AB\cos(\theta_B - \theta_A)}. \tag{2.43}$$

The positive square root is chosen arbitrarily; the negative square root would yield a negative solution for C with θ_C differing by 180°. Dividing Eq. (*c*) by Eq. (*b*), and rearranging, the angle θ_C can be written as

$$\theta_C = \tan^{-1}\frac{A\sin\theta_A + B\sin\theta_B}{A\cos\theta_A + B\cos\theta_B}, \tag{2.44}$$

where the signs of the numerator and the denominator must be considered separately in determining the proper quadrant of θ_C.* Only a single solution is found for case 1, as previously illustrated in Fig. 2.10.

* When writing computer programs, most programming languages, including both ANSI/ISO standard FORTRAN and C, provide a library function named ATAN2(*y*, *x*) that accepts the numerator and denominator separately and provides the solution (in radians) in the proper quadrant. If such a function is not available, then a solution (in radians) in either the first or the fourth quadrant is usually provided, and π radians should be added to the angle if the denominator is negative.

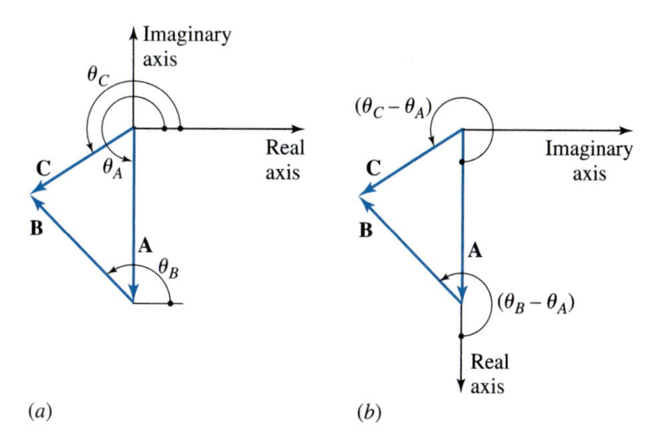

Figure 2.22 Rotation of axes by division of complex polar equations by $e^{j\theta_A}$: (*a*) original axes and (*b*) rotated axes.

The graphical solution of case 2 was illustrated in Fig. 2.11. The two unknowns are A and θ_B. One convenient way of solving this case, in complex polar form, is first to divide Eq. (2.42) by $e^{j\theta_A}$. The resulting equation can be written as

$$Ce^{j(\theta_C-\theta_A)} = A + Be^{j(\theta_B-\theta_A)}. \tag{d}$$

Comparing this equation with Fig. 2.22, we see that division by the complex polar form of a unit vector $e^{j\theta_A}$ has the effect of rotating the real and imaginary axes by the angle θ_A such that the real axis lies along the vector **A**.

We can now use Euler's equation, Eq. (2.40), to separate the real and imaginary components, that is,

$$C\cos(\theta_C - \theta_A) = A + B\cos(\theta_B - \theta_A) \tag{e}$$

$$C\sin(\theta_C - \theta_A) = B\sin(\theta_B - \theta_A). \tag{f}$$

The solutions are then obtained directly from Eqs. (*f*) and (*e*), respectively, as

$$\theta_B = \theta_A + \sin^{-1}\frac{C\sin(\theta_C - \theta_A)}{B} \tag{2.45}$$

$$A = C\cos(\theta_C - \theta_A) - B\cos(\theta_B - \theta_A). \tag{2.46}$$

The solutions given by Eqs. (2.45) and (2.46) are intentionally presented in this order because Eq. (2.46) cannot be evaluated numerically until after θ_B is found. We also note that the arc sine term in Eq. (2.45) is double valued. Therefore, case 2 has two distinct solutions, A, θ_B and A', θ'_B; these are illustrated in Fig. 2.11.

For case 3, the two unknowns of Eq. (2.42) are the two magnitudes A and B. The graphical solution to this case was illustrated in Fig. 2.12. Again, we begin by aligning the real axis along vector **A** by dividing Eq. (2.42) by $e^{j\theta_A}$ as was done to obtain Eq. (*d*) above. Separating the real and imaginary components leads again to Eqs. (*e*) and (*f*), where we note that the vector **A**, now real, has been totally eliminated from Eq. (*f*). The solution for

B is now easily obtained from the resulting relation

$$B = C\frac{\sin(\theta_C - \theta_A)}{\sin(\theta_B - \theta_A)}. \tag{2.47}$$

The solution for the other unknown magnitude A is found in completely analogous fashion. Dividing Eq. (2.42) by $e^{j\theta_B}$ aligns the real axis along the vector **B**. That equation is then separated into real and imaginary parts and yields

$$A = C\frac{\sin(\theta_C - \theta_B)}{\sin(\theta_A - \theta_B)}. \tag{2.48}$$

As before, case 3 yields a unique solution.

Case 4 has the two angles θ_A and θ_B as unknowns. The graphical solution was illustrated in Fig. 2.13. In this case we align the real axis along vector **C**:

$$C = Ae^{j(\theta_A - \theta_C)} + Be^{j(\theta_B - \theta_C)}. \tag{g}$$

Using Euler's equation to separate components, and then rearranging terms, we obtain

$$A\cos(\theta_A - \theta_C) = C - B\cos(\theta_B - \theta_C) \tag{h}$$

$$A\sin(\theta_A - \theta_C) = -B\sin(\theta_B - \theta_C). \tag{i}$$

Squaring both equations and adding results in

$$A^2 = C^2 + B^2 - 2BC\cos(\theta_B - \theta_C). \tag{j}$$

We recognize this as the law of cosines for the vector triangle. Rearranging this equation, the angle θ_B can be written as

$$\theta_B = \theta_C \pm \cos^{-1}\frac{C^2 + B^2 - A^2}{2BC}. \tag{2.49}$$

Moving C to the left-hand side of Eq. (h) before squaring and adding results in another form of the law of cosines, from which

$$\theta_A = \theta_C \mp \cos^{-1}\frac{C^2 + A^2 - B^2}{2AC}. \tag{2.50}$$

The plus-or-minus signs in these two equations are a reminder that the arc cosines are each double valued and therefore θ_B and θ_A each have two solutions. These two pairs of angles can be paired naturally together as θ_A, θ_B and θ'_A, θ'_B under the restriction of Eq. (i). Therefore, case 4 has two distinct solutions, as illustrated in Fig. 2.13.

2.11 THE CHACE SOLUTIONS TO PLANAR VECTOR EQUATIONS

As we observed in the previous section, the algebra involved in solving even simple planar vector equations can become cumbersome. Chace[2] took advantage of the brevity of vector notation in obtaining explicit closed-form solutions to both two- and three-dimensional vector equations. In this section we will study the Chace solutions to planar equations in terms of the four cases of Table 2.2 of the loop-closure equation.

We again recall Eq. (2.16), the typical planar vector equation. In terms of magnitudes and unit vectors it can be written

$$C\hat{\mathbf{C}} = A\hat{\mathbf{A}} + B\hat{\mathbf{B}} \tag{2.51}$$

and may contain two unknowns consisting of two magnitudes, two directions, or one magnitude and one direction.

Case 1 is the situation where the magnitude and direction of the same vector, say, C and $\hat{\mathbf{C}}$, form the two unknowns. The method of solution for this case was presented in Example 2.3. The general form of the solution is

$$\mathbf{C} = (\mathbf{A}\cdot\hat{\mathbf{i}} + \mathbf{B}\cdot\hat{\mathbf{i}})\hat{\mathbf{i}} + (\mathbf{A}\cdot\hat{\mathbf{j}} + \mathbf{B}\cdot\hat{\mathbf{j}})\hat{\mathbf{j}}. \tag{2.52}$$

For case 2, the two unknowns are the magnitude of one vector and the orientation of another, say, A and $\hat{\mathbf{B}}$. The Chace approach for this case consists of eliminating one of the unknowns by taking the dot product of every vector with a new vector chosen so that one of the unknowns is eliminated. We can eliminate the unknown A by taking the dot product of every term of the equation with $\hat{\mathbf{A}} \times \hat{\mathbf{k}}$, that is,

$$\mathbf{C}\cdot(\hat{\mathbf{A}} \times \hat{\mathbf{k}}) = A\hat{\mathbf{A}}\cdot(\hat{\mathbf{A}} \times \hat{\mathbf{k}}) + B\hat{\mathbf{B}}\cdot(\hat{\mathbf{A}} \times \hat{\mathbf{k}}). \tag{a}$$

Thus, because $\hat{\mathbf{A}} \times \hat{\mathbf{k}}$ is perpendicular to $\hat{\mathbf{A}}$, $\hat{\mathbf{A}}\cdot(\hat{\mathbf{A}} \times \hat{\mathbf{k}}) = \mathbf{0}$, then

$$\mathbf{C}\cdot(\hat{\mathbf{A}} \times \hat{\mathbf{k}}) = B\hat{\mathbf{B}}\cdot(\hat{\mathbf{A}} \times \hat{\mathbf{k}}). \tag{b}$$

Now, from the definition of the dot product of two vectors, namely,

$$\mathbf{P}\cdot\mathbf{Q} = PQ\cos\phi,$$

we note that

$$B\hat{\mathbf{B}}\cdot(\hat{\mathbf{A}} \times \hat{\mathbf{k}}) = B\cos\phi, \tag{c}$$

where ϕ is the angle between the two vectors $\hat{\mathbf{B}}$ and $(\hat{\mathbf{A}} \times \hat{\mathbf{k}})$. Rearranging Eq. ($c$) gives

$$\cos\phi = \hat{\mathbf{B}}\cdot(\hat{\mathbf{A}} \times \hat{\mathbf{k}}). \tag{d}$$

The vectors $\hat{\mathbf{A}}$ and $\hat{\mathbf{A}} \times \hat{\mathbf{k}}$ are perpendicular to each other; hence, we are free to choose another coordinate system $\hat{\lambda}\hat{\mu}$ having the orientations $\hat{\lambda} = \hat{\mathbf{A}} \times \hat{\mathbf{k}}$ and $\hat{\mu} = \hat{\mathbf{A}}$. In this reference system, the unknown unit vector $\hat{\mathbf{B}}$ can be written as

$$\hat{\mathbf{B}} = \cos\phi(\hat{\mathbf{A}} \times \hat{\mathbf{k}}) + \sin\phi\hat{\mathbf{A}}. \tag{e}$$

Substituting Eq. (*d*) into Eq. (*b*) and solving for cos ϕ gives

$$\cos\phi = \frac{\mathbf{C}\cdot(\hat{\mathbf{A}}\times\hat{\mathbf{k}})}{B}. \tag{f}$$

Then solving for sinϕ gives

$$\sin\phi = \pm\sqrt{1-\cos^2\phi} = \pm\frac{1}{B}\sqrt{B^2-[C\cdot(\hat{\mathbf{A}}\times\hat{\mathbf{k}})]^2}. \tag{g}$$

Substituting Eqs. (*f*) and (*g*) into Eq. (*e*) and multiplying both sides by the known magnitude B gives

$$\mathbf{B} = [\mathbf{C}\cdot(\hat{\mathbf{A}}\times\hat{\mathbf{k}})](\hat{\mathbf{A}}\times\hat{\mathbf{k}}) \pm \sqrt{B^2-[\mathbf{C}\cdot(\hat{\mathbf{A}}\times\hat{\mathbf{k}})]^2}\hat{\mathbf{A}}. \tag{2.53}$$

To obtain the vector **A** we may wish to use Eq. (2.51) directly and perform the vector subtraction. Alternatively, if we substitute Eq. (2.53) into Eq. (2.16) and rearrange, we obtain

$$\mathbf{A} = \mathbf{C} - [\mathbf{C}\cdot(\hat{\mathbf{A}}\times\hat{\mathbf{k}})](\hat{\mathbf{A}}\times\hat{\mathbf{k}}) \mp \sqrt{B^2-[\mathbf{C}\cdot(\hat{\mathbf{A}}\times\hat{\mathbf{k}})]^2}\hat{\mathbf{A}}. \tag{h}$$

The first two terms of this equation can be simplified, as illustrated in Fig. 2.23*a*. The $\hat{\mathbf{A}}\times\hat{\mathbf{k}}$ orientation is located 90° clockwise from the $\hat{\mathbf{A}}$ orientation. The magnitude $\mathbf{C}\cdot(\hat{\mathbf{A}}\times\hat{\mathbf{k}})$ is the projection of **C** in the $\hat{\mathbf{A}}\times\hat{\mathbf{k}}$ direction. Therefore, when $[\mathbf{C}\cdot(\hat{\mathbf{A}}\times\hat{\mathbf{k}})](\hat{\mathbf{A}}\times\hat{\mathbf{k}})$ is subtracted from **C**, the result is a vector of magnitude $\mathbf{C}\cdot\hat{\mathbf{A}}$ in the $\hat{\mathbf{A}}$ direction. With this substitution, Eq. (*h*) becomes

$$\mathbf{A} = \left[\mathbf{C}\cdot\hat{\mathbf{A}} \mp \sqrt{B^2-[\mathbf{C}\cdot(\hat{\mathbf{A}}\times\hat{\mathbf{k}})]^2}\right]\hat{\mathbf{A}}. \tag{2.54}$$

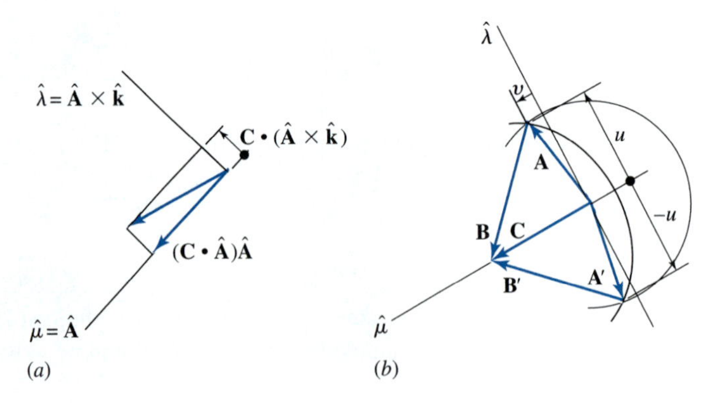

Figure 2.23

In case 3 the magnitudes of two different vectors—say, A and B—are unknown. We begin the solution by eliminating B from Eq. (2.51)

$$\mathbf{C} \cdot (\hat{\mathbf{B}} \times \hat{\mathbf{k}}) = A\hat{\mathbf{A}} \cdot (\hat{\mathbf{B}} \times \hat{\mathbf{k}}) + B\hat{\mathbf{B}} \cdot (\hat{\mathbf{B}} \times \hat{\mathbf{k}}).$$

Thus, because $\hat{\mathbf{B}} \times \hat{\mathbf{k}}$ is perpendicular to $\hat{\mathbf{B}}$, we have $\hat{\mathbf{B}} \cdot (\hat{\mathbf{B}} \times \hat{\mathbf{k}}) = \mathbf{0}$; hence, we obtain the magnitude

$$A = \frac{\mathbf{C} \cdot (\hat{\mathbf{B}} \times \hat{\mathbf{k}})}{\hat{\mathbf{A}} \cdot (\hat{\mathbf{B}} \times \hat{\mathbf{k}})}. \tag{2.55}$$

In a similar manner, we obtain the unknown magnitude B.

$$B = \frac{\mathbf{C} \cdot (\hat{\mathbf{A}} \times \hat{\mathbf{k}})}{\hat{\mathbf{B}} \cdot (\hat{\mathbf{A}} \times \hat{\mathbf{k}})}. \tag{2.56}$$

Finally, in case 4 the unknowns are the orientations of two different vectors—say, $\hat{\mathbf{A}}$ and $\hat{\mathbf{B}}$. This case is illustrated in Fig. 2.23*b*, where the vector $\mathbf{C}$ and the two magnitudes A and B are given. The problem is solved by finding the intersections of two circles of radii A and B. We begin by defining a new coordinate system $\hat{\lambda}\hat{\mu}$ whose axes are directed so that $\hat{\lambda} = \hat{\mathbf{C}} \times \hat{\mathbf{k}}$ and $\hat{\mu} = \hat{\mathbf{C}}$, as shown in the figure. If the coordinates of one of the intersections in the $\hat{\lambda}\hat{\mu}$ coordinate system are designated u and v, then

$$\mathbf{A} = u\hat{\lambda} + v\hat{\mu} \quad \text{and} \quad \mathbf{B} = -u\hat{\lambda} + (C - v)\hat{\mu}. \tag{i}$$

The equation of the circle of radius A is

$$u^2 + v^2 = A^2 \tag{j}$$

and the equation of the circle of radius B is

$$u^2 + (v - C)^2 = B^2$$

or

$$u^2 + v^2 - 2Cv + C^2 = B^2. \tag{k}$$

Substituting Eq. (*j*) into Eq. (*k*) and rearranging gives

$$v = \frac{A^2 - B^2 + C^2}{2C}. \tag{l}$$

Then substituting this result into Eq. (*j*) and solving for u gives

$$u = \pm\sqrt{A^2 - \left(\frac{A^2 - B^2 + C^2}{2C}\right)^2}. \tag{m}$$

The final step is to substitute these values of u and v into Eqs. (i) and to replace $\hat{\lambda}$ and $\hat{\mu}$ according to their definitions. The results are

$$\mathbf{A} = \pm\sqrt{A^2 - \left(\frac{A^2 - B^2 + C^2}{2C}\right)^2}(\hat{\mathbf{C}} \times \hat{\mathbf{k}}) + \frac{A^2 - B^2 + C^2}{2C}\hat{\mathbf{C}} \tag{2.57}$$

$$\mathbf{B} = \mp\sqrt{A^2 - \left(\frac{A^2 - B^2 + C^2}{2C}\right)^2}(\hat{\mathbf{C}} \times \hat{\mathbf{k}}) + \frac{B^2 - A^2 + C^2}{2C}\hat{\mathbf{C}}. \tag{2.58}$$

For an example application of the Chace equations, let us begin with the slider-crank linkage of Fig. 2.14a. From Fig. 2.14b, the loop-closure equation is written in the form

$$R_B\hat{\mathbf{R}}_B = R_A\hat{\mathbf{R}}_A + R_{BA}\hat{\mathbf{R}}_{BA}. \tag{n}$$

With θ_2 given, the unknowns in Eq. (n) are the magnitude R_B and the orientation $\hat{\mathbf{R}}_{BA}$. The solution corresponds to case 2 and is found by making appropriate substitutions in Eqs. (2.53) and (2.54).

$$\mathbf{R}_{BA} = -[\mathbf{R}_A \cdot (\hat{\mathbf{R}}_B \times \hat{\mathbf{k}})](\hat{\mathbf{R}}_B \times \hat{\mathbf{k}}) + \sqrt{R_{BA}^2 - [\mathbf{R}_A \cdot (\hat{\mathbf{R}}_B \times \hat{\mathbf{k}})]^2}\hat{R}_B \tag{2.59}$$

$$\mathbf{R}_B = (\mathbf{R}_A \cdot \hat{\mathbf{R}}_B + \sqrt{R_{BA}^2 - [\mathbf{R}_A \cdot (\hat{\mathbf{R}}_B \times \hat{\mathbf{k}})]^2})\hat{\mathbf{R}}_B. \tag{2.60}$$

EXAMPLE 2.4

Use the Chace equations to find the position of the slider of Fig. 2.14 with $R_A = 25$ mm, $R_{BA} = 75$ mm, and $\theta_2 = 150°$.

SOLUTION

Writing the given information in vector form gives

$$\mathbf{R}_A = 25\ \text{mm}\angle 150° = -21.7\ \text{mm}\hat{\mathbf{i}} + 12.5\ \text{mm}\hat{\mathbf{j}},$$

$$R_{BA} = 75\ \text{mm}, \quad \hat{\mathbf{R}}_B = \hat{\mathbf{i}}, \quad \hat{R}_B \times \hat{\mathbf{k}} = -\hat{\mathbf{j}}.$$

Substituting these values into Eq. (2.59) gives

$$\begin{aligned}\mathbf{R}_{BA} &= -[(-21.7\ \text{mm}\hat{\mathbf{i}} + 12.5\ \text{mm}\hat{\mathbf{j}}) \cdot (-\hat{\mathbf{j}})](-\hat{\mathbf{j}}) \\ &\quad + \sqrt{(75\ \text{mm})^2 - [(-21.7\ \text{mm}\hat{\mathbf{i}} + 12.5\ \text{mm}\hat{\mathbf{j}}) \cdot (-\hat{\mathbf{j}})]^2}\hat{\mathbf{i}} \\ &= -12.5\ \text{mm}\hat{\mathbf{j}} + \sqrt{(75\ \text{mm})^2 - (-12.5\ \text{mm})^2}\hat{\mathbf{i}} = 73.9\ \text{mm}\hat{\mathbf{i}} - 12.5\ \text{mm}\hat{\mathbf{j}}.\end{aligned}$$

Therefore,

$$\mathbf{R}_{BA} = 75\ \text{mm}\angle -9.60° = 75\ \text{mm}\angle 350.4°. \qquad \textit{Ans.}$$

Then using Eq. (2.60) gives

$$\mathbf{R}_B = \left\{(-21.7\text{ mm}\hat{\mathbf{i}} + 12.5\text{ mm}\hat{\mathbf{j}}) \cdot \hat{\mathbf{i}}\right\} \hat{\mathbf{i}} + \left\{\sqrt{(75\text{ mm})^2 - [(-21.7\text{ mm}\hat{\mathbf{i}} + 12.5\text{ mm}\hat{\mathbf{j}}) \cdot (-\hat{\mathbf{j}})]^2}\right\} \hat{\mathbf{i}}.$$

$$= 52.2\text{ mm}\hat{\mathbf{i}}. \qquad \textit{Ans.}$$

2.12 POSITION ANALYSIS TECHNIQUES

EXAMPLE 2.5

Perform a position analysis of the sliding-block linkage illustrated in Fig. 2.24 by finding θ_4 and the distance R_{AO_4}.

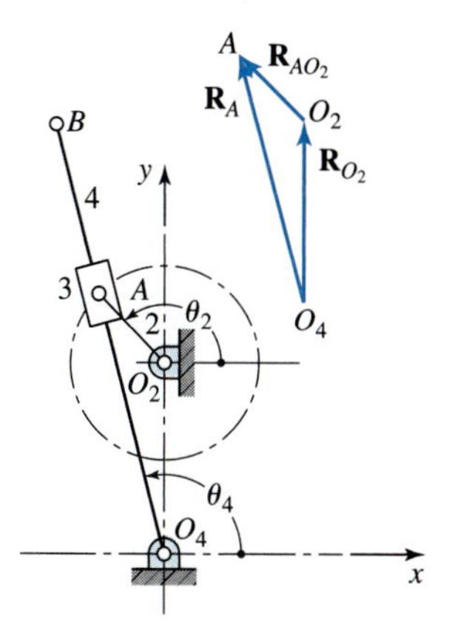

Figure 2.24 Example 2.5. A sliding-block linkage, $R_{O_2O_4} = 9.0$ in, $R_{AO_2} = 4.5$ in, $\theta_2 = 135°$.

SOLUTION

For illustrative purposes, we choose to employ five different approaches here to solve the problem. The five approaches shown here are as follows:

(*a*) Graphic approach.
(*b*) Analytic approach.
(*c*) Complex algebraic approach.
(*d*) Vector algebraic approach.
(*e*) Numeric approach.

First, using the vector diagram in Fig. 2.24 we recognize this as a case 1 problem and so

$$\overset{??}{\mathbf{R}_A} = \overset{\surd\surd}{\mathbf{R}_{O_2}} + \overset{\surd\surd}{\mathbf{R}_{AO_2}}. \qquad (1)$$

(a) Graphic approach Figure 2.24 is drawn at a scale of 6 in./in. Then by direct measurement of this figure we find that

$$\theta_4 = 105.3° \qquad \textit{Ans.}$$

$$R_{AO_2} = 2.08 \text{ in}(6 \text{ in/in}) = 12.48 \text{ in.} \qquad \textit{Ans.}$$

(b) Analytic approach For case 1 we employ Eqs. (2.43) and (2.44); therefore,

$$\begin{aligned} R_A &= \sqrt{R_{O_2}^2 R_{AO_2}^2 + 2R_{O_2}R_{AO_2}\cos(\theta_2 - 90°)} \\ &= \sqrt{(9 \text{ in})^2 + (4.5 \text{ in})^2 + 2(9 \text{ in})(4.5 \text{ in})\cos(135° - 90°)} \\ &= 12.59 \text{ in} \qquad \textit{Ans.} \end{aligned}$$

$$\begin{aligned} \theta_4 &= \tan^{-1}\frac{R_{O_2}\sin 90° + R_{AO_2}\sin\theta_2}{R_{O_2}\cos 90° + R_{AO_2}\cos\theta_2} \\ &= \tan^{-1}\frac{(9 \text{ in})\sin 90° + (4.5 \text{ in})\sin 135°}{(9 \text{ in})\cos 90° + (4.5 \text{ in})\cos 135°} = \tan^{-1}\left(\frac{12.182 \text{ in}}{-3.182 \text{ in}}\right) \\ &= -75.36°. \end{aligned}$$

However, the calculator used here does not recognize that the negative sign in the denominator indicates an angle in the second quadrant. Therefore, the angle is

$$\theta_4 = 180° - 75.36° = 104.64°. \qquad \textit{Ans.}$$

(c) Complex algebra approach The terms for Eq. (1) are

$$\begin{aligned} \mathbf{R}_A &= R_A\angle\theta_4 = R_A\cos\theta_4 + jR_A\sin\theta_4 \\ \mathbf{R}_{O_2} &= 9 \text{ in}\angle 90° = (9 \text{ in})\cos 90° + j(9 \text{ in})\sin 90° = 0 + j9 \text{ in.} \\ \mathbf{R}_{AO_2} &= 4.5 \text{ in}\angle 135° = (4.5 \text{ in})\cos 135° + j(4.5 \text{ in})\sin 135° \\ &= -3.182 \text{ in} + j3.182 \text{ in.} \end{aligned}$$

Substituting these into Eq. (1) yields

$$\mathbf{R}_A = (0 + j9 \text{ in}) + (-3.182 \text{ in} + j3.182 \text{ in}) = -3.182 \text{ in} + j12.182 \text{ in.}$$

Thus,

$$R_A = \sqrt{(-3.182 \text{ in})^2 + (12.182 \text{ in})^2} = 12.59 \text{ in} \qquad \textit{Ans.}$$

and

$$\theta_4 = \tan^{-1}\frac{y}{x} = \tan^{-1}\frac{12.182 \text{ in}}{-3.182 \text{ in}} = 104.64°. \qquad \textit{Ans.}$$

However, note the comment in (*b*) above regarding arc tangent results from the calculator.

(d) Vector algebraic approach Here we employ a scientific hand calculator that will add and subtract complex numbers in rectangular notation and will convert from rectangular to polar notation or vice-versa. Thus,

$$\mathbf{R}_{O_2} = 9 \text{ in}\angle 90^\circ = 0 + j9 \text{ in}$$

$$\mathbf{R}_{AO_2} = 4.5 \text{ in}\angle 135^\circ = -3.182 \text{ in} + j3.182 \text{ in}$$

$$\mathbf{R}_{A} = \mathbf{R}_{O_2} + \mathbf{R}_{AO_2} = (0 + j9 \text{ in}) + (-3.182 \text{ in} + j3.182)$$

$$= -3.182 \text{ in} + j12.182 \text{ in} = 12.59 \text{ in}\angle 104.64^\circ. \qquad \textit{Ans.}$$

(e) Numeric approach Some scientific hand calculators permit equations containing complex numbers to be solved in rectangular or polar form or both, with the results displayed in either mode. Using such a calculator, we enter

$$\mathbf{R}_{A} = 9 \text{ in}\angle 90^\circ + 4.5 \text{ in}\angle 135^\circ = 12.59 \text{ in}\angle 104.64^\circ. \qquad \textit{Ans.}$$

EXAMPLE 2.6

Perform a position analysis of the four-bar linkage illustrated in Fig. 2.25 by finding θ_3 and θ_4 for the two positions shown.

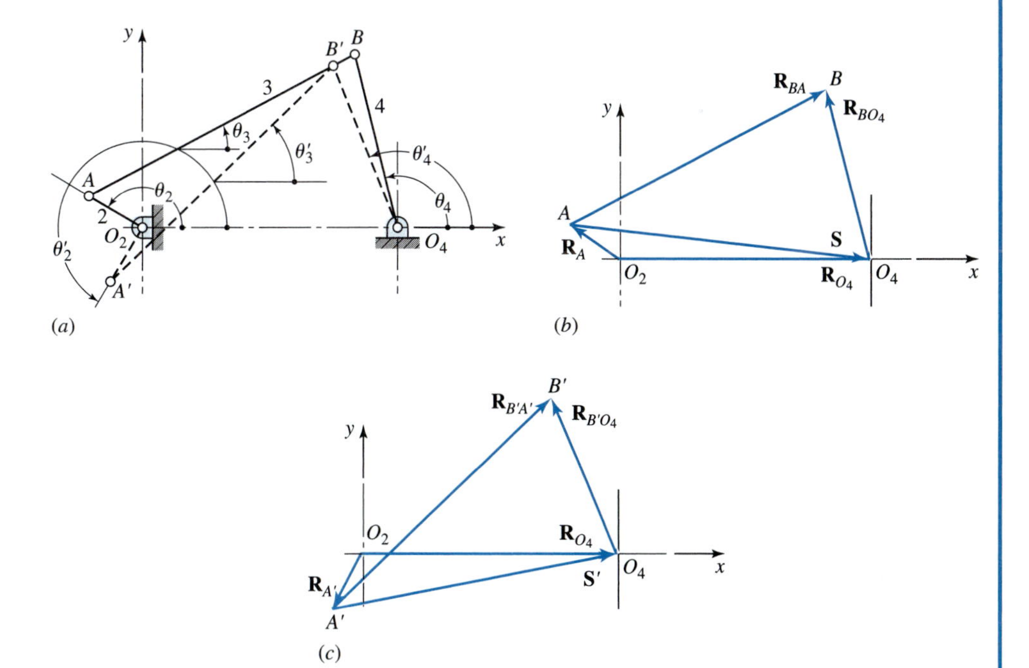

Figure 2.25 Example 2.6. Crank-and-rocker linkage illustrated in two different input positions. $R_{O_4O_2} = 600$ mm, $R_{AO_2} = 140$ mm, $R_{BA} = 690$ mm, $R_{BO_4} = 400$ mm, $\theta_2 = 150^\circ$, and $\theta_2' = 240^\circ$.

SOLUTION

For the first input angle $\theta_2 = 150°$, we refer to Fig. 2.25*b* and observe that

$$\mathbf{R}_A = 0.140 \text{ m}\angle 150°, \quad \text{and} \quad \mathbf{R}_{O_4} = 0.600 \text{ m}\angle 0°.$$

Therefore,

$$\mathbf{S} = \mathbf{R}_{O_4} - \mathbf{R}_A = 0.600 \text{ m}\angle 0° - 0.140 \text{ m}\angle 150° = 0.725 \text{ m}\angle - 5.54°.$$

Referring again to Fig. 2.25*b*, we note that the vectors in the triangle ABO_4 are related by the equation

$$\overset{\surd\surd}{\mathbf{S}} = \overset{\surd ?}{\mathbf{R}_{BA}} - \overset{\surd ?}{\mathbf{R}_{BO_4}}. \tag{a}$$

There are two unknown orientations in this equation, and so we identify this as case 4. Using Eq. (2.49) and substituting $\mathbf{S}$ for $\mathbf{C}$, R_{BA} for A, $-R_{BO_4}$ for B, θ_S for θ_C, and θ_4 for θ_B gives

$$\begin{aligned}\theta_4 &= \theta_S \pm \cos^{-1}\frac{S^2 + R_{BO_4}^2 - R_{BA}^2}{2SR_{BO_4}}\\ &= -5.54° \pm \cos^{-1}\frac{(0.725 \text{ m})^2 + (-0.400 \text{ m})^2 - (0.690 \text{ m})^2}{2(0.725 \text{ m})(-0.400 \text{ m})}\\ &= -5.54° \pm 111.18° = 105.64° \text{ or } -116.72°. \qquad \textit{Ans.}\end{aligned}$$

Note that we could have substituted $R_{BO_4} = +0.400$ m for B and we would have obtained $\theta_4 + 180°$ for the final result.

Next, using Eq. (2.50) and substituting θ_3 for θ_A gives

$$\begin{aligned}\theta_3 &= \theta_S \mp \cos^{-1}\frac{S^2 + R_{BA}^2 - R_{BO_4}^2}{2SR_{BA}}\\ &= -5.54° \mp \cos^{-1}\frac{(0.725 \text{ m})^2 + (0.690 \text{ m})^2 - (-0.400 \text{ m})^2}{2(0.725 \text{ m})(0.690 \text{ m})}\\ &= -5.54° \mp 32.72° = -38.26° \text{ or } 27.18°. \qquad \textit{Ans.}\end{aligned}$$

We follow the same procedure for the second input angle $\theta_2' = 240°$. Using Fig. 2.25*c* yields

$$\mathbf{S}' = \mathbf{R}_{O_4} - \mathbf{R}_A' = 0.600 \text{ m}\angle 0° - 0.140 \text{ m}\angle 240° = 0.681 \text{ m}\angle 10.26° \text{ m}$$

$$\theta_4' = 10.26° \pm \cos^{-1} \frac{(0.681\text{ m})^2 + (-0.400\text{ m})^2 - (0.690\text{ m})^2}{2(0.681\text{ m})(-0.400\text{ m})}$$

$$= 10.26° \pm 105.73° = 115.99° \text{ or } -95.47° \qquad \textit{Ans.}$$

$$\theta_3' = 10.26° \mp \cos^{-1} \frac{(0.681\text{ m})^2 + (0.690\text{ m})^2 - (-0.400\text{ m})^2}{2(0.681\text{ m})(0.690\text{ m})}$$

$$= 10.26° \mp 33.92° = -23.66° \text{ or } 44.18°. \qquad \textit{Ans.}$$

We recognize the positive angles as the solution of interest in all cases.

2.13 COUPLER-CURVE GENERATION

In Chapter 1, Fig. 1.23, we learned of the vast variety of useful coupler curves that are capable of being generated by the planar four-bar linkage. These curves are quite easy to obtain graphically, but computer-generated curves can be obtained more rapidly and are easier to vary to obtain the desired curve characteristics. Here we present the basic equations but omit the computer programming details required for screen display.

Applying the Chace approach to the four-bar linkage of Fig. 2.15, we first form the vector

$$\mathbf{S} = \mathbf{R}_C - \mathbf{R}_A.$$

Then we also note that

$$\mathbf{S} = R_{BA}\hat{\mathbf{R}}_{BA} - R_{BC}\hat{\mathbf{R}}_{BC},$$

where the two orientations $\hat{\mathbf{R}}_{BA}$ and $\hat{\mathbf{R}}_{BC}$ are the unknowns. This is case 4, and the solutions are given by Eqs. (2.57) and (2.58). Substituting gives

$$\mathbf{R}_{BC} = \pm\sqrt{R_{BA}^2 - \left(\frac{R_{BA}^2 - R_{BC}^2 + S^2}{2S}\right)^2}(\hat{\mathbf{S}} \times \hat{\mathbf{k}}) + \frac{R_{BC}^2 - R_{BA}^2 + S^2}{2S}\hat{\mathbf{S}} \qquad (2.61)$$

$$\mathbf{R}_{BA} = \pm\sqrt{R_{BA}^2 - \left(\frac{R_{BA}^2 - R_{BC}^2 + S^2}{2S}\right)^2}(\hat{\mathbf{S}} \times \hat{\mathbf{k}}) + \frac{R_{BA}^2 - R_{BC}^2 + S^2}{2S}\hat{\mathbf{S}}. \qquad (2.62)$$

The upper set of signs gives the solution for the crossed configuration of the linkage; the lower set thus applies to the open configuration of the linkage, as illustrated in Fig. 2.15*a* and Fig. 2.15*b*.

In Fig. 2.15, point P represents a coupler point that generates a curve called the coupler curve when crank 2 is rotated. From Fig. 2.15b we see that

$$\mathbf{R}_P = R_P e^{j\theta_6} = R_A e^{j\theta_2} + R_{PA} e^{j(\theta_3+\alpha)}. \tag{2.63}$$

We recognize this as a case 1 vector equation because R_p and θ_6 are two unknowns. The solutions can be found directly by applying Eqs. (2.43) and (2.44).

$$R_P = \sqrt{R_A^2 + R_{PA}^2 + 2R_A R_{PA}\cos(\theta_3 + \alpha - \theta_2)} \tag{2.64}$$

$$\theta_6 = \tan^{-1}\frac{R_A\sin\theta_2 + R_{PA}\sin(\theta_3 + \alpha)}{R_A\cos\theta_2 + R_{PA}\cos(\theta_3 + \alpha)}. \tag{2.65}$$

Note that both of these equations give double values coming from the double values for θ_3 and corresponding to the two closures of the linkage.

EXAMPLE 2.7

Calculate and plot points on the coupler curve of a four-bar linkage having the following dimensions: $R_C = 200$ mm, $R_A = 100$ mm, $R_{BA} = 250$ mm, $R_{BC} = 300$ mm, $R_{PA} = 150$ mm, and $\alpha = -45°$. This notation corresponds to that of Fig. 2.15b.

SOLUTION

For each value of the crank angle θ_2, the transmission angle γ is evaluated from Eq. (2.33). Next, Eqs. (2.29) and (2.31) or Eq. (2.50) are applied to give θ_3. Finally, the coupler-point positions are calculated from Eqs. (2.64) and (2.65). The solutions for crank angles from 0° to 90° are displayed in Table 2.3, and the full coupler curve is illustrated in Fig. 2.26. Only one of the two solutions is calculated and plotted.

TABLE 2.3 Coupler-curve points obtained for Example 2.7.

θ_2, deg	γ, deg	θ_3, deg	R_P, mm	θ_6, deg	R_P^x, mm	R_P^y, mm
0.0	18.2	110.5	212	40.1	162	136
10.0	18.9	99.4	232	36.9	186	139
20.0	20.9	87.8	245	33.7	204	136
30.0	23.9	77.5	250	31.5	213	131
40.0	27.4	69.2	248	30.5	213	126
50.0	31.3	62.9	241	30.7	207	123
60.0	35.2	58.3	230	31.7	196	121
70.0	39.2	55.1	218	33.5	182	120
80.0	43.1	53.0	204	35.8	166	119
90.0	46.9	51.8	190	38.3	149	118

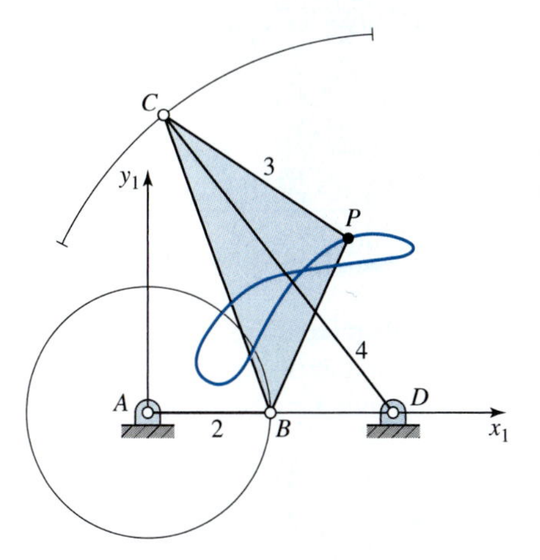

Figure 2.26 Plot of coupler curve of Example 2.7.

2.14 DISPLACEMENT OF A MOVING POINT

We have concerned ourselves so far with only a single instantaneous position of a point, but because we wish to study motion, we must be concerned with the relationship between a succession of positions.

In Fig. 2.27 a particle, originally at point P, is moving along the path shown and, some time later, arrives at the position P'. The *displacement* of the point $\Delta\mathbf{R}_P$ during the time interval is defined as the *net change in position*.

$$\Delta\mathbf{R}_P = \mathbf{R}'_P - \mathbf{R}_P. \tag{2.66}$$

Displacement is a vector quantity having the magnitude and direction of the vector from point P to point P'.

It is important to note that the displacement $\Delta\mathbf{R}_P$ is the net change in position and does not depend on the particular path taken between points P and P'. The magnitude of this vector is *not* necessarily equal to the length of the path (the distance traveled), and its direction is *not* necessarily along the tangent to the path, although both are true when the displacement is infinitesimally small. Knowledge of the path actually traveled from P

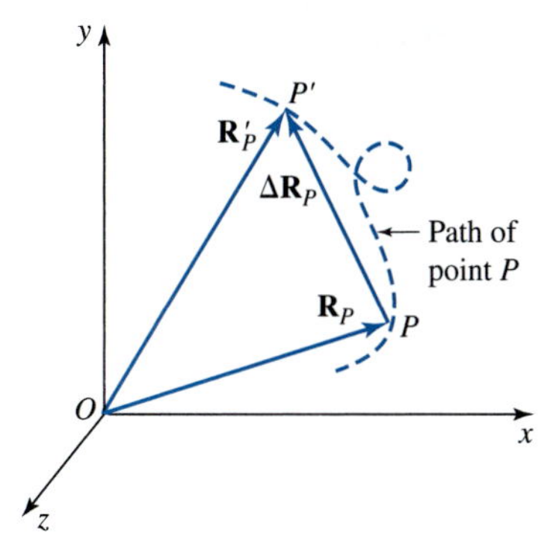

Figure 2.27 Displacement of a moving point.

to P' is not even necessary to find the displacement vector, providing the initial and final positions of the point are known.

2.15 DISPLACEMENT DIFFERENCE BETWEEN TWO POINTS

In this section we consider the difference in the displacements of two moving points. In particular, we are concerned with the case where the two moving points are both fixed in the same rigid body. The situation is illustrated in Fig. 2.28, where rigid body 2 moves from an initial position defined by $x_2y_2z_2$ to a later position defined by $x'_2y'_2z'_2$.

From Eq. (2.6), the position difference between the two points P and Q of body 2 at the initial instant is

$$\mathbf{R}_{PQ} = \mathbf{R}_P - \mathbf{R}_Q. \tag{a}$$

After the displacement of body 2, the two points are located at P' and Q'. At that time the position difference is

$$\mathbf{R}'_{PQ} = \mathbf{R}'_P - \mathbf{R}'_Q. \tag{b}$$

During the time interval of the movement the two points have undergone individual displacements of $\Delta\mathbf{R}_P$ and $\Delta\mathbf{R}_Q$, respectively.

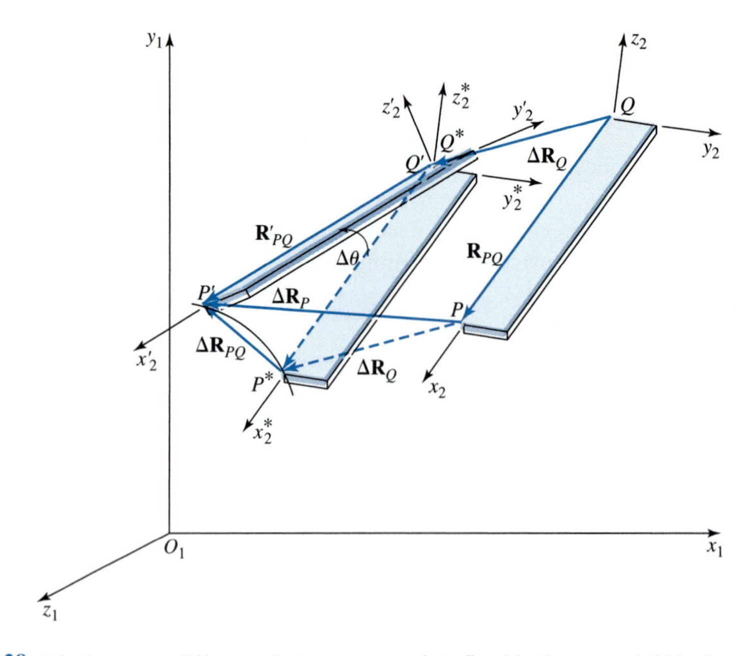

Figure 2.28 Displacement difference between two points fixed in the same rigid body.

As the name implies, the *displacement difference* between the two points is defined as the net difference between their respective displacements and can be written as

$$\Delta\mathbf{R}_{PQ} = \Delta\mathbf{R}_P - \Delta\mathbf{R}_Q. \tag{2.67}$$

Note that this equation corresponds to the vector triangle PP^*P' in Fig. 2.28. As stated in the previous section, the displacement depends only on the net change in position and not on the path by which it was achieved. Therefore, no matter how the body (containing points P and Q) was *actually* displaced, we are free to visualize the path as we choose. Equation (2.67) leads us to visualize the displacement as taking place in two stages. First, the body translates (slides without rotation) from $x_2y_2z_2$ to $x_2^*y_2^*z_2^*$; during this movement all particles, including P and Q, have the same displacement $\Delta\mathbf{R}_Q$. Second, the body rotates about point Q' through the angle $\Delta\theta$ to the final position $x_2'y_2'z_2'$.

A different interpretation can be obtained by manipulating Eq. (2.67) as

$$\begin{aligned}\Delta\mathbf{R}_{PQ} &= (\mathbf{R}_P' - \mathbf{R}_P) - (\mathbf{R}_Q' - \mathbf{R}_Q)\\ &= (\mathbf{R}_P' - \mathbf{R}_Q') - (\mathbf{R}_P - \mathbf{R}_Q)\end{aligned} \tag{c}$$

and then, from Eqs. (a) and (b),

$$\Delta\mathbf{R}_{PQ} = \mathbf{R}_{PQ}' - \mathbf{R}_{PQ}. \tag{2.68}$$

This equation corresponds to the vector triangle $Q'P^*P'$ in Fig. 2.28 and demonstrates that the displacement difference, defined as the difference between the two displacements, is equal to the net change between the position-difference vectors.

In either interpretation we are illustrating *Euler's theorem*, which states that *any displacement of a rigid body is equivalent to the sum of a net translation of one point and a net rotation of the body about that point*. We also see that only the rotation contributes to the displacement difference between two points fixed in the same rigid body; that is, *there is no difference between the displacements of any two points fixed in the same rigid body as the result of a translation.* (See Section 2.16 for the definition of the term *translation.*)

In view of the above discussion, we can visualize the displacement difference $\Delta\mathbf{R}_{PQ}$ as the displacement that would be seen for point P by a moving observer who travels along, always staying coincident with point Q *but not rotating* with the moving body—that is, always using the absolute coordinate axes $x_1y_1z_1$ to measure direction. It is important to understand the difference between the interpretation of an observer moving with point Q but not rotating and the case of the observer *on* the moving body. To an observer *on* body 2, both points P and Q would appear stationary; neither would be seen to have a displacement because they do not move relative to the observer, and the displacement difference seen by such an observer would be zero.

2.16 ROTATION AND TRANSLATION

Using the concept of displacement difference between two points fixed in the same rigid body, we are now able to define the terms translation and rotation.

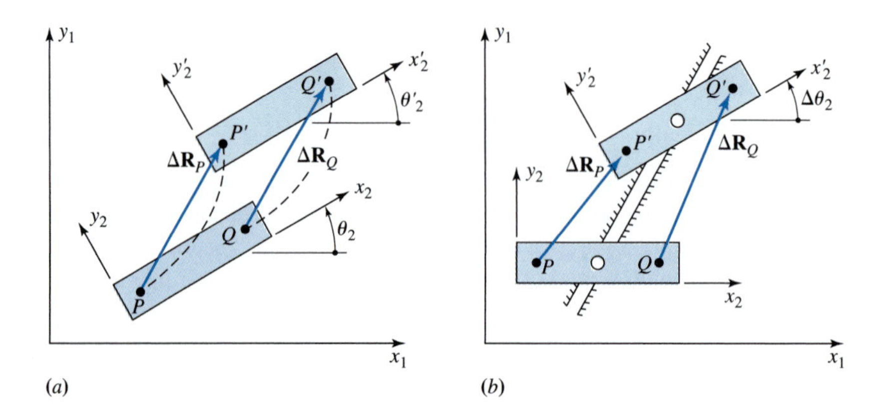

Figure 2.29 (*a*) Translation: $\Delta\mathbf{R}_P = \Delta\mathbf{R}_Q$, $\Delta\theta_2 = 0$; (*b*) rotation $\Delta\mathbf{R}_P \neq \Delta\mathbf{R}_Q$, $\Delta\theta_2 \neq 0$.

Translation is defined as *a motion of a body for which the displacement difference between any two points P and Q in the body is zero* or, from the displacement-difference equation (2.68),

$$\Delta\mathbf{R}_{PQ} = \Delta\mathbf{R}_P - \Delta\mathbf{R}_Q = \mathbf{0}$$
$$\Delta\mathbf{R}_P = \Delta\mathbf{R}_Q, \tag{2.69}$$

which states that *the displacements of any two points in the body are equal.* Rotation is a motion of the body for which different points of the body exhibit different displacements.

Figure 2.29*a* illustrates a situation where the body has moved along a curved path from position x_2y_2 to position $x'_2y'_2$. Despite the fact that the point paths are curved,* $\Delta\mathbf{R}_P$ is still equal to $\Delta\mathbf{R}_Q$ and the body has undergone a translation. Note that in translation the point paths described by any two points on the body are identical and there is no change of angular orientation between the moving coordinate system and the coordinate system of the observer; that is, $\Delta\theta_2 = \theta'_2 - \theta_2 = 0$.

In Fig. 2.29*b* the center point of the moving body is *constrained* to move along a straight-line path. Yet, as it does so, the body rotates so that $\Delta\theta_2 = \theta'_2 - \theta_2 \neq 0$ and the displacements $\Delta\mathbf{R}_P$ and $\Delta\mathbf{R}_Q$ are not equal. Although there is no obvious point on the body about which it has rotated, the coordinate system $x'_2y'_2$ has changed angular orientation relative to x_1y_1 and the body is said to have undergone a rotation. Note that the point paths described by P and Q are not equal.

We see from these two examples that rotation, or translation, of a body cannot be defined from the motion of a single point. These are characteristic motions of a body or a coordinate system. It is improper to speak of "rotation of a point" because there is no meaning for angular orientation of a point. It is also improper to associate the terms "rotation" and "translation" with the rectilinear or curvilinear characteristics of a single

* Translation in which the point paths are straight lines is called *rectilinear* translation; translation in which the point paths are *not* straight lines is referred to as *curvilinear* translation.

point path. Although it does not matter which points of the body are chosen, the motion of two or more points must be compared before meaningful definitions exist for these terms.

2.17 APPARENT DISPLACEMENT

We have already observed that the displacement of a moving point does not depend on the particular path traveled. However, because displacement is computed from the position vectors of the endpoints of the path, knowledge of the coordinate system of the observer is essential.

In Fig. 2.30 we identify three bodies. The first, body 1, is a fixed or stationary body containing the absolute reference system $x_1y_1z_1$. The second is a moving body, which we call body 2, and we fix to it the reference system $x_2y_2z_2$. Thus, $x_2y_2z_2$ is a moving system. We then identify the third as body 3 and permit it to move with respect to body 2.

We also identify two observers. We designate HE to be an observer fixed to body 1, the stationary system, and we designate SHE to be another observer on body 2, fixed to the moving system $x_2y_2z_2$. Thus, we might say HE is on the ground observing, whereas SHE is going for a ride on body 2.

Now consider a particle, or passenger, P_3, situated on body 3 and moving along a known path on body 2. HE and SHE are both observing the motion of particle P_3 and we want to discover what they see.

We must define another point P_2, which is fixed in (or rigidly attached to) body 2 and is instantaneously coincident with P_3.

Now let body 2 and $x_2y_2z_2$ be displaced to a new position $x'_2y'_2z'_2$. While this motion is taking place, let P_3 move to another position on body 2, which we identify as P'_3. But P_2, because it is fixed to body 2, moves with body 2 and is now found in the new position identified as P'_2. SHE, on body 2, reports the motion of the passenger or particle P_3 as the vector $\Delta\mathbf{R}_{P_3/2}$, which is read as the displacement of P_3 as it appears to an observer on

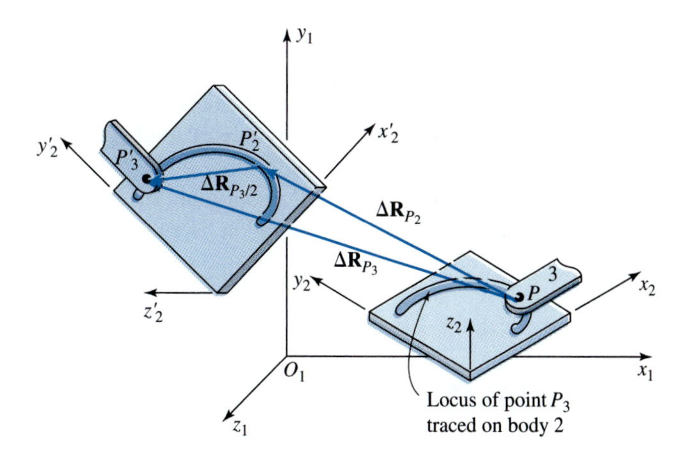

Figure 2.30 Apparent displacement of a point.

body 2. This is called the apparent-displacement vector. Note that SHE sees no motion of P_2 because it appears to HER to be stationary on body 2. Therefore, $\Delta\mathbf{R}_{P_2/2} = \mathbf{0}$.

However, HE, on the stationary body, reports the displacement of particle P_3 by the vector $\Delta\mathbf{R}_{p_3}$. Note that HE also reports the displacement of P_2 by the vector $\Delta\mathbf{R}_{P_2}$. From the vector triangle in Fig. 2.30 we see that the observations of the two observers are related by the *apparent-displacement equation*,

$$\Delta\mathbf{R}_{P_3} = \Delta\mathbf{R}_{P_2} + \Delta\mathbf{R}_{P_3/2}. \tag{2.70}$$

We can take this equation as the definition of the apparent-displacement vector, although it is important also to understand the physical concepts involved. Note that the apparent-displacement vector relates the absolute displacements of two *coincident points*, which are particles of *different moving bodies*. Note also that there is no restriction on the actual location of the observer moving with coordinate system 2, only that SHE be fixed in that coordinate system so that SHE senses no displacement for point P_2.

One primary use of the apparent displacement is to determine an absolute displacement. It is not uncommon in machines to find a point such as P_3 that is constrained to move along a known slot or path or guideway defined by the shape of another moving link 2. In such cases it may be much more convenient to measure or calculate $\Delta\mathbf{R}_{P_2}$ and $\Delta\mathbf{R}_{P_3/2}$ and to use Eq. (2.70) than to measure the absolute displacement $\Delta\mathbf{R}_{P_3}$ directly.

2.18 ABSOLUTE DISPLACEMENT

In reflecting on the definition and concept of the apparent-displacement vector, we conclude that the absolute displacement of a moving point $\Delta\mathbf{R}_{P_3/1}$ is the special case of an apparent displacement where the observer is fixed in the absolute coordinate system. As explained for the position vector, the notation is often abbreviated to read $\Delta\mathbf{R}_{P_3}$ or just $\Delta\mathbf{R}_P$ and an absolute observer is implied whenever it is not noted explicitly.

Perhaps a better physical understanding of apparent displacement can be achieved by relating it to absolute displacement. Imagine an automobile P_3 traveling along a roadway and under observation by an absolute observer some distance off to one side. Consider how this observer visually senses the motion of the car. Although HE may not be conscious of all of the following steps, the contention here is that the observer first imagines a point P_1, *coincident* with P_3, which HE defines in his mind as stationary; HE may relate to a fixed point of the roadway or nearby tree or roadsign, for example. HE then compares his later observations of the car P_3 with those of P_1 to sense displacement. Note that HE does not compare with his own location but with the initially coincident point P_1. In this instance, the apparent-displacement equation becomes an identity:

$$\Delta\mathbf{R}_{P_3} = \overset{0}{\cancel{\Delta\mathbf{R}_{P_1}}} + \Delta\mathbf{R}_{P_3/1}.$$

2.19 APPARENT ANGULAR DISPLACEMENT

In general, rotations cannot be treated as vectors (this will be explained in some detail in Section 3.2). However, for planar motion, the idea of apparent displacement also extends

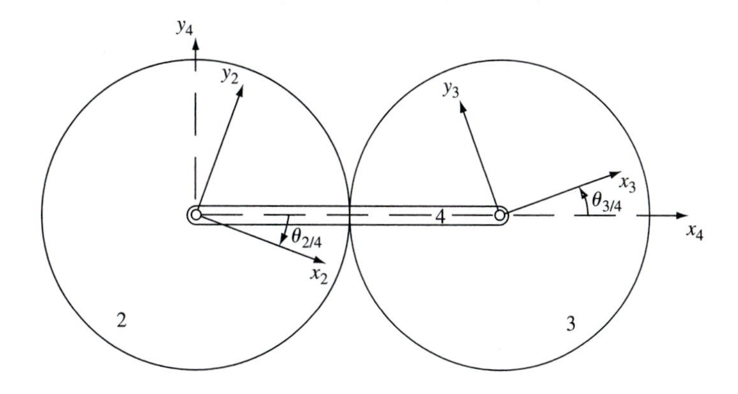

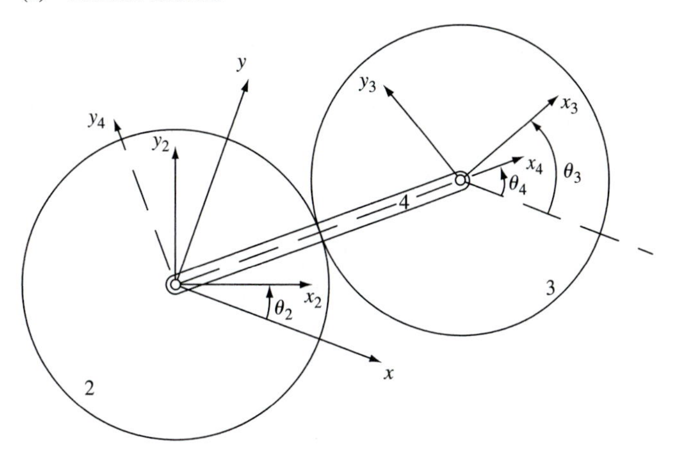

Figure 2.31 (*a*) Apparent rotations seen from arm 4; (*b*) absolute rotations.

to include rotation. For example, suppose we consider the rotations of the two gears 2 and 3 connected by arm 4 illustrated in Fig. 2.31. In Fig. 2.31*a*, we see the rotations as they would appear in HER coordinate system attached to the arm, body 4. In Fig. 2.31*b*, we see the same situation, but as it might appear to HIM in the absolute frame of reference.

Although it may not be easy to write equations relating the absolute rotations of the gears illustrated in Fig. 2.31, it is quite easy to relate their apparent rotations if we take HER point of view as an observer riding on the moving coordinate system attached to arm 4. From HER vantage point it is obvious that, if there is no slipping between bodies 2 and 3, then the arc lengths of the two gears that pass arm 4 at the point of contact must be equal. That is, if ρ_2 is the radius of gear 2 and ρ_3 is the radius of gear 3, then

$$\rho_2 \Delta\theta_{2/4} = -\rho_3 \Delta\theta_{3/4}, \tag{a}$$

where $\Delta\theta_{2/4}$ and $\Delta\theta_{3/4}$ are the angular displacements of gears 2 and 3 as they appear to HER in coordinate system 4, and the minus sign accounts for the difference in the senses of the two apparent rotations.

When these angular displacements are replaced by those seen by HIM from the absolute coordinate system, then this equation becomes

$$\rho_2(\Delta\theta_2 - \Delta\theta_4) = -\rho_3(\Delta\theta_3 - \Delta\theta_4). \tag{b}$$

Using these ideas, the vector loop-closure equations for many problems having rolling contact without slipping can be solved. The following two examples will make the procedure clear.

EXAMPLE 2.8

For the mechanism illustrated in Fig. 2.32, define a set of vectors that is suitable for a complete kinematic analysis. Label and show the sense and orientation of each vector on the mechanism. Write the vector loop equation(s) for the mechanism and identify suitable input(s) for the mechanism. Identify the known quantities, the unknown variables, and any constraints. If you have identified constraints, then write the constraint equation(s).

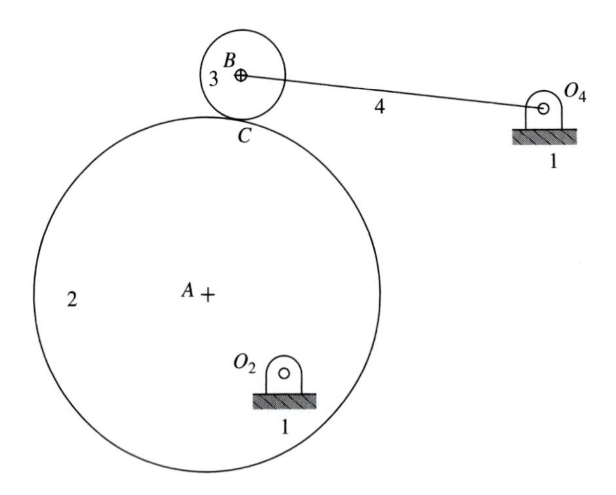

Figure 2.32 Example 2.8. Original mechanism.

SOLUTION

Because the circular cam, link 2, and the roller follower, link 3, are to remain in contact, let us consider them linked together by a (fictitious) arm, link 23. This leads us to consider the set of vectors illustrated in Fig. 2.33.

The vector loop equation for the mechanism can be written as

$$\overset{\surd\surd}{\mathbf{R}_2} + \overset{\surd ?}{\mathbf{R}_{23}} - \overset{\surd ?}{\mathbf{R}_4} - \overset{\surd\surd}{\mathbf{R}_{12}} - \overset{\surd\surd}{\mathbf{R}_{11}} = \mathbf{0}. \qquad \textit{Ans.} \ (1)$$

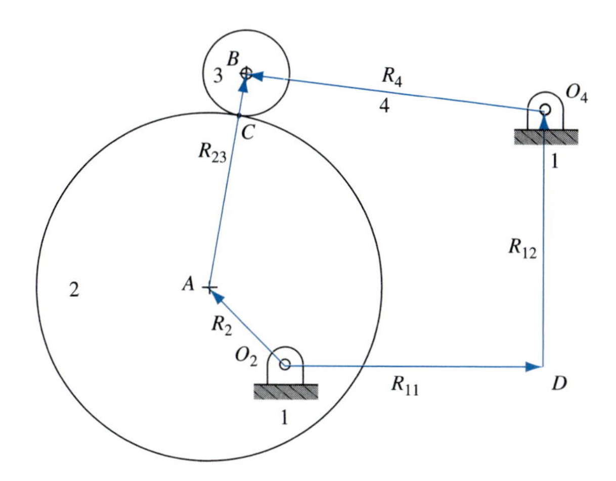

Figure 2.33 Vectors chosen for Example 2.8.

This is a valid loop-closure equation and, given θ_2 as input, it can be solved for the directions of link 4 and the fictitious link 23. However, this does not yet allow a solution for the rotation of the roller, link 3.

Because there is rolling contact between cam 2 and roller 3 at point C, however, let us consider the apparent angular displacements of links 2 and 3 as they would appear to an observer riding on the fictitious arm, link 23. Such an observer would observe

$$\rho_3 \Delta\theta_{3/23} = -\rho_2 \Delta\theta_{2/23}, \tag{2}$$

which gives the following relationship between the absolute angular displacements:

$$\rho_3(\Delta\theta_3 - \Delta\theta_{23}) = -\rho_2(\Delta\theta_2 - \Delta\theta_{23}). \tag{3}$$

Rearranging this equation gives

$$\rho_3 \Delta\theta_3 = (\rho_2 + \rho_3)\Delta\theta_{23} - \rho_2 \Delta\theta_2. \qquad \textit{Ans.} \ (4)$$

Because the angular displacement of links 2 is known as input and $\Delta\theta_{23}$ is determined from Eq. (1), the angular displacement $\Delta\theta_3$ can be determined from Eq. (4). It is worth noting that we have a solution for the rotation of link 3 although no vector is attached to or rotates with link 3.

EXAMPLE 2.9

For the mechanism illustrated in Fig. 2.34, define a set of vectors that is suitable for a complete kinematic analysis of the mechanism. Label and show the sense and orientation of each vector on the mechanism. Write the vector loop equation(s) for the mechanism and identify suitable input(s) for the mechanism. Identify the known quantities, the unknown variables, and any constraints. If you have identified constraints, then write the constraint equation(s).

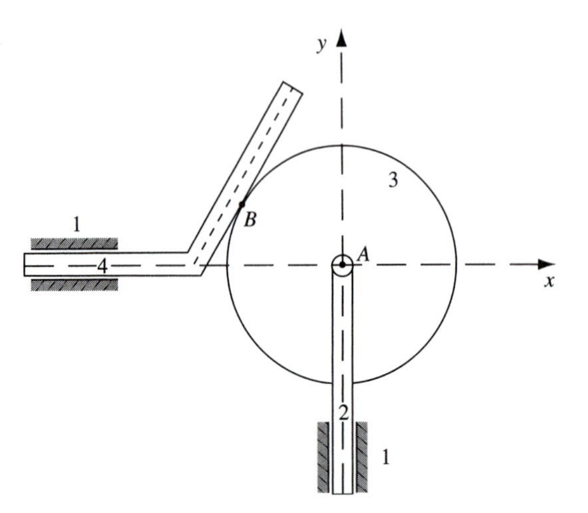

Figure 2.34 Example 2.9. Original mechanism.

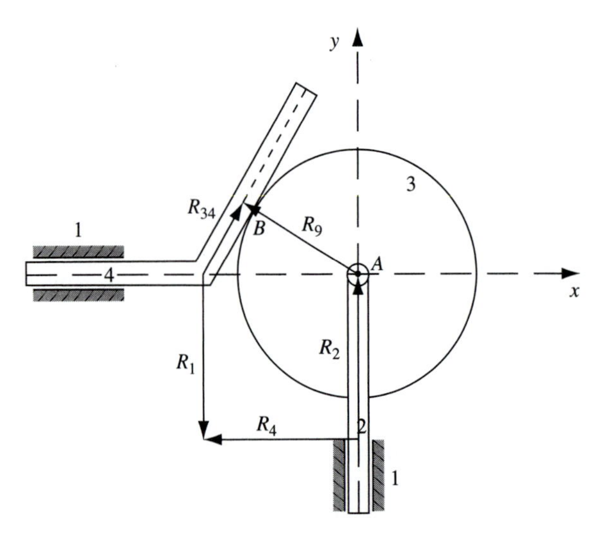

Figure 2.35 Vectors chosen for Example 2.9.

SOLUTION

Let us consider the set of vectors illustrated in Fig. 2.35.

The length of the vector $\mathbf{R}_2$ represents the input motion and the length of the vector $\mathbf{R}_4$ indicates the output motion. The loop-closure equation is

$$\overset{I\surd}{\mathbf{R}_2} + \overset{\surd C}{\mathbf{R}_9} - \overset{?\surd}{\mathbf{R}_{34}} + \overset{\surd\surd}{\mathbf{R}_1} - \overset{?\surd}{\mathbf{R}_4} = \mathbf{0}. \qquad \textit{Ans.}\ (1)$$

The constraint that gives the orientation of vector $\mathbf{R}_9$ is

$$\theta_9 = \theta_{34} + 90^\circ. \qquad \textit{Ans.}\ (2)$$

The angular displacement of roller 3 can be determined by considering how it appears to an observer riding on link 4. As such, an observer watches the changing position of the point of contact B; SHE must see the same increment of surface on both link 3 and link 4. The constraint relationship can be written as

$$\Delta R_{34} = \rho_3 \Delta\theta_{3/4} = \rho_3(\Delta\theta_3 - \Delta\not{\theta}_4^0) = \rho_3 \Delta\theta_3.$$

Therefore, the angular displacement of roller 3 is

$$\Delta\theta_3 = \Delta R_{34}/\rho_3. \qquad \textit{Ans.} \ (3)$$

2.20 REFERENCES

[1] See, for example, C. R. Mischke, *Mathematical Model Building*. Ames, IA: Iowa State University Press, 1980, 86.

[2] Chace, M. A. 1963. Vector analysis of linkages. *J. Eng. Ind. ASME Trans. B* 55: 289–97.

PROBLEMS*

†**2.1** Describe and sketch the locus of a point A that moves according to the equations $R_A^x = a t\cos(2\pi t)$, $R_A^y = a t\sin(2\pi t)$, and $R_A^z = 0$.

†**2.2** Find the position difference from point P to point Q on the curve $y = x^2 + x - 16$, where $R_P^x = 2$ and $R_Q^x = 4$.

†**2.3** The path of a moving point is defined by the equation $y = 2x^2 - 28$. Find the position difference from point P to point Q if $R_P^x = 4$ and $R_Q^x = -3$.

†**2.4** The path of a moving point P is defined by the equation $y = 60 - x^3/3$. What is the displacement of the point if its motion begins when $R_P^x = 0$ and ends when $R_P^x = 3$?

†**2.5** If point A moves on the locus of Problem 2.1, find its displacement from $t = 2$ to $t = 2.5$.

†**2.6** The position of a point is given by the equation $\mathbf{R} = 100e^{j2\pi t}$. What is the path of the point? Determine the displacement of the point from $t = 0.10$ to $t = 0.40$.

†**2.7** The equation $\mathbf{R} = (t^2 + 4)e^{-j\pi t/10}$ defines the position of a point. In which direction is the position vector rotating? Where is the point located when $t = 0$? What is the next value t can have if the orientation of the position vector is to be the same as it is when $t = 0$? What is the displacement from the first position of the point to the second?

†**2.8** The location of a point is defined by the equation $R = (4t + 2)e^{j\pi t^2/30}$, where t is time in seconds. Motion of the point is initiated when $t = 0$. What is the displacement during the first 3 s? Find the change in angular orientation of the position vector during the same time interval.

†**2.9** Link 2 in Fig. P2.9 rotates according to the equation $\theta = \pi t/4$. Block 3 slides outward on link 2 according to the equation $r = t^2 + 2$. What is the absolute displacement $\Delta\mathbf{R}_{P_3}$ from $t = 1$ to $t = 2$? What is the apparent displacement $\Delta\mathbf{R}_{P_{3/2}}$?

---

* When assigning problems, the instructor may wish to specify the method of solution to be used, because a variety of approaches have been presented in the text.

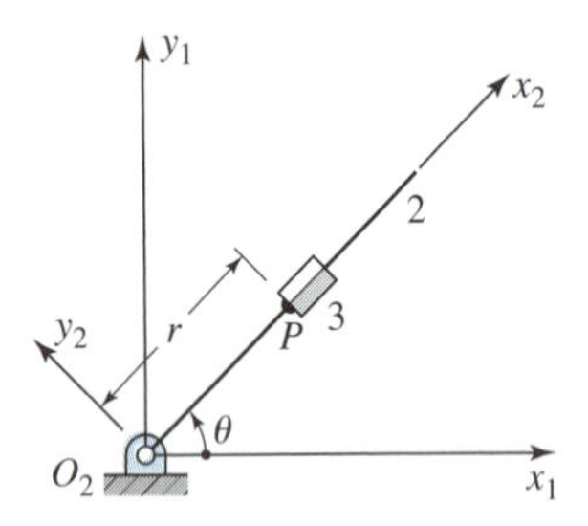

Figure P2.9

†2.10 A wheel with center at O rolls without slipping so that its center is displaced 10 in. to the right. What is the displacement of point P on the periphery during this interval?

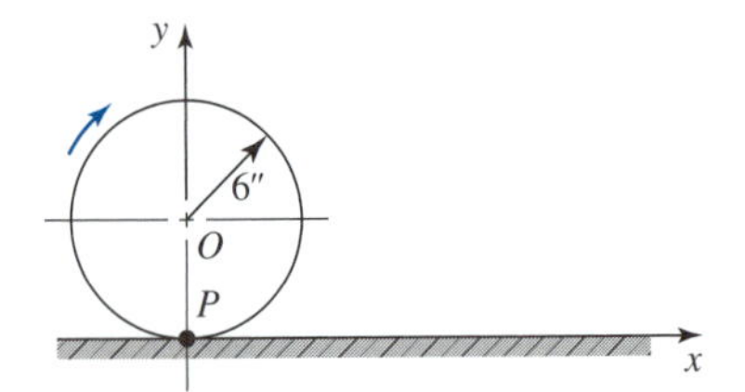

Figure P2.10 Rolling wheel.

†2.11 A point Q moves from A to B along link 3, whereas link 2 rotates from $\theta_2 = 30°$ to $\theta_2' = 120°$. Find the absolute displacement of Q.

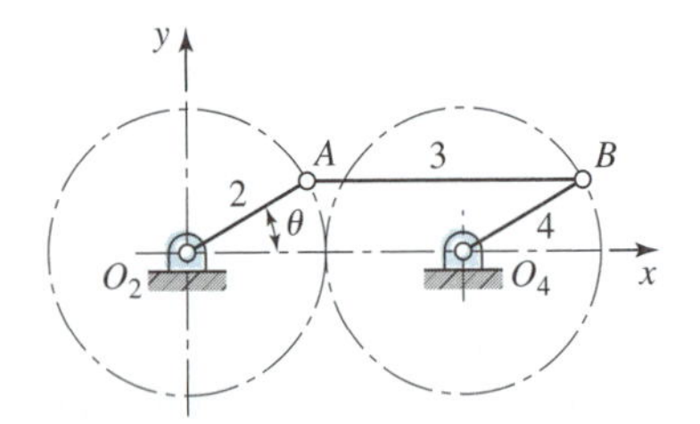

Figure P2.11 $R_{AO_2} = R_{BO_4} = 3$ in, $R_{BA} = R_{O_4O_2} = 6$ in.

†2.12 The linkage is driven by moving sliding block 2. Write the loop-closure equation. Solve analytically for the position of sliding block 4. Check the result graphically for the position where $\phi = -45°$.

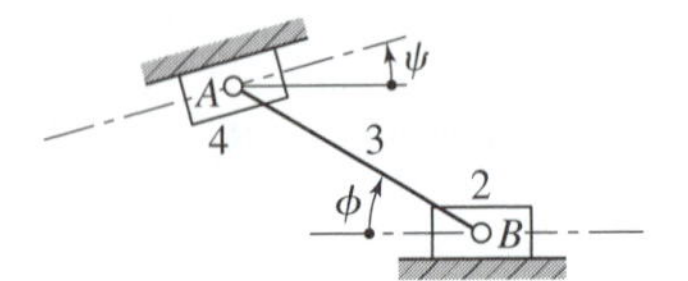

Figure P2.12 $R_{AB} = 200$ mm, $\psi = 15°$.

†2.13 The offset slider-crank mechanism is driven by rotating crank 2. Write the loop-closure equation. Solve for the position of slider 4 as a function of θ_2.

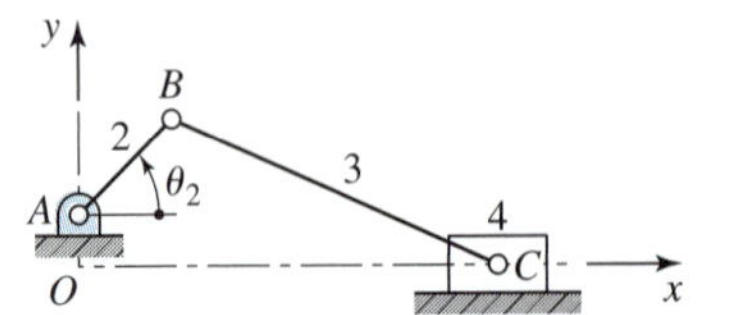

Figure P2.13 $R_{AO} = 1$ in, $R_{BA} = 2.5$ in, and $R_{CB} = 7$ in.

2.14 For the mechanism shown in Fig. P2.14, define a set of vectors that is suitable for a complete kinematic analysis of the mechanism. Label and show the sense and orientation of each vector in Fig. P2.14. Write the vector loop equation(s) for the mechanism. Identify suitable input(s) for the mechanism. Identify the known quantities, the unknown variables, and any constraints. If you have identified constraints, then write the constraint equation(s).

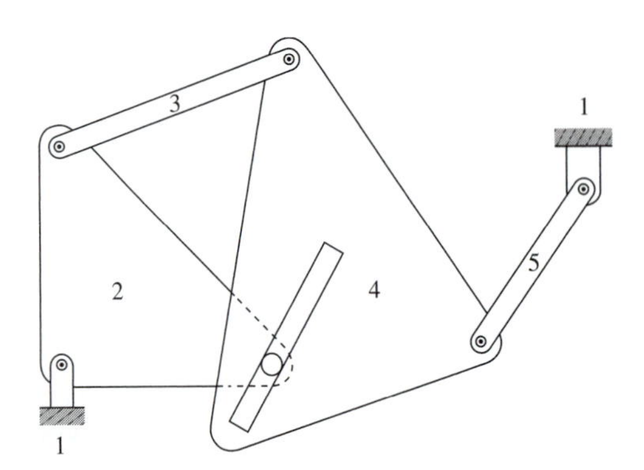

Figure P2.14

2.15 Assume rolling with no slip between pinion 5 and rack 4 in the mechanism illustrated in Fig. P2.15. Define a set of vectors that is suitable for a complete kinematic analysis of the mechanism. Label and show the sense and orientation of each vector in Fig. P2.15. Write the vector loop equation(s) for the mechanism. Identify suitable input(s) for the mechanism. Identify the known quantities, the unknown variables, and any constraints. If you have identified constraints, then write the constraint equation(s).

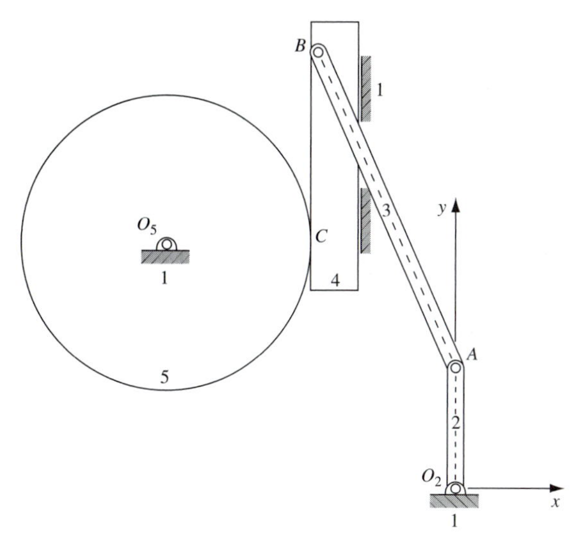

Figure P2.15 A rack-and-pinion mechanism.

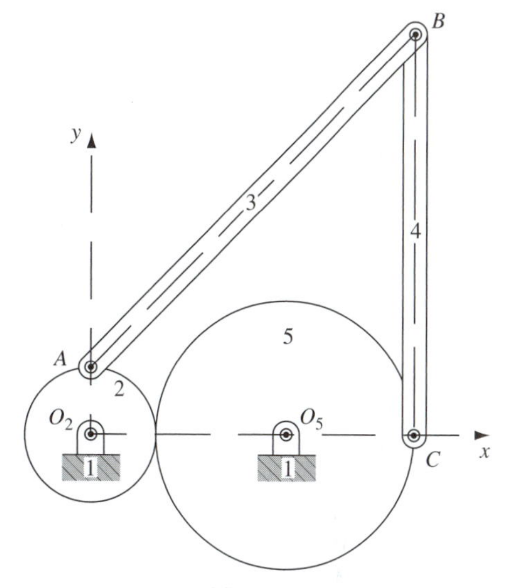

Figure P2.16 Geared five-bar mechanism.

2.16 For the geared five-bar mechanism illustrated in Fig. P2.16, there is rolling with no slipping between gears 2 and 5. Define a set of vectors that is suitable for a complete kinematic analysis of the mechanism. Label and show the sense and orientation of each vector in Fig. P2.16. Write the vector loop equation(s) for the mechanism. Identify suitable input(s) for the mechanism. Identify the known quantities, the unknown variables, and any constraints. If you have identified constraints, then write the constraint equation(s).

2.17 For the mechanism illustrated in Fig. P2.17, gear 3 is pinned to link 4 at point B and is rolling without slipping on semicircular ground link 1. The radius of the semicircular ground link is ρ_1 and the radius of gear 3 is ρ_3. Define a set of vectors that is suitable for a complete kinematic analysis of the mechanism. Label and show the sense and orientation of each vector in Fig. P2.17. Write the vector loop equation(s) for the mechanism. Identify suitable input(s) for the mechanism. Identify the known quantities, the unknown variables, and

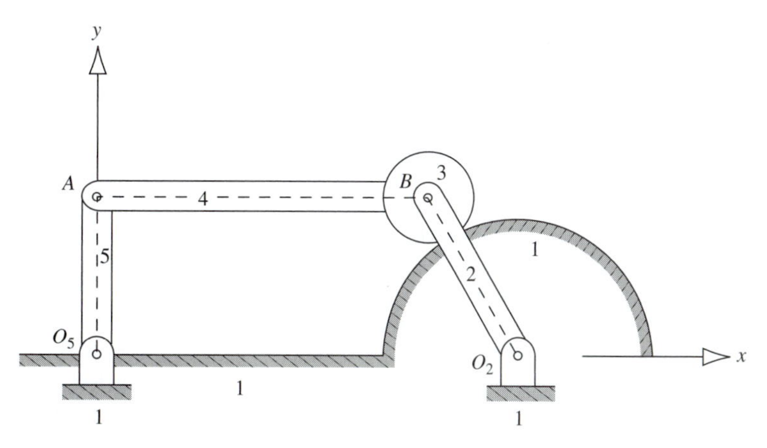

Figure P2.17

any constraints. If you have identified constraints, then write the constraint equation(s).

2.18 For the mechanism illustrated in Fig. P1.6, define a set of vectors that is suitable for a complete kinematic analysis of the mechanism. Label and show the sense and orientation of each vector in Fig. P1.6. Write the vector loop equation(s) for the mechanism. Identify suitable input(s) for the mechanism. Identify the known quantities, the unknown variables, and any constraints. If you have identified constraints, then write the constraint equation(s).

2.19 For the mechanism illustrated in Fig. P1.8, define a set of vectors that is suitable for a complete kinematic analysis of the mechanism. Label and show the sense and orientation of each vector in Fig. P1.8. Write the vector loop equation(s) for the mechanism. Identify suitable input(s) for the mechanism. Identify the known quantities, the unknown variables, and any constraints. If you have identified constraints, then write the constraint equation(s).

2.20 For the mechanism illustrated in Fig. P1.9, define a set of vectors that is suitable for a complete kinematic analysis of the mechanism. Label and show the sense and orientation of each vector in Fig. P1.9. Write the vector loop equation(s) for the mechanism. Identify suitable input(s) for the mechanism. Identify the known quantities, the unknown variables, and any constraints. If you have identified constraints, then write the constraint equation(s).

2.21 For the mechanism illustrated in Fig. P1.10, define a set of vectors that is suitable for a complete kinematic analysis of the mechanism. Label and show the sense and orientation of each vector on Fig. P1.10. Write the vector loop equation(s) for the mechanism. Identify suitable input(s) for the mechanism. Identify the known quantities, the unknown variables, and any constraints. If you have identified constraints, then write the constraint equation(s).

†2.22 Write a calculator program to find the sum of any number of two-dimensional vectors expressed in mixed rectangular or polar forms. The result should be obtainable in either form with the magnitude and angle of the polar form having only positive values.

†2.23 Write a computer program to plot the coupler curve of any crank–rocker or double-crank form of the four-bar linkage. The program should accept four link lengths and either rectangular or polar coordinates of the coupler point relative to the coupler.

†2.24 For each linkage illustrated in Fig. P2.24, find the path of point P: (*a*) inverted slider-crank mechanism; (*b*) second inversion of the slider-crank mechanism; (*c*) straight-line mechanism; and (*d*) drag–link mechanism.

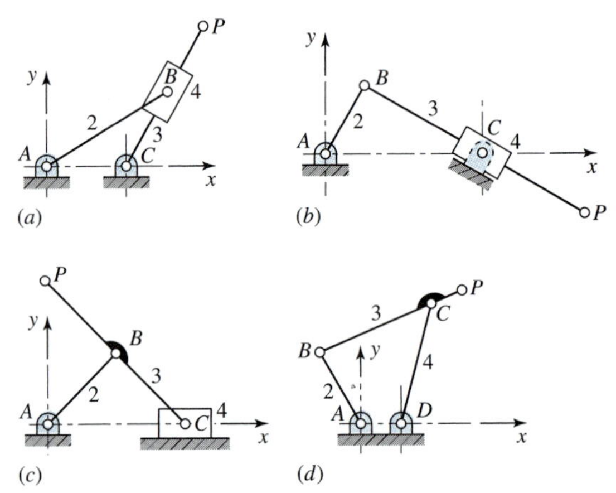

Figure P2.24 (*a*) $R_{CA} = 2$ in, $R_{BA} = 3.5$ in, and $R_{PC} = 4$ in; (*b*) $R_{CA} = 40$ mm, $R_{BA} = 20$ mm, and $R_{PB} = 65$ mm; (*c*) $R_{BA} = R_{CB} = R_{PB} = 25$ mm; (*d*) $R_{DA} = 1$ in, $R_{BA} = 2$ in, $R_{CC} = R_{DC} = 3$ in, and $R_{PB} = 4$ in.

†2.25 Using the offset slider-crank mechanism in Fig. P2.13, find the crank angles corresponding to the extreme values of the transmission angle.

2.26 In Section 1.10 it is pointed out that the transmission angle reaches an extreme value for the four-bar linkage when the crank lies on the line between the fixed pivots. Referring to Fig. 2.19, this means that γ reaches a maximum or minimum when crank 2 is collinear with the line O_2O_4. Show analytically that this statement is true.

†2.27 Figure P2.27 illustrates a crank-and-rocker four-bar linkage in the first of its two limit positions. In a limit position, points O_2, A, and B lie on a straight line; that is, links 2 and 3 form a straight line. The two limit positions of a crank–rocker describe the extreme positions of the rocking angle. Suppose that such a linkage has $r_1 = 400$ mm, $r_2 = 200$ mm, $r_3 = 500$ mm, and $r_4 = 400$ mm.

(a) Find θ_2 and θ_4 corresponding to each limit position.

(b) What is the total rocking angle of link 4?

(c) What are the transmission angles at the extremes?

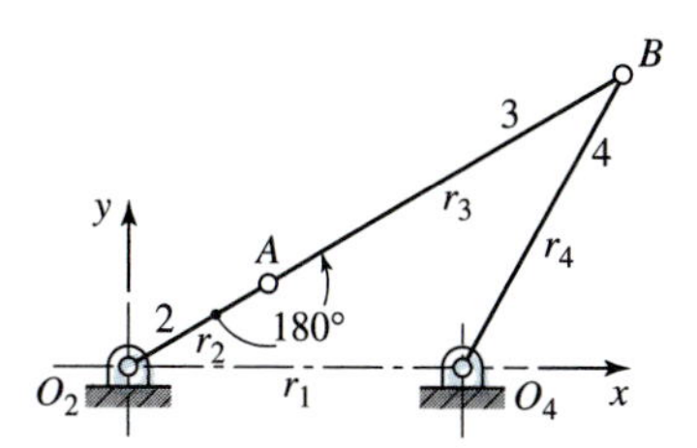

Figure P2.27 A limit position of a four-bar linkage.

†2.28 A double-rocker four-bar linkage has a dead-center position and may also have a limit position (see Prob. 2.27). These positions occur when links 3 and 4 in Fig. P2.28 lie along a straight line. In the dead-center position the transmission angle is 180° and the mechanism is locked. The designer must either avoid such positions or provide the external force, such as a spring, to unlock the linkage. Suppose, for the linkage illustrated in Fig. P2.28, that $r_1 = 14$ in, $r_2 = 5.5$ in, $r_3 = 5$ in, and $r_4 = 12$ in. Find θ_2 and θ_4 corresponding to the dead-center position. Is there a limit position?

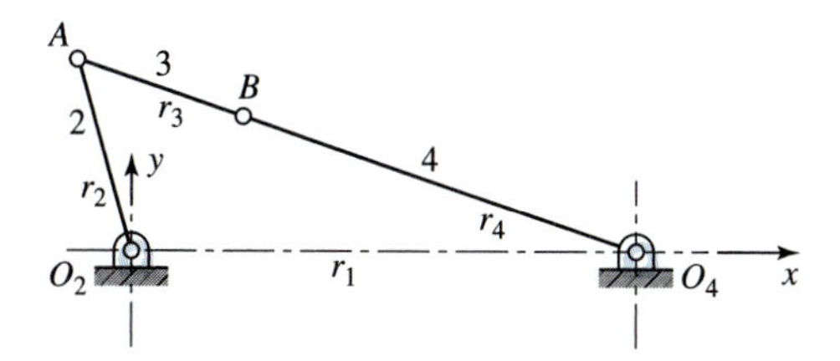

Figure P2.28 A double-rocker four-bar linkage in its dead-center position.

†2.29 Figure P2.29 illustrates a slider-crank linkage that has an offset e and that is placed in one of its limiting positions. By changing the offset e, it is possible to cause the angle that crank 2 makes in traversing between the two limiting positions to vary in such a manner that the driving or forward stroke of the slider takes place over a larger angle than the angle used for the return stroke. Such a linkage is then called a quick-return mechanism. The problem here is to develop a formula for the crank angle traversed during the forward stroke and also to develop a similar formula for the angle traversed during the return stroke. The ratio of these two angles would then constitute a time ratio of the drive to return strokes. Also, determine in which direction the crank should rotate.

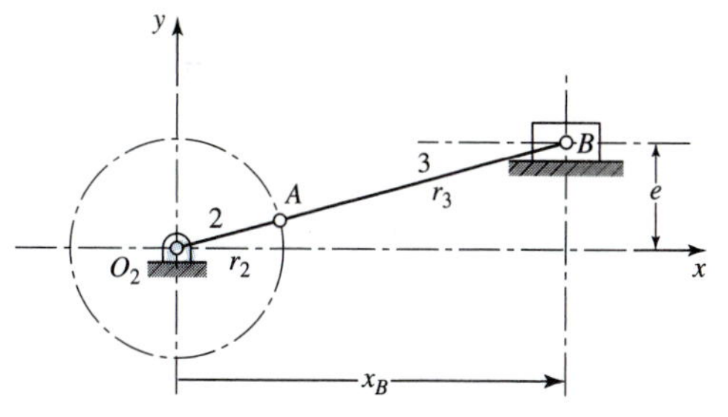

Figure P2.29 An offset slider-crank linkage in one of its limiting positions.

3 Velocity

3.1 DEFINITION OF VELOCITY

In Fig. 3.1, a moving point is first observed at location P defined by the absolute position vector $\mathbf{R}_P$. After a short time interval Δt, its location is observed to have changed to P', defined by $\mathbf{R}'_P$. From Eq. (2.66) we recall that the displacement during this time interval is defined as

$$\Delta\mathbf{R}_P = \mathbf{R}'_P - \mathbf{R}_P.$$

The *average velocity* of the point during the time interval Δt is defined by the ratio $\Delta\mathbf{R}_P/\Delta t$. The *instantaneous velocity* (hereafter called simply *velocity*) is defined by the limit of this ratio for an infinitesimally small time interval and is given by

$$\mathbf{V}_P = \lim_{\Delta t\to 0}\frac{\Delta\mathbf{R}_P}{\Delta t} = \frac{dR_P}{dt}. \tag{3.1}$$

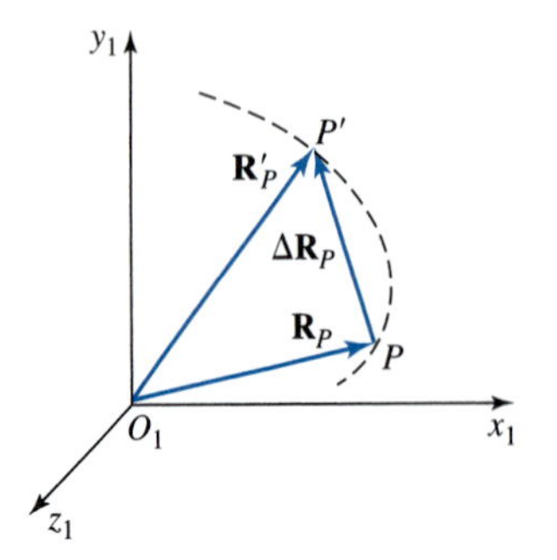

Figure 3.1 Displacement of a moving particle.

Because $\Delta\mathbf{R}_P$ is a vector, there are two convergences in taking this limit, the magnitude and the direction. Therefore, the velocity of a point is a vector quantity equal to the time rate of change of its position. Like the position and displacement vectors, the velocity vector is defined for a specific point. Velocity should not be applied to a line, coordinate system, volume, or other collection of points, because the velocity of each point may be different.

We recall that, for their definitions, the position vectors $\mathbf{R}_P$ and $\mathbf{R}'_P$ depend on the location and orientation of the coordinate system of the observer. The displacement vector $\Delta\mathbf{R}_P$ and the velocity vector $\mathbf{V}_P$, on the other hand, are independent of the initial location of the coordinate system or the location of the observer within the coordinate system. However, the velocity vector $\mathbf{V}_P$ does depend on the motion, if any, of the observer or the coordinate system during the time interval; it is for this reason that the observer is assumed to remain stationary within the coordinate system. If the coordinate system involved is the absolute coordinate system, the velocity is referred to as an *absolute velocity* and is denoted $\mathbf{V}_{P/1}$ or simply $\mathbf{V}_P$. This is consistent with the notation used for absolute displacement.

3.2 ROTATION OF A RIGID BODY

When a rigid body translates (see Section 2.16), the motion of any particular particle is equal to the motion of every other particle of the body. When a rigid body rotates, however, two arbitrarily chosen particles P and Q do not undergo the same motion and a coordinate system attached to the body does not remain parallel to its initial orientation; that is, the body undergoes some angular displacement $\Delta\theta$.

Angular displacements were not discussed in detail in Chapter 2 because, in general, they cannot be treated as vectors. The reason is that they do not obey the usual laws of vector addition. If a rigid body undergoes multiple finite angular displacements in succession, in three dimensions, the result depends on the order in which the displacements take place. To illustrate this, consider the rectangle $ABCO$ in Fig. 3.2a. The rectangular body is first

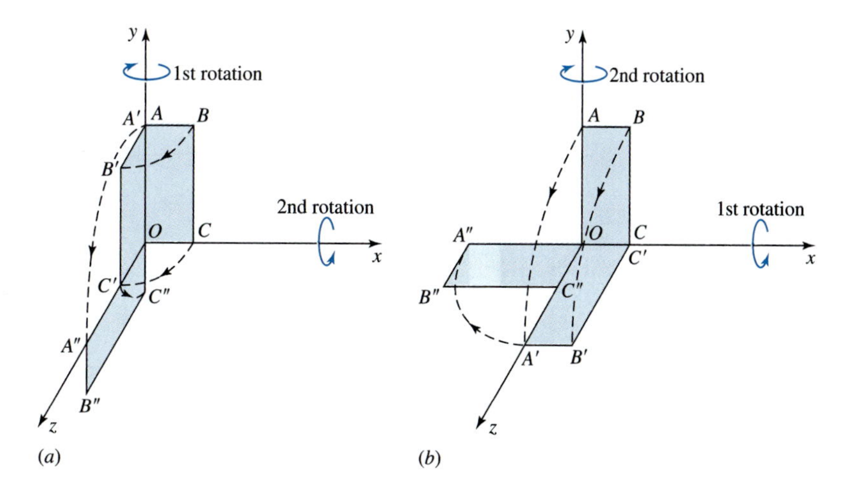

Figure 3.2 Angular displacements cannot be added vectorially because the result depends on the order in which they are added.

rotated by $-90°$ about the y axis and then rotated by $+90°$ about the x axis. The final position of the body is seen to be in the yz plane. In Fig. 3.2b the body occupies the same starting position and is again rotated about the same axes, through the same angles, and in the same directions; however, in this case, the first rotation is about the x axis and the second is about the y axis. The order of the rotations is reversed, and the final position of the rectangle is now seen to be in the zx plane rather than in the yz plane, as it was before. Because this characteristic does not correspond to the commutative law of vector addition, three-dimensional angular displacements cannot be treated as vectors.

Angular displacements that occur about the same axis or parallel axes, on the other hand, do follow the commutative law. Also, infinitesimally small angular displacements are commutative. To avoid confusion, we will treat all finite angular displacements as scalar quantities. However, we will have occasion to treat infinitesimal angular displacements as vectors.

In Fig. 3.3 we recall the definition of the displacement difference between two points, P and Q, both attached to the same rigid body. As indicated in Section 2.16, the displacement-difference vector is entirely attributable to the rotation of the body; there is no displacement difference between points in a body undergoing a translation. We reached this conclusion by picturing the displacement as occurring in two steps. First, the body was assumed to translate through the displacement $\Delta\mathbf{R}_Q$ to the position $x_2^* y_2^* z_2^*$. Next, the body was assumed to rotate about point Q^* to the position $x_2' y_2' z_2'$.

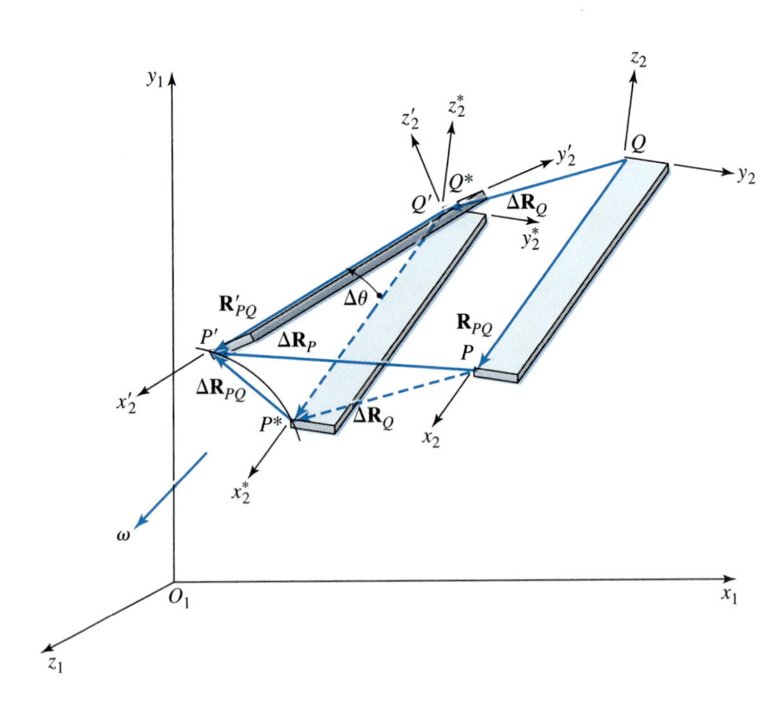

Figure 3.3 Displacement difference between two points on the same rigid body.

Another way to picture the displacement difference $\Delta\mathbf{R}_{PQ}$ is to conceive of a moving coordinate system whose origin travels along with point Q but whose axes remain parallel to the absolute axes $x_1y_1z_1$. Note that this coordinate system does not rotate. An observer in this moving coordinate system observes no motion for point Q because it remains at the origin of HER coordinate system. For the displacement of point P, SHE will observe the displacement difference vector $\Delta\mathbf{R}_{PQ}$. It seems to such an observer that point Q remains fixed and the body rotates about this fixed point, as illustrated in Fig. 3.4.

No matter whether the observer is in the ground coordinate system or in the moving coordinate system described, the body appears to rotate through some total angle $\Delta\theta$ in its displacement from $x_2y_2z_2$ to $x_2'y_2'z_2'$. If we take the point of view of the fixed observer, the location of the axis of rotation is not obvious. As seen by the translating observer, the axis passes through the apparently stationary point Q; all points in the body appear to travel in circular paths about this axis, and any line in the body whose direction is normal to this axis appears to undergo an identical angular displacement of $\Delta\theta$.

The *angular velocity* of a rotating body is now defined as a vector quantity $\boldsymbol{\omega}$ having a direction along the instantaneous axis of rotation. The magnitude of the angular velocity is defined as the time rate of change of the angular orientation of any line in the body whose direction is normal to the axis of rotation. If we designate the angular displacement of any of these lines $\Delta\theta$ and the time interval Δt, the magnitude of the angular velocity vector $\boldsymbol{\omega}$ is

$$\omega = \lim_{\Delta t \to 0} \frac{\Delta\theta}{\Delta t} = \frac{d\theta}{dt}. \tag{3.2}$$

Because we have agreed to treat counterclockwise rotations as positive, the sense of the angular velocity vector along the axis of rotation is in accordance with the right-hand rule.

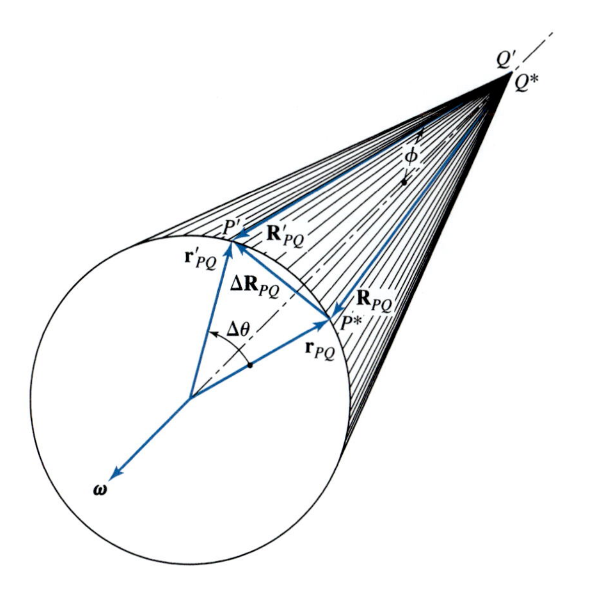

Figure 3.4 Displacement difference $\Delta\mathbf{R}_{PQ}$ as seen by a translating observer.

3.3 VELOCITY DIFFERENCE BETWEEN POINTS OF A RIGID BODY

Figure 3.5*a* illustrates another view of the same rigid-body displacement which was pictured in Fig. 3.3. This is the view seen by an observer in the absolute coordinate system looking directly along the axis of rotation of the moving body, from the tip of the angular velocity vector. In this view, the angular displacement $\Delta\theta$ is observed in true size, and *all* lines in the body rotate through this same angle during the displacement. The displacement vectors and the position difference vectors illustrated are not necessarily seen in true size; they may appear foreshortened under this viewing angle.

Figure 3.5*b* illustrates the same rigid-body rotation from the same viewing angle, but this time from the point of view of the translating observer. Thus, Fig. 3.5*b* corresponds to the base of the cone in Fig. 3.4. We note that the two vectors labeled $\mathbf{r}_{PQ}$ and $\mathbf{r}'_{PQ}$ are the foreshortened views of $\mathbf{R}_{PQ}$ and $\mathbf{R}'_{PQ}$, and from Fig. 3.4 we observe that their magnitudes are

$$r_{PQ} = r'_{PQ} = R_{PQ} \sin\phi, \tag{a}$$

where ϕ is the constant angle from the angular velocity vector $\boldsymbol{\omega}$ to the rotating position-difference vector $\mathbf{R}_{PQ}$ as it traverses the cone.

Looking again at Fig. 3.5*b*, we see that it can also be interpreted as a scale drawing corresponding to Eq. (2.68). This figure indicates that the displacement-difference vector $\Delta\mathbf{R}_{PQ}$ is equal to the vector change in the absolute position difference $\mathbf{R}_{PQ}$ produced during

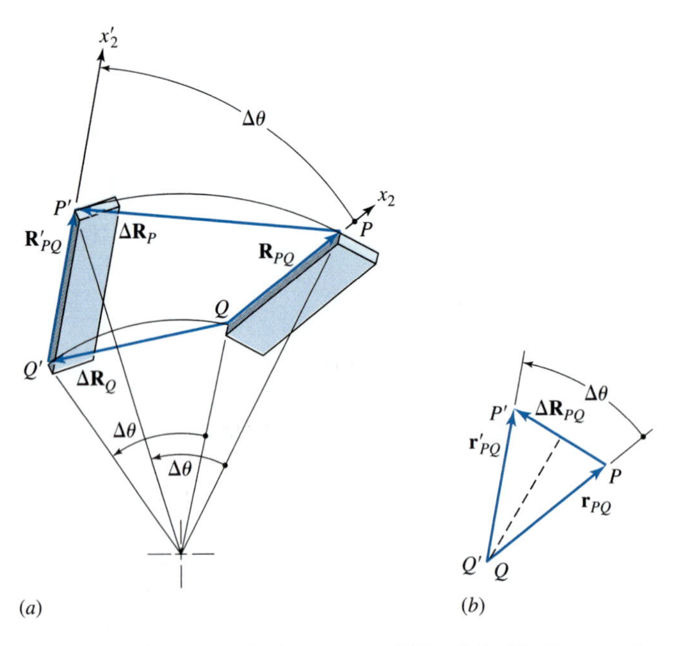

Figure 3.5 (*a*) True view of angular displacements of Fig. 3.3. (*b*) Vector subtraction to form displacement difference $\Delta\mathbf{R}_{PQ}$.

the displacement:

$$\Delta\mathbf{R}_{PQ} = \mathbf{R}'_{PQ} - \mathbf{R}_{PQ}. \tag{b}$$

We are now ready to calculate the magnitude of the displacement-difference vector ΔR_{PQ}. In Fig. 3.5*b*, where it appears in true size, we construct its perpendicular bisector, from which we see that

$$\Delta R_{PQ} = 2r_{PQ} \sin \frac{\Delta\theta}{2} \tag{c}$$

and, from Eq. (*a*),

$$\Delta R_{PQ} = 2(R_{PQ} \sin\phi) \sin \frac{\Delta\theta}{2}. \tag{d}$$

If we now limit ourselves to small motions, the sine of the angular displacement can be approximated by the angle itself, that is,

$$\Delta R_{PQ} = 2(R_{PQ} \sin\phi)\frac{\Delta\theta}{2} = \Delta\theta R_{PQ} \sin\phi. \tag{e}$$

Dividing by the small time increment Δt, noting that the magnitude R_{PQ} and the angle ϕ are constant during the interval, and taking the limit gives

$$\lim_{\Delta t \to 0}\left(\frac{\Delta R_{PQ}}{\Delta t}\right) = \lim_{\Delta t \to 0}\left(\frac{\Delta\theta}{\Delta t}\right) R_{PQ} \sin\phi = \omega R_{PQ} \sin\phi. \tag{f}$$

Recalling the definition of ϕ as the angle between the $\boldsymbol{\omega}$ and $\mathbf{R}_{PQ}$ vectors, we can restore the vector attributes of the above equation by recognizing it as the form of a vector cross-product. Therefore,

$$\lim_{\Delta t \to 0}\left(\frac{\Delta\mathbf{R}_{PQ}}{\Delta t}\right) = \frac{d\mathbf{R}_{PQ}}{dt} = \boldsymbol{\omega} \times \mathbf{R}_{PQ}. \tag{g}$$

This form is so important and so useful that it is given its own name and symbol; it is called the *velocity-difference* vector and is denoted $\mathbf{V}_{PQ}$, that is,

$$\mathbf{V}_{PQ} = \frac{d\mathbf{R}_{PQ}}{dt} = \boldsymbol{\omega} \times \mathbf{R}_{PQ}. \tag{3.3}$$

Let us now recall the displacement-difference equation (2.67), namely,

$$\Delta\mathbf{R}_{P} = \Delta\mathbf{R}_{Q} + \Delta\mathbf{R}_{PQ}. \tag{h}$$

Dividing this equation by Δt and taking the limit as the time increment goes to zero gives

$$\lim_{\Delta t \to 0}\left(\frac{\Delta\mathbf{R}_{P}}{\Delta t}\right) = \lim_{\Delta t \to 0}\left(\frac{\Delta\mathbf{R}_{Q}}{\Delta t}\right) + \lim_{\Delta t \to 0}\left(\frac{\Delta\mathbf{R}_{PQ}}{\Delta t}\right), \tag{i}$$

which from Eqs. (3.1) and (3.3) can be written as

$$\mathbf{V}_P = \mathbf{V}_Q + \mathbf{V}_{PQ}. \tag{3.4}$$

This important equation is called the *velocity-difference equation*; together with Eq. (3.3), it forms one of the primary bases of all velocity-analysis techniques. Equation (3.4) can be written for any two points with no restriction. However, as will be recognized by reviewing the above derivation, Eq. (3.3) cannot be applied to any arbitrary pair of points. *This form is valid only if the two points are both attached to the same rigid body.* This restriction* can perhaps be better remembered if all subscripts are written explicitly, that is,

$$\mathbf{V}_{P_2Q_2} = \boldsymbol{\omega}_2 \times \mathbf{R}_{P_2Q_2}. \tag{j}$$

In the interest of brevity, however, the link-number subscripts are often suppressed in this equation. Note that the link-number subscripts are the same throughout Eq. (*j*). If a mistaken attempt is made to apply Eq. (3.3) when points P and Q are not part of the same link, the error should be discovered because it will not be clear which angular velocity vector $\boldsymbol{\omega}$ should be used.

3.4 GRAPHIC METHODS; VELOCITY POLYGONS

One important approach to velocity analysis is graphical. As observed in graphic position analysis (see Section 2.7), it is primarily of use in two-dimensional problems when only a single position requires a solution. The major advantages are that a solution can be achieved quickly and that visualization of and insight into the problem are enhanced by the graphical approach.

As a first example of graphical velocity analysis, let us consider the two-dimensional motion of the unconstrained link ABC illustrated in Fig. 3.6*a*. Suppose that we know the velocities of points A and B and wish to determine the velocity of point C and the angular velocity of the link. We assume that a scale diagram of the link, Fig. 3.6*a*, has already been drawn for the instant considered, that is, that a position analysis has been completed and that position-difference vectors can be measured from the diagram.

Next we consider the velocity difference equation (3.4) relating points A and B, that is,

$$\overset{\surd\surd}{\mathbf{V}_B} = \overset{\surd\surd}{\mathbf{V}_A} + \overset{??}{\mathbf{V}_{BA}}, \tag{a}$$

where the two unknowns are the magnitude and the direction of the velocity-difference vector $\mathbf{V}_{BA}$, as indicated above this symbol in Eq. (*a*). Figure 3.6*b* illustrates the graphical solution to the equation. After a scale is chosen to represent velocity vectors, the vectors $\mathbf{V}_A$ and $\mathbf{V}_B$ are both drawn to scale, starting from a common origin and in the given directions. The vector spanning the termini of $\mathbf{V}_A$ and $\mathbf{V}_B$ is the velocity-difference vector $\mathbf{V}_{BA}$ and is correct, within graphical accuracy, in both magnitude and direction.

---

* More precisely, the restriction is the requirement that the distance R_{PQ} must remain constant. However, in application, the above wording fits most real situations.

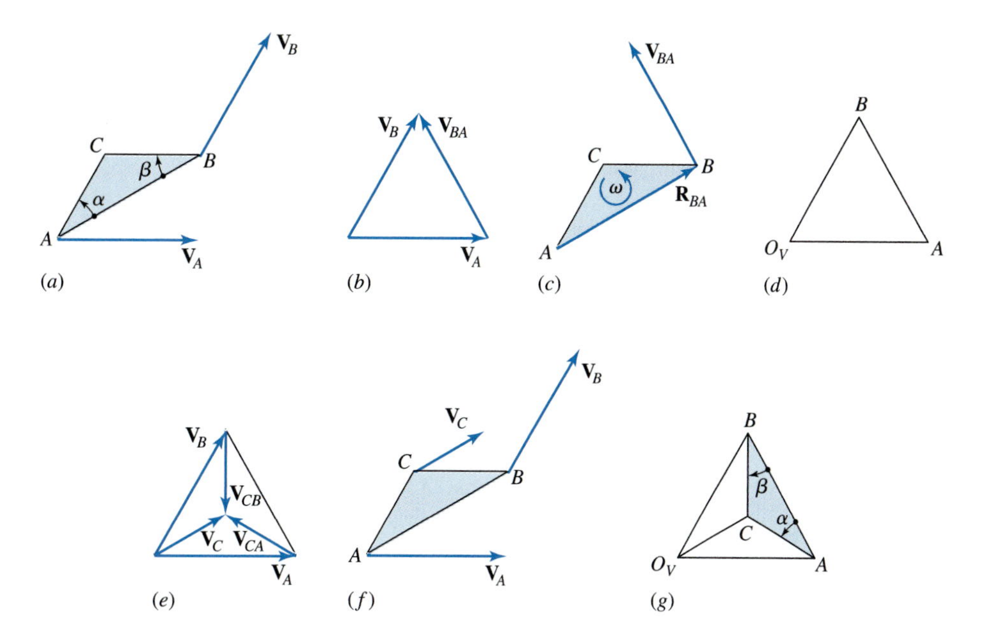

Figure 3.6

The angular velocity $\boldsymbol{\omega}$ for the link can now be found from Eq. (3.3), that is,

$$\mathbf{V}_{BA} = \boldsymbol{\omega} \times \mathbf{R}_{BA}. \tag{b}$$

Because the link is in planar motion, the $\boldsymbol{\omega}$ vector lies perpendicular to the plane of motion, that is, perpendicular to the vectors $\mathbf{V}_{BA}$ and $\mathbf{R}_{BA}$. Therefore, considering only the magnitudes in Eq. (*b*) gives

$$V_{BA} = \omega R_{BA}$$

or

$$\omega = \frac{V_{BA}}{R_{BA}}. \tag{c}$$

The numerical magnitude ω can, therefore, be found by scaling V_{BA} from Fig. 3.6*b* and R_{BA} from Fig. 3.6*a*, being careful to properly apply the scale factors for units; it is usual practice to evaluate ω in units of radians per second.

The magnitude ω is not a complete solution for the angular velocity vector; the direction must also be determined. As observed above, the $\boldsymbol{\omega}$ vector is perpendicular to the plane of the link itself because the motion is planar. However, this does not indicate whether $\boldsymbol{\omega}$ is directed into or out of the plane of the figure. Taking the point of view of a translating observer, moving with point A but not rotating, we can picture the link as rotating about point A; this is illustrated in Fig. 3.6*c*. The velocity difference $\mathbf{V}_{BA}$ is the only velocity seen by such an observer. Therefore, interpreting $\mathbf{V}_{BA}$ to indicate the direction of rotation of point B about point A, we find the direction of $\boldsymbol{\omega}$, counterclockwise in this example. Although it is not strict vector notation, it is common practice in two-dimensional problems to indicate the final solution in the form $\boldsymbol{\omega} = 15$ rad/s ccw, which indicates both magnitude and direction.

The practice of constructing vector diagrams using thick dark lines, such as in Fig. 3.6*b*, makes them easy to read, but when the diagram is the graphical solution of an equation, it is not very accurate. For this reason it is customary to construct the graphical solution with thin, sharp lines, made with a hard drawing pencil, as illustrated in Fig. 3.6*d*. The solution begins by choosing a scale and a point labeled O_V to represent zero velocity. Absolute velocities such as $\mathbf{V}_A$ and $\mathbf{V}_B$ are constructed with their origins at O_V,and their termini are labeled as points A and B. The line *from A to B* then represents the velocity difference $\mathbf{V}_{BA}$. As we continue, we observe that these labels at the vertices are sufficient to determine the precise notation of all velocity differences represented by lines in the diagram. Note, for example, that $\mathbf{V}_{BA}$ is represented by the vector *to point B from point A*. With this labeling convention, no arrowheads or additional notation are necessary and do not clutter the diagram. Such a diagram is called a *velocity polygon* and, as we will see, adds considerable convenience to the graphical solution technique.

A danger of this convention, however, is that the analyst will begin to think of the technique as a series of graphical "tricks" and lose sight of the fact that each line drawn can and should be fully justified by a corresponding vector equation. The graphics are merely a convenient solution technique and not a substitute for a sound theoretical basis.

Returning to Fig. 3.6*c*, it may have appeared coincidental that the vector $\mathbf{V}_{BA}$ was perpendicular to $\mathbf{R}_{BA}$. Looking back to Eq. (*b*), however, we see that it was a necessary outcome, resulting from the cross-product with the $\boldsymbol{\omega}$ vector. We will take advantage of this property in the next step.

Now that $\boldsymbol{\omega}$ has been found, let us determine the absolute velocity of point C. We can relate this by two velocity-difference equations to the absolute velocities of both points A and B, that is,

$$\mathbf{V}_C = \overset{\surd\surd}{\mathbf{V}_A} + \overset{?\surd}{\mathbf{V}_{CA}} = \overset{\surd\surd}{\mathbf{V}_B} + \overset{?\surd}{\mathbf{V}_{CB}}. \qquad (d)$$

Because points A, B, and C are all fixed in the same rigid link, each of the velocity difference vectors, $\mathbf{V}_{CA}$ and $\mathbf{V}_{CB}$, is of the form $\boldsymbol{\omega} \times \mathbf{R}$ using $\mathbf{R}_{CA}$ and $\mathbf{R}_{CB}$, respectively. As a result, $\mathbf{V}_{CA}$ is perpendicular to $\mathbf{R}_{CA}$ and $\mathbf{V}_{CB}$ is perpendicular to $\mathbf{R}_{CB}$. The directions of these two terms are therefore indicated as known in Eq. (*d*).

Because $\boldsymbol{\omega}$ has already been determined, it is easy to calculate the magnitudes of $\mathbf{V}_{CA}$ and $\mathbf{V}_{CB}$ using equations similar to Eq. (*c*); however, we will not do this here. Instead, we will form the graphical solution to Eq. (*d*). Equation (d) states that a vector that is perpendicular to $\mathbf{R}_{CA}$ must be added to $\mathbf{V}_A$ and that the result will equal the sum of $\mathbf{V}_B$ and a vector perpendicular to R_{CB}. The solution is illustrated in Fig. 3.6*e*. In practice, the solution is commonly continued on the same diagram as Fig. 3.6*d* and results in Fig. 3.6*g*. A line perpendicular to $\mathbf{R}_{CA}$ (representing $\mathbf{V}_{CA}$) is drawn starting at point A (representing the addition to $\mathbf{V}_A$); similarly, a line is drawn perpendicular to $\mathbf{R}_{CB}$ starting at point B. The point of intersection of these two lines is labeled C and represents the solution to Eq. (*d*).

The line from O_V to point C now represents the absolute velocity $\mathbf{V}_C$. This velocity can be transferred back to the link and interpreted as $\mathbf{V}_C$ in both magnitude and direction, as illustrated in Fig. 3.6*f*.

In seeing the shading and the labeled angles α and β in Figs. 3.6*g* and 3.6*a*, we are led to investigate whether the two triangles labeled ABC in each of these figures are similar in shape, as they appear to be. In reviewing the construction steps we see that indeed they are similar, because the velocity-difference vectors $\mathbf{V}_{BA}$, $\mathbf{V}_{CA}$, and $\mathbf{V}_{CB}$ are each perpendicular to the respective position-difference vectors, $\mathbf{R}_{BA}$, $\mathbf{R}_{CA}$, and $\mathbf{R}_{CB}$. This property will be true no matter what the shape of the moving link; a similarly shaped figure will appear in the velocity polygon. The sides of the polygon are always scaled up or down by a factor equal to the angular velocity of the link, and they are always rotated by 90° in the directions of their angular velocities. These properties result from the fact that each velocity-difference vector between two points on the link is of the form of a cross-product of the same $\boldsymbol{\omega}$ vector with the corresponding position-difference vector. This similarly shaped figure in the velocity polygon is commonly referred to as the *velocity image* of the link, and any moving link will have a corresponding velocity image in the velocity polygon.

If the concept of the velocity image had been known initially, the solution process could have been speeded up considerably. Once the solution has progressed to the state of Fig. 3.6*d*, the velocity-image points A and B are known. One can use these two points as the base of a triangle similar to the link shape and label the image point C directly, without explicitly writing Eq. (*d*). Care must be taken not to allow the triangle to be flipped over between the position diagram and the velocity image, but the solution can proceed quickly, accurately, and naturally, resulting in Fig. 3.6*g*. Here again, the caution is repeated that all steps in the solution are based on strictly derived vector equations and are not tricks. It is wise to continue to write the corresponding vector equations until one is thoroughly familiar with the procedure and its vector basis.

To increase familiarity with graphical velocity analysis techniques, we will analyze two typical example problems.

EXAMPLE 3.1

The four-bar linkage, drawn to scale in Fig. 3.7*a* with all the necessary dimensions, is driven by crank 2 at a constant angular velocity of $\omega_2 = 900$ rev/min ccw. Find the instantaneous velocities of point E in link 3 and point F in link 4 and the angular velocities of links 3 and 4 at the position shown.

SOLUTION

To obtain a graphical solution, we first calculate the angular velocity of link 2 in radians per second. This is

$$\omega_2 = \left(900 \frac{\text{rev}}{\text{min}}\right)\left(2\pi \frac{\text{rad}}{\text{rev}}\right)\left(\frac{1 \text{ min}}{60 \text{ s}}\right) = 94.2 \text{ rad/s ccw}. \tag{1}$$

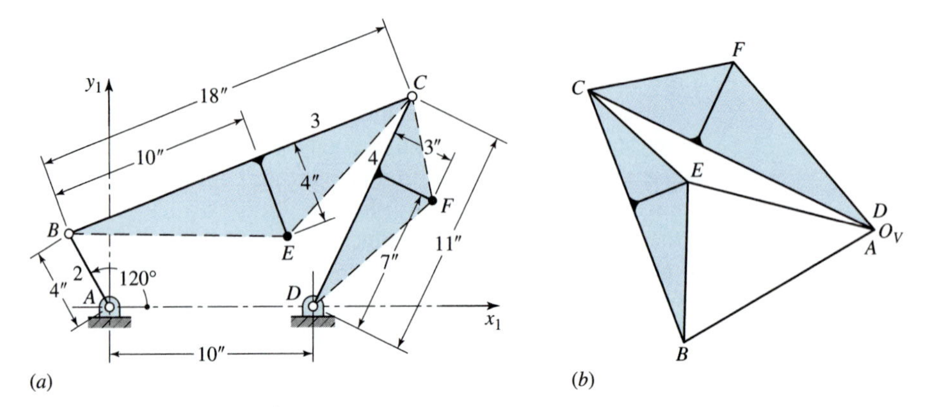

Figure 3.7 Example 3.1: graphical velocity analysis of a four-bar linkage. (*a*) Scale diagram; (*b*) velocity polygon.

Then we note that point A remains fixed and calculate the velocity of point B:

$$\mathbf{V}_B = \cancel{\mathbf{V}_A^0} + \mathbf{V}_{BA} = \boldsymbol{\omega}_2 \times \mathbf{R}_{BA}$$

$$V_B = (94.2 \text{ rad/s})\left(\frac{4}{12}\text{ft}\right) = 31.4 \text{ ft/s}. \tag{2}$$

We note that the form "$\boldsymbol{\omega} \times \mathbf{R}$" was used for the velocity difference and not for the absolute velocity $\mathbf{V}_B$ directly. In Fig. 3.7*b* we choose the point O_V and a velocity scale factor. We note that the image point A is coincident with O_V and construct the line AB perpendicular to $\mathbf{R}_{BA}$ and toward the lower left because of the counterclockwise sense of $\boldsymbol{\omega}_2$; this line represents $\mathbf{V}_{BA}$.

If we attempt at this time to write an equation directly for the velocity of point E, we find by counting the unknowns that it cannot be solved yet. Therefore, we next write two equations for the velocity of point C. Because the velocities of points C_3 and C_4 must be equal (links 3 and 4 are pinned together at C), we have

$$\mathbf{V}_C = \overset{\surd\surd}{\mathbf{V}_B} + \overset{?\surd}{\mathbf{V}_{CB}} = \cancel{\mathbf{V}_D^0} + \overset{?\surd}{\mathbf{V}_{CD}}. \tag{3}$$

We now construct two lines in the velocity polygon: the line BC is drawn from B perpendicular to $\mathbf{R}_{CB}$, and the line DC is drawn from D (coincident with O_V because $\mathbf{V}_D = \mathbf{0}$) perpendicular to $\mathbf{R}_{CD}$. We label the point of intersection of these two lines point C. When the lengths of these two lines are scaled, we find that $V_{CB} = 38.4$ ft/s and $V_C = V_{CD} = 45.5$ ft/s. The angular velocities of links 3 and 4 can now be found,

$$\omega_3 = \frac{V_{CB}}{R_{CB}} = \frac{38.4 \text{ ft/s}}{18/12 \text{ ft}} = 25.6 \text{ rad/s ccw}, \qquad \textit{Ans.} \ (4)$$

$$\omega_4 = \frac{V_{CD}}{R_{CD}} = \frac{45.5 \text{ ft/s}}{11/12 \text{ ft}} = 49.6 \text{ rad/s ccw}, \qquad \textit{Ans.} \ (5)$$

where the directions of $\boldsymbol{\omega}_3$ and $\boldsymbol{\omega}_4$ were determined by the technique illustrated in Fig. 3.6*c*.

There are now several methods of finding the velocity of point E, that is, $\mathbf{V}_E$. In one method we measure $\mathbf{R}_{EB}$ from the scale drawing of Fig. 3.7*a* and then, because points B and E are both attached to link 3, we can calculate*

$$V_{EB} = \omega_3 R_{EB} = (25.6 \text{ rad/s})\left(\frac{10.8}{12}\text{ft}\right) = 23.0 \text{ ft/s}. \tag{6}$$

We can now construct the line BE in the velocity polygon, drawn to the proper scale and perpendicular to $\mathbf{R}_{EB}$, thus solving† the velocity-difference equation,

$$\mathbf{V}_E = \mathbf{V}_B + \mathbf{V}_{EB}. \tag{7}$$

The result is

$$V_E = 27.6 \text{ ft/s}, \qquad \textit{Ans.}$$

as scaled from the velocity polygon.

Alternatively, the velocity of point E can be obtained from the velocity-difference equation,

$$\mathbf{V}_E = \mathbf{V}_C + \mathbf{V}_{EC}, \tag{8}$$

by a procedure identical to that used for Eq. (7). This solution would produce the triangle $O_V EC$ in the velocity polygon.

Suppose we wish to find $\mathbf{V}_E$ without the intermediate step of calculating ω_3. In this case, we write Eqs. (7) and (8) simultaneously, that is

$$\mathbf{V}_E = \overset{\surd\surd}{\mathbf{V}_B} + \overset{?\surd}{\mathbf{V}_{EB}} = \overset{\surd\surd}{\mathbf{V}_C} + \overset{?\surd}{\mathbf{V}_{EC}}. \tag{9}$$

Drawing lines EB (perpendicular to $\mathbf{R}_{EB}$) and EC (perpendicular to $\mathbf{R}_{EC}$) in the velocity polygon, we find their intersection and thus solve Eq. (9).

Perhaps the easiest method of solving for $\mathbf{V}_E$, however, is to take advantage of the concept of the velocity image of link 3. Recognizing that the velocity-image points B and C have already been found, we can construct the triangle BEC in the velocity polygon, similar in shape to the triangle BEC in the scale diagram of link 3. This locates point E in the velocity polygon and, therefore, gives a solution for the velocity of point E.

The velocity of point F can also be found by any of the above methods using points C, D, and F of link 4. The result is

$$V_F = 31.8 \text{ ft/s}. \qquad \textit{Ans.}$$

---

* There is no restriction in our derivation that requires that $\mathbf{R}_{EB}$ lie along the material portion of link 3 to use Eq. (6), only that points E and B remain a constant distance apart.

† Note that numerical values should not be substituted into Eq. (7) directly: this is a vector equation and requires vector addition, not scalar; this is precisely the purpose of constructing the velocity polygon.

EXAMPLE 3.2

The offset slider-crank mechanism illustrated in Fig. 3.8*a* is driven by slider 4 at a velocity $\mathbf{V}_C = -10\hat{\mathbf{i}}$ m/s at the position shown. Determine the instantaneous velocity of point *D* and the angular velocities of links 2 and 3.

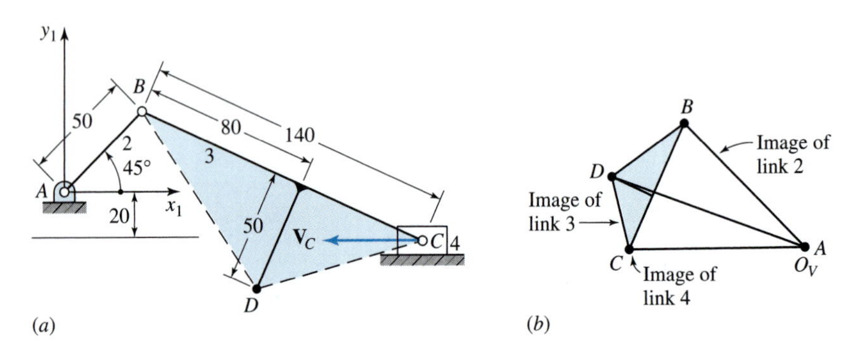

Figure 3.8 Example 3.2: (*a*) scale diagram of offset slider-crank mechanism (dimensions in millimeters); (*b*) velocity polygon.

SOLUTION

The velocity scale and pole O_V are chosen and $\mathbf{V}_C$ is drawn, thus locating point *C* as illustrated in Fig. 3.8*b*. Simultaneous equations are then written for the velocity of point *B*,

$$\mathbf{V}_B = \overset{\surd\surd}{\mathbf{V}_C} + \overset{?\surd}{\mathbf{V}_{BC}} = \cancel{\mathbf{V}}_A^0 + \overset{?\surd}{\mathbf{V}_{BA}}, \tag{10}$$

and solved for the location of point *B* in the velocity polygon.

Having found points *B* and *C*, we can construct the velocity image of link 3 as illustrated to locate point *D*; we then scale the line $O_V D$, which gives

$$V_D = 12.0 \text{ m/s} \qquad \textit{Ans.}$$

with the direction shown in the velocity polygon.

The angular velocities of links 2 and 3, respectively, are

$$\omega_2 = \frac{V_{BA}}{R_{BA}} = \frac{10.0 \text{ m/s}}{0.050 \text{ m}} = 200 \text{ rad/s ccw} \qquad \textit{Ans.} \ (11)$$

$$\omega_3 = \frac{V_{BC}}{R_{BC}} = \frac{7.5 \text{ m/s}}{0.140 \text{ m}} = 53.6 \text{ rad/s cw.} \qquad \textit{Ans.} \ (12)$$

In this example, Fig. 3.8*b*, the velocity image of each link is indicated in the polygon. Once the analysis of any problem is carried through completely, there will be a velocity image for each link of the mechanism. The following points are true, in general, and can be verified in the above two examples.

1. The velocity image of each link is a scale reproduction of the shape of the link in the velocity polygon.
2. The velocity image of each link is rotated 90° in the direction of the angular velocity of that link.
3. The letters identifying the vertices of each link are the same as those in the velocity polygon and progress around the velocity image in the same order and in the same angular direction as around the link.
4. The ratio of the size of the velocity image of a link to the size of the link itself is equal to the magnitude of the angular velocity of the link. In general, it is not the same for different links in the same mechanism.
5. The velocities of all points on a translating link are equal, and the angular velocity of the link is zero. Therefore, the velocity image of a link that is translating shrinks to a single point in the velocity polygon.
6. The point O_V in the velocity polygon is the image of all points with zero absolute velocity; it is the velocity image of the fixed link.
7. The absolute velocity of any point on any link is represented by the line from O_V to the image of the point. The velocity-difference vector between any two points, say P and Q, is represented by the line from image point P to image point Q.

3.5 APPARENT VELOCITY OF A POINT IN A MOVING COORDINATE SYSTEM

In analyzing the velocities of various machine components, we frequently encounter situations in which it is convenient to describe how a point moves with respect to another moving link but not at all convenient to describe the absolute motion of the point. An example of this occurs when a rotating link contains a slot along which a point of another link is constrained to slide. With the motion of the link containing the slot and the relative sliding motion taking place in the slot as known quantities, we may wish to find the absolute motion of the sliding member. It was for problems of this type that the apparent-displacement vector was defined in Section 2.17, and we now wish to extend this concept to velocity.

In Fig. 3.9 we recall the definition of the apparent-displacement vector. A rigid link having some general motion carries a coordinate system $x_2y_2z_2$ attached to it. After a time increment Δt, the coordinate system lies at $x_2'y_2'z_2'$. All points of link 2 move with the coordinate system.

Also, during the same time interval, another point P_3 of another link 3 is constrained in some manner to move along a known path with respect to link 2. In Fig. 3.9 this constraint is depicted as a slot carrying a pin from link 3; the center of the pin is the point P_3. Although it is pictured in this way, the constraint may occur in a variety of different forms. The only assumption here is that the path the moving point P_3 traces in coordinate system $x_2y_2z_2$, that is, the locus of the tip of the apparent-position vector $\mathbf{R}_{P_3/2}$, is known.

Recall the apparent-displacement equation, see Eq. (2.70), that is,

$$\Delta\mathbf{R}_{P_3} = \Delta\mathbf{R}_{P_2} + \Delta\mathbf{R}_{P_3/2}.$$

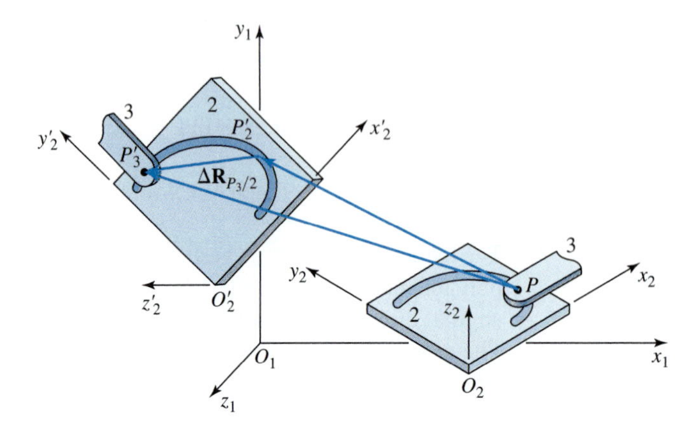

Figure 3.9 Apparent displacement.

Dividing this equation by Δt and taking the limit gives

$$\lim_{\Delta t \to 0}\left(\frac{\Delta\mathbf{R}_{P_3}}{\Delta t}\right) = \lim_{\Delta t \to 0}\left(\frac{\Delta\mathbf{R}_{P_2}}{\Delta t}\right) + \lim_{\Delta t \to 0}\left(\frac{\Delta\mathbf{R}_{P_3/2}}{\Delta t}\right). \tag{a}$$

Now we define the *apparent-velocity* vector as

$$\mathbf{V}_{P_3/2} = \lim_{\Delta t \to 0}\left(\frac{\Delta\mathbf{R}_{P_3/2}}{\Delta t}\right) = \frac{d\mathbf{R}_{P_3/2}}{dt} \tag{3.5}$$

and, in the limit, Eq. (a) becomes

$$\mathbf{V}_{P_3} = \mathbf{V}_{P_2} + \mathbf{V}_{P_3/2}, \tag{3.6}$$

called the *apparent-velocity equation.*

We note from its definition, Eq. (3.5), that the apparent velocity resembles the absolute velocity except that it comes from the apparent displacement rather than the absolute displacement. Thus, in concept, it is the velocity of the moving point P_3 *as it would appear to an observer attached to the moving link* 2 and making observations in coordinate system $x_2y_2z_2$. This concept accounts for its name. We also note that the absolute velocity is a special case of the apparent velocity where the observer happens to be fixed to the $x_1y_1z_1$ coordinate system.

We can get further insight into the nature of the apparent-velocity vector by studying Fig. 3.10. This figure illustrates the view of the moving point P_3 as it would be seen by the moving observer. To HER, the path traced on link 2 appears stationary and the moving point moves along this path from P_3 to P'_3. Working in this coordinate system, suppose we locate point C as the center of curvature of the path of point P_3. For small distances from P_2, the path follows the circular arc $P_3P'_3$ with center C and radius of curvature ρ. We now define the unit-vector tangent to the path $\hat{\mathbf{u}}^t$ with positive sense in the direction of positive movement. The plane defined by this unit tangent vector $\hat{\mathbf{u}}^t$ and the center of curvature C

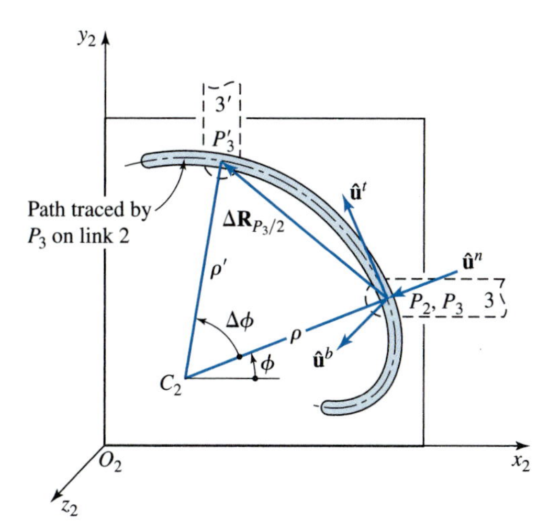

Figure 3.10 Apparent displacement of point P_3 as seen by an observer on link 2.

of the path is called the *osculating plane.* If we choose a preferred side of this plane as the positive side and denote it by the positive unit vector $\hat{\mathbf{u}}^b$ (commonly referred to as the *binormal* vector), we can complete a right-hand Cartesian coordinate system by defining the unit vector normal to the path

$$\hat{\mathbf{u}}^n = \hat{\mathbf{u}}^b \times \hat{\mathbf{u}}^t. \tag{3.7}$$

Therefore, a rule to remember is that the unit normal vector $\hat{\mathbf{u}}^n$ is always 90° counterclockwise from the unit tangent vector $\hat{\mathbf{u}}^t$. This implies that the radius of curvature ρ of the path has a positive value when $\hat{\mathbf{u}}^n$ points from point P_3 toward the center of curvature C of the path and a negative value when $\hat{\mathbf{u}}^n$ points away from the center of curvature.

This coordinate system $\hat{\mathbf{u}}^t\hat{\mathbf{u}}^n\hat{\mathbf{u}}^b$ (commonly referred to as *path coordinates*) moves with its origin tracking the motion of point P_3. However, it rotates with the radius of curvature (through the angle $\Delta\phi$) as the motion progresses, *not* the same rotation as either link 2 or link 3. Note that if positive movement had been chosen in the opposite direction along this curve, the sense of both the tangent vector $\hat{\mathbf{u}}^t$ and the normal vector $\hat{\mathbf{u}}^n$ would have been reversed. This would mean that the radius of curvature ρ of the path would have had a negative value; however, the unit vector $\hat{\mathbf{u}}^n$ would still have come from Eq. (3.7) rather than from the sense of the radius of curvature. Because the movement would now have been in the negative direction, the angle $\Delta\phi$ would still have been counterclockwise (as seen from the positive $\hat{\mathbf{u}}^b$ side of the plane) and would still have had a positive value.

We now define the scalar Δs as the distance along the curve from P_3 *to* P'_3 and note that $\Delta\mathbf{R}_{P_3/2}$ is a chord of the same arc. However, for a very short time interval Δt, the magnitude of the chord and the arc distance approach equality. Therefore,

$$\lim_{\Delta t\to 0}\left(\frac{\Delta\mathbf{R}_{P_3/2}}{\Delta s}\right) = \frac{d\mathbf{R}_{P_3/2}}{ds} = \hat{\mathbf{u}}^t. \tag{3.8}$$

Here, both $\Delta\mathbf{R}_{P_3/2}$ and Δs are considered functions of time. Therefore, from Eq. (3.5), the *apparent-velocity* vector can be written as

$$\mathbf{V}_{P_3/2} = \lim_{\Delta t \to 0} \left(\frac{\Delta\mathbf{R}_{P_3/2}}{\Delta s} \frac{\Delta s}{\Delta t} \right) = \frac{d\mathbf{R}_{P_3/2}}{ds} \frac{ds}{dt} = \frac{ds}{dt} \hat{\mathbf{u}}^t$$

or as

$$\mathbf{V}_{P_3/2} = \dot{s}\hat{\mathbf{u}}^t, \tag{3.9}$$

where $\dot{s}$ is referred to as the instantaneous speed of P_3 along the path. There are two important conclusions from this result: (*a*) the magnitude of the apparent velocity is equal to the speed with which point P_3 progresses along the path, and (*b*) the apparent-velocity vector is always tangent to the path traced by the point in the coordinate system of the observer.

The first of these two results is seldom useful in the solution of problems, although it is an important concept. The second result is extremely useful, because the apparent path traced by the point can often be visualized from the nature of the constraints and thus the direction of the apparent-velocity vector becomes known. Note that only the tangent to the path is used in Chapter 3; the radius of curvature ρ of the path is not needed until we attempt acceleration analysis in Chapter 4.

EXAMPLE 3.3

An inversion of the slider-crank mechanism is illustrated in Fig. 3.11*a*. Link 2, the crank, is driven at an angular velocity of 36 rad/s cw. Link 3 slides on link 4 and is pivoted to the crank at A. Find the angular velocity of link 4.

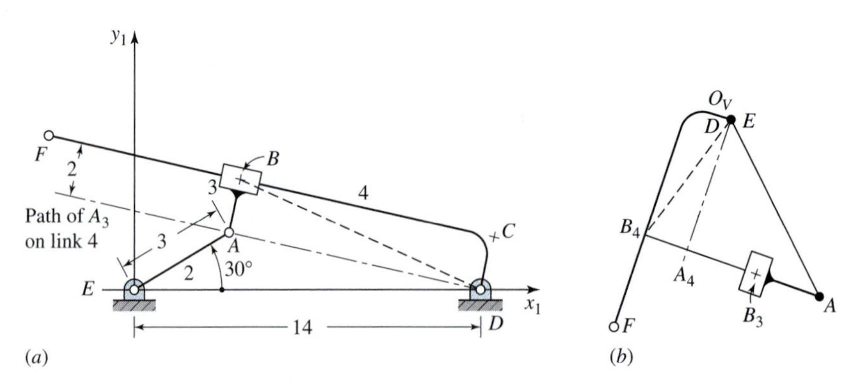

Figure 3.11 Example 3.3: (*a*) inverted slider-crank mechanism with dimensions in inches; (*b*) velocity polygon.

SOLUTION

We first calculate the velocity of point A:

$$\mathbf{V}_A = \cancel{\mathbf{V}}_E^{\mathbf{0}} + \mathbf{V}_{AE} = \boldsymbol{\omega}_2 \times \mathbf{R}_{AE}$$

$$V_A = (36 \text{ rad/s})\left(\frac{3}{12}\text{ft}\right) = 9 \text{ ft/s} \tag{1}$$

and we plot this from the pole O_V to locate point A in the velocity polygon (see Fig. 3.11*b*).

Next, we distinguish two different points, B_3 and B_4, at the location of sliding. Point B_3 is part of link 3, and point B_4 is attached to link 4, but at the instant shown the two points are coincident. Note that, as seen by an observer on link 4, point B_3 seems to slide along link 4, thus defining a straight path along the line CF. Thus, we can write the apparent-velocity equation

$$\mathbf{V}_{B_3} = \mathbf{V}_{B_4} + \mathbf{V}_{B_3/4}. \tag{2}$$

When point B_3 is related to A and point B_4 is related to D by velocity differences, expansion of Eq. (2) gives

$$\overset{\surd\surd}{\mathbf{V}_A} + \overset{?\surd}{\mathbf{V}_{B_3A}} = \cancel{\mathbf{V}}_D^{\mathbf{0}} + \overset{?\surd}{\mathbf{V}_{B_4D}} + \overset{?\surd}{\mathbf{V}_{B_3/4}}, \tag{3}$$

where $\mathbf{V}_{B_3A}$ is perpendicular to $\mathbf{R}_{BA}$, $\mathbf{V}_{B_4D}$ (dashed line in Fig. 3.11*b*) is perpendicular to $\mathbf{R}_{BD}$, and $\mathbf{V}_{B_3/4}$ has a direction defined by the tangent to the path of sliding at B.

Although Eq. (3) appears to have three unknowns, if we note that $\mathbf{V}_{B_3A}$ and $\mathbf{V}_{B_3/4}$ have identical orientations, the equation can be rearranged as

$$\overset{\surd\surd}{\mathbf{V}_A} + \overbrace{(\mathbf{V}_{B_3A} - \mathbf{V}_{B_3/4})}^{?\surd} = \overset{?\surd}{\mathbf{V}_{B_4D}} \tag{4}$$

and the vector difference indicated in parentheses can be treated as a single vector of known orientation. The equation is thereby reduced to two unknowns and can be solved graphically to locate point B_4 in the velocity polygon.

The magnitude $\mathbf{R}_{BD}$ can be computed or measured from the diagram, and $\mathbf{V}_{B_4D}$ can be scaled from the velocity polygon (the dashed line from O_V to B_4). Therefore,

$$\omega_4 = \frac{V_{B_4D}}{R_{BD}} = \frac{7.3 \text{ ft/s}}{11.6/12 \text{ ft}} = 7.55 \text{ rad/s ccw.} \qquad \textit{Ans.} \tag{5}$$

Although the problem as stated is now complete, the velocity polygon has been extended to illustrate the images of links 2, 3, and 4. In so doing, it was necessary to note that because links 3 and 4 always remain perpendicular to each other, they must rotate at the same rate. The angular velocities of the two links are the same; $\boldsymbol{\omega}_3 = \boldsymbol{\omega}_4$. This allows the calculation of $\mathbf{V}_{BA} = \boldsymbol{\omega}_3 \times \mathbf{R}_{BA}$ and the plotting of the velocity-image point B_3. We also note that the velocity images of links 3 and 4 are of comparable size because $\boldsymbol{\omega}_3 = \boldsymbol{\omega}_4$. However, they have quite a different scale than the velocity image of link 2, the line $O_V A$, because ω_2 is a larger angular velocity than either $\boldsymbol{\omega}_3$ or $\boldsymbol{\omega}_4$.

Another approach to the same problem avoids the need to combine terms as in Eq. (4). If we consider an observer riding on link 4 and ask what SHE would see for the path of point A in HER coordinate system, we find that this path is a straight line parallel to the line CF, as indicated in Fig. 3.11a. Let us now define one point of this path as A_4. At the instant shown, the point A_4 is located coincident with points A_2 and A_3. However, A_4 *does not move with the pin; it is attached to link 4 and rotates with the path around the fixed point D.* Because we can identify the path traced by A_2 and A_3 on link 4, we can write the apparent-velocity equation

$$\mathbf{V}_{A_2} = \mathbf{V}_{A_4} + \mathbf{V}_{A_2/4} \tag{6}$$

and, because point A_4 is part of link 4, we have

$$\mathbf{V}_{A_4} = \cancel{\mathbf{V}_D}^{0} + \mathbf{V}_{A_4 D}. \tag{7}$$

Substituting Eq. (7) into Eq. (6) gives

$$\overset{\surd\surd}{\mathbf{V}_{A_2}} = \overset{?\surd}{\mathbf{V}_{A_4 D}} + \overset{?\surd}{\mathbf{V}_{A_2/4}}, \tag{8}$$

where $\mathbf{V}_{A_4 D}$ is perpendicular to $\mathbf{R}_{AD}$ and $\mathbf{V}_{A_2/4}$ is tangent to the path. Solving Eq. (8) locates image point A_4 in the velocity polygon and allows a solution for $\omega_4 = V_{A_4 D}/R_{AD}$. The remainder of the velocity polygon can then be determined, as demonstrated above.

It would, however, indicate a faulty concept to attempt to use the equation

$$\mathbf{V}_{A_4} = \mathbf{V}_{A_2} + \mathbf{V}_{A_4/2}$$

rather than Eq. (6) because the *path* traced by point A_4 in a coordinate system attached to link 2 *is not known*.*

Another insight into the nature and use of the apparent-velocity equation is provided by the following example.

EXAMPLE 3.4

An airplane is traveling at a speed of 300 km/h on a circular path of radius 5 km with center at C, as illustrated in Fig. 3.12. A rocket 30 km away from the plane is traveling on a straight course at a speed of 2000 km/h. Determine the velocity of the rocket as observed by the pilot of the plane.

---

* Although the use of this equation suggests a faulty understanding, it still yields a correct solution. If the corresponding path were found, it would be tangent to the path used for point A_2on link 4. Because the *tangents* to the two paths are the same although the paths are not, the two solutions both yield the same correct numeric result. This is *not* true in acceleration analysis in Chapter 4. Therefore, the concept should be studied and this "backward" use should be avoided.

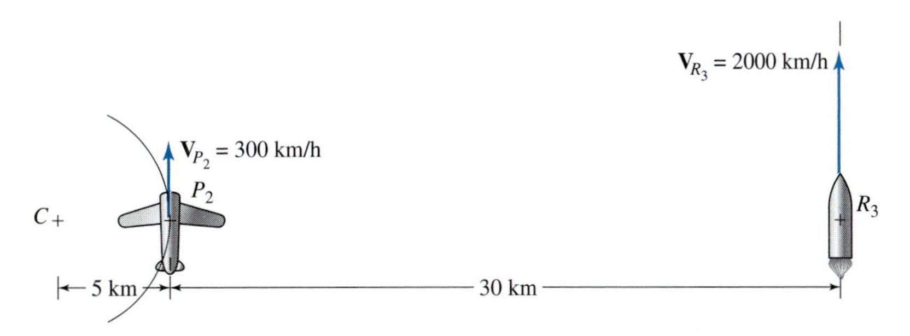

Figure 3.12 Example 3.4.

SOLUTION

Because the plane is on a circular course, the point C_2, attached to the coordinate system of the plane but coincident with C, has no motion. Therefore, the angular velocity of the plane is

$$\omega_2 = \frac{V_{PC}}{R_{PC}} = \frac{V_P}{R_{PC}} = \frac{300 \text{ km/h}}{5 \text{ km}} = 60 \text{ rad/h ccw.} \tag{9}$$

The posed question obviously requires the calculation of the apparent velocity $\mathbf{V}_{R_3/2}$, but this equation applies *only between coincident points*. Therefore, we define another point R_2, attached to the rotating coordinate system of the plane but located coincident with the rocket R_3 at the instant shown. As part of the plane, the velocity of this point is

$$\mathbf{V}_{R_2} = \mathbf{V}_P + \boldsymbol{\omega}_2 \times \mathbf{R}_{RP}$$

$$V_{R_2} = 300\frac{\text{km}}{\text{h}} + \left(60\frac{\text{rad}}{\text{h}}\right)(30 \text{ km}) = 2\ 100 \text{ km/h}, \tag{10}$$

where the values are added algebraically because the vectors are parallel. The apparent velocity can now be calculated, that is

$$\mathbf{V}_{R_3/2} = \mathbf{V}_{R_3} - \mathbf{V}_{R_2}$$

$$V_{R_3/2} = 2\ 000 \text{ km/h} - 2\ 100 \text{ km/h} = -100 \text{ km/h.} \qquad \textit{Ans.}\ (11)$$

Thus, as seen by the pilot of the plane, the rocket appears to be *backing up* at a speed of 100 km/h. This result becomes better understood as we consider the motion of point R_2. This point is being treated as *attached to the plane* and, therefore, seems stationary to the pilot. Yet, in the absolute coordinate system, this point is traveling faster than the rocket; the rocket is not keeping up with this point and therefore appears to the pilot to be backing up.

3.6 APPARENT ANGULAR VELOCITY

When two rigid bodies rotate with different angular velocities, the vector difference between the two is defined as the *apparent angular velocity*. Thus,

$$\omega_{3/2} = \omega_3 - \omega_2, \tag{3.10}$$

which can also be written as

$$\omega_3 = \omega_2 + \omega_{3/2}. \tag{3.11}$$

Note that $\omega_{3/2}$ is the angular velocity of body 3 as it would appear to an observer attached to, and rotating with, body 2. Compare this equation with Eq. (3.6) for the apparent velocity of a point.

3.7 DIRECT CONTACT AND ROLLING CONTACT

Two elements of a mechanism that are in direct contact with each other have relative motion that may or may not involve sliding between the links at the point of direct contact. In the cam-and-follower system illustrated in Fig. 3.13*a*, the cam, link 2, drives the follower, link 3, by direct contact. We see that if slip were not possible between links 2 and 3 at point P, the triangle PAB would form a truss; therefore, sliding as well as rotation must take place between the cam and the follower.

Let us distinguish between the two points P_2 attached to link 2 and P_3 attached to link 3. They are coincident points, both located at P at the instant shown; therefore, we can write the apparent-velocity equation

$$\mathbf{V}_{P_{3/2}} = \mathbf{V}_{P_3} - \mathbf{V}_{P_2}. \tag{3.12}$$

If the two absolute velocities $\mathbf{V}_{P_3}$ and $\mathbf{V}_{P_2}$ were both known, they could be subtracted to find $\mathbf{V}_{P_{3/2}}$. Components could then be taken along directions defined by the common normal and the common tangent to the surfaces at the point of direct contact. The components of $\mathbf{V}_{P_3}$ and $\mathbf{V}_{P_2}$ along the common normal must be equal, and this component of $\mathbf{V}_{P_{3/2}}$ must be zero. Otherwise, the two links would either separate or they would interfere, both contrary to

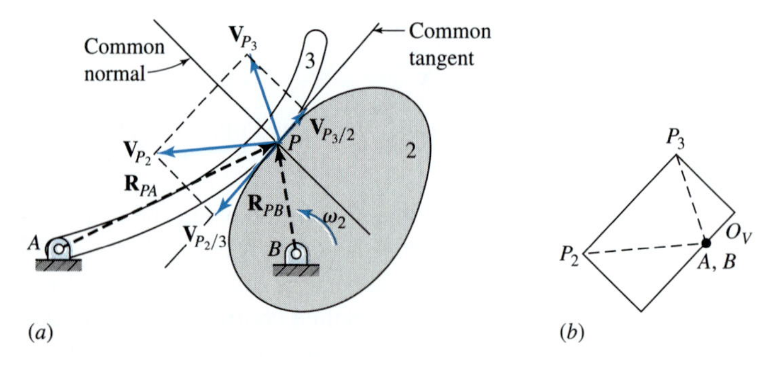

Figure 3.13 Apparent sliding velocity at a point of direct contact.

our basic assumption that contact persists. The total apparent velocity $\mathbf{V}_{P_3/2}$ must therefore lie along the common tangent and is the velocity of the relative sliding motion within the direct-contact interface. The velocity polygon for this system is illustrated in Fig. 3.13*b*.

It is possible in other mechanisms for there to be direct contact between links without slip between the links. The cam–follower system of Fig. 3.14, for example, might have high friction between the roller, link 3, and the cam surface, link 2, and restrain the wheel to roll against the cam without slip. Henceforth, we will restrict our use of the term *rolling contact* to situations where *no slip* takes place. By "no slip" we imply that the apparent "slipping" velocity of Eq. (3.12) is zero:

$$\mathbf{V}_{P_3/2} = \mathbf{0}. \tag{3.13}$$

This equation is sometimes referred to as the rolling contact condition for velocity. From Eq. (3.12), the rolling contact condition can also be written as

$$\mathbf{V}_{P_3} = \mathbf{V}_{P_2}. \tag{3.14}$$

This says that *the absolute velocities of two points in rolling contact are equal.*

The graphical solution of the problem of Fig. 3.14 is also illustrated here. Given $\boldsymbol{\omega}_2$, the velocity difference $\mathbf{V}_{P_2B}$ can be calculated and plotted, thus locating point P_2 in the velocity polygon. Using Eq. (3.13), the rolling contact condition, we also label this point P_3. Next, writing simultaneous equations for $\mathbf{V}_C$, using $\mathbf{V}_{CP_3}$ and $\mathbf{V}_{CA}$, we can find the velocity-image point C. Then ω_3 and ω_4 can be found from $\mathbf{V}_{CP}$ and $\mathbf{V}_{CA}$, respectively.

Another approach to the solution of the same problem involves defining a fictitious point C_2, which is located instantaneously coincident with points C_3 and C_4 but is understood to be attached to, and moving with, link 2, as illustrated by the shaded triangle BPC_2. When the velocity-image concept is used for link 2, the velocity-image point C_2 can be located. Noting that point C_4 (and C_3) traces a known path on link 2, we can write and solve the apparent-velocity equation involving $\mathbf{V}_{C_4/2}$. Then the velocity $\mathbf{V}_{C_4}$ (and ω_4, if desired) is obtained without dealing with the point of direct contact. This second approach would be necessary if we had not assumed rolling contact (no slip) at P.

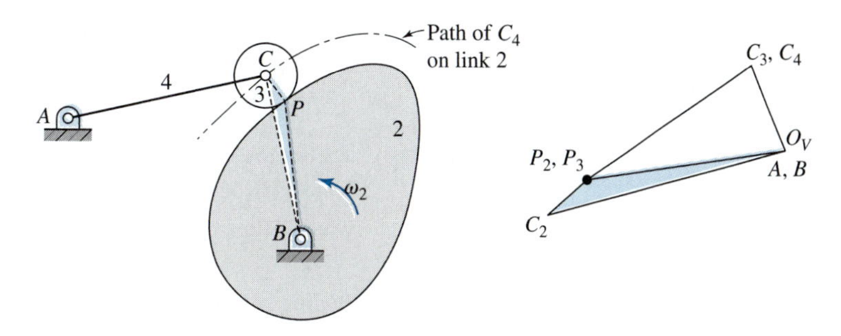

Figure 3.14 Cam–follower system with rolling contact between links 2 and 3.

3.8 SYSTEMATIC STRATEGY FOR VELOCITY ANALYSIS

Review of the preceding sections and their example problems indicates that we have now developed sufficient tools for dealing with those situations that normally arise in the analysis of rigid-body mechanical systems. It will also be noted that the word "relative" velocity has been carefully avoided. Instead, we note that whenever the desire for using relative velocity arises, there are always two points whose velocities are to be "related." Also, these two points are attached either to the same rigid body or to two different rigid bodies. Therefore, we can organize all situations into the four cases shown in Table 3.1.

In Table 3.1 we can see that, when the two points are separated by a distance, only the velocity difference equation is appropriate for use and two points on the same link should be used. However, when it is desirable to switch to a point of another link, then coincident points should be chosen and the apparent velocity equation should be used.

The notation has been made different to continually remind us that these are two totally different situations and the formulae are not interchangeable between the two. We should not try to use an "$\boldsymbol{\omega} \times \mathbf{R}$" formula when using the apparent velocity; if we do try, then we will not find a single $\boldsymbol{\omega}$ or a useful $\mathbf{R}$. Similarly, when using the velocity difference, there will be no question of which $\boldsymbol{\omega}$ to use because only one link will pertain. Further advantages will become clear when we study accelerations in Chapter 4.

According to Beyer[1], Mehmke published the following theorem in 1893[2]:

> The end points of the velocity vectors of the points of a rigid plane body, when plotted from a common origin, produce a figure which is geometrically similar to the original figure (image diagram).

It is this theorem that results in the velocity polygon presented in Section 3.4. It is also a result of this theorem that allows clarity in the velocity polygon despite the very minimal labeling that has been used in the preceding examples and will continue throughout the text. Review of the discussion in Section 3.4 may be appropriate.

Careful review of Example 3.1 through Example 3.4 indicates how the strategy suggested in Table 3.1, and the labeling of the velocity polygons, is applied to problems of velocity analysis.

TABLE 3.1 "Relative" Velocity Equations

Points are	Coincident	Separated
On same body	*Trivial case;* $\mathbf{V}_P = \mathbf{V}_Q$.	*Velocity difference;* $\mathbf{V}_P = \mathbf{V}_Q + \mathbf{V}_{PQ}$ $\mathbf{V}_{PQ} = \omega_j \times \mathbf{R}_{PQ}$
On different bodies	*Apparent velocity;* $\mathbf{V}_{P_i} = \mathbf{V}_{P_j} + \mathbf{V}_{P_i/j}$ where path $P_{i/j}$ is known. ---------------- *Rolling contact velocity;* $\mathbf{V}_{P_i} = \mathbf{V}_{P_j}$ and $\mathbf{V}_{P_i/j} = 0$.	*Too general; use two steps.*

3.9 ANALYTIC METHODS

In this section we will consider simple algebraic methods of velocity analysis. For some types of linkages, a numerical solution is often more convenient by calculator or computer, especially when the mechanism must be analyzed for multiple input positions. When solutions for multiple positions are required, graphical methods become very cumbersome.

The central (or in-line) slider-crank mechanism illustrated in Fig. 3.15 is the engine mechanism and, for this reason, we designate the lengths of the crank and the connecting rod r and l, respectively. From the geometry of the mechanism we see that

$$r\sin\theta = l sin\phi \tag{a}$$

$$x = r\cos\theta + l cos\phi. \tag{b}$$

The angle ϕ can be eliminated from Eq. (*b*) by the following procedure. Squaring Eq. (*a*) gives

$$l^2 sin^2\phi = r^2\sin^2\theta. \tag{c}$$

Therefore,

$$l cos\phi = l\sqrt{1-\sin^2\phi} = \sqrt{l^2 - r^2\sin^2\theta}. \tag{d}$$

Then substituting Eq. (*d*) into Eq. (*b*) gives

$$x = r\cos\theta + \sqrt{l^2 - r^2\sin^2\theta}. \tag{3.15}$$

Now differentiating this equation with respect to time and rearranging, the slider (or piston) velocity can be written as

$$\dot{x} = -r\omega\left[sin\theta + \frac{r\ sin2\theta}{2\sqrt{l^2 - r^2\sin^2\theta}}\right] \tag{e}$$

or as

$$\dot{x} = -r\omega\left[sin\theta + \frac{\eta\ sin2\theta}{2\sqrt{1 - \eta^2\sin^2\theta}}\right], \tag{3.16}$$

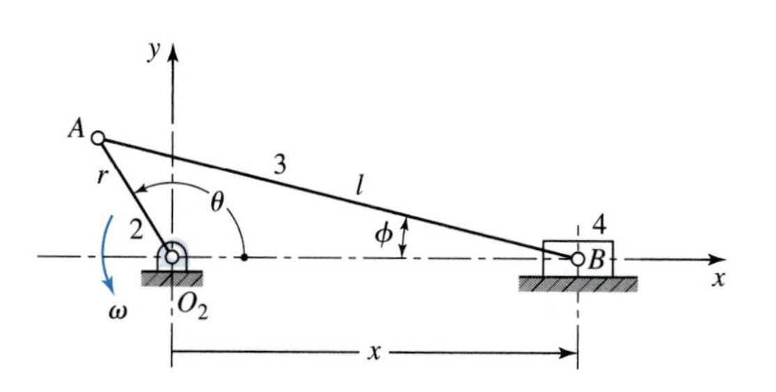

Figure 3.15 Central slider-crank mechanism.

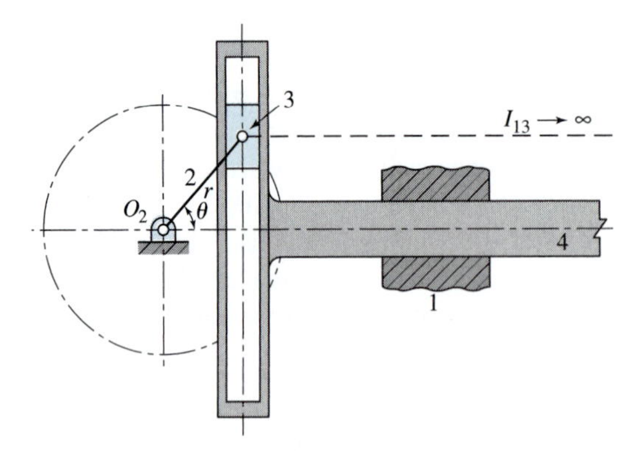

Figure 3.16 Scotch-yoke linkage.

where $\eta = r/l$ is a number less than 1. Typically, the range of values for η for most engines is $0.1 \leq \eta \leq 0.25$ (see Chapter 18).

The *Scotch yoke* illustrated in Fig. 3.16 is an interesting variation of the slider-crank mechanism. Here, link 3 rotates about a point called an instant center (denoted I_{13}) located at infinity (see Section 3.13). This has the effect of a connecting rod of infinite length, and the second terms of Eqs. (3.15) and (3.16) become zero. Hence, for the Scotch yoke we have

$$x = r\cos\theta \qquad \text{and} \qquad \dot{x} = -r\omega\sin\theta. \tag{3.17}$$

Thus, the slider moves with simple harmonic motion. It is for this reason that the deviation of the kinematics of the slider-crank motion from simple harmonic motion is sometimes said to be caused by "the angularity of the connecting rod."

3.10 COMPLEX-ALGEBRA METHODS

We recall from Section 2.9 that complex algebra provides an alternative algebraic formulation for two-dimensional kinematics problems. As we saw, the complex-algebra formulation provides the advantage of increased accuracy over graphical methods, and it is amenable to solution by digital computer at a large number of positions once the program is written. On the other hand, the solution of the loop-closure equation for the unknown position variables is a nonlinear problem and can lead to tedious algebraic manipulations. Fortunately, as we will see, the extension of the complex-algebra approach to velocity analysis leads to a set of *linear* equations, and solution is quite straightforward.

Recalling the complex polar form of a two-dimensional vector from Eq. (2.41), that is,

$$\mathbf{R} = Re^{j\theta},$$

we find the general form of its time derivative,

$$\dot{\mathbf{R}} = \frac{d\mathbf{R}}{dt} = \dot{R}e^{j\theta} + j\dot{\theta}Re^{j\theta}, \tag{3.18}$$

where $\dot{R}$ and $\dot{\theta}$ denote, respectively, the time rates of change of the magnitude and angle of the vector $\mathbf{R}$. We will see in the following examples that the first term of this equation

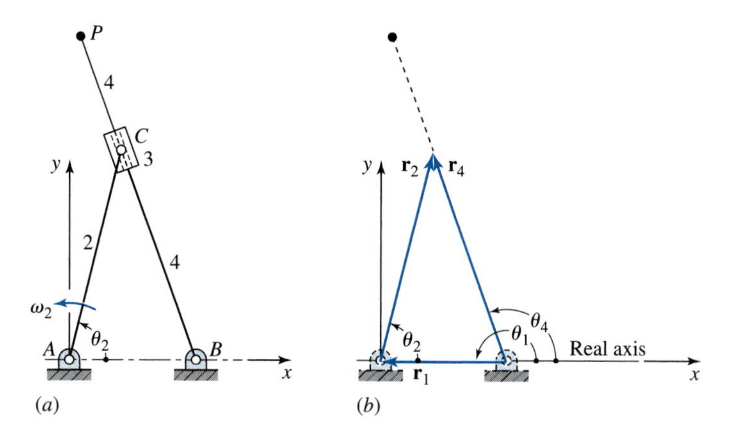

Figure 3.17 Inverted slider-crank mechanism.

often represents an apparent velocity and the second term usually represents a velocity difference. The methods illustrated in these examples were developed by Raven.[3] Although the original work provides methods applicable to both planar and spatial mechanisms, only the planar aspects are illustrated here.

To illustrate Raven's approach, let us analyze the inversion of the slider-crank mechanism in Fig. 3.17*a*. We will consider link 2, the driver, to have a known angular position θ_2 and a known angular velocity ω_2 at the instant under consideration. We wish to derive expressions for the angular velocity of link 4 and the absolute velocity of point *P*.

To simplify the notation in this example we will use the symbolism illustrated in Fig. 3.17*b* for the position-difference vectors; thus, $\mathbf{R}_{AB}$ is denoted $\mathbf{r}_1$, $\mathbf{R}_{C_2A}$ is denoted $\mathbf{r}_2$, and $\mathbf{R}_{C_4B}$ is denoted $\mathbf{r}_4$. In terms of these symbols, the loop-closure equation is

$$\mathbf{r}_1 + \mathbf{r}_2 = \mathbf{r}_4, \tag{a}$$

where the vector $\mathbf{r}_1$ has constant magnitude and direction.* The vector $\mathbf{r}_2$ has constant magnitude and its direction θ_2 varies, but θ_2 is the input angle. We assume that θ_2 is given or, more precisely, that the unknown variables will be solved as functions of θ_2. The vector $\mathbf{r}_4$ has unknown magnitude and direction.

Recognizing this as case 1 (Section 2.7), we obtain the position solution from Eqs. (2.43) and (2.44):

$$r_4 = \sqrt{r_1^2 + r_2^2 - 2r_1r_2\cos\theta_2} \tag{b}$$

$$\theta_4 = \tan^{-1}\left(\frac{r_2\sin\theta_2}{r_2\cos\theta_2 - r_1}\right). \tag{c}$$

The velocity solution is initiated by differentiating the loop-closure equation (*a*) with respect to time. Applying the general form, Eq. (3.18), to each term of this equation in turn

* Note particularly that the angle of $\mathbf{r}_1$ *is* $\theta_1 = 180°$, not zero.

and, keeping in mind that r_1, θ_1, and r_2 are constants, we obtain

$$j\dot{\theta}_2 r_2 e^{j\theta_2} = \dot{r}_4 e^{j\theta_4} + j\dot{\theta}_4 r_4 e^{j\theta_4}. \tag{d}$$

Because $\dot{\theta}_2$ and $\dot{\theta}_4$ are the same as ω_2 and ω_4, respectively, and because we recognize that

$$\dot{\theta}_2 r_2 = V_{C_2}, \qquad \dot{r}_4 = V_{C_2/4}, \qquad \dot{\theta}_4 r_4 = V_{C_4},$$

we see that Eq. (*d*) is, in fact, the complex polar form of the apparent-velocity equation

$$\mathbf{V}_{C_2} = \mathbf{V}_{C_4} + \mathbf{V}_{C_2/4}.$$

(This is pointed out for comparison only and is not a necessary step in the solution process.)

The velocity solution is performed using Euler's formula, Eq. (2.40), to separate Eq. (*d*) into its real and imaginary components. This gives

$$-\omega_2 r_2 \sin\theta_2 = \dot{r}_4 \cos\theta_4 - \omega_4 r_4 \sin\theta_4 \tag{e}$$

$$\omega_2 r_2 \cos\theta_2 = \dot{r}_4 \sin\theta_4 + \omega_4 r_4 \cos\theta_4. \tag{f}$$

Solving these two equations simultaneously for the two unknowns $\dot{r}_4$ and ω_4 yields

$$\dot{r}_4 = \omega_2 r_2 \sin(\theta_4 - \theta_2) \tag{3.19}$$

$$\omega_4 = \omega_2 \frac{r_2}{r_4} \cos(\theta_4 - \theta_2). \tag{3.20}$$

Although the variables r_4 and θ_4 could be substituted from Eqs. (*b*) and (*c*) to reduce these results to functions of θ_2 and ω_2 alone, the above forms are considered sufficient. The reason is that in writing a computer program, numeric values are generally found first for r_4 and θ_4 while performing the position analysis, and these numeric values are then used in finding $\dot{r}_4$ and ω_4 at each input angle θ_2.

To find the velocity of point P, we write

$$\mathbf{R}_P = R_{PB} e^{j\theta_4} \tag{g}$$

and we use Eq. (3.18) to differentiate with respect to time, remembering that R_{PB} is a constant length. This yields

$$\mathbf{V}_P = j\omega_4 R_{PB} e^{j\theta_4}. \tag{h}$$

Then, substituting Eq. (3.20) into this equation gives

$$\mathbf{V}_P = j\omega_2 R_{PB} \frac{r_2}{r_4} \cos(\theta_4 - \theta_2)\, e^{j\theta_4}. \tag{3.21}$$

Therefore, the horizontal and vertical components of the velocity of point P are

$$V_P^x = -\omega_2 R_{PB} \frac{r_2}{r_4} \cos(\theta_4 - \theta_2) \sin\theta_4 \tag{i}$$

$$V_P^y = -\omega_2 R_{PB} \frac{r_2}{r_4} \cos(\theta_4 - \theta_2) \cos\theta_4. \tag{j}$$

As another illustration of Raven's approach, consider the following example problem.

EXAMPLE 3.5

Develop an equation for the relationship between the angular velocities of the input and output cranks of the four-bar linkage illustrated in Fig. 3.18*a*.

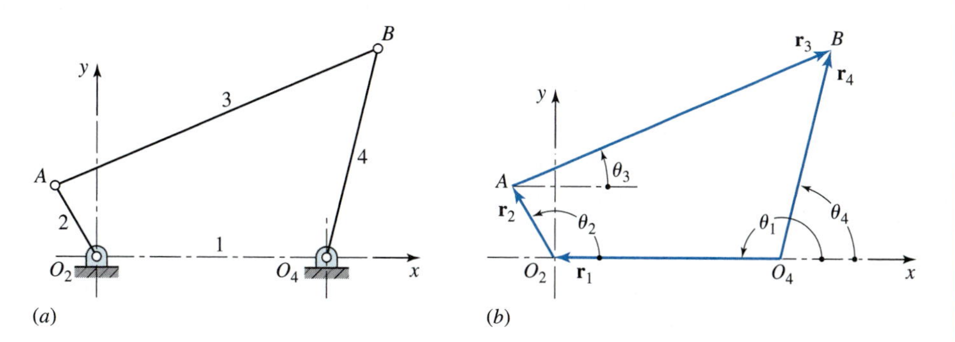

Figure 3.18 Example 3.5. (*a*) The four-bar linkage; (*b*) links 1, 2, 3, and 4 are replaced by vectors.

SOLUTION

First, replace each link of Fig. 3.18*a* by a vector as illustrated in Fig. 3.18*b*. Then, writing the loop-closure equation, we have the vector equation

$$\mathbf{r}_1 + \mathbf{r}_2 + \mathbf{r}_3 - \mathbf{r}_4 = 0. \tag{1}$$

Now, replacing each vector in Eq. (1) with a complex number in polar form, the result is

$$r_1 e^{j\theta_1} + r_2 e^{j\theta_2} + r_3 e^{j\theta_3} - r_4 e^{j\theta_4} = 0. \tag{2}$$

Because link 1 is the fixed link, the first time derivative of Eq. (2) is

$$jr_2\dot{\theta}_2 e^{j\theta_2} + jr_3\dot{\theta}_3 e^{j\theta_3} - jr_4\dot{\theta}_4 e^{j\theta_4} = 0. \tag{3}$$

Now we transform Eq. (3) into rectangular form and separate the real and the imaginary terms. Noting that $\omega_2 = \dot{\theta}_2$, $\omega_3 = \dot{\theta}_3$, and $\omega_4 = \dot{\theta}_4$, we have

$$\begin{aligned} r_2\omega_2 \cos\theta_2 + r_3\omega_3 \cos\theta_3 - r_4\omega_4 \cos\theta_4 &= 0 \\ r_2\omega_2 \sin\theta_2 + r_3\omega_3 \sin\theta_3 - r_4\omega_4 \sin\theta_4 &= 0. \end{aligned} \tag{4}$$

With the input angular velocity ω_2 given, the angular velocities ω_3 and ω_4 are the two unknown variables. Solving Eqs. (4) simultaneously yields

$$\omega_3 = \frac{r_2 \sin(\theta_2 - \theta_4)}{r_3 \sin(\theta_4 - \theta_3)}\omega_2 \quad \text{and} \quad \omega_4 = \frac{r_2 \sin(\theta_2 - \theta_3)}{r_4 \sin(\theta_4 - \theta_3)}\omega_2. \qquad \textit{Ans.} \ (3.22)$$

Note that in both of the above problems, the simultaneous equations that were solved were linear equations. This was not a coincidence but is true in all velocity solutions. It results from the fact that the general form, in Eq. (3.18), is linear in the velocity variables.

When real and imaginary components are taken, the *coefficients* may become complicated, but the *equations* remain linear with respect to the velocity unknowns. Therefore, their solution is straightforward.

Another symptom of the linearity of velocity relationships can be observed by recalling that in the graphical velocity solutions of previous sections it was possible to pick an arbitrary scale factor for a velocity polygon. If the input speed of a mechanism is doubled, the scale factor of the velocity polygon could also be doubled and the same polygon would be valid. This is another indication of linear equations.

It is also worth noting that both equations in Eqs. (3.22) include the term $\sin(\theta_4 - \theta_3)$ in their denominators. In general, any velocity-analysis problem will have similar denominators in the solutions for each of the velocity unknowns; these denominators are the determinant of the matrix of coefficients of the unknowns of the linear equations, as will be recognized by recalling Cramer's rule. In the four-bar linkage it can be seen from Fig. 2.15 that the term $(\theta_4 - \theta_3)$ is the transmission angle γ. Recall that when the transmission angle becomes small, the ratio of the output speed to the input speed becomes very large and difficulty results (Section 1.10).

3.11 THE VECTOR METHOD

Chace's approach to position analysis was discussed in Section 2.11. It will be demonstrated here how this approach is applied to the velocity analysis of linkages. The method is illustrated by again solving the inverted slider-crank mechanism of Fig. 3.17.

The procedure begins by writing the loop-closure equation,

$$\mathbf{r}_1 + \mathbf{r}_2 = \mathbf{r}_4. \tag{a}$$

The velocity relationships are determined by differentiating this equation with respect to time. The derivative of a typical term is

$$\dot{\mathbf{R}}_j = \frac{d}{dt}\left(R_j\hat{\mathbf{R}}_j\right) = \dot{R}_j\hat{\mathbf{R}}_j + R_j\dot{\hat{\mathbf{R}}}_j. \tag{b}$$

However, because $\hat{\mathbf{R}}_j$ is of constant length and generally rotates with one of the links, then $\dot{\hat{\mathbf{R}}}_j$ can be expressed as

$$\dot{\hat{\mathbf{R}}}_j = \boldsymbol{\omega}_j \times \hat{\mathbf{R}}_j = \omega_j(\hat{\mathbf{k}} \times \hat{\mathbf{R}}_j), \tag{3.23}$$

from which Eq. (b) becomes

$$\dot{R}_j = \dot{R}_j\hat{\mathbf{R}}_j + \omega_j R_j\left(\hat{\mathbf{k}} \times \hat{\mathbf{R}}_j\right). \tag{3.24}$$

Using this general form and recognizing that the magnitudes r_1 and r_2 and the direction $\hat{\mathbf{r}}_1$ are constant, we can take the time derivative of the loop-closure equation (a). This yields

$$\omega_2 r_2\left(\hat{\mathbf{k}} \times \hat{\mathbf{r}}_2\right) = \dot{r}_4\hat{\mathbf{r}}_4 + \omega_4 r_4\left(\hat{\mathbf{k}} \times \hat{\mathbf{r}}_4\right). \tag{c}$$

Because it is assumed that r_4 and $\hat{\mathbf{r}}_4$ are known from a previous position analysis, perhaps following the Chace approach of Section 2.11, and because ω_2 is the known input velocity, the only two unknown variables of this equation are the velocities $\dot{r}_4$ and ω_4.

Rather than taking components of Eq. (*c*) in the horizontal and vertical directions, which leads to two simultaneous equations in two unknowns, Chace's approach is to eliminate an unknown by careful choice of the direction along which components are taken. In Eq. (*c*), for example, we note that the unit vector $\hat{\mathbf{r}}_4$ is perpendicular to $\hat{\mathbf{k}} \times \hat{\mathbf{r}}_4$ and, therefore,

$$\hat{\mathbf{r}}_4 \cdot \left(\hat{\mathbf{k}} \times \hat{\mathbf{r}}_4\right) = \mathbf{0}. \qquad (d)$$

We take advantage of this to eliminate the unknown velocity $\dot{r}_4$. Taking the dot product of each term of Eq. (*c*) with $\left(\hat{\mathbf{k}} \times \hat{\mathbf{r}}_4\right)$, we obtain

$$\omega_2 r_2 \left(\hat{\mathbf{k}} \times \hat{\mathbf{r}}_2\right) \cdot \left(\hat{\mathbf{k}} \times \hat{\mathbf{r}}_4\right) = \omega_4 r_4.$$

Therefore, the angular velocity of link 4 can be written as

$$\omega_4 = \omega_2 \frac{r_2}{r_4} \left(\hat{\mathbf{k}} \times \hat{\mathbf{r}}_2\right) \cdot \left(\hat{\mathbf{k}} \times \hat{\mathbf{r}}_4\right). \qquad (e)$$

Similarly, we can take the dot product of Eq. (*c*) with the unit vector $\hat{\mathbf{r}}_4$ and thus eliminate the unknown angular velocity ω_4. This gives

$$\dot{r}_4 = \omega_2 r_2 \left(\hat{\mathbf{k}} \times \hat{\mathbf{r}}_2\right) \cdot \hat{\mathbf{r}}_4. \qquad (f)$$

It is easy to demonstrate that these solutions are indeed the same as those obtained by Raven's (complex-algebra) method. From Eq. (*e*) we can write

$$\hat{\mathbf{k}} \times \hat{\mathbf{r}}_2 = \begin{vmatrix} \hat{\mathbf{i}} & \hat{\mathbf{j}} & \hat{\mathbf{k}} \\ 0 & 0 & 1 \\ \cos\theta_2 & \sin\theta_2 & 0 \end{vmatrix} = -\sin\theta_2 \hat{\mathbf{i}} + \cos\theta_2 \hat{\mathbf{j}}$$

and, similarly,

$$\hat{\mathbf{k}} \times \hat{\mathbf{r}}_4 = -\sin\theta_4 \hat{\mathbf{i}} + \cos\theta_4 \hat{\mathbf{j}}.$$

Then

$$\begin{aligned} \left(\hat{\mathbf{k}} \times \hat{\mathbf{r}}_2\right) \cdot \left(\hat{\mathbf{k}} \times \hat{\mathbf{r}}_4\right) &= \left(-\sin\theta_2 \hat{\mathbf{i}} + \cos\theta_2 \hat{\mathbf{j}}\right) \cdot \left(-\sin\theta_4 \hat{\mathbf{i}} + \cos\theta_4 \hat{\mathbf{j}}\right) \\ &= \sin\theta_2 \sin\theta_4 + \cos\theta_2 \cos\theta_4 = \cos\left(\theta_4 - \theta_2\right) \end{aligned} \qquad (g)$$

and, in a similar manner,

$$\left(\hat{\mathbf{k}} \times \hat{\mathbf{r}}_2\right) \cdot \hat{\mathbf{r}}_4 = \sin(\theta_4 - \theta_2). \qquad (h)$$

When Eqs. (*g*) and (*h*) are substituted into Eqs. (*e*) and (*f*), the results match identically with those obtained from Raven's approach, Eqs. (3.19) and (3.20).

3.12 THE METHOD OF KINEMATIC COEFFICIENTS

A useful method that provides geometric insight into the motion of a linkage is to differentiate the x and y components of the loop-closure equation with respect to the input position variable, rather than differentiating directly with respect to time. This analytical approach is referred to as the method of kinematic coefficients. The numeric values of the first-order kinematic coefficients can also be checked with the graphical approach of finding the locations of the instantaneous centers of velocity (see Section 3.13). This method is illustrated here by again solving the four-bar linkage problem that was presented in Example 3.1 and the offset slider-crank mechanism that was presented in Example 3.2.

EXAMPLE 3.6

The four-bar linkage illustrated in Fig. 3.7*a* is driven by crank 2 at a constant angular velocity of $\omega_2 = 900$ rev/ min ccw. Find the angular velocities of the coupler link and the output link and the instantaneous velocities of points E and F for the position shown.

SOLUTION

First, we replace each link of Fig. 3.7*a* by a vector (see Fig. 3.18*b*). Then the vector loop-closure equation can be written as

$$\mathbf{r}_1 + \mathbf{r}_2 + \mathbf{r}_3 - \mathbf{r}_4 = \mathbf{0}. \tag{1}$$

The two scalar component equations are

$$\begin{aligned} r_1 \cos\theta_1 + r_2 \cos\theta_2 + r_3 \cos\theta_3 - r_4 \cos\theta_4 &= 0 \\ r_1 \sin\theta_1 + r_2 \sin\theta_2 + r_3 \sin\theta_3 - r_4 \sin\theta_4 &= 0. \end{aligned} \tag{2}$$

Because the input is the joint displacement θ_2, the method of kinematic coefficients implies differentiating Eqs. (2) with respect to the position variable θ_2. Rearranging the result gives

$$-r_3 \sin\theta_3 \theta_3' + r_4 \sin\theta_4 \theta_4' = r_2 \sin\theta_2$$

and

$$r_3 \cos\theta_3 \theta_3' - r_4 \cos\theta_4 \theta_4' = -r_2 \cos\theta_2, \tag{3}$$

where

$$\theta_3' = \frac{d\theta_3}{d\theta_2} \quad \text{and} \quad \theta_4' = \frac{d\theta_4}{d\theta_2} \tag{4}$$

are referred to as the *first-order kinematic coefficients* of links 3 and 4, respectively. Algebraic forms for the kinematic coefficients can be obtained using Cramer's rule. Writing Eqs. (3) in matrix form gives

$$\begin{bmatrix} -r_3 \sin\theta_3 & r_4 \sin\theta_4 \\ r_3 \cos\theta_3 & -r_4 \cos\theta_4 \end{bmatrix} \begin{bmatrix} \theta_3' \\ \theta_4' \end{bmatrix} = \begin{bmatrix} r_2 \sin\theta_2 \\ -r_2 \cos\theta_2 \end{bmatrix}. \tag{5}$$

The determinant of the (2×2) coefficient matrix can be written as

$$\Delta = -r_3 r_4 \sin(\theta_4 - \theta_3) \tag{6}$$

and provides geometric insight into special positions of the mechanism (see Example 3.5). For example, when this determinant tends toward zero, the kinematic coefficients tend toward infinity. Note that the determinant is zero (that is, the transmission angle $\theta_4 - \theta_3$ is zero) when (*i*) $\theta_3 = \theta_4$ or when (*ii*) $\theta_3 = \theta_4 \pm 180°$, that is, when links 3 and 4 are fully extended or folded on top of each other.

From Eq. (5), the first-order kinematic coefficients of links 3 and 4 are

$$\theta_3' = \frac{r_2 \sin(\theta_2 - \theta_4)}{r_3 \sin(\theta_4 - \theta_3)} \quad \text{and} \quad \theta_4' = \frac{r_2 \sin(\theta_2 - \theta_3)}{r_4 \sin(\theta_4 - \theta_3)}. \tag{7}$$

The angular velocities of links 3 and 4, obtained from the chain rule, are

$$\omega_3 = \theta_3' \omega_2 \quad \text{and} \quad \omega_4 = \theta_4' \omega_2. \tag{8}$$

Substituting Eqs. (7) into Eqs. (8) gives the same results as Eqs. (3.22). For the given link dimensions and the specified input angle θ_2, the angular positions θ_3 and θ_4 are known from the position analysis. The answers can be demonstrated to be $\theta_3 = 20.92°$ and $\theta_4 = 64.05°$. Substituting the known data into Eqs. (7), the first-order kinematic coefficients of links 3 and 4 are

$$\theta_3' = +0.271\,8 \text{ rad/rad} \quad \text{and} \quad \theta_4' = +0.526\,5 \text{ rad/rad}. \tag{9}$$

Substituting these values and the input angular velocity into Eqs. (8), the angular velocities of links 3 and 4 are

$$\omega_3 = 25.6 \text{ rad/s ccw} \quad \text{and} \quad \omega_4 = 49.6 \text{ rad/s ccw.} \qquad \textit{Ans.} \ (10)$$

These answers agree with the results obtained from the velocity polygon method (see Eqs. (4) and (5) of Example 3.1).

The position of point E with respect to the ground pivot A (see Fig. 3.7*a*), can be written as

$$\mathbf{r}_E = \mathbf{r}_2 + \mathbf{r}_{EB}. \tag{11}$$

The x and y components of this vector equation are

$$\begin{aligned} x_E &= r_2 \cos\theta_2 + r_{EB} \cos(\theta_3 - \phi) \\ y_E &= r_2 \sin\theta_2 + r_{EB} \sin(\theta_3 - \phi), \end{aligned} \tag{12}$$

where $\phi = \tan^{-1}(4 \text{ in}/10 \text{ in}) = 21.80°$.

Differentiating Eqs. (12) with respect to the input joint displacement θ_2, the first-order kinematic coefficients for point E are

$$\begin{aligned} x_E' &= -r_2 \sin\theta_2 - r_{EB} \sin(\theta_3 - \phi)\,\theta_3' \\ y_E' &= r_2 \cos\theta_2 + r_{EB} \cos(\theta_3 - \phi)\,\theta_3'. \end{aligned} \tag{13}$$

Substituting the known values into these equations, the first-order kinematic coefficients are

$$x'_E = -3.419\,1 \text{ in/rad} \quad \text{and} \quad y'_E = +0.926\,9 \text{ in/rad}. \tag{14}$$

The velocity of point E can be written from the chain rule as

$$\mathbf{V}_E = (x'_E\hat{\mathbf{i}} + y'_E\hat{\mathbf{j}})\omega_2. \tag{15}$$

Substituting Eqs. (14) and the input angular velocity into this equation, the velocity of point E is

$$\mathbf{V}_E = -26.85\hat{\mathbf{i}} + 7.28\hat{\mathbf{j}} \text{ ft/s.} \qquad \textit{Ans.}$$

Therefore, the instantaneous speed of point E is $V_E = 27.8$ ft/s, which agrees closely with the result obtained from the graphical method in Example 3.1 ($V_E = 27.6$ ft/s).

The velocity of point F can be obtained in a similar manner, and the result is

$$\mathbf{V}_F = -20.60\hat{\mathbf{i}} + 23.82\hat{\mathbf{j}} \text{ ft/s.} \qquad \textit{Ans.}$$

Therefore, the instantaneous speed of point F is $V_F = 31.5$ ft/s, which agrees closely with the result obtained in Example 3.1 ($V_F = 31.8$ ft/s).

EXAMPLE 3.7

The offset slider-crank linkage illustrated in Fig. 3.8 is driven by slider 4 at a speed of $V_C = 10$ m/s to the left at the position indicated. Determine the angular velocity of links 2 and 3 and the instantaneous velocity of point D for this position.

SOLUTION

The loop-closure equation in complex polar form for the offset slider-crank linkage is

$$jr_1 + r_2e^{j\theta_2} + r_3e^{j\theta_3} - r_4 = 0. \tag{1}$$

The two scalar equations are

$$\begin{aligned} r_2\cos\theta_2 + r_3\cos\theta_3 - r_4 &= 0 \\ r_1 + r_2\sin\theta_2 + r_3\sin\theta_3 &= 0. \end{aligned} \tag{2}$$

Because the input is the linear displacement of link 4 then differentiating Eqs. (2) with respect to r_4 gives

$$\begin{aligned} -r_2\sin\theta_2\theta'_2 - r_3\sin\theta_3\theta'_3 &= 1 \\ r_2\cos\theta_2\theta'_2 + r_3\cos\theta_3\theta'_3 &= 0, \end{aligned} \tag{3}$$

where now

$$\theta'_2 = \frac{d\theta_2}{dr_4} \quad \text{and} \quad \theta'_3 = \frac{d\theta_3}{dr_4} \tag{4}$$

are the first-order kinematic coefficients of links 2 and 3. Symbolic equations for the first-order kinematic coefficients are obtained using Cramer's rule. Writing Eqs. (3) in matrix form gives

$$\begin{bmatrix} -r_2 \sin\theta_2 & -r_3 \sin\theta_3 \\ r_2 \cos\theta_2 & r_3 \cos\theta_3 \end{bmatrix} \begin{bmatrix} \theta_2' \\ \theta_3' \end{bmatrix} = \begin{bmatrix} 1 \\ 0 \end{bmatrix}. \tag{5}$$

The determinant of the (2×2) coefficient matrix in Eq. (5) can be written as

$$\Delta = r_2 r_3 \sin(\theta_3 - \theta_2). \tag{6}$$

Note that the determinant is zero when (*i*) $\theta_2 = \theta_3$ or when (*ii*) $\theta_2 = \theta_3 \pm 180°$, that is, when links 2 and 3 are fully extended or folded on top of each other.

From Eq. (5), the first-order kinematic coefficients of links 2 and 3 can be written as

$$\theta_2' = \frac{\cos\theta_3}{r_2 \sin(\theta_3 - \theta_2)} \quad \text{and} \quad \theta_3' = \frac{-\cos\theta_2}{r_3 \sin(\theta_3 - \theta_2)}. \tag{7}$$

For the given link dimensions and the given input position $r_4 = 164$ mm, the angular positions of links 2 and 3 are $\theta_2 = 45°$ and $\theta_3 = 337°$, respectively. Substituting these values into Eqs. (7), the first-order kinematic coefficients of links 2 and 3 are

$$\theta_2' = -19.856 \text{ rad/m} \quad \text{and} \quad \theta_3' = 5.447 \text{ rad/m}. \tag{8}$$

The angular velocities of links 2 and 3 can be written from the chain rule as

$$\omega_2 = \theta_2' \dot{r}_4 \quad \text{and} \quad \omega_3 = \theta_3' \dot{r}_4, \tag{9}$$

where the linear velocity of the slider is $\dot{r}_4 = -10$ m/s. Substituting Eqs. (8) and the velocity of the slider into Eqs. (9), the angular velocities of the two links are

$$\omega_2 = 198.6 \text{ rad/s ccw} \quad \text{and} \quad \omega_3 = 54.5 \text{ rad/s cw}. \qquad \textit{Ans.}$$

These answers agree closely with the results obtained from the velocity polygon method in Example 3.2 ($\omega_2 = 200$ rad/s ccw and $\omega_3 = 53.6$ rad/s cw).

The position of point D with respect to the ground pivot A (see Fig. 3.8*a*) can be written as

$$\mathbf{r}_D = \mathbf{r}_2 + \mathbf{r}_{DB}. \tag{10}$$

The x and y components of this vector equation are

$$x_D = r_2 \cos\theta_2 + r_{DB} \cos(\theta_3 - \beta) \tag{11a}$$

$$y_D = r_2 \sin\theta_2 + r_{DB} \sin(\theta_3 - \beta), \tag{11b}$$

where $\beta = \tan^{-1}(50 \text{ mm}/80 \text{ mm}) = 32.01°$.

Differentiating Eqs. (11) with respect to the input displacement r_4, the first-order kinematic coefficients for point D are

$$x'_D = -r_2 \sin\theta_2\theta'_2 - r_{DB}\sin(\theta_3 - \beta)\,\theta'_3 \tag{12a}$$

$$y'_D = r_2 \cos\theta_2\theta'_2 + r_{DB}\cos(\theta_3 - \beta)\,\theta'_3. \tag{12b}$$

Substituting the known values into these equations, the first-order kinematic coefficients are

$$x'_D = 1.123 \text{ m/m} \quad \text{and} \quad y'_D = -0.407 \text{ m/m}. \tag{13}$$

The velocity of the point D can be written as

$$\mathbf{V}_D = (x'_D\hat{\mathbf{i}} + y'_D\hat{\mathbf{j}})\dot{r}_4. \tag{14}$$

Substituting Eqs. (13) and the input velocity $\dot{r}_4 = -10$ m/s into this equation, the velocity of point D is

$$\mathbf{V}_D = -11.23\hat{\mathbf{i}} + 4.07\hat{\mathbf{j}} \text{ m/s.} \qquad \textit{Ans.}\ (15)$$

Therefore, the instantaneous speed of point D is 11.94 m/s, which agrees closely with the result obtained from the velocity polygon method in Example 3.2 ($V_D = 12$ m/s).

The method of kinematic coefficients also provides geometric insight into the motion of mechanisms with links that are in rolling contact. The following two examples will help to illustrate this important concept.

EXAMPLE 3.8

For the mechanism illustrated in Fig. 3.19, the wheel is rolling without slipping on the ground. The radius of the wheel is 100 mm. The length of the input crank 2 is 1.5 times the radius of the wheel. The distance from pin B to the center of the wheel G is one-half the radius of the wheel, and the distance from pin O_2 to the point of contact C is twice the radius of the wheel. For the position illustrated in Fig. 3.19, that is, with $\theta_2 = 90°$, determine the following:

a. The first-order kinematic coefficients of links 3 and 4;
b. The angular velocities of links 3 and 4 if the input is rotating with a constant angular velocity of 10 rad/s ccw; and
c. The velocity of the center of the wheel.

SOLUTION

The loop-closure equation for the mechanism is

$$\mathbf{r}_2 + \mathbf{r}_3 + \mathbf{r}_4 + \mathbf{r}_7 - \mathbf{r}_9 = \mathbf{0}. \tag{1}$$

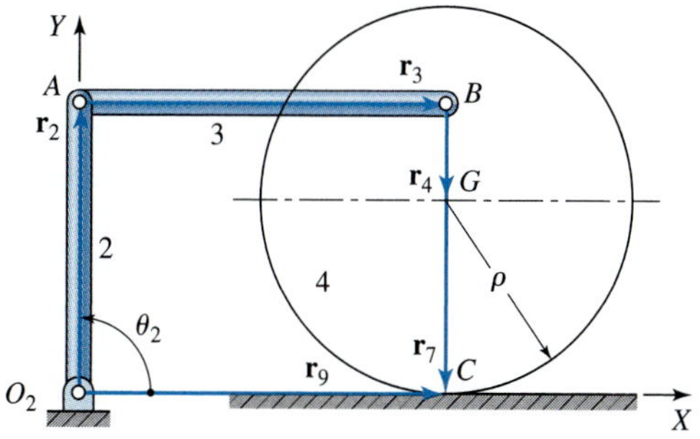

Figure 3.19 Example 3.8.

The horizontal and vertical component scalar equations are

$$r_2 \cos\theta_2 + r_3 \cos\theta_3 + r_4 \cos\theta_4 + r_7 \cos\theta_7 - r_9 \cos\theta_9 = 0 \tag{2a}$$

$$r_2 \sin\theta_2 + r_3 \sin\theta_3 + r_4 \sin\theta_4 + r_7 \sin\theta_7 - r_9 \sin\theta_9 = 0. \tag{2b}$$

Because the input is the angular displacement of link 2, we differentiate Eqs. (2) with respect to θ_2 and, by setting $\theta_9 = 0°$, we obtain

$$-r_2 \sin\theta_2 - r_3 \sin\theta_3\theta_3' - r_4 \sin\theta_4\theta_4' - r_9' = 0 \tag{3a}$$

and

$$r_2 \cos\theta_2 + r_3 \cos\theta_3\theta_3' + r_4 \cos\theta_4\theta_4' = 0, \tag{3b}$$

where we have taken note that both r_7 and θ_7 are constants and where

$$\theta_3' = \frac{d\theta_3}{d\theta_2}, \quad \theta_4' = \frac{d\theta_4}{d\theta_2}, \quad \text{and} \quad r_9' = \frac{dr_9}{d\theta_2} \tag{4}$$

are the first-order kinematic coefficients of links 3 and 4 and the position of the contact point.

We note that the rotation of the wheel, θ_4, and the change in the distance r_9 are not independent because the wheel is rolling on the ground without slipping. This rolling contact (no slip) condition can be written as

$$-\Delta r_9 = \rho\, \Delta\left(\theta_4 - \theta_7\right), \tag{5}$$

where ρ is the radius of the wheel, $\theta_7 = 270°$ is constant for all positions of the wheel, and $\Delta\theta_7 = 0$; the factor $\Delta\left(\theta_4 - \theta_7\right)$ is the angular displacement from the position illustrated for link 4 with respect to the vertical orientation of vector 7. The negative sign on the left of Eq. (5) is the result of the decreasing magnitude of r_9 for a positive change in the input variable $\Delta\theta_2$.

In the limit, for infinitesimal displacements, the constraint Eq. (5) can be written in terms of first-order kinematic coefficients as

$$-r_9' = \rho\theta_4'. \tag{6}$$

Substituting Eq. (6) into Eqs. (3) and writing the resulting equations in matrix form gives

$$\begin{bmatrix} -r_3 \sin\theta_3 & \rho - r_4 \sin\theta_4 \\ r_3 \cos\theta_3 & r_4 \cos\theta_4 \end{bmatrix} \begin{bmatrix} \theta_3' \\ \theta_4' \end{bmatrix} = \begin{bmatrix} +r_2 \sin\theta_2 \\ -r_2 \cos\theta_2 \end{bmatrix}. \tag{7}$$

The determinant of the (2×2) coefficient matrix in Eq. (7) can be written as

$$\Delta = r_3 \left[r_4 \sin(\theta_4 - \theta_3) - \rho \cos\theta_3 \right]. \tag{8}$$

From Eq. (7), using Cramer's rule, the symbolic equations for the first-order kinematic coefficients of links 3 and 4, respectively, are

$$\theta_3' = \frac{r_2 \left[\rho \cos\theta_2 - r_4 \sin(\theta_4 - \theta_2) \right]}{\Delta} \tag{9a}$$

$$\theta_4' = \frac{r_2 r_3 \sin(\theta_3 - \theta_2)}{\Delta}. \tag{9b}$$

For the position illustrated in Fig. 3.19, we have $\theta_2 = 90°$, $\theta_3 = 0°$, and $\theta_4 = -90°$. Therefore, the determinant in Eq. (8) can be written as

$$\Delta = -r_3 (r_4 + \rho). \tag{10}$$

Substituting the known data and Eq. (10) into Eqs. (9) gives

$$\theta_3' = 0 \quad \text{and} \quad \theta_4' = \frac{r_2}{r_4 + \rho} = 1 \text{ rad/rad}. \qquad \textit{Ans.} \ (11)$$

Note that these results are in agreement with our intuition; in this position link 3 is not rotating and link 4 is rotating at the same angular velocity as the input. Therefore, the angular velocity of link 3 is

$$\omega_3 = \theta_3' \omega_2 = 0 \qquad \textit{Ans.}$$

and the angular velocity of the wheel, link 4, is

$$\omega_4 = \theta_4' \omega_2 = 10 \text{ rad/s ccw}. \qquad \textit{Ans.}$$

The velocity of the center of the wheel, point G, is

$$\mathbf{V}_G = r_9' \omega_2 \angle \theta_9. \tag{12}$$

Substituting Eq. (11) into Eq. (6) gives $r_9' = -\rho = -100$ mm, and substituting this result into Eq. (12) gives

$$\mathbf{V}_G = r_9' \omega_2 \angle \theta_9 = (-100 \text{ mm/rad})\,(10 \text{ rad/s}) \angle 0° = -1.0 \text{ m/s} \angle 0°. \qquad \textit{Ans.}$$

This result can be verified by direct calculation by noting that

$$\mathbf{V}_G = \boldsymbol{\omega}_4 \times \left(\rho \hat{\mathbf{j}} \right) = (10 \text{ rad/s})\, \hat{\mathbf{k}} \times (100 \text{ mm})\, \hat{\mathbf{j}} = -1.0 \hat{\mathbf{i}} \text{ m/s}. \qquad \textit{Ans.}$$

EXAMPLE 3.9

For the mechanism illustrated in Fig. 3.20, gear 3 is rolling without slip on input gear 2. The radius of gear 3 is half the radius of gear 2, the length of link 5 is three times the radius of the input gear, and the distance from pin A to pin B is half the radius of gear 3. For the position illustrated—that is, with arm 4 vertical and link 5 horizontal—determine (*i*) the first-order kinematic coefficients of links 3, 4, and 5 and (*ii*) the angular velocities of links 3, 4, and 5 if the input gear is rotating with a constant angular velocity of 10 rad/s ccw.

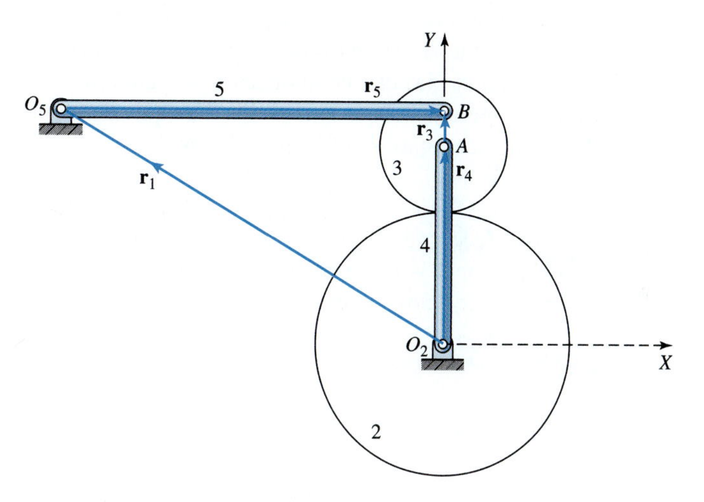

Figure 3.20 Example 3.9.

SOLUTION

The loop-closure equation for the mechanism is

$$\mathbf{r}_4 + \mathbf{r}_3 - \mathbf{r}_5 - \mathbf{r}_1 = \mathbf{0}. \tag{1}$$

The two scalar equations are

$$r_4 \cos\theta_4 + r_3 \cos\theta_3 - r_5 \cos\theta_5 - r_1 \cos\theta_1 = 0 \tag{2a}$$

$$r_4 \sin\theta_4 + r_3 \sin\theta_3 - r_5 \sin\theta_5 - r_1 \sin\theta_1 = 0. \tag{2b}$$

Because the input is the angular displacement of gear 2, differentiating Eqs. (2) with respect to θ_2 gives

$$-r_4 \sin\theta_4\theta_4' - r_3 \sin\theta_3\theta_3' + r_5 \sin\theta_5\theta_5' = 0 \tag{3a}$$

$$r_4 \cos\theta_4\theta_4' + r_3 \cos\theta_3\theta_3' - r_5 \cos\theta_5\theta_5' = 0, \tag{3b}$$

where

$$\theta_3' = \frac{d\theta_3}{d\theta_2}, \quad \theta_4' = \frac{d\theta_4}{d\theta_2}, \quad \text{and} \quad \theta_5' = \frac{d\theta_5}{d\theta_2} \tag{4}$$

are the first-order kinematic coefficients of links 3, 4, and 5.

Note that the joint variables θ_2, θ_3, and θ_4 are not independent because gear 3 is constrained to roll on gear 2. The rolling contact condition can be written to demonstrate that the arc lengths, passed during a small movement, are equal on the two surfaces. This rolling contact (no slip) condition can be written as

$$\rho_2\left(\Delta\theta_2-\Delta\theta_4\right)=-\rho_3\left(\Delta\theta_3-\Delta\theta_4\right), \tag{5a}$$

where $\Delta\theta_2$, $\Delta\theta_3$, and $\Delta\theta_4$ are angular displacements of links 2, 3, and 4 from the position illustrated in Fig. 3.20. The negative sign, in front of the right-hand term, arises from the reversal in the direction of the angular displacements.

Equation (5a) can be divided by a change in the input motion $\Delta\theta_2$ and the limit for small increments taken. This yields the equivalent constraint equation written in terms of first-order kinematic coefficients, that is

$$\rho_2\left(\theta_2'-\theta_4'\right)=-\rho_3\left(\theta_3'-\theta_4'\right). \tag{5b}$$

Since the input is the rotation of gear 2, then, by definition, $\theta_2'=1$ and by substituting $\rho_2=2\rho_3$, Eq. (5*b*) can be written as

$$\theta_3'=3\theta_4'-2. \tag{6}$$

Substituting Eq. (6) into Eqs. (3) and writing the resulting equations in matrix form gives

$$\begin{bmatrix}-r_4\sin\theta_4-3r_3\sin\theta_3 & r_5\sin\theta_5\\ r_4\cos\theta_4+3r_3\cos\theta_3 & -r_5\cos\theta_5\end{bmatrix}\begin{bmatrix}\theta_4'\\ \theta_5'\end{bmatrix}=\begin{bmatrix}-2r_3\sin\theta_3\\ 2r_3\cos\theta_3\end{bmatrix}. \tag{7}$$

The determinant of the (2×2) coefficient matrix in Eq. (7) can be written as

$$\Delta=r_5\left[r_4\sin\left(\theta_4-\theta_5\right)+3r_3\sin\left(\theta_3-\theta_5\right)\right]. \tag{8}$$

From Eq. (7), using Cramer's rule, the symbolic equations for the first-order kinematic coefficients of links 4 and 5 are

$$\theta_4'=\frac{2r_3r_5\sin\left(\theta_3-\theta_5\right)}{\Delta} \tag{9a}$$

$$\theta_5'=\frac{2r_3r_4\sin\left(\theta_3-\theta_4\right)}{\Delta}. \tag{9b}$$

For the position illustrated in Fig. 3.20, we have $\theta_3=90°$, $\theta_4=90°$, and $\theta_5=0°$; therefore, Eq. (8) can be written as

$$\Delta=r_5\left(r_4+3r_3\right). \tag{10}$$

Also, Eqs. (9) reduce to

$$\theta_4'=\frac{2\,r_3}{r_4+3r_3}=\frac{2}{9}\text{ rad/rad}\quad\text{and}\quad\theta_5'=0. \tag{11}$$

It is interesting to note that link 5 is not rotating in this position; that is, its angular velocity is $\omega_5 = 0$. By substituting Eq. (11) into Eq. (6) the first-order kinematic coefficient of gear 3 becomes

$$\theta_3' = -\frac{4}{3} \text{ rad/rad.} \tag{12}$$

The angular velocity of link j can now be written as

$$\omega_j = \theta_j' \omega_2. \tag{13}$$

Therefore, by substituting Eqs. (11) and (12) into Eq. (13) and setting $\omega_2 = 10$ rad/s ccw, the angular velocities of gear 3 and links 4 and 5 become

$$\omega_3 = -13.33 \text{ rad/s}, \quad \omega_4 = +2.22 \text{ rad/s}, \quad \text{and} \quad \omega_5 = 0, \qquad \textit{Ans.} \tag{14}$$

where the negative sign indicates clockwise rotation and the positive sign indicates counterclockwise rotation.

Note that kinematic coefficients are functions of *position only*; that is, they are not directly functions of time. Also, note that the units of the first-order kinematic coefficients depend on the specified input and the variable under consideration. The units may be nondimensional (rad/rad or length/length), length (length/rad), or the reciprocal of length (rad/length). Table 3.2 summarizes the first-order kinematic coefficients that are related to link j of a mechanism (that is, the vector r_j) having (*i*) unknown angle θ_j and/or (*ii*) unknown magnitude r_j.

TABLE 3.2 Summary of first-order kinematic coefficients

	Variable of interest Angle θ_j (Use symbol θ_j' for kinematic coefficient regardless of input)	Variable of interest Magnitude r_j (Use symbol r_j' for kinematic coefficient regardless of input)
Input ψ = angle θ_i	$\omega_j = \theta_j' \dot{\psi}$ $\theta_j' = \frac{d\theta_j}{d\psi}$ (dimensionless, rad/rad)	$\dot{r}_j = r_j' \dot{\psi}$ $r_j' = \frac{dr_j}{d\psi}$ (length, length/rad)
Input ψ = distance r_i	$\omega_j = \theta_j' \dot{\psi}$ $\theta_j' = \frac{d\theta_j}{d\psi}$ (1/length, rad/length)	$\dot{r}_j = r_j' \dot{\psi}$ $r_j' = \frac{dr_j}{d\psi}$ (dimensionless, length/length)

3.13 INSTANTANEOUS CENTER OF VELOCITY

One of the more interesting concepts in kinematics is that of an instantaneous velocity axis for a pair of rigid bodies that move with respect to one another. In particular, we shall find that an axis exists that is common to both bodies and about which each body is rotating with respect to the other.

Because our study of these axes is restricted to planar motions,* each axis is perpendicular to the plane of the motion and, therefore, reduces to a point in that plane. We shall refer to these points as *instant centers of velocity*, commonly abbreviated as *instant centers*. (Some texts prefer the name *velocity poles*; see Section 3.21). Chapter 3 only considers instant centers for mechanisms with mobility $m = 1$. However, instant centers can also be used for velocity analysis of mechanisms with a mobility greater than 1 (see Chapter 5).

For planar motion, an instant center can be regarded as a pair of coincident points, one attached to each body, about which one body may have a rotational velocity, but no translational velocity, with respect to the other. This property is true only instantaneously, and a new pair of coincident points becomes the instant center at the next instant. It is not correct, therefore, to speak of an instant center as the center of rotation, because it is generally not located at the center of curvature of the apparent point path that a point of one body generates with respect to the coordinate system of the other. Even with this restriction, however, we will find that instant centers contribute substantially to understanding the kinematics of planar motion.

The instantaneous center of velocity is defined as *the instantaneous location of a pair of coincident points of two different rigid bodies for which the absolute velocities of the two points* are equal. The instantaneous center of velocity may also be defined as the location of a pair of coincident points of two different rigid bodies for which the apparent velocity of one of the points is zero, as seen by an observer on the other body.

Let us consider a rigid body 2 having some general motion with respect to the x_1y_1 plane; the motion might be translation, rotation, or a combination of both. As illustrated in Fig. 3.21, suppose that some point A of the body has a known velocity $\mathbf{V}_A$ and that the body has a known angular velocity $\boldsymbol{\omega}_2$. With these two quantities known, the velocity of any other point of the body can be determined from the velocity-difference equation. Suppose we define a point I of the same body, for example, whose position difference $\mathbf{R}_{IA}$ from the

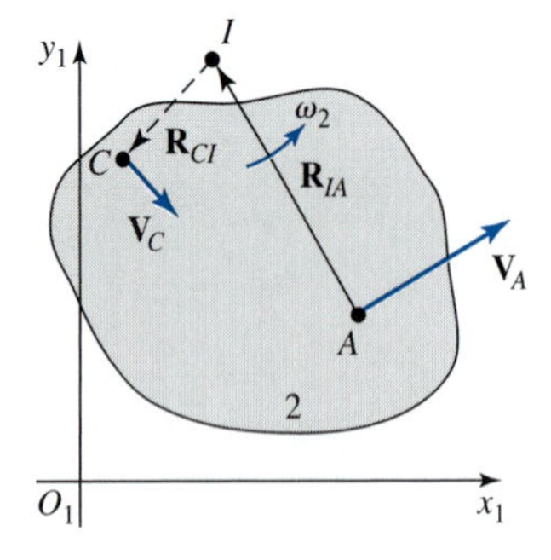

Figure 3.21

* For three-dimensional motion, this axis is referred to as the *instantaneous screw axis*. The classic work covering its properties is R. S. Ball, *A Treatise on the Theory of Screws*, Cambridge University Press, Cambridge, 1900, reprinted in paperback in 1998.

point A is chosen to be

$$\mathbf{R}_{IA} = \frac{\boldsymbol{\omega}_2 \times \mathbf{V}_A}{\omega_2^2}. \tag{3.25}$$

Because of the cross-product, we see that point I is located perpendicular to $\mathbf{V}_A$ and vector $\mathbf{R}_{IA}$ is rotated from the direction of $\mathbf{V}_A$ in the direction of $\boldsymbol{\omega}_2$, as illustrated in Fig. 3.21. The length of $\mathbf{R}_{IA}$ can be calculated from the above equation, and point I can be located. The velocity of point I can be written as

$$\mathbf{V}_I = \mathbf{V}_A + \mathbf{V}_{IA} = \mathbf{V}_A + \boldsymbol{\omega}_2 \times \mathbf{R}_{IA} = \mathbf{V}_A + \boldsymbol{\omega}_2 \times \frac{\boldsymbol{\omega}_2 \times \mathbf{V}_A}{\omega_2^2}.$$

Recalling the vector triple product $\boldsymbol{a} \times (\boldsymbol{b} \times \boldsymbol{c}) = \boldsymbol{b}(\boldsymbol{c}.\boldsymbol{a}) - \boldsymbol{c}(\boldsymbol{a}.\boldsymbol{b})$ and substituting this identity into the above equation gives

$$\mathbf{V}_I = \mathbf{V}_A + \frac{\overbrace{(\boldsymbol{\omega}_2 \cdot \mathbf{V}_A)}^{0}\,\boldsymbol{\omega}_2 - \overbrace{(\boldsymbol{\omega}_2 \cdot \boldsymbol{\omega}_2)}^{\omega_2^2}\,\mathbf{V}_A}{\omega_2^2} = \mathbf{V}_A - \mathbf{V}_A = \mathbf{0}. \tag{a}$$

Because the absolute velocity of the particular point I chosen is zero, the same as the velocity of the coincident point of the fixed link, then point I is the instant center between links 1 and 2.

The velocity of any other point of the moving body can now be determined using point I. For example, for point C illustrated in Fig. 3.21, the velocity can be written as

$$\mathbf{V}_C = \cancelto{\mathbf{0}}{\mathbf{V}_I} + \mathbf{V}_{CI} = \boldsymbol{\omega}_2 \times \mathbf{R}_{CI}. \tag{b}$$

The direction of the velocity of this point is as illustrated in Fig. 3.21.

The instant center can be located more easily when the absolute velocities of two points are given. In Fig. 3.22a, suppose that the points A and C have known velocities $\mathbf{V}_A$ and $\mathbf{V}_C$.

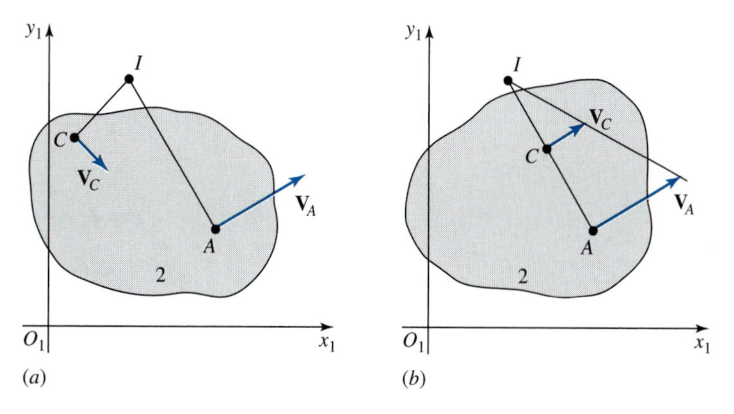

Figure 3.22 Locating an instant center from two known velocities.

The perpendiculars to $\mathbf{V}_A$ and $\mathbf{V}_C$ intersect at I, the instant center. Figure 3.22*b* illustrates how to locate the instant center I when points A, C, and I happen to fall on the same straight line.

The instant center between two bodies, in general, is not a stationary point. It changes its location with respect to both bodies as the motion progresses and describes a path or locus on each. These paths of the instant centers, called centrodes, will be discussed in Section 3.21.

Because we have adopted the convention of numbering the links of a mechanism, it is convenient to designate an instant center using the numbers of the two links associated with it. Thus, I_{32}, for example, identifies the instant centers between links 3 and 2. This same instant center could be identified as I_{23} because the order of the numbers has no significance. A mechanism has as many instant centers as there are ways of pairing the link numbers. Thus, the number of instant centers in an n-link mechanism is

$$N = \frac{n\,(n-1)}{2}. \tag{3.26}$$

3.14 THE ARONHOLD–KENNEDY THEOREM OF THREE CENTERS

According to Eq. (3.26), the number of instant centers in a four-bar linkage is 6. As illustrated in Fig. 3.23*a*, we can identify four of the instant centers by inspection (called primary instant centers). The four pins can be identified as primary instant centers I_{12}, I_{23}, I_{34}, and I_{14}, because each satisfies the definition. Point I_{23}, for example, is a point of link 2 about which link 3 appears to rotate; it is a point of link 3 that has no apparent velocity as seen from link 2; it is a pair of coincident points of links 2 and 3 that have the same absolute velocities.

A good method of keeping track of which instant centers have been identified is to arrange the link numbers around the perimeter of a circle (called the Kennedy circle), as illustrated in Fig. 3.23*b*. Then, as each instant center is identified, a line is drawn connecting the corresponding pair of link numbers. Figure 3.23*b* illustrates that I_{12}, I_{23}, I_{34}, and I_{14} have been identified; it also includes dashed lines indicating that I_{13} and I_{24} (called secondary instant centers) have not yet been identified. The locations of the secondary instant centers cannot be determined simply by applying the definition visually. After locating the primary instant centers by inspection, the secondary instants are located by applying the

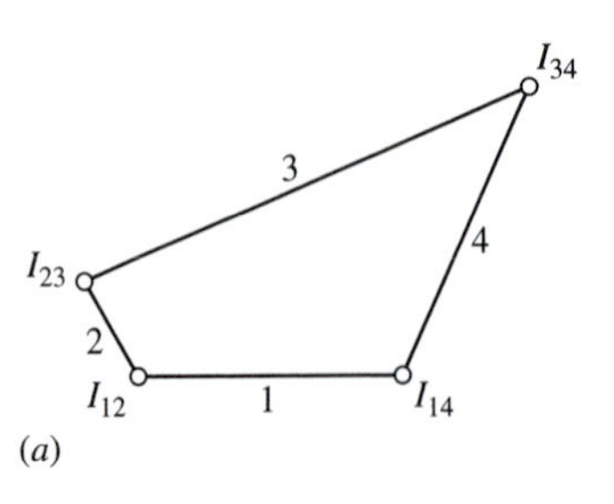

(*a*)

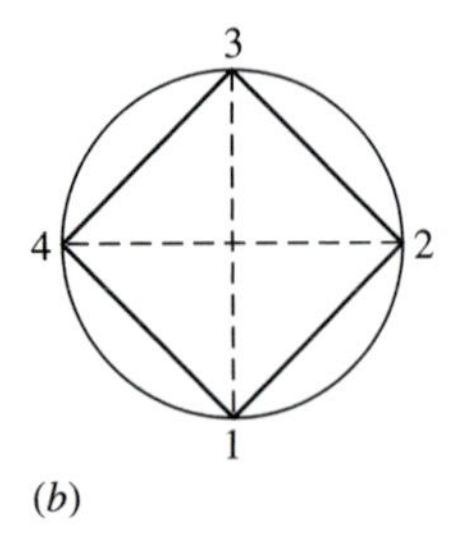

(*b*)

Figure 3.23

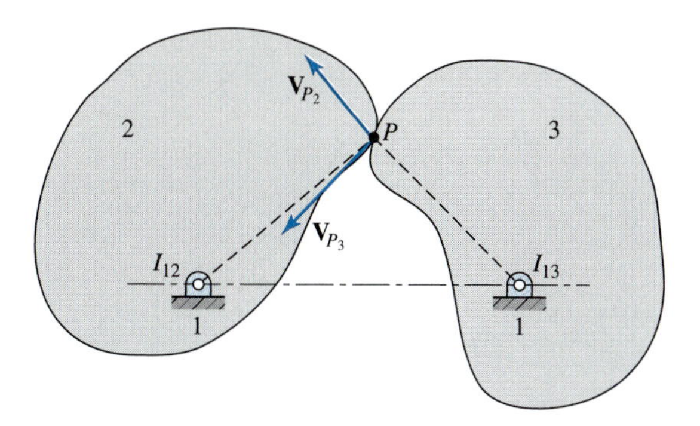

Figure 3.24 The Aronhold–Kennedy theorem.

Aronhold–Kennedy theorem of three centers (often just called Kennedy's theorem*). This theorem states that *the three instant centers shared by three rigid bodies in relative motion to one another (whether or not they are connected) all lie on the same straight line.*

The theorem can be proven by contradiction, as illustrated in Fig. 3.24. Link 1 is a stationary frame, and instant center I_{12} is located where link 2 is pin connected to it. Similarly, I_{13} is located at the pin connecting links 1 and 3. The shapes of links 2 and 3 are arbitrary. The Aronhold–Kennedy theorem states that the three instant centers I_{12}, I_{13}, and I_{23} must all lie on the same straight line, the line connecting the two pins. Let us suppose that this were not true; in fact, let us suppose that I_{23} were located at the point labeled P in Fig. 3.24. Then the velocity of P as a point of link 2 would have the orientation $\mathbf{V}_{P_2}$, perpendicular to $\mathbf{R}_{PI_{12}}$. However, the velocity of P as a point of link 3 would have the orientation V_{P_3}, perpendicular to $\mathbf{R}_{PI_{13}}$. The orientations of these two velocities are inconsistent with the definition that an instant center must have equal absolute velocities as a part of either link. The point P chosen, therefore, cannot be the instant center I_{23}. This same contradiction in the directions of $\mathbf{V}_{P_2}$ and $\mathbf{V}_{P_3}$ occurs for any location chosen for point P unless the point is chosen on the straight line through the instant centers I_{12} and I_{13}.

3.15 LOCATING INSTANT CENTERS OF VELOCITY

In the previous two sections we have considered several methods of locating instant centers of velocity. They can often be located by inspecting the figure of a mechanism and visually seeking out coincident point pairs that fit the definition, such as pin-joint centers. Also, after these primary instant centers are identified, the secondary instant centers can be determined using the theorem of three centers. Section 3.14 demonstrated that an instant center between a moving body and the fixed link can be identified if the directions of the absolute velocities of two points of the body are known or if the absolute velocity of one point and the angular velocity of the body are known. The purpose of this section is to expand the list of techniques and to present examples.

* The Aronhold–Kennedy theorem is named after its two independent discoverers, S. H. Aronhold in 1872 and A. B. W. Kennedy in 1886. It is known as the Aronhold theorem in German-speaking countries and is called Kennedy's theorem in English-speaking countries.

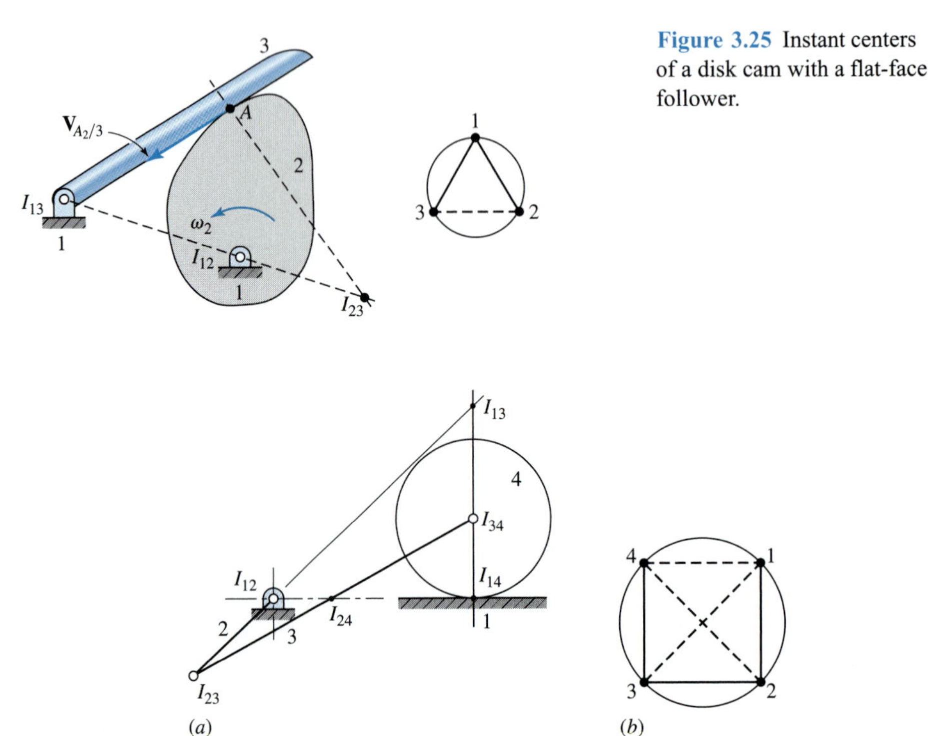

Figure 3.25 Instant centers of a disk cam with a flat-face follower.

Figure 3.26 Instant center at a point of rolling contact.

Consider the cam–follower system of Fig. 3.25. The instant centers I_{12} and I_{13} are primary instant centers since, by inspection, they are located at the two pin centers. However, the location of the secondary instant center, I_{23} (indicated by the dashed line in the Kennedy circle), is not immediately obvious. According to the Aronhold–Kennedy theorem, it must lie on the straight line connecting I_{12} and I_{13}, but where on this line? After some reflection we see that the orientation of the apparent velocity $\mathbf{V}_{A_2/3}$ must be along the common tangent to the two moving links at the point of contact. As seen by an observer on link 3, this velocity must result from the apparent relative rotation of body 2 about the instant center I_{23}. Therefore, I_{23} must lie on the line that is perpendicular to $\mathbf{V}_{A_2/3}$. This line now locates I_{23} as shown in the figure. The concept illustrated in this example should be remembered because it is often useful in locating the instant centers of mechanisms involving direct contact.

A special case of direct contact, as we have seen before, is rolling contact with no slip. Considering the mechanism of Fig. 3.26, we can immediately locate the instant centers I_{12}, I_{23}, and I_{34}. As demonstrated by the previous example, if the contact between links 1 and 4 involves any slippage, we can only say that the instant center I_{14} is located on the vertical line through the point of contact. However, if we also know that there is no slip—that is, *if there is rolling contact—then the instant center must be located at the point of contact*. This is also a general principle, as can be seen by comparing the definition of rolling contact, Eq. (3.14), and the definition of an instant center; they are equivalent.

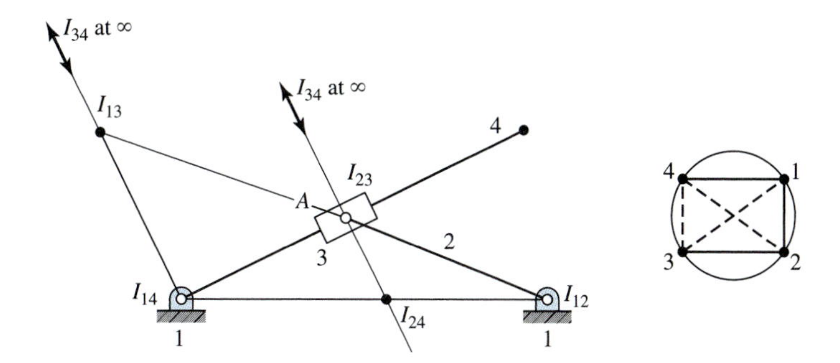

Figure 3.27 Instant centers of an inverted slider-crank mechanism.

Another special case of direct contact is evident between links 3 and 4 in Fig. 3.27. In this case there *is* an apparent (slip) velocity $\mathbf{V}_{A_3/4}$ between points A of links 3 and 4, but there is *no apparent rotation between the links*. Here, as in Fig. 3.25, the instant center I_{34} lies along a common perpendicular to the known line of sliding, but now it is located infinitely far away, in the direction defined by this perpendicular line. This infinite distance can be demonstrated by considering the kinematic inversion of the mechanism in which link 4 becomes stationary. Writing Eq. (3.25) for the inverted mechanism, we see that

$$\left|\mathbf{R}_{I_{34}A}\right| = \left|\frac{\boldsymbol{\omega}_{3/4} \times \mathbf{V}_{A_3/4}}{\omega_{3/4}^2}\right| = \left|\hat{\boldsymbol{\omega}}_{3/4} \times \left(\frac{\mathbf{V}_{A_3/4}}{\omega_{3/4}}\right)\right| = \infty. \tag{3.27}$$

The direction stated earlier is confirmed by the cross-product of Eq. (3.27). We also see that, because there is no relative rotation between links 3 and 4, the denominator is zero and the distance from I_{23} to I_{34} is infinite. The remaining instant centers of Fig. 3.27 can be identified by inspection or by the Aronhold–Kennedy theorem. Note that the line through I_{14} is drawn parallel to the line through the instant centers I_{23} and I_{34}. On this line must lie the instant center I_{13} (in agreement with the Aronhold–Kennedy theorem).

One final example will be presented here to again illustrate the above concepts.

EXAMPLE 3.10

Locate all instant centers of the mechanism illustrated in Fig. 3.28 assuming there is rolling contact between links 1 and 2.

SOLUTION

The three pin joints are the instant centers I_{13}, I_{34}, and I_{15}, and the point of rolling contact between links 1 and 2 is the instant center I_{12}. The instant center I_{24} could be located by drawing perpendicular lines to the directions of the apparent velocities at two of the corners of link 4. However, from observation, this instant center is the center of the apparent rotation between links 2 and 4. One line for the instant center I_{25} is perpendicular to the direction of slipping between links 2 and 5 and the other line is the line through the instant centers $I_{12}I_{15}$.

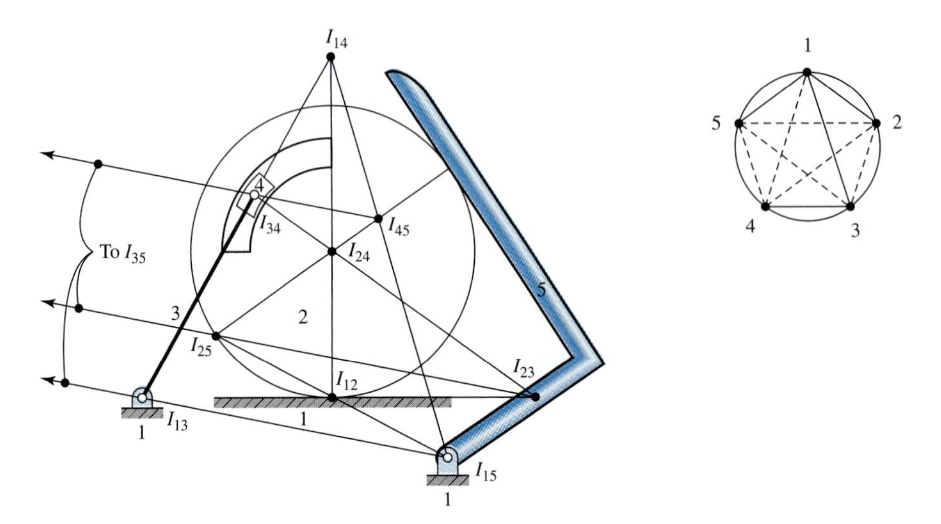

Figure 3.28 Example 3.10.

The locations of the remaining instant centers can be identified by repeated applications of the theorem of three centers.

It should be noted before closing this section that in the above examples the locations of all instant centers were identified without requiring knowledge of the actual operating speed of the mechanism. This is another indication of the linearity of the equations relating velocities, as indicated in Section 3.9. *For a single-degree-of-freedom mechanism, the locations of all instant centers are uniquely determined by the geometry alone and do not depend on the operating speed.*

3.16 VELOCITY ANALYSIS USING INSTANT CENTERS

In the previous section we learned how to locate the instant centers of planar mechanisms. Now we will demonstrate how the properties of instant centers can also provide a simple graphical approach for their velocity analysis.

EXAMPLE 3.11

Assume that the angular velocity ω_2 of crank 2 of the four-bar linkage illustrated in Fig. 3.29*a* is known. The problem is to determine the velocities of points B, D, and E at the position illustrated.

SOLUTION

Consider the line defined by the instant centers I_{12}, I_{14}, and I_{24}. This must be a straight line according to the Aronhold–Kennedy theorem and is commonly referred to as the line

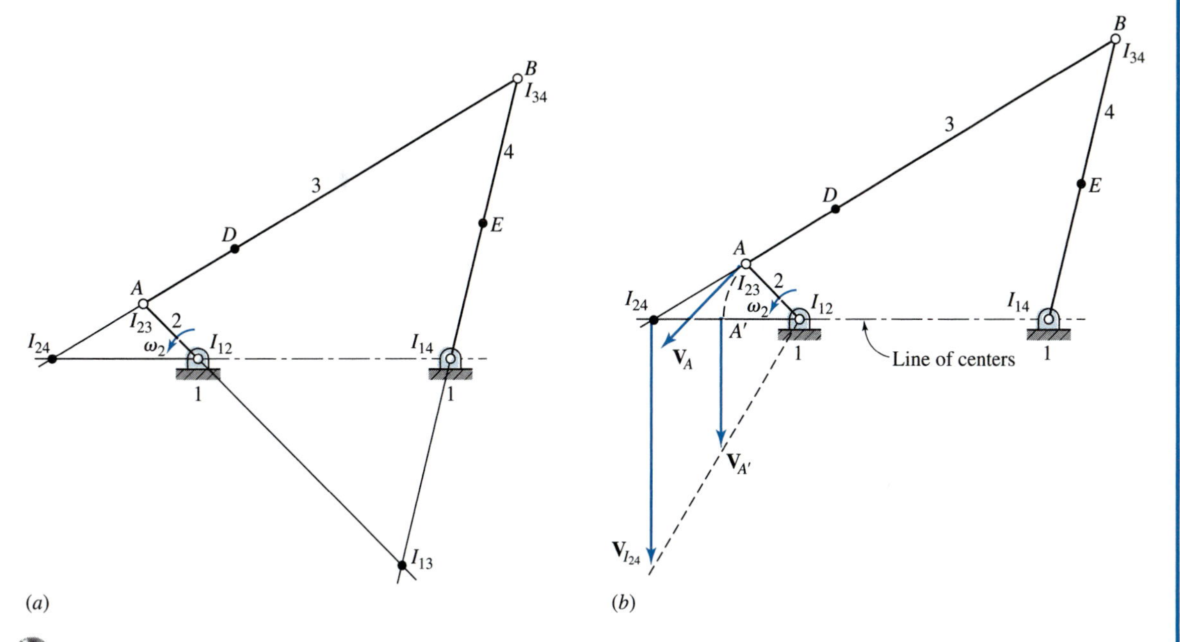

Figure 3.29 Example 3.11. Graphical velocity determination using instant centers.

of centers. According to the definition, the instant center I_{24} is a pair of points common to both links 2 and 4 and has equal absolute velocities in both links.

First consider instant center I_{24} as a point of link 2. The velocity $\mathbf{V}_A$ can be determined from $\boldsymbol{\omega}_2$ using the velocity-difference equation about I_{12}, and the velocity of I_{24} is known from it; the graphical construction is illustrated in Fig. 3.29*b*. When point A' of link 2 is located on the line of centers at an equal distance from I_{12}, its absolute velocity $\mathbf{V}_{A'}$ is equal in magnitude to the velocity V_A. Now the magnitude of $\mathbf{V}_{I_{24}}$ can be determined* by constructing a line from I_{12} through the terminus of $\mathbf{V}_{A'}$ as shown.

Next, consider I_{24} a point of link 4 rotating about I_{14}. Knowing the velocity $\mathbf{V}_{I_{24}}$, we can find the velocity of any other point of link 4, such as B' or E' (see Fig. 3.30*a*), using the reverse construction. Because B' and E' were chosen to have the same radii from I_{14} as B and E, their velocities have magnitudes equal to those of $\mathbf{V}_B$ and $\mathbf{V}_E$, respectively, and these can be laid out in their proper directions as illustrated in Fig. 3.30*a*.

To obtain the velocity $\mathbf{V}_D$ we note that D is in link 3; the known velocity $\boldsymbol{\omega}_2$ (or $\mathbf{V}_A$) is for link 2, and the reference link is link 1. Therefore, a new line of centers $I_{12}I_{13}I_{23}$ is chosen, as shown in Fig. 3.30*b*. Using ω_2 and I_{12}, we find the absolute velocity of the common instant center I_{23}. Here this step is trivial, because $\mathbf{V}_{I_{23}} = \mathbf{V}_A$. Locating point D' on the new line of centers, we find $\mathbf{V}_{D'}$ as illustrated and use its magnitude to find the desired velocity $\mathbf{V}_D$. Note that, according to its definition, the instant center I_{13}, as part of

* Note that $\mathbf{V}_{I_{24}}$ could have been determined directly from its velocity difference from I_{12}. This construction was demonstrated to illustrate the approach of the graphical method.

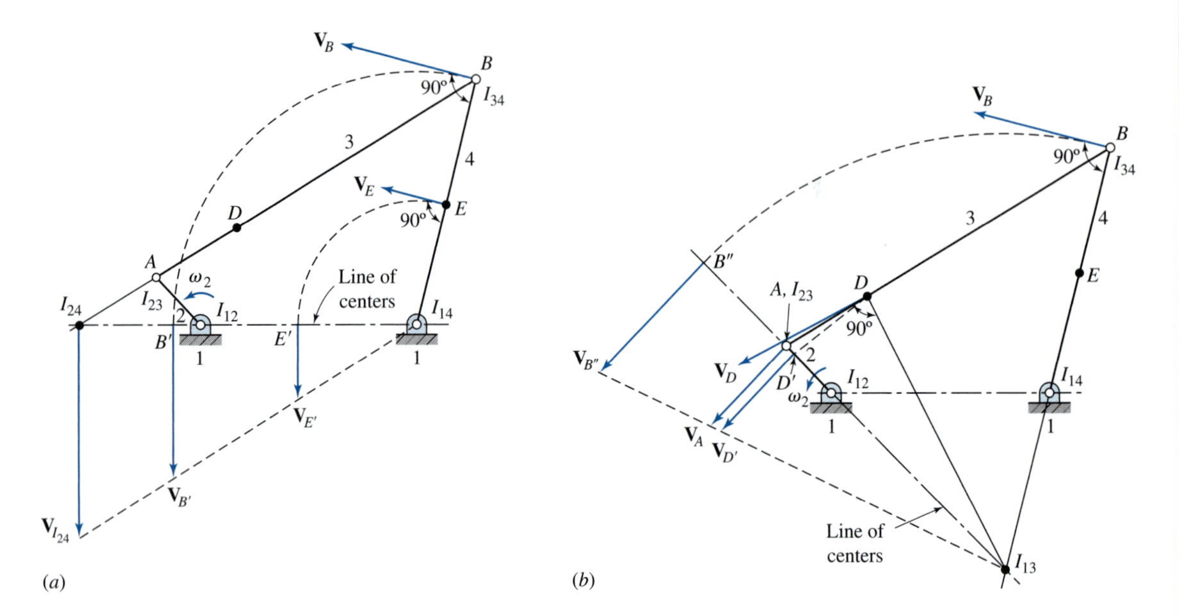

Figure 3.30 Example 3.11 continued.

link 3, has zero velocity at this instant. Because B can also be considered a point of link 3, its velocity can be determined in a similar manner by determining $V_{B''}$ as demonstrated.

The *line-of-centers* method of velocity analysis using instant centers can be summarized as follows:

1. Identify the three link numbers associated with the given velocity, the velocity to be determined, and the reference link. The reference link is usually link 1, because absolute velocity information is usually given and requested.
2. Locate the three instant centers defined by the links of step 1 and draw the line of centers.
3. Find the velocity of the common instant center by treating it as a point of the link whose velocity is given.
4. With the velocity of the common instant center known, consider it a point of the link whose velocity is to be determined. The velocity of any other point in that link can now be determined.

Another example will help to illustrate the procedure and will also demonstrate how to treat instant centers located at infinity.

EXAMPLE 3.12

For the device illustrated in Fig. 3.31, some of the links can be seen, whereas other links are enclosed in a housing. However, the location of the instant center I_{25} is known, see the

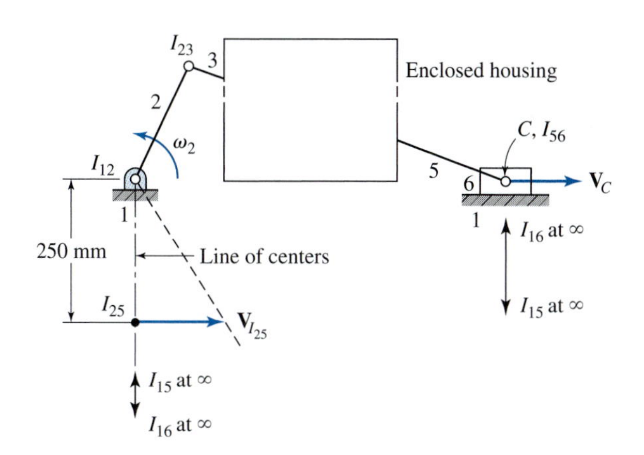

Figure 3.31 Example 3.12.

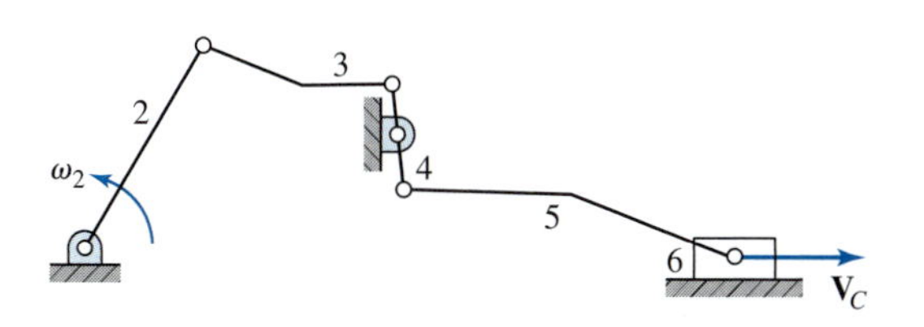

Figure 3.32

figure. If the velocity of point C in link 6 is $\mathbf{V}_C = 10$ m/s to the right, determine the angular velocity $\boldsymbol{\omega}_2$ of crank 2.

SOLUTION

Because we are given $\mathbf{V}_{C_5/1}$ and are required to find $\boldsymbol{\omega}_{2/1}$, we need to use the instant centers I_{15}, I_{12}, and I_{25}. Because I_{16} is at infinity as illustrated, then by inspection and applying the theorem of three centers we locate I_{15} (which is also at infinity). We can draw the line of centers $I_{12}I_{25}I_{15}$.

Considering I_{25} a point in link 5, we wish to find the velocity of this point from the given $\mathbf{V}_C$. We have difficulty in locating a point C' on the line of centers at the same radius from I_{15} as C because I_{15} is at infinity. How can we proceed?

Recalling the discussion of Section 3.15 and Eq. (3.27), we see that because I_{15} is at infinity, the relative motion between links 5 and 1 is translation and $\omega_{5/1} = 0$. Because this is true, *every* point of link 5 has the same absolute velocity, including $\mathbf{V}_{I_{25}} = \mathbf{V}_C$. Thus, we lay out $\mathbf{V}_{I_{25}}$ on Fig. 3.31.

Next, we turn our attention to link 2. We treat I_{25} as a point of link 2, rotating about I_{15}, and solve for the angular velocity of link 2:

$$\omega_2 = \frac{V_{I_{25}}}{R_{I_{25}I_{12}}} = \frac{10 \text{ m/s}}{0.25 \text{ m}} = 40 \text{ rad/s ccw.} \qquad \textit{Ans.}$$

Noting the apparent paradox between the directions of $\mathbf{V}_C$ and $\boldsymbol{\omega}_2$, we may speculate on the validity of our solution. This would be resolved, however, by opening the enclosed housing and discovering the linkage illustrated in Fig. 3.32.

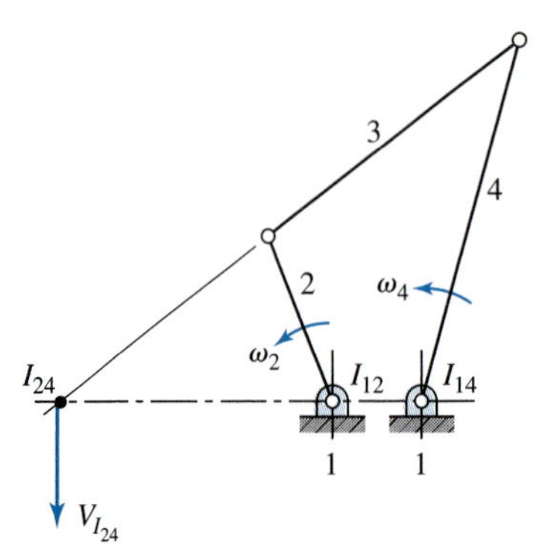

Figure 3.33 The angular-velocity-ratio theorem.

3.17 THE ANGULAR VELOCITY RATIO THEOREM

In Fig. 3.33, the instant center I_{24} is common to both links 2 and 4. The absolute velocity $\mathbf{V}_{I_{24}}$ is the same whether I_{24} is considered a point in link 2 or a point in link 4. Considering it each way, we can write

$$\mathbf{V}_{I_{24}} = \cancel{\mathbf{V}}^{\mathbf{0}}_{I_{12}} + \boldsymbol{\omega}_{2/1} \times \mathbf{R}_{I_{24}I_{12}} = \cancel{\mathbf{V}}^{\mathbf{0}}_{I_{14}} + \boldsymbol{\omega}_{4/1} \times \mathbf{R}_{I_{24}I_{14}}, \tag{a}$$

where $\boldsymbol{\omega}_{2/1}$ and $\boldsymbol{\omega}_{4/1}$ are the same as $\boldsymbol{\omega}_2$ and $\boldsymbol{\omega}_4$, respectively, but the additional subscript has been written to emphasize the presence of the third link (the frame).

Considering only the magnitudes of Eq. (a), the equation can be rewritten as

$$\frac{\omega_{4/1}}{\omega_{2/1}} = \frac{R_{I_{24}I_{12}}}{R_{I_{24}I_{14}}}. \tag{b}$$

This system illustrates the *angular-velocity-ratio theorem*. The theorem states that *the angular-velocity ratio of any two bodies in planar motion with respect to a third body is inversely proportional to the segments into which the common instant center cuts the line of centers.* Written in general notation for the motion of bodies j and k with respect to body i, the equation is

$$\frac{\omega_{k/i}}{\omega_{j/i}} = \frac{R_{I_{jk}I_{ij}}}{R_{I_{jk}I_{ik}}}. \tag{3.28}$$

Choosing an arbitrary positive direction along the line of centers, you should prove for yourself that *the angular-velocity ratio is positive when the common instant center falls outside the other two centers and negative when it falls between them.*

3.18 RELATIONSHIPS BETWEEN FIRST-ORDER KINEMATIC COEFFICIENTS AND INSTANT CENTERS

The first-order kinematic coefficients (see Section 3.12) can be expressed in terms of the locations of the instantaneous centers of velocity. For a planar linkage, the angular velocity of link j can be written from the chain rule as

$$\omega_j = \theta_j' \omega_i, \tag{3.29}$$

where ω_i is the input angular velocity. Therefore, the first-order kinematic coefficient of link j can be written as

$$\theta_j' = \frac{\omega_j}{\omega_i}.$$

Consistent with Eq. (3.28), the first-order kinematic coefficient of link j can be written as

$$\theta_j' = \frac{R_{I_{ij}I_{1i}}}{R_{I_{ij}I_{1j}}}, \tag{3.30}$$

where I_{1i} and I_{1j} are the absolute instant centers of link i and link j, respectively, and I_{ij} is the relative instant center of links i and j.

For the four-bar linkage in Example 3.11, the first-order kinematic coefficients of link 3 and 4 can be written from Eqs. (3.30) as

$$\theta_3' = \frac{R_{I_{23}I_{12}}}{R_{I_{23}I_{13}}} \quad \text{and} \quad \theta_4' = \frac{R_{I_{24}I_{12}}}{R_{I_{24}I_{14}}}.$$

The lengths of the input crank and the frame are $R_{I_{12}I_{23}} = 4$ in and $R_{I_{12}I_{14}} = 10$ in, respectively. From the scaled drawing of the linkage (see Fig. 3.29*a*), we measure $R_{I_{23}I_{13}} =$ 15 in and $R_{I_{24}I_{12}} = 11.2$ in. Therefore, the first-order kinematic coefficients of the two links are

$$\theta_3' = \frac{4 \text{ in}}{15 \text{ in}} = 0.267 \text{ rad/rad} \quad \text{and} \quad \theta_4' = \frac{11.2 \text{ in}}{21.2 \text{ in}} = 0.528 \text{ rad/rad}.$$

Substituting these values and the input angular velocity $\omega_2 = 94.2$ rad/s ccw into Eqs. (3.29), the angular velocity of the coupler link and the output link, respectively, are

$$\omega_3 = 25.15 \text{ rad/s ccw} \quad \text{and} \quad \omega_4 = 49.74 \text{ rad/s ccw}.$$

These answers agree closely with the results obtained from the velocity polygon method of Example 3.1 ($\omega_3 = 25.6$ rad/s ccw and $\omega_4 = 49.6$ rad/s ccw).

For a slider-crank linkage with the displacement of the slider (denoted as link i) regarded as the input, the angular velocity of link j can be written from the chain rule as

$$\theta_j' = \frac{\omega_j}{\dot{r}_i}, \tag{3.31}$$

where $\dot{r}_i$ is the velocity of the slider. When $R_{I_{ij}I_{1i}}$ becomes infinite, Eq. (3.30) for the first-order kinematic coefficient of link j becomes

$$\theta_j' = \frac{\pm 1}{R_{I_{ij}I_{1j}}}, \tag{3.32}$$

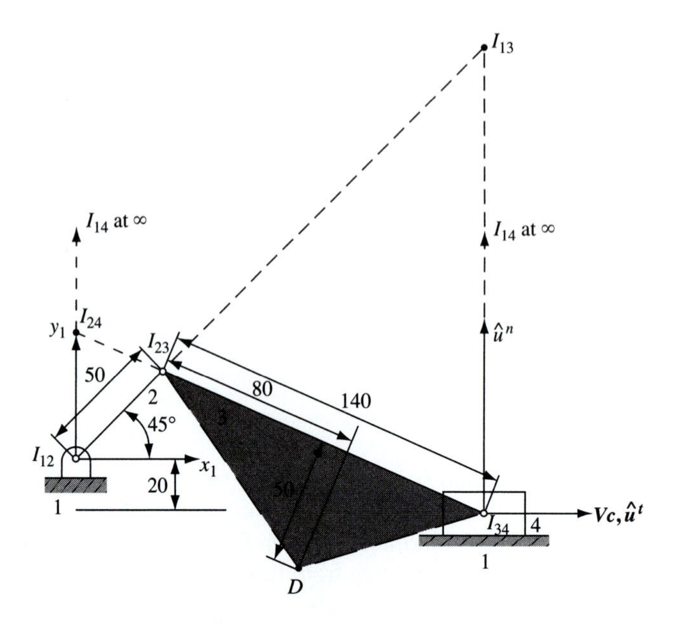

Figure 3.34 The offset slider-crank mechanism of Example 3.2.

where the choice of the plus or minus sign depends on the positive direction along the line of centers. Consistent with the statement after Eq. (3.28), the angular-velocity ratio is positive when the relative instant center falls outside the two absolute centers and negative when it falls between them.

The instant centers for the offset slider-crank mechanism of Example 3.2 are illustrated in Fig. 3.34.

For this example, the first-order kinematic coefficients of links 2 and 3 can be written from Eq. (3.32) as

$$\theta_2' = \frac{\pm 1}{R_{I_{24}I_{12}}} \quad \text{and} \quad \theta_3' = \frac{\pm 1}{R_{I_{34}I_{13}}}.$$

If we choose the velocity of slider 4 as positive to the right, then we have $\hat{\mathbf{u}}^t = \hat{\mathbf{i}}$ for the positive input direction and then [by Eq. (3.7)] we choose $\hat{\mathbf{u}}^n = \hat{\mathbf{u}}^b \times \hat{\mathbf{u}}^t = \hat{\mathbf{k}} \times \hat{\mathbf{i}} = \hat{\mathbf{j}}$ (upward) as the positive direction along the line of centers.* From the scaled drawing of the offset slider-crank linkage (Fig. 3.34), we measure $R_{I_{24}I_{12}} = +51$ mm and $R_{I_{34}I_{13}} = -185$ mm and find that the kinematic coefficients are

$$\theta_2' = \frac{\pm 1}{0.051 \text{ m}} = -19.6 \text{ rad/m} \quad \text{and} \quad \theta_3' = \frac{\pm 1}{-0.185 \text{ m}} = +5.41 \text{ rad/m},$$

* Note that this sign convention is the same as saying that positive input motion of slider 4 is the same as a positive (counterclockwise) rotation of link 4 about instant center I_{14} now located at infinity in the positive $\hat{\mathbf{u}}^n$ direction.

where θ_2' is negative because I_{24} lies between I_{12} and I_{14} (which is at infinity upward in the positive $\hat{\mathbf{u}}^n$ direction along the line of centers). Similarly, θ_3' is positive because I_{34} is outside of I_{13} and I_{14}.

Substituting these values and the input velocity, $\dot{r}_4 = -10$ m/s, into Eqs. (3.31), the angular velocities of links 2 and 3, respectively, are

$$\omega_2 = 196 \text{ rad/s ccw} \quad \text{and} \quad \omega_3 = 54.1 \text{ rad/s cw}.$$

These answers agree closely with the results obtained from the velocity polygon method in Example 3.2 ($\omega_2 = 200$ rad/s ccw and $\omega_3 = 53.6$ rad/s cw). However, because graphical measurements are used for finding the positions of the instant centers, these results are probably still not as accurate as those of Example 3.7 ($\omega_2 = 197.7$ rad/s ccw and $\omega_3 = 54.4$ rad/s cw).

The velocity of a point, say P, fixed in a link of a mechanism can be written as

$$\mathbf{V}_P = V_P \hat{\mathbf{u}}_P^t \tag{3.33}$$

or as

$$\mathbf{V}_P = (x_P'\hat{\mathbf{i}} + y_P'\hat{\mathbf{j}})\dot{\psi}, \tag{3.34}$$

where $\dot{\psi}$ is the generalized input velocity to the mechanism. The magnitude of the velocity, commonly referred to as the speed, can be written from the chain rule as

$$V_P = r_P'\dot{\psi}, \tag{3.35}$$

where the first-order kinematic coefficient is defined as

$$r_P' = \pm\sqrt{x_P'^2 + y_P'^2}. \tag{3.36}$$

Here the sign convention is as follows: We use the positive sign if the instantaneous change in the input position is positive, and we use the negative sign if the instantaneous change in the input position is negative.

Rearranging Eq. (3.33), the unit tangent vector to the point trajectory can be written as

$$\hat{\mathbf{u}}_P^t = \frac{\mathbf{V}_P}{V_P}. \tag{3.37}$$

Then, substituting Eqs. (3.34) and (3.35) into this relation gives

$$\hat{\mathbf{u}}_P^t = \left(\frac{x_P'}{r_P'}\right)\hat{\mathbf{i}} + \left(\frac{y_P'}{r_P'}\right)\hat{\mathbf{j}}. \tag{3.38}$$

Consistent with Eq. (3.7), the unit normal vector to the point trajectory can now be written as $\hat{\mathbf{u}}_P^n = \hat{\mathbf{u}}^b \times \hat{\mathbf{u}}_P^t$. Substituting Eq. (3.38) into this relation, the unit normal vector can be written

$$\hat{\mathbf{u}}_P^n = \left(\frac{-y_P'}{r_P'}\right)\hat{\mathbf{i}} + \left(\frac{x_P'}{r_P'}\right)\hat{\mathbf{j}}. \tag{3.39}$$

3.19 FREUDENSTEIN'S THEOREM

In the analysis and design of linkages it is often important to know the positions of the linkage at which the extreme values of the output velocity occur or, more precisely, the input position values at which the ratio of the output and input velocities reaches its extremes.

The earliest work in determining extreme values is apparently that of Krause,[4] who stated that the velocity ratio ω_2/ω_4 of the drag–link mechanism (Fig. 3.35) reaches an extreme value when the connecting rod and follower, links 3 and 4, become perpendicular to each other. Rosenauer, however, demonstrated that this is not strictly true.[5] Following Krause, Freudenstein developed a simple graphical method for determining the positions of the four-bar linkage at which the extreme values of the output velocity do occur.[6]

Freudenstein's theorem makes use of the line connecting instant centers I_{13} and I_{24} (Fig. 3.36), called the *collineation axis*. The theorem states that *at an extreme of the output to input angular velocity ratio of a four-bar linkage, the collineation axis is perpendicular to the coupler link.*[7]

Using the angular-velocity-ratio theorem, Eq. (3.28), we write

$$\frac{\omega_4}{\omega_2} = \frac{R_{I_{24}I_{12}}}{R_{I_{24}I_{12}} + R_{I_{12}I_{14}}}.$$

Because $R_{I_{12}I_{14}}$ is the fixed length of the frame, the extremes of the velocity ratio occur when $R_{I_{24}I_{12}}$ is either a maximum or a minimum. Such positions may occur on either or both sides of the instant center I_{12}. Thus, the problem reduces to finding the geometry of the linkage for which $R_{I_{24}I_{12}}$ is an extremum.

During the motion of the linkage, the instant center I_{24} travels along the line $I_{12}I_{14}$ as seen by the theorem of three centers but, at an extreme value of the velocity ratio, I_{24} must instantaneously be at rest (its direction of travel on this line must be reversing). This occurs when the velocity of I_{24}, considered a point of link 3, is directed along the coupler

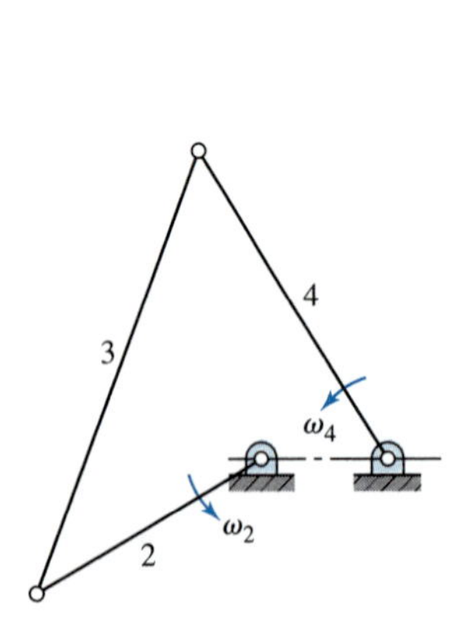

Figure 3.35 The drag–link mechanism.

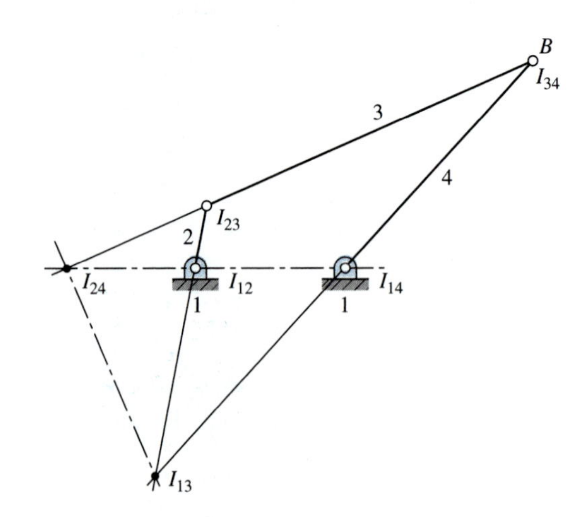

Figure 3.36 The collineation axis.

link. This will be true only when the coupler link is perpendicular to the collineation axis, because I_{13} is the instant center of link 3.

An inversion of the theorem (treating link 2 as fixed) states that an *extreme value of the velocity ratio* ω_3/ω_2 *of a four-bar linkage occurs when the collineation axis is perpendicular to the follower* (*link 4*).

3.20 INDICES OF MERIT; MECHANICAL ADVANTAGE

In this section we will study some of the various ratios, angles, and other parameters of mechanisms that tell us whether a mechanism is a good one or a poor one (see Chapter 1). Many such parameters have been defined by various authors over the years, and there is no common agreement on a single "index of merit" for all mechanisms. Yet the many used have a number of features in common, including the fact that most can be related to the velocity ratios of the mechanism and, therefore, can be determined solely from its geometry. In addition, most depend on some knowledge of the application of the mechanism, especially the input and output variables. It is often desirable in the analysis or synthesis of a mechanism to plot the index of merit for a revolution of the input crank and to note in particular the minimum and maximum values when evaluating the design or its suitability for a given application.

In Section 3.17 we learned that the ratio of the angular velocity of the output link to the input velocity of a mechanism is inversely proportional to the segments into which the common instant center cuts the line of centers. Thus, in the four-bar linkage illustrated in Fig. 3.37, if links 2 and 4 are the input and output cranks, respectively, then

$$\frac{\omega_4}{\omega_2} = \frac{R_{IA}}{R_{ID}}$$

is the equation for the output-to-input-velocity ratio. We also learned in Section 3.19 that the extremes of this ratio occur when the collineation axis is perpendicular to the coupler, link 3.

If we now assume that the linkage of Fig. 3.37 has no friction or inertia forces during its operation or that these are negligible compared with the input torque T_2, applied to link 2, and the output torque T_4, the resisting load torque on link 4, then we can derive a relation between T_2 and T_4. Because friction and inertia forces are negligible, the input

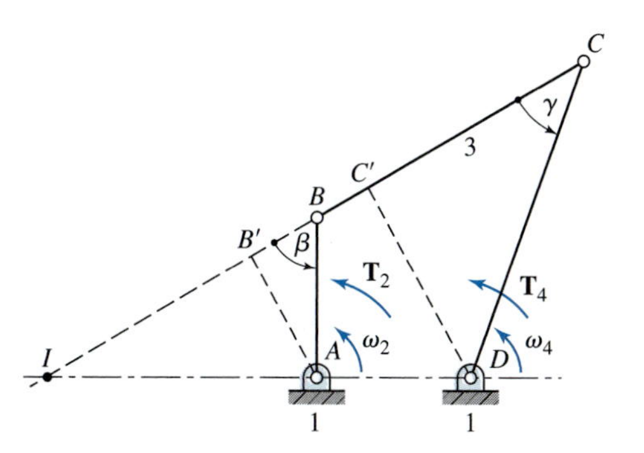

Figure 3.37 Four-bar linkage.

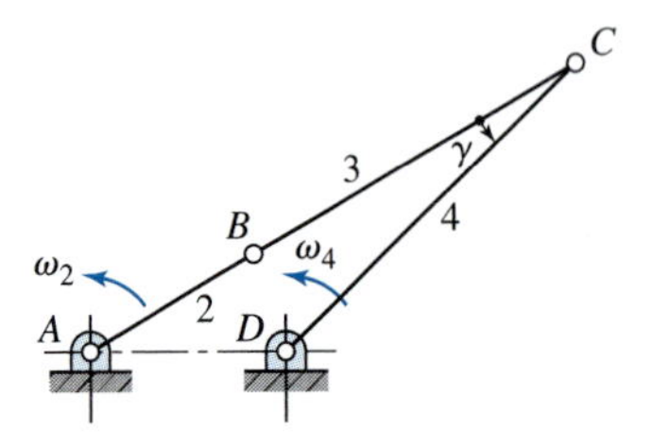

Figure 3.38 Four-bar linkage in toggle.

power applied to link 2 is the negative of the power applied to link 4 by the load; hence,

$$T_2\omega_2 = -T_4\omega_4 \tag{a}$$

or

$$\frac{T_4}{T_2} = -\frac{\omega_2}{\omega_4} = -\frac{R_{ID}}{R_{IA}}. \tag{3.40}$$

The *mechanical advantage* of a mechanism is the instantaneous ratio of the output force (torque) to the input force (torque). Here we see that the mechanical advantage is the negative reciprocal of the velocity ratio. Either can be used as an index of merit in judging the ability of a mechanism to transmit force or power.

The mechanism is redrawn in Fig. 3.38 at the position where links 2 and 3 are aligned. At this position, R_{IA} and ω_4 are passing through zero; hence, an extreme value (infinity) of the mechanical advantage is obtained. A mechanism in such a position is said to be *in toggle*. Such toggle positions are often used to produce a high mechanical advantage; an example is the clamping mechanism of Fig. 2.8.

Proceeding further, we construct $B'A$ and $C'D$ perpendicular to the line IBC in Fig. 3.37. Also, we assign labels β and γ to the acute angles made by the coupler or its extension and the input and output links, respectively. Then, by similar triangles,

$$\frac{R_{ID}}{R_{IA}} = \frac{R_{C'D}}{R_{B'A}} = \frac{R_{CD}\sin\gamma}{R_{BA}\sin\beta}. \tag{b}$$

Then, using Eq. (3.40), we see that another expression for mechanical advantage is

$$\frac{T_4}{T_2} = -\frac{\omega_2}{\omega_4} = -\frac{R_{CD}\sin\gamma}{R_{BA}\sin\beta}. \tag{3.41}$$

This equation demonstrates that the mechanical advantage is infinite whenever the angle β is 0 or 180°—that is, whenever the mechanism is in a toggle position.

In Sections 1.10 and 2.8 we defined the angle γ between the coupler link and the follower link as the *transmission angle*. This angle is also often used as an index of merit for a four-bar linkage. Equation (3.41) indicates that the mechanical advantage diminishes when the transmission angle is much less than a right angle. If the transmission angle becomes too small, the mechanical advantage becomes small and even a very small amount of friction may cause a mechanism to lock or jam. To avoid this, a common rule of thumb is that a four-bar linkage should not be used in a region where the transmission angle is less than, say, 45° or 50°. The best four-bar linkage, based on the quality of its force transmission, will have a transmission angle that deviates from 90° by the smallest amount.

In other mechanisms—for example, meshing gear teeth or a cam–follower system—the *pressure angle* is used as an index of merit. The pressure angle is defined as the acute angle between the direction of the output force and the direction of the velocity of the point where the output force is applied. Pressure angles are discussed more thoroughly in Chapters 6, 7, and 8. In the four-bar linkage, the pressure angle is the complement of the transmission angle.

Another index of merit which has been proposed[8] is the determinant of the matrix of coefficients of the simultaneous equations relating the dependent velocities of a mechanism. In Example 3.5, for example, we saw that the dependent velocities of a four-bar linkage are related by

$$(r_3 \sin\theta_3)\omega_3 - (r_4 \sin\theta_4)\omega_4 = -(r_2 \sin\theta_2)\omega_2$$

$$(r_3 \cos\theta_3)\omega_3 - (r_4 \cos\theta_4)\omega_4 = -(r_2 \cos\theta_2)\omega_2.$$

The matrix of coefficients is called the Jacobian of the system, and its determinant is

$$\Delta = \begin{vmatrix} r_3 \sin\theta_3 & -r_4 \sin\theta_4 \\ r_3 \cos\theta_3 & -r_4 \cos\theta_4 \end{vmatrix} = r_3 r_4 \sin(\theta_4 - \theta_3).$$

As is obvious from Cramer's rule, the solutions for the dependent velocities, in this case ω_3 and ω_4, must include this determinant in the denominator. This is borne out in the solution of the four-bar linkage, Eqs. (3.22). Although the form of this determinant changes for different mechanisms, such a determinant can always be defined and always appears in the denominators of all dependent velocity solutions.

If this determinant becomes small, however, the mechanical advantage also becomes small and the usefulness of the mechanism is reduced in such regions. We have not seen it yet, but it is also true that this same determinant appears in the denominator of the dependent accelerations (see Section 4.12) and all other quantities that require taking derivatives of the loop-closure equation. If this determinant is small, the mechanism will function poorly in all respects—force transmission, motion transformation, sensitivity to manufacturing errors, and so on.

3.21 CENTRODES

We noted in Section 3.13 that the location of an instant center of velocity is defined only instantaneously and changes as the mechanism moves. When the changing locations of an instant center are identified for all possible positions of a single-degree-of-freedom mechanism, they describe a curve or locus, called a *centrode*.* In Fig. 3.39, the instant

* Opinion seems divided on whether these loci should be termed *centrodes* or *polodes*. Generally, those preferring the name *instant center* call them *centrodes* and those preferring the name *velocity*

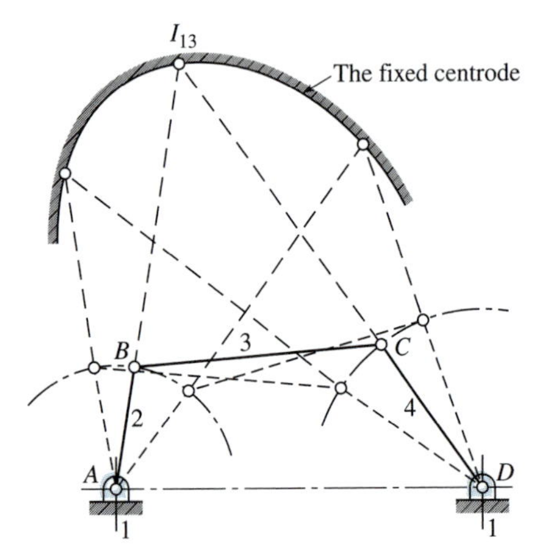

Figure 3.39 The fixed centrode.

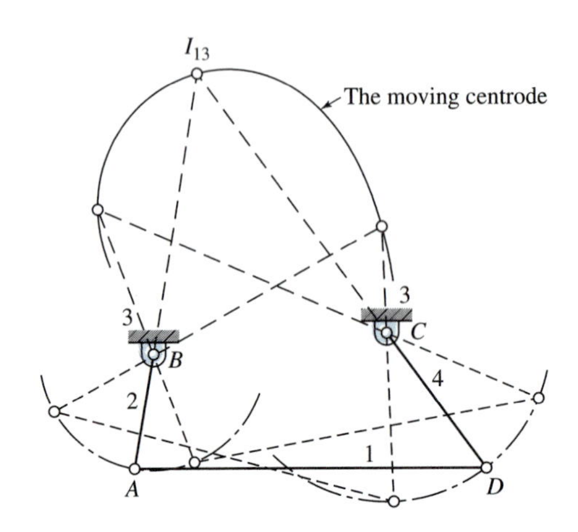

Figure 3.40 The moving centrode.

center I_{13} is located at the intersection of the extension of links 2 and 4. As the linkage moves through all possible positions, I_{13} traces out the curve called the *fixed centrode* on link 1.

Figure 3.40 illustrates the inversion of the same linkage in which link 3 is fixed and link 1 is movable. When this inversion moves through all possible input positions, I_{13} traces a *different* curve on link 3. For the original linkage, with link 1 fixed, this is the curve traced by I_{13} on the coordinate system of the moving link 3; it is called the *moving centrode*.

Figure 3.41 illustrates the moving centrode, attached to link 3, and the fixed centrode, attached to link 1. It is imagined here that links 1 and 3 have been machined to the actual shapes of the respective centrodes and that links 2 and 4 have been removed entirely. If the moving centrode is now permitted to roll on the fixed centrode without slip, link 3 has exactly the same motion as it had in the original linkage. This remarkable property, which stems from the fact that a point of rolling contact is an instant center, turns out to be quite useful in the synthesis of linkages.

We can restate this property as follows: *the plane motion of one rigid body with respect to another is completely equivalent to the rolling motion of one centrode on the other*. The instantaneous point of rolling contact is the instant center, as illustrated in Fig. 3.41. Also illustrated are the common tangent to the two centrodes and the common normal, called the *centrode tangent* and the *centrode normal*; they are sometimes used as the axes of a coordinate system for developing equations for a coupler curve or other properties of the motion.

The centrodes of Fig. 3.41 were generated by the instant center I_{13} on links 1 and 3. Another set of centrodes, both moving, is generated on links 2 and 4 when instant center

pole call them *polodes*. The French name *roulettes* has also been applied. The three-dimensional equivalents are ruled surfaces and are referred to as *axodes*.

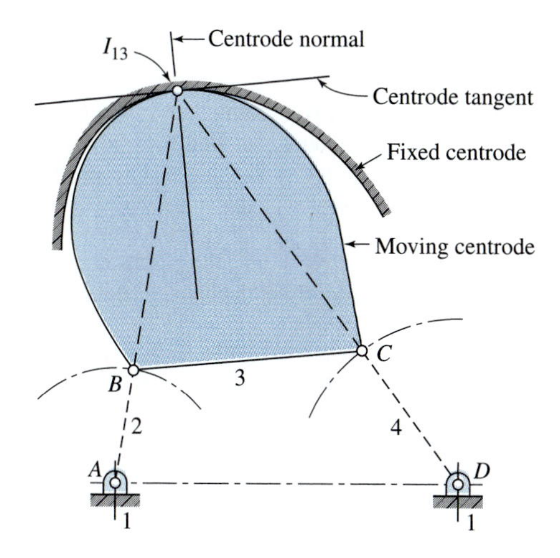

Figure 3.41 Rolling contact between centrodes.

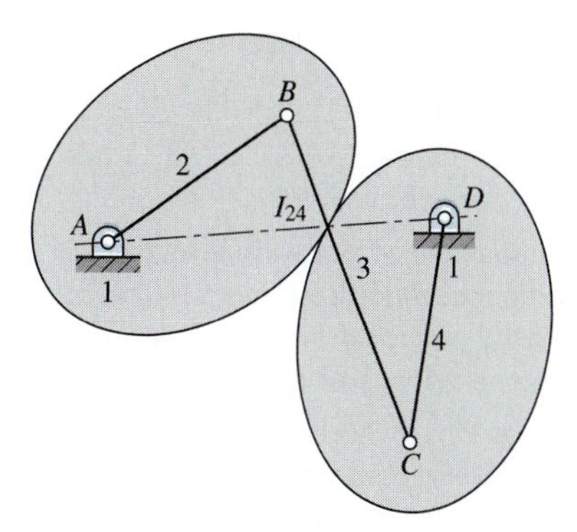

Figure 3.42 Elliptical gears.

I_{24} is considered. Figure 3.42 illustrates these centrodes as two ellipses for the case of a crossed double-crank linkage with equal cranks. The two centrodes roll upon each other and describe the identical motion between links 2 and 4 that would result from the operation of the original four-bar linkage. This construction can be used as the basis for the development of a pair of elliptical gears.

3.22 REFERENCES

[1] Beyer, R. *The Kinematic Synthesis of Mechanisms*, McGraw-Hill, 1963 (Trans. H. Kuenzel, Dept. of Mechanical Engineering, University of Alabama).

[2] Mehmke, R. Über die Geschwindigkeiten beliebiger Ordung eines in seiner Ebene bewegten ähnlich veränddерlichen Systems, *Civil Inginieur*, 1883.

[3] Raven, F. H. 1958. Velocity and acceleration analysis of plane and space mechanisms by means of independent-position equations. *J. Appl. Mech. ASME Trans. E* 80:1–6.

[4] Krause, R. 1939. Die Doppelkurbel und Ihre Geschwindigkeitssgrenzen. *Maschinenbau/Getriebetechnik* 18:37–41; 1939. Zur Synthese der Doppelkurbel. *Maschinenbau/Getriebetechnik* 18:93–4.

[5] Rosenauer, N. 1957. Synthesis of drag–link mechanisms for producing nonuniform rotational motion with prescribed reduction ratio limits. *Aus. J. Appl. Sci.* 8:1–6.

[6] Freudenstein, F. 1956. On the maximum and minimum velocities and accelerations in four-link mechanisms. *Trans. ASME* 78:779–87.

[7] A. S. Hall, Jr. contributed a rigorous proof of this theorem in an appendix to Freudenstein's paper.

[8] Denavit, J., et al. 1965. Velocity, acceleration, and static force analysis of spatial linkages. *J. Appl. Mech. ASME Trans. E* 87:903–10.

PROBLEMS*

†3.1 The position vector of a point is given by the equation $\mathbf{R} = 100e^{j\pi t}$, where R is in inches. Find the velocity of the point at $t = 0.40$ s.

†3.2 The equation $R = \left(t^2 + 4\right) e^{-j\pi t/10}$ defines the path of a particle. If R is in meters, find the velocity of the particle at $t = 20$ s.

†3.3 If automobile A is traveling south at 55 mi/h and automobile B north 60° east at 40 mi/h, what is the velocity difference between B and A? What is the apparent velocity of B to the driver of A?

†3.4 In Fig. P3.4, wheel 2 rotates at 600 rev/min and drives wheel 3 without slipping. Find the velocity difference between points B and A.

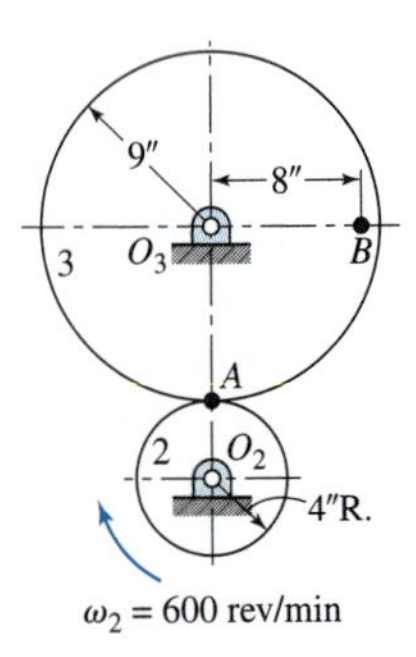

Figure P3.4

†3.5 Two points A and B, located along the radius of a wheel (see Fig. P3.5), have speeds of 80 and 140 m/s, respectively. The distance between the points is $R_{BA} = 300$ mm. (*a*) What is the diameter of the wheel? (*b*) Find $\mathbf{V}_{AB}, \mathbf{V}_{BA}$, and the angular velocity of the wheel.

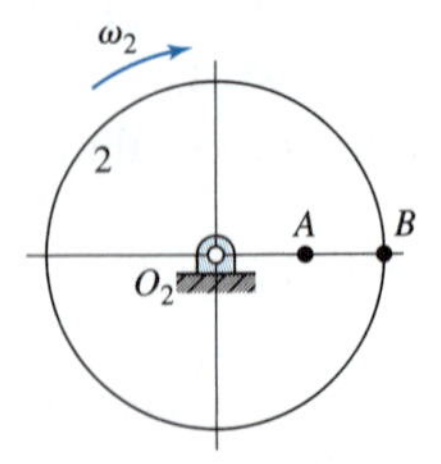

Figure P3.5

†3.6 A plane leaves point B and flies east at 350 mi/h. Simultaneously, at point A, 200 miles southeast (see Fig. P3.6), a plane leaves and flies northeast at 390 mi/h.

(a) How close will the planes come to each other if they fly at the same altitude?

(*b*) If they both leave at 6:00 p.m., at what time will this occur?

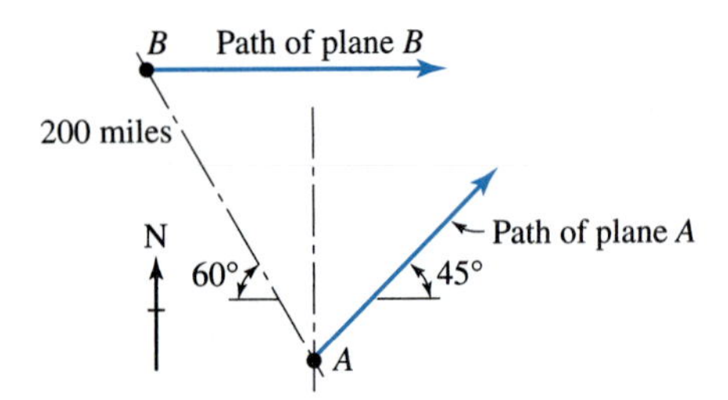

Figure P3.6 $R_{AB} = 200$ mi.

†3.7 To the data of Problem 3.6, add a wind of 30 mi/h from the west. (a) If A flies the same heading, what is its new path? (b) What change does the wind make in the results of Problem 3.6?

†3.8 The velocity of point B on the linkage shown in Fig. P3.8 is 40 m/s. Find the velocity of point A and the angular velocity of link 3.

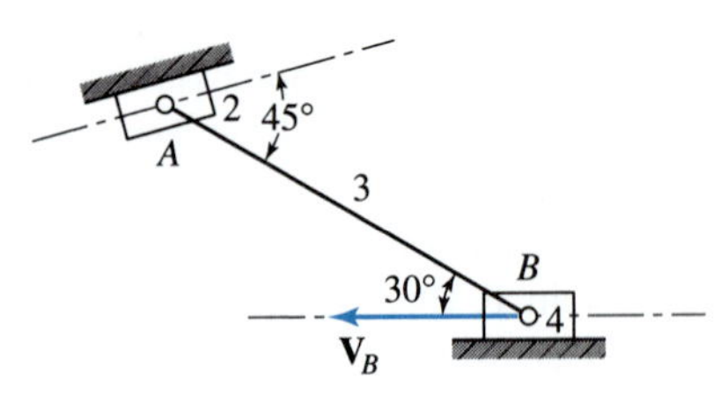

Figure P3.8 $R_{AB} = 400$ mm.

†3.9 The mechanism illustrated in Fig. P3.9 is driven by link 2 at $\omega_2 = 45$ rad/s ccw. Find the angular velocities of links 3 and 4.

* When assigning problems, the instructor may wish to specify the method of solution to be used because a variety of approaches are presented in the text.

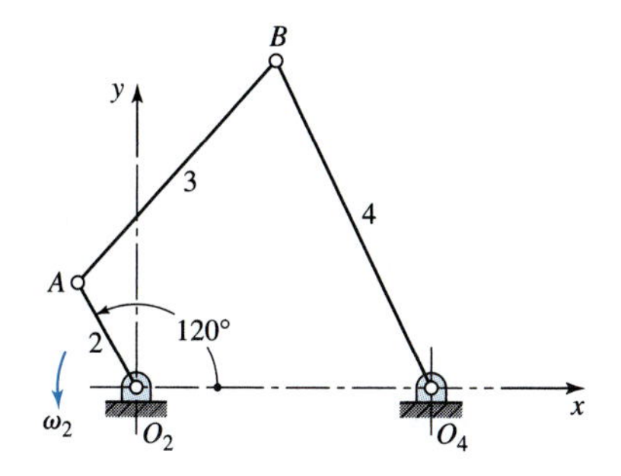

Figure P3.9 $R_{AO_2} = 4$ in, $R_{BA} = 10$ in, $R_{O_4O_2} = 10$ in, and $R_{BO_4} = 12$ in.

†3.10 Crank 2 of the push–link mechanism illustrated in Fig. P3.10 is $\omega_2 = 60$ rad/s cw. Find the velocities of points B and C and the angular velocities of links 3 and 4.

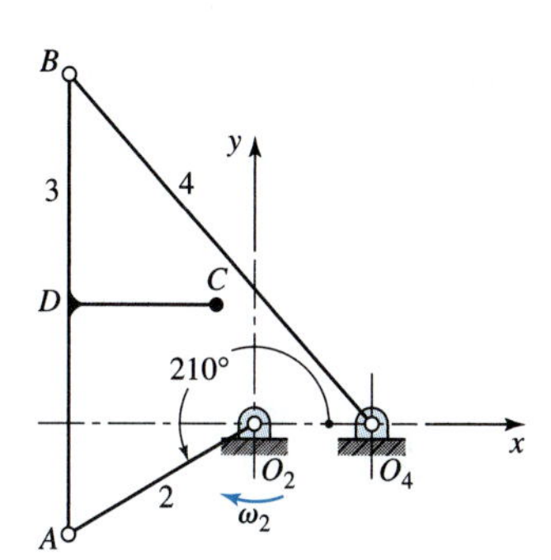

Figure P3.10 $R_{AO_2} = 150$ mm, $R_{BA} = 300$ mm, $R_{O_4O_2} = 75$ mm, $R_{BO_4} = 300$ mm, $R_{DA} = 150$ mm, and $R_{CD} = 100$ mm.

†3.11 Find the velocity of point C on link 4 of the mechanism P3.11 if crank 2 is driven at $\omega_2 = 48$ rad/s ccw. What is the angular velocity of link 3?

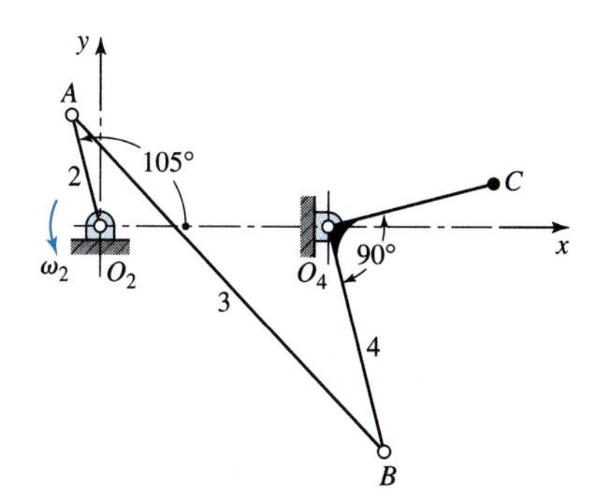

Figure P3.11 $R_{AO_2} = 8$ in, $R_{BA} = 32$ in, $R_{O_4O_2} = 16$ in, $R_{BO_4} = 16$ in, and $R_{CO_4} = 12$ in.

†3.12 Figure P3.12 illustrates a parallel-bar linkage in which opposite links have equal lengths. For this linkage, demonstrate that ω_3 is always zero and that $\omega_4 = \omega_2$. How would you describe the motion of link 4 with respect to link 2?

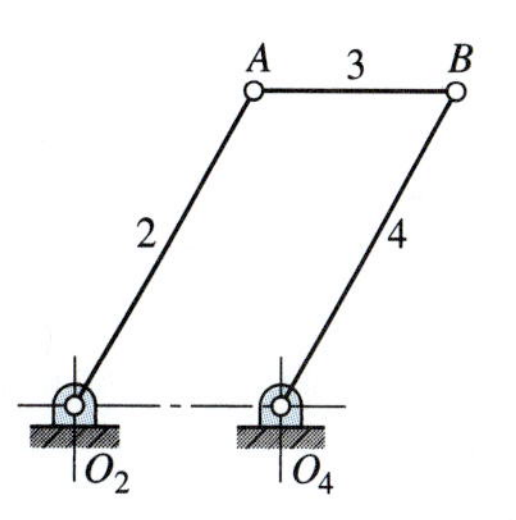

Figure P3.12

†3.13 Figure P3.13 illustrates the antiparallel or crossed-bar linkage. If link 2 is driven at $\omega_2 = 1$ rad/s ccw, find the velocities of points C and D.

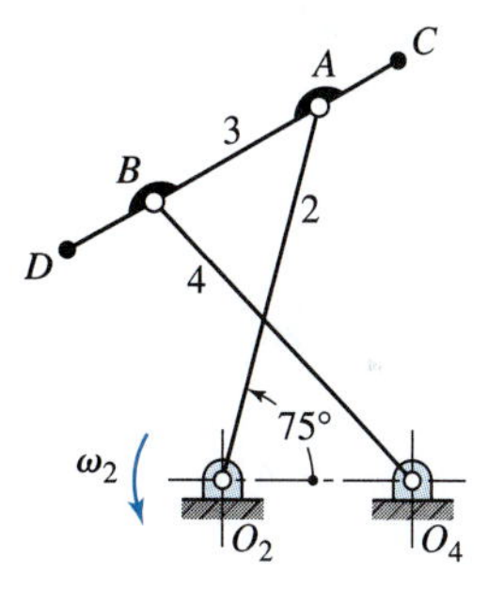

Figure P3.13 $R_{AO_2} = R_{BO_4} = 300$ mm, $R_{BA} = R_{O_4O_2} = 150$ mm, and $R_{CA} = R_{DB} = 75$ mm.

†3.14 Find the velocity of point C of the linkage illustrated in Fig. P3.14 assuming that link 2 has an angular velocity of 60 rad/s ccw. Also find the angular velocities of links 3 and 4.

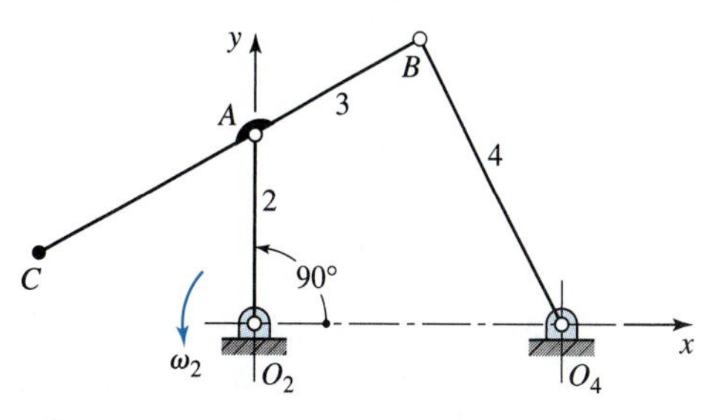

Figure P3.14 $R_{AO_2} = R_{BA} = 6$ in, $R_{O_4O_2} = R_{BO_4} = 10$ in, and $R_{CA} = 8$ in.

†**3.15** The inversion of the slider-crank mechanism illustrated in Fig. P3.15 is driven by link 2 at $\omega_2 =$ 60 rad/s ccw. Find the velocity of point B and the angular velocities of links 3 and 4.

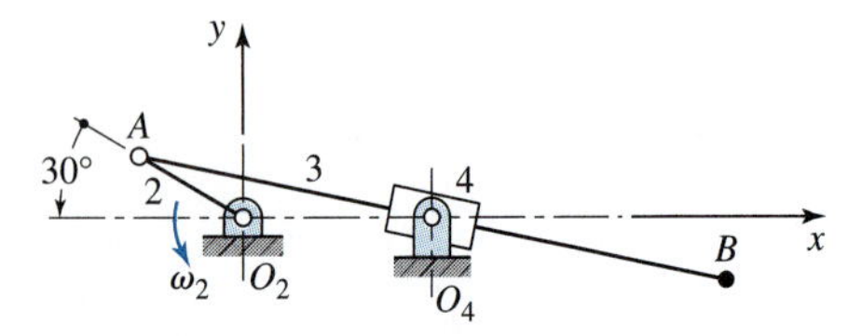

Figure P3.15 $R_{AO_2} = 75$ mm, $R_{BA} = 400$ mm, and $R_{O_4O_2} = 125$ mm.

†**3.16** Find the velocity of the coupler point C and the angular velocities of links 3 and 4 of the mechanism illustrated if crank 2 has an angular velocity of 30 rad/s cw.

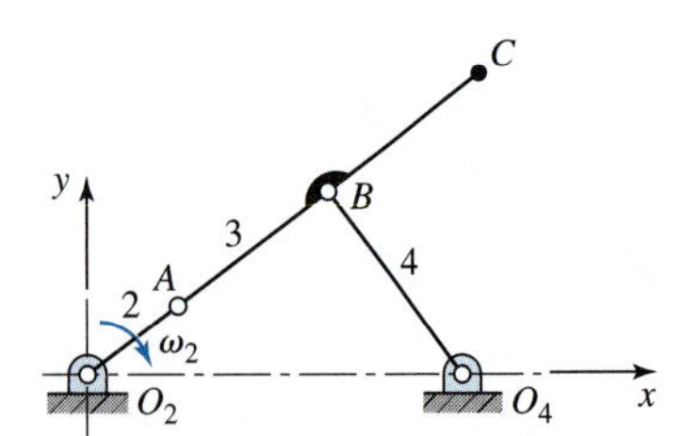

Figure P3.16 $R_{AO_2} = 3$ in, $R_{BA} = R_{CB} = 5$ in, $R_{O_4O_2} = 10$ in, and $R_{BO_4} = 6$ in.

†**3.17** Link 2 of the linkage illustrated in Fig. P3.17 has an angular velocity of 10 rad/s ccw. Find the angular velocity of link 6 and the velocities of points B, C, and D.

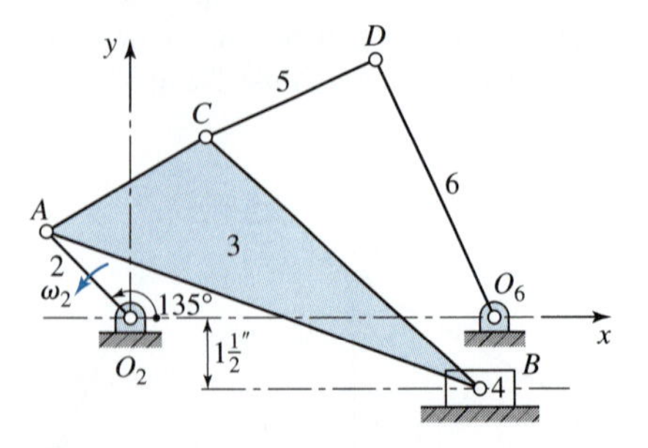

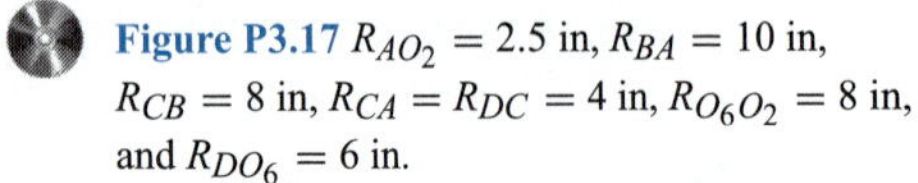

Figure P3.17 $R_{AO_2} = 2.5$ in, $R_{BA} = 10$ in, $R_{CB} = 8$ in, $R_{CA} = R_{DC} = 4$ in, $R_{O_6O_2} = 8$ in, and $R_{DO_6} = 6$ in.

†**3.18** The angular velocity of link 2 of the drag–link mechanism illustrated in Fig. P3.18 is 16 rad/s cw. Plot a polar velocity diagram for the velocity of point B for all crank positions. Check the positions of maximum and minimum velocities by using Freudenstein's theorem.

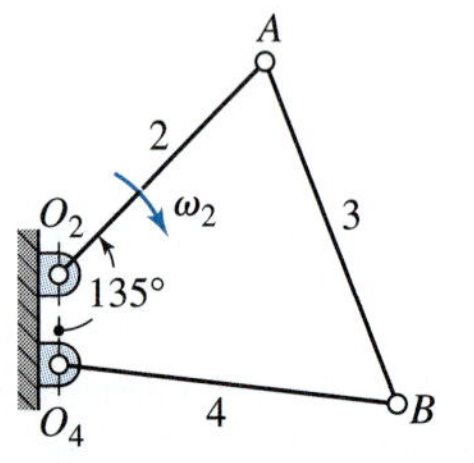

Figure P3.18 $R_{AO_2} = 350$ mm, $R_{BA} = 425$ mm, $R_{O_4O_2} = 100$ mm, and $R_{BO_4} = 400$ mm.

†**3.19** Link 2 of the mechanism illustrated in Fig. P3.19 is driven at $\omega_2 = 36$ rad/s cw. Find the angular velocity of link 3 and the velocity of point B.

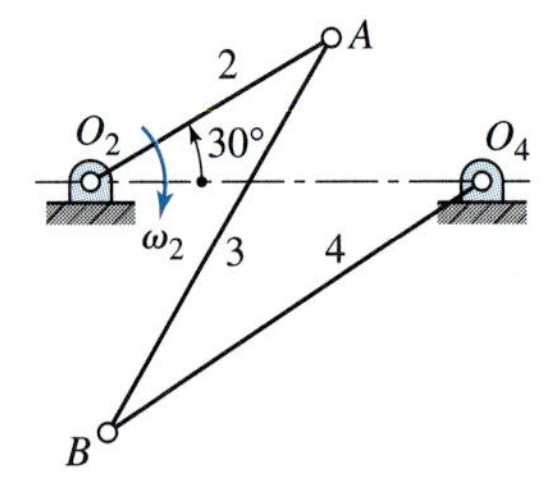

Figure P3.19 $R_{AO_2} = 5$ in, $R_{BA} = R_{BO_4} = 8$ in, and $R_{O_4O_2} = 7$ in.

†**3.20** Find the velocity of point C and the angular velocity of link 3 of the push–link mechanism illustrated in Fig. P3.20. Link 2 is the driver and rotates at 8 rad/s ccw.

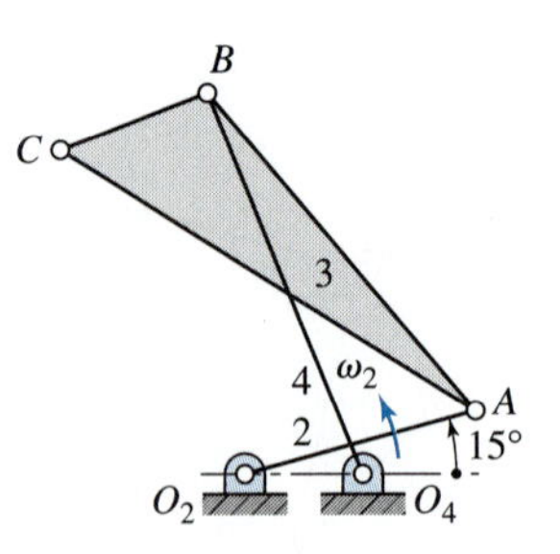

Figure P3.20 $R_{AO_2} = 150$ mm, $R_{BA} = R_{BO_4} =$ 250 mm, $R_{O_4O_2} = 75$ mm, $R_{CA} = 300$ mm, and $R_{CB} = 100$ mm.

†3.21 Link 2 of the mechanism illustrated in Fig. P3.21 has an angular velocity of 56 rad/s ccw. Find the velocity of point C.

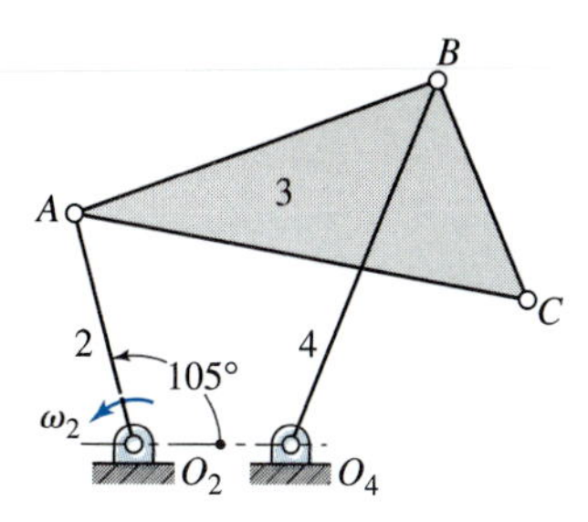

Figure P3.21 $R_{AO_2} = 150$ mm, $R_{BA} = R_{BO_4} = 250$ mm, $R_{O_4O_2} = 100$ mm, and $R_{CA} = 300$ mm.

†3.22 Find the velocities of points B, C, and D of the double-slider mechanism illustrated in Fig. P3.22 if crank 2 rotates at 42 rad/s cw.

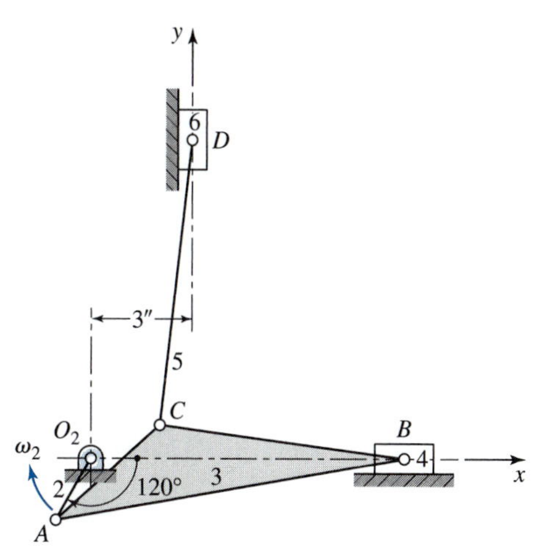

Figure P3.22 $R_{AO_2} = 2$ in, $R_{BA} = 10$ in, $R_{CA} = 4$ in, $R_{CB} = 7$ in, and $R_{DC} = 8$ in.

†3.23 Figure P3.23 illustrates the mechanism used in a two-cylinder 60° V engine consisting, in part, of an articulated connecting rod. Crank 2 rotates at 2000 rev/min cw. Find the velocities of points B, C, and D.

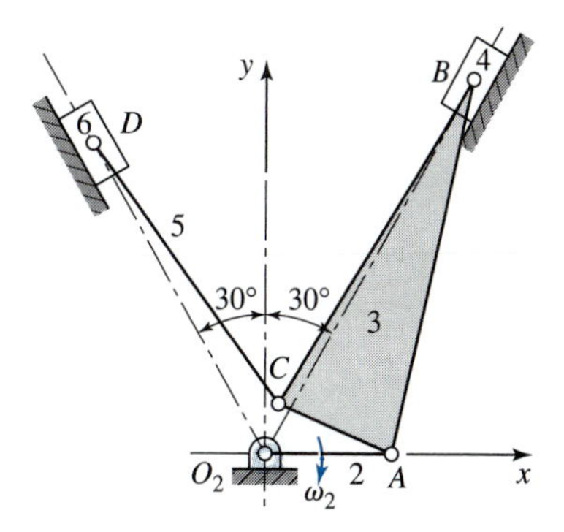

Figure P3.23 $R_{AO_2} = 2$ in, $R_{BA} = R_{CB} = 6$ in, $R_{CA} = 2$ in, and $R_{DC} = 5$ in.

3.24 Make a complete velocity analysis of the linkage illustrated in Fig. P3.24 given that $\omega_2 = 24$ rad/s cw. What is the absolute velocity of point B? What is its apparent velocity to an observer moving with link 4?

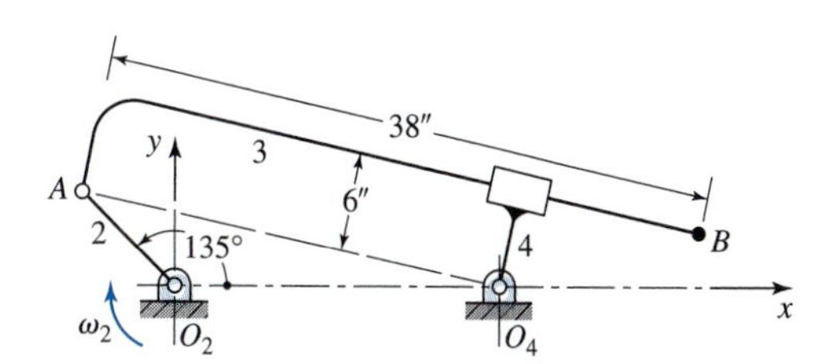

Figure P3.24 $R_{AO_2} = 8$ in, $R_{O_4O_2} = 20$ in.

3.25 Find $\mathbf{V}_B$ for the linkage illustrated in Fig. P3.25 if $V_A = 1$ ft/s.

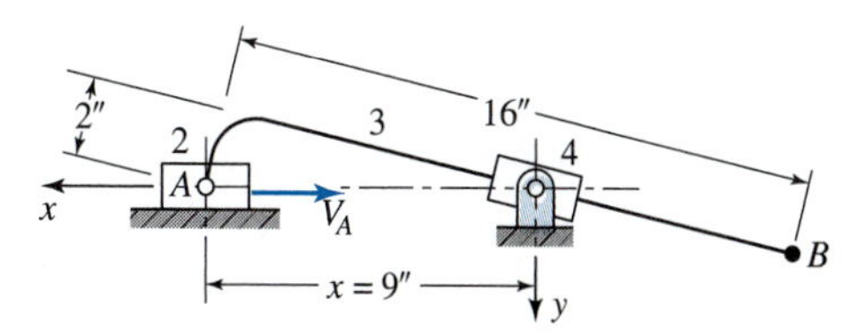

Figure P3.25

†3.26 Figure P3.26 illustrates a variation of the Scotch-yoke mechanism. The mechanism is driven by crank 2 at $\omega_2 = 36$ rad/s ccw. Find the velocity of the crosshead, link 4.

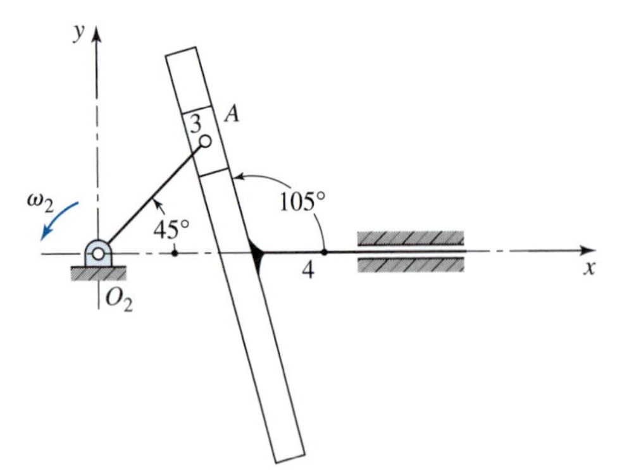

Figure P3.26 $R_{AO_2} = 250$ mm.

3.27 Make a complete velocity analysis of the linkage illustrated in Fig. P3.27 for $\omega_2 = 72$ rad/s ccw.

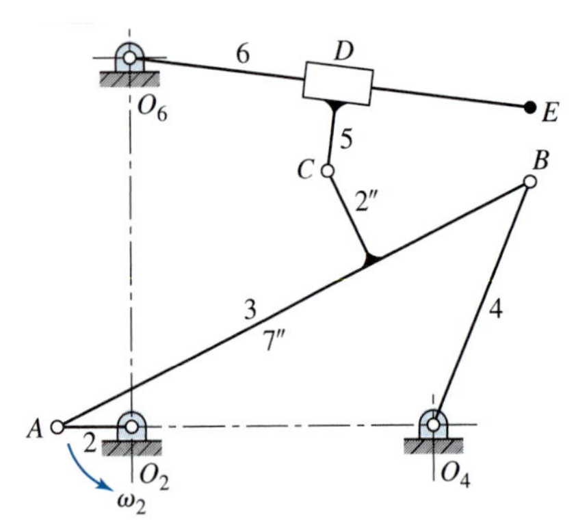

Figure P3.27 $R_{AO_2} = R_{DC} = 1.5$ in, $R_{BA} = 10.5$ in, $R_{O_4O_2} = 6$ in, $R_{BO_4} = 5$ in, $R_{O_6O_2} = 7$ in, and $R_{EO_6} = 8$ in.

3.28 The mechanism illustrated in Fig. 3.28 is driven such that $V_C = 10$ in/s to the right. Rolling contact is assumed between links 1 and 2, but slip is possible between links 2 and 3. Determine the angular velocity of link 3.

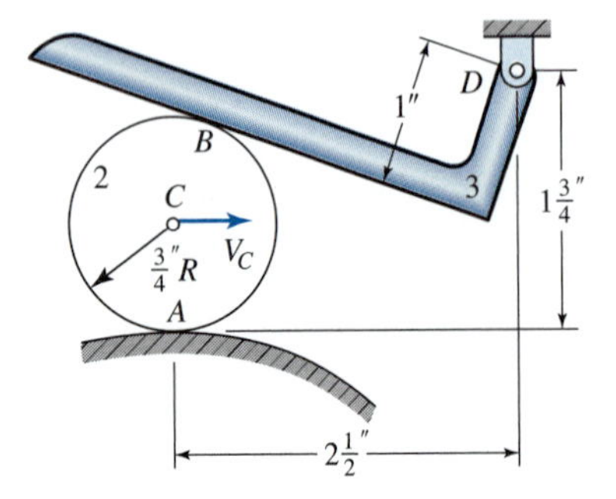

Figure P3.28

3.29 The circular cam illustrated in Fig. 3.29 is driven at an angular velocity of $\omega_2 = 15$ rad/s ccw. There is rolling contact between the cam and the roller, link 3. Find the angular velocity of the oscillating follower, link 4.

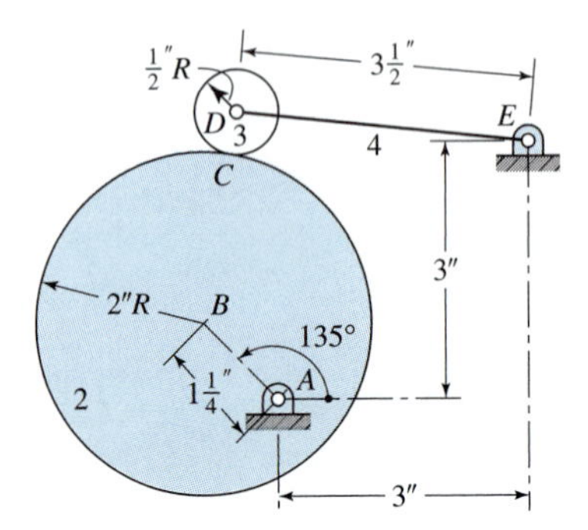

Figure P3.29

3.30 The mechanism illustrated in Fig. 3.30 is driven by link 2 at 10 rad/s ccw. There is rolling contact at point F. Determine the velocity of points E and G and the angular velocities of links 3, 4, 5, and 6.

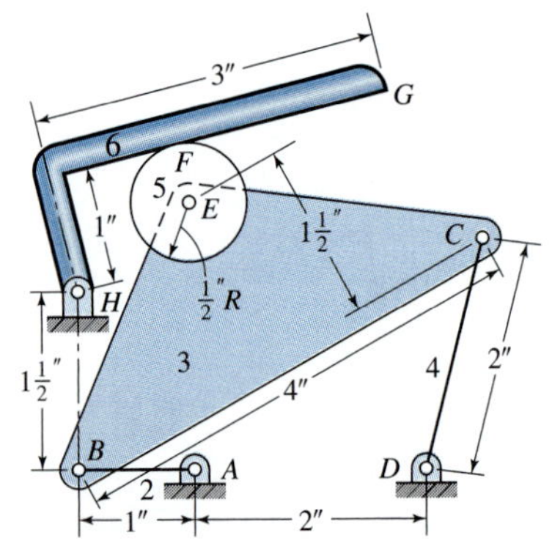

Figure P3.30

3.31 Figure 3.31 is a schematic diagram for a two-piston pump. The pump is driven by a circular eccentric, link 2, at $\omega_2 = 25$ rad/s ccw. Find the velocities of the two pistons, links 6 and 7.

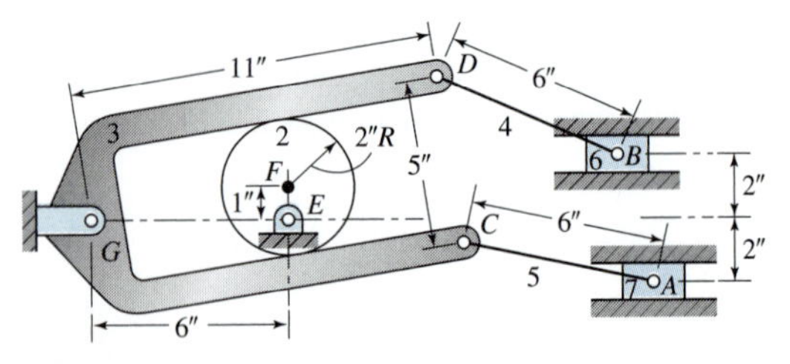

Figure P3.31

†**3.32** The epicyclic gear train illustrated in Fig. 3.32 is driven by the arm, link 2, at $\omega_2 = 10$ rad/s cw. Determine the angular velocity of the output shaft, attached to gear 3.

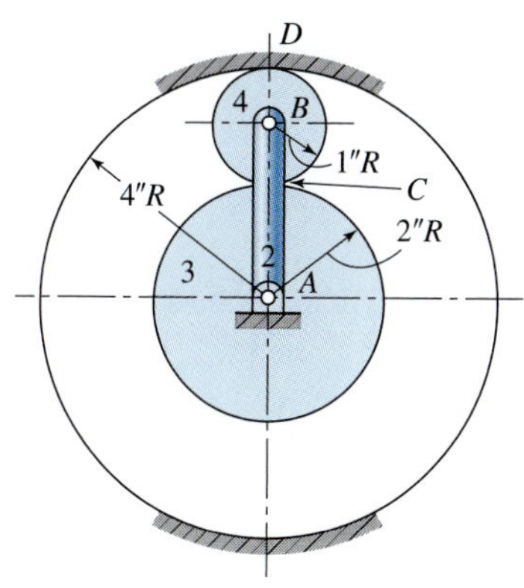

Figure P3.32

3.33 The diagram in Fig. P3.33 illustrates a planar schematic approximation of an automotive front suspension. The roll center is the term used by the industry to describe the point about which the auto body seems to rotate with respect to the ground. The assumption is made that there is pivoting but no slip between the tires and the road. After making a sketch, use the concepts of instant centers to find a technique to locate the roll center.

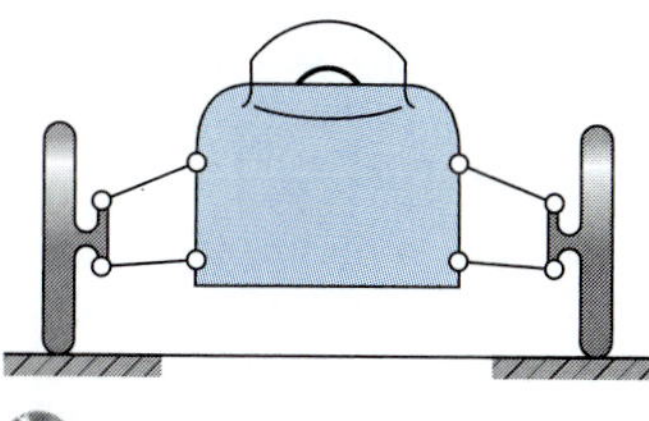

Figure P3.33

3.34 Locate all instant centers for the linkage of Problem 3.22.

3.35 Locate all instant centers for the mechanism of Problem 3.25.

3.36 Locate all instant centers for the mechanism of Problem 3.26.

3.37 Locate all instant centers for the mechanism of Problem 3.27.

3.38 Locate all instant centers for the mechanism of Problem 3.28.

3.39 Locate all instant centers for the mechanism of Problem 3.29.

3.40 For the mechanism illustrated in Fig. P3.40, the input link 2 is in the position $R_{AO_4} = 120$ mm and is moving to the right at a velocity of $V_A = 15.0$ m/s. Determine the first-order kinematic coefficients for the mechanism. Find the angular velocities of links 3 and 4.

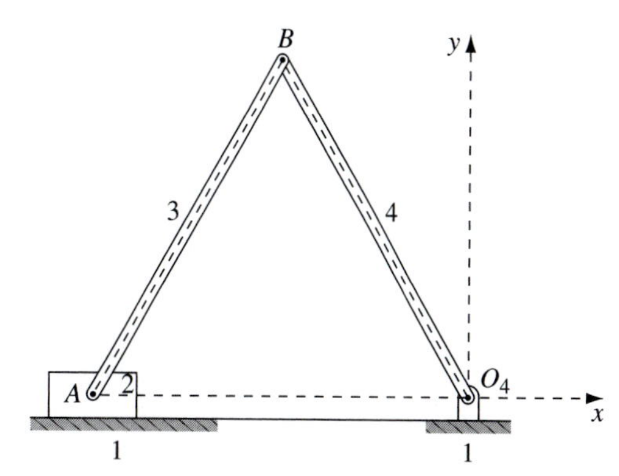

Figure P3.40 $R_{BO_4} = R_{BA} = 120$ mm.

3.41 For the mechanism illustrated in Fig. P3.41, pinion 3 is rolling without slipping on rack 4 at point D. Input link 2 is in the position $R_{GO_4} = 10$ in., and the input velocity is $\mathbf{V}_G = 3\hat{\mathbf{i}}$ in/s. Determine the first-order kinematic coefficients of the mechanism. Find the angular velocities of both rack 4 and pinion 3.

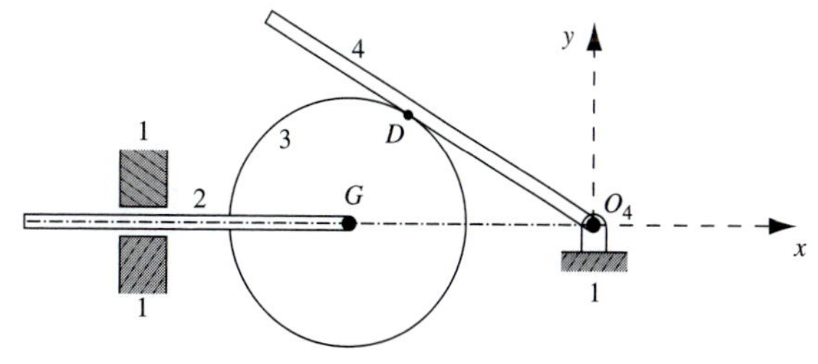

Figure P3.41 Rack-and-pinion mechanism. $R_{DG} = \rho_3 = 5$ in.

3.42 For the mechanism of Example 2.9, see Fig. 2.34, the dimensions are $R_1 = 800$ mm, $R_9 = 550$ mm, and $\rho_3 = 500$ mm. In the position where $R_2 = 750$ mm, the input link 2 has a velocity of $\mathbf{V}_A = 0.150\hat{\mathbf{j}}$ m/s. Determine the first-order kinematic coefficients for the mechanism. Find the velocity of rack 4 and the angular velocity of pinion 3.

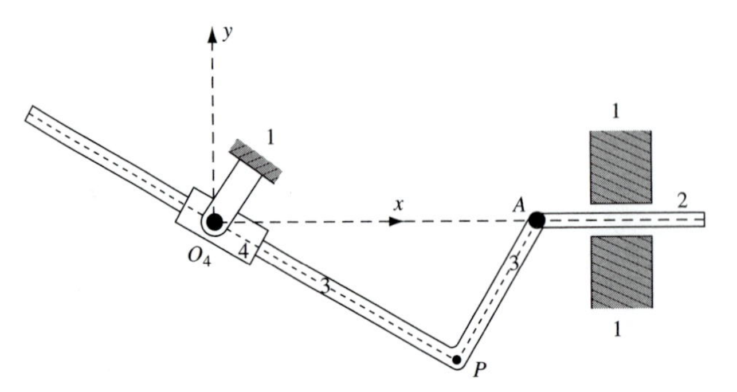

Figure P3.43 $R_{PA} = 5$ in and $\angle APO_4 = \angle 90°$.

3.43 For the mechanism illustrated in Fig. P3.43, in the current position $R_{AO_4} = 10$ in., and the input velocity is $\mathbf{V}_A = -5\hat{\mathbf{i}}$ in/s. Determine the first-order kinematic coefficients of the mechanism. Find the angular velocity of link 3 and the slipping velocity between links 3 and 4.

3.44 For the mechanism illustrated in Fig. 3.30, the input link has an angular velocity $\omega_2 = 10$ rad/s ccw and there is rolling contact between links 5 and 6 at point F. Determine the first-order kinematic coefficients for links 3, 4, 5, and 6. Find the angular velocities for links 3, 4, 5, and 6 and the velocities of points E and G.

3.45 For the mechanism illustrated in Fig. P3.45, input link 2 is moving vertically upward with a velocity of $V_A = 0.150$ m/s. Pinion 4 has a radius of 20 mm and is rolling without slipping on rack 3 at point B. The distance from point E to point B is equal to the distance from point B to pin A. The distance from O_4 to A is 40 mm. Determine the first-order kinematic coefficients for rack 3 and pinion 4. Find the angular velocity of rack 3 and pinion 4 and the velocity of point E. Also, find the velocity along rack 3 of the point of contact between links 3 and 4 (that is, point B).

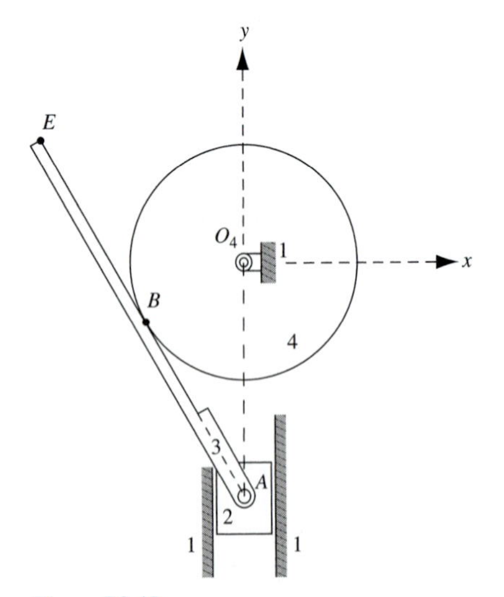

Figure P3.45

3.46 For the mechanism illustrated in Fig. P3.46, the dimensions are $R_{AO_2} = 10$ in. and $R_{PO_4} = 20$ in. At the position illustrated, where $\angle O_4O_2A = 30°$, $R_{PA} = R_{AO_4}$, and $R_{PB} = R_{BA}$, the angular velocity of the input link 2 is ω_2= 5 rad/s cw. Determine the first-order kinematic coefficients for links 3, 4, and 5. Find: (*i*) the angular velocities of links 3 and 4, (*ii*) the velocity of link 5, and (*iii*) the velocity of point P fixed in link 4.

3.47 For the mechanism illustrated in Fig. P3.47, the input link 2 is moving parallel to the X axis with a constant velocity $V_B = 15$ in/s to the right. At the instant indicated, the angle $\theta_4 = 60^o$. (*i*) Determine the first-order kinematic coefficients for links 3 and 4 and find the angular velocities of links 3 and 4. (*ii*) Determine the conditions for the determinant of the coefficient matrix of part (*i*) to be zero; then sketch the mechanism in the position when the determinant is zero.

3.48 For the mechanism illustrated in Fig. P2.15, the dimensions are $R_{AO_2} = 2$ in., $R_{BA} = 6$ in., and $R_{CO_5} = 2.5$ in. In the position indicated, the angle $\angle BAO_2$ is 150° and the distance $R_{BC} = 3.20$ in. The input link 2 is vertical and the angular velocity

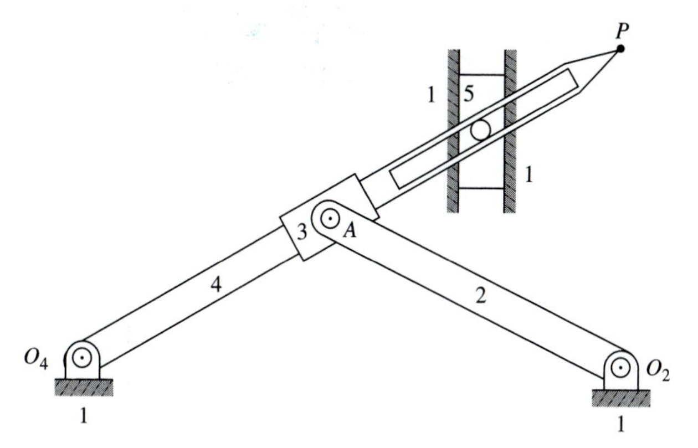

Figure P3.46 A planar linkage.

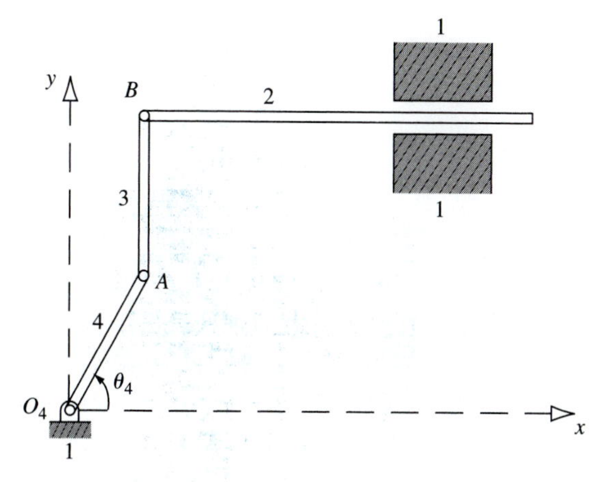

Figure P3.47 A planar mechanism. $R_{BA} = R_{AO_4} = 4$ in.

is $\omega_2 = 10$ rad/s cw. (*i*) Show the locations of all instant centers. (*ii*) Using instant centers, determine the first-order kinematic coefficients of link 3, rack 4, and pinion 5. (*iii*) Determine the angular velocity of link 3, the velocity of rack 4, and the angular velocity of pinion 5.

3.49 For the mechanism illustrated in Fig. P2.16, the dimensions are $R_{BA} = 7.071$ in. and $R_{BC} = 6$ in. The radius of gear 2 is $\rho_2 = 1$ in. and the radius of gear 5 is $\rho_5 = 2$ in. In the position indicated, the angular velocity $\omega_2 = 5$ rad/s ccw. Determine the first-order kinematic coefficients of links 3, 4, and 5. Find the angular velocities of links 3, 4, and 5.

3.50 For the mechanism illustrated in Fig. P2.17, the radius of wheel 3 is $\rho_3 = 0.75$ in. and the other dimensions are $R_{O_2O_5} = 7.0$ in., $R_{BA} = 5.5$ in., and $R_{AO_5} = 2.6$ in. For the given position, link 4 is parallel to the X axis and link 5 is coincident with the Y axis. Also, the input link 2 has an angular velocity of $\omega_2 = 15$ rad/s cw. Determine the first-order kinematic coefficients for links 3, 4, and 5. Find the angular velocities of links 3, 4 and 5.

4 Acceleration

4.1 DEFINITION OF ACCELERATION

Consider a moving point, first observed at location P where it has a velocity $\mathbf{V}_P$, as illustrated in Fig. 4.1*a*. After a short time interval Δt, the point is observed to have moved along some path to a new location P', where it has a velocity $\mathbf{V}'_P$, which may be different from $\mathbf{V}_P$ in both magnitude and direction. The change in the velocity of the point between the two positions can be evaluated as illustrated in Fig. 4.1*b*; namely,

$$\Delta \mathbf{V}_P = \mathbf{V}'_P - \mathbf{V}_P.$$

The *average acceleration* of the point P during the given time interval is defined as $\Delta \mathbf{V}_P / \Delta t$. The *instantaneous acceleration* (henceforth referred to simply as the *acceleration*) of point P is defined as the time rate of change of its velocity, that is, the limit of the average acceleration for an infinitesimally small time interval:

$$\mathbf{A}_P = \lim_{\Delta t \to 0} \left(\frac{\Delta \mathbf{V}_P}{\Delta t} \right) = \frac{d\mathbf{V}_P}{dt} = \frac{d^2 \mathbf{R}_P}{dt^2}. \tag{4.1}$$

Because velocity is a vector quantity, the change in velocity $\Delta \mathbf{V}_P$ and the acceleration $\mathbf{A}_P$ are also vector quantities, that is, they have both magnitude and direction. Also, like velocity, the acceleration vector is properly defined only for a point; the term should not be applied to a line, a coordinate system, a volume, or any other collection of points, because the accelerations of these points may be different.

As in the case of velocity (Chapter 3), the acceleration of a moving point will appear differently to different observers. Acceleration does not depend on the actual location of the observer but does depend critically on the motion of the observer or, more precisely, on

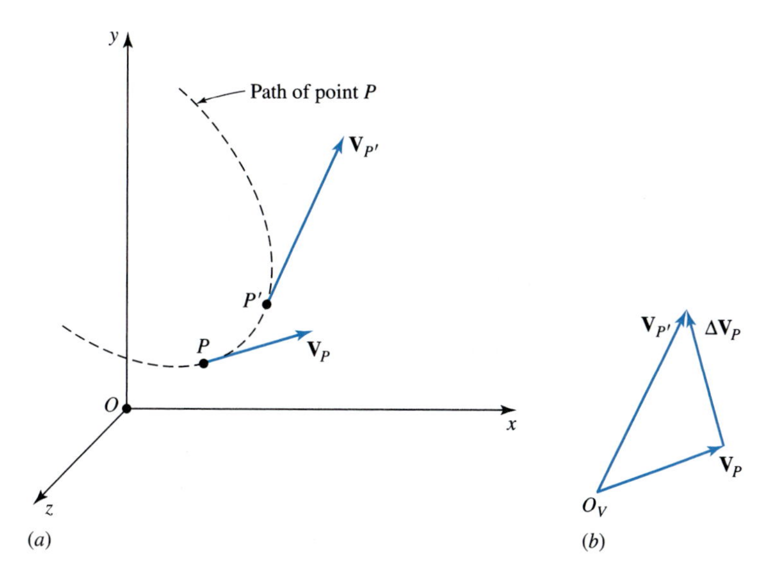

Figure 4.1 Change in velocity of a moving point.

the motion of the observer's coordinate system. If the acceleration is sensed by an observer using the absolute coordinate system, it is referred to as *absolute acceleration* and is denoted by the symbol $\mathbf{A}_{P/1}$ or simply $\mathbf{A}_P$, which is consistent with the notation used for position, displacement, and velocity.

Recall that the velocity of point P, moving along the path AB (see Fig. 4.2a), can be written from Eq. (3.9) as

$$\mathbf{V}_P = \dot{s}\hat{\mathbf{u}}^t, \tag{a}$$

where $\dot{s}$ is the instantaneous speed of P along the path and $\hat{\mathbf{u}}^t$ is the unit vector tangent to the path of P and with positive sense in the direction of positive motion for s (see Section 3.5). Identifying the osculating plane defined by $\hat{\mathbf{u}}^t$ and the center of curvature C and designating its preferred positive side by the unit vector $\hat{\mathbf{u}}^b$, we complete the right-hand vector triad $\hat{\mathbf{u}}^b\hat{\mathbf{u}}^t\hat{\mathbf{u}}^n$ by defining $\hat{\mathbf{u}}^n = \hat{\mathbf{u}}^b \times \hat{\mathbf{u}}^t$, consistent with Eq. (3.7). Thus, $\hat{\mathbf{u}}^n$ and $\hat{\mathbf{u}}^t$ are normal and tangent, respectively, to the path at the instantaneous position of point P.

The acceleration of point P is obtained by differentiating Eq. (a) with respect to time, that is,

$$\mathbf{A}_P = \dot{s}\dot{\hat{\mathbf{u}}}^t + \ddot{s}\hat{\mathbf{u}}^t, \tag{b}$$

where $\ddot{s}$ is the instantaneous rate of change in the speed of P along the path. The first term on the right-hand side of Eq. (b) requires additional clarification. Let ϕ represent the inclination angle of $\hat{\mathbf{u}}^t$ to any axis selected in the osculating plane defined by $\hat{\mathbf{u}}^n$, $\hat{\mathbf{u}}^t$, and point C the instantaneous center of curvature of the path at P (see Fig. 4.2a). As point P moves along the path AB, both $\hat{\mathbf{u}}^n$ and $\hat{\mathbf{u}}^t$ are functions of ϕ. In Fig. 4.2b, the unit vectors $\hat{\mathbf{u}}^t$ and $\hat{\mathbf{u}}^t + \Delta\hat{\mathbf{u}}^t$ have been transferred to a common origin. Then the limit of the ratio

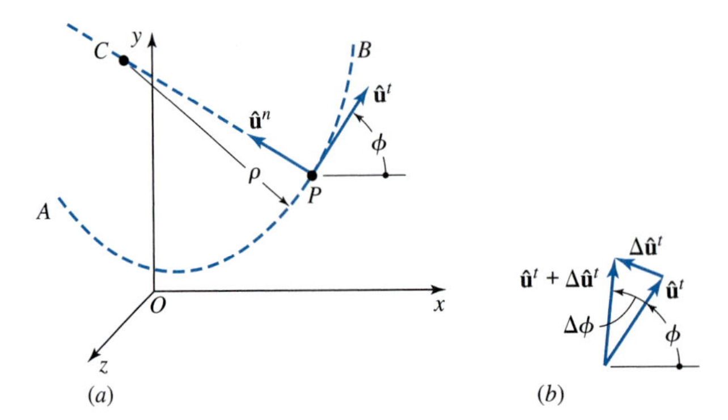

Figure 4.2 The motion of point P generates the spatial path AB.

$\Delta\hat{\mathbf{u}}^t/\Delta\phi$ is

$$\frac{d\hat{\mathbf{u}}^t}{d\phi} = \lim_{\Delta\phi\to 0}\left(\frac{\Delta\hat{\mathbf{u}}^t}{\Delta\phi}\right) = \lim_{\Delta\phi\to 0}\left(\frac{2\sin(\Delta\phi/2)\hat{\mathbf{u}}^n}{\Delta\phi}\right) = \hat{\mathbf{u}}^n. \qquad (c)$$

The first term on the right-hand side of Eq. (b) can be arranged in the form

$$\dot{s}\dot{\hat{\mathbf{u}}}^t = \frac{ds}{dt}\left(\frac{d\hat{\mathbf{u}}^t}{d\phi}\frac{d\phi}{ds}\frac{ds}{dt}\right). \qquad (d)$$

The term $d\phi/ds$ is the rate with respect to the change in distance s along the path with which the inclination of the tangent to the path changes. This quantity is called the *curvature* of the path, and its reciprocal is the *radius of curvature* ρ; that is,

$$\frac{d\phi}{ds} = \frac{1}{\rho}. \qquad (e)$$

Substituting Eqs. (c) and (e) into Eq. (d) and rearranging gives

$$\dot{s}\dot{\hat{\mathbf{u}}}^t = \frac{\dot{s}^2}{\rho}\hat{\mathbf{u}}^n. \qquad (f)$$

Then substituting this relation into Eq. (b), the acceleration of point P can be written as

$$\mathbf{A}_P = \frac{\dot{s}^2}{\rho}\hat{\mathbf{u}}^n + \ddot{s}\hat{\mathbf{u}}^t. \qquad (4.2)$$

Therefore, the acceleration vector $\mathbf{A}_P$ has two perpendicular components, a normal component of magnitude $\dot{s}^2/\rho$ (called normal because it is oriented along the $\hat{\mathbf{u}}^n$axis*) and a tangential component of magnitude $\ddot{s}$ along the $\hat{\mathbf{u}}^t$ axis. Hence, Eq. (4.2) can be written as

$$\mathbf{A}_P = \mathbf{A}_P^n + \mathbf{A}_P^t, \qquad (4.3)$$

---

* The reader should verify that the normal acceleration component is always oriented toward the center of curvature of the path no matter which orientations are positive for $\hat{\mathbf{u}}^t$ or $\hat{\mathbf{u}}^n$.

where the superscript n is for the normal direction and the superscript t is for the tangential direction.

4.2 ANGULAR ACCELERATION

The previous section demonstrated that the acceleration of a point is a vector quantity having a magnitude and a direction. But a point has no dimensions, so we cannot speak of the angular acceleration of a point. Thus, *angular acceleration deals with the motion of a rigid body*, whereas *rectilinear acceleration*, or simply acceleration, *deals with the motion of a point.*

Suppose, at one instant in time, a rigid body has an angular velocity of $\boldsymbol{\omega}$ and an instant later an angular velocity of $\boldsymbol{\omega}'$. The difference,

$$\Delta\boldsymbol{\omega} = \boldsymbol{\omega}' - \boldsymbol{\omega}, \tag{a}$$

is also a vector quantity. The angular velocities $\boldsymbol{\omega}$ and $\boldsymbol{\omega}'$ may have different magnitudes as well as different directions. Thus, we define *angular acceleration* as *the time rate of change of the angular velocity of a rigid body* and designate it by the symbol $\boldsymbol{\alpha}$, that is,

$$\boldsymbol{\alpha} = \lim_{\Delta t \to 0} \left(\frac{\Delta\boldsymbol{\omega}}{\Delta t} \right) = \frac{d\boldsymbol{\omega}}{dt} = \dot{\boldsymbol{\omega}}. \tag{4.4}$$

As is the case with $\Delta\boldsymbol{\omega}$, there is no reason to believe that $\boldsymbol{\alpha}$ has the same direction as either $\boldsymbol{\omega}$ or $\boldsymbol{\omega}'$; it may have an entirely new direction.

We further note that the angular acceleration vector $\boldsymbol{\alpha}$ applies to the absolute rotation of the entire rigid body and hence may be subscripted by the number of the coordinate system of the rigid body, for example, $\boldsymbol{\alpha}_2$ or $\boldsymbol{\alpha}_{2/1}$.

4.3 ACCELERATION DIFFERENCE BETWEEN POINTS OF A RIGID BODY

In Section 3.3 we determined the velocity difference between two points of a rigid body moving with both translation and rotation. We determined that the velocity of a point in a rigid body could be obtained as the sum of the velocity of *any* reference point of the body plus a rotational component, called the velocity difference, caused by the angular velocity $\boldsymbol{\omega}$ of the body. Thus, we saw that the velocity of any point P in a rigid body can be obtained from the velocity-difference equation [see Eq. (3.4)], that is,

$$\mathbf{V}_P = \mathbf{V}_Q + \mathbf{V}_{PQ}, \tag{a}$$

where $\mathbf{V}_Q$ is the velocity of the reference point Q, and $\mathbf{V}_{PQ}$ is the velocity difference and is given by the equation

$$\mathbf{V}_{PQ} = \boldsymbol{\omega} \times \mathbf{R}_{PQ}. \tag{b}$$

Here $\boldsymbol{\omega}$ is the angular velocity of the body and $\mathbf{R}_{PQ}$ is the position difference vector that defines the position of point P with respect to the reference point Q. Substituting Eq. (*b*)

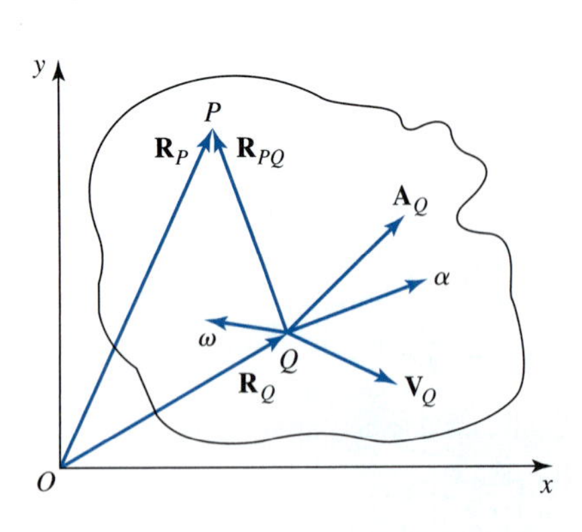

Figure 4.3

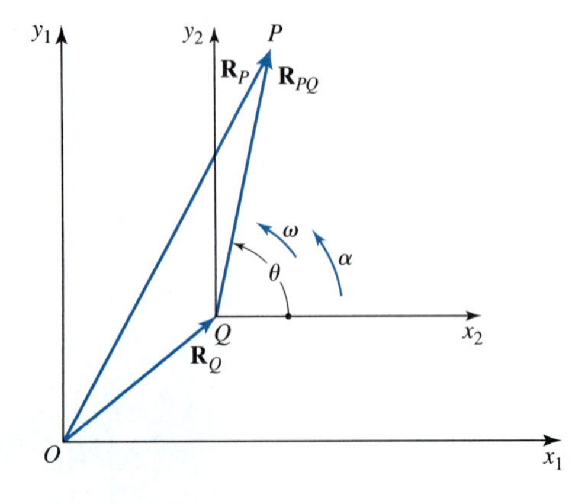

Figure 4.4

into Eq. (*a*) gives

$$\mathbf{V}_P = \mathbf{V}_Q + \boldsymbol{\omega} \times \mathbf{R}_{PQ}. \tag{c}$$

We employ a similar nomenclature in Fig. 4.3 and specify that a reference point Q of a rigid body has an acceleration $\mathbf{A}_Q$ and that the body has an angular acceleration $\boldsymbol{\alpha}$ in addition to its angular velocity $\boldsymbol{\omega}$, as illustrated. Note that, in general, $\boldsymbol{\alpha}$ may not have the same orientation as $\boldsymbol{\omega}$.

The acceleration of point P is obtained by taking the derivative of Eq. (*c*):

$$\dot{\mathbf{V}}_P = \dot{\mathbf{V}}_Q + \boldsymbol{\omega} \times \dot{\mathbf{R}}_{PQ} + \dot{\boldsymbol{\omega}} \times \mathbf{R}_{PQ}. \tag{d}$$

However, we already know that $\dot{\mathbf{V}}_P = \mathbf{A}_P, \dot{\mathbf{V}}_Q = \mathbf{A}_Q$, $\dot{\mathbf{R}}_{PQ} = \boldsymbol{\omega} \times \mathbf{R}_{PQ}$, and $\dot{\boldsymbol{\omega}} = \boldsymbol{\alpha}$. Therefore, Eq. (*d*) can be written as

$$\mathbf{A}_P = \mathbf{A}_Q + \boldsymbol{\omega} \times (\boldsymbol{\omega} \times \mathbf{R}_{PQ}) + \boldsymbol{\alpha} \times \mathbf{R}_{PQ}. \tag{4.5}$$

In this expression $\mathbf{A}_Q$ is the acceleration of the reference and is one component of the total acceleration of P. The other two components are caused by the rotation of the body. To picture the directions of these *components*, let us first study them in terms of a two-dimensional problem.

In Fig. 4.4, let P and Q be two points in a rigid body that has a combined motion of translation and rotation with respect to the ground reference plane x_1y_1. We can also define a moving system x_2y_2 with origin at Q, but restrict this system to only translation; thus, x_2 must remain parallel to x_1. As given quantities we specify the velocity and acceleration of reference point Q. We also specify the angular velocity $\boldsymbol{\omega}$ and the angular acceleration $\boldsymbol{\alpha}$ of the rigid body. For plane motion, these angular rates can be treated as scalars because the corresponding vectors always have axes perpendicular to the plane of the motion. They may have different senses, however, and the scalar quantities can be either positive or negative.

The location of point P in Fig. 4.4 can now be specified by the position-difference equation,

$$\mathbf{R}_P = \mathbf{R}_Q + \mathbf{R}_{PQ}. \tag{e}$$

This equation can also be written in the alternative form,

$$\mathbf{R}_p = \mathbf{R}_Q + R_{PQ}\angle\theta, \tag{f}$$

or, for two-dimensional problems, it can be written in complex polar form as

$$\mathbf{R}_P = \mathbf{R}_Q + R_{PQ}e^{j\theta}. \tag{g}$$

Differentiating Eq. (g) gives the velocity of point P, that is,

$$\mathbf{V}_P = \mathbf{V}_Q + j\omega R_{PQ}e^{j\theta}. \tag{h}$$

Note that this is the complex polar form of Eq. (c) for plane motion where the second term on the right-hand side corresponds to the velocity difference vector $\mathbf{V}_{PQ}$. The magnitude is ωR_{PQ} and the direction is $je^{j\theta}$, which is perpendicular to $\mathbf{R}_{PQ}$ in the sense of $\boldsymbol{\omega}$, as illustrated in Fig. 4.5.

Differentiating Eq. (h) gives the acceleration of point P, which can be written as

$$\mathbf{A}_P = \mathbf{A}_Q - \omega^2 R_{PQ}e^{j\theta} + j\alpha R_{PQ}e^{j\theta}. \tag{i}$$

The second and third components on the right-hand side of this equation correspond exactly with the second and third components of Eq. (4.5). The second term of Eq. (i) is called the *normal* or *centripetal* component of acceleration. For two-dimensional motion, the magnitude of this component is

$$\omega^2 R_{PQ} = \frac{V_{PQ}^2}{R_{PQ}}$$

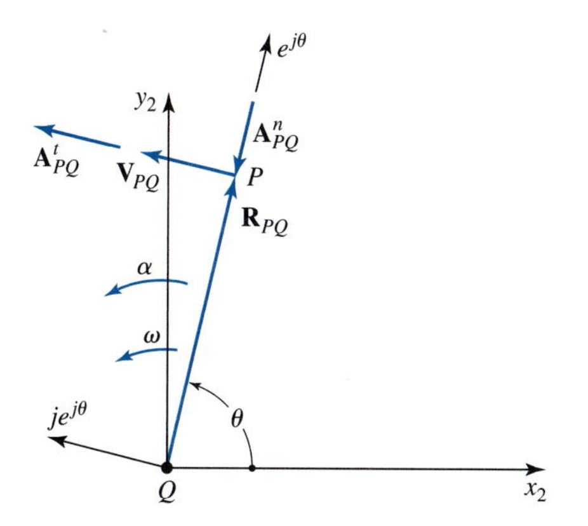

Figure 4.5
$A^t_{PQ} = \alpha R_{PQ}; V_{PQ} = \omega R_{PQ}.$

and the direction $-e^{j\theta}$ is opposite to the position-difference vector R_{PQ}. This component is also illustrated in Fig. 4.5. The superscript n is used to denote the normal component, that is,

$$A^n_{PQ} = \omega^2 R_{PQ} = \frac{V^2_{PQ}}{R_{PQ}}. \tag{4.6}$$

Note that the third term of Eq. (i), that is, $j\alpha R_{PQ}e^{j\theta}$, is associated with the angular acceleration of the body. The magnitude is αR_{PQ} and the direction is $je^{j\theta}$, which is along the same line as the velocity difference $\mathbf{V}_{PQ}$. Note that P traces out a circle in its motion about the translating reference point Q. Because the term $j\alpha R_{PQ}e^{j\theta}$ is perpendicular to the position-difference vector $\mathbf{R}_{PQ}$ and hence tangent to the circle, it is convenient to call it the *tangential* component of acceleration and designate the magnitude

$$A^t_{PQ} = \alpha R_{PQ}. \tag{4.7}$$

Let us now examine the last two terms of Eq. (4.5) again, but this time in three-dimensional space. The direction of the normal component of acceleration,

$$\mathbf{A}^n_{PQ} = \boldsymbol{\omega} \times (\boldsymbol{\omega} \times \mathbf{R}_{PQ}), \tag{4.8}$$

is illustrated in Fig. 4.6. This component is in the plane containing $\boldsymbol{\omega}$ and $\mathbf{R}_{PQ}$ and it is perpendicular to $\boldsymbol{\omega}$. The magnitude is

$$\left|\boldsymbol{\omega} \times (\boldsymbol{\omega} \times \mathbf{R}_{PQ})\right| = \omega^2 R_{PQ} \sin\phi = \frac{V^2_{PQ}}{R_{PQ}\sin\phi},$$

where $R_{PQ}\sin\phi$ is the radius of the circle in Fig. 4.6.

According to the definition of the vector cross-product, the tangential component of acceleration,

$$\mathbf{A}^t_{PQ} = \boldsymbol{\alpha} \times \mathbf{R}_{PQ}, \tag{4.9}$$

is perpendicular to the plane containing $\boldsymbol{\alpha}$ and $\mathbf{R}_{PQ}$ with a sense indicated by the right-hand rule. Because of $\boldsymbol{\alpha}$ we visualize P as accelerating around a circle, as illustrated in Fig. 4.7.

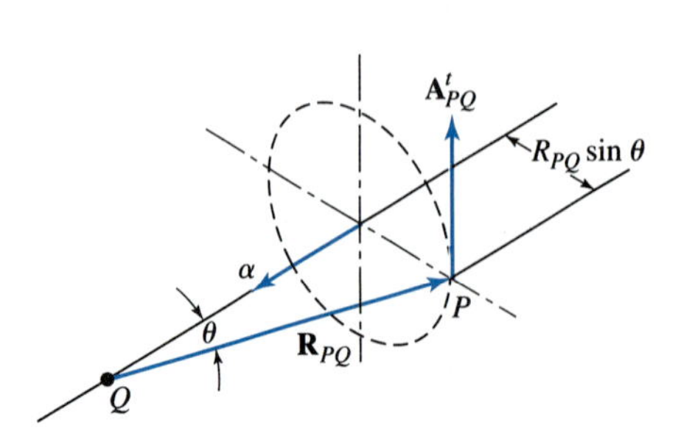

Figure 4.6
$V_{PQ} = \boldsymbol{\omega} \times R_{PQ}$; $A^n_{PQ} = \boldsymbol{\omega} \times (\boldsymbol{\omega} \times R_{PQ})$.

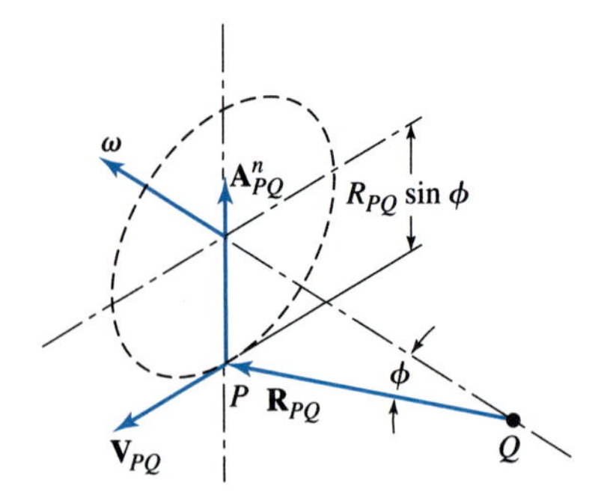

Figure 4.7 $A^t_{PQ} = \boldsymbol{\alpha} \times R_{PQ}$.

The plane of this circle is normal to the plane containing $\boldsymbol{\alpha}$ and $\mathbf{R}_{PQ}$. Using the definition of the cross-product, the magnitude of this tangential component of acceleration can be written as

$$\left|\boldsymbol{\alpha} \times \mathbf{R}_{PQ}\right| = \alpha R_{PQ} \sin\theta,$$

where $R_{PQ} \sin\theta$, as illustrated in Fig. 4.7, is the radius of the circle.

Again, we emphasize that in general $\boldsymbol{\alpha}$ and $\boldsymbol{\omega}$ may not have the same directions in three-dimensional space.

Let us now summarize the results of this section. The acceleration of a point fixed in a rigid body can be determined from the sum of three components. The first of these is the acceleration of a reference point (point Q in Fig. 4.3). This is only one component of the acceleration, and its value depends on the motion, if any, of the particular reference point selected. There are two additional components of acceleration caused by the rotation of the body. One of these is the normal component and comes solely from the angular velocity. The other is the tangential component and is caused by the rate of change of the angular velocity of the body.

Equation (4.5) can now be written as

$$\mathbf{A}_P = \mathbf{A}_Q + \mathbf{A}_{PQ}, \tag{4.10}$$

which is called the *acceleration-difference equation*. It is also convenient to designate the components of the acceleration-difference equation as

$$\mathbf{A}_{PQ} = \mathbf{A}^n_{PQ} + \mathbf{A}^t_{PQ}. \tag{4.11}$$

The acceleration-difference equation can also be dealt with by most of the methods that were presented in Chapter 3.

EXAMPLE 4.1

For the four-bar linkage in the position illustrated in Fig. 4.8*a*, determine the acceleration of points A and B and the angular acceleration of links 3 and 4. The input crank 2 is rotating counterclockwise with a constant angular velocity $\omega_2 = 200$ rad/s.

SOLUTION

Figures 4.8*b* and 4.8*c* illustrate the velocity and acceleration polygons, respectively, with the circled numbers indicating the order of the steps in the construction; the method of finding the directions of each vector is also indicated, using one symbol (||) to indicate parallelism and another (⊥) to indicate perpendicularity. The velocity polygon must, of course, be completed first, because the velocities of points and angular velocities of links 3 and 4 are needed in the acceleration analysis.

The magnitude of the velocity of point A is

$$V_A = \omega_2 R_{AO_2} = (200 \text{ rad/s}) \left(6/12\text{ft}\right) = 100 \text{ ft/s}.$$

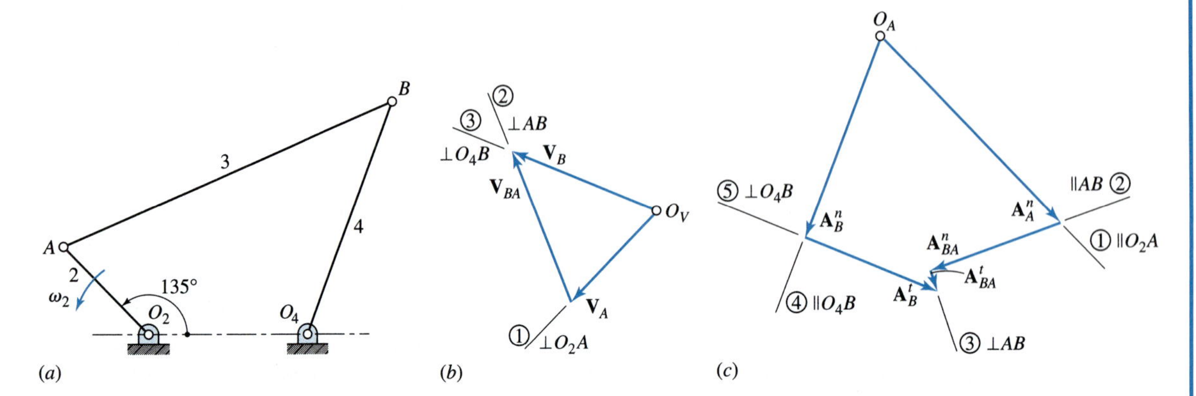

Figure 4.8 Example 4.1. (*a*) The four-bar linkage; $R_{O_4O_2} = 8$ in, $R_{AO_2} = 6$ in, $R_{BA} = 18$ in, $R_{BO_4} = 12$ in. (*b*) The velocity polygon. (*c*) The acceleration polygon.

Now the velocity polygon (Fig. 4.8*b*) can be drawn. From the polygon we obtain

$$V_{BA} = 128 \text{ ft/s} \quad \text{and} \quad V_B = 129 \text{ ft/s}.$$

Therefore, the angular velocities of links 3 and 4, respectively, are

$$\omega_3 = \frac{V_{BA}}{R_{BA}} = \frac{128 \text{ ft/s}}{18/12 \text{ ft}} = 85.3 \text{ rad/s ccw}$$

$$\omega_4 = \frac{V_{BO_4}}{R_{BO_4}} = \frac{129 \text{ ft/s}}{12/12 \text{ ft}} = 129 \text{ rad/s ccw},$$

where the directions are obtained from an examination of the velocity polygon.

The next step is to write the acceleration-difference equation in terms of its components. Because

$$\mathbf{A}_B = \mathbf{A}_A + \mathbf{A}_{BA},$$

then

$$\overset{\surd\surd}{\mathbf{A}^n_{BO_4}} + \overset{?\surd}{\mathbf{A}^t_{BO_4}} = \overset{\surd\surd}{\mathbf{A}^n_{AO_2}} + \overset{\mathbf{0}}{\cancel{\mathbf{A}^t_{AO_2}}} + \overset{\surd\surd}{\mathbf{A}^n_{BA}} + \overset{?\surd}{\mathbf{A}^t_{BA}}.$$

We note that there are two unknowns, that is, the magnitudes of the two tangential-component vectors.

We have sufficient information to calculate some of the components as follows.

$$A^t_{AO_2} = \alpha_2 R_{AO_2} = 0$$

$$A^n_{AO_2} = \omega_2^2 R_{AO_2} = (200 \text{ rad/s})^2\,(6/12 \text{ ft}) = 20\,000 \text{ ft/s}^2 = A_A \qquad \textit{Ans.}$$

$$A_{BA}^{n} = \frac{V_{BA}^{2}}{R_{BA}} = \frac{(128 \text{ ft/s})^{2}}{18/12 \text{ ft}} = 10\,923 \text{ ft/s}^{2}$$

$$A_{BO_4}^{n} = \frac{V_{BO4}^{2}}{R_{BO_4}} = \frac{(129 \text{ ft/s})^{2}}{(12/12 \text{ ft})} = 16\,641 \text{ ft/s}^{2}.$$

Beginning with the right-hand side of the acceleration-difference equation, the acceleration polygon is drawn by choosing an acceleration origin O_A and scale and constructing $\mathbf{A}_{AO_2}^{n}$, then $\mathbf{A}_{BA}^{n}$, and then $\mathbf{A}_{BA}^{t}$, temporarily of indefinite length because its magnitude is unknown. Beginning again at the acceleration pole O_A and using the left-hand side of the equation, we now construct $\mathbf{A}_{BO_4}^{n}$ and then $\mathbf{A}_{BO_4}^{t}$ with magnitude also unknown. Where the lines for the two unknowns intersect, this completes the polygon; we label it as demonstrated and measure the following results:

$$A_{BO_4}^{t} = 11\,900 \text{ ft/s}^{2}, A_{BA}^{t} = 2\,000 \text{ ft/s}^{2} \text{ and } A_B = 20\,500 \text{ ft/s}^{2}. \qquad \textit{Ans.}$$

The angular accelerations of links 3 and 4 are then computed as follows,

$$\alpha_3 = \frac{A_{BA}^{t}}{R_{BA}} = \frac{2\,000 \text{ ft/s}^{2}}{18/12 \text{ ft}} = 1\,333 \text{ rad/s}^{2} \text{ cw} \qquad \textit{Ans.}$$

$$\alpha_4 = \frac{A_{BO4}^{t}}{R_{BO_4}} = \frac{11\,900 \text{ ft/s}^{2}}{12/12 \text{ ft}} = 11\,900 \text{ rad/s}^{2} \text{ cw}, \qquad \textit{Ans.}$$

where the directions are obtained from an examination of the acceleration polygon.

EXAMPLE 4.2

Solve Example 4.1 using the direct analytical approach.

SOLUTION

The first two steps are to perform a position analysis and a velocity analysis of the linkage. Because the procedure has been demonstrated in earlier chapters, we omit the analysis here and display only the results.

In vector form, the position vectors corresponding to the links (see Fig. 4.9) are as follows:

$$\mathbf{R}_{AO_2} = \left(\frac{6}{12}\right) \text{ ft } \angle 135^\circ = -0.353\,55\hat{\mathbf{i}} + 0.353\,55\hat{\mathbf{j}} \text{ ft}$$

$$\mathbf{R}_{BA} = \left(\frac{18}{12}\right) \text{ ft } \angle 22.4^\circ = 1.386\,82\hat{\mathbf{i}} + 0.571\,61\hat{\mathbf{j}} \text{ ft}$$

$$\mathbf{R}_{BO_4} = \left(\frac{12}{12}\right) \text{ ft } \angle 68.4^\circ = 0.368\,12\hat{\mathbf{i}} + 0.929\,78\hat{\mathbf{j}} \text{ ft}.$$

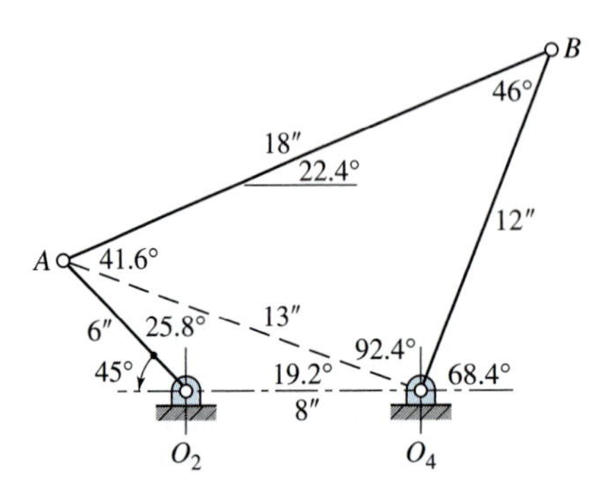

Figure 4.9

The results of the velocity analysis are

$$\boldsymbol{\omega}_2 = 200\hat{\mathbf{k}} \text{ rad/s}, \ \boldsymbol{\omega}_3 = 84.253\hat{\mathbf{k}} \text{ rad/s, and } \boldsymbol{\omega}_4 = 129.393\hat{\mathbf{k}} \text{ rad/s.}$$

The known acceleration components are computed next as

$$\mathbf{A}^n_{AO_2} = \mathbf{A}_A = \boldsymbol{\omega}_2 \times (\boldsymbol{\omega}_2 \times \mathbf{R}_{AO_2}) = 14\,142\hat{\mathbf{i}} - 14\,142\hat{\mathbf{j}} \text{ ft/s}^2 \qquad \textit{Ans.} \ (1)$$

$$\mathbf{A}^n_{BA} = \boldsymbol{\omega}_3 \times (\boldsymbol{\omega}_3 \times \mathbf{R}_{BA}) = -9\,844\hat{\mathbf{i}} - 4\,058\hat{\mathbf{j}} \text{ ft/s}^2 \qquad (2)$$

$$\mathbf{A}^n_{BO_4} = \boldsymbol{\omega}_4 \times (\boldsymbol{\omega}_4 \times \mathbf{R}_{BO_4}) = -6\,163\hat{\mathbf{i}} - 15\,567\hat{\mathbf{j}} \text{ ft/s}^2. \qquad (3)$$

Although the angular accelerations $\boldsymbol{\alpha}_3$ and $\boldsymbol{\alpha}_4$ both have unknown magnitudes, we can incorporate them into the solution in the following manner:

$$\mathbf{A}^t_{BA} = \boldsymbol{\alpha}_3 \times \mathbf{R}_{BA} = \begin{vmatrix} \hat{\mathbf{i}} & \hat{\mathbf{j}} & \hat{\mathbf{k}} \\ 0 & 0 & \alpha_3 \\ 1.386\,82 \text{ ft} & 0.571\,61 \text{ ft} & 0 \end{vmatrix} = -0.571\,61\alpha_3\hat{\mathbf{i}} + 1.386\,82\alpha_3\hat{\mathbf{j}} \text{ ft/s}^2 \qquad (4)$$

$$\mathbf{A}^t_{BO_4} = \boldsymbol{\alpha}_4 \times \mathbf{R}_{BO_4} = \begin{vmatrix} \hat{\mathbf{i}} & \hat{\mathbf{j}} & \hat{\mathbf{k}} \\ 0 & 0 & \alpha_4 \\ 0.368\,12 \text{ ft} & 0.929\,78 \text{ ft} & 0 \end{vmatrix} = -0.929\,78\alpha_4\hat{\mathbf{i}} + 0.368\,12\alpha_4\hat{\mathbf{j}} \text{ ft/s}^2. \qquad (5)$$

Writing the acceleration-difference equation for point B and noting that $\mathbf{A}^t_{AO_2} = \mathbf{0}$ gives

$$\mathbf{A}^n_{BO_4} + \mathbf{A}^t_{BO_4} = \mathbf{A}^n_{AO_2} + \mathbf{A}^n_{BA} + \mathbf{A}^t_{BA} \qquad (6)$$

Then substituting Eqs. (1) through (5) into Eq. (6) and separating the $\hat{\mathbf{i}}$ and $\hat{\mathbf{j}}$ components, we obtain the following pair of simultaneous equations:

$$0.571\,61 \text{ ft}\,\alpha_3 - 0.929\,78 \text{ ft}\,\alpha_4 = 10\,461 \text{ ft/s}^2 \qquad (7)$$

$$-1.386\,82 \text{ ft}\,\alpha_3 + 0.368\,12 \text{ ft}\,\alpha_4 = -2\,633 \text{ ft/s}^2. \qquad (8)$$

When these equations are solved simultaneously, the results are found to be

$$\boldsymbol{\alpha}_3 = -1\,300\hat{\mathbf{k}} \text{ rad/s}^2 \quad \text{and} \quad \boldsymbol{\alpha}_4 = -12\,050\hat{\mathbf{k}} \text{ rad/s}^2, \qquad \textit{Ans.}$$

where the negative signs indicate that both are clockwise.

4.4 ACCELERATION POLYGONS

The acceleration image of a link can be obtained in much the same manner as the velocity image of the link. Figure 4.10*a* illustrates a slider-crank mechanism in which links 2 and 3 have been given triangular shapes to emphasize their images. Figure 4.10*b* illustrates the velocity images, and the acceleration images are illustrated in Fig. 4.10*c*. The angular

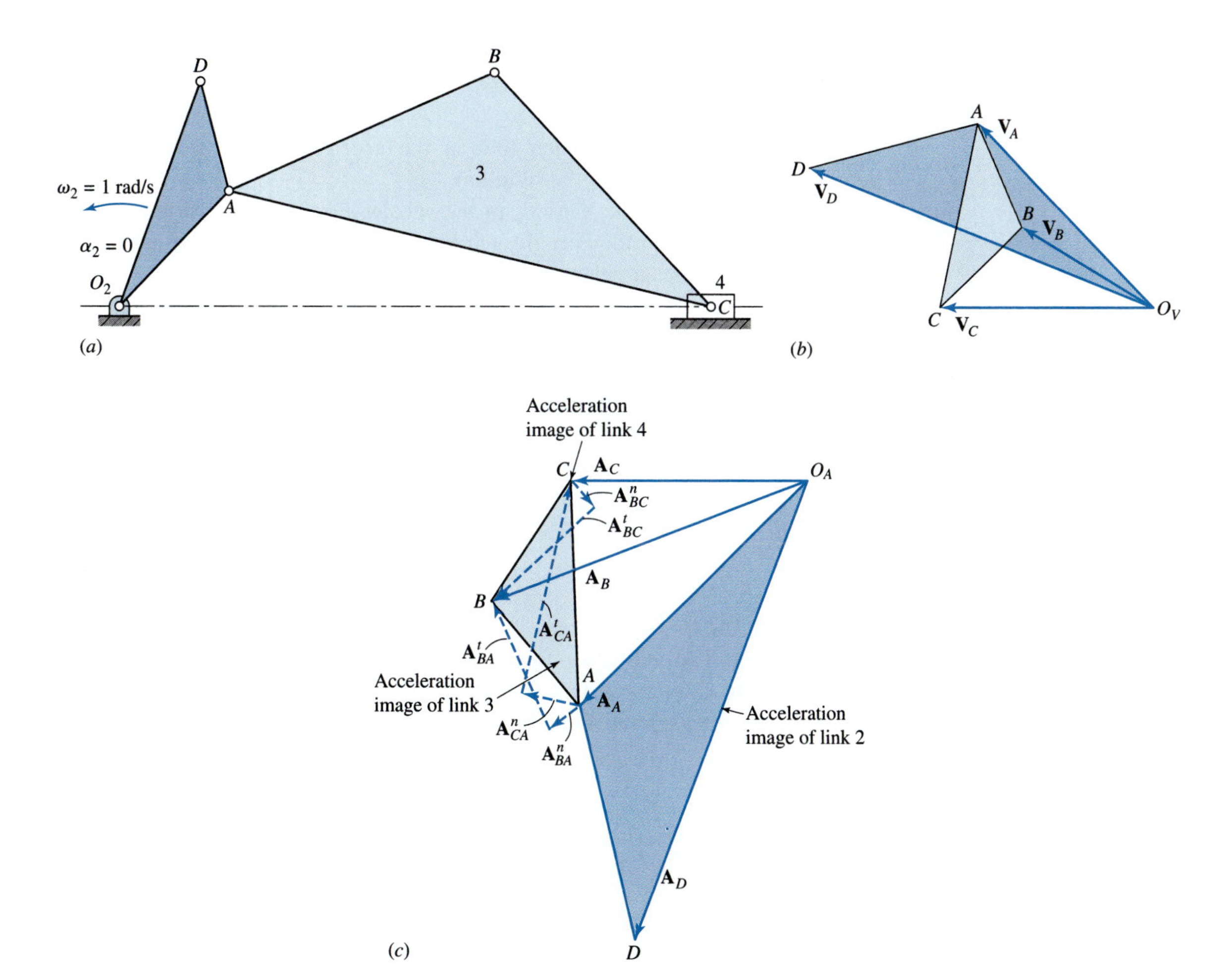

Figure 4.10 (*a*) Slider-crank mechanism; (*b*) velocity polygon; (*c*) acceleration polygon.

acceleration of the crank (link 2) is zero and note that its corresponding acceleration image is turned 180° from the orientation of the link itself. On the other hand, note that link 3 has a counterclockwise angular acceleration and that its image is oriented less than 180° from the orientation of the link itself. Thus, the orientation of the acceleration image depends on both the angular velocity and the angular acceleration of the link in question. It can be demonstrated from the geometry of Fig. 4.10 that the orientation of an acceleration image is given by the equation

$$\delta = 180^\circ - \tan^{-1}\frac{\alpha}{\omega^2}, \tag{4.12}$$

where δ is the angle in degrees, measured in the (positive) counterclockwise direction, from the orientation of the link itself to its acceleration image.

Also, the acceleration image is determined in exactly the same manner as was the velocity image. Note particularly that the acceleration image is formed from the *total acceleration difference* vectors, not the component vectors. We also note the following properties of the acceleration image.

1. The acceleration image of each link in the acceleration polygon is a scale reproduction of the shape of the rigid link.
2. The letters identifying the vertices in the acceleration polygon are the same as those of each link, and they progress around the acceleration image in the same order and in the same angular sense as around the link itself.
3. The point O_A in the acceleration polygon is the image of all points with zero absolute acceleration. It is the acceleration image of the fixed link.
4. The absolute acceleration of any point on any link is represented by the line from O_A to the image of the point in the acceleration polygon. The acceleration difference between two points, say P and Q, is represented by the line to the acceleration image point P from the acceleration image point Q.

EXAMPLE 4.3

A velocity analysis of the four-bar linkage in the position illustrated in Fig. 4.11 was presented in Example 3.1. The velocity polygon is illustrated in Fig. 3.7*b*. Given that the angular velocity of link 2 is a constant 900 rev/min = 94.25 rad/s ccw, determine the angular accelerations of links 3 and 4 and the accelerations of points E and F.

SOLUTION

Because the angular acceleration of link 2 is zero, there remains only the normal component of the acceleration of B and, hence,

$$A_B = A^n_{BA} = \omega_2^2 R_{BA} = (94.25\ \text{rad/s})^2\,(4/12\ \text{ft}) = 2\,961\ \text{ft/s}^2.$$

Point O_A and an acceleration scale are chosen and $\mathbf{A}_B$ is constructed (opposite in direction to the vector $\mathbf{R}_{BA}$) to locate the acceleration image point B, as illustrated in Fig. 4.12.

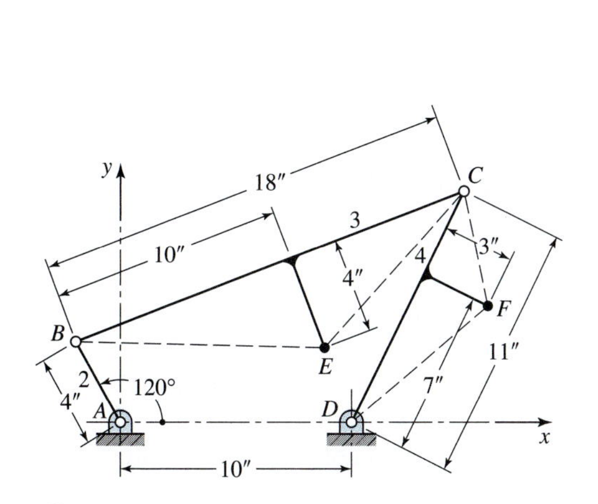

Figure 4.11 Example 4.3.

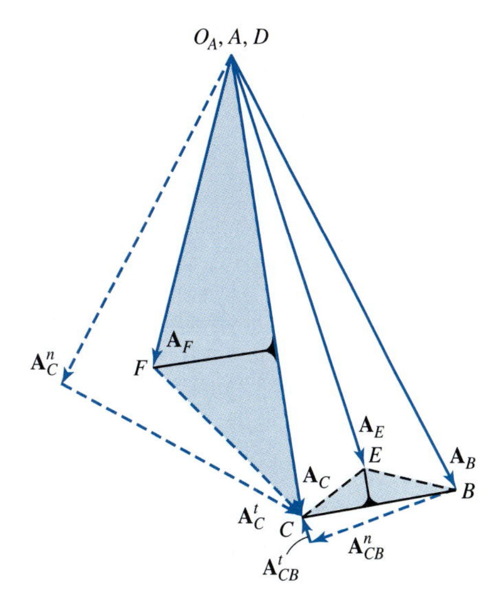

Figure 4.12 Example 4.3. Acceleration polygon.

Next, the acceleration-difference equation is written to relate point C to point B and to point D in the acceleration polygon. Thus,

$$\mathbf{A}_C = \overset{\surd\surd}{\mathbf{A}_B} + \overset{\surd\surd}{\mathbf{A}_{CB}^n} + \overset{?\surd}{\mathbf{A}_{CB}^t} = \overset{\surd\surd}{\mathbf{A}_{CD}^n} + \overset{?\surd}{\mathbf{A}_{CD}^t}. \tag{1}$$

Using information scaled from the velocity polygon in Fig. 3.7b, we calculate the magnitude of the two normal components of Eq. (1). Thus,

$$A_{CB}^n = \frac{V_{CB}^2}{R_{CB}} = \frac{(38.4\ \text{ft/s})^2}{18/12\ \text{ft}} = 983\ \text{ft/s}^2$$

$$A_{CD}^n = \frac{V_{CD}^2}{R_{CD}} = \frac{(45.5\ \text{ft/s})^2}{11/12\ \text{ft}} = 2\,258\ \text{ft/s}^2.$$

These two normal components are constructed with orientations opposite to the vectors $\mathbf{R}_{CB}$ and $\mathbf{R}_{CD}$, respectively. As required by Eq. (1), they are added to the acceleration polygon originating from points B and D, respectively, and are indicated by dashed lines in Fig. 4.12. Perpendicular dashed lines are then drawn through the termini of these two normal components; these represent the addition of the two tangential components $\mathbf{A}_{CB}^t$ and $\mathbf{A}_{CD}^t$, as required to complete Eq. (1). Their intersection is labeled acceleration-image point C.

The angular accelerations of links 3 and 4 are now determined from the two tangential components that are scaled from the acceleration polygon

$$\alpha_3 = \frac{A^t_{CB}}{R_{CB}} = \frac{170 \text{ ft/s}^2}{18/12 \text{ ft}} = 113 \text{ rad/s}^2 \text{ ccw} \qquad \textit{Ans.}$$

$$\alpha_4 = \frac{A^t_{CD}}{R_{CD}} = \frac{1\,670 \text{ ft/s}^2}{11/12 \text{ ft}} = 1\,822 \text{ rad/s}^2 \text{ cw.} \qquad \textit{Ans.}$$

There are several approaches for finding the acceleration of point E. One approach is to relate it through acceleration-difference equations to points B and C, which are also in link 3. Thus,

$$\mathbf{A}_E = \mathbf{A}_B + \mathbf{A}^n_{EB} + \mathbf{A}^t_{EB} = \mathbf{A}_C + \mathbf{A}^n_{EC} + \mathbf{A}^t_{EC}. \qquad (2)$$

The solution of these equations can follow the same methods used for Eq. (1) if desired. A second approach is to use the value of the angular acceleration $\boldsymbol{\alpha}_3$, now known, to calculate one or both of the tangential components in Eq. (2). Probably the easiest approach, however, and the one used in Fig. 4.12, is to form the acceleration-image triangle BCE for link 3, using A_{CB} and the shape of link 3 as the basis.* Any of these approaches leads to the location of the acceleration-image point E illustrated in Fig. 4.12. The magnitude of the acceleration of point E is then scaled and found to be

$$A_E = 2\,580 \text{ ft/s}^2. \qquad \textit{Ans.}$$

The same approach can be used to find the magnitude of the acceleration of point F. The result is found to be

$$A_F = 1\,960 \text{ ft/s}^2. \qquad \textit{Ans.}$$

4.5 APPARENT ACCELERATION OF A POINT IN A MOVING COORDINATE SYSTEM

In Section 3.5 we found it helpful to develop the apparent-velocity equation for situations where it was convenient to describe the path along which a point moves relative to another moving link but where it was not convenient to describe the absolute motion of the same point. Let us now investigate the acceleration of such a point.

To review, Fig. 4.13 illustrates a point P_3 of link 3 that moves along a known path (the slot) relative to the moving reference frame $x_2y_2z_2$. Point P_2 is fixed in the moving link 2 and is instantaneously coincident with P_3. The problem is to find an equation relating the accelerations of points P_3 and P_2 in terms of meaningful parameters that can be calculated (or measured) in a typical mechanical system.

---

* We must be extremely careful that the shape of the acceleration image is not "flipped over" with respect to the original shape. A convenient test is to note that, for link 3 of this example, because the labels BCE are in clockwise order for the original link shape, they are still clockwise in the velocity and the acceleration images.

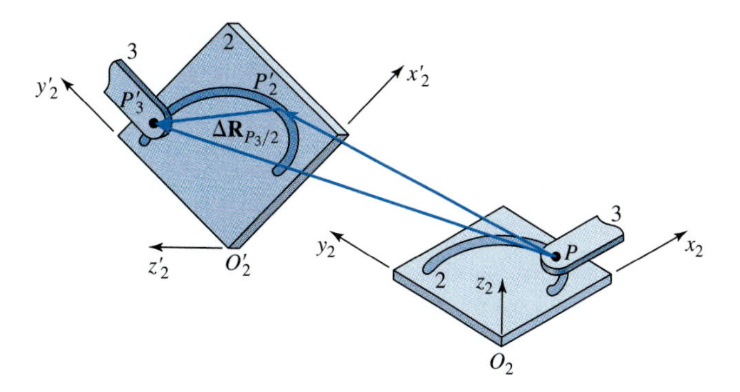

Figure 4.13 Apparent displacement.

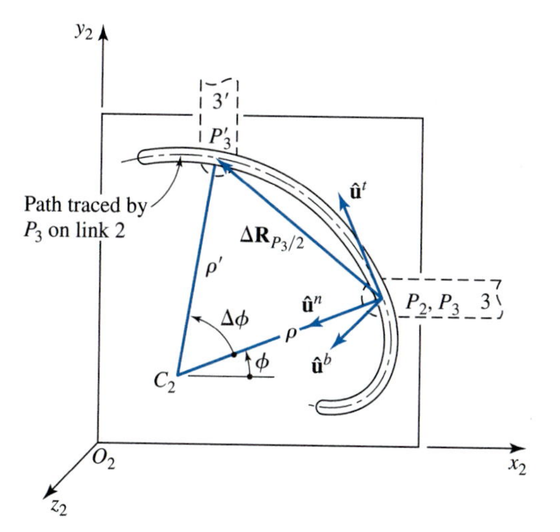

Figure 4.14 Apparent displacement of point P_3 as seen by an observer on link 2.

In Fig. 4.14 we recall how the same situation would be perceived by a moving observer attached to link 2. To HER, the path of P_3, the slot, would appear stationary and the point P_3 would appear to move along its tangent with the apparent velocity $\mathbf{V}_{P_3/2}$.

In Section 3.5 (see Fig. 3.10), we defined another moving coordinate system $\hat{\mathbf{u}}^t\hat{\mathbf{u}}^n\hat{\mathbf{u}}^b$, where $\hat{\mathbf{u}}^t$ is defined as the unit tangent vector to the path of P in the direction of positive movement, and $\hat{\mathbf{u}}^b$ is normal to the plane containing $\hat{\mathbf{u}}^t$ and the center of curvature C and is directed positive to the preferred side of the plane. A third unit vector $\hat{\mathbf{u}}^n$ is then obtained from Eq. (3.7), that is, $\hat{\mathbf{u}}^n = \hat{\mathbf{u}}^b \times \hat{\mathbf{u}}^t$, thus completing a right-hand Cartesian coordinate system. After defining s as the scalar arc distance measuring the travel of P_3 along its curved path, we derived Eq. (3.9) for the apparent velocity

$$\mathbf{V}_{P_3/2} = \frac{ds}{dt}\hat{\mathbf{u}}^t. \tag{a}$$

Consider the rotation of the radius of curvature ρ; it sweeps through some small angle $\Delta\phi$ as P_3 travels the small arc distance Δs during a short time interval Δt. The angle and arc distance are related by

$$\Delta\phi = \frac{\Delta s}{\rho}. \tag{b}$$

Note here that the center of curvature C can lie along either the positive or the negative extension of $\hat{\mathbf{u}}^n$. Therefore, the radius of curvature ρ, measured *from P to C*, can have either a positive or a negative value according to the sense of $\hat{\mathbf{u}}^n$. Also, this implies that the angle $\Delta\phi$ is positive when counterclockwise as seen from positive $\hat{\mathbf{u}}^b$.

Dividing Eq. (*b*) by Δt and taking the limit for infinitesimally small Δt, we find

$$\frac{d\phi}{dt} = \frac{1}{\rho}\frac{ds}{dt} = \frac{V_{P_3/2}}{\rho}. \tag{c}$$

This is the angular rate at which the radius of curvature ρ (and also the unit vectors $\hat{\mathbf{u}}^t$ and $\hat{\mathbf{u}}^n$) appears to rotate as seen by an observer in coordinate system 2 as point P_3 moves along its path. We can give this rotational speed its proper vector properties as an apparent angular velocity by noting that the axis of this rotation is parallel to $\hat{\mathbf{u}}^b$. Thus, we define the apparent angular velocity vector as

$$\dot{\boldsymbol{\phi}} = \frac{d\phi}{dt}\hat{\mathbf{u}}^b = \frac{V_{P_3/2}}{\rho}\hat{\mathbf{u}}^b. \tag{d}$$

Next, we seek to find the time derivative of the unit tangent vector $\hat{\mathbf{u}}^t$ so that we can differentiate Eq. (*a*). Because $\hat{\mathbf{u}}^t$ is a unit vector, its length does not change; however, it does have a derivative because of its change in direction, its rotation. In the absolute coordinate system, $\hat{\mathbf{u}}^t$ is subject to the apparent angular velocity $\dot{\boldsymbol{\phi}}$ and also to the angular velocity $\boldsymbol{\omega}$, with which the moving coordinate system 2 is rotating. Therefore, the time derivative of $\hat{\mathbf{u}}^t$ is written as

$$\frac{d\hat{\mathbf{u}}^t}{dt} = (\boldsymbol{\omega} + \dot{\boldsymbol{\phi}})\times\hat{\mathbf{u}}^t = \boldsymbol{\omega}\times\hat{\mathbf{u}}^t + \dot{\boldsymbol{\phi}}\times\hat{\mathbf{u}}^t. \tag{e}$$

Substituting Eq. (*d*) into this equation gives

$$\frac{d\hat{\mathbf{u}}^t}{dt} = \boldsymbol{\omega}\times\hat{\mathbf{u}}^t + \frac{V_{P_3/2}}{\rho}\hat{\mathbf{u}}^b\times\hat{\mathbf{u}}^t = \boldsymbol{\omega}\times\hat{\mathbf{u}}^t + \frac{V_{P_3/2}}{\rho}\hat{\mathbf{u}}^n. \tag{f}$$

Similarly, the time derivative of $\hat{\mathbf{u}}^n$ is written as

$$\begin{aligned}\frac{d\hat{\mathbf{u}}^n}{dt} &= (\boldsymbol{\omega} + \dot{\boldsymbol{\phi}})\times\hat{\mathbf{u}}^n = \boldsymbol{\omega}\times\hat{\mathbf{u}}^n + \dot{\boldsymbol{\phi}}\times\hat{\mathbf{u}}^n \\ &= \boldsymbol{\omega}\times\hat{\mathbf{u}}^n + \frac{V_{P_3/2}}{\rho}\hat{\mathbf{u}}^b\times\hat{\mathbf{u}}^n = \boldsymbol{\omega}\times\hat{\mathbf{u}}^n - \frac{V_{P_3/2}}{\rho}\hat{\mathbf{u}}^t.\end{aligned} \tag{g}$$

Now, taking the time derivative of Eq. (*a*) and using Eq. (*f*), we find that

$$\frac{d\mathbf{V}_{P_3/2}}{dt} = \frac{ds}{dt}\frac{d\hat{\mathbf{u}}^t}{dt} + \frac{d^2s}{dt^2}\hat{\mathbf{u}}^t = \frac{ds}{dt}\hat{\boldsymbol{\omega}} \times \hat{\mathbf{u}}^t + \frac{ds}{dt}\frac{V_{P_3/2}}{\rho}\hat{\mathbf{u}}^n + \frac{d^2s}{dt^2}\hat{\mathbf{u}}^t.$$

Finally, using Eqs. (*a*) and (*c*), this equation reduces to

$$\frac{d\mathbf{V}_{P_3/2}}{dt} = \boldsymbol{\omega}\times\mathbf{V}_{P_3/2} + \frac{\mathbf{V}^2_{P_3/2}}{\rho}\hat{\mathbf{u}}^n + \frac{d^2s}{dt^2}\hat{\mathbf{u}}^t. \qquad (h)$$

Note that the three terms on the right-hand side of Eq. (*h*) are *not* all defined as *apparent-acceleration* components. To be consistent, the term apparent acceleration should include only those components *that would be seen by an observer attached to the moving coordinate system*. Equation (*h*) is derived in the absolute coordinate system and includes the rotational effect of $\boldsymbol{\omega}$ that would not be sensed by the moving observer. The apparent acceleration, which is given the notation $\mathbf{A}_{P_3/2}$, can easily be determined, however, by setting $\boldsymbol{\omega}$ to zero in this equation. The two remaining components can then be written as

$$\mathbf{A}_{P_3/2} = \mathbf{A}^n_{P_3/2} + \mathbf{A}^t_{P_3/2}, \qquad (4.13)$$

where

$$\mathbf{A}^n_{P_3/2} = \frac{V^2_{P_3/2}}{\rho}\hat{\mathbf{u}}^n \qquad (4.14)$$

is called the *normal component*, indicating that it is always normal to the path and is always directed from point P toward the center of curvature (that is, in the $\hat{\mathbf{u}}^n$ direction when ρ is positive or in the $-\hat{\mathbf{u}}^n$ direction when ρ is negative), whereas

$$\mathbf{A}^t_{P_3/2} = \frac{d^2s}{dt^2}\hat{\mathbf{u}}^t \qquad (4.15)$$

is called the *tangential component*, indicating that it is always tangent to the path (along the $\hat{\mathbf{u}}^t$ direction, but it may be either positive or negative).

Next we write the position equation from Fig. 4.14 as

$$\mathbf{R}_{P_3} = \mathbf{R}_{C_2} - \rho\hat{\mathbf{u}}^n$$

and, with the help of Eq. (*g*), its time derivative can be written as*

$$\mathbf{V}_{P_3} = \mathbf{V}_{C_2} - \boldsymbol{\omega}\times\left(\rho\hat{\mathbf{u}}^n\right) + \rho\frac{V_{P_3/2}}{\rho}\hat{\mathbf{u}}^t = \mathbf{V}_{C_2} - \boldsymbol{\omega}\times\left(\rho\hat{\mathbf{u}}^n\right) + \mathbf{V}_{P_3/2}. \qquad (i)$$

* The first two terms on the right of Eq. (*i*) are equal to $\mathbf{V}_{P_2}$; thus, this is equivalent to the apparent-velocity equation. Note, however, that although $\rho\hat{\mathbf{u}}^n = \mathbf{R}_{C_2P_2}$ is true instantaneously, their derivatives are not equal; they do not rotate at the same rate. Thus, some of the terms of the next equation might be missed if the apparent-velocity equation were differentiated instead.

Then, differentiating this equation, again with respect to time, gives

$$\mathbf{A}_{P_3} = \mathbf{A}_{C_2} - \boldsymbol{\alpha}\times\left(\rho\hat{\mathbf{u}}^n\right) - \boldsymbol{\omega}\times\frac{d\left(\rho\hat{\mathbf{u}}^n\right)}{dt} + \frac{d\mathbf{V}_{P_3/2}}{dt}$$

and, with the help of Eqs. (g) and (h), this becomes

$$\mathbf{A}_{P_3} = \mathbf{A}_{C_2} + \boldsymbol{\alpha}\times\left(-\rho\hat{\mathbf{u}}^n\right) + \boldsymbol{\omega}\times\left[\boldsymbol{\omega}\times\left(-\rho\hat{\mathbf{u}}^n\right)\right] + 2\boldsymbol{\omega}\times\mathbf{V}_{P_3/2} + \frac{V_{P_3/2}^2}{\rho}\hat{\mathbf{u}}^n + \frac{d^2s}{dt^2}\hat{\mathbf{u}}^t. \quad (j)$$

The first three terms on the right-hand side of this equation are the components of $\mathbf{A}_{P_2}$ [see Eq. (4.5)], and the last two terms are the components of the apparent acceleration $\mathbf{A}_{P_3/2}$ [see Eqs. (4.13)–(4.15)]. Therefore, we define the following symbol for the fourth term,

$$\mathbf{A}_{P_3P_2}^c = 2\boldsymbol{\omega}\times\mathbf{V}_{P_3/2}. \tag{4.16}$$

This term is called the *Coriolis component of acceleration*. We note that it is one term of the total apparent acceleration equation. However, unlike the components of $\mathbf{A}_{P_3/2}$, it is not sensed by an observer attached to moving coordinate system 2. Still, it is a necessary term in Eq. (j) and it is a part of the difference between $\mathbf{A}_{P_3}$ and $\mathbf{A}_{P_2}$ sensed by an absolute observer.

With the definition of Eq. (4.16), Eq. (j) can be written in the following form, called the *apparent-acceleration equation*,

$$\mathbf{A}_{P_3} = \mathbf{A}_{P_2} + \mathbf{A}_{P_3P_2}^c + \mathbf{A}_{P_3/2}^n + \mathbf{A}_{P_3/2}^t, \tag{4.17}$$

where the definitions of the individual components are given in Eqs. (4.14) through (4.16).

It is extremely important to recognize certain features of Eq. (4.17).

1. It serves the objectives of this section because it relates the accelerations of two *coincident points on different links* in a meaningful way.
2. There is only *one unknown* among the three new components defined. The normal component and the Coriolis component can be calculated from Eqs. (4.14) and (4.16) from velocity information; they do not contribute any new unknowns. The tangential component $\mathbf{A}_{P_3/2}^t$, given by Eq. (4.15), however, will almost always have an unknown magnitude in application, because d^2s/dt^2 will not be known.
3. It is important in each application to note the dependence of Eq. (4.17) on the ability to recognize the point path that P_3 traces on coordinate system 2. This path is the basis for the axes of the normal and tangential components and is also necessary for the determination of ρ for Eq. (4.14).

Finally, a word of warning: *The path described by P_3 on link 2 is not necessarily the same as the path described by P_2 on link 3.* In Fig. 4.14, the path of P_3 on link 2 is clear; it is the curved slot. The path of P_2 on link 3 is not at all clear. As a result, there is a natural "right" and "wrong" way to write the apparent-acceleration equation for this situation. The equation

$$\mathbf{A}_{P_2} = \mathbf{A}_{P_3} + \mathbf{A}_{P_2P_3}^c + \mathbf{A}_{P_2/3}^n + \mathbf{A}_{P_2/3}^t$$

is a perfectly valid equation, but it is *useless* because the path and ρ are not known for the normal component. Note also that $\mathbf{A}^c_{P_3P_2}$ makes use of $\boldsymbol{\omega}_2$, whereas $\mathbf{A}^c_{P_2P_3}$ would make use of $\boldsymbol{\omega}_3$. *We must be extremely careful to write the appropriate equation for each application, recognizing which path is known.*

EXAMPLE 4.4

For the mechanism in the position illustrated in Fig. 4.15*a*, block 3 is sliding outward on link 2 at a uniform rate of 30 m/s, while link 2 is rotating at a constant angular velocity of 50 rad/s ccw. Determine the acceleration of point A of the sliding block.

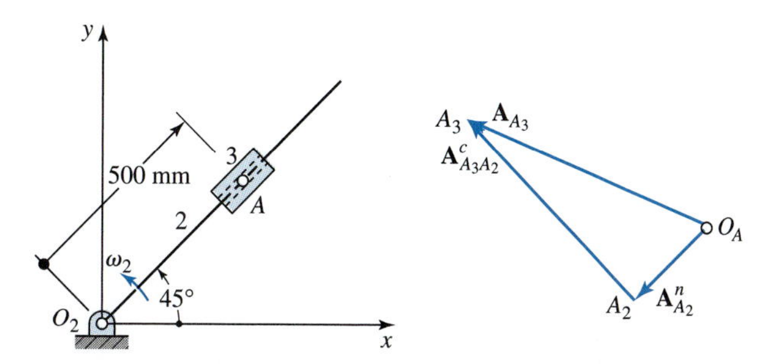

Figure 4.15 Example 4.4.

SOLUTION

We first calculate the acceleration of the coincident point A_2, immediately under the block but attached to link 2:

$$\mathbf{A}_{A_2} = \overset{0}{\cancel{\mathbf{A}_{O_2}}} + \mathbf{A}^n_{A_2O_2} + \overset{0}{\cancel{\mathbf{A}^t_{A_2O_2}}}$$

$$A^n_{A_2O_2} = \omega_2^2 R_{A_2O_2} = (50\text{ rad/s})^2(0.500\text{ m}) = 1\,250\text{ m/s}^2.$$

We draw this to scale, determining the acceleration-image point A_2. Next, we recognize that point A_3 is constrained to travel only along the axis of link 2. This provides a path for which we can write the apparent-acceleration equation,

$$\mathbf{A}_{A_3} = \mathbf{A}_{A_2} + \mathbf{A}^c_{A_3A_2} + \mathbf{A}^n_{A_3/2} + \mathbf{A}^t_{A_3/2}.$$

The terms for this equation are computed as follows and graphically added into the acceleration polygon:

$$A^c_{A_3A_2} = 2\omega_2 V_{A_3/2} = 2(50\text{ rad/s})(30\text{ m/s}) = 3\,000\text{ m/s}^2$$

$$A^n_{A_3/2} = \frac{V^2_{A_3/2}}{\rho} = \frac{(30\text{ m/s})^2}{\infty} = 0$$

$$A^t_{A_3/2} = \frac{d^2s}{dt^2} = 0\text{ (uniform rate along path).}$$

Noting that the sense of $\mathbf{A}^c_{A_3A_2}$ comes from the sense of the vector cross-product of Eq. (4.16), this locates the acceleration-image point A_3, and the result is

$$A_{A_3} = 3\,250 \text{ m/s}^2. \qquad \textit{Ans.}$$

EXAMPLE 4.5

Perform an acceleration analysis of the linkage in the position illustrated in Fig. 4.16 for a constant input speed $\omega_2 = 18$ rad/s cw.

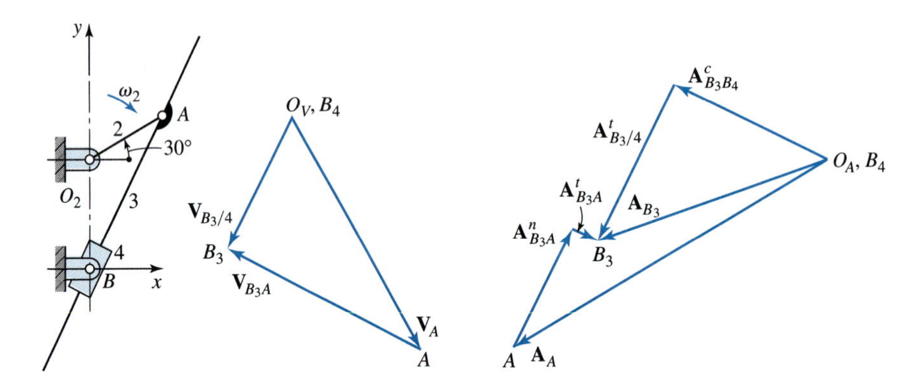

Figure 4.16 Example 4.5; $R_{AO_2} = 8$ in, $R_{BO_2} = 10$ in.

SOLUTION

A complete velocity analysis is performed first, as illustrated in Fig. 4.16. This yields

$$V_A = 12.0 \text{ ft/s}, \; V_{B_3A} = 10.0 \text{ ft/s, and } V_{B_3/4} = 6.7 \text{ ft/s}$$

$$\omega_3 = \omega_4 = 7.67 \text{ rad/s cw.}$$

To solve for accelerations we first find

$$\mathbf{A}_A = \overset{\mathbf{0}}{\cancel{\mathbf{A}_{O_2}}} + \mathbf{A}^n_{AO_2} + \overset{\mathbf{0}}{\cancel{\mathbf{A}^t_{AO_2}}}$$

$$A^n_{AO_2} = \omega_2^2 R_{AO_2} = (18 \text{ rad/s})^2 (8/12 \text{ ft}) = 216.0 \text{ ft/s}^2 \qquad \textit{Ans.}$$

and plot this to locate the acceleration image point A. Next we write the acceleration-difference equation,

$$\overset{??}{\mathbf{A}_{B_3}} = \overset{\surd\surd}{\mathbf{A}_A} + \overset{\surd\surd}{\mathbf{A}^n_{B_3A}} + \overset{?\surd}{\mathbf{A}^t_{B_3A}} \tag{1}$$

$$A^n_{B_3A} = \frac{V^2_{B_3A}}{R_{BA}} = \frac{(10.0 \text{ ft/s})^2}{(15.6/12) \text{ ft}} = 76.9 \text{ ft/s}^2.$$

The term $\mathbf{A}^n_{B_3A}$ is directed from B toward A and is added to the acceleration polygon as illustrated in Fig. 4.16. The term $\mathbf{A}^t_{B_3A}$ has an unknown magnitude and sense but is perpendicular to $\mathbf{R}_{BA}$.

Because Eq. (1) has three unknowns, it cannot be solved by itself. Therefore, we seek a second equation for $\mathbf{A}_{B_3}$. Consider the view of an observer located on link 4; SHE would see point B_3 moving on a straight-line path along the centerline of the block. Using this path, we can write the apparent-acceleration equation

$$\mathbf{A}_{B_3} = \overset{0}{\cancel{\mathbf{A}_{B_4}}} + \mathbf{A}^c_{B_3B_4} + \mathbf{A}^n_{B_3/4} + \mathbf{A}^t_{B_3/4}. \tag{2}$$

Because point B_4 is pinned to the ground link, it has zero acceleration. The other components of Eq. (2) are

$$A^c_{B_3B_4} = 2\omega_4 V_{B_3/4} = 2\,(7.67\text{ rad/s})\,(6.7\text{ ft/s}) = 103\text{ ft/s}^2$$

$$A^n_{B_3/4} = \frac{V^2_{B_3/4}}{\rho} = \frac{(6.5\text{ ft/s})^2}{\infty} = 0.$$

The Coriolis component is added to the acceleration polygon, originating at point B_4 (O_A) as indicated. Finally, $\mathbf{A}^t_{B_3/4}$, whose magnitude and sense are unknown, is graphically added in the direction defined by the path tangent. It crosses the unknown-length line of $\mathbf{A}^t_{B_3A}$, Eq. (1), locating the acceleration-image point B_3. When the polygon is scaled, the results are found to be

$$A^t_{B_3/4} = 103\text{ ft/s}^2,\ A^t_{B_3A} = 17\text{ ft/s}^2,\ A_{B_3} = 145\text{ ft/s}^2. \qquad \textit{Ans.}$$

The angular accelerations of links 3 and 4 are

$$\alpha_4 = \alpha_3 = \frac{A^t_{B_3A}}{R_{BA}} = \frac{17\text{ ft/s}^2}{15.6/12\text{ ft}} = 13.1\text{ rad/s}^2\text{ ccw}. \qquad \textit{Ans.}$$

We note that, in this example, the path of B_3 on link 4 and the path of B_4 on link 3 can both be visualized and either could have been used in deciding the approach. However, although B_4 is pinned to the ground (link 1), the path of point B_3 on link 1 is not known. Therefore, the term $\mathbf{A}^n_{B_3/1}$ cannot be calculated directly.

EXAMPLE 4.6

The velocity analysis of the inverted slider-crank mechanism in the position illustrated in Fig. 4.17 was performed in Example 3.3 (Fig. 3.11). Determine the angular acceleration of link 4 assuming that the angular velocity of link 2 is a constant $\omega_2 = 36$ rad/s cw.

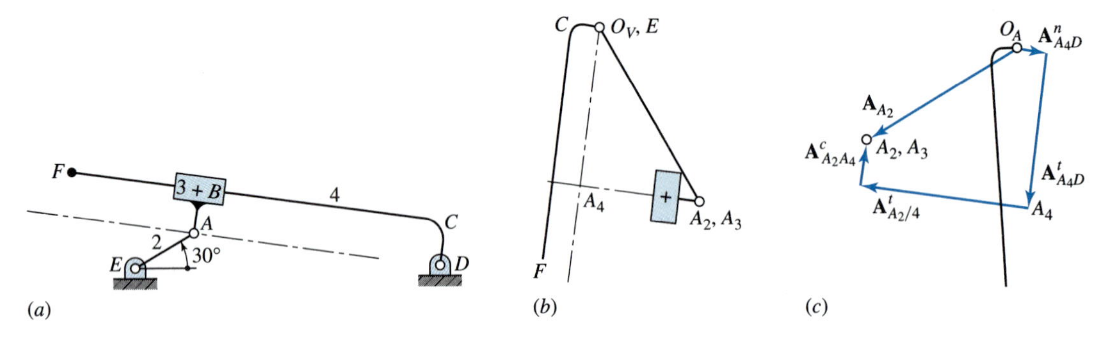

Figure 4.17 Example 4.6.

SOLUTION

From Example 3.3 for the given angular velocity $\omega_2 = 36$ rad/s cw, the answers for the velocity analysis are

$$V_{A_2} = 9 \text{ ft/s}, \quad V_{A_4D} = 7.17 \text{ ft/s}, \quad V_{A_2/4} = 5.48 \text{ ft/s} \quad \text{and} \quad \omega_3 = \omega_4 = 7.55 \text{ rad/s ccw}.$$

To analyze for accelerations, we start by writing

$$\mathbf{A}_{A_2} = \overset{0}{\cancel{\mathbf{A}_E}} + \mathbf{A}^n_{AE} + \overset{0}{\cancel{\mathbf{A}^t_{AE}}}$$

$$A^n_{AE} = \omega_2^2 R_{AE} = (36 \text{ rad/s})^2 \left(3/12 \text{ ft}\right) = 324 \text{ ft/s}^2$$

and plot this as illustrated in Fig. 4.17*c*.

Next, we note that point A_2 travels along the straight-line path indicated, as seen by an observer on link 4. Knowing this path, we write

$$\overset{\surd\surd}{\mathbf{A}_{A_2}} = \overset{?\surd}{\mathbf{A}_{A_4}} + \overset{\surd\surd}{\mathbf{A}^c_{A_2A_4}} + \overset{0}{\cancel{\mathbf{A}^n_{A_2/4}}} + \overset{?\surd}{\mathbf{A}^t_{A_2/4}} \tag{3}$$

$$A^c_{A_2A_4} = 2\omega_4 V_{A_2/4} = 2\,(7.55 \text{ rad/s})\,(5.48 \text{ ft/s}) = 82.7 \text{ ft/s}^2 \tag{4}$$

and $\mathbf{A}^n_{A_2/4} = 0$ because $\rho = \infty$. In Eq. (3), the acceleration $\mathbf{A}_{A_4}$ was marked as having only one unknown because

$$\mathbf{A}_{A_4} = \overset{0}{\cancel{\mathbf{A}_D}} + \overset{\surd\surd}{\mathbf{A}^n_{A_4D}} + \overset{?\surd}{\mathbf{A}^t_{A_4D}}$$

$$A^n_{A_4D} = \frac{V^2_{A_4D}}{R_{A_4D}} = \frac{(7.17 \text{ ft/s})^2}{11.5/12 \text{ ft}} = 53.6 \text{ ft/s}^2. \tag{5}$$

The term $\mathbf{A}^n_{A_4D}$ is added from O_A followed by a line of unknown length for $\mathbf{A}^t_{A_4D}$. Because the image point A_4 is not yet known, the terms $\mathbf{A}^c_{A_2A_4}$ and $\mathbf{A}^t_{A_2/4}$ cannot be added as directed by Eq. (3). However, these two terms can be transferred to the other side of Eq. (3) and graphically subtracted from image point A_2, thus completing the acceleration polygon. The angular acceleration of link 4 can now be found:

$$\alpha_4 = \frac{A^t_{A_4D}}{R_{AD}} = \frac{279.9\ \text{ft/s}^2}{11.5/12\ \text{ft}} = 292\ \text{rad/s}^2\ \text{ccw}. \qquad \textit{Ans.}$$

This need to subtract vectors is common in acceleration problems involving the Coriolis component and should be studied carefully. Note that the opposite equation involving $\mathbf{A}_{A_4/2}$ cannot be used, because ρ, and therefore $A^n_{A_4/2}$, would become an additional (third) unknown.

EXAMPLE 4.7

The angular velocity and angular acceleration of the direct contact circular cam, link 2, with the oscillating flat-face follower, link 3, at the position illustrated in Fig. 4.18a are

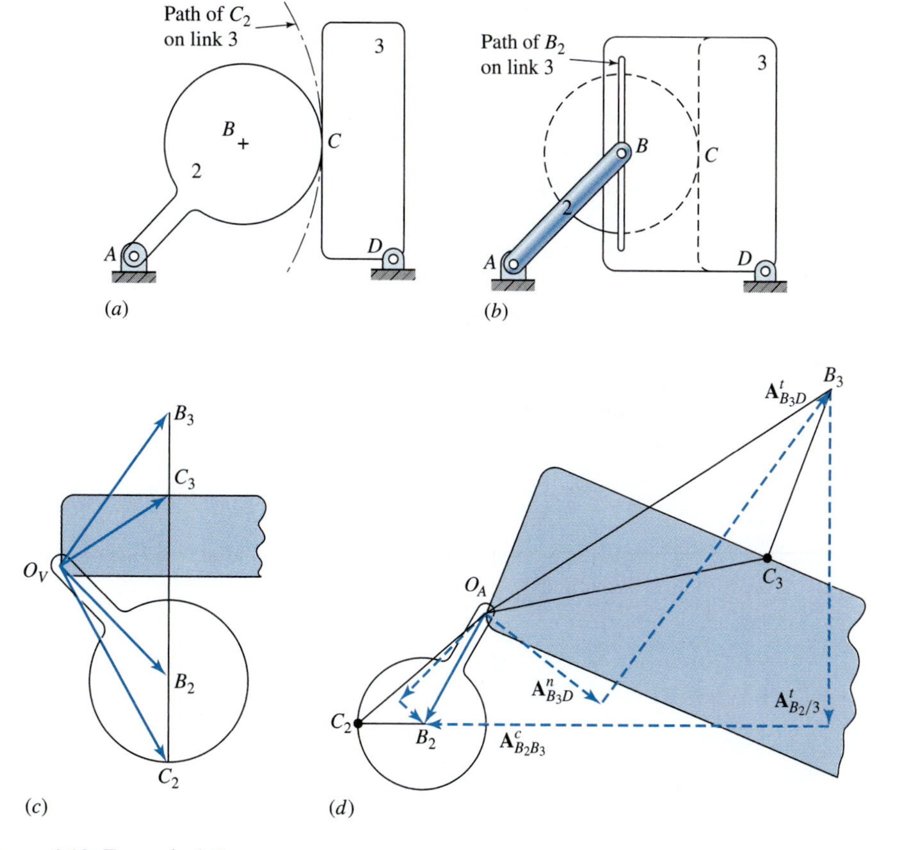

Figure 4.18 Example 4.7.

$\omega_2 = 10$ rad/s cw and $\alpha_2 = 25$ rad/s^2 cw, respectively. Determine the angular acceleration of the follower for this position.

SOLUTION

We are given the scaled drawing of the cam-and-follower system illustrated in Fig. 4.18*a* and the velocity polygon illustrated in Fig. 4.18*c*.

We begin the acceleration polygon (see Fig. 4.18*d*) by calculating and plotting the acceleration of point B_2 relative to point A:

$$\mathbf{A}_{B_2} = \overset{0}{\cancel{\mathbf{A}_A}} + \mathbf{A}^n_{B_2A} + \mathbf{A}^t_{B_2A}$$

$$A^n_{B_2A} = \omega_2^2 R_{B_2A} = (10 \text{ rad/s})^2 \,(3 \text{ in}) = 300 \text{ in/s}^2$$

$$A^t_{B_2A} = \alpha_2 R_{B_2A} = (25 \text{ rad/s}^2)(3 \text{ in}) = 75 \text{ in/s}^2.$$

These are plotted as illustrated in Fig. 4.18*d*, and then the acceleration image-point C_2 is determined by constructing the acceleration image of the triangle AB_2C_2.

Proceeding as in Section 3.7, we would next like to find the velocity of point C_3. If we attempt this approach to acceleration analysis, however, the equation is

$$\mathbf{A}_{C_2} = \mathbf{A}_{C_3} + \mathbf{A}^c_{C_2C_3} + \mathbf{A}^n_{C_2/3} + \mathbf{A}^t_{C_2/3}$$

and is based on the path that point C_2 traces on link 3, sketched approximately in Fig. 4.18*a*. However, this approach is useless here because the radius of curvature of this path is not known and, therefore, the magnitude of $A^n_{C_2/3}$ cannot be calculated. To avoid this problem, we take a different approach; we look for another pair of coincident points where the curvature of the path is known.

If we consider the path traced by point B_2 on the (extended) link 3, we see that it remains a constant distance from the surface; this path is a straight line. Our new approach is better visualized if we consider the mechanism of Fig. 4.18*b* and note that it has motion equivalent to the original mechanism. Having thus visualized a "slot" in the equivalent mechanism that can be used for a path, it becomes clear how to proceed; the appropriate equation is

$$\mathbf{A}_{B_2} = \mathbf{A}_{B_3} + \mathbf{A}^c_{B_2B_3} + \mathbf{A}^n_{B_2/3} + \mathbf{A}^t_{B_2/3}, \tag{1}$$

where B_3 is a point coincident with B_2 but attached to link 3. Because the path is a straight line, we obtain

$$A^n_{B_2/3} = \frac{V^2_{B_2/3}}{\rho} = \frac{(50 \text{ in/s})^2}{\infty} = 0$$

$$A^c_{B_2B_3} = 2\omega_3 V_{B_2/3} = 2\,(10 \text{ rad/s})\,(50 \text{ in/s}) = 1\,000 \text{ in/s}^2.$$

The acceleration of B_3 can be determined from

$$\mathbf{A}_{B_3} = \overset{0}{\cancel{\mathbf{A}_D}} + \mathbf{A}^n_{B_3D} + \mathbf{A}^t_{B_3D}, \tag{2}$$

where

$$A^n_{B_3D} = \frac{V^2_{B_3D}}{R_{B_3D}} = \frac{(35.4 \text{ in/s})^2}{3.58 \text{ in}} = 350 \text{ in/s}^2.$$

Substituting Eq. (2) into Eq. (1) and rearranging terms, we arrive at an equation with only two unknowns:

$$\overset{\surd\surd}{\mathbf{A}_{B_2}} - \overset{\surd\surd}{A^c_{B_2B_3}} - \overset{?\surd}{\mathbf{A}^t_{B_2/3}} = \overset{\surd\surd}{\mathbf{A}^n_{B_3D}} + \overset{?\surd}{\mathbf{A}^t_{B_3D}}. \tag{3}$$

This equation is solved graphically as illustrated in Fig. 4.18*d*. Once image point B_3 has been determined, image point C_3 is determined by constructing the acceleration image of triangle DB_3C_3, all on link 3. Fig. 4.18*d* has been extended to illustrate the acceleration images of links 2 and 3 to aid in visualization and illustrate once again that there is no obvious relation between the final locations of image points C_2 and C_3, as suggested by Eq. (2).

Finally, the angular acceleration of link 3 is determined as follows:

$$\alpha_3 = \frac{A^t_{B_3D}}{R_{B_3D}} = \frac{938 \text{ in/s}^2}{3.58 \text{ in}} = 262 \text{ rad/s}^2 \text{ cw.} \qquad \textit{Ans.}$$

4.6 APPARENT ANGULAR ACCELERATION

Although it is seldom useful, completeness suggests that we should also define the term *apparent angular acceleration*. When two rigid bodies rotate with different angular accelerations, the vector difference between them is defined as the apparent-angular acceleration,

$$\boldsymbol{\alpha}_{3/2} = \boldsymbol{\alpha}_3 - \boldsymbol{\alpha}_2.$$

The apparent-angular-acceleration equation can also be written as

$$\boldsymbol{\alpha}_3 = \boldsymbol{\alpha}_2 + \boldsymbol{\alpha}_{3/2}. \tag{4.18}$$

We can recognize that $\boldsymbol{\alpha}_{3/2}$ is the angular acceleration of body 3 as it would appear to an observer attached to, and rotating with, body 2.

4.7 DIRECT CONTACT AND ROLLING CONTACT

We recall from Section 3.7 that the relative motion between two bodies in direct contact at a point can be either of two different kinds; there may be an apparent slipping velocity between the bodies or there may be no such slip. We defined the term *rolling contact* to imply that no slip is in progress and developed the rolling contact condition, Eq. (3.13), to indicate that the apparent velocity at such a point is zero. Here we will investigate the apparent acceleration at a point of rolling contact.

Consider the case of circular wheel 3 in rolling contact with fixed straight link 2, as illustrated in Fig. 4.19. Although this is admittedly a very simplified case, the arguments

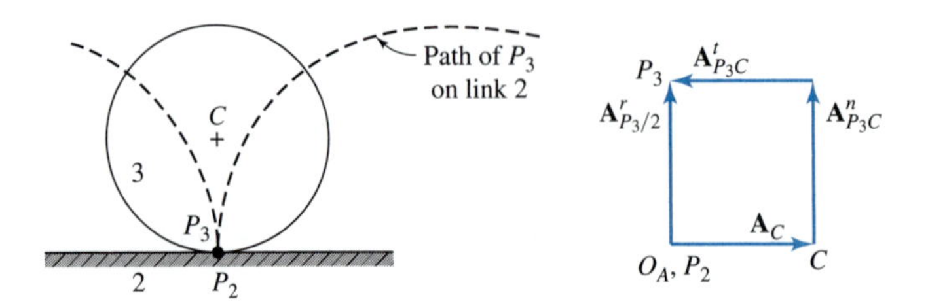

Figure 4.19 Rolling contact acceleration.

made and the conclusions reached are completely general and apply to any rolling contact situation, no matter what the shapes of the two bodies or whether either is the ground link. To keep this clear in our minds, the ground link has been numbered 2 for this example.

Once the acceleration A_C of the center point of the wheel is given, the pole O_A can be chosen and the acceleration polygon can be started by plotting A_C. In relating the accelerations of points P_3 and P_2 at the rolling contact point, however, we are dealing with two coincident points of different bodies. Therefore, it is appropriate to use the apparent acceleration equation, Eq. (4.17). To do this we must identify a path that one of these points traces on the other body. The path that point P_3 traces on link 2 is sketched in Fig. 4.19.* Although the precise shape of this path depends on the shapes of the two contacting links, provided that there is no slip, there is always a cusp at the point of rolling contact and the tangent to this cusp-shape path is always normal to the surfaces that are in contact.

Because this path is known, we are free to write the apparent acceleration equation:

$$\mathbf{A}_{P_3} = \mathbf{A}_{P_2} + \mathbf{A}^c_{P_3P_2} + \mathbf{A}^n_{P_3/2} + \mathbf{A}^t_{P_3/2}.$$

In evaluating the components, we keep in mind the rolling contact velocity condition, that is, $\mathbf{V}_{P_3/2} = \mathbf{0}$. Therefore,

$$\mathbf{A}^c_{P_3P_2} = 2\omega_2 \times \mathbf{V}_{P_3/2} = \mathbf{0} \quad \text{and} \quad A^n_{P_3/2} = \frac{V^2_{P_3/2}}{\rho} = 0.$$

Thus, only one component of the apparent acceleration equation, $\mathbf{A}^t_{P_3/2}$, may be nonzero.

Because of possible confusion in calling this nonzero term a tangential component (tangent to the cusp-shape path) while its direction is *normal* to the rolling surfaces, we will adopt a new superscript and refer to it as the *rolling contact acceleration* $\mathbf{A}^r_{P_3/2}$. Thus, for rolling contact with no slip, the apparent acceleration equation becomes

$$\mathbf{A}_{P_3} = \mathbf{A}_{P_2} + \mathbf{A}^r_{P_3/2} \tag{4.19}$$

and the term $\mathbf{A}^r_{P_3/2}$ is known to always have its line of action normal to the surfaces at the point of rolling contact.

---

* This particular curve is called a *cycloid*.

EXAMPLE 4.8

Consider circular roller 4 rolling without slip on oscillating flat-face follower 3 as illustrated in Fig. 4.20. The angular velocity and angular acceleration of input link 2 are $\omega_2 = 10$ rad/s cw and $\alpha_2 = 25$ rad/s^2 cw, respectively. Given the scale drawing and the velocity analysis of the mechanism, determine the angular accelerations of the roller and the follower at the instant indicated.

SOLUTION

Noting the similarity between this problem and that of Example 4.7, not only in their geometries, but also in their input motions, allows a quick completion of the velocity

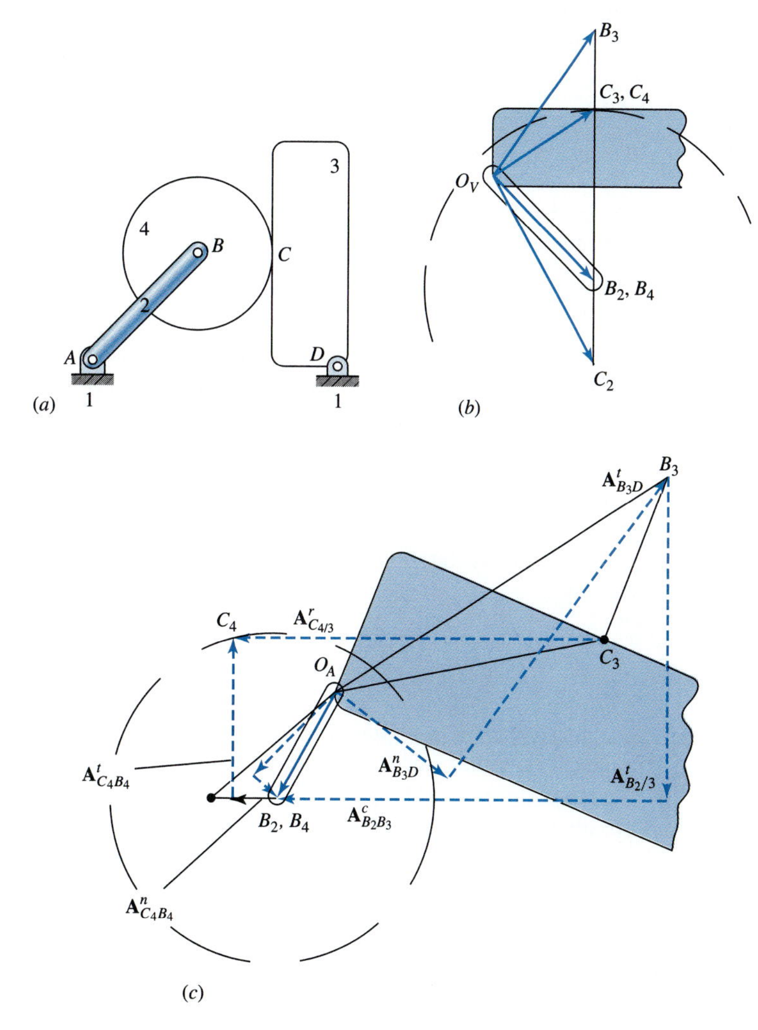

 **Figure 4.20** Example 4.8.

polygon illustrated in Fig. 4.20*b*. The rolling contact condition for velocity implies that $\mathbf{V}_{C_4} = \mathbf{V}_{C_3}$ which allows drawing the velocity image of link 4 as indicated.

Next, we might be tempted to write the following equation for the acceleration of point C_4:

$$\mathbf{A}_{C_4} = \overset{\surd\surd}{\mathbf{A}^n_{B_2A_2}} + \overset{\surd\surd}{\mathbf{A}^t_{B_2A_2}} + \overset{\surd\surd}{\mathbf{A}^n_{C_4B_4}} + \overset{?\surd}{\mathbf{A}^t_{C_4B_4}} = \overset{\surd\surd}{\mathbf{A}^n_{C_3D_3}} + \overset{?\surd}{\mathbf{A}^t_{C_3D_3}} + \overset{?\surd}{\mathbf{A}^r_{C_4/3}}. \tag{1}$$

Unfortunately, this equation cannot be solved because it contains three unknowns, namely, the two angular accelerations α_3 and α_4 and the rolling contact acceleration $A^r_{C_4/3}$. A different solution strategy must be determined.

Let us first approach the problem exactly as we did with Example 4.7. We can proceed as we did in Fig. 4.18*d*, all the way to the solution for $\boldsymbol{\alpha}_3$ and to finding $\mathbf{A}_{C_3}$. From this, we will have constructed a good portion of Fig. 4.20*c*.

Now we can relate the acceleration of point C_4 to that of B_4, that is,

$$\mathbf{A}_{C_4} = \mathbf{A}_{B_4} + \mathbf{A}^n_{C_4B_4} + \mathbf{A}^t_{C_4B_4} \tag{2}$$

$$A^n_{C_4B_4} = \frac{V^2_{C_4B_4}}{R_{CB}} = \frac{(14.8 \text{ in/s})^2}{1.50 \text{ in}} = 146 \text{ in/s}^2. \tag{3}$$

We can also write the rolling contact acceleration condition [see Eq. (4.19)] as

$$\mathbf{A}_{C_4} = \mathbf{A}_{C_3} + \mathbf{A}^r_{C_4/3}. \tag{4}$$

Remembering that $\mathbf{A}^r_{C_4/3}$ is perpendicular to the surfaces at C, we can graphically construct the simultaneous solution to Eqs. (2) and (4) as illustrated in Fig. 4.20*c*. Finally, the angular acceleration of the roller 4 can be obtained as follows:

$$\alpha_4 = \frac{A^t_{C_4B_4}}{R_{CB}} = \frac{406 \text{ in/s}^2}{1.50 \text{ in}} = 271 \text{ rad/s}^2 \text{ ccw}. \qquad \textit{Ans.}$$

The angular acceleration of the oscillating follower is identical to that in Example 4.7, namely,

$$\alpha_3 = 262 \text{ rad/s}^2 \text{ cw}. \qquad \textit{Ans.}$$

In general, it is almost always necessary to determine the motions of the two links adjacent on either side of the point of rolling contact (links 2 and 3 in Example 4.8) and then to perform the rolling contact computation working from both sides. It is almost never possible to work straight through the point of contact, as tried in Eq. (1) above. This almost always requires the visualization of an equivalent mechanism and the solution of a Coriolis equation. However, with a little practice, it is usually quite straightforward.

4.8 SYSTEMATIC STRATEGY FOR ACCELERATION ANALYSIS

Review of the preceding sections and example problems will demonstrate that we have now developed sufficient tools for dealing with those situations that normally arise in the

TABLE 4.1 "Relative" Acceleration Equations

Points are	Coincident	Separated
On same body	*Trivial Case:* $\mathbf{A}_P = \mathbf{A}_Q$	*Acceleration difference:* $\mathbf{A}_P = \mathbf{A}_Q + \mathbf{A}^n_{PQ} + \mathbf{A}^t_{PQ}$ $\mathbf{A}^n_{PQ} = \boldsymbol{\omega} \times (\boldsymbol{\omega} \times \mathbf{R}_{PQ})$ $\mathbf{A}^t_{PQ} = \boldsymbol{\alpha} \times \mathbf{R}_{PQ}$
On different bodies	*Apparent Acceleration:* $\mathbf{A}_{P_i} = \mathbf{A}_{P_j} + \mathbf{A}^c_{P_iP_j} + \mathbf{A}^n_{P_i/j} + \mathbf{A}^t_{P_i/j}$ *where path $P_{i/j}$ is known and* $\mathbf{A}^c_{P_iP_j} = 2\boldsymbol{\omega}_j \times \mathbf{V}_{P_i/j}$ $\mathbf{A}^n_{P_i/j} = \frac{V^2_{P_i/j}}{\rho}\hat{\mathbf{u}}^n$ $\mathbf{A}^t_{P_i/j} = \frac{d^2s}{dt^2}\hat{\mathbf{u}}^t$ - *Rolling Contact Acceleration:* $\mathbf{A}_{P_i} = \mathbf{A}_{P_j} + \mathbf{A}^r_{P_i/j}$ *where path $\mathbf{A}^r_{P_i/j}$ is normal to surfaces at point of contact.*	*Too general; use two steps.*

acceleration analysis of rigid-body mechanical systems. It will also be noted that the word "relative" acceleration has been carefully avoided. Instead we note that, as with velocity analysis, whenever the desire for using relative acceleration arises, there are always two points whose accelerations are to be "related"; also, these two points are fixed either to the same rigid body or to two different rigid bodies. Therefore, similar to velocity analysis (see Table 3.1), we can organize all situations into the four cases illustrated in Table 4.1.

In Table 4.1 we see that, when the two points are separated by a distance, only the acceleration-difference equation is appropriate for use, and two points on the *same link* should be used. When it is desirable to switch to a point of another link, then *coincident points* should be chosen and the apparent-acceleration equation should be used. The path of one of these points on the other link is then required.

Even the notation has been made different to continually remind us that these are two totally different situations and the formulae are not interchangeable between the two. We should not try to use an "$\boldsymbol{\alpha} \times \mathbf{R}$" formula when the apparent acceleration is called for; however, if we do try then we will not find an appropriate $\boldsymbol{\alpha}$ or $\mathbf{R}$. Similarly, when using the acceleration difference, there is no question of which $\boldsymbol{\alpha}$ to use because only one link pertains. The two questions that always arise with respect to apparent acceleration are: (*a*) When should we include the Coriolis term and when do we only use normal and tangential components? and (*b*) Which $\boldsymbol{\omega}$ should we use in the Coriolis term? The answer to the first question is relatively straightfoward: Whenever we use the apparent-acceleration equation, the Coriolis term should be included; if it should not be there, such as when the "path" is not rotating, it should be included anyway and the calculation will give an answer of zero.

The answer to the second question is also straightforward: Whenever we use the apparent-acceleration equation, we must always visualize a path that a point P_i makes on another link j; then $\boldsymbol{\omega}_j$ for the link that contains the path is used. As indicated in Section 4.7 and Table 4.1, rolling contact acceleration is a special case of apparent acceleration.

According to Rosenauer and Willis,[1] the theorem of Mehmke can be written as follows:

> The end points of the acceleration vectors of the points of a rigid plane body, when plotted from a common origin, produce a figure which is geometrically similar to the original figure (image diagram).

It is this theorem that results in the acceleration image presented in Section 4.4. Also, it is a result of this theorem that allows clarity in the acceleration polygon despite the very minimal labeling that has been used in the preceding examples and will continue throughout the text.

Careful review of Examples 4.3 through 4.8 demonstrates how the strategy suggested in Table 4.1 and the labeling strategy are applied in a variety of problems of acceleration analysis.

4.9 ANALYTIC METHODS

In this section we shall continue some of the analytic approaches that we began in Chapter 3. One mechanism studied there was the central (or in-line) slider-crank linkage of Fig. 4.21.

The acceleration of the slider can be obtained by differentiating Eq. (*c*) of Section 3.9. The result, after considerable manipulation, is

$$\ddot{x} = -r\omega^2\left(\cos\theta + \frac{r\cos 2\theta}{l\cos\phi} + \frac{r^3\sin^2 2\theta}{4l^3\cos^3\phi}\right) - r\alpha\left(\sin\theta + \frac{r\sin 2\theta}{2l\cos\phi}\right). \tag{4.20}$$

This is an involved expression, and it becomes even more involved when ϕ is eliminated by the substitution:

$$\cos\phi = \sqrt{1 - \left(\frac{r}{l}\sin\theta\right)^2}. \tag{4.21}$$

An approximate expression is sometimes used for the acceleration of the slider (see Chapter 16 for one approach that uses the binomial theorem). For small values of (r/l), $\cos\phi$ is

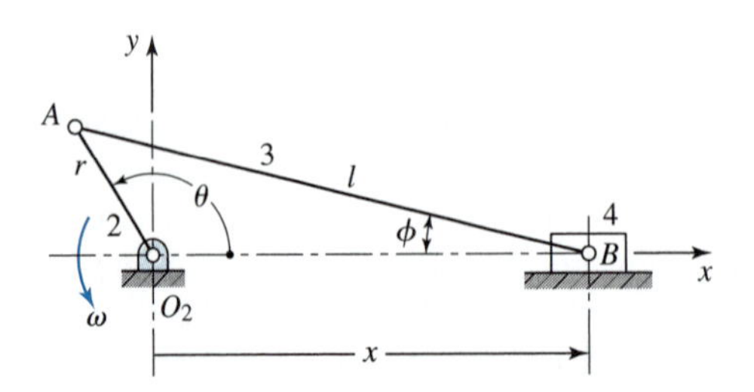

Figure 4.21 Central slider-crank mechanism.

near unity and the last term in the first bracket of Eq. (4.20) can be neglected. Using these approximations, Eq. (4.20) can be written as

$$\ddot{x} = -r\omega^2\left(\cos\theta + \frac{r}{l}\cos 2\theta\right) - r\alpha\left(\sin\theta + \frac{r}{2l}\sin 2\theta\right). \tag{4.22}$$

These expressions are still complex and, if a computer is to be used in the analysis, the use of Raven's method may be easier. In the special case that the angular velocity is constant, that is, $\alpha = 0$, then the acceleration of the slider is

$$\ddot{x} = -r\omega^2\left(\cos\theta + \frac{r}{l}\cos 2\theta\right).$$

4.10 COMPLEX-ALGEBRA METHODS

Let us now see how Raven's method is extended for the analysis of accelerations. The general method is illustrated here for the offset slider-crank linkage illustrated in Fig. 4.22.

In complex polar form the loop-closure equation is

$$r_2e^{j\theta_2} + r_3e^{j\theta_3} - r_1e^{-j(\pi/2)} - r_4e^{j0} = 0. \tag{a}$$

If we separate this equation into its real and imaginary components, we obtain the two position equations,

$$r_2\cos\theta_2 + r_3\cos\theta_3 - r_4 = 0 \tag{b}$$

$$r_2\sin\theta_2 + r_3\sin\theta_3 + r_1 = 0. \tag{c}$$

With r_1, r_2, r_3, and θ_2 given, these two equations can be solved for the two position unknowns. In this example, Eq. (*c*) can be solved for θ_3 and the result is

$$\theta_3 = \sin^{-1}\left(\frac{-r_1 - r_2\sin\theta_2}{r_3}\right). \tag{4.23}$$

Then Eq. (*b*) can be solved for r_4, that is,

$$r_4 = r_2\cos\theta_2 + r_3\cos\theta_3. \tag{4.24}$$

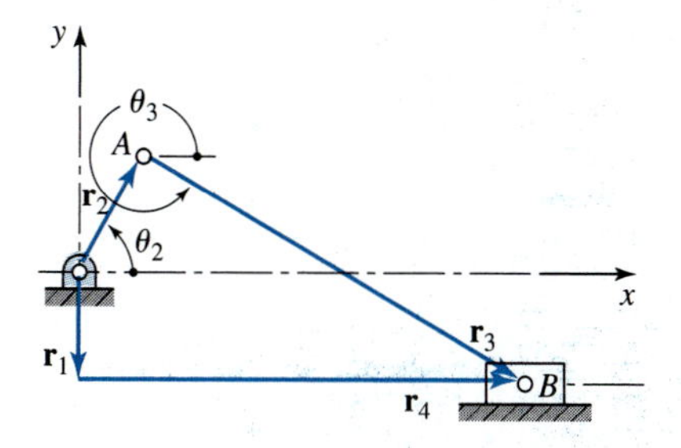

Figure 4.22 Offset slider-crank mechanism.

Differentiating Eq. (*a*) with respect to time gives the velocity equation, which is really the equation of the velocity polygon in exponential form. After replacing $\dot{\theta}$ with ω, the result is

$$jr_2\omega_2 e^{j\theta_2} + jr_3\omega_3 e^{j\theta_3} - \dot{r}_4 e^{j0} = 0. \tag{d}$$

Raven's method, as we have seen, consists of making the trigonometric transformation and separating the result into real and imaginary terms as was performed in obtaining Eqs. (*b*) and (*c*) from Eq. (*a*). When this is carried out for the velocity equation, Eq. (*d*), we find

$$\omega_3 = -\frac{r_2 \cos\theta_2}{r_3 \cos\theta_3}\omega_2 \tag{4.25}$$

and

$$\dot{r}_4 = -r_2 \sin\theta_2\omega_2 - r_3 \sin\theta_3\omega_3. \tag{4.26}$$

We use the same approach to obtain the acceleration equations by first differentiating Eq. (*d*). Then separating the result into its real and imaginary components gives two equations that can be solved for the two acceleration unknowns; namely

$$\alpha_3 = \frac{r_2 \sin\theta_2\omega_2^2 - r_2 \cos\theta_2\alpha_2 + r_3 \sin\theta_3\omega_3^2}{r_3 \cos\theta_3} \tag{4.27}$$

$$\ddot{r}_4 = -r_2 \cos\theta_2\omega_2^2 - r_2 \sin\theta_2\alpha_2 - r_3 \cos\theta_3\omega_3^2 - r_3 \sin\theta_3\alpha_3. \tag{4.28}$$

If the input angular velocity is constant, that is, if $\alpha_2 = 0$, then the acceleration results are

$$\alpha_3 = \frac{r_2 \sin\theta_2\omega_2^2 + r_3 \sin\theta_3\omega_3^2}{r_3 \cos\theta_3} \tag{4.29}$$

$$\ddot{r}_4 = -r_2 \cos\theta_2\omega_2^2 - r_3 \cos\theta_3\omega_3^2 - r_3 \sin\theta_3\alpha_3. \tag{4.30}$$

The same procedure, when carried out for the four-bar linkage, gives results that can be used for computer solutions of both the crank–rocker and the drag–link mechanisms. Unfortunately, they cannot be used for other four-bar linkages unless an arrangement is built into the program to cause the solution to stop when an extreme position is reached.

If the vector loop-closure equation for the four-bar linkage is written as

$$\mathbf{r}_1 + \mathbf{r}_2 + \mathbf{r}_3 - \mathbf{r}_4 = \mathbf{0}, \tag{e}$$

where the subscripts are the link numbers and where link 2 is the driver having a constant input angular velocity, then the acceleration relations that are obtained by Raven's method are

$$\alpha_3 = \frac{r_2 \cos(\theta_2 - \theta_4)\,\omega_2^2 + r_3 \cos(\theta_3 - \theta_4)\,\omega_3^2 - r_4\omega_4^2}{r_3 \sin(\theta_4 - \theta_3)} \tag{4.31}$$

$$\alpha_4 = \frac{r_2 \cos(\theta_2 - \theta_3)\,\omega_2^2 - r_4 \cos(\theta_3 - \theta_4)\,\omega_4^2 + r_3\omega_3^2}{r_4 \sin(\theta_4 - \theta_3)}. \tag{4.32}$$

4.11 THE CHACE SOLUTIONS

Here we employ the four-bar linkage as an example of Chace's method of solution. First, we write the loop-closure equation,

$$\mathbf{r}_1 + \mathbf{r}_2 + \mathbf{r}_3 - \mathbf{r}_4 = \mathbf{0}. \tag{a}$$

Here again, the subscripts designate the link numbers. In using Chace's approach, we first differentiate Eq. (a) to obtain the velocities. Because, r_2, r_3, and r_4 are constants for the four-bar linkage, this yields

$$\boldsymbol{\omega}_2 \times \mathbf{r}_2 + \boldsymbol{\omega}_3 \times \mathbf{r}_3 - \boldsymbol{\omega}_4 \times \mathbf{r}_4 = \mathbf{0}. \tag{b}$$

In this example we are taking link 2 as the driver, ω_2 as a known constant, and $\boldsymbol{\alpha}_2 = \mathbf{0}$. Now, in a planar mechanism all the angular velocities are in the $\hat{\mathbf{k}}$ direction. Therefore, we can write Eq. (b) as

$$r_2\omega_2(\hat{\mathbf{k}} \times \hat{\mathbf{r}}_2) + r_3\omega_3(\hat{\mathbf{k}} \times \hat{\mathbf{r}}_3) - r_4\omega_4(\hat{\mathbf{k}} \times \hat{\mathbf{r}}_4) = \mathbf{0}. \tag{c}$$

To solve this equation, we take the dot product with $\hat{\mathbf{r}}_4$ and then with $\hat{\mathbf{r}}_3$. Each of these operations will reduce Eq. (c) to a scalar equation with only a single unknown. The dot product of Eq. (c) with $\hat{\mathbf{r}}_4$ gives

$$r_2\omega_2(\hat{\mathbf{k}} \times \hat{\mathbf{r}}_2)\cdot\hat{\mathbf{r}}_4 + r_3\omega_3(\hat{\mathbf{k}} \times \hat{\mathbf{r}}_3)\cdot\hat{\mathbf{r}}_4 = 0 \tag{d}$$

because

$$r_4\omega_4(\hat{\mathbf{k}} \times \hat{\mathbf{r}}_4)\cdot\hat{\mathbf{r}}_4 = 0.$$

Solving Eq. (d) for ω_3 then yields

$$\omega_3 = -\frac{r_2}{r_3}\frac{(\hat{\mathbf{k}} \times \hat{\mathbf{r}}_2)\cdot\hat{\mathbf{r}}_4}{(\hat{\mathbf{k}} \times \hat{\mathbf{r}}_3)\cdot\hat{\mathbf{r}}_4}\omega_2. \tag{4.33}$$

We then take the dot product of Eq. (c) with $\hat{\mathbf{r}}_3$ and solving gives

$$\omega_4 = \frac{r_2}{r_4}\frac{(\hat{\mathbf{k}} \times \hat{\mathbf{r}}_2)\cdot\hat{\mathbf{r}}_3}{(\hat{\mathbf{k}} \times \hat{\mathbf{r}}_4)\cdot\hat{\mathbf{r}}_3}\omega_2. \tag{4.34}$$

Next, we differentiate Eq. (b) to get the acceleration relations, remembering that the input angular acceleration $\boldsymbol{\alpha}_2 = \mathbf{0}$. Thus, in vector form, we have

$$\boldsymbol{\omega}_2 \times (\boldsymbol{\omega}_2 \times \mathbf{r}_2) + \boldsymbol{\omega}_3 \times (\boldsymbol{\omega}_3 \times \mathbf{r}_3) + \boldsymbol{\alpha}_3 \times \mathbf{r}_3 - \boldsymbol{\omega}_4 \times (\boldsymbol{\omega}_4 \times \mathbf{r}_4) - \boldsymbol{\alpha}_4 \times \mathbf{r}_4 = \mathbf{0}. \tag{e}$$

The unknowns are α_3 and α_4. For plane motion we can write Eq. (e), as before, in the form

$$-r_2\omega_2^2\hat{\mathbf{r}}_2 - r_3\omega_3^2\hat{\mathbf{r}}_3 + r_3\alpha_3(\hat{\mathbf{k}} \times \hat{\mathbf{r}}_3) + r_4\omega_4^2\hat{\mathbf{r}}_4 - r_4\alpha_4(\hat{\mathbf{k}} \times \hat{\mathbf{r}}_4) = \mathbf{0}. \tag{f}$$

Taking the dot product of this equation with the unit vector $\hat{\mathbf{r}}_3$, we note that

$$r_3\omega_3^2\hat{\mathbf{r}}_3\cdot\hat{\mathbf{r}}_3 = r_3\omega_3^2.$$

Also,

$$r_3\alpha_3(\hat{\mathbf{k}}\times\hat{\mathbf{r}}_3)\cdot\hat{\mathbf{r}}_3 = 0$$

because $\hat{\mathbf{k}}\times\hat{\mathbf{r}}_3$ is perpendicular to $\hat{\mathbf{r}}_3$. Then Eq. (*f*) becomes

$$-r_2\omega_2^2\hat{\mathbf{r}}_2\cdot\hat{\mathbf{r}}_3 - r_3\omega_3^2 + r_4\omega_4^2\hat{\mathbf{r}}_4\cdot\hat{\mathbf{r}}_3 - r_4\alpha_4\left(\hat{\mathbf{k}}\times\hat{\mathbf{r}}_4\right)\cdot\hat{\mathbf{r}}_3 = 0. \tag{g}$$

The only unknown in this scalar equation is α_4. Therefore, rearranging the equation gives

$$\alpha_4 = \frac{r_2\omega_2^2\hat{\mathbf{r}}_2\cdot\hat{\mathbf{r}}_3 + r_3\omega_3^2 - r_4\omega_4^2\hat{\mathbf{r}}_4\cdot\hat{\mathbf{r}}_3}{-r_4\left(\hat{\mathbf{k}}\times\hat{\mathbf{r}}_4\right)\cdot\hat{\mathbf{r}}_3}. \tag{4.35}$$

By taking the dot product of Eq. (*f*) throughout with the unit vector $\hat{\mathbf{r}}_4$ and using a similar procedure, the angular acceleration of link 3 is

$$\alpha_3 = \frac{r_2\omega_2^2\hat{\mathbf{r}}_2\cdot\hat{\mathbf{r}}_4 + r_3\omega_3^2\hat{\mathbf{r}}_3\cdot\hat{\mathbf{r}}_4 - r_4\omega_4^2}{r_3\left(\hat{\mathbf{k}}\times\hat{\mathbf{r}}_3\right)\cdot\hat{\mathbf{r}}_4}. \tag{4.36}$$

You should prove for yourself that Eqs. (4.35) and (4.36) are identical to Eqs. (4.32) and (4.31), respectively, determined using Raven's method.

4.12 THE METHOD OF KINEMATIC COEFFICIENTS

Again, we employ the four-bar linkage as an example of this method of solution. The angular accelerations of links 3 and 4 can be obtained by differentiating Eqs. (3) of Example 3.6 with respect to the input position θ_2, that is,

$$-r_3\cos\theta_3\theta_3'^2 - r_3\sin\theta_3\theta_3'' + r_4\cos\theta_4\theta_4'^2 + r_4\sin\theta_4\theta_4'' = r_2\cos\theta_2 \tag{4.37}$$

$$-r_3\sin\theta_3\theta_3'^2 + r_3\cos\theta_3\theta_3'' + r_4\sin\theta_4\theta_4'^2 - r_4\cos\theta_4\theta_4'' = r_2\sin\theta_2, \tag{4.38}$$

where, by definition,

$$\theta_3'' = \frac{d^2\theta_3}{d\theta_2^2} \text{ and } \theta_4'' = \frac{d^2\theta_4}{d\theta_2^2}$$

and are referred to as the *second-order kinematic coefficients* of links 3 and 4. Writing Eqs. (4.37) and (4.38) in matrix form gives

$$\begin{bmatrix} -r_3\sin\theta_3 & r_4\sin\theta_4 \\ r_3\cos\theta_3 & -r_4\cos\theta_4 \end{bmatrix}\begin{bmatrix}\theta_3''\\ \theta_4''\end{bmatrix} = \begin{bmatrix}B_1\\ B_2\end{bmatrix}. \tag{4.39}$$

where

$$B_1 = r_2 \cos\theta_2 + r_3 \cos\theta_3 \theta_3'^2 - r_4 \cos\theta_4 \theta_4'^2$$
$$B_2 = r_2 \sin\theta_2 + r_3 \sin\theta_3 \theta_3'^2 - r_4 \sin\theta_4 \theta_4'^2. \tag{4.40}$$

and are known from position and velocity analysis. Also, note that the (2×2) coefficient matrix on the left-hand side of Eq. (4.39) is the same as the (2×2) coefficient matrix in the velocity analysis* (see Eq. (5) in Example 3.6). This is not a coincidence. The coefficient matrix is the same for all positional derivatives, that is, first-order, second-order, third-order kinematic coefficients, and so on. Therefore, this matrix provides a built-in check of all higher-order differentiation. The determinant of the coefficient matrix is as given by Eq. (6) in Example 3.6; that is,

$$\Delta = -r_3 r_4 \sin(\theta_4 - \theta_3). \tag{4.41}$$

The second-order kinematic coefficients for links 3 and 4 can be obtained from Eq. (4.38), using Cramer's rule, and the results are

$$\theta_3'' = \frac{B_1 \cos\theta_4 + B_2 \sin\theta_4}{r_3 \sin(\theta_4 - \theta_3)} \text{ and } \theta_4'' = \frac{B_1 \cos\theta_3 + B_2 \sin\theta_3}{r_4 \sin(\theta_4 - \theta_3)}. \tag{4.42}$$

Note that the second-order kinematic coefficients for the four-bar linkage are nondimensional (rad/rad^2). The angular accelerations of links 3 and 4, obtained from the chain rule, are

$$\alpha_3 = \theta_3'' \omega_2^2 + \theta_3' \alpha_2 \quad \text{and} \quad \alpha_4 = \theta_4'' \omega_2^2 + \theta_4' \alpha_2. \tag{4.43}$$

TABLE 4.2 Summary of Second-Order Kinematic Coefficients.

	Variable of interest Angle θ_j (Use symbol θ_j'' for kinematic coefficient regardless of input)	Variable of interest Magnitude r_j (Use symbol r_j'' for kinematic coefficient regardless of input)
Input	$\alpha_j = \theta_j'' \dot{\psi}^2 + \theta_j' \ddot{\psi}$	$\ddot{r}_j = r_j'' \dot{\psi}^2 + r_j' \ddot{\psi}$
ψ = angle θ_i	$\theta_j' = \frac{d\theta_j}{d\psi}$ (dimensionless)	$r_j' = \frac{dr_j}{d\psi}$ (length)
	$\theta_j'' = \frac{d^2\theta_j}{d\psi^2}$ (dimensionless)	$r_j'' = \frac{d^2 r_j}{d\psi^2}$ (length)
Input	$\alpha_j = \theta_j'' \dot{\psi}^2 + \theta_j' \ddot{\psi}$	$\ddot{r}_j = r_j'' \dot{\psi}^2 + r_j' \ddot{\psi}$
ψ = magnitude r_i	$\theta_j' = \frac{d\theta_j}{d\psi}$ (1/length)	$r_j' = \frac{dr_j}{d\psi}$ (dimensionless)
	$\theta_j'' = \frac{d^2\theta_j}{d\psi^2}$ (1/length)^2	$r_j'' = \frac{d^2 r_j}{d\psi^2}$ (1/length)

* This coefficient matrix is called the *Jacobian* of the system.

Table 4.2 summarizes the second-order kinematic coefficients that are related to link j of a planar mechanism (that is, vector $\mathbf{r}_j$) having (i) variable angular displacement θ_j and/or (ii) variable magnitude r_j.

EXAMPLE 4.9

To illustrate the method of kinematic coefficients for determining the angular acceleration of a link and the rectilinear acceleration of a point fixed in a link, we will revisit the four-bar linkage in Example 4.3. Recall that link 2 is driven at a constant angular velocity of 900 rev/min = 94.2 rad/s ccw. The problem is to determine the angular accelerations of links 3 and 4 and the absolute accelerations of points E and F.

SOLUTION

Recall that the first-order kinematic coefficients, from Eq. (9) in Example 3.6, are

$$\theta_3' = 0.271\,8 \text{ rad/rad} \quad \text{and} \quad \theta_4' = 0.526\,5 \text{ rad/rad}. \tag{1}$$

Substituting Eqs. (1) and the known data into Eqs. (4.40) and substituting the results into Eq. (4.42), the second-order kinematic coefficients of links 3 and 4 are

$$\theta_3'' = 0.013\,1 \text{ rad/rad}^2 \quad \text{and} \quad \theta_4'' = -0.203\,0 \text{ rad/rad}^2. \tag{2}$$

Then, substituting Eqs. (1) and (2) and the angular velocity and acceleration of the input into Eqs. (4.43), the angular accelerations of links 3 and 4 are

$$\alpha_3 = 116 \text{ rad/s}^2 \text{ ccw} \quad \text{and} \quad \alpha_4 = 1\,801 \text{ rad/s}^2 \text{ cw.} \qquad \textit{Ans.}$$

These answers are in good agreement with the answers obtained from the graphical analysis of Example 4.3; that is, $\alpha_3 = 113$ rad/s^2 ccw and $\alpha_4 = 1\,822$ rad/s^2 cw.

To obtain the acceleration of point E, we differentiate Eqs. (13) in Example 3.6 with respect to the input angular displacement θ_2. Therefore, the second-order kinematic coefficients for point E are

$$\begin{aligned} x_E'' &= -r_2\cos\theta_2 - r_{EB}\cos(\theta_3-\phi)\theta_3'^2 - r_{EB}\sin(\theta_3-\phi)\theta_3'' \\ y_E'' &= -r_2\sin\theta_2 - r_{EB}\sin(\theta_3-\phi)\theta_3'^2 + r_{EB}\cos(\theta_3-\phi)\theta_3''. \end{aligned} \tag{3}$$

Substituting Eqs. (1) and (2) and the known data into Eqs. (3) gives

$$x_E'' = 1.206\,7 \text{ in/rad}^2 \text{ and } y_E'' = -3.310\,8 \text{ in/rad}^2. \tag{4}$$

Differentiating the velocity of point E, from Eq. (15) in Example 3.6, with respect to time, the acceleration of point E can be written as

$$\mathbf{A}_E = (x_E''\omega_2^2 + x_E'\alpha_2)\hat{\mathbf{i}} + (y_E''\omega_2^2 + y_E'\alpha_2)\hat{\mathbf{j}}. \tag{5}$$

Recall that the first-order kinematic coefficients for point E, from Eq. (14) in Example 3.6, are

$$x'_E = -3.419\,1 \text{ in/rad} \quad \text{and} \quad y'_E = 0.926\,9 \text{ in/rad}. \tag{6}$$

Substituting Eqs. (4) and (6), and the angular velocity and acceleration of the input into Eq. (5) gives

$$\mathbf{A}_E = 892.32\hat{\mathbf{i}} - 2\,448.24\hat{\mathbf{j}} \text{ ft/s}^2. \qquad \textit{Ans.}$$

Therefore, the magnitude of the acceleration of point E is $A_E = 2\,606$ ft/s^2. This result is in good agreement with the answer in Example 4.3; that is, $A_E = 2\,580$ ft/s^2.

The acceleration of point F can be obtained in a similar manner. The first-order and second-order kinematic coefficients for point F are

$$x'_F = -2.626\,2 \text{ in/rad} \quad \text{and} \quad y'_F = 3.068\,6 \text{ in/rad}$$
$$x''_F = -0.586\,0 \text{ in/rad}^2 \quad \text{and} \quad y''_F = -2.551\,7 \text{ in/rad}^2. \tag{7}$$

Substituting Eqs. (7) and the angular velocity and acceleration of the input into the equation for the acceleration of point F, namely,

$$\mathbf{A}_F = (x''_F\omega_2^2 + x'_F\alpha_2)\hat{\mathbf{i}} + (y''_F\omega_2^2 + y'_F\alpha_2)\hat{\mathbf{j}},$$

we obtain

$$\mathbf{A}_F = -433\hat{\mathbf{i}} - 1\,886\hat{\mathbf{j}} \text{ ft/s}^2. \qquad \textit{Ans.}$$

Therefore, the magnitude of the acceleration of point F is $A_F = 1\,935$ ft/s^2. This result is in good agreement with the answer of Example 4.3; that is, $A_F = 1\,960$ ft/s^2.

EXAMPLE 4.10

Link 4 of the offset slider-crank linkage in the position illustrated in Fig. 3.8 is driven at a constant speed of 10 m/s to the left. Determine the angular acceleration of links 2 and 3 and the instantaneous acceleration of point D.

SOLUTION

The angular acceleration of links 2 and 3 is obtained by differentiating Eqs. (3) in Example 3.7, with respect to the input displacement r_4, that is,

$$-r_2\cos\theta_2\theta_2'^2 - r_2\sin\theta_2\theta_2'' - r_3\cos\theta_3\theta_3'^2 - r_3\sin\theta_3\theta_3'' = 0$$
$$-r_2\sin\theta_2\theta_2'^2 + r_2\cos\theta_2\theta_2'' - r_3\sin\theta_3\theta_3'^2 + r_3\cos\theta_3\theta_3'' = 0, \tag{1}$$

where the second-order kinematic coefficients of links 2 and 3 are defined as

$$\theta_2'' = \frac{d^2\theta_2}{dr_4^2} \text{ and } \theta_3'' = \frac{d^2\theta_3}{dr_4^2}.$$

We recall the first-order kinematic coefficients from Eq. (8) in Example 3.7; they are

$$\theta_2' = -19.856 \text{ rad/m} \quad \text{and} \quad \theta_3' = 5.447 \text{ rad/m}. \tag{2}$$

Substituting Eqs. (2) and the known data into Eqs. (1), the second-order kinematic coefficients of links 2 and 3 are

$$\theta_2'' = -246.25 \text{ rad/m}^2 \quad \text{and} \quad \theta_3'' = 162.73 \text{ rad/m}^2. \tag{3}$$

The angular acceleration of links 2 and 3 are now written from Table 4.2 as

$$\alpha_2 = \theta_2''\dot{r}_4^2 + \theta_2'\ddot{r}_4 \quad \text{and} \quad \alpha_3 = \theta_3''\dot{r}_4^2 + \theta_3'\ddot{r}_4. \tag{4}$$

Substituting Eqs. (2) and (3) and the constant input speed $\dot{r}_4 = -10$ m/s into Eq. (4), the angular accelerations of links 2 and 3 are

$$\alpha_2 = -24\,625 \text{ rad/s}^2 \text{ (cw)} \quad \text{and} \quad \alpha_3 = 16\,273 \text{ rad/s}^2 \text{ ccw}. \qquad \textit{Ans.}$$

To obtain the acceleration of point D, we differentiate Eqs. (12) in Example 3.7 with respect to the input displacement r_4. The second-order kinematic coefficients for point D are

$$\begin{aligned} x_D'' &= -r_2\cos\theta_2\theta_2'^2 - r_2\sin\theta_2\theta_2'' - r_{DB}\cos(\theta_3-\beta)\theta_3'^2 - r_{DB}\sin(\theta_3-\beta)\theta_3'' \\ y_D'' &= -r_2\sin\theta_2\theta_2'^2 + r_2\cos\theta_2\theta_2'' - r_{DB}\sin(\theta_3-\beta)\theta_3'^2 + r_{DB}\cos(\theta_3-\beta)\theta_3''. \end{aligned} \tag{5}$$

Substituting Eqs. (2) and (3) and the known data into Eqs. (5), we find

$$x_D'' = 6.52 \text{ m/m}^2 \quad \text{and} \quad y_D'' = -12.31 \text{ m/m}^2. \tag{6}$$

Differentiating the velocity of point D, Eq. (14) in Example 3.7, with respect to time, the acceleration of point D can be written as

$$\mathbf{A}_D = (x_D''\dot{r}_4^2 + x_D'\ddot{r}_4)\hat{\mathbf{i}} + (y_D''\dot{r}_4^2 + y_D'\ddot{r}_4)\hat{\mathbf{j}}. \tag{7}$$

Next, we recall the first-order kinematic coefficients for point D, from Eq. (13) in Example 3.7, as

$$x_D' = 1.123 \text{ m/m} \quad \text{and} \quad y_D' = -0.407 \text{ m/m}. \tag{8}$$

Substituting Eqs. (8) and the speed and acceleration of the input link 4 into Eq. (7) we get

$$\mathbf{A}_D = 652\hat{\mathbf{i}} - 1\,231\hat{\mathbf{j}} \text{ m/s}^2. \qquad \textit{Ans.}$$

Therefore, the magnitude of the acceleration of point D is $A_D = 1\,393$ m/s^2.

EXAMPLE 4.11

The mechanism illustrated in Fig. 4.23 is a marine steering gear called Rapson's slide. Link O_4B is the tiller, and link AC is the actuating rod. For the position shown in the figure, the

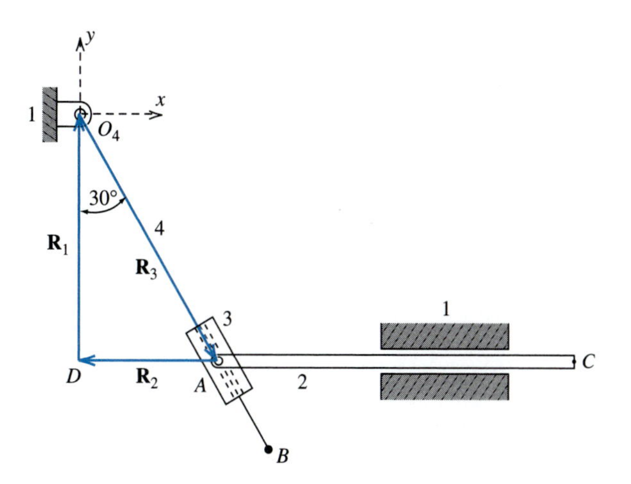

Figure 4.23 Example 4.11. $R_{O_4D} = R_1 = 8$ ft, and $R_{BO_4} = 11$ ft.

velocity of link AC is a constant 15 ft/s to the left. Determine the angular acceleration of the tiller. Also, find the velocity and acceleration of point B.

SOLUTION

The vectors that are chosen for the kinematic analysis of the mechanism are illustrated in Fig. 4.23. Note that point D is a point on link 1 (treated as fixed) such that it lies vertically below point O_4; thus, the input is the magnitude of vector $\mathbf{R}_2$ from point A on link 2 to point D on the ground link. Also, note that the change in the input is negative; that is, the magnitude of the input is decreasing.

With the vectors illustrated in Fig. 4.23, the vector loop equation for the mechanism is

$$\overset{I\surd}{\mathbf{R}_2} + \overset{\surd\surd}{\mathbf{R}_1} + \overset{??}{\mathbf{R}_3} = \mathbf{0}, \tag{1}$$

where $\theta_1 = 90^o$, $\theta_2 = 180^o$ and $\theta_3 = 300^o$. The X and Y components of Eq. (1) are

$$R_2 \cos\theta_2 + R_1 \cos\theta_1 + R_3 \cos\theta_3 = 0 \tag{2a}$$

$$R_2 \sin\theta_2 + R_1 \sin\theta_1 + R_3 \sin\theta_3 = 0. \tag{2b}$$

Differentiating Eqs. (2) with respect to the input position R_2 gives

$$\cos\theta_2 - R_3 \sin\theta_3 \theta_3' + R_3' \cos\theta_3 = 0 \tag{3a}$$

$$\sin\theta_2 + R_3 \cos\theta_3 \theta_3' + R_3' \sin\theta_3 = 0. \tag{3b}$$

In matrix form, Eqs. (3) can be written as

$$\begin{bmatrix} -R_3 \sin\theta_3 & \cos\theta_3 \\ R_3 \cos\theta_3 & \sin\theta_3 \end{bmatrix} \begin{bmatrix} \theta_3' \\ R_3' \end{bmatrix} = \begin{bmatrix} -\cos\theta_2 \\ -\sin\theta_2 \end{bmatrix}. \tag{4}$$

The determinant of the coefficient matrix in Eq. (4) is

$$\Delta = \begin{vmatrix} -R_3 \sin\theta_3 & \cos\theta_3 \\ R_3 \cos\theta_3 & \sin\theta_3 \end{vmatrix} = -R_3 \sin^2\theta_3 - R_3 \cos^2\theta_3 = -R_3. \tag{5}$$

From Eq. (2*a*), $R_3 = (8 \text{ ft}) / \cos 30^o = 9.237\,6$ ft. Therefore, $\Delta = -9.237\,6$ ft.

Differentiating Eqs. (3) with respect to the input position R_2 gives

$$-R_3 \cos\theta_3\theta_3'^2 - R_3 \sin\theta_3\theta_3'' - 2R_3' \sin\theta_3\theta_3' + R_3'' \cos\theta_3 = 0 \tag{6a}$$

$$-R_3 \sin\theta_3\theta_3'^2 + R_3 \cos\theta_3\theta_3'' + 2R_3' \cos\theta_3\theta_3' + R_3'' \sin\theta_3 = 0. \tag{6b}$$

In matrix form, Eqs. (6) can be written as

$$\begin{bmatrix} -R_3 \sin\theta_3 & \cos\theta_3 \\ R_3 \cos\theta_3 & \sin\theta_3 \end{bmatrix} \begin{bmatrix} \theta_3'' \\ R_3'' \end{bmatrix} = \begin{bmatrix} 2R_3' \sin\theta_3\theta_3' + R_3 \cos\theta_3\theta_3'^2 \\ -2R_3' \cos\theta_3\theta_3' + R_3 \sin\theta_3\theta_3'^2 \end{bmatrix}. \tag{7}$$

Note that the coefficient matrix in Eq. (7) is the same as the coefficient matrix in Eq. (4). Recall that the coefficient matrix does not change with differentiation of the unknown variables.

Substituting the known data into Eq. (4) and using Cramer's rule, the first-order kinematic coefficients are

$$\theta_3' = \frac{\begin{vmatrix} -\cos\theta_2 & \cos\theta_3 \\ -\sin\theta_2 & \sin\theta_3 \end{vmatrix}}{\Delta} = \frac{\sin(\theta_2 - \theta_3)}{-R_3} = +0.093\,75 \text{ rad/ft} \tag{8a}$$

and

$$R_3' = \frac{\begin{vmatrix} -R_3 \sin\theta_3 & -\cos\theta_2 \\ R_3 \cos\theta_3 & -\sin\theta_2 \end{vmatrix}}{\Delta} = \frac{R_3 \cos(\theta_2 - \theta_3)}{-R_3} = +0.500 \text{ ft/ft}. \tag{8b}$$

Substituting the known data into Eqs. (7) and using Cramer's rule, the second-order kinematic coefficients are

$$\theta_3'' = \frac{\begin{vmatrix} 2R_3' \sin\theta_3\theta_3' + R_3 \cos\theta_3\theta_3'^2 & \cos\theta_3 \\ -2R_3' \cos\theta_3\theta_3' + R_3 \sin\theta_3\theta_3'^2 & \sin\theta_3 \end{vmatrix}}{\Delta} = \frac{2R_3'\theta_3'}{-R_3} = -0.010\,1 \text{ rad/ft}^2 \tag{9a}$$

and

$$R_3'' = \frac{\begin{vmatrix} -R_3 \sin\theta_3 & 2R_3' \sin\theta_3\theta_3' + R_3 \cos\theta_3\theta_3'^2 \\ R_3 \cos\theta_3 & -2R_3' \cos\theta_3\theta_3' + R_3 \sin\theta_3\theta_3'^2 \end{vmatrix}}{\Delta} = \frac{-\left(R_3\theta_3'\right)^2}{-R_3} = +0.081\,2 \text{ ft/ft}^2. \tag{9b}$$

The angular velocity of link 3 is

$$\omega_3 = \theta_3'\dot{R}_2 = (+0.093\,75 \text{ rad/ft})(-15 \text{ ft/s}) = -1.41 \text{ rad/s}. \tag{10}$$

The negative sign indicates that the direction of the angular velocity of link 3 is clockwise. Note that $\dot{R}_2 = -15$ ft/s because link 2 is moving to the left, that is, the input magnitude R_2 is decreasing. Also, note that link 4 is constrained to rotate with the same angular velocity as link 3. Therefore, the angular velocity of link 4 is

$$\omega_4 = \omega_3 = -1.41 \text{ rad/s (cw)}. \tag{11}$$

The angular acceleration of link 3 is

$$\begin{aligned}\alpha_3 &= \theta_3'\ddot{R}_2 + \theta_3''\dot{R}_2^2 \\ &= (+0.093\,75 \text{ rad/ft})(0) + (-0.010\,1 \text{ rad/ft}^2)(-15 \text{ ft/s})^2. \\ &= -2.27 \text{ rad/s}^2\end{aligned} \tag{12}$$

The negative sign indicates that link 3 is accelerating in the clockwise direction. Because link 4 is constrained to rotate with link 3, then the angular acceleration of link 4 is

$$\alpha_4 = \alpha_3 = -2.27 \text{ rad/s}^2 \text{ (cw)}. \qquad \textit{Ans.} \tag{13}$$

The sliding velocity of link 3 relative to link 4 is

$$V_{A_3/4} = R_3'\dot{R}_2 = (+0.5 \text{ ft/ft})\,(-15 \text{ ft/s}) = -7.5 \text{ ft/s}. \tag{14}$$

The negative sign indicates that the direction of the relative velocity is along link 4 pointing toward the pin O_4; that is, the direction is $\angle \mathbf{V}_{A_3/4} = 120°$.

The sliding acceleration of link 3 relative to link 4 can be written as

$$\begin{aligned}A_{A_3/4} &= R_3'\ddot{R}_2 + R_3''\dot{R}_2^2 \\ &= (+0.5 \text{ ft/ft})\,(0) + \left(+0.081\,2 \text{ ft/ft}^2\right)(-15 \text{ ft/s})^2 \\ &= +18.27 \text{ ft/s}^2.\end{aligned} \tag{15}$$

The positive sign indicates that the direction of the relative acceleration is along link 4 pointing away from the pin O_4; that is, the direction is $\angle \mathbf{A}_{A_3/4} = -60°$.

Figure 4.24 illustrates the vector $\mathbf{R}_B$, which represents the motion of point B of link 4. The X and Y components of the position of point B are

$$X_B = R_B \cos\theta_3 = 5.50 \text{ ft} \tag{16a}$$

$$Y_B = R_B \sin\theta_3 = -9.53 \text{ ft}. \tag{16b}$$

Differentiating Eqs. (16) with respect to the input position R_2, the first-order coefficients of point B are

$$X_B' = -R_B \sin\theta_3\theta_3' = +0.893\,09 \text{ ft/ft} \tag{17a}$$

$$Y_B' = R_B \cos\theta_3\theta_3' = +0.515\,63 \text{ ft/ft}. \tag{17b}$$

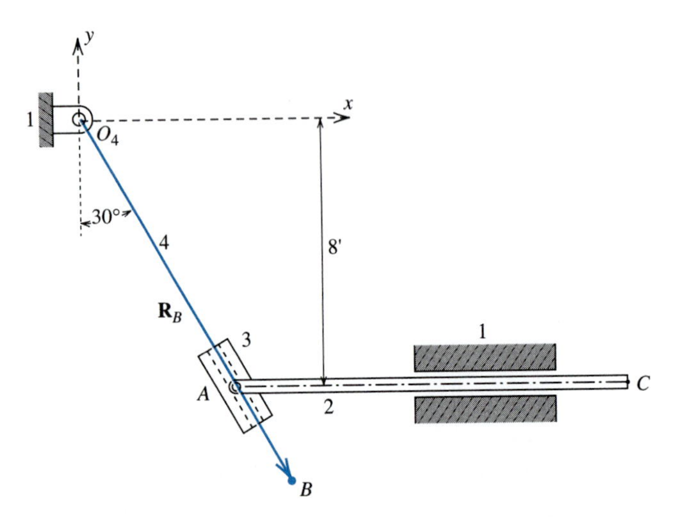

Figure 4.24 Example 4.11. The vector for the motion of point B.

Then, differentiating Eqs. (17) with respect to the input position R_2 gives the second-order coefficients of point B

$$X''_B = -R_B \cos\theta_3 \theta_3'^2 - R_B \sin\theta_3 \theta''_3 = -0.145 \text{ ft/ft}^2 \tag{18a}$$

$$Y''_B = -R_B \sin\theta_3 \theta_3'^2 + R_B \cos\theta_3 \theta''_3 = +0.028 \text{ ft/ft}^2. \tag{18b}$$

The velocity of point B can be written as

$$\begin{aligned} \mathbf{V}_B &= \left(X'_B\hat{\mathbf{i}} + Y'_B\hat{\mathbf{j}}\right)\dot{R}_2 \\ &= \left(0.893\,09\,\hat{\mathbf{i}} + 0.515\,63\,\hat{\mathbf{j}}\right)(-15 \text{ ft/s}) = -13.40\,\hat{\mathbf{i}} - 7.73\,\hat{\mathbf{j}} \text{ ft/s} \\ &= 15.47 \text{ ft/s}\angle 210°. \end{aligned} \qquad \textit{Ans.}\ (19)$$

As a check, the velocity of point B can also be written as

$$V_B = \omega_4 R_{BO_4} = (1.41 \text{ rad/s})\,(11 \text{ ft}) = 15.51 \text{ ft/s},$$

where the difference is a result of truncation in the value of ω_4 in Eq. (10), [ω_4 = 1.406 25 rad/s].

The acceleration of point B can be written as

$$\begin{aligned} \mathbf{A}_B &= \left(X'_B\hat{\mathbf{i}} + Y'_B\hat{\mathbf{j}}\right)\ddot{R}_2 + \left(X''_B\hat{\mathbf{i}} + Y''_B\,\hat{\mathbf{j}}\right)\dot{R}_2^2 \\ &= \left(0.893\hat{\mathbf{i}} + 0.516\hat{\mathbf{j}} \text{ ft/ft}\right)(0) + (-0.145\hat{\mathbf{i}} + 0.028\hat{\mathbf{j}} \text{ ft/ft}^2)(-15 \text{ ft/s})^2 \\ &= -32.6\hat{\mathbf{i}} + 6.3\hat{\mathbf{j}} \text{ ft/s}^2 \\ &= 33.2 \text{ ft/s}^2\angle 169.1°. \end{aligned} \qquad \textit{Ans.}\ (20)$$

Note that the inverse tangent function on many calculators initially indicates an angle of $-10.9°$ for $\mathbf{A}_B$. However, this angle must be transposed into the second quadrant; that is, $\angle\mathbf{A}_B = -10.9° + 180° = 169.1°$.

4.13 THE EULER–SAVARY EQUATION[2]

In Section 4.5 we developed the apparent-acceleration equation, Eq. (4.17). Then, in the examples that followed, we reported that it is very important to carefully choose a point whose apparent path is known so that the radius of curvature of the path, needed for the normal component in Eq. (4.14), can be determined. This need to know the radius of curvature of the path often dictates the method of approach to such a problem and sometimes even requires the visualization of an equivalent mechanism. It would be convenient if an arbitrary point could be chosen and the radius of curvature of its path could be calculated. In planar mechanisms this can be accomplished by the methods that will be presented here.

When two rigid bodies move relative to each other with planar motion, any arbitrarily chosen point A of one describes a path, or locus, relative to a coordinate system fixed to the other. At any given instant, there is a point A', attached to the other body, that is the center of curvature of the locus of A. If we take the kinematic inversion of this motion, A' also describes a locus relative to the body containing A, and it so happens that A is the center of curvature of this locus. Each point, therefore, acts as the center of curvature of the path traced by the other, and the two points are called *conjugates* of each other. The distance between these two conjugate points is the magnitude of the radius of curvature of either locus.

Figure 4.25 illustrates two circles with centers at C and C'. Let us think of the circle with center C' as the fixed centrode and think of the circle with center C as the moving centrode of two bodies experiencing some particular relative planar motion. In actuality, the fixed centrode need not be fixed but is attached to the body that contains the path whose curvature is sought. Also, it is not necessary that the two centrodes be circles; we are interested only in instantaneous values and, for convenience, we will think of the centrodes as circles matching the curvatures of the two actual centrodes in the region near their point of contact I. When the bodies containing the two centrodes move relative to each other, the centrodes appear to roll against each other without slip (see Section 3.21). The point of contact I is, of course, the instantaneous center of velocity (also called the velocity pole). Because of these properties, we can think of the two circular centrodes as actually representing the shapes of the two moving bodies, if this helps in visualizing the motion.

If the moving centrode has an angular velocity $\boldsymbol{\omega}$, relative to the fixed centrode, the instantaneous velocity* of point C is

$$V_C = \omega R_{CI}. \tag{a}$$

Similarly, the arbitrary point A, whose conjugate point A' we wish to find, has a velocity of

$$V_A = \omega R_{AI}. \tag{b}$$

* All velocities used in this section are actually apparent velocities, relative to the coordinate system of the fixed centrode; they are written as absolute velocities to simplify the notation.

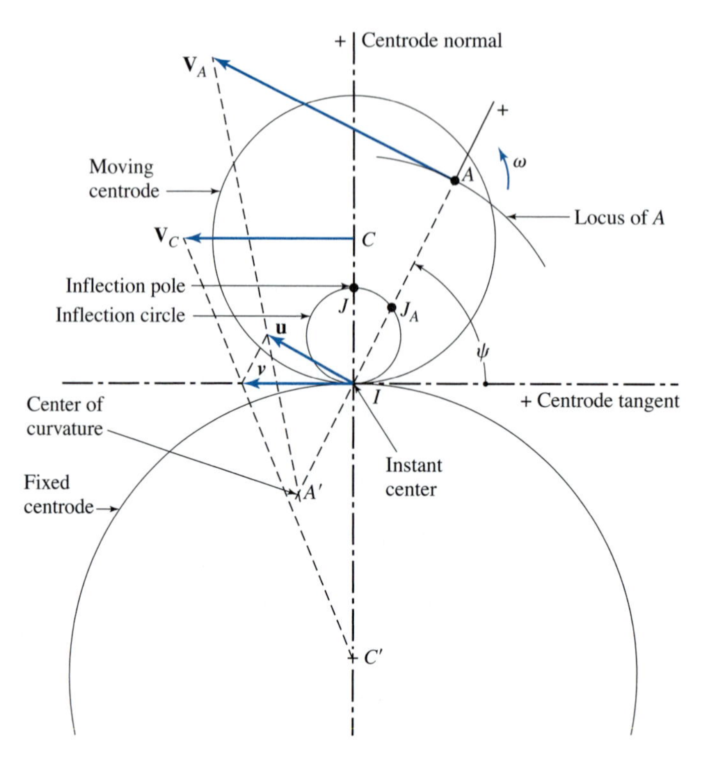

Figure 4.25 The Hartmann construction.

As the motion progresses, the point of contact of the two centrodes, and therefore the location of the instantaneous center of velocity, I, moves along both centrodes with some velocity (or rate) $\boldsymbol{v}$. As illustrated in Fig. 4.25, $\boldsymbol{v}$ can be determined by connecting a straight line from the terminus of V_C to point C'. Alternatively, the magnitude of this velocity can be obtained from

$$v = \frac{R_{IC'}}{R_{CC'}} V_C. \tag{c}$$

A graphical construction for A', the center of curvature of the locus of the arbitrary point A, is illustrated in Fig. 4.25 and is called the *Hartmann construction*. First, we find the component $\mathbf{u}$ of the instant center's velocity $\boldsymbol{v}$ as that component parallel to $\mathbf{V}_A$ or perpendicular to $\mathbf{R}_{AI}$. Then the intersection of the line AI and a line connecting the termini of the velocities $\mathbf{V}_A$ and $\mathbf{u}$ gives the location of the conjugate point A'. The radius of curvature ρ of the locus of point A is $\rho = R_{AA'}$.

An analytical expression for locating point A' would also be desirable and can be derived from the Hartmann construction. The magnitude of the velocity $\mathbf{u}$ is given by

$$u = v \sin \psi, \tag{d}$$

where ψ is the positive counterclockwise angle measured from the centrode tangent to the line of R_{AI}. Then, noting the similar triangles in Fig. 4.25, we can write

$$u = \frac{R_{IA'}}{R_{AA'}} V_A. \tag{e}$$

Now, equating the expressions of Eqs. (*d*) and (*e*) and substituting Eqs. (*a*), (*b*), and (*c*) into the resulting equation gives

$$u = \frac{R_{IC'} R_{CI}}{R_{CC'}} \omega \sin \psi = \frac{R_{IA'} R_{AI}}{R_{AA'}} \omega. \tag{f}$$

Dividing this equation by $\omega \sin \psi$ and inverting, we obtain

$$\frac{R_{AA'}}{R_{AI} R_{IA'}} \sin \psi = \frac{R_{CC'}}{R_{CI} R_{IC'}} = \frac{\omega}{v}. \tag{g}$$

Next, upon noting that $R_{AA'} = R_{AI} - R_{A'I}$ and $R_{CC'} = R_{CI} - R_{C'I}$, we can reduce this equation to the form

$$\left(\frac{1}{R_{AI}} - \frac{1}{R_{A'I}} \right) \sin \psi = \left(\frac{1}{R_{CI}} - \frac{1}{R_{C'I}} \right). \tag{4.44}$$

This important equation is one form of the *Euler–Savary equation*. If we assume that the radii of curvature of the two centrodes R_{CI} and $R_{C'I}$ are known, then this equation can be used to determine the position of one of the two conjugate points (A or A') from the position of the other relative to the instantaneous center of velocity, I.

Before continuing, an explanation of the sign convention is important. In using the Euler–Savary equation, we may arbitrarily choose a positive sense for the centrode tangent; the positive centrode normal is then 90° counterclockwise from it. This establishes a positive direction for the line CC', which may be used in assigning appropriate signs to R_{CI} and $R_{C'I}$. Similarly, an arbitrary positive direction can be chosen for the line AA'. The angle ψ is then taken as positive counterclockwise from the positive centrode tangent to the positive sense of line AA'. The sense of the line AA' also gives the appropriate signs for R_{AI} and $R_{A'I}$ for Eq. (4.44).

There is a major disadvantage to the above form of the Euler–Savary equation in that the radii of curvature of both centrodes, R_{CI} and $R_{C'I}$, must be known. Usually these are not known, any more than the curvature of the locus itself is known. However, this difficulty can be overcome by seeking another form of the equation.

Let us consider the particular point labeled J in Fig. 4.25. This point is located on the centrode normal at a location defined by

$$\frac{1}{R_{JI}} = \frac{1}{R_{CI}} - \frac{1}{R_{C'I}}. \tag{h}$$

If this particular point is chosen for A in Eq. (4.44), we find that its conjugate point J' must be located at infinity. The radius of curvature of the path of point J is infinite, and the locus of J therefore has an inflection point at J. The point J is called the *inflection pole*.

Let us now consider whether there are any other points J_A of the moving body that also have infinite radii of curvature at the instant considered. If so, then for each of these points $R_{AA'} = \infty$. Substituting this condition into Eq. (4.44) and with the aid of Eq. (h), we obtain

$$R_{J_A I} = R_{JI} \sin \psi. \tag{4.45}$$

This is the equation of a circle in polar coordinates whose diameter is R_{JI}, as illustrated in Fig. 4.25. This circle is called the *inflection circle*. Every point on this circle is an inflection point; it has its conjugate point at infinity, and each point therefore has an infinite radius of curvature at the instant indicated. These points are instantaneously moving on straight lines.

Now, with the help of Eq. (4.45), the Euler–Savary equation can be written in the form

$$\frac{1}{R_{AI}} - \frac{1}{R_{A'I}} = \frac{1}{R_{J_A I}}. \tag{4.46}$$

Also, after some further manipulation, the radius off curvature of the path of point A can be written as

$$\rho = R_{AA'} = \frac{R_{AI}^2}{R_{AJ_A}}. \tag{4.47}$$

Either of these two forms of the Euler–Savary equation, Eqs. (4.46) and (4.47), is more useful in practice than Eq. (4.44), because they do not require knowledge of the curvatures of the two centrodes. They do require finding the inflection circle, but the following example will illustrate the graphical procedure for drawing the inflection circle.

EXAMPLE 4.12

Draw the inflection circle for the coupler link 3 of the slider-crank linkage in the position illustrated in Fig. 4.26. Then use the inflection circle and the Euler–Savary equation to determine the radius of curvature of the path of the coupler point C.

SOLUTION

We begin by locating the instantaneous center of velocity, I, at the intersection of the line O_2A and a line through B perpendicular to its direction of travel (see Fig. 4.27). By definition, points B and I must both lie on the inflection circle; hence, we need only one additional point to construct the circle.

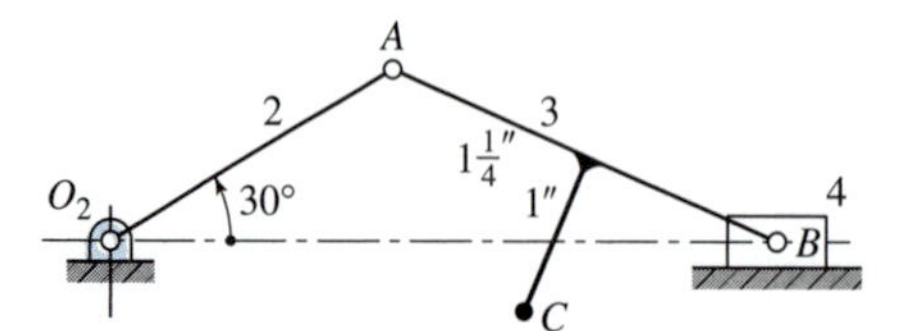

Figure 4.26 Example 4.12: $R_{AO_2} = 2$ in, $R_{BA} = 2.5$ in.

The center of curvature of point A is, of course, at O_2, which we now call A'. Taking the positive sense of the line AI as being downward and to the left, we measure $R_{AI} = 2.64$ in and $R_{AA'} = -2.00$ in. Then, substituting these values into Eq. (4.47), we obtain

$$R_{AJ_A} = \frac{R_{AI}^2}{R_{AA'}} = \frac{(2.64 \text{ in})^2}{-2.00 \text{ in}} = -3.48 \text{ in.}$$

With this, we lay off 3.48 in from A to locate J_A, which is a third point on the inflection circle. The inflection circle for the motion 3/1 can now be constructed through the three points B, I, and J_A, and the diameter of the circle is measured as

$$R_{JI} = 6.28 \text{ in.} \qquad \textit{Ans.}$$

The centrode normal and the centrode tangent can also be drawn, if desired, as illustrated in Fig. 4.27.

Drawing the ray R_{CI} and continuing to take the positive sense as downward and to the left, we measure $R_{CI} = 3.10$ in and $R_{CJ_C} = -1.75$ in. Substituting these values into Eq.

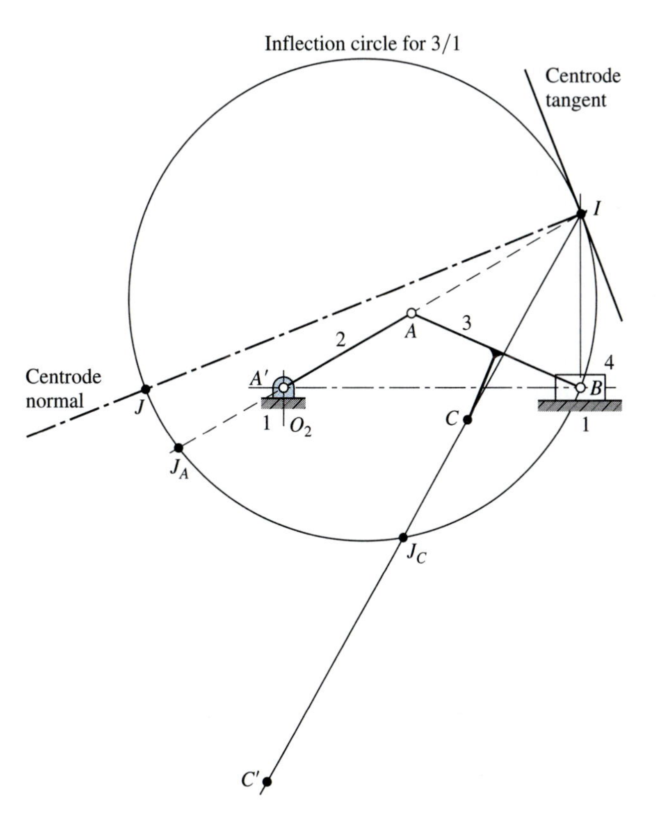

Figure 4.27 The inflection circle.

(4.47), the radius of curvature of the path of point C is

$$\rho = R_{CC'} = \frac{R_{CI}^2}{R_{CJ_C}} = \frac{(3.10\ \text{in})^2}{-1.75\ \text{in}} = -5.49\ \text{in}, \qquad \textit{Ans.}$$

where the negative sign indicates that C is above C' on the line ICJ_CC'.

4.14 THE BOBILLIER CONSTRUCTIONS

The Hartmann construction provides one graphical method of finding the conjugate point and the radius of curvature of the path of a moving point, but it requires knowledge of the curvature of the fixed and moving centrodes. It would be desirable to have *graphical* methods for obtaining the inflection circle and the conjugate of a given point without requiring the curvature of the centrodes. Such graphical solutions are presented in this section and are called the *Bobillier constructions.*

To understand these constructions, consider the inflection circle, the centrode normal (denoted N), and the centrode tangent (denoted T), as illustrated in Fig. 4.28. Let us select any two points A and B of the moving body that do not lie on a straight line through the instantaneous center of velocity, I. Using the Euler–Savary equation, we can find the two corresponding conjugate points A' and B'. The intersection of the lines AB and $A'B'$ is labeled Q. Then, the straight line drawn through I and Q is called the *collineation axis*. This axis applies only to the two lines AA' and BB' and so is said to belong to these two rays; also, point Q will be located differently on the collineation axis if another pair of points A and B is chosen on the same rays. Nevertheless, there is a unique relationship between the collineation axis and the two rays used to define it. This relationship is expressed in *Bobillier's theorem*, which states that *the angle from the collineation axis to the first ray is equal to the angle from the second ray to the centrode tangent.*

In applying the Euler–Savary equation to a planar mechanism, we can usually find two pairs of conjugate points by inspection, and from these we wish to graphically determine the inflection circle. For example, a four-bar linkage with a crank O_2A and a follower O_4B has A and O_2 as one set of conjugate points and B and O_4 as another when we are interested in the motion of the coupler relative to the frame (3 relative to 1). Given these two pairs of conjugate points, how do we use the Bobillier theorem to find the inflection circle?

In Fig. 4.29*a*, let A and A' and B and B' represent the known pairs of conjugate points. Rays constructed through each pair intersect at the instantaneous center of velocity, I, giving one point on the inflection circle. Point Q is located next by the intersection of a ray through A and B with a ray through A' and B'. Then the collineation axis can be drawn as the line IQ.

The next step is illustrated in Fig. 4.29b. Drawing a straight line through I parallel to $A'B'$, we identify the point W as the intersection of this line with the line AB. Now, through W we draw a second line parallel to the collineation axis. This line intersects AA' at J_A and BB' at J_B, two additional points on the inflection circle for which we are searching.

We could now construct a circle through the three points J_A, J_B, and I, but there is an easy way to do this. Remembering that a triangle inscribed in a semicircle is a right triangle having the diameter as its hypotenuse, we erect a perpendicular to AI at J_A and another perpendicular to BI at J_B. The intersection of these two perpendiculars gives point the *inflection pole*, J, as illustrated in Fig. 4.29*c*. Because IJ is the diameter, then the inflection circle, the centrode normal N, and the centrode tangent T can all be easily constructed.

To demonstrate that this construction satisfies the Bobillier theorem, note that the arc from I to J_A is inscribed by the angle that J_AI makes with the centrode tangent. But this same arc is also inscribed by the angle IJ_BJ_A. Therefore, these two angles are equal. But the line J_BJ_A was originally constructed parallel to the collineation axis. Therefore, the line IJ_B also makes the same angle β with the collineation axis.

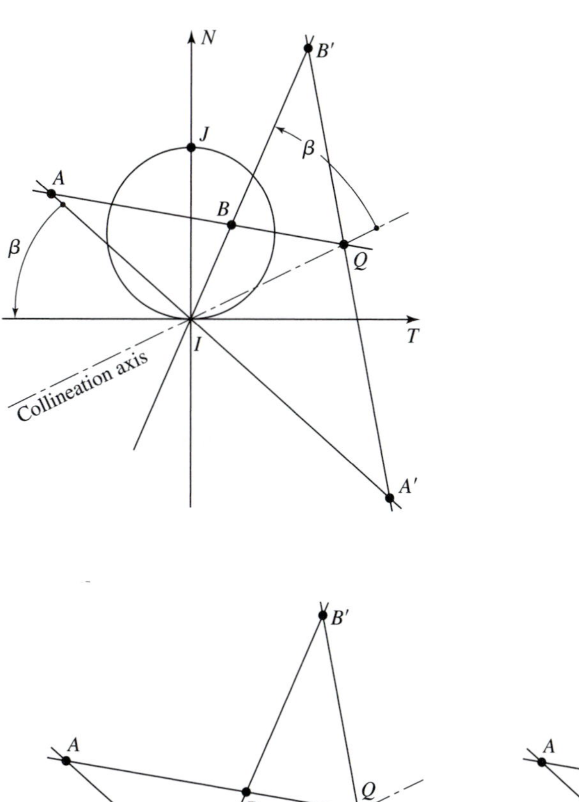

Figure 4.28 The Bobillier theorem.

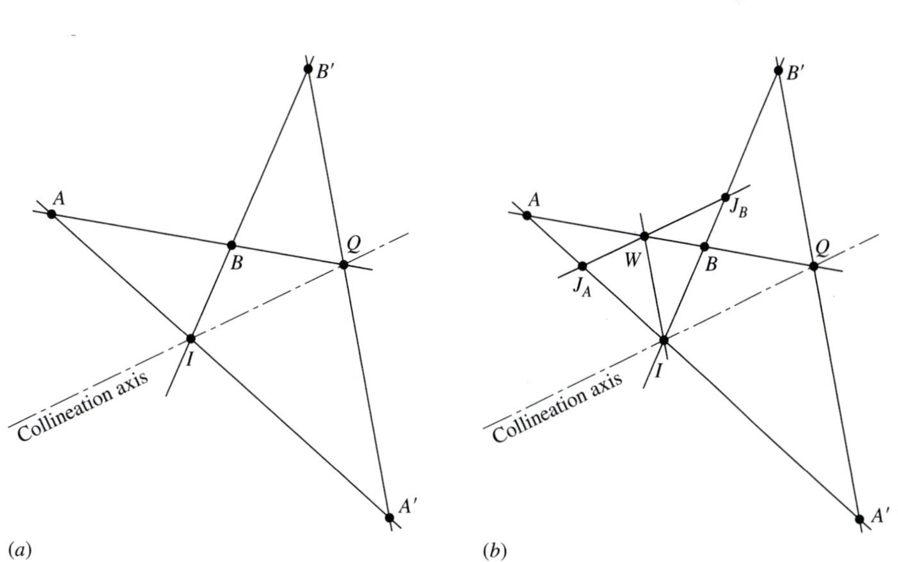

Figure 4.29 Continued

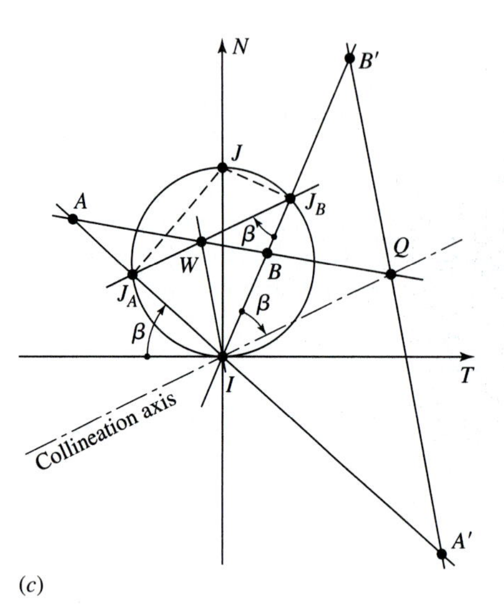

Figure 4.29 The Bobillier construction for locating the inflection circle.

Our final problem is to learn how to use the Bobillier theorem to find the conjugate of another arbitrary point, say C, when the inflection circle is given. In Fig. 4.30 we draw a line through point C and the instantaneousness center of velocity, I, and locate the point of intersection of this line with the inflection circle (point J_C). This ray serves as one of the two necessary rays to locate the collineation axis. For the other ray, we can use the centrode normal, because J and its conjugate point J', at infinity, are both known. For these two rays, the collineation axis is a line through I parallel to the line $J_C J$, as we learned in Fig. 4.29c. The balance of the construction is similar to that of Fig. 4.27. Point Q is located by the intersection of a line through J and C with the collineation axis. Then a line through Q and J' at infinity intersects the ray IC at C', the conjugate point for C.

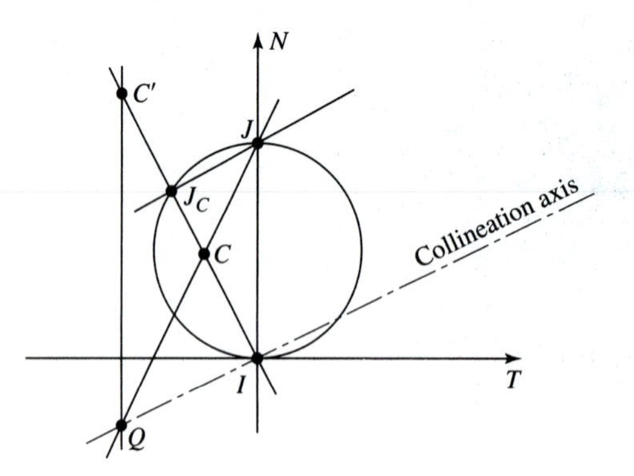

Figure 4.30 The Bobillier construction for locating the conjugate point C'.

EXAMPLE 4.13

For the four-bar linkage $A'ABB$ illustrated in Fig. 4.31, use the Bobillier theorem to find the center of curvature of the coupler curve of point C.

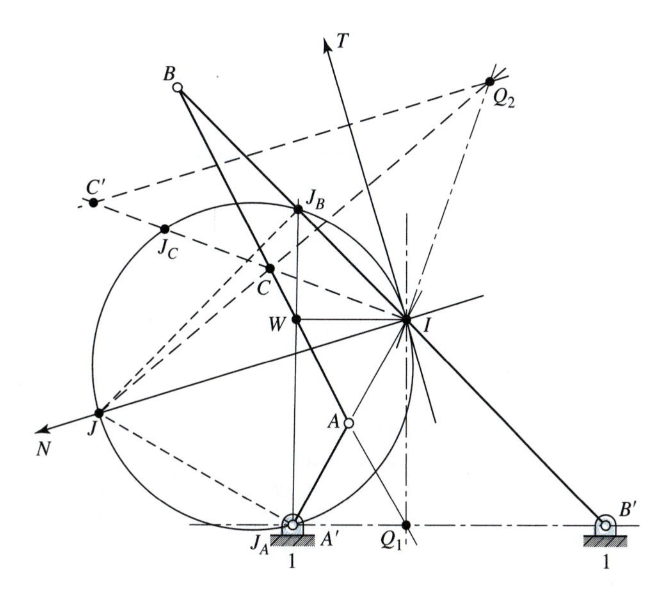

Figure 4.31 Example 4.13.

SOLUTION

Locate the instantantaneous center of velocity, I, at the intersection of AA' and BB'; also locate Q_1 at the intersection of AB and $A'B'$. IQ_1 is the first collineation axis. Through I, draw a line parallel to $A'B'$ to locate W on AB. Through W, draw a line parallel to IQ_1 to locate J_A on AA' and J_B on BB'. Then, through J_A, draw a perpendicular to AA' and through J_B draw a perpendicular to BB'. These perpendiculars intersect at the inflection pole J and define the inflection circle, the centrode normal N, and the centrode tangent T.

To obtain the conjugate point of C, we draw the ray IC and locate J_C on the inflection circle. The second collineation axis IQ_2, belonging to the pair of rays IC and IJ, is a line through I parallel to a line (not shown) from J to J_C. The point Q_2 is obtained as the intersection of this collineation axis and a line JC. Now, through Q_2, we draw a line parallel to the centrode normal; its intersection with the ray IC yields C', the center of curvature of the path of C.

4.15 INSTANTANEOUS CENTER OF ACCELERATION

This section defines the *instantaneous center of acceleration*, or the *acceleration pole*, for a planar mechanism. It is important to note that, in general, the acceleration pole is not

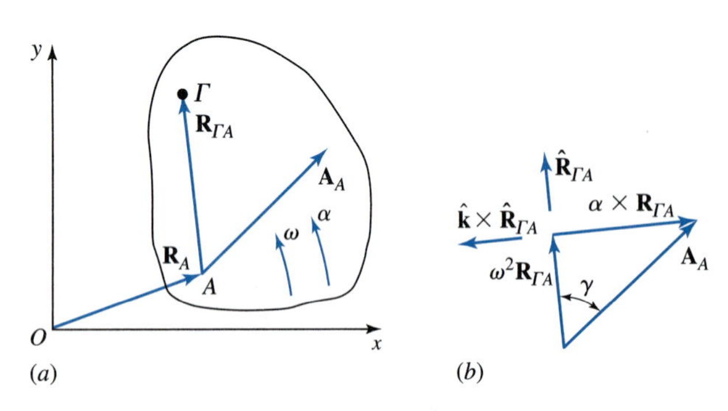

Figure 4.32 The instantaneous center of acceleration.

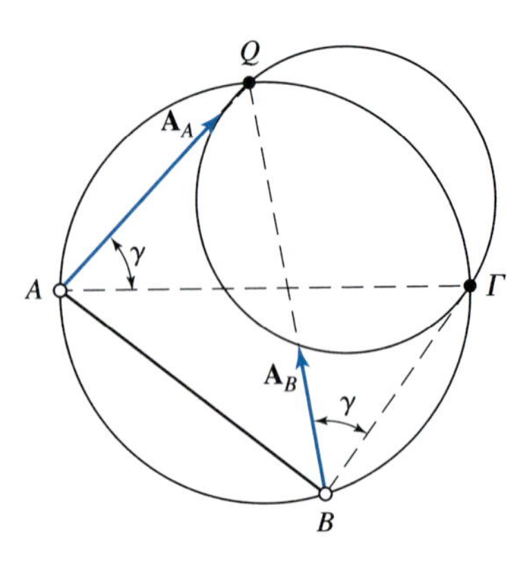

Figure 4.33 The four-circle method of locating the instantaneous center of acceleration Γ.

coincident with the instantaneous center of velocity. In other words, the instantaneous center of velocity does have acceleration. The acceleration pole is defined as *the instantaneous location of a pair of coincident points of two different rigid bodies where the absolute accelerations of the two bodies are identical.* If we consider a fixed and a moving body, the acceleration pole is the point of the moving body that has zero absolute acceleration at the instant under consideration.

For the moving plane illustrated in Fig. 4.32*a*, assume that $\boldsymbol{\omega}$ and $\boldsymbol{\alpha}$ are known and that point A in the plane has a known acceleration $\mathbf{A}_A$. Let Γ denote the acceleration pole, a point of zero absolute acceleration whose location is unknown. The acceleration difference equation can then be written as

$$\mathbf{A}_\Gamma = \mathbf{A}_A - \omega^2\mathbf{R}_{\Gamma A} + \boldsymbol{\alpha} \times \mathbf{R}_{\Gamma A} = \mathbf{0}. \tag{a}$$

Solving for $\mathbf{A}_A$ gives

$$\mathbf{A}_A = \omega^2 R_{\Gamma A}\hat{\mathbf{R}}_{\Gamma A} - \alpha R_{\Gamma A}(\hat{\mathbf{k}} \times \hat{\mathbf{R}}_{\Gamma A}). \tag{b}$$

Now, because $\hat{\mathbf{R}}_{\Gamma A}$ is perpendicular to $\hat{\mathbf{k}} \times \hat{\mathbf{R}}_{\Gamma A}$, the two terms on the right of Eq. (*b*) are rectangular components of $\mathbf{A}_A$, as illustrated in Fig. 4.32*b*. From Fig. 4.32*b* we can solve for the direction and magnitude of $\mathbf{R}_{\Gamma A}$, that is.

$$\gamma = \tan^{-1}\frac{\alpha}{\omega^2} \tag{4.48}$$

$$R_{\Gamma A} = \frac{A_A}{\sqrt{\omega^4 + \alpha^2}} = \frac{A_A \cos\gamma}{\omega^2}. \tag{4.49}$$

Equation (4.49) states that the distance $R_{\Gamma A}$ from point A to the instantaneous center of acceleration can be determined from the magnitude of the acceleration A_A of any point of the moving plane. Our convention is that the angle γ is defined *from* the line $I\Gamma$ *to* the positive centrode normal.

There are many graphical methods of locating the instantaneous center of acceleration.[3] Here, without proof, we present one method. In Fig. 4.33 we are given points A and B and their absolute accelerations, $\mathbf{A}_A$ and $\mathbf{A}_B$. We extend $\mathbf{A}_A$ and $\mathbf{A}_B$ until they intersect at Q; then we construct a circle through points A, B, and Q. Next, we draw another circle through the termini of $\mathbf{A}_A$ and $\mathbf{A}_B$ and point Q. The second intersection of the two circles locates point Γ, the instantaneous center of acceleration.

4.16 THE BRESSE CIRCLE (OR DE LA HIRE CIRCLE)

Another graphical method of locating the instantaneous center of acceleration is by drawing a circle called the Bresse circle. In Section 4.13 we defined the inflection circle as the locus of points that have their conjugate points at infinity, and each therefore has an infinite radius of curvature at the instant under consideration. Therefore, the inflection circle can also be defined as the locus of points with zero normal acceleration. The locus of points with zero tangential acceleration also defines a circle, illustrated in Fig. 4.33, referred to as the *Bresse circle* or the *de La Hire circle.*

The inflection circle and the Bresse circle intersect at two points, as illustrated in Fig. 4.34; one point is the velocity pole I and the other point is the acceleration pole Γ. Because I will have acceleration, in general, then it must be discounted as a possible solution. In fact, the acceleration of I is directed along the centrode normal N and can be written as

$$A_I = \omega \nu, \tag{4.50}$$

where ν is the velocity of I (see Eq. (*b*) in Section 4.13) and can be written as

$$\nu = \omega R_{JI}. \tag{4.51}$$

Substituting Eq. (4.51) into Eq. (4.50), the acceleration of I can also be written as

$$A_I = \omega^2 R_{JI}.$$

The angle from the line ΓI (that is, the line connecting the acceleration pole and the velocity pole) to the centrode normal is the angle γ. This angle was defined in Section 4.15, Eq. (4.48), as

$$\gamma = \tan^{-1} \frac{\alpha}{\omega^2}.$$

The diameter of the Bresse circle can be written as

$$b = R_{JI} \frac{\omega^2}{\alpha}. \tag{4.52}$$

Sign convention If the angular acceleration of the moving plane is positive (that is, counterclockwise), then the Bresse circle will lie on the negative side of the centrode tangent T, as illustrated in Fig. 4.34.

In the special case that the angular velocity of the moving plane is a constant (that is, $\alpha = 0$), then the diameter of the Bresse circle is infinite (the Bresse circle tends to the centrode normal N) and the acceleration pole is coincident with the inflection pole.

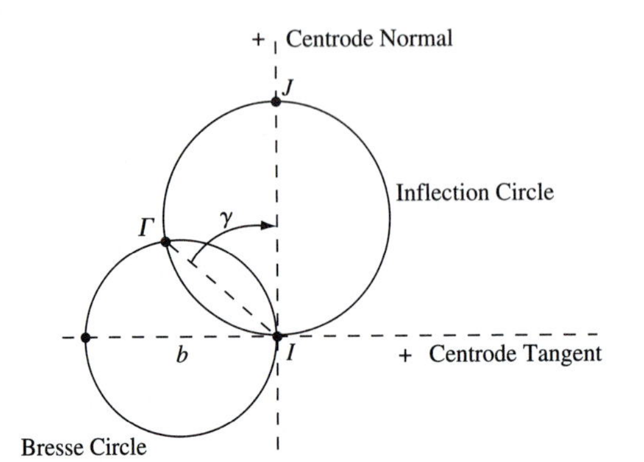

Figure 4.34 The Bresse circle and the instantaneous center of acceleration.

EXAMPLE 4.14

Consider the planar four-bar linkage of Example 4.1, illustrated in Fig. 4.8*a* and redrawn here as Fig. 4.35. For the specified position (that is, the crank angle is 135° ccw from the ground link O_2O_4), the angular velocity and angular acceleration of the coupler link AB are $\omega_3 = 5$ rad/s ccw and $\alpha_3 = 20$ rad/s^2 cw, respectively. For the instantaneous motion of the coupler link, show: (*a*) the velocity pole I, centrode tangent T, and centrode normal N; (*b*) the inflection circle and Bresse circle; and (*c*) the acceleration center of the coupler link AB. Then determine: (*d*) the radius of curvature of the trajectory of coupler point C where $R_{CB} = 6$ in., (*e*) the magnitude and direction of the velocity of C, (*f*) the magnitude and direction of the angular velocity of the crank, (*g*) the magnitude and direction of the velocity of I, (*h*) the magnitude and direction of the acceleration of I, and (*i*) the magnitude and direction of the acceleration of C.

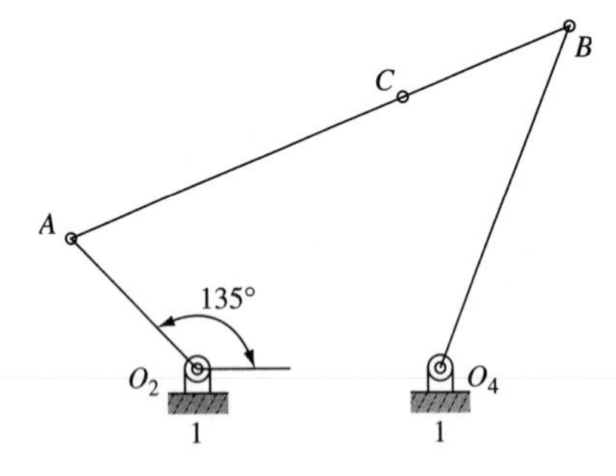

Figure 4.35 Example 4.14. Planar four-bar linkage; $R_{O_4O_2} = 8$ in., $R_{AO_2} = 6$ in., $R_{BA} = 18$ in., and $R_{BO_4} = 12$ in.

SOLUTION

a. In this problem, the velocity pole I is coincident with the instant center I_{13}, as illustrated in Fig. 4.36. The instant center I_{24} (labeled point Q) and the collineation axis IQ are also illustrated in Fig. 4.36. From Bobillier's theorem, the angle from the collineation axis to the first ray (say link 2) is obtained from a scaled drawing as

$$\alpha = 29.08^\circ \text{ cw.} \qquad \textit{Ans.} \ (1)$$

This is equal to the angle from the second ray (say link 4) to the pole tangent T. Therefore, we draw the centrode tangent T and the centrode normal N, which is 90° counterclockwise from T, as illustrated in Fig. 4.36.

b. Now, by measuring the ray $R_{AI} = 14.14$ in., we can use Eq. (4.47) to locate the inflection point J_A,

$$R_{AJ_A} = \frac{R_{AI}^2}{R_{AA'}} = \frac{(14.14 \text{ in})^2}{6.00 \text{ in}} = 33.31 \text{ in.}$$

Similarly, measuring the ray $R_{BI} = 18.20$ in, we locate the inflection point J_B,

$$R_{BJ_B} = \frac{R_{BI}^2}{R_{BB'}} = \frac{(18.20 \text{ in})^2}{12.00 \text{ in}} = 27.60 \text{ in.}$$

Erecting perpendiculars from these two rays, we locate the inflection pole J and draw the inflection circle as illustrated in Fig. 4.36.
The diameter of the inflection circle is

$$R_{JI} = 19.26 \text{ in.} \qquad \textit{Ans.} \ (2)$$

The diameter of the Bresse circle, from Eq. (4.52), is

$$b = R_{JI}\frac{\omega_3^2}{\alpha_3} = (19.26 \text{ in})\frac{(5 \text{ rad/s})^2}{-20 \text{ rad/s}^2} = -24.08 \text{ in}, \qquad \textit{Ans.} \ (3)$$

where the negative sign indicates that the Bresse circle lies on the positive side of the centrode tangent.

c. The acceleration center Γ of the coupler link is as illustrated in Fig. 4.36. *Ans.*

d. The radius of curvature of the path of coupler point C (where $R_{CB} = 6$ in) is

$$\rho_C = R_{CC'} = \frac{R_{CI}^2}{R_{CJ_C}} = \frac{(14.67 \text{ in})^2}{28.62 \text{ in}} = 7.52 \text{ in.} \qquad \textit{Ans.} \ (4)$$

The center of curvature of the path traced by point C, that is C', is illustrated in Fig. 4.36.

e. The velocity of coupler point C is

$$V_C = \omega_3 R_{CI} = (5 \text{ rad/s})\ (14.65 \text{ in}) = 73.25 \text{ in/s.} \qquad \textit{Ans.} \ (5)$$

The orientation of the velocity vector is as illustrated in Fig. 4.36. *Ans.*

f. The angular velocity of the crank can be written as

$$\omega_2 = \frac{R_{I_{23}I_{13}}}{R_{I_{23}I_{12}}}\omega_3 = \frac{14.14 \text{ in}}{6.00 \text{ in}}(5 \text{ rad/s}) = 11.78 \text{ rad/s ccw.} \qquad \textit{Ans.} \ (6)$$

g. The velocity of I, from Eq. (4.51), is

$$\nu = \omega_3 R_{JI} = (5 \text{ rad/s})\ (19.26 \text{ in}) = 96.3 \text{ in/s.} \qquad \textit{Ans.} \ (7)$$

The direction of the velocity of I is as illustrated in Fig. 4.36. *Ans.*

h. The acceleration of the velocity pole I, from Eq. (4.50), is

$$A_I = \omega_3 \nu = (5\text{ rad/s})\,(96.3\text{ in/s}) = 481.5\text{ in/s}^2. \qquad \textit{Ans.}\ (8)$$

The orientation of the acceleration of I is as illustrated in Fig. 4.36. *Ans.*

i. The acceleration of coupler point C is found by rearranging Eq. (4.49), that is,

$$A_C = R_{\Gamma C}\sqrt{\omega_3^4 + \alpha_3^2} = (29.64\text{in})\sqrt{(5\text{rad/s})^4 + \left(20\text{rad/s}^2\right)^2} = 948.94\text{ in/s}^2. \qquad \textit{Ans.}\ (9)$$

The orientation of the acceleration of point C, from Eq. (4.48), is

$$\gamma = \tan^{-1}\frac{\alpha_3}{\omega_3^2} = \tan^{-1}\frac{-20\text{ rad/s}^2}{(5\text{ rad/s})^2} = -38.66°. \qquad \textit{Ans.}\ (10)$$

The orientation of the acceleration vector of point C is illustrated in Fig. 4.36.

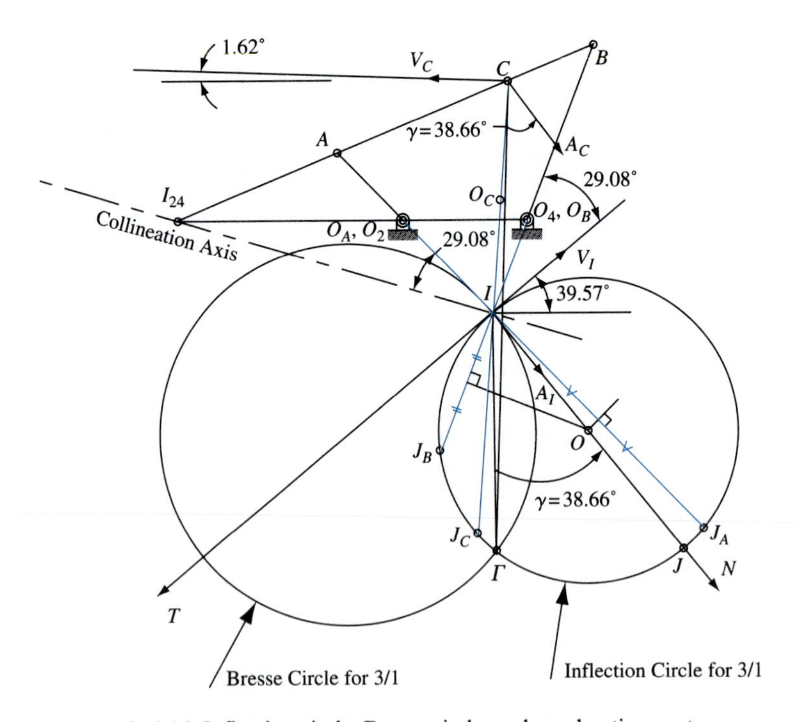

Figure 4.36 Example 4.14. Inflection circle, Bresse circle, and acceleration center.

4.17 RADIUS OF CURVATURE OF A POINT TRAJECTORY USING KINEMATIC COEFFICIENTS

The radius of curvature of a point trajectory (say point P), at the instant considered, can be written from Eqs. (4.2) and (4.3) or from Eq. (4.14) as

$$\rho = \frac{V_P^2}{A_P^n}. \tag{4.53}$$

From Eq. (3.34*a*), the speed of point P can be written as

$$V_P = r'_P \dot{\psi}. \tag{a}$$

Also, the normal component of the acceleration of point P can be written as

$$A_P^n = \mathbf{A}_P \cdot \hat{\mathbf{u}}^n, \tag{4.54}$$

where the unit normal vector to the point trajectory at the position considered [see Eq. (3.39)] is

$$\hat{\mathbf{u}}^n = \left(\frac{-y'_P}{r'_P}\right)\hat{\mathbf{i}} + \left(\frac{x'_P}{r'_P}\right)\hat{\mathbf{j}}. \tag{b}$$

Taking the time derivative of Eq. (3.33*b*), the acceleration of point P can be written as

$$\mathbf{A}_P = (x''_P \dot{\psi}^2 + x'_P \ddot{\psi})\hat{\mathbf{i}} + (y''_P \dot{\psi}^2 + y'_P \ddot{\psi})\hat{\mathbf{j}}. \tag{4.55}$$

Substituting Eqs. (*b*) and (4.55) into Eq. (4.54), the normal component of the acceleration of point P can be written as

$$A_P^n = \left(\frac{x'_P y''_P - y'_P x''_P}{r'_P}\right)\dot{\psi}^2. \tag{4.56}$$

Then, substituting Eqs. (*a*) and (4.56) into Eq. (4.53), the radius of curvature of the point trajectory at the position considered can be written as

$$\rho = \frac{r'^3_P}{x'_P y''_P - y'_P x''_P}. \tag{4.57}$$

Sign convention If the unit normal vector to the point trajectory $\hat{\mathbf{u}}^n$ points toward the center of curvature of the path, then the radius of curvature has a positive value. If the unit normal vector to the point trajectory points away from the center of curvature of the path, then the radius of curvature has a negative value.

The coordinates of the center of curvature of the point trajectory, at the position under investigation, can be written as

$$x_C = x_P - \rho\left(\frac{y'_P}{r'_P}\right) \quad \text{and} \quad y_C = y_P + \rho\left(\frac{x'_P}{r'_P}\right). \tag{4.58}$$

EXAMPLE 4.15

Determine the radius of curvature and the center of curvature of the path of point B for the Rapson slide mechanism presented in Example 4.11.

SOLUTION

Here we continue on from Example 4.11 in all respects, including equation numbers. Therefore, the unit tangent vector to the path of point B is the unit vector pointing in the direction of the velocity vector of point B. The unit tangent vector can be written as

$$\hat{\mathbf{u}}^t = \frac{X'_B\hat{\mathbf{i}} + Y'_B\hat{\mathbf{j}}}{r'_B}, \tag{21}$$

where, using the data of Eqs. (17),

$$r'_B = \pm\sqrt{X_B'^2 + Y_B'^2} = -\sqrt{(0.893\ 09\ \text{ft/ft})^2 + (0.515\ 63\ \text{ft/ft})^2} = -1.031\ 25\ \text{ft/ft}. \tag{22}$$

Note that the negative sign was chosen because the input vector R_2 is becoming shorter and, therefore, the input is negative. Substituting Eqs. (17) and (22) into Eq. (21) gives

$$\hat{\mathbf{u}}^t = \frac{X'_B\hat{\mathbf{i}} + Y'_B\hat{\mathbf{j}}}{r'_B} = \frac{(0.893\ 09\ \text{ft/ft})\hat{\mathbf{i}} + (0.515\ 63\ \text{ft/ft})\hat{\mathbf{j}}}{-1.031\ 25\ \text{ft/ft}} = -0.866\ 02\hat{\mathbf{i}} - 0.500\ 00\hat{\mathbf{j}}. \tag{23}$$

As a check, using Eq. (19), the unit tangent vector can also be written as

$$\hat{\mathbf{u}}^t = \frac{\mathbf{V}_B}{V_B} = \frac{-13.40\,\hat{\mathbf{i}} - 7.73\,\hat{\mathbf{j}}\ \text{ft/s}}{15.47\ \text{ft/s}} = -0.866\ 19\hat{\mathbf{i}} - 0.499\ 68\hat{\mathbf{j}}.$$

The unit normal vector is 90° counterclockwise from the unit tangent vector; that is, using Eqs. (17) and (22),

$$\hat{\mathbf{u}}^n = \hat{\mathbf{k}} \times \hat{\mathbf{u}}^t = \hat{\mathbf{k}}\times\frac{X'_B\hat{\mathbf{i}} + Y'_B\hat{\mathbf{j}}}{r'_B} = \frac{-Y'_B\hat{\mathbf{i}} + X'_B\hat{\mathbf{j}}}{r'_B} = 0.500\ 00\hat{\mathbf{i}} - 0.866\ 02\hat{\mathbf{j}}. \tag{24}$$

The direction of the unit tangent vector and the unit normal vector are illustrated in Fig. 4.37.

Using Eqs. (17), (18), and (22), the radius of curvature of the path of point B can be found from Eq. (4.57) as

$$\begin{aligned}\rho_B &= \frac{r_B'^3}{X'_BY''_B - Y'_BX''_B}\\ &= \frac{(-1.031\ 25\ \text{ft/ft})^3}{(0.893\ 09\ \text{ft/ft})\left(0.028\ \text{ft/ft}^2\right) - (0.515\ 63\ \text{ft/ft})\left(-0.145\ \text{ft/ft}^2\right)}. \qquad \textit{Ans.}\ (25)\\ &= -11.0\ \text{ft}\end{aligned}$$

The negative sign indicates that the unit normal vector points away from the center of curvature of the path of point B (see Fig. 4.37).

The coordinates of the center of curvature of the path of point B can be found from Eqs. (4.58). Using known values gives

$$X_C = X_B - \rho_B\left[\frac{Y'_B}{r'_B}\right] = 5.50\text{ ft} - (-11.0\text{ ft})\left[\frac{(0.515\,63\text{ ft/ft})}{(-1.031\,25\text{ ft/ft})}\right] = 0.00\text{ ft} \qquad (26a)$$

$$Y_C = Y_B + \rho_B\left[\frac{X'_B}{r'_B}\right] = -9.53\text{ ft} + (-11.0\text{ ft})\left[\frac{0.893\,09\text{ ft/ft}}{-1.031\,25\text{ ft/ft}}\right] = 0.00\text{ ft.} \qquad (26b)$$

Note that Eqs. (25), (26*a*), and (26*b*) make intuitive sense because point B is located on link 4 and link 4 is pinned at the origin O_4, and, therefore, must rotate about the origin. Figure 4.37 illustrates the directions of the unit tangent vector $\hat{\mathbf{u}}^t$ and the unit normal vector $\hat{\mathbf{u}}^n$. Note that the unit normal vector points away from the center of curvature of the path of point B. The center of curvature of the path of point B is coincident with the ground pin O_4. From observation, these answers are corroborated.

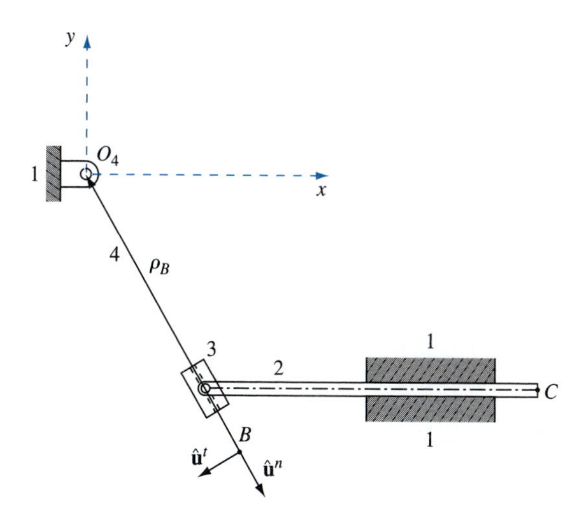

Figure 4.37 Example 4.15. The unit tangent and unit normal vectors and the center of curvature of the path of point B.

4.18 THE CUBIC OF STATIONARY CURVATURE

Now let us consider a point fixed in the coupler link of a planar four-bar linkage that generates a path relative to the frame whose radius of curvature, at the instant considered, is ρ. Because the coupler curve is, in general, a tricircular sextic[4], this radius of curvature changes continuously as the point moves. In certain situations, however, the path will have stationary curvature, which means that

$$\frac{d\rho}{ds} = 0, \qquad (a)$$

where s is the increment traveled along the path. The locus of all points in the coupler, or moving plane, which have stationary curvature at the instant considered is called the *cubic of stationary curvature* or sometimes the *circling-point curve*. We should note that stationary curvature does not necessarily mean constant curvature, but rather that the continually varying radius of curvature is passing through a maximum or minimum.

Here we present a quick and easy graphical method for obtaining the cubic of stationary curvature for a coupler link, as described by Hain.[5] Consider the four-bar linkage $A'ABB'$ illustrated in Fig. 4.38, where A' and B' are the frame pivots. Note that the loci of points A and B have stationary curvature (in fact, they have constant curvature about centers at A' and B'); hence, A and B must lie on the cubic.

The first step of the construction is to obtain the centrode normal and the centrode tangent. Because the inflection circle is not needed, we locate the collineation axis IQ as indicated and draw the centrode tangent T at the angle ψ from the line IB', equal to the angle from the collineation axis to the line IA'. This construction follows directly from Bobillier's theorem. We also construct the centrode normal N. At this time it may be convenient to reorient the drawing on the working surface so that the centrode normal lies along the horizontal axis.

Next, we construct a line through A perpendicular to IA and another line through B perpendicular to IB. These lines intersect the centrode normal and centrode tangent at A_N, A_T and B_N, B_T, respectively, as illustrated in Fig. 4.38. Now we draw the two rectangles $IA_NA_GA_T$ and $IB_NB_GB_T$; the points A_G and B_G define an auxiliary line G that we will use to obtain other points on the cubic of stationary curvature.

Next, we choose any point S_G on the line G. A ray parallel to N locates S_T, and another ray parallel to T locates S_N. Connecting S_T with S_N and drawing a perpendicular to this line through I locates point S, another point on the cubic of stationary curvature. We now repeat this process as often as desired by choosing different points on the line G, and we draw the cubic as a smooth curve through all the points S obtained.

Note that the cubic of stationary curvature has two tangents at the velocity pole I: one is the *centrode-tangent tangent* and the other is the *centrode-normal tangent*. The radius of curvature of the cubic at these tangents is obtained as follows: Extend the line G to intersect T at G_T and N at G_N as indicated. Then, half the distance IG_T is the radius of curvature of the cubic tangent to the centrode normal, and half the distance IG_N is the radius of curvature of the cubic tangent to the centrode tangent.

A point with interesting properties is the point of intersection of the cubic of stationary curvature with the inflection circle (other than the velocity pole I); this point is called *Ball's point*. The point fixed in the coupler that is coincident with Ball's point describes a path that is practically rectilinear for a considerable distance[6]; that is, it describes an excellent straight line near the design position because it is located at an inflection point of its path and has stationary curvature.

The equation of the cubic of stationary curvature[7] can be written in polar coordinates as

$$\frac{1}{r} = \frac{1}{M \sin \psi} - \frac{1}{N \cos \psi}, \tag{4.59}$$

where r is the distance from the velocity pole I to the point on the cubic, measured at an angle ψ from the centrode tangent. The constant parameters M and N are determined using any two points known to lie on the cubic, such as points A and B of Fig. 4.38. Equations for M and N can be written as

$$\frac{1}{M} = \frac{1}{3}\left(\frac{1}{R_{JI}} - \frac{1}{R_{IO_M}}\right) \quad \text{and} \quad \frac{1}{N} = \frac{1}{3R_{JI}}\left(\frac{dR_{JI}}{ds}\right). \tag{4.60}$$

It so happens[8] that M and N are, respectively, the diameters IG_T and IG_N of the circles centered on the centrode tangent and centrode normal whose radii represent the two curvatures of the cubic at the instant center.

The equation of the cubic of stationary curvature can also be expressed in Cartesian coordinates as

$$(x^2 + y^2)\left(\frac{x}{M} - \frac{y}{N}\right) = xy. \tag{4.61}$$

Degenerate forms. From Eq. (4.59) or Eq. (4.61), we observe that the cubic of stationary curvature will degenerate to a circle and a straight line when either: (*i*) N tends to infinity, that is, $1/N$ approaches zero; or (*ii*) M tends to infinity, that is, $1/M$ approaches zero. Consider the following example.

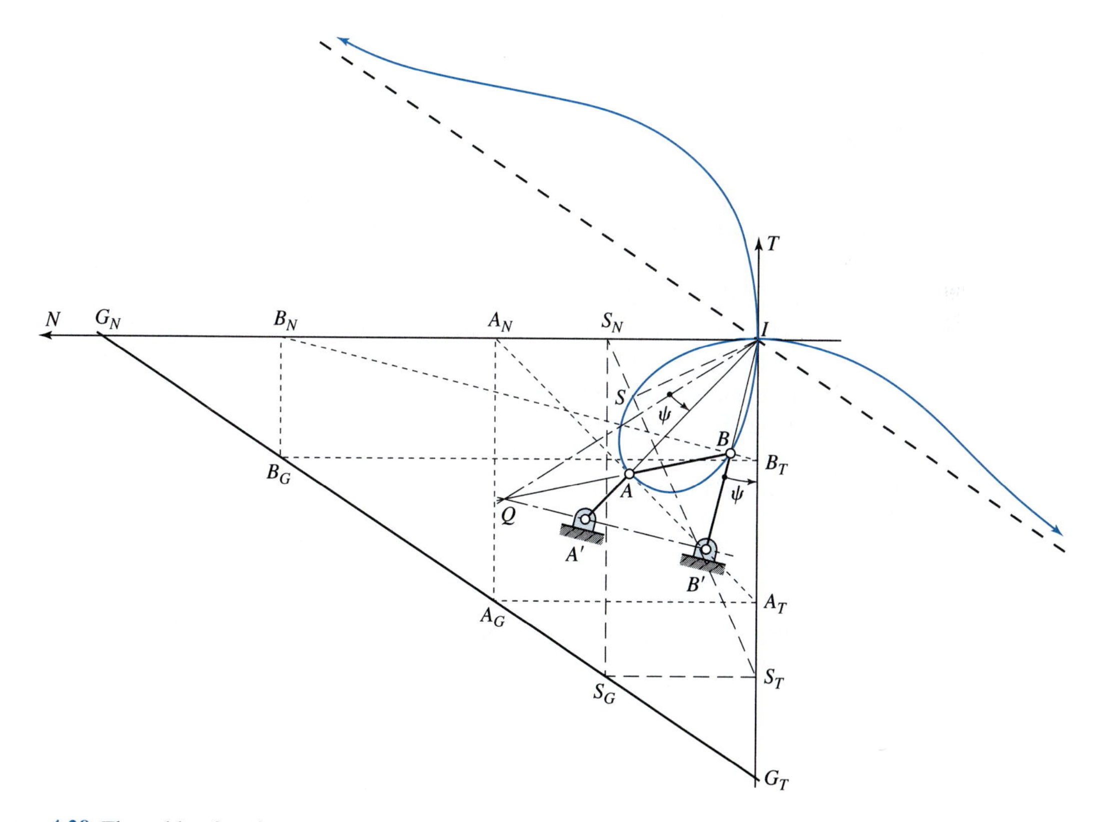

Figure 4.38 The cubic of stationary curvature.

EXAMPLE 4.16

For the four-bar linkage illustrated in Fig. 4.39, the link dimensions are $R_{O_2O_4} = 1$ in., $R_{O_2A} = 1$ in., and $R_{AB} = 2$ in. In the design position, link 2 is perpendicular to the frame O_2O_4 and the coupler link AB is parallel to the frame.

For the absolute motion of the coupler link, in the specified position, determine: (*i*) the diameter of the inflection circle; (*ii*) the cubic of stationary curvature, the values of the parameters M and N; clearly indicate the location of Ball's point; (*iii*) the radius of curvature and center of curvature of the moving centrode; (*iv*) the radius of curvature and the center of curvature of the fixed centrode; and (*v*) the radius of curvature and center of curvature of coupler point C, which is midway between pins A and B.

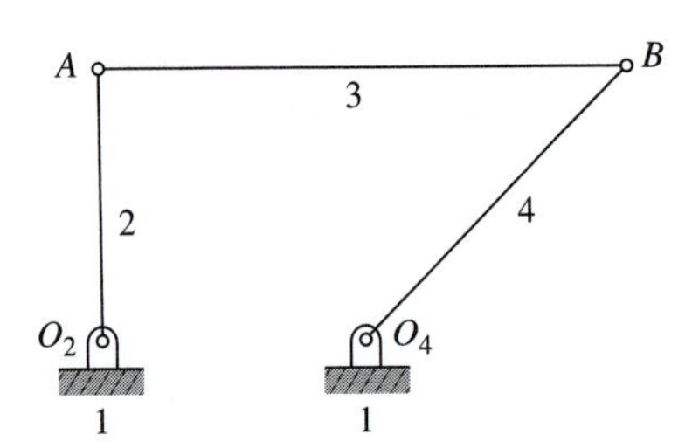

Figure 4.39 Example 4.16. A planar four-bar linkage.

SOLUTION

i. The velocity pole I is coincident with the instant center I_{13}, as illustrated in Fig. 4.39. The instant center I_{24} (with label Q) lies at infinity (say to the right) and the collineation axis IQ is parallel to the coupler link. From Bobillier's theorem, the angle from the collineation axis to the first ray (link 2) is $\alpha = 90°$ ccw. This is equal to the angle from the second ray (link 4) to the centrode tangent T. Therefore, the centrode tangent T and the centrode normal N (which is 90° ccw from the centrode tangent T) are as illustrated in Fig. 4.40. Note that link 4 lies along the centrode normal N.
The location of the inflection point J_A for point A on the coupler link is obtained from the Euler–Savary equation, Eq. (4.47); that is,

$$R_{AJ_A} = \frac{R_{AI}^2}{R_{AA'}} = \frac{(2\text{ in})^2}{1\text{ in}} = 4\text{ in.} \tag{1}$$

The inflection point J_B for point B on the coupler link is obtained in a similar manner, that is,

$$R_{BJ_B} = \frac{R_{BI}^2}{R_{BB'}} = \frac{\left(2\sqrt{2}\text{ in}\right)^2}{\sqrt{2}\text{ in}} = 5.66\text{ in.} \tag{2}$$

The location of the inflection points J_A and J_B are illustrated in Fig. 4.39. Note that J_B lies on the centrode normal N; therefore, this inflection point is coincident with the inflection pole J.

Knowing the centrode normal N and the two inflection points, the inflection circle for the motion of link 3 with respect to 1 can now be drawn. The diameter of the inflection circle for the motion 3/1 is

$$R_{JI} = \frac{1}{2}R_{JB} = 2\sqrt{2}\text{ in} = 2.83\text{ in} \qquad \textit{Ans.}\ (3)$$

The inflection circle, the inflection pole J, and the center of the inflection circle (denoted point O) are illustrated in Fig. 4.40. Note that the centrode normal N is directed from the velocity pole I toward the inflection pole J and the centrode tangent T is 90° clockwise from the centrode normal N.

ii. The cubic of stationary curvature equation for the coupler link, including pins A and B, can be written from Eq. (4.59) as

$$\frac{1}{R_{AI}} = \frac{1}{M\sin\psi_A} - \frac{1}{N\cos\psi_A} \qquad (4a)$$

and also as

$$\frac{1}{R_{BI}} = \frac{1}{M\sin\psi_B} - \frac{1}{N\cos\psi_B}, \qquad (4b)$$

where the parameters M and N are as given by Eqs. (4.60). The angle ψ_A is the counterclockwise angle from the centrode tangent T to the ray containing point A and the angle ψ_B is the counterclockwise angle from the centrode tangent T to the ray containing point B. From the scale drawing, these angles are measured as

$$\psi_A = -45° \quad \text{and} \quad \psi_B = -90° \qquad (5a)$$

Note that these measurements can easily be verified from trigonometry.
The distances to pin A from the velocity pole I and to pin B from the velocity pole I are measured consistently as

$$R_{AI} = 2\text{ in} \quad \text{and} \quad R_{BI} = 2\sqrt{2}\text{ in}. \qquad (5b)$$

Substituting Eqs. (5*a*) into Eq. (4*b*) gives

$$\frac{1}{R_{BI}} = \frac{1}{-M}. \qquad (6)$$

Therefore, using Eq. (5*b*), we find the parameter

$$M = -R_{BI} = R_{JI} = -2\sqrt{2}\text{ in}. \qquad \textit{Ans.}\ (7)$$

Note that substituting $\psi_A = -45°$ into Eq. (4*a*) indicates that $1/N = 0$ and the parameter N tends to infinity. *Ans.*
This is a degenerate form of the cubic of stationary curvature. The cubic of stationary curvature degenerates to a straight line (that is, the centrode normal) and a circle of diameter M (the center must lie on the centrode normal). The circle

must pass through pins A and B on link 3. Therefore, the center of the circle is coincident with the ground pin O_4. *Ans.*

Substituting the condition $N = \infty$ into Eq. (4.60) gives

$$\frac{1}{3R_{JI}}\left(\frac{dR_{JI}}{ds}\right) = 0. \tag{8a}$$

Because the diameter of the inflection circle is infinite [see Eq. (3)], the rate of change in the inflection circle diameter must be zero, that is,

$$\frac{dR_{JI}}{ds} = 0. \tag{8b}$$

In other words, the diameter of the inflection circle is at a maximum or a minimum as the four-bar linkage passes through this position.

Recall that Ball's point is located at the intersection of the inflection circle and the cubic of stationary curvature. Figure 4.39 illustrates two apparent points of intersection, that is, the velocity pole I and the inflection pole J. Since the velocity pole is not a point fixed in the coupler link then it is not a solution. Therefore, Ball's point is coincident with the inflection pole J. *Ans.*

iii. Rearranging Eq. (4.60), the radius of curvature of the moving centrode can be written as

$$\frac{1}{R_{IO_M}} = \frac{1}{R_{JI}} - \frac{3}{M}. \tag{9}$$

Then substituting Eq. (7) into this equation gives

$$\frac{1}{R_{IO_M}} = \frac{1}{R_{JI}} - \frac{3}{R_{JI}} = -\frac{2}{R_{JI}}. \tag{10a}$$

Therefore, the radius of curvature of the moving centrode is

$$R_{IO_M} = -\frac{R_{JI}}{2} = \sqrt{2}\text{ in.} \qquad \textit{Ans.} \tag{10b}$$

The negative signs in these last two equations indicate that the direction from O_M to I is opposite to the direction from I to J. Therefore, the moving centrode is coincident with (or coalesces with) the inflection circle (see Fig. 4.40).

iv. The Euler–Savary equation can be written as

$$\frac{1}{R_{JI}} = \frac{1}{R_{IO_F}} - \frac{1}{R_{IO_M}}. \tag{11}$$

Substituting Eq. (10b) into this equation and rearranging, the radius of curvature of the fixed centrode can be written as

$$R_{IO_F} = -R_{JI} = 2\sqrt{2}\text{ in.} \qquad \textit{Ans.} \tag{12}$$

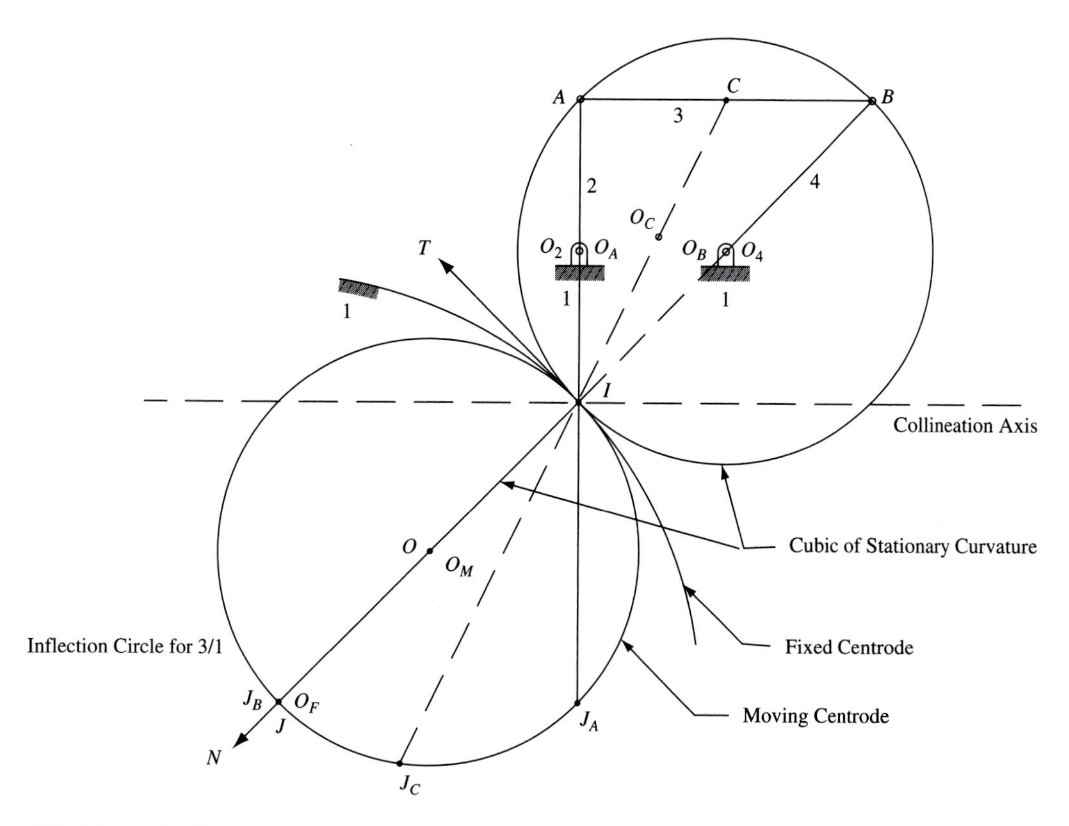

Figure 4.40 The cubic of stationary curvature for the coupler link.

The negative sign indicates that the direction from O_F to I is opposite to the direction from I to J. The radius of the osculating circle for the fixed centrode is the same as the diameter of the inflection circle, that is, O_F is coincident with inflection pole J (see Fig. 4.40).

v. The position of the coupler point C is illustrated in Fig. 4.40. Using complex polar notation, this can be written as

$$\mathbf{R}_{CI} = x_{CI}\hat{\mathbf{i}} + y_{CI}\hat{\mathbf{j}} = jR_{O_2I} + R_{AO_2}e^{j\theta_2} + R_{CA}e^{j\theta_3}, \tag{13}$$

which has real and imaginary components of

$$x_{CI} = R_{AO_2}\cos\theta_2 + R_{CA}\cos\theta_3 \quad \text{and} \quad y_{CI} = R_{O_2I} + R_{AO_2}\sin\theta_2 + R_{CA}\sin\theta_3. \tag{14}$$

Substituting the know data $R_{O_2I} = 1$ in., $R_{AO_2} = 1$ in., $R_{CA} = 1$ in. into Eq. (14) gives

$$x_{CI} = (1\text{ in})\cos\theta_2 + (1\text{ in})\cos\theta_3 \quad \text{and}$$
$$y_{CI} = (1\text{ in}) + (1\text{ in})\sin\theta_2 + (1\text{ in})\sin\theta_3. \tag{15}$$

For this position, $\theta_2 = 90°$ and $\theta_3 = 0$; $x_{CI} = 1$ in. and $y_{CI} = 2$ in. Taking the derivative of Eqs. (14) with respect to the input angle θ_2 gives

$$x'_{CI} = -R_{AO_2} \sin\theta_2 - \theta'_3 R_{CA} \sin\theta_3 \quad \text{and} \quad y'_{CI} = R_{AO_2} \cos\theta_2 + \theta'_3 R_{CA} \cos\theta_3. \tag{16}$$

Next we use the angular velocity ratio theorem, Eq. (3.30), to find the first-order kinematic coefficient,

$$\theta'_3 = \frac{R_{I_{23}I_{12}}}{R_{I_{23}I_{13}}} = \frac{1 \text{ in}}{2 \text{ in}} = 0.5 \text{ rad/rad}. \tag{17}$$

Then, using the dimensions given above and evaluating at the same position gives

$$x'_{CI} = -1 \text{ in/rad} \quad \text{and} \quad y'_{CI} = 0.5 \text{ in/rad} \tag{18}$$

and, from Eq. (3.34*b*),

$$r'_{CI} = \sqrt{(x'_{CI})^2 + (y'_{CI})^2} = \sqrt{(-1 \text{ in/rad})^2 + (0.5 \text{ in/rad})^2} = 1.12 \text{ in/rad}. \tag{19}$$

Measuring from the centrode tangent and centrode normal axes, the polar coordinates of coupler point C are

$$R_{CI} = \sqrt{(2 \text{ in})^2 + (1 \text{ in})^2} = \sqrt{5} \text{ in} \quad \text{and} \quad \psi_C = -71.57°. \tag{20}$$

The inflection point for point C, from Eq. (4.46), is

$$R_{J_C I} = R_{JI} \sin\psi_C = -2.68 \text{ in}. \tag{21}$$

Then, from Eq. (4.47), the radius of curvature of the path of coupler point C is

$$\rho_C = R_{CC'} = \frac{R^2_{CI}}{R_{CI} - R_{J_C I}} = \frac{\left(\sqrt{5} \text{ in}\right)^2}{\left(\sqrt{5} \text{ in}\right) - (-2.68 \text{ in})} = 1.02 \text{ in}. \quad \textit{Ans.} \tag{22}$$

Finally, measuring from point I, Eqs. (4.58) give the center of curvature of the coupler curve as

$$x_{C'I} = x_{CI} - \rho_C \left(\frac{y'_C}{r'_C}\right) = (1 \text{ in}) - (1.02 \text{ in}) \left(\frac{0.5 \text{ in/rad}}{1.12 \text{ in/rad}}\right) = 0.54 \text{ in} \quad \textit{Ans.} \tag{23}$$

$$y_{C'I} = y_{CI} - \rho_C \left(\frac{x'_C}{r'_C}\right) = (2 \text{ in}) - (1.02 \text{ in}) \left(\frac{1 \text{ in/rad}}{1.12 \text{ in/rad}}\right) = 1.09 \text{ in}. \quad \textit{Ans.} \tag{24}$$

These results are in good agreement with measurements from Fig. 4.40.

4.19 REFERENCES

[1] Rosenauer, N., and A. H. Willis. *Kinematics of Mechanisms.* Sydney, Australia: Associated General Publications, 1953, 161; reprinted New York: Dover, 1967.

[2] Among the most important and most useful references on this subject are N. Rosenauer and A. H. Willis, *op. cit.*, Chap. 4; A. E. R. deJonge, A Brief Account of Modern Kinematics, *ASME Trans.* 65 (1943): 663–83; R. S. Hartenberg and J. Denavit, *Kinematic Synthesis of Linkages*, New York: McGraw–Hill, 1964, Chap. 7; A. S. Hall, Jr., *Kinematics and Linkage Design*, Englewood Cliffs, NJ: Prentice–Hall, 1961, Chap. 5 (this book is a classic on the theory of mechanisms and contains many useful examples); K. Hain (translated by T. P. Goodman et al.), *Applied Kinematics*, 2nd ed., New York: McGraw–Hill, 1967, Chap. 4.

[3] Rosenauer, N., and A. H. Willis, *op. cit.*, pp. 145–67; K. Hain, op. cit., pp. 148–58.

[4] R. Beyer, *Technische Kinematik*, Johann Ambrosius Barth, Leipzig, 1931 (Lithoprint publication by J. W. Edwards, Ann Arbor, MI, 1948), pp. 313–4.

[5] K. Hain, *op. cit.*, pp. 498–502.

[6] J. Hirschhorn, *Kinematics and Dynamics of Plane Mechanisms*, New York: McGraw–Hill, 1962, 357.

[7] For a derivation of this equation see A. S. Hall, Jr., *op. cit.*, p. 98, or R. S. Hartenberg and J. Denavit, *op. cit.*, p. 206.

[8] Tao, D. C. *Applied Linkage Synthesis*, Reading, MA: Addison–Wesley, 1964, 111.

PROBLEMS*

†4.1 The position vector of a point is defined by the equation

$$\mathbf{R} = \left(4t - \frac{t^3}{3}\right)\hat{\mathbf{i}} + 10\hat{\mathbf{j}},$$

where R is in inches and t is in seconds. Find the acceleration of the point at $t = 2$ s.

†4.2 Find the acceleration at $t = 3$s of a point that moves according to the equation

$$\mathbf{R} = \left(t^2 - \frac{t^3}{6}\right)\hat{\mathbf{i}} + \frac{t^3}{3}\hat{\mathbf{j}}.$$

The units are meters and seconds.

4.3 The path of a point is described by the equation

$$\mathbf{R} = (t^2 + 4)e^{-j\pi t/10},$$

where R is in millimeters and t is in seconds. For $t = 20$ s, find the unit tangent vector for the path, the normal and tangential components of the point's absolute acceleration, and the radius of curvature of the path.

4.4 The motion of a point is described by the equations

$$x = 4t\, cos\pi t^3 \text{ and } y = \frac{t^3 \sin 2\pi t}{6},$$

where x and y are in feet and t is in seconds. Find the acceleration of the point at $t = 1.40$ s.

†4.5 Link 2 in Fig. P4.5 has an angular velocity of $\omega_2 = 120$ rad/s ccw and an angular acceleration of 4800 rad/s^2 ccw at the instant indicated. Determine the absolute acceleration of point A.

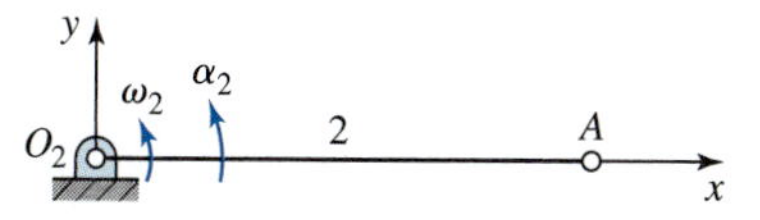

Figure P4.5 $R_{AO_2} = 500$ mm.

* When assigning problems, the instructor may wish to specify the method of solution to be used because a variety of approaches are presented in the text.

4.6 Link 2 is rotating clockwise as illustrated in Fig. P4.6. Find its angular velocity and acceleration and the acceleration of its midpoint C.

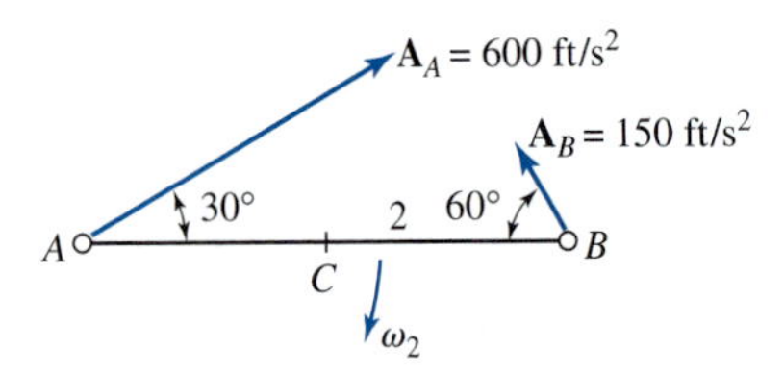

Figure P4.6 $R_{BA} = 20$ in.

4.7 For the data given in Fig. P4.7, find the velocity and acceleration of points B and C.

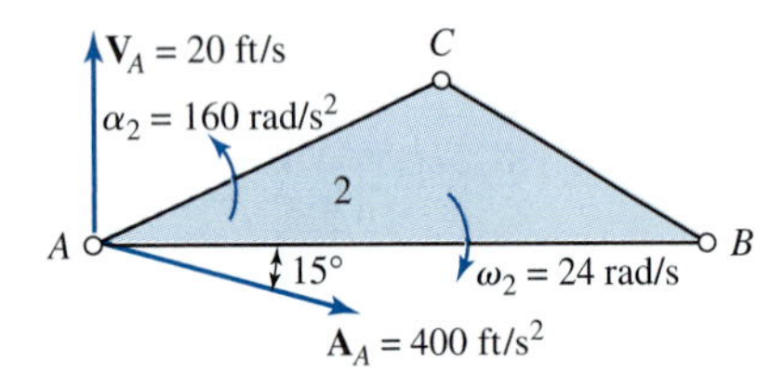

Figure P4.7 $R_{BA} = 16$ in, $R_{CA} = 10$ in, $R_{CB} = 8$ in.

†4.8 For the straight-line mechanism illustrated in Fig. P4.8, $\omega_2 = 20$ rad/s cw and $\alpha_2 = 140$ rad/s^2 cw. Determine the velocity and acceleration of point B and the angular acceleration of link 3.

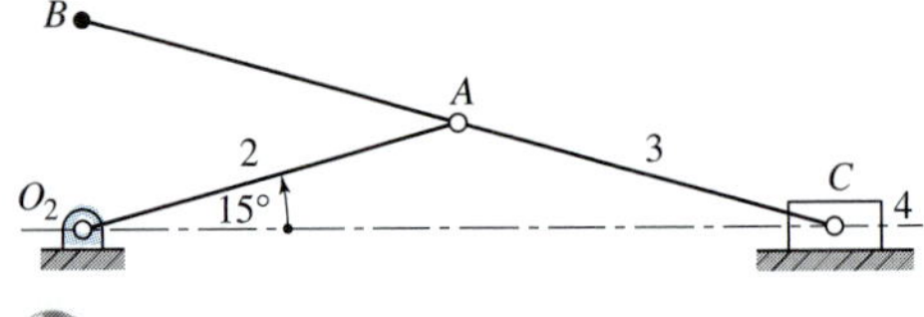

Figure P4.8 $R_{AO_2} = R_{CA} = R_{BA} = 100$ mm.

†4.9 In Fig. P4.8, the slider 4 is moving to the left with a constant velocity of 2.0 m/s. Find the angular velocity and angular acceleration of link 2.

†4.10 Solve Problem 3.8 using constant input velocity for the acceleration of point A and the angular acceleration of link 3.

†4.11 For Problem 3.9, using constant input velocity, find the angular accelerations of links 3 and 4.

†4.12 For Problem 3.10, using constant input velocity, find the acceleration of point C and the angular accelerations of links 3 and 4.

†4.13 For Problem 3.11, using constant input velocity, find the acceleration of point C and the angular accelerations of links 3 and 4.

†4.14 Using the data of Problem 3.13 and assuming constant input velocity, solve for the accelerations of points C and D and the angular acceleration of link 4.

†4.15 For Problem 3.14, using constant input velocity, find the acceleration of point C and the angular acceleration of link 4.

†4.16 Solve Problem 3.16 using constant input velocity for the acceleration of point C and the angular acceleration of link 4.

4.17 For Problem 3.17, using constant input velocity, find the acceleration of point B and the angular accelerations of links 3 and 6.

4.18 For the data of Problem 3.18, what angular acceleration must be given to link 2 for the position indicated to make the angular acceleration of link 4 zero?

4.19 For the data of Problem 3.19, what angular acceleration must be given to link 2 for the angular acceleration of link 4 to be 100 rad/s^2 cw at the instant indicated?

†4.20 Solve Problem 3.20 using constant input velocity for the acceleration of point C and the angular acceleration of link 3.

†4.21 For Problem 3.21, using constant input velocity, find the acceleration of point C and the angular acceleration of link 3.

4.22 Find the accelerations of points B and D of Problem 3.22 using constant input velocity.

4.23 Find the accelerations of points B and D of Problem 3.23 using constant input velocity.

†4.24 to 4.30 The nomenclature for this group of problems is illustrated in Fig. P4.24, and the dimensions and data are given in Table P4.24 to P4.30. For each problem, determine θ_3, θ_4, ω_3, ω_4, α_3, and α_4. The angular velocity ω_2 is constant for each problem, and a negative sign is used to indicate the clockwise direction. The dimensions of even-numbered problems are given in inches and odd-numbered problems are given in millimeters.

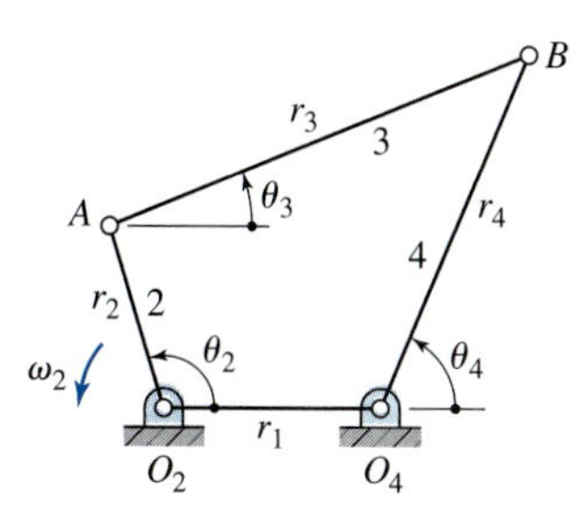

Figure P4.24

Table P4.24 to P4.30

Prob.	r_1	r_2	r_3	r_4	θ_2, deg	ω_2, rad/s
P4.24	4	6	9	10	240	1
P4.25	100	150	250	250	−45	56
P4.26	14	4	14	10	0	10
P4.27	250	100	500	400	70	−6
P4.28	8	2	10	6	40	12
P4.29	400	125	300	300	210	−18
P4.30	16	5	12	12	315	−18

4.31 Crank 2 of the system illustrated in Fig. P4.31 has a constant speed of 60 rev/min ccw. Find the velocity and acceleration of point B and the angular velocity and acceleration of link 4.

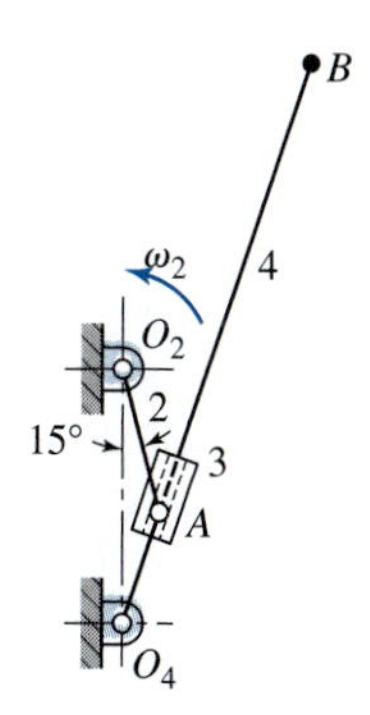

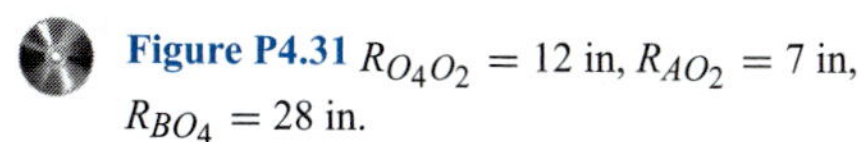
Figure P4.31 $R_{O_4O_2} = 12$ in, $R_{AO_2} = 7$ in, $R_{BO_4} = 28$ in.

4.32 Determine the acceleration of link 4 of Problem 3.26 assuming constant input velocity.

4.33 For Problem 3.27, using constant input velocity, find the acceleration of point E.

4.34 Find the acceleration of point B and the angular acceleration of link 4 of Problem 3.24 using constant input velocity.

4.35 For Problem 3.25, using constant input velocity, find the acceleration of point B and the angular acceleration of link 3.

4.36 Solve Problem 3.31 for the accelerations of points A and B assuming constant input velocity.

4.37 For Problem 3.32, determine the acceleration of point C_4 and the angular acceleration of link 3 if crank 2 is given an angular acceleration of 2 rad/s^2 ccw.

4.38 Determine the angular accelerations of links 3 and 4 of Problem 3.29 assuming constant input velocity.

4.39 For Problem 3.30, using constant input velocity, determine the acceleration of point G and the angular accelerations of links 5 and 6.

4.40 Continue with Problem 3.40 and find the second-order kinematic coefficients of links 3 and 4. Assuming an input acceleration of $A_{A_2} = 5$ m/s^2, find the angular accelerations of links 3 and 4.

4.41 Continue with Problem 3.49 and find the second-order kinematic coefficients of links 3, 4, and 5. Assuming constant angular velocity for link 2, find the angular accelerations of links 3, 4, and 5.

4.42 Continue with Problem 3.50 and find the second-order kinematic coefficients of links 3, 4, and 5. Assuming constant angular velocity for link 2, find the angular accelerations of links 3, 4, and 5.

4.43 Find the inflection circle for motion of the coupler of the double-slider mechanism illustrated in Fig. P4.43. Select several points on the centrode normal and find their conjugate points. Plot portions of the paths of these points to demonstrate for yourself that the conjugates are indeed the centers of curvature.

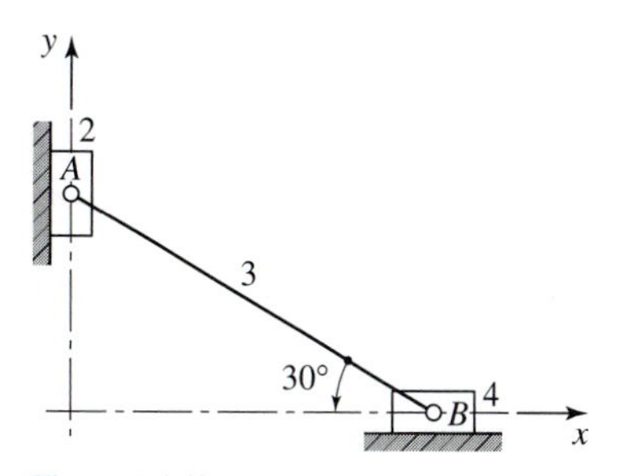

Figure P4.43 $R_{BA} = 125$ mm.

4.44 Find the inflection circle for motion of the coupler relative to the frame of the linkage illustrated in Fig. P4.44. Find the center of curvature of the coupler curve of point C and generate a portion of the path of C to verify your findings.

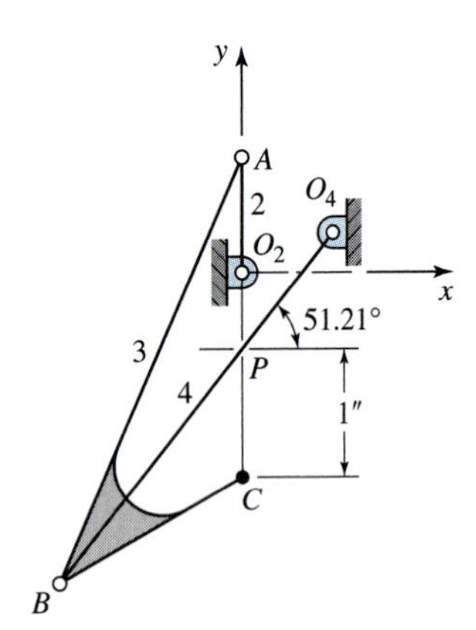

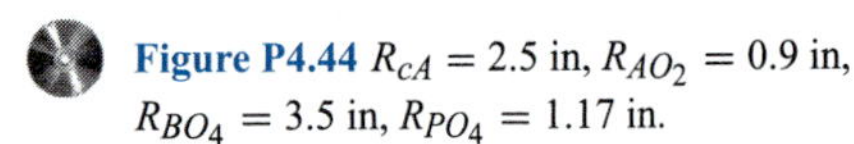
Figure P4.44 $R_{cA} = 2.5$ in, $R_{AO_2} = 0.9$ in, $R_{BO_4} = 3.5$ in, $R_{PO_4} = 1.17$ in.

4.45 For the motion of the coupler relative to the frame, find the inflection circle, the centrode normal, the centrode tangent, and the centers of curvature of points C and D of the linkage of Problem 3.13. Choose points on the coupler coincident with the instant center and inflection pole and plot nearby portions of their paths.

4.46 The planar four-bar linkage illustrated in Fig. P4.46 has link dimensions $R_{O_4O_2} = 2.5$ in, $R_{AO_2} = 1$ in, $R_{BA} = 3.15$ in, and $R_{BO_4} = 1.5$ in. For the position indicated, link 2 is 30° counterclockwise from the ground link O_2O_4 and the angular velocity and angular acceleration of the coupler link AB are $\omega_3 = 5$ rad/s ccw and $\alpha_3 = 20$ rad/s^2 cw, respectively. For the instantaneous motion of the coupler link AB, show: (*a*) the velocity pole I, pole tangent T, and pole normal N; (*b*) the inflection circle and the Bresse circle; and (*c*) the acceleration center of the coupler link AB. Then determine: (*d*) the radius of curvature of the path of coupler point C where $R_{CB} = 1$ in, (*e*) the magnitude and direction of the velocity of point C, (*f*) the magnitude and direction of the angular velocity of link 2, (*g*) the magnitude and direction of the velocity of the pole, (*h*) the magnitude and direction of the acceleration of point C, and (*i*) the magnitude and direction of the acceleration of the velocity pole.

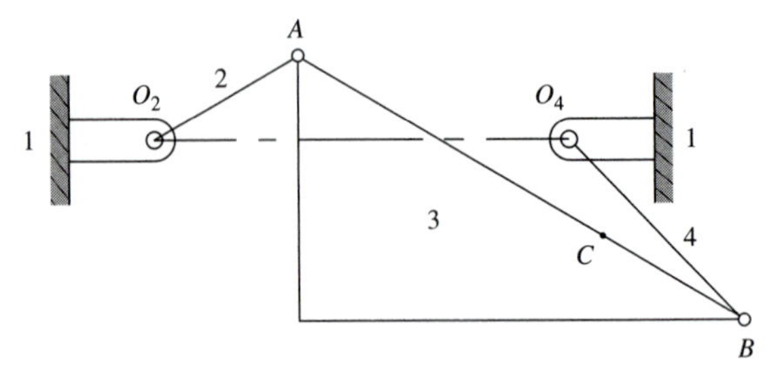

Figure P4.46

4.47 Consider the double-slider mechanism in the position given in Problem 3.8. Point B moves with a constant velocity $V_B = 40$ m/s to the left, as illustrated in the figure. The angular velocity and angular acceleration of coupler link AB are $\omega_3 = 36.6$ rad/s ccw and $\alpha_3 = 1\,340$ rad/s^2 cw, respectively. For the absolute motion of coupler link AB in the specified position, draw the inflection circle and the Bresse circle. Then determine: (*a*) the radius of curvature of the path of point C, which is a point in link 3 midway between points A and B; and (*b*) the magnitude and direction of the velocity of the velocity pole I. Using the acceleration pole, determine: (*c*) the magnitude and direction of the acceleration of I; (*d*) the magnitude and direction of the velocity of points A and C; and (*e*) the magnitude and direction of the acceleration of points A and C.

4.48 For the mechanism of Problem 3.17, link 2 is rotating with an angular velocity $\omega_2 = 15$ rad/s ccw and an angular acceleration $\alpha_2 = 320.93$ rad/s^2 cw. For the instantaneous motion of the connecting rod 3, find: (i) the inflection circle and the Bresse circle; (ii) the location of the acceleration pole; (iii) the center of curvature of the path traced by the coupler point C; (iv) the magnitude and direction of the velocity and acceleration of points A, B, and C; and (v) the magnitude and direction of the velocity and acceleration of inflection pole J.

4.49 Figure P3.32 illustrates an epicyclic gear train driven by the arm, link 2, with an angular velocity $\omega_2 = 3.33$ rad/s cw and an angular acceleration $\alpha_2 = 15$ rad/s^2 ccw. Define point E as a point on the circumference of the planet gear 4 horizontal to the right of point B such that the angle $\angle DBE = 90^o$. For the absolute motion of the planet gear 4, draw the inflection circle and the

Bresse circle on a scaled drawing of the epicyclic gear train. Then determine: (*a*) the location of the acceleration center of the planet gear; (*b*) the radii of curvature of the paths of points B and E; (*c*) the locations of the centers of curvature of the paths of points B and E; (*d*) the magnitudes and the directions of the velocities of points B and E and pole I; and (*e*) the magnitudes and directions of the accelerations of points B and E and pole I.

4.50 On 18 × 24 in. paper, draw the linkage illustrated in Fig. P4.50 in full size, placing A' 6 in. from the lower edge and 7 in. from the right edge. Better utilization of the paper is obtained by tilting the frame through about 15°, as indicated. (*a*) Find the inflection circle. (*b*) Draw the cubic of stationary curvature. (*c*) Choose a coupler point C coincident with the cubic and plot a portion of its coupler curve in the vicinity of the cubic. (*d*) Find the conjugate point C'. Draw a circle through C with center at C' and compare this circle with the actual path of C. (*e*) Find Ball's point. Locate a point D on the coupler at Ball's point and plot a portion of its path. Compare the result with *a* straight line.

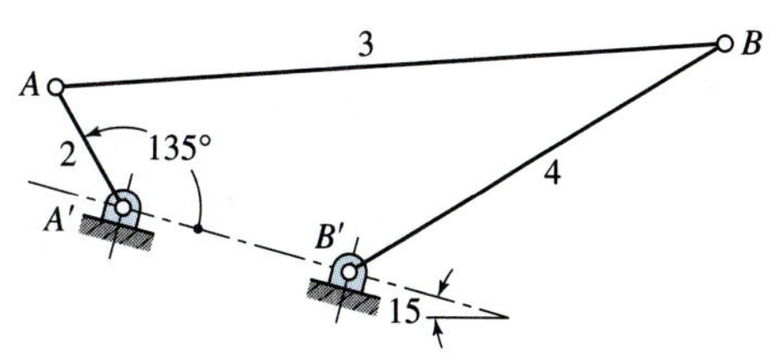

Figure P4.50 $R_{AA'} = 1$ in, $R_{BA} = 5$ in, $R_{B'A'} = 1.75$ in, $R_{BB'} = 3.25$ in.

5 Multi-Degree-of-Freedom Planar Linkages

5.1 INTRODUCTION

In Chapters 2, 3, and 4 we concentrated totally on problems that exhibit a single degree of freedom and can be analyzed by giving the motion of a single input variable. This was justifiable because, by far, the large majority of practical linkages are designed to have only one degree of freedom so that they can be driven by a single power source. However, there are mechanisms that have multiple degrees of freedom and can only be analyzed if more than one input motion is given. In this chapter we will look at how our methods can be used to find the positions, velocities, and accelerations of these mechanisms.

The Kutzbach criterion, Eq. (1.1), demonstrates that the planar five-bar linkage illustrated in Fig. 5.1 has a mobility of 2 and requires two input motions to provide a unique output motion. This mechanism is operated by rotating cranks 2 and 3 independently and can, therefore, produce a wide variety of motions for the two coupler links 4 and 5.

A practical application of the five-bar linkage is to position the end effector of an industrial robotic manipulator, for example, the General Electric Model P80 robotic manipulator illustrated in Fig. 5.2.

Another more common practical application of the five-bar linkage is the pantograph linkage illustrated in Fig. 1.26. A variation of the pantograph is the linkage illustrated in Fig. 5.3, where the path of point P is a magnified copy of the path of point C.

If the two input rotations of Fig. 5.1 are interconnected by gears that have rolling contact with each other, as illustrated in Fig. 5.4, then the resulting mechanism has only

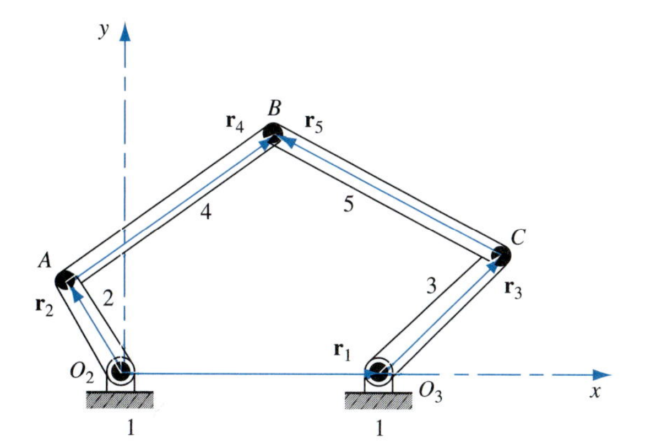

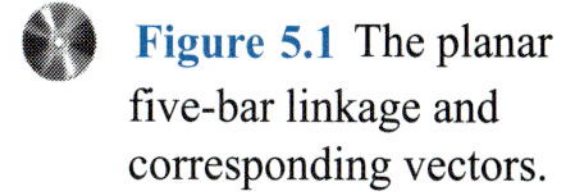

Figure 5.1 The planar five-bar linkage and corresponding vectors.

Figure 5.2 General Electric Model P80 robotic manipulator.

a single degree of freedom and is commonly referred to as a *geared five-bar linkage*. Such a linkage is commonly found in machinery because it can provide more complex motions than the well-known planar four-bar linkage that was investigated in previous chapters.

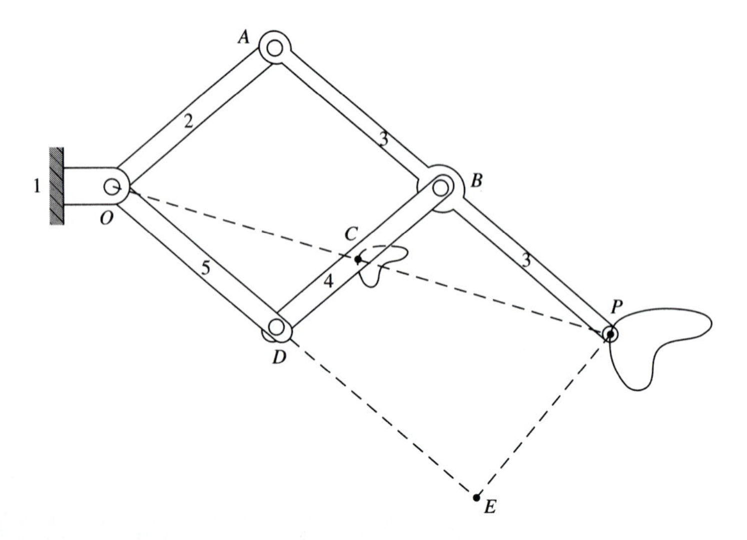

Figure 5.3 A pantograph linkage.

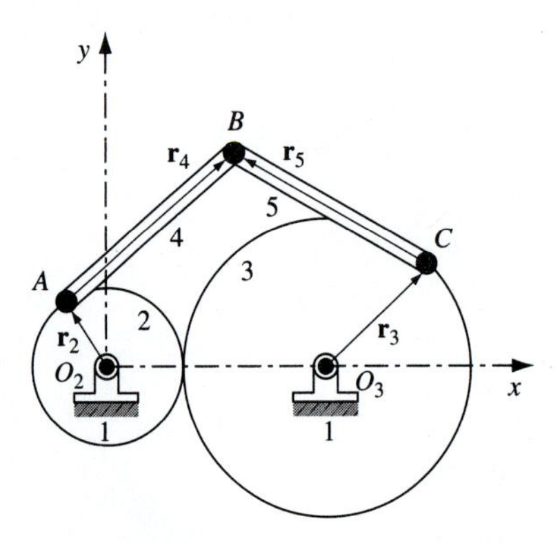

Figure 5.4 The geared five-bar linkage.

EXAMPLE 5.1

The vectors for the kinematic analysis of a five-bar linkage are shown in Fig. 5.5. The lengths of the links are as follows: link 1 is $r_1 = R_{O_3O_2} = 6$ in., link 2 is $r_2 = R_{AO_2} = 2.5$ in., link 3 is $r_3 = R_{CO_3} = 4$ in., link 4 is $r_4 = R_{BA} = 6$ in., and link 5 is $r_5 = R_{BC} = 6$ in. For the position under investigation, the orientations of the two input links are $\theta_2 = 120°$ and $\theta_3 = 45°$. The problem is to determine the orientations of the two coupler links, that is, θ_4 and θ_5.

SOLUTION

Once links 2 and 3 are drawn in their correct orientations, Fig. 5.5 can be completed. The orientations of links 4 and 5 can then be measured from the drawing and are found to be

$$\theta_4 = 36.5° \quad \text{and} \quad \theta_5 = 151.1°. \qquad \textit{Ans.}$$

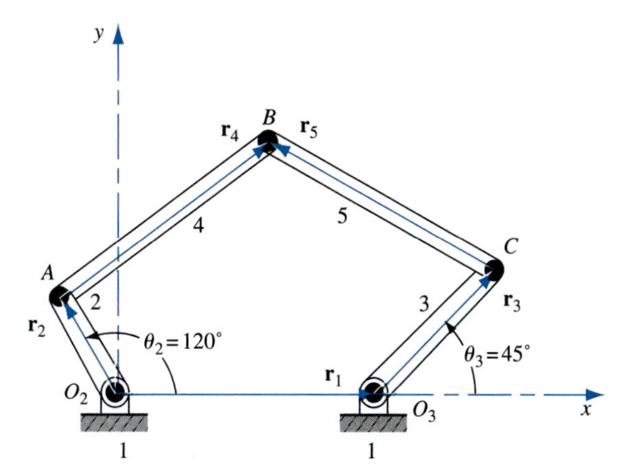

Figure 5.5 Example 5.1. The specified configuration of the five-bar linkage.

Control of the two independent input motions is required for a point fixed in coupler link 4, or coupler link 5, to follow a unique path. Still, there is an extremely wide variety of curves that can be generated by a coupler point of the five-bar linkage. For purposes of illustration, consider the arbitrary point P, shown in Fig. 5.6, that is fixed in coupler link 4.

EXAMPLE 5.2

Consider the continuation of Example 5.1 to include the coupler point P illustrated in Fig. 5.6. The location of point P is given by $r_7 = R_{PA} = 8$ in and $\beta = 30°$. For the given configuration of the linkage, the problem is to determine the absolute position coordinates of point P.

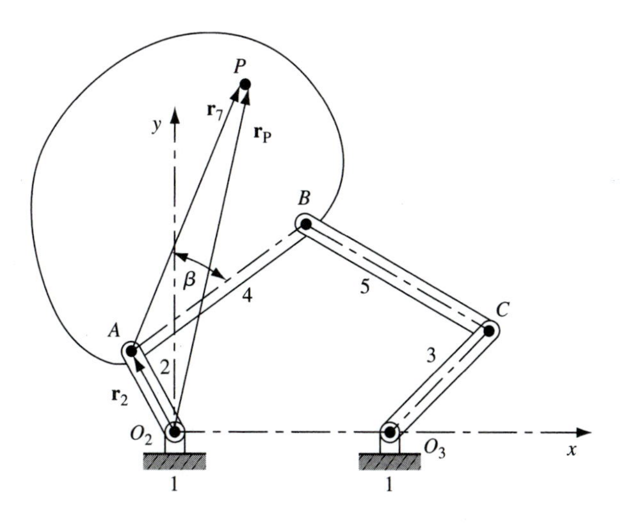

Figure 5.6 Example 5.2. Vectors for coupler point P fixed in link 4.

SOLUTION

After Fig. 5.5 of Example 5.1 is completed to scale, the vector $\mathbf{r}_7$ can be drawn at the orientation $\theta_7 = \theta_4 + \beta = 66.5°$. The absolute coordinates of point P can then be measured and the answers are

$$X_P = 1.95 \text{ in.} \quad \text{and} \quad Y_P = 9.50 \text{ in.} \qquad \textit{Ans.}$$

5.2 POSITION ANALYSIS; ALGEBRAIC SOLUTION

The analytical approaches for the solution of the position equations of multi-degree-of-freedom systems are parallel to those in Chapter 2. When sufficient input data are given to represent the input positions of all degrees of freedom, the loop-closure equations can be formulated. For planar systems, these equations can be separated into real and imaginary parts (or horizontal and vertical components), and these allow the solution of two unknowns per loop. We will demonstrate this by continuing with the planar five-bar linkage of Examples 5.1 and 5.2.

EXAMPLE 5.3

For the planar five-bar linkage of Fig. 5.5, the loop-closure equation is

$$\overset{\surd\surd}{\mathbf{r}_2} + \overset{\surd ?}{\mathbf{r}_4} - \overset{\surd ?}{\mathbf{r}_5} - \overset{\surd\surd}{\mathbf{r}_3} - \overset{\surd\surd}{\mathbf{r}_1} = \mathbf{0}, \tag{1}$$

where θ_2 and θ_3 are the input angles, checked as known above the appropriate terms in Eq. (1). For the given configuration of the linkage, the problem is to determine: (i) the unknown coupler angles θ_4 and θ_5; and (ii) the absolute position coordinates of point P.

SOLUTION

In complex polar notation, Eq. (1) becomes

$$r_2 e^{j\theta_2} + r_4 e^{j\theta_4} - r_5 e^{j\theta_5} - r_3 e^{j\theta_3} - r_1 = 0. \tag{2}$$

Separating the real and imaginary parts of this equation gives

$$r_2 \cos\theta_2 + r_4 \cos\theta_4 - r_5 \cos\theta_5 - r_3 \cos\theta_3 - r_1 = 0 \tag{3}$$

$$r_2 \sin\theta_2 + r_4 \sin\theta_4 - r_5 \sin\theta_5 - r_3 \sin\theta_3 = 0. \tag{4}$$

The solution to these equations can be determined either analytically or numerically. For example, using the Newton–Raphson iterative procedure (see Section 2.8), the angular positions of the two coupler links are evaluated as

$$\theta_4 = 36.447° \quad \text{and} \quad \theta_5 = 151.084°. \qquad \textit{Ans.}$$

These verify the results determined graphically in Example 5.1 and have higher accuracy.

The vectors defining the location of coupler point P are illustrated in Fig. 5.6. The vector equation for this point can be written as

$$\overset{??}{\mathbf{r}_P} = \overset{\surd\surd}{\mathbf{r}_2} + \overset{\surd C}{\mathbf{r}_7}, \tag{5}$$

where the constraint equation for the direction of vector $\mathbf{r}_7$ is

$$\theta_7 = \theta_4 + \beta. \tag{6}$$

From Eq. (5), the absolute position coordinates of point P are

$$X_P = r_2 \cos\theta_2 + r_7 \cos\theta_7 \tag{7}$$

$$Y_P = r_2 \sin\theta_2 + r_7 \sin\theta_7. \tag{8}$$

Substituting the given data into these equations, the position of point P (for the given configuration of the linkage) is

$$X_P = 2.5 \cos 120^\circ + 8 \cos 66.447^\circ = 1.947 \text{ in.} \qquad \textit{Ans.}$$

$$Y_P = 2.5 \sin 120^\circ + 8 \sin 66.447^\circ = 9.499 \text{ in.} \qquad \textit{Ans.}$$

These verify the results determined graphically in Example 5.2 and have higher accuracy.

5.3 GRAPHICAL METHODS; VELOCITY POLYGONS

No new methods are required for the development of velocity polygons when a mechanism has more than one degree of freedom. However, the input motions must be given for each degree of freedom. An example is presented here to demonstrate how the methods can be extended. A velocity analysis can be performed quite easily by the graphical velocity polygon methods of Sections 3.3 through 3.8 when the velocities of each of the independent inputs are specified.

EXAMPLE 5.4

For the five-bar linkage of Example 5.1, in the configuration shown in Fig. 5.5, suppose that the constant angular velocities of links 2 and 3 are $\omega_2 = 10$ rad/s ccw and $\omega_3 = 5$ rad/s cw, respectively. The problem is to determine: (i) the angular velocities of coupler links 4 and 5; and (ii) the velocity of the coupler point P.

SOLUTION

First, the velocities of points A and C, respectively, are

$$V_A = V_{AO_2} = \omega_2 R_{AO_2} = (10 \text{ rad/s})\,(2.5 \text{ in}) = 25.0 \text{ in/s}$$

$$V_C = V_{CO_3} = \omega_3 R_{CO_3} = (5 \text{ rad/s})\,(4.0 \text{ in}) = 20.0 \text{ in/s}.$$

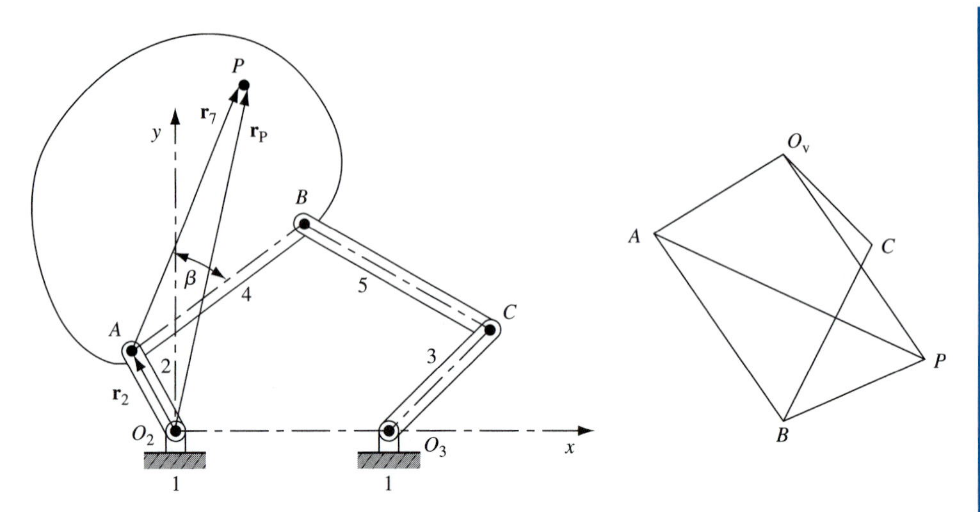

Figure 5.7 Example 5.4. Velocity polygon.

From these, we can write simultaneous equations for the velocity of point B,

$$\mathbf{V}_B = \overset{\surd\surd}{\mathbf{V}_A} + \overset{?\surd}{\mathbf{V}_{BA}} = \overset{\surd\surd}{\mathbf{V}_C} + \overset{?\surd}{\mathbf{V}_{BC}}.$$

Solving this with the velocity polygon illustrated in Fig. 5.7, we can locate the velocity image point B. Then we can measure the velocities V_{BA} and V_{BC} and obtain the angular velocities of coupler links 4 and 5, that is

$$\omega_4 = \frac{V_{BA}}{R_{BA}} = \frac{35.34 \text{ in/s}}{6.0 \text{ in}} = 5.89 \text{ rad/s cw} \qquad \textit{Ans.}$$

$$\omega_5 = \frac{V_{BC}}{R_{BC}} = \frac{30.60 \text{ in/s}}{6.0 \text{ in}} = 5.10 \text{ rad/s ccw}. \qquad \textit{Ans.}$$

Once the velocity of point B is known, we can construct the velocity image of coupler link 4 as explained in Section 3.4 and obtain the velocity of coupler point P, that is

$$\mathbf{V}_P = 38 \text{ in/s} \angle -55.5^\circ. \qquad \textit{Ans.}$$

5.4 INSTANT CENTERS OF VELOCITY

If we attempt to use the method of instant centers for planar mechanisms with multiple degrees of freedom, we encounter a difficulty. We find that the locations of the secondary instant centers cannot be determined by the methods of Sections. 3.13 through 3.15 alone. In fact, the locations of these secondary instant centers cannot be determined using only the geometry of the mechanism; their locations depend on the ratios of the independent input velocities[1]. One method for finding the secondary instant centers is demonstrated in the following example.

EXAMPLE 5.5

For the five-bar linkage in the configuration of Example 5.1, see Fig. 5.5, suppose that the input links 2 and 3 are rotating with constant angular velocities $\omega_2 = 10$ rad/s ccw and $\omega_3 = 5$ rad/s cw, respectively. Using the method of instant centers, determine the angular velocities of the coupler links 4 and 5 and the velocity of coupler point P.

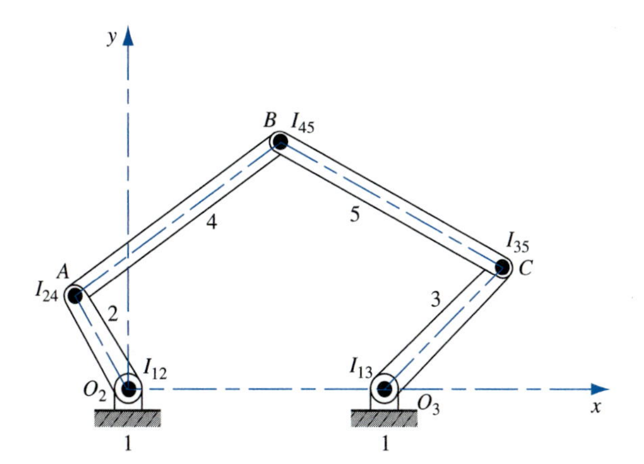

Figure 5.8 Example 5.5. The primary instant centers.

SOLUTION

The five primary instant centers for the five-bar linkage, denoted $I_{12}, I_{24}, I_{45}, I_{35}$, and I_{13}, are the five pin joint centers and are shown in Fig. 5.8. From the Aronhold–Kennedy theorem (Section 3.14), we know that the secondary instant center I_{23}, which relates the two input velocities, must lie on the line containing the instant centers I_{12} and I_{13}.

Also, from the angular velocity ratio theorem, Eq. (3.28), we know that the ratio of the angular velocity of link 2 to the angular velocity of link 3 can be written as

$$\frac{\omega_2}{\omega_3} = \frac{R_{I_{23}I_{13}}}{R_{I_{23}I_{12}}} = \frac{10.0 \text{ rad/s}}{-5.0 \text{ rad/s}} = -2.0. \tag{9}$$

From Section 3.17 we know that if the instant center I_{23} is located between the two absolute instant centers I_{12} and I_{13}, then the angular velocity ratio is negative. Similarly, if the instant center I_{23} is outside the two absolute instant centers I_{12} and I_{13} then the angular velocity ratio is positive. Note, however, that a sign convention is not necessary in Eq. (9) when directed lines are used for measuring the locations of the instant centers.

In our example, Eq. (9) gives

$$R_{I_{23}I_{13}} = -2.0 R_{I_{23}I_{12}}. \tag{10}$$

But we also know from Fig. 5.8, using positive distances to the right, that

$$R_{I_{23}I_{13}} = R_{I_{23}I_{12}} + R_{I_{12}I_{13}} = R_{I_{23}I_{12}} - 6.0 \text{ in.} \tag{11}$$

and solving Eqs. (10) and (11) simultaneously gives

$$-2.0R_{I_{23}I_{12}} = R_{I_{23}I_{12}} - 6.0 \text{ in.}$$

$$R_{I_{23}I_{12}} = 2.0 \text{ in.} \tag{12}$$

Therefore, the instant center I_{23} is located 2.0 in. to the right of instant center I_{12}.

The location of instant center I_{23} is indicated on a scaled drawing of the linkage in Fig. 5.9. The Kennedy circle, which is used to help locate the secondary instant centers, is also shown in the figure. The remaining secondary instant centers can now be obtained directly from the Aronhold–Kennedy theorem. The secondary instant center I_{14}, for example, must lie on the line containing the instant centers I_{12} and I_{24}. Similarly, the secondary instant center I_{15} must lie on the line containing the instant centers I_{13} and I_{35}. The locations

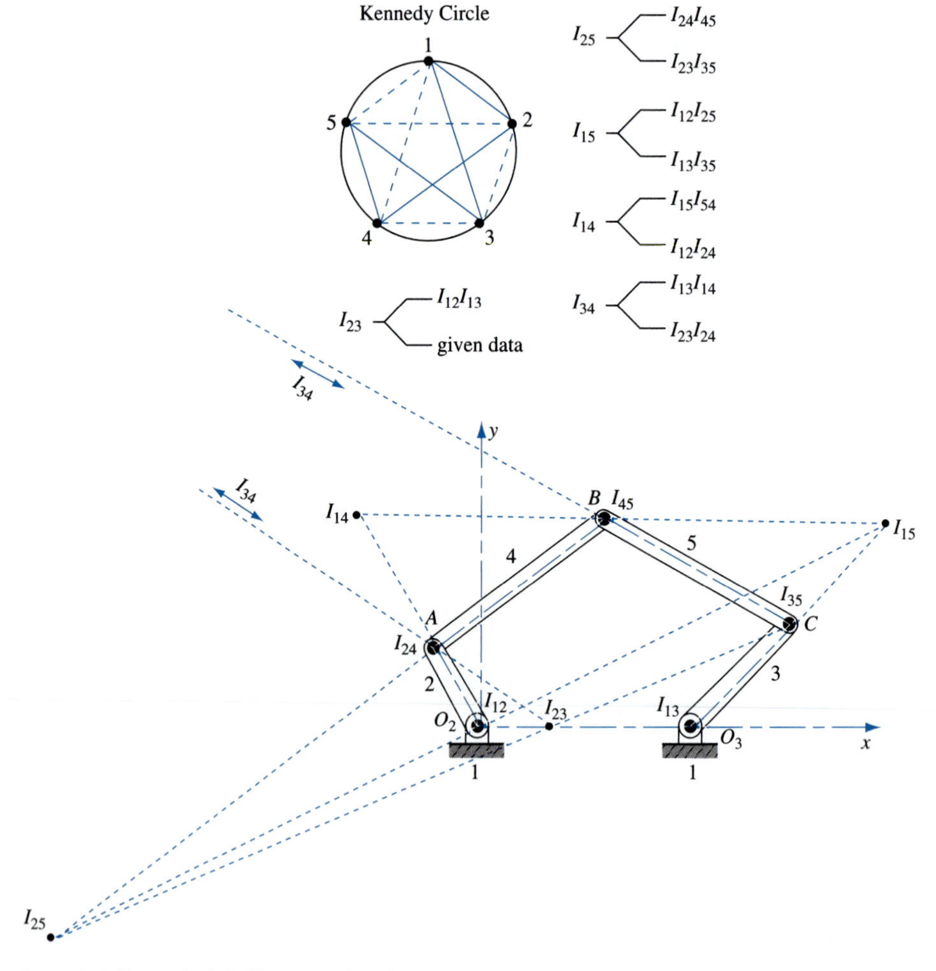

Figure 5.9 Example 5.5. The secondary instant centers.

of all secondary instant centers that lie on the page are illustrated in Fig. 5.9. The location of the instant center I_{34} does not lie on the page, but it is indicated by the two dashed lines. Note that it is not necessary to find the locations of all secondary instant centers in order to determine the unknown angular velocities ω_4 and ω_5.

Choosing the upward direction along each line of centers as positive, the distances between instant centers I_{12} and I_{24} and between instant centers I_{13} and I_{35} are known from the lengths of the links 2 and 3, respectively; these are

$$R_{I_{24}I_{12}} = 2.50 \text{ in.} \quad \text{and} \quad R_{I_{35}I_{13}} = 4.00 \text{ in.} \tag{13}$$

Distances between other instant centers are also measured from the scaled drawing as

$$R_{I_{24}I_{14}} = -4.24 \text{ in.}, \quad R_{I_{25}I_{12}} = -13.41 \text{ in.}, \quad R_{I_{25}I_{15}} = -26.29 \text{ in.}, \quad \text{and}$$
$$R_{I_{35}I_{15}} = -3.92 \text{ in.}$$

Substituting these distances into Eq. (3.28), angular velocity ratios can be written as

$$\frac{\omega_4}{\omega_2} = \frac{R_{I_{24}I_{12}}}{R_{I_{24}I_{14}}} = \frac{2.50 \text{ in}}{-4.24 \text{ in}} = -0.590 \tag{14}$$

$$\frac{\omega_5}{\omega_2} = \frac{R_{I_{25}I_{12}}}{R_{I_{25}I_{15}}} = \frac{-13.41 \text{ in}}{-26.29 \text{ in}} = 0.510 \tag{15}$$

$$\frac{\omega_5}{\omega_3} = \frac{R_{I_{35}I_{13}}}{R_{I_{35}I_{15}}} = \frac{4.00 \text{ in}}{-3.92 \text{ in}} = -1.020. \tag{16}$$

Then, substituting the angular velocity of link 2 into Eqs. (14) and (15), the angular velocities of links 4 and 5 are

$$\omega_4 = -0.590\omega_2 = -0.590\,(10 \text{ rad/s}) = -5.90 \text{ rad/s (cw)} \qquad \textit{Ans.}$$

$$\omega_5 = 0.510\omega_2 = 0.510\,(10 \text{ rad/s}) = 5.10 \text{ rad/s (ccw).} \qquad \textit{Ans.}$$

As a check, we can substitute the angular velocity of link 3 into Eq. (16) to verify that the angular velocity of link 5 is

$$\omega_5 = -1.020\omega_3 = -1.020(-5 \text{ rad/s}) = 5.10 \text{ rad/s (ccw).} \qquad \textit{Ans.}$$

Using the method of instant centers, the magnitude of the velocity of point P can be written as

$$V_P = \omega_4 R_{PI_{14}}. \tag{17}$$

From the scaled drawing (see Fig. 5.10), the distance from the absolute instant center of link 4 (I_{14}) to point P is measured as $R_{PI_{14}} = 6.46$ in. Substituting this value into Eq. (17), the magnitude of the velocity of point P is

$$V_P = (|-5.90 \text{ rad/s}|)\,(6.46 \text{ in}) = 38.11 \text{ in/sec.} \qquad \textit{Ans.}$$

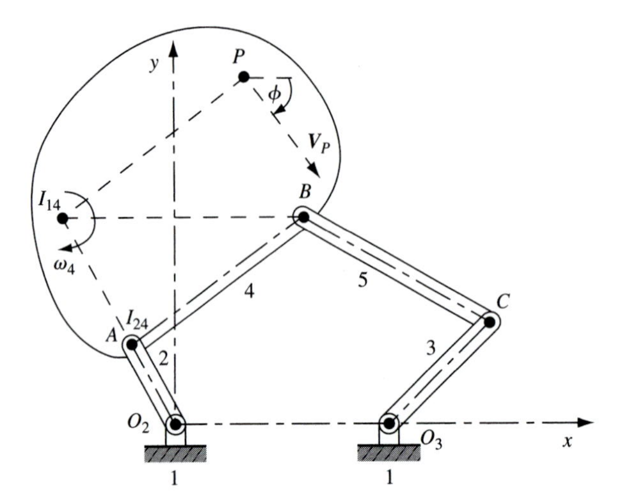

Figure 5.10 Example 5.5. The absolute instant center for coupler link 4.

The direction of the velocity of point P is measured from Fig. 5.10 as

$$\phi = -55.5^\circ. \qquad \textit{Ans.}$$

Note that these results are in good agreement with those obtained in Example 5.4.

A special case of the five-bar linkage is when the instant center I_{23} is stationary on the line of centers $I_{12}I_{13}$, then links 2 and 3 can be replaced by two circular gears as illustrated in Fig. 5.4. This mechanism, referred to as a geared five-bar linkage, has a mobility of 1 (the input is either link 2 or link 3). The two gears; that is, links 2 and 3, are in rolling contact at instant center I_{23}. The ratio of the angular velocity of gear 3 to the angular velocity of the gear 2 (also referred to as the gear ratio) is the reciprocal of Eq. (9).

5.5 FIRST-ORDER KINEMATIC COEFFICIENTS

If we require higher accuracy in the solution for velocities, we look for an analytical or numerical technique. The method of kinematic coefficients, which was presented in Section 3.12, provides such a technique.

When dealing with mechanisms having multiple degrees of freedom, however, the method of kinematic coefficients raises a new complexity in notation. It requires an additional subscript to keep track of the multiple input motions. For example, the time rate of change of a dependent variable such as θ_i, the angular velocity of link i, can be written as $\omega_i = d\theta_i/dt$. However, because there are multiple degrees of freedom, the dependent variable θ_i is a function of two or more independent variables, say θ_j and θ_k; that is, $\theta_i = \theta_i\left(\theta_j, \theta_k\right)$. In this case, the time rate of change of the dependent variable θ_i, can be written as

$$\omega_i = d\theta_i/dt = (\partial\theta_i/\partial\theta_j)(d\theta_j/dt) + (\partial\theta_i/\partial\theta_k)(d\theta_k/dt). \tag{5.1}$$

Similarly, a point P on one of the moving links has a position that is a function of both independent variables, $\mathbf{r}_P = \mathbf{r}_P\left(\theta_j, \theta_k\right)$. Therefore, the velocity of point P can be written as

$$\mathbf{V}_P = d\mathbf{r}_P/dt = (\partial\mathbf{r}_P/\partial\theta_j)(d\theta_j/dt) + (\partial\mathbf{r}_P/\partial\theta_k)(d\theta_k/dt). \tag{5.2}$$

Also, because $\mathbf{r}_P\left(\theta_j, \theta_k\right)$ has coordinates such as $\mathbf{r}_P = X_P\hat{\mathbf{i}} + Y_P\hat{\mathbf{j}} + Z_P\hat{\mathbf{k}}$, then the velocity of point P can be written as

$$\begin{aligned}\mathbf{V}_P &= \left(\frac{\partial X_P}{\partial\theta_j}\hat{\mathbf{i}} + \frac{\partial Y_P}{\partial\theta_j}\hat{\mathbf{j}} + \frac{\partial Z_P}{\partial\theta_j}\hat{\mathbf{k}}\right)\frac{d\theta_j}{dt} + \left(\frac{\partial X_P}{\partial\theta_k}\hat{\mathbf{i}} + \frac{\partial Y_P}{\partial\theta_k}\hat{\mathbf{j}} + \frac{\partial Z_P}{\partial\theta_k}\hat{\mathbf{k}}\right)\frac{d\theta_k}{dt} \\ &= \left(\frac{\partial X_P}{\partial\theta_j}\frac{d\theta_j}{dt} + \frac{\partial X_P}{\partial\theta_k}\frac{d\theta_k}{dt}\right)\hat{\mathbf{i}} + \left(\frac{\partial Y_P}{\partial\theta_j}\frac{d\theta_j}{dt} + \frac{\partial Y_P}{\partial\theta_k}\frac{d\theta_k}{dt}\right)\hat{\mathbf{j}} + \left(\frac{\partial Z_P}{\partial\theta_j}\frac{d\theta_j}{dt} + \frac{\partial Z_P}{\partial\theta_k}\frac{d\theta_k}{dt}\right)\hat{\mathbf{k}}.\end{aligned} \tag{5.3}$$

If we wish to continue to use the prime notation, where a first-order kinematic coefficient carries the symbol $\theta'_{ij} = \partial\theta_i/\partial\theta_j$ or $r'_{Pj} = \partial r_P/\partial\theta_j$, we see that a second subscript is required to indicate with respect to which independent variable the derivative is taken. With this additional subscript, Eq. (5.1) becomes

$$\omega_i = d\theta_i/dt = \theta'_{ij}\omega_j + \theta'_{ik}\omega_k. \tag{5.4}$$

Similarly, Eq. (5.2) becomes

$$\mathbf{V}_P = d\mathbf{r}_P/dt = \mathbf{r}'_{Pj}\omega_j + \mathbf{r}'_{Pk}\omega_k \tag{5.5}$$

and Eq. (5.3) becomes

$$\mathbf{V}_P = \left(X'_{Pj}\hat{\mathbf{i}} + Y'_{Pj}\hat{\mathbf{j}} + Z'_{Pj}\hat{\mathbf{k}}\right)\omega_j + \left(X'_{Pk}\hat{\mathbf{i}} + Y'_{Pk}\hat{\mathbf{j}} + Z'_{Pk}\hat{\mathbf{k}}\right)\omega_k$$

or

$$\mathbf{V}_P = \left(X'_{Pj}\omega_j + X'_{Pk}\omega_k\right)\hat{\mathbf{i}} + \left(Y'_{Pj}\omega_j + Y'_{Pk}\omega_k\right)\hat{\mathbf{j}} + \left(Z'_{Pj}\omega_j + Z'_{Pk}\omega_k\right)\hat{\mathbf{k}}. \tag{5.6}$$

Although we find need for this additional subscript, first-order kinematic coefficients do provide insight into the velocity analysis of multi-degree-of-freedom mechanisms, just as they did in Section 3.12 for single-degree-of-freedom mechanisms. This will be illustrated in the following example.

EXAMPLE 5.6

Consider the five-bar linkage in the configuration of Example 5.1 (see Fig. 5.5), and with links 2 and 3 rotating with the same constant angular velocities as Example 5.5. Using the method of kinematic coefficients, determine the angular velocities of the coupler links 4 and 5 and the velocity of coupler point P.

SOLUTION

We begin the analysis by writing the loop-closure equation as we did in Example 5.3 and separating the real and imaginary parts, as in Eqs. (3) and (4). As we demonstrated in that example, those equations can be solved for the position variables θ_4 and θ_5.

Next, we take the partial derivatives of Eqs. (3) and (4) with respect to input position variable θ_2, that is

$$-r_2 \sin\theta_2 - r_4 \sin\theta_4 \theta'_{42} + r_5 \sin\theta_5 \theta'_{52} = 0 \tag{18a}$$

$$r_2 \cos\theta_2 + r_4 \cos\theta_4 \theta'_{42} - r_5 \cos\theta_5 \theta'_{52} = 0. \tag{18b}$$

Rearranging these two equations and expressing them in matrix form gives

$$\begin{bmatrix} -r_4 \sin\theta_4 & r_5 \sin\theta_5 \\ r_4 \cos\theta_4 & -r_5 \cos\theta_5 \end{bmatrix} \begin{bmatrix} \theta'_{42} \\ \theta'_{52} \end{bmatrix} = \begin{bmatrix} r_2 \sin\theta_2 \\ -r_2 \cos\theta_2 \end{bmatrix}.$$

Now we can solve this matrix equation for the first-order kinematic coefficients θ'_{42} and θ'_{52}. Using the data obtained so far, the answers are $\theta'_{42} = -0.237$ and $\theta'_{52} = 0.456$, which are both dimensionless (rad/rad).

Similarly, taking the partial derivative of Eqs. (3) and (4) with respect to input position variable θ_3, and rearranging the resulting equations in matrix form, gives

$$\begin{bmatrix} -r_4 \sin\theta_4 & r_5 \sin\theta_5 \\ r_4 \cos\theta_4 & -r_5 \cos\theta_5 \end{bmatrix} \begin{bmatrix} \theta'_{43} \\ \theta'_{53} \end{bmatrix} = \begin{bmatrix} -r_3 \sin\theta_3 \\ r_3 \cos\theta_3 \end{bmatrix}. \tag{19}$$

We can solve this matrix equation for the first-order kinematic coefficients θ'_{43} and θ'_{53}. The answers are $\theta'_{43} = 0.705$ rad/rad and $\theta'_{53} = -0.109$ rad/rad.

Table 5.1 presents the angular positions of the two coupler links and the first-order kinematic coefficients of the coupler links, θ_4 and θ_5, for input angles in the range $40° \leq \theta_2 \leq 150°$ and $85° \geq \theta_3 \geq 30°$.

As indicated in Section 4.12, it is not a coincidence that the (2×2) coefficient matrices in Eqs. (18) and (19) are identical. This is always true for all sets of derivative equations determined by differentiating the loop-closure constraint equations. This matrix is called the Jacobian of the system. In this example, the determinant of the Jacobian is

$$\Delta = r_4 r_5 \sin(\theta_4 - \theta_5). \tag{20}$$

At the current position, this determinant has a value of $\Delta = -32.72\ \text{in}^2$ and does not cause any numeric difficulty. However, we note that this determinant becomes zero when $\theta_5 = \theta_4$ or when $\theta_5 = \theta_4 \pm 180°$. These occur when the coupler links are either fully extended or folded on top of each other, that is, when the five-bar linkage is in the configuration of a quadrilateral with the two coupler links aligned.

TABLE 5.1 Input angles, coupler angles, first-order kinematic coefficients.

θ_2 deg	θ_3 deg	θ_4 deg	θ_5 deg	θ'_{42} —	θ'_{43} —	θ'_{52} —	θ'_{53} —
40	85	93.419	142.993	−0.533	0.743	−0.440	0.128
50	80	84.547	138.843	−0.513	0.703	−0.291	0.065
60	75	76.200	136.526	−0.466	0.674	−0.134	0.016
70	70	68.479	135.944	−0.412	0.659	0.012	−0.019
80	65	61.343	136.872	−0.360	0.654	0.138	−0.044
90	60	54.690	139.054	−0.316	0.658	0.242	−0.062
100	55	48.401	142.263	−0.281	0.667	0.327	−0.077
110	50	42.360	146.315	−0.254	0.683	0.397	−0.091
120	45	36.447	151.084	−0.237	0.705	0.456	−0.109
130	40	30.526	156.510	−0.230	0.737	0.508	−0.136
140	35	24.383	162.646	−0.241	0.793	0.564	−0.185
150	30	17.496	169.884	−0.306	0.927	0.663	−0.311

The angular velocities of links 4 and 5 can be determined from Eq. (5.4). At the position illustrated in Fig. 5.1, the values are

$$\omega_4 = \theta'_{42}\omega_2 + \theta'_{43}\omega_3 = -0.237\,(10.0\text{ rad/s}) + 0.705\,(-5.0\text{ rad/s}) = -5.90\text{ rad/s (cw)}$$

Ans.

$$\omega_5 = \theta'_{52}\omega_2 + \theta'_{53}\omega_3 = 0.456\,(10.0\text{ rad/s}) - 0.109\,(-5.0\text{ rad/s}) = 5.10\text{ rad/s (ccw)}.$$

Ans.

These confirm the results in both Examples 5.4 and 5.5.

Differentiating constraint Eq. (6) with respect to the two independent inputs, the first-order kinematic coefficients of vector $\mathbf{r}_7$ are

$$\theta'_{72} = \theta'_{42} \text{ and } \theta'_{73} = \theta'_{43}. \tag{21}$$

Partially differentiating Eqs. (7) and (8) with respect to the input position θ_2 and using Eq. (21), the first-order kinematic coefficients of point P are

$$X'_{P2} = -r_2\sin\theta_2 - r_7\sin\theta_7\theta'_{42} \tag{22}$$

$$Y'_{P2} = r_2\cos\theta_2 + r_7\cos\theta_7\theta'_{42}. \tag{23}$$

Similarly, partially differentiating Eqs. (7) and (8) with respect to the input position θ_3 and using Eq. (21), the first-order kinematic coefficients of point P are

$$X'_{P3} = -r_7\sin\theta_7\theta'_{43} \tag{24}$$

$$Y'_{P3} = r_7\cos\theta_7\theta'_{43}. \tag{25}$$

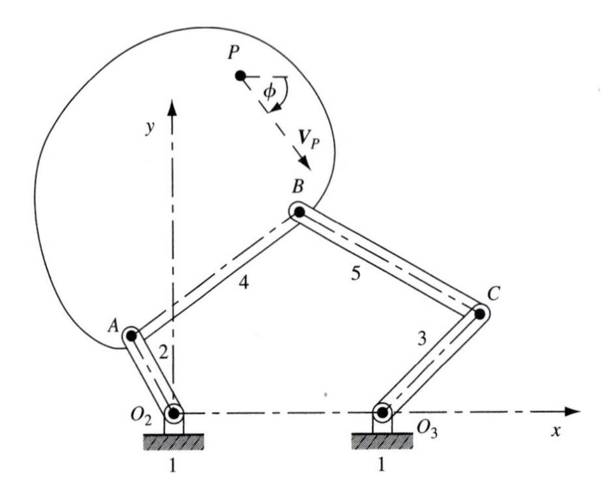

Figure 5.11 Example 5.6. The velocity of coupler point P.

TABLE 5.2 The first-order kinematic coefficients of coupler point P.

θ_2 deg	θ_3 deg	X'_{P2} in.	Y'_{P2} in.	X'_{P3} in.	Y'_{P3} in.	V_P in./s	ϕ deg
40	85	1.9543	4.2650	−4.9588	−3.2721	73.811	53.081
50	80	1.8180	3.3120	−5.1124	−2.3350	62.609	45.681
60	75	1.4177	2.2909	−5.1816	−1.5054	50.330	37.209
70	70	0.9102	1.3410	−5.2151	−0.7774	39.200	26.183
80	65	0.4201	0.5017	−5.2332	−0.1227	30.885	10.503
90	60	0.0191	−0.2342	−5.2391	0.4870	26.816	−10.260
100	55	−0.2610	−0.8859	−5.2303	1.0735	27.506	−31.146
110	50	−0.4107	−1.4715	−5.2054	1.6553	31.766	−46.367
120	45	−0.4294	−2.0066	−5.1681	2.2528	38.024	−55.483
130	40	−0.3144	−2.5116	−5.1347	2.9020	45.583	−60.380
140	35	−0.0397	−3.0379	−5.1568	3.6942	55.053	−62.539
150	30	0.5534	−3.8178	−5.4662	5.0096	71.258	−62.535

Substituting the specified data into Eqs. (22) through (25), the first-order kinematic coefficients of point P are

$$X'_{P2} = -0.429\,4 \text{ in.}, \quad Y'_{P2} = -2.006\,6 \text{ in.}, \quad X'_{P3} = -5.168\,1 \text{ in.}, \quad \text{and}$$
$$Y'_{P3} = 2.252\,8 \text{ in.}$$

Finally, substituting these values and the given input angular velocities into Eq. (5.6), the velocity of point P is

$$\mathbf{V}_P = [(-0.429\,4 \text{ in})(10 \text{ rad/s}) + (-5.168\,14 \text{ in})(-5 \text{ rad/s})]\hat{\mathbf{i}}$$
$$+ [(-2.006\,6 \text{ in})(10 \text{ rad/s}) + (2.252\,84 \text{ in})(-5 \text{ rad/s})]\hat{\mathbf{j}}.$$

That is,

$$\mathbf{V}_P = 21.546 \text{ in/s}\,\hat{\mathbf{i}} - 31.330 \text{ in/s}\,\hat{\mathbf{j}} = 38.024 \text{ in/s}\angle - 55.483^\circ. \qquad \textit{Ans.}$$

The velocity of point P is shown in Fig. 5.11. Note that these results are in good agreement with those obtained in both Examples 5.4 and 5.5.

Table 5.2 indicates the first-order kinematic coefficients of point P and the magnitude and direction of the velocity of point P for the range of input angles used in Table 5.1.

5.6 THE METHOD OF SUPERPOSITION

Equations (5.4), (5.5), and (5.6) suggest that for mechanisms with more than one degree of freedom, another method of solution for velocities is to solve the problem multiple times, with all except one of the inputs considered inactive (frozen, or locked) during each solution, and then to sum the results to find the total solution with all inputs active. This method is called the method of superposition, and it is valid for velocity analysis because the velocity equations are linear, that is, because all dependent velocities are linear combinations of the input velocities. The procedure may be best understood through an example.

EXAMPLE 5.7

Let us again solve the velocity analysis of Example 5.3, this time using the method of superposition. There are two cases that must be considered: case (*i*), where link 3 is instantaneously frozen whereas link 2 has an angular velocity of $\omega_2 = 10$ rad/s ccw, and case (*ii*), where link 2 is instantaneously frozen whereas link 3 has an angular velocity of $\omega_3 = 5$ rad/s cw.

SOLUTION

The angular velocity of link 4 can be written as a linear combination of case (*i*) and case (*ii*). Using the angular velocity ratio theorem, this gives

$$\omega_4 = \left(\frac{R_{I_{12}^2 I_{24}^2}}{R_{I_{14}^2 I_{24}^2}}\right)\omega_2 + \left(\frac{R_{I_{13}^3 I_{34}^3}}{R_{I_{14}^3 I_{34}^3}}\right)\omega_3, \tag{26}$$

where the superscripts in the instant center labels denote the input variable that is moving, whereas all others are instantaneously frozen. Similarly, the angular velocity of link 5 can be written as

$$\omega_5 = \left(\frac{R_{I_{12}^2 I_{25}^2}}{R_{I_{15}^2 I_{25}^2}}\right)\omega_2 + \left(\frac{R_{I_{13}^3 I_{35}^3}}{R_{I_{15}^3 I_{35}^3}}\right)\omega_3. \tag{27}$$

Comparing Eqs. (26) and (27) with Eqs. (5.4), the first-order kinematic coefficients of links 4 and 5 can be written as

$$\theta'_{42} = \frac{R_{I^2_{12}I^2_{24}}}{R_{I^2_{14}I^2_{24}}}, \quad \theta'_{43} = \frac{R_{I^3_{13}I^3_{34}}}{R_{I^3_{14}I^3_{34}}}, \quad \theta'_{52} = \frac{R_{I^2_{12}I^2_{25}}}{R_{I^2_{15}I^2_{25}}}, \quad \text{and} \quad \theta'_{53} = \frac{R_{I^3_{13}I^3_{35}}}{R_{I^3_{15}I^3_{35}}}. \tag{28}$$

Note that a sign convention is not necessary if directed lines are used when measuring the relative locations of the instant centers. However, if a sign convention is preferred, then each first-order kinematic coefficient is negative if the relative instant center I^k_{ij} is between the absolute instant centers I^k_{1i} and I^k_{1j} and positive if the relative instant center I^k_{ij} is outside the absolute instant centers I^k_{1i} and I^k_{1j}.

Case (*i*): Link 3 is instantaneously frozen. The linkage can be regarded instantaneously as a four-bar linkage, as illustrated in Fig. 5.12 with input from the rotation of link 2 alone.

From the Aronhold–Kennedy theorem, the secondary instant center I^2_{14} is the point of intersection of the line containing the instant centers I^2_{15} and I^2_{45} and the line containing the instant centers I^2_{12} and I^2_{24}. Similarly, the secondary instant center I^2_{25} is the point of

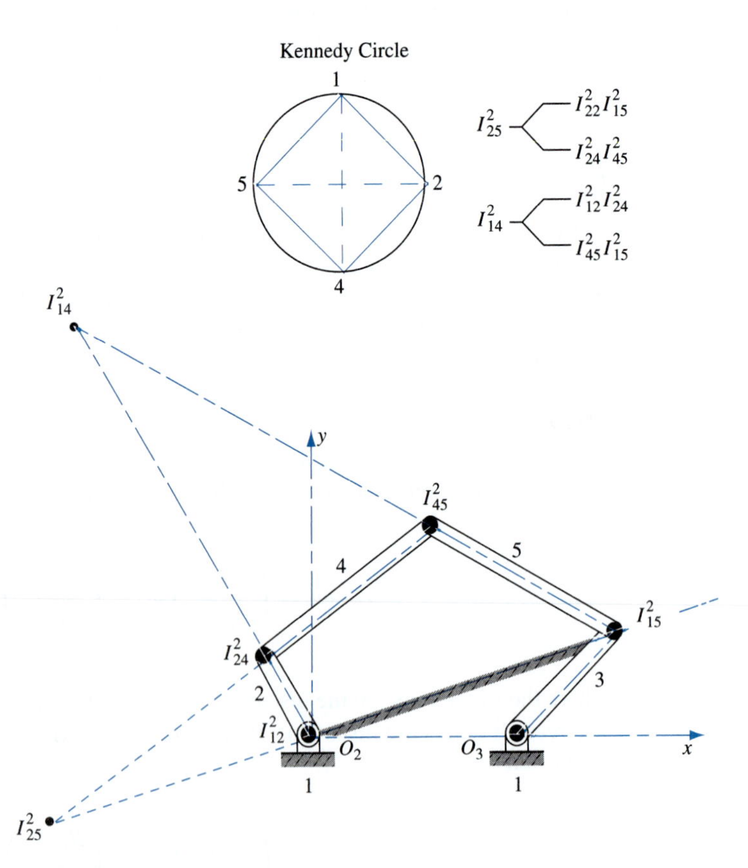

Figure 5.12 Example 5.7. Case (*i*). The instantaneous four-bar linkage with link 3 frozen.

intersection of the line containing the instant centers I_{24}^2 and I_{45}^2 and the line containing the instant centers I_{12}^2 and I_{15}^2.

Using the convention that lines proceed positive to the right, the distances between the instant centers, measured from the scaled drawing of Fig. 5.12, are

$$R_{I_{12}^2 I_{24}^2} = 2.50 \text{ in.}, \quad R_{I_{14}^2 I_{24}^2} = -10.56 \text{ in.}, \quad R_{I_{12}^2 I_{25}^2} = 7.76 \text{ in.}, \quad \text{and}$$
$$R_{I_{15}^2 I_{25}^2} = 17.03 \text{ in.} \tag{29}$$

Case (*ii*): Link 2 is instantaneously frozen. The linkage can be regarded instantaneously as a four-bar linkage, as illustrated in Fig. 5.13 with input from the rotation of link 3 alone.

From the Aronhold–Kennedy theorem, the secondary instant center I_{34}^3 is the point of intersection of the line containing the instant centers I_{13}^3 and I_{14}^3 and the line containing the instant centers I_{35}^3 and I_{45}^3. Similarly, the secondary instant center I_{15}^3 is the point of intersection of the line containing the instant centers I_{14}^3 and I_{45}^3 and the line containing the instant centers I_{13}^3 and I_{35}^3.

Continuing with the convention that lines proceed positive to the right, the distances between the instant centers, measured from the scaled drawing of Fig. 5.13, are

$$R_{I_{13}^3 I_{34}^3} = -18.06 \text{ in.}, \quad R_{I_{14}^3 I_{34}^3} = -25.62 \text{ in.}, \quad R_{I_{13}^3 I_{35}^3} = -4.00 \text{ in.}, \text{ and}$$
$$R_{I_{15}^3 I_{35}^3} = 36.67 \text{ in.} \tag{30}$$

Substituting Eqs. (29) and (30) into Eqs. (28), the first-order kinematic coefficients of links 4 and 5 are

$$\theta'_{42} = \frac{R_{I_{12}^2 I_{24}^2}}{R_{I_{14}^2 I_{24}^2}} = \frac{2.50 \text{ in}}{-10.56 \text{ in}} = -0.237, \quad \theta'_{43} = \frac{R_{I_{13}^3 I_{34}^3}}{R_{I_{14}^3 I_{34}^3}} = \frac{-18.06 \text{ in}}{-25.62 \text{ in}} = 0.705 \tag{31a}$$

$$\theta'_{52} = \frac{R_{I_{12}^2 I_{25}^2}}{R_{I_{15}^2 I_{25}^2}} = \frac{7.76 \text{ in}}{17.03 \text{ in}} = 0.455, \quad \theta'_{53} = \frac{R_{I_{13}^3 I_{35}^3}}{R_{I_{15}^3 I_{35}^3}} = \frac{-4.00 \text{ in}}{36.67 \text{ in}} = -0.109. \tag{31b}$$

Note that the values of the first-order kinematic coefficients of links 4 and 5, given by Eqs. (31), are all nondimensional and in good agreement with the values determined analytically in Example 5.6.

Substituting Eqs. (31) into Eq. (5.4), the angular velocities of links 4 and link 5, respectively, are

$$\omega_4 = -0.237(10 \text{ rad/s}) + 0.705(-5 \text{ rad/s}) = -5.90 \text{ rad/s (cw)} \qquad \textit{Ans.}$$

$$\omega_5 = 0.455(10 \text{ rad/s}) - 0.109(-5 \text{ rad/s}) = 5.10 \text{ rad/s ccw.} \qquad \textit{Ans.}$$

Note that these answers for the angular velocities of coupler links 4 and 5 also agree with the results obtained from the analytic approach and the method of instant centers; see Examples 5.4 and 5.5.

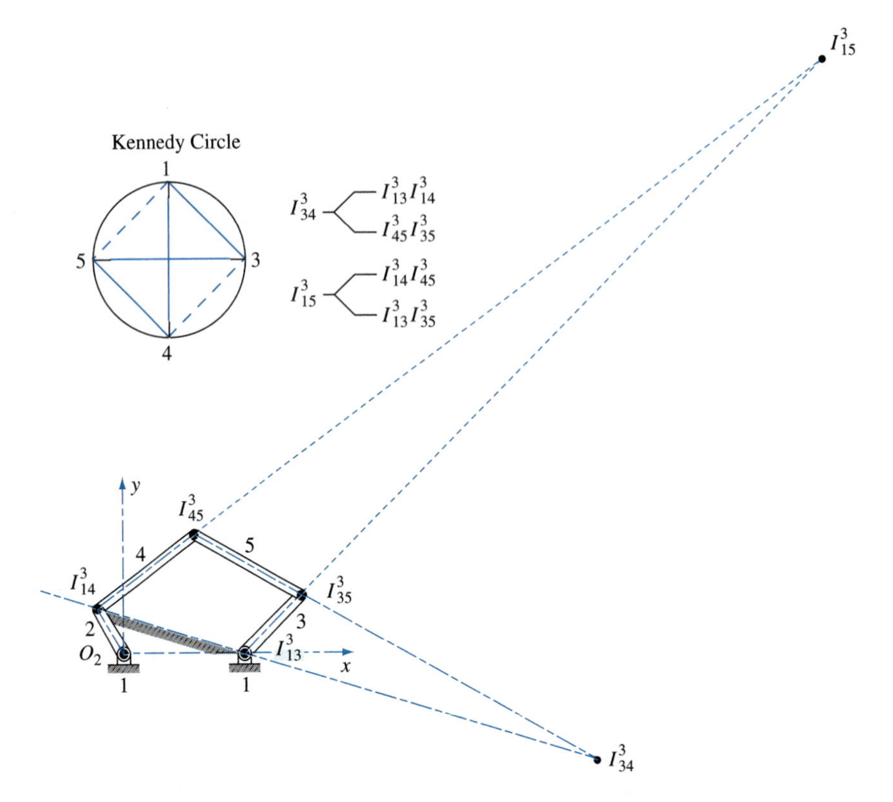

Figure 5.13 Example 5.7. Case (*ii*). The instantaneous four-bar linkage with link 2 frozen.

5.7 GRAPHICAL METHOD; ACCELERATION POLYGONS

As with velocity analysis, acceleration analysis requires that accelerations must be known for the input motions of each degree of freedom. An example is presented here to demonstrate how the graphical methods can be extended. An acceleration analysis can be performed quite easily by the graphical acceleration polygon method of Sections 4.3 through 4.8 once accelerations are given for each of the independent input motions.

EXAMPLE 5.8

For the five-bar linkage of Example 5.1 in the configuration illustrated in Fig. 5.5, suppose that links 2 and 3 are rotating with constant angular velocities of $\omega_2 = 10$ rad/s ccw and $\omega_3 = 5$ rad/s cw, respectively. The velocities were determined, using this same input data, in the velocity polygon of Example 5.4. The problem now is to determine the angular accelerations of coupler links 4 and 5.

SOLUTION

Because the angular velocities of links 2 and 3 are constant, we have $\alpha_2 = \alpha_3 = 0$ and there remain only the normal components of acceleration for points A and C. Hence,

$$A_A = A^n_{AO_2} = \omega_2^2 R_{AO_2} = (10\ \text{rad/s})^2\ (2.5\ \text{in}) = 250\ \text{in/s}^2$$

$$A_C = A^n_{CO_3} = \omega_3^2 R_{CO_3} = (5\ \text{rad/s})^2\ (4\ \text{in}) = 100\ \text{in/s}^2.$$

Next, the acceleration-difference equation is written to relate the acceleration of point B to that of points A and C, that is,

$$\mathbf{A}_B = \overset{\surd\surd}{\mathbf{A}_A} + \overset{\surd\surd}{\mathbf{A}^n_{BA}} + \overset{?\surd}{\mathbf{A}^t_{BA}} = \overset{\surd\surd}{\mathbf{A}_C} + \overset{\surd\surd}{\mathbf{A}^n_{BC}} + \overset{?\surd}{\mathbf{A}^t_{BC}}. \tag{32}$$

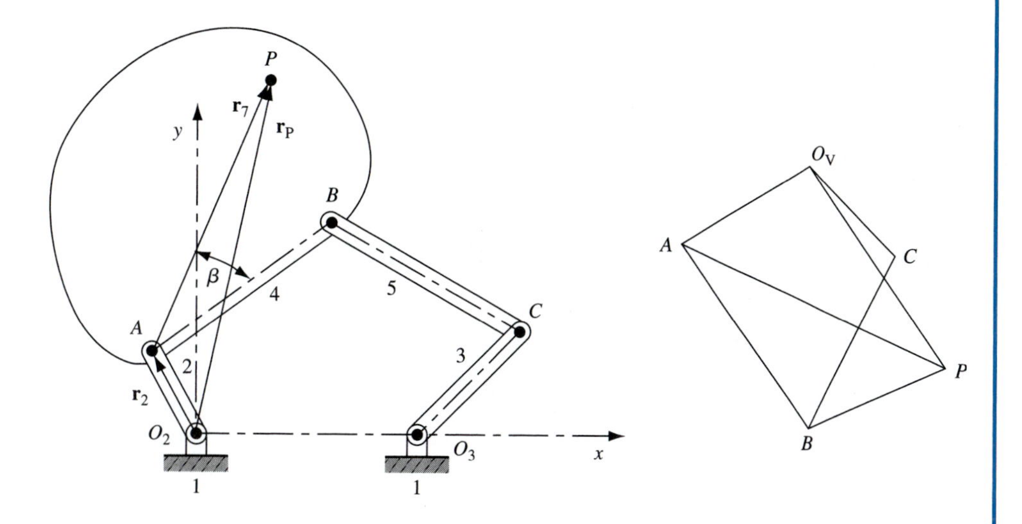

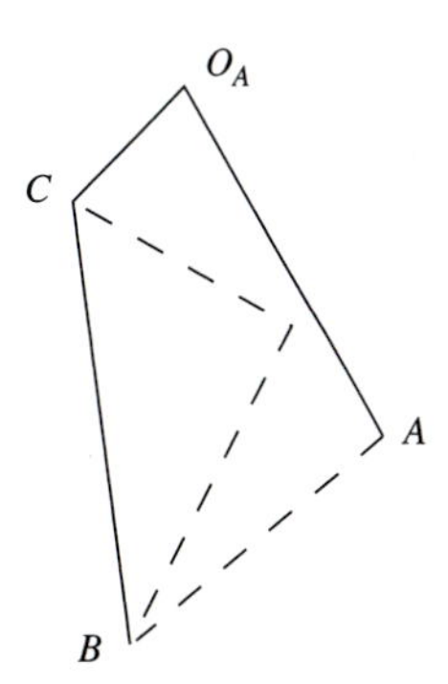

Figure 5.14 Example 5.8. Acceleration polygon.

Using information scaled from the velocity polygon determined in Example 5.4, we calculate the magnitudes of the two normal components of Eq. (32). Thus,

$$A_{BA}^n = \frac{V_{BA}^2}{R_{BA}} = \frac{(35.34 \text{ in/s})^2}{6 \text{ in}} = 208.15 \text{ in/s}^2$$

and

$$A_{BC}^n = \frac{V_{BC}^2}{R_{BC}} = \frac{(30.60 \text{ in/s})^2}{6 \text{ in}} = 156.06 \text{ in/s}^2.$$

With these, the terms of Eq. (32) can be constructed graphically, as illustrated in Fig. 5.14.

The angular accelerations of links 4 and 5 can now be determined from the two tangential components scaled from the acceleration polygon, namely,

$$\alpha_4 = \frac{A_{BA}^t}{R_{BA}} = \frac{1.2 \text{ in/s}^2}{6 \text{ in}} = 0.2 \text{ rad/s}^2 \text{ cw} \qquad \textit{Ans.}$$

$$\alpha_5 = \frac{A_{BC}^t}{R_{BC}} = \frac{222.7 \text{ in/s}^2}{6 \text{ in}} = 37.1 \text{ rad/s}^2 \text{ ccw.} \qquad \textit{Ans.}$$

5.8 SECOND-ORDER KINEMATIC COEFFICIENTS

For higher accuracy in the solution to the acceleration problems we look for an analytic or numeric technique. The method of kinematic coefficients, which was presented in Section 4.12, provides such a technique.

As with first-order kinematic coefficients, we will use subscripts to signify each independent variable when we have multiple degrees of freedom. For example, when we have the second-order kinematic coefficient of the dependent variable θ_i with respect to the two independent input variables, θ_j and θ_k, we will use the notation

$$\theta_{ijk}'' = \frac{\partial^2 \theta_i}{\partial \theta_j \partial \theta_k}. \tag{5.7}$$

Using this notation, the derivative of Eq. (5.4) with respect to time (that is the angular acceleration of link i) can be written as

$$\alpha_i = \frac{d^2\theta_i}{dt^2} = \frac{d\omega_i}{dt} = \theta_{ijj}''\omega_j^2 + 2\theta_{ijk}''\omega_j\omega_k + \theta_{ikk}''\omega_k^2 + \theta_{ij}'\alpha_j + \theta_{ik}'\alpha_k. \tag{5.8}$$

Differentiating Eq. (5.6) with respect to time, the acceleration of point P can be written as

$$\begin{aligned}\mathbf{A}_P = &\left(X_{Pjj}''\omega_j^2 + 2X_{Pjk}''\omega_j\omega_k + X_{Pkk}''\omega_k^2 + X_{Pj}'\alpha_j + X_{Pk}'\alpha_k\right)\hat{\mathbf{i}} \\ &+ \left(Y_{Pjj}''\omega_j^2 + 2Y_{Pjk}''\omega_j\omega_k + Y_{Pkk}''\omega_k^2 + Y_{Pj}'\alpha_j + Y_{Pk}'\alpha_k\right)\hat{\mathbf{j}}.\end{aligned} \tag{5.9}$$

The angle for the direction of the acceleration of point P can be written as

$$\psi = \tan^{-1}\left(\frac{A_P^y}{A_P^x}\right). \tag{5.10}$$

Substituting the components of Eq. (5.9) into this equation, the direction of $\mathbf{A}_P$ can be written as

$$\psi = \tan^{-1}\left(\frac{Y''_{Pjj}\omega_j^2 + 2Y''_{Pjk}\omega_j\omega_k + Y''_{Pkk}\omega_k^2 + Y'_{Pj}\alpha_j + Y'_{Pk}\alpha_k}{X''_{Pjj}\omega_j^2 + 2X''_{Pjk}\omega_j\omega_k + X''_{Pkk}\omega_k^2 + X''_{Pj}\alpha_j + X''_{Pk}\alpha_k}\right). \tag{5.11}$$

For the special case where the input velocities ω_j and ω_k are constant, that is, when the input accelerations $\alpha_j = 0$ and $\alpha_k = 0$, then Eq. (5.11) can be written as

$$\psi = \tan^{-1}\left(\frac{Y''_{Pjj}\omega_j^2 + 2Y''_{Pjk}\omega_j\omega_k + Y''_{Pkk}\omega_k^2}{X''_{Pjj}\omega_j^2 + 2X''_{Pjk}\omega_j\omega_k + X''_{Pkk}\omega_k^2}\right). \tag{5.12}$$

Even with these additional subscripts, second-order kinematic coefficients can be used for acceleration analysis of multi-degree-of-freedom mechanisms, almost identically as they were in Section 4.12 with a single degree of freedom. Let us demonstrate this by again solving Example 5.8 using second-order kinematic coefficients.

EXAMPLE 5.9

For the five-bar linkage of Example 5.1, in the configuration illustrated in Fig. 5.5, suppose that links 2 and 3 are rotating with constant angular velocities of $\omega_2 = 10$ rad/s ccw and $\omega_3 = 5$ rad/s cw, respectively. The velocities were determined in Example 5.6 using the first-order kinematic coefficients. The problem now is to determine the angular accelerations of coupler links 4 and 5 and the acceleration of coupler point P.

SOLUTION

We continue from our work in Example 5.6 by taking the partial derivative of Eqs. (18) with respect to the input position θ_2; writing this in matrix form, gives

$$\begin{bmatrix} -r_4\sin\theta_4 & r_5\sin\theta_5 \\ r_4\cos\theta_4 & -r_5\cos\theta_5 \end{bmatrix}\begin{bmatrix} \theta''_{422} \\ \theta''_{522} \end{bmatrix} = \begin{bmatrix} r_2\cos\theta_2 + r_4\cos\theta_4\theta'^2_{42} - r_5\cos\theta_5\theta'^2_{52} \\ r_2\sin\theta_2 + r_4\sin\theta_4\theta'^2_{42} - r_5\sin\theta_5\theta'^2_{52} \end{bmatrix}. \tag{33}$$

We can solve this matrix equation for the second-order kinematic coefficients θ''_{422} and θ''_{522}. Using the data obtained so far, the answers are $\theta''_{422} = 0.139$ rad/rad^2 and $\theta''_{522} = 0.208$ rad/rad^2.

We also partially differentiate Eqs. (18) with respect to the input position θ_3; writing this in martrix form, gives

$$\begin{bmatrix} -r_4\sin\theta_4 & r_5\sin\theta_5 \\ r_4\cos\theta_4 & -r_5\cos\theta_5 \end{bmatrix}\begin{bmatrix} \theta''_{423} \\ \theta''_{523} \end{bmatrix} = \begin{bmatrix} r_4\cos\theta_4\theta'_{42}\theta'_{43} - r_5\cos\theta_5\theta'_{52}\theta'_{53} \\ r_4\sin\theta_4\theta'_{42}\theta'_{43} - r_5\sin\theta_5\theta'_{52}\theta'_{53} \end{bmatrix}. \tag{34}$$

Then solving this matrix equation will give the second-order kinematic coefficients θ''_{423} and θ''_{523}. Note that taking the partial derivative of Eqs. (19) with respect to the input position θ_2 gives the same result. For the current geometry and configuration of the linkage, the results are $\theta''_{423} = 0.131$ rad/rad^2 and $\theta''_{523} = -0.206$ rad/rad^2.

Next, we partially differentiate Eqs. (19) with respect to the input position θ_3, which gives

$$\begin{bmatrix} -r_4 \sin\theta_4 & r_5 \sin\theta_5 \\ r_4 \cos\theta_4 & -r_5 \cos\theta_5 \end{bmatrix} \begin{bmatrix} \theta''_{433} \\ \theta''_{533} \end{bmatrix} = \begin{bmatrix} -r_3 \cos\theta_3 + r_4 \cos\theta_4 \theta'^2_{43} - r_5 \cos\theta_5 \theta'^2_{53} \\ -r_3 \sin\theta_3 + r_4 \sin\theta_4 \theta'^2_{43} - r_5 \sin\theta_5 \theta'^2_{53} \end{bmatrix}. \quad (35)$$

Then we can solve this matrix equation for the second-order kinematic coefficients θ''_{433} and θ''_{533}. For the current geometry and configuration of the linkage, the results are $\theta''_{433} = -0.038$ rad/rad^2 and $\theta''_{533} = -0.173$ rad/rad^2.

Table 5.3 presents the second-order kinematic coefficients of the two coupler links for the same range of input angles given in Table 5.1.

Substituting the numerical values of the first-order and second-order kinematic coefficients into Eqs. (5.8), the angular accelerations of the coupler links (for the position illustrated in Fig. 5.5) are

$$\alpha_4 = -0.150 \text{ rad/s}^2 \text{ (cw)} \quad \text{and} \quad \alpha_5 = 37.075 \text{ rad/s}^2 \text{ (ccw)}. \qquad \textit{Ans.}$$

Note that these answers are in good agreement with the results obtained from the acceleration polygon; see Example 5.8.

Differentiating Eq. (21) with respect to the two independent inputs, the second-order kinematic coefficients of vector $\mathbf{r}_7$ are

$$\theta''_{722} = \theta''_{422},\ \theta''_{723} = \theta''_{423}, \quad \text{and} \quad \theta''_{733} = \theta''_{433}. \quad (36)$$

TABLE 5.3 Input angles and second-order kinematic coefficients of coupler links.

θ_2 deg	θ_3 deg	θ''_{422} —	θ''_{423} —	θ''_{433} —	θ''_{522} —	θ''_{523} —	θ''_{533} —
40	85	−0.135	−0.263	−0.016	0.535	−0.472	−0.156
50	80	0.095	−0.236	−0.075	0.686	−0.430	−0.214
60	75	0.215	−0.177	−0.107	0.701	−0.361	−0.244
70	70	0.254	−0.112	−0.114	0.635	−0.294	−0.251
80	65	0.249	−0.055	−0.106	0.537	−0.242	−0.245
90	60	0.225	−0.005	−0.088	0.436	−0.208	−0.233
100	55	0.196	0.038	−0.068	0.346	−0.190	−0.217
110	50	0.167	0.081	−0.049	0.269	−0.188	−0.198
120	45	0.139	0.131	−0.038	0.208	−0.206	−0.173
130	40	0.104	0.208	−0.050	0.168	−0.259	−0.128
140	35	0.034	0.371	−0.144	0.174	−0.404	−0.002
150	30	−0.281	0.987	−0.752	0.434	−1.006	0.635

Differentiating Eqs. (22) and (23) with respect to the input position θ_2 and using Eqs. (21) and (36), the second-order kinematic coefficients of point P are

$$X''_{P22} = -r_2 \cos\theta_2 - r_7 \cos\theta_7 \theta'^{2}_{42} - r_7 \sin\theta_7 \theta''_{422} \tag{37}$$

$$Y''_{P22} = -r_2 \sin\theta_2 - r_7 \sin\theta_7 \theta'^{2}_{42} + r_7 \cos\theta_7 \theta''_{422}. \tag{38}$$

Similarly, differentiating Eqs. (22) to (23) with respect to the input position θ_3 and using Eqs. (21) and (36), the second-order kinematic coefficients of point P are

$$X''_{P23} = -r_7 \cos\theta_7 \theta'_{42}\theta'_{43} - r_7 \sin\theta_7 \theta''_{423} \tag{39}$$

$$Y''_{P23} = -r_7 \sin\theta_7 \theta'_{42}\theta'_{43} + r_7 \cos\theta_7 \theta''_{423} \tag{40}$$

$$X''_{P33} = -r_7 \cos\theta_7 \theta'^{2}_{43} - r_7 \sin\theta_7 \theta''_{433} \tag{41}$$

$$Y''_{P33} = -r_7 \sin\theta_7 \theta'^{2}_{43} + r_7 \cos\theta_7 \theta''_{433}. \tag{42}$$

Substituting the results of Eqs. (18) and (19) and Eqs. (33) to (35) into Eqs. (37) to (42), the second-order kinematic coefficients of point P are

$$X''_{P22} = -2.5 \cos 120° - 8 \cos 66.447°(-0.237)^2 - 8 \sin 66.447°(+0.139) = +0.054 \text{ in/rad}^2$$

$$Y''_{P22} = -2.5 \sin 120° - 8 \sin 66.447°(-0.237)^2 + 8 \cos 66.447°(+0.139) = -2.133 \text{ in/rad}^2$$

$$X''_{P23} = -8 \cos 66.447°(-0.237)(+0.705) - 8 \sin 66.447°(0.131) = -0.429 \text{ in/rad}^2$$

$$Y''_{P23} = -8 \sin 66.447°(-0.237)(+0.705) + 8 \cos 66.447°(0.131) = +1.642 \text{ in/rad}^2$$

$$X''_{P33} = -8 \cos 66.447°(+0.705)^2 - 8 \sin 66.447°(-0.038) = -1.311 \text{ in/rad}^2$$

$$Y''_{P33} = -8 \sin 66.447°(+0.705)^2 + 8 \cos 66.447°(-0.038) = -3.762 \text{ in/rad}^2.$$

Then, substituting these results and the specified input angular velocities and accelerations into Eq. (5.9), the acceleration of point P is

$$\mathbf{A}_P = 15.51\hat{\mathbf{i}} - 471.57\hat{\mathbf{j}} \text{ in/s}^2.$$

Therefore, the magnitude and direction of the acceleration of point P, respectively, are

$$A_P = \sqrt{(15.51 \text{ in/s}^2)^2 + (-471.57 \text{ in/s}^2)^2} = 471.82 \text{ in/s}^2$$

and

$$\psi = \tan^{-1}\left(\frac{-471.57 \text{ in/s}^2}{15.51 \text{ in/s}^2}\right) = -88.117°.$$

The direction of the acceleration of point P is illustrated in Fig. 5.15.

Table 5.4 presents the second-order kinematic coefficients of point P and the magnitude and direction of the acceleration of point P for the range of input angles given in Table 5.1.

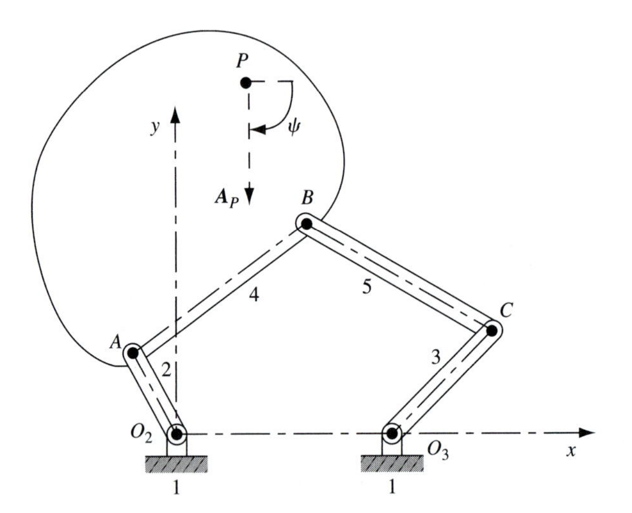

Figure 5.15 Example 5.9. The acceleration of coupler point P.

TABLE 5.4 The second-order kinematic coefficients of coupler point P.

θ_2 deg	θ_3 deg	X''_{P22} in.	Y''_{P22} in.	X''_{P23} in.	Y''_{P23} in.	X''_{P33} in.	Y''_{P33} in.	A_P in./s^2	ψ deg
40	85	0.236	−2.914	0.013	3.805	2.537	−3.612	766.98	−83.583
50	80	−1.425	−4.147	0.517	3.406	2.188	−3.342	850.35	−99.446
60	75	−2.417	−4.316	0.656	2.811	1.837	−3.256	836.00	−108.217
70	70	−2.666	−3.992	0.569	2.281	1.417	−3.302	766.03	−112.089
80	65	−2.402	−3.547	0.393	1.896	0.926	−3.404	679.65	−112.157
90	60	−1.870	−3.130	0.197	1.653	0.384	−3.511	599.38	−109.199
100	55	−1.232	−2.764	0.005	1.530	−0.184	−3.600	535.02	−103.873
110	50	−0.578	−2.436	−0.193	1.519	−0.758	−3.673	490.69	−96.725
120	45	0.054	−2.133	−0.429	1.642	−1.311	−3.762	471.82	−88.117
130	40	0.677	−1.875	−0.783	1.999	−1.794	−3.981	497.37	−78.265
140	35	1.423	−1.826	−1.520	2.969	−1.991	−4.761	646.57	−67.780
150	30	3.318	−3.321	−4.290	7.007	−0.210	−9.128	1470.05	−59.072

5.9 PATH CURVATURE OF A COUPLER POINT

After the first- and second-order kinematic coefficients of a multi-degree-of-freedom linkage are known, a study of the curvature of the path traced by a coupler point, which was discussed in Section 4.17, can be undertaken. The radius of curvature of the path traced by coupler point P was given by Eq. (4.53). However, a modified version of this equation is required for multi-degree-of-freedom linkages.

The position of point P can be written as

$$\mathbf{r}_P = X_P\hat{\mathbf{i}} + Y_P\hat{\mathbf{j}} \tag{5.13}$$

and the velocity of point P, $\mathbf{V}_P$, can be determined by differentiating this equation with respect to time. Recall that the velocity of point P can be written as in Eq. (5.6).

The unit vector tangent to the path of point P is given by Eq. (3.37), that is,

$$\hat{\mathbf{u}}_P^t = \frac{\mathbf{V}_P}{V_P}. \tag{5.14}$$

For the case where X_P and Y_P are functions of two input variables θ_j and θ_k, by substituting Eq. (5.6) into this equation, the unit tangent vector can be written as

$$\hat{\mathbf{u}}_P^t = \frac{\left(X'_{Pj}\omega_j + X'_{Pk}\omega_k\right)\hat{\mathbf{i}} + \left(Y'_{Pj}\omega_j + Y'_{Pk}\omega_k\right)\hat{\mathbf{j}}}{\sqrt{\left(X'_{Pj}\omega_j + X'_{Pk}\omega_k\right)^2 + \left(Y'_{Pj}\omega_j + Y'_{Pk}\omega_k\right)^2}}. \tag{5.15}$$

Then, the unit normal vector is 90° counterclockwise from the unit tangent vector; that is,

$$\hat{\mathbf{u}}_P^n = \hat{\mathbf{k}} \times \hat{\mathbf{u}}_P^t = \frac{-\left(Y'_{Pj}\omega_j + Y'_{Pk}\omega_k\right)\hat{\mathbf{i}} + \left(X'_{Pj}\omega_j + X'_{Pk}\omega_k\right)\hat{\mathbf{j}}}{\sqrt{\left(X'_{Pj}\omega_j + X'_{Pk}\omega_k\right)^2 + \left(Y'_{Pj}\omega_j + Y'_{Pk}\omega_k\right)^2}}. \tag{5.16}$$

The normal acceleration of point P is

$$A_P^n = \mathbf{A}_P \cdot \hat{\mathbf{u}}_P^n = A_P^x \hat{u}_P^{nx} + A_P^y \hat{u}_P^{ny}. \tag{5.17}$$

Substituting Eqs. (5.9) and (5.16) into Eq. (5.17) and performing the dot product operation, the normal acceleration of point P can be written as

$$A_P^n = \frac{\begin{array}{c}\left(Y''_{Pjj}\omega_j^2 + 2Y''_{Pjk}\omega_j\omega_k + Y''_{Pkk}\omega_k^2 + Y'_{Pj}\alpha_j + Y'_{Pk}\alpha_k\right)\left(X'_{Pj}\omega_j + X'_{Pk}\omega_k\right) \\ -\left(X''_{Pjj}\omega_j^2 + 2X''_{Pjk}\omega_j\omega_k + X''_{Pkk}\omega_k^2 + X'_{Pj}\alpha_j + X'_{Pk}\alpha_k\right)\left(Y'_{Pj}\omega_j + Y'_{Pk}\omega_k\right)\end{array}}{\sqrt{\left(X'_{Pj}\omega_j + X'_{Pk}\omega_k\right)^2 + \left(Y'_{Pj}\omega_j + Y'_{Pk}\omega_k\right)^2}}. \tag{5.18}$$

For the special case where links 2 and 3 have constant angular velocities, then Eq. (5.18) can be written as

$$A_P^n = \frac{\begin{array}{c}\left(Y''_{Pjj}\omega_j^2 + 2Y''_{Pjk}\omega_j\omega_k + Y''_{Pkk}\omega_k^2\right)\left(X'_{Pj}\omega_j + X'_{Pk}\omega_k\right) \\ -\left(X''_{Pjj}\omega_j^2 + 2X''_{Pjk}\omega_j\omega_k + X''_{Pkk}\omega_k^2\right)\left(Y'_{Pj}\omega_j + Y'_{Pk}\omega_k\right)\end{array}}{\sqrt{\left(X'_{Pj}\omega_j + X'_{Pk}\omega_k\right)^2 + \left(Y'_{Pj}\omega_j + Y'_{Pk}\omega_k\right)^2}}. \tag{5.19}$$

Substituting Eqs. (5.6) and (5.19) into Eq. (4.53) and simplifying, the radius of curvature of the path traced by coupler point P can be written as

$$\rho_P = \frac{\left[\left(X'_{Pj}\omega_j + X'_{Pk}\omega_k\right)^2 + \left(Y'_{Pj}\omega_j + Y'_{Pk}\omega_k\right)^2\right]^{3/2}}{\left(Y''_{Pjj}\omega_j^2 + 2Y''_{Pjk}\omega_j\omega_k + Y''_{Pkk}\omega_k^2\right)\left(X'_{Pj}\omega_j + X'_{Pk}\omega_k\right) - \left(X''_{Pjj}\omega_j^2 + 2X''_{Pjk}\omega_j\omega_k + X''_{Pkk}\omega_k^2\right)\left(Y'_{Pj}\omega_j + Y'_{Pk}\omega_k\right)}. \tag{5.20}$$

If we let the angular velocity ratio $\eta = \omega_k/\omega_j$, then Eq. (5.20) can be written as

$$\rho_P = \frac{\left[\left(X'_{Pj} + X'_{Pk}\eta\right)^2 + \left(Y'_{Pj} + Y'_{Pk}\eta\right)^2\right]^{3/2}}{\left(Y''_{Pjj} + 2Y''_{Pjk}\eta + Y''_{Pkk}\eta^2\right)\left(X'_{Pj} + X'_{Pk}\eta\right) - \left(X''_{Pjj} + 2X''_{Pjk}\eta + X''_{Pkk}\eta^2\right)\left(Y'_{Pj} + Y'_{Pk}\eta\right)}. \tag{5.21}$$

The sign convention here is the same as it has been in earlier chapters: a negative value for the radius of curvature indicates that the unit normal vector is pointing away from the center of curvature of the path traced by point P.

The coordinates of the center of curvature of the path traced by point P, as given by Eq. (4.58), are written here as

$$X_C = X_P + \rho_P u_P^{nx} \text{ and } Y_C = Y_P + \rho_P u_P^{ny}, \tag{5.22}$$

where X_P and Y_P are determined as indicated in Example 5.3 and u_P^{nx} and u_P^{ny} are given by the $\hat{\mathbf{i}}$ and $\hat{\mathbf{j}}$ components of Eq. (5.16).

EXAMPLE 5.10

Continuing the five-bar linkage of Examples 5.1 through 5.9 in the configuration illustrated in Fig. 5.5, suppose that links 2 and 3 are rotating with constant angular velocities of $\omega_2 =$ 10 rad/s ccw and $\omega_3 = 5$ rad/s cw, respectively. The velocity of coupler point P, obtained from the first-order kinematic coefficients with the same input data in Example 5.6, is

$$\mathbf{V}_P = 21.546\hat{\mathbf{i}} - 31.330\hat{\mathbf{j}} \text{ in/s} = 38.024 \text{ in/s} \angle -55.483^\circ.$$

The acceleration of point P, obtained from the second-order kinematic coefficients with the same input data in Example 5.9, is

$$\mathbf{A}_P = 15.508\hat{\mathbf{i}} - 471.570\hat{\mathbf{j}} \text{ in/s}^2 = 471.824 \text{ in/s}^2 \angle -88.117^\circ.$$

Find the unit tangent vector and unit normal vector to the path of point P, the radius of curvature, and the Cartesian coordinates of the center of curvature of the path of point P.

SOLUTION

Substituting the velocity components into Eq. (5.14), the unit tangent vector is

$$\hat{\mathbf{u}}^t = \frac{21.546 \text{ in/s } \hat{\mathbf{i}} - 31.330 \text{ in/s } \hat{\mathbf{j}}}{38.02 \text{ in/s}} = 0.567\hat{\mathbf{i}} - 0.824\hat{\mathbf{j}}.$$

Then from Eq. (5.16), the unit normal vector is

$$\hat{\mathbf{u}}^n = 0.824\hat{\mathbf{i}} + 0.567\hat{\mathbf{j}}. \tag{43}$$

The directions of the unit tangent vector and unit normal vector are illustrated in Fig. 5.16.

Substituting data from Examples 5.6 and 5.9 for the first- and second-order kinematic coefficients and the specified input angular velocities into Eq. (5.20), the radius of curvature of the path of point P is

$$\rho_P = \frac{1445.81 \text{ in}^3/\text{s}^3}{-254.44 \text{ in}^2/\text{s}^3} = -5.682 \text{ in}. \tag{44}$$

The negative sign indicates that the center of curvature is located in the direction of the negative unit normal vector $\hat{\mathbf{u}}^n$ from point P (see Fig. 5.17).

Finally, substituting the answers from Example 5.3 and Eqs. (43) and (44) into Eqs. (5.22), the coordinates of the center of curvature of the path of point P are

$$X_C = 1.947 \text{ in} + (-5.682 \text{ in})(0.824) = -2.735 \text{ in}.$$

$$Y_C = 9.499 \text{ in} + (-5.682 \text{ in})(0.567) = 6.277 \text{ in}.$$

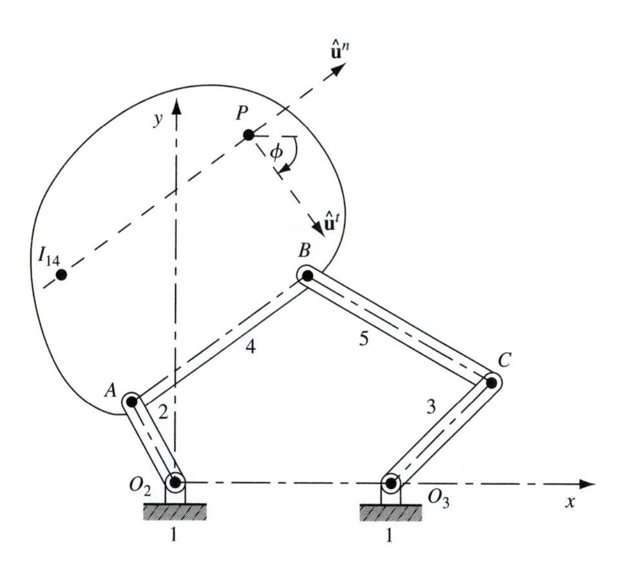

Figure 5.16 Example 5.10. The unit tangent and the unit normal vectors.

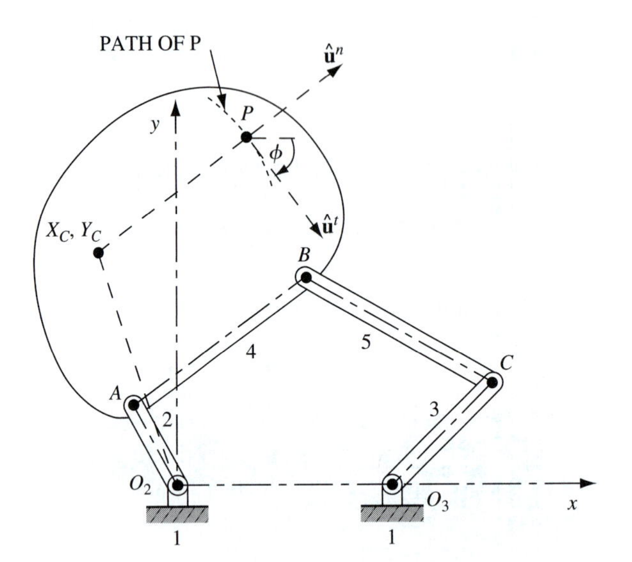

Figure 5.17 Example 5.10. The center of curvature of the path traced by point P.

TABLE 5.5 The radius of curvature and the coordinates of the center of curvature.

θ_2 deg	θ_3 deg	ρ in.	X_C in.	Y_C in.
40	85	−10.350	5.784	2.067
50	80	−8.062	4.052	3.559
60	75	−5.340	2.247	5.595
70	70	−3.014	1.005	7.557
80	65	−1.667	0.551	8.821
90	60	−1.214	0.524	9.271
100	55	−1.481	0.408	9.031
110	50	−2.671	−0.364	8.130
120	45	−5.682	−2.735	6.279
130	40	−13.603	−9.496	2.157
140	35	−51.321	−42.794	−15.556
150	30	57.190	53.984	33.525

The center of curvature of the path of point P is illustrated in Fig. 5.17. This figure also illustrates the unit tangent vector $\hat{\mathbf{u}}^t$ and the unit normal vector $\hat{\mathbf{u}}^n$ to the path traced by point P.

Table 5.5 presents the radius of curvature of the path traced by point P and the Cartesian coordinates of the center of curvature of this path for the range of input angles given in Table 5.1.

5.10 THE FINITE DIFFERENCE METHOD

The finite difference method can be used to check: (*i*) the first-order kinematic coefficients of links 4 and 5; and (*ii*) the second-order kinematic coefficients of links 4 and 5.

(*i*) In general, the first-order kinematic coefficient of link *i* with respect to link *j* can be written as

$$\theta'_{ij} = \partial\theta_i/\partial\theta_j \approx \Delta\theta_i/\Delta\theta_j.$$

Keep in mind that the definition of a *partial* derivative implies that $\Delta\theta_j$ in the denominator corresponds to the change of an independent input while all other independent inputs are held frozen or locked during the change of $\Delta\theta_j$. However, data such as those in Tables 5.1 and 5.2, for example, do not correspond to this restriction because they were derived with both $\Delta\theta_2$ and $\Delta\theta_3$ changing simultaneously. Therefore, the first-order kinematic coefficient of link *i* with respect to link *j* *cannot* be directly approximated by the following formula,

$$\theta'_{ij} \neq \frac{\Delta\theta_i}{\Delta\theta_j} = \frac{(\theta_i)_A - (\theta_i)_B}{(\theta_j)_A - (\theta_j)_B},$$

where subscript *A* denotes the value after and *B* denotes the value before the value that is to be checked by the finite difference method. In situations such as Tables 5.1 and 5.2, however, where the total change of $\Delta\theta_i$ is the result of separate changes of more than one independent variable, say $\Delta\theta_j$ and $\Delta\theta_k$, the procedure is to write the finite change in the angular displacement of link *i* as

$$\Delta\theta_i \approx \theta'_{ij}\Delta\theta_j + \theta'_{ik}\Delta\theta_k$$

or as

$$[(\theta_i)_A - (\theta_i)_B] \approx \theta'_{ij}\left[(\theta_j)_A - (\theta_j)_B\right] + \theta'_{ik}\left[(\theta_k)_A - (\theta_k)_B\right]. \tag{5.23}$$

The following example will illustrate the use of these equations to check the first-order kinematic coefficients.

EXAMPLE 5.11

Use finite differences to check the first-order kinematic coefficients of links 4 and 5 of the five-bar linkage of Example 5.6.

SOLUTION

By Eq. (5.23), the finite change in the angular displacement of link 4 can be written as

$$\Delta\theta_4 = [(\theta_4)_A - (\theta_4)_B] \approx \theta'_{42}\,[(\theta_2)_A - (\theta_2)_B] + \theta'_{43}\,[(\theta_3)_A - (\theta_3)_B] \tag{45}$$

and the finite change in the angular displacement of link 5 can be written as

$$\Delta\theta_5 = [(\theta_5)_A - (\theta_5)_B] \approx \theta'_{52}\,[(\theta_2)_A - (\theta_2)_B] + \theta'_{53}\,[(\theta_3)_A - (\theta_3)_B]\,. \tag{46}$$

For the input position angles $\theta_2 = 120°$ and $\theta_3 = 45°$ (see row nine of Table 5.1), the first-order kinematic coefficients are

$$\theta'_{42} = -0.237, \theta'_{43} = 0.705, \theta'_{52} = 0.455, \text{ and } \theta'_{53} = -0.109.$$

Substituting these values into Eqs. (45) and (46), the finite change in the angular displacements of links 4 and 5, respectively, are

$$\Delta\theta_4 = \left[30.526° - 42.360°\right] \approx -0.237\left[130° - 110°\right] + 0.705\left[40° - 50°\right] \tag{47}$$

$$\Delta\theta_5 = \left[156.510° - 146.315°\right] \approx 0.455\left[130° - 110°\right] - 0.109\left[40° - 50°\right]. \tag{48}$$

Note that the answers given by the two separate calculations in Eqs. (47) and (48) are in reasonable agreement with each other. If the increments of the two input angular displacements are decreased, then even closer agreement between the two sets of calculations is obtained. The general conclusion, therefore, is that the first-order kinematic coefficients of the five-bar linkage as indicated by Table 5.1 are correct.

From Eqs. (5.7), the second-order kinematic coefficients of link i with respect to the input angles θ_j and θ_k can be written as

$$\theta''_{ijk} = \partial^2\theta_i/\partial\theta_j\partial\theta_k = \partial\theta'_{ij}/\partial\theta_k.$$

Remembering the directional nature of finite differences, then the finite change in the first-order kinematic coefficients of link i with respect to link j can be written as

$$\Delta\theta'_{ij} = \left[(\theta'_{ij})_A - (\theta'_{ij})_B\right] \approx \theta''_{ijj}\,[(\theta_j)_A - (\theta_j)_B] + \theta''_{ijk}\,[(\theta_k)_A - (\theta_k)_B], \tag{5.24}$$

where subscript A denotes the value after and B denotes the value before the value that is to be checked by the finite difference method.

The following example will illustrate the value of these equations to check the second-order kinematic coefficients.

EXAMPLE 5.12

Use finite differences to check the second-order kinematic coefficients of links 4 and 5 of the five-bar linkage of Example 5.9.

SOLUTION

The second-order kinematic coefficients can be checked by writing the finite change in the first-order kinematic coefficients of links 4 and 5 with respect to the input links 2 and 3 [see Eq. (5.24)] as

$$\Delta\theta'_{42} = \left[(\theta'_{42})_A - (\theta'_{42})_B\right] \approx \theta''_{422}\left[(\theta_2)_A - (\theta_2)_B\right] + \theta''_{423}\left[(\theta_3)_A - (\theta_3)_B\right]$$

$$\Delta\theta'_{43} = \left[(\theta'_{43})_A - (\theta'_{43})_B\right] \approx \theta''_{432}\left[(\theta_2)_A - (\theta_2)_B\right] + \theta''_{433}\left[(\theta_3)_A - (\theta_3)_B\right]$$

$$\Delta\theta'_{52} = \left[(\theta'_{52})_A - (\theta'_{52})_B\right] \approx \theta''_{522}\left[(\theta_2)_A - (\theta_2)_B\right] + \theta''_{523}\left[(\theta_3)_A - (\theta_3)_B\right]$$

and

$$\Delta\theta'_{53} = \left[(\theta'_{53})_A - (\theta'_{53})_B\right] \approx \theta''_{532}\left[(\theta_2)_A - (\theta_2)_B\right] + \theta''_{533}\left[(\theta_3)_A - (\theta_3)_B\right],$$

where $\theta''_{432} = \theta''_{423}$ and $\theta''_{532} = \theta''_{523}$. From row nine of Tables 5.1 and 5.3, the values in these equations are

$$\Delta\theta'_{42} = [(-0.230) - (-0.254)] \approx 0.139\left[\frac{130^\circ - 110^\circ}{57.296\ ^\circ/\text{rad}}\right] + 0.131\left[\frac{40^\circ - 50^\circ}{57.296\ ^\circ/\text{rad}}\right]$$

$$\Delta\theta'_{43} = [0.737 - 0.683] \approx 0.131\left[\frac{130^\circ - 110^\circ}{57.296\ ^\circ/\text{rad}}\right] + (-0.038)\left[\frac{40^\circ - 50^\circ}{57.296\ ^\circ/\text{rad}}\right]$$

$$\Delta\theta'_{52} = [0.508 - 0.397] \approx 0.208\left[\frac{130^\circ - 110^\circ}{57.296\ ^\circ/\text{rad}}\right] + (-0.206)\left[\frac{40^\circ - 50^\circ}{57.296\ ^\circ/\text{rad}}\right]$$

$$\Delta\theta'_{53} = [(-0.136) - (-0.091)] \approx (-0.206)\left[\frac{130^\circ - 110^\circ}{57.296\ ^\circ/\text{rad}}\right] + (-0.173)\left[\frac{40^\circ - 50^\circ}{57.296\ ^\circ/\text{rad}}\right].$$

Note that the answers given by the two separate calculations in each of these equations are in reasonable agreement with each other. The general conclusion, therefore, is that the second-order kinematic coefficients of links 4 and 5 of the five-bar linkage as given by Table 5.3 are correct.

5.11 REFERENCES

[1] Pennock, G. R. 2008. Curvature Theory for a Two-Degree-of-Freedom Planar Linkage. *Mechanism and Machine Theory* 43:525–48.

PROBLEMS*

5.1 The slotted links 2 and 3 are driven independently at constant speeds of $\omega_2 = 30$ rad/s cw and $\omega_3 = 20$ rad/s cw, respectively. Find the absolute velocity and acceleration of the center of the pin P carried in the two slots.

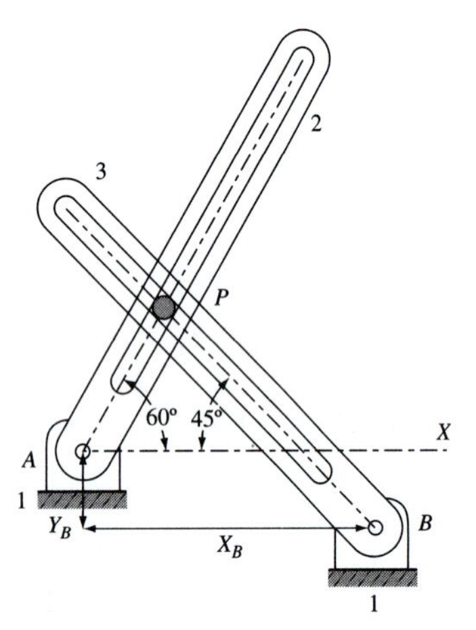

Figure P5.1 Dimensions in millimeters.

5.2 For the five-bar linkage in the position illustrated in Fig. P5.2, the angular velocity of link 2 is 15 rad/s cw and the angular velocity of link 5 is 15 rad/s cw. Determine the angular velocity of link 3 and the apparent velocity $V_{B4/5}$.

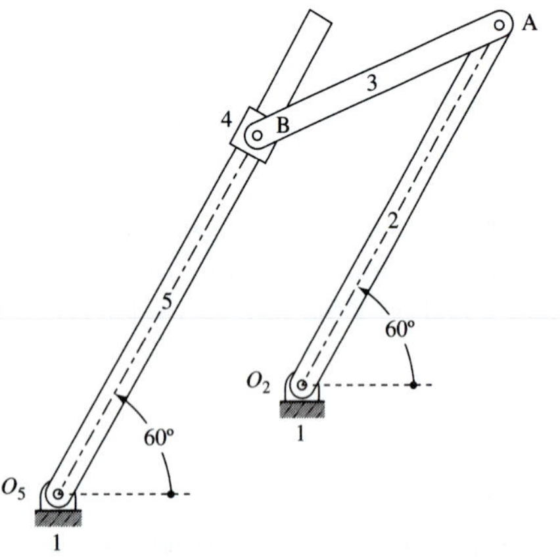

Figure P5.2 The dimensions are $\mathbf{R}_{O_2O_5} = 200$ mm $\angle 23.1°$, $R_{AO_2} = 300$ mm, and $R_{BA} = 200$ mm.

5.3 For the five-bar linkage in the position illustrated in Fig. P5.2, the angular velocity of link 2 is $\omega_2 =$ 25 rad/s ccw and the apparent velocity $V_{B4/5}$ is 5 m/s upward along link 5. Determine the angular velocities of links 3 and 5.

5.4 For Problem 5.2, assuming that the two given input velocities are constant, determine the angular acceleration of link 3 at the instant indicated.

5.5 Figure P5.5 illustrates link 2 rotating at a constant angular velocity of 10 rad/s ccw while the sliding block 3 slides toward point A on link 2 at the constant rate of 5 in./s. At the instant indicated, $R_{PA} = 4.0$ in. Find the absolute velocity and absolute acceleration of point P of block 3.

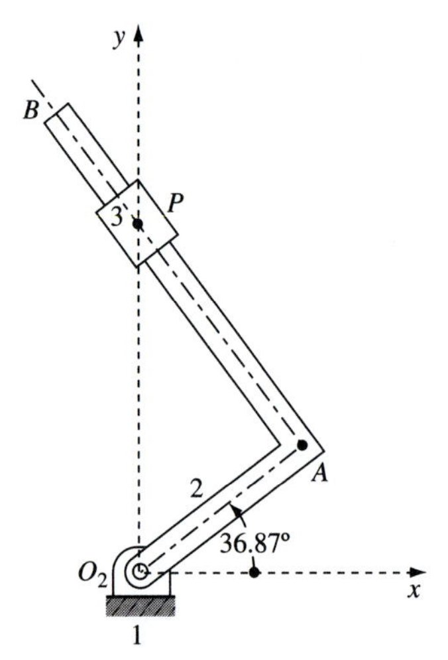

Figure P5.5 The dimensions are $R_{AO_2} = 3.0$ in. and $R_{BA} = 6.0$ in.

5.6 For Problem 5.5, determine the value of the sliding velocity $V_{P_3/2}$ that minimizes the absolute velocity of point P of block 3. Find the value of $V_{P_3/2}$ that minimizes the absolute acceleration of point P of block 3.

5.7 The two-link planar robots shown in Fig. P5.7 have the link lengths $R_{AO_2} = R_{AO_4} = 0.3$ m and $R_{PA} = R_{PB} = 0.4$ m. The two robots are carrying a small object labeled P. At the instant indicated, the angular positions are $\theta_2 = 45°$ and $\theta_{3/2} = -15°$. (Note that the angle $\theta_{3/2} = \theta_3 - \theta_2$ is given because that is the angle controlled by the motor in joint A.) A second robot having identical dimensions is stationed

* When assigning problems, the instructor may wish to specify the method of solution to be used because a variety of approaches are presented in the text.

at position O_4, 1 m to the right. What angular positions must the two joints θ_4 and $\theta_{5/4}$ of the second robot have at this moment to allow it to take over possession of the object P?

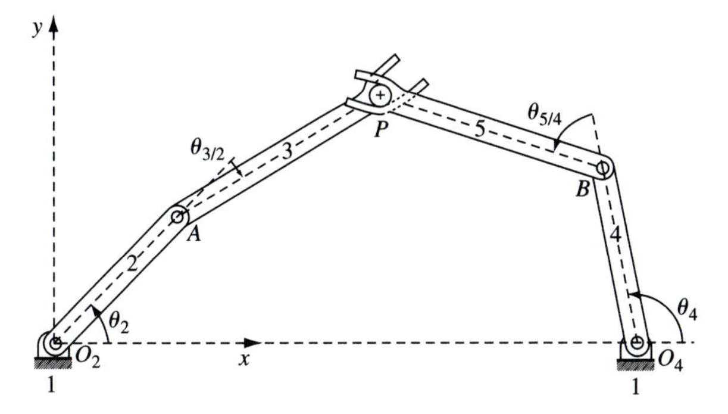

Figure P5.7 The dimensions are $R_{AO_2} = R_{AO_4} = 0.3$ m and $R_{PA} = R_{PB} = 0.4$ m.

5.8 For the transfer of the object described in Problem 5.7, it is necessary that the velocities of point P of the two robots match. If the two input velocities of the first robot are $\omega_2 = 10$ rad/s cw and $\omega_{3/2} = 15$ rad/s ccw, what angular velocities must be used for ω_4 and $\omega_{5/4}$?

5.9 For the transfer of the object described in Problem 5.7, it is necessary that the velocities of point P of the two robots match. If the two input velocities of the first robot are $\omega_2 = 10$ rad/s cw and $\omega_{3/2} = 10$ rad/s ccw, what angular velocities must be used for ω_4 and $\omega_{5/4}$?

5.10 For the transfer of the object described in Problem 5.7, it is necessary that the velocities of point P of the two robots match. If the two input velocities of the first robot are $\omega_2 = 10$ rad/s cw and $\omega_{3/2} = 0$, what angular velocities must be used for ω_4 and $\omega_{5/4}$?

5.11 To successfully transfer an object between two robots, as described in Problems 5.7 and 5.8, it is helpful if the accelerations are also matched at point P. Assuming that the two input accelerations are $\alpha_2 = \alpha_3 = 0$ at this instant for the robot on the left, what angular accelerations must be given to the two input joints of the robot on the right to achieve this?

PART 2
Design of Mechanisms

6 Cam Design

6.1 INTRODUCTION

In the previous chapters we have been learning how to analyze the kinematic characteristics of a given mechanism. We were given the design of a mechanism and we studied ways to determine its mobility, its position, its velocity, and its acceleration, and we even discussed its suitability for given types of tasks. However, we have said little about how the mechanism is designed, that is, how the sizes and the shapes of the links are chosen by the designer.

The next several chapters will introduce this *design* point of view as it relates to mechanisms. We will find ourselves looking more at individual types of machine components and learning when and why such components are used and how they are sized. In Chapter 6, which is devoted to the design of cams, for example, we will assume that we know the task to be accomplished. However, we will not know but will look for techniques to help discover the size and the shape of the cam to perform this task.

Of course, there is the creative process of deciding whether a cam should be used in the first place, as opposed to a gear train, a linkage, or some other mechanical device. This question often cannot be answered on the basis of scientific principles alone; it requires experience and imagination and involves such factors as economics, marketability, reliability, maintenance, esthetics, ergonomics, ability to manufacture, and suitability to the task. These aspects are not well studied by a general scientific approach; they require human judgment of factors that are often not easily reduced to numbers or formulae. There is usually not a single "right" answer, and generally these questions cannot be answered by this or any other text or reference book.

On the other hand, this is not to say that there is no place for a general science-based approach in design situations. Most mechanical design is based on repetitive analysis. Therefore, in Chapter 6 and in several of the upcoming chapters we will use the principles of analysis presented in Part 1 of the book. Also, we will use the governing analysis equations to help in our choice of part sizes and shapes and to help us assess the quality of our design as we proceed. It is important to point out that the upcoming chapters are still based on the laws of mechanics. The primary shift for Part 2 is that the component dimensions are often the unknowns of the problem, whereas the input and output speeds, for example, may be given information. In Chapter 6 we will discover how to determine a cam contour, or profile, which delivers a specified motion characteristic.

6.2 CLASSIFICATION OF CAMS AND FOLLOWERS

A *cam* is a machine element used to drive another element, called a *follower*, through a specified motion by direct contact. Cam-and-follower mechanisms are simple and inexpensive, have few moving parts, and occupy very little space. Furthermore, follower motions having almost any desired characteristics are not difficult to design. For these reasons, cam mechanisms are used extensively in modern machinery.

The versatility and flexibility in the design of cam systems are among their more attractive features, yet this also leads to a wide variety of shapes and forms and the need for terminology to distinguish them.

Sometimes cams are classified according to their basic shapes. Figure 6.1 illustrates four different types of cams:

(a) A *plate cam*, also called a *disk cam* or a *radial cam*;
(b) A *wedge* cam;
(c) A *cylindric cam* or *barrel cam*; and
(d) An *end cam* or *face cam*.

The least common of these in practical applications is the wedge cam because of its need for a reciprocating motion rather than a continuous input motion. By far the most common is the plate cam. For this reason, most of the remainder of Chapter 6 specifically addresses plate cams, although the concepts presented pertain universally.

Cam systems can also be classified according to the basic shape of the follower. Figure 6.2 illustrates plate cams actuating four different types of followers:

(a) A *knife-edge* follower;
(b) A *flat-face* follower;
(c) A *roller* follower; and
(d) A *spherical-face* or *curved-shoe* follower.

Note that the follower face is usually chosen to have a simple geometric shape and the motion is achieved by careful design of the shape of the cam to mate with it. This is not always the case, and examples of *inverse cams*, where the output element is machined to a complex shape, can be found.

Another method of classifying cams is according to the characteristic output motion allowed between the follower and the frame. Thus, some cams have *reciprocating* (translating) followers, as in Figs. 6.1*a*, *b*, and *d* and Figs. 6.2*a* and *b*, whereas others have *oscillating* (rotating) followers, as in Fig. 6.1*c*, Fig. 6.2*c*, and Fig. 6.2*d*. Further classification of

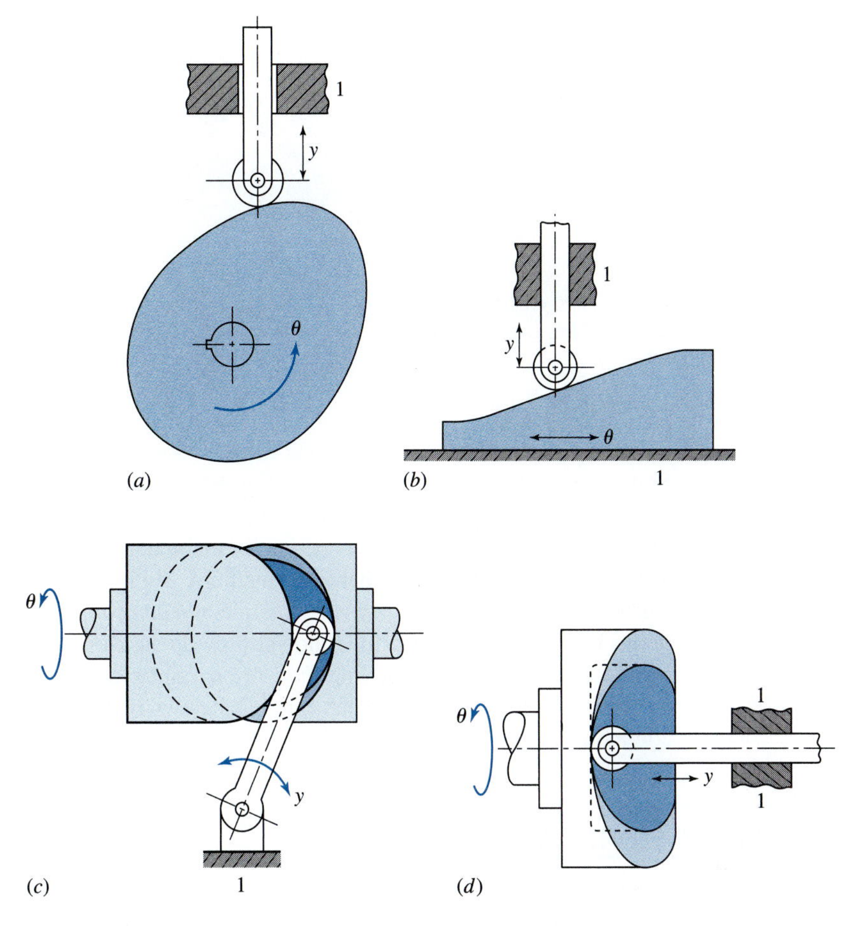

Figure 6.1 Types of cams: (*a*) plate cam, (*b*) wedge cam, (*c*) barrel cam, and (*d*) face cam.

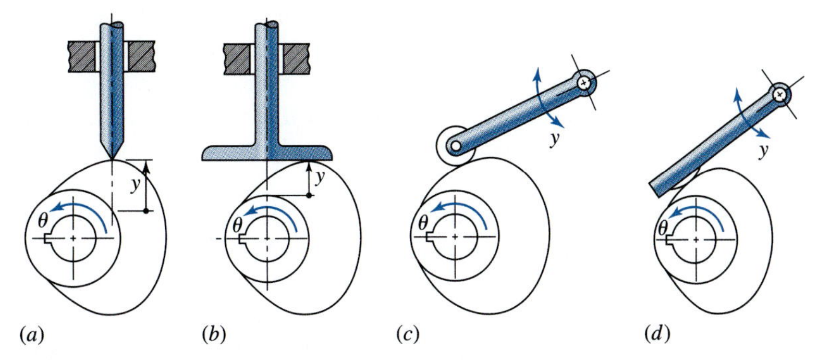

Figure 6.2 Plate cams with (*a*) an offset reciprocating knife-edge follower, (*b*) a reciprocating flat-face follower, (*c*) an oscillating roller follower, and (*d*) an oscillating curved-shoe follower.

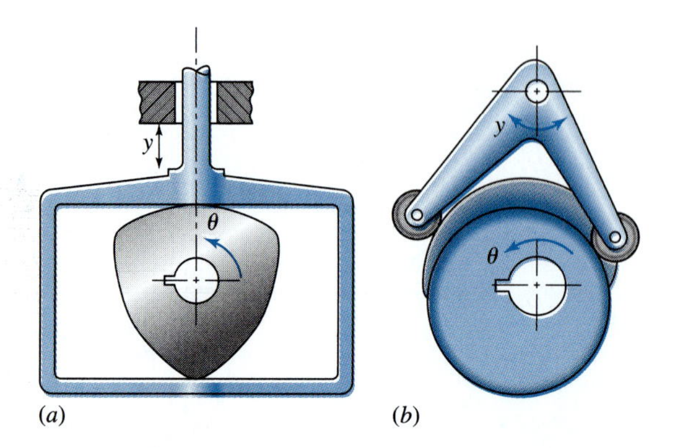

Figure 6.3
(*a*) Constant-breadth cam with a reciprocating flat-face follower. (*b*) Conjugate cams with an oscillating roller follower.

reciprocating followers distinguishes whether the centerline of the follower stem relative to the center of the cam is *offset*, as in Fig. 6.2*a*, or *radial*, as in Fig. 6.2*b*.

In all cam systems, the designer must ensure that the follower maintains contact with the cam at all times. This can be accomplished by depending on gravity, by the inclusion of a suitable spring, or by a mechanical constraint. In Fig. 6.1*c*, the follower is constrained by a groove. Figure 6.3*a* illustrates an example of a *constant-breadth* cam, where two contact points between the cam and the follower provide the constraint. Mechanical constraint can also be introduced by employing *dual* or *conjugate* cams in an arrangement such as that illustrated in Fig. 6.3*b*. Here each cam has its own roller, but the rollers are mounted on a common follower.

6.3 DISPLACEMENT DIAGRAMS

Despite the wide variety of cam types used and their differences in form, they also have certain features in common that allow a systematic approach to their design. Usually a cam system is a single-degree-of-freedom device. It is driven by a known input motion, usually a shaft that rotates at constant speed, and it is intended to produce a certain desired periodic output motion for the follower.

To investigate the design of cams in general, we will denote the known input motion by $\theta\,(t)$ and the output motion by y. A review of Fig. 6.1 to Fig. 6.3 will demonstrate the definitions of θ and y for various types of cams. Figs. 6.1 to 6.3 illustrate that the input θ is an angle for most cams, but it can be a distance, as in Fig. 6.1*b*. Also, the output y is a translational distance for a reciprocating follower but it is an angle for an oscillating follower.

During the rotation of the cam through one cycle of input motion, the follower executes a series of events as demonstrated in graphical form in the *displacement diagram* of Fig. 6.4. In such a diagram, the abscissa represents one cycle of the input motion θ (usually, one revolution of the cam) and is drawn to any convenient scale. The ordinate represents the follower travel y and, for a reciprocating follower, is usually drawn at full scale to help in the layout of the cam. On a displacement diagram it is possible to identify a portion of the graph called the *rise*, where the motion of the follower is away from the cam center. The maximum rise is called the *lift*. The *return* is the portion in which the motion of the

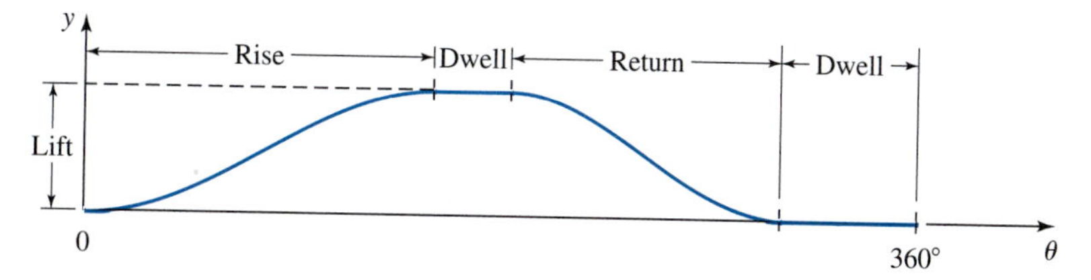

Figure 6.4 Displacement diagram for a cam.

follower is toward the cam center. Portions of the cycle during which the follower is at rest are referred to as *dwells*.

Many of the essential features of a displacement diagram, such as the total lift or the placement and duration of dwells, are usually dictated by the requirements of the application. There are, however, many possible choices of follower motions that might be used for the rise and return, and some are preferable to others depending on the situation. One of the key steps in the design of a cam is the choice of suitable forms for these motions. Once the motions have been chosen, that is, once the exact relationship is set between the input θ and the output y, the displacement diagram can be constructed precisely and is a graphical representation of the functional relationship

$$y = y(\theta).$$

This equation has stored within it the exact nature of the shape of the final cam, the necessary information for its layout and manufacture, and also the important characteristics that determine the quality of its dynamic performance. Before looking further at these topics, however, we will display graphical methods of constructing the displacement diagrams for various rise and return motions.

The displacement diagram for *uniform motion* is a straight line with a constant slope. Thus, for constant input speed, the velocity of the follower is also constant. This motion is not useful for the full lift because of the sharp corners produced at the boundaries with other segments of the displacement diagram. It is often used, however, between other curve segments that eliminate the corners.

The displacement diagram for a *modified uniform motion* is illustrated in Fig. 6.5*a*. The central portion of the diagram, subtended by the cam angle β_2 and the lift L_2, is uniform motion. The ends, with angles β_1 and β_3 and with corresponding lifts L_1 and L_3, are shaped to deliver *parabolic motion* to the follower. Soon we shall learn that these produce constant acceleration of the follower. The diagram illustrates how to match the slopes of the parabolic motion with that of the uniform motion. With β_1, β_2, β_3, and the total lift L known, the individual lifts L_1, L_2, and L_3 are determined by locating the midpoints of the β_1 and β_3 segments and constructing a straight line as indicated. Figure 6.5*b* illustrates a graphical construction for a parabola to fit within a given rectangular boundary defined by L_1 and β_1. The abscissa and ordinate are first divided into a convenient but equal number of divisions and numbered as indicated. The construction of each point of the parabola then follows that indicated by dashed lines for division 3.

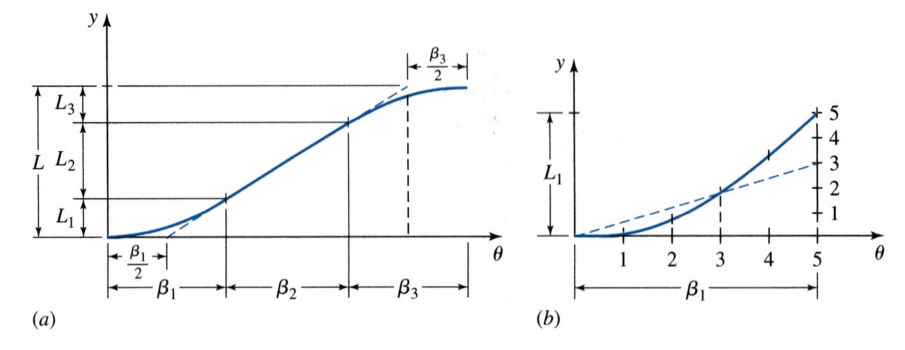

Figure 6.5 Parabolic motion displacement diagram: (*a*) interfaces with uniform motion and (*b*) graphical construction.

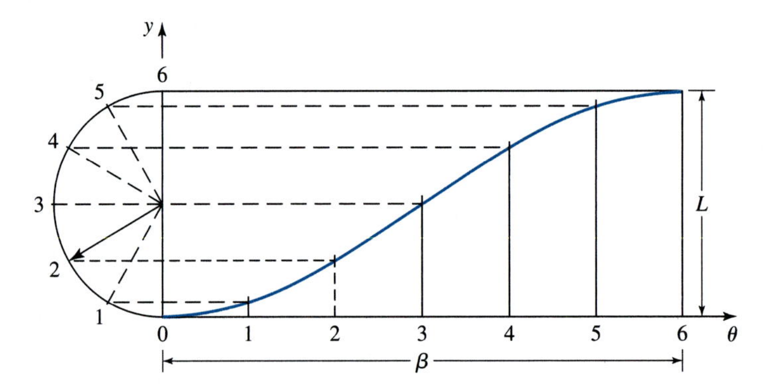

Figure 6.6 Simple harmonic motion displacement diagram; graphical construction.

In the layout of an actual cam, if this is done graphically, a great many divisions are used to obtain good accuracy. At the same time, the drawing is made to a large scale, perhaps 10 times full size, and then reduced to actual size by a pantographic method. However, for clarity in reading, the figures in this text are shown with a lesser number of divisions to define the curves and illustrate the graphic techniques.

The displacement diagram for *simple harmonic motion* is illustrated in Fig. 6.6. The graphical construction makes use of a semicircle having a diameter equal to the lift L. The semicircle and abscissa are divided into an equal number of divisions, and the construction then follows that indicated by dashed lines for division number 2.

Cycloidal motion obtains its name from the geometric curve called a cycloid. As illustrated on the left-hand side of Fig. 6.7, a circle of radius $L/2\pi$, where L is the lift, makes exactly one revolution by rolling along the ordinate from the origin to $y = L$. A point P of the circle, originally located at the origin, traces a cycloid, as demonstrated. As the circle rolls without slip, the graph of the vertical position y of the point versus the rotation angle θ gives the displacement diagram illustrated at the right of Fig. 6.7. We find it much more convenient for graphical purposes to draw the circle only once, using point B as the center.

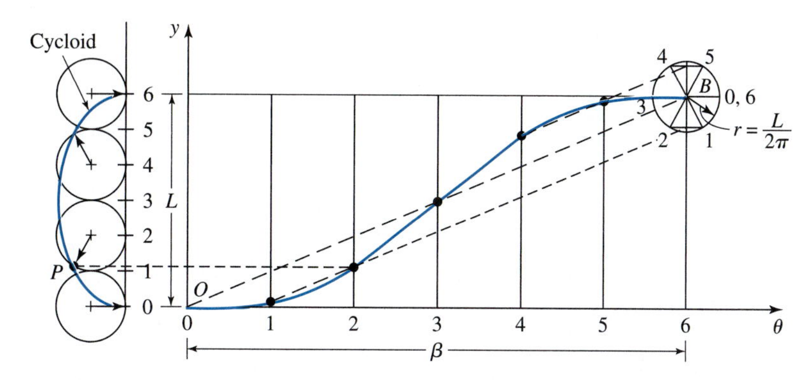

Figure 6.7 Cycloidal motion displacement diagram; graphical construction.

After dividing the circle and the abscissa into an equal number of divisions and numbering them as indicated, we can project each point of the circle horizontally until it intersects the ordinate; next, from the ordinate, we project parallel to the diagonal *OB* to obtain the corresponding point on the displacement diagram.

6.4 GRAPHICAL LAYOUT OF CAM PROFILES

Let us now examine the problem of determining the exact shape of the cam surface required to deliver a specified follower motion. We assume here that the required motion has been completely defined—graphically, analytically, or numerically—as discussed in later sections. Thus, a complete displacement diagram can be drawn to scale for the entire cam rotation. The problem now is to lay out the proper cam shape to achieve the follower motion represented by this displacement diagram.

We illustrate the procedure using the case of a plate cam as illustrated in Fig. 6.8. Let us first note some additional nomenclature illustrated in Fig. 6.8.

The *trace point* is a theoretical point of the follower; it corresponds to the tip of a fictitious knife-edge follower. It is located at the center of a roller follower or along the surface of a flat-face follower.

The *pitch curve* is the locus generated by the trace point as the follower moves relative to the cam. For a knife-edge follower, the pitch curve and cam profile are identical. For a roller follower they are separated by the radius of the roller.

The *prime circle* is the smallest circle that can be drawn with center at the cam rotation axis and tangent to the pitch curve. The radius of this circle is denoted R_0.

The *base circle* is the smallest circle centered on the cam rotation axis and tangent to the cam profile. For a roller follower it is smaller than the prime circle by the radius of the roller, and for a flat-face follower it is identical to the prime circle.

In constructing the cam profile, we employ the principle of kinematic inversion. We imagine the sheet of paper on which we are working to be fixed to the cam, and we note that the follower appears to rotate *opposite to the actual direction of cam rotation*. As illustrated in Fig. 6.8, we divide the prime circle into a number of segments and assign station numbers to the boundaries of these segments. Dividing the displacement-diagram

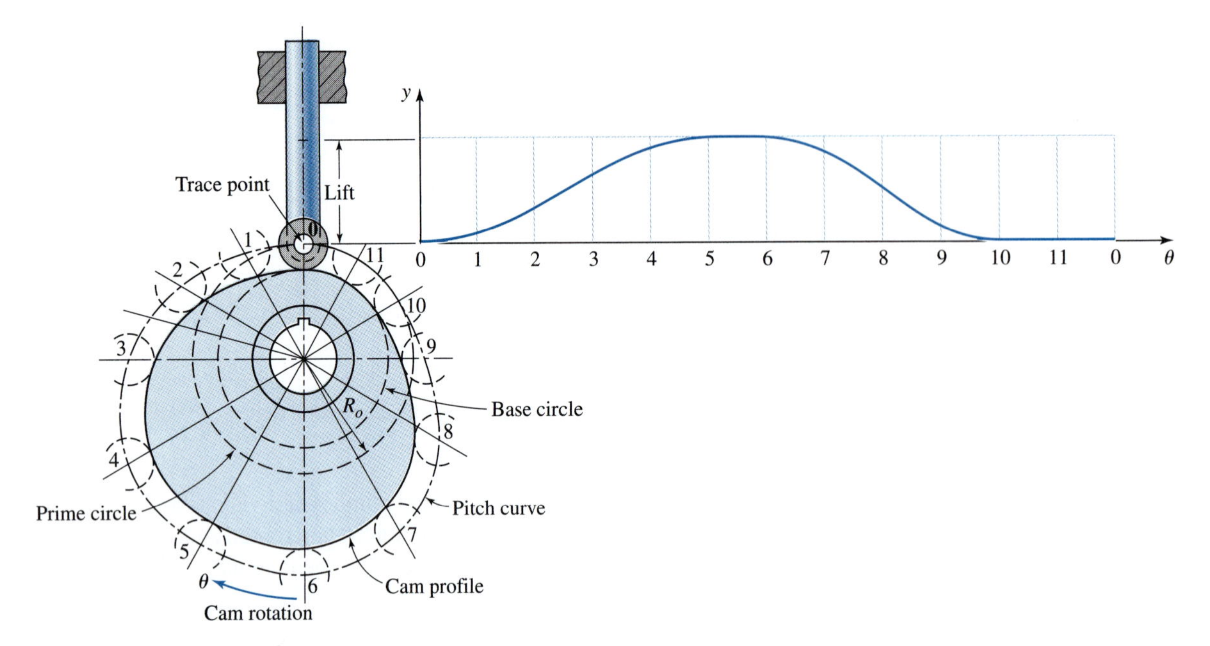

Figure 6.8 Cam nomenclature. The cam surface is developed by holding the cam stationary and rotating the follower from station 0 through stations 1, 2, 3, etc.

abscissa into corresponding segments, we transfer distances, by means of dividers, from the displacement diagram directly onto the cam layout to locate the corresponding locations of the trace point. The smooth curve through these points is the *pitch curve*. For the case of a roller follower, as in this example, we simply draw the roller in its proper location at each station and then construct the cam profile as a smooth curve tangent to all of these roller locations.

Figure 6.9 illustrates how the method of construction is modified for an offset roller follower. We begin by constructing an *offset circle*, using a radius equal to the offset distance. After identifying station numbers around the prime circle, the centerline of the follower is constructed for each station, making it tangent to the offset circle. The roller centers for each station are established by transferring distances from the displacement diagram directly to these follower centerlines, always measuring positive outward from the prime circle. An alternative procedure is to identify the points $0'$, $1'$, $2'$, and so on, on a single follower centerline and then to rotate them about the cam center to the corresponding follower centerline locations. In either case, the roller circles are drawn next and a smooth curve tangent to all roller locations is the required cam profile.

Figure 6.10 illustrates the construction for a plate cam with a reciprocating flat-face follower. The pitch curve is constructed using a method similar to that used for the roller follower in Fig. 6.8. A line representing the flat face of the follower is then constructed at each station. The cam profile is a smooth curve drawn tangent to all the follower locations. It may be helpful to extend each straight line representing a location of the follower face to form a series of triangles. If these triangles are lightly shaded, as suggested in the illustration,

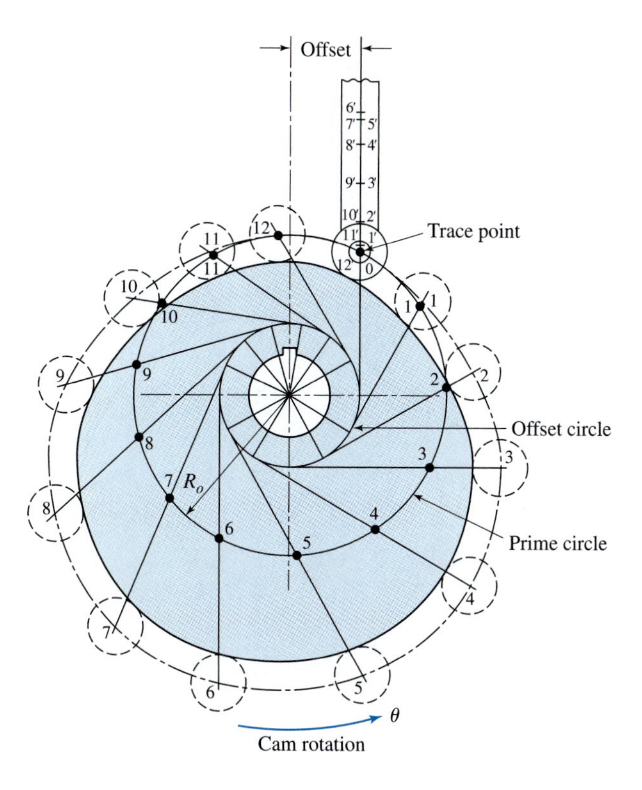

Figure 6.9 Graphical layout of a plate cam profile with an offset reciprocating roller follower.

Figure 6.10 Graphical layout of a plate cam profile with a reciprocating flat-face follower.

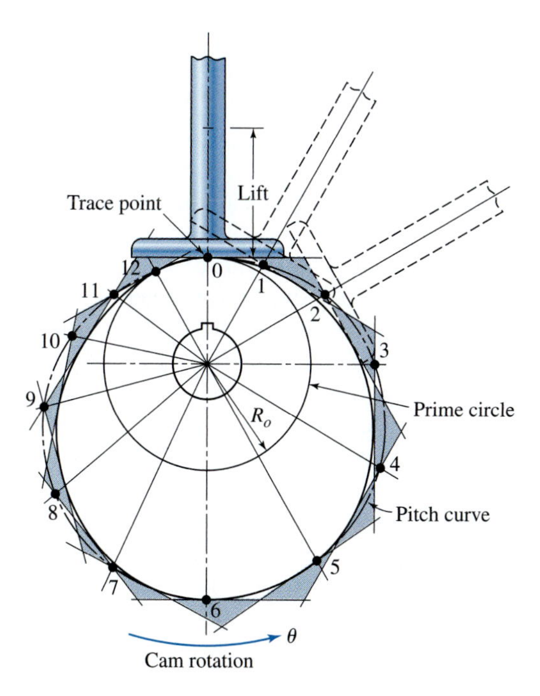

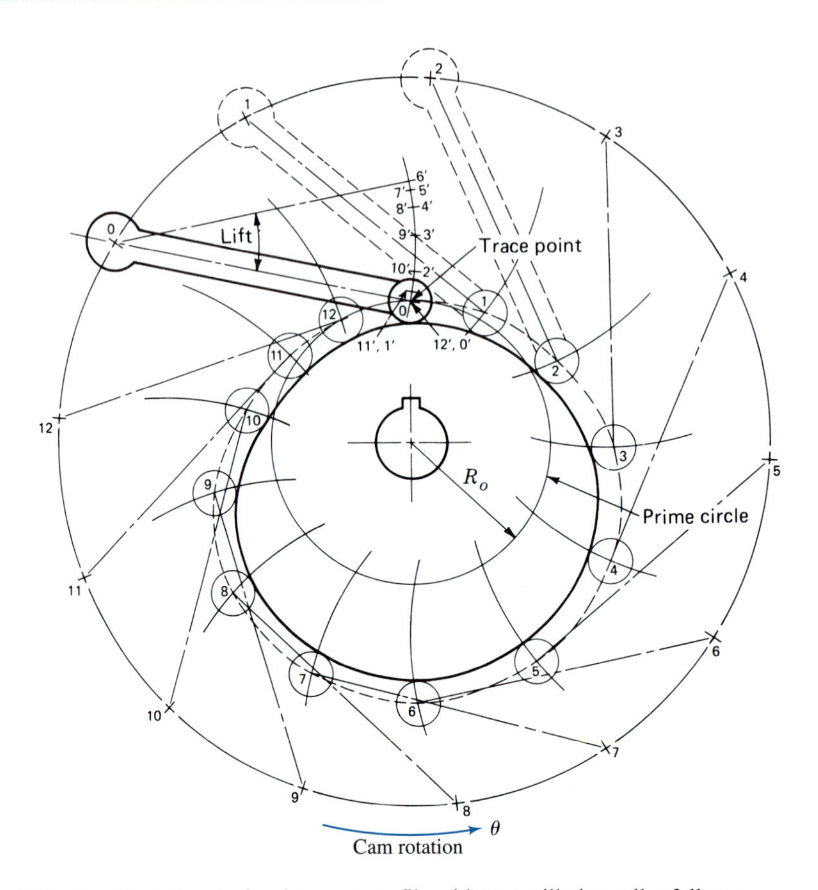

Figure 6.11 Graphical layout of a plate cam profile with an oscillating roller follower.

it may be easier to draw the cam profile inside all the shaded triangles and tangent to the inner sides of the triangles.

Figure 6.11 illustrates the layout of the profile of a plate cam with an oscillating roller follower. In this case we must rotate the fixed pivot center of the follower opposite to the direction of cam rotation to develop the cam profile. To perform this inversion, first a circle is drawn about the camshaft center through the fixed pivot of the follower. This circle is then divided and given station numbers to correspond to the displacement diagram. Next, arcs are drawn about each of these centers, all with equal radii corresponding to the length of the follower.

In the case of an oscillating follower, the ordinate values of the displacement diagram represent angular movements of the follower. If the ordinate scale of the displacement diagram is properly chosen initially, however, and if the total lift of the follower is a reasonably small angle, then ordinate distances from the displacement diagram at each station can be transferred directly to the corresponding arc traveled by the roller using dividers and measuring positive outward along the arc from the prime circle to locate the center of the roller for that station. Finally, circles representing the roller locations are drawn at each station, and the cam profile is constructed as a smooth curve tangent to each of these roller locations.

From the examples presented in this section, it should be clear that each different type of cam-and-follower system requires its own method of construction to determine the cam profile graphically from the displacement diagram. The examples presented here are not intended to be exhaustive of those possible, but they illustrate the general approach. They should also serve to illustrate and reinforce the discussion of the previous section; it should now be clear that much of the detailed shape of the cam itself results directly from the shape of the displacement diagram. Although different types of cams and followers have different shapes for the same displacement diagram, once a few parameters (such as the prime-circle radius) are chosen to determine the size of a cam, the remainder of its shape results directly from the motion requirements specified in the displacement diagram.

6.5 KINEMATIC COEFFICIENTS OF THE FOLLOWER MOTION

We have seen that the displacement diagram is plotted with the cam input motion θ as the abscissa and the follower output motion y as the ordinate regardless of the type of cam or the type of follower. The displacement diagram is, therefore, a graph representing some mathematical function relating the input and output motions of the cam system. In general terms, this relationship is

$$y = y(\theta). \tag{6.1}$$

Additional graphs can be plotted representing the derivatives of y with respect to the input position θ, that is, the kinematic coefficients of the follower motion. The first-order kinematic coefficient is denoted

$$y'(\theta) = \frac{dy}{d\theta} \tag{6.2}$$

and represents the slope of the displacement diagram at each input position θ. The first-order kinematic coefficient, although it may seem to be of little practical value, is a measure of the "steepness" of the displacement diagram. We will find later that it is closely related to the mechanical advantage of the cam system and manifests itself in such things as the pressure angle (see Section 6.10). If we consider a wedge cam (Fig. 6.1*b*) with a knife-edge follower (Fig. 6.2*a*), the displacement diagram itself is of the same shape as the corresponding cam. In such a case we can visualize that difficulties will occur if the cam is too steep, that is, if the first-order kinematic coefficient y' has too high a value.

The second-order kinematic coefficient (that is, the second derivative of y with respect to the input position θ) is also significant. The second-order kinematic coefficient is denoted

$$y''(\theta) = \frac{d^2y}{d\theta^2}. \tag{6.3}$$

Although it is not as easy to visualize the reason, the second-order kinematic coefficient is very closely related to the curvature of the cam at locations along its profile. Recall that curvature is the reciprocal of the radius of curvature, see Eq. (e) in Section 4.1. Therefore, as y'' becomes large, the radius of curvature becomes small. For example, if the second-order kinematic coefficient becomes infinite, then the radius of curvature becomes zero, that is, the cam profile at such a position becomes pointed. This would be a highly unsatisfactory condition from the point of view of contact stresses between the cam and the follower and would very quickly cause surface damage.

The third-order kinematic coefficient denoted

$$y'''(\theta) = \frac{d^3y}{d\theta^3} \tag{6.4}$$

can also be plotted if desired. Although it is not easy to describe geometrically, this demonstrates the rate of change of y'' with respect to the input position θ. We will see below that the third-order kinematic coefficient can also be controlled when choosing the detailed shape of the displacement diagram.

EXAMPLE 6.1

Derive equations to describe the displacement diagram of a plate cam that rises with parabolic motion from a dwell to another dwell such that the total lift is L and the total cam rotation angle is β. Plot the displacement diagram and the first-, second-, and third-order kinematic coefficients with respect to cam rotation.

SOLUTION

As illustrated in Fig. 6.5*a*, two parabolas are required, meeting at a common tangent taken here at midrange. For the first half of the motion we choose the general equation of a parabola; that is,

$$y = A\theta^2 + B\theta + C. \tag{a}$$

The first three derivatives of Eq. (*a*), with respect to the input position θ, are

$$y' = 2A\theta + B \tag{b}$$

$$y'' = 2A \tag{c}$$

$$y''' = 0. \tag{d}$$

To properly match the position and the slope with those of the preceding dwell, at $\theta = 0$ we must have $y(0) = y'(0) = 0$. Thus, Eqs. (*a*) and (*b*) demonstrate that $B = C = 0$. Looking next at the inflection point, at $\theta = \beta/2$, we want $y = L/2$. Substituting these conditions into Eq. (*a*) and rearranging gives

$$A = \frac{2L}{\beta^2}.$$

Therefore, the displacement equation for the first half of the parabolic motion becomes

$$y = 2L\left(\frac{\theta}{\beta}\right)^2. \tag{6.5a}$$

Differentiating this equation with respect to the input position θ, the first-, second-, and third-order kinematic coefficients, respectively, are

$$y' = \frac{4L}{\beta}\left(\frac{\theta}{\beta}\right) \tag{6.5b}$$

$$y'' = \frac{4L}{\beta^2} \tag{6.5c}$$

$$y''' = 0. \tag{6.5d}$$

The maximum value for the first-order kinematic coefficient (that is, the maximum slope) occurs at the midpoint, where $\theta = \beta/2$. Substituting this value into Eq. (6.5*b*), the maximum value for the first-order kinematic coefficient is

$$y'_{\max} = \frac{2L}{\beta}. \tag{e}$$

For the second half of the parabolic motion we return to the general equations (*a*) through (*d*) for a parabola. Substituting the conditions that $y = L$, and $y' = 0$ at $\theta = \beta$ into Eqs. (*a*) and (*b*) gives

$$L = A\beta^2 + B\beta + C \tag{f}$$

$$0 = 2A\beta + B. \tag{g}$$

Because the slope must match that of the first parabola at $\theta = \beta/2$, then from Eqs. (*b*) and (*e*) we have

$$\frac{2L}{\beta} = 2A\frac{\beta}{2} + B. \tag{h}$$

Solving Eqs. (*f*), (*g*), and (*h*) simultaneously gives

$$A = -\frac{2L}{\beta^2}, \quad B = \frac{4L}{\beta}, \quad C = -L. \tag{i}$$

Substituting these constraints into Eq. (*a*), the displacement equation for the second half of the parabolic motion can be written as

$$y = L\left[1 - 2\left(1 - \frac{\theta}{\beta}\right)^2\right]. \tag{6.6a}$$

Also, substituting Eqs. (*i*) into Eqs. (*b*), (*c*), and (*d*), the first-, second-, and third-order kinematic coefficients for the second half of the parabolic motion, respectively, are

$$y' = \frac{4L}{\beta}\left(1 - \frac{\theta}{\beta}\right) \tag{6.6b}$$

$$y'' = -\frac{4L}{\beta^2} \tag{6.6c}$$

$$y''' = 0. \tag{6.6d}$$

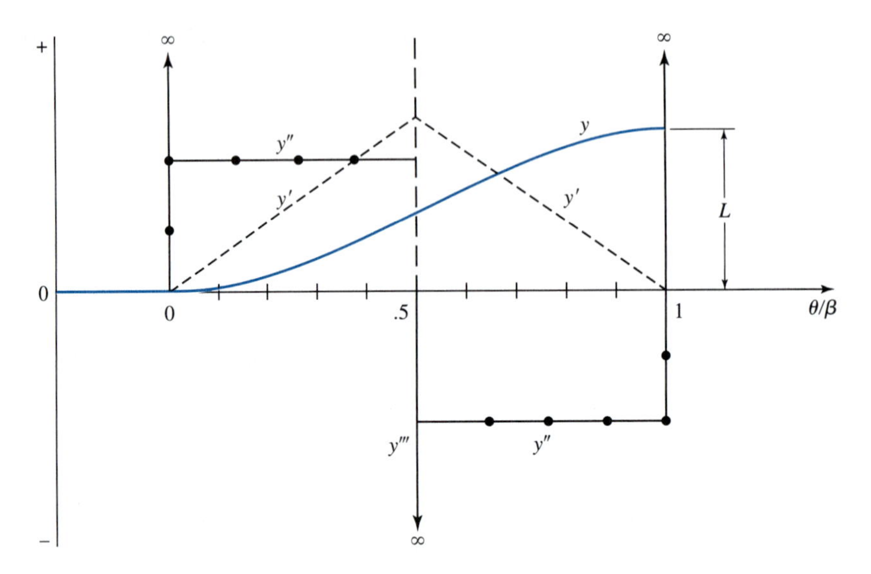

Figure 6.12 Displacement diagram and derivatives for full-rise parabolic motion.

The displacement diagram and the first-, and second-order kinematic coefficients for this example of parabolic motion are illustrated in Fig. 6.12.

The above discussion relates to the kinematic coefficients of the follower motion. These coefficients are derivatives with respect to the input position θ and relate to the geometry of the cam system. Let us now consider the derivatives of the follower motions with respect to time. First, we assume that the time variation of the input motion $\theta(t)$ is known. The angular velocity $\omega = d\theta/dt$, the angular acceleration $\alpha = d^2\theta/dt^2$, and the next derivative (often called angular jerk or second angular acceleration) $\dot{\alpha} = d^3\theta/dt^3$ are all assumed to be known. Usually, a plate cam is driven by a constant-speed input shaft. In this case, ω is a known constant, $\theta = \omega t$, and $\alpha = \dot{\alpha} = 0$. During start-up of the cam system, however, this is not the case, and we will consider the more general situation first.

From the general equation of the displacement diagram we have

$$y = y(\theta) \quad \text{and} \quad \theta = \theta(t). \tag{6.7}$$

Therefore, we can differentiate to find the time derivatives of the follower motion. The velocity of the follower, for example, is given by

$$\dot{y} = \frac{dy}{dt} = \left(\frac{dy}{d\theta}\right)\left(\frac{d\theta}{dt}\right),$$

which, using the first-order kinematic coefficient, can be written as

$$\dot{y} = y'\omega. \tag{6.8}$$

Similarly, the acceleration and the jerk of the follower can be written, respectively, as

$$\ddot{y} = \frac{d^2y}{dt^2} = y''\omega^2 + y'\alpha \tag{6.9}$$

and

$$\dddot{y} = \frac{d^3y}{dt^3} = y'''\omega^3 + 3y''\omega\alpha + y'\dot{\alpha}. \tag{6.10}$$

When the camshaft speed is constant, then Eqs. (6.8) through (6.10) reduce to

$$\dot{y} = y'\omega, \quad \ddot{y} = y''\omega^2, \quad \dddot{y} = y'''\omega^3. \tag{6.11}$$

For this reason, it has become somewhat common to refer to the graphs of the kinematic coefficients y', y'', y''', as the "velocity," "acceleration," and "jerk" curves for a given motion. These are only appropriate names for a constant-speed cam and then only when scaled by ω, ω^2, and ω^3, respectively.* However, it is helpful to use these names for the kinematic coefficients when considering the physical implications of a certain choice of displacement diagram. For example, consider the parabolic motion of Fig. 6.12, it is intuitively meaningful to say that the "velocity" of the follower rises linearly to a maximum at the midpoint $\theta = \beta/2$ and then decreases linearly to zero. The "acceleration" of the follower is zero during the initial dwell and changes abruptly (that is, a step change) to a constant positive value upon beginning the rise. There are two more step changes in the "acceleration" of the follower, namely, at the midpoint and at the end of the rise. At each of these three step changes in the "acceleration" of the follower, the "jerk" of the follower becomes infinite.

6.6 HIGH-SPEED CAMS

Continuing with our discussion of parabolic motion, let us consider briefly the implications of the "acceleration" curve of Fig. 6.12 on the dynamic performance of the cam system. Any real follower must, of course, have at least some mass and, when multiplied by acceleration, will exert an inertia force (see Chapter 14). Therefore, the "acceleration" curve of Fig. 6.12 can also be thought of as indicating the inertia force of the follower, which, in turn, will be felt at the follower bearings and at the contact point with the cam surface. An "acceleration" curve with abrupt changes (that is, the "jerk" becomes infinite), such as those demonstrated for parabolic motion, will exert abruptly changing contact stresses at the bearings and on the cam surface and will lead to noise, surface wear, and early failure. Thus, it is very important in choosing a displacement diagram to ensure that the first- and second-order kinematic coefficients (that is, the "velocity" and "acceleration" curves) are smooth and continuous, that is, they contain no step changes.

---

* Accepting the word "velocity" literally, for example, leads to consternation when it is discovered that, for a plate cam with a reciprocating follower, the units of "velocity" y' are length per radian. Multiplying these units by radians per second, the units of ω, give units of length per second for $\dot{y}$, however.

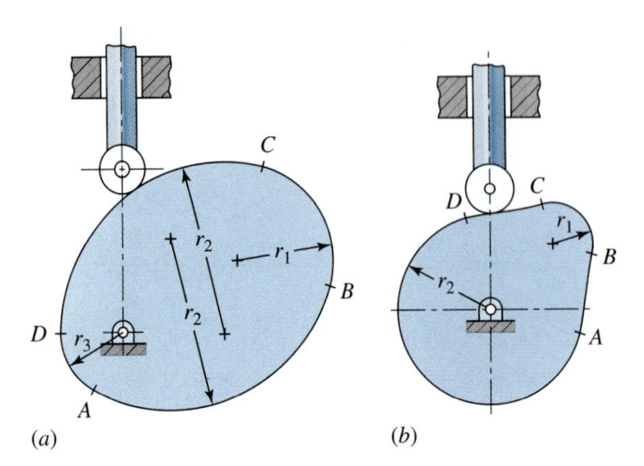

Figure 6.13 (*a*) Circle-arc cam and (*b*) tangent cam.

Sometimes in low-speed cam applications compromises are made with the "velocity" and "acceleration" relationships. It is sometimes simpler to employ a reverse procedure and design the cam shape first, obtaining the displacement diagram as a second step. Such cams are sometimes composed of a combination of curves, such as straight lines and circular arcs, which are readily produced by machine tools. Two examples are the *circle-arc cam* and the *tangent cam* illustrated in Fig. 6.13. The design approach is by iteration. A trial cam is designed and its kinematic characteristics are computed. The process is then repeated until a cam with acceptable characteristics is obtained. Points A, B, C, and D of the circle-arc cam and the tangent cam are points of tangency or blending points. It is worth noting, as with the parabolic-motion example above, that the acceleration changes abruptly at each of the tangency points because of the instantaneous change in the radius of curvature of the cam profile.

Although cams with discontinuous acceleration characteristics have sometimes been accepted to save cost in low-speed applications, such cams have invariably exhibited major problems at some later time when the speed was raised to increase the productivity of the application. For any high-speed cam application, it is extremely important that not only the displacement and "velocity" curves but also the "acceleration" curve be made continuous for the entire motion cycle. No discontinuities should be allowed at the boundaries of different portions of the cam.

As demonstrated by Eq. (6.11), the importance of continuous derivatives becomes more serious as the camshaft speed is increased. The higher the speed, the greater the need for smooth curves. At very high speeds it might also be desirable to require that jerk, which is related to rate of change of force, and perhaps even higher derivatives, be made continuous as well. In many applications, however, this is not necessary.

There is no simple answer as to how high a speed one must have before considering the application to require high-speed design techniques. The answer depends not only on the mass of the follower but also on the stiffness of the return spring, the materials used, the flexibility of the follower, and many other factors.[1] Further analysis techniques on cam dynamics are presented in Chapter 18. Still, with the methods presented below, it is not difficult to achieve continuous derivative displacement diagrams. Therefore, it is recommended that this be undertaken as standard practice. Cycloidal-motion cams, for example, are no more difficult to manufacture than parabolic-motion cams and there is no good reason for use of the latter. The circle-arc cam and the tangent cam may be easier to produce, but with modern machining methods cutting more complex cam shapes is not expensive and is recommended.

6.7 STANDARD CAM MOTIONS

Example 6.1 in Section 6.5 gave a detailed derivation of the equations for parabolic motion and the first three derivatives. Then, in Section 6.6, reasons were provided for avoiding the use of parabolic motion in high-speed cam systems. The purpose of this section is to present equations for a number of standard types of displacement curves that can be used to address most high-speed cam-motion requirements. The derivations parallel those of Example 6.1 and are not presented in this text.

The displacement equation and the first-, second-, and third-order kinematic coefficients for *full-rise simple harmonic motion* are

$$y = \frac{L}{2}\left(1 - \cos\frac{\pi\theta}{\beta}\right) \tag{6.12a}$$

$$y' = \frac{\pi L}{2\beta}\sin\frac{\pi\theta}{\beta} \tag{6.12b}$$

$$y'' = \frac{\pi^2 L}{2\beta^2}\cos\frac{\pi\theta}{\beta} \tag{6.12c}$$

$$y''' = -\frac{\pi^3 L}{2\beta^3}\sin\frac{\pi\theta}{\beta}. \tag{6.12d}$$

The displacement diagram and the first-, second-, and third-order kinematic coefficients for full-rise simple harmonic motion are illustrated in Fig. 6.14. Unlike parabolic motion, simple harmonic motion shows no discontinuity at the inflection point.

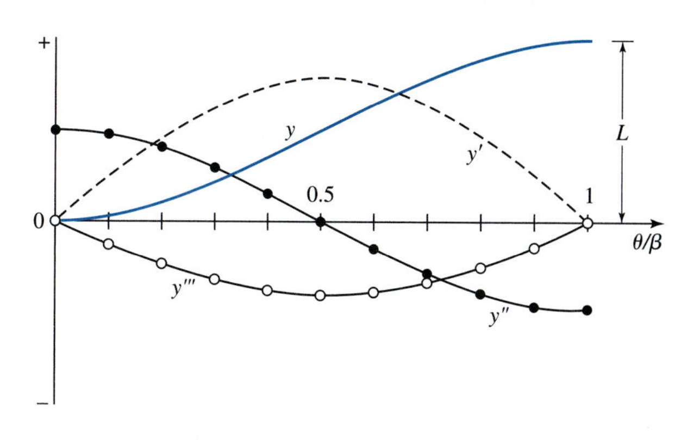

Figure 6.14 Displacement diagram and derivatives for full-rise simple harmonic motion, Eqs. (6.12).

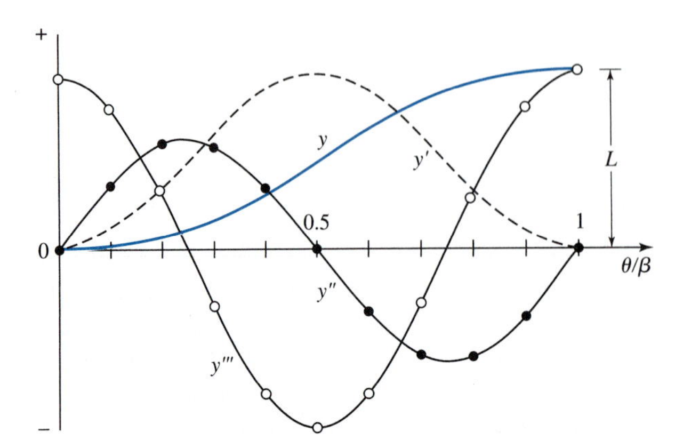

Figure 6.15 Displacement diagram and derivatives for full-rise cycloidal motion, Eqs. (6.13).

The displacement equation and the first-, second-, and third-order kinematic coefficients for *full-rise cycloidal motion* are

$$y = L\left(\frac{\theta}{\beta} - \frac{1}{2\pi}\sin\frac{2\pi\theta}{\beta}\right) \tag{6.13a}$$

$$y' = \frac{L}{\beta}\left(1 - \cos\frac{2\pi\theta}{\beta}\right) \tag{6.13b}$$

$$y'' = \frac{2\pi L}{\beta^2}\sin\frac{2\pi\theta}{\beta} \tag{6.13c}$$

$$y''' = \frac{4\pi^2 L}{\beta^3}\cos\frac{2\pi\theta}{\beta}. \tag{6.13d}$$

The displacement diagram and the first-, second-, and third-order kinematic coefficients for full-rise cycloidal motion are illustrated in Fig. 6.15.

Unlike the simple harmonic motion curves, the full-rise cycloidal motion has zero values for derivatives at the beginning and end of its span, including the second-order kinematic coefficient ("acceleration" curve). The peak "velocity," "acceleration," and "jerk" values, however, are higher than those for simple harmonic motion.

The displacement equation and the first-, second-, and third-order kinematic coefficients for a *full-rise eighth-order polynomial motion* are

$$y = L\left[6.097\,55\left(\frac{\theta}{\beta}\right)^3 - 20.780\,40\left(\frac{\theta}{\beta}\right)^5 + 26.731\,55\left(\frac{\theta}{\beta}\right)^6 - 13.609\,65\left(\frac{\theta}{\beta}\right)^7 + 2.560\,95\left(\frac{\theta}{\beta}\right)^8\right] \tag{6.14a}$$

$$y' = \frac{L}{\beta}\left[18.292\,65\left(\frac{\theta}{\beta}\right)^2 - 103.902\,00\left(\frac{\theta}{\beta}\right)^4 + 160.389\,30\left(\frac{\theta}{\beta}\right)^5 - 95.267\,55\left(\frac{\theta}{\beta}\right)^6 + 20.487\,60\left(\frac{\theta}{\beta}\right)^7\right] \tag{6.14b}$$

$$y'' = \frac{L}{\beta^2}\left[36.585\,30\left(\frac{\theta}{\beta}\right) - 415.608\,00\left(\frac{\theta}{\beta}\right)^3 + 801.946\,50\left(\frac{\theta}{\beta}\right)^4 - 571.605\,30\left(\frac{\theta}{\beta}\right)^5 + 143.413\,20\left(\frac{\theta}{\beta}\right)^6\right] \tag{6.14c}$$

$$y''' = \frac{L}{\beta^3}\left[36.585\,30 - 1246.824\,00\left(\frac{\theta}{\beta}\right)^2 + 3207.786\,00\left(\frac{\theta}{\beta}\right)^3 - 2858.026\,50\left(\frac{\theta}{\beta}\right)^4 + 860.479\,20\left(\frac{\theta}{\beta}\right)^5\right]. \tag{6.14d}$$

The displacement diagram and the first-, second-, and third-order kinematic coefficients for the full-rise motion formed from this eighth-order polynomial are illustrated in Fig. 6.16. Equations have seemingly strange coefficients because they have been specially derived to have many "nice" properties.[2] Among these, Fig. 6.16 illustrates that several of the kinematic coefficients are zero at both ends of the range but that the "acceleration" characteristics are nonsymmetric, whereas the peak values of "acceleration" are kept as small as possible (that is, the magnitudes of the positive and negative peak "accelerations" are equal).

The displacement diagrams of simple harmonic, cycloidal, and eighth-order polynomial motions look quite similar at first glance. Each rises through a lift of L in a total cam rotation angle of β, and each begins and ends with a horizontal slope. For these reasons, they are all referred to as *full-rise* motions. However, their "acceleration" curves are quite different.

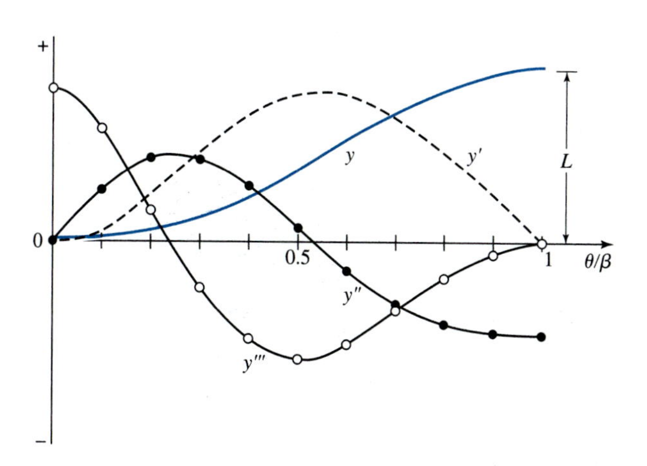

Figure 6.16 Displacement diagram and derivatives for full-rise eighth-order polynomial motion, Eqs. (6.14).

LIVERPOOL JOHN MOORES UNIVERSITY
LEARNING SERVICES

Simple harmonic motion has nonzero "acceleration" at the two ends of the range; cycloidal motion has zero "acceleration" at both boundaries; and eighth-order polynomial motion has one zero and one nonzero "acceleration" at its two ends. This variety provides the selections necessary when matching these curves with neighboring curves of different types.

Full-return motions of the same three types are illustrated in Figs. 6.17 through 6.19.

The displacement equation and the first-, second-, and third-order kinematic coefficients for *full-return simple harmonic motion* are

$$y = \frac{L}{2}\left(1 + \cos\frac{\pi\theta}{\beta}\right) \tag{6.15a}$$

$$y' = -\frac{\pi L}{2\beta}\sin\frac{\pi\theta}{\beta} \tag{6.15b}$$

$$y'' = -\frac{\pi^2 L}{2\beta^2}\cos\frac{\pi\theta}{\beta} \tag{6.15c}$$

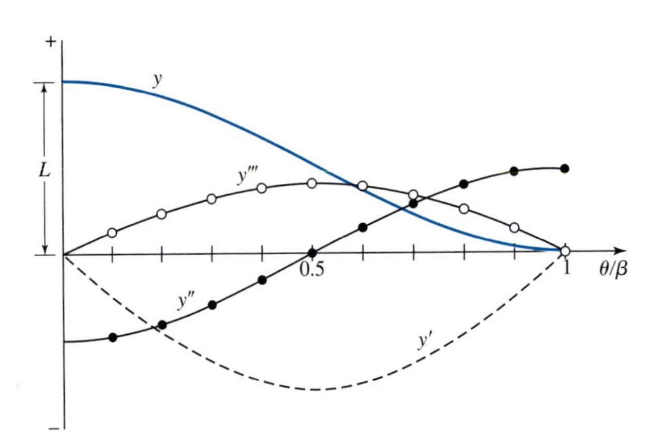

Figure 6.17 Displacement diagram and derivatives for full-return simple harmonic motion, Eqs. (6.15).

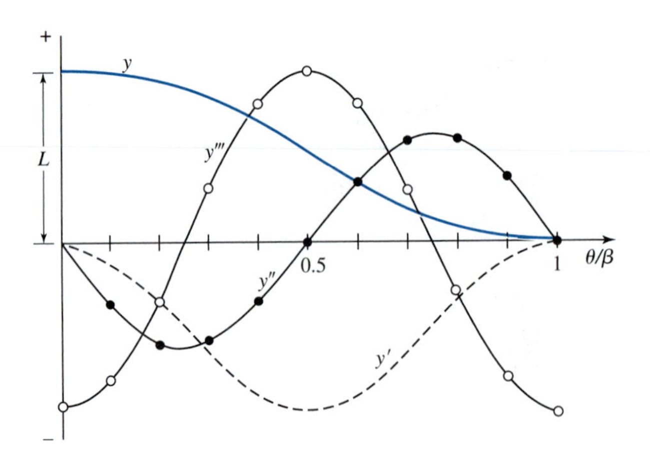

Figure 6.18 Displacement diagram and derivatives for full-return cycloidal motion, Eqs. (6.16).

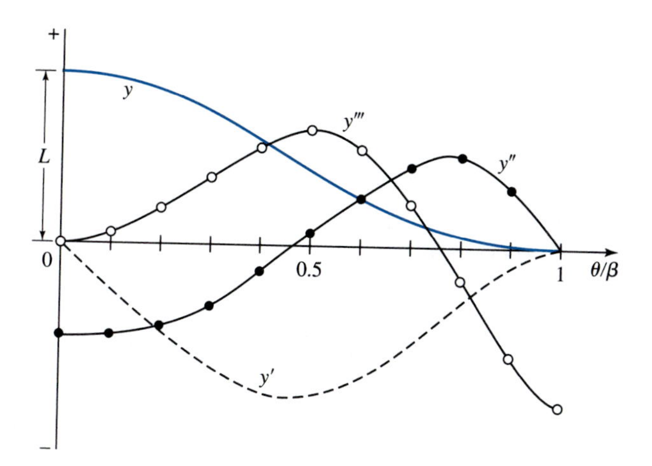

Figure 6.19 Displacement diagram and derivatives for full-return eighth-order polynomial motion, Eqs. (6.17).

$$y''' = \frac{\pi^3 L}{2\beta^3} \sin \frac{\pi\theta}{\beta}. \tag{6.15d}$$

For *full-return cycloidal motion*, the displacement equation and the first-, second-, and third-order kinematic coefficients are

$$y = L\left(1 - \frac{\theta}{\beta} + \frac{1}{2\pi} \sin \frac{2\pi\theta}{\beta}\right) \tag{6.16a}$$

$$y' = -\frac{L}{\beta}\left(1 - \cos \frac{2\pi\theta}{\beta}\right) \tag{6.16b}$$

$$y'' = -\frac{2\pi L}{\beta^2} \sin \frac{2\pi\theta}{\beta} \tag{6.16c}$$

$$y''' = -\frac{4\pi^2 L}{\beta^3} \cos \frac{2\pi\theta}{\beta}. \tag{6.16d}$$

For *full-return eighth-order-polynomial motion*, the displacement equation and the first-, second-, and third-order kinematic coefficients are

$$y = L\left[1.000\,00 - 2.634\,15\left(\frac{\theta}{\beta}\right)^2 + 2.780\,55\left(\frac{\theta}{\beta}\right)^5 + 3.170\,60\left(\frac{\theta}{\beta}\right)^6 - 6.877\,95\left(\frac{\theta}{\beta}\right)^7 + 2.560\,95\left(\frac{\theta}{\beta}\right)^8\right] \tag{6.17a}$$

$$y' = -\frac{L}{\beta}\left[5.268\,30\frac{\theta}{\beta} - 13.902\,75\left(\frac{\theta}{\beta}\right)^4 - 19.023\,60\left(\frac{\theta}{\beta}\right)^5 + 48.145\,65\left(\frac{\theta}{\beta}\right)^6 - 20.487\,60\left(\frac{\theta}{\beta}\right)^7\right] \tag{6.17b}$$

$$y'' = -\frac{L}{\beta^2}\left[5.268\,30 - 55.611\,00\left(\frac{\theta}{\beta}\right)^3 - 95.118\,00\left(\frac{\theta}{\beta}\right)^4 + 288.873\,90\left(\frac{\theta}{\beta}\right)^5 - 143.413\,20\left(\frac{\theta}{\beta}\right)^6\right] \tag{6.17c}$$

$$y''' = \frac{L}{\beta^3}\left[166.833\,00\left(\frac{\theta}{\beta}\right)^2 + 380.472\,00\left(\frac{\theta}{\beta}\right)^3 - 1444.369\,50\left(\frac{\theta}{\beta}\right)^4 + 860.479\,20\left(\frac{\theta}{\beta}\right)^5\right]. \tag{6.17d}$$

Polynomial displacement equations of much higher order and meeting many more conditions than those presented here are also in common use. Automated procedures for determining the coefficients have been developed by Stoddart,[3] who also indicates how the choice of coefficients can be made to compensate for elastic deformation of the follower system under dynamic conditions. Such cams are referred to as *polydyne cams*.

In addition to the full-rise and full-return motions presented above, it is often useful to have a selection of standard *half-rise* or *half-return* motions available. These are curves for which one boundary has a nonzero slope and can be used to blend with uniform motion. The displacement diagrams and the first-, second-, and third-order kinematic coefficients for *half-rise simple harmonic motions*, sometimes called half-harmonic rise motions, are illustrated in Fig. 6.20. The equations corresponding to Fig. 6.20*a* are

$$y = L\left(1 - \cos\frac{\pi\theta}{2\beta}\right) \tag{6.18a}$$

$$y' = \frac{\pi L}{2\beta}\sin\frac{\pi\theta}{2\beta} \tag{6.18b}$$

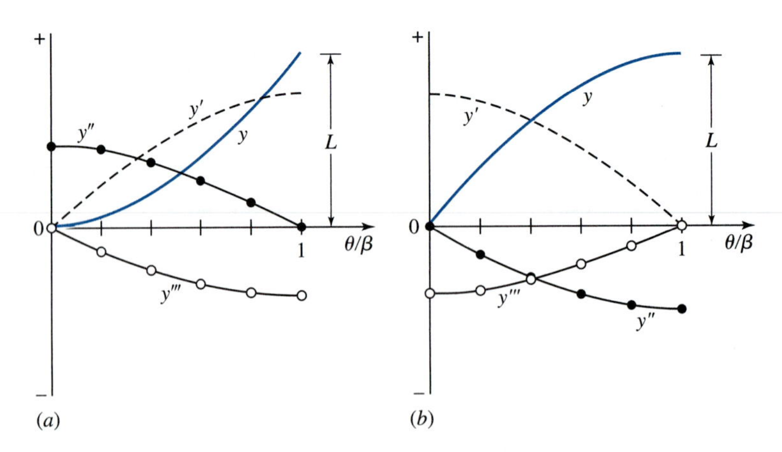

Figure 6.20 Displacement diagram and derivatives for half-rise simple harmonic motions: (*a*) Eqs. (6.18) and (*b*) Eqs. (6.19).

$$y'' = \frac{\pi^2 L}{4\beta^2} \cos \frac{\pi\theta}{2\beta} \tag{6.18c}$$

$$y''' = -\frac{\pi^3 L}{8\beta^3} \sin \frac{\pi\theta}{2\beta}. \tag{6.18d}$$

The displacement equation and the first-, second-, and third-order kinematic coefficients corresponding to the half-rise simple harmonic motions of Fig. 6.20*b* are

$$y = L \sin \frac{\pi\theta}{2\beta} \tag{6.19a}$$

$$y' = \frac{\pi L}{2\beta} \cos \frac{\pi\theta}{2\beta} \tag{6.19b}$$

$$y'' = -\frac{\pi^2 L}{4\beta^2} \sin \frac{\pi\theta}{2\beta} \tag{6.19c}$$

$$y''' = -\frac{\pi^3 L}{8\beta^3} \cos \frac{\pi\theta}{2\beta}. \tag{6.19d}$$

The curves for *half-return simple-harmonic motions* are illustrated in Fig. 6.21. The equations corresponding to Fig. 6.21*a* are

$$y = L \cos \frac{\pi\theta}{2\beta} \tag{6.20a}$$

$$y' = -\frac{\pi L}{2\beta} \sin \frac{\pi\theta}{2\beta} \tag{6.20b}$$

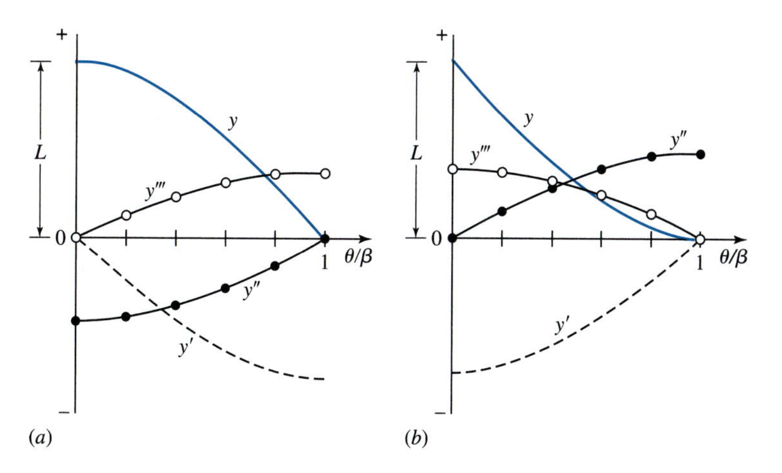

Figure 6.21 Displacement diagram and derivatives for half-return simple harmonic motions: (*a*) Eqs. (6.20) and (*b*) Eqs. (6.21).

$$y'' = -\frac{\pi^2 L}{4\beta^2} \cos \frac{\pi\theta}{2\beta} \tag{6.20c}$$

$$y''' = \frac{\pi^3 L}{8\beta^3} \sin \frac{\pi\theta}{2\beta}. \tag{6.20d}$$

The equations corresponding to Fig. 6.21*b* are

$$y = L\left(1 - \sin \frac{\pi\theta}{2\beta}\right) \tag{6.21a}$$

$$y' = -\frac{\pi L}{2\beta} \cos \frac{\pi\theta}{2\beta} \tag{6.21b}$$

$$y'' = \frac{\pi^2 L}{4\beta^2} \sin \frac{\pi\theta}{2\beta} \tag{6.21c}$$

$$y''' = \frac{\pi^3 L}{8\beta^3} \cos \frac{\pi\theta}{2\beta}. \tag{6.21d}$$

In addition to the half-harmonics, half-cycloidal motions are also useful, because their "accelerations" are zero at both boundaries. The displacement diagrams and first-, second-, and third-order kinematic coefficients for *half-rise cycloidal motions* are illustrated in Fig. 6.22. The equations corresponding to Fig. 6.22*a* are

$$y = L\left(\frac{\theta}{\beta} - \frac{1}{\pi} \sin \frac{\pi\theta}{\beta}\right) \tag{6.22a}$$

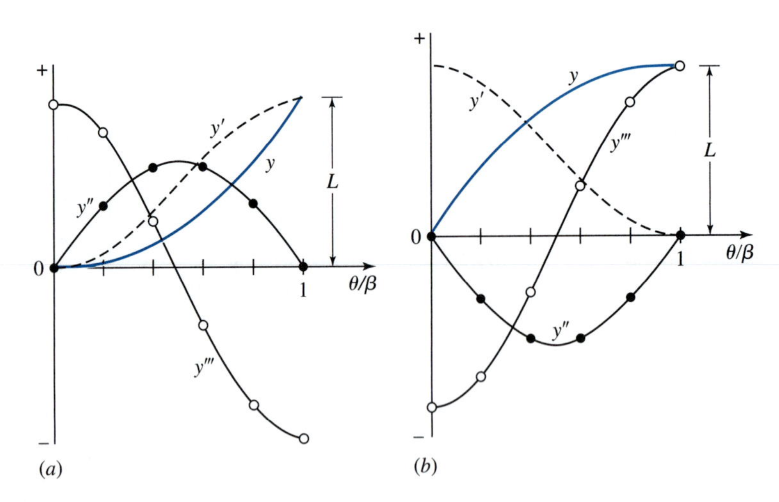

Figure 6.22 Displacement diagram and derivatives for half-rise cycloidal motions: (*a*) Eqs. (6.22) and (*b*) Eqs. (6.23).

$$y' = \frac{L}{\beta}\left(1 - \cos\frac{\pi\theta}{\beta}\right) \tag{6.22b}$$

$$y'' = \frac{\pi L}{\beta^2}\sin\frac{\pi\theta}{\beta} \tag{6.22c}$$

$$y''' = \frac{\pi^2 L}{\beta^3}\cos\frac{\pi\theta}{\beta}. \tag{6.22d}$$

The equations corresponding to Fig. 6.22*b* are

$$y = L\left(\frac{\theta}{\beta} + \frac{1}{\pi}\sin\frac{\pi\theta}{\beta}\right) \tag{6.23a}$$

$$y' = \frac{L}{\beta}\left(1 + \cos\frac{\pi\theta}{\beta}\right) \tag{6.23b}$$

$$y'' = -\frac{\pi L}{\beta^2}\sin\frac{\pi\theta}{\beta} \tag{6.23c}$$

$$y''' = -\frac{\pi^2 L}{\beta^3}\cos\frac{\pi\theta}{\beta}. \tag{6.23d}$$

The curves for *half-return cycloidal motions* are illustrated in Fig. 6.23. The equations corresponding to Fig. 6.23*a* are

$$y = L\left(1 - \frac{\theta}{\beta} + \frac{1}{\pi}\sin\frac{\pi\theta}{\beta}\right) \tag{6.24a}$$

$$y' = -\frac{L}{\beta}\left(1 - \cos\frac{\pi\theta}{\beta}\right) \tag{6.24b}$$

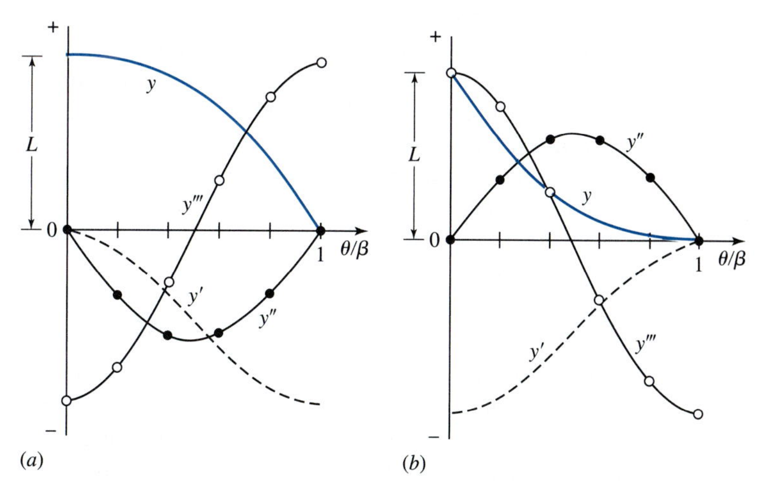

Figure 6.23 Displacement diagram and derivatives for half-return cycloidal motions: (*a*) Eqs. (6.24) and (*b*) Eqs. (6.25).

$$y'' = -\frac{\pi L}{\beta^2} \sin \frac{\pi \theta}{\beta} \tag{6.24c}$$

$$y''' = -\frac{\pi^2 L}{\beta^3} \cos \frac{\pi \theta}{\beta}. \tag{6.24d}$$

The equations corresponding to Fig. 6.23*b* are

$$y = L\left(1 - \frac{\theta}{\beta} - \frac{1}{\pi} \sin \frac{\pi \theta}{\beta}\right) \tag{6.25a}$$

$$y' = -\frac{L}{\beta}\left(1 + \cos \frac{\pi \theta}{\beta}\right) \tag{6.25b}$$

$$y'' = \frac{\pi L}{\beta^2} \sin \frac{\pi \theta}{\beta} \tag{6.25c}$$

$$y''' = \frac{\pi^2 L}{\beta^3} \cos \frac{\pi \theta}{\beta}. \tag{6.25d}$$

We shall soon see how the "standard" graphs and equations presented in this section can greatly reduce the analytical effort involved in designing the full displacement diagram for a high-speed cam. First, however, we should note several important features of the graphs of Fig. 6.14 through Fig. 6.23.

Each graph illustrates only one segment of a full displacement diagram; the total lift for that segment is labeled L for each, and the total cam travel is labeled β. The abscissa of each graph is normalized so that the ratio θ/β ranges from $\theta/\beta = 0$ at the left end to $\theta/\beta = 1$ at the right end.

The scales used in plotting the graphs are not depicted but are consistent for all full-rise and full-return curves and for all half-rise and half-return curves. Thus, in judging the suitability of one curve compared with another, the magnitudes of the "accelerations," for example, can be compared. For this reason, when other factors are equivalent, simple harmonic motion should be used where possible and, in order to keep "accelerations" small, cycloidal motion should be avoided except where necessary.

Finally, it should be noted that the standard cam motions presented in this section do not form an exhaustive set. The set presented here is sufficient for most practical applications. However, cams with good dynamic characteristics can also be formed from a wide variety of other possible motion curves. A much more extensive set can be found, for example, in the text by F. Y. Chen.[4]

6.8 MATCHING DERIVATIVES OF DISPLACEMENT DIAGRAMS

In the previous section, a great many equations were presented that might be used to represent different segments of the displacement diagram of a cam. In this section we will study how they can be joined together to form the motion specification for a complete cam. The procedure is one of solving for proper values of L and β for each segment so that:

1. The motion requirements of the particular application are met.

2. The displacement diagram, as well as the diagrams of the first- and second-order kinematic coefficients, is continuous across the boundaries of the merged segments. The diagram of the third-order kinematic coefficient may be allowed discontinuities if necessary, but must not become infinite; that is, the "acceleration" curve may contain corners but not discontinuities.
3. The maximum magnitudes of the "velocity" and "acceleration" peaks are kept as low as possible, consistent with the first two conditions.

The procedure may best be understood through an example.

EXAMPLE 6.2

A plate cam with a reciprocating follower is to be driven by a constant-speed motor at 150 rpm. The follower is to start from a dwell, accelerate to a uniform velocity of 25 in/s, maintain this velocity for 1.25 in of rise, decelerate to the top of the lift, return, and then dwell for 0.10 s. The total lift is to be 3.00 in. Determine the complete specifications of the displacement diagram.

SOLUTION

The speed of the input shaft is

$$\omega = 150 \text{ rev/min} = 15.707\,96 \text{ rad/s}. \tag{1}$$

Using Eq. (6.8), the first-order kinematic coefficient (that is, the slope of the uniform "velocity" segment) is

$$y' = \frac{\dot{y}}{\omega} = \frac{25 \text{ in/s}}{15.707\,96 \text{ rad/s}} = 1.591\,55 \text{ in/rad}. \tag{2}$$

Because this "velocity" is held constant for 1.25 in of rise, the cam rotation in this segment is*

$$\beta_2 = \frac{L_2}{y'} = \frac{1.25 \text{ in}}{1.591\,55 \text{ in/rad}} = 0.785\,40 \text{ rad} = 45.000°. \tag{3}$$

Similarly, from Eq. (1), the cam rotation during the final dwell is

$$\beta_5 = 0.10\text{s}\,(15.707\,96 \text{ rad/s}) = 1.570\,796 \text{ rad} = 90.000°. \tag{4}$$

From these results and the given information, we can sketch the beginnings of the displacement diagram, not necessarily working to scale, but to visualize the motion requirements. This gives the general shapes illustrated by the heavy curves of Fig. 6.24*a*. The lighter

* Note that several digits of accuracy higher than usual are used here and are recommended as standard practice when matching cam motion derivatives. Any inaccuracy in the L and β values results in discontinuities in the smoothness of derivatives at the junctures of the segments and discontinuities in force, as previously explained.

segments of the displacement curve are not yet accurately known, but can be sketched by lightly outlining a smooth curve for visualization. Working from this curve, we can also sketch the general nature of the derivative curves. From the changing slope of the displacement diagram we sketch the "velocity" curve (Fig. 6.24*b*), and from the changing slope of this curve we sketch the "acceleration" curve (Fig. 6.24*c*). At this stage, no attempt is made to produce accurate curves drawn to scale, but only to provide an idea of the desired curve shapes.

Next, using the sketches of Fig. 6.24, we compare these desired motion curves with the various standard curves of Fig. 6.14 through Fig. 6.23 to choose an appropriate set of equations for each segment of the cam. In segment *AB*, for example, we find that Fig. 6.22*a* is the only standard motion curve available with half-rise characteristics, an appropriate slope curve, and the necessary zero "acceleration" at both ends of the segment. Thus, we choose the cycloidal half-rise motion of Eq. (6.22) for that portion of the cam. There are two sets of choices possible for segments *CD* and *DE*. One set might be the choice of Fig. 6.22*b*

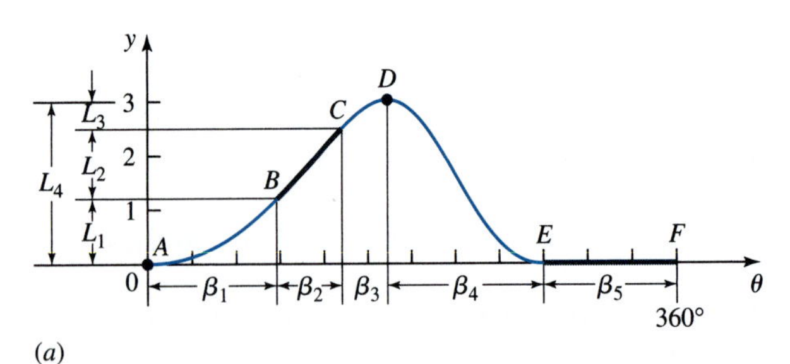

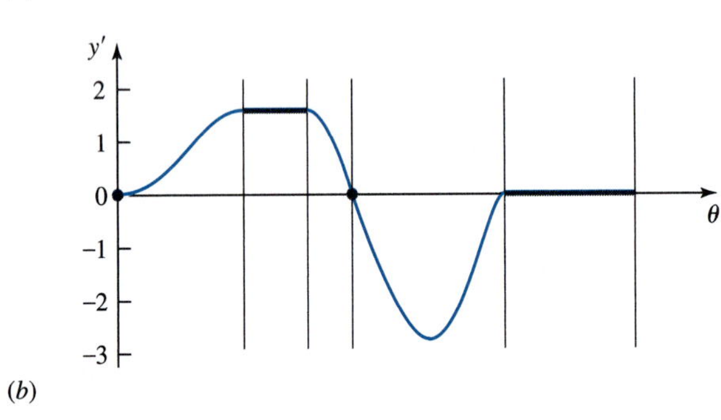

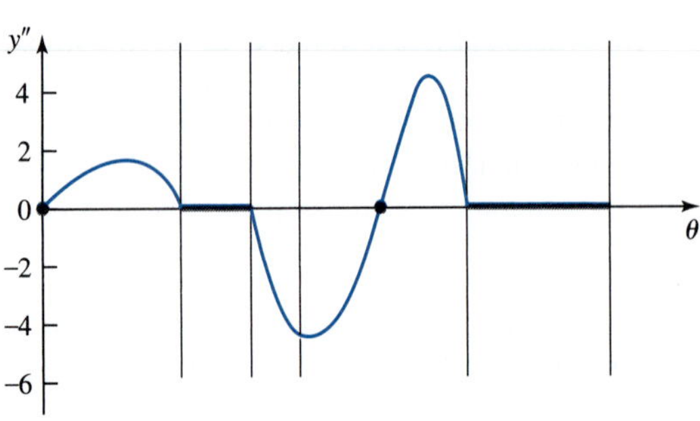

Figure 6.24 Example 6.2: (*a*) displacement diagram, in; (*b*) "velocity" diagram, in/rad; and (*c*) "acceleration" diagram, in/rad^2.

matched with Fig. 6.18. However, to keep the peak "accelerations" low and to keep the "jerk" curves as smooth as possible, we will choose Fig. 6.20*b* matched with Fig. 6.19. Thus, for segment *CD* we will use the harmonic half-rise motion of Eq. (6.19), and for segment *DE* we will choose the eighth-order-polynomial return motion of Eq. (6.17).

Choosing the motion curve types, however, is not sufficient to fully specify the motion characteristics. We must also find values for the unknown parameters of the motion equations; these are $L_1, L_3, \beta_1, \beta_3$, and β_4. We do this by equating the kinematic coefficients at each nonzero boundary. For example, to match the "velocities" at point *B* we must equate the first-order kinematic coefficient from Eq. (6.22*b*) at the right end (that is, at $\theta/\beta = 1$) with the first-order kinematic coefficient of the *BC* segment; that is,

$$y'_B = \frac{2L_1}{\beta_1} = \frac{L_2}{\beta_2} = \frac{1.25 \text{ in}}{0.785\ 40 \text{ rad}} = 1.591\ 55 \text{ in/rad}$$

or

$$L_1 = 0.795\ 77\beta_1. \tag{5}$$

Similarly, to match the "velocities" at point *C*, we equate the first-order kinematic coefficient of the *BC* segment with the first-order kinematic coefficient of Eq. (6.19*b*) at its left end (that is, at $\theta/\beta = 0$). This gives

$$y'_C = \frac{L_2}{\beta_2} = \frac{\pi L_3}{2\beta_3} = 1.591\ 55 \text{ in/rad}$$

or

$$L_3 = 1.013\ 21\beta_3. \tag{6}$$

To match the "accelerations" (that is, the curvatures) at point *D*, we equate the second-order kinematic coefficient of Eq. (6.19*c*) at its right end (that is, at $\theta/\beta = 1$) with the second-order kinematic coefficient of Eq. (6.17*c*) at its left end (that is, at $\theta/\beta = 0$). This gives

$$y''_D = -\frac{\pi^2 L_3}{4\beta_3^2} = -5.268\ 30\frac{L_4}{\beta_4^2},$$

where the total lift is $L_4 = 3$ in. Substituting Eq. (6) and the total lift into this result and rearranging gives

$$\beta_3 = 0.158\ 18\beta_4^2. \tag{7}$$

Finally, for geometric compatibility, we have the constraints

$$L_1 + L_3 = L_4 - L_2 = 1.750 \text{ in}. \tag{8}$$

and, considering Eqs. (3) and (4),

$$\beta_1 + \beta_3 + \beta_4 = 2\pi - \beta_2 - \beta_5 = 3.926\ 99 \text{ rad}. \tag{9}$$

Solving the five equations—that is, Eqs. (5) through (9)—simultaneously for the five unknowns, $L_1, L_3, \beta_1, \beta_3$, and β_4, provides the proper values of the remaining parameters. In summary, the results are

$$L_1 = 1.183\,1 \text{ in} \qquad \beta_1 = 1.486\,74 \text{ rad} = 85.184^\circ$$

$$L_2 = 1.250\,0 \text{ in} \qquad \beta_2 = 0.785\,40 \text{ rad} = 45.000^\circ$$

$$L_3 = 0.566\,9 \text{ in} \qquad \beta_3 = 0.559\,51 \text{ rad} = 32.058^\circ$$

$$L_4 = 3.000\,0 \text{ in} \qquad \beta_4 = 1.880\,74 \text{ rad} = 107.758^\circ$$

$$L_5 = 0.000\,0 \text{ in} \qquad \beta_5 = 1.570\,80 \text{ rad} = 90.000^\circ. \qquad \textit{Ans.}$$

At this time, an accurate layout of the displacement diagram and, if desired, the kinematic coefficients can be made to replace the sketches. The curves of Fig. 6.24 have been drawn to scale using these values.

6.9 PLATE CAM WITH RECIPROCATING FLAT-FACE FOLLOWER

Once the displacement diagram of a cam system has been completely determined, as described in Section 6.8, the layout of the actual cam shape can be attempted, as demonstrated in Section 6.4. In laying out the cam, however, we recall the need for a few more parameters, depending on the type of cam and follower—for example, the prime-circle radius, any offset distance, roller radius, and so on. Also, as we will see, each different type of cam can be subject to certain further problems unless these remaining parameters are carefully chosen.

In this section we study the problems that may be encountered in the design of a plate cam with a reciprocating flat-face follower. The geometric parameters of such a system that must yet be chosen are the prime-circle radius R_0, the offset (or eccentricity) ε of the follower stem, and the width of the follower face.

Figure 6.25 illustrates the layout of a plate cam with a radial reciprocating flat-face follower. In this case, the displacement chosen was a full-rise cycloidal motion with $L = 100$ mm during $\beta_1 = 90^\circ$ of cam rotation, followed by a full-return cycloidal motion during the remaining $\beta_2 = 270^\circ$ of cam rotation. The layout procedure of Fig. 6.10 was followed to develop the cam shape, and the radius that was chosen for the prime circle was $R_0 = 25$ mm. Obviously, there is a problem because the cam profile intersects itself. During machining, part of the cam shape will be lost and, when in operation, the intended cycloidal motion will not be achieved. Such a cam is said to be *undercut*.

Why did undercutting occur in this example and how can it be avoided? It resulted from attempting to achieve too great a lift in too little cam rotation with too small a cam. One possible cure for this problem is to decrease the desired lift L or to increase the cam rotation angle, β_1. However, this is not possible while still achieving the original design objectives. Another cure is to continue with the same displacement characteristics but to increase the prime-circle radius R_0 to avoid undercutting. This does produce a larger cam, but with sufficient increase it will overcome the undercutting problem.

The minimum value of R_0 to avoid undercutting can be determined by developing an equation for the radius of curvature of the cam profile. We start by writing the loop-closure

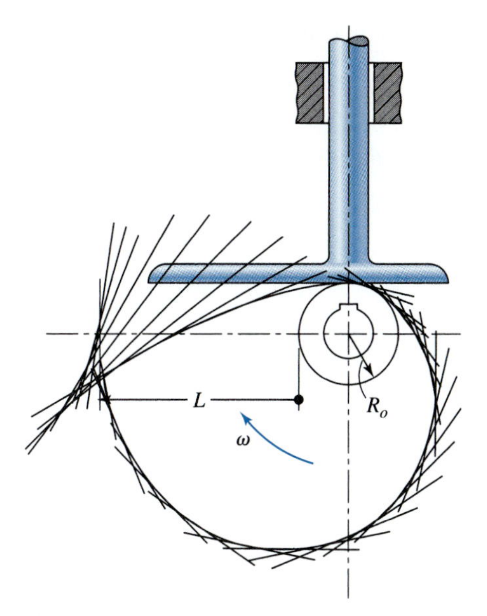

Figure 6.25 Undercut plate-cam profile layout with flat-face follower.

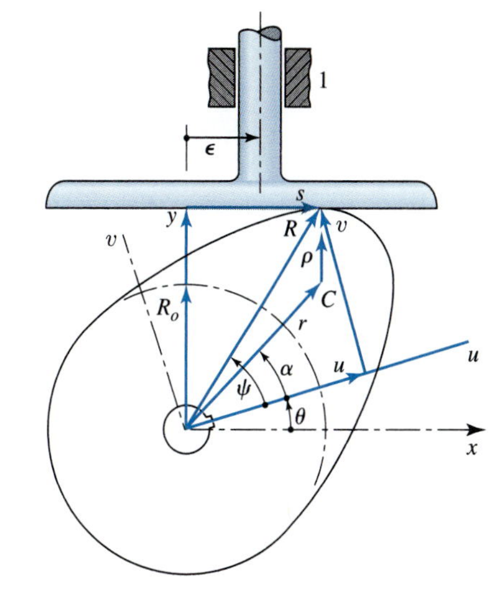

Figure 6.26 Plate-cam profile with flat-face follower.

equation using the vectors illustrated in Fig. 6.26. Using complex polar notation, the loop-closure equation is

$$\mathbf{R} = re^{j(\theta+\alpha)} + j\rho = j(R_0 + y) + s. \qquad (a)$$

Here we have carefully chosen the vectors so that point C is located at the instantaneous center of curvature and ρ is the radius of curvature corresponding to the current contact point. The line along the u axis that separates the angles θ and α is fixed on the cam and is horizontal for the cam position $\theta = 0$. The angle θ changes as the cam rotates, and the u and v axes rotate with the cam.

Separating Eq. (a) into real and imaginary parts, respectively, gives

$$r\cos(\theta + \alpha) = s \qquad (b)$$

$$r\sin(\theta + \alpha) + \rho = R_0 + y. \qquad (c)$$

Because point C is the center of curvature, the magnitudes of r, α, and ρ do not change for small increments in cam rotation*; that is,

$$\frac{dr}{d\theta} = \frac{d\alpha}{d\theta} = \frac{d\rho}{d\theta} = 0.$$

* The values of r, α, and ρ are not truly constant, but are currently at stationary values; their higher derivatives are nonzero.

Therefore, differentiating Eq. (*a*) with respect to the cam rotation angle θ gives

$$jre^{j(\theta+\alpha)} = jy' + s', \tag{d}$$

where $s' = ds/d\theta$. Separating Eq. (*d*) into real and imaginary parts, the first-order kinematic coefficients are

$$s' = -r\sin(\theta + \alpha) \tag{e}$$

$$y' = r\cos(\theta + \alpha). \tag{f}$$

Equating Eqs. (*b*) and (*f*), we find the location of the trace point along the surface of the follower as

$$s = y'. \tag{6.26}$$

Differentiating this equation with respect to the cam rotation angle θ, we find

$$s' = y''. \tag{g}$$

Substituting Eq. (*g*) into Eq. (*e*) and then substituting the result into Eq. (*c*), the radius of curvature of the cam profile can be written as

$$\rho = R_0 + y + y''. \tag{6.27}$$

We should carefully note the importance of Eq. (6.27); it states that the radius of curvature of the cam profile can be obtained for each cam rotation angle θ directly from the displacement equations, *before laying out the cam profile*. All that is needed is the choice of the prime circle radius R_0 and values for the displacement y and the second-order kinematic coefficient y''.

We can use Eq. (6.27) to select a value for R_0 that will avoid undercutting. When undercutting occurs, the radius of curvature of the cam profile switches sign from positive to negative. On the verge of undercutting, the cam will come to a point and the radius of curvature will be zero for some value of the cam rotation angle θ. We can choose R_0 large enough that this is never the case. In fact, to avoid high contact stresses, we may wish to ensure that ρ is everywhere larger than some specified value $\rho_{\min}$. To do this, from Eq. (6.27), we must require that

$$\rho = R_0 + y + y'' > \rho_{\min}.$$

Because R_0 and y are always positive, the critical situation occurs at or near the position where the second-order kinematic coefficient y'' has its largest negative value. Denoting this minimum value of y'' as $y''_{\min}$ and remembering that y corresponds to the same position, defined by cam angle θ, we have the condition

$$R_0 > \rho_{\min} - y - y''_{\min}, \tag{6.28}$$

which must be satisfied. This can easily be checked once the displacement equations have been established, and an appropriate value of R_0 can be chosen before the cam layout is attempted.

Returning now to Fig. 6.26, we see that Eq. (6.26) can also be of value. This equation states that the distance of travel of the point of contact on either side of the cam rotation center corresponds precisely with the plot of the first-order kinematic coefficient. Thus, the minimum face width for a flat-face follower must extend at least y'_{max} to the right and $-y'_{min}$ to the left of the camshaft center to maintain contact; that is,

$$\text{Face width} > y'_{max} - y'_{min}. \tag{6.29}$$

EXAMPLE 6.3

Assuming that the displacement characteristics in Example 6.2 are to be achieved by a plate cam with a reciprocating flat-face follower, determine the minimum face width and the minimum prime-circle radius to ensure that the radius of curvature of the cam is everywhere greater than 0.25 in.

SOLUTION

From Fig. 6.24*b* we see that the maximum "velocity" (that is, the maximum value of the first-order kinematic coefficient) occurs in the segment *BC* and is

$$y'_{max} = \frac{L_2}{\beta_2} = \frac{1.250\,0 \text{ in}}{0.785\,40 \text{ rad}} = 1.592 \text{ in/rad}. \tag{1}$$

The minimum "velocity" occurs in segment *DE* at approximately $\theta/\beta_4 = 0.5$. From Eq. (6.17*b*), the minimum value of the first-order kinematic coefficient is approximately

$$y'_{min} \approx y'\left(\theta/\beta = 0.5\right) = -2.812 \text{ in/rad}. \tag{2}$$

Substituting Eqs. (1) and (2) into Eq. (6.29), the minimum face width is

$$\text{Face width} > (1.592 \text{ in}) - (-2.812 \text{ in}) = 4.404 \text{ in}. \qquad \textit{Ans.}$$

Therefore, the follower would be positioned with 1.592 in. to the right and 2.812 in. to the left of the cam rotation axis, and some appropriate additional allowance would be added on each side.

The maximum negative "acceleration" (the minimum value of the second-order kinematic coefficient) occurs at *D* and can be obtained from Eq. (6.19*c*) at $\theta/\beta_3 = 1$. That is,

$$y''_{min} = -\frac{\pi^2 L_3}{4\beta_3^2} = -\frac{\pi^2(0.5669 \text{ in})}{4(0.559\,51 \text{ rad})^2} = -4.468\,18 \text{ in/rad}^2.$$

Substituting this result and the known parameters into Eq. (6.28), the minimum prime-circle radius is

$$R_0 > 0.250 \text{ in} - (-4.468 \text{ in}) - 3.000 \text{ in} = 1.718 \text{ in}. \qquad \textit{Ans.}$$

From this calculation we would choose the actual prime-circle radius as, say, $R_0 = 1.75$ in.

We can see that the eccentricity of the flat-face follower stem does not affect the geometry of the cam. This eccentricity is usually chosen to relieve high bending stresses in the follower. Also, there may be a higher load in the follower during the working stroke, say the lift stroke, than during the return motion. In such a case, the eccentricity may be chosen to locate the follower stem more centrally over the contact point during the lift portion of the motion cycle.

Looking again at Fig. 6.26, we can write another loop-closure equation; that is,

$$ue^{j\theta} + ve^{j(\theta+\pi/2)} = j\,(R_0 + y) + s,$$

where we recall that u and v denote the coordinates of the contact point in a coordinate system attached to the cam. Dividing this equation by $e^{j\theta}$ gives

$$u + jv = j\,(R_0 + y)\,e^{-j\theta} + se^{-j\theta}.$$

Using Eq. (6.26), the real and imaginary parts of this equation can be written as

$$u = (R_0 + y)\sin\theta + y'\cos\theta \tag{6.30a}$$

$$v = (R_0 + y)\cos\theta - y'\sin\theta. \tag{6.30b}$$

These two equations give the coordinates of the cam profile and provide an alternative to the graphical layout procedure of Fig. 6.10. They can be used to generate a table of numeric rectangular coordinate data from which the cam can be machined. Polar coordinate equations for this same curve are

$$R = \sqrt{(R_0 + y)^2 + (y')^2} \tag{6.31a}$$

and

$$\psi = \frac{\pi}{2} - \theta - \tan^{-1}\frac{y'}{R_0 + y}. \tag{6.31b}$$

6.10 PLATE CAM WITH RECIPROCATING ROLLER FOLLOWER

Figure 6.27 illustrates a plate cam with a reciprocating roller follower. We see that three geometric parameters remain to be chosen after the displacement diagram is completed and before the cam layout can be finalized. These three parameters are the radius of the prime circle R_0, the eccentricity ε, and the radius of the roller R_r. There are also two potential problems to be considered when choosing these parameters. One problem is undercutting and the other is an excessive pressure angle.

Pressure angle is the name used for the angle between the axis of motion of the follower stem and the line of action of the force exerted by the cam onto the roller follower, that is, the normal to the pitch curve through the trace point. The pressure angle is labeled ϕ in Fig. 6.27. Only the component of force along the line of motion of the follower is useful in overcoming the output load; the perpendicular component should be kept low to reduce sliding friction between the follower and its guideway and to reduce bending of the follower

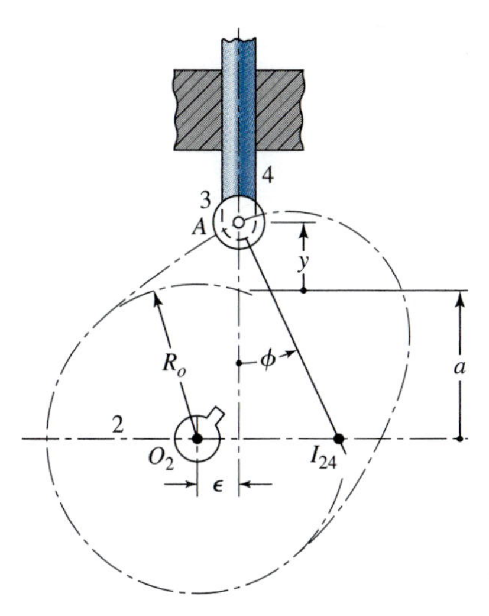

Figure 6.27 Plate cam with a reciprocating roller follower.

stem. Too high a pressure angle increases the deleterious effect of friction and may cause the translating follower to chatter or perhaps even to jam. Cam pressure angles of up to about 30° or 35° are about the largest that can be used without difficulties.

In Fig. 6.27 we see that the normal to the pitch curve intersects the horizontal axis at point I_{24}, that is, at the instantaneous center of velocity between the cam 2 and the follower stem 4. Because the follower stem is translating, all points of the follower stem have velocities equal to that of the instant center I_{24}. This velocity must also be equal to the velocity of the coincident point of link 2; that is,

$$V_{I_{24}} = \dot{y} = \omega R_{I_{24}O_2}.$$

Dividing this equation by the angular velocity of the cam ω [see Eq. (6.11)], the first-order kinematic coefficient is

$$y' = \frac{\dot{y}}{\omega} = R_{I_{24}O_2}.$$

This kinematic coefficient can also be expressed in terms of the eccentricity of the follower stem and the pressure angle of the cam as

$$y' = \varepsilon + (a + y)\tan\phi, \tag{a}$$

where, as illustrated in Fig. 6.27, the vertical distance from the cam axis to the prime circle is

$$a = \sqrt{R_o^2 - \varepsilon^2}. \tag{b}$$

Substituting Eq. (b) into Eq. (a) and rearranging, the pressure angle of the cam can be written as

$$\phi = \tan^{-1}\left(\frac{y' - \varepsilon}{\sqrt{R_0^2 - \varepsilon^2} + y}\right). \tag{6.32}$$

From this equation we observe that, once the displacement equations and the first-order kinematic coefficient have been determined, the two parameters, R_0 and ε, can be adjusted to obtain a suitable pressure angle. We also note that the pressure angle is continuously changing as the cam rotates, and therefore we are particularly interested in studying its extreme values.

Let us first consider the effect of eccentricity. From Eq. (6.32) we observe that increasing ε either increases or decreases the magnitude of the numerator, depending on the sign of the first-order kinematic coefficient y'. Thus, a small eccentricity ε can be used to reduce the pressure angle ϕ during the rise motion when y' is positive, but only at the expense of an increased pressure angle during the return motion when y' is negative. Still, because the magnitudes of the forces are usually greater during rise, it is common practice to offset the follower to take advantage of this reduction in pressure angle.

A much more significant effect can be made in reducing the pressure angle by increasing the prime-circle radius R_0. To study this effect, let us take the conservative approach and assume a radial follower, that is, there is no eccentricity. Substituting $\varepsilon = 0$ into Eq. (6.32), the pressure angle reduces to

$$\phi = \tan^{-1}\left(\frac{y'}{R_0 + y}\right). \tag{6.33}$$

To find the extremum values of the pressure angle, it is possible to differentiate this equation with respect to the cam rotation angle and equate it to zero, thus finding the values of the rotation angle θ that yield the maximum and the minimum pressure angles. This

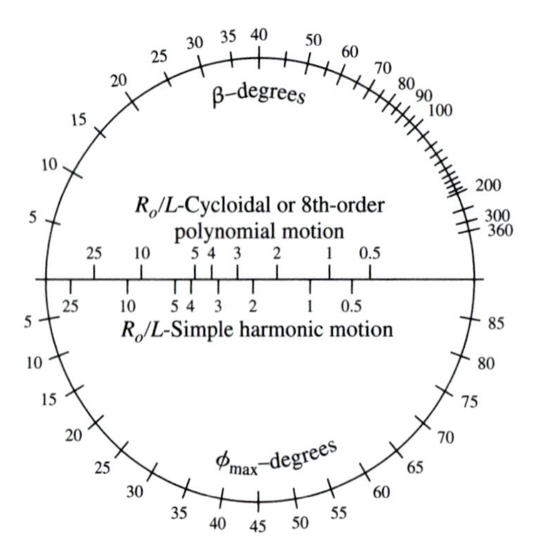

Figure 6.28 Nomogram relating maximum pressure angle ϕ_{max} to prime-circle radius R_0, lift L, and active cam angle β for radial reciprocating roller-follower cams with full-rise or full-return simple harmonic, cycloidal, or eighth-order-polynomial motion.

is a tedious mathematical process, however, and can be avoided by using the nomogram of Fig. 6.28. This nomogram was produced by searching out on a digital computer the maximum value of ϕ from Eq. (6.33) for each of the standard full-rise and full-return motion curves of Section 6.7. With the nomogram it is possible to use the known values of L and β for each segment of the displacement diagram and to read directly the maximum pressure angle occurring in that motion segment for a particular choice of R_0. Alternatively, a desired maximum pressure angle can be chosen and an appropriate value of R_0 can be determined. The process is best illustrated by an example.

EXAMPLE 6.4

Assuming that the displacements determined in Example 6.2 are to be achieved by a plate cam with a reciprocating radial roller follower, determine the minimum prime-circle radius that ensures that the pressure angle is everywhere less than 30°.

SOLUTION

Each segment of the displacement diagram can be checked in succession using the nomogram of Fig. 6.28.

For segment AB of Fig. 6.24, we have half-rise cycloidal motion with $L_1 = 1.1831$ in. and $\beta_1 = 85.184°$. Because this is a half-rise curve, whereas the nomogram of Fig. 6.28 is made only for full-rise curves, it is necessary to double both L_1 and β_1, thus pretending that the curve is full-rise. This gives $L = 2.37$ in. and $\beta = 170°$. Next, connecting a straight line from $\beta = 170°$ to $\phi_{max} = 30°$, we read from the upper scale on the center axis of the nomogram a value of $R_0/L \geq 0.75$, from which

$$R_0 \geq 0.75(2.37 \text{ in}) = 1.78 \text{ in.} \tag{1}$$

The segment BC need not be checked because its maximum pressure angle for this segment will occur at the boundary B and cannot be greater than that for segment AB.

Segment CD has half-rise harmonic motion with $L_3 = 0.5669$ in. and $\beta_3 = 32.058°$. Again, because this is a half-rise curve, these values are doubled and $L = 1.13$ in. and $\beta = 64°$ are used instead. Then, from the nomogram, we find $R_0/L \geq 2.45$, from which

$$R_0 \geq 2.45(1.13 \text{ in}) = 2.77 \text{ in.}$$

However, here we must be careful. This value is the radius of a fictitious prime circle for which the horizontal axis of our doubled "full-rise" harmonic curve would have $y = 0$. It is not the R_0 we seek, because our full-harmonic curve has a nonzero y value at its base of

$$y = 3.00 - 1.13 = 1.87 \text{ in.}$$

The appropriate value for R_0 for this situation is

$$R_0 \geq 2.77 - 1.87 = 0.90 \text{ in.} \tag{2}$$

Next we check the segment DE, which has eighth-order polynomial motion with $L_4 = 3.0000$ in and $\beta_4 = 107.758°$. Because this is a full-return motion curve with $y = 0$ at its base, no adjustments are necessary for use of the nomogram. We find $R_0/L \geq 1.3$ and

$$R_0 \geq 1.3(3.00 \text{ in}) = 3.90 \text{ in}. \tag{3}$$

To ensure that the pressure angle does not exceed 30° throughout all segments of the cam motion, we must chose the prime-circle radius to be at least as large as the maximum of these discovered values. Remembering the inability to read the nomogram with great precision, we might choose a yet larger value, such as

$$R_0 = 4.00 \text{ in}. \qquad \textit{Ans.}$$

Once a final value has been chosen, we can use Fig. 6.28 again to find the actual maximum pressure angle in each segment of the motion:

$$AB: \quad \frac{R_0}{L} = \frac{4.00}{2.37} = 1.69 \quad \phi_{\text{max}} = 18°$$

$$CD: \quad \frac{R_0}{L} = \frac{5.87}{1.13} = 5.19 \quad \phi_{\text{max}} = 14°$$

$$DE: \quad \frac{R_0}{L} = \frac{4.00}{3.00} = 1.33 \quad \phi_{\text{max}} = 29°.$$

Although the prime circle has been sized to give a satisfactory pressure angle, the follower may still not complete the desired motion. It is still possible that the curvature of the pitch curve is too sharp and that the cam profile may be undercut. Figure 6.29*a* illustrates a portion of a cam pitch curve and two cam profiles generated by two different size rollers. It is clear from Fig. 6.29*a* that a small roller moving on the given pitch curve generates a satisfactory cam profile. The cam profile generated by the larger roller, however, intersects itself and is said to be *undercut*. The result, after machining, is a pointed cam that does not produce the desired motion. Still, if the prime circle and thus the cam size is increased enough, even the larger roller will generate a cam profile that will operate satisfactorily.

In Fig. 6.29*b* we see that the cam profile will become pointed when the roller radius R_r is equal to the magnitude* of the radius of curvature of the pitch curve. Therefore, to achieve some chosen minimum size $|\rho|_{\text{min}}$ for the radius of curvature of the cam profile, the radius of curvature of the pitch curve must be of greater magnitude than this value by the radius of the roller, that is,

$$|\rho|_{\text{pitch}} \geq |\rho|_{\text{min}} + R_r. \tag{c}$$

Recall Section 4.17 where the radius of curvature of a point trajectory was given by Eq. (4.56).

In concept, it is possible to search out the minimum value of $|\rho|_{\text{pitch}}$ for a particular choice of the displacement equation y and a particular prime-circle radius R_0. However,

* Remember that the radius of curvature of the pitch curve can have either a positive or a negative value (Section 4.17). However, here we are only concerned with its size, that is, its absolute value.

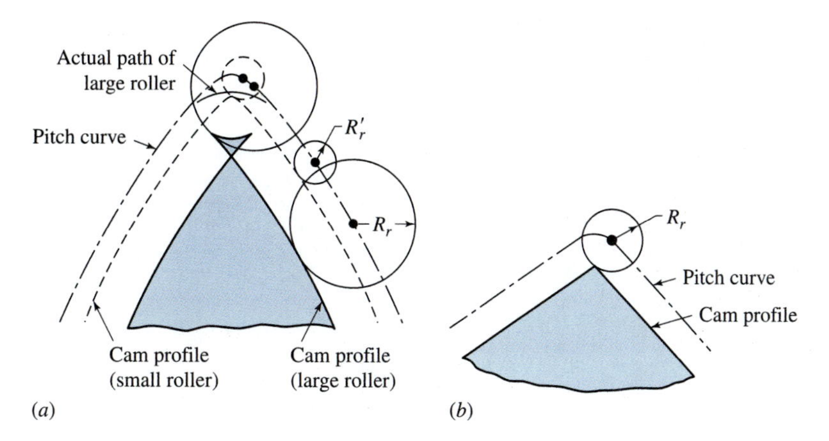

Figure 6.29

because it would be burdensome to perform such a search for each new cam design, the minimum-size radius of curvature (normalized with respect to R_0) has been sought out by a digital computer program for each of the standard cam motions of Section 6.7. The results are presented graphically in Fig. 6.30 through Fig. 6.34. Each of these figures illustrates a graph of $(|\rho|_{\min} + R_r)/R_0$ versus β for one type of standard-motion curve with various ratios of R_0/L. Because we have already chosen the displacement equations and have found a suitable value of R_0, each segment of the cam can now be checked to find its minimum size radius of curvature.

Saving even more effort, it is not necessary to check those segments of the cam where the second-order kinematic coefficient y'' remains positive throughout the segment, such as the half-rise motions of Eqs. (6.18) and (6.22) or the half-return motions of Eqs. (6.21) and (6.25). Assuming that the "acceleration" curve has been made continuous, the minimum size radius of curvature of the cam cannot occur in these segments. For each of these segments,

$$|\rho|_{\min} = R_0 - R_r. \tag{6.34}$$

EXAMPLE 6.5

Assuming that the displacement characteristics of Example 6.2 are to be achieved by a plate cam with a reciprocating roller follower, determine the minimum size of the radius of curvature of the cam profile if a prime-circle radius of $R_0 = 4.00$ in (from Example 6.4) and a roller radius of $R_r = 0.50$ in are used.

SOLUTION

For the segment AB of Fig. 6.24, we find from Eq. (6.34) that

$$|\rho|_{\min} = R_0 - R_r = 4.00 \text{ in} - 0.50 \text{ in} = 3.50 \text{ in}. \tag{1}$$

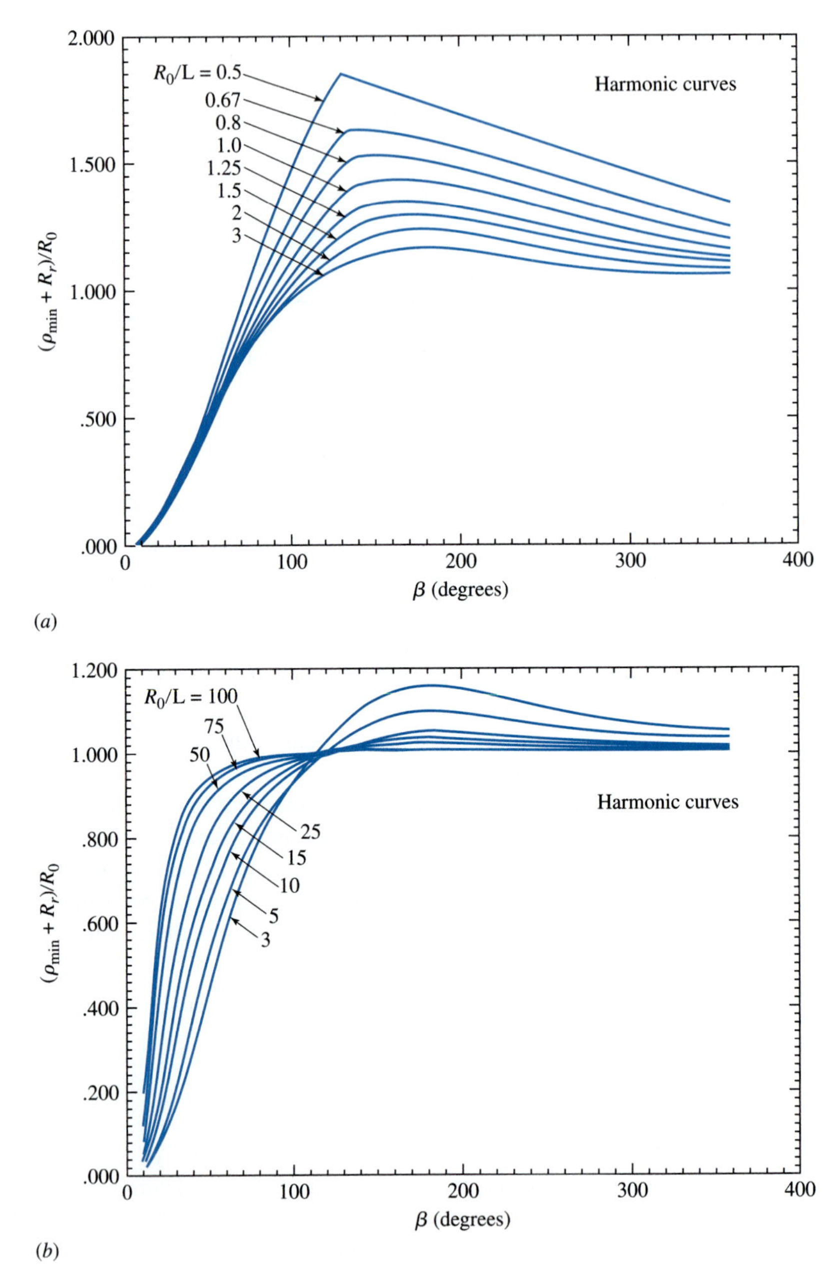

Figure 6.30 Minimum size radius-of-curvature of radial reciprocating roller follower cams with full-rise or full-return simple harmonic motion, Eqs. (6.12) or (6.15). (From M. A. Ganter and J. J. Uicker, Jr., *J. Mech. Des. ASME Trans. B* 101 (1979): 465–70.)

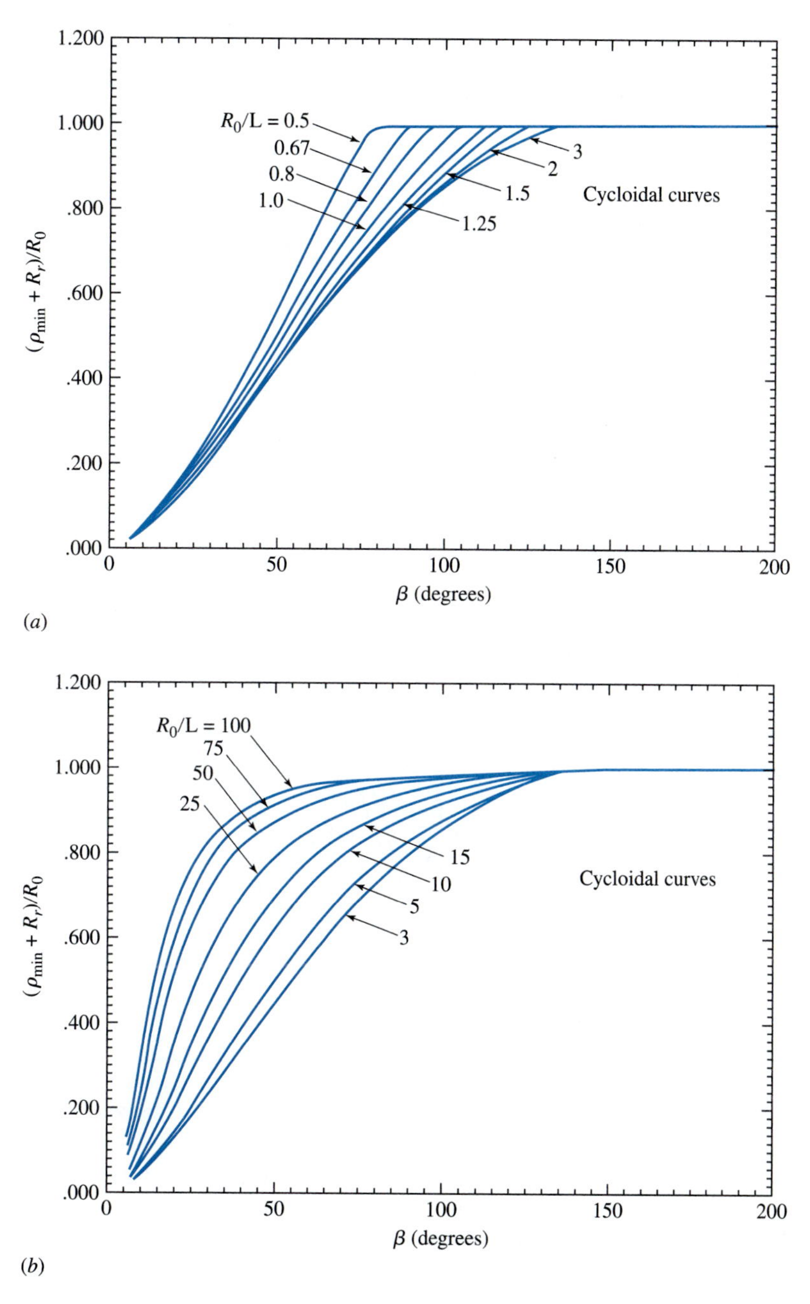

Figure 6.31 Minimum size radius-of-curvature of radial reciprocating roller follower cams with full-rise or full-return cycloidal motion, Eqs. (6.13) or (6.16). (From M. A. Ganter and J. J. Uicker, Jr., *J. Mech. Des. ASME Trans. B* 101 (1979): 465–70.)

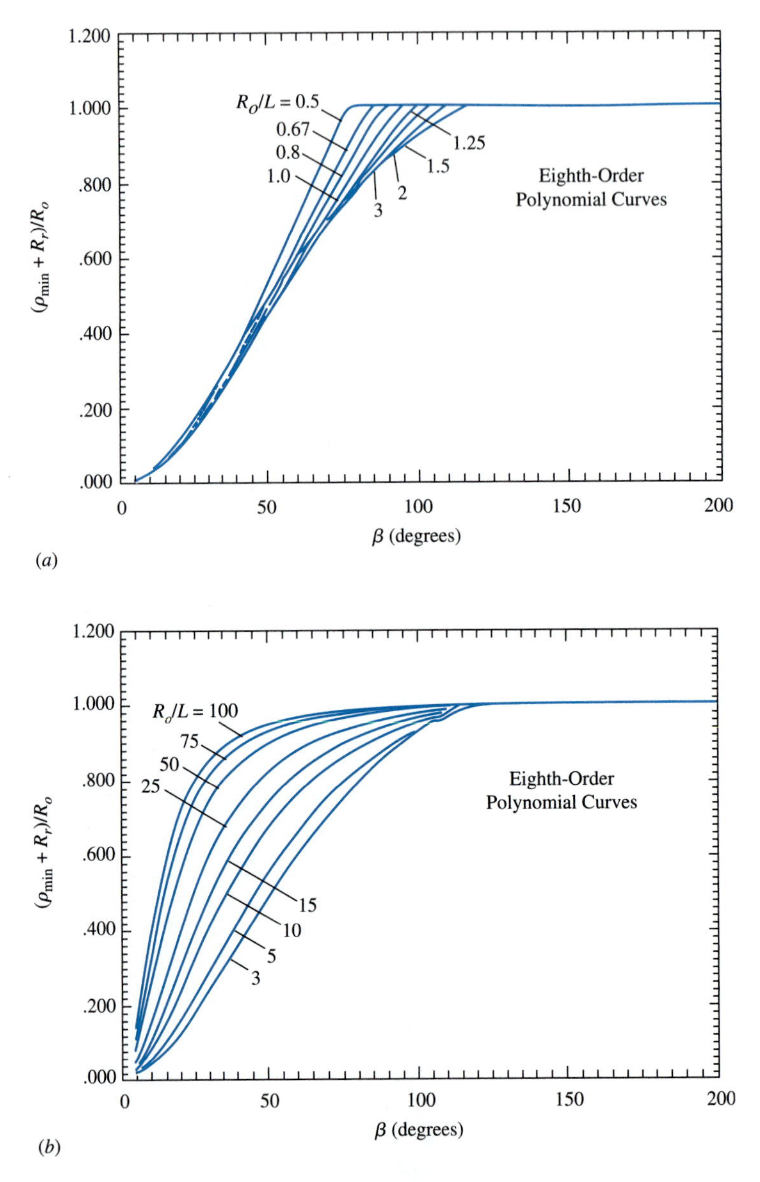

Figure 6.32 Minimum size radius-of-curvature of radial reciprocating roller follower cams with full-rise or full-return eighth-order polynomial motion, Eqs. (6.14) or (6.17). (From M. A. Ganter and J. J. Uicker, Jr., *J. Mech. Des. ASME Trans. B* 101 (1979): 465–70.)

For the segment *CD* we have half-harmonic rise motion with $L_3 = 0.5669$ in. and $\beta_3 = 32.058°$, from which we find

$$\frac{R_0}{L} = \frac{4.00 + (L_1 + L_2)}{L_3} = \frac{6.433\,1 \text{ in}}{0.566\,9 \text{ in}} = 11.35,$$

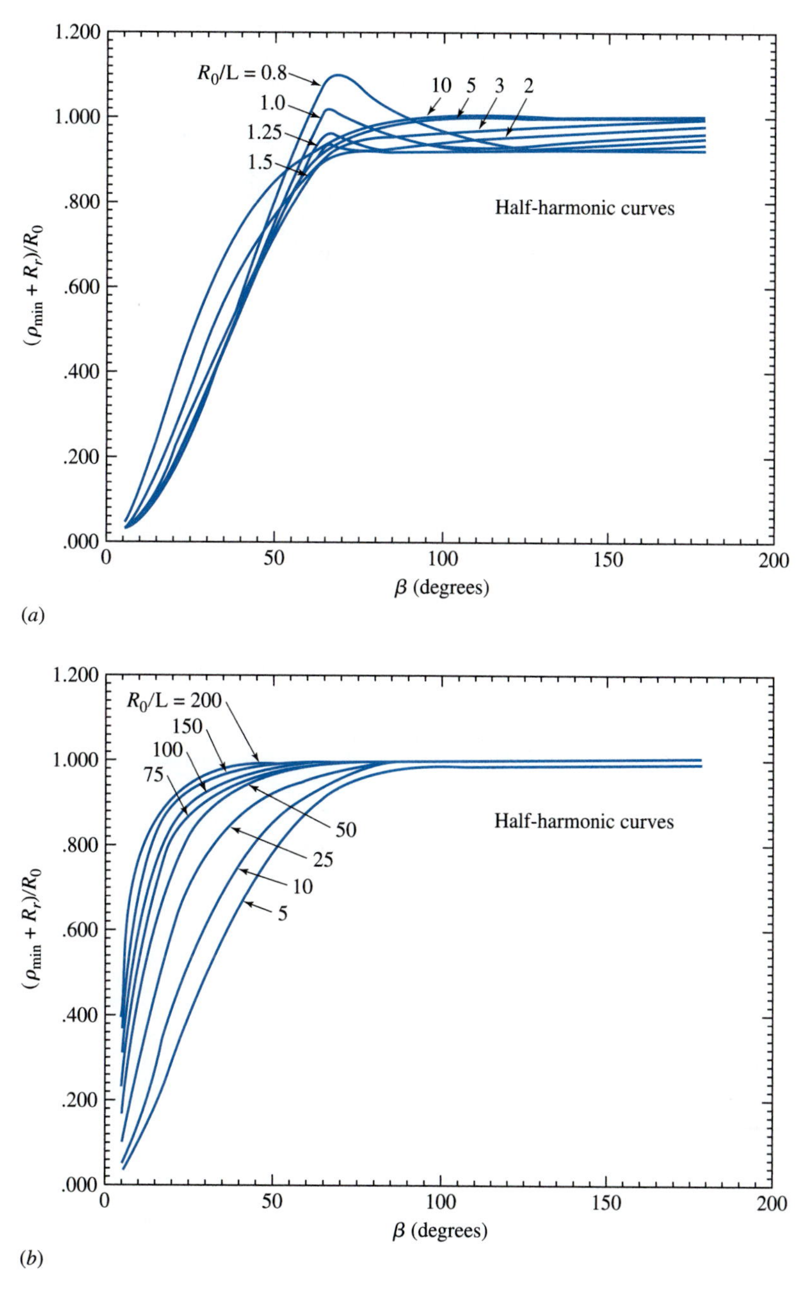

Figure 6.33 Minimum size radius-of-curvature of radial reciprocating roller follower cams with half-harmonic motion, Eqs. (6.19) or (6.20). (From M. A. Ganter and J. J. Uicker, Jr., *J. Mech. Des. ASME Trans. B* 101 (1979): 465–70.)

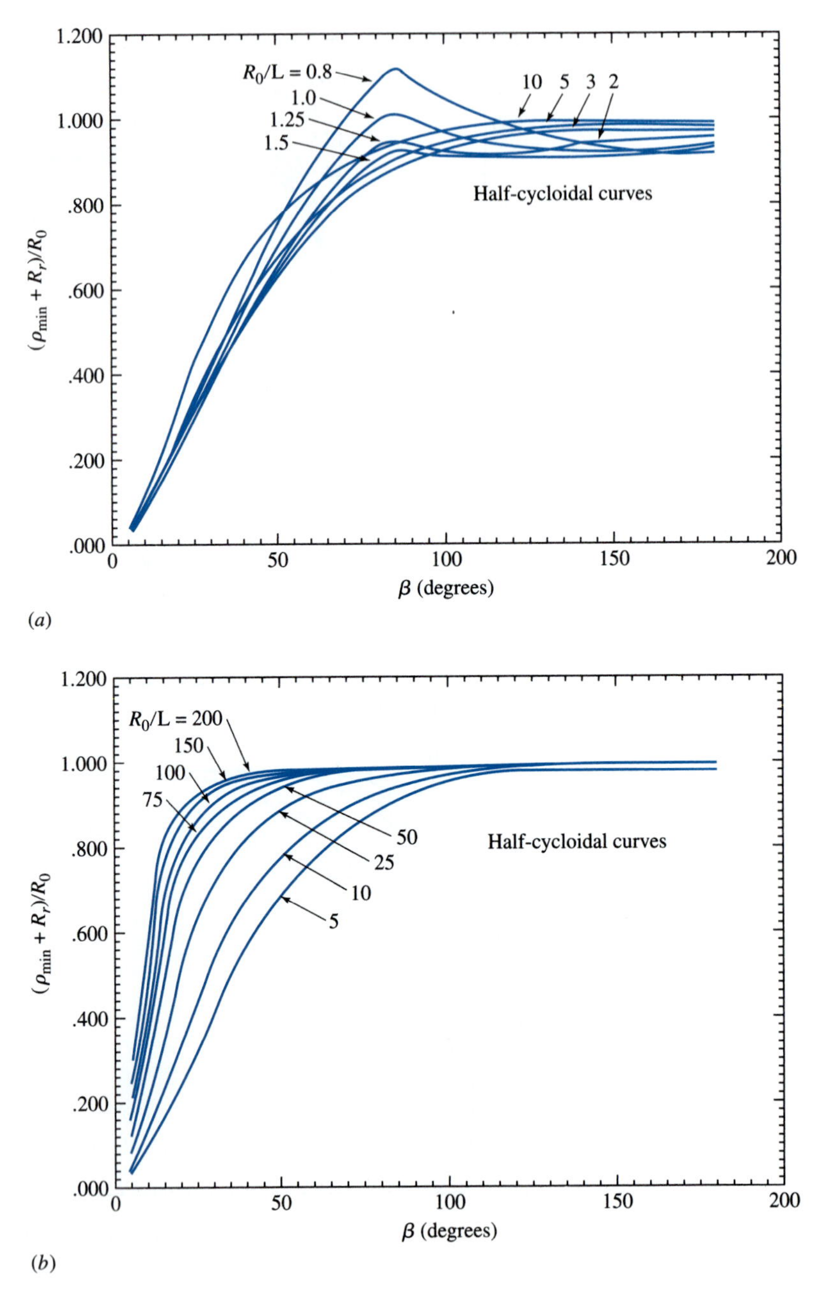

Figure 6.34 Minimum size radius-of-curvature of radial reciprocating roller follower cams with half-cycloidal motion, Eqs. (6.23) or (6.24). (From M. A. Ganter and J. J. Uicker, Jr., *J. Mech. Des. ASME Trans. B* 101 (1979): 465–70.)

where R_0 was adjusted by $(L_1 + L_2)$ because the curves of Fig. 6.33 were plotted for $y = 0$ at the base of that motion segment. Now, using Fig. 6.33*b*, we find $(|\rho|_{\min} + R_r)/R_0 = 0.66$. Therefore,

$$|\rho|_{\min} = 0.66R_0 - R_r = 0.66\,(6.433\,1\text{ in}) - 0.50\text{ in} = 3.75\text{ in.} \tag{2}$$

where, again, the adjusted value of R_0 was used.

For the segment *DE* we have eighth-order polynomial motion with $L_4 = 3.000\,0$ in. and $\beta_4 = 107.758°$, from which $R_0/L = 1.33$. Using Fig. 6.32*a*, we find $(|\rho|_{\min} + R_r)/R_0 = 0.80$, and

$$|\rho|_{\min} = 0.80R_0 - R_r = 0.80\,(4.00\text{ in}) - 0.50\text{ in} = 2.70\text{ in.} \tag{3}$$

Choosing the smallest value among Eqs. (1), (2), and (3), the minimum size of the radius of curvature of the entire cam profile is taken as

$$|\rho|_{\min} = 2.70\text{ in.} \qquad \textit{Ans.}$$

To machine the cam to the proper shape we will need the coordinates of the cam surface in the moving *uv* coordinate system attached to the cam.

The rectangular coordinates of the follower center (the pitch curve) of a plate cam with a reciprocating roller follower in the moving *uv* coordinate system are

$$u = \left(\sqrt{R_0^2 - \varepsilon^2} + y\right)\sin\theta + \varepsilon\cos\theta \tag{d}$$

$$v = \left(\sqrt{R_0^2 - \varepsilon^2} + y\right)\cos\theta - \varepsilon\sin\theta. \tag{e}$$

Differentiating these with respect to the rotation of the cam, the first-order kinematic coefficients of the follower center are

$$u' = y'\sin\theta + \left(\sqrt{R_0^2 - \varepsilon^2} + y\right)\cos\theta - \varepsilon\sin\theta \tag{f}$$

$$v' = y'\cos\theta - \left(\sqrt{R_0^2 - \varepsilon^2} + y\right)\sin\theta - \varepsilon\cos\theta \tag{g}$$

and, differentiating again, the second-order kinematic coefficients of the follower center are

$$u'' = y''\sin\theta + 2y'\cos\theta - \left(\sqrt{R_0^2 - \varepsilon^2} + y\right)\sin\theta - \varepsilon\cos\theta \tag{h}$$

$$v'' = y''\cos\theta - 2y'\sin\theta - \left(\sqrt{R_0^2 - \varepsilon^2} + y\right)\cos\theta + \varepsilon\sin\theta \tag{i}$$

We note that positive (counterclockwise) rotation of the cam causes the center point of the follower to increment in the clockwise direction around the pitch curve of the cam in the moving *uv* coordinate system. This defines the positive sense of the unit tangent vector $\hat{\mathbf{u}}^t$ for the pitch curve, and the unit normal vector is then given by $\hat{\mathbf{u}}^n = \hat{\mathbf{k}} \times \hat{\mathbf{u}}^t$ in the moving coordinate system. Because this unit normal points away from the center of curvature of the pitch curve then the sign convention is that the radius of curvature of the pitch curve will be a negative value.

To normalize Eqs. (f) and (g), we define

$$w' = +\sqrt{u'^2 + v'^2} \tag{6.35}$$

or

$$w' = +\left[(y' - \varepsilon)^2 + \left(\sqrt{R_0^2 - \varepsilon^2} + y\right)^2\right]^{1/2} \tag{j}$$

and with this we find the unit tangent

$$\hat{\mathbf{u}}^t = \left(\frac{u'}{w'}\right)\hat{\mathbf{i}} + \left(\frac{v'}{w'}\right)\hat{\mathbf{j}} \tag{k}$$

and unit normal

$$\hat{\mathbf{u}}^n = \left(\frac{-v'}{w'}\right)\hat{\mathbf{i}} + \left(\frac{u'}{w'}\right)\hat{\mathbf{j}} \tag{l}$$

in the moving uv coordinate system attached to the cam.

The coordinates of the point of contact between the cam and the roller follower, in the moving coordinate system of the cam, can now be written as

$$u_{\text{cam}} = u + R_r\left(\frac{v'}{w'}\right) \quad \text{and} \quad v_{\text{cam}} = v - R_r\left(\frac{u'}{w'}\right). \tag{6.36}$$

The radius of curvature of the pitch curve can be written from Eq. (4.57) as

$$\rho = \frac{w'^3}{u'v'' - v'u''} \tag{6.37}$$

Remembering that the radius of curvature of the pitch curve is expected to have a negative value, as explained above, then the radius of curvature of the cam profile is

$$\rho_{\text{cam}} = \rho + R_r, \tag{6.38}$$

Note that Eq. (6.38) will still yield a negative value, although smaller, than Eq. (6.37) because the unit normal vector will still point away from the center of curvature of the cam.

The unit vector in the direction of motion of the follower (the y direction), when expressed in the moving uv coordinate system, is

$$\hat{\mathbf{u}}^y = \sin\theta\hat{\mathbf{i}} + \cos\theta\hat{\mathbf{j}}. \tag{m}$$

The pressure angle can be written from Fig. 6.27, and the aid of Eqs. (l) amd (m), as

$$\cos\phi = \hat{\mathbf{u}}^n \cdot \hat{\mathbf{u}}^y = -\left(\frac{v'}{w'}\right)\sin\theta + \left(\frac{u'}{w'}\right)\cos\theta. \tag{6.39}$$

EXAMPLE 6.6

A plate cam with a reciprocating radial roller follower is to be designed such that the displacement of the follower is

$$y = 15\,(1 - \cos 2\theta)\ \text{mm}.$$

The prime-circle radius is to be $R_0 = 40$ mm and the roller is to have a radius of $R_r = 12$ mm. The cam is to rotate counterclockwise. For the cam rotation angle $\theta = 30°$, determine the following:

(a) The coordinates of the point of contact between the cam and the follower in the moving coordinate system;
(b) The radius of curvature of the cam profile; and
(c) The pressure angle of the cam.

SOLUTION

At the cam rotation angle $\theta = 30°$, the lift of the follower from the specified displacement equation is

$$y = 15\,(1 - \cos 2\theta)\ \text{mm} = 7.500\ \text{mm}. \tag{1}$$

By differentiating the displacement equation with respect to cam rotation angle we find

$$y' = 30 \sin 2\theta\ \text{mm/rad} = 25.981\ \text{mm/rad} \tag{2a}$$

$$y'' = 60 \cos 2\theta\ \text{mm/rad}^2 = 30.000\ \text{mm/rad}^2. \tag{2b}$$

From Eqs. (*d*) and (*e*), with offset $\varepsilon = 0$, the coordinates of the roller center in the moving coordinate system attached to the cam are

$$u = (R_0 + y) \sin\theta = (55 - 15 \cos 2\theta) \sin\theta\ \text{mm} = 23.750\ \text{mm} \tag{3a}$$

$$v = (R_0 + y) \cos\theta = (55 - 15 \cos 2\theta) \cos\theta\ \text{mm} = 41.136\ \text{mm}. \tag{3b}$$

Then, from Eqs. (*f*) and (*g*), the first-order kinematic coefficients of the follower center are

$$\begin{aligned} u' &= y' \sin\theta + (R_0 + y)\cos\theta \\ &= (25.981\ \text{mm/rad}) \sin\theta + (47.500\ \text{mm}) \cos\theta = 54.127\ \text{mm/rad} \end{aligned} \tag{4a}$$

$$\begin{aligned} v' &= y' \cos\theta - (R_0 + y)\sin\theta \\ &= (25.981\ \text{mm/rad}) \cos\theta - (47.500\ \text{mm}) \sin\theta = -1.250\ \text{mm/rad} \end{aligned} \tag{4b}$$

and, from Eqs. (*h*) and (*i*), the second-order kinematic coefficients of the follower center are

$$\begin{aligned} u'' &= y'' \sin\theta + 2y' \cos\theta - (R_0 + y)\sin\theta \\ &= \left(30.000\ \text{mm/rad}^2\right) \sin\theta + (51.962\ \text{mm/rad}) \cos\theta - (47.500\ \text{mm}) \sin\theta \\ &= 36.250\ \text{mm/rad}^2 \end{aligned} \tag{5a}$$

$$v'' = y'' \cos\theta - 2y' \sin\theta - (R_0 + y)\cos\theta$$
$$= \left(30.000 \text{ mm/rad}^2\right)\cos\theta - (51.962 \text{ mm/rad})\sin\theta - (47.500 \text{ mm})\cos\theta$$
$$= -41.136 \text{ mm/rad}^2. \tag{5b}$$

From Eq. (6.35) we have

$$w' = +\sqrt{u'^2 + v'^2}$$
$$= +\sqrt{(54.127 \text{ mm/rad})^2 + (-1.250 \text{ mm/rad})^2} = +54.141 \text{ mm/rad}. \tag{6}$$

(a) Substituting Eqs. (3), (4), and (6) and the given dimensions into Eqs. (6.36), the coordinates of the point of contact between the cam and the roller follower, in the moving coordinate system, are

$$u_{\text{cam}} = 23.750 \text{ mm} + (12 \text{ mm})\left(\frac{-1.250 \text{ mm/rad}}{54.141 \text{ mm/rad}}\right) = 23.413 \text{ mm} \qquad \textit{Ans.}$$

$$v_{\text{cam}} = 41.136 \text{ mm} + (12 \text{ mm})\left(\frac{-54.127 \text{ mm/rad}}{54.141 \text{ mm/rad}}\right) = 29.139 \text{ mm}. \qquad \textit{Ans.}$$

(b) From Eq. (6.37), using Eqs. (4)–(6), the radius of curvature of the pitch curve is

$$\rho = \frac{(54.141 \text{ mm/rad})^3}{(54.127 \text{ mm/rad})(-41.136 \text{ mm/rad}^2) - (-1.250 \text{ mm/rad})(36.250 \text{ mm/rad}^2)}$$
$$\rho = -72.775 \text{ mm}$$

and we note that this value is negative, as explained above. With this, Eq. (6.38), gives the radius of curvature of the cam profile as

$$\rho_{\text{cam}} = \rho + R_r = -72.775 \text{ mm} + 12.0 \text{ mm} = -60.775 \text{ mm}, \qquad \textit{Ans.}$$

which we note is still negative. This confirms that the center of curvature is still further inward and that there is no undercutting at this location on the cam profile.

(c) From Eq. (6.39) we find the pressure angle for this position,

$$\cos\phi = -\left(\frac{-1.250 \text{ mm/rad}}{54.141 \text{ mm/rad}}\right)\sin 30^\circ + \left(\frac{54.127 \text{ mm/rad}}{54.141 \text{ mm/rad}}\right)\cos 30^\circ;\ \phi = 28.68^\circ.$$

Ans.

In this and the previous section, we have considered problems that result from poor choice of the prime-circle radius for a plate cam with a reciprocating follower. Although the equations are different for oscillating followers or other types of cams, a similar approach can be used to guard against undercutting[5] and severe pressure angles.[6] Similar equations can also be developed for cam profile data.[7] An extensive survey of the cam design literature has been compiled by Chen.[8] We will present one more example here to illustrate a general approach for the design of cams that requires a vector analysis in addition to the above equations.

EXAMPLE 6.7

A plate cam with an oscillating roller follower illustrated in Fig. 6.35a is to be designed such that the displacement of the follower will follow the equation

$$y = 0.5(1 - \cos 2\theta) \text{ rad}$$

and the cam is to rotate counterclockwise. The prime-circle radius is $R_0 = 1.50$ in., the roller radius is $R_r = 0.50$ in., the distance between the center of the camshaft and the follower pivot is $R_{MO} = r_1 = 3.00$ in., and the length of the follower arm is $R_{CM} = r_4 = 2.598$ in. For the cam rotation angle of $\theta_2 = \theta = 45°$, determine the following:

(a) The coordinates of the point of contact between the cam and the follower;
(b) The radius of curvature of the cam profile; and
(c) The pressure angle of the cam.

SOLUTION

Let us designate the rotation of the oscillating follower by the angle θ_4 as illustrated in Fig. 6.35b. With the specified dimensions, the angle of the follower with zero displacement must be $\theta_4 = 150° = 5\pi/6$ rad. Combining this with the specified displacement gives the complete equation for the rotation of the follower.

$$\theta_4 = 5\pi/6 - 0.5(1 - \cos 2\theta) \text{ rad} \tag{1}$$

Note that a negative sign was used because the follower displacement, as illustrated by Fig. 6.35b, is in the clockwise direction.

From Fig. 6.35b, the coordinates of the trace point C in the fixed coordinate system are

$$x = r_1 + r_4 \cos \theta_4 \tag{2a}$$

$$y = r_4 \sin \theta_4 \tag{2b}$$

These coordinates can be transformed into the moving coordinate system of the cam by using the relations

$$u = x \cos \theta + y \sin \theta \tag{3a}$$

$$v = -x \sin \theta + y \cos \theta \tag{3b}$$

Therefore, substituting Eqs. (2) into Eqs. (3), the coordinates of the trace point C in the moving coordinate system are

$$u = (r_4 \sin \theta_4) \sin \theta + (r_1 + r_4 \cos \theta_4) \cos \theta$$

$$v = (r_4 \sin \theta_4) \cos \theta - (r_1 + r_4 \cos \theta_4) \sin \theta,$$

Figure 6.35 Example 6.7. Plate cam with oscillating roller follower.

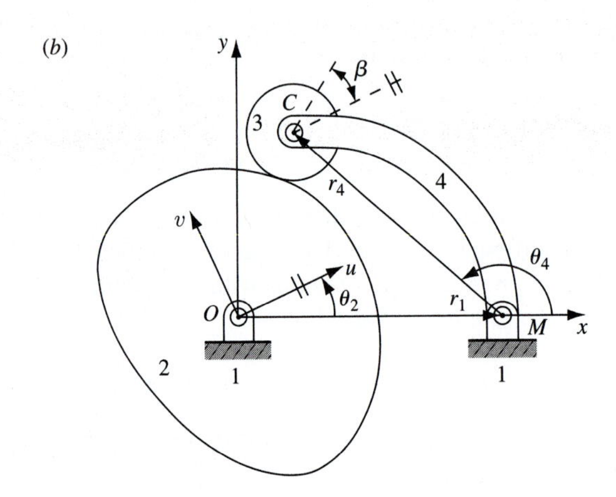

which can be written as

$$u = r_1 \cos\theta + r_4 \cos(\theta_4 - \theta) \tag{4a}$$

$$v = -r_1 \sin\theta + r_4 \sin(\theta_4 - \theta). \tag{4b}$$

Differentiating Eqs. (4) with respect to the rotation angle of the cam θ, the first- and second-order kinematic coefficients of the trace point C are

$$u' = -r_1 \sin\theta - r_4 \sin(\theta_4 - \theta)[\theta_4' - 1] \tag{5a}$$

$$v' = -r_1 \cos\theta + r_4 \cos(\theta_4 - \theta)[\theta_4' - 1] \tag{5b}$$

$$u'' = -r_1 \cos\theta - r_4 \cos(\theta_4 - \theta)[\theta_4' - 1]^2 - r_4\theta_4'' \sin(\theta_4 - \theta) \tag{6a}$$

$$v'' = +r_1 \sin\theta - r_4 \sin(\theta_4 - \theta)[\theta_4' - 1]^2 + r_4\theta_4'' \cos(\theta_4 - \theta), \tag{6b}$$

where the first- and second-order kinematic coefficients of the follower arm, from Eq. (1), are

$$\theta_4' = -\sin 2\theta \quad \text{and} \quad \theta_4'' = -2\cos 2\theta. \tag{7}$$

The coordinates of the point of contact between the cam and the follower, in the moving coordinate system, are determined in the same way as for Eqs. (6.35) and (6.36).

At the cam rotation angle $\theta = 45°$, from Eqs. (1) and (7), we have

$$\theta_4 = 2.118 \text{ rad} = 121.35°, \quad \theta_4' = -1.000, \quad \text{and} \quad \theta_4'' = 0.000. \tag{8}$$

Substituting Eqs. (8) into Eqs. (4), the coordinates of the trace point C are

$$u = (3.000 \text{ in}) \cos 45° + (2.598 \text{ in}) \cos(121.35° - 45°) = 2.734 \text{ in} \tag{9a}$$

$$v = -(3.000 \text{ in}) \sin 45° + (2.598 \text{ in}) \sin(121.35° - 45°) = 0.403 \text{ in} \tag{9b}$$

Substituting Eqs. (8) and the given geometry data into Eqs. (5), the first-order kinematic coefficients of the trace point are

$$u' = -(3.000 \text{ in}) \sin 45° - (2.598 \text{ in}) \sin(121.35° - 45°)[-2.000] = 2.928 \text{ in/rad} \tag{10a}$$

$$v' = -(3.000 \text{ in}) \cos 45° + (2.598 \text{ in}) \cos(121.35° - 45°)[-2.000] = -3.347 \text{ in/rad} \tag{10b}$$

and from Eq. (6.35) we have

$$w' = +\sqrt{(2.928 \text{ in/rad})^2 + (-3.347 \text{ in/rad})^2} = 4.447 \text{ in/rad} \tag{11}$$

(a) Then substituting Eqs. (9)–(11) into Eqs. (6.36), the coordinates of the contact point between the cam and the follower, in the moving coordinate system, are

$$u_{\text{cam}} = (2.734 \text{ in}) + (0.5 \text{ in}) \left(\frac{-3.347 \text{ in/rad}}{4.447 \text{ in/rad}} \right) = 2.358 \text{ in} \qquad \textit{Ans.}$$

$$v_{\text{cam}} = (0.403 \text{ in}) - (0.5 \text{ in}) \left(\frac{2.928 \text{ in/rad}}{4.447 \text{ in/rad}} \right) = 0.074 \text{ in}. \qquad \textit{Ans.}$$

(b) In the moving coordinate system of the cam, the direction of motion of the roller center C is defined by the angle β [see Fig. 6.35b)], which can be written as

$$\beta = \theta_4 - 90° - \theta = 121.35° - 90° - 45° = -13.65°. \tag{12}$$

Therefore, from Eq. (6.39), the pressure angle of the cam can be written as

$$\cos\phi = \left(-\frac{v'}{w'}\right) \cos\beta + \left(\frac{u'}{w'}\right) \sin\beta.$$

Substituting Eqs. (10), (11), and (12) into this equation, the pressure angle (at the cam rotation angle $\theta = 45°$) is

$$\phi = 54.83°. \qquad \textit{Ans.}$$

(c) Substituting Eqs. (8) and the given geometry into Eqs. (6), the second-order kinematic coefficients of the follower center are

$$u'' = -(3.000 \text{ in}) \cos 45° - (2.598 \text{ in}) \cos(121.35° - 45°)[-2.000]^2$$
$$= -4.573 \text{ in/rad}^2 \qquad (13a)$$
$$v'' = (3.000 \text{ in}) \sin 45° - (2.598 \text{ in}) \sin(121.35° - 45°)[-2.000]^2$$
$$= -7.977 \text{ in/rad}^2. \qquad (13b)$$

Then, substituting Eqs. (10), (11), and (13) into Eq. (6.37), the radius of curvature of the pitch curve at this location is

$$\rho = \frac{(4.447 \text{ in/rad})^3}{(2.928 \text{ in/rad})(-7.977 \text{ in/rad}^2) - (-3.347 \text{ in/rad})(-4.573 \text{ in/rad}^2)}$$
$$= -2.275 \text{ in.}$$

The negative sign indicates that the unit normal vector to the pitch curve, through the follower center, points away from the center of curvature as explained above. Therefore, from Eq. (6.38), the radius of curvature of the cam profile is

$$\rho_{\text{cam}} = \rho + R_r = -2.275 \text{ in} + 0.500 \text{ in} = -1.775 \text{ in.} \qquad \textit{Ans.}$$

6.11 REFERENCES

[1] A good analysis of this subject is presented in Tesar, D., and G. K. Matthew. *The Dynamic Synthesis, Analysis, and Design of Modeled Cam Systems*. Lexington, MA: Heath, 1976.

[2] Kloomak, M., and R. V. Muffley.1955. Plate Cam Design—With Emphasis on Dynamic Effects. *Prod. Eng.* 26; 178–82.

[3] Stoddart, D. A. 1953. Polydyne Cam Design. *Mach. Design* 25, no. 1:121–35; no. 2:146–54; no. 3:149–64.

[4] Chen, F. Y. *The Mechanics and Design of Cam Mechanisms*. Oxford: Pergamon Press, 1982.

[5] Kloomak, M., and R. V. Mufley. 1955. Plate Cam Design: Radius of Curvature. *Prod. Eng.* 26, no. 9:186–201.

[6] Kloomak, M., and R. V. Mufley. 1955. late Cam Design: Pressure Angle Analysis. *Prod. Eng.* 26, no. 5:155–71.

[7] Molian, S. *The Design of Cam Mechanisms and Linkages*. London: Constable, 1968.

[8] Chen, F. Y. 1977. A Survey of the State of the Art of Cam System Dynamics. *Mech. Mach. Theory* 12, no. 3:201–24.

PROBLEMS

6.1 The reciprocating radial roller follower of a plate cam is to rise 2 in. with simple harmonic motion in 180° of cam rotation and return with simple harmonic motion in the remaining 180°. If the roller radius is 0.375 in and the prime-circle radius is 2 in, construct the displacement diagram, the pitch curve, and the cam profile for clockwise cam rotation.

6.2 A plate cam with a reciprocating flat-face follower has the same motion as in Problem 6.1. The prime-circle radius is 2 in, and the cam rotates counterclockwise. Construct the displacement diagram and the cam profile, offsetting the follower stem by 0.75 in in the direction that reduces the bending in the follower during rise.

6.3 Construct the displacement diagram and the cam profile for a plate cam with an oscillating radial flat-face follower that rises through 30° with cycloidal motion in 150° of counterclockwise cam rotation, then dwells for 30°, returns with cycloidal motion in 120°, and dwells for 60°. Determine the necessary length for the follower face, allowing 5 mm clearance at the free end. The prime-circle radius is 30 mm, and the follower pivot is 120 mm to the right.

6.4 A plate cam with an oscillating roller follower is to produce the same motion as in Problem 6.3. The prime-circle radius is 60 mm, the roller radius is 10 mm, the length of the follower is 100 mm, and it is pivoted at 125 mm to the right of the cam rotation axis. The cam rotation is clockwise. Determine the maximum pressure angle.

6.5 For full-rise simple harmonic motion, write the equations for the velocity and the jerk at the midpoint of the motion. Also, determine the acceleration at the beginning and the end of the motion.

6.6 For full-rise cycloidal motion, determine the values of θ for which the acceleration is maximum and minimum. What is the formula for the acceleration at these points? Find the equations for the velocity and the jerk at the midpoint of the motion.

6.7 A plate cam with a reciprocating follower is to rotate clockwise at 400 rev/min. The follower is to dwell for 60° of cam rotation, after which it is to rise to a lift of 2.5 in. During 1 in. of its return stroke, it must have a constant velocity of −40 in/s. Recommend standard cam motions from Section 6.7 to be used for high-speed operation and determine the corresponding lifts and cam rotation angles for each segment of the cam.

6.8 Repeat Problem 6.7 except with a dwell for 20° of cam rotation.

6.9 If the cam of Problem 6.7 is driven at constant speed, determine the time of the dwell and the maximum and minimum velocity and acceleration of the follower for the cam cycle.

6.10 A plate cam with an oscillating follower is to rise through 20° in 60° of cam rotation, dwell for 45°, then rise through an additional 20°, return and dwell for 60° of cam rotation. Assuming high-speed operation, recommend standard cam motions from Section 6.7 to be used and determine the lifts and cam-rotation angles for each segment of the cam.

6.11 Determine the maximum velocity and acceleration of the follower for Problem 6.10, assuming that the cam is driven at a constant speed of 600 rev/min.

6.12 The boundary conditions for a polynomial cam motion are as follows: for $\theta = 0$, $y = 0$, and $y' = 0$; for $\theta = \beta$, $y = L$, and $y' = 0$. Determine the appropriate displacement equation and the first three derivatives of this equation with respect to the cam rotation angle. Sketch the corresponding diagrams.

6.13 Determine the minimum face width using 0.1-in. allowances at each end and determine the minimum radius of curvature for the cam of Problem 6.2.

6.14 Determine the maximum pressure angle and the minimum radius of curvature for the cam of Problem 6.1.

6.15 A radial reciprocating flat-face follower is to have the motion described in Problem 6.7. Determine the minimum prime-circle radius if the radius of curvature of the cam is not to be less than 0.5 in. Using this prime-circle radius, what is the minimum length of the follower face using allowances of 0.15 in on each side?

6.16 Graphically construct the cam profile of Problem 6.15 for clockwise cam rotation.

6.17 A radial reciprocating roller follower is to have the motion described in Problem 6.7. Using a prime-circle radius of 20 in., determine the maximum pressure angle and the maximum roller radius that can be used without producing undercutting.

6.18 Graphically construct the cam profile of Problem 6.17 using a roller radius of 0.75 in. The cam rotation is to be clockwise.

6.19 A plate cam rotates at 300 rev/min and drives a reciprocating radial roller follower through a full rise of 75 mm in 180° of cam rotation. Find the minimum radius of the prime circle if simple harmonic motion is used and the pressure angle is not to exceed 25°. Find the maximum acceleration of the follower.

6.20 Repeat Problem 6.19 except that the motion is cycloidal.

6.21 Repeat Problem 6.19 except that the motion is eighth-order polynomial.

6.22 Using a roller diameter of 20 mm, determine whether the cam of Problem 6.19 will be undercut.

6.23 Equations (6.30) and (6.31) describe the profile of a plate cam with a reciprocating flat-face follower. If such a cam is to be cut on a milling machine with cutter radius R_C, determine similar equations for the center of the cutter.

6.24 Write computer programs for each of the displacement equations of Section 6.7.

6.25 Write a computer program to plot the cam profile for Problem 6.2.

6.26 A plate cam with an offset reciprocating roller follower is to have a dwell of 60° and then rise in 90° to another dwell of 120°, after which it is to return in 90° of cam rotation. The radius of the base circle is 40 mm, the radius of the roller follower is 15 mm, and the follower offset is 20 mm. For the rise motion $60° \leq \theta \leq 150°$, the equation of the displacement (the lift) is to be

$$y = 40\left(\frac{\varphi}{\pi} + \sin\varphi\right),$$

where y is in millimeters and φ is the cam rotation angle in radians. (*i*) Find equations for the first- and second-order kinematic coefficients of the lift y for this rise motion. (*ii*) Sketch the displacement diagram and the first- and second-order kinematic coefficients for the follower motion described. Comment on the suitability of this rise motion in the context of the other displacements specified. At the cam angle $\theta = 120°$, determine the following: (*iii*) the location of the point of contact between the cam and follower, expressed in the moving Cartesian coordinate system attached to the cam; (*iv*) the radius of the curvature of the pitch curve and the radius of curvature of the cam surface; and (*v*) the pressure angle of the cam. Is this pressure angle acceptable?

6.27 A plate cam with an offset reciprocating roller follower is to be designed using the input, the rise and fall, and the output motion shown in Table P6.27. The radius of the base circle is to be 30 mm, the radius of the roller follower is 12.5 mm, and the follower offset (or eccentricity) is to be 15 mm.

TABLE P6.27 Displacement information for plate cam with reciprocating roller follower.

Cam Angle (deg)	Rise or Fall (mm)	Follower Motion
0°–20°	0	Dwell
20°–110°	+25	Full-rise simple harmonic motion
110°–120°	0	Dwell
120°–200°	+5	Full-rise cycloidal motion
200°–270°	0	Dwell
270°–360°	−30	Full-return cycloidal motion

Comment on the suitability of the motions specified. At the cam angle $\theta = 50°$, determine the following: (*i*) the first-, second-, and third-order kinematic coefficients of the lift curve; (*ii*) the coordinates of the point of contact between the roller follower and the cam surface, expressed in the Cartesian coordinate system rotating with the cam; (*iii*) the radius of curvature of the pitch curve; (*iv*) the unit tangent and the unit normal vectors to the pitch curve; and (*v*) the pressure angle of the cam.

6.28 A plate cam with a radial reciprocating roller follower is to be designed using the input, the rise and fall, and the output motion shown in Table P6.28. The base circle diameter is 3 in and the diameter of the roller is 1 in. Displacements are specified as follows.

TABLE P6.28 Displacement information for plate cam with reciprocating roller follower.

Input θ (deg)	Lift L (in.)	Output y
$0° - 90°$	3.0	Cycloidal rise
$90° - 105°$	0	Dwell
$105° - 195°$	-3.0	Cycloidal fall
$195° - 210°$	0	Dwell
$210° - 270°$	2.0	Simple harmonic rise
$270° - 285°$	0	Dwell
$285° - 345°$	-2.0	Simple harmonic fall
$345° - 360°$	0	Dwell

Plot the lift curve (the displacement diagram) and the profile of the cam. (*i*) Comment on the lift curves at appropriate positions of the cam (for example, when the cam angle is $\theta = 0°$, $\theta = 45°$, $\theta = 180°$, $\theta = 210°$, $\theta = 225°$, and $\theta = 300°$). (*ii*) Identify on your cam profile the location(s) and the value of the largest pressure angle. Would this pressure angle cause difficulties for a practical cam-follower system? (*iii*) Identify on your cam profile the location(s) of discontinuities in position, velocity, acceleration, and/or jerk. Are these discontinuities acceptable (why or why not)? (*iv*) Identify on your cam profile any regions of positive radius of curvature of the cam profile. Are these regions acceptable (why or why not)? (*v*) For the values given in Table P6.28, what design changes would you suggest to improve the cam design?

6.29 Continue using the same displacement information and the same design parameters as in Problem 6.28. Use a spreadsheet to determine and plot the following for a complete rotation of the cam: (*i*) the first-order kinematic coefficients of the follower center, (*ii*) the second-order kinematic coefficients of the follower center, (*iii*) the third-order kinematic coefficients of the follower center, (*iv*) the lift curve (displacement diagram), (*v*) the radius of curvature of the cam surface, and (*vi*) the pressure angle of the cam-follower system. Is the pressure angle suitable for a practical cam-follower system?

7 Spur Gears

Gears are machine elements used to transmit rotary motion between two shafts, usually with a constant speed ratio. In this chapter we will discuss the case where the axes of the two shafts are parallel and the teeth are straight and parallel to the axes of rotation of the shafts; such gears are called *spur gears*.

7.1 TERMINOLOGY AND DEFINITIONS

A pair of spur gears in mesh is illustrated in Fig. 7.1. The *pinion* is a name given to the smaller of the two mating gears; the larger is often called the *gear* or the *wheel*. The pair of gears, chosen to work together, is often called a *gearset*.

The terminology of gear teeth is illustrated in Fig. 7.2, where most of the following definitions are given.

The *pitch circle* is a theoretical circle on which all calculations are based. The pitch circles of a pair of mating gears are tangent to each other, and it is these pitch circles that were pictured in earlier chapters as rolling against each other without slip.

The *diametral pitch P* is the ratio of the number of teeth on the gear to its pitch diameter, that is,

$$P = \frac{N}{2R}, \tag{7.1}$$

where N is the number of teeth and R is the pitch circle radius. Note that the diametral pitch cannot be directly measured on the gear itself. Also, note that, as the value of the diametral pitch becomes larger, the teeth become smaller; this is illustrated clearly in Fig. 7.3. In addition, a pair of mating gears have the same diametral pitch. The diametral pitch is used to indicate the tooth size in U.S. customary units and usually has units of teeth per inch.

Figure 7.1 Pair of spur gears in mesh. (Courtesy of Gleason Works, Rochester, NY.)

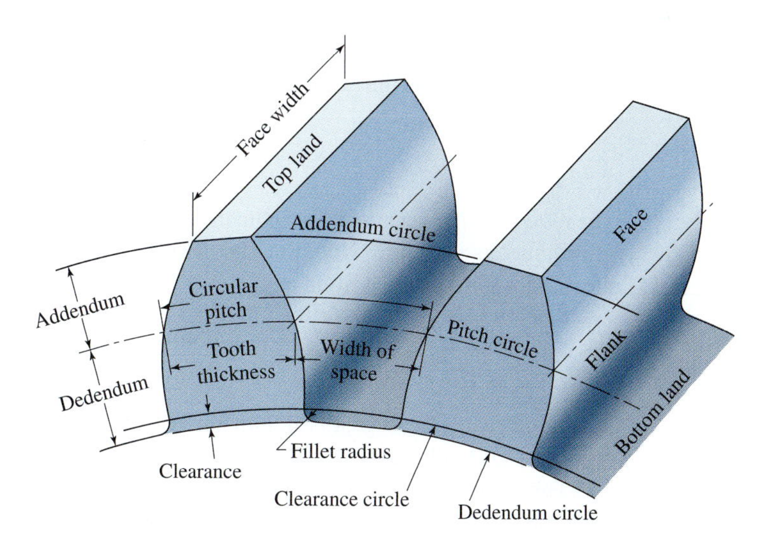

Figure 7.2 Gear tooth terminology.

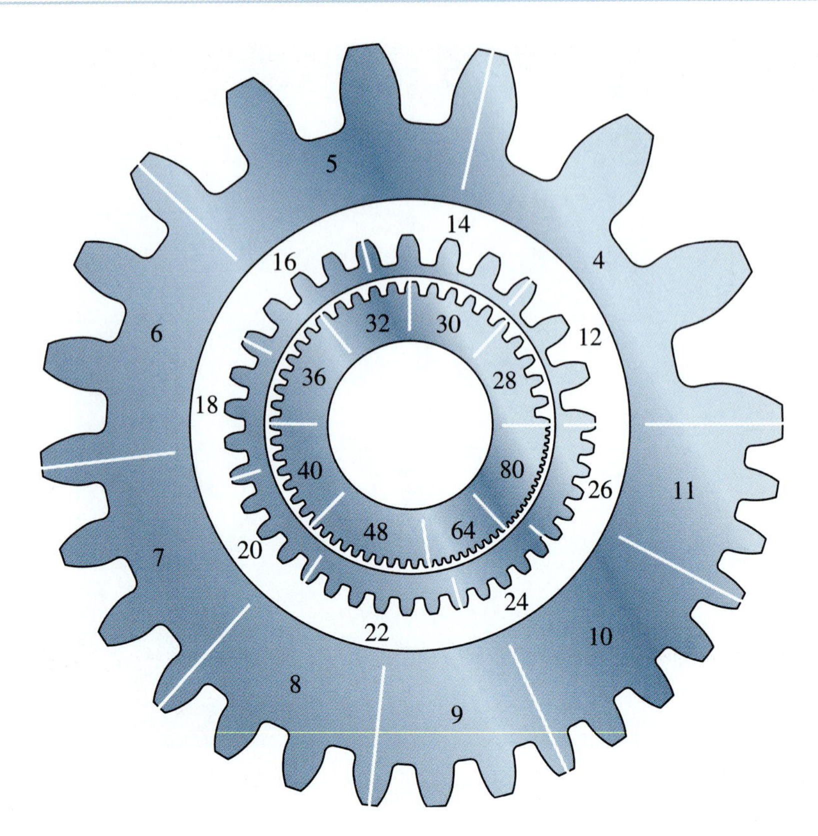

Figure 7.3 Tooth sizes for various diametral pitches in teeth per inch. (Courtesy of Gleason Cutting Tools Corp., Loves Park, IL.)

The *module m* is the ratio of the pitch diameter of the gear to its number of teeth, that is

$$m = \frac{2R}{N}. \tag{7.2}$$

The module is the usual unit for indicating tooth size in SI units and it customarily has units of millimeters per tooth. Note that the module is the reciprocal of the diametral pitch, and the relationship can be written as

$$m = \frac{25.4\ \text{(mm/in)}}{P\ \text{(teeth/in)}} = \frac{25.4}{P}\ \text{mm/tooth}.$$

Also note that metric gears should not be interchanged with U.S. gears because their standards for tooth sizes are not the same.

The *circular pitch p* is the distance from one tooth to the adjacent tooth measured along the pitch circle. Therefore, it can be determined from

$$p = \frac{2\pi R}{N}. \tag{7.3}$$

Circular pitch is related to the previous definitions, depending on the units, by

$$p = \frac{\pi}{P} = \pi m. \tag{7.4}$$

The *addendum* a is the radial distance between the pitch circle and the top land of each tooth.

The *dedendum* d is the radial distance from the pitch circle to the bottom land of each tooth.

The *whole depth* is the sum of the addendum and dedendum.

The *clearance* c is the amount by which the dedendum of a gear exceeds the addendum of the mating gear.

The *backlash* is the amount by which the width of a tooth space exceeds the thickness of the engaging tooth measured along the pitch circles.

7.2 FUNDAMENTAL LAW OF TOOTHED GEARING

Gear teeth mating with each other to produce rotary motion are similar to a cam and follower. When the tooth profiles (or cam and follower profiles) are shaped so as to produce a constant angular velocity ratio between the two shafts, then the two mating surfaces are said to be *conjugate*. It is possible to specify an arbitrary profile for one tooth and then to find a profile for the mating tooth so that the two surfaces are conjugate. One possible choice for such conjugate solutions is the *involute* profile, which, with few exceptions, is in universal use for gear teeth.

A single pair of mating gear teeth as they pass through their entire period of contact must be shaped such that the ratio of the angular velocity of the driven gear to that of the driving gear, that is, *the first-order kinematic coefficient, must remain constant*. This is the fundamental criterion that governs the choice of the tooth profiles. If this were not true in gearing, very serious vibration and impact problems would result, even at low speeds.

In Section 3.17 we learned that the angular velocity ratio theorem states that the first-order kinematic coefficient of any mechanism is inversely proportional to the segments into which the common instant center cuts the line of centers. In Fig. 7.4, two profiles are in contact at point T; let profile 2 represent the driver and profile 3 represent the driven. The normal to the surfaces CD is called the *line of action*. The normal to the profiles at the point of contact T intersects the line of centers O_2O_3 at the instant center of velocity. In gearing, this instant center is generally referred to as the *pitch point* and usually carries the label P.

Designating the pitch circle radii of the two gear profiles R_2 and R_3, from the angular velocity ratio theorem, Eq. (3.28), we see that

$$\frac{\omega_2}{\omega_3} = \frac{R_3}{R_2}. \tag{7.5}$$

This equation is frequently used to define what is called the fundamental law of gearing, which states that as gears go through their mesh, *the pitch point must remain stationary on the line of centers* so that the speed ratio remains constant. This means that the line of

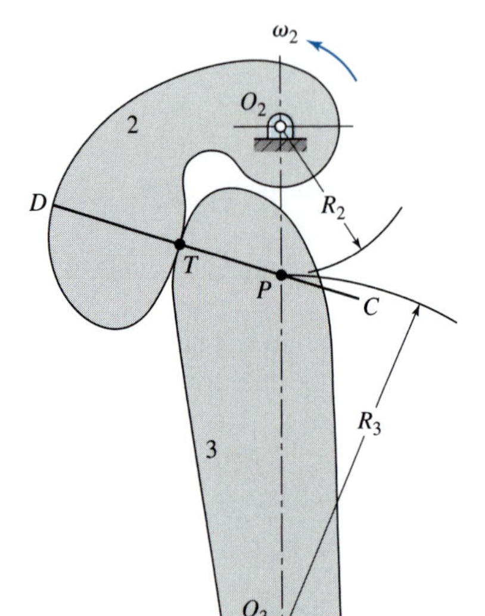

Figure 7.4

action for every new instantaneous point of contact must always pass through the stationary pitch point P. Thus, the problem of finding a conjugate profile for a given shape is to find a mating shape that satisfies the fundamental law of gearing.

It should not be assumed that any shape or profile is satisfactory just because a conjugate profile can be found. Although theoretically conjugate curves might be found, the practical problems of reproducing these curves from steel gear blanks or other materials while using existing machinery still exist. In addition, the sensitivity of the law of gearing to small dimensional changes of the shaft center distance caused either by misalignment or by large forces must also be considered. Finally, the tooth profile selected must be one that can be reproduced quickly and economically in very large quantities. A major portion of this chapter is devoted to illustrating how the involute curve profile fulfills these requirements.

7.3 INVOLUTE PROPERTIES

An *involute* curve is the path generated by a tracing point on a cord as the cord is unwrapped from a cylinder called the *base cylinder*. This is illustrated in Fig. 7.5, where T is the tracing point. Note that the cord AT is normal to the involute at T, and the distance AT is the instantaneous value of the radius of curvature. As the involute is generated from its origin T_0 to T_1, the radius of curvature varies continuously; it is zero at T_0 and increases continuously to T_1. Thus, the cord is the generating line, and it is always normal to the involute.

If the two mating tooth profiles both have the shapes of involute curves, the condition that the pitch point P remain stationary is satisfied. This is illustrated in Fig. 7.6, where two gear blanks with fixed centers O_2 and O_3 are illustrated having base cylinders with respective radii of O_2A and O_3B. We now imagine that a cord is wound clockwise around the base cylinder of gear 2, pulled tightly between points A and B, and wound counterclockwise

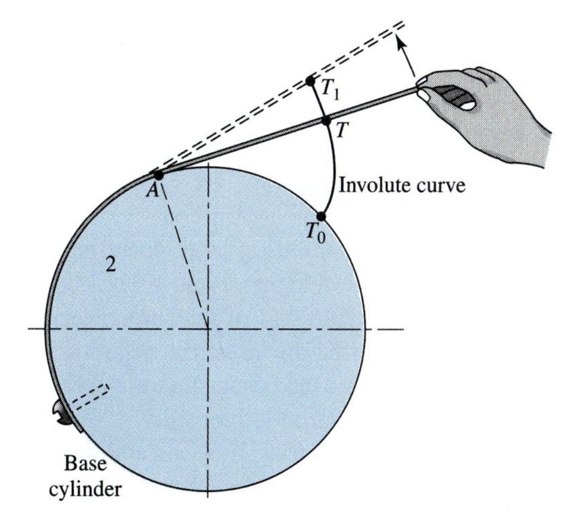

Figure 7.5 Involute curve.

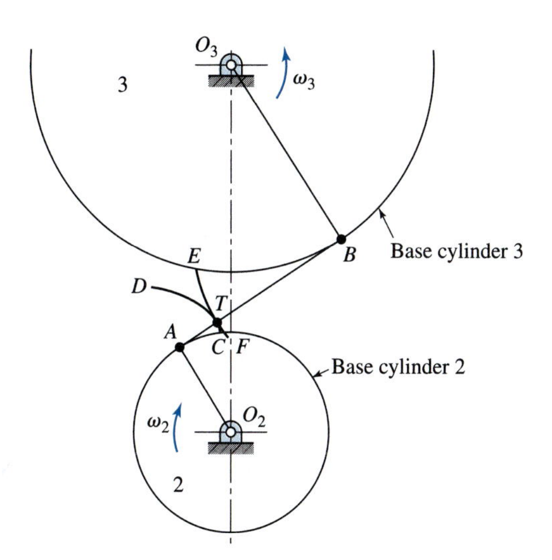

Figure 7.6 Conjugate involute curves.

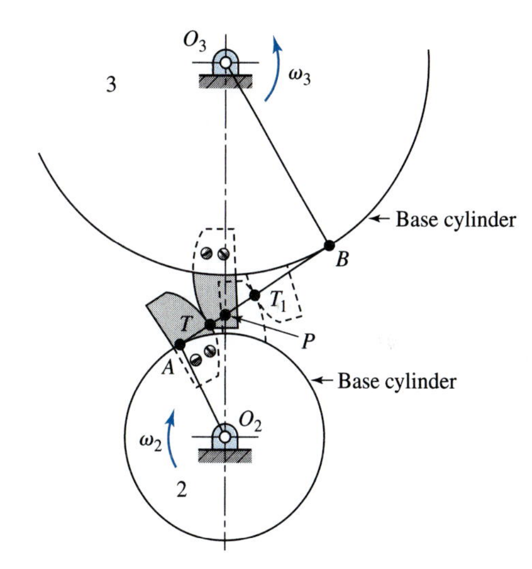

Figure 7.7 Involute action.

around the base cylinder of gear 3. If the two base cylinders are now rotated in opposite directions so as to keep the cord tight, a tracing point T traces out the involutes CD on gear 2 and EF on gear 3. The two involutes thus generated simultaneously by the single tracing point T are conjugate profiles.

Next imagine that the involutes of Fig. 7.6 are scribed on plates and that the plates are cut along the scribed curves and then bolted to the respective cylinders in the same positions. The result is illustrated in Fig. 7.7. The cord can now be removed and, if gear 2 is moved clockwise, gear 3 is caused to move counterclockwise by the camlike action of the two curved plates. The path of contact is the line AB formerly occupied by the cord. Because the line AB is the generating line for each involute, it is normal to both profiles at all points of contact. Also, it always occupies the same position because it is always tangent to both base cylinders. Therefore, point P is the pitch point. Point P does not move; therefore, the involute curves are conjugate curves and satisfy the fundamental law of gearing.

7.4 INTERCHANGEABLE GEARS; AGMA STANDARDS

A *tooth system* is the name given to a *standard*[1] that specifies the relationships among addendum, dedendum, clearance, tooth thickness, and fillet radius to attain interchangeability of gears of different tooth numbers but of the same pressure angle and the same diametral pitch or module. We should be aware of the advantages and disadvantages of such a tooth system so that we can choose the best gears for a given design and have a basis for comparison if we depart from a standard tooth profile.

For a pair of spur gears to properly mesh, they must share the same pressure angle and the same tooth size as specified by the choice of the diametral pitch or module. The numbers of teeth and the pitch diameters of the two gears in mesh need not match, but are chosen to give the desired speed ratio, as demonstrated in Eq. (7.5).

The sizes of the teeth used are chosen by selecting the diametral pitch P or module m. Standard cutters are generally available for the sizes listed in Table 7.1. Once the diametral pitch or module is chosen, the remaining dimensions of the tooth are set by the standards in Table 7.2. Tables 7.1 and 7.2 contain the standards for the spur gears most in use today, and they include the values for both SI and U.S. customary units.

Let us illustrate the design choices by an example.

TABLE 7.1 Standard gear tooth sizes

	Standard diametral pitches P, US customary, teeth/in.
Coarse	1, $1\frac{1}{4}$, $1\frac{1}{2}$, $1\frac{3}{4}$, 2, $2\frac{1}{2}$, 3, 4, 5, 6, 8, 10, 12, 14, 16, 18
Fine	20, 24, 32, 40, 48, 64, 72, 80, 96, 120, 150, 200
	Standard modules m, SI, mm/tooth
Preferred	1, 1.25, 1.5, 2, 2.5, 3, 4, 5, 6, 8, 10, 12, 16, 20, 25, 32, 40, 50
Next choice	1.125, 1.375, 1.75, 2.25, 2.75, 3.5, 4.5, 5.5, 7, 9, 11, 14, 18, 22, 28, 36, 45

TABLE 7.2 Standard tooth systems for spur gears

System	Pressure angle, ϕ (deg)	Addendum, a	Dedendum, d
Full depth	20°	$1/P$ or $1m$	$1.25/P$ or $1.25m$
Full depth	$22\frac{1}{2}$°	$1/P$ or $1m$	$1.25/P$ or $1.25m$
Full depth	25°	$1/P$ or $1m$	$1.25/P$ or $1.25m$
Stub teeth	20°	$0.8/P$ or $0.8m$	$1/P$ or $1m$

EXAMPLE 7.1

Two parallel shafts, separated by a distance (commonly referred to as the center distance) of 3.5 in, are to be connected by a gear set so that the output shaft rotates at 40% of the speed of the input shaft. Design a gearset to fit this situation.

SOLUTION

The center distance can be written as $R_2 + R_3 = 3.5$ in, and substituting the given information into Eq. (7.5) we have $\omega_3/\omega_2 = R_2/R_3 = 0.40$. Then substituting the first equation into the second equation, and rearranging, we find that $R_2 = 1.0$ in and $R_3 = 2.5$ in. Next, we must choose the size of the teeth by picking a value for the diametral pitch or module. From Eq. (7.1) we find the numbers of teeth on the two gears to be $N_2 = 2PR_2$ and $N_3 = 2PR_3$. This choice of P or m for tooth size is often iterative. First, we might choose a value of $P = 6$ teeth/in; this gives the numbers of teeth as $N_2 = 12$ teeth and $N_3 = 30$ teeth; if we choose $P = 10$ teeth/in, then we get $N_2 = 20$ teeth and $N_3 = 50$ teeth. At this time, either choice appears acceptable, and we choose $P = 10$ teeth/in. However, this choice of P (or m) must later be checked for possible undercutting, as we will study in Section 7.7, and for contact ratio, which we will study in Section 7.8, and for strength and wear of the teeth.[2]

7.5 FUNDAMENTALS OF GEAR–TOOTH ACTION

To illustrate the fundamentals we now proceed, step by step, through the actual graphical layout of a pair of spur gears. The dimensions used here are those of Example 7.1 assuming standard 20° full-depth involute tooth form as specified in Table 7.2. The various steps, in the correct order, are illustrated in Figs. 7.8 and 7.9 and are as follows.

STEP 1 Calculate the two pitch circle radii, R_2 and R_3, as in Example 7.1 and draw the two pitch circles tangent to each other, identifying O_2 and O_3 as the two shaft centers (Fig. 7.8).
STEP 2 Draw the common tangent to the pitch circles perpendicular to the line of centers and through the pitch point P (Fig. 7.8). Draw the *line of action* at an angle equal to the *pressure angle* $\phi = 20°$ from the common tangent. This line of action corresponds to the generating line discussed in Section 7.3; it is always normal to the involute curves and always passes through the pitch point. It is called the line of action or the pressure line because, assuming no friction, the resultant tooth force acts along this line.
STEP 3 Through the centers of the two gears, draw the two perpendiculars to the line of action O_2A and O_3B (Fig. 7.8). Draw the two *base circles* with radii of $r_2 = O_2A$ and $r_3 = O_3B$; these correspond to the base cylinders of Section 7.3.
STEP 4 From Table 7.2, continuing with $P = 10$ teeth/in., the addendum for both of the gears is found to be

$$a = \frac{1}{P} = \frac{1}{10 \text{ teeth/in}} = 0.10 \text{ in.}$$

Adding this to each of the pitch circle radii, draw the two addendum circles that define the top lands of the teeth on each gear. Carefully identify and label point C where the addendum circle of gear 3 intersects the line of action (Fig. 7.9). Similarly, identify and label point D where the addendum circle of gear 2 intersects the line of action. Visualizing the rotation

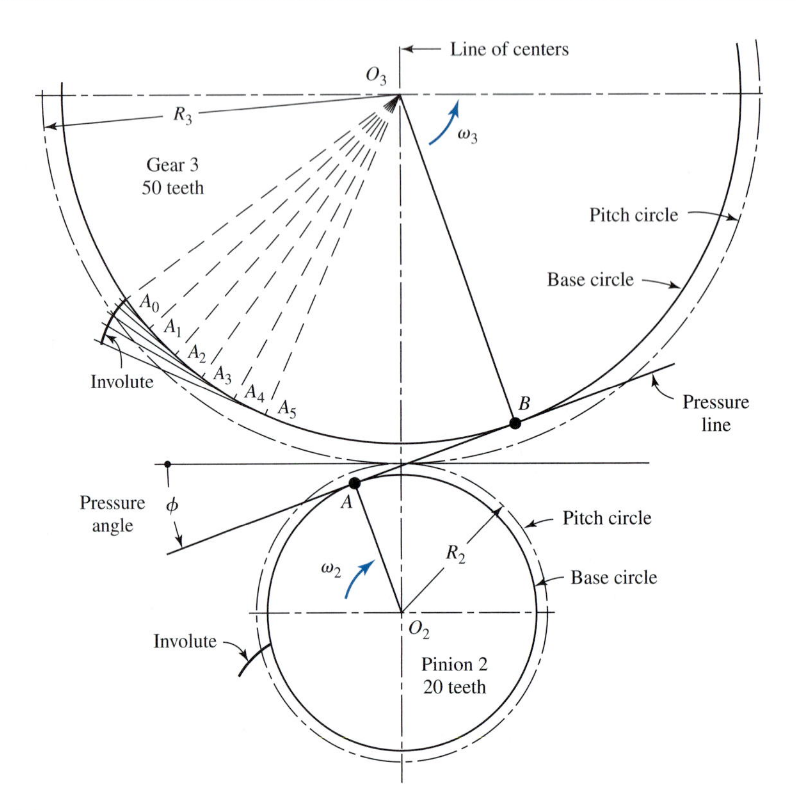

Figure 7.8 Partial gear pair layout.

of the two gears in the directions given, we see that contact is not possible before point C because the teeth of gear 3 are not of sufficient height; thus, C is the first point of contact between the teeth. Similarly, the teeth of gear 2 are too short to allow further contact after reaching point D; thus, contact between one or more pairs of mating teeth continues between C and D and then ceases.

Steps 1 through 4 are critical for verifying the choice of any gear pair. We will continue with the diagram illustrated in Fig. 7.9 when we check for interference, undercutting, and contact ratio in later sections. However, to complete our visualization of gear tooth action, let us first proceed to the construction of the complete involute tooth shapes as illustrated in Fig. 7.8.

STEP 5 From Table 7.2, the dedendum for each gear is found to be

$$d = \frac{1.25}{P} = \frac{1.25}{10 \text{ teeth/in}} = 0.125 \text{ in.}$$

Subtracting this from each of the pitch circle radii, draw the two dedendum circles that define the bottom lands of the teeth on each gear (Fig. 7.9). Note that the dedendum circles often lie quite close to the base circles; however, they have distinctly different meanings. In this example, the dedendum circle of gear 3 is larger than its base circle and the dedendum circle of the pinion 2 is smaller than its base circle. However, this is not always the case.

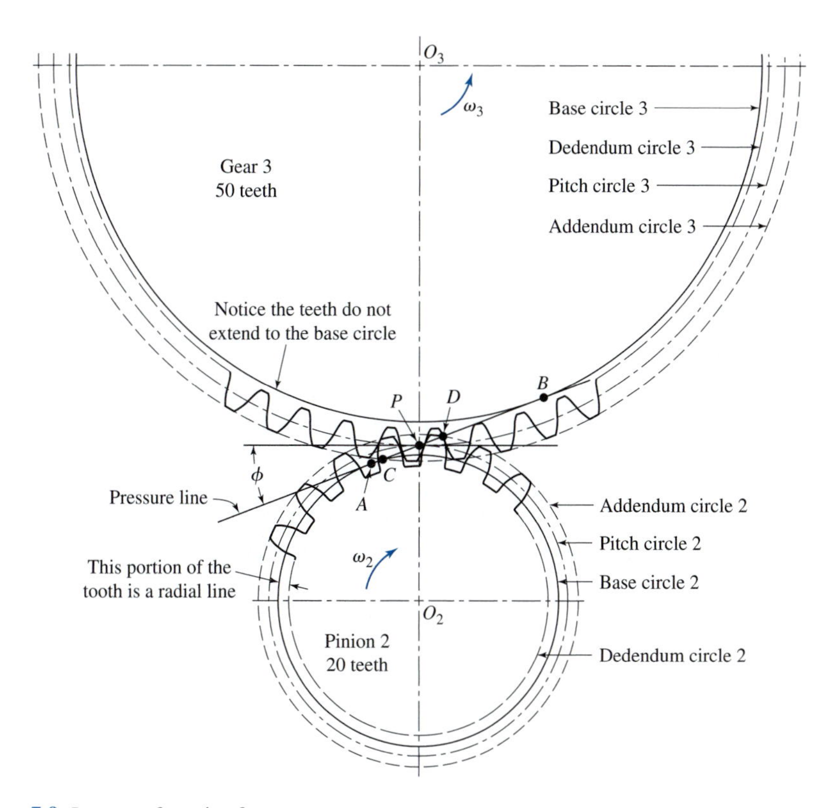

Figure 7.9 Layout of a pair of spur gears.

STEP 6 Generate an involute curve on each base circle as illustrated for gear 3 in Fig. 7.8. This is done by first dividing a portion of the base circle into a series of equal small parts, A_0, A_1, A_2, and so on. Next the radial lines O_3A_0, O_3A_1, O_3A_2, and so on are constructed, and tangents to the base circle are drawn perpendicular to each of these. The involute begins at A_0. The second point is obtained by striking an arc, with center A_1 and radius A_0A_1, up to the tangent line through A_1. The next point is found by striking a similar arc with center at A_2 and so on. This construction is continued until the involute curve is generated far enough to meet the addendum circle of gear 3. If the dedendum circle lies inside of the base circle, as is true for pinion 2 of this example, then, except for the fillet, the curve is extended inward to the dedendum circle by a radial line; this portion of the curve is not involute.

STEP 7 Using cardboard or preferably a sheet of clear plastic, cut a template for the involute curve and mark on it the center point of the corresponding gear. Note that two templates are needed because the involute curves are different for gears 2 and 3.

STEP 8 Calculate the circular pitch using Eq. (7.4).

$$p = \frac{\pi}{P} = \frac{\pi}{10 \text{ teeth/in}} = 0.314\,16 \text{ in/tooth}.$$

This distance from one tooth to the next is now marked along the pitch circle and the template is used to draw the involute portion of each tooth (Fig. 7.9). The width of a tooth and that of a tooth space are each equal to half of the circular pitch or (0.314 16 in/tooth)/2 = 0.157 08 in/tooth. These distances are marked along the pitch circle, and the same template is turned over and used to draw the opposite sides of the teeth. The portion of the tooth space between the clearance and the dedendum may be used for a fillet radius. The top and bottom lands are now drawn as circular arcs along the addendum and dedendum circles to complete the tooth profiles. The same process is performed on the other gear using the other template.

Remember that steps 5 through 8 are not necessary for the proper design of a gear set. They are only included here to help us visualize the relation between real tooth shapes and the theoretical properties of the involute curve.

Involute Rack. We may imagine a *rack* as a spur gear having an infinitely large pitch diameter. Therefore, in theory, a rack has an infinite number of teeth and its base circle is located an infinite distance from the pitch point. For involute teeth, the curves on the sides of the teeth of a rack become straight lines making an angle with the line of centers equal to the pressure angle. The addendum and dedendum distances are the same as those given in Table 7.2. Figure 7.10 illustrates an involute rack in mesh with the pinion of the previous example.

Base Pitch. Corresponding sides of involute teeth are parallel curves. The *base pitch* is the constant and fundamental distance between these curves, that is, the distance from one tooth to the next, measured along the common normal to the tooth profiles, which is the line of action (Fig. 7.10). The base pitch p_b and the circular pitch p are related as follows:

$$p_b = p \cos \phi. \tag{7.6}$$

The base pitch is a much more fundamental measurement, as we will see below.

Internal Gear. Figure 7.11 depicts the pinion of the preceding example in mesh with an *internal*, or *annular*, gear. With internal contact, both centers are on the same side of the pitch point. Thus, the positions of the addendum and dedendum circles of an internal gear are reversed with respect to the pitch circle; the addendum circle of the internal gear lies *inside* the pitch circle, whereas the dedendum circle lies *outside* the pitch circle. The base

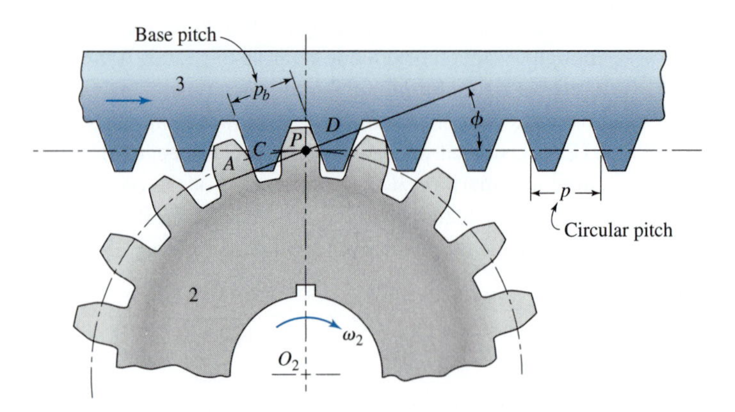

Figure 7.10 Involute pinion and rack.

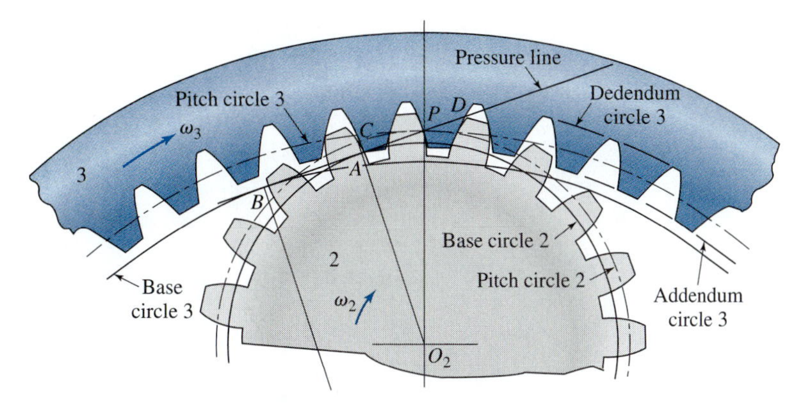

Figure 7.11 Involute pinion and internal gear.

circle lies inside the pitch circle as with an external gear, but is now near the addendum circle. Otherwise, Fig. 7.11 is constructed in the same manner as Fig. 7.9.

7.6 THE MANUFACTURE OF GEAR TEETH

There are many ways of manufacturing the teeth of gears; for example, they can be made by sand casting, shell molding, investment casting, permanent-mold casting, die casting, or centrifugal casting. They can be formed by the powder-metallurgy process, or a single bar of aluminum can be formed by extrusion and then sliced into gears. Gears that carry large loads in comparison with their sizes are usually made of steel and are cut with either *form cutters* or *generating cutters*. In form cutting, the cutter is of the exact shape of the tooth space. With generating cutters, a tool having a shape different from the tooth space is moved through several cuts relative to the gear blank to obtain the proper shape for the teeth.

Probably the oldest method of cutting gear teeth is *milling*. A form milling cutter corresponding to the shape of the tooth space, such as that illustrated in Fig. 7.12*a*, is used to machine one tooth space at a time, as illustrated in Fig. 7.12*b*, after which the gear is indexed through one circular pitch to the next position. Theoretically, with this method, a different cutter is required for each gear to be cut because, for example, the shape of the tooth space in a 25-tooth gear is different from the shape of the tooth space in, say, a 24-tooth gear. Actually, the change in tooth space shape is not very large, and eight form cutters can be used to cut any gear in the range from 12 teeth to a rack with reasonable accuracy. Of course, a separate set of form cutters is required for each pitch.

Shaping is a highly favored method of *generating* gear teeth. The cutting tool may be either a rack cutter or a pinion cutter. The operation is explained by reference to Fig. 7.13. For shaping, the reciprocating cutter is first fed into the gear blank until the pitch circles are tangent. Then, after each cutting stroke, the gear blank and the cutter roll slightly on their pitch circles. When the blank and cutter have rolled by a total distance equal to the circular pitch, one tooth has been generated and the cutting continues with the next tooth until all teeth have been cut. Shaping of an internal gear with a pinion cutter is illustrated in Fig. 7.14.

Hobbing is another method of generating gear teeth, which is quite similar to shaping them with a rack cutter. However, hobbing is done with a special tool called a *hob,* a

Figure 7.12 Manufacture of gear teeth by a form cutter. (a) A single-tooth involute hob. (b) Machining of a single tooth space. (Courtesy of Gleason Works, Rochester, NY.)

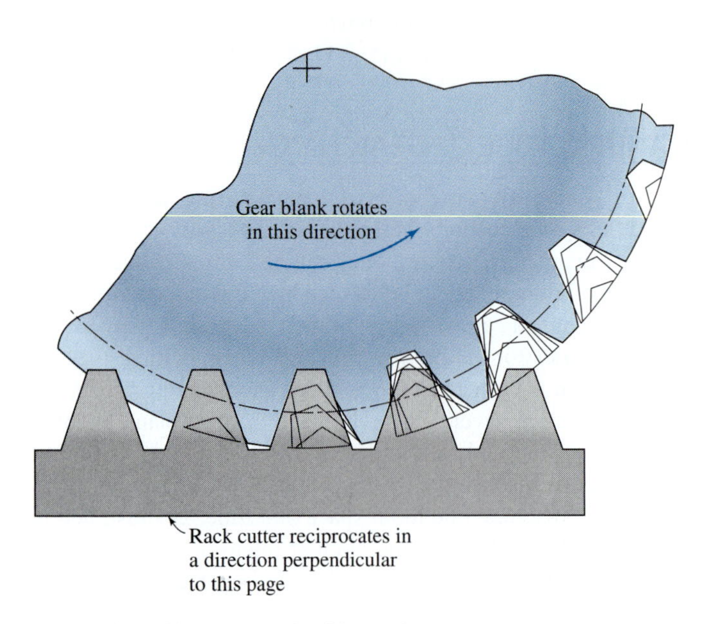

 **Figure 7.13** Shaping of involute teeth with a rack cutter.

cylindrical cutter with one or more helical threads quite like a screw-thread tap; the threads have straight sides like a rack. A number of different gear hobs are displayed in Fig. 7.15. A view of the hobbing of a gear is illustrated in Fig. 7.16. The hob and the gear blank are both rotated continuously at the proper angular velocity ratio, and the hob is fed slowly across the face of the blank to cut the full thickness of the teeth.

Following the cutting process, grinding, lapping, shaving, and burnishing are often used as final finishing processes when tooth profiles of very good accuracy and surface finish are desired.

Figure 7.14 Shaping of an internal gear with a pinion cutter. (Courtesy of Gleason Works, Rochester, NY.)

Figure 7.15 A variety of involute gear hobs. (Courtesy of Gleason Works, Rochester, NY.)

Figure 7.16 The hobbing of a gear. (Courtesy of Gleason Works, Rochester, NY.)

7.7 INTERFERENCE AND UNDERCUTTING

Figure 7.17 illustrates the pitch circles of the same gears used for discussion in Section 7.5. Let us assume that the pinion is the driver and that it is rotating clockwise.

We saw in Section 7.5 that for involute teeth, contact always takes place along the line of action AB. Contact begins at point C where the addendum circle of the driven gear crosses the line of action. Thus, initial contact is on the tip of the driven gear tooth and on the flank of the pinion tooth.

As the pinion tooth drives the gear tooth, contact approaches the pitch point P. Near the pitch point, contact slides *up* the flank of the pinion tooth and *down* the face of the gear tooth. At the pitch point, contact is at the pitch circles; note that P is the instant center and therefore the motion must be rolling with no slip at that point. Note also that this is the only location where the motion can be true rolling.

As the teeth recede from the pitch point, the point of contact continues to travel in the same direction as before along the line of action. Contact continues to slide *up* the face of the pinion tooth and *down* the flank of the gear tooth. The last point of contact occurs at the tip of the pinion and the flank of the gear tooth, at the intersection D of the line of action and the addendum circle of the pinion.

The *approach* phase of the motion is the period between the initial contact at point C and the pitch point P. The *angles of approach* are the angles through which the two gears rotate as the point of contact progresses from C to P. However, reflecting on the unwrapping

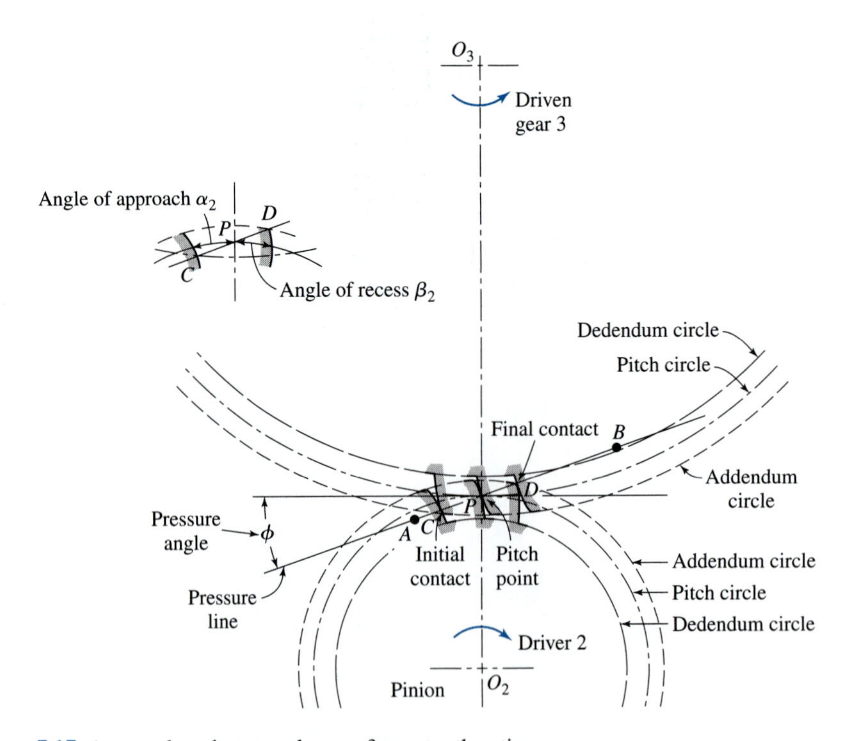

Figure 7.17 Approach and recess phases of gear tooth action.

cord analogy of Fig. 7.6, we see that the distance CP is equal to a length of cord unwrapped from the base circle of the pinion during the approach phase of the motion. Similarly, an equal amount of cord has wrapped onto the driven gear during that same phase. Thus, the angles of approach for the pinion and the gear, in radians, are

$$\alpha_2 = \frac{CP}{r_2} \quad \text{and} \quad \alpha_3 = \frac{CP}{r_3}. \tag{7.7}$$

The *recess* phase of the motion is the period during which contact progresses from the pitch point P to final contact at point D. The *angles of recess* are the angles through which the two gears rotate as the point of contact progresses from P to D. Again, from the unwrapping cord analogy, we find these angles, in radians, to be

$$\beta_2 = \frac{PD}{r_2} \quad \text{and} \quad \beta_3 = \frac{PD}{r_3}. \tag{7.8}$$

If the teeth come into contact such that they are not conjugate, this is called *interference*. Consider Fig. 7.18; illustrated here are two 16-tooth $14\frac{1}{2}^\circ$ pressure angle gears* with

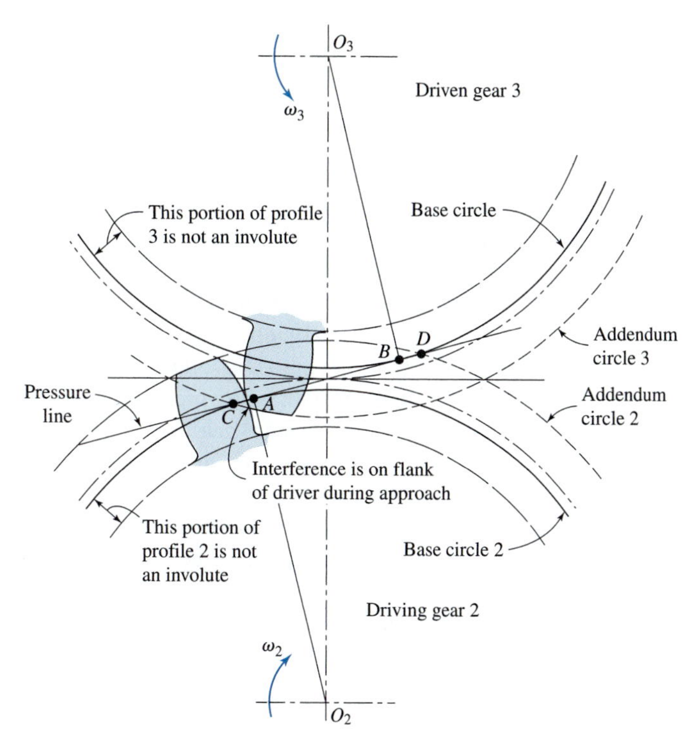

Figure 7.18 Interference in gear tooth action.

* Such gears were part of an older standard and are now obsolete. They are chosen here only to illustrate an example of interference.

full-depth involute teeth. The driver, gear 2, turns clockwise. As with previous figures, the points labeled A and B indicate the points of tangency of the line of action with the two base circles, whereas the points labeled C and D indicate the initial and final points of contact. Note that the points C and D are now *outside* of points A and B. This indicates interference.

The interference is explained as follows. Contact begins when the tip of the driven gear 3 contacts the flank of the driving tooth. In this case the flank of the driving tooth first tries to make contact with the driven tooth at point C, and this occurs *before* the involute portion of the driving tooth comes within range. In other words, contact occurs before the two teeth become tangent. The actual effect is that the nontangent tip of the driven gear interferes with and digs into the flank of the driver.

In this example a similar effect occurs again as the teeth leave contact. Contact should end at or before point B. Because for this example contact does not end until point D, the effect is for the nontangent tip of the driving tooth to interfere with and dig into the flank of the driven tooth.

When gear teeth are produced by a generating process, interference is automatically eliminated because the cutting tool removes the interfering portion of the flank. This effect is called *undercutting*. If undercutting is at all pronounced, the undercut tooth can be considerably weakened. Thus, the effect of eliminating interference by a generation process is merely to substitute another problem for the original.

The importance of the problem of teeth that have been weakened by undercutting cannot be emphasized too strongly. Of course, interference can be eliminated by using more teeth on the gears. However, if the gears are to transmit a given amount of power, more teeth can be used only by increasing the pitch diameter. This makes the gears larger, which is seldom desirable. It also increases the pitch-line velocity, which makes the gears noisier and somewhat reduces the power transmission, although not in direct proportion. In general, however, the use of more teeth to eliminate interference or undercutting is seldom an acceptable solution.

Another method of reducing interference and the resulting undercutting is to employ a larger pressure angle. The larger pressure angle creates smaller base circles, so that a greater portion of the tooth profile has an involute shape. In effect, this means that fewer teeth can be used; as a result, gears with larger pressure angle are often smaller.

Of course, the use of standard gears is far less expensive than manufacturing specially made nonstandard gears. However, as indicated in Table 7.2, gears with larger pressure angles can be found without deviating from the standards.

One more way to eliminate interference is to use gears with shorter teeth. If the addendum distance is reduced, then points C and D move inward. One way to do this is to purchase standard gears and then grind the tops of the teeth to a new addendum distance. This, of course, makes the gears nonstandard and causes concern about repair or replacement, but it can be effective in eliminating interference. Again, careful study of Table 7.2 indicates that this is possible by use of standard 20° *stub tooth* gears.

7.8 CONTACT RATIO

The zone of action of meshing gear teeth is illustrated in Fig. 7.19, where tooth contact begins and ends at the intersections of the two addendum circles with the line of action. As always, initial contact occurs at C and final contact at D. Tooth profiles drawn through

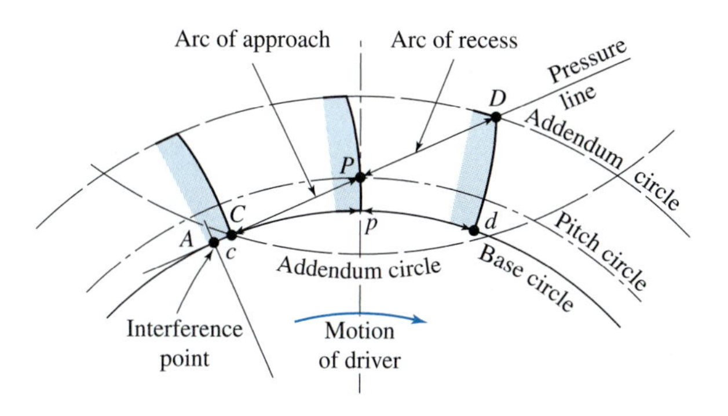

Figure 7.19

these points intersect the base circle at points c and d. Thinking back to our analogy of the unwrapping cord of Fig. 7.6, the linear distance CD, measured along the line of action, is equal to the arc length cd, measured along the base circle.

Consider a situation in which the arc length cd, or distance CD, is exactly equal to the base pitch p_b of Eq. (7.6). This means that one tooth and its space spans the entire arc cd. In other words, when a tooth is just beginning contact at C, the tooth ahead of it is just ending its contact at D. Therefore, during the tooth action from C to D there is exactly one pair of teeth in contact.

Next, consider a situation for which the arc length cd, or distance CD, is greater than the base pitch, but not much greater, say $cd = 1.1\, p_b$. This means that when one pair of teeth is just entering contact at C, the previous pair, already in contact, has not yet reached D. Thus, for a short time, there are two pairs of teeth in contact, one in the vicinity of C and the other nearing D. As meshing proceeds, the previous pair reaches D and ceases contact, leaving only a single pair of teeth in contact again, until the situation repeats itself with the next pair of teeth.

Because of the nature of this tooth action, with one, two, or even more pairs of teeth in contact simultaneously, it is convenient to define the term contact ratio m_c as

$$m_c = \frac{CD}{p_b}. \tag{7.9}$$

This is a value for which the next lower integer indicates the average number of pairs of teeth in contact. Thus, a contact ratio of $m_c = 1.35$, for example, implies that there is always at least one tooth in contact and there are two teeth in contact 35% of the time.

The minimum acceptable value of the contact ratio for smooth operation of meshing gears is $1.2 \leq m_c \leq 1.4$ and the recommended range of the contact ratio for most spur gearsets is $m_c > 1.4$.

The distance CD is quite convenient to measure if we are working graphically by making a drawing like Fig. 7.20 or Fig. 7.9. However, the distances CP and PD can also be determined analytically. From triangles O_3BC and O_3BP we can write

$$CP = \sqrt{(R_3 + a)^2 - (R_3 \cos\phi)^2} - R_3 \sin\phi. \tag{7.10}$$

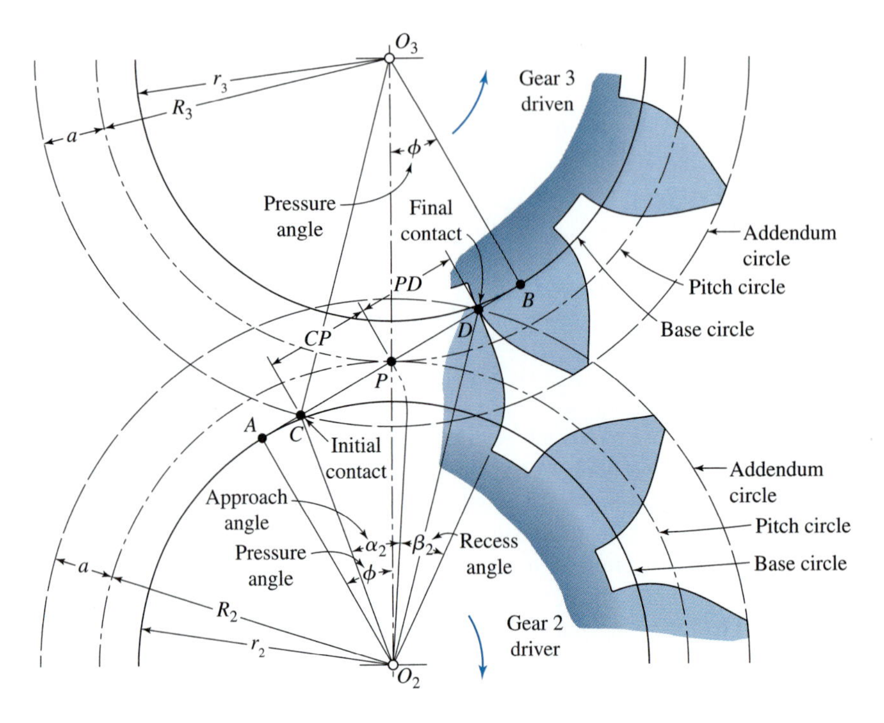

Figure 7.20

Similarly, from triangles O_2AD and O_2AP we have

$$PD = \sqrt{(R_2 + a)^2 - (R_2 \cos\phi)^2} - R_2 \sin\phi. \tag{7.11}$$

The contact ratio is then obtained by substituting the sum of Eqs. (7.10) and (7.11) into Eq. (7.9).

We should note, however, that Eqs. (7.10) and (7.11) are only valid for the conditions where

$$CP \leq R_2 \sin\phi \quad \text{and} \quad PD \leq R_3 \sin\phi \tag{7.12}$$

because proper contact cannot begin before point A or end after point B. If either of these inequalities is not satisfied, then the gear teeth have interference and undercutting results.

7.9 VARYING THE CENTER DISTANCE

Figure 7.21*a* illustrates a pair of meshing gears having 20° full-depth involute teeth. Because both sides of the teeth are in contact, the center distance $R_{O_3O_2}$ cannot be reduced without jamming or deforming the teeth. However, Fig. 7.21*b* illustrates the same pair of gears, but

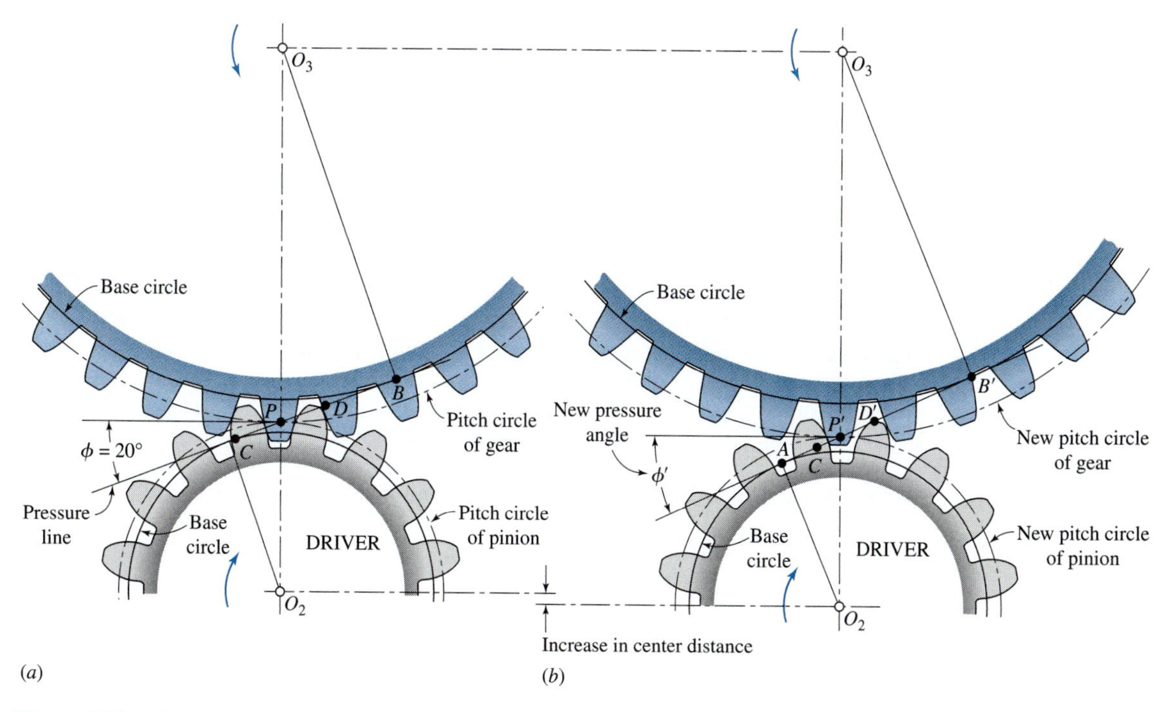

Figure 7.21 Effect of increased center distance on the action of involute gearing; mounting the gears at (*a*) normal center distance and (*b*) increased center distance.

mounted with a slightly increased distance $R_{O_3O_2}$ between the shaft centers as might happen through the accumulation of tolerances of surrounding parts. Clearance, or *backlash*, now exists between the teeth, as illustrated.

When the center distance is increased, the base circles of the two gears do not change; they are fundamental to the shapes of the gears, once manufactured. However, review of Fig. 7.6 indicates that the same involute tooth shapes still touch as conjugate curves and the fundamental law of gearing is still satisfied. However, the larger center distance results in an increase of the pressure angle and larger pitch circles passing through a new adjusted pitch point.

In Fig. 7.21*b* we can see that the triangles $O_2A'P'$ and $O_3B'P'$ are still similar to each other, although they are both modified by the change in pressure angle. Also, the distances O_2A' and O_3B' are the base circle radii and have not changed. Therefore, the ratio of the new pitch radii, O_2P' and O_3P', and the new velocity ratio remain the same as in the original design.

Another effect of increasing the center distance, observable in Fig. 7.21, is the shortening of the path of contact. The original path of contact CD in Fig. 7.21*a* is shortened to $C'D'$ in Fig. 7.21*b*. The contact ratio, Eq. (7.9), is also reduced when the path of contact $C'D'$ is shortened. Because a contact ratio of less than unity would imply periods during which no teeth would be in contact at all, the center distance must never be increased larger than that corresponding to a contact ratio of unity.

7.10 INVOLUTOMETRY

The study of the geometry of the involute curve is called *involutometry*. In Fig. 7.22 a base circle with center at O is used to generate the involute BC. AT is the generating line, ρ is the instantaneous radius of curvature of the involute, and r is the radius to point T on the curve. If we designate the radius of the base circle r_b, the line AT has the same length as the arc distance AB and so

$$\rho = r_b(\alpha + \varphi), \tag{a}$$

where α is the angle between the origin of the involute OB and the radius AT, and φ is the angle between the radius of the base circle OA and the radius OT. Because OAT is a right triangle,

$$\rho = r_b \tan\varphi. \tag{7.13}$$

Solving Eqs. (a) and (7.13) simultaneously to eliminate ρ and r_b gives

$$\alpha = \tan\varphi - \varphi,$$

which can be written

$$\text{inv}\,\varphi = \tan\varphi - \varphi. \tag{7.14}$$

and defines the *involute function*. The angle φ in this equation is the variable involute angle, given in radians. Once φ is known, inv φ can readily be determined from Eq. (7.14). The inverse problem, when inv φ is given and φ is to be found, is more difficult. One approach is to expand Eq. (7.14) in an infinite series and to employ the first several terms to obtain a numerical approximation. Another approach is to use a root-finding technique.* Here, we

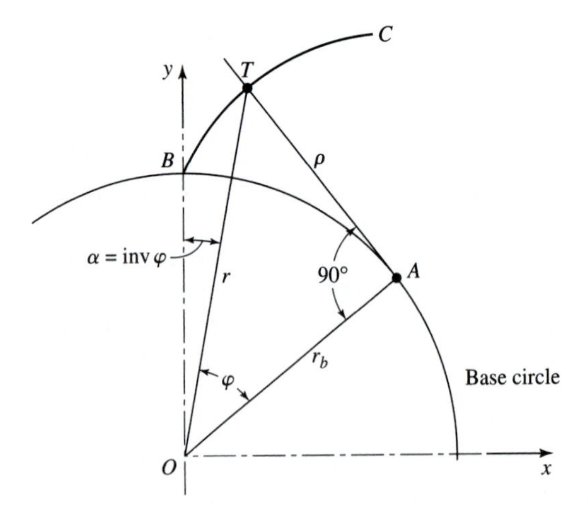

Figure 7.22

* See, for example, C. R. Mischke, *Mathematical Model Building*. Ames, IA: Iowa State University Press, 1980, chap. 4.

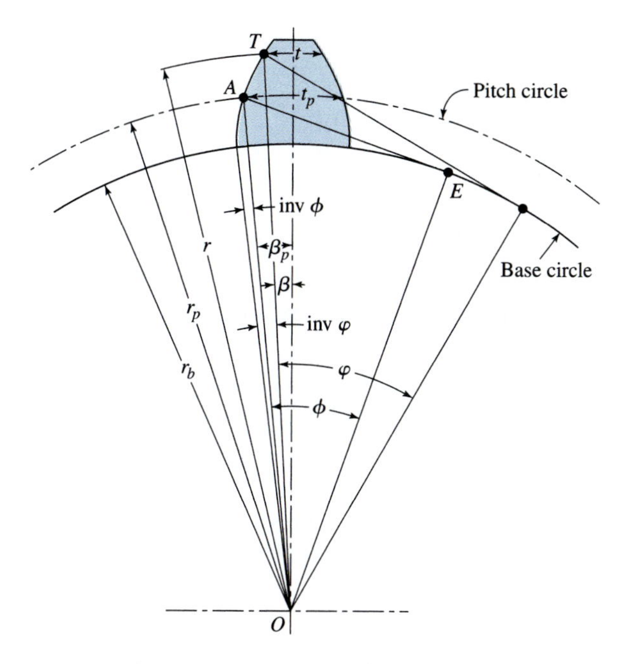

Figure 7.23

refer to Table 6 of Appendix A, in which the value of the involute function is tabulated and the angle φ can be determined directly, in degrees.

Referring again to Fig. 7.22, we see that

$$r = \frac{r_b}{\cos\varphi}. \tag{7.15}$$

To illustrate the use of the relations obtained above, let us determine the tooth dimensions of Fig. 7.23. Here, only the portion of the tooth extending above the base circle has been drawn, and the thickness of the tooth t_p at the pitch circle (point A), equal to half of the circular pitch, is given. The problem is to determine the tooth thickness at some other point, say point T. The various quantities illustrated in Fig. 7.23 are identified as follows:

r_b = radius of the base circle;
r_p = radius of the pitch circle;
r = radius at which the tooth thickness is to be determined;
t_p = tooth thickness at the pitch circle;
t = tooth thickness to be determined;
ϕ = pressure angle corresponding to the pitch circle radius r_p;
φ = involute angle corresponding to point T;
β_p = angular half-tooth thickness at the pitch circle; and
β = angular half-tooth thickness at point T.

The half-tooth thicknesses at points A and T are

$$\frac{t_p}{2} = \beta_p r_p \quad \text{and} \quad \frac{t}{2} = \beta r \tag{b}$$

so that

$$\beta_p = \frac{t_p}{2r_p} \quad \text{and} \quad \beta = \frac{t}{2r}. \tag{c}$$

From these we can write

$$\text{inv}\,\varphi - \text{inv}\,\phi = \beta_p - \beta = \frac{t_p}{2r_p} - \frac{t}{2r}. \tag{d}$$

The tooth thickness at point T is obtained by solving Eq. (d) for t:

$$t = 2r\left(\frac{t_p}{2r_p} + \text{inv}\,\phi - \text{inv}\,\varphi\right). \tag{7.16}$$

EXAMPLE 7.2

A gear has 22 teeth cut full-depth with pressure angle $\phi = 20°$, and a diametral pitch $P = 2$ teeth/in. Find the thickness of the teeth at the base circle and at the addendum circle.

SOLUTION

By the equations of Section 7.1 and Table 7.2 we find the radius of the pitch circle $r_P = 5.500$ in., the circular pitch $p = 1.571$ in/tooth, the addendum $a = 0.500$ in, and the dedendum $d = 0.625$ in.

From the right-angled triangle OEA in Fig. 7.23, the radius of the base circle can be written as

$$r_b = r_P \cos\phi = (5.500 \text{ in}) \cos 20° = 5.168 \text{ in}.$$

The thickness of the tooth at the pitch circle is

$$t_p = \frac{p}{2} = \frac{1.571 \text{ in/tooth}}{2} = 0.785\,5 \text{ in}.$$

Converting the tooth pressure angle into radians gives $\phi = 20° = 0.349$ rad. Then the involute function from Eq. (7.14) is

$$\text{inv}\,\phi = \tan 0.349 - 0.349 = 0.014\,9 \text{ rad}.$$

The involute angle at the base circle, from Eq. (7.15), is $\varphi_b = 0$. Therefore, the involute function is

$$\text{inv}\,\varphi_b = 0.$$

Substituting these results into Eq. (7.16), the tooth thickness at the base circle is

$$t_b = 2r_b\left[\frac{t_p}{2r_p} + \text{inv}\,\phi - \text{inv}\varphi_b\right] = 2(5.168 \text{ in})\left[\frac{0.785\,5 \text{ in}}{2(5.500 \text{ in})} + 0.014\,9 - 0\right] = 0.892 \text{ in}.$$

Ans.

The radius of the addendum circle is $r_a = r_P + a = 5.500 + 0.500 = 6.000$ in. Therefore, the involute angle corresponding to this radius, from Eq. (7.15), is

$$\varphi = \cos^{-1}\left(\frac{r_b}{r}\right) = \cos^{-1}\left(\frac{5.168 \text{ in}}{6.000 \text{ in}}\right) = 30.53° = 0.533 \text{ rad.}$$

Thus, the involute function is

$$\text{inv}\,\varphi = \tan 0.533 - 0.533 = 0.056\,9 \text{ rad.}$$

Substituting these results into Eq. (7.16), the tooth thickness at the addendum circle is

$$t_a = 2r_a\left[\frac{t_P}{2r_P} + \text{inv}\phi - \text{inv}\varphi\right] = 2(6.000 \text{ in})\left[\frac{0.785\,5 \text{ in}}{2(5.500 \text{ in})} + 0.014\,9 - 0.056\,9\right]$$

$$= 0.353 \text{ in.} \qquad \textit{Ans.}$$

Note that the tooth thickness at the base circle is more than double the tooth thickness at the addendum circle.

7.11 NONSTANDARD GEAR TEETH

In this section we will examine the effects obtained by deviating from the specified standards and modifying such things as pressure angle, tooth depth, addendum, or center distance. Some of these modifications do not eliminate interchangeability; all of them are discussed with the intent of obtaining improved performance. Still, making such modifications probably means increased cost because modified gears will not be available and will need to be specially machined for the particular application. Of course, this will also be necessary at the time of any future repair or design modification.

The designer is often under great pressure to produce a design using gears that is small and yet will transmit a large amount of power. Consider, for example, a gearset that must have a 4:1 velocity ratio. If the smallest pinion that will carry the load has a pitch diameter of 2 in, the mating gear will have a pitch diameter of 8 in, making the overall space required for the two gears more than 10 in. On the other hand, if the pitch diameter of the pinion can be reduced by only $\frac{1}{4}$ in, the pitch diameter of the gear is reduced by a full 1 in and the overall size of the gearset is reduced by $1\frac{1}{4}$ in. This reduction assumes considerable importance when it is realized that the associated machine elements, such as shafts, bearings, and enclosure, are also reduced in size.

If a tooth of a certain pitch is required to carry the load, the only method of decreasing the pinion diameter is to use fewer teeth. However, we have already seen that problems involving interference, undercutting, and contact ratio are encountered when the tooth numbers are made too small. Thus, three principal reasons for employing nonstandard gears are to: (*i*) eliminate undercutting, (*ii*) prevent interference, and (*iii*) maintain a reasonable contact ratio. It should be noted too that if a pair of gears are manufactured of the same material, the pinion is the weaker and is subject to greater wear because each of its teeth is in contact a greater portion of the time. Therefore, any undercutting weakens the tooth that is already weaker. Thus, another objective of nonstandard gears is to gain a better balance of strength between the pinion and the gear.

As an involute curve is generated from its base circle, its radius of curvature becomes larger and larger. Near the base circle the radius of curvature is quite small, being theoretically zero at the base circle. Contact near this region of sharp curvature should be avoided if possible because of the difficulty in obtaining good cutting accuracy in areas of small curvature and because contact stresses are likely to be very high. Nonstandard gears present the opportunity of designing to avoid these sensitive areas.

Clearance Modification. A larger fillet radius at the root of the tooth increases the fatigue strength of the tooth and provides extra depth for shaving the tooth profile. Because interchangeability is not lost, the dedendum is sometimes increased to $1.300/P$ or $1.400/P$ to obtain space for a larger fillet radius.

Center-Distance Modification. When gears of low tooth numbers are paired with each other, or with larger gears, reduction in interference and improvement in the contact ratio can be obtained by increasing the center distance to greater than standard. Although such a system changes the tooth proportions and the pressure angle of the gears, the resulting tooth shapes can be generated with rack cutters (or hobs) of standard pressure angles or with standard pinion shapers. Before introducing this system, however, it will be of value to develop certain additional relations about the geometry of gears.

The first new relation is for finding the thickness of a tooth that is cut by a rack cutter (or hob) when the pitch line of the rack cutter is displaced or offset a distance e from the pitch circle of the gear being cut. What we are doing here is moving the rack cutter away from the center of the gear being cut. Stated another way, suppose the rack cutter does not cut as deeply into the gear blank and the teeth are not cut to full depth. This produces teeth that are thicker than the standard, and this thickness will now be determined. Figure 7.24*a* illustrates the problem, and Fig. 7.24*b* illustrates the solution. The increase of tooth thickness at the pitch circle is $2e \tan \phi$, so that

$$t = 2e \tan \phi + \frac{p}{2}, \tag{7.17}$$

where ϕ is the pressure angle of the rack cutter and t is the thickness of the modified gear tooth measured on its pitch circle.

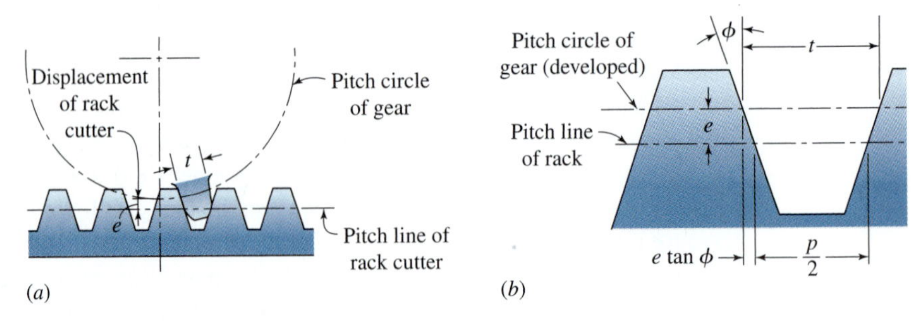

Figure 7.24

Now suppose that two gears of different tooth numbers have both been cut with the cutter offset from their pitch circles as in the previous paragraph. Because the teeth of both have been cut with offset cutters, they will mate at a modified pressure angle and with modified pitch circles and consequently modified center distances. The word modified is used here in the sense of being nonstandard. Our problem is to determine the radii of these modified pitch circles and the value of the modified pressure angle.

In the following notation, the word *standard* refers to values that would have been obtained had the usual, or standard, systems been employed to obtain the dimensions:

ϕ = pressure angle of generating rack cutter;
ϕ' = modified pressure angle at which gears will mate;
R_2 = standard pitch radius of pinion;
R'_2 = modified pitch radius of pinion when meshing with given gear;
R_3 = standard pitch radius of gear;
R'_3 = modified pitch radius of gear when meshing with given pinion;
t_2 = thickness of pinion tooth at standard pitch radius R_2;
t'_2 = thickness of pinion tooth at modified pitch radius R'_2;
t_3 = thickness of gear tooth at standard pitch radius R_3; and
t'_3 = thickness of gear tooth at modified pitch radius R'_3.

From Eq. (7.16), the thickness of a gear tooth at the standard pitch radius and at the modified pitch radius can be written, respectively, as

$$t'_2 = 2R'_2\left(\frac{t_2}{2R_2} + \text{inv}\,\phi - \text{inv}\,\phi'\right) \tag{a}$$

and

$$t'_3 = 2R'_3\left(\frac{t_3}{2R_3} + \text{inv}\,\phi - \text{inv}\,\phi'\right). \tag{b}$$

Note that the sum of these two thicknesses must be the new circular pitch. Therefore, using Eq. (7.3), we can write

$$t'_2 + t'_3 = p' = \frac{2\pi R'_2}{N_2}. \tag{c}$$

Since the pitch diameters of a pair of mating gears are proportional to their tooth numbers then

$$R_3 = \frac{N_3}{N_2}R_2 \quad \text{and} \quad R'_3 = \frac{N_3}{N_2}R'_2. \tag{d}$$

Substituting Eqs. (a), (b), and (d) into Eq. (c) and rearranging gives

$$\text{inv}\,\phi' = \frac{N_2\left(t'_2 + t'_3\right) - 2\pi R_2}{2R_2\left(N_2 + N_3\right)} + \text{inv}\,\phi. \tag{7.18}$$

This equation gives the modified pressure angle ϕ' at which a pair of gears will operate when the tooth thicknesses on their standard pitch circles are modified to t'_2 and t'_3.

Although the base circle of a gear is fundamental to its shape and fixed once the gear is generated, gears have no pitch circles until a pair of them is brought into contact. Bringing a pair of gears into contact creates a pair of pitch circles that are tangent to each other at the modified pitch point. Throughout this discussion, the idea of a pair of so-called standard pitch circles has been used to define a certain point on the involute curves. These standard pitch circles, as we have seen, are the ones that would come into existence when the gears are paired *if the gears are not modified from the standard dimensions*. On the other hand, the base circles are fixed circles that are not changed by tooth modifications. The base circle remains the same whether the tooth dimensions are changed or not, so we can determine the base circle radius using either the standard pitch circle or the new pitch circle. Thus, from Eq. (7.15), we can write

$$R_2 \cos\varphi = R_2' \cos\varphi'$$

Therefore, the modified pitch radius of the pinion can be written as

$$R_2' = \frac{R_2 \cos\phi}{\cos\phi'}. \tag{7.19}$$

Similarly, the modified pitch radius of the gear can be written as

$$R_3' = \frac{R_3 \cos\phi}{\cos\phi'}. \tag{7.20}$$

Equations (7.19) and (7.20) give the values of the actual pitch radii when the two gears with modified teeth are brought into mesh without backlash. The new center distance is, of course, the sum of these two radii.

All necessary relations have now been developed to create nonstandard gears with changes in the center distance. The use of these relations is now illustrated by an example.

Figure 7.25 is a drawing of a 20° pressure angle, 1 tooth/in diametral pitch, 12-tooth pinion generated with a rack cutter to full depth with a standard clearance of $0.250/P$. In

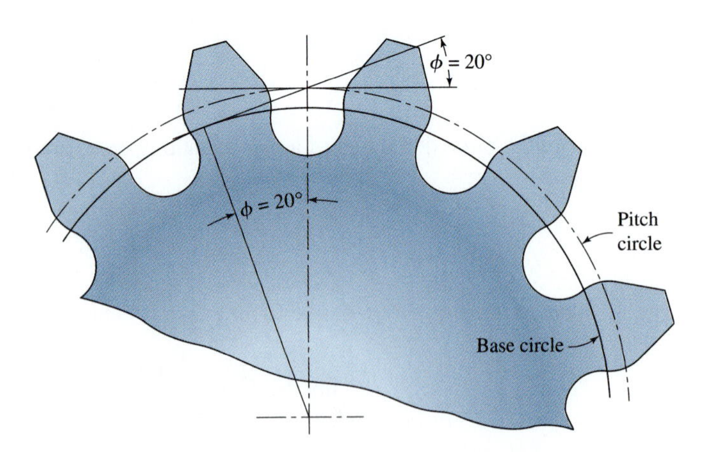

Figure 7.25 Standard 20° pressure angle, 1-tooth/in diametral pitch, 12-tooth full-depth involute gear showing undercut.

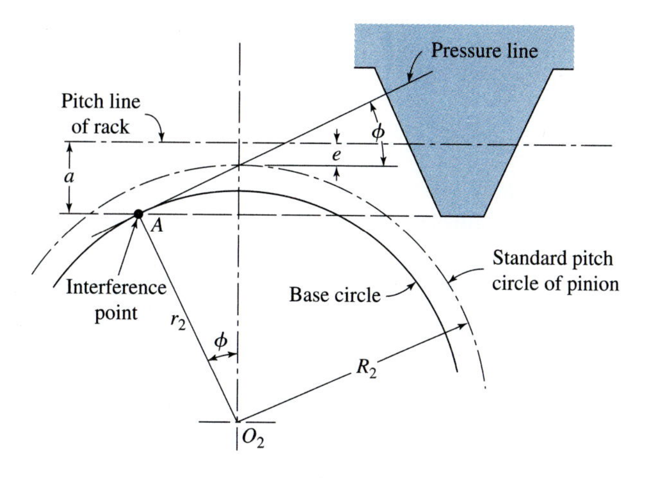

Figure 7.26 Offset of a rack cutter to cause its addendum line to pass through the interference point.

the 20° full-depth system, interference is severe when the number of teeth is less than 14. The resulting undercutting is evident in the drawing.

In an attempt to eliminate undercutting, improve the tooth action, and increase the contact ratio, suppose that this pinion were not cut to full depth; suppose instead that the rack cutter were only allowed to cut to a depth for which its addendum passes through the interference point A of the pinion being cut—that is, the point of tangency of the 20° line of action and the base circle—as illustrated in Fig. 7.26. From Eq. (7.15) we know that

$$r_2 = R_2 \cos\phi. \tag{e}$$

Then, from Fig. 7.26, the depth of the cut would be offset from the standard by

$$e = a + r_2 \cos\phi - R_2. \tag{f}$$

Substituting Eq. (e) into Eq. (f), the offset can be written as

$$e = a + R_2 \cos^2\phi - R_2 = a - R_2 \sin^2\phi. \tag{7.21}$$

If the offset is any less than this, then the rack will cut below the interference point A and will result in undercutting.

EXAMPLE 7.3

A 12-tooth pinion with pressure angle $\varphi = 20°$ and diametral pitch $P = 1$ tooth/in is to be mated with a standard 40-tooth gear. If the pinion were cut to full depth, then Eq. (7.9). demonstrates that the contact ratio would be 1.41, but there would be undercutting as indicated in Fig. 7.25. Instead, let the 12-tooth pinion be cut from a larger blank using center-distance modifications. Determine the cutter offset, the modified pressure angle, the modified pitch radii of the pinion and gear, the modified center distance, the modified outside radii of the pinion and gear, and the contact ratio. Has the contact ratio increased significantly?

SOLUTION

Designating the pinion as subscript 2 and the gear as 3, then with $P = 1$ tooth/in and $\phi = 20°$, the following values are determined:

$$p = 3.142 \text{ in/tooth}, \; R_2 = 6 \text{ in}, \; R_3 = 20 \text{ in}, \; N_2 = 12 \text{ teeth},$$
$$N_3 = 40 \text{ teeth}, \quad \text{and} \quad t_3 = 1.571 \text{ in}.$$

For a standard rack cutter, from Table 7.2, the addendum is $a = 1/P = 1.0$ in.

From Eq. (7.21) the rack cutter will be offset by

$$e = 1.0 - 6.0 \sin^2 20° = 0.298 \text{ in.} \qquad \textit{Ans.}$$

Then the thickness of the pinion tooth at the 6 in pitch circle, using Eq. (7.17), is

$$t_2' = 2e \tan\phi + \frac{p}{2} = 2\,(0.298 \text{ in}) \tan 20° + \frac{3.142 \text{ in}}{2} = 1.788 \text{ in.}$$

The pressure angle at which this (and only this) gearset will operate is determined from Eq. (7.18), that is

$$\begin{aligned} \text{inv}\,\phi' &= \frac{N_2\left(t'_2 + t'_3\right) - 2\pi R_2}{2R_2\,(N_2 + N_3)} + \text{inv}\phi \\ &= \frac{12(1.788 \text{ in} + 1.571 \text{ in}) - 2\pi(6.0 \text{ in})}{2(6.0 \text{ in})\,(12 + 40)} + \text{inv } 20° = 0.019\,08 \text{ rad.} \end{aligned}$$

From Appendix A, Table 6, we find that the new pressure angle is $\varphi' = 21.65°$. *Ans.*

Using Eqs. (7.19) and (7.20), the modified pitch radii are determined to be

$$R_2' = \frac{R_2 \cos\phi}{\cos\phi'} = \frac{(6.0 \text{ in}) \cos 20°}{\cos 21.65°} = 6.066 \text{ in.} \qquad \textit{Ans.}$$

$$R_3' = \frac{R_3 \cos\phi}{\cos\phi'} = \frac{(20.0 \text{ in}) \cos 20°}{\cos 21.65°} = 20.220 \text{ in.} \qquad \textit{Ans.}$$

So the modified center distance is

$$R_2' + R_3' = 6.066 + 20.220 = 26.286 \text{ in.} \qquad \textit{Ans.}$$

Note that the center distance has not increased as much as the offset of the rack cutter.

Standard clearance of $0.25/P$ results from the standard dedendums equal to $1.25/P$ as indicated in Table 7.2. So the root radii of the two gears are

$$\text{Root radius of pinion} = 6.298 - 1.250 = 5.048 \text{ in.}$$

$$\text{Root radius of gear} = 20.000 - 1.250 = 18.750 \text{ in.}$$

$$\text{Sum of root radii} = 23.798 \text{ in.}$$

The difference between this sum and the center distance is the working depth plus twice the clearance. Because the clearance is 0.25 in for each gear, the working depth is

$$\text{Working depth} = 26.286 - 23.798 - 2(0.250) = 1.988 \text{ in.}$$

The outside radius of each gear is the sum of the root radius, the clearance, and the working depth, that is

$$\text{Outside radius of pinion} = 5.048 + 0.250 + 1.988 = 7.286 \text{ in.} \qquad \textit{Ans.}$$

$$\text{Outside radius of gear} = 18.750 + 0.250 + 1.988 = 20.988 \text{ in.} \qquad \textit{Ans.}$$

The result is illustrated in Fig. 7.27, and the pinion is seen to have a stronger looking form than the one of Fig. 7.25. Undercutting has been completely eliminated.

The contact ratio can be obtained from Eqs. (7.9) through (7.11). The following quantities are needed:

$$\text{Outside radius of pinion} = R_2' + a = 7.286 \text{ in.}$$

$$\text{Outside radius of gear} = R_3' + a = 20.988 \text{ in.}$$

$$r_2 = R_2 \cos\phi = (6.000 \text{ in}) \cos 20° = 5.638 \text{ in.}$$

$$r_3 = R_3 \cos\phi = (20.000 \text{ in}) \cos 20° = 18.794 \text{ in.}$$

$$p_b = p \cos\phi = (3.141\,6 \text{ in / tooth}) \cos 20° = 2.952 \text{ in/tooth.}$$

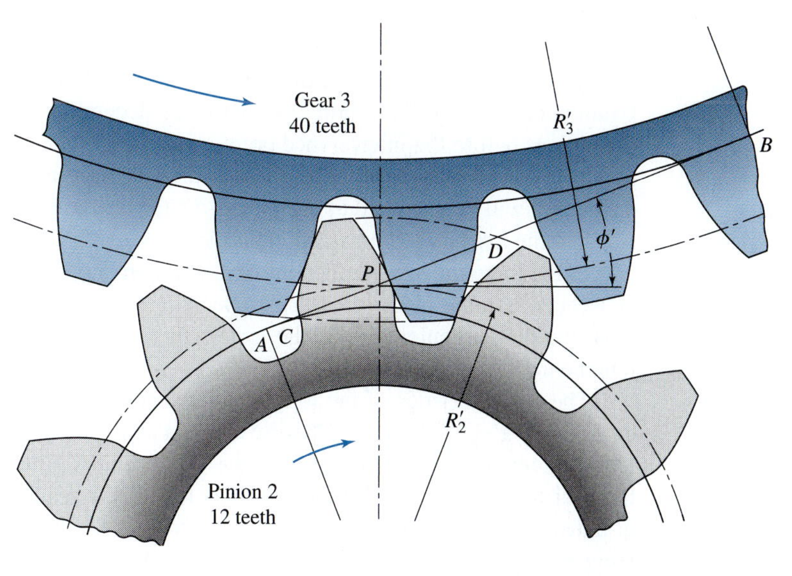

Figure 7.27

Therefore, for Eqs. (7.10) and (7.11) we have

$$
\begin{aligned}
CP &= \sqrt{(R_3' + a)^2 - r_3^2} - R_3' \sin\phi \\
&= \sqrt{(20.988\text{ in})^2 - (18.794\text{ in})^2} - (20.220\text{ in})\sin 21.65^\circ = 1.883\text{ in} \\
PD &= \sqrt{(R_2' + a)^2 - r_2^2} - R_2' \sin\phi \\
&= \sqrt{(7.286\text{ in})^2 - (5.638\text{ in})^2} - (6.066\text{ in})\sin 21.65^\circ \\
&= 2.377\text{ in}
\end{aligned}
$$

Finally, from Eq. (7.9), the contact ratio is

$$
m_c = \frac{CP + PD}{p_b} = \frac{1.883\text{ in} + 2.377\text{ in}}{2.952\text{ in/tooth}} = 1.443\text{ teeth avg.} \qquad \textit{Ans.}
$$

Therefore, the contact ratio has increased only slightly (approximately a 2% increase). The modification, however, is justified because of the elimination of undercutting, which results in a substantial improvement in the strength of the teeth.

Long-and-Short-Addendum Systems It often happens in the design of machinery that the center distance between a pair of gears is fixed by some other design consideration or feature of the machine. In such a case, modifications to obtain improved performance cannot be made by varying the center distance.

In the previous section we saw that improved action and tooth shape can be obtained by backing the rack cutter away from the gear blank during forming of the teeth. The effect of this withdrawal is to create the active tooth profile farther away from the base circle. Examination of Fig. 7.27 indicates that more dedendum could be used on the gear (not the pinion) before the interference point is reached. If the rack cutter is advanced into the gear blank by a distance equal to the withdrawal from the pinion blank, more of the gear dedendum will be used and at the same time the center distance will not be changed. This is called the *long-and-short-addendum system.*

In the long-and-short-addendum system there are no changes in the pitch circles and consequently none in the pressure angle. The effect is to move the contact region away from the pinion center toward the gear center, thus shortening the approach action and lengthening the recess action.

The characteristics of the long-and-short-addendum system can be explained by reference to Fig. 7.28. Figure 7.28*a* illustrates a conventional (standard) set of gears having a dedendum equal to the addendum plus the clearance. Interference exists, and the tip of the gear tooth will have to be relieved as illustrated or the pinion will be undercut. This is indicated because the addendum circle crosses the line of action at C, outside of the tangency or interference point A; hence, the distance AC is a measure of the degree of interference.

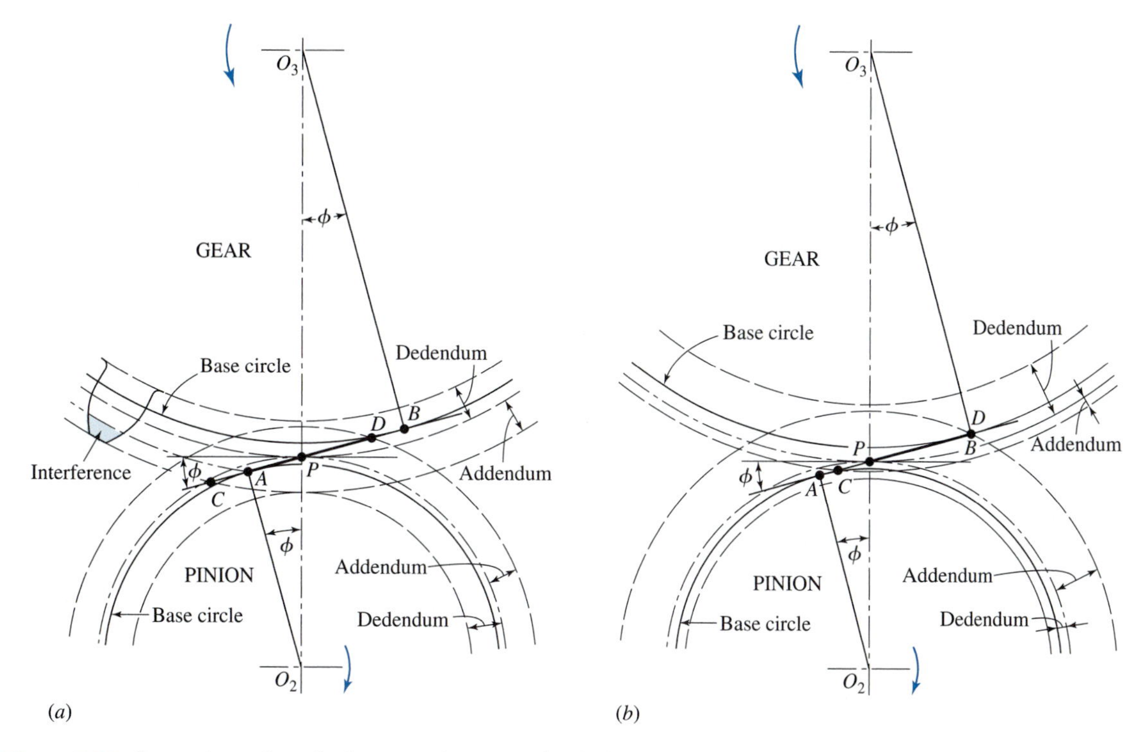

Figure 7.28 Comparison of standard gears and gears cut by the long-and-short-addendum system: (*a*) gear and pinion with standard addendum and dedendum; (*b*) gear and pinion with long-and-short addendum.

To eliminate the undercutting or interference, the pinion addendum may be enlarged, as in Fig. 7.28*b*, until the addendum circle of the pinion passes through the interference point (point *B*) of the gear. In this manner, we shall be using all of the gear–tooth profile. The same whole depth may be retained; hence, the dedendum of the pinion may be reduced by the same amount that the addendum is increased. This means that we must also lengthen the gear dedendum and shorten the dedendum of the mating pinion. With these changes the path of contact is the line *CD* of Fig. 7.28*b*. It is longer than the path *AD* of Fig. 7.28*a*, and so the contact ratio is higher. Note too that the base circles, the pitch circles, the pressure angle, and the center distance have not changed. Both gears can be cut with standard cutters by advancing the cutter into the gear blank, for this modification by a distance equal to the amount of withdrawal from the pinion blank. Finally, note that the blanks from which the pinion and gear are cut must now be of different diameters than the standard blanks.

The tooth dimensions for the long-and-short-addendum system can be determined using the equations developed in the previous sections.

A less obvious advantage of the long-and-short-addendum system is that more recess action than approach action is obtained. The approach action of gear teeth is analogous to pushing a piece of chalk across a blackboard; the chalk screeches. But when the chalk is pulled across a blackboard, analogous to the recess action, it glides smoothly. Thus, recess action is always preferable because of the smoothness and the lower frictional forces.

7.12 REFERENCES

[1] Standards are defined by the American Gear Manufacturers Association (AGMA) and the American National Standards Institute (ANSI). The AGMA standards may be quoted or extracted in their entirety, provided that an appropriate credit line is included—for example, "Extracted from AGMA Information Sheet—Strength of Spur, Helical, Herringbone, and Bevel Gear Teeth (AGMA 225.01) with permission of the publisher, the American Gear Manufacturers Association, 1500 King Street, Suite 201, Alexandria, VA 22314." These standards have been used extensively in Chapter 7 and in Chapter 8.

[2] The strength and wear of gears are covered in texts such as *Shigley's Mechanical Engineering Design,* 8th ed., R.G. Budynas and J.K. Nisbett, New York: McGraw–Hill, 2008.

PROBLEMS

7.1 Find the diametral pitch of a pair of gears having 32 and 84 teeth, respectively, whose center distance is 3.625 in.

7.2 Find the number of teeth and the circular pitch of a 6-in pitch diameter gear whose diametral pitch is 9 teeth/in.

7.3 Determine the module of a pair of gears having 18 and 40 teeth, respectively, whose center distance is 58 mm.

7.4 Find the number of teeth and the circular pitch of a gear whose pitch diameter is 200 mm if the module is 8 mm/tooth.

7.5 Find the diametral pitch and the pitch diameter of a 40-tooth gear whose circular pitch is 3.50 in/tooth.

7.6 The pitch diameters of a pair of mating gears are 3.50 and 8.25 in, respectively. If the diametral pitch is 16 teeth/in, how many teeth are there on each gear?

7.7 Find the module and the pitch diameter of a gear whose circular pitch is 40 mm/tooth if the gear has 36 teeth.

7.8 The pitch diameters of a pair of gears are 60 and 100 mm, respectively. If their module is 2.5 mm/tooth, how many teeth are there on each gear?

7.9 What is the diameter of a 33-tooth gear if its circular pitch is 0.875 in/tooth?

7.10 A shaft carries a 30-tooth, 3-teeth/in diametral pitch gear that drives another gear at a speed of 480 rev/min. How fast does the 30-tooth gear rotate if the shaft center distance is 9 in?

7.11 Two gears having an angular velocity ratio of 3:1 are mounted on shafts whose centers are 136 mm apart. If the module of the gears is 4 mm/tooth, how many teeth are there on each gear?

7.12 A gear having a module of 4 mm/tooth and 21 teeth drives another gear at a speed of 240 rev/min. How fast is the 21-tooth gear rotating if the shaft center distance is 156 mm?

7.13 A 4-tooth/in diametral pitch, 24-tooth pinion is to drive a 36-tooth gear. The gears are cut on the 20° full-depth involute system. Find and tabulate the addendum, dedendum, clearance, circular pitch, base pitch, tooth thickness, pitch circle radii, base circle radii, length of paths of approach and recess, and contact ratio.

7.14 A 5-tooth/in diametral pitch, 15-tooth pinion is to mate with a 30-tooth internal gear. The gears are 20° full-depth involute. Make a drawing of the gears showing several teeth on each gear. Can these gears be assembled in a radial direction? If not, what remedy should be used?

7.15 A $2\frac{1}{2}$-teeth/in diametral pitch 17-tooth pinion and a 50-tooth gear are paired. The gears are cut on the 20° full-depth involute system. Find the angles of approach and recess of each gear and the contact ratio.

7.16 A gearset with a module of 5 mm/tooth has involute teeth with $22\frac{1}{2}°$ pressure angle and 19 and

31 teeth, respectively. They have 1.0m for the addendum and 1.25m for the dedendum.* Tabulate the addendum, dedendum, clearance, circular pitch, base pitch, tooth thickness, base circle radius, and contact ratio.

7.17 A gear with a module of 8 mm/tooth and 22 teeth is in mesh with a rack; the pressure angle is 25°. The addendum and dedendum are 1.0m and 1.25m, respectively.* Find the lengths of the paths of approach and recess and determine the contact ratio.

7.18 Repeat Problem 7.15 using the 25° full-depth system.

7.19 Draw a 2-tooth/in diametral pitch, 26-tooth, 20° full-depth involute gear in mesh with a rack.

(a) Find the lengths of the paths of approach and recess and the contact ratio.

(b) Draw a second rack in mesh with the same gear but offset $^{1}/_{8}$ in further away from the gear center. Determine the new contact ratio. Has the pressure angle changed?

7.20 through 7.24 Shaper gear cutters have the advantage that they can be used for either external or internal gears and also that only a small amount of runout is necessary at the end of the stroke. The generating action of a pinion shaper cutter can easily be simulated by employing a sheet of clear plastic. Figure P7.20 illustrates one tooth of a 16-tooth pinion cutter with 20° pressure angle as it can be cut from a plastic sheet. To construct the cutter, lay out the tooth on a sheet of drawing paper. Be sure to include the clearance at the top of the tooth. Draw radial lines through the pitch circle spaced at distances equal to one fourth of the tooth thickness, as illustrated in Fig. P7.20. Next, fasten the sheet of plastic to the drawing and scribe the cutout, the pitch circle, and the radial lines onto the sheet. Then remove the sheet and trim the tooth outline with a razor blade. Then use a small piece of fine sandpaper to remove any burrs.

To generate a gear with the cutter, only the pitch circle and the addendum circle need be drawn. Divide the pitch circle into spaces equal to those used on the template and construct radial lines through them. The tooth outlines are then obtained by rolling the template pitch circle upon that of the gear and drawing the cutter tooth lightly for each position. The resulting generated tooth upon the gear will be evident. The following problems all employ a standard 1-tooth/in diametral pitch 20° full-depth template constructed as described above. In each case you should generate a few teeth and estimate the amount of undercutting.

Table P7.20 to P7.24

Problem no.	P7.20	P7.21	P7.22	P7.23	P7.24
No. of teeth	10	12	14	20	36

7.25 A 10-mm/tooth module gear has 17 teeth, a 20° pressure angle, an addendum of 1.0m, and a dedendum of 1.25m.* Find the thickness of the teeth at the base circle and at the addendum circle. What is the pressure angle corresponding to the addendum circle?

7.26 A 15-tooth pinion has $1^{1}/_{2}$-tooth/in diametral pitch 20° full-depth involute teeth. Calculate the thickness of the teeth at the base circle. What are the tooth thickness and the pressure angle at the addendum circle?

7.27 A tooth is 0.785 in thick at a pitch circle radius of 8 in. and a pressure angle of 25°. What is the thickness at the base circle?

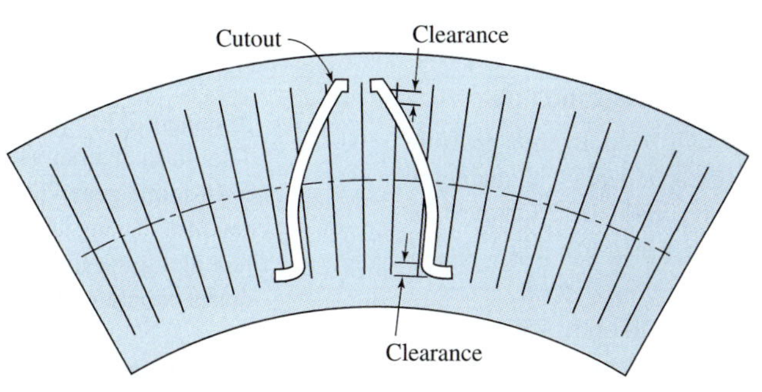

Figure P7.20

* In SI, tooth sizes are given in modules, m, and $a = 1.0\,\mathrm{m}$ means 1 module, not 1 meter.

7.28 A tooth is 1.571 in thick at the pitch radius of 16 in and a pressure angle of 20°. At what radius does the tooth become pointed?

7.29 A 25° full-depth involute, 12-tooth/in diametral pitch pinion has 18 teeth. Calculate the tooth thickness at the base circle. What are the tooth thickness and pressure angle at the addendum circle?

7.30 A nonstandard 10-tooth 8-tooth/in diametral pitch involute pinion is to be cut with a $22^1/_2$° pressure angle. What maximum addendum can be used before the teeth become pointed?

7.31 The accuracy of cutting gear teeth can be measured by fitting hardened and ground pins in diametrically opposite tooth spaces and measuring the distance over the pins. For a 10-tooth/in diametral pitch 20° full-depth involute system 96 tooth gear:

(a) Calculate the pin diameter that will contact the teeth at the pitch lines if there is to be no backlash.

(b) What should be the distance measured over the pins if the gears are cut accurately?

7.32 A set of interchangeable gears with 4-tooth/in diametral pitch is cut on the 20° full-depth involute system. The gears have tooth numbers of 24, 32, 48, and 96. For each gear, calculate the radius of curvature of the tooth profile at the pitch circle and at the addendum circle.

7.33 Calculate the contact ratio of a 17-tooth pinion that drives a 73-tooth gear. The gears are 96-tooth/in diametral pitch and cut on the 20° full-depth involute system.

7.34 A 25° pressure angle 11-tooth pinion is to drive a 23-tooth gear. The gears have a diametral pitch of 8 teeth/in and have involute stub teeth. What is the contact ratio?

7.35 A 22-tooth pinion mates with a 42-tooth gear. The gears have full-depth involute teeth, have a diametral pitch of 16 teeth/in, and are cut with a $17^1/_2$° pressure angle.* Find the contact ratio.

7.36 The center distance of two 24-tooth, 20° pressure angle, full-depth involute spur gears with diametral pitch of 2 teeth/in is increased by 0.125 in over the standard distance. At what pressure angle do the gears operate?

7.37 The center distance of two 18-tooth, 25° pressure angle, full-depth involute spur gears with diametral pitch of 3 teeth/in is increased by 0.0625 in over the standard distance. At what pressure angle do the gears operate?

7.38 A pair of mating gears have 24 teeth/in. diametral pitch and are generated on the 20° full-depth involute system. If the tooth numbers are 15 and 50, what maximum addendums may they have if interference is not to occur?

7.39 A set of gears is cut with a $4^1/_2$-in/tooth circular pitch and a $17^1/_2$° pressure angle.* The pinion has 20 full-depth teeth. If the gear has 240 teeth, what maximum addendum may it have to avoid interference?

7.40 Using the method described for Problems 7.20 through 7.24, cut a 1-tooth/in diametral pitch 20° pressure angle full-depth involute rack tooth from a sheet of clear plastic. Use a nonstandard clearance of 0.35/P to obtain a stronger fillet. This template can be used to simulate the generating action of a hob. Now, using the variable-center-distance system, generate an 11-tooth pinion to mesh with a 25-tooth gear without interference. Record the values found for center distance, pitch radii, pressure angle, gear blank diameters, cutter offset, and contact ratio. Note that more than one satisfactory solution exists.

7.41 Using the template cut in Problem 7.40, generate an 11-tooth pinion to mesh with a 44-tooth gear with the long-and-short-addendum system. Determine and record suitable values for gear and pinion addendum and dedendum and for the cutter offset and contact ratio. Compare the contact ratio with that of standard gears.

7.42 A pair of involute spur gears with 9 and 36 teeth are to be cut with a 20° full-depth cutter with diametral pitch of 3 teeth/in.

(a) Determine the amount that the addendum of the gear must be decreased to avoid interference.

(b) If the addendum of the pinion is increased by the same amount, determine the contact ratio.

7.43 A standard 20° pressure angle full-depth involute 1-tooth/in diametral pitch 20-tooth pinion drives a 48-tooth gear. The speed of the pinion is 500 rev/min. Using the position of the point of contact along the line of action as the abscissa, plot a curve indicating the sliding velocity at all points of contact. Note that the sliding velocity changes sign when the point of contact passes through the pitch point.

---

* Such gears came from an older standard and are now obsolete.

8 Helical Gears, Bevel Gears, Worms, and Worm Gears

When rotational motion is to be transmitted between parallel shafts, engineers often prefer to use spur gears because they are easy to design and very economical to manufacture. However, sometimes the design requirements are such that helical gears are a better choice. This is especially true when the loads are heavy, the speeds are high, or the noise level must be kept low.

When motion is to be transmitted between shafts that are not parallel, spur gear cannot be used; the designer must then choose between crossed-helical, bevel, hypoid, or worm gears. Bevel gears have straight teeth, line contact, and high efficiency. Crossed-helical and worm gears have a much lower efficiency because of their increased sliding action; however, if good engineering is used, crossed-helical and worm gears may be designed with quite acceptable values of efficiency. Bevel and hypoid gears are used for similar applications and, although hypoid gears have inherently stronger teeth, their efficiency is often much less. Worm gears are used when a very small velocity ratio (first-order kinematic coefficient) is required.

8.1 PARALLEL-AXIS HELICAL GEARS

The shape of the tooth of a helical gear is illustrated in Fig. 8.1. If a piece of paper is cut into the shape of a parallelogram and wrapped around a cylinder, the angular edge of the paper wraps into a helix. The cylinder plays the same role as the base cylinder of a spur gear in Chapter 7. If the paper is unwound, each point on the angular edge generates an involute curve as indicated in Section 7.3 for spur gears. The surface obtained when every point on the angular edge of the paper generates an involute is called an *involute helicoid*. If we imagine the strip of paper as unwrapping from a base cylinder on one gear and wrapping

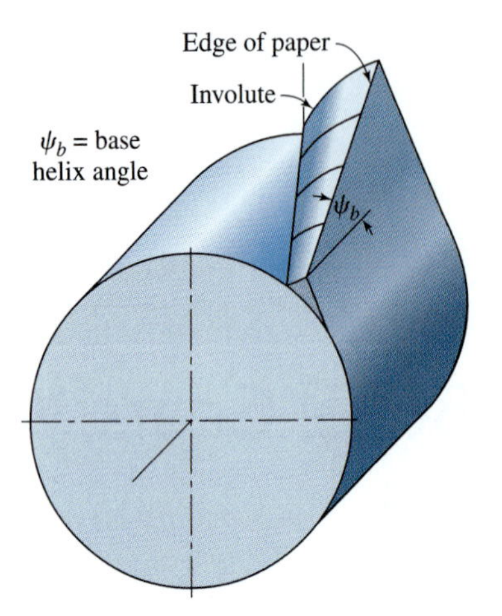

Figure 8.1 An involute helicoid.

up onto the base cylinder of another, then a slanted line on this strip of paper generates two involute helicoids meshing as two tangent tooth shapes.

The initial contact of spur gear teeth, as we saw in Chapter 7, is a line extending across the face of the tooth. The initial contact of helical gear teeth starts as a point and changes into a line as the teeth come into more engagement; in helical gears, however, the line is diagonal across the face of the tooth. It is this gradual engagement of the teeth and the smooth transfer of load from one tooth to another that give helical gears the ability to quietly transmit heavy loads at high speeds.

8.2 HELICAL GEAR TOOTH RELATIONS

As illustrated in Fig. 8.2, to mesh properly, two parallel shaft helical gears must have equal pitches and equal helix angles, but must be of opposite hand. Helical gears with the same hand, for example, a right-hand driver and a right-hand driven gear, can be meshed with their axes skewed and are commonly referred to as crossed-axis gears (see Section 8.7).

Figure 8.3 represents a portion of the top view of a helical rack. Lines AB and CD are the centerlines of two adjacent helical teeth taken on the pitch plane. The angle ψ is the helix angle and is measured at the pitch diameter unless otherwise specified. The distance AC, in the plane of rotation of the gear, is the *transverse circular pitch* p_t. The distance AE is the *normal circular pitch* p_n and is related to the transverse circular pitch as follows:

$$p_n = p_t \cos\psi. \tag{8.1}$$

The distance AD is called *axial pitch* p_x and can be written as

$$p_x = \frac{p_t}{\tan\psi}. \tag{8.2}$$

Figure 8.2 Pair of helical gears in mesh. Note the opposite hand of the two gears. (Courtesy of Gleason Works, Rochester, NY.)

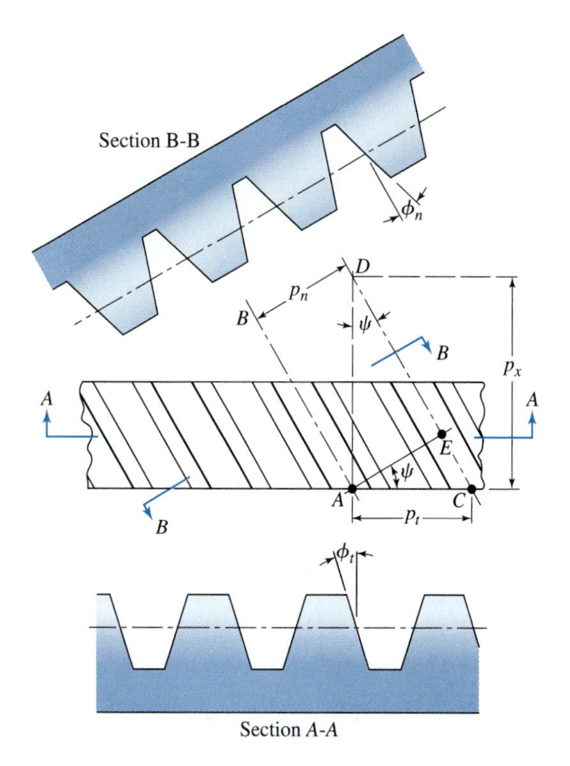

Figure 8.3 Helical gear tooth relations.

The *normal diametral pitch* P_n can be written as

$$P_n = \frac{\pi}{p_n} = \frac{\pi}{p_t \cos\psi} = \frac{P_t}{\cos\psi}, \tag{8.3}$$

where P_t is the *transverse diametral pitch.*

Because of the angularity of the teeth, we must define two different pressure angles, the *normal pressure angle* ϕ_n and the *transverse pressure angle* ϕ_t, both illustrated in Fig. 8.3. The two pressure angles are related by

$$\tan\phi_n = \tan\phi_t \cos\psi. \tag{8.4}$$

In applying these equations it is convenient to remember that all equations and relations that are valid for a spur gear apply equally for the transverse plane of a helical gear.

A better picture of the tooth relations can be obtained by an examination of Fig. 8.4. To obtain the geometric relations, a helical gear has been cut by the oblique plane AA at an angle ψ to a right section. For convenience, only the pitch cylinder of radius R is given. Figure 8.4 illustrates that the intersection of the AA plane and the pitch cylinder is an ellipse whose radius at the pitch point P is R_e. This is called the *equivalent pitch radius*, and it is the radius of curvature of the pitch surface in the normal cross-section. For the condition

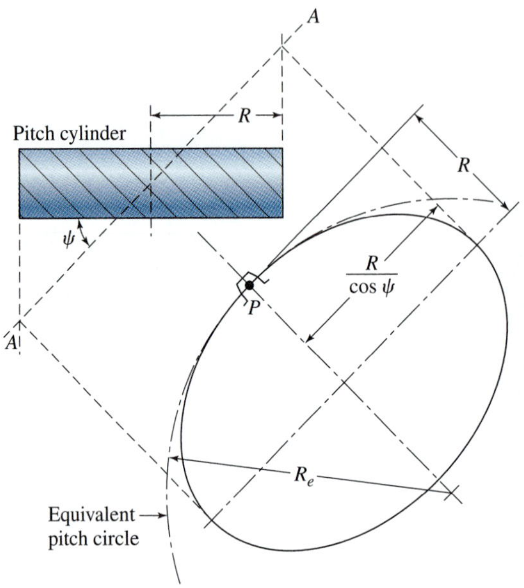

Figure 8.4

that $\psi = 0$, this radius of curvature is $R_e = R$. If we imagine the angle ψ to be gradually increased from 0 to 90°, we see that R_e begins at a value of $R_e = R$ and increases until, when $\psi = 90°$, the value of $R_e = \infty$.

It is demonstrated in a note at the end of this chapter[1] that

$$R_e = \frac{R}{\cos^2 \psi}, \tag{8.5}$$

where R is the pitch radius of the helical gear and R_e is the pitch radius of an equivalent spur gear. This equivalence is taken on the normal section of the helical gear.

Let us define the number of teeth on the helical gear as N and that on the equivalent spur gear as N_e. Then,

$$N_e = 2R_e P_n.$$

Using Eqs. (8.3) and (8.5) we can write this as

$$N_e = 2\frac{R}{\cos^2 \psi}\frac{P_t}{\cos \psi} = \frac{N}{\cos^3 \psi}. \tag{8.6}$$

8.3 HELICAL GEAR TOOTH PROPORTIONS

Except for fine pitch (normal diametral pitch of 20 teeth/in. and finer), there is no generally accepted standard for the proportions of helical gear teeth.

In determining the tooth proportions for helical gears, it is necessary to consider the manner in which the teeth are formed. If the helical gear is hobbed, then tooth proportions are calculated in a plane normal to the tooth. As a general guide, tooth proportions are then often based on a normal pressure angle of $\phi_n = 20°$. Most of the proportions used for spur gears, given in Table 7.2, can then be used. The tooth proportions are calculated using the normal diametral pitch P_n. These proportions are suitable for helix angles from 0° to 30°, and all helix angles can be cut with the same hob. Of course, the normal diametral pitch of the hob and the gear will be the same.

If the gear is to be cut by a shaper, an alternative set of tooth proportions is used based on a transverse pressure angle of $\phi_t = 20°$ and the transverse diametral pitch P_t. For these gears the helix angles are generally restricted to 15°, 23°, 30°, or 45°; helix angles greater than 45° are not recommended. The normal diametral pitch P_n must be used to compute the tooth dimensions; the proportions given in Table 7.2 are usually satisfactory. If the shaper method is used, however, the same cutter cannot be used to cut both spur and helical gears.

8.4 CONTACT OF HELICAL GEAR TEETH

For spur gears, contact between meshing teeth occurs along a line that is parallel to their axes of rotation. As illustrated in Fig. 8.5, contact between meshing helical gear teeth occurs along a diagonal line.

Several kinds of contact ratios are used in evaluating the performance of helical gearsets. The *transverse contact ratio* is designated m_t and is the contact ratio in the transverse plane. It is obtained exactly as was m_c for spur gears.

The *normal contact ratio* m_n is the contact ratio in the normal section. It is also determined exactly as was the contact ratio m_c for spur gears, but the dimensions of equivalent spur gears must be used in the determination. The base helix angle ψ_b and the pitch helix angle ψ, for helical gears, are related by

$$\tan \psi_b = \tan \psi \cos \phi_t, \tag{8.7}$$

Then the transverse and normal contact ratios are related by

$$m_n = \frac{m_t}{\cos^2 \psi_b}. \tag{8.8}$$

The *axial contact ratio* m_x, also called the *face contact ratio*, is the ratio of the face width of the gear to the axial pitch, determined from Eq. (8.2). It is given by

$$m_x = \frac{F}{p_x} = \frac{F \tan \psi}{p_t}, \tag{8.9}$$

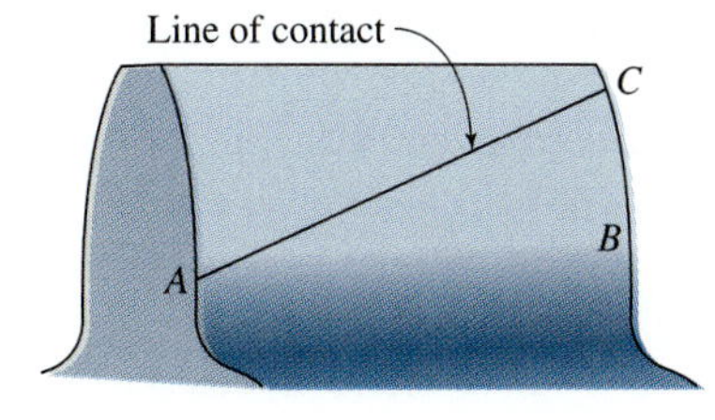

Figure 8.5 When contact of another tooth is just beginning at A, contact at the other end of the tooth may have already progressed from B to C.

where F is the face width of the helical gear. Figure 8.5 illustrates that this contact ratio is greater than unity when another tooth is beginning contact solely because of the helix angle of the teeth before the previous tooth contact has finished. Note that this face contact ratio, also called *overlap*, has no parallel for spur gears and note that, because of the helix angle, this face contact ratio can be made greater than unity for helical gears by the choice of face width despite the choice of tooth size. If the face width is made greater than the axial pitch, continuous contact of at least one tooth is assured. This means that fewer teeth may be used on helical pinions than on spur pinions. The overlapping action also results in smoother operation of the gears. Note also that the face contact ratio depends solely on the geometry of a single gear, whereas the transverse and normal contact ratios depend upon the geometry of a pair of mating gears.

The *total contact ratio* is the sum of the face contact ratio m_x and the transverse contact ratio m_t. In a sense this sum gives the average total number of teeth in contact.

8.5 REPLACING SPUR GEARS WITH HELICAL GEARS

Because of their ability to carry heavy loads at high speed with little noise, it is sometimes desirable to replace a pair of spur gears by parallel shaft helical gears although the cost may be slightly more. An example illustrates the calculations.

EXAMPLE 8.1

A pair of 20° full-depth involute spur gears with 32 and 80 teeth, diametral pitch of 16 teeth/in, and face width of 0.75 in are to be replaced by helical gears. The same hob used for the spur gears is to be used for the helical gears. The shaft center distance and the angular velocity ratio must remain the same. The helix angle is to be as small as possible, and the overlap is to be 1.5 or greater. Determine the helix angle, the numbers of teeth, and the face width of the new helical gears.

SOLUTION

From the spur gear data and Eq. (7.1), the center distance is

$$R_2 + R_3 = \frac{N_2 + N_3}{2P} = \frac{32\text{ teeth} + 80\text{ teeth}}{2(16\text{ teeth/in})} = 3.5\text{ in.} \tag{1}$$

From Eq. (7.5), the first-order kinematic coefficient, the angular velocity ratio, is

$$\left|\theta'_{3/2}\right| = \left|\frac{\omega_3}{\omega_2}\right| = \frac{R_2}{R_3} = \frac{N_2}{N_3} = \frac{32\text{ teeth}}{80\text{ teeth}} = 0.4. \tag{2}$$

Because the same hob is to be used, the normal diametral pitch P_n for the helical gears must also be 16 teeth/in. Because the shaft center distance must remain the same, that is,

$$R_2 + R_3 = \frac{N_2 + N_3}{2P_n \cos\psi} = \frac{N_2 + N_3}{2(16\text{ teeth/in})\cos\psi} = 3.5\text{ in},$$

or

$$\cos\psi = \frac{N_2 + N_3}{112 \text{ teeth}} = \frac{(N_2/N_3)N_3 + N_3}{112 \text{ teeth}} = \frac{1.4N_3}{112 \text{ teeth}} = \frac{N_3}{80 \text{ teeth}}. \quad (3)$$

This implies that N_3 must be less than 80 teeth, whereas Eq. (2) requires that the ratio N_2/N_3 must remain 0.4 or $N_2 = 0.4N_3$.

Because $N_3 = 79$ teeth does not give an integer solution for N_2, the next smallest integer solution (which gives the smallest nonzero helix angle ψ) is $N_3 = 75$ teeth and $N_2 = 30$ teeth, giving a helix angle of $\psi = 20.364°$. The transverse circular pitch is

$$p_t = \frac{\pi}{P_n \cos\psi} = \frac{\pi}{(16 \text{ teeth/in}) \cos 20.364°} = 0.209 \text{ in/tooth},$$

for which Eq. (8.9) indicates a face width of $F \geq 0.845$ in. Unfortunately, the space available will not allow this increase in face width. Therefore, this solution is not acceptable.

The next integer solution is $N_3 = 70$ teeth and $N_2 = 28$ teeth, giving a helix angle of $\psi = 28.955°$. The transverse circular pitch is $p_t = \pi/(P_n \cos\psi) = 0.224$ in/tooth and the face width is $F \geq 0.607$ in. This is an acceptable solution with face width of less than the original spur gears.

8.6 HERRINGBONE GEARS

Double-helical or *herringbone gears* comprise teeth having both a right- and a left-hand helix cut on the same gear blank, as illustrated schematically in Fig. 8.6. One of the primary disadvantages of the single helical gear is the axial thrust loads that must be accounted for in the design of the bearings. In addition, the desire to obtain a good overlap without an excessively large face width may lead to the use of a comparatively larger helix angle, thus producing even higher axial thrust loads. These thrust loads are eliminated by the herringbone configuration because the axial force of the right-hand half is balanced by that of the left-hand half. Thus, with the absence of thrust reactions, helix angles are usually larger for herringbone gears than for single-helical gears. However, one of the members of a herringbone gearset should always be mounted with some axial play or float to accommodate slight tooth errors and mounting tolerances.

For the efficient transmission of large amounts of power at high speeds, herringbone gears are almost universally employed.

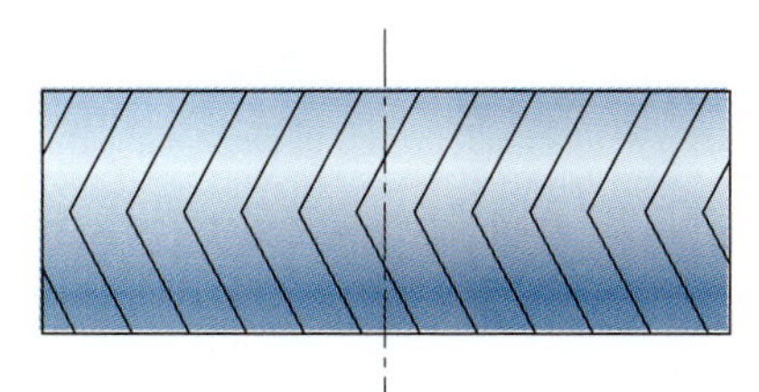

Figure 8.6 Schematic drawing of the pitch cylinder of a herringbone gear.

8.7 CROSSED-AXIS HELICAL GEARS

Crossed-axis helical or *spiral gears* are sometimes used when the shaft centerlines are neither parallel nor intersecting. These are essentially nonenveloping worm gears (see Section 8.13) because the gear blanks have a cylindrical form with the two cylinder axes skew to each other.

The tooth action of crossed-axis helical gears is quite different from that of parallel-axis helical gears. The teeth of crossed-axis helical gears have only *point contact*. In addition, there is much greater sliding action along the tooth surfaces than for parallel-axis helical gears. For these reasons they are chosen only to carry small loads. Because of the point contact, however, they need not be mounted accurately; either the center distance or the shaft angle may also vary slightly without affecting the amount of contact.

There is no difference between crossed-axis helical gears and other helical gears until they are mounted in mesh. They are manufactured in the same way. Two meshing crossed-axis helical gears usually have the same hand; that is, a right-hand driver goes with a right-hand driven gear. The relation among thrust, hand, and rotation for crossed-axis helical gears is illustrated in Fig. 8.7.

For crossed-axis helical gears to mesh properly, they must share the same normal pitch. When tooth sizes are specified, the normal pitch should always be used. The reason for this is that when different helix angles are used for the driver and the driven gear, the transverse pitches are not the same. The relation between the shaft and helix angles is

$$\Sigma = \psi_2 \pm \psi_3. \tag{8.10}$$

The positive sign is used when both helix angles are of the same hand, and the negative sign is used when they are of opposite hand. Opposite-hand crossed-axis helical gears are used when the shaft angle Σ is small. The first-order kinematic coefficient, the angular velocity

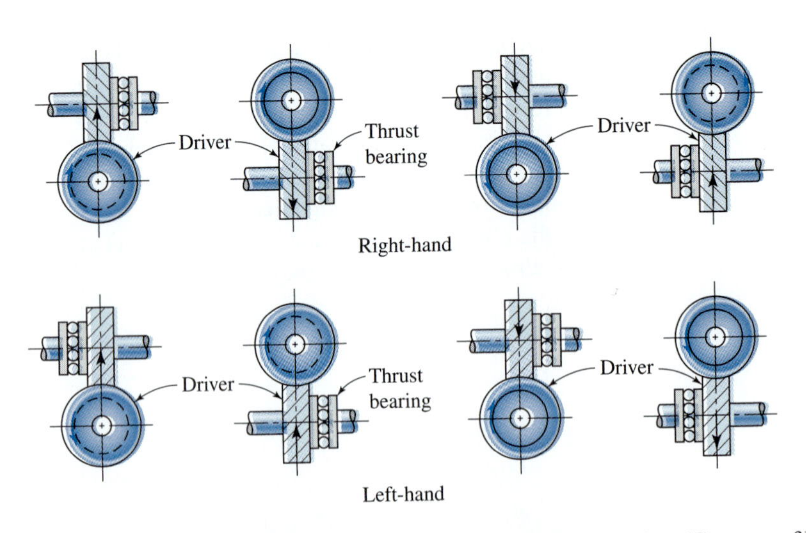

Figure 8.7 Thrust, rotation, and hand relations for crossed-axis helical gearing. (Courtesy of Boston Gear Works, Inc., North Quincy, MA.)

TABLE 8.1 Tooth proportions for crossed-axis helical gears Normal diametral pitch $P_n = 1$ teeth/in.; working depth = 2.400 in; whole depth = 2.650 in; addendum $a = 1.200$ in.

Driver		Driven	Both
Helix angle	Minimum tooth number	Helix angle	Normal pressure angle
ψ_2	N_2	ψ_3	ϕ_n
45°	20	45°	$14\frac{1}{2}$°
60°	9	30°	$17\frac{1}{2}$°
75°	4	15°	$19\frac{1}{2}$°
86°	1	4°	20°

ratio between the shafts, is

$$\left|\theta'_{3/2}\right| = \left|\frac{\omega_3}{\omega_2}\right| = \frac{N_2}{N_3} = \frac{R_2 \cos \psi_2}{R_3 \cos \psi_3}. \tag{8.11}$$

Crossed-axis helical gears have the least sliding at the point of contact when the two helix angles are equal. If the two helix angles are not equal, the larger helix angle should be used with the driver if both gears have the same hand.

There is no widely accepted standard for crossed-axis helical gear tooth proportions. Many different combinations of proportions give good tooth action. Because the teeth are in point contact, an effort should be made to obtain a contact ratio of 2 or more. For this reason, crossed-axis helical gears are usually cut with a low pressure angle and a deep tooth. The tooth proportions given in Table 8.1 are representative of good design. The driver tooth numbers indicated are the minimum required to avoid undercut. The driven gear should have 20 teeth or more if a contact ratio of 2 is to be obtained.

To illustrate the calculations for a pair of crossed-axis helical gears, consider the following example.

EXAMPLE 8.2

Two shafts at an angle of 60° are to have a velocity ratio of 1:1.5. The center distance between the shafts is 8.63 in. Design a pair of crossed-axis helical gears for this application.

SOLUTION

Choosing $\psi_2 = 35°$ for the pinion, then Eq. (8.10) gives $\psi_3 = 25°$ for the gear. Substituting these angles into Eq. (8.11), the first-order kinematic coefficient can be written as

$$\left|\theta'_{3/2}\right| = \left|\frac{\omega_3}{\omega_2}\right| = \frac{R_2 \cos 35°}{R_3 \cos 25°} = \frac{1}{1.5}.$$

Therefore, the pitch radius of the pinion is

$$R_2 = 0.737\,6 R_3.$$

This, along with the given shaft center distance, $R_2 + R_3 = 8.63$ in., gives the pitch radius of the pinion $R_2 = 3.663$ in. and the pitch radius of the gear $R_3 = 4.967$ in. Choosing a normal diametral pitch of $P_n = 6$ teeth/in., the numbers of teeth on the pinion and the gear, respectively, are

$$N_2 = 2P_n R_2 \cos\psi_2 = 2(6\ \text{teeth/in})(3.663\ \text{in})\cos 35^\circ = 36\ \text{teeth} \qquad \textit{Ans.}$$

and

$$N_3 = 2P_n R_3 \cos\psi_3 = 2(6\ \text{teeth/in})(4.967\ \text{in})\cos 25^\circ = 54\ \text{teeth.} \qquad \textit{Ans.}$$

8.8 STRAIGHT-TOOTH BEVEL GEARS

When rotational motion is transmitted between shafts whose axes intersect, some form of bevel gears is usually used. Bevel gears have pitch surfaces that are cones, with their cone axes matching the two shaft rotation axes, as illustrated in Fig. 8.8. The gears are mounted so that the apexes of the two pitch cones are coincident with the point of intersection of the shaft axes. These pitch cones roll together without slipping.

Although bevel gears are often made for an angle of 90° between the shafts, they can be designed for almost any angle. When the shaft intersection angle is other than 90°, the gears are called *angular bevel gears*. For the special case where the shaft intersection angle is 90° and both gears are of equal size, such bevel gears are called *miter gears*. A pair of miter gears is illustrated in Fig. 8.9.

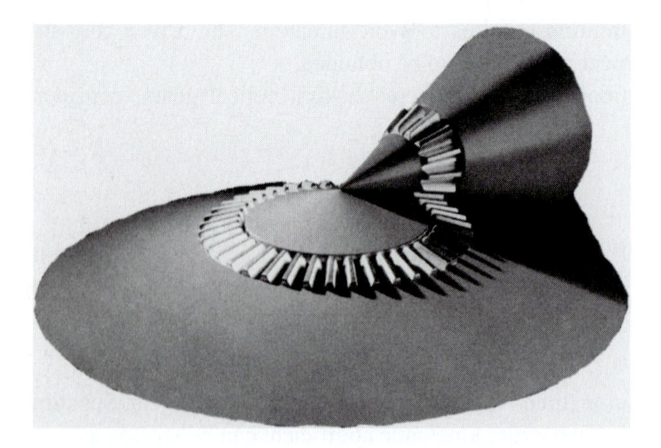

Figure 8.8 The pitch surfaces of bevel gears are cones that have only rolling contact. (Courtesy of Gleason Works, Rochester, NY.)

Figure 8.9 A pair of miter gears in mesh. (Courtesy of Gleason Works, Rochester, NY.)

For straight-tooth bevel gears, the true shape of a tooth is obtained by taking a spherical section through the tooth, where the center of the sphere is at the common apex, as illustrated in Fig. 8.8. As the radius of the sphere increases, the same number of teeth is projected onto a larger surface; therefore, the size of the teeth increases as larger spherical sections are taken. We have seen that the action and contact conditions for spur gear teeth may be viewed in a plane taken at right angles to the axes of the spur gears. For bevel gear teeth, the action and contact conditions should properly be viewed on a spherical surface (instead of a plane). We can even think of spur gears as a special case of bevel gears in which the spherical radius is infinite, thus producing a plane surface on which the tooth action is viewed. Figure 8.10 is typical of many straight-tooth bevel gear sets.

It is standard practice to specify the pitch diameter of a bevel gear at the large end of the teeth. In Fig. 8.11, the pitch cones of a pair of bevel gears are drawn and the pitch radii are given as R_2 and R_3, respectively, for the pinion and the gear. The cone angles γ_2 and γ_3 are defined as the pitch angles, and their sum is equal to the shaft intersection angle Σ, that is

$$\Sigma = \gamma_2 + \gamma_3.$$

The first-order kinematic coefficient, the angular velocity ratio between the shafts, is obtained in the same manner as for spur gears and is

$$\left|\theta'_{3/2}\right| = \left|\frac{\omega_3}{\omega_2}\right| = \frac{R_2}{R_3} = \frac{N_2}{N_3}. \tag{8.12}$$

In the kinematic design of bevel gears, the tooth numbers of the two gears and the shaft angle are usually given, and the corresponding pitch angles are to be determined. Although

Figure 8.10 A pair of straight-tooth bevel gears. (Courtesy of Gleason Works, Rochester, NY.)

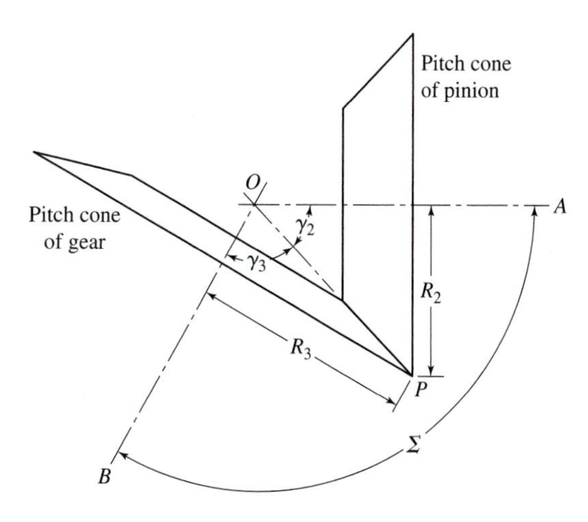

Figure 8.11 Pitch cones of bevel gears.

they can be determined graphically, the analytical approach gives more exact values. From Fig. 8.11 the distance OP may be written as

$$OP = \frac{R_2}{\sin\gamma_2} = \frac{R_3}{\sin\gamma_3}, \tag{a}$$

so that

$$\sin\gamma_2 = \frac{R_2}{R_3}\sin\gamma_3 = \frac{R_2}{R_3}\sin(\Sigma - \gamma_2) \tag{b}$$

or

$$\sin\gamma_2 = \frac{R_2}{R_3}(\sin\Sigma\cos\gamma_2 - \cos\Sigma\sin\gamma_2). \tag{c}$$

Dividing both sides of this equation by $\cos\gamma_2$ and rearranging gives

$$\tan\gamma_2 = \frac{R_2}{R_3}(\sin\Sigma - \cos\Sigma\tan\gamma_2).$$

Then, rearranging this equation gives

$$\tan\gamma_2 = \frac{\sin\Sigma}{(R_3/R_2) + \cos\Sigma} = \frac{\sin\Sigma}{(N_3/N_2) + \cos\Sigma}. \tag{8.13}$$

Similarly,

$$\tan\gamma_3 = \frac{\sin\Sigma}{(N_2/N_3) + \cos\Sigma}. \tag{8.14}$$

For a shaft angle of $\Sigma = 90°$, the above expressions reduce to

$$\tan\gamma_2 = \frac{N_2}{N_3} \tag{8.15}$$

and

$$\tan\gamma_3 = \frac{N_3}{N_2}. \tag{8.16}$$

The projection of bevel gear teeth onto the surface of a sphere would indeed be a difficult and time-consuming task. Fortunately, an approximation that reduces the problem to that of ordinary spur gears is common. This approximation is called *Tredgold's approximation* and, as long as the gear has eight or more teeth, it is accurate enough for practical purposes. It is in almost universal use, and the terminology of bevel gear teeth has evolved around it.

In Tredgold's method, a *back cone* is formed of elements that are perpendicular to the elements of the pitch cone at the large end of the teeth, as illustrated in Fig. 8.12. The length of a back-cone element is called the back-cone radius. Now an equivalent spur gear is constructed whose pitch radius R_e is equal to the back-cone radius. Thus, from a pair of bevel gears, using Tredgold's approximation, we can obtain a pair of equivalent spur

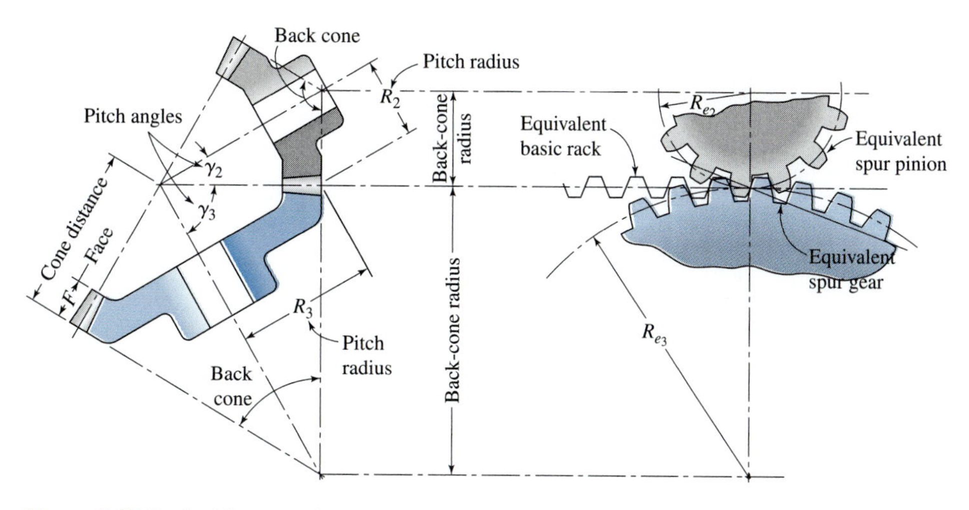

Figure 8.12 Tredgold's approximation.

gears that are then used to define the tooth profiles. They can also be used to determine the tooth action and the contact conditions, just as for ordinary spur gears, and the results will correspond closely to those for the bevel gears.

From the geometry of Fig. 8.12, the equivalent pitch radii are

$$R_{e2} = \frac{R_2}{\cos \gamma_2}, \quad R_{e3} = \frac{R_3}{\cos \gamma_3}. \tag{8.17}$$

The number of teeth on each of the equivalent spur gears is

$$N_e = \frac{2\pi R_e}{p}, \tag{8.18}$$

where p is the circular pitch of the bevel gear measured at the large end of the teeth. Usually the equivalent spur gears will *not* have integral numbers of teeth.

8.9 TOOTH PROPORTIONS FOR BEVEL GEARS

Practically all straight-tooth bevel gears manufactured today use a 20° pressure angle. It is not necessary to use an interchangeable tooth form because bevel gears cannot be interchanged. For this reason, the long-and-short-addendum system, described in Section 7.11, is used. The proportions are tabulated in Table 8.2.

Bevel gears are usually mounted on the outboard side of the bearings because the shaft axes intersect, which means that the effect of shaft deflection is to pull the small end of the teeth away from mesh, causing the larger end to take more of the load. Thus, the load across the tooth is variable; for this reason, it is desirable to design a fairly short tooth. As indicated in Table 8.2, the face width is usually limited to about one third of the cone distance. We note also that a short face width simplifies the tooling problems in cutting bevel gear teeth.

TABLE 8.2 Tooth proportions for 20° straight-tooth bevel gears

Item	Formula
Working depth	$h_k = 2.0/P$
Clearance	$c = 0.188/P + 0.002$ in.
Addendum of gear	$a_G = \dfrac{0.540}{P} + \dfrac{0.460}{P(m_{90})^2}$
Gear ratio	$m_G = N_G/N_P$
Equivalent 90° ratio	$m_{90} = \begin{cases} m_G & \text{when } \Sigma = 90^\circ \\ \sqrt{m_G \dfrac{\cos \gamma_P}{\cos \gamma_G}} & \text{when } \Sigma \neq 90^\circ \end{cases}$
Face width	$F = \dfrac{1}{3}$ or $F = \dfrac{10}{P}$ whichever is smaller
Minimum number of teeth	Pinion: 16, 15, 14, 13; Gear: 16, 17, 20, 30

Figure 8.13 defines additional terminology characteristic of bevel gears. Note that a constant clearance is maintained by making the elements of the face cone parallel to the elements of the root cone of the mating gear. This explains why the face cone apex is not coincident with the pitch cone apex in Fig. 8.13. This permits a larger fillet than would otherwise be possible.

8.10 CROWN AND FACE GEARS

If the pitch angle of one of a pair of bevel gears is made equal to 90°, the pitch cone becomes a flat surface and the resulting gear is called a *crown gear*. Figure 8.14 illustrates a crown gear in mesh with a bevel pinion. Note that a crown gear is the counterpart of a rack in spur gearing. The back cone of a crown gear is a cylinder, and the resulting involute teeth have straight sides, as indicated in Fig. 8.12.

A pseudo-bevel gearset can be obtained using a cylindrical spur gear for a pinion in mesh with a gear having a planar pitch surface (similar to a crown gear) called a *face gear*. When the axes of the pinion and gear intersect, the face gear is called *on center*; when the axes do not intersect, the face gear is called *off center*.

To understand how a spur pinion, with a cylindrical rather than conical pitch surface, can properly mesh with a face gear we must consider how the face gear is formed; it is generated by a reciprocating cutter that is a replica of the spur pinion. Because the cutter and the gear blank are rotated as if in mesh, the resulting face gear is conjugate to the cutter and, therefore, to the spur pinion. The face width of the teeth on the face gear must be held quite short, however; otherwise the top land will become pointed.

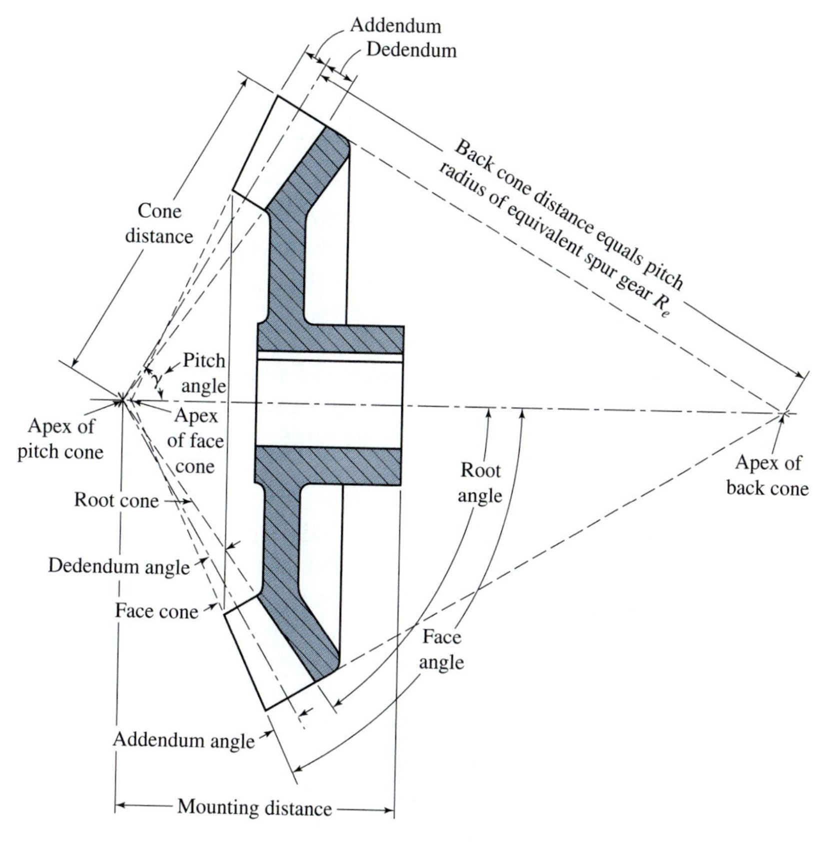

Figure 8.13

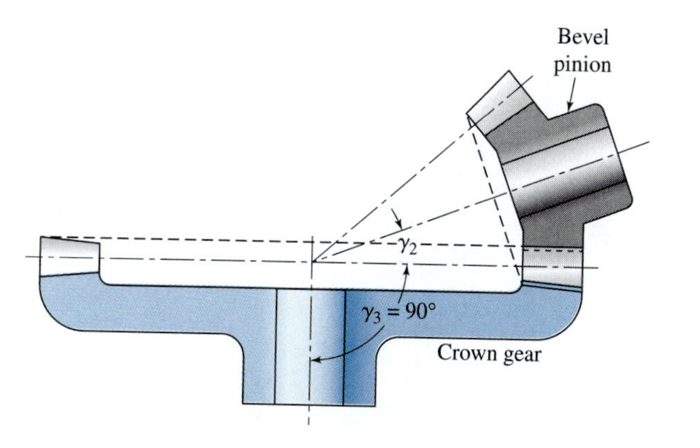

Figure 8.14 A crown gear and bevel pinion.

Face gears are not capable of carrying heavy loads, but because the axial mounting position of the pinion is not critical, they are sometimes more suitable for angular drives than bevel gears.

8.11 SPIRAL BEVEL GEARS

Straight-tooth bevel gears are easy to design and simple to manufacture and give very good results in service if they are mounted accurately and positively. As in the case of spur gears, however, they become noisy at higher pitch-line velocities. In such cases it is often good design practice to use *spiral bevel gears*, which are the bevel counterparts of helical gears. A mating pair of spiral bevel gears is illustrated in Fig. 8.15. The pitch surfaces and the nature of contact are the same as for straight-tooth bevel gears except for the differences brought about by the spiral-shape teeth.

Spiral bevel gear teeth are conjugate to a basic crown rack, which can be generated as illustrated in Fig. 8.16 using a circular cutter. The spiral angle ψ is measured at the mean radius of the gear. As with helical gears, spiral bevel gears give much smoother tooth action than straight-tooth bevel gears and hence are useful where high speeds are encountered. To obtain true spiral tooth action, the face contact ratio should be at least 1.25.

Pressure angles used with spiral bevel gears are generally $14\frac{1}{2}°$ to $20°$, whereas the spiral angle is about $30°$ or $35°$. As far as tooth action is concerned, the spiral may be either right- or left-handed; it makes no difference. However, looseness in the bearings might result in jamming or separating of the teeth, depending on the direction of rotation and the hand of the spiral. Because jamming of the teeth would do the most damage, the hand of the spiral should be such that the teeth tend to separate.

Figure 8.15 Spiral bevel gears. (Courtesy of Gleason Works, Rochester, NY.)

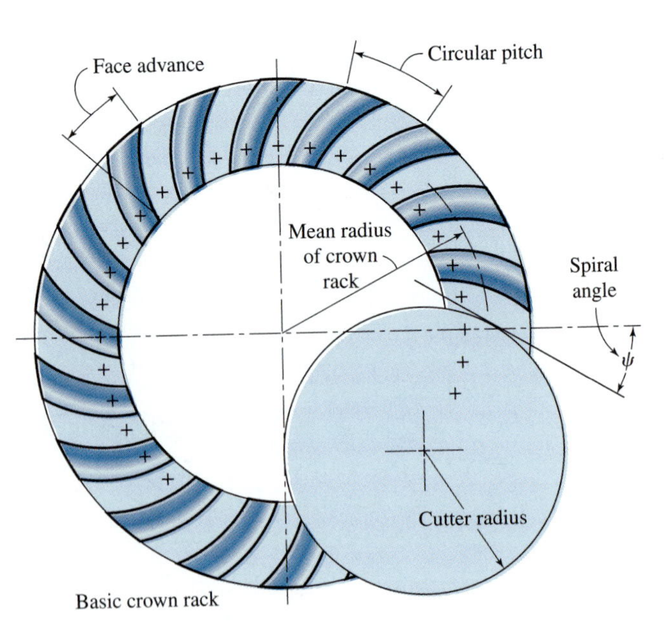

Figure 8.16 Cutting spiral bevel gear teeth on a basic crown rack.

Figure 8.17 Zerol bevel gears. (Courtesy of Gleason Works, Rochester, NY.)

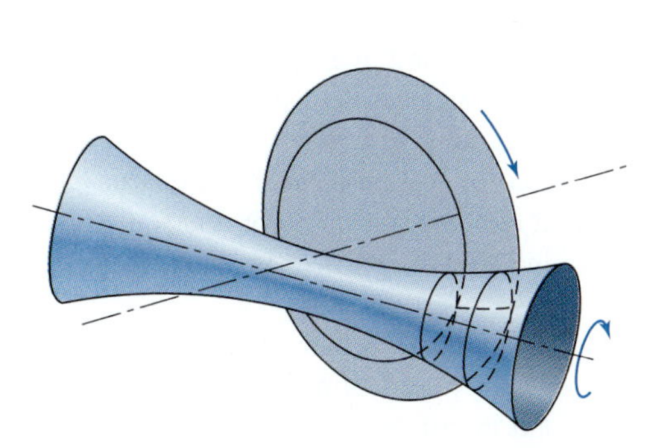

Figure 8.18 The pitch surfaces for hypoid gears are hyperboloids of revolution.

Figure 8.19 Hypoid gears. (Courtesy of Gleason Works, Rochester, NY.)

Zerol Bevel Gears The Zerol bevel gear is a patented gear that has curved teeth but a zero-degree spiral angle. An example is illustrated in Fig. 8.17. It has no advantage in tooth action over the straight-tooth bevel gear and is designed simply to take advantage of the cutting machinery used for cutting spiral bevel gears.

8.12 HYPOID GEARS

It is frequently desirable, as in the case of rear-wheel drive automotive differential applications, to have a gearset similar to bevel gears but where the shafts do not intersect. Such gears are called *hypoid gears* because, as illustrated in Fig. 8.18, their pitch surfaces are hyperboloids of revolution. Figure 8.19 illustrates a pair of hypoid gears in mesh. The tooth action between these gears is a combination of rolling and sliding along a straight line and has much in common with that of worm gears (see Section 8.13).

8.13 WORMS AND WORM GEARS

A *worm* is a machine member having a screw-like thread, and worm teeth are frequently spoken of as threads. A worm meshes with a conjugate gear-like member called a *worm wheel* or a *worm gear*. Figure 8.20 illustrates a worm and worm gear in an application. These gears are used with nonintersecting shafts that are usually at a shaft angle of 90°, but there is no reason why shaft angles other than 90° cannot be used if a design demands it.

Worms in common use have from one to four teeth and are said to be *single-threaded*, *double-threaded*, and so on. As we will see, there is no definite relation between the number of teeth and the pitch diameter of a worm. The number of teeth on a worm gear is usually much higher and, therefore, the angular velocity of the worm gear is usually much lower than that of the worm. In fact, often, one primary application for a worm and worm gear is to obtain a very large angular velocity reduction, that is, a very low first-order kinematic coefficient. In keeping with this low velocity ratio, the worm gear is usually the driven member of the pair and the worm is usually the driving member.

A worm gear, unlike a spur or helical gear, has a face that is made concave so that it partially wraps around, or envelops, the worm, as illustrated in Fig. 8.21. Worms are sometimes designed with a cylindrical pitch surface or they may have an hourglass shape, such that the worm also wraps around or partially encloses the worm gear. If an enveloping worm gear is mated with a cylindrical worm, the set is said to be *single enveloping*. When the worm is hourglass shaped, the worm and worm gearset is said to be *double enveloping* because each member partially wraps around the other; such a worm is sometimes called a *Hindley worm*. The nomenclature of a single-enveloping worm and worm gearset is illustrated in Fig. 8.21.

A worm and worm gear combination is similar to a pair of mating crossed-helical gears except that the worm gear partially envelops the worm. For this reason, they have line contact instead of the point contact found in crossed-helical gears and are thus able to transmit more power. When a double-enveloping worm and worm gearset is used, even more power can be transmitted, at least in theory, because contact is distributed over an area on both tooth surfaces.

Figure 8.20 A single-enveloping worm and worm gear set. (Courtesy of Gleason Works, Rochester, NY.)

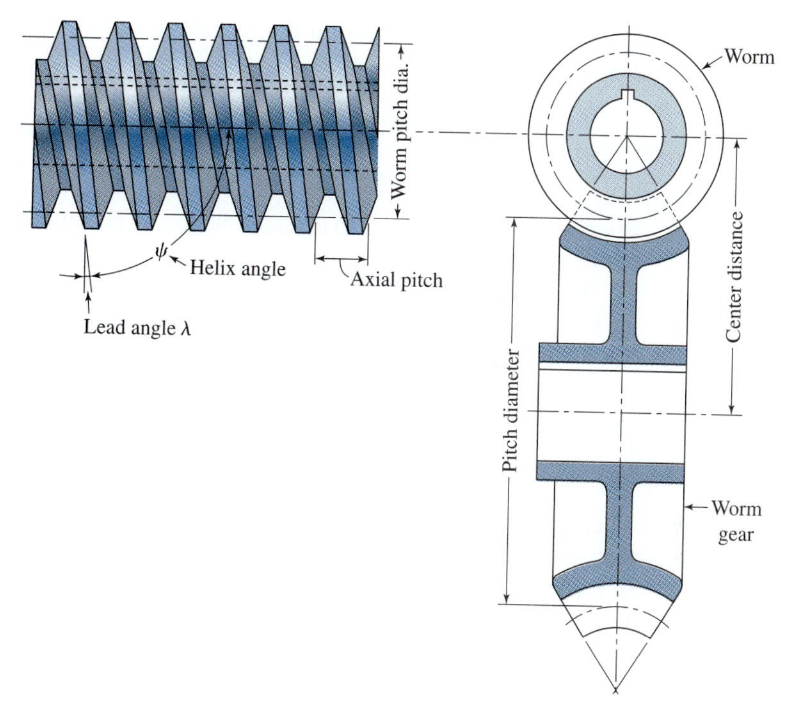

Figure 8.21 Nomenclature of a single-enveloping worm and worm gearset.

In a single-enveloping worm and worm gearset it makes no difference whether the worm rotates on its own axis and drives the gear by a screwing action or whether the worm is translating along its axis and drives the worm gear through rack action. The resulting motion and contact are the same. For this reason, a single-enveloping worm need not be accurately mounted along its axis. However, the worm gear should be accurately mounted along its rotation axis; otherwise, its pitch surface is not properly aligned with the worm axis. In a double-enveloping worm and worm gearset, both members are throated and therefore both must be accurately mounted in all directions to obtain correct contact.

A mating worm and worm gear with a 90° shaft angle have the same hand of helix, but the helix angles are usually very different. The helix angle on the worm is usually quite large (at least for one or two teeth) and quite small on the worm gear. On the worm, the *lead angle* is the complement of the helix angle, as illustrated in Fig. 8.21. Because of this, it is customary to specify the lead angle for the worm and specify the helix angle for the worm gear. This is convenient because these two angles are equal for a 90° shaft angle.

In specifying the pitch of a worm and worm gearset, it is usual to specify the axial pitch of the worm and the circular pitch of the worm gear. These are equal if the shaft angle is 90°. It is common to employ even fractions, such as $1/4$, $3/8$, $1/2$, $3/4$, 1, and $1\,1/4$ in/tooth, for the circular pitch of the worm gear; there is no reason, however, why the AGMA standard diametral pitches used for spur gears (Table 7.1) should not also be used for worm gears.

The pitch radius of a worm gear is determined in the same manner as that of a spur gear, that is,

$$R_3 = \frac{N_3 p}{2\pi}, \tag{8.19}$$

where all values are defined in the same manner as for spur gears, but refer to the parameters of the worm gear.

The pitch radius of the worm may have any value, but it should be the same as that of the hob used to cut the worm gear teeth. The relation between the pitch radius of the worm and the center distance, as recommended by AGMA, is

$$R_2 = \frac{(R_2 + R_3)^{0.875}}{4.4}, \tag{8.20}$$

where the quantity (R_2+R_3) is the center distance in inches. This equation gives proportions that result in good power capacity. The AGMA standard also states that the denominator of Eq. (8.20) may vary from 3.4 to 6.0 without appreciably affecting the power capacity. Equation (8.20) is not required, however; other proportions will also serve well and, in fact, power capacity may not always be the primary consideration. However, there are a lot of variables in worm gear design, and the equation is helpful in obtaining trial dimensions.

The *lead* of a worm has the same meaning as for a screw thread and is the axial distance through which a point on the helix will move when the worm is turned through one revolution. Thus, in equation form, the lead of the worm is given by

$$l = p_x N_2, \tag{8.21}$$

where p_x is the axial pitch and N_2 is the number of teeth (threads) on the worm. The lead and the *lead angle* are related as follows,

$$\lambda = \tan^{-1}\left(\frac{l}{2\pi R_2}\right), \tag{8.22}$$

where λ is the lead angle, as illustrated in Fig. 8.21.

The teeth on a worm are usually cut in a milling machine or on a lathe. Worm gear teeth are most often produced by hobbing. Except for clearance at the top of the hob teeth, the worm should be an exact duplicate of the hob to obtain conjugate action. This also means that, where possible, the worm should be designed using the dimensions of existing hobs.

The pressure angles used on worms and worm gearsets vary widely and should depend approximately on the value of the lead angle. Good tooth action is obtained if the pressure angle is made large enough to eliminate undercutting of the worm gear tooth on the side at which the contact ends. Recommended values are given in Table 8.3.

A satisfactory tooth depth that has about the right relation to the lead angle is obtained by making the depth a proportion of the normal circular pitch. Using an addendum of $1/P = p_n/\pi$, as for full-depth spur gears, we obtain the following proportions for worms

TABLE 8.3 Recommended pressure angles for worm and worm gear sets

Lead angle λ	Pressure angle ϕ
0°–16°	$14\frac{1}{2}$°
16°–25°	20°
25°–35°	25°
35°–45°	30°

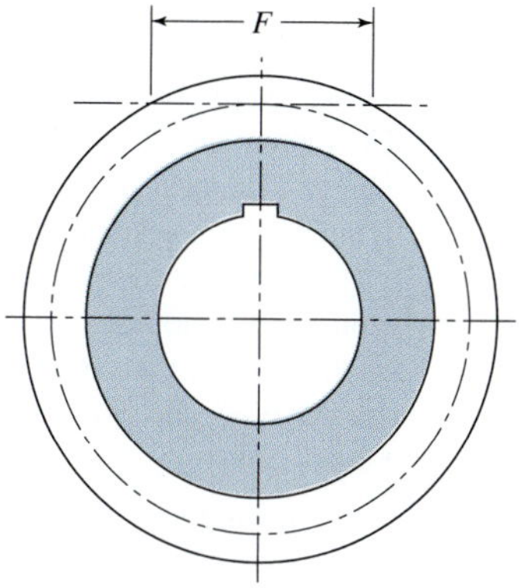

Figure 8.22 Face width of a worm gear.

and worm gears.

$$\text{Addendum} = 1.000/\text{P} = 0.318\,3p_n$$

$$\text{Whole depth} = 2.000/\text{P} = 0.636\,6p_n$$

$$\text{Clearance} = 0.157/\text{P} = 0.050\,7p_n$$

The face width of the worm gear should be obtained as illustrated in Fig. 8.22. This makes the face of the worm gear equal to the length of a tangent to the worm pitch circle between its points of intersection with the addendum circle.

8.14 NOTES

1. The equation of an ellipse with its center at the origin of an xy coordinate system with a and b as its semimajor and semiminor axes, respectively, is

$$\frac{x^2}{a^2} + \frac{y^2}{b^2} = 1. \tag{a}$$

Also, the formula for radius of curvature is

$$\rho = \frac{\left[1 + (dy/dx)^2\right]^{3/2}}{d^2y/dx^2}. \tag{b}$$

Using these two equations, it is not difficult to find the radius of curvature corresponding to $x = 0, y = b$. The result is

$$\rho = a^2/b. \tag{c}$$

Then, referring to Fig. 8.3, we substitute $a = R/\cos\psi$ and $b = R$ into Eq. (c) and obtain Eq. (8.5).

PROBLEMS

8.1 A pair of parallel-axis helical gears has $14\frac{1}{2}°$ normal pressure angle, diametral pitch of 6 teeth/in, and 45° helix angle. The pinion has 15 teeth, and the gear has 24 teeth. Calculate the transverse and normal circular pitch, the normal diametral pitch, the pitch radii, and the equivalent tooth numbers.

8.2 A set of parallel-axis helical gears are cut with a 20° normal pressure angle and a 30° helix angle. They have diametral pitch of 16 teeth/in and have 16 and 40 teeth, respectively. Find the transverse pressure angle, the normal circular pitch, the axial pitch, and the pitch radii of the equivalent spur gears.

8.3 A parallel-axis helical gearset is made with a 20° transverse pressure angle and a 35° helix angle. The gears have diametral pitch of 10 teeth/in and have 15 and 25 teeth, respectively. If the face width is 0.75 in, calculate the base helix angle and the axial contact ratio.

8.4 A set of helical gears is to be cut for parallel shafts whose center distance is to be about 3.5 in to give a velocity ratio of approximately 1.8. The gears are to be cut with a standard 20° pressure angle hob whose diametral pitch is 8 teeth/in. Using a helix angle of 30°, determine the transverse values of the diametral and circular pitches and the tooth numbers, pitch radii, and center distance.

8.5 A 16-tooth helical pinion is to run at 1 800 rev/min and drive a helical gear on a parallel shaft at 400 rev/min. The centers of the shafts are to be spaced 11.0 in apart. Using a helix angle of 23° and a pressure angle of 20°, determine the values for the tooth numbers, pitch radii, normal circular and diametral pitch, and face width.

8.6 The catalog description of a pair of helical gears is as follows: $14\frac{1}{2}°$ normal pressure angle, 45° helix angle, diametral pitch of 8 teeth/in, 1.0-in face width, and normal diametral pitch of 11.31 teeth/in. The pinion has 12 teeth and a 1.500-in pitch diameter, and the gear has 32 teeth and a 4.000-in pitch diameter. Both gears have full-depth teeth, and they may be purchased either right or left handed. If a right-hand pinion and left-hand gear are placed in mesh, find the transverse contact ratio, the normal contact ratio, the axial contact ratio, and the total contact ratio.

8.7 In a medium-size truck transmission a 22 tooth clutch-stem gear meshes continuously with a 41-tooth countershaft gear. The data indicate normal diametral pitch of 7.6 teeth/in, $18\frac{1}{2}°$ normal pressure angle, $23\frac{1}{2}°$ helix angle, and a 1.12-in face width. The clutch-stem gear is cut with a left-hand helix, and the countershaft gear is cut with a right-hand helix. Determine the normal and total contact ratio if the teeth are cut full depth with respect to the normal diametral pitch.

8.8 A helical pinion is right handed, has 12 teeth, has a 60° helix angle, and is to drive another gear at a velocity ratio of 3.0. The shafts are at a 90° angle, and the normal diametral pitch of the gears is 8 teeth/in. Find the helix angle and the number of teeth on the mating gear. What is the shaft center distance?

8.9 A right-hand helical pinion is to drive a gear at a shaft angle of 90°. The pinion has 6 teeth and a 75° helix angle and is to drive the gear at a velocity ratio of 6.5. The normal diametral pitch of the gear is 12 teeth/in. Calculate the helix angle and the number of teeth on the mating gear. Also determine the pitch radius of each gear.

8.10 Gear 2 in Fig. P8.10 is to rotate clockwise and drive gear 3 counterclockwise at a velocity ratio of 2. Use a normal diametral pitch of 5 teeth/in, a shaft center distance of about 10 in, and the same helix angle for

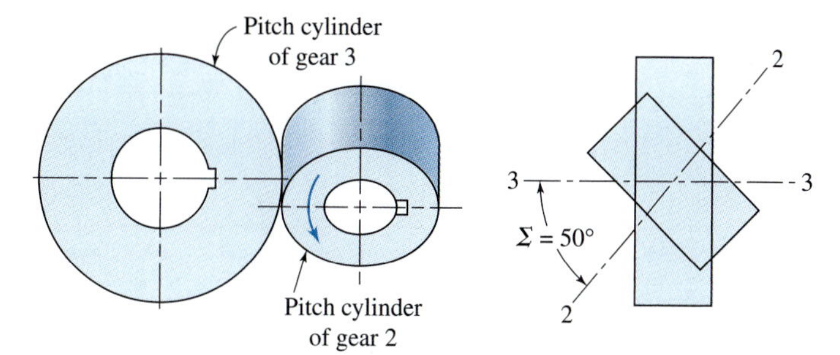

Figure P8.10

both gears. Find the tooth numbers, the helix angles, and the exact shaft center distance.

8.11 A pair of straight-tooth bevel gears is to be manufactured for a shaft angle of 90°. If the driver is to have 18 teeth and the velocity ratio is to be 3:1, what are the pitch angles?

8.12 A pair of straight-tooth bevel gears has a velocity ratio of 1.5 and a shaft angle of 75°. What are the pitch angles?

8.13 A pair of straight-tooth bevel gears is to be mounted at a shaft angle of 120°. The pinion and gear are to have 15 and 33 teeth, respectively. What are the pitch angles?

8.14 A pair of straight-tooth bevel gears with diametral pitch of 2 teeth/in have 19 and 28 teeth, respectively. The shaft angle is 90°. Determine the pitch diameters, pitch angles, addendum, dedendum, face width, and pitch diameters of the equivalent spur gears.

8.15 A pair of straight-tooth bevel gears with diametral pitch of 8 teeth/in have 17 and 28 teeth, respectively, and a shaft angle of 105°. For each gear, calculate the pitch radius, pitch angle, addendum, dedendum, face width, and equivalent tooth numbers. Make a sketch of the two gears in mesh. Use standard tooth proportions as for a 90° shaft angle.

8.16 A worm having 4 teeth and a lead of 1.0 in drives a worm gear at a velocity ratio of 7.5. Determine the pitch diameters of the worm and worm gear for a center distance of 1.75 in.

8.17 Specify a suitable worm and worm gear combination for a velocity ratio of 60 and a center distance of 6.50 in. Use an axial pitch of 0.500 in/tooth.

8.18 A triple-threaded worm drives a worm gear having 40 teeth. The axial pitch is 1.25 in, and the pitch diameter of the worm is 1.75 in. Calculate the lead and lead angle of the worm. Find the helix angle and pitch diameter of the worm gear.

8.19 A triple-threaded worm with a lead angle of 20° and an axial pitch of 0.400 in/tooth drives a worm gear with a velocity reduction of 15 to 1. Determine the following for the worm gear: (*a*) the number of teeth, (*b*) the pitch radius, and (*c*) the helix angle. (*d*) Determine the pitch radius of the worm. (*e*) Compute the center distance.

9 Mechanism Trains

Mechanisms arranged in combinations so that the driven member of one mechanism is the driver for another mechanism are called *mechanism trains*. With certain exceptions, to be explored here, the analysis of such trains can proceed in serial fashion using the methods developed in the previous chapters.

9.1 PARALLEL-AXIS GEAR TRAINS

In Chapter 3, we learned that the first-order *kinematic coefficient* is the term used to describe the ratio of the angular velocity of the driven member to that of the driving member. Thus, for example, in a four-bar linkage with link 2 as the driving or input member and link 4 as the driven or output member, we have

$$\theta'_{42} = \frac{\omega_4}{\omega_2} = \frac{d\theta_4/dt}{d\theta_2/dt} = \frac{d\theta_4}{d\theta_2} \tag{a}$$

where it is noted that, as in Chapter 5, we adopt the second subscript to explicitly indicate the number of the driving or input member. This second subscript is important in Chapter 9 because many mechanism trains have more than one degree of freedom.

In this section, where we deal with serially connected gear trains, we prefer to write Eq. (*a*) as

$$\theta'_{LF} = \frac{\omega_L}{\omega_F} = \frac{d\theta_L/dt}{d\theta_F/dt} = \frac{d\theta_L}{d\theta_F}, \tag{9.1}$$

where ω_L is the angular velocity of the *last* gear and ω_F is the angular velocity of the *first* gear in the train because, usually, the last gear is the output and is the driven gear and the first is the input and driving gear.

The term θ'_{LF} in Eq. (9.1) is the first-order kinematic coefficient, called the *speed ratio* by some or the *train value* by others. Equation (9.1) is often written in the more convenient form:

$$\omega_L = \theta'_{LF}\omega_F. \tag{9.2}$$

Next, we consider a pinion 2 driving a gear 3. The speed of the driven gear is

$$\omega_3 = \pm\frac{R_2}{R_3}\omega_2 = \pm\frac{N_2}{N_3}\omega_2, \tag{b}$$

where, for each gear, R is the radius of the pitch circle, N is the number of teeth, and ω is either the angular velocity or the angular displacement completed during a chosen time interval.

For parallel-shaft gearing, the directions can be kept track of by following the vector sense, that is, by specifying that angular velocity is positive when counterclockwise as seen from a chosen side. For parallel-shaft gearing we shall use the following sign convention: If the last gear of a parallel-shaft gear train rotates in the same sense as the first gear, then θ'_{LF} is positive; if the last gear rotates in the opposite sense to the first gear, then θ'_{LF} is negative. This sign convention approach is not as easy, however, when the gear shafts are not parallel, as in bevel, crossed-helical, or worm gearing. In such cases, it is often simpler to track the directions by visually inspecting a sketch of the train.

The gear train illustrated in Fig. 9.1 is made up of five gears in series. Applying Eq. (*b*) three times, we find the speed of gear 6 to be

$$\omega_6 = -\frac{N_5}{N_6}\frac{N_4}{N_5}\frac{N_2}{N_3}\omega_2.$$

Here we note that gear 5 is an idler; that is, its tooth numbers cancel in Eq. (*c*) and hence the only purpose served by gear 5 is to change the direction of rotation of gear 6. We further note that gears 5, 4, and 2 are drivers, whereas gears 6, 5, and 3 are driven members. Thus, Eq. (9.1) can also be written

$$\theta'_{LF} = \pm\frac{\text{product of driving tooth numbers}}{\text{product of driven tooth numbers}}. \tag{9.3}$$

Note also that because they are proportional, pitch radii can be used in Eq. (9.3) just as well as tooth numbers.

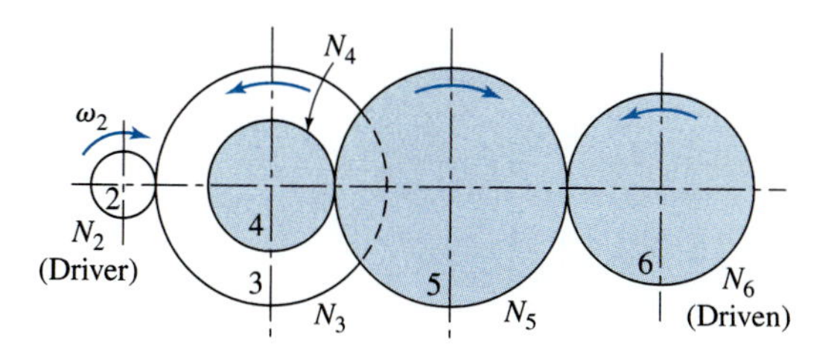

Figure 9.1

9.2 EXAMPLES OF GEAR TRAINS

In speaking of gear trains it is convenient to describe a train having only one gear on each axis as a *simple* gear train. A *compound* gear train then is one that has two or more gears on one or more axes, such as the train illustrated in Fig. 9.1. Another example of a compound gear train is illustrated in Fig. 9.2. Figure 9.2 illustrates a transmission for a small- or medium-size truck, which has four speeds forward and one in reverse.

The compound gear train illustrated in Fig. 9.3 is composed of bevel, helical, and spur gears. The helical gears are crossed, and so their direction of rotation depends upon their hand.

A *reverted* gear train is one in which the first and last gears have collinear axes of rotation, such as the one illustrated in Fig. 9.4. This produces a compact arrangement and is

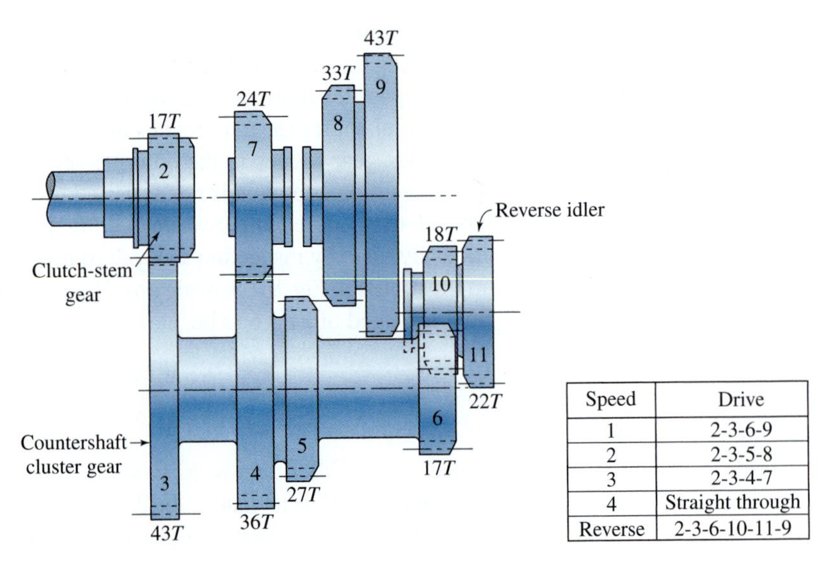

Speed	Drive
1	2-3-6-9
2	2-3-5-8
3	2-3-4-7
4	Straight through
Reverse	2-3-6-10-11-9

Figure 9.2 A truck transmission with gears having diametral pitch of 7 teeth/in. and pressure angle of 22.5°.

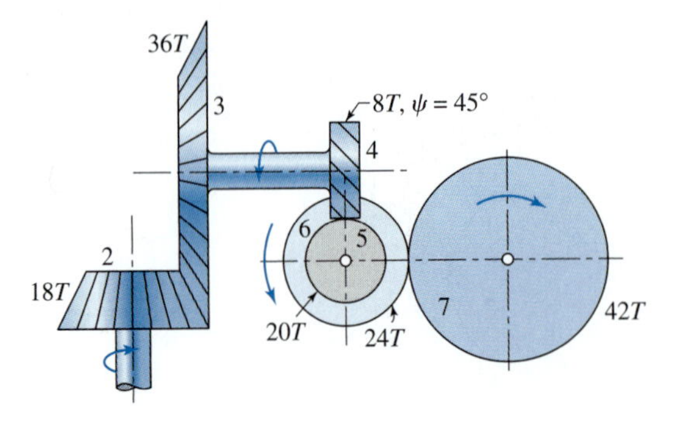

Figure 9.3 A gear train composed of bevel, crossed-helical, and spur gears.

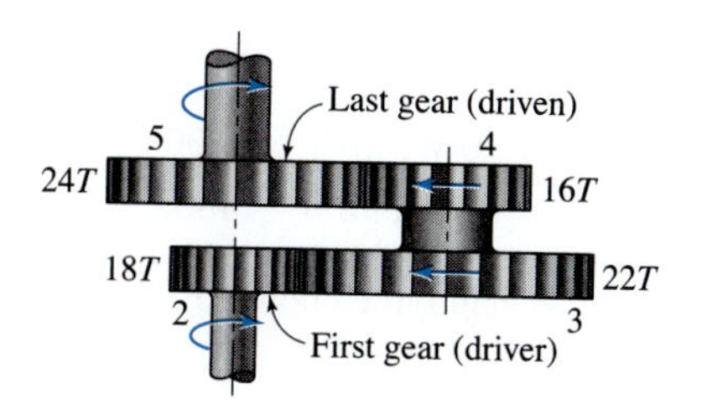

Figure 9.4 A reverted gear train.

used in such applications as speed reducers, clocks (to connect the hour hand to the minute hand), and machine tools. As an exercise, it is suggested that you seek out a suitable set of diametral pitches for each pair of gears illustrated in Fig. 9.4 so that the first and last gears have the same axis of rotation with all gears properly engaged.

9.3 DETERMINING TOOTH NUMBERS

When notable power is transmitted through a speed reduction unit, the speed ratio of the last pair of meshing gears is usually chosen larger than that of the first gear pair because the torque is greater at the low-speed end. In a given amount of space, more teeth can be used on gears of lesser pitch; hence, a greater speed reduction can be obtained at the high-speed end.

Without examining the problem of tooth strength, suppose we wish to use two pairs of gears in a train to obtain an overall kinematic coefficient of $\theta'_{LF} = 1/12$. Let us also impose the restriction that the tooth numbers must not be less than 15 and that the reduction in the first pair of gears should be about twice that of the second pair. This means that

$$\theta'_{52} = \frac{N_4}{N_5}\frac{N_2}{N_3} = \frac{1}{12}, \tag{a}$$

where N_2/N_3 is the kinematic coefficient of the first gear pair and N_4/N_5 is that of the second pair. Because the kinematic coefficient of the first pair should be half that of the second, Eq. (a) can be written as

$$\left(\frac{N_4}{N_5}\right)\left(\frac{N_4}{2N_5}\right) = \frac{1}{12} \tag{b}$$

or

$$\frac{N_4}{N_5} = \sqrt{\frac{1}{6}} = 0.408\,248 \tag{c}$$

to six decimal places. The following tooth numbers are seen to be close:

$$\frac{15}{37}\ \frac{16}{39}\ \frac{18}{44}\ \frac{20}{49}\ \frac{22}{54}\ \frac{24}{59}.$$

Of these, $N_4/N_5 = 20/49$ is the closest approximation, but note that

$$\theta'_{52} = \left(\frac{N_4}{N_5}\right)\left(\frac{N_2}{N_3}\right) = \left(\frac{20}{49}\right)\left(\frac{20}{98}\right) = \frac{400}{4802} = \frac{1}{12.005},$$

which is very close to 1/12. On the other hand, the choice of $N_4/N_5 = 18/44$ gives exactly

$$\theta'_{52} = \left(\frac{N_4}{N_5}\right)\left(\frac{N_2}{N_3}\right) = \left(\frac{18}{44}\right)\left(\frac{22}{108}\right) = \frac{396}{4752} = \frac{1}{12}.$$

In this case, the reduction in the first gear pair is not exactly twice the reduction in the second gear pair. However, this consideration is usually of only minor importance.

The problem of specifying tooth numbers and the number of pairs of gears to give a kinematic coefficient with a specified degree of accuracy has interested many people. Consider, for instance, the problem of specifying a set of gears to have a kinematic coefficient of $\theta'_{LF} = \pi/10$ accurate to eight decimal places, while we know that π is an irrational number and cannot be expressed as a ratio of integers.

9.4 EPICYCLIC GEAR TRAINS

Figure 9.5 illustrates an elementary epicyclic gear train together with its schematic diagram as suggested by Lévai.* The train consists of a *central gear* 2 and an *epicyclic gear* 4, which produces epicyclic motion for its points by rolling around the periphery of the central gear. A *crank arm* 3 contains the bearings for the epicyclic gear to maintain the gears in mesh.

Epicyclic trains are also called *planetary* or *sun-and-planet* gear trains. In this nomenclature, gear 2 of Fig. 9.5 is called the *sun gear*, gear 4 is called the *planet gear*, and crank 3 is called the *planet carrier*. Figure 9.6 illustrates the train of Fig. 9.5 with two redundant planet gears added. This produces better force balance; also, adding more planet gears

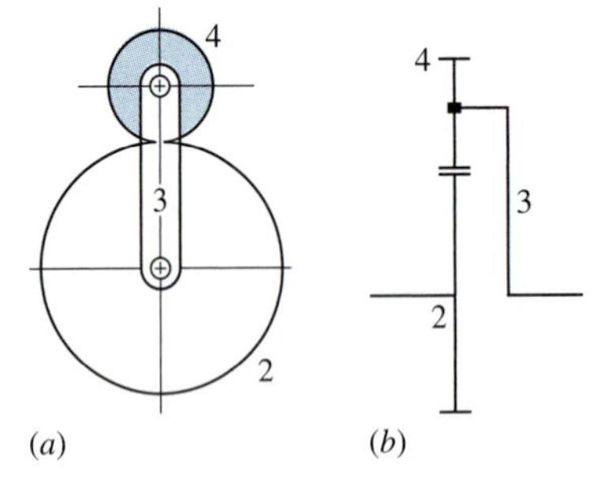

Figure 9.5 (*a*) The elementary epicyclic gear train; (*b*) its schematic diagram.

* Literature devoted to epicyclic gear trains is rather scarce; however, see G. R. Pennock and J. J. Alwerdt, "Duality Between the Kinematics of Gear Trains and the Statics of Beam Systems," *Mechanism and Machine Theory*, 42, no. 11 (2007):1527–46. For a comprehensive study in the English language, see Z. L. Lévai, *Theory of Epicyclic Gears and Epicyclic Change-Speed Gears*, Budapest: Technical University of Building, Civil, and Transport Engineering, 1966. This book lists 104 references.

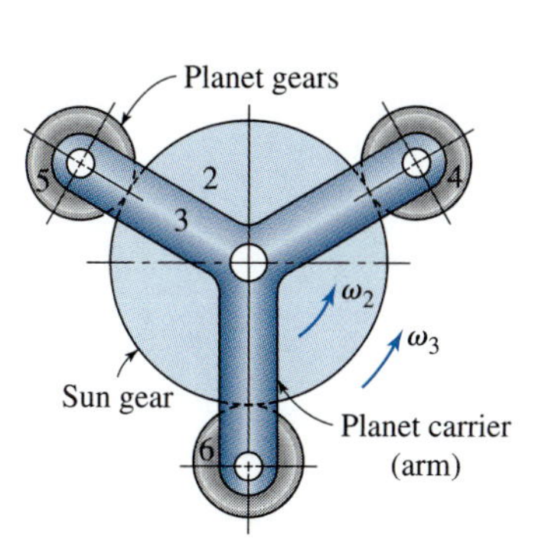

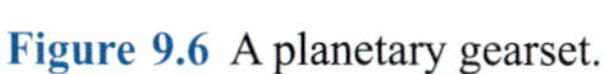
Figure 9.6 A planetary gearset.

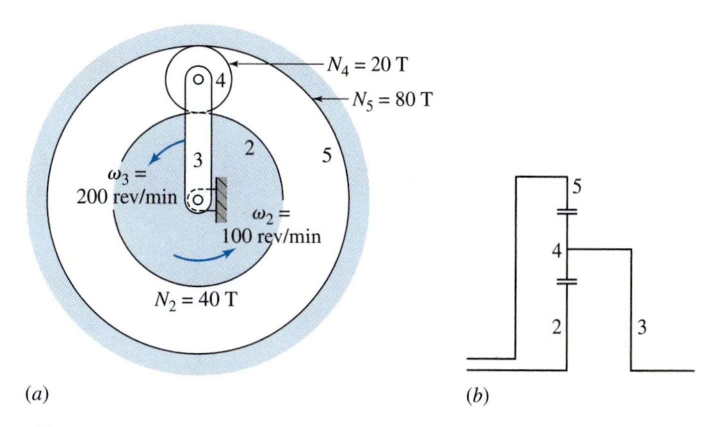

Figure 9.7 (*a*) A simple epicyclic gear train; (*b*) its schematic diagram.

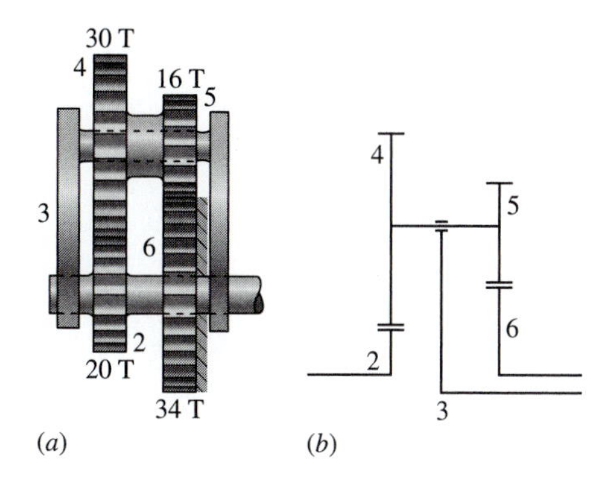

Figure 9.8 A simple epicyclic gear train with double-planet gears.

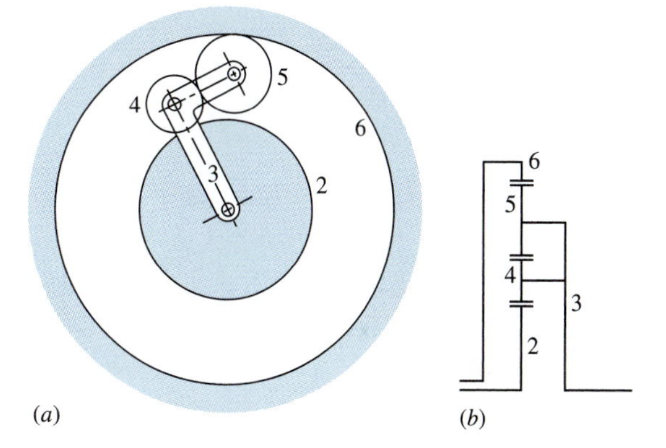

Figure 9.9 An epicyclic gear train with two planet gears.

allows lower forces by more force sharing. However, these additional planet gears do not change the kinematic characteristics at all. For this reason, we generally indicate only a single planet in the illustrations and problems in Chapter 9, although an actual machine would probably be designed with planets in trios.

The simple epicyclic gear train together with its schematic designation in Fig. 9.7 illustrates how the motion of the planet gear can be transmitted to another central gear. The second central gear in this case is gear 5, an internal gear. In Fig. 9.7*a*, internal gear 5 is stationary, but this is not a requirement, as illustrated in Fig. 9.7*b*. Figure 9.8 illustrates a similar arrangement with the difference that both central gears are external gears. Note, in Fig. 9.8, that the double-planet gears are mounted on a single-planet shaft and that each planet gear is in mesh with a separate sun gear rotating at a different speed.

In any case, no matter how many planets are used, only one planet carrier or arm may be used. This principle is illustrated in Fig. 9.6, in which redundant planets are used, and in Fig. 9.9, where two planets are used to alter the kinematic performance.

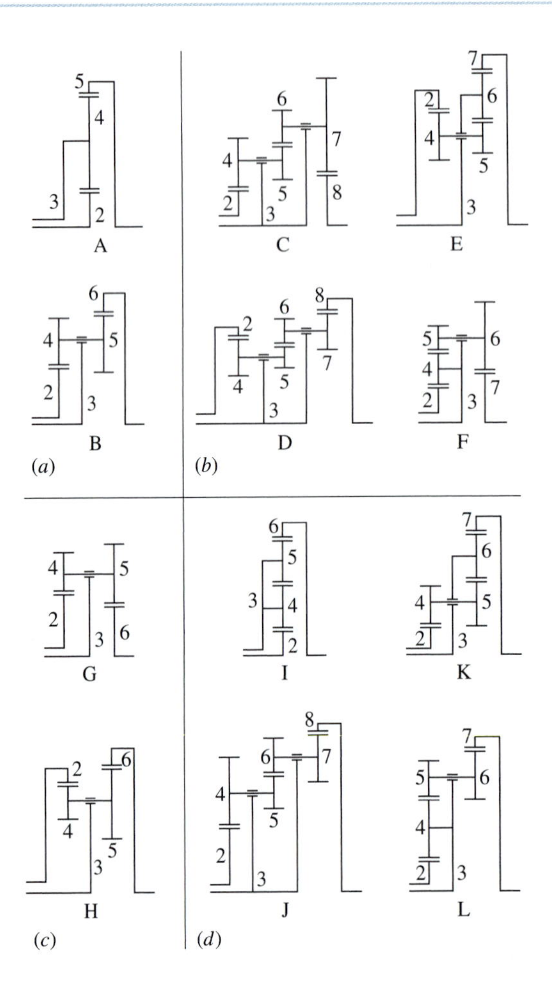

Figure 9.10 All 12 possible epicyclic gear train types according to Lévai. (Reproduced with permission of the author.)

According to Lévai, only 12 variations of epicyclic gear trains are possible; they are all illustrated in schematic form in Fig. 9.10 as Lévai arranged them. In all variations, the arm (the planet carrier) is shown as link number 3. The trains in Fig. 9.10*a* and Fig. 9.10*c* are simple trains in which the planet gears mesh with both sun gears. The trains illustrated in Fig. 9.10*b* and Fig. 9.10*d* have planet gear pairs that are partly in mesh with each other and partly in mesh with the sun gears.

9.5 BEVEL GEAR EPICYCLIC TRAINS

The bevel gear train illustrated in Fig. 9.11 is called *Humpage's reduction gear*. Bevel gear epicyclic trains are used quite frequently, and they are similar to spur gear epicyclic trains except that their axes of rotation are not all on parallel shafts. The train of Fig. 9.11 is, in fact, a double epicyclic train, and the spur gear counterpart of each can be found in Fig. 9.10. We will find in the next section that the analysis of such trains can be done the same as for spur gear trains.

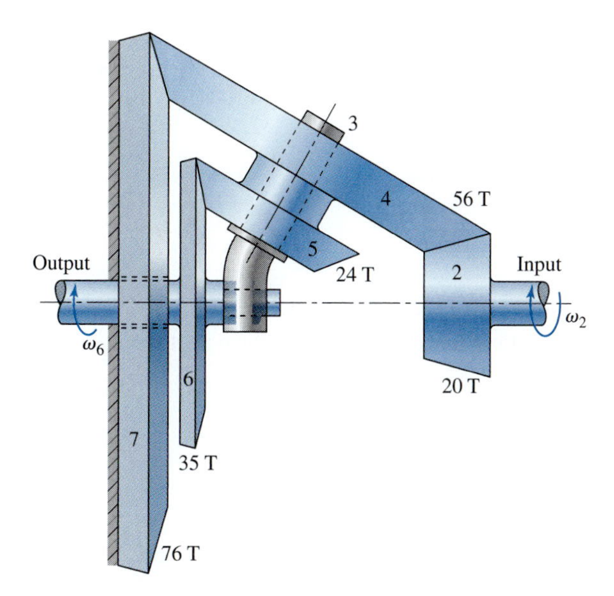

Figure 9.11 Humpage's reduction gear.

9.6 ANALYSIS OF EPICYCLIC GEAR TRAINS BY FORMULA

Figure 9.12 illustrates a planetary gear train composed of a sun gear 2, an arm or planet carrier 3, and planet gears 4 and 5. Using the apparent angular velocity equation, Eq. (3.10), we can write that the angular velocity of gear 2 as it would appear from a coordinate system fixed to the arm 3 is

$$\omega_{2/3} = \omega_2 - \omega_3. \tag{a}$$

Also, the angular velocity of gear 5 as it would appear from the coordinate system fixed to arm 3 is

$$\omega_{5/3} = \omega_5 - \omega_3. \tag{b}$$

Dividing Eq. (*b*) by Eq. (*a*) gives

$$\frac{\omega_{5/3}}{\omega_{2/3}} = \frac{\omega_5 - \omega_3}{\omega_2 - \omega_3}. \tag{c}$$

Equation (*c*) expresses the ratio of the apparent angular velocity of gear 5 to that of gear 2 with both taken as they would appear from arm 3. This ratio, which is proportional to the tooth numbers, appears the same whether arm 3 is rotating or not; it is the first-order kinematic coefficient of the gear train. Therefore, the first-order kinematic coefficient can be expressed as

$$\theta'_{52/3} = \frac{\omega_5 - \omega_3}{\omega_2 - \omega_3}. \tag{d}$$

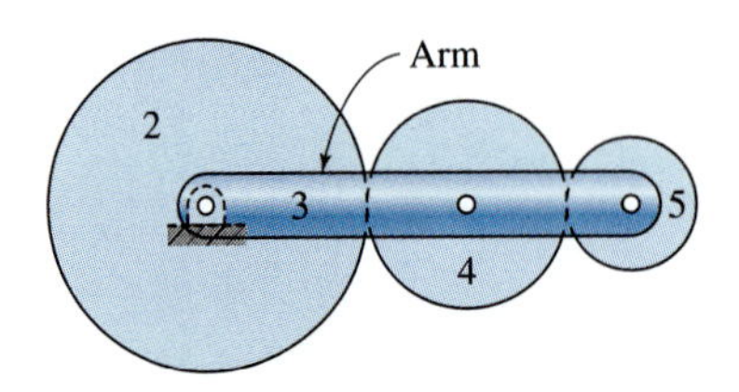

Figure 9.12

An equation similar to Eq. (*d*) is all that we need to find the angular velocities in a planetary gear train. It is convenient to express it in the form as it would appear from the arm,

$$\theta'_{LF/A} = \frac{\omega_L - \omega_A}{\omega_F - \omega_A}, \tag{9.4}$$

where

ω_F = angular velocity of the first gear in the train;
ω_L = angular velocity of the last gear in the train; and
ω_A = angular velocity of the arm.

Note that Eq. (*d*) and Eq. (9.4) can be expressed entirely in terms of kinematic coefficients, that is,

$$\theta'_{52/3} = \frac{\theta'_5 - \theta'_3}{\theta'_2 - \theta'_3} \tag{e}$$

and

$$\theta'_{LF/A} = \frac{\theta'_L - \theta'_A}{\theta'_F - \theta'_A}, \tag{9.5}$$

where

θ'_F = kinematic coefficient of the first gear in the train;
θ'_L = kinematic coefficient of the last gear in the train;
θ'_A = kinematic coefficient of the arm.

The following examples help to illustrate the use of Eqs. (9.4) and (9.5).

EXAMPLE 9.1

Figure 9.8 illustrates a reverted planetary gear train. Gear 2 is fastened to its shaft and is driven at 250 rev/min in a clockwise direction. Gears 4 and 5 are planet gears that are joined but are free to turn on the shaft carried by the arm. Gear 6 is stationary. Find the speed and direction of rotation of the arm.

SOLUTION

We must first decide which gears to designate the first and last members of the train. Because the speeds of gears 2 and 6 are both given, then either gear may be chosen as the first. The choice makes no difference to the results, but once the decision is made, it may not be changed. Here we choose gear 2 as first; therefore, gear 6 is last. Thus, choosing counterclockwise as positive gives

$$\omega_F = \omega_2 = -250 \text{ rev/min} \quad \text{and} \quad \omega_L = \omega_6 = 0 \text{ rev/min}$$

and, according to Eq. (9.3), the first-order kinematic coefficient is

$$\theta'_{LF} = \theta'_{62} = \left(\frac{16}{34}\right)\left(\frac{20}{30}\right) = \frac{16}{51},$$

where the positive sign is chosen because there are two external contacts.

Substituting the values into Eq. (9.4) gives

$$\theta'_{LF/A} = \frac{16}{51} = \frac{0 - \omega_3}{-250 \text{ rev/min} - \omega_3}$$

$$\omega_A = \omega_3 = 114.3 \text{ rev/min ccw.} \qquad \textit{Ans.}$$

EXAMPLE 9.2

In the bevel gear train in Fig. 9.11, the input is gear 2 and the output is gear 6, which is connected to the output shaft. The arm 3 turns freely on the output shaft and carries planets 4 and 5. Gear 7 is fixed to the frame. What is the output shaft speed if gear 2 rotates at 2000 rev/min?

SOLUTION

The problem is solved in two steps. In the first step we consider the train to be made up of gears 2, 4, and 7 and calculate the rotational speed of the arm. Thus,

$$\omega_F = \omega_2 = 2\,000 \text{ rev/min} \quad \text{and} \quad \omega_L = \omega_7 = 0 \text{ rev/min}$$

and, according to Eq. (9.3),

$$\theta'_{LF} = \theta'_{72} = -\left(\frac{56}{76}\right)\left(\frac{20}{56}\right) = -\frac{5}{19}.$$

The negative sign is chosen because, if gear 7 were not fixed, it would appear to rotate in the direction opposite to that of gear 2 when viewed from a coordinate system fixed to arm 3.

Substituting into Eq. (9.4) and solving for the angular velocity of arm 3 gives

$$\theta'_{LF/A} = -\frac{5}{19} = \frac{0 - \omega_3}{2\,000 \text{ rev/min} - \omega_3}$$

$$\omega_A = \omega_3 = 416.67 \text{ rev/min.}$$

Next we consider the train as composed of gears 2, 4, 5, and 6. Then $\omega_F = \omega_2 =$ 2000 rev/min, as before, and $\omega_L = \omega_6$, which is to be determined. The first-order kinematic coefficient of the train is

$$\theta'_{LF} = \theta'_{62} = \left(\frac{24}{35}\right)\left(-\frac{20}{56}\right) = -\frac{12}{49}.$$

Again, the negative sign is chosen because gear 6 would appear to rotate in the direction opposite to that of gear 2 when viewed from a coordinate system fixed to arm 3.

Substituting into Eq. (9.4) again and solving for ω_6, with ω_3 known above, now gives

$$\theta'_{LF/A} = -\frac{12}{49} = \frac{\omega_L - 416.67 \text{ rev/min}}{2\,000 \text{ rev/min} - 416.67 \text{ rev/min}}.$$

Rearranging this equation, the speed of gear 6 (and the output shaft) is

$$\omega_L = \omega_6 = 28.91 \text{ rev/min}. \qquad \textit{Ans.}$$

Because the result is positive, we conclude that the output shaft rotates in the same direction as the input shaft 2 with a speed reduction of 2000:28.91 or 69.18:1.

EXAMPLE 9.3

Consider the simple planetary gear train illustrated in Fig. 9.13 in which the ring gear h is locked (that is, the angular velocity of the ring gear is $\omega_h = 0$) and the input is the planet carrier (the arm). The dimensions of the ring gear h, the planet gear j, the sun gear k, and the arm A are $R_h = 200$ mm, $R_j = 50$ mm, $R_k = 100$ mm, and $R_A = 150$ mm, respectively. If the input angular velocity is $\omega_A = \omega_i = 10$ rad/s ccw, determine the angular velocities of the planet gear j and the sun gear k.

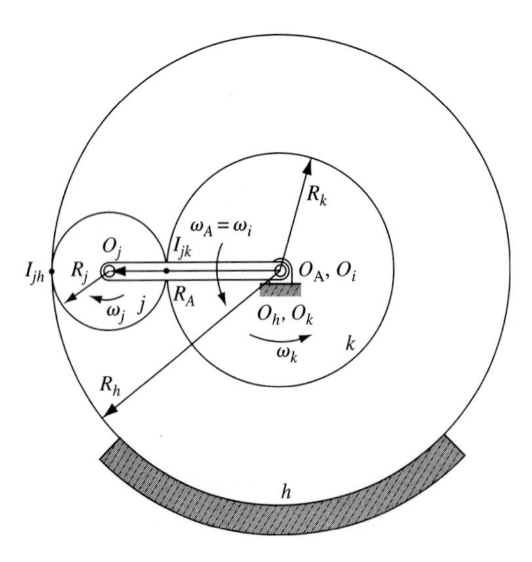

Figure 9.13 Example 9.3. A simple planetary gear train with the ring gear locked.

SOLUTION

Using Eq. (9.4), the first-order kinematic coefficient of the planet gear j in rolling contact with rotating ring gear h can be written as

$$\theta'_{jh/A} = \pm\frac{R_h}{R_j} = \frac{\omega_j - \omega_A}{\omega_h - \omega_A}. \tag{1a}$$

Similarly, the first-order kinematic coefficient of the planet gear j in rolling contact with the sun gear k can be written as

$$\theta'_{jk/A} = \pm\frac{R_k}{R_j} = \frac{\omega_j - \omega_A}{\omega_k - \omega_A}. \tag{1b}$$

Note that if the ring gear is removed and the arm is locked (that is, $\omega_A = 0$), then the planetary gear train reduces to an ordinary gear train, that is, Eq. (1b) reduces to

$$\theta'_{jk/A} = \pm\frac{R_k}{R_j} = \frac{\omega_j}{\omega_k}. \tag{2}$$

Because there is internal contact between the planet gear and the fixed ring gear, then the positive sign is used in Eq. (1a). Then, rearranging the equation and substituting $\omega_h = 0$ gives

$$R_j\omega_j = -(R_h - R_j)\omega_A = -R_A\omega_A. \tag{3}$$

Therefore, the angular velocity of the planet gear is

$$\omega_j = -\frac{R_A\omega_A}{R_j} = -\frac{(150\text{ mm})(10\text{ rad/s})}{50\text{ mm}} = -30\text{ rad/s}. \qquad \textit{Ans.}$$

The negative sign indicates that the direction of the angular velocity of the planet gear is opposite to the direction of the arm and, therefore, clockwise.

Similarly, because there is external contact between the planet gear and the sun gear, the negative sign is used in Eq. (1b) and the equation can be written as

$$R_k\omega_k = -R_j\omega_j + (R_k + R_j)\omega_A. \tag{4a}$$

The angular velocity of the sun gear can be expressed in terms of the input angular velocity of the arm by substituting Eq. (3) into Eq. (4a) and simplifying, that is,

$$R_k\omega_k = (R_k + R_h)\omega_A = 2R_A\omega_A. \tag{4b}$$

Therefore, the angular velocity of the sun gear is

$$\omega_k = \frac{2R_A\omega_A}{R_k} = \frac{2(150\text{ mm})(10\text{ rad/s})}{100\text{ mm}} = 30\text{ rad/s}. \qquad \textit{Ans.}$$

The positive result indicates that the direction of the angular velocity of the sun gear is the same as the arm and, therefore, counterclockwise.

EXAMPLE 9.4

Consider the compound planetary gear train that is commonly used in a drill or an electric screwdriver, illustrated in the schematic diagram in Fig. 9.14. The gear and the arm dimensions are $R_1 = 150$ mm, $R_{2A} = R_{4A} = 100$ mm, and $R_3 = R_4 = R_5 = R_6 = 50$ mm. The angular velocity of the input shaft (and arm 2) is $\omega_{2A} = 10$ rad/s ccw (looking from the left) and the ring gear 1 is locked, that is, the angular velocity of the ring gear is $\omega_1 = 0$. Determine the angular velocities of gears 3, 4, 5, and 6.

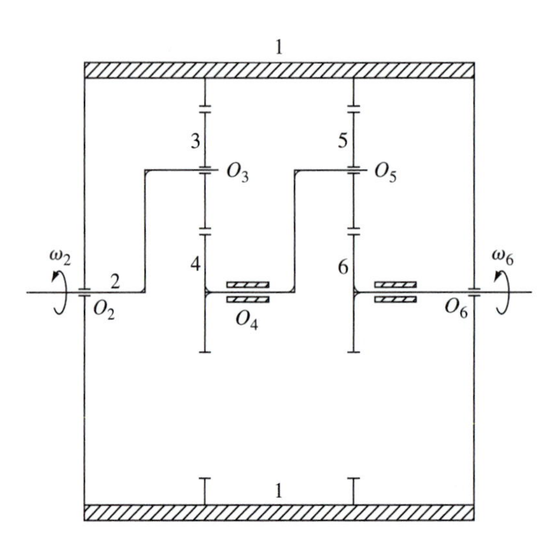

Figure 9.14 Example 9.4. Schematic diagram of a compound planetary gear train.

SOLUTION

This gear train is composed of two simple planetary gear trains in series; that is, the first planetary gear train is composed of ring gear 1, input arm 2, planet gear 3, and sun gear 4 and the second planetary gear train is composed of ring gear 1, arm 4, planet gear 5, and output sun gear 6. Note that the angular velocities of gear 4 and arm 4 are equal in magnitude and direction because they are rigidly attached. Also, note that gear 4 acts as both the output of the first planetary gear train and the input to the second planetary gear train. Using Eq. (9.4), the first-order kinematic coefficient of gear 4 with respect to ring gear 1 can be written as

$$\theta'_{41/2A} = -\frac{R_1}{R_4} = \frac{\omega_4 - \omega_{2A}}{\omega_1 - \omega_{2A}}. \tag{1a}$$

The correct sign is negative because gears 4 and 3 are in external contact and gears 3 and 1 are in internal contact. Substituting the given data into Eq. (1a) gives

$$-\frac{150 \text{ mm}}{50 \text{ mm}} = \frac{\omega_4 - 10 \text{ rad/s}}{0 - 10 \text{ rad/s}}. \tag{1b}$$

Therefore, the angular velocity of gear 4 is

$$\omega_4 = 40 \text{ rad/s}, \qquad \textit{Ans.}$$

where the positive result implies that gear 4 is rotating in the same direction as the input gear 2, that is, counterclockwise.

Using Eq. (9.4), the first-order kinematic coefficient of gear 3 with respect to gear 4 can be written as

$$\theta'_{34/2A} = -\frac{R_4}{R_3} = \frac{\omega_3 - \omega_{2A}}{\omega_4 - \omega_{2A}}, \tag{2a}$$

where the negative sign denotes external contact between gears 3 and 4. Therefore, the first-order kinematic coefficient is

$$\theta'_{34/2A} = -\frac{50 \text{ mm}}{50 \text{ mm}} = \frac{\omega_3 - 10 \text{ rad/s}}{40 \text{ rad/s} - 10 \text{ rad/s}}. \tag{2b}$$

Therefore, the angular velocity of gear 3 is

$$\omega_3 = -20 \text{ rad/s}, \qquad \textit{Ans.}$$

where the negative sign indicates that gear 3 is rotating in the direction opposite to the input gear 2, that is, clockwise.

Using Eq. (9.4), the first-order kinematic coefficient of gear 6 with respect to the ring gear 1 can be written as

$$\theta'_{61/4A} = -\frac{R_1}{R_6} = \frac{\omega_6 - \omega_{4A}}{\omega_1 - \omega_{4A}}. \tag{3a}$$

The correct sign is negative because gears 6 and 5 are in external contact and gears 5 and 1 are in internal contact. Substituting the given data into Eq. (3a) gives

$$-\frac{150 \text{ mm}}{50 \text{ mm}} = \frac{\omega_6 - 40 \text{ rad/s}}{0 - 40 \text{ rad/s}}. \tag{3b}$$

Therefore, the angular velocity of gear 6 is

$$\omega_6 = 160 \text{ rad/s}, \qquad \textit{Ans.}$$

where the positive result implies that gear 4 is rotating in the same direction as the input gear 2, that is, counterclockwise.

Using Eq. (9.4), the first-order kinematic coefficient of gear 5 with respect to gear 6 can be written as

$$\theta'_{56/4A} = -\frac{R_6}{R_5} = \frac{\omega_5 - \omega_{4A}}{\omega_6 - \omega_{4A}}, \tag{4a}$$

where the negative sign denotes external contact between gears 5 and 6. Substituting the known data, the first-order kinematic coefficient is

$$\theta'_{56/4A} = -\frac{50 \text{ mm}}{50 \text{ mm}} = \frac{\omega_5 - 40 \text{ rad/s}}{160 \text{ rad/s} - 40 \text{ rad/s}}. \tag{4b}$$

Therefore, the angular velocity of gear 5 is

$$\omega_5 = -80 \text{ rad/s}, \qquad \textit{Ans.}$$

where the negative sign implies that gear 5 is rotating in the direction opposite to the input gear 2, that is, clockwise.

ALTERNATE SOLUTION

The rolling contact constraint between planet gear 3 and ring gear 1 (see Section 3.12) can be written as

$$\frac{\theta'_{12} - \theta'_{22}}{\theta'_{32} - \theta'_{22}} = \frac{R_3}{R_1} = \frac{50 \text{ mm}}{150 \text{ mm}} = \frac{1}{3}, \tag{1}$$

where the positive sign is chosen because planet gear 3 has internal contact with ring gear 1.

Because ring gear 1 is fixed, the kinematic coefficient of the ring gear is $\theta'_{12} = 0$ and, by definition, the kinematic coefficient of input arm 2 is $\theta'_{22} = 1$. Substituting these into Eq. (1), the kinematic coefficient of planet gear 3 with respect to input gear 2 is

$$\theta'_{32} = -2 \text{ rad/rad}, \tag{2}$$

where the negative sign indicates that gear 3 is rotating in the direction opposite to input gear 2, that is, clockwise. Therefore, the angular velocity of gear 3 is

$$\omega_3 = \theta'_{32}\omega_2 = -20 \text{ rad/s}. \qquad \textit{Ans.}$$

The rolling contact constraint between sun gear 4 and planet gear 3 can be written from Eq. (1) as

$$\frac{\theta'_{32} - \theta'_{22}}{\theta'_{42} - \theta'_{22}} = -\frac{R_4}{R_3} = -1, \tag{3}$$

where the negative sign is chosen because planet gear 3 has external contact with sun gear 4.

Substituting Eq. (2) and $\theta'_{22} = 1$ and the known geometry into Eq. (3), the kinematic coefficient of the sun gear 4 with respect to gear 2 is

$$\theta'_{42} = 4 \text{ rad/rad}, \tag{4}$$

where the positive result implies that gear 4 is rotating in the same direction as the input gear 2, that is, counterclockwise. Therefore, the angular velocity of gear 4 is

$$\omega_4 = \theta'_{42}\omega_2 = 40 \text{ rad/s ccw}. \qquad \textit{Ans.}$$

The rolling contact constraint between planet gear 5 and ring gear 1 can be written from Eq. (1) as

$$\frac{\theta'_{12} - \theta'_{42}}{\theta'_{52} - \theta'_{42}} = \frac{R_5}{R_1} = \frac{50 \text{ mm}}{150 \text{ mm}} = \frac{1}{3}, \tag{5}$$

where the positive sign is used because planet gear 5 has internal contact with ring gear 1.

Substituting Eq. (4) and $\theta'_{12} = 0$, the kinematic coefficient of planet gear 5 with respect to gear 2 is

$$\theta'_{52} = -8 \text{ rad/rad}, \tag{6}$$

where the negative sign indicates that gear 5 is rotating in the direction opposite to the input gear 2, that is, clockwise. Therefore, the angular velocity of gear 5 is

$$\omega_5 = \theta'_{52}\omega_2 = -80 \text{ rad/s.} \qquad \textit{Ans.}$$

The rolling contact constraint between sun gear 6 and planet gear 5 can be written from Eq. (5) as

$$\frac{\theta'_{52} - \theta'_{42}}{\theta'_{62} - \theta'_{42}} = -\frac{R_6}{R_5} = -1, \tag{7}$$

where the negative sign is used because of the external contact between planet gear 5 and sun gear 6.

Substituting Eqs. (4) and (6) and the known geometry into Eq. (7), the kinematic coefficient of sun gear 6 with respect to gear 2 is

$$\theta'_{62} = 16 \text{ rad/rad}, \tag{8}$$

where the positive result implies that gear 6 is rotating in the same direction as input gear 2. Therefore, the angular velocity of gear 6 is

$$\omega_6 = \theta'_{62}\omega_2 = 160 \text{ rad/s.} \qquad \textit{Ans.}$$

EXAMPLE 9.5

Consider the planetary gear train illustrated in the schematic diagram in Fig. 9.15. The radius of the fixed sun gear is $R_1 = 100$ mm, the radius of the input gear 2 is $R_2 = 100$ mm, the radius of the ring gear 4 is $R_4 = 200$ mm, and the radii of gears 3 and 5 are $R_3 = R_5 = 50$ mm. The angular velocity of the input shaft 2 is $\omega_2 = 10$ rad/s ccw (looking from the left). Determine the angular velocities of gears 3, 4, and 5.

SOLUTION

The rolling contact constraint between input gear 2 and gear 5 can be written as

$$\frac{\theta'_{52}}{\theta'_{22}} = -\frac{R_2}{R_5} = -\frac{100 \text{ mm}}{50 \text{ mm}} = -2. \tag{1}$$

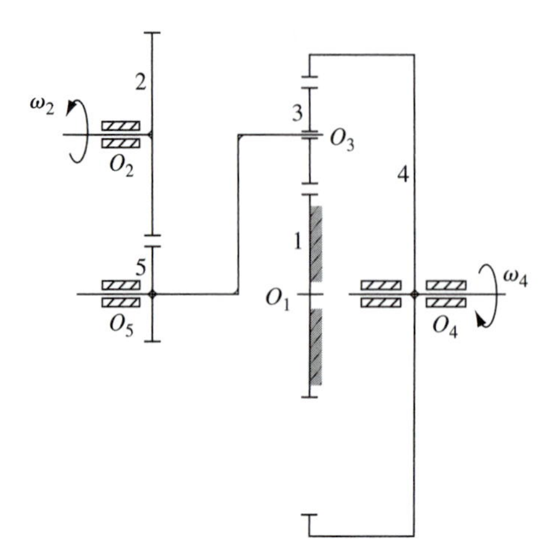

Figure 9.15 Example 9.5. Schematic diagram of a planetary gear train.

Because the kinematic coefficient of the input shaft 2 is $\theta'_{22} = 1$ rad/rad, the kinematic coefficient of gear 5 is

$$\theta'_{52} = -2 \text{ rad/rad}, \tag{2}$$

where the negative sign implies that gear 5 is rotating in the opposite direction to input gear 2, that is, clockwise. Therefore, the angular velocity of gear 5 is

$$\omega_5 = \theta'_{52}\omega_2 = -20 \text{ rad/s}. \qquad \textit{Ans.}$$

The rolling contact constraint between planet gear 3 and fixed sun gear 1 can be written as

$$-\frac{R_3}{R_1} = \frac{\theta'_{12} - \theta'_{5A2}}{\theta'_{32} - \theta'_{5A2}}, \tag{3}$$

which, for the given conditions, becomes

$$-\frac{50 \text{ mm}}{100 \text{ mm}} = \frac{0 + 2 \text{ rad/rad}}{\theta'_{32} + 2 \text{ rad/rad}}.$$

Therefore, the kinematic coefficient of gear 3 with respect to fixed sun gear 1 is

$$\theta'_{32} = -6 \text{ rad/rad} \tag{4}$$

and the angular velocity of gear 3 is

$$\omega_3 = \theta'_{32}\omega_2 = -60 \text{ rad/s}, \qquad \textit{Ans.}$$

where the negative sign indicates that the direction of ω_3 is clockwise.

The rolling contact constraint between ring gear 4 and planet gear 3 can be written as

$$\frac{R_3}{R_4} = \frac{\theta'_{42} - \theta'_{5A2}}{\theta'_{32} - \theta'_{5A2}}, \tag{5}$$

which, using Eqs. (2) and (4) and the given dimensions, can be written as

$$\frac{50\text{ mm}}{200\text{ mm}} = \frac{\theta'_{42} + 2\text{ rad/rad}}{-6\text{ rad/rad} + 2\text{ rad/rad}}.$$

Therefore, the kinematic coefficient of gear 4 with respect to fixed sun gear 1 is

$$\theta'_{42} = -3\text{ rad/rad} \tag{6}$$

and the angular velocity of gear 4 is

$$\omega_4 = \theta'_{42}\omega_2 = -30\text{ rad/s}, \qquad \textit{Ans.}$$

where the negative sign indicates that the direction of ω_4 is clockwise.

The kinematic coefficients in Eqs. (2), (4), and (6) define the geometry of the gear train and provide the capability for obtaining the angular velocities (magnitudes and directions) in terms of the input angular velocity ω_2.

9.7 TABULAR ANALYSIS OF EPICYCLIC GEAR TRAINS

Another method of determining the rotational speeds of epicyclic gear trains uses the principle of superposition. The total analysis is carried out by finding the apparent rotations of the components with respect to the arm or planet carrier and then summing with the rotation of the components as if all the gears are fixed to the arm. The process is easily carried out in a tabular procedure.

Figure 9.7 illustrates a planetary gear train composed of sun gear 2, planet carrier (arm) 3, planet gear 4, and internal gear 5 that is in mesh with the planet gear. Because this gear train has two degrees of freedom, we might reasonably specify the angular velocities of the sun gear and of the arm and wish to determine the angular velocity of the internal gear.

The analysis can be carried out in the following three steps.

1. Consider all gears (including a fixed gear, if any) to be locked to the arm and allow the arm to rotate with angular velocity ω_A. Tabulate the angular velocities of all components under this condition as also equal to ω_A.
2. Free all constraints of step 1, fix the arm, and allow some other gear B (such as the sun gear) to rotate with angular velocity $\omega_{B/A}$. Tabulate the apparent angular velocities of all other gears with respect to the arm as multiples of $\omega_{B/A}$.
3. Add the angular velocities of each gear from steps 1 and 2 and apply the given input velocities to find numeric values for ω_A and $\omega_{B/A}$.

A few worked examples will clarify how this can be done in a convenient tabular procedure.

EXAMPLE 9.6

Let us assume the tooth numbers of Fig. 9.7, and let the angular velocities of the sun gear and the arm be $\omega_2 = 100$ rev/min and $\omega_3 = 200$ rev/min, respectively, both in the ccw direction, chosen positive. What is the angular velocity of internal ring gear 5?

SOLUTION

The solution process described by the three steps explained above is demonstrated in Table 9.1.

TABLE 9.1 Tabular analysis for Examples 9.6 and 9.7

Step Number	Gear 2	Arm 3	Gear 4	Gear 5
1. Gears fixed to arm	ω_3	ω_3	ω_3	ω_3
2. Arm fixed	$\omega_{2/3}$	0	$(-40/20)\omega_{2/3}$	$(20/80)\ (-40/20)\omega_{2/3}$
3. Total	$\omega_3 + \omega_{2/3}$	ω_3	$\omega_3 + (-40/20)\omega_{2/3}$	$\omega_3 + (20/80)\ (-40/20)\omega_{2/3}$

Next, comparing the given input velocities with the bottom row of columns 2 and 3, we see that because $\omega_2 = \omega_3 + \omega_{2/3} = 100$ rev/min and $\omega_3 = 200$ rev/min, then $\omega_{2/3} = -100$ rev/min. Therefore, from the bottom row of column 5, the angular velocity of internal ring gear 5 can be written as

$$\omega_5 = \omega_3 + (20/80)(-40/20)\omega_{2/3},$$

that is,

$$\omega_5 = (200 \text{ rev/min}) + (-40/80)(-100 \text{ rev/min}) = 250 \text{ rev/min},$$

where the positive result implies that gear 5 is rotating in the same direction as the positive inputs, that is, counterclockwise.

EXAMPLE 9.7

What is the angular velocity of internal gear 5 of Fig. 9.7 if gear 2 rotates at 100 rev/min clockwise while arm 3 rotates at 200 rev/min counterclockwise?

SOLUTION

The analysis is identical to that performed in Table 9.1 for Example 9.6. However, the input velocities have now changed. Still, taking counterclockwise as positive, the bottom row of columns 2 and 3 indicate that because

$$\omega_2 = \omega_3 + \omega_{2/3} = -100 \text{ rev/min} \quad \text{and} \quad \omega_3 = 200 \text{ rev/min},$$

then $\omega_{2/3} = -300$ rev/min. Therefore, the bottom row of column 5 indicates

$$\omega_5 = \omega_3 + (20/80)(-40/20)\omega_{2/3} = (200 \text{ rev/min}) + (-40/80)(-300 \text{ rev/min})$$
$$= 350 \text{ rev/min ccw.} \qquad \textit{Ans.}$$

EXAMPLE 9.8

The planetary gear train illustrated in Fig. 9.16 is called Ferguson's paradox.* Gear 2 is fixed to the frame. Arm 3 and gears 4 and 5 are free to turn upon the shaft. Gears 2, 4, and 5 have tooth numbers of 100, 101, and 99, respectively, all cut with the same circular pitch (but with slightly different pitch circle radii) so that planet gear 6 meshes with all of them. Find the angular rotations of gears 4 and 5 when arm 3 is given one counterclockwise turn.

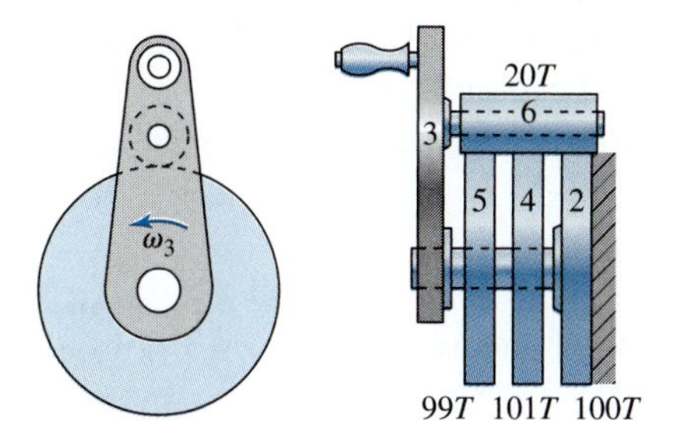

Figure 9.16 Example 9.8. Ferguson's paradox.

SOLUTION

The solution process is indicated in Table 9.2.

TABLE 9.2 Tabular analysis for Example 9.8

Step Number	Gear 2	Arm 3	Gear 4	Gear 5	Gear 6
1. Gears fixed to arm	$\Delta\theta_3$	$\Delta\theta_3$	$\Delta\theta_3$	$\Delta\theta_3$	$\Delta\theta_3$
2. Arm fixed	$\Delta\theta_{2/3}$	0	$(-20/101)(-100/20)\Delta\theta_{2/3}$	$(-20/99)(-100/20)\Delta\theta_{2/3}$	$(-100/20)\Delta\theta_{2/3}$
3. Total	$\Delta\theta_3 + \Delta\theta_{2/3}$	$\Delta\theta_3$	$\Delta\theta_3 + (-20/101)(-100/20)\Delta\theta_{2/3}$	$\Delta\theta_3 + (-20/99)(-100/20)\Delta\theta_{2/3}$	$\Delta\theta_3 + (-100/20)\Delta\theta_{2/3}$

It should be noted that angular displacements are indicated in Table 9.2 instead of angular velocities. This comes from the nature of the question asked and recognizing that, during a chosen time interval, $\Delta\theta = \omega\Delta t$ for each of the elements. According to the problem statement, we choose the time interval Δt such that for the arm $\Delta\theta_3 = 1$ rev ccw. Then, for gear 2 to remain stationary, column 2 indicates that $\Delta\theta_3 + \Delta\theta_{2/3} = (1 \text{ rev}) + \Delta\theta_{2/3} = 0$

* James Ferguson (1710–1776), Scottish physicist and astronomer, first published this device under the title *The Description and Use of a New Machine Called the Mechanical Paradox,* London, 1764.

and, therefore, $\Delta\theta_{2/3} = -1$ rev. Finally, from the bottom row of columns 4 and 5 we determine that

$$\Delta\theta_4 = \Delta\theta_3 + (-20/101)(-100/20)\Delta\theta_{2/3} = (1 \text{ rev}) + (100/101)(-1 \text{ rev}) = 1/101 \text{ rev}$$

and

$$\Delta\theta_5 = \Delta\theta_3 + (-20/99)(-100/20)\Delta\theta_{2/3} = (1 \text{ rev}) + (100/99) - 1 \text{ rev}) = -1/99 \text{ rev}.$$

Thus, when arm 3 is turned 1 rev ccw, gear 4 rotates 1/101 rev ccw, whereas gear 5 turns 1/99 rev cw. *Ans.*

EXAMPLE 9.9

The overdrive unit illustrated in Fig. 9.17 is sometimes used to follow a standard automotive transmission to further reduce engine speed. The engine speed (after the transmission) corresponds to the speed of planet carrier 3, and the drive shaft speed corresponds to that of gear 5; sun gear 2 is held stationary. Determine the percentage reduction in engine speed obtained when the overdrive is active.

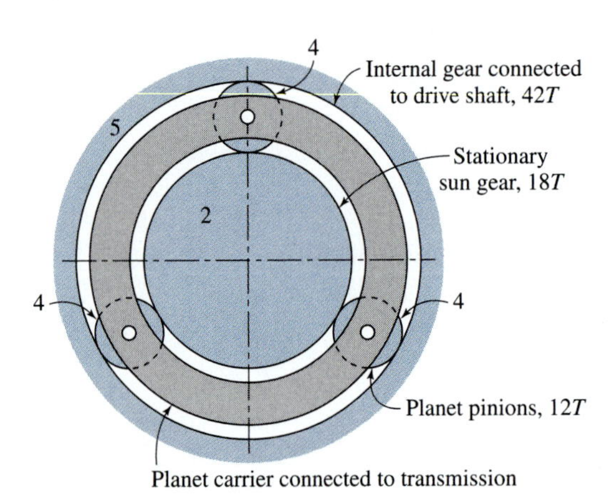

Figure 9.17 Example 9.9. Overdrive unit.

SOLUTION

The analysis for this problem is demonstrated in Table 9.3. For gear 2 to remain stationary, column 2 indicates that $\omega_3 + \omega_{2/3} = 0$; therefore, $\omega_{2/3} = -\omega_3$. Putting this into column 5 indicates that

$$\omega_5 = \omega_3 + (12/42)(-18/12)\omega_{2/3} = \omega_3 + (-18/42)(-\omega_3) = 1.429\omega_3.$$

Therefore, the percentage reduction in engine speed is

$$(1.429\omega_3 - 1.0\omega_3)/(1.429\omega_3) = 0.300 = 30\%. \qquad \textit{Ans.}$$

TABLE 9.3 Tabular analysis for Example 9.9

Step Number	Gear 2	Arm 3	Gear 4	Gear 5
1. Gears fixed to arm	ω_3	ω_3	ω_3	ω_3
2. Arm fixed	$\omega_{2/3}$	0	$(-18/12)\omega_{2/3}$	$(12/42)(-18/12)\omega_{2/3}$
3. Total	$\omega_3 + \omega_{2/3}$	ω_3	$\omega_3 + (-18/12)\omega_{2/3}$	$\omega_3 + (12/42)(-18/12)\omega_{2/3}$

9.8 SUMMERS AND DIFFERENTIALS

Figure 9.18 illustrates a variety of mechanisms used as computing devices. Because each of these is a two-degree-of-freedom mechanism, two input motion variables must be defined so that the motions of the remaining elements of the system are determined. The equation below each of these mechanisms indicates that the output position is a direct measure of the sum of the two input positions. For this reason, such a mechanism is referred to as an *adder* or a *summer*.

The spur gear differential of Fig. 9.19 helps in visualizing its action. If planet carrier 2 is held stationary and gear 3 is turned by some amount $\Delta\theta_3$, then gear 8 turns in the opposite direction by an amount $\Delta\theta_8 = -\Delta\theta_3$. If the planet carrier is also allowed to turn, then $\Delta\theta_8 = \Delta\theta_2 - \Delta\theta_3$. It is for this reason that this two-degree-of-freedom mechanism is called a *differential*. Of course, a better force balance is obtained by employing several sets of planets equally spaced about the sun gears; three sets is usual. Also, you will note in Fig. 9.19 that planets 4, 5, 6, and 7 are identical; by making planets 4 and 7 longer (thicker), they meet with each other and eliminate the need for planets 5 and 6.

If a spur gear differential were used on the driving axles of an automobile, then shafts *A* and *B* of Fig. 9.19 would drive the right and left wheels, respectively, whereas arm 2 receives power from a main drive shaft connected to the transmission.

It is interesting that a differential was used in China, long before the invention of the magnetic compass, to indicate geographic direction. In Fig. 9.20, each wheel of a carriage drives a vertical shaft through pin wheels. The right-hand shaft drives the upper pin wheel and the left-hand shaft drives the lower pin wheel of the differential illustrated in Fig. 9.21. When the cart is driven in a straight line, the upper and lower pin wheels rotate at the same speed but in opposite directions. Thus, the planet gear turns about its own center, but the axle to which it is mounted remains stationary and so the character continues to point in a constant direction. However, when the cart makes a turn, one of the pin wheels rotates faster than the other, causing the planet axle to turn just enough to cause the character to continue to point in the same geographic direction as before.

Figure 9.22 is a schematic drawing of the ordinary bevel gear automotive differential. The drive shaft pinion and the ring gear are normally hypoid gears (see Chapter 8). The ring gear acts as the planet carrier, and its speed can be calculated as for a simple gear train when the speed of the drive shaft is given. Gears 5 and 6 are connected, respectively, to each of the rear wheels and, when the car is traveling in a straight line, these two gears rotate in the same direction with exactly the same speed. Thus, for straight-line motion of the car, there is no relative motion between the planet gears and gears 5 and 6. The planet gears, in effect, serve only as keys to transmit motion from the planet carrier to both wheels.

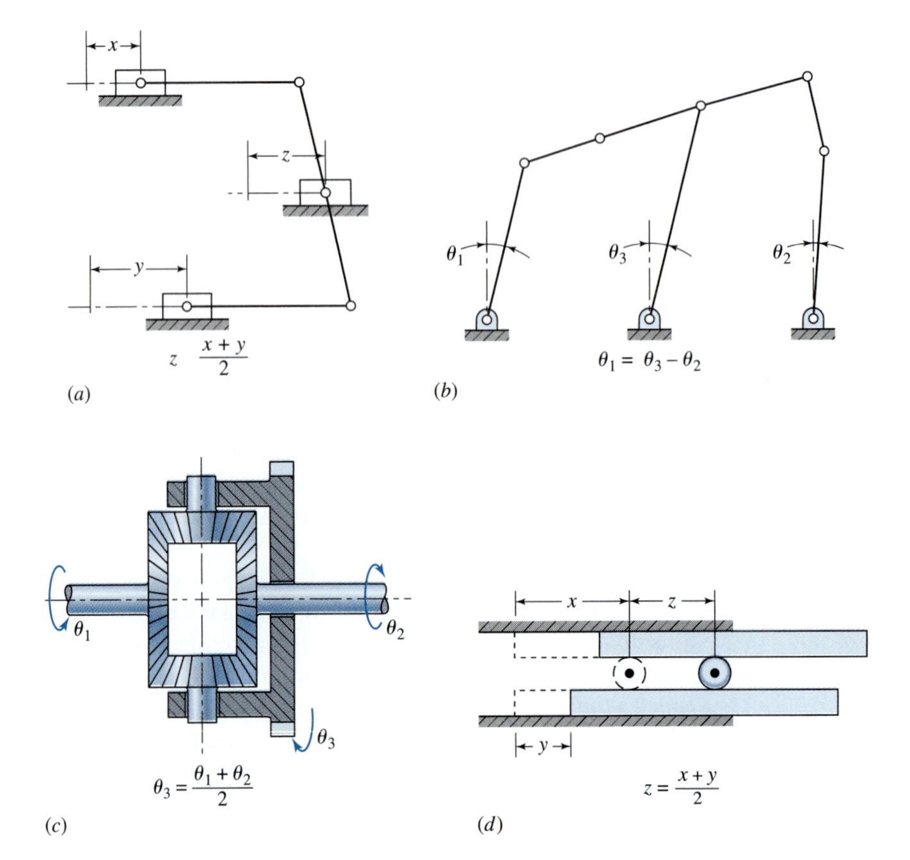

Figure 9.18 Differential mechanisms used for (*a*) adding, (*b*) subtracting, and (*c*, *d*) averaging two quantities.

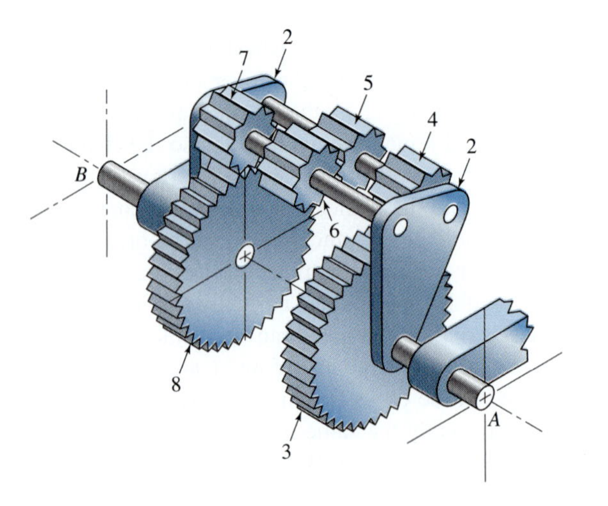

Figure 9.19 A spur gear differential.

Figure 9.20 The character on this chariot continues to point in a constant geographic direction. (Smithsonian Institute, photo P63158-B.)

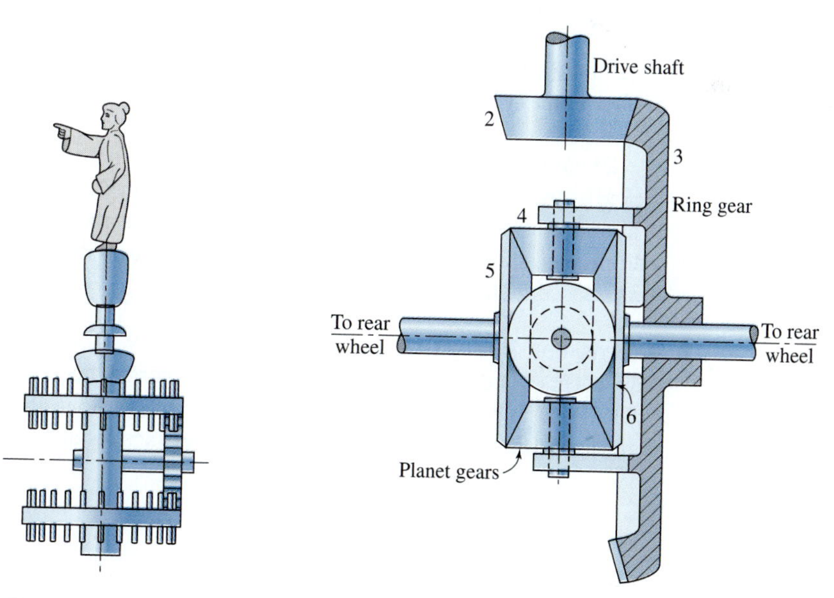

Figure 9.21 The Chinese differential.

Figure 9.22 Schematic drawing of a bevel gear automotive rear axle differential.

When the vehicle is making a turn, the wheel on the inside of the curve makes fewer revolutions than the wheel with the larger turning radius. Unless this difference in speed is accommodated in some manner, one or both of the tires must slip to make the turn. The differential permits the two wheels to rotate at different angular velocities, while at the same time delivering power to both. During a turn, the planet gears rotate about their own axes, thus permitting gears 5 and 6 to revolve at different angular velocities.

The purpose of the differential is to allow different speeds for the two driving wheels. In the usual differential of a rear-wheel-drive passenger car, the torque is divided approximately equally whether the car is traveling in a straight line or on a curve. Sometimes, however, the road conditions are such that the tractive effort developed by the two wheels is unequal. In such a case, the total tractive effort is only twice that at the wheel having the least traction, because the differential divides the torque equally. If one wheel happens to be traveling on snow or ice, the total tractive effort possible at that wheel is very small because only a small torque is required to cause the wheel to slip. Thus, the car remains stationary with one wheel spinning and the other having only trivial tractive effort. If the car is in motion and encounters a slippery surface, then all traction as well as control is lost!

Limited-Slip Differential It is possible to overcome this disadvantage of the simple bevel gear differential by adding a coupling unit that is sensitive to wheel speeds. The object of such a unit is to cause more of the torque to be directed to the slower moving wheel. Such a combination is then called a *limited-slip differential.*

Mitsubishi, for example, utilizes a viscous coupling unit, called a VCU, which is torque sensitive to wheel speeds. A difference in wheel speeds causes more torque to be delivered to the slower moving wheel. A large difference, perhaps caused by the spinning of one wheel on ice, causes a much larger share of the torque to be delivered to the nonspinning wheel. The arrangement, as used on the rear axle of an automobile, is illustrated in Fig. 9.23.

Another approach is to employ Coulomb friction, or clutching action, in the coupling. Such a unit, as with the VCU, is engaged whenever a significant difference in wheel speeds occurs.

Of course, it is also possible to design a bevel gear differential that is capable of being locked by the driver whenever dangerous road conditions are encountered. This is equivalent to a solid axle and forces both wheels to move at the same speed. It seems obvious that such a differential should not be locked when the tires are on dry pavement because of excessive wear caused by tire slipping.

Worm Gear Differential If gears 3 and 8 in Fig. 9.19 were replaced with worm gears and planet gears 4 and 7 were replaced with mating worm wheels, then the result is a *worm gear differential.* Of course, planet carrier 2 would have to rotate about a new axis perpendicular to the axle AB, because worm and worm wheel axes act at right angles to each other. Such

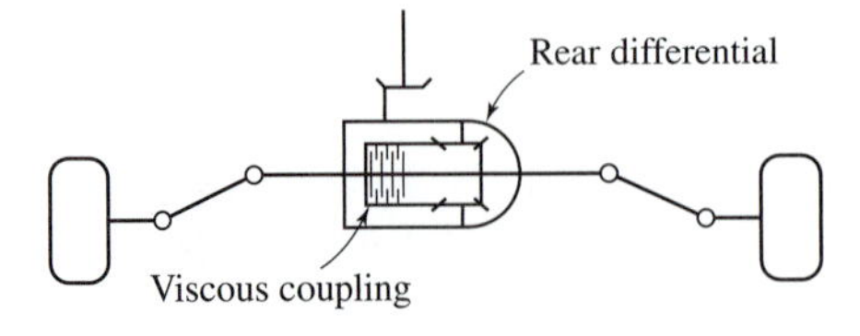

Figure 9.23 Viscous coupling used on the rear axle of the Mitsubishi Galant and Eclipse GSX automobiles.

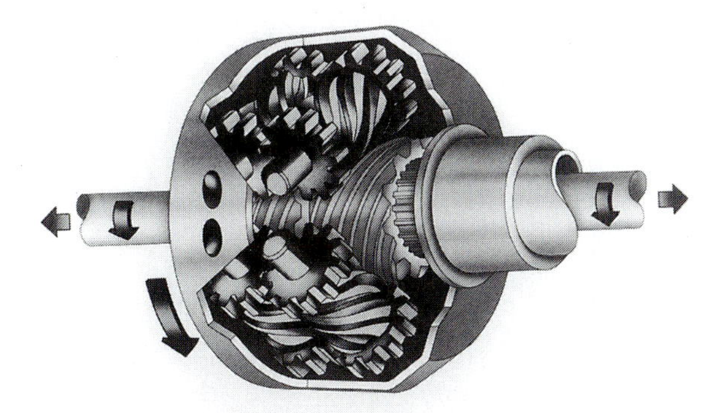

Figure 9.24 The TORSEN differential used on the drive shaft of Audi automobiles. This is the invention of Mr. Vernon Gleasman of Pittsford, NY. (Courtesy of Audi of America, Inc., Troy, MI.)

an arrangement can provide the traction of a locked differential or solid axle without the penalty of restricting small differential movement between the wheels.

The worm gear differential was invented by Mr. Vernon Gleasman and developed by Gleason Works as the TORSEN differential, a registered trademark now owned by JTEKT Torsen North America Inc. The word TORSEN is an acronym for the words "torque-sensing" because the differential can be designed to provide any desired locking value by varying the lead angle of the worm. Figure 9.24 illustrates a TORSEN differential as used on Audi automobiles.

9.9 ALL-WHEEL DRIVE TRAIN

As illustrated in Fig. 9.25, an all-wheel drive train consists of a center differential, geared to the transmission, driving the ring gears on both the front- and the rear-axle differentials. Dividing the thrusting force between all four wheels instead of only two is itself an advantage, but it also allows easier handling on curves and in crosswinds.

Dr. Herbert H. Dobbs, Colonel (Ret.), an automotive engineer, stated:

> One of the major improvements in automotive design is antilocking braking. This provides stability and directional control when stopping by ensuring a balanced transfer of momentum from the car to the road through all wheels. As too many have found out, loss of traction at one of the wheels when braking produces unbalanced forces on the car which can throw it out of control.
>
> It is equally important to provide such control during starting and acceleration. As with braking, this is not a problem when a car is driven prudently and driving conditions are good. When driving conditions are not good, the problems quickly become manifold. The antilock braking system provides the answer for the deceleration portion of the driving cycle, but the devices generally available to help during the remainder of the cycle are much less satisfactory.*

---

* Dr. Herbert H. Dobbs, Rochester Hills, MI, personal communication.

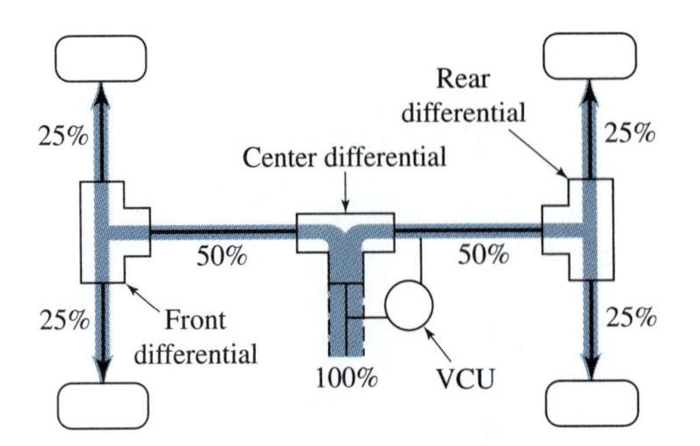

Figure 9.25 All-wheel drive (AWD) system used on the Mitsubishi Galant, demonstrating the power distribution for straight-ahead operation.

An early solution to this problem used by Audi is to electrically lock the center or the rear differential, or both, when driving conditions deteriorate. Locking only the center differential causes one half of the power to be delivered to the rear wheels and one half to the front wheels. If one of the rear wheels, say, rests on slippery ice, the other rear wheel has no traction. But the front wheels still provide traction. So the car has two-wheel drive. If the rear differential is then also locked, the car has three-wheel drive because the rear-wheel drive is then 50–50 distributed.

Another solution is to use a limited slip differential as the center differential on an all-wheel drive (AWD) vehicle. This then has the effect of distributing most of the driving torque to the front or rear axle depending on which is moving the slowest. An even better solution is to use limited slip differentials on both the center and the rear differentials.

Unfortunately, both locking differentials and limited slip differentials interfere with antilock braking systems. However, they are quite effective during low-speed winter operation.

The most effective solution seems to be the use of TORSEN differentials in an AWD vehicle. Here is what Dr. Dobbs has to say about their use:

> If they are cut to preclude any slip, the TORSEN distributes torque proportional to available traction at the driven wheels under all conditions just like a solid axle does, but it never locks up under any circumstances. Both of the driven wheels are always free to follow the separate paths dictated for them by the vehicle's motion, but are constrained by the balancing gears to stay synchronized with each other. All this adds up to a true "TORque SENsing and proportioning" differential, which of course is where the name came from.
>
> The result, particularly with a high performance front-wheel drive vehicle, is remarkable to say the least. The Army has TORSENS in the High Mobility Multipurpose Wheeled Vehicle (HMMWV) or "Hummer," which has replaced the Jeep. The only machines in production more mobile off-road than this one have tracks, and it is very capable on highway as well. It is fun to drive. The troops love it. Beyond that, Teledyne has an experimental "FAst Attack Vehicle" with TORSENS front, center, and rear. I've driven that machine over 50 mi/h on loose sand washes at Hank Hodges' Nevada Automotive Center, and it handled like it was running on dry pavement. The constant redistribution of torque to where traction was available kept all wheels driving and none digging!

PROBLEMS

9.1 Find the speed and direction of gear 8 in Fig. P9.1. What is the first-order kinematic coefficient of the train?

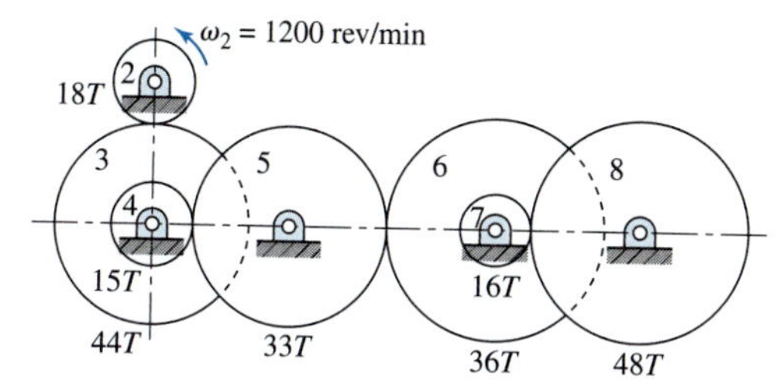

Figure P9.1

9.2 Figure P9.2 gives the pitch diameters of a set of spur gears forming a train. Compute the first-order kinematic coefficient of the train. Determine the speed and direction of rotation of gears 5 and 7.

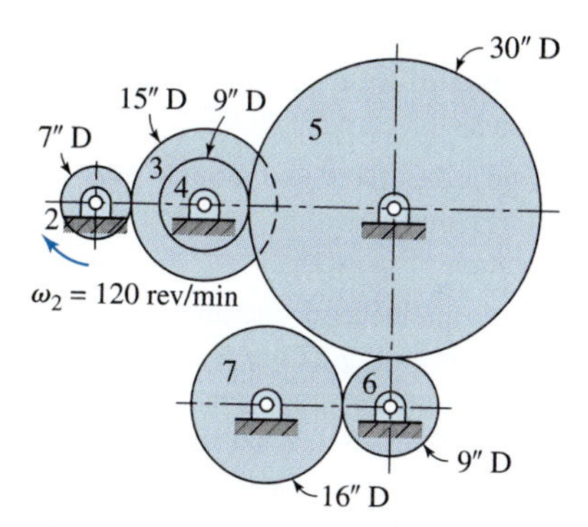

Figure P9.2

9.3 Figure P9.3 illustrates a gear train consisting of bevel gears, spur gears, and a worm and worm gear. The bevel pinion is mounted on a shaft that is driven by a V-belt on pulleys. If pulley 2 rotates at 1200 rev/min in the direction indicated, find the speed and direction of rotation of gear 9.

9.4 Use the truck transmission of Fig. 9.2 and an input speed of 3000 rev/min to find the drive shaft speed for each forward gear and for the reverse gear.

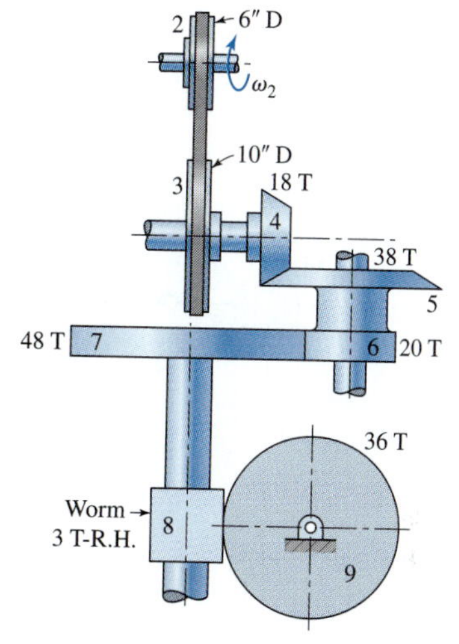

Figure P9.3

9.5 Figure P9.5 illustrates the gears in a speed-change gearbox used in machine tool applications. By sliding the cluster gears on shafts *B* and *C*, nine speed changes can be obtained. The problem of the machine tool designer is to select tooth numbers for the various gears so as to produce a reasonable distribution of speeds for the output shaft. The smallest and largest gears are gears 2 and 9, respectively. Using 20 and 45 teeth for these gears, determine a set of suitable tooth numbers for the remaining gears. What are the corresponding speeds of the output shaft? Note that the problem has many solutions.

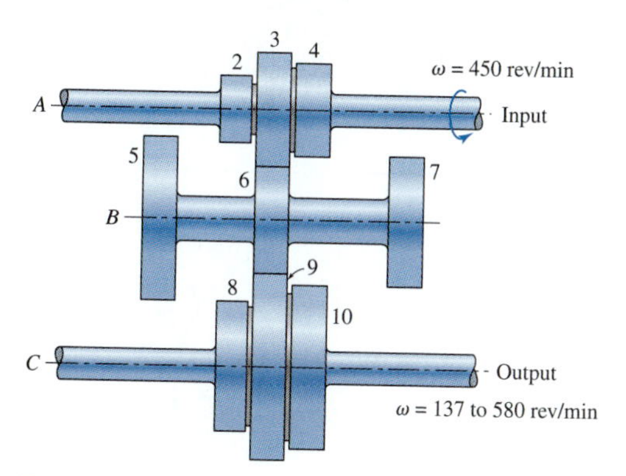

Figure P9.5

9.6 The internal gear (gear 7) in Fig. P9.6 turns at 60 rev/min ccw. What are the speed and direction of rotation of arm 3?

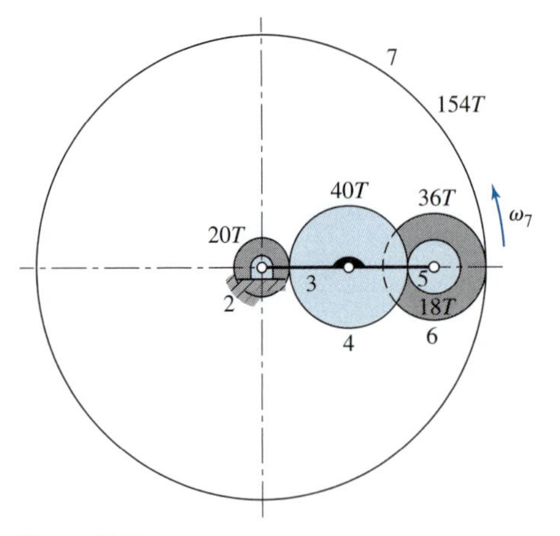

Figure P9.6

9.7 If the arm in Fig. P9.6 rotates at 300 rev/min ccw, find the speed and direction of rotation of internal gear 7.

9.8 In Fig. P9.8*a*, shaft *C* is stationary. If gear 2 rotates at 800 rev/min ccw, what are the speed and direction of rotation of shaft *B*?

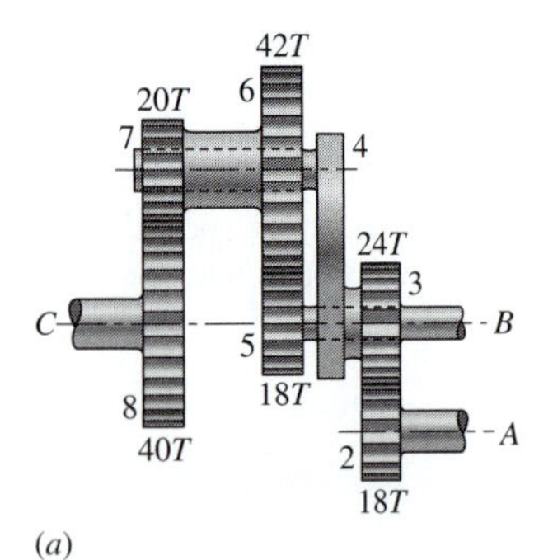

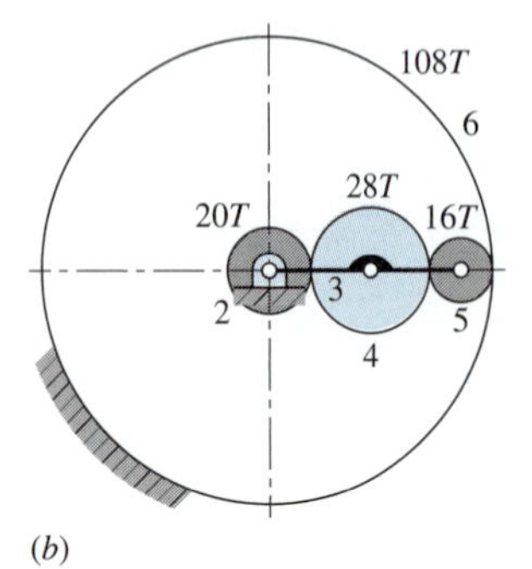

Figure P9.8

9.9 In Fig. 9.8*a*, consider shaft *B* as stationary. If shaft *C* is driven at 380 rev/min ccw, what are the speed and direction of rotation of shaft *A*?

9.10 In Fig. 9.8*a*, determine the speed and direction of rotation of shaft *C* under the following conditions:

(a) Shafts *A* and *B* both rotate at 360 rev/min ccw; and

(b) Shaft *A* rotates at 360 rev/min cw and shaft *B* rotates at 360 rev/min ccw.

9.11 In Fig. 9.8*b*, gear 2 is connected to the input shaft. If arm 3 is connected to the output shaft, what speed reduction can be obtained? What is the sense of rotation of the output shaft? What changes could be made in the train to produce the opposite sense of rotation for the output shaft?

9.12 The Lévai type-L train illustrated in Fig. 9.10 has $N_2 = 16\text{T}$, $N_4 = 19\text{T}$, $N_5 = 17\text{T}$, $N_6 = 24\text{T}$, and $N_7 = 95\text{T}$. Internal gear 7 is fixed. Find the speed and direction of rotation of the arm if gear 2 is driven at 100 rev/min cw.

9.13 The Lévai type-A train of Fig. 9.10 has $N_2 = 20\text{T}$ and $N_4 = 32\text{T}$.

(a) If the module is $m = 6$ mm/tooth, find the number of teeth on gear 5 and the crank arm radius.

(b) If gear 2 is fixed and internal gear 5 rotates at 10 rev/min ccw, find the speed and direction of rotation of the arm.

9.14 The tooth numbers for the automotive differential illustrated in Fig. 9.22 are $N_2 = 17\text{T}$, $N_3 = 54\text{T}$, $N_4 = 11\text{T}$, and $N_5 = N_6 = 16\text{T}$. The drive shaft turns at 1200 rev/min. What is the speed of the right wheel if it is jacked up and the left wheel is resting on the road surface?

9.15 A vehicle using the differential illustrated in Fig. 9.22 turns to the right at a speed of 30 mi/h on a curve of 80-ft radius. Use the same tooth numbers as in Problem 9.14. The tire diameter is 15 in. Use 60 in. as the distance between treads.

(a) Calculate the speed of each rear wheel.

(b) Find the rotational speed of the ring gear.

9.16 Figure P9.16 illustrates a possible arrangement of gears in a lathe headstock. Shaft *A* is driven by a motor at a speed of 720 rev/min. The three pinions can slide along shaft *A* so as to yield the meshes 2 with 5, 3 with 6, or 4 with 8. The gears on shaft *C* can also slide so as to mesh either 7 with 9 or 8 with 10. Shaft *C* is the mandrel shaft.

(a) Make a table demonstrating all possible gear arrangements, beginning with the slowest speed for shaft C and ending with the highest, and enter in this table the speeds of shafts B and C.

(b) If the gears all have a module of $m = 5$ mm/tooth, what must be the shaft center distances?

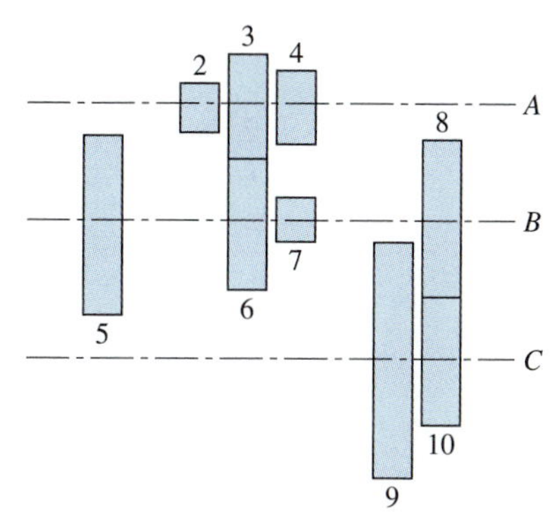

Figure P9.16 $N_2 = 16$T, $N_3 = 36$T, $N_4 = 25$T, $N_5 = 64$T, $N_6 = 66$T, $N_7 = 17$T, $N_8 = 55$T, $N_9 = 79$T, $N_{10} = 41$T.

9.17 Shaft A in Fig. P9.17 is the output and is connected to the arm. If shaft B is the input and drives gear 2, what is the speed ratio? Can you identify the Lévai type for this train?

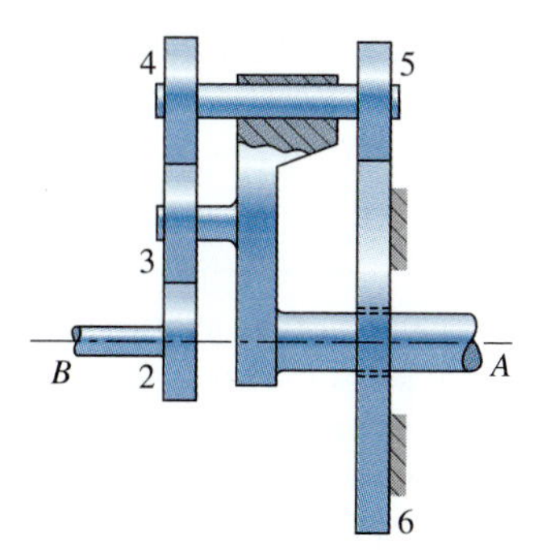

Figure P9.17 $N_2 = 16$T, $N_3 = 18$T, $N_4 = 16$T, $N_5 = 18$T, $N_6 = 50$T.

9.18 In Problem 9.17, shaft B rotates at 100 rev/min cw. Find the speed of shaft A and of gears 3 and 4 about their own axes.

9.19 Bevel gear 2 is driven by the engine in the reduction unit illustrated in Fig. P9.19. Bevel planets 3 mesh with crown gear 4 and are pivoted on the spider (arm), which is connected to propeller shaft B. Find the percentage speed reduction.

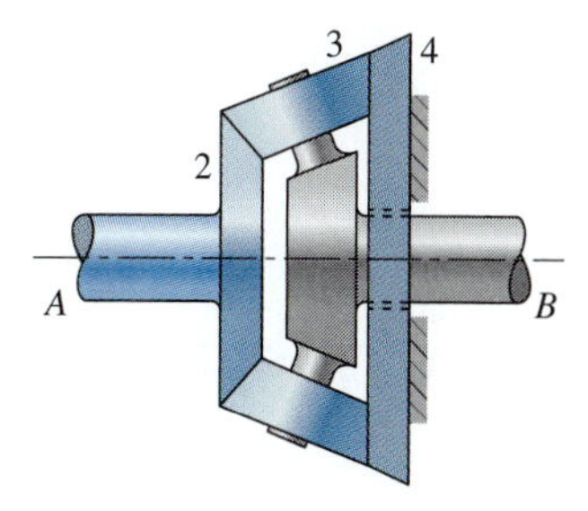

Figure P9.19 A marine reduction differential; $N_2 = 36$T, $N_3 = 21$T, $N_4 = 52$T; crown gear 4 is fixed.

9.20 In the clock mechanism illustrated in Fig. P9.20, a pendulum on shaft A drives an anchor (see Fig. 1.12c). The pendulum period is such that one tooth of the 30T escapement wheel on shaft B is released every 2 s, causing shaft B to rotate once every minute. In Fig. P9.20, note that the second (to the right) 64T gear is pivoted loosely on shaft D and is connected by a tubular shaft to the hour hand.

(a) Show that the train values are such that the minute hand rotates once every hour and that the hour hand rotates once every 12 hours.

(b) How many turns does the drum on shaft F make every day?

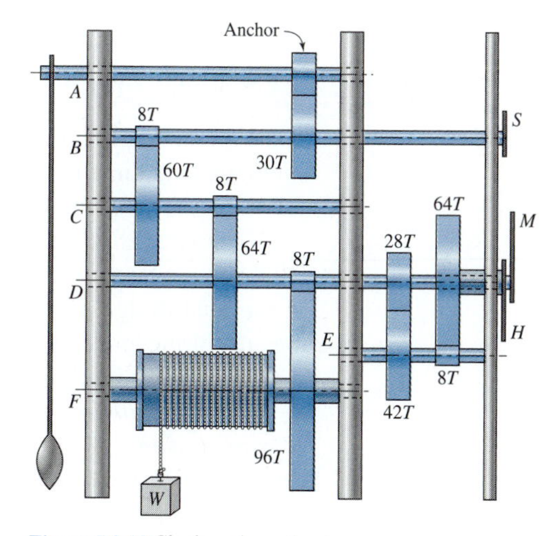

Figure P9.20 Clockwork mechanism.

10 Synthesis of Linkages[1]

In previous chapters we have concentrated on the analysis of mechanisms, where the dimensions of the links are known. By kinematic *synthesis* we mean the design or creation of a new mechanism to yield a desired set of motion characteristics. Because of the very large number of techniques available, some of which may be quite frustrating, we present here only a few of the more useful approaches to illustrate applications of the theory.[2]

10.1 TYPE, NUMBER, AND DIMENSIONAL SYNTHESIS

There are three general stages in the design of a new mechanism, type synthesis, number synthesis, and dimensional synthesis. Although these three steps may not be consciously followed, they are always present in the creation of any new device.

Type synthesis refers to the choice of the kind of mechanism to be used; it might be a linkage, a geared system, belts and pulleys, or even a cam system. This beginning stage of the total design process usually involves the consideration of design factors such as manufacturing processes, materials, space, safety, and economics. The study of kinematics is usually only slightly involved in type synthesis.

Number synthesis deals with finding a satisfactory number of links and number of joints or pairs to obtain the desired mobility (see Section 1.6). Number synthesis is the second stage in the design process and follows type synthesis.

The third stage in design, determining the detailed dimensions of the individual links, is called *dimensional synthesis*. This is the subject of the balance of Chapter 10.

10.2 FUNCTION GENERATION, PATH GENERATION, AND BODY GUIDANCE

A frequent requirement in mechanism design is that of causing an output member to rotate, oscillate, or reciprocate according to a specified function of time or function of the input motion. This is called *function generation*. A simple example is that of synthesizing a four-bar linkage to generate the function $y = f(x)$. In this case, x represents the motion (crank angle) of the input crank, and the linkage is to be designed so that the motion (angle) of the output rocker approximates the desired function y. A few other examples of function generation are as follows.

1. In a conveyor line the output member of a mechanism must move at the constant velocity of the conveyer while some operation is performed—for example, capping a bottle—and then it must return, pick up the next cap, and repeat.
2. The output member of a mechanism must pause or stop during its motion cycle to provide time for another event. The second event might be a sealing, stapling, or fastening operation of some kind.
3. The output member of a mechanism must rotate according to a specified nonuniform velocity function because it must be synchronized with another mechanism that requires such a rotating motion.

A second type of synthesis problem is called *path generation*. This refers to a problem in which a coupler point must follow a path having a prescribed shape. Common requirements are that a portion of the path be a circular arc, elliptical, or a straight line. Sometimes it is required that the path cross over itself, as in a figure eight.

The third general class of synthesis problems is called *body guidance*. Here we are interested in moving an object from one position to another. The problem may call for a simple translation or a combination of translation and rotation. In the construction industry, for example, a heavy part such as a scoop or a bulldozer blade must be moved through a series of prescribed positions.

The general problem of dimensional synthesis is to design a mechanism that will guide a moving rigid body through N finitely separated positions in a single plane (where $N = 2, 3, 4, \ldots$). In general, the N positions are specified; that is, the design problem dictates the N positions to which the rigid body must move. The desire is to synthesize a planar, single-degree-of-freedom mechanism that will achieve the location and orientation of the rigid body for each of the N positions. For example, the synthesis of a planar four-bar linkage could be the first attempt. If the rigid body that is to be guided through the N positions is the coupler link, then the synthesis problem is one of body guidance. If the rigid body is the input or output crank, then the synthesis problem is one of function generation.

Here we present a general geometric approach that can be used to synthesize a four-bar linkage for $N = 2, 3, 4$, or 5 finitely separated positions of the coupler link. We will see that the number of four-bar linkages that can be found to guide a rigid body through two finitely separated positions is ∞^6; for three finitely separated positions this number is ∞^4; for four finitely separated positions there are ∞^1 solutions; and through five finitely separated positions there are a small finite number of solutions. In general, it is not possible to synthesize a four-bar linkage for N greater than five finitely separated positions of a rigid body unless the positions have special geometric relationships.

10.3 TWO FINITELY SEPARATED POSITIONS OF A RIGID BODY ($N = 2$)

Let us start with the study of simple two-position synthesis methods.

Two-Position Synthesis of a Slider-Crank Mechanism The central slider-crank mechanism of Fig. 10.1a has a stroke B_1B_2 equal to twice the crank radius r_2. As illustrated, the extreme positions of B_1 and B_2, also called limiting positions of the slider, are determined by constructing circular arcs through O_2 of length $r_3 - r_2$ and $r_3 + r_2$, respectively. The two dimensions r_2 and r_3 can be determined from these two measured lengths.

In general, the central slider-crank mechanism must have r_3 larger than r_2. However, the special case of $r_3 = r_2$ results in the *isosceles slider-crank mechanism*, in which the slider reciprocates through O_2 and the stroke is four times the crank radius. All points on the coupler of the isosceles slider-crank generate elliptic paths. The paths generated by points on the coupler of the central slider-crank of Fig. 10.1a, which is not isosceles, are not elliptical; however, they are always symmetric about the sliding axis O_2B.

The linkage of Fig. 10.1b is called the *general* or *offset slider-crank mechanism.* Note that the limiting positions B_1 and B_2 of the slider can be determined in the same manner as explained above for the central slider-crank mechanism. Certain special effects can be obtained by changing the offset distance e. For example, stroke B_1B_2 is always greater than twice the crank radius. Also, the crank angle required to execute the forward stroke is different from that for the return stroke. This feature can be used to synthesize quick-return mechanisms where a slower working stroke may be desired to reduce power requirements (see Section 1.7).

Two-Position Synthesis of a Crank-and-Rocker Mechanism The limiting positions of the rocker in a crank-and-rocker mechanism are illustrated as points B_1 and B_2 in Fig. 10.2. Note that the crank and the coupler form a single straight line at each of

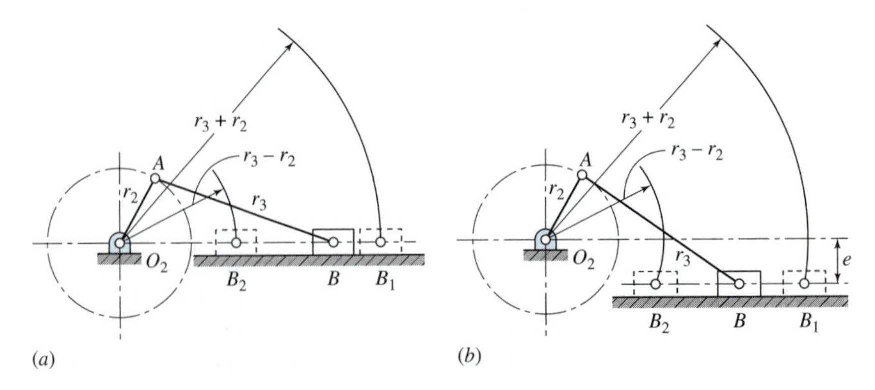

Figure 10.1 (a) Central slider-crank mechanism. (b) General or offset slider-crank mechanism.

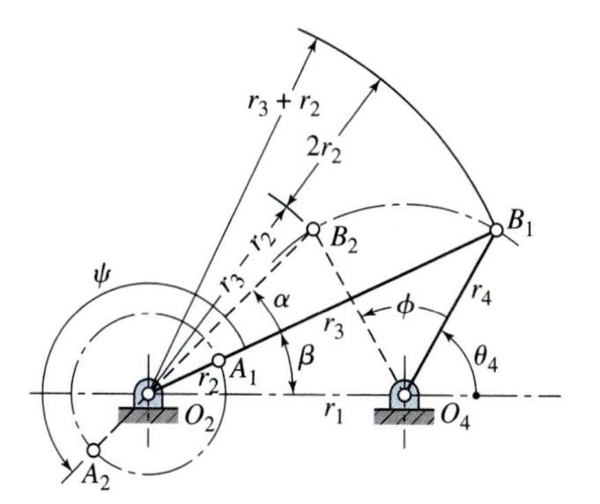

Figure 10.2 The two extreme positions of the crank-and-rocker mechanism.

the extreme positions. Also, note that the two dimensions r_2 and r_3 can be determined from measurements made in these two positions in the same way as for the slider-crank linkage.

In this case, the crank executes the angle ψ while the rocker moves from B_1 to B_2 through the angle ϕ. Note on the return stroke that the rocker swings from B_2 back to B_1 through the same angle $-\phi$ but the crank moves through the angle $360° - \psi$.

There are many cases in which a crank-and-rocker mechanism is superior to a cam-and-follower system. Among the advantages over cam systems are the smaller forces involved, the elimination of the retaining spring, and the smaller clearances obtainable by the use of revolute pairs.

If the direction of rotation of the input crank is chosen so that $\psi > 180°$ in Fig. 10.2, then we can define $\alpha = \psi - 180°$, and an equation for the advance to return time ratio (see Section 1.7) can be written as

$$Q = \frac{180° + \alpha}{180° - \alpha}. \tag{10.1}$$

A problem that frequently arises in the synthesis of crank-and-rocker linkages is to obtain the dimensions or geometry that will cause the mechanism to generate a specified output displacement ϕ when the time ratio is specified.[3]

To synthesize a crank-and-rocker mechanism for specified values of ϕ and α, we locate an arbitrary point O_4 in Fig. 10.3*a* and choose any desired rocker length r_4; we note that this choice does not change the solution, but only sets the scale of the figure. Then we draw the two positions O_4B_1 and O_4B_2 of link 4 separated by the given angle ϕ. Through B_1 we construct any line X. Then, through B_2, we construct the line Y at the given angle α to the line X. The intersection of these two lines defines the location of the crank pivot O_2. Because the orientation of line X was chosen arbitrarily, there are an infinite number of solutions to this problem.

Next, as illustrated in Fig. 10.3*a*, the distance B_2C is $2r_2$, or twice the crank length. So, we can bisect this distance to find r_2. Then the coupler length is found from $r_3 = O_2B_1 - r_2$. The completed linkage is illustrated in Fig. 10.3*b*.

The Pole of a Finite Displacement

For two finitely separated positions of a rigid body in planar motion, the moving body can be regarded as a line AB of constant length, as

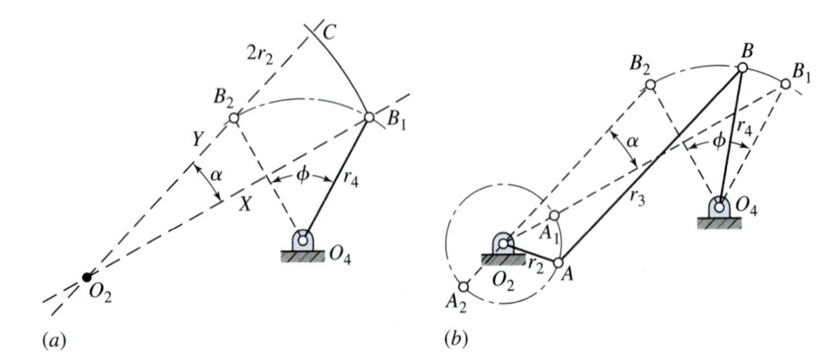

Figure 10.3 Synthesis of a four-bar linkage to generate a specified rocker angle ϕ.

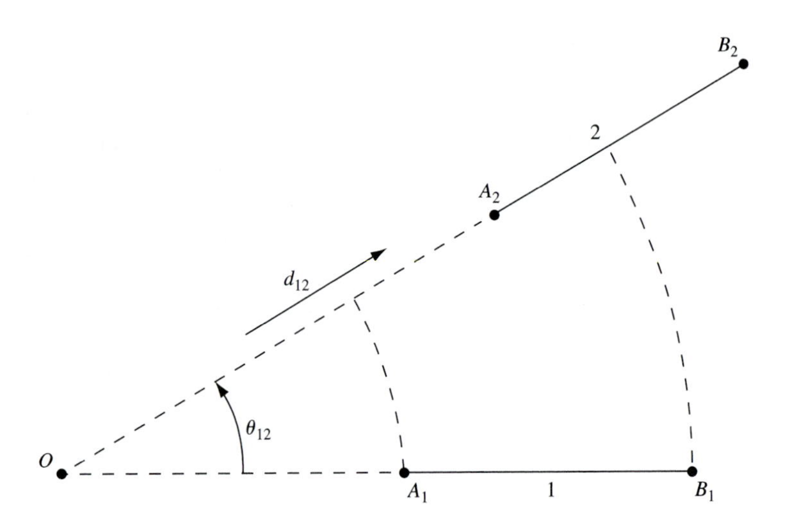

Figure 10.4 Two finitely separated positions of a rigid body.

illustrated in Fig. 10.4. In the first position, the body is denoted A_1B_1 and in the second position the body is denoted A_2B_2. The displacement of the body from position 1 to position 2 can be described by a rotation θ_{12} and a translation d_{12}, as illustrated in Fig. 10.4.

Note that a single pivot can be used as the center of rotation of the rigid body displacement. The center of rotation is the point of intersection of the perpendicular bisector of the line connecting A_1 to A_2 and the perpendicular bisector of the line connecting B_1 to B_2. The center of rotation will henceforth be referred to as the *pole* for the finite displacement and denoted P_{12} as illustrated in Fig. 10.5.

The angle of rotation of the body AB is denoted

$$\angle A_1P_{12}A_2 = \angle B_1P_{12}B_2 = \theta_{12} = 2\phi_{12}. \tag{10.2}$$

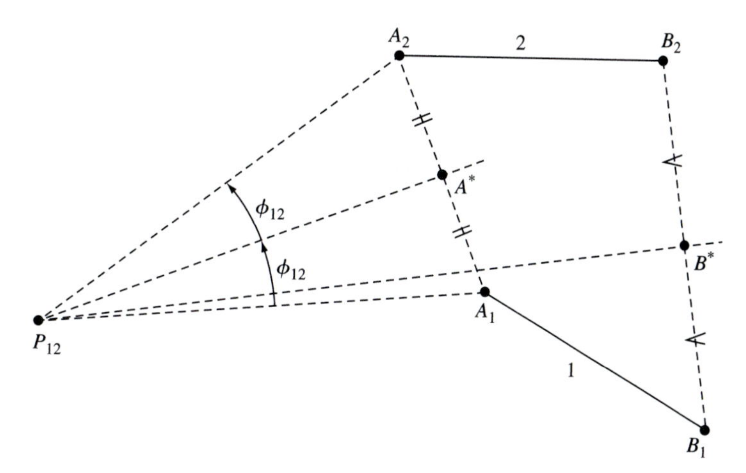

Figure 10.5 A single pivot to guide the rigid body through a finite displacement.

Therefore, the angles

$$\begin{aligned} \angle A_1P_{12}A^* &= \angle A^*P_{12}A_2 = \phi_{12} \\ \angle B_1P_{12}B^* &= \angle B^*P_{12}B_2 = \phi_{12}, \end{aligned} \tag{10.3}$$

where A^* and B^* are the midpoints of the lines A_1A_2 and B_1B_2, respectively. Note that the pole P_{ij} for a finite displacement from position i to position j is analogous to the instantaneous center of velocity for an infinitesimal displacement of a rigid body (see Section 3.13). The following two statements are important: (*i*) the pole for a finite displacement is located at that point of the moving body having zero displacement and (*ii*) for the purpose of calculating displacements, the finite displacement of the body can be considered a rotation around the pole.

A better solution than a single pivot is to use a four-bar linkage to guide the moving body between two positions because a four-bar linkage is more stable and can support a larger load than a single pivot. A possible four-bar linkage solution is illustrated in Fig. 10.6. The body AB would be attached to the coupler link; the previous or new points A and B could be used as the two coupler pivots; the ground pivot O_A could be chosen as any point on the perpendicular bisector A^*P_{12}, and the ground pivot O_B could be chosen as any point on the perpendicular bisector B^*P_{12}.

The total number of possible four-bar linkages that can guide the body through two finitely separated positions can be obtained from the fact that there are ∞^2 locations for pin A, ∞^2 locations for pin B, ∞^1 choices for the ground pivot O_A, and ∞^1 choices for the ground pivot O_B along the perpendicular bisector lines. Therefore, there are a total of ∞^6 possible four-bar linkages that can guide the body through the two specified finitely separated positions.

It is important to note the relationships between the signs of the rotation angles and the order of their subscripts. The rotation angle through which the body turns in moving from position 1 to position 2 is denoted $2\phi_{12}$ [see Eq. (10.2)]. Similarly, the rotation angle

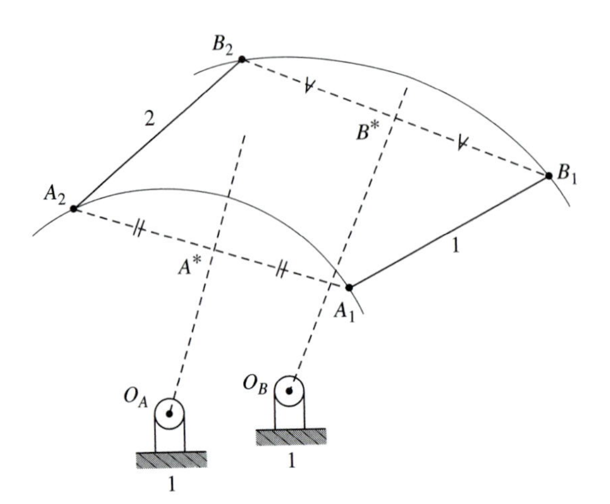

Figure 10.6 A possible four-bar linkage.

through which the body turns in moving from position 2 back to position 1 is denoted $2\phi_{21}$. Therefore, we can write that

$$2\phi_{12} + 2\phi_{21} = 0°, 360°$$

or

$$\phi_{12} + \phi_{21} = 0°, 180°. \tag{10.4}$$

Note that Eqs. (10.4) are valid when the two angles are measured in the same direction. If the angles are measured in opposite directions, then

$$\phi_{21} = -\phi_{12}. \tag{10.5}$$

An expression for the average velocity of an arbitrary point fixed in the body, say point A, can be written as

$$(V_A)_{\text{avg}} = \frac{\Delta s_A}{\Delta t}, \tag{10.6}$$

where the distance $\Delta s_A = A_1A_2$. The relationship between the distance Δs_A and the angle ϕ_{12} can be obtained as follows. Consider the triangle $A_1P_{12}A^*$ of Fig. 10.5; then,

$$\sin\phi_{12} = \frac{A_1A^*}{P_{12}A_1},$$

which can be written as

$$A_1A^* = (P_{12}A_1)\sin\phi_{12}. \tag{10.7}$$

Note that

$$A_1A^* = \frac{A_1A_2}{2} = \frac{1}{2}\Delta s_A. \tag{10.8}$$

Setting Eq. (10.8) equal to Eq. (10.7) and rearranging, the displacement can be written as

$$\Delta s_A = 2A_1A^* = 2(P_{12}A_1)\sin\phi_{12}. \tag{10.9}$$

Then substituting Eq. (10.9) into Eq. (10.6) and rearranging, the average velocity of point A can be written as

$$(V_A)_{\text{avg}} = \frac{2\sin\phi_{12}}{\Delta t}(P_{12}A_1). \tag{10.10}$$

From this equation, the magnitude of the instantaneous velocity of point A in the body can be written as

$$V_A = \lim_{\Delta t\to 0}\left(\frac{2\sin\phi_{12}}{\Delta t}\right)P_{12}A_1. \tag{10.11}$$

Recall that the magnitude of the instantaneous velocity of point A (see Chapter 3) can also be written as

$$V_A = \omega R, \tag{10.12}$$

where ω is the angular velocity of the body and $R = P_{12}A_1$ is the distance of point A from the pole (center of rotation). Comparing Eqs. (10.11) and (10.12), we note that the angular velocity of the body can be written as

$$\omega = \lim_{\Delta t\to 0}\left(\frac{2\sin\phi_{12}}{\Delta t}\right). \tag{10.13}$$

10.4 THREE FINITELY SEPARATED POSITIONS OF A RIGID BODY ($N = 3$)

Three finitely separated positions of a body, denoted positions 1, 2, and 3, can be specified by: (*i*) the poles P_{12}, P_{23} and P_{31}; and (*ii*) the rotation angle from position 1 to position 2 denoted $2\phi_{12}$, the rotation angle from position 2 to position 3 denoted $2\phi_{23}$, and the rotation angle from position 3 to position 1 denoted $2\phi_{31}$.

We can obtain the locations of the three poles in the same manner as before. We choose an arbitrary line in the body, say the line DE illustrated in Fig. 10.7. For the body in its first position, the line is denoted D_1E_1; in the second position, the line is denoted D_2E_2; and in the third position, the line is denoted D_3E_3. The intersection of the perpendicular bisectors of the lines D_1D_2 and E_1E_2 is the pole P_{12}. The intersection of the perpendicular bisectors of the lines D_2D_3 and E_2E_3 is the pole P_{23}. Finally, the intersection of the perpendicular bisectors of the lines D_3D_1 and E_3E_1 is the pole P_{31}. The three poles for the three finitely separated positions of the body are illustrated in Fig. 10.7.

We connect poles P_{12} and P_{23} by a straight line; we connect poles P_{23} and P_{31} by a straight line; and we connect poles P_{31} and P_{12} by a straight line. These three lines form a triangle called the *pole triangle*. An understanding of the geometry of the pole triangle

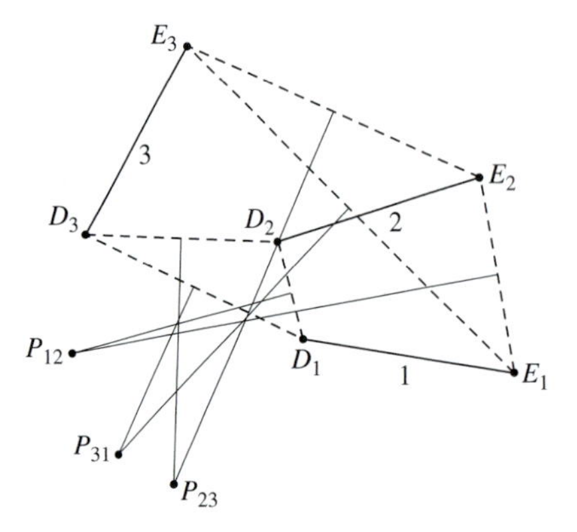

Figure 10.7 The three poles for three positions of rigid body DE.

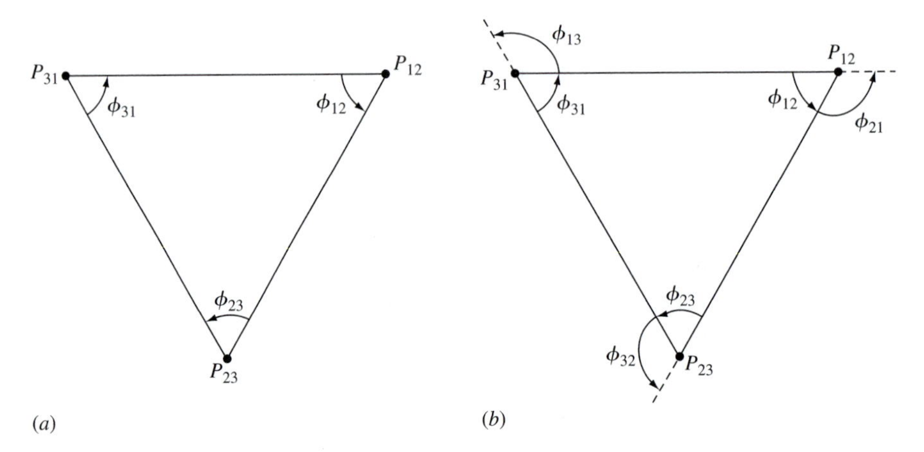

Figure 10.8 (*a*) The interior angles of a pole triangle. (*b*) The exterior angles of a pole triangle.

(for example, the interior angles and the exterior angles) is essential for synthesizing a mechanism that can guide the body through the three finitely separated positions.

For convenience, the finite displacements of the rigid body in Fig. 10.8*a* are illustrated counterclockwise. For the pole triangle illustrated, the sides are labeled so that side 1 is the side with the common subscript 1 for the two poles ($P_{31}P_{12}$), side 2 is the side with the common subscript 2 for the two poles ($P_{12}P_{23}$), and side 3 is the side with the common subscript 3 for the two poles ($P_{23}P_{31}$).

Interior Angles of a Pole Triangle The order of the subscripts for the interior angles of a pole triangle are defined as follows: (*i*) the interior angle about the pole P_{12} from side 1 to side 2 of the pole triangle is the angle ϕ_{12} (positive ccw), (*ii*) the interior angle about the pole P_{23} from side 2 to side 3 of the pole triangle is the angle ϕ_{23} (positive ccw), and (*iii*) the interior angle about the pole P_{31} from side 3 to side 1 of the pole triangle is the angle ϕ_{31} (positive ccw).

Recall that these interior angles are half the rotation angles for the body; that is,

$$\phi_{12} = \frac{\theta_{12}}{2}, \quad \phi_{23} = \frac{\theta_{23}}{2}, \quad \text{and} \quad \phi_{31} = \frac{\theta_{31}}{2}. \tag{10.14}$$

Also note that, consistent with Eq. (10.14), the sum of the three interior angles of the pole triangle must be

$$\phi_{12} + \phi_{23} + \phi_{31} = 180^\circ. \tag{10.15}$$

Exterior Angles of a Pole Triangle Note the following relationships, again consistent with Eq. (10.14),

$$\phi_{12} + \phi_{21} = 180^\circ, \quad \phi_{23} + \phi_{32} = 180^\circ, \quad \text{and} \quad \phi_{31} + \phi_{13} = 180^\circ, \tag{10.16}$$

which are valid when the angles are measured in the same direction, as illustrated in Fig. 10.8*b*. Also, note the relationships

$$\phi_{12} = -\phi_{21}, \quad \phi_{23} = -\phi_{32}, \quad \text{and} \quad \phi_{31} = -\phi_{13}, \tag{10.17}$$

which are valid when the angles are measured in opposite directions, consistent with Eq. (10.5).

EXAMPLE 10.1

Given the pole triangle illustrated in Fig. 10.9, where the length of side 1 is $P_{31}P_{12} = 2$ in. and the interior angles are $\phi_{12} = \theta_{12}/2 = 53.13^\circ$ ccw and $\phi_{31} = \theta_{31}/2 = 36.87^\circ$ ccw, and also given that the location of point D fixed in the moving body in position 1, that is, point D_1, is midway between the poles P_{31} and P_{12} and 1.5 in vertically below this line, then determine the location of point D when the rigid body is in position 2 and in position 3; that is, find points and D_2 and D_3.

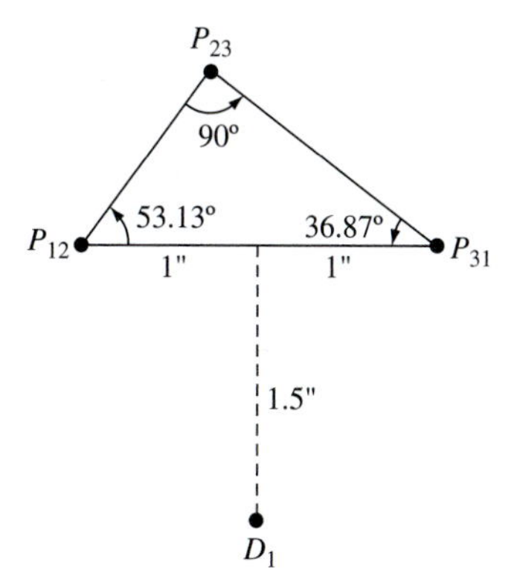

Figure 10.9 The pole triangle and point D_1.

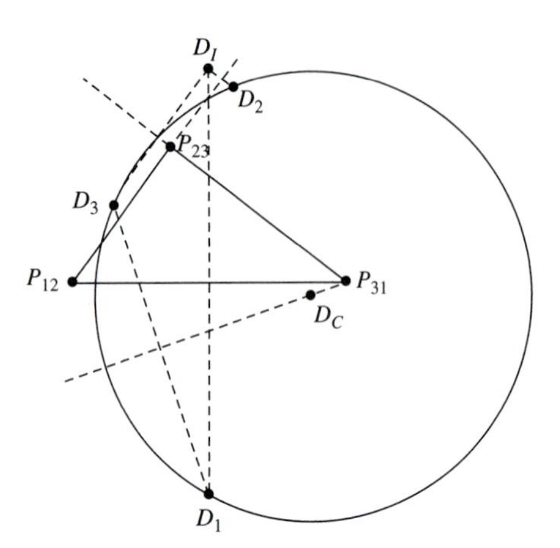

Figure 10.10 Example 10.1. The location of point D in the three positions (that is, D_1, D_2, and D_3).

SOLUTION

Note that the interior angle $\phi_{23} = \theta_{23}/2 = 90°$ ccw; therefore, the pole triangle is a 3, 4, 5 right-angled triangle and the lengths of the other two sides are $P_{12}P_{23} = 1.2$ in for side 2 and $P_{23}P_{31} = 1.6$ in. for side 3.

A straightforward approach for finding the locations of points D_2 and D_3 is to use the so-called *image point* (also referred to as the *image pole*). The image point, denoted here D_I, is defined as that point that will give D_i $(i = 1, 2, 3)$ when reflected about the side (or the extended side) of the pole triangle with the common subscript i. In other words, D_1 is the reflection of D_I about side 1 $(P_{31}P_{12})$, D_2 is the reflection of D_I about side 2 $(P_{12}P_{23})$, and D_3 is the reflection of D_I about side 3 $(P_{12}P_{12})$. The points $D_1 D_2$, and D_3 are illustrated in Fig. 10.10.

The three points D_1, D_2, and D_3 must also lie on the circumference of a circle, that is, we know that it is always possible to find a circle that passes through the three locations defined by any arbitrary point fixed in the body as it travels through three finitely separated positions. This makes it possible to design a four-bar linkage to guide a rigid body through three finitely separated positions because pins A and B on the coupler link must be located on circular arcs.

The center of the circle that passes through the three points D_1, D_2, and D_3 is called a *center point* and is denoted D_C. All points such as D_C make up the well-known center point system (or center system), which will be explained in more detail later. The center point D_C is the point of intersection of the perpendicular bisectors of the lines D_1D_3 and D_2D_3. Because the perpendicular bisector of the line D_1D_2 must also pass through the center point D_C, this can be used to check the accuracy of the previous constructions. Also note that, consistent with the definition of a pole, the perpendicular bisector of the line D_iD_j must pass through the pole P_{ij}.

The geometric relationships can be written as follows.

$$\begin{aligned}\angle D_1P_{12}D_C &= \angle D_CP_{12}D_2 = \phi_{12}\\ \angle D_2P_{23}D_C &= \angle D_CP_{23}D_3 = \phi_{23}\\ \angle D_3P_{31}D_C &= \angle D_CP_{31}D_1 = \phi_{31}\end{aligned} \tag{10.18}$$

EXAMPLE 10.2

Assume we are given the pole triangle and the center point E_C, as illustrated in Fig. 10.11. The length of the side $P_{31}P_{12}$ is 2 in., the interior angles $\phi_{12} = \theta_{12}/2 = 25°$ ccw and $\phi_{31} = \theta_{31}/2 = 40°$ ccw, the distance from P_{12} to the center point E_C is 1 in, and the angle $\angle P_{31}P_{12}E_C = 45°$ cw as illustrated in Fig. 10.11. Find the locations of point E when the body is in positions 1, 2, and 3 (that is, find E_1E_2, and E_3) and specify the radius of the circle that passes through these three points.

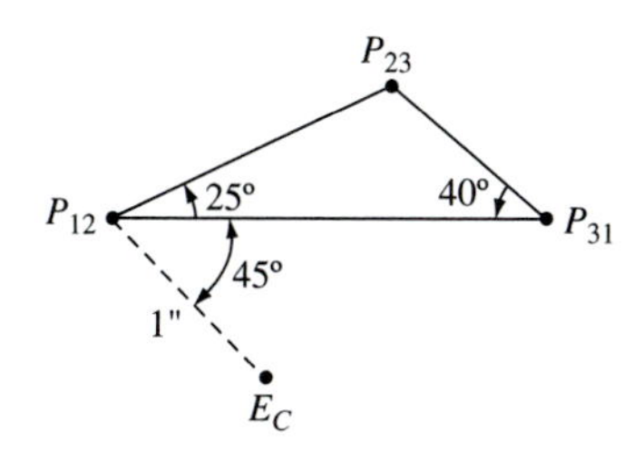

Figure 10.11 Example 10.2. The pole triangle and the center point E_C.

SOLUTION

The interior angle from position 1 to position 2 can be written as

$$\phi_{12} = \angle P_{31}P_{12}P_{23} = 25° \text{ ccw} \tag{1a}$$

and the interior angle from position 3 to position 1 can be written as

$$\phi_{31} = \angle P_{23}P_{31}P_{12} = 40° \text{ ccw.} \tag{1b}$$

Therefore, the interior angle from position 2 to position 3 is

$$\phi_{23} = \angle P_{12}P_{23}P_{31} = 115° \text{ ccw.} \tag{1c}$$

The geometric relationships are

$$\begin{aligned} \angle E_1P_{12}E_C &= \angle E_CP_{12}E_2 = \phi_{12} \\ \angle E_2P_{23}E_C &= \angle E_CP_{23}E_3 = \phi_{23} \quad . \\ \angle E_3P_{31}E_C &= \angle E_CP_{31}E_1 = \phi_{31}. \end{aligned} \tag{2}$$

Two of these three equations can be used to locate E_1, that is, the first and the third equations can be written as

$$\begin{aligned} \angle E_CP_{12}E_1 &= -\phi_{12} = 25° \text{cw} \\ \angle E_CP_{31}E_1 &= \phi_{31} = 40° \text{cww.} \end{aligned} \tag{3}$$

Therefore, the point of intersection of these two lines is E_1, as illustrated in Fig. 10.12. The location of point E in positions 2 and 3 can be obtained from a similar procedure; that is, to locate E_2 use the first two of Eqs. (2) and to locate E_3 use the second and third equations. An alternative procedure is to find the image point E_I and then reflect this point about the common 2 side and the common 3 side of the pole triangle to find E_2 and E_3, respectively.

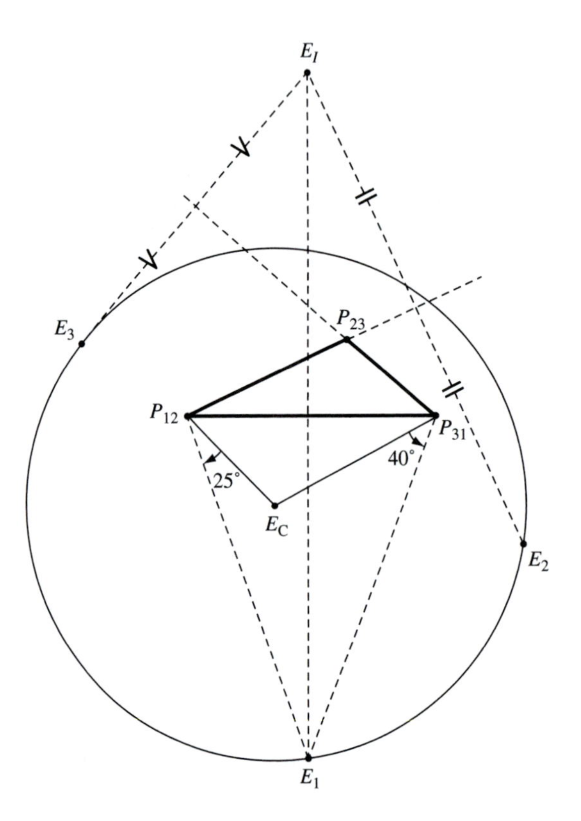

Figure 10.12 Example 10.2. The pole triangle and points E_1, E_2, and E_3.

The circle can be drawn with center E_C and passing through points E_1, E_2, and E_3. The radius of this circle is measured as

$$E_C\mathrm{E}_1 = E_C\mathrm{E}_2 = E_C\mathrm{E}_3 = \ 1.9 \text{ in.}$$

A special case, which is important in synthesis, is when the radius of the circle is infinite, that is when the circle degenerates to a straight line. This implies that point E is moving on a straight line through the three finitely separated positions. In such a case, point E would be suitable for a prismatic joint. An example of this special case is presented next.

EXAMPLE 10.3

We are given the pole triangle and the point D_1, as illustrated in Fig. 10.13. The length of the side $P_{31}P_{12}$ is 2 in, and the interior angles are $\phi_{12} = 30°$ ccw, $\phi_{23} = 90°$ ccw, and $\phi_{31} = 60°$ ccw. The location of point D_1 is midway between the poles P_{31} and P_{12} and 1 in below this line, as illustrated in Fig. 10.13. Find the location of point D when the rigid body is in positions 2 and 3; that is, find the locations of points D_2 and D_3.

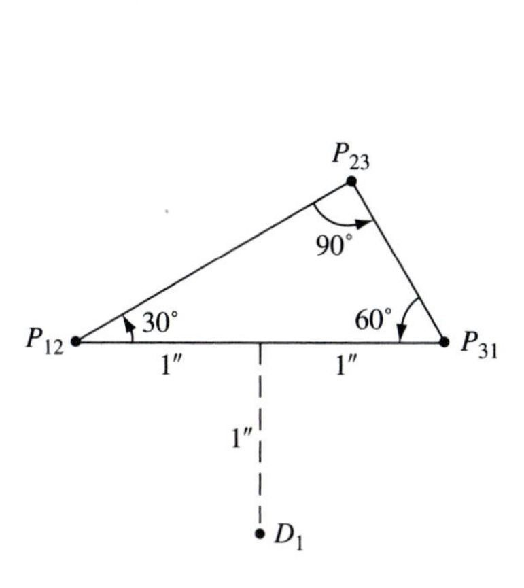

Figure 10.13 Example 10.3. The pole triangle and the point D_1.

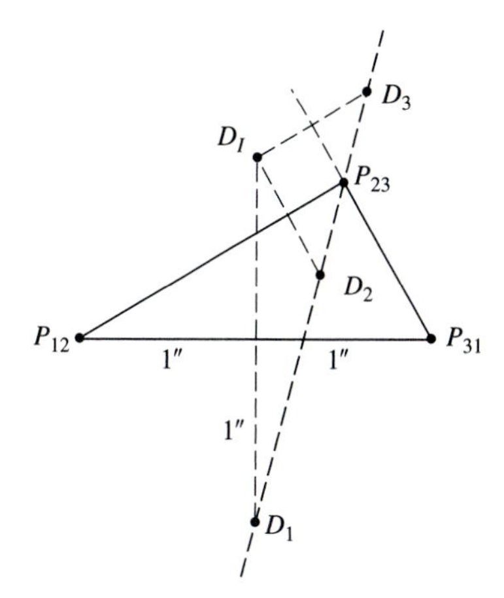

Figure 10.14 Example 10.3. The locations of points D_1, D_2, and D_3.

SOLUTION

The pole triangle is a 1, 2, $\sqrt{3}$ right-angled triangle; therefore, the lengths of the remaining two sides are $P_{12}P_{23} = \sqrt{3}$ in and $P_{23}P_{31} = 1$ in. Reflecting point D_1 about the side $P_{31}P_{12}$ gives the image point D_I. Then reflecting D_I about the side $P_{12}P_{23}$ gives point D_2 and, finally, reflecting D_I about the side $P_{23}P_{31}$ gives point D_3. The points D_1, D_2, and D_3 are illustrated in Fig. 10.14.

Note that the three points D_1, D_2, and D_3 lie on a straight line. Also note that this line passes through the pole P_{23}, that is, the rotation angle

$$\angle D_2P_{23}D_3 = 2\phi_{23} = 2\angle P_{12}P_{23}P_{31} = 180^\circ \text{ ccw}.$$

The following will clarify why the result, in this example, is a straight line.

The Circumscribing Circle

An important geometric property of a triangle is its circumscribing circle. A circle can always be drawn such that the three vertices of the triangle lie on the circumference of this circle, referred to as the circumscribing circle. The center of the circumscribing circle is the intersection of the perpendicular bisectors of the three sides of the triangle. The center of the circumscribing circle is denoted point O, as illustrated in Fig. 10.15.

The reflection of point O about side 1 ($P_{31}P_{12}$) is denoted O_1, that about side 2 ($P_{12}P_{23}$) is denoted O_2, and that about side 3 ($P_{23}P_{31}$) is denoted O_3. Poles P_{31} and P_{12} lie on the circumference of a circle with the same radius as the circumscribing circle and with center O_1; poles P_{12} and P_{23} lie on the circumference of a circle with the same radius as the circumscribing circle and with center O_2; and poles P_{23} and P_{31} lie on the circumference of a circle with the same radius as the circumscribing circle and with center O_3. These three

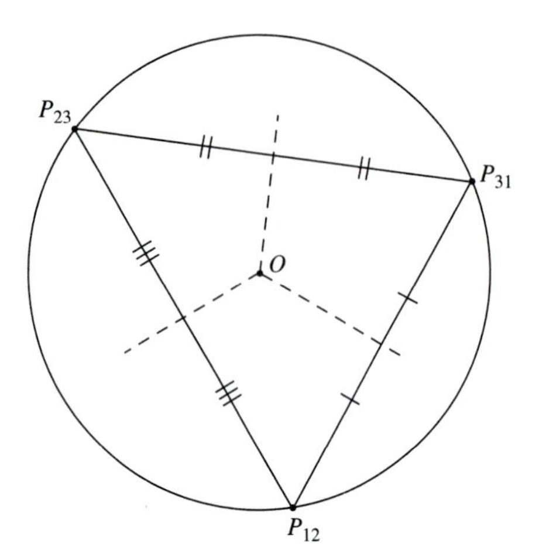

Figure 10.15 The circumscribing circle of a pole triangle.

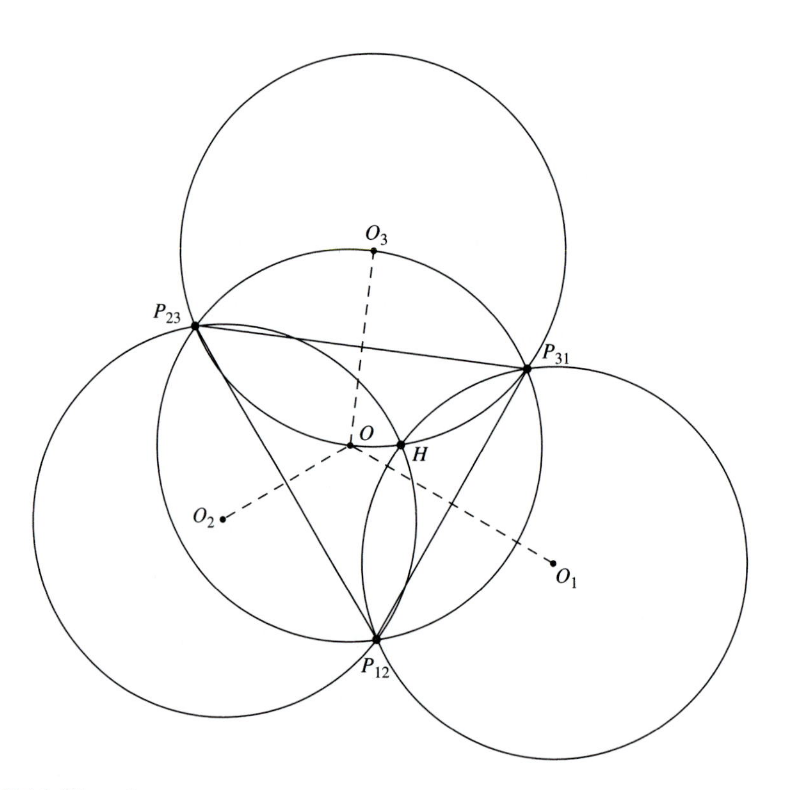

Figure 10.16 The orthocenter.

circles also intersect at a unique point, which is referred to as the *orthocenter* and is denoted point H, as illustrated in Fig. 10.16.

The orthocenter of a triangle is defined as the point of intersection of the bisectors of the three angles of the triangle. Therefore, if the pole triangle is an acute-angled triangle,

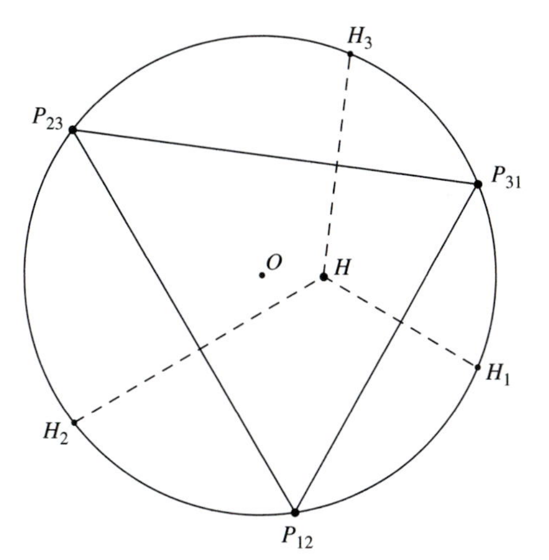

Figure 10.17 Reflections of the orthocenter.

then the orthocenter lies inside the triangle. If the pole triangle is a right-angled triangle, then the orthocenter is coincident with the pole and is located at the apex of the right angle. If the pole triangle has an obtuse angle, then the orthocenter lies outside the triangle.

The reflection of the orthocenter H about side 1 ($P_{31}P_{12}$) is denoted H_1, the reflection of point H about side 2 ($P_{12}P_{23}$) is denoted H_2, and the reflection of point H about side 3 ($P_{23}P_{31}$) is denoted H_3. Note that the reflections of the orthocenter lie on the circumscribing circle of the pole triangle, as illustrated in Fig. 10.17.

EXAMPLE 10.4

For the pole triangle given in Example 10.3, find the center of the circumscribing circle, point O, and draw the circumscribing circle. Then locate the points O_1, O_2, and O_3 and draw the circles with the same radius as the circumscribing circle and with centers at O_1, O_2, and O_3. Finally, determine the locus of all the points in the body having three positions on straight lines.

SOLUTION

Because the pole triangle from Example 10.3 is a right-angled triangle, the center of the circumscribing circle, point O, must lie on the hypotenuse. Note that the hypotenuse is side 1 of the pole triangle, with common subscript 1; therefore, point O_1 is coincident with O. Reflecting point O about side 2 of the pole triangle with the common 2 subscript gives O_2 and reflecting point O about side 3 of the pole triangle with the common 3 subscript gives O_3, as illustrated in Fig. 10.18.

Because point O_1 is coincident with O, then the circle with center at O_1 and the same radius as the circumscribing circle is coincident with the circumscribing circle. Also note that the given point D_1 lies on this circle; therefore, points D_2 and D_3 must lie on the circles with centers at O_2 and O_3. The conclusion is that the locus of points having three positions

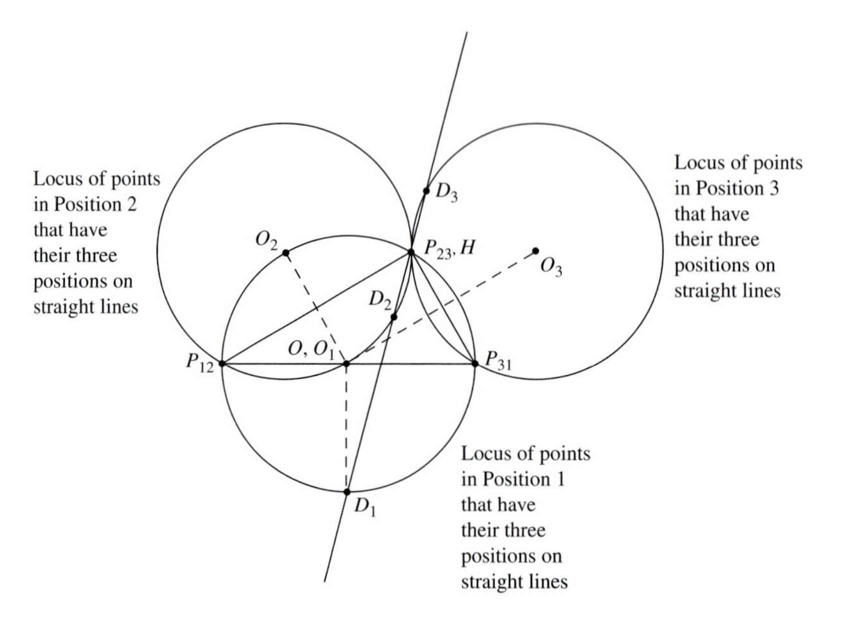

Figure 10.18 Example 10.4. The locus of points having three positions on a straight line.

on a straight line are the circles with the same radii as the circumscribing circle and centers at O_1, O_2, and O_3. These are the only points in the body that can travel on straight lines through the three finitely separated positions.

It is interesting to note that, for three infinitesimally separated positions of the body, the three circles coalesce and become the inflection circle (see Section 4.13). Recall that the inflection circle is the locus of all inflection points in the body, that is, points that instantaneously travel on straight lines.

In designing a four-bar linkage to guide the coupler link through three finitely separated positions, there are ∞^4 possible solutions, that is, ∞^2 choices for the crankpin A in the coupler link and ∞^2 choices for the crankpin B in the coupler link. Based on the choices of A and B, then the locations of the ground pivots O_A and O_B are uniquely determined.

10.5 FOUR FINITELY SEPARATED POSITIONS OF A RIGID BODY ($N = 4$)

For four finitely separated positions of a rigid body there are six poles: (*i*) pole P_{12} for the finite displacement from position 1 to position 2, (*ii*) pole P_{13} for the finite displacement from position 1 to position 3, (*iii*) pole P_{14} for the finite displacement from position 1 to position 4, (*iv*) pole P_{23} for the finite displacement from position 2 to position 3, (*v*) pole P_{24} for the finite displacement from position 2 to position 4, and (*vi*) pole P_{34} for the finite displacement from position 3 to position 4.

The four finitely separated positions of an arbitrary point fixed in the moving body will not, in general, lie on the circumference of a circle. However, there are some points whose four positions will lie on the circumference of a circle; these points are important in kinematic synthesis of a mechanism to guide the body through four given positions and are referred to as *circle points*. Circle points are suitable for the crankpins of a four-bar linkage with the ground pivots at the center of the circle; these are called *center points*. For four finitely separated positions of a plane, all possible circle points lie on a curve that is referred to as the *circle point curve* and all center points lie on a curve that is referred to as the *center point curve*. These two curves are third-order polynomial curves, that is, cubic curves.

To design a four-bar linkage to guide a rigid body through four finitely separated positions we only need to focus on four of the six poles. In other words, the circle point curve and the center point curve can be obtained by considering only four of the six poles. If the solution is not satisfactory then we can choose a different combination of four poles and repeat the synthesis procedure. The geometry involved is that of a pole quadrilateral referred to as an *opposite pole quadrilateral*.

Two poles whose subscripts do not contain a common numeric subscript are referred to as *opposite poles*. There are three pairs of opposite poles, namely, (P_{12}, P_{34}), (P_{13}, P_{24}), and (P_{14}, P_{23}). Two poles whose subscripts do contain a common numeric subscript are referred to as *adjacent poles*. There are twelve pairs of adjacent poles, namely, (P_{12}, P_{13}), (P_{12}, P_{14}), (P_{12}, P_{23}), (P_{12}, P_{24}), (P_{13}, P_{14}), (P_{13}, P_{23}), (P_{13}, P_{34}), (P_{14}, P_{24}), (P_{14}, P_{34}), (P_{23}, P_{24}), (P_{23}, P_{34}), and (P_{24}, P_{34}). The sides of an opposite pole quadrilateral are lines connecting adjacent poles. Therefore, there a total of three opposite pole quadrilaterals; they have diagonals that connect pairs of opposite poles: (*i*) (P_{13}, P_{12}), (P_{12}, P_{24}), (P_{24}, P_{34}), and (P_{34}, P_{13}); (*ii*) (P_{14}, P_{12}), (P_{12}, P_{23}), (P_{23}, P_{34}), and (P_{34}, P_{14}); and (*iii*) (P_{14}, P_{13}), (P_{13}, P_{23}), (P_{23}, P_{24}), and (P_{24}, P_{14}). These three opposite pole quadrilaterals are illustrated in Fig. 10.19.

Theorem *If four positions of a point fixed in a rigid body are to lie on the circumference of a circle, then the center of the circle (that is, a center point) must view the opposite sides of an opposite pole quadrilateral under angles that are equal or differ by* 180°.

The converse is also true, namely: *If a point (say E_C) views the opposite sides of an opposite pole quadrilateral under angles that are equal, or differ by* 180°, *then that point is the center of a circle that passes through the four positions of point E of the body.*

The procedure to find circle points is illustrated in Fig. 10.20; it is as follows. Consider one of the three opposite pole quadrilaterals, say the opposite pole quadrilateral (P_{13}, P_{12}), (P_{12}, P_{24}), (P_{24}, P_{34}), and (P_{34}, P_{13}). Then choose one pair of opposite sides of this quadrilateral, say the opposite sides (P_{12}, P_{24}) and (P_{13}, P_{34}). Draw a circle with the side (P_{12}, P_{24})

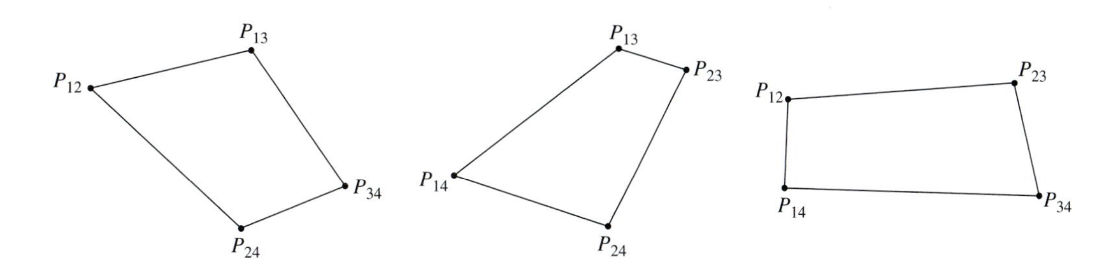

Figure 10.19 The three possible opposite pole quadrilaterals.

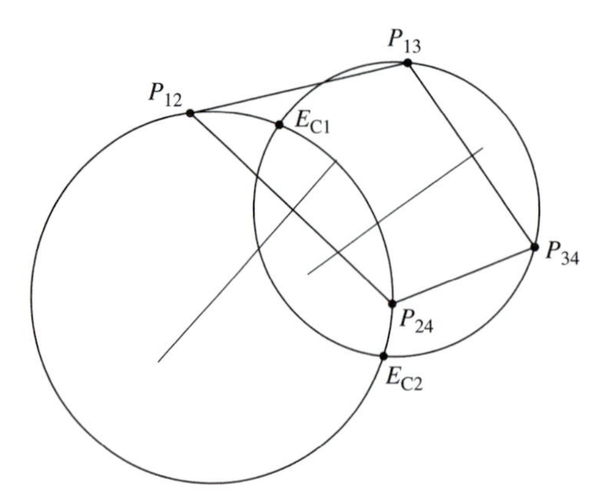

Figure 10.20 A circle with side (P_{12}, P_{24}) as a chord.

as a chord; that is, perpendicularly bisect the side (P_{12}, P_{24}) and choose the center of the circle as any point on this bisector. Choose a convenient radius R and draw the circle as illustrated in Fig. 10.20. The center of this circle can be chosen on the left side of the line (P_{12}, P_{24}) or on the right side of the line. Left and right are defined as follows: if we stand on pole P_{12}, for example, and look toward pole P_{24}, that is, we are looking from 1 to 4, we ignore the common 2 subscript. So, as illustrated in Fig. 10.20, the center of the circle has been chosen on the right side of the line.

Now we draw a circle with the opposite side (P_{13}, P_{34}) as a chord. The radius r of this circle must be chosen in the ratio

$$r = \frac{P_{13}P_{34}}{P_{12}P_{24}} R. \tag{10.19}$$

The center of this circle must be chosen on the same side of the line (P_{13}, P_{34}) as the center of the first circle was chosen relative to the line (P_{12}, P_{24}). So we stand on pole P_{13} and look toward the pole P_{34}, that is, looking from 1 to 4, we ignore the common 3 subscript, consistent with the previous procedure. The center must again be chosen on the right of the line, as illustrated in Fig. 10.20.

In general, these two circles will intersect in two points, denoted here as E_{C_1} and E_{C_2}, both are possible center points. These points both satisfy the theorem, that is, they both view the opposite sides of the opposite pole quadrilateral under angles that are equal or differ by 180°. The relationships can be written as

$$\begin{aligned} \angle P_{12}E_{C_1}P_{24} &= \angle P_{13}E_{C_1}P_{34} \\ \angle P_{12}E_{C_2}P_{24} &= \angle P_{13}E_{C_2}P_{34} \end{aligned} \tag{10.20}$$

or

$$\begin{aligned} \angle P_{12}E_{C_1}P_{24} &= \angle P_{13}E_{C_1}P_{34} \pm 180^\circ \\ \angle P_{12}E_{C_2}P_{24} &= \angle P_{13}E_{C_2}P_{34} \pm 180^\circ \end{aligned} \tag{10.21}$$

The second of Eqs. (10.20) and the first of Eqs. (10.21) can be verified to be true for the example illustrated in Fig. 10.20.

By choosing different values for R (and r) and following the above procedure, a set of center points can be obtained. A curve can then be drawn through these points and is called the *center point curve*. Points on this curve are all suitable candidates for fixed pivots of the four-bar linkage. Note that all six poles, by definition, must lie on the center point curve.

EXAMPLE 10.5

The locations of five of the six poles for four finitely separated positions of the coupler link of a planar four-bar linkage are as illustrated in Fig. 10.21. Draw an opposite pole quadrilateral. Draw the center point curve, which can be used in the synthesis of a four-bar linkage. For the specified fixed pivots of a four-bar linkage, O_A and O_B, which are illustrated in Fig. 10.21, perform the following: (*i*) locate the corresponding circle points A and B in the first three positions of the four-bar linkage (that is, A_1B_1, A_2B_2, and A_3B_3). (*ii*) Specify the lengths of the two links O_AA and O_BB and the length of the coupler link AB of the synthesized four-bar linkage O_A, A, B, and O_B. (*iii*) Is the synthesized four-bar linkage a Grashof four-bar linkage or a non-Grashof four-bar linkage?

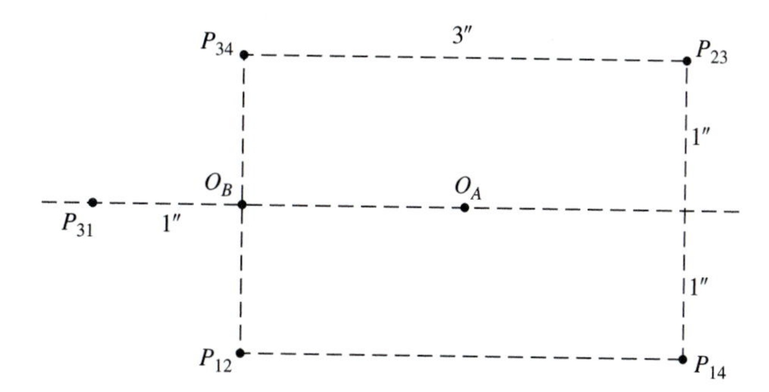

Figure 10.21 Example 10.5. Four finitely separated positions of a coupler link.

SOLUTION

First, an opposite pole quadrilateral is drawn. The four sides of the opposite pole quadrilateral are (P_{12}, P_{14}), (P_{14}, P_{34}), (P_{34}, P_{23}), and (P_{23}, P_{12}), as illustrated in Fig. 10.22. Note that this is only one of three possible opposite pole quadrilaterals. However, it is the only one that can be drawn from the given five poles. The pole P_{24} is not known; therefore, pole P_{31} cannot be used to draw an opposite pole quadrilateral.

The procedure to draw the center point curve is as follows.

1. Choose a side (P_{12}, P_{14}) and the opposite side (P_{34}, P_{23}) of the opposite pole quadrilateral. From the property of similar triangles,

$$\frac{R}{r} = \frac{P_{12}P_{14}}{P_{23}P_{34}} = \frac{3\text{ in}}{3\text{ in}} = 1, \quad \text{that is,} \quad R = r, \tag{1}$$

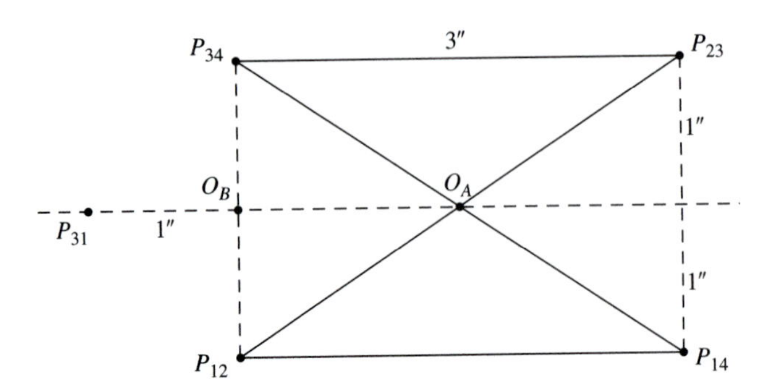

Figure 10.22 Example 10.5. An opposite pole quadrilateral.

where R is the radius of the circle with center on the perpendicular bisector of $P_{12}P_{14}$, and r is the radius of the circle with center on the perpendicular bisector of $P_{34}P_{23}$.

2. Consider the side (P_{12}, P_{14}) of the opposite pole quadrilateral. Standing at pole P_{12}, we look at pole P_{14}; we are looking from 2 to 4, ignoring the common subscript 1. We draw a circle with radius R and with center to our right on the perpendicular bisector of (P_{12}, P_{14}).
3. Now we consider the opposite side (P_{23}, P_{34}) of the opposite pole quadrilateral. Standing at pole P_{23}, we look at pole P_{34}, looking from 2 to 4, ignoring the common subscript 3. We draw a circle with radius $r = R$ and with center again to our right on the perpendicular bisector of (P_{23}, P_{34}).
4. The two points of intersection of the two circles give us two center points. Next, we choose different values for R (and r) and follow the above procedure (steps 2 to 4) to obtain another set of center points. Finally, after doing this repeatedly, we draw a curve through these center points. This is the center point curve and any two points on this curve are suitable as fixed pivots of the four-bar linkage.

As illustrated in Fig. 10.23, the center point curve for this example is a straight horizontal line passing through O_A and O_B and a circle circumscribing the pole quadrilateral. This is overly simplified and a degenerate case; the cubic curve here consists of a straight line and a circle. Still, we note that the center point curve passes through the four poles of the opposite pole quadrilateral (and the given fifth pole). As mentioned previously, the poles (by definition) must always lie on the center point curve; this helps to confirm the graphical construction.

To locate circle points A and B corresponding to the given center points O_A and O_B, consider the pole triangle formed by poles P_{12}, P_{23} and P_{31}; the interior angles are

$$\phi_{12} = 101.3° \text{ cw}, \quad \phi_{23} = 19.6° \text{ cw}, \quad \text{and} \quad \phi_{31} = 59.1° \text{ cw}. \tag{2}$$

Consider the center point, that is, the fixed pivot O_A; then we find the angles

$$\angle O_A P_{12} A_1 = -\phi_{12} = 101.3° \text{ ccw} \quad \text{and} \quad \angle O_A P_{31} A_1 = +\phi_{31} = 59.1° \text{ cw}. \tag{3}$$

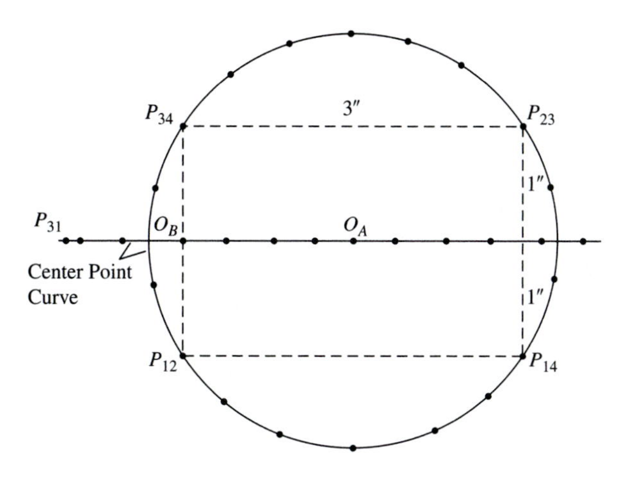

Figure 10.23 Example 10.5. The center point curve.

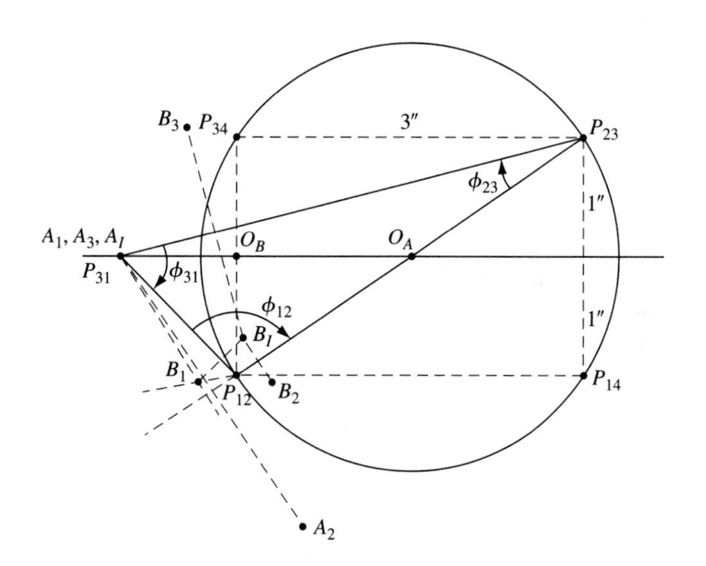

Figure 10.24 Example 10.5. The four-bar linkage.

We draw a line through pole P_{12} such that it makes an angle of 101.3° ccw from the line O_AP_{12}. Then we draw a line through pole P_{31} such that it makes an angle of 59.1° cw from the line O_AP_{31}. The intersection of these two lines gives point A_1. Note that point A_1 is coincident with pole P_{31}. Next, we reflect A_1 about the side 1 ($P_{12}P_{31}$) to find image point A_I. Then we reflect image point A_I about side 2 ($P_{12}P_{23}$) and side 3 ($P_{23}P_{31}$) to give points A_2 and A_3, respectively. The locations of points A_1, A_2, and A_3 are as illustrated in Fig. 10.24. Note that image point A_I is coincident with point A_1.

Similarly, we consider the center point, that is, the fixed pivot O_B where the angles are

$$\angle O_B P_{12} B_1 = -\phi_{12} = 101.3^\circ \text{ ccw} \quad \text{and} \quad \angle O_B P_{31} B_1 = +\phi_{31} = 59.1^\circ \text{ cw}. \tag{4}$$

Then we follow the same procedure as above to obtain points B_1, B_2 and B_3. The locations of points B_1, B_2 and B_3 are as illustrated in Fig. 10.24.

The lengths of the links of the synthesized four-bar linkage $O_A ABO_B$ are measured and found to be

$$AB = 1.25 \text{ in} = p, \quad O_A O_B = 1.5 \text{ in} = q, \quad O_A A = 2.5 \text{ in} = l \quad \text{and} \quad O_B B = 1.2 \text{ in} = s.$$

From the Grashof criterion, Eq. (1.6), a planar four-bar linkage has a crank if and only if

$$s + l < p + q,$$

where $l =$ the length of the longest link, $s =$ the length of the shortest link, and p and q are the lengths of the remaining two links. From the measurements,

$$s + l = 1.2 \text{ in} + 2.5 \text{ in} = 3.7 \text{ in} \quad \text{and} \quad p + q = 1.25 \text{ in} + 1.2 \text{ in} = 2.45 \text{ in}.$$

Therefore, the synthesized planar four-bar linkage is a non-Grashof four-bar linkage and cannot have a continuously rotating crank.

EXAMPLE 10.6

For a rigid body in plane motion, position 2 coincides with position 1 and position 4 coincides with position 3. The locations of the six poles for four finitely separated positions of the moving body are as illustrated in Fig. 10.25. Draw the center point curve that can be used in the synthesis of a four-bar linkage to guide the rigid body through the four finitely separated positions.

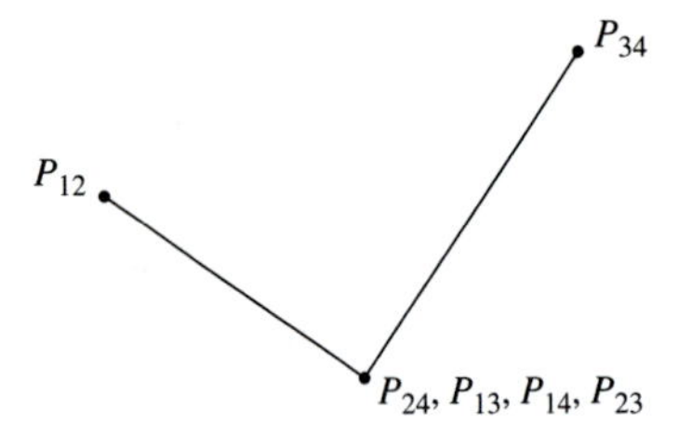

Figure 10.25 Example 10.6. Four finitely separated positions of a rigid body.

SOLUTION

Following the procedure that was outlined in the previous example, the center point curve is as illustrated in Fig. 10.26.

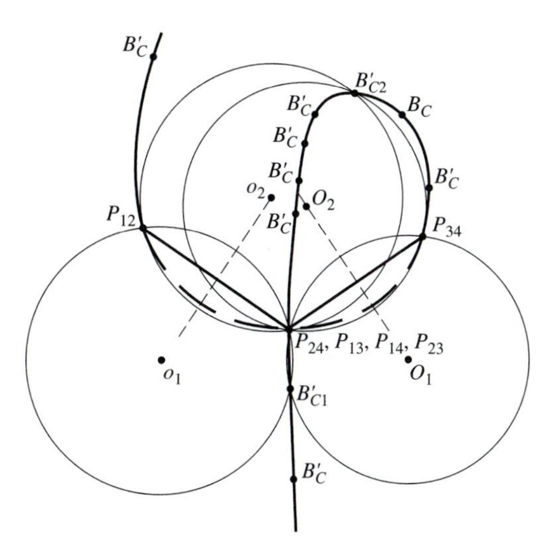

Figure 10.26 Example 10.6. The center point curve.

10.6 FIVE FINITELY SEPARATED POSITIONS OF A RIGID BODY ($N = 5$)

It may be possible to guide a rigid body through five finitely separated positions using a four-bar linkage. However, to do this, we must find two points belonging to the rigid body and having their five positions on the circumferences of two circles. If such circle points exist, they must lie at the intersection of two circle point curves. For example, one could plot the circle point curve that is associated with positions 1, 2, 3, and 4 and then plot the circle point curve that is associated with positions 1, 2, 3, and 5. The points of intersection of these two curves are called *Burmester points* and they are possible locations for the coupler pivots of the four-bar linkage. If the two center points corresponding to these points of intersection lie in locations where there are no practical obstructions to establishing ground pivots for the linkage, then a workable design may be possible.

We probably agree that it is not likely that these particular points will also happen to fall on the circle point curve of yet another group of four positions of the body. Therefore, it is generally *not* possible to synthesize a planar four-bar linkage that passes through six or more arbitrarily prescribed positions.

It is worth noting that a mechanism designed using the methods presented in this section may not be workable although the theory has been applied correctly. Because we are dealing with finitely separated positions of a body and we are not exerting any control over intermediate positions, it is possible that the designed mechanism may not pass through the desired motion because of intermediate limiting positions.

10.7 PRECISION POSITIONS; STRUCTURAL ERROR; CHEBYCHEV SPACING

The synthesis examples in the preceding sections are of the body guidance type. However, it was pointed out that linkages of the function generation type can also be synthesized by kinematic inversion if we consider the motion of the input crank with respect to the output crank. That is, if x is the angular position of the input crank and y is the position of the output link, then, for function generation, we are trying to find the dimensions of a linkage for which the input/output relationship fits a given functional relationship:

$$y = f(x). \tag{a}$$

In general, however, a mechanism has only a limited number of design parameters, a few link lengths, starting angles for the input and output links, and a few more. Therefore, except for very special cases, a linkage synthesis problem usually has no exact solution over its entire range of travel.

In the preceding sections we have chosen to work with two, three, four, or even five positions of the linkage, called *precision positions*, and to seek a linkage that exactly satisfies the desired requirements at these few chosen positions. Our implicit assumption has been that if the design fits the specifications at these few positions, then it will probably deviate only slightly from the desired motion between the precision positions and the deviation will probably be acceptably small. *Structural error* is defined as the theoretical difference between the function produced by the synthesized linkage and the function originally prescribed. For many function generation problems, the structural error in a four-bar linkage solution can be held to less than 4%. We should note, however, that structural error usually exists even with no *graphical error* resulting from a graphical solution process and even with no *mechanical error*, which stems from imperfect manufacturing tolerances.

Of course, the amount of structural error in the solution can be affected by the choice of the precision positions. One problem of linkage design is to select a set of precision positions for use in the synthesis procedure that will minimize this structural error.

Although it is not perfect, a very good trial for the distribution of these precision positions is called *Chebychev* spacing. For N precision positions in the range $x_0 \leq x \leq x_{N+1}$, *Chebychev* spacing, according to Freudenstein and Sandor,[4] is given by

$$x_j = \frac{1}{2}(x_{N+1} + x_0) - \frac{1}{2}(x_{N+1} - x_0)\cos\frac{(2j-1)\pi}{2N} \qquad j = 1, 2, \ldots, N. \tag{10.22}$$

As an example, suppose we wish to devise a linkage to generate the function

$$y = x^{0.8} \tag{b}$$

over the range $1 \leq x \leq 3$ using three precision positions. Then, from Eq. (10.22), the three values of x_j are

$$x_1 = \frac{1}{2}(3+1) - \frac{1}{2}(3-1)\cos\frac{(2-1)\pi}{2(3)} = 2 - \cos\frac{\pi}{6} = 1.134$$

$$x_2 = 2 - \cos\frac{3\pi}{6} = 2.000$$

$$x_3 = 2 - \cos\frac{5\pi}{6} = 2.866.$$

From Eq. (*b*), we find the corresponding values of y to be

$$y_1 = 1.106, \quad y_2 = 1.741, \quad y_3 = 2.322.$$

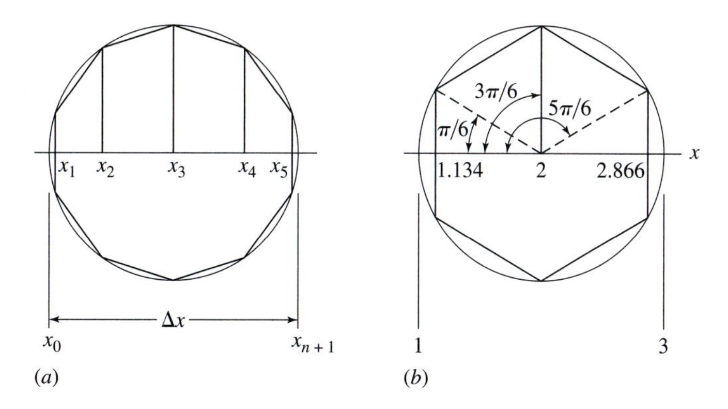

Figure 10.27 Graphical determination of Chebychev spacing.

Chebychev spacing of the precision positions is also easily determined using the graphical approach illustrated in Fig. 10.27. As illustrated in Fig. 10.27*a*, a circle is first constructed whose diameter is equal to the range Δx, given by

$$\Delta x = x_{N+1} - x_0. \tag{c}$$

Next we inscribe a regular polygon having $2N$ sides in this circle, with its first side placed perpendicular to the x axis. Perpendiculars dropped from each jth vertex now intersect the diameter Δx at the precision position value of x_j. Figure 10.27*b* illustrates the construction for the numerical example above.

It should be noted that Chebychev spacing is a good approximation of precision positions that reduce structural error in a design. Depending on the accuracy requirements of the problem, it may be satisfactory. If additional accuracy is required, then by plotting a curve of structural error versus x we can usually determine visually the adjustments to be made in the choice of precision positions for another trial.

Before closing this section, however, we should note two more problems that can arise to confound the designer in choosing precision positions for synthesis. These are called *branch defect* and *order defect*. Branch defect refers to a possible completed design that meets all of the prescribed requirements at each of the precision positions, but which cannot be moved continuously between these positions without being taken apart and reassembled. Order defect refers to a completed linkage design that can reach all of the precision positions, but not in the desired order.[5]

10.8 THE OVERLAY METHOD

Synthesis of a function generator mechanism, say, using the overlay method, is the easiest and quickest of all methods in use. It is not always possible to obtain a solution, and sometimes the accuracy may be rather poor. Theoretically, however, one can employ as many precision positions as are desired in the process.

TABLE 10.1

Position	x	ψ, deg	y	ϕ, deg
1	1	0	1	0
2	1.366	22.0	1.284	14.2
3	1.756	45.4	1.568	28.4
4	2.16	69.5	1.852	42.6
5	2.58	94.8	2.136	56.8
6	3.02	121.0	2.420	71.0

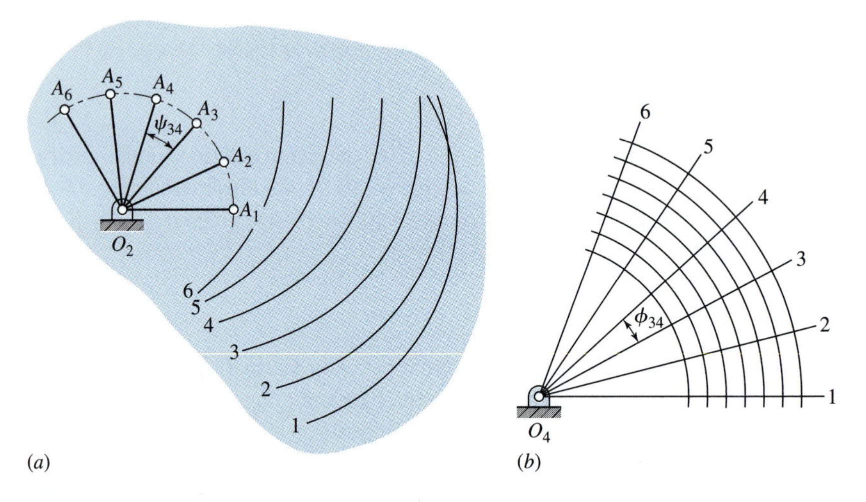

Figure 10.28

Let us design a function generator linkage to generate the equation

$$y = x^{0.8}, \quad 1 \leq x \leq 3.$$

Suppose we choose six precision positions of the linkage for this example and use uniform spacing of the output rocker. Table 10.1 indicates the values of x and y, rounded, and the corresponding angles selected for the input and output rockers.

The first step in the synthesis is illustrated in Fig. 10.28*a*. We use a sheet of tracing paper and construct the input rocker O_2A in all of its positions. This requires an arbitrary choice for the length of crank O_2A. Also on this sheet, we choose another arbitrary length for the coupler AB and draw arcs numbered 1 to 6 using points A_1 to A_6, respectively, as centers.

Now, on another sheet of paper, we construct the output rocker, whose length is unknown, in all of its angular positions, as illustrated in Fig. 10.28*b*. Through point O_4 we draw a number of arbitrarily spaced arcs intersecting the lines O_41, O_42, and so on; these represent possible lengths of the output rocker.

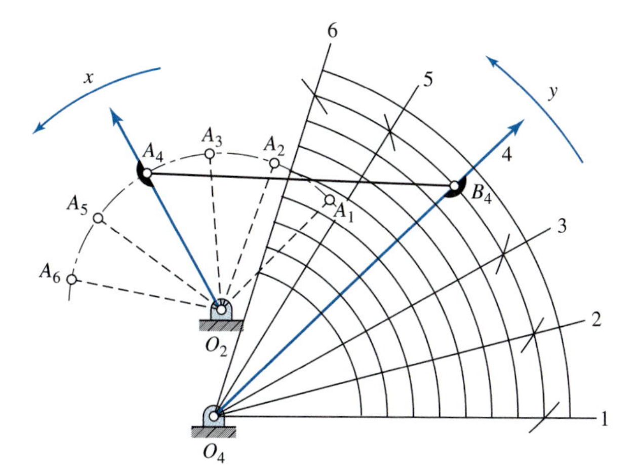

Figure 10.29

As the final step, we lay the tracing over the drawing and manipulate it in an effort to find a fit. In this case a fit is found, and the result is illustrated in Fig. 10.29.

10.9 COUPLER-CURVE SYNTHESIS[6]

In this section we synthesize a four-bar linkage so that a tracing point on the coupler traces a specified path when the linkage is moved. Then, in sections to follow we discover that paths having certain characteristics are particularly useful, for example, in synthesizing linkages having dwells of the output member for certain periods of rotation of the input member.

In synthesizing a linkage to generate a path, we can choose up to six precision positions along the path. If the synthesis is successful, the tracing point will pass through each precision position. The final result may or, because of the branch or order defects, may not approximate the desired path.

Two positions of a four-bar linkage are illustrated in Fig. 10.30. Link 2 is the input member; it is connected at A to coupler 3, containing the tracing point C, and connected to output link 4 at B. Two postures of the linkage are illustrated by subscripts 1 and 3. Points C_1 and C_3 are two positions of the tracing point on the path to be generated. In this example, C_1 and C_3 have been especially selected so that the perpendicular bisector c_{13} passes through O_4. Note, for the selection of points, that the angle $\angle C_1O_4C_3$ is equal to the angle $\angle A_1O_4A_3$, as indicated in Fig. 10.30.

The advantage of making these two angles equal is that when the linkage is finally synthesized, the triangles $C_3A_3O_4$ and $C_1A_1O_4$ are congruent. Thus, if the tracing point is made to pass through C_1 on the path, it will also pass through C_3.

To synthesize a linkage so that the coupler will pass through four precision positions, we locate any four points C_1, C_2, C_3, and C_4 on the desired path (see Fig. 10.31). Choosing C_1 and C_3, say, we first locate O_4 anywhere on the perpendicular bisector c_{13}. Then, with O_4 as a center and using any radius R, we construct a circular arc. Next, with centers at C_1 and C_3, and any other radius r, we strike arcs to intersect the arc of radius R. These two intersections define points A_1 and A_3 on the input link. We construct the perpendicular

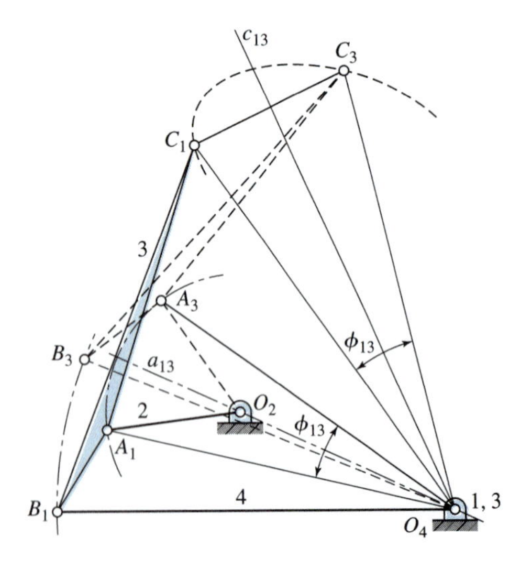

Figure 10.30

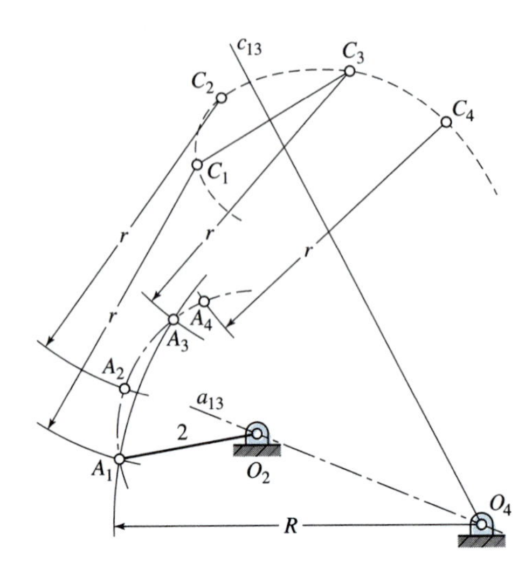

Figure 10.31

bisector a_{13} to A_1A_3 and note that it passes through O_4. We locate O_2 anywhere on a_{13}. This provides an opportunity to choose a convenient length for the input link. Now we use O_2 as a center and draw the crank circle through A_1 and A_3. Points A_2 and A_4 on this circle are obtained by striking arcs of radius r again about C_2 and C_4. This completes the first phase of the synthesis; we have located O_2 and O_4 relative to the desired path and hence defined the distance O_2O_4. We have also defined the length of the input link and located its positions relative to the four precision points on the path.

Our next task is to locate point B, the point of attachment of the coupler and output link. Any one of the four locations of B can be used; in this example we use the B_1 position.

Before beginning the final step, we note that the linkage is now defined. Four arbitrary decisions were made: the location of O_4, the radii R and r, and the location of O_2. Thus, ∞^4 solutions are possible.

Referring to Fig. 10.32, we locate point 2 by making triangles $C_2A_2O_4$ and C_1A_12 congruent. We locate point 4 by making $C_4A_1O_4$ and C_1A_14 congruent. Points 4, 2, and O_4 lie on a circle whose center is B_1. So we find B_1 at the intersection of the perpendicular bisectors of O_42 and O_44. Note that the procedure used causes points 1 and 3 to coincide with O_4. With B_1 located, the links can be drawn in place and the mechanism tested to see how well it traces the desired path.

To synthesize a linkage to generate a path through five precision points, we can make two point reductions. We begin by choosing five points, C_1 to C_5, on the path to be traced. We choose two pairs of these for reduction purposes. In Fig. 10.33, we choose the pairs C_1C_5 and C_2C_3. Other pairs that could have been chosen are

$$C_1C_5,\ C_2C_4; \quad C_1C_5,\ C_3C_4; \quad C_1C_4,\ C_2C_3; \quad C_2C_5,\ C_3C_4.$$

We construct the perpendicular bisectors c_{23} and c_{15} of the lines connecting each pair. These intersect at point O_4. Note that O_4 can, therefore, be located conveniently by a judicious

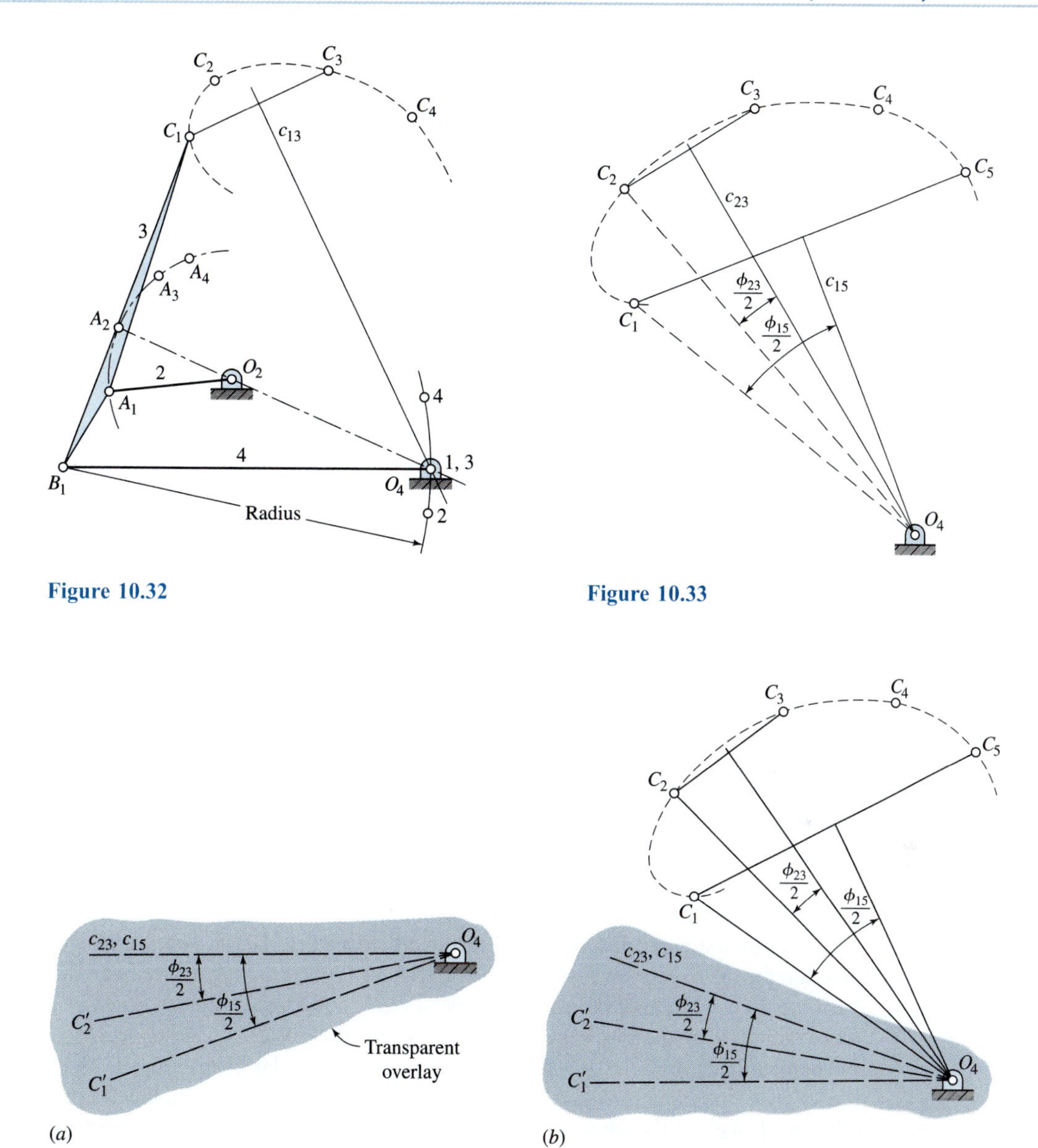

Figure 10.32

Figure 10.33

Figure 10.34

choice of the pairs to be used as well as by the choice of the positions of the points C_i on the path.

The next step is best performed using a sheet of tracing paper as an overlay. We secure the tracing paper to the drawing and mark upon it the center O_4, the perpendicular bisector c_{23}, and another line from O_4 to C_2. Such an overlay is illustrated in Fig. 10.34*a* with the line O_4C_2 designated $O_4C'_2$. This defines the angle $\phi_{23}/2$. Now we rotate the overlay about

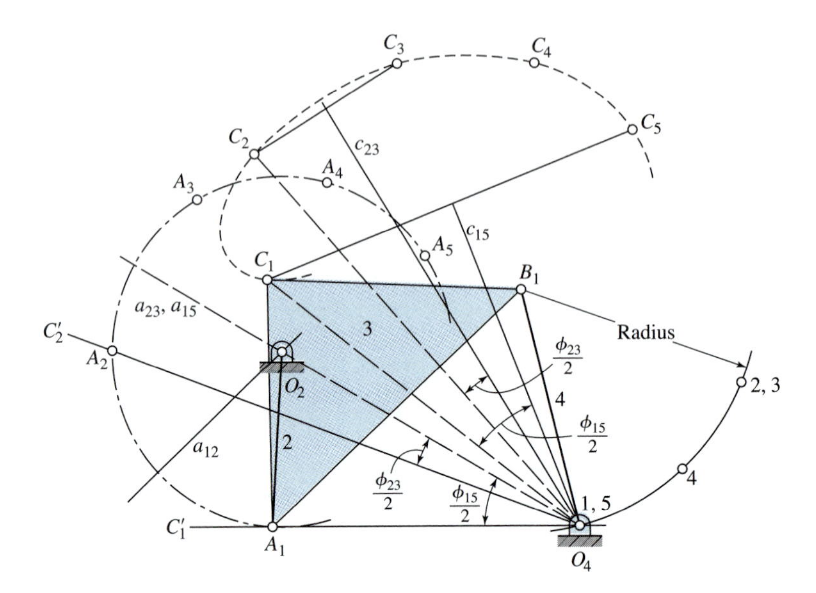

Figure 10.35

the point O_4 until the perpendicular bisector coincides with c_{15} and repeat for point C_1. This defines the angle $\phi_{15}/2$ and the corresponding line O_4C_1'.

Now we pin the overlay at point O_4, using a thumbtack, and rotate it until a good position is found. It is helpful to set the compass for some convenient radius r and draw circles about each point C_i. The intersection of these circles with the lines O_4C_1' and O_4C_2' on the overlay, and with each other, will reveal which areas will be profitable to investigate (see Fig. 10.34*b*).

The final steps in the solution are illustrated in Fig. 10.35. Having located a good position for the overlay, we transfer the three lines to the drawing and remove the overlay. Now we draw a circle of radius r to intersect O_4C_1' and locate point A_1. Another arc of the same radius r from point C_2 intersects O_4C_2' at point A_2. With A_1 and A_2 located, we draw the perpendicular bisector a_{12}; it intersects the perpendicular bisector a_{23} at O_2, giving us the length of the input rocker. A circle through A_1 about O_2 contains all the design positions of point A; we use the same radius r and locate A_3, A_4, and A_5 on arcs about and C_3, C_4, and C_5.

We have now located everything except point B_1, and this is determined as before. A double point 2, 3 exists because of the choice of point O_4 on the perpendicular bisector c_{23}. To locate this point, we strike an arc from C_1 of radius C_2O_4. Then we strike another arc from A_1 of radius A_2O_4. These intersect at the double point 2, 3. To locate point 4, we strike an arc from C_1 of radius C_4O_4 and another from A_1 of radius A_4O_4. Note that point O_4 and the double point 1, 5 are coincident because the synthesis is based on inversion on the O_4B_1 position. Points O_4, 4 and double point 2, 3 lie on a circle whose center is B_1, as illustrated in Fig. 10.35. The linkage is completed by drawing the coupler link and the follower link in the first design position.

10.10 COGNATE LINKAGES; THE ROBERTS–CHEBYCHEV THEOREM

One of the remarkable properties of the planar four-bar linkage is that there is not just one but three four-bar linkages that generate the same coupler curve. This was discovered by Roberts* in 1875 and by Chebychev in 1878 and hence is known as the Roberts–Chebychev theorem. Although mentioned in an English publication in 1954,[7] it did not appear in the American literature until 1958[8], when it was presented, independently and almost simultaneously, by R. S. Hartenberg and J. Denavit of Northwestern University and by R. T. Hinkle of Michigan State University.

In Fig. 10.36, let O_1ABO_2 be the original four-bar linkage with coupler point P attached to AB. The remaining two linkages defined by the Roberts–Chebychev theorem were termed *cognate linkages* by Hartenberg and Denavit. Each of the cognate linkages is illustrated in Fig. 10.36; one is $O_1A_1C_1O_3$ and uses short dashes for showing the links, and the other is $O_2B_2C_2O_3$ and uses long dashes. The construction is evident by observing that there are four similar triangles, each containing the angles α, β, and γ, and three different parallelograms.

A good way to obtain the dimensions of the two cognate linkages is to imagine that the frame connections O_1, O_2, and O_3 can be unfastened. Then, imagine that O_1, O_2, and O_3 are "pulled" away from each other until a straight line is formed by the crank, coupler, and follower of each linkage. If we were to do this for Fig. 10.36, then we would obtain Fig. 10.37. Note that the frame distances are now incorrect, but all the movable links are of the correct lengths. Given any four-bar linkage and its coupler point, one can create a

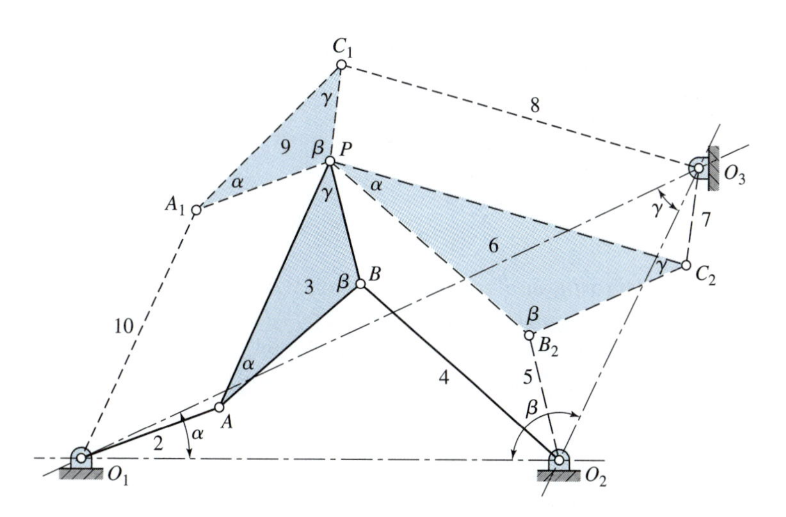

Figure 10.36

* Samuel Roberts (1827–1913), a mathematician; this was not the same Roberts of the approximate-straight-line generator (Fig. 1.19*b*).

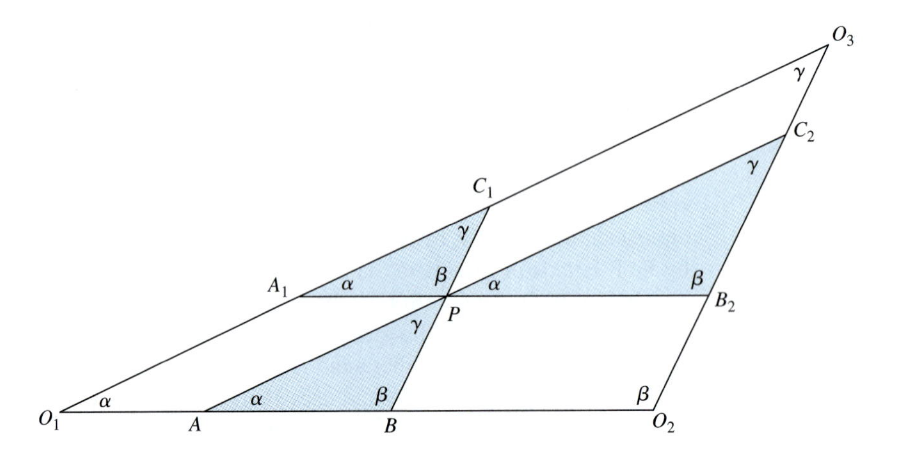

Figure 10.37 The Cayley diagram.

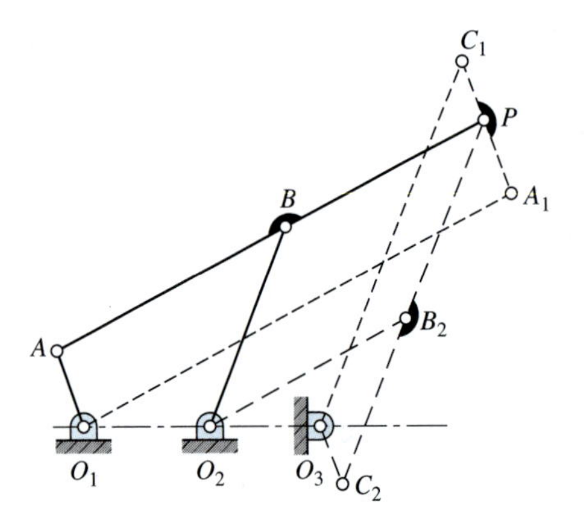

Figure 10.38

drawing similar to Fig. 10.36 and obtain the dimensions of the other two cognate linkages. This approach was discovered by A. Cayley and is called the *Cayley diagram*.*

If the tracing point P is on the straight line AB or its extensions, a figure like Fig. 10.37 is of little help because all three linkages are compressed into a single straight line. An example is illustrated in Fig. 10.38, where O_1ABO_2 is the original linkage having a coupler point P on an extension of AB. To find the cognate linkages, locate O_3 on an extension of O_1O_2 in the same ratio as AB is to BP. Then construct, in order, the parallelograms O_1A_1PA, O_2B_2PB, and $O_3C_1PC_2$.

* Arthur Cayley (1821–1895), "On Three-Bar Motion," *Proc. Lond. Math. Soc.* 7 (1876):136–66. In Cayley's time, a four-bar linkage was described as a three-bar mechanism because the idea of a kinematic chain had not yet been conceived.

Hartenberg and Denavit demonstrated that the angular-velocity relations between the links in Fig. 10.36 are

$$\omega_9 = \omega_2 = \omega_7, \quad \omega_{10} = \omega_3 = \omega_5, \quad \omega_8 = \omega_4 = \omega_6. \tag{11.6}$$

They also observed that if crank 2 is driven at a constant angular velocity and if the velocity relationships are to be preserved during generation of the coupler curve, the cognate mechanisms must be driven at variable angular velocities.

10.11 FREUDENSTEIN'S EQUATION[9]

In Fig. 10.39, we replace the links of a four-bar linkage by position vectors and write the vector loop-closure equation

$$\mathbf{r}_1 + \mathbf{r}_2 + \mathbf{r}_3 + \mathbf{r}_4 = \mathbf{0}. \tag{a}$$

In complex polar notation Eq. (a) is written as

$$r_1 e^{j\theta_1} + r_2 e^{j\theta_2} + r_3 e^{j\theta_3} + r_4 e^{j\theta_4} = 0. \tag{b}$$

From Fig. 10.39, we see that $\theta_1 = 180° = \pi$ radians, from which $e^{j\theta_1} = -1$. Therefore, if Eq. (b) is transformed into complex rectangular form, and if the real and the imaginary components are separated, we obtain the two algebraic equations

$$-r_1 + r_2 \cos\theta_2 + r_3 \cos\theta_3 + r_4 \cos\theta_4 = 0 \tag{c}$$

$$r_2 \sin\theta_2 + r_3 \sin\theta_3 + r_4 \sin\theta_4 = 0. \tag{d}$$

The coupler angle θ_3 can be deleted and the output angle θ_4 can be expressed in terms of the input angle θ_2 by the following procedure. Moving all terms except those involving r_3 to the right-hand side and squaring both sides gives

$$r_3^2 \cos^2\theta_3 = (r_1 - r_2\cos\theta_2 - r_4\cos\theta_4)^2 \tag{e}$$

$$r_3^2 \sin^2\theta_3 = (-r_2\sin\theta_2 - r_4\sin\theta_4)^2. \tag{f}$$

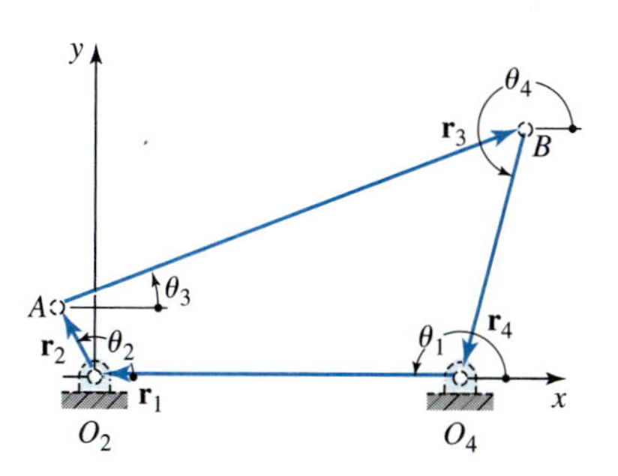

Figure 10.39

Now expanding the right-hand sides and adding the two equations gives

$$\begin{aligned} r_3^2 &= r_1^2 + r_2^2 + r_4^2 - 2r_1r_2\cos\theta_2 - 2r_1r_4\cos\theta_4 \\ &\quad + 2r_2r_4\left(\cos\theta_2\cos\theta_4 + \sin\theta_2\sin\theta_4\right). \end{aligned} \tag{g}$$

Making the substitution $(\cos\theta_2\cos\theta_4 + \sin\theta_2\sin\theta_4) = \cos(\theta_2 - \theta_4)$, and then dividing by the factor $2r_2r_4$, and rearranging again gives

$$\frac{r_3^2 - r_1^2 - r_2^2 - r_4^2}{2r_2r_4} + \frac{r_1}{r_4}\cos\theta_2 + \frac{r_1}{r_2}\cos\theta_4 = \cos(\theta_2 - \theta_4). \tag{h}$$

Freudenstein writes this equation in the form

$$K_1\cos\theta_2 + K_2\cos\theta_4 + K_3 = \cos(\theta_2 - \theta_4), \tag{10.23}$$

where

$$K_1 = \frac{r_1}{r_4} \tag{10.24}$$

$$K_2 = \frac{r_1}{r_2} \tag{10.25}$$

$$K_3 = \frac{r_3^2 - r_1^2 - r_2^2 - r_4^2}{2r_2r_4}. \tag{10.26}$$

We have already learned graphical methods for synthesizing a four-bar linkage so that the motion of the output member is coordinated with that of the input member. Freudenstein's equation, Eq. (10.23), enables us to perform this same task by analytic means. Thus, suppose we wish the output link of a four-bar linkage to occupy the positions ψ_1, ψ_2, and ψ_3 corresponding to the angular positions ϕ_1, ϕ_2, and ϕ_3 of the input link. In Eq. (10.23), we simply replace θ_2 with ϕ_i and θ_4 with ψ_i, and write the equation three times, once for each position. This gives

$$\begin{aligned} K_1\cos\phi_1 + K_2\cos\psi_1 + K_3 &= \cos(\phi_1 - \psi_1) \\ K_1\cos\phi_2 + K_2\cos\psi_2 + K_3 &= \cos(\phi_2 - \psi_2) \\ K_1\cos\phi_3 + K_2\cos\psi_3 + K_3 &= \cos(\phi_3 - \psi_3) \end{aligned} \tag{i}$$

Equations (*i*) are then solved simultaneously for the three unknowns, K_1, K_2, and K_3. Then a length, say r_1, is selected for one of the links and Eqs. (10.24) through (10.26) are solved for the dimensions of the other three links. The method is best illustrated by an example.

EXAMPLE 10.7

Synthesize a function generator to solve the equation

$$y = \frac{1}{x} \quad \text{over the range} \quad 1 \le x \le 2$$

using three precision positions.

SOLUTION

Choosing Chebychev spacing, we find, from Eq. (10.22), the values of x and corresponding values of y to be

$$\begin{aligned} x_1 &= 1.067 \qquad y_1 = 0.937 \\ x_2 &= 1.500 \qquad y_2 = 0.667\ . \\ x_3 &= 1.933 \qquad y_3 = 0.517 \end{aligned}$$

We must now choose starting angles for the input and output links and also total swing angles for each. These are arbitrary decisions and may or may not result in a good linkage design in the sense that the structural errors between the precision points may be large or the transmission angles may be poor. Sometimes, in such a synthesis, it is even determined that one of the pivots must be disconnected to move from one precision position to another. Generally, some trial-and-error work may be necessary to discover the best choices of starting angles and swing angles.

Here, for the input link, we choose a starting angle of $\phi_{\text{min}} = 30°$ and a total swing angle of $\Delta\phi = \phi_{\text{max}} - \phi_{\text{min}} = 90°$. For the output link, we choose a starting angle of $\psi_{\text{min}} = 240°$ and again choose a range of $\Delta\psi = \psi_{\text{max}} - \psi_{\text{min}} = 90°$ total travel. With these choices made, the first and last rows of Table 10.2 can be completed.

Next, to obtain the values of ϕ and ψ corresponding to the precision points, we write

$$\phi = ax + b \quad \psi = cy + d \tag{1}$$

and use the data in the first and last rows of Table 10.2 to evaluate the constants a, b, c, and d. When this is done, we find Eqs. (1) are

$$\phi = 90°x - 60° \quad \psi = -180°y + 420°. \tag{2}$$

These equations can now be used to compute the data for the remaining rows in Table 10.2 and to determine the scales of the input and output links of the synthesized linkage.

Now we take the values of ϕ and ψ from the second line of Table 10.2 and substitute them for θ_2 and θ_4 in Eq. (10.23). Then, if we repeat this for the third and fourth lines, we have the three equations

$$\begin{aligned} K_1 \cos 36.03° + K_2 \cos 251.34° + K_3 &= \cos\left(36.03° - 251.34°\right) \\ K_1 \cos 75.00° + K_2 \cos 300.00° + K_3 &= \cos\left(75.00° - 300.00°\right) \\ K_1 \cos 113.97° + K_2 \cos 326.94° + K_3 &= \cos\left(113.97° - 326.94°\right) \end{aligned} \tag{3}$$

TABLE 10.2 Example 10.7: Accuracy positions

Position	x	ψ, deg	y	ϕ, deg
–	1.000	30.00	1.000	240.00
1	1.067	36.03	0.937	251.34
2	1.500	75.00	0.667	300.00
3	1.933	113.97	0.517	326.94
–	2.000	120.00	0.500	330.00

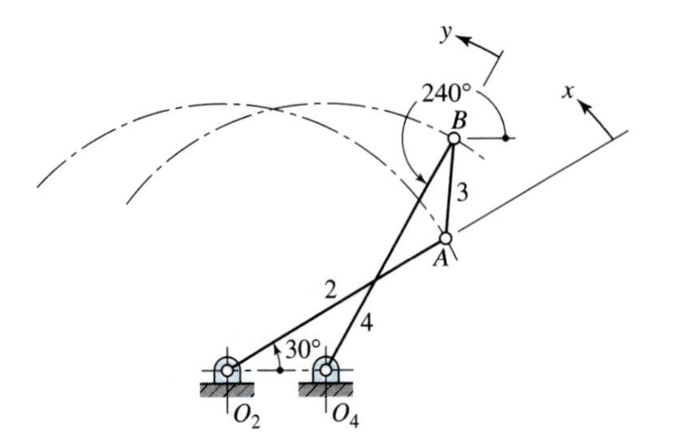

Figure 10.40

When the trigonometric operations are carried out, we have

$$\begin{aligned} 0.8087K_1 - 0.3200K_2 + K_3 &= -0.8160 \\ 0.2588K_1 + 0.5000K_2 + K_3 &= -0.7071 \\ -0.4062K_1 + 0.8381K_2 + K_3 &= -0.8389 \end{aligned} \tag{4}$$

Upon solving Eqs. (4), we obtain

$$K_1 = 0.4032, \quad K_2 = 0.4032, \quad K_3 = -1.0130.$$

Using $r_1 = 1.000$ units, we obtain, from Eq. (10.24),

$$r_4 = \frac{r_1}{K_1} = \frac{1.000}{0.4032} = 2.480 \text{ units.} \qquad \textit{Ans.}$$

Similarly, from Eqs. (10.25) and (10.26), we learn that

$$r_2 = 2.480 \text{ units} \quad \text{and} \quad r_3 = 0.917 \text{ units.} \qquad \textit{Ans.}$$

The result is the crossed linkage illustrated in Fig. 10.40.

Freudenstein offers the following suggestions, which are helpful in synthesizing such function generators.

1. The total swing angles of the input and output links should be less than 120°.
2. Avoid the generation of symmetric functions such as $y = x^2$ over a symmetric range, such as $-1 \leq x \leq 1$.
3. Avoid the generation of functions having abrupt changes in slope.

10.12 ANALYTICAL SYNTHESIS USING COMPLEX ALGEBRA

Another very powerful approach to the synthesis of planar linkages takes advantage of the concept of precision positions and the operations available through the use of complex algebra. Basically, as with Freudenstein's equation in the previous section, the idea is to write complex algebra equations describing the final linkage in each of its precision positions.

Because links do not change lengths during the motion, the magnitudes of these complex vectors do not change from one position to the next, but their angles vary. By writing equations at several precision positions, we obtain a set of simultaneous equations that may be solved for the unknown magnitudes and angles.

The method is very flexible and much more general than is illustrated here. More complete coverage is given in texts such as that by Erdman, Sandor, and Kota.[10] However, the fundamental ideas and some of the operations are illustrated here by an example.

EXAMPLE 10.8

In this example we wish to design a mechanical strip-chart recorder. The concept of the final design is illustrated in Fig. 10.41. We assume that the signal to be recorded is available as a shaft rotation having a range of $0 \leq \phi \leq 90°$ clockwise. This rotation is to be converted into a straight-line motion of the recorder pen over a range of $0 \leq s \leq 4$ in to the right, with a linear relationship between changes of ϕ and s.[11]

SOLUTION

We choose three accuracy positions for our design approach. Using Chebychev spacing over the range to reduce structural error and taking counterclockwise rotation as positive,

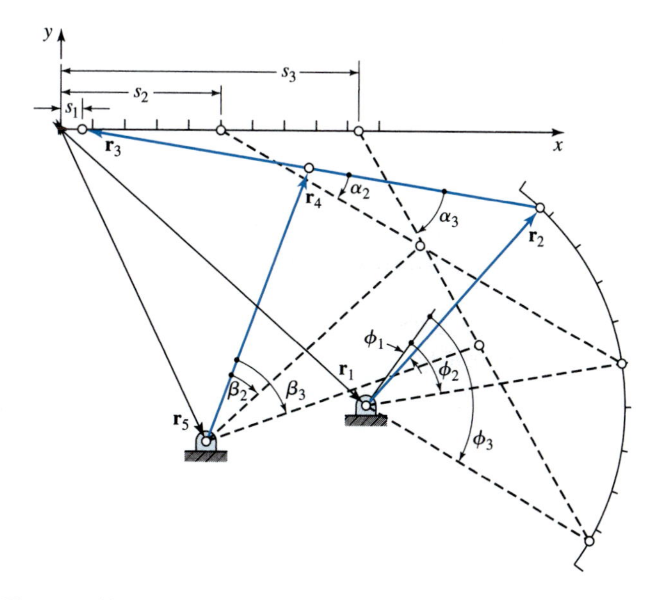

Figure 10.41 Three-position synthesis of chart-recorder linkage using complex algebra.

the three accuracy positions are given by Eq. (10.22):

$$\begin{aligned} \phi_1 &= -6^\circ = -0.104\,72 \text{ rad} & s_1 &= 0.267\,95 \text{ in} \\ \phi_2 &= -45^\circ = -0.785\,40 \text{ rad} & s_2 &= 2.000\,00 \text{ in} \\ \phi_3 &= -84^\circ = -1.466\,08 \text{ rad} & s_3 &= 3.732\,05 \text{ in} \end{aligned} \quad . \tag{a}$$

First, we tackle the design of the dyad consisting of the input crank $\mathbf{r}_2$ and the coupler link $\mathbf{r}_3$. Taking these to be complex vectors for the mechanism in its first accuracy position and designating the unknown position of the fixed pivot by the complex vector $\mathbf{r}_1$, we write a loop-closure equation at each of the three precision positions:

$$\begin{aligned} \mathbf{r}_1 + \mathbf{r}_2 + \mathbf{r}_3 &= s_1 \\ \mathbf{r}_1 + \mathbf{r}_2 e^{j(\phi_2 - \phi_1)} + \mathbf{r}_3 e^{j\alpha_2} &= s_2 \\ \mathbf{r}_1 + \mathbf{r}_2 e^{j(\phi_3 - \phi_1)} + \mathbf{r}_3 e^{j\alpha_3} &= s_3 \end{aligned} \; , \tag{b}$$

where the angles α_j represent the angular displacements of the coupler link from its first position. Next, by subtracting the first of these equations from each of the others and rearranging, we obtain

$$\begin{aligned} \left[e^{j(\phi_2-\phi_1)} - 1\right]\mathbf{r}_2 + \left[e^{j\alpha_2} - 1\right]\mathbf{r}_3 &= s_2 - s_1 \\ \left[e^{j(\phi_3-\phi_1)} - 1\right]\mathbf{r}_2 + \left[e^{j\alpha_3} - 1\right]\mathbf{r}_3 &= s_3 - s_1 \end{aligned} \; . \tag{c}$$

Here we note that we have two complex equations in two complex unknowns, $\mathbf{r}_2$ and $\mathbf{r}_3$, except we note that the coupler displacement angles α_j, which appear in the coefficients, are also unknowns. Thus, we have more unknowns than equations and are free to specify additional conditions or additional data for the problem. Making estimates based on crude sketches of our contemplated design, therefore, we make the following arbitrary decisions:

$$\begin{aligned} \alpha_2 &= -20^\circ = -0.349\,07 \text{ rad} \\ \alpha_3 &= -50^\circ = -0.872\,66 \text{ rad} \end{aligned} \; . \tag{d}$$

Collecting the data from Eqs. (a) and (d), substituting into Eqs. (c), and evaluating, we find

$$\begin{aligned} -(0.222\,85 + j0.629\,32)\mathbf{r}_2 - (0.060\,31 + j0.342\,02)\mathbf{r}_3 &= 1.732\,05 \\ -(0.792\,09 + j0.978\,15)\mathbf{r}_2 - (0.357\,21 + j0.766\,04)\mathbf{r}_3 &= 3.464\,10 \end{aligned} \; ,$$

which we can now solve for the two unknowns:

$$\mathbf{r}_2 = 2.153\,26 + j2.448\,60 = 3.261 \text{ in}\angle 48.67^\circ \qquad \textit{Ans.}$$

$$\mathbf{r}_3 = -5.725\,48 + j0.952\,04 = 5.804 \text{ in}\angle 170.56^\circ. \qquad \textit{Ans.}$$

Then, using the first of Eqs. (b), we solve for the position of the fixed pivot:

$$\begin{aligned} \mathbf{r}_1 &= s_1 - \mathbf{r}_2 - \mathbf{r}_3 \\ &= 3.840\,17 - j3.400\,64 = 5.129 \text{ in}\angle -41.53^\circ \end{aligned} \; . \qquad \textit{Ans.}$$

Thus far, we have completed the design of the dyad which includes the input crank. Before proceeding, we should note that an identical procedure could have been used for the design of a slider-crank mechanism, for one dyad of a four-bar linkage used for any path generation or motion generation problem, or for a variety of other applications. Our total design is not yet completed, but we should note the general applicability of the procedures covered to other linkage synthesis problems.

Continuing with our design of the recording instrument, however, we now must find the location and dimensions of the dyad, $\mathbf{r}_4$ and $\mathbf{r}_5$ of Fig. 10.41. As illustrated in Fig. 10.41, we choose to connect the moving pivot of the output crank at the midpoint of the coupler link to minimize its mass and to keep dynamic forces low. Thus, we can write another loop-closure equation including the rocker at each of the three precision positions:

$$\begin{aligned} \mathbf{r}_5 + \mathbf{r}_4 + 0.5\mathbf{r}_3 &= s_1 \\ \mathbf{r}_5 + \mathbf{r}_4 e^{j\beta_2} + 0.5\mathbf{r}_3 e^{j\alpha_2} &= s_2 \, . \\ \mathbf{r}_5 + \mathbf{r}_4 e^{j\beta_3} + 0.5\mathbf{r}_3 e^{j\alpha_3} &= s_3 \end{aligned} \tag{e}$$

Substituting the known data and rearranging these equations, we obtain

$$\begin{aligned} \mathbf{r}_5 + \mathbf{r}_4 + (-3.130\,69 + j0.476\,02) &= \mathbf{0} \\ \mathbf{r}_5 + \left(e^{j\beta_2}\right)\mathbf{r}_4 + (-4.527\,25 + j1.426\,52) &= \mathbf{0}\, . \\ \mathbf{r}_5 + \left(e^{j\beta_3}\right)\mathbf{r}_4 + (-5.207\,45 + j2.499\,03) &= \mathbf{0} \end{aligned} \tag{f}$$

These appear to be three simultaneous complex equations in only two complex unknowns, $\mathbf{r}_4$ and $\mathbf{r}_5$, and thus we are not free to choose the rotation angles β_2 and β_3 arbitrarily. For Eqs. (f) to have consistent nontrivial solutions, it is necessary that the determinant of the matrix of coefficients be zero. Thus, β_2 and β_3 must be chosen such that

$$\begin{vmatrix} 1 & 1 & (-3.130\,69 + j0.476\,02) \\ 1 & e^{j\beta_2} & (-4.527\,25 + j1.426\,52) \\ 1 & e^{j\beta_3} & (-5.207\,45 + j2.499\,03) \end{vmatrix} = 0, \tag{g}$$

which expands to

$$\begin{aligned} (-2.076\,76 + j2.023\,01)\, e^{j\beta_2} + (1.396\,56 - j0.950\,50)\, e^{j\beta_3} \\ + (0.680\,20 - j1.072\,51) = 0. \end{aligned}$$

Solving this for $e^{j\beta_2}$ gives

$$e^{j\beta_2} = (0.573\,82 + j0.101\,28)\, e^{j\beta_3} + (0.426\,19 - j0.101\,28)$$

and equating the real and imaginary parts, respectively, gives

$$\begin{aligned} \cos\beta_2 &= 0.573\,82\cos\beta_3 - 0.101\,28\sin\beta_3 + 0.426\,19 \\ \sin\beta_2 &= 0.101\,28\cos\beta_3 + 0.573\,82\sin\beta_3 - 0.101\,28 \end{aligned} \, . \tag{h}$$

These equations can now be squared and added to eliminate the unknown β_2. The result, after rearrangement, is a single equation in β_3:

$$0.468\,59\cos\beta_3 - 0.202\,56\sin\beta_3 - 0.468\,57 = 0. \tag{i}$$

This equation can be solved by substituting the tangent of the half-angle identities,

$$x = \tan\frac{\beta_3}{2}, \quad \cos\beta_3 = \frac{1-x^2}{1+x^2}, \quad \sin\beta_3 = \frac{2x}{1+x^2} \tag{j}$$

$$0.468\,59\left(1-x^2\right) - 0.202\,56\,(2x) - 0.468\,57\left(1+x^2\right) = 0.$$

This reduces to the quadratic equation

$$-0.937\,16x^2 - 0.405\,12x + 0.000\,02 = 0,$$

for which the roots are

$$x = -0.432\,66 \quad \text{or} \quad x = 0.000\,27$$

and, from Eq. (*j*),

$$\beta_3 = -46.79° \quad \text{or} \quad \beta_3 = 0.03°.$$

Guided by our sketch of the desired design, we choose the first of these roots, $\beta_3 = -46.79°$. Then, returning to Eqs. (*h*), we find the value $\beta_2 = -26.76°$, and finally, from Eqs. (*f*), we find the final solution:

$$\mathbf{r}_4 = 1.299\,71 + j3.410\,92 = 3.650 \text{ in}\angle 69.14° \qquad \textit{Ans.}$$

$$\mathbf{r}_5 = 1.830\,98 - j3.886\,94 = 4.297 \text{ in}\angle -64.78°. \qquad \textit{Ans.}$$

Note that this second part of the solution, solving for the rocker $\mathbf{r}_4$, is also a very general approach that could be used to design a crank to go through three given precision positions in a variety of other problems. Although only a specific case is presented here, the approach arises repeatedly in linkage design.

Of course, before we finish we should evaluate the quality of our solution by analysis of the linkage we have designed. This has been done here using the equations of Chapter 2 to find the locations of the coupler point for 20 equally spaced crank increments spanning the given range of motion. As expected, there is structural error; the coupler curve of the recording pen tip is not exactly straight, and the displacement increments are not perfectly linear over the range of travel of the pen. However, the solution is quite good; the deviation from a straight line is less than 0.020 in., or 0.5% of the travel, and the linearity between the input crank rotation and coupler point travel is better than 1% of the travel. As expected, the structural error follows a regular pattern and vanishes at the three precision positions. The transmission angle remains larger than 70° throughout the range; thus, no problems with force transmission are expected. Although the design might be improved slightly using additional precision positions, the present solution seems excellent and the additional effort does not appear worthwhile.

10.13 SYNTHESIS OF DWELL MECHANISMS

One of the most interesting uses of coupler curves having straight-line or circular-arc segments is in the synthesis of mechanisms having a substantial dwell during a portion of their

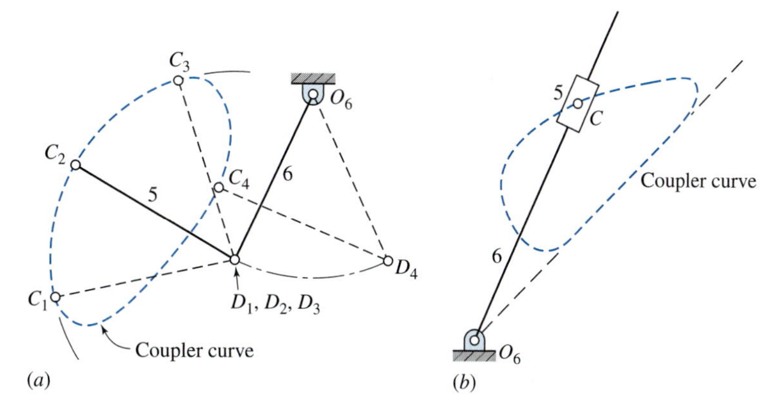

Figure 10.42 Synthesis of dwell mechanisms; in each case, the four-bar linkage that generates the coupler curve is not illustrated. (*a*) Link 6 dwells, as point C travels the circular-arc path $C_1C_2C_3$. (*b*) Link 6 dwells, as point C travels along the straight portion of the coupler curve.

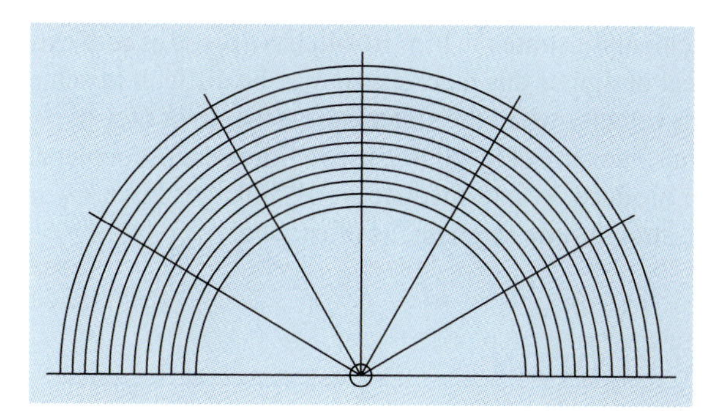

Figure 10.43 Overlay for use with the Hrones and Nelson atlas.

operating cycle. Using segments of coupler curves, it is not difficult to synthesize linkages having a dwell at either or both of the extremes of their motion or at an intermediate position.

In Fig. 10.42*a*, a coupler curve having an approximately elliptical shape is selected from the Hrones and Nelson atlas so that a substantial portion of the curve approximates a circular arc. Connecting link 5 is then given a length equal to the radius of this arc. Thus, in Fig. 10.42*a*, points D_1, D_2, and D_3 are stationary, as coupler point C moves through positions C_1, C_2, and C_3. The length of output link 6 and the location of the frame pivot O_6 depend upon the desired angle of oscillation of this link. The frame pivot should also be positioned for a desirable transmission angle.

When segments of circular arcs are desired for a coupler curve, an organized method of searching the Hrones and Nelson atlas can be employed. The overlay, illustrated in Fig. 10.43, is made on a sheet of tracing paper and can be fitted over the paths in the atlas very quickly. It reveals immediately the radius of curvature of the segment, the location of pivot point D, and the swing angle of the connecting link.

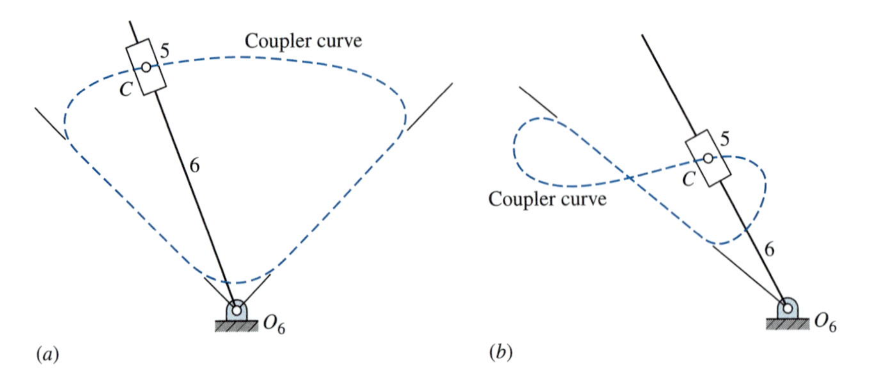

Figure 10.44 (*a*) Link 6 dwells at each end of its swing; (*b*) link 6 dwells in the central portion of its swing.

Figure 10.44 illustrates two ideas for dwell mechanisms employing a slider. A coupler curve having a straight-line segment is used in each, and the pivot point O_6 is placed on an extension of this line.

The arrangement illustrated in Fig. 10.44*a* has a dwell at both extremes of the motion of link 6. A practical design of this mechanism may be difficult to achieve, however, because link 6 has a high velocity when the slider is near the pivot O_6.

The slider mechanism of Fig. 10.44*b* uses a figure-eight coupler curve having a straight-line segment to produce an intermediate dwell linkage. Pivot O_6 must be located on an extension of the straight-line segment, as illustrated.

10.14 INTERMITTENT ROTARY MOTION

The *Geneva wheel*, or *Maltese cross*, is a cam-like mechanism that provides intermittent rotary motion and is widely used in both low-speed and high-speed machinery. Although originally developed as a stop to prevent overwinding of watches, it is now used extensively in automatic machinery, for example, where a spindle, turret, or worktable must be indexed. It is also used in motion-picture projectors to provide the intermittent advance of the film.

A drawing of a six-slot Geneva mechanism is illustrated in Fig. 10.45. Note that the centerlines of the slot and crank are mutually perpendicular at engagement and at disengagement. The crank, which usually rotates at a uniform angular velocity, carries a roller to engage with the slots. During one revolution of the crank, the Geneva wheel rotates a fractional part of a revolution, the amount of which is dependent upon the number of slots. The circular segment attached to the crank effectively locks the wheel against rotation when the roller is not in engagement and also positions the wheel for correct engagement of the roller with the next slot.

The design of a Geneva mechanism is initiated by specifying the crank radius, the roller diameter, and the number of slots. At least three slots are necessary, but most problems can be solved with wheels having from 4 to 12 slots. The design procedure is illustrated in

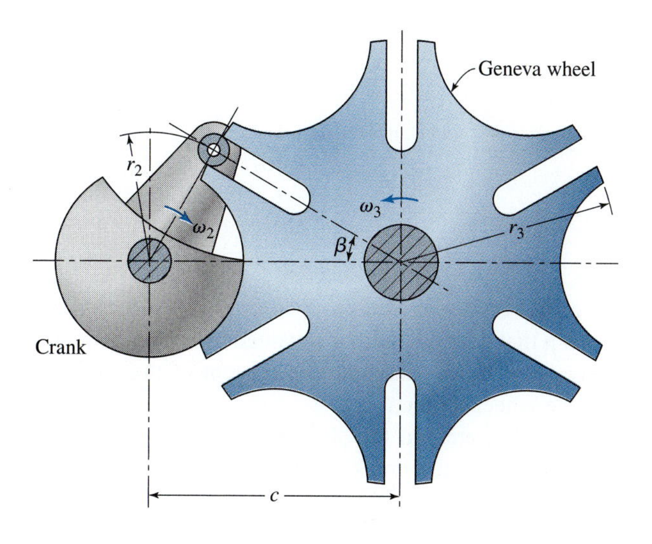

Figure 10.45 The Geneva mechanism.

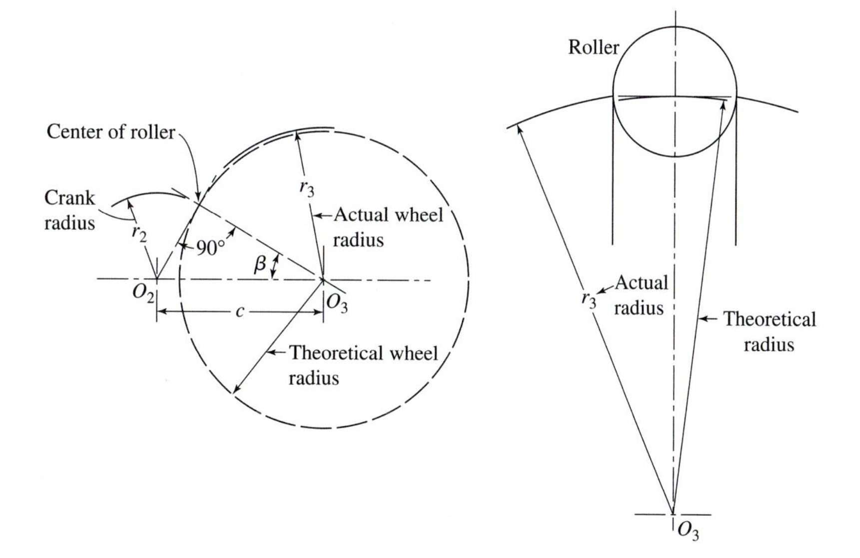

Figure 10.46 Design of a Geneva wheel.

Fig. 10.46. The angle β is half the angle subtended by adjacent slots; that is,

$$\beta = \frac{360^\circ}{2n}, \qquad (a)$$

where n is the number of slots in the wheel. Then, defining r_2 as the crank radius, we have

$$c = \frac{r_2}{\sin\beta}, \qquad (b)$$

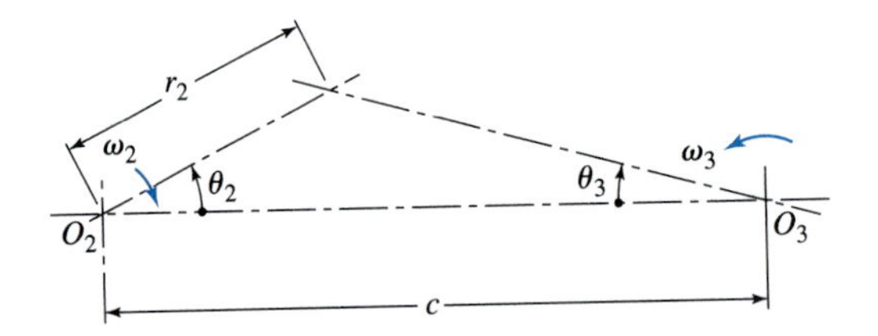

Figure 10.47

where c is the center distance. Note, too, from Fig. 10.46, that the actual Geneva-wheel radius is more than that which would be obtained by a zero-diameter roller. This is because of the difference between the sine and the tangent of the angle subtended by the roller, measured from the wheel center.

After the roller has entered the slot and is driving the wheel, the geometry is that of Fig. 10.47. Here, θ_2 is the crank angle and θ_3 is the wheel angle. They are related trigonometrically by

$$\tan\theta_3 = \frac{\sin\theta_2}{(c/r_2) - \cos\theta_2}. \tag{c}$$

We can determine the angular velocity of the wheel for any value of θ_2 by differentiating Eq. (c) with respect to time. This produces

$$\omega_3 = \frac{(c/r_2)\cos\theta_2 - 1}{1 + \left(c^2/r_2^2\right) - 2\,(c/r_2)\cos\theta_2}\,\omega_2. \tag{10.27}$$

The maximum wheel velocity occurs when the crank angle is zero. Substituting $\theta_2 = 0$ therefore gives

$$(\omega_3)_{\max} = \frac{r_2}{c - r_2}\,\omega_2. \tag{10.28}$$

The angular acceleration with constant input crank speed, obtained by differentiating Eq. (10.27) with respect to time, is

$$\alpha_3 = \frac{(c/r_2)\sin\theta_2\left(1 - c^2/r_2^2\right)}{\left[1 + (c/r_2)^2 - 2\,(c/r_2)\cos\theta_2\right]^2}\,\omega_2^2. \tag{10.29}$$

The angular acceleration reaches a maximum where

$$\theta_2 = \cos^{-1}\left\{\pm\sqrt{\left[\frac{1 + \left(c^2/r_2^2\right)}{4\,(4/r_2)}\right]^2 + 2} - \frac{1 + (c/r_2)^2}{4\,(c/r_2)}\right\}. \tag{10.30}$$

This occurs when the roller has advanced about 30% into the slot.

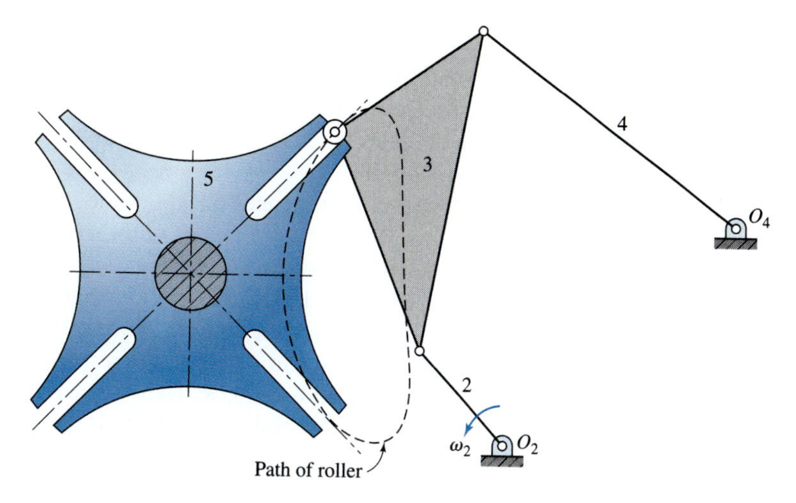

Figure 10.48 Geneva wheel driven by a four-bar linkage synthesized by the Hrones and Nelson atlas. Link 2 is the driving crank.

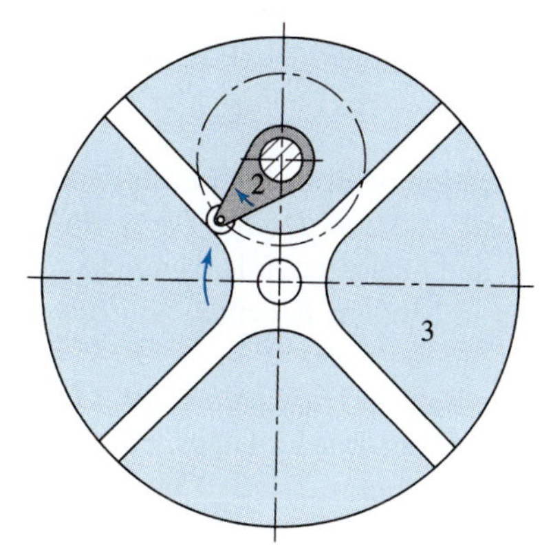

Figure 10.49 The inverse Geneva mechanism.

Several methods have been employed to reduce the wheel acceleration to reduce inertia forces and the consequent wear on the sides of the slot. Among these is the idea of using a curved slot. This can reduce the acceleration, but also increases the deceleration and consequently the wear on the other side of the slot.

Another method uses the Hrones and Nelson atlas for synthesis. The idea is to place the roller on the coupler of a four-bar linkage. During the period in which it drives the wheel, the path of the roller should be curved and should have a low value of acceleration. Figure 10.48 illustrates one solution and includes the path taken by the roller. This is the type of path that is sought while leafing through the atlas.

The inverse Geneva mechanism of Fig. 40.49 enables the wheel to rotate in the same direction as the crank and requires less radial space. The locking device is not illustrated, but this can be a circular segment attached to the crank, as before, which locks by wiping against a built-up rim on the periphery of the wheel.

10.15 REFERENCES

[1] Extensive bibliographies may be found in K. Hain (translated by H. Kuenzel, T. P. Goodman, et al.), *Applied Kinematics*, 2nd ed., New York: McGraw–Hill, 1967, 639–727; and F. Freudenstein and G. N. Sandor, Kinematics of Mechanism, in H. A. Rothbart (ed.), *Mechanical Design and Systems Handbook*, 2nd ed. New York: McGraw–Hill, 1985, 4-56–4-68.

[2] The following are some of the most useful references on kinematic synthesis in the English language: R. Beyer (translated by H. Kuenzel) *Kinematic Synthesis of Linkages*. New York: McGraw–Hill, 1964; A. G. Erdman, G. N. Sandor, and S. Kota, *Mechanism Design: Analysis and Synthesis*, vol. I, 4th ed. Englewood Cliffs, NJ: Prentice Hall, 2001; R. E. Gustavson, "Linkages", Chap. 41 in J. E. Shigley and C. R. Mischke (eds.), *Standard Handbook of Machine Design*. New York: McGraw–Hill, 1986; R. E. Gustavson, "Linkages," Chap. 3 in J. E. Shigley and C. R. Mischke (eds.); *Mechanisms—A Mechanical Designer's Workbook*. New York: McGraw–Hill, 1990; K. Hain, *op. cit.*; A. S. Hall, Jr., *Kinematics and Linkage Design*. Englewood Cliffs, NJ: Prentice Hall, 1961; R. S. Hartenberg and J. Denavit, *Kinematic Synthesis of Linkages*. New York: McGraw–Hill, 1964; J. Hirschhorn, *Kinematics and Dynamics of Plane Mechanisms*. New York: McGraw–Hill 1962; K. H. Hunt, *Kinematic Geometry of Mechanisms*. Oxford: Oxford University Press, 1978; A. H. Soni, *Mechanism Synthesis and Analysis*. New York: McGraw–Hill, 1974; C. H. Suh and C. W. Radcliffe, *Kinematics and Mechanism Design* New York: Wiley, 1978; D. C. Tao, *Fundamentals of Applied Kinematics*. Reading, MA: Addison–Wesley, 1967.

[3] The method described here appears in A. S. Hall, Jr., *op. cit.*, p. 33, and A. H. Soni, *op. cit.*, p. 257. Both D. C. Tao, *op. cit.*, p. 241, and K. Hain, *op. cit.*, p. 317, describe another method that gives different results.

[4] Freudenstein and Sandor, *op. cit.*, p. 14–27.

[5] See K. J. Waldron and E. N. Stephensen, Jr., Elimination of Branch, Grashof, and Order Defects in Path-Angle Generation and Function Generation Synthesis. *J. Mech. Design Trans. ASME* 101 (1979):428–37.

[6] The methods presented here were devised by Hain and presented in K. Hain, *op. cit.*, Chap. 17.

[7] P. Grodzinski and E. M'Ewan, Link Mechanisms in Modern Kinematics. *Proc. Inst. Mech. Eng.* 168, no. 37 (1954):877–96.

[8] R. S. Hartenberg and J. Denavit, The Fecund Four-Bar, in *Trans. 5th Conf. Mech.* West Lafayette, IN: Purdue University, 1958, 194. R. T. Hinkle, Alternate Four-Bar Linkages, *Prod. Eng.* 29 (1958):4.

[9] F. Freudenstien, Approximate Synthesis of Four-Bar Linkages. *Trans. ASME* 77 (1955):853–61.

[10] Erdman, Sandor and Kota, *op. cit.*, Chap. 8.

[11] This example is adapted from a similar problem solved graphically by R. S. Hartenberg and J. Denavit, *op. cit.*, pp. 244–8 and pp. 274–8.

PROBLEMS

10.1 A function varies from 0 to 10. Find the Chebychev spacing for two, three, four, five, and six precision positions.

10.2 Determine the link lengths of a slider-crank linkage to have a stroke of 600 mm and a time ratio of 1.20.

10.3 Determine a set of link lengths for a slider-crank linkage such that the stroke is 16 in and the time ratio is 1.25.

10.4 The rocker of a crank–rocker linkage is to have a length of 500 mm and swing through a total angle of 45° with a time radio of 1.25. Determine a suitable set of dimensions for r_1, r_2, and r_3.

10.5 A crank-and-rocker mechanism is to have a rocker of 6 ft length and a rocking angle of 75°. If the time ratio is to be 1.32, what are a suitable set of link lengths for the remaining three links?

10.6 Design a crank and coupler to drive rocker 4 in Fig. P10.6 such that slider 6 will reciprocate through a distance of 16 in with a time radio of 1.20. Use $a = r_4 = 16$ in and $r_5 = 24$ in with $\mathbf{r}_4$ vertical at midstroke. Record the location of O_2 and dimensions r_2 and r_3.

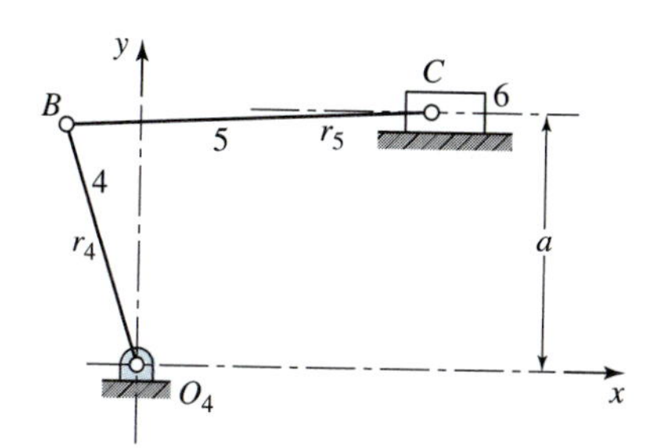

Figure P10.6

10.7 Design a crank and rocker for a six-link mechanism such that the slider in Fig. P10.6 for Problem 10.6 reciprocates through a distance of 800 mm with a time ratio of 1.12; use $a = r_4 = 1\,200$ mm and $r_5 = 1\,800$ mm. Locate O_4 such that rocker 4 is vertical when the slider is at midstroke. Find suitable coordinates for O_2 and lengths for r_2 and r_3.

10.8 Figure P10.8 illustrates two positions of a folding seat used in the aisles of buses to accommodate extra passengers. Design a four-bar linkage to support the seat so that it will lock in the open position and fold to a stable closing position along the side of the aisle.

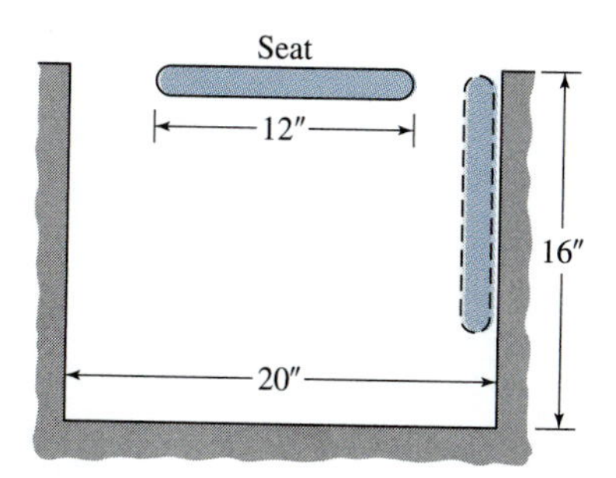

Figure P10.8

10.9 Design a spring-operated four-bar linkage to support a heavy lid like the trunk lid of an automobile. The lid is to swing through an angle of 80° from the closed to the open position. The springs are to be mounted so that the lid will be held closed against a stop, and they should also hold the lid in a stable open position without the use of a stop.

10.10 For Fig. P10.10, synthesize a linkage to move AB from position 1 to position 2 and return.

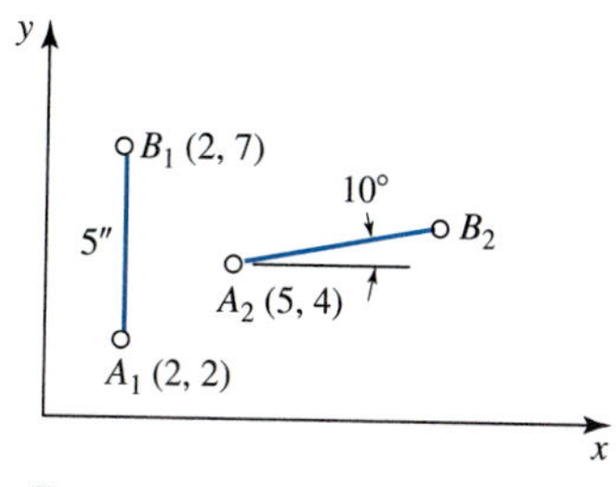

Figure P10.10

10.11 Synthesize a mechanism to move AB successively through positions 1, 2, and 3 of Fig. P10.11.

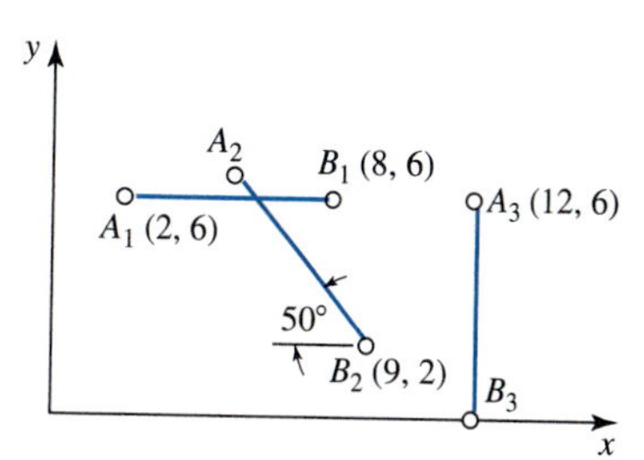

Figure P10.11

10.12 through 10.21* Figure P10.12 illustrates a function-generator linkage in which the motion of rocker 2 corresponds to x and the motion of rocker 4 to the function $y = f(x)$. Use four precision points with Chebychev spacing and synthesize a linkage to generate the functions illustrated in Table P10.12 through 10.31. Plot a curve of the desired function and a curve of the actual function that the linkage generates. Compute the maximum error between them in percent.

* Solutions for these problems were among the earliest computer work in kinematic synthesis and results are illustrated in F. Freudenstein, "Four-Bar Function Generators," *Machine Design*, 30, no. 24 (1958):119–23.

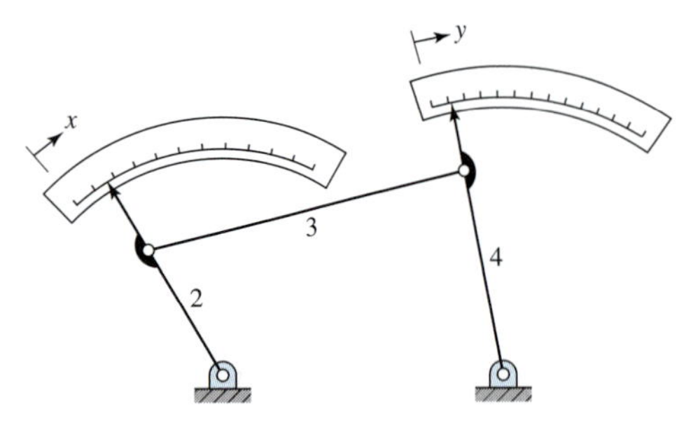

Figure P10.12

TABLE P10.12 to P10.31.

Problem number	Function, $y =$	Range
P10.12, P10.22	$\log_{10} x$	$1 \le x \le 2$
P10.13, P10.23	$\sin x$	$0 \le x \le \pi/2$
P10.14, P10.24	$\tan x$	$0 \le x \le \pi/4$
P10.15, P10.25	e^x	$0 \le x \le 1$
P10.16, P10.26	$1/x$	$1 \le x \le 2$
P10.17, P10.27	$x^{1.5}$	$0 \le x \le 1$
P10.18, P10.28	x^2	$0 \le x \le 1$
P10.19, P10.29	$x^{2.5}$	$0 \le x \le 1$
P10.20, P10.30	x^3	$0 \le x \le 1$
P10.21, P10.31	x^2	$-1 \le x \le 1$

10.22 through 10.31 Repeat Problems 10.12 through 10.21 using the overlay method.

10.32 Figure P10.32 illustrates a coupler curve that can be generated by a four-bar linkage (not illustrated). Link 5 is to be attached to the coupler point, and link 6 is to be a rotating member with O_6 as the frame connection. In this problem we wish to find a coupler curve from the Hrones and Nelson atlas or by precision positions, such that, for an appreciable distance, point C moves through an arc of a circle. Link 5 is then proportioned so that D lies at the center of curvature of this arc. The result is then called a *hesitation motion* because link 6 will hesitate in its rotation for the period during which point C transverses the approximate circular arc. Make a drawing of the complete linkage and plot the velocity- displacement diagram for 360° of displacement of the input link.

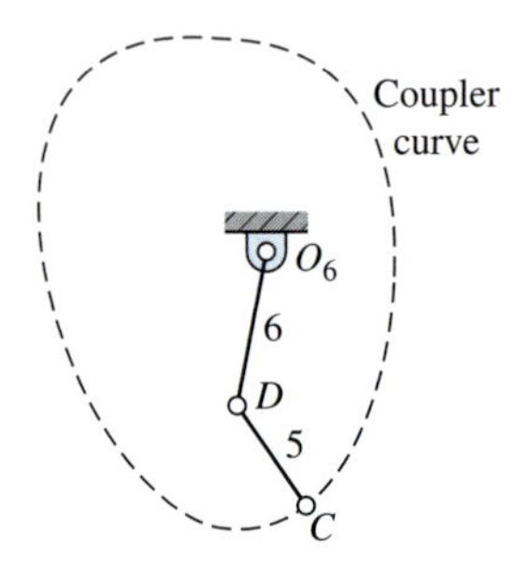

Figure P10.32

10.33 Synthesize a four-bar linkage to obtain a coupler curve having an approximate straight-line segment. Then, using the suggestion included in Fig. 10.42*b* or Fig. 10.44*b*, syjnthesize a dwell motion. Using an input crank angular velocity of unity, plot the velocity of rocker 6 versus the input crank displacement.

10.34 Synthesize a dwell mechanism using the idea suggested in Fig. 10.42*a* and the Hrones and Nelson atlas. Rocker 6 is to have a total angular displacement of 60°. Using this displacement as the abscissa, plot a velocity diagram of the motion of the rocker to illustrate the dwell motion.

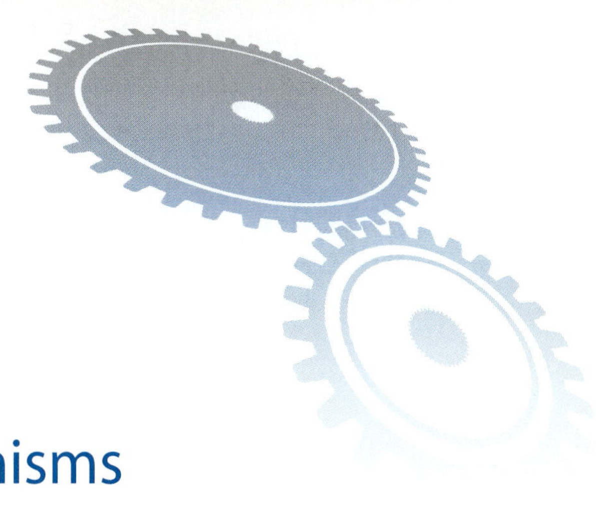

11 Spatial Mechanisms

11.1 INTRODUCTION

The large majority of mechanisms in use today have *planar* motion, that is, the motions of all points produce paths that lie in a single plane or in parallel planes. This means that all motions can be seen in true size and shape from a single viewing direction and that graphical methods of solution require only a single view. If the coordinate system is chosen with the x and y axes parallel to the plane(s) of motion, then all z values remain constant and the problem can be solved, either graphically or analytically, with only two-dimensional methods. Although this is usually the case, it is not a necessity. Mechanisms having three-dimensional point paths do exist and are called *spatial mechanisms*. Another special category, called *spherical mechanisms*, has point paths that lie on concentric spherical surfaces.

Recall that these definitions were raised in Chapter 1, however, almost all of the examples presented in the previous chapters have dealt only with planar mechanisms. This is justified because of their very extensive use in practical applications. Although a few nonplanar mechanisms such as universal shaft couplings and bevel gears have been known for centuries, it is only relatively recently that kinematicians have made substantial progress in developing design procedures for spatial mechanisms. It is probably not a coincidence, given the greater difficulty of the mathematical manipulations, that the emergence of such tools awaited the development and availability of computers.

Although we have concentrated so far on examples with planar motion, a brief review will demonstrate that most of the previous theory has been derived in sufficient generality for both planar and spatial motion. Examples have been planar because they can be more easily visualized and require less tedious computations than the three-dimensional case. Still, most of the theory extends directly to spatial mechanisms. This chapter will review some of the previous techniques, giving examples with spatial motion. The chapter will also introduce a few new mathematical tools and solution techniques that were not required for planar motion.

In Section 1.6 we learned that the mobility of a kinematic chain can be obtained from the Kutzbach criterion. The three-dimensional form of this criterion was given in Eq. (1.3), that is

$$m = 6\,(n-1) - 5j_1 - 4j_2 - 3j_3 - 2j_4 - j_5, \tag{11.1}$$

where m is the mobility of the mechanism, n is the number of links, and each j_k is the number of joints having k degrees of freedom.

One of the numerical solutions to Eq. (11.1) is $m = 1$, when $n = 7$, $j_1 = 7$, and $j_2 = j_3 = j_4 = j_5 = 0$. Harrisberger called such a solution a mechanism type[1]; in particular, he called this the $7j_1$ type. Other combinations of j_k's produce different types of mechanisms. For example, with a mobility $m = 1$, the $3j_1 + 2j_2$ type has $n = 5$ links and the $1j_1 + 2j_3$ type has $n = 3$ links.

Each mechanism type contains a finite number of *kinds* of mechanisms; there are as many kinds of mechanisms of each type as there are ways of arranging the different types of

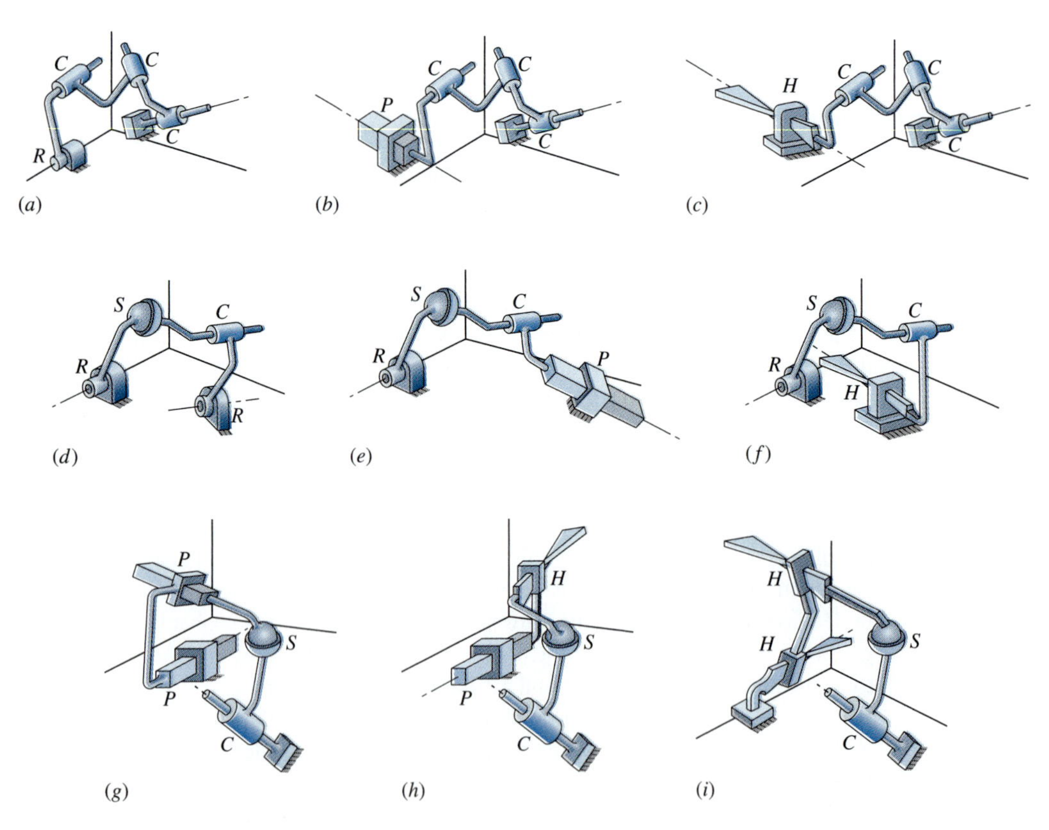

Figure 11.1 Spatial four-bar linkages with a mobility of $m = 1$: (*a*) *RCCC*, (*b*) *PCCC*, (*c*) *HCCC*, (*d*) *RSCR*, (*e*) *RSCP*, (*f*) *RSCH*, (*g*) *PPSC*, (*h*) *PHSC*, and (*i*) *HHSC*.[1]

joints between the links. In Table 1.1 we saw that three of the six lower pairs have one degree of freedom, the revolute, R, the prismatic, P, and the helical, H. Thus, using any seven of these lower pairs, we obtain 36 kinds of type $7j_1$ mechanisms. All together, Harrisberger lists 435 kinds of mechanisms that satisfy the Kutzbach criterion with a mobility of $m = 1$. Not all of these types, or kinds, however, are likely to have much practical value. Consider, for example, the $7j_1$ type with all revolute joints, each connected in series in a single loop.*

For mechanisms having mobility $m = 1$, Harrisberger selected nine kinds from two types that appeared to him to be useful; these are illustrated in Fig. 11.1. They are all spatial four-bar linkages having four joints with either rotating or sliding input and output members. The designations in the legend, such as *RSCH* for Fig. 11.1*f*, identify the kinematic pair types (see Table 1.1) beginning with the input link and proceeding through the coupler and the output link back to the frame. Thus, for the *RSCH*, the input crank is pivoted to the frame by a revolute pair R and to the coupler by a spheric pair S, and the motion of the output member is determined by the helical pair H. The freedoms of these pairs from Table 1.1 are $R = 1$, $S = 3$, $C = 2$, and $H = 1$.

The mechanisms of Fig. 11.1*a* through Fig. 11.1*c* are described by Harrisberger as type 1, or of the $1j_1 + 3j_2$ type. The remaining linkages of Fig. 11.1 are described as type 2, or of the $2j_1 + 1j_2 + 1j_3$ type. All have $n = 4$ links and have a mobility of $m = 1$.

11.2 EXCEPTIONS TO THE MOBILITY OF MECHANISMS

Curiously enough, the most common and most useful spatial mechanisms that have been discovered date back many years and are, in fact, exceptions to the Kutzbach criterion. As indicated in Fig. 1.6, certain geometric conditions sometimes occur that are not included in the Kutzbach criterion and lead to apparent exceptions. As a case in point, consider that every planar mechanism, once constructed, truly exists in three dimensions. Yet a planar four-bar linkage has $n = 4$ and is of type $4j_1$; thus, Eq. (11.1) predicts a mobility of $m = 6(4 - 1) - 5(4) = -2$ (implies redundant constraints). However, we know that the mobility is, in fact, $m = 1$. The special geometric conditions in this case lie in the fact that all revolute axes remain parallel and are all perpendicular to the plane of motion. The Kutzbach criterion does not consider such conditions and can result in false predictions.

At least three more exceptions to the Kutzbach criterion are also four-link *RRRR* linkages. Thus, as with the planar four-bar linkage, the Kutzbach criterion predicts $m = -2$, and yet they are truly of mobility $m = 1$. One of these is the spherical four-bar linkage, illustrated in Fig. 11.2. The axes of all four revolute joints intersect at the center of a sphere. The links may be regarded as great-circle arcs existing on the surface of the sphere; what would be link lengths are now spherical angles. By properly proportioning these angles, it is possible to design all of the spherical counterparts of the planar four-bar linkage such as the spherical crank-rocker linkage and the spherical drag-link linkage. The spherical four-bar linkage is easy to design and manufacture and hence is one of the most useful of all spatial mechanisms. The Hooke or Cardan joint, which is the basis of the universal shaft coupling,

* The only application known to the authors for this type is its use in the front landing gear mechanism of the Boeing 727 aircraft.

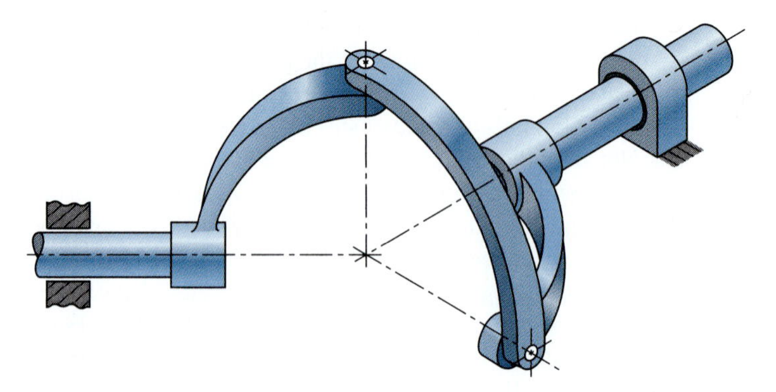

Figure 11.2 Spherical four-bar linkage.

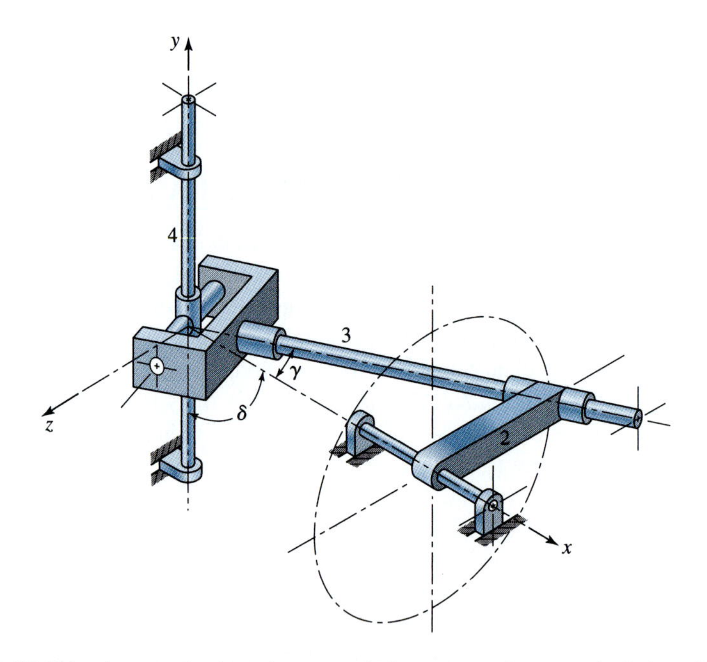

Figure 11.3 Wobble-plate mechanism; input crank 2 rotates and output shaft 4 oscillates. When $\delta = 90°$ the mechanism is called a spherical-slide oscillator. If $\gamma > \delta$, the output shaft rotates.

is a special case of a spherical four-bar linkage with input and output cranks that subtend equal spherical angles.

The wobble-plate mechanism, illustrated in Fig. 11.3, is also a special case. It is another four-bar spherical *RRRR* linkage, which is an exception to the Kutzbach criterion but is movable. Note again that all of the revolute axes intersect at the origin; thus, it is a spherical mechanism.

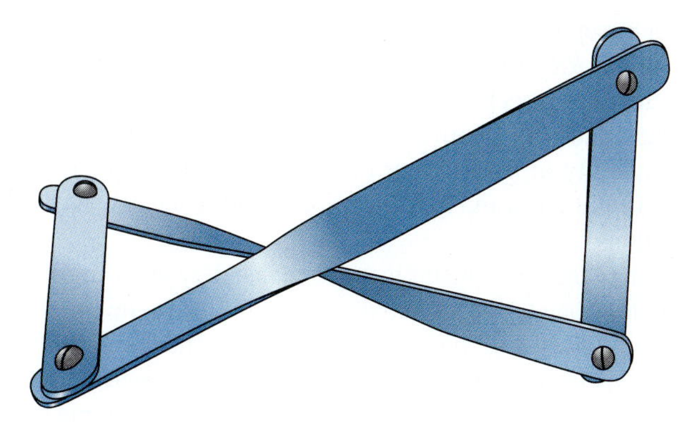

Figure 11.4 The Bennett *RRRR* mechanism.

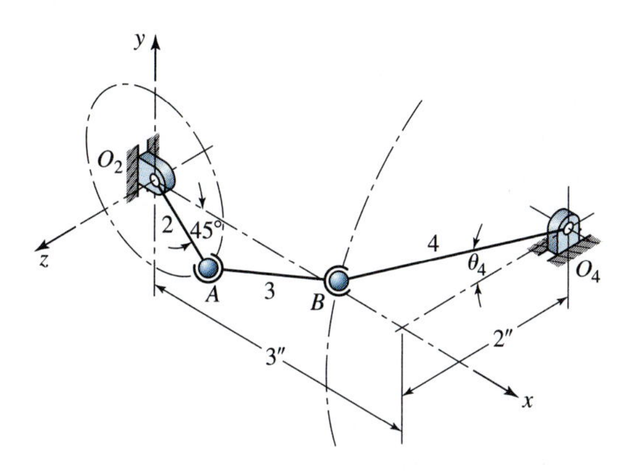

Figure 11.5 Four-link *RSSR* linkage for Example 11.1; $R_{AO_2} = 1$ in, $R_{BA} = 3.5$ in, and $R_{BO_4} = 4$ in.

The Bennett four-bar *RRRR* mechanism[2], illustrated in Fig. 11.4, is probably one of the least practical of the known spatial four-bar linkages. However, it has stimulated the development of kinematic theory for spatial mechanisms[3]. In this mechanism, opposite links are of equal lengths and are twisted by equal amounts. The sines of the twist angles α_1 and α_2 are proportional to the link lengths a_1 and a_2 according to the equation

$$\frac{\sin\alpha_1}{a_1} = \pm\frac{\sin\alpha_2}{a_2}$$

The spatial four-bar *RSSR* linkage, illustrated in Fig. 11.5, is an important and useful linkage. Because $n = 4, j_1 = 2$, and $j_3 = 2$, the Kutzbach criterion of Eq. (11.1) predicts a mobility of $m = 2$. Although this might appear at first glance to be another exception, upon closer examination we find that the second degree of freedom actually exists; it is the freedom of the coupler link to spin about its own axis between the two spheric joints. Because this degree of freedom does not affect the input–output kinematic relationship, it is called an *idle* freedom. This extra freedom does no harm if the mass of the coupler link is truly distributed along its axis; in fact, it may be an advantage because the rotation of the coupler about its axis may equalize wear on the two ball-and-socket joints. If the mass center lies off-axis, however, then this second freedom is not idle dynamically and can cause quite erratic performance of the mechanism at high speed.

Still other exceptions to the Kutzbach mobility criterion are the Goldberg (Michael, not Rube!) five-bar *RRRRR* linkage and the Bricard six-bar *RRRRRR* linkage.[4] Again, it is doubtful whether these mechanisms have much practical value.

Harrisberger and Soni identified all spatial linkages having one general constraint, that is, that have mobility of $m = 1$, but for which the Kutzbach criterion predicts $m = 0$.[5] They identified eight types and 212 kinds and found seven new mechanisms that may have useful applications.

Note that all of the mechanisms mentioned that defy the Kutzbach criterion always predict a mobility less than the actual mobility. This is always the case; the Kutzbach criterion always predicts a lower limit on the mobility, even when the criterion shows exceptions. The reason for this is mentioned in Section 1.6. The argument for the development of the Kutzbach equation, Eq. (11.1), came from counting the freedoms for motion of all of the bodies before any connections are made less the numbers of these presumably eliminated by connecting various types of joints. Yet, when there are special geometric conditions such as intersections or parallelism between joint axes, the criterion counts each joint as eliminating its own share of freedoms although two (or more) of them may eliminate the same freedom(s). Thus, the exceptions arise from the false assumption of independence among the constraints of the joints.

Let us carry this thought one step further. When two (or more) of the constraint conditions eliminate the same motion freedom, the problem is said to have *redundant constraints*. Under these conditions, the same redundant constraints also determine how the forces are shared where the motion freedom is eliminated. Thus, when we come to analyzing forces, we find that there are too many constraints (equations) relating the number of unknown forces. The force analysis problem is then said to be *overconstrained*, and we find that there are statically indeterminate forces in the same number as the error in the predicted mobility.

There is an important lesson buried in this argument*: whenever there are redundant constraints on the motion, there are an equal number of statically indeterminate force components in the mechanism. Despite the higher simplicity in the design equations of planar linkages, for example, we should consider the force effects of these redundant constraints. All of the out-of-plane force and moment components become statically indeterminate. Slight machining tolerance errors or misalignments of axes can cause indeterminate stresses with cyclic loading as the mechanism is operated. The engineer needs to address the question: What effects will these stresses have on the fatigue life of the members of the mechanism?

On the other hand, as pointed out by Phillips†, when motion is only occasional and loads are not very high, this might be an ideal design decision. If errors are small, the additional indeterminate forces may be small and such designs are tolerated although they may seem to exhibit friction effects and wear in the joints. As errors become larger, however, we may find binding in no uncertain terms.

The very existence of the large number of planar linkages in the world testify that such effects can be tolerated with appropriate tolerances on manufacturing errors, good

---

* An extremely detailed discussion of this entire topic forms one of the main themes of an excellent two volume set: J. Phillips, *Freedom of Machinery, Volume 1, Introducing Screw Theory*, Cambridge University Press, 1984, and *Volume 2, Screw Theory Exemplified*, Cambridge University Press, 1990.
† Ibid, Section 20.16, The advantages of overconstraint, p. 151.

lubrication, and proper fits between mating joint elements. Still, too few mechanical designers truly understand that the root of the problem can be eliminated by removing the redundant constraints in the first place. Thus, even in planar motion mechanisms, for example, we can make use of ball-and-socket or cylindric joints that, when properly located, can help to relieve statically indeterminate force and moment components.

11.3 THE SPATIAL POSITION-ANALYSIS PROBLEM

Like planar mechanisms, a spatial mechanism is often connected to form one or more closed loops. Thus, following methods similar to those of Section 2.6, loop-closure equations can be written that define the kinematic relationships of the mechanism. A number of different mathematical forms can be used, including vectors,[6] dual numbers, quaternions,[7] and transformation matrices.[8] In vector notation, the closure of a spatial linkage such as the mechanism of Fig. 11.5 can be defined by a loop-closure equation of the form

$$\mathbf{r} + \mathbf{s} + \mathbf{t} + \mathbf{C} = \mathbf{0}. \tag{11.2}$$

This equation is called the *vector tetrahedron equation* because the individual vectors can be thought of as defining four of the six edges of a tetrahedron.

The vector tetrahedron equation is three-dimensional and hence can be solved for three scalar unknowns. These can be either magnitudes or angles and can exist in any combination in vectors **r**, **s**, and **t**. The vector **C** is the sum of all known vectors in the loop. Using spherical coordinates, each of the vectors **r**, **s**, and **t** can be expressed as a magnitude and two angles. Vector **r**, for example, is defined once its magnitude r and two angles θ_r and ϕ_r are known. Thus, in Eq. (11.2), any three of the nine quantities r, θ_r, ϕ_r, s, θ_s, ϕ_s, t, θ_t, and ϕ_t can be unknowns that must be found from the vector equation. Chace has solved these nine cases by first reducing each to a polynomial.[9] He classifies the solutions depending upon whether the three unknowns occur in one, two, or three separate vectors, and he tabulates the forms of the solutions as shown in Table 11.1.

In Table 11.1, the unit vectors $\hat{\omega}_r$, $\hat{\omega}_s$, and $\hat{\omega}_t$ are axes about which the angles ϕ_r, ϕ_s, and ϕ_t are measured. In case 1, vectors **s** and **t** are not needed, set to zero, and dropped from the equation; the three unknowns are all in vector **r**. In cases 2*a*, 2*b*, 2*c*, and 2*d*, vector **t** is not needed and dropped; the three unknowns are shared by vectors **r** and **s**. Cases 3*a*, 3*b*, 3*c*, and 3*d* have single unknowns in each of the vectors **r**, **s**, and **t**.

One advantage of the Chace vector tetrahedron solutions is that, because they provide known forms for the solutions of the nine cases, we can write a set of nine program modules for computer or calculator evaluation. All nine cases have been reduced to explicit closed-form solutions for the unknowns and therefore can be quickly evaluated. Only case 3*d*, involving the solution of an eighth-order polynomial, must be solved by numerical iteration.

Although the vector tetrahedron equation and its nine case solutions can be used to solve most practical spatial problems, we recall from Section 11.1 that the Kutzbach criterion predicts the existence of up to seven j_1 joints in a single-loop, one-degree-of-freedom mechanism. For example, the seven-link 7*R* mechanism[10] has one input and six unknown joint variables. Using the vector form of the loop-closure equation, it is not possible to determine the values for the six unknowns because the vector equation is equivalent to only three scalar equations and does not fully describe three-dimensional rotation. Therefore,

TABLE 11.1 Classification of solutions of the vector tetrahedron equation

Case number	Unknowns	Known quantities: Vectors	Known quantities: Scalars	Degree of polynomial
1	r, θ_r, ϕ_r	$\mathbf{C}$		1
2a	r, θ_r, s	$\mathbf{C},\ \hat{\boldsymbol{\omega}}_r, \hat{\mathbf{s}}$	ϕ_r	2
2b	r, θ_r, θ_s	$\mathbf{C}, \hat{\boldsymbol{\omega}}_r, \hat{\boldsymbol{\omega}}_s$	$\phi_r, s\phi_s$	4
2c	θ_r, ϕ_r, s	$\mathbf{C}, \hat{\mathbf{s}}$	r	2
2d	$\theta_r, \phi_r, \theta_s$	$\mathbf{C}, \hat{\boldsymbol{\omega}}_s$	r, s, ϕ_s	2
3a	r, s, t	$\mathbf{C}, \hat{\mathbf{r}}, \hat{\mathbf{s}}, \hat{\mathbf{t}}$		1
3b	r, s, θ_t	$\mathbf{C}, \hat{\mathbf{r}}, \hat{\mathbf{s}}, \hat{\boldsymbol{\omega}}_t$	t, ϕ_t	2
3c	r, θ_s, θ_t	$\mathbf{C}, \hat{\mathbf{r}}, \hat{\boldsymbol{\omega}}_s, \hat{\boldsymbol{\omega}}_t$	s, ϕ_s, t, ϕ_t	4
3d	$\theta_r, \theta_s, \theta_t$	$\mathbf{C}, \hat{\boldsymbol{\omega}}_r, \hat{\boldsymbol{\omega}}_s, \hat{\boldsymbol{\omega}}_t$	$r, \phi_r, s, \phi_s, t,\ \phi_t$	8

other mathematical tools such as dual numbers, quaternions, dual quaternions, or transformation matrices must be used. In general, such problems require numerical techniques for their evaluation. Anyone attempting the solution of such problems by hand-algebraic techniques quickly appreciates that *position* analysis, not velocity or acceleration analysis, is the most difficult problem in kinematics.

Solving the polynomials of the Chace vector tetrahedron equation turns out to be equivalent to finding the intersections of straight lines or circles with various surfaces of revolution. In general, such problems can easily and quickly be solved by the methods of descriptive geometry. The graphical approach has the additional advantage that the geometric nature of the problem is not concealed in a multiplicity of mathematical operations. Graphic methods performed by hand are not nearly as accurate as an analytic or a numeric approach. However, graphic methods performed by a computer software package can give the same accuracy as an analytic or numeric approach.

EXAMPLE 11.1

A spatial four-bar *RSSR* crank–rocker linkage is illustrated in Fig. 11.5. The knowns are the dimensions of the links, the plane of rotation and the angle of the input crank, and the plane of rotation of the output crank. For the input crank angle $\theta_2 = -45°$ as shown in the figure, determine the positions of the coupler and the output link, that is, links 3 and 4.

GRAPHIC SOLUTION

If we replace the output link 4 by the vector $\mathbf{R}_{BO_4}$, then the only unknown is the angle θ_4 because the magnitude of the vector and the plane of rotation are given. Similarly, if the coupler link 3 is replaced by the vector $\mathbf{R}_{BA}$, then the magnitude is known and there are only two unknown spherical coordinate angular direction variables for this vector. The loop-closure equation is

$$(\mathbf{R}_{BA}) + (-\mathbf{R}_{BO_4}) + \left(\mathbf{R}_{AO_2} - \mathbf{R}_{O_4O_2}\right) = \mathbf{0}. \tag{1}$$

We identify this as case $2d$ in Table 11.1, requiring the solution of a second-order polynomial (a quadratic) and hence yielding two solutions.

This problem is solved graphically using two orthographic views, the frontal and profile views. If we imagine, in Fig. 11.5, that the coupler is disconnected from the output crank at B and allowed to occupy any possible position with respect to A, then B of link 3 must lie on the surface of a sphere of known radius with its center at A. With joint B still disconnected, the locus of B of link 4 is a circle of known radius about O_4 in a plane parallel to the yz plane. Therefore, to solve this problem, we need only find the two points of intersection of a circle and a sphere.

The solution is shown in Fig. 11.6, where the subscripts F and P denote projections on the frontal and profile planes, respectively. First we locate points O_2, A, and O_4 in both views. In the profile view we draw a circle of radius $R_{BO_4} = 4$ in about center O_{4P}; this is the locus of point B_4. This circle appears in the frontal view as the vertical line $M_F O_{4F} N_F$.

Next, in the frontal view, we construct the outline of a sphere with A_F as its center and the coupler length $R_{BA} = 3.5$ in as its radius. If $M_F O_{4F} N_F$ is regarded as the trace of a plane normal to the frontal view plane, the intersection of this plane with the sphere appears as the full circle in the profile view, having diameter $M_P N_P = M_F N_F$. The circular arc of radius R_{BO_4} intersects the circle at two points, yielding two solutions. One of the points is

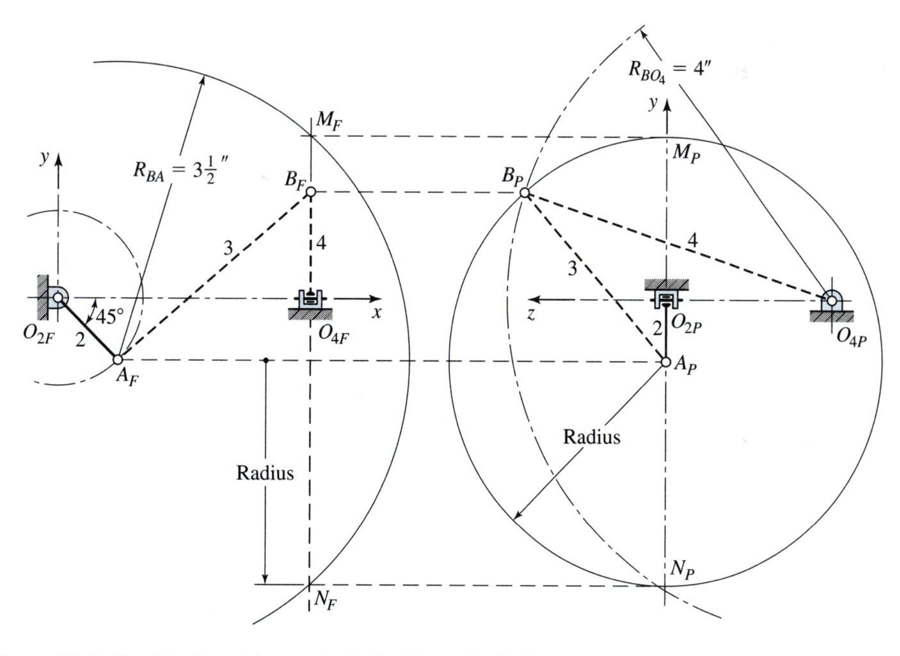

Figure 11.6 Graphical position analysis for Example 11.1.

chosen as B_P and is projected back to the frontal view to locate B_F. Links 3 and 4 are drawn in next as the dashed lines illustrated in the profile and frontal views of Fig. 11.6.

Measuring the x, y, and z projections from the graphic solution, we can write the vectors for each link:

$$\mathbf{R}_{O_4O_2} = 3.00\hat{\mathbf{i}} - 2.00\hat{\mathbf{k}} \text{ in.} \qquad \textit{Ans.}$$

$$\mathbf{R}_{AO_2} = 0.71\hat{\mathbf{i}} - 0.71\hat{\mathbf{j}} \text{ in.} \qquad \textit{Ans.}$$

$$\mathbf{R}_{BA} = 2.30\hat{\mathbf{i}} + 1.95\hat{\mathbf{j}} + 1.77\hat{\mathbf{k}} \text{ in.} \qquad \textit{Ans.}$$

$$\mathbf{R}_{BO_4} = 1.22\hat{\mathbf{j}} + 3.81\hat{\mathbf{k}} \text{ in.} \qquad \textit{Ans.}$$

These components were obtained from a true-size graphical solution; better accuracy would be possible, of course, by making the drawings two or four times actual size.

ANALYTIC SOLUTION

The given information establishes that

$$\begin{aligned}
\mathbf{R}_{O_4O_2} &= 3.000\hat{\mathbf{i}} - 2.000\hat{\mathbf{k}} \text{ in.} \\
\mathbf{R}_{AO_2} &= \cos(-45^\circ) + \sin(-45^\circ) = 0.707\hat{\mathbf{i}} - 0.707\hat{\mathbf{j}} \text{ in.} \\
\mathbf{R}_{BA} &= R^x_{BA}\hat{\mathbf{i}} + R^y_{BA}\hat{\mathbf{j}} + R^z_{BA}\hat{\mathbf{k}} \\
\mathbf{R}_{BO_4} &= -4.000\sin\theta_4\hat{\mathbf{j}} + 4.000\cos\theta_4\hat{\mathbf{k}} \text{ in.}
\end{aligned}$$

The same vector loop-closure equation as in Eq. (1) above must be solved. Substituting the given information gives

$$\begin{aligned}
&\left(R^x_{BA}\hat{\mathbf{i}} + R^y_{BA}\hat{\mathbf{j}} + R^z_{BA}\hat{\mathbf{k}}\right) + \left(4.000\sin\theta_4\hat{\mathbf{j}} - 4.000\cos\theta_4\hat{\mathbf{k}}\right) \\
&\quad + \left(0.707\hat{\mathbf{i}} - 0.707\hat{\mathbf{j}} - 3.000\hat{\mathbf{i}} + 2.000\hat{\mathbf{k}}\right) = \mathbf{0}.
\end{aligned}$$

Then separating this equation into its $\hat{\mathbf{i}}$, $\hat{\mathbf{j}}$, and $\hat{\mathbf{k}}$ components and rearranging gives the following three equations:

$$\begin{aligned}
\text{R}^x_{BA} &= 2.293 \text{ in.} \\
\text{R}^y_{BA} &= -4.000\sin\theta_4 + 0.707 \text{ in.} \\
\text{R}^z_{BA} &= 4.000\cos\theta_4 - 2.000 \text{ in.}
\end{aligned}$$

Next, we square and add these equations and, remembering that $R_{BA} = 3.500$ in., we obtain

$$(2.293 \text{ in})^2 + (-4.000\sin\theta_4 + 0.707 \text{ in})^2 + (4.000\cos\theta_4 - 2.000 \text{ in})^2 = (3.500 \text{ in})^2$$

Expanding this equation and rearranging gives

$$-5.656\sin\theta_4 - 16.000\cos\theta_4 + 13.508 \text{ in}^2 = 0$$

We now have a single equation with only one unknown, namely, θ_4. However, it contains a mixture of sine and cosine terms (a transcendental equation). A standard technique for dealing with this situation is to convert the equation into a second-order polynomial (a quadratic) using the tangent of the half-angle identities:

$$\sin\theta_4 = \frac{2\tan(\theta_4/2)}{1+\tan^2(\theta_4/2)} \quad \text{and} \quad \cos\theta_4 = \frac{1-\tan^2(\theta_4/2)}{1+\tan^2(\theta_4/2)}.$$

Substituting these identities into the above equation and multiplying by the common denominator gives a single quadratic equation in one unknown, $\tan(\theta_4/2)$, that is,

$$-11.312\tan(\theta_4/2) - 16.000\left[1-\tan^2(\theta_4/2)\right] + 13.508\left[1+\tan^2(\theta_4/2)\right] = 0,$$

or

$$29.508\tan^2(\theta_4/2) - 11.312\tan(\theta_4/2) - 2.492 = 0.$$

This quadratic equation can now be solved for two roots, which are

$$\tan\frac{\theta_4}{2} = \begin{cases} 0.5398; & \theta_4 = -123.3^\circ \\ -0.1565; & \theta_4 = 162.2^\circ \end{cases}.$$

Of these two roots, we choose the second as the solution of interest. Back-substituting into the previous conditions, we find numeric values for the four vectors, namely:

$$\mathbf{R}_{O_4O_2} = 3.000\hat{\mathbf{i}} - 2.000\hat{\mathbf{k}} \text{ in.} \qquad \textit{Ans.}$$

$$\mathbf{R}_{AO_2} = 0.707\hat{\mathbf{i}} - 0.707\hat{\mathbf{j}} \text{ in.} \qquad \textit{Ans.}$$

$$\mathbf{R}_{BA} = 2.293\hat{\mathbf{i}} + 1.930\hat{\mathbf{j}} + 1.809\hat{\mathbf{k}} \text{ in.} \qquad \textit{Ans.}$$

$$\mathbf{R}_{BO_4} = 1.223\hat{\mathbf{j}} + 3.809\hat{\mathbf{k}} \text{ in.} \qquad \textit{Ans.}$$

These results match those of the graphic solution within graphical error.

We should note from this example that we have solved for the output angle θ_4 and the current values of the four vectors for the given position of the input crank angle θ_2. However, have we really solved for the positions of all links? The answer is no! As indicated above, we have not determined how the coupler link may have rotated about its axis $\mathbf{R}_{BA}$ between the two spheric joints. This is still unknown and explains how we were able to solve this two-degree-of-freedom system without specifying another input angle for the idle freedom. Depending upon our motivation, the above solution may be sufficient; however, it is important to note that the above vectors are not sufficient to solve for this additional unknown variable.

The spherical four-bar *RRRR* linkage illustrated in Fig. 11.2 is also case 2*d* of the vector tetrahedron equation and, for a specified position of the input link, can be solved in the same manner as the example given.

11.4 SPATIAL VELOCITY AND ACCELERATION ANALYSES

When the positions of all the links of a spatial mechanism are known, the velocities and accelerations can then be determined using the methods of Chapters 3 and 4. In planar mechanisms, the angular velocity and acceleration vectors are always perpendicular to the plane of motion. This considerably reduces the effort required in the solution process for both graphical and analytical approaches. In spatial problems, however, these vectors may be skewed in space. Otherwise, the equations and the methods of analysis are the same. The following example will illustrate the differences.

EXAMPLE 11.2

For the spatial four-bar *RSSR* linkage illustrated in Fig. 11.7, the constant angular velocity of link 2 is $\omega_2 = 40\hat{\mathbf{k}}$ rad/s. Find the angular velocities and angular accelerations of links 3 and 4 and the velocity and acceleration of point *B* for the specified position.

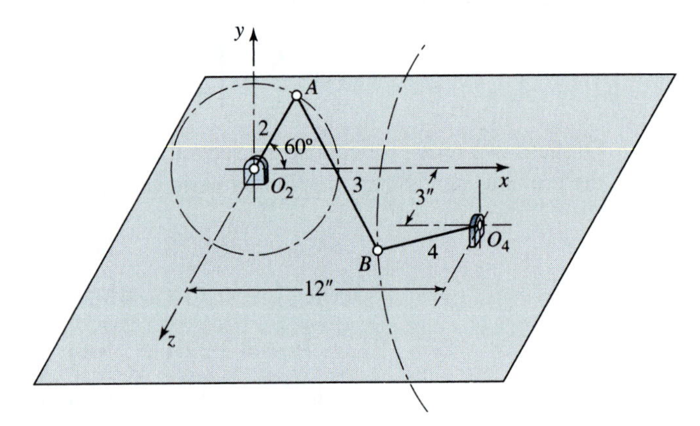

Figure 11.7 Spatial four-bar *RSSR* linkage for Example 11.2; $R_{AO_2} = 4$ in, $R_{BA} = 15$ in, and $R_{BO_4} = 10$ in.

ANALYTIC SOLUTION

The position analysis follows exactly the procedure given in Section 11.3 and Example 11.1; the graphic solution is illustrated in Fig. 11.8. For the given crank angle, the results are:

$$\mathbf{R}_{O_4O_2} = 12.000\hat{\mathbf{i}} + 3.000\hat{\mathbf{k}} \text{ in.}$$

$$\mathbf{R}_{AO_2} = 2.000\hat{\mathbf{i}} + 3.464\hat{\mathbf{j}} \text{ in.}$$

$$\mathbf{R}_{BA} = 10.000\hat{\mathbf{i}} + 2.746\hat{\mathbf{j}} + 10.838\hat{\mathbf{k}} \text{ in.}$$

$$\mathbf{R}_{BO_4} = 6.210\hat{\mathbf{j}} + 7.838\hat{\mathbf{k}} \text{ in.}$$

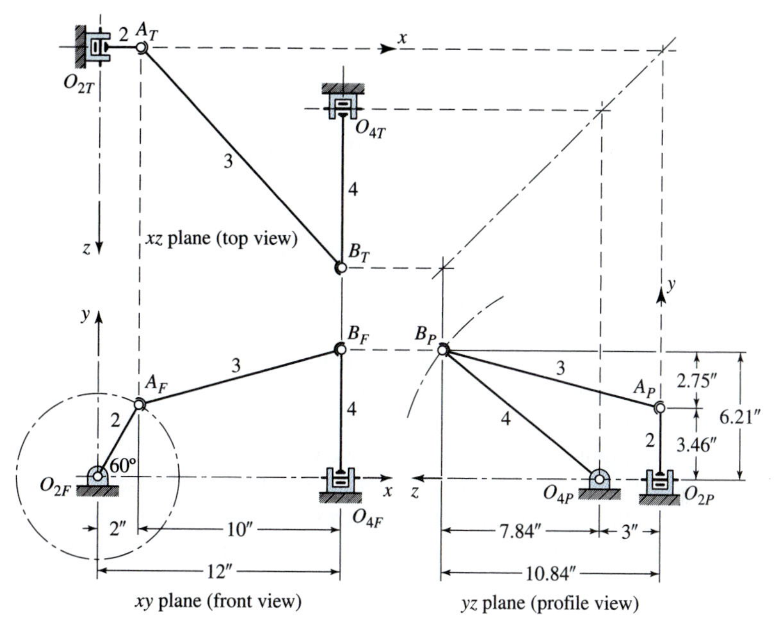

Figure 11.8 Graphical position analysis for Example 11.2.

From the given information and the constraints, we see that the angular velocities and accelerations can be written as

$$\begin{aligned} \boldsymbol{\omega}_2 &= 40\hat{\mathbf{k}} \text{ rad/s} & \boldsymbol{\alpha}_2 &= \mathbf{0} \\ \boldsymbol{\omega}_3 &= \omega_3^x\hat{\mathbf{i}} + \omega_3^y\hat{\mathbf{j}} + \omega_3^z\hat{\mathbf{k}} & \boldsymbol{\alpha}_3 &= \alpha_3^x\hat{\mathbf{i}} + \alpha_3^y\hat{\mathbf{j}} + \alpha_3^z\hat{\mathbf{k}} \\ \boldsymbol{\omega}_4 &= \omega_4\hat{\mathbf{i}} & \boldsymbol{\alpha}_4 &= \alpha_4\hat{\mathbf{i}}. \end{aligned}$$

First we find the velocity of point A as the velocity difference from point O_2. Thus,

$$\begin{aligned} \mathbf{V}_A = \boldsymbol{\omega}_2 \times \mathbf{R}_{AO_2} &= \left(40\hat{\mathbf{k}} \text{ rad/s}\right) \times \left(2.000\hat{\mathbf{i}} + 3.464\hat{\mathbf{j}} \text{ in}\right) \\ &= -138.560\hat{\mathbf{i}} + 80.000\hat{\mathbf{j}} \text{ in/s} \end{aligned} \tag{1}$$

Similarly, for link 3, the velocity difference is

$$\begin{aligned} \mathbf{V}_{BA} &= \left(\omega_3^x\hat{\mathbf{i}} + \omega_3^y\hat{\mathbf{j}} + \omega_3^z\hat{\mathbf{k}}\right) \times \left(10.000\hat{\mathbf{i}} + 2.746\hat{\mathbf{j}} + 10.838\hat{\mathbf{k}} \text{ in}\right) \\ &= \left(10.838\omega_3^y - 2.746\omega_3^z\right)\hat{\mathbf{i}} + \left(10.000\omega_3^z - 10.838\omega_3^x\right)\hat{\mathbf{j}} \\ &\quad + \left(2.746\omega_3^x - 10.000\omega_3^y\right)\hat{\mathbf{k}} \text{ in} \end{aligned} \tag{2}$$

and for link 4 the velocity of point B is

$$\begin{aligned} \mathbf{V}_B = \boldsymbol{\omega}_4 \times \mathbf{R}_{BO_4} &= \left(\omega_4\hat{\mathbf{i}}\right) \times \left(6.210\hat{\mathbf{j}} + 7.838\hat{\mathbf{k}} \text{ in}\right) \\ &= -\left(7.838 \text{ in}\right)\omega_4\hat{\mathbf{j}} + \left(6.210 \text{ in}\right)\omega_4\hat{\mathbf{k}}. \end{aligned} \tag{3}$$

The next step is to substitute these results into the velocity difference equation,

$$\mathbf{V}_B = \mathbf{V}_A + \mathbf{V}_{BA}. \tag{4}$$

After this, we can separate the $\hat{\mathbf{i}}$, $\hat{\mathbf{j}}$, and $\hat{\mathbf{k}}$ components to obtain the three algebraic equations:

$$10.838 \text{ in } \omega_3^y - 2.746 \text{ in } \omega_3^z = 138.560 \text{ in/s} \tag{5}$$

$$-10.838 \text{ in } \omega_3^x + 10.000 \text{ in } \omega_3^z + 7.838 \text{ in } \omega_4 = -80.000 \text{ in/s} \tag{6}$$

$$2.746 \text{ in } \omega_3^x - 10.000 \text{ in } \omega_3^y - 6.210 \text{ in } \omega_4 = 0.000. \tag{7}$$

We now have three equations; however, we note that there are four unknowns, ω_3^x, ω_3^y, ω_3^z, and ω_4. This would not occur in most problems, but does here because of the idle freedom of the coupler to spin about its own axis. Because this spin does not affect the input–output relationship, we will find the same result for ω_4 regardless of the rate of spin. One way to proceed, therefore, would be to choose one component of $\boldsymbol{\omega}_3$ and give it a value, thus setting the rate of spin, and then solve for the other unknowns. Another approach is to set the rate of spin about the coupler link axis to zero by requiring that

$$\boldsymbol{\omega}_3 \cdot \mathbf{R}_{BA} = 0$$

$$10.000 \text{ in } \omega_3^x + 2.746 \text{ in } \omega_3^y + 10.838 \text{ in } \omega_3^z = 0. \tag{8}$$

Equations (5) through (8) can now be solved simultaneously for the four unknowns. The result is

$$\boldsymbol{\omega}_3 = -7.692\hat{\mathbf{i}} + 13.704\hat{\mathbf{j}} + 3.625\hat{\mathbf{k}} \text{ rad/s} \qquad \textit{Ans.}$$

$$\boldsymbol{\omega}_4 = -25.468\hat{\mathbf{i}} \text{ rad/s}. \qquad \textit{Ans.}$$

Substituting these angular velocities into Eq. (3), the velocity of point B is

$$\mathbf{V}_B = 199.618\hat{\mathbf{j}} - 158.156\hat{\mathbf{k}} \text{ in/s} \qquad \textit{Ans.}$$

Turning next to acceleration analysis, we compute the following components:

$$\mathbf{A}_{AO_2}^n = \boldsymbol{\omega}_2 \times \left(\boldsymbol{\omega}_2 \times \mathbf{R}_{AO_2}\right) = -3\,200.00\hat{\mathbf{i}} - 5\,542.56\hat{\mathbf{j}} \text{ in/s}^2 \tag{9}$$

$$\mathbf{A}_{AO_2}^t = \boldsymbol{\alpha}_2 \times \mathbf{R}_{AO_2} = \mathbf{0} \tag{10}$$

$$\mathbf{A}_{BA}^n = \boldsymbol{\omega}_3 \times (\boldsymbol{\omega}_3 \times \mathbf{R}_{BA}) = -2\,601.01\hat{\mathbf{i}} - 714.26\hat{\mathbf{j}} - 2\,818.99\hat{\mathbf{k}} \text{ in/s}^2 \tag{11}$$

$$\begin{aligned} \mathbf{A}_{BA}^t &= \boldsymbol{\alpha}_3 \times \mathbf{R}_{BA} = \left(\alpha_3^x\hat{\mathbf{i}} + \alpha_3^y\hat{\mathbf{j}} + \alpha_3^z\hat{\mathbf{k}}\right) \times \left(10.000 \text{ in } \hat{\mathbf{i}} + 2.746 \text{ in } \hat{\mathbf{j}} + 10.838 \text{ in } \hat{\mathbf{k}}\right) \\ &= \left(10.838 \text{ in } \alpha_3^y - 2.746 \text{ in } \alpha_3^z\right)\hat{\mathbf{i}} + \left(10.000 \text{ in } \alpha_3^z - 10.838 \text{ in } \alpha_3^x\right)\hat{\mathbf{j}} \\ &\quad + \left(2.746 \text{ in } \alpha_3^x - 10.000 \text{ in } \alpha_3^y\right)\hat{\mathbf{k}} \end{aligned} \tag{12}$$

$$\mathbf{A}_{BO_4}^n = \boldsymbol{\omega}_4 \times \left(\boldsymbol{\omega}_4 \times \mathbf{R}_{BO_4}\right) = -4\,028.05\hat{\mathbf{j}} - 5\,084.03\hat{\mathbf{k}} \text{ in/s}^2 \tag{13}$$

$$\mathbf{A}_{BO_4}^t = \boldsymbol{\alpha}_4 \times \mathbf{R}_{BO_4} = -7.838 \text{ in } \alpha_4\hat{\mathbf{j}} + 6.210 \text{ in } \alpha_4\hat{\mathbf{k}}. \tag{14}$$

These quantities are now substituted into the acceleration difference equation,

$$\mathbf{A}^{n}_{BO_4} + \mathbf{A}^{t}_{BO_4} = \mathbf{A}^{n}_{AO_2} + \mathbf{A}^{t}_{AO_2} + \mathbf{A}^{n}_{BA} + \mathbf{A}^{t}_{BA}, \tag{15}$$

and separated into $\hat{\mathbf{i}}$, $\hat{\mathbf{j}}$, and $\hat{\mathbf{k}}$ components. Along with the condition $\boldsymbol{\alpha}_3 \cdot \mathbf{R}_{BA} = 0$ for the spin of the idle freedom, this results in four equations in four unknowns as follows.

$$10.838 \text{ in } \alpha_3^y - 2.746 \text{ in } \alpha_3^z = 5\,801.01 \text{ in/s}^2$$

$$-10.838 \text{ in } \alpha_3^x + 10.000 \text{ in } \alpha_3^z + 7.838 \text{ in } \alpha_4 = 2\,228.61 \text{ in/s}^2$$

$$2.746 \text{ in } \alpha_3^x - 10.000 \text{ in } \alpha_3^y - 6.210 \text{ in } \alpha_4 = -2\,265.04 \text{ in/s}^2$$

$$10.000 \text{ in } \alpha_3^x + 2.746 \text{ in } \alpha_3^y + 10.838 \text{ in } \alpha_3^z = 0.00.$$

The coefficients (on the left-hand side) of these acceleration equations are identical to those of the velocity equations [see Eqs. (5)–(8)]. This will always be the case. These equations are now solved to give the desired accelerations.

$$\boldsymbol{\alpha}_3 = -526.94\hat{\mathbf{i}} + 618.71\hat{\mathbf{j}} + 329.43\hat{\mathbf{k}} \text{ rad/s}^2 \qquad \textit{Ans.}$$

$$\boldsymbol{\alpha}_4 = -864.59\hat{\mathbf{i}} \text{ rad/s}^2 \qquad \textit{Ans.}$$

$$\mathbf{A}_B = \mathbf{A}^{n}_{BO_4} + \mathbf{A}^{t}_{BO_4} = 2\,748.61\hat{\mathbf{j}} - 10\,453.10\hat{\mathbf{k}} \text{ in/s}^2. \qquad \textit{Ans.}$$

GRAPHIC SOLUTION

The determination of the velocities and accelerations of a spatial mechanism by graphic means can be conducted in the same manner as for a planar mechanism. However, the position information as well as the velocity and acceleration vectors often do not appear in their true lengths in the front, top, and profile views, but are foreshortened. This means that spatial motion problems usually require the use of auxiliary views where the vectors appear in true lengths.

The velocity solution for this example is illustrated in Fig. 11.9 with notation corresponding to that used in many texts on descriptive geometry. The letters F, T, and P designate the front, top, and profile views, and the numbers 1 and 2 indicate the first and second auxiliary views, respectively. Points projected into these views contain subscripts F, T, P, and so on. The steps in obtaining the velocity are as follows.

1. We first construct to scale the front, top, and profile views of the linkage and designate each point.
2. Next, we calculate $\mathbf{V}_A$ as above and place this vector in position with its origin at A in the three views. The velocity $\mathbf{V}_A$ shows in true length in the frontal view. We designate its terminus as point a_F and project this point to the top and profile views to find a_T and a_P.
3. The magnitude of the velocity $\mathbf{V}_B$ is unknown, but its direction is perpendicular to $\mathbf{R}_{BO_4}$ and, once the problem is solved, it will show in true size in the profile view. Therefore, we construct a line in the profile view that originates at point A_P (which

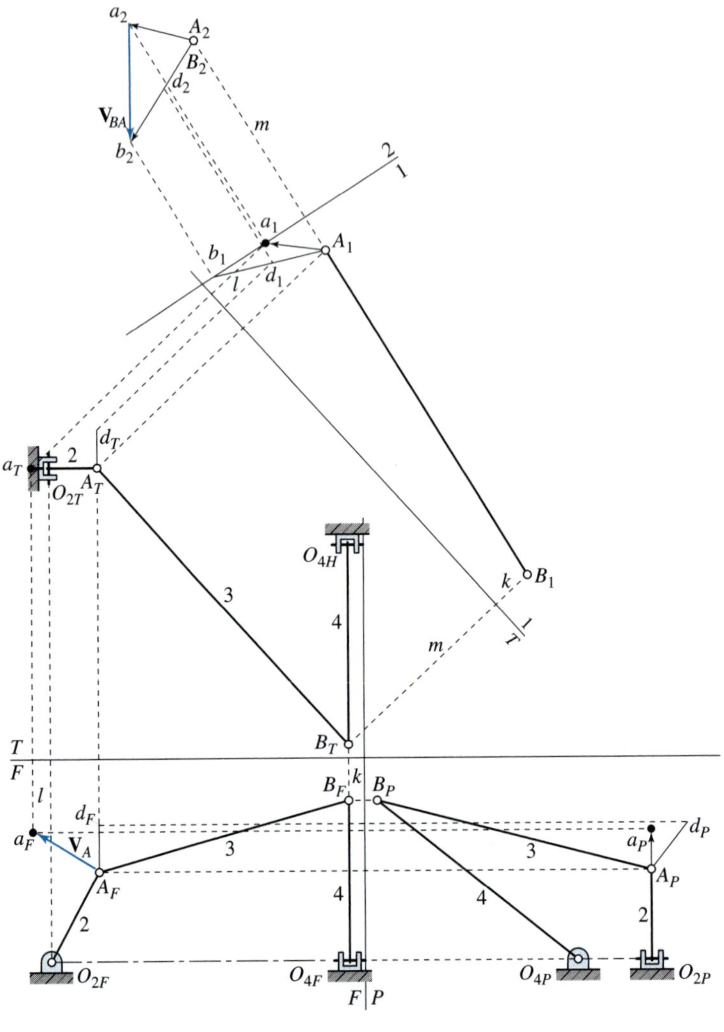

Figure 11.9 Graphical velocity analysis for Example 11.2.

will become the origin of our velocity polygon) and corresponds in direction to that of $\mathbf{V}_B$ (perpendicular to $R_{B_P O_{4P}}$). We then choose any point d_P on this line and project it to the front view (d_F) and top view (d_T) to establish the line of action of $\mathbf{V}_B$ in those views.

4. The equation to be solved is

$$\mathbf{V}_B = \mathbf{V}_A + \mathbf{V}_{BA}, \tag{16}$$

where $\mathbf{V}_A$ and the directions of $\mathbf{V}_B$ and $\mathbf{V}_{BA}$ are known. We note that $\mathbf{V}_{BA}$ must lie in a plane perpendicular (in space) to $\mathbf{R}_{BA}$, but its magnitude and that of $\mathbf{V}_B$ are unknown. The vector $\mathbf{V}_{BA}$ must originate at the terminus of $\mathbf{V}_A$ and must lie in a plane perpendicular to $\mathbf{R}_{BA}$; it terminates by intersecting the line Ad or its extension. To find the plane perpendicular to $\mathbf{R}_{BA}$, we start by constructing the first auxiliary view, which shows vector $\mathbf{R}_{BA}$ in true length, so we construct the edge view of plane 1 parallel to $A_T B_T$ and project $\mathbf{R}_{BA}$ to this plane. In this projection we note that the distances k and l are the same in this first auxiliary view as in the frontal view. The first auxiliary view of AB is $A_1 B_1$, which is true length. We also project points a and d to this view, but the remaining links need not be projected.

5. In this step we construct a second auxiliary view, plane 2, such that the projection of AB upon it is a point. Then all lines drawn parallel to this plane will be perpendicular to $\mathbf{R}_{BA}$. The edge view of such a plane must be perpendicular to $A_1 B_1$ extended. In this example it is convenient to choose this plane so that it contains point a; therefore, we construct the edge view of plane 2 through point a_1 perpendicular to $A_1 B_1$ extended. Now we project points A, B, a, and d onto this plane. Note that the distances—m, for example—of points from plane 1 must be the same in the top view and in the second auxiliary view.
6. We now extend the line $A_1 d_1$ until it intersects the edge view of plane 2 at b_1, and we find the projection b_2 of this point in plane 2 where the projector meets the line $A_2 d_2$ extended. Now both points a and b lie in plane 2, and any line drawn in plane 2 is perpendicular to $\mathbf{R}_{BA}$. Therefore, line ab is $\mathbf{V}_{BA}$ and the second auxiliary view of that line shows its true length. Line $A_2 b_2$ is the projection of $\mathbf{V}_B$ on the second auxiliary plane, but not in its true length because point A is not in plane 2.
7. To simplify the reading of the drawing, step 7 is not illustrated; those who follow the first six steps carefully should have no difficulty with the seventh. We can project the three vectors back to the top, front, and profile views. The velocity V_B can be measured from the profile view where it appears in true length. The result is

$$\mathbf{V}_B = 200\hat{\mathbf{j}} - 158\hat{\mathbf{k}} \text{ in/s.} \qquad \textit{Ans.}$$

When all vectors have been projected back to these three views, we can measure their x, y, and z projections directly.

8. If we assume that the idle freedom of spin is not active and, therefore, that the angular velocity vector $\boldsymbol{\omega}_3$ is perpendicular to $\mathbf{R}_{BA}$ and appears in true length in the second auxiliary view,* then the magnitudes of the angular velocities can be determined from the equations

$$\omega_3 = \frac{V_{BA}}{R_{BA}} = \frac{242 \text{ in/s}}{15 \text{ in}} = 16.13 \text{ rad/s}$$

$$\omega_4 = \frac{V_{BO_4}}{R_{BO_4}} = \frac{255 \text{ in/s}}{10 \text{ in}} = 25.50 \text{ rad/s.}$$

Therefore, we can draw the angular velocity vectors in the views where they appear in their true length and project them into the views where their vector components

* This is the same assumption we used in the analytic solution when we wrote the equation $\boldsymbol{\omega}_3 \cdot \mathbf{R}_{BA} = \mathbf{0}$.

can be measured. The results are

$$\omega_3 = -7.69\hat{\mathbf{i}} + 13.71\hat{\mathbf{j}} + 3.63\hat{\mathbf{k}} \text{ rad/s} \qquad \textit{Ans.}$$

$$\omega_4 = 25.50\hat{\mathbf{i}} \text{ rad/s}. \qquad \textit{Ans.}$$

9. The solution to the acceleration problem, also not shown, is obtained in identical manner, using the same two auxiliary planes. The equation to be solved is

$$\mathbf{A}^n_{BO_4} + \mathbf{A}^t_{BO_4} = \mathbf{A}^n_{AO_2} + \mathbf{A}^t_{AO_2} + \mathbf{A}^n_{BA} + \mathbf{A}^t_{BA}, \tag{17}$$

where the vectors $\mathbf{A}^n_{BO_4}$, $\mathbf{A}^n_{AO_2}$, $\mathbf{A}^t_{AO_2}$, and $\mathbf{A}^n_{BA}$ can be determined now that the velocity analysis has been completed. Also, we see that $\mathbf{A}^t_{BO_4}$ and $\mathbf{A}^t_{BA}$ have known directions. Therefore, the solution can proceed exactly as for the velocity polygon; the only difference in approach is that there are more known vectors. The final results are

$$\boldsymbol{\alpha}_3 = -527\hat{\mathbf{i}} + 619\hat{\mathbf{j}} + 329\hat{\mathbf{k}} \text{ rad/s}^2 \qquad \textit{Ans.}$$

$$\boldsymbol{\alpha}_4 = -865\hat{\mathbf{i}} \text{ rad/s}^2 \qquad \textit{Ans.}$$

$$\mathbf{A}_B = 2\,750\hat{\mathbf{j}} - 10\,450\hat{\mathbf{k}} \text{ in/s}^2 \qquad \textit{Ans.}$$

11.5 EULER ANGLES

We saw in Section 3.2 that angular velocity is a vector quantity; hence, like all vector quantities, it has components along any set of rectilinear axes,

$$\boldsymbol{\omega} = \omega^x\hat{\mathbf{i}} + \omega^y\hat{\mathbf{j}} + \omega^z\hat{\mathbf{k}},$$

and it obeys the laws of vector algebra. Unfortunately, we also saw in Fig. 3.2 that three-dimensional angular displacements do not behave as vectors. Their order of summation is not arbitrary; that is, they are not commutative in addition. Therefore, they do not follow the rules of vector algebra. The inescapable conclusion is that we cannot find a set of three angles, or angular displacements, that specify the three-dimensional orientation of a rigid body and that also have ω^x, ω^y, and ω^z as their time derivatives.

To clarify the problem further, we visualize a rigid body rotating in space about a fixed point O that we take as the origin of a grounded or absolute reference frame xyz. We also visualize a moving reference system $x'y'z'$ sharing the same origin, but attached to and moving with the rotating body. The $x'y'z'$ system is called *body-fixed axes*. Our problem here is to determine some way to specify the orientation of the body-fixed axes with respect to the absolute reference axes, a method that is convenient for use with three-dimensional finite rotations.

Let us assume that the stationary axes are aligned along unit vector directions $\hat{\mathbf{i}}$, $\hat{\mathbf{j}}$, and $\hat{\mathbf{k}}$ and that the rotated body-fixed axes have directions denoted by $\hat{\mathbf{i}}'$, $\hat{\mathbf{j}}'$, and $\hat{\mathbf{k}}'$. Then any point in the absolute coordinate system is located by the position vector $\mathbf{R}$ and in the rotated system by $\mathbf{R}'$. Since both vectors are descriptions of the position of the same point, then we conclude that $\mathbf{R} = \mathbf{R}'$ and

$$R^x\hat{\mathbf{i}} + R^y\hat{\mathbf{j}} + R^z\hat{\mathbf{k}} = R^{x'}\hat{\mathbf{i}}' + R^{y'}\hat{\mathbf{j}}' + R^{z'}\hat{\mathbf{k}}'.$$

Now taking the vector dot products of this equation with $\hat{\mathbf{i}}$, $\hat{\mathbf{j}}$, and $\hat{\mathbf{k}}$, respectively, gives the transformation equations between the two sets of axes, namely

$$\begin{aligned} R^x &= \left(\hat{\mathbf{i}}\cdot\hat{\mathbf{i}}'\right)R^{x'} + \left(\hat{\mathbf{i}}\cdot\hat{\mathbf{j}}'\right)R^{y'} + \left(\hat{\mathbf{i}}\cdot\hat{\mathbf{k}}'\right)R^{z'} \\ R^y &= \left(\hat{\mathbf{j}}\cdot\hat{\mathbf{i}}'\right)R^{x'} + \left(\hat{\mathbf{j}}\cdot\hat{\mathbf{j}}'\right)R^{y'} + \left(\hat{\mathbf{j}}\cdot\hat{\mathbf{k}}'\right)R^{z'} \\ R^z &= \left(\hat{\mathbf{k}}\cdot\hat{\mathbf{i}}'\right)R^{x'} + \left(\hat{\mathbf{k}}\cdot\hat{\mathbf{j}}'\right)R^{y'} + \left(\hat{\mathbf{k}}\cdot\hat{\mathbf{k}}'\right)R^{z'}. \end{aligned} \qquad (a)$$

But we remember that the dot product of two unit vectors is equal to the cosine of the angle between them. For example, if the angle between the unit vectors $\hat{\mathbf{i}}$ and $\hat{\mathbf{j}}'$ is denoted by $\theta_{ij'}$, then

$$(\hat{\mathbf{i}}\cdot\hat{\mathbf{j}}') = \cos\theta_{ij'}. \qquad (b)$$

Thus, the above set of equations becomes

$$\begin{aligned} R^x &= \cos\theta_{ii'}R^{x'} + \cos\theta_{ij'}R^{y'} + \cos\theta_{ik'}R^{z'} \\ R^y &= \cos\theta_{ji'}R^{x'} + \cos\theta_{jj'}R^{y'} + \cos\theta_{jk'}R^{z'} \\ R^z &= \cos\theta_{ki'}R^{x'} + \cos\theta_{kj'}R^{y'} + \cos\theta_{kk'}R^{z'} \end{aligned}. \qquad (11.3)$$

These equations can be written, in matrix notation, as

$$\begin{bmatrix} R^x \\ R^y \\ R^z \end{bmatrix} = \begin{bmatrix} \cos\theta_{ii'} & \cos\theta_{ij'} & \cos\theta_{ik'} \\ \cos\theta_{ji'} & \cos\theta_{jj'} & \cos\theta_{jk'} \\ \cos\theta_{ki'} & \cos\theta_{kj'} & \cos\theta_{kk'} \end{bmatrix} \begin{bmatrix} R^{x'} \\ R^{y'} \\ R^{z'} \end{bmatrix}. \qquad (11.4)$$

This means of defining the orientation of the $x'y'z'$ axes with respect to the xyz axes uses what are called *direction cosines* as coefficients in a set of *transformation equations* between the two coordinate systems. Although we will have use for direction cosines, we note the disadvantage that there are nine of them, whereas only three variables can be independent for spatial rotation. The nine direction cosines are not all independent, but are related by six *orthogonality conditions*. We would prefer a technique that had only three independent variables, preferably all angles.

Three angles, called *Euler angles*, can be used to specify the orientation of the body-fixed axes with respect to the reference axes, as illustrated in Fig. 11.10. To explain the Euler angles, we begin with the body-fixed axes coincident with the reference axes. We then specify three successive rotations, θ, ϕ, and ψ, which must occur in the specified order and about the specified axes, to move the body-fixed axes into their final orientation.

The first Euler angle describes a rotation through the angle θ and is taken positive counterclockwise about the positive z axis, as illustrated in Fig. 11.10a. This rotation goes

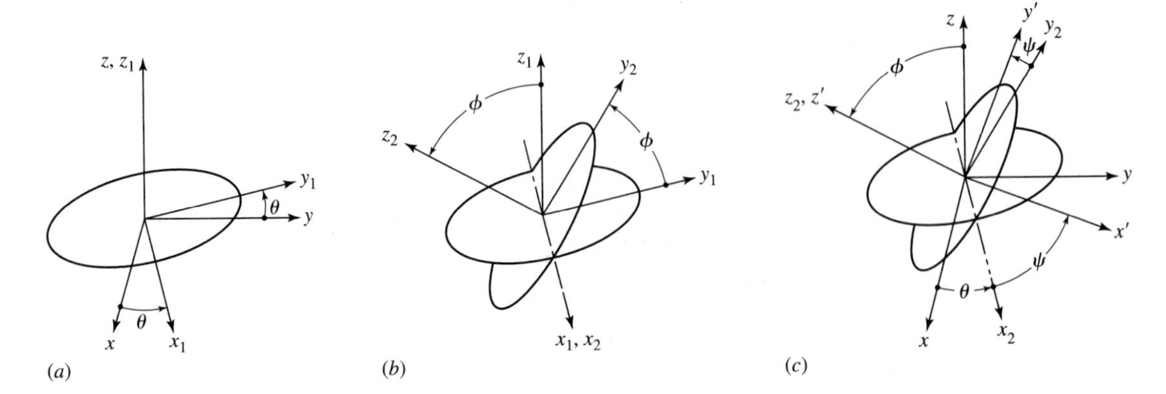

Figure 11.10 Euler angles.

from the xy axes by the angle θ to their new x_1y_1 locations shown, whereas z and z_1 remain fixed and collinear. The transformation equations for this first rotation are

$$\begin{bmatrix} R^x \\ R^y \\ R^z \end{bmatrix} = \begin{bmatrix} \cos\theta & -\sin\theta & 0 \\ \sin\theta & \cos\theta & 0 \\ 0 & 0 & 1 \end{bmatrix} \begin{bmatrix} R^{x_1} \\ R^{y_1} \\ R^{z_1} \end{bmatrix}. \tag{11.5}$$

The second Euler angle describes a rotation through the angle ϕ and is taken positive counterclockwise about the positive x_1 axis, as illustrated in Fig. 11.10*b*. This rotation goes from the displaced y_1z_1 axes by the angle ϕ to their new y_2z_2 locations shown, whereas x_1 and x_2 remain fixed. The transformation equations for this second rotation are

$$\begin{bmatrix} R^{x_1} \\ R^{y_1} \\ R^{z_1} \end{bmatrix} = \begin{bmatrix} 1 & 0 & 0 \\ 0 & \cos\phi & -\sin\phi \\ 0 & \sin\phi & \cos\phi \end{bmatrix} \begin{bmatrix} R^{x_2} \\ R^{y_2} \\ R^{z_2} \end{bmatrix}. \tag{11.6}$$

The third Euler angle describes a rotation through the angle ψ and is taken positive counterclockwise about the positive z_2 axis, as illustrated in Fig. 11.10*c*. This rotation goes from the x_2y_2 axes by the angle ψ to their final $x'y'$ locations shown, whereas z_2 and z' remain fixed. The transformation equations for this third rotation are

$$\begin{bmatrix} R^{x_2} \\ R^{y_2} \\ R^{z_2} \end{bmatrix} = \begin{bmatrix} \cos\psi & -\sin\psi & 0 \\ \sin\psi & \cos\psi & 0 \\ 0 & 0 & 1 \end{bmatrix} \begin{bmatrix} R^{x'} \\ R^{y'} \\ R^{z'} \end{bmatrix}. \tag{11.7}$$

Now, substituting Eq. (11.6) into Eq. (11.5) will give the transformation from the xyz axes to the $x_2y_2z_2$ axes:

$$\begin{bmatrix} R^x \\ R^y \\ R^z \end{bmatrix} = \begin{bmatrix} \cos\theta & -\sin\theta\cos\phi & \sin\theta\sin\phi \\ \sin\theta & \cos\theta\cos\phi & -\cos\theta\sin\phi \\ 0 & \sin\phi & \cos\phi \end{bmatrix} \begin{bmatrix} R^{x_2} \\ R^{y_2} \\ R^{z_2} \end{bmatrix}. \tag{11.8}$$

Finally, substituting Eq. (11.7) into Eq. (11.8) will give the total transformation from the xyz axes to the $x'y'z'$ axes:

$$\begin{bmatrix} R^x \\ R^y \\ R^z \end{bmatrix} = \begin{bmatrix} \cos\theta\cos\psi - \sin\theta\cos\phi\sin\psi & -\cos\theta\sin\psi - \sin\theta\cos\phi\cos\psi & \sin\theta\sin\phi \\ \sin\theta\cos\psi + \cos\theta\cos\phi\sin\psi & -\sin\theta\sin\psi + \cos\theta\cos\phi\cos\psi & -\cos\theta\sin\phi \\ \sin\phi\sin\psi & \sin\phi\cos\psi & \cos\phi \end{bmatrix} \begin{bmatrix} R^{x'} \\ R^{y'} \\ R^{z'} \end{bmatrix}. \tag{11.9}$$

Because the transformation is expressed as a function of only three Euler angles, θ, ϕ, and ψ, it represents the same information as Eqs. (11.3) and (11.4), but without the difficulty of having nine variables (direction cosines) related by six (orthogonality) conditions. Thus, Euler angles have become a useful tool in treating problems involving three-dimensional rotation.

Note, however, that the form of the transformation equations given here depends on using precisely the conventions for the Euler angles defined in Fig. 11.10. Unfortunately, there appears to be little agreement among different authors regarding how these angles should be defined. A wide variety of other definitions, differing in the axes about which the rotations are to be measured, or in their order, or in the sign conventions for positive values of the angles, are to be found in other references. The differences are not great, but are sufficient to frustrate easy comparison of the formulae derived.

We must also remember that the time derivatives of the Euler angles, $\dot{\theta}$, $\dot{\phi}$, and $\dot{\psi}$, are *not* the components of the angular velocity $\boldsymbol{\omega}$ of the body-fixed axes. We see that each of these rotations acts about a different axis and they are expressed in inconsistent coordinate systems. From Fig. 11.10 we can see that the angular velocity of the body-fixed axes is

$$\boldsymbol{\omega} = \dot{\theta}\hat{\mathbf{k}} + \dot{\phi}\hat{\mathbf{i}}_1 + \dot{\psi}\hat{\mathbf{k}}'.$$

When these different unit vectors are all transformed into the fixed reference directions and then added, the three components of the angular velocity vector can be written as

$$\begin{aligned} \omega^x &= \dot{\phi}\cos\theta + \dot{\psi}\sin\theta\sin\phi \\ \omega^y &= \dot{\phi}\sin\theta - \dot{\psi}\cos\theta\sin\phi \\ \omega^z &= \dot{\theta} + \dot{\psi}\cos\phi. \end{aligned} \tag{11.10}$$

On the other hand, we can also transform the unit vectors into the body-fixed axes; then

$$\begin{aligned} \omega^{x'} &= \dot{\theta}\sin\phi\sin\psi + \dot{\phi}\cos\psi \\ \omega^{y'} &= \dot{\theta}\sin\phi\cos\psi - \dot{\phi}\sin\psi \\ \omega^{z'} &= \dot{\theta}\cos\phi + \dot{\psi}. \end{aligned} \tag{11.11}$$

11.6 THE DENAVIT–HARTENBERG PARAMETERS

The transformation equations of the previous section dealt only with three-dimensional rotations about a fixed point. Yet the same approach can be generalized to include both

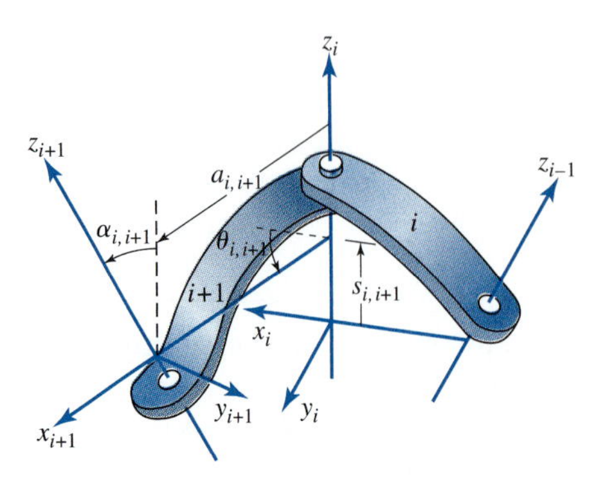

Figure 11.11 Definitions of Denavit–Hartenberg parameters.

translations and rotations and thus to treat general spatial motion. This has been done, and much literature dealing with these techniques exists, usually under titles such as "matrix methods". Most modern publications on this approach stems from the work of Denavit and Hartenberg.[11] They developed a notation for labeling all single-loop lower pair linkages and also devised a transformation matrix technique for their analysis.

In the Denavit–Hartenberg approach we start by numbering the joints of the linkage, usually starting with the input joint and numbering consecutively around a kinematic loop to the output joint. If we assume that the linkage has only a single loop with only j_1 joints and a mobility of $m = 1$, then the Kutzbach criterion indicates that there will be $n = 7$ binary links and $j_1 = 7$ joints numbered from 1 through 7.* Figure 11.11 illustrates a typical joint of this loop, revolute joint number i in this case, and the two links that it joins.

Next we identify the motion axis of each of the joints, choose a positive orientation on each, and label each as a z_i axis. Although the z_i axes may be skew in space, it is possible to find a common perpendicular between each consecutive pair and label these as x_i axes such that each x_i axis is the common perpendicular between z_{i-1} and z_i, choosing an arbitrary positive orientation.†

Having done this, we can now identify and label y_i axes such that there is a right-hand Cartesian coordinate system $x_i y_i z_i$ associated with each joint of the loop. Careful study of Fig. 11.11 and visualization of the motion allowed by the joint indicates that coordinate system $x_i y_i z_i$ remains fixed to the link carrying joint $i-1$ and joint i, whereas $x_{i+1} y_{i+1} z_{i+1}$ moves with and remains fixed to the link carrying joint i and joint $i+1$. Thus, we have a basis for assigning numbers to the links that correspond to the numbers of the coordinate system attached to each link.

The key to the Denavit–Hartenberg approach comes in the standard method they defined for determining the dimensions of the links. The relative position and orientation of any

* The treatment presented here deals only with j_1 joints; multi-degree-of-freedom lower pairs may be represented as combinations of revolute joints and prismatic joints, however, with fictitious links between them.

† When $i = 1$, then $i-1$ must be taken as n because, in the loop, the last joint is also the joint before the first. Similarly, when $i = n$, then $i+1$ must be taken as 1.

two consecutive coordinate systems placed as described above can be defined by four parameters, labeled a, α, θ, and s, illustrated in Fig. 11.11, and defined as follows.

$a_{i,i+1}$ = the distance along x_{i+1} from z_i to z_{i+1} with sign taken from the sense of x_{i+1}
$\alpha_{i,i+1}$ = the angle from positive z_i to positive z_{i+1} taken positive counterclockwise as seen from positive x_{i+1}
$\theta_{i,i+1}$ = the angle from positive x_i to positive x_{i+1} taken positive counterclockwise as seen from positive z_i
$s_{i,i+1}$ = the distance along z_i from x_i to x_{i+1} with sign taken from the sense of z_i

We see that, when joint i is a revolute, as depicted in Fig. 11.11, then the $a_{i,i+1}$, $\alpha_{i,i+1}$, and $s_{i,i+1}$ parameters are constants defining the shape of link $i+1$, but $\theta_{i,i+1}$ is a variable; in fact, it serves to measure the joint variable of joint i. When joint i is a prismatic joint, then the $a_{i,i+1}$, $\alpha_{i,i+1}$, and $\theta_{i,i+1}$ parameters are constants and $s_{i,i+1}$ is the joint variable.

EXAMPLE 11.3

Find the Denavit–Hartenberg parameters for the Hooke, or Cardan, universal joint illustrated in Fig. 11.12.

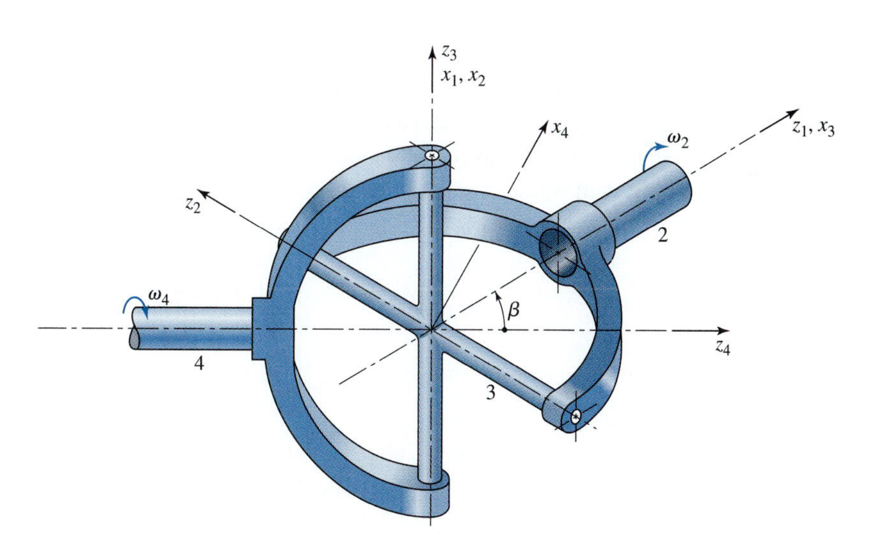

Figure 11.12 The Hooke, or Cardan, universal joint.

SOLUTION

The mechanism has four links and four revolute joints (it is identical to the spherical four-bar linkage illustrated in Fig. 11.2). First, we label the axes of the four revolutes joints as z_1, z_2, z_3, and z_4. Next, we identify and label the common perpendiculars between the joint axes, choosing positive orientations for these, and label them x_1, x_2, x_3, and x_4 as shown.

Note that, at the instant shown, x_1 and x_3 appear to lie along z_3 and z_1, respectively; this is only a temporary coincidence and will change as the mechanism moves.

Now, following the definitions given above, we find the values for the parameters.

$$\begin{array}{llll} a_{12} = 0 & a_{23} = 0 & a_{34} = 0 & a_{41} = 0 \\ \alpha_{12} = 90^\circ & \alpha_{23} = 90^\circ & \alpha_{34} = 90^\circ & \alpha_{41} = \beta \\ \theta_{12} = \phi_1 & \theta_{23} = \phi_2 & \theta_{34} = \phi_3 & \theta_{41} = \phi_4 \\ s_{12} = 0 & s_{23} = 0 & s_{34} = 0 & s_{41} = 0 \end{array} \qquad \begin{array}{r} \textit{Ans.} \\ \textit{Ans.} \\ \textit{Ans.} \\ \textit{Ans.} \end{array}$$

where β is the angle between the two shafts.

We note that all a and s distance parameters are zero; this signifies that the mechanism is, indeed, spherical with all motion axes intersecting at a common center. Also, we see that all θ parameters are joint variables because this is an *RRRR* linkage. They have been given the symbols ϕ_i rather than numeric values because they are variables; solutions for their values will be determined in the next section.

11.7 TRANSFORMATION-MATRIX POSITION ANALYSIS

The Denavit–Hartenberg parameters provide a standard method for measuring the important geometric characteristics of a linkage, but the value of this approach does not stop at that. Having standardized the placement of the coordinate systems on each link, Denavit and Hartenberg also demonstrated that the transformation equations between successive coordinate systems can be written in a standard matrix format that uses these parameters. If we know the position coordinates of a point measured in one of the coordinate systems, say R_{i+1}, then we can find the position coordinates of the same point in the previous coordinate system R_i as follows,

$$R_i = T_{i,i+1} R_{i+1}, \tag{11.12}$$

where the transformation matrix $T_{i,i+1}$ has the standard form*

$$T_{i,i+1} = \begin{bmatrix} \cos\theta_{i,i+1} & -\cos\alpha_{i,i+1}\sin\theta_{i,i+1} & \sin\alpha_{i,i+1}\sin\theta_{i,i+1} & a_{i,i+1}\cos\theta_{i,i+1} \\ \sin\theta_{i,i+1} & \cos\alpha_{i,i+1}\cos\theta_{i,i+1} & -\sin\alpha_{i,i+1}\cos\theta_{i,i+1} & a_{i,i+1}\sin\theta_{i,i+1} \\ 0 & \sin\alpha_{i,i+1} & \cos\alpha_{i,i+1} & s_{i,i+1} \\ 0 & 0 & 0 & 1 \end{bmatrix} \tag{11.13}$$

* Note that the first three rows and columns of the T matrix are the direction cosines needed for the rotation between the coordinate systems [for purposes of comparison see Eq. (11.8)]. The fourth column adds the translation terms for the separation of the origins. The fourth row represents a dummy equation, $1 = 1$, and keeps the T matrix square and nonsingular.

and each point position vector is given by

$$R = \begin{bmatrix} x \\ y \\ z \\ 1 \end{bmatrix}. \tag{11.14}$$

Now, by applying Eq. (11.12) recursively from one link to the next, we see that, for an n-link single-loop mechanism,

$$R_1 = T_{1,2}R_2$$
$$R_1 = T_{1,2}T_{2,3}R_3$$
$$R_1 = T_{1,2}T_{2,3}T_{3,4}R_4$$

and, in general,

$$R_1 = T_{1,2}T_{2,3}\cdots T_{i-1,i}R_{i}. \tag{11.15}$$

If we agree on a notation in which a product of these T matrices is still denoted a T matrix,

$$T_{i,j} = T_{i,i+1}T_{i+1,i+2}\cdots T_{j-1,j}, \tag{11.16}$$

then Eq. (11.15) becomes

$$R_1 = T_{1,i}R_i. \tag{11.17}$$

Finally, because link 1 follows link n at the end of the loop,

$$R_1 = T_{1,2}T_{2,3}\cdots T_{n-1,n}T_{n,1}R_1,$$

and, because this equation must be true no matter what point we choose for R_1, we see that

$$T_{1,2}T_{2,3}\cdots T_{n-1,n}T_{n,1} = I, \tag{11.18}$$

where I is the 4×4 identity transformation matrix.

This important equation is the transformation matrix form of the *loop-closure equation*. Just as the vector tetrahedron equation, Eq. (11.2), states that the sum of vectors around a kinematic loop must equal zero for the loop to close, Eq. (11.18) states that the product of transformation matrices around a kinematic loop must equal the identity transformation. Whereas the vector sum ensures that the loop returns to its starting location, the transformation-matrix product also ensures that the loop returns to its starting angular orientation. This becomes critical, for example, in spherical motion problems and cannot be shown by the vector-tetrahedron equation. The following example illustrates this point.

EXAMPLE 11.4

Find equations for the positions of all the joint variables of the Hooke universal joint illustrated in Fig. 11.12 when the input shaft angle ϕ_1 is specified.

SOLUTION

The Denavit–Hartenberg parameters for this mechanism were determined in Example 11.3. Substituting these parameters into Eq. (11.13) gives the individual transformation matrices for each link, namely

$$T_{1,2} = \begin{bmatrix} \cos\phi_1 & 0 & \sin\phi_1 & 0 \\ \sin\phi_1 & 0 & -\cos\phi_1 & 0 \\ 0 & 1 & 0 & 0 \\ 0 & 0 & 0 & 1 \end{bmatrix} \tag{1}$$

$$T_{2,3} = \begin{bmatrix} \cos\phi_2 & 0 & \sin\phi_2 & 0 \\ \sin\phi_2 & 0 & -\cos\phi_2 & 0 \\ 0 & 1 & 0 & 0 \\ 0 & 0 & 0 & 1 \end{bmatrix} \tag{2}$$

$$T_{3,4} = \begin{bmatrix} \cos\phi_3 & 0 & \sin\phi_3 & 0 \\ \sin\phi_3 & 0 & -\cos\phi_3 & 0 \\ 0 & 1 & 0 & 0 \\ 0 & 0 & 0 & 1 \end{bmatrix} \tag{3}$$

$$T_{4,1} = \begin{bmatrix} \cos\phi_4 & -\cos\beta\sin\phi_4 & \sin\beta\sin\phi_4 & 0 \\ \sin\phi_4 & \cos\beta\cos\phi_4 & -\sin\beta\cos\phi_4 & 0 \\ 0 & \sin\beta & \cos\beta & 0 \\ 0 & 0 & 0 & 1 \end{bmatrix}. \tag{4}$$

Although we could now use Eq. (11.18) directly, that is,

$$T_{1,2}T_{2,3}T_{3,4}T_{4,1} = I,$$

the number of computations can be reduced if we first rearrange the equation as

$$T_{1,2}T_{2,3}T_{3,4} = T_{4,1}^{-1}. \tag{5}$$

The inverse matrix on the right-hand side can easily be found here by simply transposing the matrix, that is, by switching rows and columns.* Therefore, substituting and carrying

* Although this is true in this special case, where all translation terms are zero, it is not true for all 4×4 transformation matrices.

out the matrix computations, Eq. (5) becomes

$$\begin{bmatrix} \cos\phi_1\cos\phi_2\cos\phi_3 + \sin\phi_1\sin\phi_3 & \cos\phi_1\sin\phi_2 & -\cos\phi_1\cos\phi_2\sin\phi_3 - \sin\phi_1\cos\phi_3 & 0 \\ \sin\phi_1\cos\phi_2\cos\phi_3 - \cos\phi_1\sin\phi_3 & \sin\phi_1\sin\phi_2 & \sin\phi_1\cos\phi_2\sin\phi_3 + \cos\phi_1\cos\phi_3 & 0 \\ \sin\phi_2\cos\phi_3 & -\cos\phi_2 & \sin\phi_2\sin\phi_3 & 0 \\ 0 & 0 & 0 & 1 \end{bmatrix}$$

$$= \begin{bmatrix} \cos\phi_4 & \sin\phi_4 & 0 & 0 \\ -\cos\beta\sin\phi_4 & \cos\beta\cos\phi_4 & \sin\beta & 0 \\ \sin\beta\sin\phi_4 & -\sin\beta\cos\phi_4 & \cos\beta & 0 \\ 0 & 0 & 0 & 1 \end{bmatrix}. \tag{6}$$

Because corresponding row and column elements on both sides of this equation must be equal, then using the second-column elements in the first and second rows we have

$$\sin\phi_4 = \cos\phi_1\sin\phi_2 \quad \text{and} \quad \cos\beta\cos\phi_4 = \sin\phi_1\sin\phi_2.$$

Then dividing the first equation by the second equation and rearranging gives

$$\phi_4 = \tan^{-1}\left(\frac{\cos\beta}{\tan\phi_1}\right). \qquad \textit{Ans.}$$

Once we have solved for the value of ϕ_4, we can equate the ratios of the third-row, third-column elements to the third-row, first-column elements. This gives

$$\phi_3 = \tan^{-1}\left(\frac{1}{\tan\beta\sin\phi_4}\right). \qquad \textit{Ans.}$$

Finally, equating the elements in the second column of the third row gives

$$\phi_2 = \cos^{-1}(\sin\beta\cos\phi_4). \qquad \textit{Ans.}$$

11.8 MATRIX VELOCITY AND ACCELERATION ANALYSES

The power of the matrix method in position analysis was demonstrated in the previous section. However, the method is not limited to position analysis. The same standardized approach can be extended to velocity and acceleration analyses. To see this, let us start by noting that, of the four Denavit–Hartenberg parameters, three are constants describing the link shape and one is the joint variable. Therefore, in the basic transformation of Eq. (11.13), that is

$$T = \begin{bmatrix} \cos\theta & -\cos\alpha\sin\theta & \sin\alpha\sin\theta & a\cos\theta \\ \sin\theta & \cos\alpha\cos\theta & -\sin\alpha\cos\theta & a\sin\theta \\ 0 & \sin\alpha & \cos\alpha & s \\ 0 & 0 & 0 & 1 \end{bmatrix}, \tag{11.19}$$

there is only one variable and it is either the θ or the s parameter depending on the type of joint.

If we consider the case where the joint variable is the angle θ, then the derivative of T with respect to its own variable is

$$\frac{dT}{d\theta} = \begin{bmatrix} -\sin\theta & -\cos\alpha\cos\theta & \sin\alpha\cos\theta & -a\sin\theta \\ \cos\theta & -\cos\alpha\sin\theta & \sin\alpha\sin\theta & a\cos\theta \\ 0 & 0 & 0 & 0 \\ 0 & 0 & 0 & 0 \end{bmatrix}. \tag{11.20}$$

On the other hand, if the joint variable is the distance s, as in a prismatic joint, then the derivative of T with respect to s is

$$\frac{dT}{ds} = \begin{bmatrix} 0 & 0 & 0 & 0 \\ 0 & 0 & 0 & 0 \\ 0 & 0 & 0 & 1 \\ 0 & 0 & 0 & 0 \end{bmatrix}. \tag{11.21}$$

It is interesting that both of these derivatives can be expressed by the same formula,

$$\frac{dT_{i,i+1}}{d\phi_i} = Q_i T_{i,i+1}, \tag{11.22}$$

where we understand that when joint i is a revolute, then $\phi_i = \theta_{i,i+1}$ and we use

$$Q_i = \begin{bmatrix} 0 & -1 & 0 & 0 \\ 1 & 0 & 0 & 0 \\ 0 & 0 & 0 & 0 \\ 0 & 0 & 0 & 0 \end{bmatrix}, \tag{11.23}$$

and when joint i is prismatic, then $\phi_i = s_{i,i+1}$ and we use

$$Q_i = \begin{bmatrix} 0 & 0 & 0 & 0 \\ 0 & 0 & 0 & 0 \\ 0 & 0 & 0 & 1 \\ 0 & 0 & 0 & 0 \end{bmatrix}. \tag{11.24}$$

For velocity analysis we need derivatives with respect to time rather than with respect to the joint variables. Therefore, using Eq. (11.22), we determine that

$$\frac{dT_{i,i+1}}{dt} = Q_i T_{i,i+1}\dot{\phi}_i, \tag{11.25}$$

where $\dot{\phi}_i = d\phi_i/dt$. These are usually not known values, but we will discover below how they can be determined. We start by differentiating the loop-closure conditions, Eq. (11.18), with respect to time. Using the chain rule with Eq. (11.25) to differentiate each factor, we get

$$\sum_{i=1}^{n} T_{1,2}T_{2,3}\cdots T_{i-1,i}Q_i T_{i,i+1}\cdots T_{n-1,n}T_{n1}\dot{\phi}_i = 0$$

and using the more condensed notation of Eq. (11.16), this becomes

$$\sum_{i=1}^{n} T_{1,i} Q_i T_{i,1} \dot{\phi}_i = 0. \tag{11.26}$$

If we now define the symbol

$$D_i = T_{1,i} Q_i T_{i,1} \tag{11.27a}$$

and take note that the loop-closure condition indicates that this is the same as

$$D_i = T_{1,i} Q_i T_{1,i}^{-1}, \tag{11.27b}$$

then Eq. (11.26) can be written as

$$\sum_{i=1}^{n} D_i \dot{\phi}_i = 0. \tag{11.28}$$

This equation contains all the conditions that the pair variable velocities must obey to fit the mechanism in question and be compatible with each other. Note that once the position analysis is completed, the D_i matrices can be evaluated from known information. Some of the pair variable velocities will be given as input information, namely, the m input variables; all other $\dot{\phi}_i$ values may then be solved for from Eq. (11.28). A continuation of Example 11.4 will make the procedure clear.

EXAMPLE 11.5

Determine the angular velocity of the output shaft for the Hooke universal joint of Example 11.4.

SOLUTION

From the results of the previous example, we find the following relationships between the input shaft angle ϕ_1 and the other pair variables,

$$\begin{aligned} \sin\phi_2 &= \frac{\cos\beta}{\sigma} & \cos\phi_2 &= \frac{\sin\beta\sin\phi_1}{\sigma} \\ \sin\phi_3 &= \sigma & \cos\phi_3 &= \sin\beta\cos\phi_1, \\ \sin\phi_4 &= \frac{\cos\beta\cos\phi_1}{\sigma} & \cos\phi_4 &= \frac{\sin\phi_1}{\sigma} \end{aligned}$$

where

$$\sigma = \sqrt{1 - \sin^2\beta\cos^2\phi_1},$$

Substituting these relationships into Eq. (11.13) gives the transformation matrices as functions of ϕ_1 alone. These, in turn, can be substituted into Eq. (11.27) to give the derivative operator matrices D_i, which become

$$D_1 = \begin{bmatrix} 0 & -1 & 0 & 0 \\ 1 & 0 & 0 & 0 \\ 0 & 0 & 0 & 0 \\ 0 & 0 & 0 & 0 \end{bmatrix}$$

$$D_2 = \begin{bmatrix} 0 & 0 & -\cos\phi_1 & 0 \\ 0 & 0 & -\sin\phi_1 & 0 \\ \cos\phi_1 & \sin\phi_1 & 0 & 0 \\ 0 & 0 & 0 & 0 \end{bmatrix}$$

$$D_3 = \begin{bmatrix} 0 & \dfrac{\sin\beta\sin\phi_1}{\sigma} & \dfrac{\cos\beta\sin\phi_1}{\sigma} & 0 \\ -\dfrac{\sin\beta\sin\phi_1}{\sigma} & 0 & \dfrac{-\cos\beta\cos\phi_1}{\sigma} & 0 \\ -\dfrac{\cos\beta\sin\phi_1}{\sigma} & \dfrac{\cos\beta\cos\phi_1}{\sigma} & 0 & 0 \\ 0 & 0 & 0 & 0 \end{bmatrix}$$

$$D_4 = \begin{bmatrix} 0 & -\cos\beta & \sin\beta & 0 \\ \cos\beta & 0 & 0 & 0 \\ -\sin\beta & 0 & 0 & 0 \\ 0 & 0 & 0 & 0 \end{bmatrix}.$$

These can now be used in Eq. (11.28). Taking the elements from row 3, column 2, then from row 1, column 3, and finally from row 2, column 1, gives the three equations:

$$(\sin\phi_1)\,\dot{\phi}_2 + \left(\frac{\cos\beta\cos\phi_1}{\sigma}\right)\dot{\phi}_3 = 0$$

$$(-\cos\phi_1)\,\dot{\phi}_2 + \left(\frac{\cos\beta\sin\phi_1}{\sigma}\right)\dot{\phi}_3 + (\sin\beta)\,\dot{\phi}_4 = 0$$

$$\left(-\frac{\sin\beta\sin\phi_1}{\sigma}\right)\dot{\phi}_3 + (\cos\beta)\,\dot{\phi}_4 = -\dot{\phi}_1.$$

Solving these equations simultaneously gives

$$\dot{\phi}_2 = \frac{-\sin\beta\cos\beta\cos\phi_1}{1-\sin^2\beta\cos^2\phi_1}\dot{\phi}_1$$

$$\dot{\phi}_3 = \frac{\sin\beta\sin\phi_1}{\sqrt{1-\sin^2\beta\cos^2\phi_1}}\dot{\phi}_1$$

$$\dot{\phi}_4 = \frac{-\cos\beta}{1-\sin^2\beta\cos^2\phi_1}\dot{\phi}_1.$$

Ans.

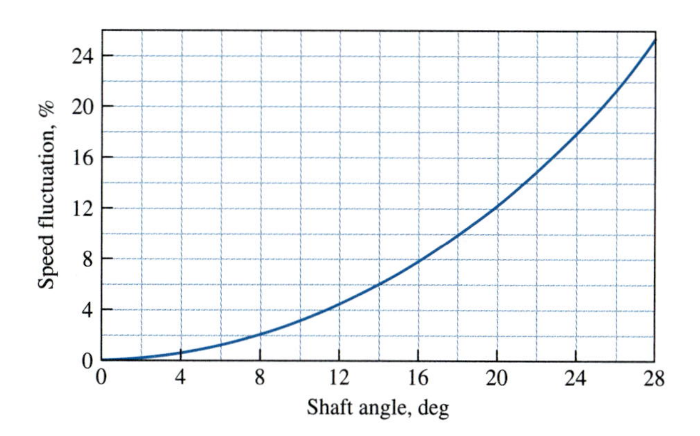

Figure 11.13 Relationship between shaft angle and speed fluctuation for Hooke universal shaft couplings.

Note that when the input shaft has constant angular velocity $\dot{\phi}_1$, the angular velocity of the output shaft $\dot{\phi}_4$ is not constant unless the shafts are in line (that is, unless the shaft angle $\beta = 0$). If β is not equal to zero, the output angular velocity fluctuates as the shaft rotates. Because the shaft angle β is constant, but usually not zero, the maximum value of the output/input velocity ratio $\dot{\phi}_4/\dot{\phi}_1$ occurs when $\cos\phi_1 = 1$, that is, when $\phi_1 = 0°, 180°, 360°, 540°$, and so on; the minimum value of this velocity ratio occurs when $\cos\phi_1 = 0$. If the difference between the maximum and minimum velocity ratio is expressed in percentages and plotted against the shaft angle β, the graph illustrated in Fig. 11.13 is the result. This curve, showing percentage speed fluctuation, is useful in evaluating the Hooke universal shaft coupling applications.

If we wanted to find the velocity of a moving point attached to link i, we would differentiate the equation for the position of the point, Eq. (11.15), with respect to time. Using the D_i operator matrices to do this, we are led to defining a set of velocity operator matrices,

$$\omega_1 = 0 \tag{11.29a}$$

$$\omega_{i+1} = \omega_i + D_i\dot{\phi}_i, \quad i = 1, 2, \ldots, n, \tag{11.29b}$$

and then the velocity of a point on link i is given by

$$\dot{R}_i = \omega_i T_{1,i} R_i. \tag{11.30}$$

Acceleration analysis follows the same approach as for the velocity analysis. Without showing all of the details, the important equations will be presented here.[12] First, we must find a formula for taking the derivative of a D_i matrix.

$$\frac{dD_i}{dt} = (\omega_i D_i - D_i\omega_i) \quad i = 1, 2, \ldots, n \tag{11.31}$$

Differentiating Eq. (11.28) again with respect to time, by means of Eq. (11.31), and rearranging, we get a set of equations relating the pair-variable accelerations:

$$\sum_{i=1}^{n} D_i \ddot{\phi}_i = -\sum_{i=1}^{n} (\omega_i D_i - D_i \omega_i) \dot{\phi}_i. \tag{11.32}$$

These equations can be solved for the $\ddot{\phi}_i$ joint variable acceleration values once the position, velocity, and input acceleration values are known. The solution process is identical to that for Eq. (11.28); in fact, the matrix of coefficients of the unknowns is identical for both sets of equations.

This matrix of coefficients, called the *Jacobian*, is essential for the solution of any set of derivatives of the joint variables. As indicated in Section 3.20, if this matrix becomes singular, there is no unique solution for the velocities (or accelerations) of the joint variables. If this occurs, such a position of the mechanism is called a *singular* position; one example would be a dead-center position.

The derivatives of the velocity operator matrices of Eq. (11.29) give rise to a definition of a set of acceleration operator matrices, that is

$$\alpha_1 = 0 \tag{11.33a}$$

$$\alpha_{i+1} = \alpha_i + D_i \ddot{\phi}_i + (\omega_i D_i - D_i \omega_i) \dot{\phi}_i, \tag{11.33b}$$

From these, the acceleration of a point on link i is given by

$$\ddot{R}_i = (\alpha_i + \omega_i \omega_i) T_{1,i} R_i. \tag{11.34}$$

Much greater detail and more power has been developed using this transformation-matrix approach for the kinematic and dynamic analysis of rigid-body systems. These methods go far beyond the scope of this book. Further examples dealing with robotics, however, are presented in Chapter 12.

11.9 GENERALIZED MECHANISM ANALYSIS COMPUTER PROGRAMS

It can probably be observed that the methods taken for the solution of each new problem are quite similar from one problem to the next. However, particularly in three-dimensional analysis, we also see that the number and complexity of the calculations can make solution by hand a very tedious task. These characteristics indicate that a general computer program might have a broad range of applications and that the development costs for such a program might be justified through repeated use and increased accuracy, relief of human drudgery, and elimination of human errors. General computer programs for the simulation of rigid-body kinematic and dynamic systems have been under development for several decades, and many are available and currently used in industrial settings, particularly in the automotive and aircraft industries.

The first widely available computer program for mechanism analysis was named Kinematic Analysis Method (KAM) and was written and distributed by IBM. It included capabilities for position, velocity, acceleration, and force analysis of both planar and spatial mechanisms and was developed using the Chace vector-tetrahedron equation solutions discussed in Section 11.3. Released in 1964, this program was the first to recognize the need for a general program for mechanical systems exhibiting large geometric movements. Being first, however, it had limitations and has been superseded by more powerful programs, including those described below.

Powerful generalized programs have also been developed using finite element and finite difference methods; NASTRAN and ANSYS are two examples. In the realm of mechanical systems these programs have been developed primarily for stress analysis and have excellent capabilities for static- and dynamic-force analysis. These also allow the links of the simulated system to deflect under load and are capable of solving statically indeterminate force problems. They are very powerful programs with wide application in industry. Although they are sometimes used for mechanism analysis, they are limited in their ability to simulate the geometric changes typical of kinematic systems.

The most widely used generalized programs for kinematic and dynamic simulation of three-dimensional rigid-body mechanical systems are ADAMS, DADS, and IMP. The MSC ADAMS program, standing for Automatic Dynamic Analysis of Mechanical Systems, grew from the research efforts of Chace, Orlandea, and others at the University of Michigan[13] and is now available from MSC Software.* DADS, an acronym for Dynamic Analysis and Design System, was originally developed by Haug and others at the University of Iowa[14] and is now made available through LMS International†. The Integrated Mechanisms Program (IMP) was developed by Uicker, Sheth, and others at the University of Wisconsin–Madison.[15] These and other similar programs are all applicable to single- or multiple-degree-of-freedom systems in both open- and closed-loop configurations. All operate on workstations, and some operate on microprocessors; all can display results with graphic animation. All are capable of solving position, velocity, acceleration, and static and dynamic force analyses. All can formulate the dynamic equations of motion and predict the system response to a given set of initial conditions with prescribed motions or forces that may be functions of time. Some of these programs include multibody collision detection, the ability to simulate impact, elasticity, or control system effects. Other commercial software in this area include the Pro/ENGINEER Mechanism Dynamics†† and MSC Working Model§ systems.

* MSC Software Corp., 2 MacArthur Place, Santa Ana, CA 92707.

† LMS International, Researchpark Z1, Interleuvenlaan 68, 3001 Leuven, Belgium; LMS CAE Division, 2651 Crosspark Road, Coralville, IA 52241.

†† Parametric Technologies, Corp., 140 Kendrick Street, Needham, MA 02494.

§ MSC Software Corp., *ibid.*

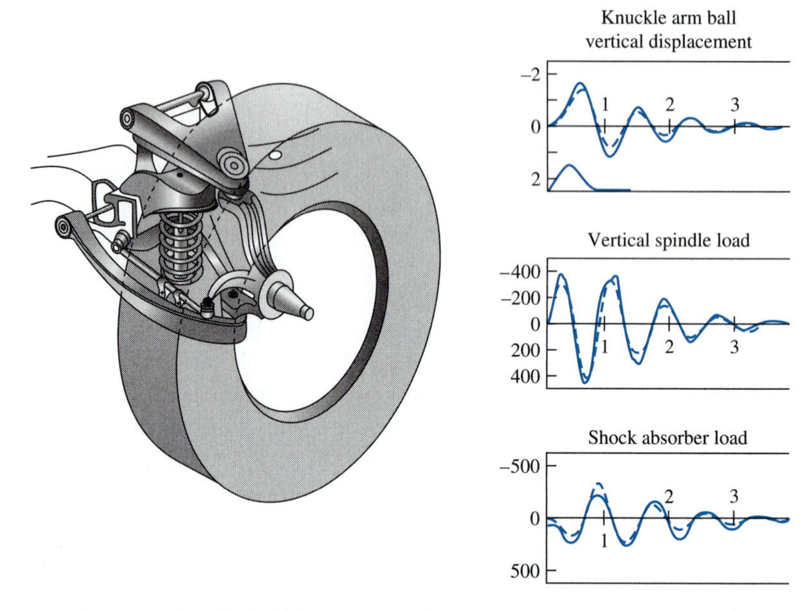

Figure 11.14 An example of a half-front automotive suspension simulated by both the ADAMS and the IMP programs. The graphs show the comparison of experimental test data and numerical simulation results as the suspension encounters a hole. (Mechanical Dynamics, Inc., Ann Arbor, MI, and JML Research, Inc., Madison, WI.)

A typical application for any of these programs is the simulation of the automotive front suspension illustrated in Fig. 11.14.‖ Simulations of this type have been performed with several of these programs and they compare well with experimental data.

Another type of generalized program available today is intended for kinematic synthesis. The earliest such program was KINSYN (KINematic SYNthesis),[16] which was followed by LINCAGES (Linkage INteractive Computer Analysis and Graphically Enhanced Synthesis package)[17,*], RECSYN (RECtified SYNthesis)[18], and others. These systems are directed toward the kinematic synthesis of planar linkages using methods analogous to those presented in Chapter 10. Users may input their motion requirements through a graphical user interface (GUI); the computer accepts the sketch and provides the required design information on the display screen. An example showing the use of KINSYN is illustrated in Fig. 11.15. A more recent system of this type is the WATT Mechanism Design Tool from Heron Technologies,† a spin-off company from Twente University in the Netherlands.

‖ Simulations of the system illustrated in Fig. 11.14 were performed using the ADAMS and IMP programs in 1974 for the Strain History Prediction Committee of the Society of Automotive Engineers. Vehicle data and experimental test results were provided by Chevrolet Division, General Motors Corp.
* LINCAGES is available through Dr. A. G. Erdman by sending e-mail to agerdman@me.umn.edu.
† Heron Technologies b.v., P.O. Box 2, 7550 AA Hengelo, The Netherlands. For availability and prices, e-mail sales@heron-technologies.com.

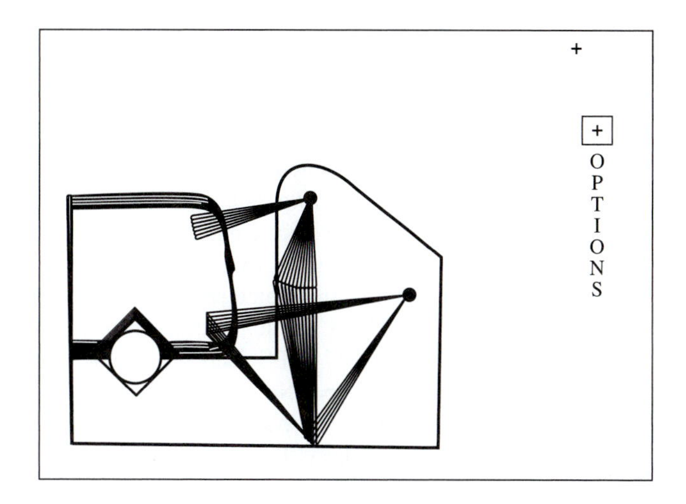

Figure 11.15 This pipe-clamp mechanism was designed in about 15 minutes using KINSYN III. KINSYN was developed at the Joint Computer Facility of the Massachusetts Institute of Technology under the direction of Dr. R. E. Kaufman, now Professor of Engineering at the George Washington University. (Courtesy of Prof. R. E. Kaufman.)

11.10 REFERENCES

[1] L. Harrisberger, A Number Synthesis Survey of Three-Dimensional Mechanisms. *J. Eng. Ind. ASME Trans.* B 87 (1965).
[2] G. T. Bennett, A New Mechanism. *Engineering* 76 (1903):777–8.
[3] J. E. Baker, The Bennett Linkage and Its Associated Quadric Surfaces. *Mech. Machine Theory* 23, no. 2 (1988):147–56.
[4] M. Goldberg, "New Five-Bar and Six-Bar Linkages in Three Dimensions," *ASME Transactions*, vol. 65, 1943, pp. 649–661. For pictures of these, see R. S. Hartenberg and J. Denavit, *Kinematic Synthesis of Linkages*, McGraw–Hill, New York, 1964, pp. 85–86.
[5] L. Harrisberger and A. H. Soni, A Survey of Three-Dimensional Mechanisms with One General Constraint. ASME Paper 66-MECH-44, October 1966. This paper contains 45 references on spatial mechanisms.
[6] M. A. Chace, Vector Analysis of Linkages. *J. Eng. Ind. ASME Trans. B* 85 (1963):289–97.
[7] A. T. Yang and F. Freudenstein, Application of Dual-Number and Quaternion Algebra to the Analysis of Spatial Mechanisms. *J. Appl. Mech. ASME Trans. E* 86 (1964): 300–8.
[8] J. J. Uicker, Jr., J. Denavit, and R. S. Hartenberg, An Iterative Method for the Displacement Analysis of Spatial Linkages. *J. Appl. Mech. ASME Trans. E* 87 (1965):309–14.
[9] M. A. Chace, *ibid.*
[10] J. Duffy and C. Crane, A Displacement Analysis of the General Spatial 7-Link, 7R Mechanisms. *Mech. Machine Theory* 15, no. 3 (1980):153–69.
[11] J. Denavit and R. S. Hartenberg, A Kinematic Notation for Lower-Pair Mechanisms Based on Matrices. *J. Appl. Mech. ASME Trans. E* 22 (1955):215–21.
[12] J. J. Uicker, Jr., B. Ravani, and P. N. Sheth, *Matrix Methods in the Design Analysis of Mechanisms and MultiBody Systems*. Cambridge, UK: Cambridge University Press, 2010.

[13] N. Orlandea, M. A. Chace, and D. A. Calahan, A Sparsity-Oriented Approach to the Dynamic Analysis and Design of Mechanical Systems, Parts I and II. *J. Eng. Ind. ASME Trans.* 99 (1977):773–84.

[14] E. J. Haug, *Computer-Aided Kinematics and Dynamics of Mechanical Systems*. Boston: Allyn & Bacon, 1989.

[15] P. N. Sheth and J. J. Uicker, Jr., "MP (Integrated Mechanisms Program), A Computer-Aided Design Analysis System for Mechanisms and Linkages. *J. Eng. Ind. ASME Trans.* 94 (1972):454–64.

[16] R. E. Kaufman, KINSYN: An Interactive Kinematic Design System. *Transactions of the Third World Congress on Theory of Machines and Mechanisms*, 1971, Dubrovnik, Yugoslavia. The paper was accompanied by the presentation of a 16-mm motion picture of 28-minute duration. The final version, KINSYN7, was developed and marketed by KINTECH, Inc., until the firm closed in 1987.

[17] A. G. Erdman and J. E. Gustafson, LINCAGES: Linkage Interactive Computer Analysis and Graphically Enhanced Synthesis Package. ASME Paper No. 77-DTC-5, 1977.

[18] J. C. Chuang, R. T. Strong, and K. J. Waldron, Implementation of Solution Rectification Techniques in an Interactive Linkage Synthesis Program. *J. Mechanical Design ASME Trans.* July (1981): 657–64.

PROBLEMS

11.1 Use the Kutzbach criterion to determine the mobility of the *SSC* linkage illustrated in Fig. P11.1. Identify any idle freedoms and state how they can be removed. What is the nature of the path described by point B?

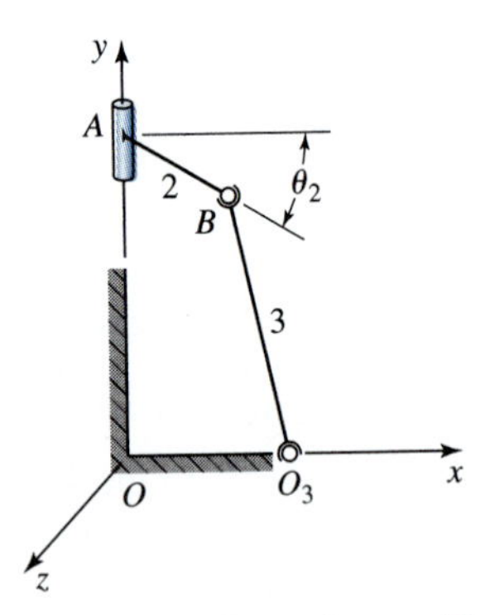

Figure P11.1 $R_{BA} = R_{O_3O} = 75\text{mm}$, $R_{BO_3} = 150\text{mm}$, and $\theta_2 = 30°$.

11.2 For the *SSC* linkage illustrated in Fig. P11.1, express the position of each link in vector form.

11.3 For the linkage of Fig. P11.1 with $V_A = -50\hat{\mathbf{j}}$ mm/s, use vector analysis to find the angular velocities of links 2 and 3 and the velocity of point B at the position specified.

11.4 Solve Problem 11.3 using graphic techniques.

11.5 For the spherical *RRRR* illustrated in Fig. P11.5, use vector algebra to make complete velocity and acceleration analyses at the position given.

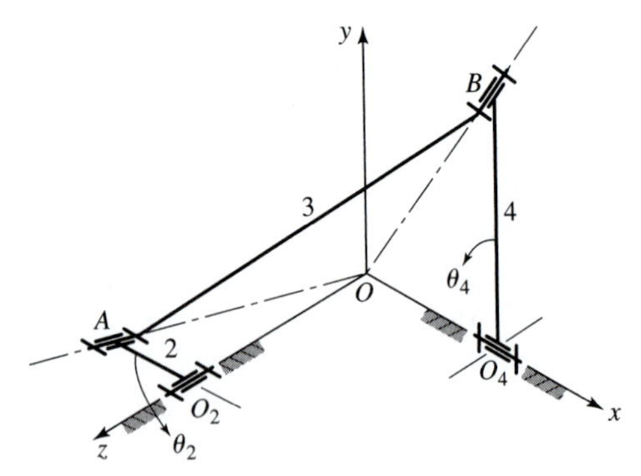

Figure P11.5 The Fig. is not drawn to scale; at the position shown, $\mathbf{R}_{O_2O} = 7\hat{\mathbf{k}}$ in., $\mathbf{R}_{O_4O} = 2\hat{\mathbf{i}}$ in., $\mathbf{R}_{AO_2} = -3\hat{\mathbf{i}}$in., $\mathbf{R}_{BO_4} = 9\hat{\mathbf{j}}$in., $\mathbf{R}_{BA} = 5\hat{\mathbf{i}} + 9\hat{\mathbf{j}} - 7\hat{\mathbf{k}}$in., and $\omega_2 = -60\hat{\mathbf{k}}$rad/s.

11.6 Solve Problem 11.5 using graphic techniques.

11.7 Solve Problem 11.5 using transformation-matrix techniques.

11.8 Solve Problem 11.5 except with $\theta_2 = 90°$.

11.9 Determine the advance-to-return time ratio for Problem 11.5. What is the total angle of oscillation of link 4?

11.10 For the spherical *RRRR* linkage illustrated in Fig. P11.10, determine whether the crank is free to turn through a complete revolution. If so, find the angle of oscillation of link 4 and the advance-to-return time ratio.

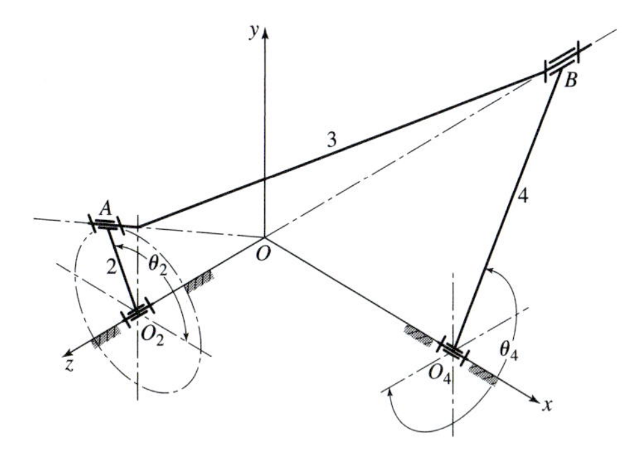

Figure P11.10 $R_{O_2O} = 150\text{mm}$, $R_{O_4O} = 225\text{mm}$, $R_{AO_2} = 37.5\text{mm}$, $R_{BO_4} = 262\text{mm}$, $R_{BA} = 412\text{mm}$, $\theta_2 = 120°$, and $\omega_2 = 30\hat{\mathbf{k}}\text{rad/s}$.

11.11 Use vector algebra to make complete velocity and acceleration analyses of the linkage of Fig. P11.10 at the position specified.

11.12 Solve Problem 11.11 using graphic techniques.

11.13 Solve Problem 11.11 using transformation-matrix techniques.

11.14 Figure P11.14 illustrates the top, front, and auxiliary views of a spatial slider–crank *RSSP* linkage. In the construction of many such mechanisms, a provision is made to vary the angle β; thus, the stroke of slider 4 becomes adjustable from zero, when $\beta = 0$, to twice the crank length, when $\beta = 90°$. With $\beta = 30°$, use vector algebra to make a complete velocity analysis of the linkage at the given position.

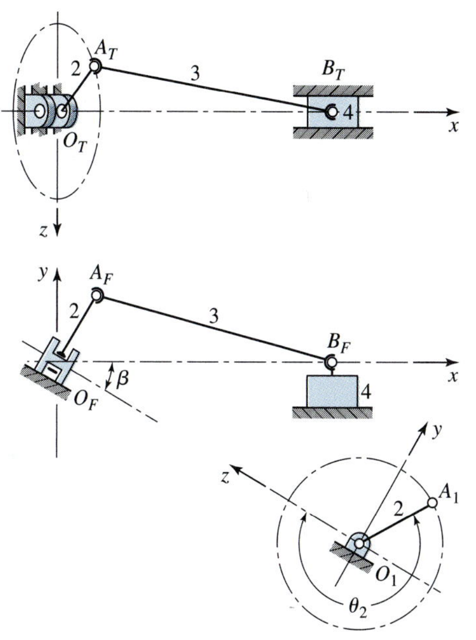

Figure P11.14 $R_{AO} = 2\text{in.}$, $R_{BA} = 6\text{in.}$, $\theta_2 = 240°$, and $\omega_2 = 24\hat{\mathbf{i}}\text{rad/s}$.

11.15 Solve Problem 11.14 using graphical techniques.

11.16 Solve Problem 11.14 using transformation-matrix techniques.

11.17 Solve Problem 11.14 with $\beta = 60°$ using vector algebra.

11.18 Solve Problem 11.14 with $\beta = 60°$ using graphical techniques.

11.19 Solve Problem 11.14 with $\beta = 60°$ using transformation-matrix techniques.

11.20 Figure P11.20 illustrates the top, front, and profile views of an *RSRC* crank and oscillating-slider linkage. Link 4, the oscillating slider, is rigidly attached to a round rod that rotates and slides in the two bearings. (*a*) Use the Kutzbach criterion to find the mobility of this linkage. (*b*) With crank 2 as the driver, find the total angular and linear travel of link 4. (*c*) Write the loop-closure equation for this

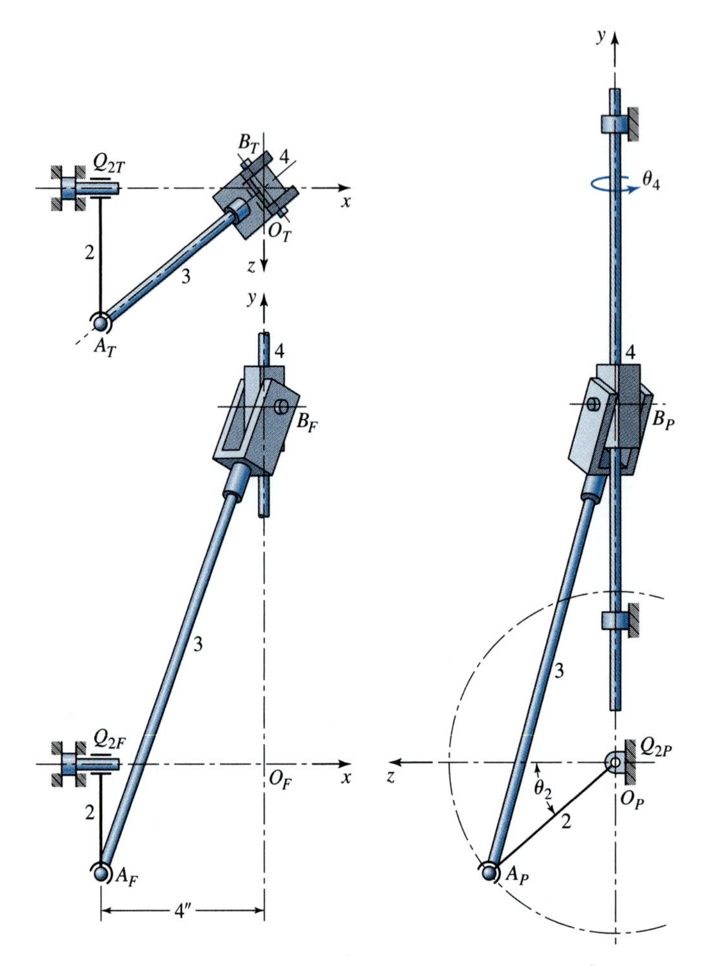

Figure P11.20 $R_{AO} = 4\text{in.}$, $R_{BA} = 12\text{in.}$, $\theta_2 = 40°$, and $\boldsymbol{\omega}_2 = -48\hat{\mathbf{i}}_{rad/s}$.

mechanism and use vector algebra to solve it for all unknown position data.

11.21 Use vector algebra to find $\boldsymbol{V}_B$, ω_3, and ω_4 for Problem 11.20.

11.22 Solve Problem 11.21 using graphical techniques.

11.23 Solve Problem 11.21 using transformation matrix techniques.

12 Robotics

12.1 INTRODUCTION

In previous chapters we have studied methods for analyzing the kinematics of machines. First we studied planar kinematics at great length, justifying this emphasis by pointing out that over 90% of all machines in use today have planar motion. Then, in Chapter 11, we demonstrated how these methods extend to problems with spatial motion. Although the algebra became lengthier with spatial problems, we determined that there was no significant new block of theory required. We also concluded that, as the mathematics became more tedious, there is a role for the computer to relieve the drudgery. Still, we reported that only a few problems have both practical application and spatial motion.

However, within the past few decades, advancing technology has given considerable attention to the development of robotic devices. Most are spatial mechanisms and require the attendant, more extensive, calculations. Fortunately, however, they also carry on-board computing capability and can deal with this added complexity. The one remaining requirement is that the engineers and designers of the robot itself have a clear understanding of their characteristics and have appropriate tools for their analysis. That is the purpose of Chapter 12.

The term *robot* is of Slavic origin, comes from the Czech word *robota*, meaning work or labor, and was introduced into the English vocabulary by Czechoslovakian dramatist Karl Capek in the early part of the 20th century. This word has been applied to a wide variety of computer-controlled electromechanical systems, from autonomous landrovers to underwater vehicles to teleoperated arms in industrial manipulators. The Robot Institute of America (RIA) defines a robot as *a reprogrammable multifunctional manipulator designed to move material, parts, tools, or specialized devices through variable programmed motions for the performance of a variety of tasks.* Many industrial manipulators bear a strong resemblance in their conceptual design to that of a human arm. However, this is not always true and is not essential; the key to the above definition is that a robot has flexibility through its programming, and its motion can be adapted to fit a variety of tasks.

12.2 TOPOLOGICAL ARRANGEMENTS OF ROBOTIC ARMS

Up to this point, we have studied machines with only a few degrees of freedom. With traditional types of machines it was usually desirable to drive the entire machine from a single source of power; thus, they were designed to have mobility of $m = 1$. In keeping with the idea of flexibility of application, however, robots must have more. We know that if a robot is to reach an arbitrary location in three-dimensional space, it must have mobility of at least $m = 3$ to adjust to the proper values of x, y, and z. In addition, if the robot is to be able to manipulate a tool or an object it carries into an arbitrary orientation, once reaching the desired location, an additional three degrees of freedom are required, giving a desired mobility of at least $m = 6$.

Along with the desire for six or more degrees of freedom, we also strive for simplicity, not only for reasons of good design and reliability, but also to minimize the computing burden in the control of the robot. Therefore, we often find that only three of the freedoms are designed into the robot arm itself and that another two or three may be included in the wrist. Because the tool, or *end effector*, usually varies with the task to be performed, this

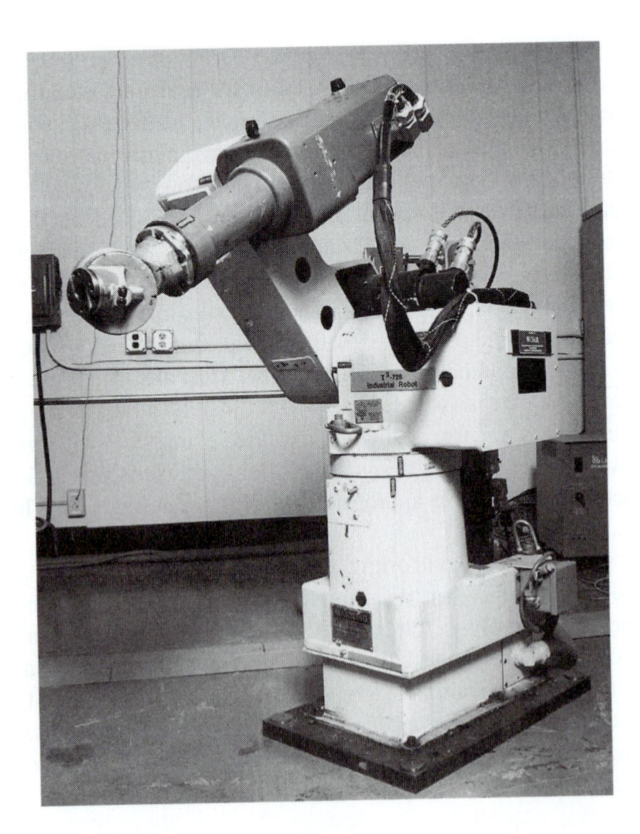

Figure 12.1 Cincinnati Milacron T^3 articulated six-axis robot. This model, T26, has a load capacity of 14 lb, horizontal reach of 40 in. from the vertical centerline, and position repeatability of 0.004 in.

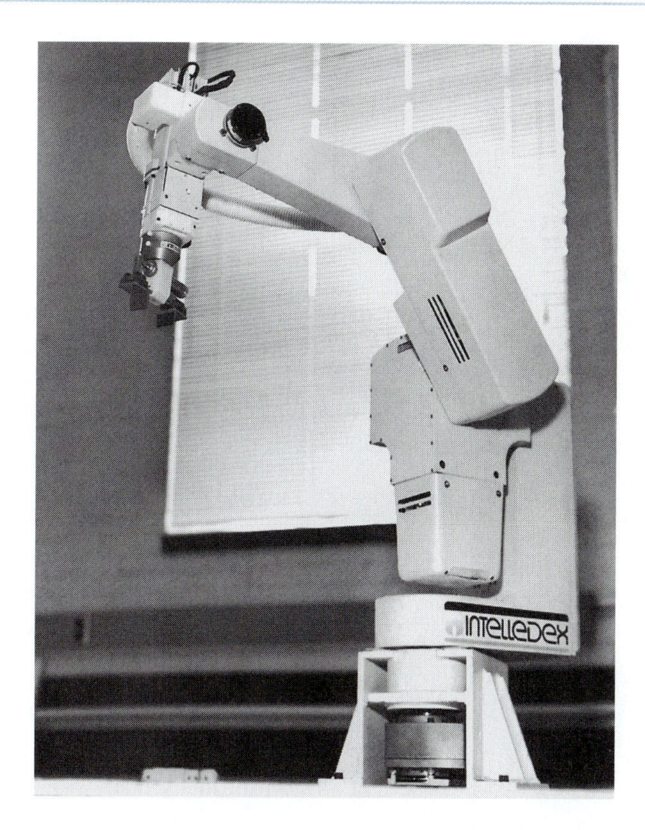

Figure 12.2 Intelledex articulated robot.

portion of the total robot may be made interchangeable; the same basic robot arm might carry any of several different end effectors specially designed for particular tasks.

Based on the first three freedoms of the arm, one common arrangement of joints for a manipulator is an *RRR* linkage, also called an *articulated* configuration. Two examples are the Cincinnati Milacron T^3 robot illustrated in Fig. 12.1 and the Intelledex articulated robot illustrated in Fig. 12.2.

The Selectively Compliant Articulated Robot for Assembly, also called the SCARA robot, is a more recent but popular configuration based on the *RRP* linkage. An example is illustrated in Fig. 12.3. Although there are other *RRP* robot configurations, note that the SCARA robot has all three joint variable axes parallel (usually pointing in the direction of gravity), restricting its freedom of movement but particularly suiting it to assembly operations.

A manipulator based on the *PPP* chain is illustrated in Fig. 12.4. It has three mutually perpendicular prismatic joints and, therefore, is called a Cartesian configuration or a *gantry* robot. The kinematic analysis of, and the programming for, this style of robot is particularly simple because of the perpendicularity of the three joint axes. It has applications in tabletop assembly and in transfer of cargo or material.

The arms depicted up to this point are simple, series-connected open kinematic chains. This is desirable because it reduces their kinematic complexity and eases their design and

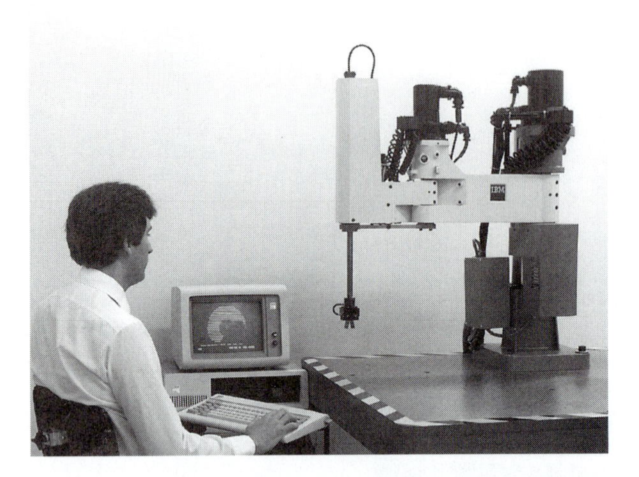

Figure 12.3 IBM model 7525 Selectively Compliant Articulated Robot for Assembly (SCARA). (Courtesy of IBM Corp., Rochester, MN.)

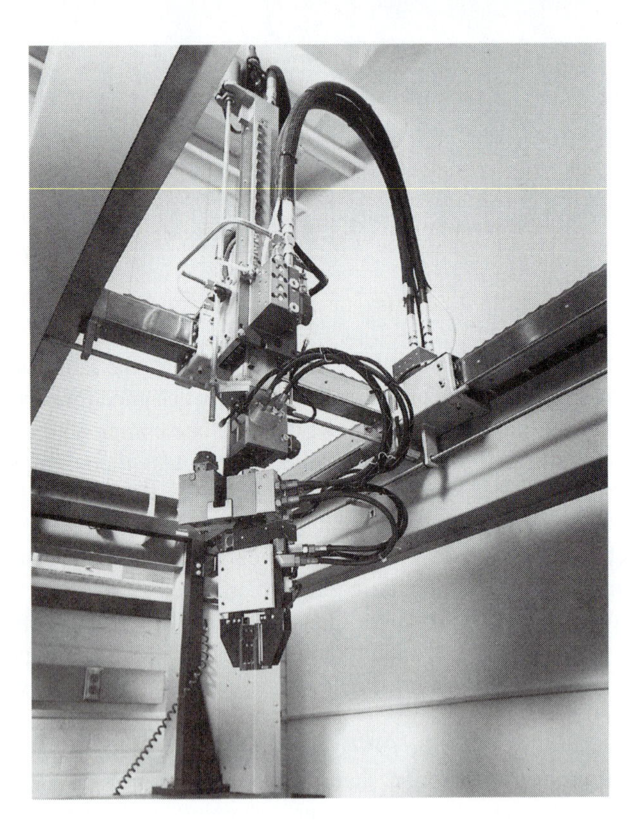

Figure 12.4 IBM model 7650 gantry robot.

Figure 12.5 General Electric model P80 robotic manipulator.

programming. However, it is not always true; there are robotic arms that include closed kinematic loops in their basic topology. One example is the robot illustrated in Fig. 12.5.

12.3 FORWARD KINEMATICS

The first kinematic analysis problem addressed here for a robot is finding the position of the tool or end effector once we are given the geometry of each component and the positions of the several actuators controlling the degrees of freedom. This can easily be done by the methods presented in Chapter 11 on spatial mechanisms. First, we should recognize three important characteristics of most robotics problems that are different from the spatial mechanisms of Chapter 11.

1. Knowledge of the position of a single point at the tip of the tool is often not enough. To have a complete knowledge of the tip of the tool, we must know the location and orientation of a coordinate system attached to the tool. This implies that vector methods are not sufficient and that the matrix methods of Section 11.7 are better suited.
2. Perhaps because of the previous observation, many robot manufacturers supply the Denavit–Hartenberg parameters (Section 11.6) for their particular robot designs. Then, use of the transformation-matrix approach is straightforward.
3. All joint variables of a serially connected chain are independent and are degrees of freedom. Thus, in the forward kinematics problem being discussed now, all joint variables are actuator variables and have given values; there are no loop-closure conditions and no dependent joint variable values to be determined.

The conclusion implied by these three observations combined is that finding the position and orientation of the end effector for given positions of the actuators is a straightforward application of Eq. (11.17). The absolute position of any chosen point R_{n+1} in the tool coordinate system attached to link $n+1$ is given by

$$R_1 = T_{1,n+1} R_{n+1}, \tag{12.1}$$

where

$$T_{1,n+1} = T_{1,2} T_{2,3} \cdots T_{n,n+1} \tag{12.2}$$

and each T matrix is given by Eq. (11.13) once the Denavit–Hartenberg parameters (including the actuator positions) are known.

Of course, if the robot has $n+1 = 7$ links ($n = 6$ joints), then symbolic multiplication of these matrices may become lengthy, unless some of the (constant) shape parameters are conveniently set to "nice" values. However, remembering that the robot will have computing capability on board, numerical evaluation of Eq. (12.2) is no great challenge for a particular set of actuator values. Still, because of speed requirements for real-time computer control, it is desirable to simplify these expressions as much as possible before programming. Toward this goal, most robot manufacturers have chosen more simplified designs (having nice shape parameters) to simplify these expressions. Many have also worked out the final expressions for their particular robots and make these available in their technical documentation.

EXAMPLE 12.1

For the Microbot model TCM five-axis robot illustrated in Fig. 12.6, find the transformation matrix T_{16} relating the position of the tool coordinate system to the ground coordinate system when the joint actuators are set to values $\phi_1 = 30°$, $\phi_2 = 60°$, $\phi_3 = -30°$, and $\phi_4 = \phi_5 = 0°$. Also, find the absolute position of the tool point that has local coordinates of $x_6 = y_6 = 0$, $z_6 = 2.5$ in.

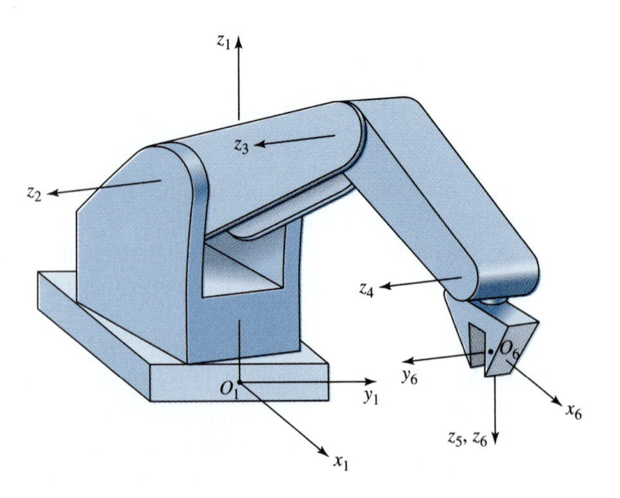

Figure 12.6 Example 12.1. The Microbot model TCM five-axis robot.

SOLUTION

The coordinate system axes are shown in Fig. 12.6. Note that the $x_6y_6z_6$ coordinate system is not based on a joint axis, but was chosen arbitrarily to form a convenient tool coordinate system. Based on these axes and the data in the manufacturer's documentation, we find the Denavit–Hartenberg parameters to be as follows:

$$
\begin{aligned}
&a_{12} = 0, && \alpha_{12} = 90°, && \theta_{12} = \phi_1 = 30°, && s_{12} = 7.68 \text{ in.} \\
&a_{23} = 7.00 \text{ in.}, && \alpha_{23} = 0°, && \theta_{23} = \phi_2 = 60°, && s_{23} = 0 \\
&a_{34} = 7.00 \text{ in.}, && \alpha_{34} = 0°, && \theta_{34} = \phi_3 = -30°, && s_{34} = 0 \\
&a_{45} = 0, && \alpha_{45} = 90°, && \theta_{45} = \phi_4 = 0°, && s_{45} = 0 \\
&a_{56} = 0, && \alpha_{56} = 0°, && \theta_{56} = \phi_5 = 0°, && s_{56} = 3.80 \text{ in}
\end{aligned}
$$

Using Eqs. (11.13) and (11.16) we can now form the following matrices and matrix products:

$$
T_{1,2} = \begin{bmatrix} 0.866 & 0 & 0.500 & 0 \\ 0.500 & 0 & -0.866 & 0 \\ 0 & 1 & 0 & 7.68 \text{ in} \\ 0 & 0 & 0 & 1 \end{bmatrix}
$$

$$
T_{2,3} = \begin{bmatrix} 0.500 & -0.866 & 0 & 3.50 \text{ in} \\ 0.866 & 0.500 & 0 & 6.06 \text{ in} \\ 0 & 0 & 1 & 0 \\ 0 & 0 & 0 & 1 \end{bmatrix}, \quad T_{1,3} = \begin{bmatrix} 0.433 & -0.750 & 0.500 & 3.03 \text{ in} \\ 0.250 & -0.433 & -0.866 & 1.75 \text{ in} \\ 0.866 & 0.500 & 0 & 13.74 \text{ in} \\ 0 & 0 & 0 & 1 \end{bmatrix}
$$

$$
T_{3,4} = \begin{bmatrix} 0.866 & 0.500 & 0 & 6.06 \text{ in} \\ -0.500 & 0.866 & 0 & -3.50 \text{ in} \\ 0 & 0 & 1 & 0 \\ 0 & 0 & 0 & 1 \end{bmatrix}, \quad T_{1,4} = \begin{bmatrix} 0.750 & -0.433 & 0.500 & 8.28 \text{ in} \\ 0.433 & -0.250 & -0.866 & 4.78 \text{ in} \\ 0.500 & 0.866 & 0 & 17.24 \text{ in} \\ 0 & 0 & 0 & 1 \end{bmatrix}
$$

$$
T_{4,5} = \begin{bmatrix} 1 & 0 & 0 & 0 \\ 0 & 0 & -1 & 0 \\ 0 & 1 & 0 & 0 \\ 0 & 0 & 0 & 1 \end{bmatrix}, \quad T_{1,5} = \begin{bmatrix} 0.750 & 0.500 & 0.433 & 8.28 \text{ in} \\ 0.433 & -0.866 & 0.250 & 4.78 \text{ in} \\ 0.500 & 0 & -0.866 & 17.24 \text{ in} \\ 0 & 0 & 0 & 1 \end{bmatrix}
$$

$$
T_{5,6} = \begin{bmatrix} 1 & 0 & 0 & 0 \\ 0 & 1 & 0 & 0 \\ 0 & 0 & 1 & 3.80 \text{ in} \\ 0 & 0 & 0 & 1 \end{bmatrix}, \quad T_{1,6} = \begin{bmatrix} 0.750 & 0.500 & 0.433 & 9.93 \text{ in} \\ 0.433 & -0.866 & 0.250 & 5.73 \text{ in} \\ 0.500 & 0 & -0.866 & 13.95 \text{ in} \\ 0 & 0 & 0 & 1 \end{bmatrix}.
$$

Ans.

The absolute position of the specified tool point is now determined from Eq. (12.1):

$$
R_1 = \begin{bmatrix} 0.750 & 0.500 & 0.433 & 9.93 \text{ in} \\ 0.433 & -0.866 & 0.250 & 5.73 \text{ in} \\ 0.500 & 0 & -0.866 & 13.95 \text{ in} \\ 0 & 0 & 0 & 1 \end{bmatrix} \begin{bmatrix} 0 \\ 0 \\ 2.5 \text{ in} \\ 1 \end{bmatrix} = \begin{bmatrix} 11.00 \text{ in} \\ 6.36 \text{ in} \\ 11.78 \text{ in} \\ 1 \end{bmatrix}. \qquad \textit{Ans.}
$$

The direct computation of velocities and accelerations of arbitrary points on the links of the robot can also be accomplished easily by the matrix methods of Section 11.8. In particular, Eqs. (11.29), (11.30), (11.33), and (11.34) are of direct use once the actuator velocities and accelerations are known. A continuation of the previous example should make this clear.

EXAMPLE 12.2

For the Microbot model TCM robot of Fig. 12.6 and in the position described in Example 12.1, find the instantaneous velocity and acceleration of the same tool point $x_6 = y_6 = 0$, $z_6 = 2.5$ in. if the actuators are given constant velocities of $\dot{\phi}_1 = 0.20$ rad/s, $\dot{\phi}_4 = -0.35$ rad/s, and $\dot{\phi}_2 = \dot{\phi}_3 = \dot{\phi}_5 = 0$.

SOLUTION

Using Eq. (11.27) and the results from Example 12.1, we find

$$D_1 = \begin{bmatrix} 0 & -1 & 0 & 0 \\ 1 & 0 & 0 & 0 \\ 0 & 0 & 0 & 0 \\ 0 & 0 & 0 & 0 \end{bmatrix}, \quad D_4 = \begin{bmatrix} 0 & 0 & -0.866 & 14.93\text{ in} \\ 0 & 0 & -0.500 & 8.62\text{ in} \\ 0.866 & 0.500 & 0 & -9.56\text{ in} \\ 0 & 0 & 0 & 0 \end{bmatrix},$$

and then, from Eq. (11.29), we obtain

$$\omega_1 = 0$$

$$\omega_2 = \omega_3 = \omega_4 = \begin{bmatrix} 0 & -0.200\text{ rad/s} & 0 & 0 \\ 0.200\text{ rad/s} & 0 & 0 & 0 \\ 0 & 0 & 0 & 0 \\ 0 & 0 & 0 & 0 \end{bmatrix}$$

$$\omega_5 = \omega_6 = \begin{bmatrix} 0 & -0.200\text{ rad/s} & 0.303\text{ rad/s} & -5.225\text{ in/s} \\ 0.200\text{ rad/s} & 0 & 0.175\text{ rad/s} & -3.017\text{ in/s} \\ -0.303\text{ rad/s} & -0.175\text{ rad/s} & 0 & 3.346\text{ in/s} \\ 0 & 0 & 0 & 0 \end{bmatrix}.$$

Now, using Eq. (11.30), we find that the velocity of the tool point is

$$\dot{R}_6 = \omega_6 T_{1,6} R_6 = \begin{bmatrix} -2.93\text{ in/s} \\ 1.24\text{ in/s} \\ -1.10\text{ in/s} \\ 0 \end{bmatrix}, \qquad \textit{Ans.}$$

which in vector notation is

$$\mathbf{V} = -2.93\hat{\mathbf{i}}_1 + 1.24\hat{\mathbf{j}}_1 - 1.10\hat{\mathbf{k}}_1\text{ in/s}. \qquad \textit{Ans.}$$

The accelerations can be obtained in similar fashion. From Eq. (11.33), we obtain

$$\alpha_1 = \alpha_2 = \alpha_3 = \alpha_4 = 0$$

$$\alpha_5 = \alpha_6 = \begin{bmatrix} 0 & 0 & -0.035 \text{ rad/s}^2 & 0.603 \text{ in/s}^2 \\ 0 & 0 & 0.152 \text{ rad/s}^2 & -1.045 \text{ in/s}^2 \\ 0.035 \text{ rad/s}^2 & -0.152 \text{ rad/s}^2 & 0 & 0 \\ 0 & 0 & 0 & 0 \end{bmatrix}.$$

Finally, using Eq. (11.34), we find that the acceleration of the tool point is

$$\ddot{R}_6 = (\alpha_6 + \omega_6\omega_6)T_{16}R_6 = \begin{bmatrix} -0.390 \text{ in/s}^2 \\ -0.033 \text{ in/s}^2 \\ 0.088 \text{ in/s}^2 \\ 0 \end{bmatrix}, \qquad \textit{Ans.}$$

which in vector notation is

$$\mathbf{A} = -0.390\hat{\mathbf{i}}_1 - 0.033\hat{\mathbf{j}}_1 + 0.088\hat{\mathbf{k}}_1 \text{ in/s}^2. \qquad \textit{Ans.}$$

12.4 INVERSE POSITION ANALYSIS

The previous section demonstrates the basic procedure for finding the positions, velocities, and accelerations of arbitrary points on the moving members of a robot once the positions, velocities, and accelerations of the joint actuators are known. The procedures are straightforward and simply applied if, as is usual, the robot is a simply connected chain and has no closed loops. In this simple case every joint variable is an independent degree of freedom, and no loop-closure equations must be solved. This avoids a major complication.

However, although the robot itself might have no closed loops, the manner in which a problem or question is presented can sometimes lead to the same complications. Consider, for example, an open-loop robot that we wish to guide to follow a specified path; we will be given the desired path, but we will not know the actuator values (or, more precisely, the time functions) required to achieve this. This problem cannot be solved by the methods of the previous section. When the joint variables (or their derivatives) are the unknowns of the problem rather than given information, the problem is called an *inverse kinematics* problem. In general, this is a more complicated problem to solve, and it does arise repeatedly in robot applications.

When inverse kinematics problems arise, we must of course find a set of equations that describe the given situation and that can be solved. In robotics this usually means that there will be one or more constraint equations, not necessarily defined by closed loops within the robot topology, but perhaps defined by the manner in which the problem is posed. For example, if we are told that the end effector of the robot is to travel along a certain path with certain timing, then we are being told the values required for the transformation $T_{1,n+1}$ as known functions of time. Then Eq. (12.2) becomes a set of required constraints that must be satisfied and that must be solved to find the joint actuator values or functions of time. An example will make the procedure clear.

EXAMPLE 12.3

The Microbot robot of Fig. 12.6 described in Example 12.1 is to be guided along a path for which the origin of the end effector O_6 follows the straight line given by

$$\mathbf{R}_{O_6}(t) = (4.0 + 1.6t)\hat{\mathbf{i}}_1 + (3.0 + 1.2t)\hat{\mathbf{j}}_1 + 2.0\hat{\mathbf{k}}_1 \text{ in},$$

with t varying from 0.0 to 5.0 s; the orientation of the end effector is to remain constant with $\hat{\mathbf{i}}_6 = \hat{\mathbf{k}}_1$ (vertical) and $\hat{\mathbf{k}}_6$ radially outward from the base of the robot. Find expressions for how each of the joint actuators must be driven, as functions of time, to achieve this motion.

SOLUTION

From the problem requirements stated, we can express the required time history of the end effector coordinate system by the transformation matrix

$$T_{1,6}(t) = \begin{bmatrix} 0 & 0.600 & 0.800 & 4.0 + 1.6t \text{ in} \\ 0 & -0.800 & 0.600 & 3.0 + 1.2t \text{ in} \\ 1 & 0 & 0 & 2.0 \text{ in} \\ 0 & 0 & 0 & 1 \end{bmatrix}. \tag{1}$$

We can also multiply out the matrix description for T_{16} from Eq. (12.2); this gives

$$T_{1,6} = \begin{bmatrix} \begin{matrix}\cos\phi_1\cos\beta\cos\phi_5 \\ -\sin\phi_1\sin\phi_5\end{matrix} & \begin{matrix}-\cos\phi_1\cos\beta\sin\phi_5 \\ +\sin\phi_1\cos\phi_5\end{matrix} & \cos\phi_1\sin\beta & \begin{matrix}7\cos\phi_1\cos\phi_2 \\ +7\cos\phi_1\cos(\phi_2+\phi_3) \\ +3.8\cos\phi_1\sin\beta\end{matrix} \\ \begin{matrix}\sin\phi_1\cos\beta\cos\phi_5 \\ -\cos\phi_1\sin\phi_5\end{matrix} & \begin{matrix}-\sin\phi_1\cos\beta\sin\phi_5 \\ -\cos\phi_1\cos\phi_5\end{matrix} & \sin\phi_1\sin\beta & \begin{matrix}7\sin\phi_1\cos\phi_2 \\ +7\sin\phi_1\sin(\phi_2+\phi_3) \\ +3.8\sin\phi_1\sin\beta\end{matrix} \\ \sin\beta\cos\phi_5 & -\sin\beta\sin\phi_5 & -\cos\beta & \begin{matrix}7\sin\phi_2 \\ +7\sin(\phi_2+\phi_3) \\ -3.8\cos\beta\end{matrix} \\ 0 & 0 & 0 & 1 \end{bmatrix}, \tag{2}$$

where, to save space, we have adopted the definition

$$\beta = \phi_2 + \phi_3 + \phi_4. \tag{3}$$

Now, to achieve the problem requirements, we must find expressions for the joint variables ϕ_i that will ensure that the elements of Eq. (2) are equal to those of Eq. (1) for all values of time; this is our required constraint condition. Equating the ratios of the elements of the second row, third column, with the elements of the first row, third column, of Eqs. (2) and (1) gives

$$\tan\phi_1 = \frac{0.600}{0.800} = 0.750.$$

Therefore, the first joint variable is

$$\phi_1 = \tan^{-1} 0.750 = 36.87°. \qquad \textit{Ans.}$$

Similarly, from the ratios of the elements of the third row, second column, with the elements of the third row, first column, we find

$$\tan \phi_5 = 0$$
$$\phi_5 = 0° \text{ or } \pm 180°.$$

Because the problem gives no preference for the orientation of the end effector in the horizontal plane, we are free to choose

$$\phi_5 = 0°. \qquad \textit{Ans.}$$

Next, from the third row, third column, elements we obtain

$$\cos \beta = 0$$
$$\beta = \phi_2 + \phi_3 + \phi_4 = \pm 90°.$$

As above, because the orientation of the end effector is not specified, we will choose

$$\beta = \phi_2 + \phi_3 + \phi_4 = 90°. \tag{4}$$

Turning now to the elements of the fourth column and simplifying according to already known results, we obtain two more independent equations:

$$7 \cos \phi_2 + 7 \cos(\phi_2 + \phi_3) = 1.2 + 2t \tag{5}$$

$$7 \sin \phi_2 + 7 \sin(\phi_2 + \phi_3) = -5.68. \tag{6}$$

Squaring and adding these, we get

$$98 + 98 \cos \phi_3 = 33.7024 + 4.8t + 4t^2, \tag{7}$$

which gives a solution for ϕ_3:

$$\phi_3 = \cos^{-1}(-0.65610 + 0.04898t + 0.040816t^2). \qquad \textit{Ans.} \ (8)$$

Because this form admits to multiple values, we will take the smallest solution with $\phi_3 \leq 0$ so that the robot elbow is kept above the path. At time $t = 0$, for example, the value of ϕ_3 is $\phi_3 = \cos^{-1}(-0.65610) = -131.00°$.

Knowing the solution for ϕ_3, we can now rewrite Eqs. (5) and (6) as follows:

$$\sin \phi_2 = \frac{-5.68}{14} - \frac{(1.2 + 2t) \sin \phi_3}{14(1 + \cos \phi_3)} \tag{9}$$

$$\cos \phi_2 = \frac{(1.2 + 2t)}{14} - \frac{5.68 \sin \phi_3}{14(1 + \cos \phi_3)}. \tag{10}$$

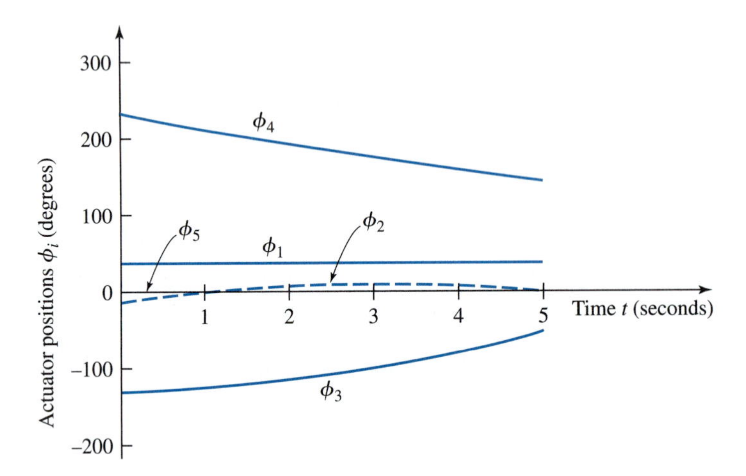

Figure 12.7 Example 12.3; graphs of inverse position solutions.

Because the third joint variable ϕ_3 now has a known solution, either of these equations can be used to find ϕ_2. Actually, both should be evaluated to ensure that the proper quadrant is found. These two equations are considered solutions although they are not explicit, closed-form functions of time.

Similarly, from Eq. (4), we can now find the fourth joint variable:

$$\phi_4 = 90^\circ - \phi_2 - \phi_3. \qquad \textit{Ans.}\ (11)$$

Graphs of the five joint variables versus time are illustrated in Fig. 12.7. Note that through the path being followed, the joint variables follow very smooth but nonlinear curves. Thus, it appears that the joint variable velocities and accelerations will not be zero, but should fall within very reasonable limits.

12.5 INVERSE VELOCITY AND ACCELERATION ANALYSES

The above example presents the general approach for inverse position analysis. However, at least for control purposes, it may also be important to find the joint variable velocities and accelerations. One way to do this would be to directly differentiate the joint variable solution equations with respect to time. However, another more general approach is to differentiate our constraint conditions.

As we saw in the above example, these were given by Eq. (12.2) as

$$T_{1,2}T_{2,3}\cdots T_{n,n+1} = T_{1,n+1}(t),$$

where, for the inverse kinematics problem, the right-hand side of this equation consists of known functions of time given by the problem specifications, the trajectory, or the path to be followed. Now, using Eq. (11.25) and the D_i differentiation operator matrices of Eq. (11.27*b*), we can differentiate the left-hand side of the equation, and by direct differentiation with respect to time we can also differentiate the right-hand side. Thus, we obtain

$$\sum_{i=1}^{n} D_i \dot{\phi}_i T_{1,n+1} = \frac{dT_{1,n+1}(t)}{dt}. \tag{12.3}$$

If we postmultiply both sides of Eq. (12.3) by $T_{1,n+1}^{-1}$, we obtain

$$\sum_{i=1}^{n} D_i \dot{\phi}_i = \frac{dT_{1,n+1}(t)}{dt} T_{1,n+1}^{-1}. \tag{12.4}$$

For convenience, let us define the right-hand side by

$$\Omega = \frac{dT_{1,n+1}(t)}{dt} T_{1,n+1}^{-1}, \tag{12.5}$$

where Ω is a new (velocity) operator matrix. Then substituting Eq. (12.5) into Eq. (12.4) gives

$$\sum_{i=1}^{n} D_i \dot{\phi}_i = \Omega, \tag{12.6}$$

which, as we will see, can be solved for the joint actuator velocities $\dot{\phi}_i$.

Let us now look more carefully at the nature of Eq. (12.6). Each of the matrices has four rows and four columns. Yet, because they represent velocities in a single chain in three dimensions, there should only be six independent equations, three for translations and three for rotations. Looking at the definitions of the Q_i differentiation operator matrices in Eq. (11.23) and reviewing the form of the D_i matrices in Examples 11.5 and 12.2, we can discover which of the elements of these matrices contain the six independent equations. Although the proof is left as an exercise, it can be demonstrated that the D_i matrices always have a bottom row of zeroes and the upper-left three rows and three columns are always antisymmetric; that is, they are always of the form[1]

$$D_i = \begin{bmatrix} 0 & -c & +b & d \\ +c & 0 & -a & e \\ -b & +a & 0 & f \\ 0 & 0 & 0 & 0 \end{bmatrix}. \tag{12.7}$$

As expected, there are only six independent values (a, b, c, d, e, and f) in each of these matrices. It will now be convenient to rearrange these into a single column and to give this column a new symbol:

$$\{D_i\} = \begin{bmatrix} a \\ b \\ c \\ d \\ e \\ f \end{bmatrix}, \tag{12.8}$$

where the braces indicate that the elements of the 4×4 matrix have been rearranged into a 6×1 column matrix. There is actually much more mathematical and physical significance to this 6×1 column vector than is explained here; they form the *Plücker coordinates* or *line coordinates* of the motion axis of the joint variable's freedom for motion.[2] The top three elements form the three-dimensional unit vector for the motion axis, whereas the bottom three uniquely locate that axis in the absolute coordinate system.

If we now reorganize all matrices in Eq. (12.6) into this column vector form, it becomes

$$\sum_{i=1}^{n} \{D_i\}\,\dot{\phi}_i = \{\Omega\}, \tag{12.9}$$

which is a set of six simultaneous algebraic equations relating the joint variable velocities. If we write the equations out explicitly, they appear as

$$\begin{aligned}
a_1\dot{\phi}_1 + a_2\dot{\phi}_2 + \cdots + a_n\dot{\phi}_n &= a \\
b_1\dot{\phi}_1 + b_2\dot{\phi}_2 + \cdots + b_n\dot{\phi}_n &= b \\
c_1\dot{\phi}_1 + c_2\dot{\phi}_2 + \cdots + c_n\dot{\phi}_n &= c \\
d_1\dot{\phi}_1 + d_2\dot{\phi}_2 + \cdots + d_n\dot{\phi}_n &= d \\
e_1\dot{\phi}_1 + e_2\dot{\phi}_2 + \cdots + e_n\dot{\phi}_n &= e \\
f_1\dot{\phi}_1 + f_2\dot{\phi}_2 + \cdots + f_n\dot{\phi}_n &= f,
\end{aligned}$$

where a, b, and so on are the elements of $\{\Omega\}$ taken in the order consistent with Eqs. (12.7) and (12.8). Regrouping the coefficients of this set of equations into matrix form, we get

$$J = \begin{bmatrix} a_1 & a_2 & \ldots & a_n \\ b_1 & b_2 & \ldots & b_n \\ c_1 & c_2 & \ldots & c_n \\ d_1 & d_2 & \ldots & d_n \\ e_1 & e_2 & \ldots & e_n \\ f_1 & f_2 & \ldots & f_n \end{bmatrix}, \tag{12.10}$$

which is called the *Jacobian* of the system. Using this matrix, Eq. (12.9) now becomes

$$J\left\{\dot{\phi}\right\} = \{\Omega\}, \tag{12.11}$$

where $\left\{\dot{\phi}\right\}$ is a column of the unknown joint actuator velocities $\dot{\phi}_i$ taken in order. This set of equations specifies the constraint conditions that must exist between the joint variable velocities to trace the specified path according to the given functions of time.

Assuming that the position analysis has already been completed, the Ω and D_i matrices are all known functions of time, and this set of equations can now be solved for the joint variable velocities as functions of time. If the robot has six independent joint variables, then the solution can be determined by inverting the Jacobian matrix. In this case, the solution is

$$\left\{\dot{\phi}\right\} = J^{-1}\{\Omega\}, \tag{12.12}$$

where J^{-1} is the inverse of the Jacobian matrix in Eq. (12.10).

If the Jacobian is not a square matrix, then two cases are possible: (1) when there are less than six joint variables, other solution methods, such as Gaussian elimination, must be used. (2) When there are more than six joint variables, the problem has no unique solution; further equations, perhaps even arbitrary conditions, must be supplied. If the J matrix is square but singular or if the equations are inconsistent, there is no finite unique solution for velocities; the problem specifications are not realistic for this robot to do the task specified. If none of these problems arises, the joint variable velocities become known functions of time.

Joint variable accelerations can be determined by completely analogous methods. Taking the next time derivative of Eq. (12.6), we see that

$$\sum_{i=1}^{n} D_i\ddot{\phi}_i = \frac{d\Omega}{dt} - \sum_{i=1}^{n}(\omega_i D_i - D_i\omega_i)\dot{\phi}_i = A, \tag{12.13}$$

where this equation defines a new matrix A and the ω_i matrices are defined as in Eqs. (11.29). Extracting the same six independent equations, in the same order as before, we get

$$\sum_{i=1}^{n}\{D_i\}\ddot{\phi}_i = \{A\}. \tag{12.14}$$

Now we note that the coefficients of the unknown $\ddot{\phi}_i$ joint variable accelerations are identical to those of Eq. (12.9); thus, we can use the same Jacobian matrix defined in Eq. (12.10). The solution to these equations now becomes

$$\{\ddot{\phi}\} = J^{-1}\{A\}, \tag{12.15}$$

where $\{\ddot{\phi}\}$ is a column of the unknown joint variable accelerations $\ddot{\phi}_i$ taken in order. This set of equations specifies the relations that must exist between the joint variable accelerations required to trace the specified path according to the given functions of time.

As with direct kinematics problems, once the joint variable velocities and accelerations are known, the velocity and acceleration of a point on one of the moving links can be determined from Eqs. (11.29) through (11.34).

EXAMPLE 12.4

Continue with Example 12.3 and find the velocities required at the actuators as functions of time to achieve the motion described.

SOLUTION

Using Eq. (11.27b) and the solutions for the position analysis from the above example, we can find the D_i matrices. These are

$$D_1 = \begin{bmatrix} 0 & -1 & 0 & 0 \\ 1 & 0 & 0 & 0 \\ 0 & 0 & 0 & 0 \\ 0 & 0 & 0 & 0 \end{bmatrix}, \quad D_2 = \begin{bmatrix} 0 & 0 & -0.8 & 6.144\text{ in} \\ 0 & 0 & -0.6 & 4.608\text{ in} \\ 0.8 & 0.6 & 0 & 0 \\ 0 & 0 & 0 & 0 \end{bmatrix}$$

$$D_3 = \begin{bmatrix} 0 & 0 & -0.8 & (5.6\sin\phi_2 + 6.144)\text{ in} \\ 0 & 0 & -0.6 & (4.2\sin\phi_2 + 4.608)\text{ in} \\ 0.8 & 0.6 & 0 & -7\cos\phi_2\text{ in} \\ 0 & 0 & 0 & 0 \end{bmatrix}$$

$$D_4 = \begin{bmatrix} 0 & 0 & -0.8 & 1.6\text{ in} \\ 0 & 0 & -0.6 & 1.2\text{ in} \\ 0.8 & 0.6 & 0 & -(2t + 1.2)\text{ in} \\ 0 & 0 & 0 & 0 \end{bmatrix}, \quad D_5 = \begin{bmatrix} 0 & 0 & 0.6 & -1.2\text{ in} \\ 0 & 0 & -0.8 & 1.6\text{ in} \\ -0.6 & 0.8 & 0 & 0 \\ 0 & 0 & 0 & 0 \end{bmatrix}. \tag{12}$$

Using Eq. (12.5) and the trajectory specified in Example 12.3, we can also find the Ω matrix.

$$\Omega = \begin{bmatrix} 0 & 0 & 0 & 1.6\text{ in/s} \\ 0 & 0 & 0 & 1.2\text{ in/s} \\ 0 & 0 & 0 & 0 \\ 0 & 0 & 0 & 0 \end{bmatrix} \tag{13}$$

From these and Eq. (12.9), we can now construct the set of equations relating the joint variable velocities:

$$\begin{bmatrix} 0 & 0.6 & 0.6 & 0.6 & 0.8 \\ 0 & -0.8 & -0.8 & -0.8 & 0.6 \\ 1 & 0 & 0 & 0 & 0 \\ 0 & 6.144\text{ in} & (5.6\sin\phi_2 + 6.144)\text{ in} & 1.6\text{ in} & -1.2\text{ in} \\ 0 & 4.608\text{ in} & (4.2\sin\phi_2 + 4.608)\text{ in} & 1.2\text{ in} & 1.6\text{ in} \\ 0 & 0 & -7\cos\phi_2\text{ in} & -(2t + 1.2)\text{ in} & 0 \end{bmatrix} \begin{bmatrix} \dot{\phi}_1 \\ \dot{\phi}_2 \\ \dot{\phi}_3 \\ \dot{\phi}_4 \\ \dot{\phi}_5 \end{bmatrix} = \begin{bmatrix} 0 \\ 0 \\ 0 \\ 1.6\text{ in/s} \\ 1.2\text{ in/s} \\ 0 \end{bmatrix}. \tag{14}$$

Here we see the Jacobian matrix for the robot following the specified trajectory. However, in this example it is not square and cannot be inverted. This happened because the robot has only five degrees of freedom and, therefore, is not capable of arbitrary spatial motion; there are six equations and only five unknowns, the five actuator velocities.

Fortunately, however, the trajectory specified is within the capability of this robot. There is a set of solutions for the five unknown actuator velocities that fits all six equations; these solutions are

$$\dot{\phi}_1 = 0 \qquad \textit{Ans.}$$

$$\dot{\phi}_2 = \frac{(2t + 1.2) - 7\cos\phi_2}{\Delta}\text{ rad/s} \qquad \textit{Ans.}$$

$$\dot{\phi}_3 = \frac{-(2t + 1.2)}{\Delta}\text{ rad/s} \qquad \textit{Ans.}$$

$$\dot{\phi}_4 = \frac{7\cos\phi_2}{\Delta}\text{ rad/s} \qquad \textit{Ans.}$$

$$\dot{\phi}_5 = 0, \qquad \textit{Ans.}$$

where Δ is defined as

$$\Delta = -3.5(2t + 1.2)\sin\phi_2 - 19.9\cos\phi_2. \qquad (15)$$

As with the position solutions, we agree that the velocity expressions shown are solutions because the parameter ϕ_2 is already known from the position analysis.

12.6 ROBOT ACTUATOR FORCE ANALYSIS

Of course, the purpose of moving the robot gripper along the planned trajectory is to perform some useful function, and this almost certainly requires the expenditure of work or power. The source of this work or power must come from the actuators at the joint variables. Therefore, it would be very helpful to find a means of knowing what size forces and/or torques must be exerted by the actuators to perform the given task. Conversely, it would be useful to know how much force is produced at the tool for a given set of forces or torques applied by the actuators.

If the force or torque capacities of the actuators is less than those demanded by the task being attempted, the robot is not capable of performing the task. It will probably not fail catastrophically, but it will deviate from the desired trajectory and perform a different motion than that desired. The purpose of this section is to find a means of evaluating the forces required of the robot actuators to perform a given trajectory with a given task loading, so that overloading of the actuators can be anticipated and avoided.

The entire subject of force analysis in mechanical systems is covered in much more depth in Chapters 13 and 14. Therefore, the treatment here is limited by simplifying assumptions. It is suited specifically to the study of robots performing specified tasks with specified loads at low speeds with no friction or other losses. If these assumptions do not apply, the more extensive treatments of the later chapters should be employed.

The interaction of the robot with the task being performed produces a set of forces and torques at the end effector or tool. Let us assume that these required forces and torques are known functions of time and are grouped into a given six-element column $\{F\}$ in the following order:

$$\{F\} = \begin{bmatrix} M^{x_1} \\ M^{y_1} \\ M^{z_1} \\ F^{x_1} \\ F^{y_1} \\ F^{z_1} \end{bmatrix}, \qquad (12.16)$$

where the first three elements are the components of the required tool torque about the x_1, y_1, z_1 axes and the next three are the components of the required force along the same axes.

The source of energy for these needs will be a set of unknown forces or torques τ_i, each one applied by the actuator at the ith joint variable location. We will arrange these into another column matrix τ in the order of numbering of the joint variables:

$$\tau = \begin{bmatrix} \tau_1 \\ \tau_2 \\ \vdots \\ \tau_n \end{bmatrix}. \tag{12.17}$$

Next, we define a small (absolute) displacement of the tool against the loads by another six-element column matrix $\{\delta R\}$, arranged in the same order as the load vector $\{F\}$,

$$\{\delta R\} = \begin{bmatrix} \delta\theta^{x_1} \\ \delta\theta^{y_2} \\ \delta\theta^{z_1} \\ \delta R^{x_1} \\ \delta R^{y_1} \\ \delta R^{z_1} \end{bmatrix} \tag{12.18}$$

and yet another column matrix $\delta\phi$ of displacements of the joint variables $\delta\phi_i$ during this small displacement.

Now, if the actual robot were to undergo the small displacement $\{\delta R\}$ against the loads $\{F\}$, the work would have to come from the torques τ of the actuators acting through their small displacements $\delta\phi$. Therefore, because work input must equal work output in the absence of other losses, we have

$$\{F\}^T \{\delta R\} = \tau^T \delta\phi, \tag{a}$$

where the superscript T signifies the transpose of the matrix. But we also know by integrating Eq. (12.11) over a short time interval δt that the task displacement $\{\delta R\}$ is related to the joint actuator displacements $\{\delta\phi\}$ by

$$\{\delta R\} = J\delta\phi. \tag{12.19}$$

Substituting this relationship into Eq. (a) and rearranging, we get

$$\left[\tau^T - \{F\}^T J\right]\delta\phi = 0. \tag{b}$$

Because we assume that the small displacement $\delta\phi$ is not zero, we can solve for the actuator torques τ by setting the leading factor to zero. This shows that

$$\tau = J^T \{F\}, \tag{12.20}$$

which is the relationship we have been seeking. Under the simplifying conditions assumed above, we have determined not only that the Jacobian matrix relates the actuator motions to the absolute tool motions, Eqs. (12.12), (12.15), and (12.19), but also that its transpose relates the actuator torques and tool loads, Eq. (12.20).

EXAMPLE 12.5

Continue with Example 12.4 and assume that the end effector is working against a time-varying force loading of $10\hat{\mathbf{i}}_1 + 5t\hat{\mathbf{k}}_1$ lb in addition to a constant torque loading of $25\hat{\mathbf{k}}_1$ in · lb. Find the torques required at the actuators as functions of time to achieve the motion described.

SOLUTION

From the data given, the tool force vector of Eq. (12.16) is

$$\{F\} = \begin{bmatrix} 0 \\ 0 \\ 25 \text{ in} \cdot \text{lb} \\ 10 \text{ lb} \\ 0 \\ 5t \text{ lb} \end{bmatrix} \tag{16}$$

Because the Jacobian matrix was already found in Example 12.4, Eq. (14), we can now use Eq. (12.20) to find the actuator torques directly. We find them to be

$$\tau_1 = 25.0 \text{ in·lb} \qquad \textit{Ans.}$$

$$\tau_2 = 61.4 \text{ in·lb} \qquad \textit{Ans.}$$

$$\tau_3 = (-35t\cos\phi_2 + 56.0\sin\phi_2 + 61.4) \text{ in·lb} \qquad \textit{Ans.}$$

$$\tau_4 = (-10t^2 - 6t + 16.0) \text{ in·lb} \qquad \textit{Ans.}$$

$$\tau_5 = -12.0 \text{ in·lb.} \qquad \textit{Ans.}$$

12.7 REFERENCES

[1] J. J. Uicker, Jr., B. Ravani, and P. N. Sheth, *Matrix Methods in the Design Analysis of Mechanisms and MultiBody Systems*. New York: Cambridge University Press, 2010.

[2] For further reading, an excellent reference is K. H. Hunt, *Kinematic Geometry of Mechanisms*. New York: Oxford University Press, 1978, especially Chapter 11.

PROBLEMS

†12.1 For the SCARA robot shown in Fig. P12.1, find the transformation matrix T_{15} relating the position of the tool coordinate system to the ground coordinate system when the joint actuators are set to the values $\phi_1 = 30°$, $\phi_2 = -60°$, $\phi_3 = 2$ in., and $\phi_4 = 0$. Also find the absolute position of the tool point that has coordinates $x_5 = y_5 = 0$, $z_5 = 1.5$ in.

†12.2 Repeat Problem 12.1 using arbitrary (symbolic) values for the joint variables.

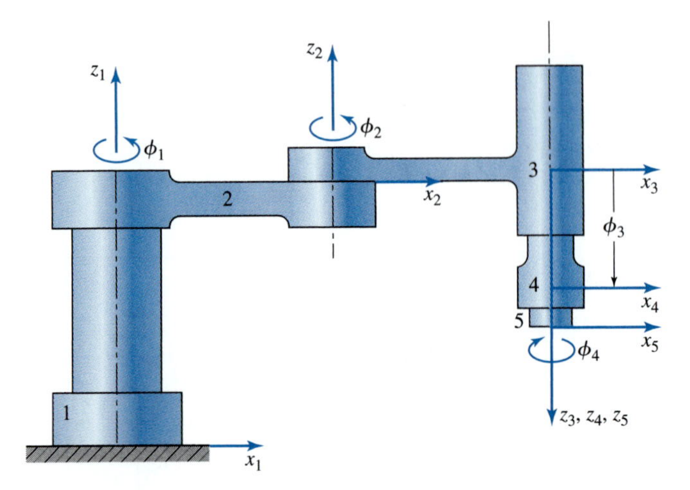

Figure P12.1 $a_{12} = a_{23} = 10$ in, $a_{34} = a_{45} = 0$, $\alpha_{12} = \alpha_{34} = \alpha_{45} = 0$, $\alpha_{23} = 180°$, $\theta_{12} = \phi_1$, $\theta_{23} = \phi_2$, $\theta_{34} = 0$, $\theta_{45} = \phi_4$, $s_{12} = 12$ in, $s_{23} = 0$, $s_{34} = \phi_3$, and $s_{45} = 2$ in.

†12.3 For the gantry robot shown in Fig. P12.3, find the transformation matrix T_{15} relating the position of the tool coordinate system to the ground coordinate system when the joint actuators are set to the values $\phi_1 = 450$ mm, $\phi_2 = 180$ mm, $\phi_3 = 50$ mm, and $\phi_4 = 0$. Also find the absolute position of the tool point that has coordinates $x_5 = y_5 = 0$, $z_5 = 45$ mm.

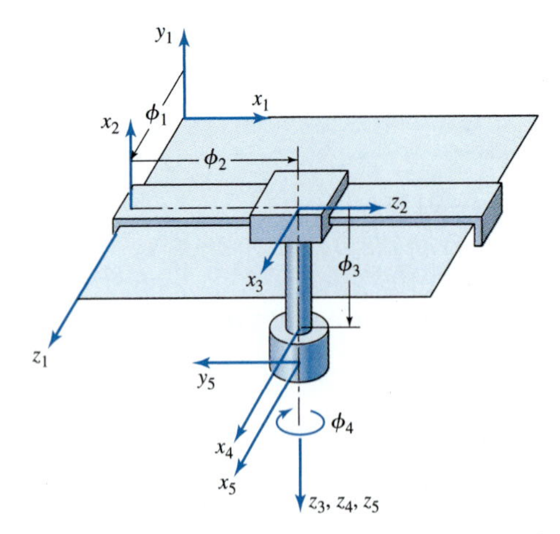

Figure P12.3 $a_{12} = a_{23} = a_{34} = a_{45} = 0$, $\alpha_{12} = 90°$, $\alpha_{23} = -90°$, $\alpha_{34} = \alpha_{45} = 0$, $\theta_{12} = \theta_{23} = 90°$, $\theta_{34} = 0$, $\theta_{45} = \phi_4$, $s_{12} = \phi_1$, $s_{23} = \phi_2$, $s_{34} = \phi_3$, and $s_{45} = 50$ mm.

†12.4 Repeat Problem 12.3 using arbitrary (symbolic) values for the joint variables.

†12.5 For the SCARA robot of Problem 12.1 in the position described, find the instantaneous velocity and acceleration of the same tool point, $x_5 = y_5 = 0$, $z_5 = 1.5$ in, if the actuators have (constant) velocities of $\dot{\phi}_1 = 0.20$ rad/s, $\dot{\phi}_2 = -0.35$ rad/s, and $\dot{\phi}_3 = \dot{\phi}_4 = 0$.

†12.6 For the gantry robot of Problem 12.3 in the position described, find the instantaneous velocity and acceleration of the same tool point, $x_5 = y_5 = 0$, $z_5 = 45$ mm, if the actuators have (constant) velocities of $\dot{\phi}_1 = \dot{\phi}_2 = 0$, $\dot{\phi}_3 = 40$ mm/s, and $\dot{\phi}_4 = 20$ rad/s.

†12.7 The SCARA robot of Problem 12.1 is to be guided along a path for which the origin of the end effector O_5 follows the straight line given by

$$\mathbf{R}_{O_5}(t) = (1.6t + 4.0)\hat{\mathbf{i}}_1 + (1.2t + 3.0)\hat{\mathbf{j}}_1 + 2.0\hat{\mathbf{k}}_1 \text{ in.},$$

with t varying from 0.0 to 5.0 s; the orientation of the end effector is to remain constant with $\hat{\mathbf{k}}_5 = -\hat{\mathbf{k}}_1$ (vertically downward) and $\hat{\mathbf{i}}_5$ radially outward from the base of the robot. Find expressions for how each of the actuators must be driven, as functions of time, to achieve this motion.

†12.8 The gantry robot of Problem 12.3 is to travel a path for which the origin of the end effector O_5 follows the straight line given by

$$\mathbf{R}_{O_5}(t) = (120t + 300)\hat{\mathbf{i}}_1 - 150\hat{\mathbf{j}}_1 + (90t + 225)\hat{\mathbf{k}}_1 \text{ mm},$$

with t varying from 0.0 to 4.0 s; the orientation of the end effector is to remain constant with $\hat{\mathbf{k}}_5 = -\hat{\mathbf{j}}_1$ (vertically downward) and $\hat{\mathbf{i}}_5 = \hat{\mathbf{i}}_1$. Find expressions for the positions of each of the actuators, as functions of time, for this motion.

†12.9 The end effector of the SCARA robot of Problem 12.1 is working against a force loading of $10\hat{\mathbf{i}}_1 + 5t\hat{\mathbf{k}}_1$ lb and a constant torque loading of $25\hat{\mathbf{k}}_1$ in · lb as it follows the trajectory described in Problem 12.7. Find the torques required at the actuators, as functions of time, to achieve the motion described.

†12.10 The end effector of the gantry robot of Problem 12.3 is working against a force loading of $20\hat{\mathbf{i}}_1 + 10t\hat{\mathbf{j}}_1$ N and a constant torque loading of $5\hat{\mathbf{j}}_1$ N · m as it follows the trajectory described in Problem 12.8. Find the torques required at the actuators, as functions of time, to achieve the motion described.

PART 3
Dynamics of Machines

13 Static Force Analysis

13.1 INTRODUCTION

In our studies of kinematic analysis in the previous chapters we limited ourselves to consideration of the geometry of the motions and of the relationships between displacement and time. The forces required to produce those motions or the motion that would result from the application of a given set of forces were not considered. We are now ready for a study of the dynamics of machines and systems. Such a study is usually simplified by starting with the statics of such systems.

In the design of a machine, consideration of only those effects that are described by units of *length* and *time* is a tremendous simplification. It frees the mind of the complicating influence of many other factors that ultimately enter into the problem, and it permits our attention to be focused on the primary goal, that of designing a mechanism to obtain a desired motion. That was the problem of *kinematics*, which was covered in the previous chapters of this book.

The fundamental units in kinematic analysis are length and time; in dynamic analysis they are *length*, *time*, and *force*.

Forces are transmitted between machine members through mating surfaces, that is, from a gear to a shaft or from one gear through meshing teeth to another gear, from a connecting rod through a bearing to a lever, from a V-belt to a pulley, from a cam to a follower, or from a brake drum to a brake shoe. It is necessary to know the magnitudes of these forces for a variety of reasons. The distribution of these forces at the boundaries of mating surfaces must be reasonable, and their intensities must remain within the working limits of the materials composing the surfaces. For example, if the force operating on a sleeve bearing becomes too high, it will squeeze out the oil film and cause metal-to-metal contact, overheating, and rapid failure of the bearing. If the forces between gear teeth are too large, the oil film may be squeezed out from between them. This could result in rough motion, noise, vibration, flaking and spalling of the metal, and eventual failure. In our study of dynamics we are interested principally in determining the magnitudes, directions, and locations of the forces, but not in sizing the members on which they act. The determination of the sizes of machine members is the subject of books usually titled machine design or mechanical design.[1]

Some of the key terms used in this phase of our study are defined as follows.

Force Our earliest ideas concerning forces arose from our desire to push, lift, or pull various objects. Thus, force is the action of one body acting on another. Our intuitive concept of force includes such ideas as *magnitude, direction*, and *point of application*, and these are commonly referred to as the *characteristics* of the force.

Matter Matter is a material substance. If it is closed and retains its shape, it is called a body.

Mass Newton defined the mass of a body as the *quantity of matter*. He recognized the fact that all bodies possess some inherent property that signifies the amount of matter and that is different from weight. Thus, a moon rock has a certain fixed amount of substance, although its moon weight is different from its earth weight. The amount of substance, or quantity of matter, is called the *mass* of the rock.

Inertia Inertia is the property of mass that causes it to resist any effort to change its motion.

Weight Weight is the force that results from gravity acting upon a mass.

Particle A particle is a body whose dimensions may be neglected. The dimensions may or may not be small; if the body is in translation then the motion of all points of the particle are equal and it may be considered located at a single point. A particle is not a point, however, in the sense that a particle can consist of matter and can have mass, whereas a point cannot.

Rigid body All real bodies are either elastic or plastic and will deform, although perhaps only slightly, when acted upon by forces. When the deformation of such a body is small enough to be neglected, such a body is frequently assumed to be *rigid*—that is, incapable of deformation—to simplify the analysis. This assumption of rigidity is the key step that allows the treatment of kinematics without consideration of forces. Without this simplifying assumption of rigidity, forces and motions are interdependent and kinematic and dynamic analyses require simultaneous solution.

Deformable body The rigid-body assumption cannot be maintained when internal stresses and strains caused by applied forces are to be analyzed. If stress is to be found, we must admit to the existence of strain; thus, we must consider the body capable of deformation, although small. If deformations are small in comparison to the gross dimensions and motion of the body, we can still treat the body as rigid while performing the kinematic (motion) analysis, but must then consider it deformable when stresses are sought. Using the additional assumption that the forces and stresses remain within the elastic range, such analysis is frequently called *elastic-body analysis*.

13.2 NEWTON'S LAWS

As stated in the *Principia*[2], Newton's three laws are as follows:

> Law I. Every body perseveres in its state of rest, or of uniform motion in a straight line, unless it is compelled to change that state by forces impressed thereon.
> Law II. The alteration of motion is proportional to the motive force impressed; and takes place in the direction of the straight line in which that force is impressed.
> Law III. To every action there is always an equal opposite reaction; or the mutual actions of two bodies upon each other are always equal, and directly opposed.

For our purposes, it convenient to restate these laws:

Law 1. If all the forces acting on a body are balanced, that is, sum to zero, then the particle either remains at rest or continues to move in a straight line at a uniform velocity.

Law 2. If the forces acting on a body are not balanced, the body experiences an acceleration proportional to and in the direction of the resultant force.

Law 3. When two particles interact, a pair of reaction forces comes into existence; these forces have equal magnitudes and opposite senses, and they act along the straight line common to the two particles.

Newton's first two laws can be summarized by the equation

$$\sum \mathbf{F}_{ij} = m_j \mathbf{A}_{G_j}, \tag{13.1}$$

which is called *the equation of motion* for body j. In this equation, $\mathbf{A}_{G_j}$ is the absolute acceleration of the center of mass of the body j that has mass m_j, and this acceleration is produced by the sum of all forces acting on the body, that is, all values of the index i. Both $\mathbf{F}_{ij}$ and $\mathbf{A}_{G_j}$ are vector quantities.

13.3 SYSTEMS OF UNITS

An important use of Eq. (13.1) occurs in the standardization of systems of units. Let us employ the following symbols to designate the units associated with a physical quantity:

Length	L
Time	T
Mass	M
Force	F

These symbols are to stand for any unit we may choose to use for that respective quantity. Thus, possible choices for L are inches, feet, miles, millimeters, meters, or kilometers. The symbols L, T, M, and F are not numbers. However, they can be substituted into any equation as if they were. The equality sign then states that the units on one side must be equivalent to those on the other. Making the indicated substitutions into Eq. (13.1) gives

$$F = MLT^{-2} \tag{13.2}$$

because the acceleration **A** has units of length divided by time squared. Equation (13.2) expresses an equivalence relationship among the four units, force, mass, length, and time. We are free to choose units for three of these, and then the units used for the fourth depend on those used for the first three. For this reason, the first three units chosen are called *basic units*, whereas the fourth is called the *derived unit*.

When force, length, and time are chosen as basic units, then mass is the derived unit, and the system is called a *gravitational system of units*. When mass, length, and time are chosen as the basic units, then force is the derived unit, and the system is called an *absolute system of units*.

In many English-speaking countries the U.S. customary foot-pound-second (fps) system and the inch-pound-second (ips) system are two gravitational systems of units that are still in use by many practicing engineers*. In the fps system the (derived) unit of mass is

$$M = FT^2/L = \text{lbf} \cdot \text{s}^2/\text{ft} = \text{slug}. \tag{13.3}$$

Thus, force, length, and time are the three basic units in the U.S. customary system. The unit of force is the pound, more properly the pound force. We sometimes, but not always, abbreviate this unit *lbf*; the abbreviation *lb* is also permissible† because we shall be dealing with U.S. customary gravitational systems.

Finally, we note in Eq. (13.3) that the derived unit of mass in the fps gravitational system is the lbf · s²/ft, sometimes referred to as a slug; there is no abbreviation for slug.

The derived unit of mass in the ips gravitational systems is

$$M = FT^2/L = \text{lbf} \cdot \text{s}^2/\text{in}. \tag{13.4}$$

Note that this unit of mass has not been given an official name.

The International System, or Systeme Internationale (SI),†† of units is an absolute system. The three basic units are the meter (m), the kilogram mass (kg), and the second (s). The unit of force is derived and is called a *Newton* to distinguish it from the kilogram, which, as indicated, is a unit of mass. The dimensions of the Newton (N) are

$$F = ML/T^2 = \text{kg} \cdot \text{m/s}^2 = \text{N}. \tag{13.5}$$

The weight of an object is the force exerted upon it by gravity. Designating the weight **W** and the acceleration caused by gravity **g**, Eq. (13.1) becomes

$$\mathbf{W} = m\mathbf{g}.$$

In the fps system, standard gravity is $g = 32.174\ 0\ \text{ft/s}^2$, which for most cases is approximated as $32.2\ \text{ft/s}^2$. Therefore, the weight of a mass of one slug under standard

* Many prefer to use gravitational systems; this helps to explain some of the resistance to the use of SI units, because the International System (SI) is an absolute system.

† The abbreviation *lb* for pound comes from the Latin word, *Libra*, the balance, the seventh sign of the zodiac, which is represented as a pair of scales.

†† SI stands for Système Internationale d'Unités (French).

gravity is

$$W = mg = (1 \text{ slug})(32.2 \text{ ft/s}^2) = 32.2 \text{ lb}.$$

In the ips system, standard gravity is 386.088 in/s^2 or approximately 386 in/s^2. Thus, in standard gravity, a unit mass weighs

$$W = mg = (1 \text{ lbf} \cdot \text{s}^2/\text{in})(386 \text{ in/s}^2) = 386 \text{ lb}.$$

In the International System of units, standard gravity is 9.806 65 m/s^2 which for most cases is approximated to 9.81 m/s^2. Therefore, the weight of a mass of 1 kg, under standard gravity, is

$$W = mg = (1 \text{ kg})(9.81 \text{ m/s}^2) = 9.81 \text{ kg} \cdot \text{m/s}^2 = 9.81 \text{ N}.$$

It is convenient to remember that the weight of a large apple is approximately 1 N.

13.4 APPLIED AND CONSTRAINT FORCES

When two or more bodies are connected to form a group or system, the action and reaction forces between any two of the connected bodies are called *constraint forces*. These forces constrain the connected bodies to behave in a specific manner defined by the nature of the connection. Forces acting on this system of bodies from outside the system are called *applied* forces.

Electrical, magnetic, and gravitational forces are examples of forces that may be applied without actual physical contact. However, a great many, if not most, of the forces with which we are concerned in mechanical equipment occur through direct physical or mechanical contact.

As we will see in the next section, the constraint forces of action and reaction at a mechanical contact occur in equal but opposite pairs and thus have no *net* force effect on the system of bodies being considered. Such pairs of constraint forces, although they clearly exist and may be large, are usually not considered further when both the action and the reaction forces act on bodies of the system being considered. However, when we try to consider a body, or a system of bodies, to be isolated from its surroundings, only one of each pair of constraint forces acts on the system being considered at any point of contact on the separation boundary. The other constraint force, the reaction, is left acting on the surroundings. When we isolate the system being considered, these constraint forces at points of separation must be clearly identified, and they are essential to further study of the dynamics of the isolated system of bodies.

As indicated earlier, the *characteristics* of a force are its *magnitude*, its *direction*, and its *point of application*. The direction of a force includes the concept of a line along which the force is acting and a *sense*. Therefore, a force may be directed either positively or negatively along its line of action.

Sometimes the position of the point of application along the line of action is not important, for example, when we are studying the equilibrium of a rigid body. Thus, in Fig. 13.1*a* it does not matter whether we diagram the force pair $\mathbf{F}_1\mathbf{F}_2$ as if they compress the link or if we diagram them as putting the link in tension, provided we are interested only in the

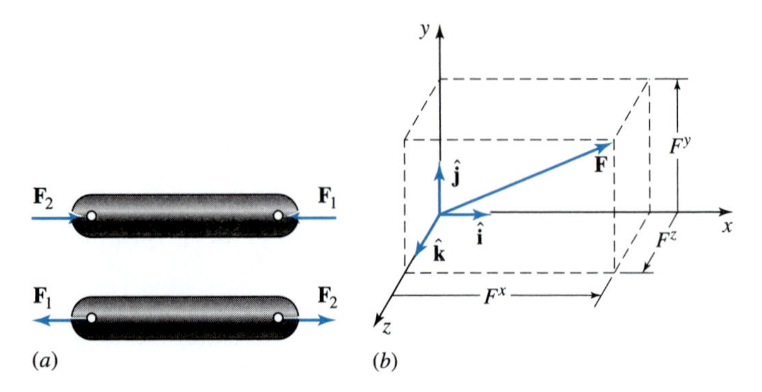

Figure 13.1 (*a*) Points of application of forces $\mathbf{F}_1$ and $\mathbf{F}_2$ to a rigid body may or may not be important. (*b*) The rectangular components of a force vector.

equilibrium of the link. Of course, if we are also interested in the internal link stresses, these forces cannot be interchanged.

The notation to be used for force vectors is illustrated in Fig. 13.1*b*. Boldface letters are used for force vectors and lightface letters for their magnitudes. Thus, the components of a force vector are

$$\mathbf{F} = F^x\hat{\mathbf{i}} + F^y\hat{\mathbf{j}} + F^z\hat{\mathbf{k}}. \tag{a}$$

Note that, as in earlier chapters of this text, the directions of components are indicated by superscripts, not subscripts.

Two equal and opposite forces along two parallel but *noncollinear* straight lines in a body cannot be combined to produce a single (null) resultant force on the body. Any two such forces acting on the body constitute a *couple*. The *arm of the couple* is the perpendicular distance between their lines of action, shown as h in Fig. 13.2*a*, and *the plane of the couple* is the plane containing the two lines of action.

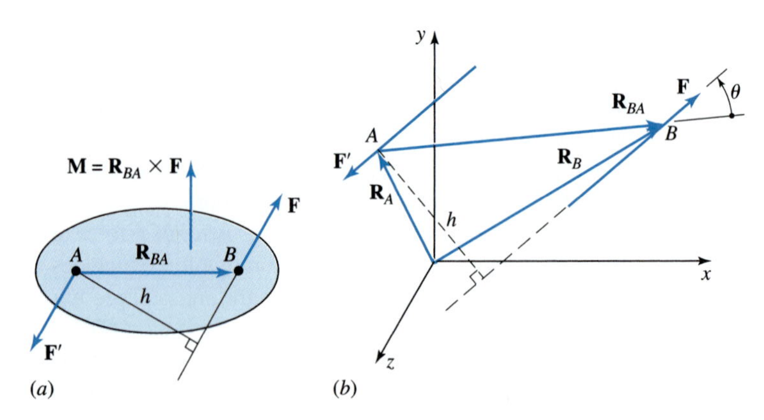

Figure 13.2 (*a*) $\mathbf{R}_{BA}$ is a position difference vector, whereas $\mathbf{F}$ and $\mathbf{F}'$ are force vectors. The free vector $\mathbf{M}$ is the moment of the couple formed by $\mathbf{F}$ and $\mathbf{F}'$. (*b*) A force couple formed by $\mathbf{F}$ and $\mathbf{F}'$.

The *moment* of a couple is another vector **M** directed normal to the plane of the couple; the sense of **M** is in accordance with the right-hand rule for rotation. The magnitude of the moment is the product of the arm h and the magnitude of one of the forces F. Thus,

$$M = hF. \tag{13.6}$$

In vector form the moment **M** is the cross-product of the position difference vector $\mathbf{R}_{BA}$ and the force vector **F**, and so it is defined by the equation

$$\mathbf{M} = \mathbf{R}_{BA} \times \mathbf{F}. \tag{13.7}$$

Some interesting properties of couples can be determined by an examination of Fig. 13.2*b*. Here **F** and $\mathbf{F}'$ are two equal-sized, opposite, and parallel forces. We can choose any point on each line of action and define the positions of these points by the vectors $\mathbf{R}_A$ and $\mathbf{R}_B$. Then the "relative" position, or position difference, vector is

$$\mathbf{R}_{BA} = \mathbf{R}_B - \mathbf{R}_A. \tag{b}$$

The moment of the couple is the sum of the moments of the two forces and is

$$\mathbf{M} = \mathbf{R}_A \times \mathbf{F}' + \mathbf{R}_B \times \mathbf{F}. \tag{c}$$

But $\mathbf{F}' = -\mathbf{F}$ and, therefore, using Eq. (*b*), Eq. (*c*) can be written as

$$\mathbf{M} = (\mathbf{R}_B - \mathbf{R}_A) \times \mathbf{F} = \mathbf{R}_{BA} \times \mathbf{F}. \tag{d}$$

Equation (*d*) indicates the following.

1. The value of the moment of the couple is independent of the choice of the reference point about which the moments are taken, because the vector $\mathbf{R}_{BA}$ is the same for all positions of the origin.
2. Because $\mathbf{R}_A$ and $\mathbf{R}_B$ define any choices of points on the lines of action of the two forces, the vector $\mathbf{R}_{BA}$ is not restricted to perpendicularity with **F** and $\mathbf{F}'$. This is a very important characteristic of the vector product because it demonstrates that the value of the moment is independent of how $\mathbf{R}_{BA}$ is chosen. The magnitude of the moment can be obtained as follows: resolve $\mathbf{R}_{BA}$ into two components $\mathbf{R}^n_{BA}$ and $\mathbf{R}^t_{BA}$, perpendicular and parallel to **F**, respectively. Then,

$$\mathbf{M} = (\mathbf{R}^n_{BA} + \mathbf{R}^t_{BA}) \times \mathbf{F}.$$

But $\mathbf{R}^n_{BA}$ is the perpendicular between the two lines of action and $\mathbf{R}^t_{BA}$ is parallel to **F**. Therefore, $\mathbf{R}^t_{BA} \times \mathbf{F} = \mathbf{0}$ and

$$\mathbf{M} = \mathbf{R}^n_{BA} \times \mathbf{F} \tag{e}$$

is the moment of the couple. Because $R^n_{BA} = R_{BA} \sin\theta$, where θ is the angle between $\mathbf{R}_{BA}$ and **F**, the magnitude of the moment is

$$M = (R_{BA} \sin\theta)F = hF. \tag{f}$$

3. The moment vector **M** is independent of any particular origin or line of application and is thus a *free vector*.
4. The forces of a couple can be rotated together within their plane, keeping their magnitudes and the distance between their lines of action constant, or they can be translated to any parallel plane, without changing the magnitude, direction, or sense of the moment vector.
5. Two couples are equal if they have the same moment vectors, regardless of the forces or moment arms. This means that it is the vector product of the two that is significant and not the individual values.

13.5 FREE-BODY DIAGRAMS

The term "body" as used here may consist of an entire machine, several connected parts of a machine, a single part, or a portion of a machine part. A *free-body diagram* is a sketch or drawing of the body or bodies, isolated from the rest of the machine and its surroundings, upon which the forces and moments are shown in action. It is often desirable to include on the diagram the known magnitudes and directions as well as other pertinent information.

The diagram so obtained is called "free" because the machine part, or parts, or portion of the system has been freed (or isolated) from the remaining machine elements, and their effects have been replaced by forces and moments. If the free-body diagram is of an entire machine part or system of parts, the forces shown on it are the external forces (applied forces) and moments exerted by adjacent or connecting parts that are not part of this free-body diagram. If the diagram is of a portion of a part, the forces and moments shown acting on the cut (separation) surface are *internal* forces and moments, that is, the summation of the internal stresses exerted on the cut surface by the remainder of the part that has been cut away.

The construction and presentation of clear, detailed, and neatly drawn free-body diagrams represent the heart of engineering communication. This is true because they represent a part of the thinking process, whether they are actually placed on paper or not, and because the construction of these diagrams is the *only* way the results of thinking can be communicated to others. It is important to acquire the habit of drawing free-body diagrams no matter how simple the problem may appear to be. They are a means of storing ideas while concentrating on the next step in the solution of a problem. Construction of or, at the very minimum, sketching of free-body diagrams speeds up the problem-solving process and dramatically decreases the chances of making mistakes.

The advantages of using free-body diagrams can be summarized as follows.

1. They make it easier for us to translate words, thoughts, and ideas into physical models.
2. They assist in seeing and understanding all facets of a problem.
3. They help in planning the approach to the problem.
4. They make mathematical relations easier to see or formulate.
5. Their use makes it easy to keep track of our progress and helps in making simplifying assumptions.
6. They are useful for storing the methods of solution for future reference.

7. They assist our memory and make it easier to present and explain our work to others.

In analyzing the forces in a machine we shall almost always need to separate the machine into its individual components or subsystems and construct free-body diagrams indicating the forces that act on each. Many of these components will have been connected to each other by kinematic pairs (joints). Accordingly, Fig. 13.3 has been prepared to illustrate

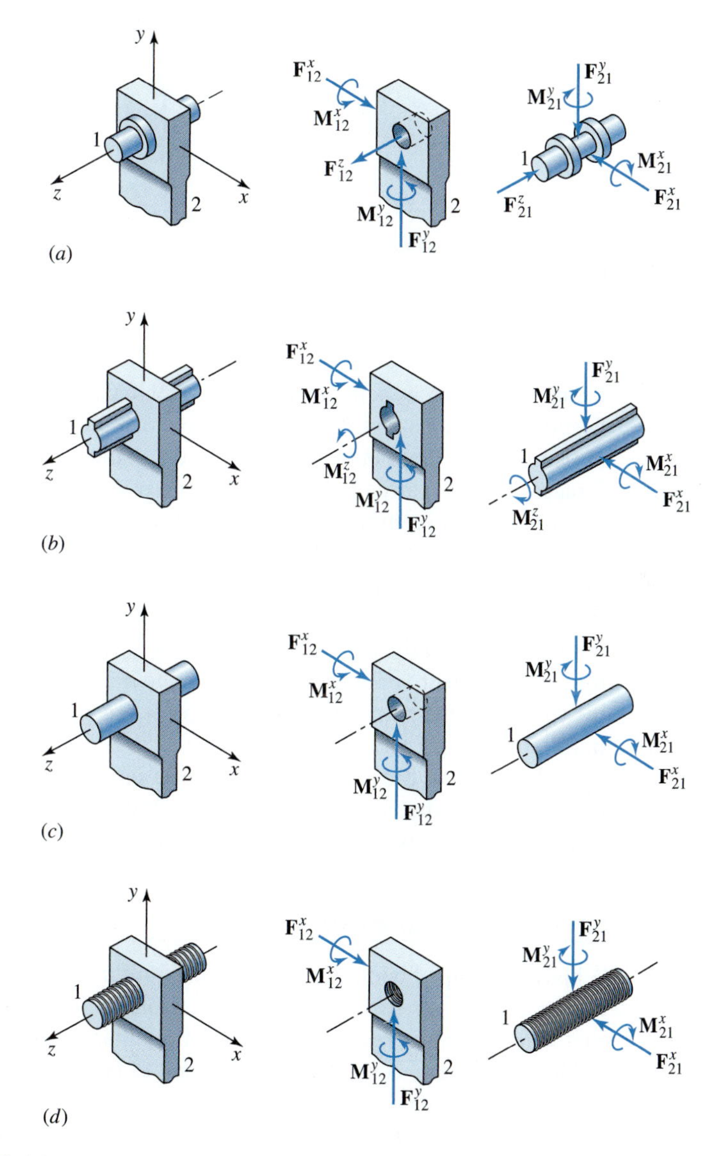

Figure 13.3 (*Continued*).

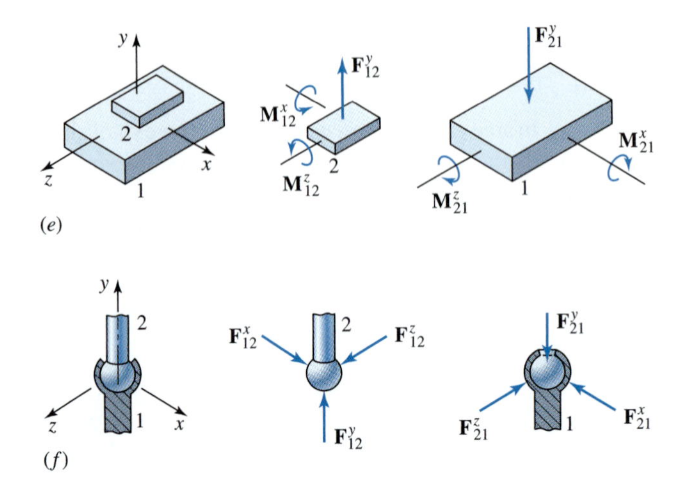

Figure 13.3 All lower pairs and their constraint forces: (*a*) revolute or turning pair with pair variable θ; (*b*) prismatic or sliding pair with pair variable z; (*c*) cylindric pair with pair variables θ and z; (*d*) helical or screw pair with pair variables θ or z; (*e*) flat or planar pair with pair variables x, z; and θ; and (*f*) spheric or globular pair with pair variables θ, ϕ, and ψ.

the constraint forces acting between the elements of the lower pairs (see Chapter 1) when friction forces are assumed to be zero. Note that these are not complete free-body diagrams in that they illustrate only the forces at the mating pair surfaces and do not illustrate the forces where the partial links have been severed from their remainders.

Upon careful examination of Fig. 13.3, we see that there is no component of a constraint force or moment transmitted along an axis where motion is possible, that is, along with the direction of a pair variable. This is consistent with our assumption of no friction. If one of these pair elements were disposed to transmit a force, or a torque, to its mating element in the direction of the pair variable, in the absence of friction, the tendency would result in motion of the pair variable rather than in the transmission of force or torque. Similarly, in the case of higher pairs, the constraint forces are always normal to the contacting surfaces in the absence of friction.

The notation illustrated in Fig. 13.3 is used consistently throughout this part of the book. The force that link i exerts onto link j is denoted $\mathbf{F}_{ij}$, whereas the reaction to this force is denoted $\mathbf{F}_{ji}$ and is the force from link j acting back onto link i.

13.6 CONDITIONS FOR EQUILIBRIUM

A body or group of bodies is said to be in equilibrium if all the forces exerted on the system are in balance. In such a situation, Newton's laws as expressed in Eq. (13.1) indicate that no acceleration results. This may imply that no motion takes place, meaning that all velocities are zero. If so, the system is said to be in *static equilibrium*. On the other hand, no acceleration may imply that velocities do exist but remain constant; the system is then said to be in *dynamic equilibrium*. In either case, Eq. (13.1) demonstrates that a system of bodies is in equilibrium if and only if

1. *The vector sum of all forces acting upon it is zero.*
2. *The vector sum of the moments of all forces acting about an arbitrary axis is zero.*

Mathematically, these two statements are expressed as

$$\sum \mathbf{F} = \mathbf{0} \quad \text{and} \quad \sum \mathbf{M} = \mathbf{0}. \tag{13.8}$$

Note how these statements are a result of Newton's first and third laws, it being understood that a body is composed of a collection of particles.

Many problems have forces acting in a single plane. When this is true, it is convenient to choose this as the xy plane. Then, Eqs. (13.8) can be simplified by taking the components in that plane as follows,

$$\sum F^x = 0, \quad \sum F^y = 0, \quad \sum M = 0,$$

where the z direction for the moment M components is implied by the fact that all forces act only in the xy plane.

13.7 TWO- AND THREE-FORCE MEMBERS

A member is the name given to a rigid body subjected to only forces, that is, there is no applied torque. The free-body diagram of Fig. 13.4*a* illustrates a two-force member, that is, an arbitrarily shaped body acted upon by two forces, $\mathbf{F}_A$ and $\mathbf{F}_B$, at points A and B, respectively. The forces have been shown with broken lines as a reminder that the magnitudes and directions of these forces are not yet known. Assuming that our free-body diagram is complete and that no other forces are active on the body, the first of Eqs. (13.8) gives

$$\sum \mathbf{F} = \mathbf{F}_A + \mathbf{F}_B = \mathbf{0}.$$

This requires that $\mathbf{F}_A$ and $\mathbf{F}_B$ have *equal magnitudes* and *opposite directions*. Using the second of Eqs. (13.8) about either point A or point B demonstrates that $\mathbf{F}_A$ and $\mathbf{F}_B$ must also have the *same line of action*; otherwise, the moments could not sum to zero. Thus, for any two-force member, we have learned that the forces must be oppositely directed along the unique line of action defined by the points A and B, as illustrated in Fig. 13.4*b*. When the magnitude of one force becomes known, the magnitude of the second force must be equal to the first force but of the opposite sense (simply stated, the two forces must be equal, opposite, and collinear).

It follows from the above that for a body with two active forces and an applied torque to be in static equilibrium, the two forces must be equal, opposite, and parallel. Also, for a four-force member to be in static equilibrium, the resultant of any two of the active forces must be equal, opposite, and collinear to the resultant of the other two active forces. In other words, a four-force member can be reduced to a two-force member (a four-force member will be discussed in more detail in the following section).

Now consider a three-force member, that is, a body with three active forces and no applied torques, as illustrated in Figs. 13.4*c* and 13.4*d*. Here we also assume that the

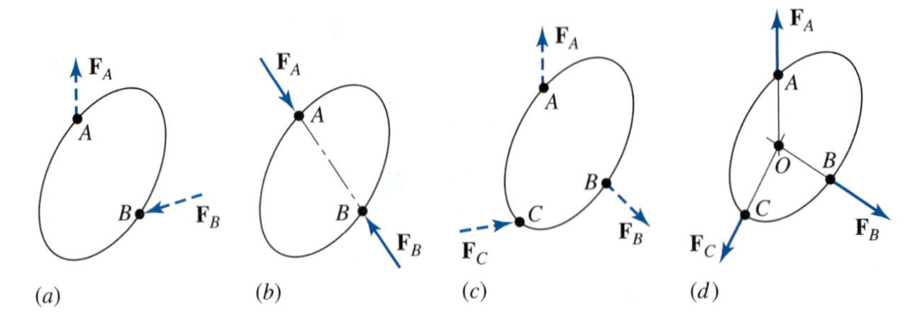

Figure 13.4 (*a*) Two-force member not in equilibrium; (*b*) two-force member in equilibrium if $\mathbf{F}_A$ and $\mathbf{F}_B$ are equal, opposite, and share the same line of action; (*c*) three-force member not in equilibrium; and (*d*) three-force member in equilibrium if $\mathbf{F}_A$, $\mathbf{F}_B$, and $\mathbf{F}_C$ are coplanar, if their lines of action intersect at a common point O, and if their vector sum is zero.

three forces, $\mathbf{F}_A$, $\mathbf{F}_B$, and $\mathbf{F}_C$, are coplanar, that is, they are all known to act in the plane defined by the three points, A, B, and C. Suppose further that we also know two of the lines of action, say, those of $\mathbf{F}_A$ and $\mathbf{F}_B$, and that they intersect at some point O. Because neither $\mathbf{F}_A$ nor $\mathbf{F}_B$ exerts any moment about point O, applying $\sum \mathbf{M} = \mathbf{0}$ for all three forces demonstrates that the moment of $\mathbf{F}_C$ about point O must also be zero. Thus, the lines of action of all three forces must intersect at the common point O; that is, the three forces acting on the body must be *concurrent*. This explains why, in two dimensions, a three-force member can be solved for only two force magnitudes, although there are three scalar equations implied in Eqs. (13.8); the moment equation has already been satisfied once the lines of action become known. If one of the three force magnitudes is known, the other two can be found using $\sum \mathbf{F} = \mathbf{0}$. This will be demonstrated in the following example.

It is also worth noting at this point that any member subjected to more than four forces can be reduced to one of the above cases.

EXAMPLE 13.1

The four-bar linkage of Fig. 13.5*a* has crank 2 driven by an input torque $\mathbf{M}_{12}$; an external load $\mathbf{P} = 120\ \text{lb}\angle 220°$ acts at point Q on link 4. For the position shown, find all the constraint forces and their reactions necessary for the linkage to be in equilibrium.

GRAPHIC SOLUTION

1. We draw the linkage and the given force or forces to scale, as illustrated in Fig. 13.5*a*. The size scale shown here is about 6.25 in/in, meaning that 1 in of the drawing represents 6.25 in of the linkage. The force scale shown is about 90 lb/in; that is, a vector shown 1 in long represents a force of 90 lb.
2. We begin drawings of the free-body diagrams of each of the moving links, as illustrated in Figs. 13.5*b*, 13.5*e*, and 13.5*f*. At this stage, all known information, such as the positions and the force **P**, should be drawn to scale. Unknown

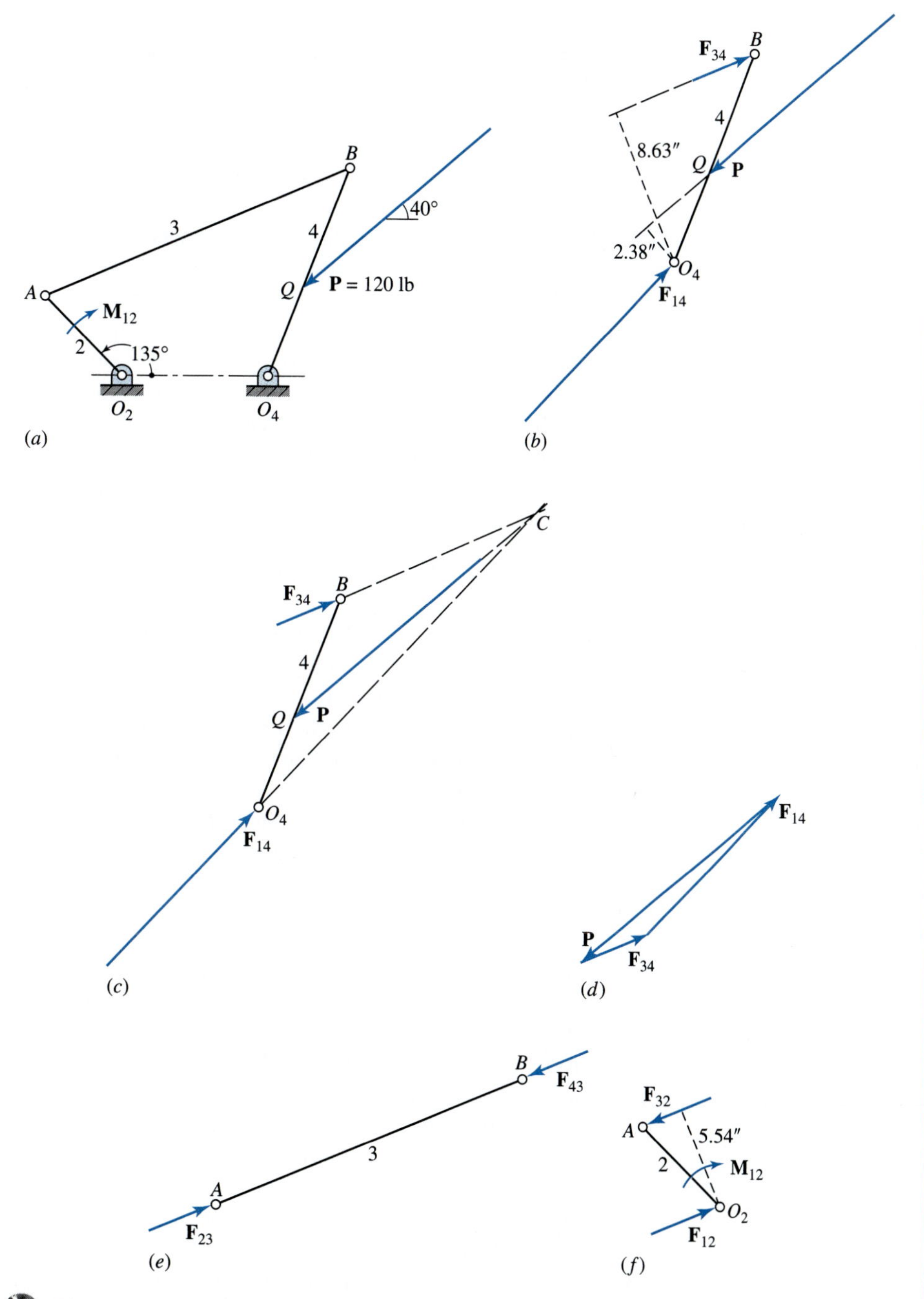

Figure 13.5 Graphic solution for Example 13.1: $R_{AO_2} = 6$ in, $R_{BA} = 18$ in, $R_{O_4O_2} = 8$ in, $R_{BO} = 12$ in, and $R_{QO_4} = 5$ in.

information, such as other applied forces on each link, are sketched lightly on the free-body diagrams, but will be redrawn to scale once they become known.

3. We ensure that all free-body diagrams are complete, that is, that all pertinent information is shown. For example, we must now either find the weights and centers of gravity of all links and the direction of gravity or, as done here, assume that the gravitational forces are small in comparison with the applied force **P** and that they can safely be ignored.
4. From the free-body diagram, we now observe that link 3 is a two-force member. Thus, according to the above discussion, we can now show the lines of action of $\mathbf{F}_{23}$ and $\mathbf{F}_{43}$ along the axis defined by points A and B in Fig. 13.5*e*. Because action and reaction forces must be equal and opposite, these also tell us the lines of action of $\mathbf{F}_{32}$ on link 2 in Fig. 13.5*f* and $\mathbf{F}_{34}$ on link 4 in Fig. 13.5*b*.

5a. We proceed next to link 4, as illustrated in Fig. 13.5*b*, where neither the magnitude nor the direction of the frame reaction $\mathbf{F}_{14}$ is known as yet. One method for continuation is to measure the moment arms **P** and $\mathbf{F}_{34}$ about O_4, which are 2.38 in and 8.63 in, respectively. Then, taking counterclockwise moments as positive, Eqs. (13.8) gives

$$\sum M_{O_4} = (2.38\text{ in})(120\text{ lb}) - (8.63\text{ in})F_{34} = 0.$$

The solution is $F_{34} = 33.1$ lb, where the fact that the result is positive confirms the sense shown for $\mathbf{F}_{34}$ in Fig. 13.5*b*.

5b. As an alternative approach to Step 5a, we could note that link 4 is a three-force member; therefore, when the lines of action of **P** and $\mathbf{F}_{34}$ are extended, they intersect at the point of concurrency C and define the line of action of $\mathbf{F}_{14}$, as illustrated in Fig. 13.5*c*.

6. No matter whether Step 5a or Step 5b is used, the force polygon illustrated in Fig. 13.5*d* can now be used to graphically solve the equation

$$\sum \mathbf{F} = \mathbf{P} + \mathbf{F}_{34} + \mathbf{F}_{14} = \mathbf{0},$$

giving $F_{34} = 33.1$ lb and $F_{14} = 89.0$ lb with directions as indicated in the force polygon. Note that this avoids the need for the moment equation of Step 5a; it is satisfied by the concurrency at point C.

7. From the free-body diagram of link 3, Fig. 13.5*e*, and the nature of action and reaction forces, we note that $\mathbf{F}_{23} = -\mathbf{F}_{43} = \mathbf{F}_{34}$. For a two-force member, it is important to note whether the link is in tension or in compression. In the case of tension, the link would fail as a result of yielding; however, in the case of compression, the link could fail by either yielding or buckling. Hence, buckling becomes an important topic in machine design and is addressed in Sections 13.14–13.17.
8. Because $\mathbf{F}_{32} = -\mathbf{F}_{23}$, this known force can be illustrated on the free-body diagram of link 2 in Fig. 13.5*f*. Because link 2 is subjected to two forces and a torque, $\mathbf{F}_{12}$ must be equal and opposite to $\mathbf{F}_{32}$. Alternatively, from the summation of forces on the free-body diagram of link 2 in Fig. 13.5*f*, we see that $\mathbf{F}_{12} = -\mathbf{F}_{32}$. When the moment arm of $\mathbf{F}_{32}$ about point O_2 is measured, it is

found to be 5.54 in. Therefore, taking counterclockwise moments as positive, we obtain

$$\sum M_{O_2} = (5.54\text{ in})F_{32} - M_{12} = 0,$$

which yields

$$M_{12} = (5.54\text{ in})(33.1\text{ lb}) = 183\text{ in}\cdot\text{lb}, \qquad \textit{Ans.}$$

where the positive sign confirms the clockwise sense illustrated in Fig. 13.5*f*.

ANALYTIC SOLUTION

1. First we make a position analysis of the linkage at the position of interest to determine the angular orientation of each link. The angles are illustrated in Fig. 13.6*a*. In addition, we find

$$\mathbf{R}_{AO_2} = 6.0\text{ in}\angle 135^\circ = -4.24\hat{\mathbf{i}} + 4.24\hat{\mathbf{j}}\text{ in.}$$

$$\mathbf{R}_{BO_4} = 12.0\text{ in}\angle 68.4^\circ = 4.42\hat{\mathbf{i}} + 11.16\hat{\mathbf{j}}\text{ in.}$$

$$\mathbf{R}_{QO_4} = 5.0\text{ in}\angle 68.4^\circ = 1.84\hat{\mathbf{i}} + 4.65\hat{\mathbf{j}}\text{ in.}$$

$$\mathbf{F}_{34} = F_{34}\angle 22.4^\circ = 0.925\text{F}_{34}\hat{\mathbf{i}} + 0.381\text{F}_{34}\hat{\mathbf{j}}\text{ lb}$$

$$\mathbf{P} = 120\text{ lb}\angle 220^\circ = -91.9\hat{\mathbf{i}} - 77.1\hat{\mathbf{j}}\text{ lb.}$$

2. Referring to Fig. 13.6*b*, we sum moments about point O_4. Thus,

$$\sum \mathbf{M}_{O_4} = \mathbf{R}_{QO_4} \times \mathbf{P} + \mathbf{R}_{BO_4} \times \mathbf{F}_{34} = \mathbf{0}.$$

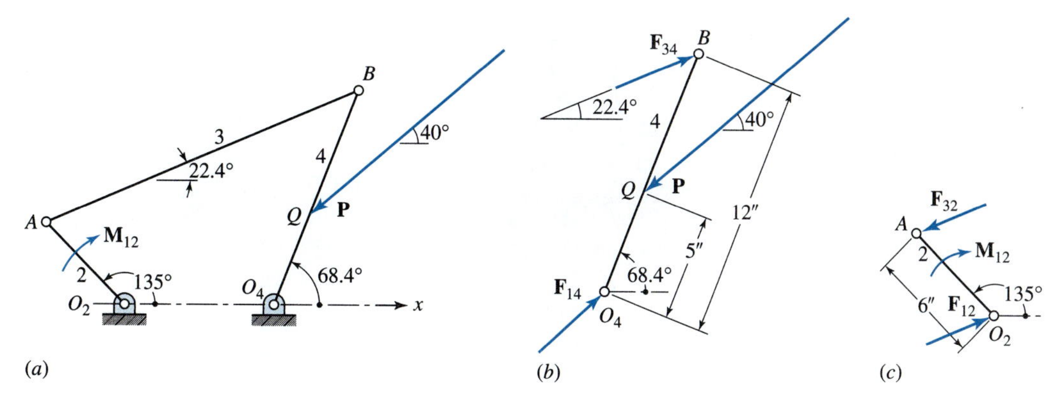

Figure 13.6 Analytic solution for Example 13.1.

Performing the cross-product operations, we find the first term to be $\mathbf{R}_{QO_4} \times \mathbf{P} = 285.5\hat{\mathbf{k}}$ in · lb, whereas the second term is $\mathbf{R}_{BO_4} \times \mathbf{F}_{34} = -8.63F_{34}\hat{\mathbf{k}}$ in · lb. Substituting these values into the above equation and solving gives

$$\mathbf{F}_{34} = 33.1 \text{ lb}\angle 22.4° = 30.6\hat{\mathbf{i}} + 12.6\hat{\mathbf{j}} \text{ lb.} \qquad \textit{Ans.}$$

3. The force $\mathbf{F}_{14}$ is determined next from the equation

$$\sum \mathbf{F} = \mathbf{P} + \mathbf{F}_{34} + \mathbf{F}_{14} = \mathbf{0}.$$

Solving gives

$$\mathbf{F}_{14} = 61.3\hat{\mathbf{i}} + 64.5\hat{\mathbf{j}} \text{ lb} = 89.0 \text{ lb}\angle 46.5°. \qquad \textit{Ans.}$$

4. Next, from the free-body diagram of link 2 (see Fig. 13.6*c*), we write

$$\sum \mathbf{M}_{O_2} = -\mathbf{M}_{12} + \mathbf{R}_{AO_2} \times \mathbf{F}_{32} = \mathbf{0}.$$

Using $\mathbf{F}_{32} = -\mathbf{F}_{34} = -30.6\hat{\mathbf{i}} - 12.6\hat{\mathbf{j}}$ lb, we find

$$\mathbf{M}_{12} = -183\hat{\mathbf{k}} \text{ in} \cdot \text{lb.} \qquad \textit{Ans.}$$

Note that in both the graphic and the analytic solutions the free-body diagram of the frame, link 1, was never drawn and is not needed for the solution. If drawn, a force $\mathbf{F}_{21} = -\mathbf{F}_{12}$ would be shown at O_2, a force $\mathbf{F}_{41} = -\mathbf{F}_{14}$ at O_4, and a moment $\mathbf{M}_{21} = -\mathbf{M}_{12}$ would be shown. In addition, a minimum of two more force components and a torque would be shown where link 1 is anchored to keep it stationary. This would introduce at least three additional unknowns and since Eqs. (13.8) would not allow the solution for more than three unknowns, then drawing the free-body diagram of the frame will be of no benefit in the

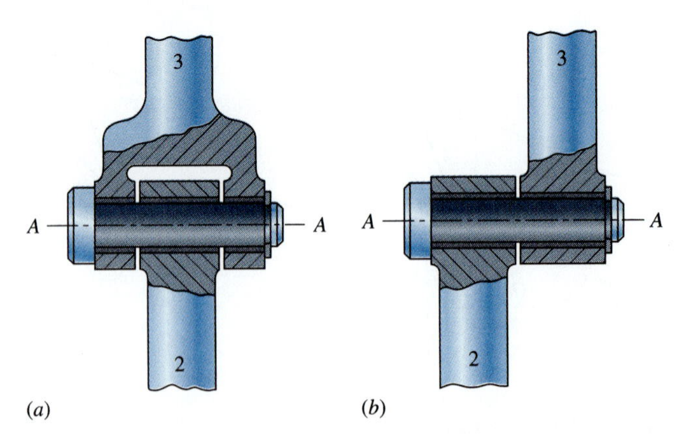

Figure 13.7 (*a*) Balanced revolute-pair connection. (*b*) An unbalanced revolute-pair connection produces a bending moment on the pin and on each link.

solution process. The only time that this is necessary is when these anchoring forces, called "shaking forces," and moments are required as part of the solution. This will be discussed in more detail in Section 14.7 of Chapter 14.

In the preceding example, it was assumed that the forces all act in the same plane. For the connecting link 3, for example, it was assumed that the line of action of the forces and the centerline of the link are coincident. A careful designer will sometimes go to extreme measures to approach these conditions as closely as possible. Note that if the pin connections are arranged as illustrated in Fig. 13.7*a*, such conditions are obtained theoretically. If, on the other hand, the connection is like the one illustrated in Fig. 13.7*b*, the pin itself as well as the link will have moments acting upon them. If the forces are not in the same plane, then moments exist proportional to the distance between the force planes.

EXAMPLE 13.2

The slider–crank mechanism of Fig. 13.8*a* has an external load $P = 100$ lb acting horizontally at point Q on link 4. Block 4 is 8 in wide by 3 in. high with the pin centrally located. Point Q is 1 in above the centerline. Determine the torque $\mathbf{M}_{12}$ that must be applied to link 2 to hold the mechanism in static equilibrium at the position illustrated in Fig. 13.8*a*. Assume that gravitational and friction forces are small in comparison with the applied force **P** and can be ignored.

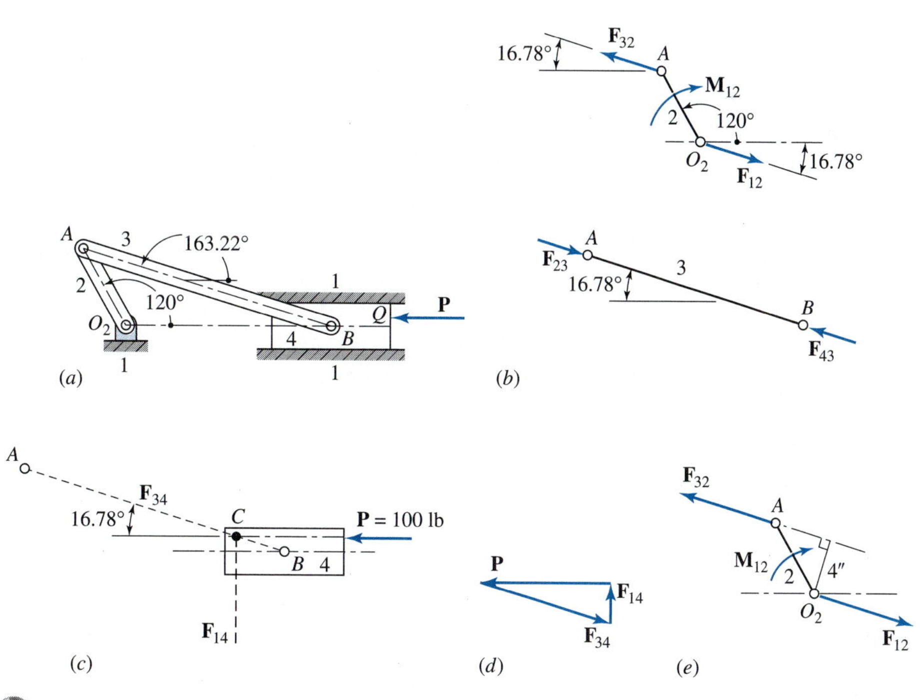

Figure 13.8 Graphic solution for Example 13.2: $R_{AO_2} = 6$ in and $R_{BA} = 18$ in.

GRAPHIC SOLUTION

1. We draw the mechanism and the given force to scale, as illustrated in Fig. 13.8*a*. The size scale used here is 15 in/in and the force scale is 140 lb/in.
2. We draw the free-body diagrams of each moving link, as illustrated in Figs. 13.8*b*, 13.8*c*, and 13.8*e*.
3. We observe that link 3 is a two-force member. Thus, we can show the lines of action of $\mathbf{F}_{23}$ and $\mathbf{F}_{43}$ along the axis defined by points A and B in Fig. 13.8*b*. These also tell us the lines of action of $\mathbf{F}_{34}$ on link 4 in Fig. 13.8*c* and $\mathbf{F}_{32}$ on link 2 in Fig. 13.8*e*.
4. We proceed to link 4 where neither the magnitude nor the location of the frame reaction force $\mathbf{F}_{14}$ is known as yet. However, the direction of this force must be perpendicular to the ground because friction is neglected. Link 4 is a three-force member; therefore, when the lines of action of $\mathbf{P}$ and $\mathbf{F}_{34}$ are extended, they intersect at the point of concurrency C and define the line of action of $\mathbf{F}_{14}$, as illustrated in Fig. 13.8*c*.
5. We use the force polygon illustrated in Fig. 13.8*d* to graphically solve the equation

$$\sum \mathbf{F} = \mathbf{P} + \mathbf{F}_{34} + \mathbf{F}_{14} = \mathbf{0},$$

giving

$$F_{34} = 105 \text{ lb} \quad \text{and} \quad F_{14} = 31 \text{ lb}.$$

6. From the free-body diagram of link 3, Fig. 13.8*b*, and the nature of action and reaction forces, we note that $\mathbf{F}_{23} = -\mathbf{F}_{43} = \mathbf{F}_{34}$. Link 3 is in compression; therefore, the problem of buckling should be investigated (see Sections 13.14–13.17).
7. From the free-body diagram of link 2, Fig. 13.8*e*, and the nature of action and reaction forces, we note that $\mathbf{F}_{32} = -\mathbf{F}_{23}$. From the summation of forces on the free-body diagram of link 2, we find $\mathbf{F}_{12} = -\mathbf{F}_{32}$. When the moment arm of $\mathbf{F}_{32}$ about point O_2 is measured, it is determined to be 4 in. Therefore, taking counterclockwise moments as positive, we obtain

$$\sum M_{O_2} = (4 \text{ in})F_{32} - M_{12} = 0,$$

which yields

$$M_{12} = (4 \text{ in})(105 \text{ lb}) = 420 \text{ in} \cdot \text{lb}, \qquad \textit{Ans.}$$

where the positive sign confirms the clockwise sense illustrated in Fig. 13.8*e*.

ANALYTIC SOLUTION

1. From a position analysis of the mechanism at the position of interest, with the angles indicated in Fig. 13.9*a*, we find

$$\mathbf{R}_{AO_2} = 6.0 \text{ in}\angle 120^\circ = -3.000\hat{\mathbf{i}} + 5.196\hat{\mathbf{j}} \text{ in.}$$

$$\mathbf{F}_{34} = F_{34}\angle 16.78^\circ = 0.957\,4F_{34}\hat{\mathbf{i}} - 0.288\,7F_{34}\hat{\mathbf{j}}.$$

2. The forces $\mathbf{F}_{34}$ and $\mathbf{F}_{14}$, see Fig. 13.9*b*, are determined from the equation

$$\sum \mathbf{F} = \mathbf{P} + \mathbf{F}_{34} + \mathbf{F}_{14} = \mathbf{0},$$

where $\mathbf{P} = -100\hat{\mathbf{i}}$ lb. This equation can be written as

$$-100 \text{ lb}\hat{\mathbf{i}} + 0.957\,4F_{34}\hat{\mathbf{i}} - 0.288\,7F_{34}\hat{\mathbf{j}} + F_{14}\hat{\mathbf{j}} = \mathbf{0}.$$

Then, equating the $\hat{\mathbf{i}}$ and $\hat{\mathbf{j}}$ components, respectively, gives

$$F_{34} = \frac{100 \text{ lb}}{0.957\,4} = 104.45 \text{ lb}$$

$$F_{14} = 0.288\,7F_{34} = 30.15 \text{ lb.}$$

These forces are written as

$$\mathbf{F}_{34} = 100 \text{ lb}\hat{\mathbf{i}} - 30.15 \text{ lb}\hat{\mathbf{j}} = 104.45 \text{ lb}\angle -16.78^\circ$$

$$\mathbf{F}_{14} = 30.15\hat{\mathbf{j}} \text{ lb.}$$

3. Next, from the free-body diagram of link 2 (see Fig. 13.9*c*), we write

$$\sum \mathbf{M}_{O_2} = -\mathbf{M}_{12} + \mathbf{R}_{AO_2} \times \mathbf{F}_{32} = \mathbf{0}.$$

Using $\mathbf{F}_{32} = -\mathbf{F}_{34} = -100\hat{\mathbf{i}} + 30.15\hat{\mathbf{j}}$ lb and solving, we find

$$\mathbf{M}_{12} = -429.15\hat{\mathbf{k}} \text{ in} \cdot \text{lb (cw).} \qquad \textit{Ans.}$$

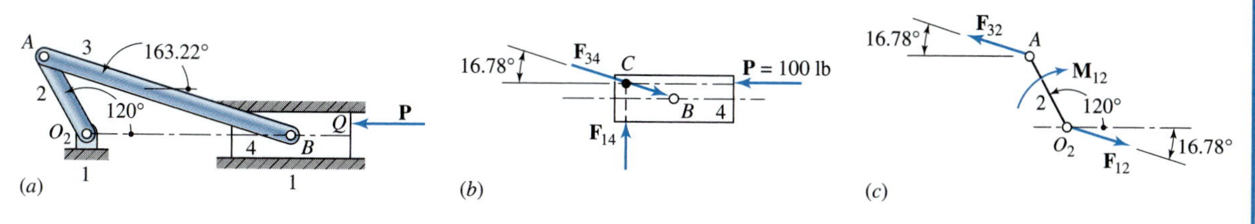

Figure 13.9 Analytic solution for Example 13.2.

13.8 FOUR-FORCE MEMBERS

When all free-body diagrams of a system have been constructed, we usually find that one or more represent either two- or three-force members without applied torques, and the techniques of the last section can be used for their solution, at least in planar problems. Note that we started by seeking out such links and using them to establish known lines of action for other unknown forces. In doing this, we were implicitly enforcing the summation-of-moments equation for such links. Then, by noting that action and reaction forces share the same line of action, we proceeded to establish more lines of action. Finally, upon reaching a link where one or more forces and all lines of action were known, we applied the summation-of-forces equations to find the magnitudes and senses of the unknown forces. Proceeding from link to link in this way, we found the solutions for the unknowns.

In some problems, however, we find that this procedure reaches a point where the solution cannot proceed in this fashion, for example, where two- or three-force members cannot be found, and lines of action remain unknown. It is then wise to count the number of unknown quantities (magnitudes and directions) to be determined for each free-body diagram. In planar problems, it is clear that Eqs. (13.8), the equilibrium conditions, cannot be solved for more than three unknowns on any single free-body diagram. Sometimes it is helpful to combine two links, or even three links, and to draw a free-body diagram of the combined assembly. This approach can sometimes be used to eliminate unknown forces on the individual links, which combine with their reactions as internal stresses of the combined system.

In some problems, however, this is still not sufficient. In such situations, it is always good practice to combine multiple forces on the same free-body diagram that are completely known by a single force representing the sum of the known forces. This sometimes simplifies the figures to reveal two- and three-force members that were otherwise not noted.

In planar problems, the most general case of a system of forces that is solvable is one in which the three unknowns are the magnitudes of three forces. If all other forces are combined into a single force, we then have a four-force system. The following example will help to demonstrate how such a system can be treated.

EXAMPLE 13.3

A cam with a reciprocating roller follower is illustrated in Fig. 13.10*a*. The follower is held in contact with the cam by a spring pushing downward at C with a spring force of $F_C = 12$ N for this particular position. Also, an external load $F_E = 35$ N acts on the follower at E in the direction indicated. For the position indicated, determine the follower pin force at A and the bearing reactions at B and D. Assume no friction and a weightless follower.

SOLUTION

A free-body diagram of the follower (link 4) is illustrated in Fig. 13.10*b*. Note that the follower can be viewed as a five-force member ($\mathbf{F}_A$, $\mathbf{F}_B$, $\mathbf{F}_C$, $\mathbf{F}_D$ and $\mathbf{F}_E$). However, the forces $\mathbf{F}_C$ and $\mathbf{F}_E$ are known, and their sum is obtained graphically in Fig. 13.10*b*. The

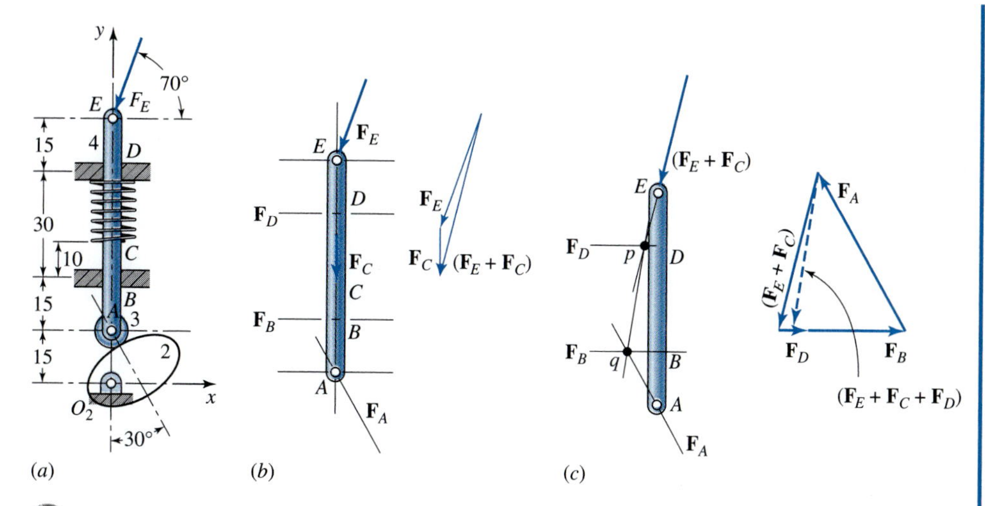

Figure 13.10 Graphic solution for Example 13.3: (*a*) A cam system with dimensions in millimeters. (*b*) Free-body diagram of link 4, a five-force member, and graphic summation of known forces $\mathbf{F}_C$ and $\mathbf{F}_E$. (*c*) Free-body diagram of link 4, reduced to a four-force member, and graphic solution.

free-body diagram is redrawn to indicate this in Fig. 13.10*c*. We show the resultant force $\mathbf{F}_E + \mathbf{F}_C$ with its point of application at E, where the original lines of action of $\mathbf{F}_E$ and $\mathbf{F}_C$ intersected, with line of action dictated by the direction of the vector sum. This placement of the point of application at E was not arbitrary, but required. Since the original forces $\mathbf{F}_E$ and $\mathbf{F}_C$ had no moments about point E then their resultant must have no moment about this point. The result of combining these two forces is that the free-body of Fig. 13.10*c* is now reduced to a four-force member with one known force, the resultant $\mathbf{F}_E + \mathbf{F}_C$, and three forces of unknown magnitudes.

In a similar manner, if the magnitude of $\mathbf{F}_D$ were known, it could be added to $\mathbf{F}_E + \mathbf{F}_C$ to produce a new resultant $\mathbf{F}_E + \mathbf{F}_C + \mathbf{F}_D$, which would act through point p.

Now consider the moment equation about point q. If we write $\sum \mathbf{M}_q = \mathbf{0}$, we see that the equation can be satisfied only if the resultant $\mathbf{F}_E + \mathbf{F}_C + \mathbf{F}_D$ has no moment about point q and thus has pq as its line of action. This is, therefore, the basis for the graphic solution illustrated in Fig. 13.10*c*. The direction of the line of action pq of the resultant $\mathbf{F}_E + \mathbf{F}_C + \mathbf{F}_D$ is used in the force polygon to determine the force $\mathbf{F}_D$; the force polygon is then completed by finding $\mathbf{F}_A$ and $\mathbf{F}_B$ whose lines of action are known. The solutions are $F_A = 51.8$ N, $F_B = 32.8$ N, and $F_D = 5.05$ N when rounded to three significant figures.

Note that this approach defines a general concept, useful in either the graphic or the analytic approach. When there are three unknown force magnitudes on a single free-body diagram, we choose a point such as q where the lines of action of two of the unknown forces meet and write the moment equation about that point, that is, $\sum \mathbf{M}_q = \mathbf{0}$. This equation will have only one unknown remaining and can be solved directly. Only then should $\sum \mathbf{F} = \mathbf{0}$ be written, because the problem has then been reduced to two unknowns.

SUMMARY

The following is a summary of the procedures for the graphical method of static force analysis that has been presented so far in Chapter 13.

1. We classify each member in the mechanism; that is, we identify two-force members, three-force members, four-force members, and so on. We draw complete free-body diagrams of each member. It is good practice to combine all known forces on a free-body diagram into a single force; we then commence with the member with the lowest number of forces. That is, we draw the two-force members first, then we draw the three-force members, and, finally, we draw any four-force members. If a member is acted upon by more than four forces, then either (*a*) it can be reduced to one of the above or (*b*) it has more than three unknowns and is not solvable.
2. Using the definitions of two-force, three-force, and four-force members, we apply the following rules.
   (a) For a two-force member, the two forces must be equal, opposite, and collinear. Note that for a link with two forces and a torque, the forces are equal, opposite, and parallel.
   (b) For a three-force member, the three forces must intersect at a single point.
   (c) For a four-force member, the resultant of any two forces must be equal, opposite, and collinear with the resultant of the other two forces.
3. We draw force polygons for three- and four-force members. We should clearly state the scale of each force polygon. To be able to draw a force polygon, we remember that we need four pieces of information (one of which must be a magnitude) for a three-force member. Five pieces of information (one of which must be a magnitude) are required for a four-force member. Note that, in general, it is less confusing if the force polygons are not superimposed on top of the mechanism diagram. If a force polygon cannot be completed for a particular member, then, in general, multiple links can be taken together and treated as a single free-body diagram.

13.9 FRICTION–FORCE MODELS

Over the years there has been much interest in the subjects of friction and wear, and many papers and books have been devoted to these subjects. It is not our purpose to explore the mechanics of friction here, but to present classical simplifications that have been used in the analysis of the performance of mechanical devices. The results of any such analysis may not be theoretically ideal, but they do correspond closely to experimental measurements, so that reliable decisions can be made from them regarding a design and its operating performance.

Consider two bodies constrained to remain in contact with each other, such as the surfaces of block 3 and link 2, as illustrated in Fig. 13.11*a*. A force $\mathbf{F}_{43}$ may be exerted on block 3 by link 4, tending to cause block 3 to slide relative to link 2. Without the presence of friction between the interface surfaces of links 2 and 3, the block cannot transmit the component of $\mathbf{F}_{43}$ tangent to the surfaces. Instead of transmitting this force, the block will slide in the direction of this unbalanced force component; without friction, equilibrium is not possible unless the line of action of $\mathbf{F}_{43}$ is normal to the surface.

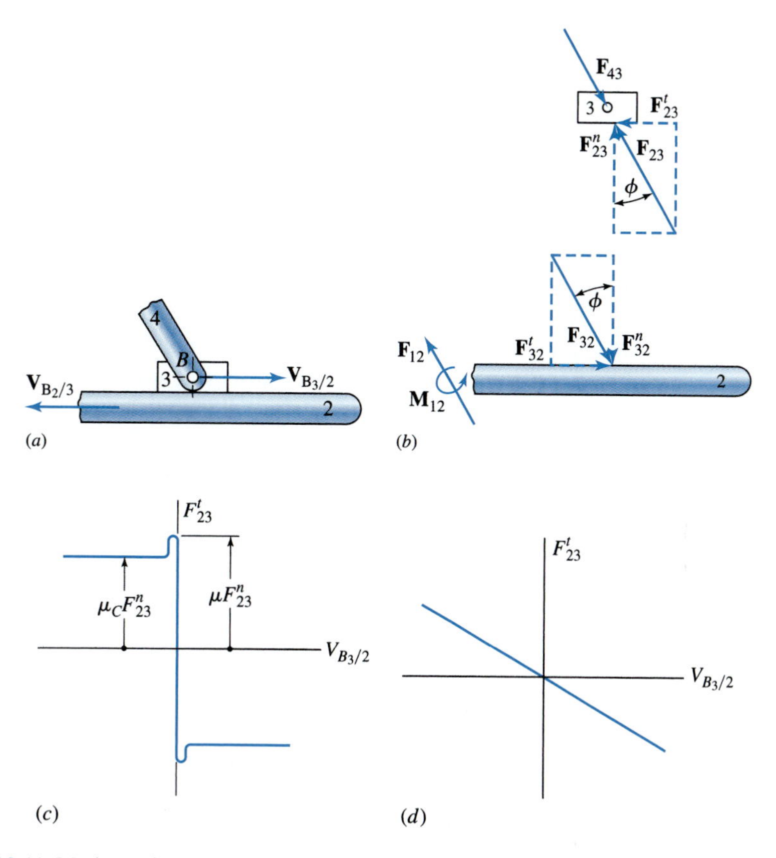

Figure 13.11 Mathematical representation of friction forces: (*a*) physical system; (*b*) free-body diagrams; (*c*) static and Coulomb friction models; and (*d*) viscous friction model.

With friction, however, a resisting force $\mathbf{F}_{23}^{t}$ is developed at the contact surface, as illustrated in the free-body diagrams of Fig. 13.11*b*. This friction force $\mathbf{F}_{23}^{t}$ acts in addition to the usual constraint force $\mathbf{F}_{23}^{n}$ across the surface of the sliding joint; together, the force components $\mathbf{F}_{23}^{n}$ and $\mathbf{F}_{23}^{t}$ form the total constraint force $\mathbf{F}_{23}$ that balances the force $\mathbf{F}_{43}$ when block 3 is in equilibrium. Of course, the reaction force components $\mathbf{F}_{23}^{n}$ and $\mathbf{F}_{23}^{t}$ (or their total) are also acting simultaneously onto link 2, as illustrated in the other free-body diagram of Fig. 13.11*b*. The force component $\mathbf{F}_{23}^{t}$ and its reaction $\mathbf{F}_{32}^{t}$ are called *friction forces*.

Depending on the materials of links 2 and 3, there is a limit to the size of the force component $\mathbf{F}_{23}^{t}$ that can be sustained by friction while still maintaining equilibrium. This limit is expressed by the relationship

$$F_{23}^{t} \leq \mu F_{23}^{n}, \tag{13.9}$$

where μ, referred to as the *coefficient of static friction*, is a characteristic property of the contacting materials. Values of the coefficient μ have been determined experimentally for many materials and can be found in many engineering handbooks.[3]

If the force $\mathbf{F}_{43}$ is tipped too much, so that its tangential component and therefore the friction force component F_{23}^t would be too large to satisfy the inequality of Eq. (13.9), equilibrium is not possible and block 3 will slide relative to link 2 with an apparent velocity $\mathbf{V}_{B_3/2}$. When sliding takes place, the friction force takes on the value

$$F_{23}^t = \mu_c F_{23}^n, \tag{13.10}$$

where μ_c is the *coefficient of sliding friction*. This manner of approximating sliding friction is called *Coulomb friction*, and we shall often use this term to refer to the relationship of Eq. (13.10). The coefficient μ_c can also be determined experimentally and for most materials is slightly less than μ, the static coefficient of friction.

To summarize what we have discussed so far, Fig. 13.11*c* illustrates a graph of the friction force F_{23}^t versus the apparent sliding velocity $V_{B_3/2}$. Here it can be seen that when the sliding velocity is zero, the friction force F_{23}^t can have any magnitude between μF_{23}^n and $-\mu F_{23}^n$. When the velocity is not zero, the magnitude of the friction force F_{23}^t drops slightly to the value $\mu_c F_{23}^n$ and has a sense opposing that of the sliding motion $V_{B_3/2}$.

Looking again at the total force $\mathbf{F}_{23}$ in Fig. 13.11*b*, we see that it is inclined at an angle to the surface normal and is equal and opposite to $\mathbf{F}_{43}$ whenever the system is in equilibrium. Thus, the angle ϕ is given by

$$\tan\phi = \frac{F_{23}^t}{F_{23}^n}$$

$$\tan\phi \le \frac{\mu F_{23}^n}{F_{23}^n} = \mu$$

$$\phi \le \tan^{-1}\mu.$$

When $\mathbf{F}_{43}$ is tipped so that the block 3 is on the verge of sliding,

$$\phi = \tan^{-1}\mu. \tag{13.11}$$

This limiting value of the angle ϕ, called the *friction angle*, defines the maximum angle through which the force $\mathbf{F}_{23}$ can tip from the surface normal before equilibrium is no longer possible and sliding begins (commonly referred to as the impending motion). Note that the limiting value of ϕ does not depend on the magnitude of the force $\mathbf{F}_{23}$ but only on the coefficient of friction of the materials involved.

We might also note that, in the above discussion, we have treated only the effects of friction between surfaces with relative sliding motion. We might ask how this can be extended to treat relative rotational motion such as in a revolute joint. Because good low-friction rotational bearings are not expensive and because they can be easily lubricated to reduce friction, this is usually not a problem. Extensions of the above ideas are known for rotational bearings.[4] However, the additional difficulty does not often warrant their use, because the results are usually not affected by more than a very small percentage.

Although the static friction or Coulomb friction models are often used and often do represent good approximations for friction forces in mechanical equipment, they are not the only models. Sometimes—for example, when representing a machine or its dynamic effects by its differential equation of motion—it is more convenient to analyze the machine's performance using another approximation for friction forces, called *viscous friction* or *viscous damping*. As illustrated in the graph of Fig. 13.11*d*, this model assumes a linear relationship between the magnitude of the friction force and the sliding velocity. This viscous friction model is especially useful when the dynamic analysis of a machine leads to the use of one or more differential equations. The nonlinear relationship of static and/or Coulomb friction, illustrated in Fig. 13.11*c*, leads to a nonlinear differential equation that is more difficult to treat.

Whether the friction effect is represented by the static, Coulomb, or viscous friction models, it is important to recognize the sense of the friction force. As a mnemonic device, the rule is often stated that "friction opposes motion," as indicated by the free-body diagram of links 3 in Fig. 13.11*b*, where the sense of $\mathbf{F}_{23}^t$ is opposite that of $\mathbf{V}_{B_3/2}$. This rule of thumb is not wrong if it is applied carefully, but it can be dangerous and even misleading. It will be noted in Fig. 13.11*a* that there are two motions that might be thought of, namely, $\mathbf{V}_{B_3/2}$ and $\mathbf{V}_{B_2/3}$; there are also two friction forces, $\mathbf{F}_{23}^t$ and $\mathbf{F}_{32}^t$. Careful examination of Fig. 13.11*b* indicates that $\mathbf{F}_{23}^t$ opposes the sense of $\mathbf{V}_{B_3/2}$, whereas $\mathbf{F}_{32}^t$ opposes the sense of $\mathbf{V}_{B_2/3}$. In machine systems, particularly where both sides of a sliding joint are in motion, it is very important to understand *which* friction force "opposes" *which* motion.

Of course, in static force analysis, the subject of Chapter 13, the assumption throughout is that forces are found under conditions of static equilibrium; therefore, there should be no motion. When we say that "friction opposes motion," we are speaking of *impending motion*; that is, we are assuming that the system is on the verge of moving and we are speaking of the movement that would start if a small change in force were to perturb the equilibrium.

13.10 STATIC FORCE ANALYSIS WITH FRICTION

We will show the effect of including friction on our previous methods of static force analysis by the following examples.

EXAMPLE 13.4

Repeat the static force analysis of the slider–crank mechanism (Fig. 13.8*a*) that was analyzed in Example 13.2, assuming that there is a coefficient of static friction of $\mu = 0.25$ between the slider and the ground. Assume that the impending motion of the slider is to the left and that friction can be neglected in the revolute joints.

SOLUTION

As is always the case, when we begin a force analysis with friction, it is necessary to solve the problem first without friction. The purpose is to find the sense of each of the normal force components, in this example $\mathbf{F}_{14}^n$. This was done in Example 13.2, where $\mathbf{F}_{14}^n$ was determined to act vertically upward.

Because the impending motion of the slider is assumed to the left—that is, $\mathbf{V}_{B_4/1}$ is to the left—the friction force $\mathbf{F}_{14}^t$ must act to the right. We can redraw the free-body diagram of link 4 (Fig. 13.8*c*) and include the friction force as illustrated in Fig. 13.12*a*. Because of

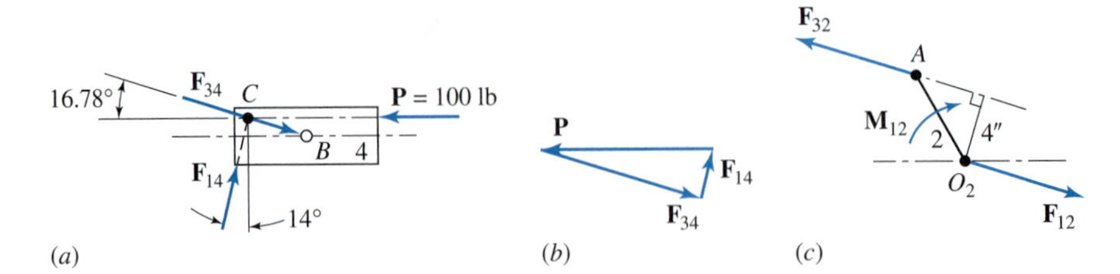

Figure 13.12 Solution for Example 13.4.

the static friction, the line of action of $\mathbf{F}_{14}$ is shown tipped through the angle ϕ, which is calculated from Eq. (13.11):

$$\phi = \tan^{-1} 0.25 = 14^\circ.$$

In deciding the direction of tip of the angle ϕ, it is necessary to know the sense of the normal force (upward) and the sense of the friction force (horizontal to the right) on link 4 at the point of contact. This explains why the solution without friction should be completed first.

Once the new line of action of the force $\mathbf{F}_{14}$ is known, the solution can proceed exactly as in Example 13.2. The graphical solution, with friction effects, is illustrated in Fig. 13.12, and it is found that

$$\mathrm{F}_{34} = 95 \text{ lb} \quad \text{and} \quad \mathrm{F}_{14} = 30 \text{ lb}. \qquad \textit{Ans.}$$

Note that the line of action of $\mathbf{F}_{34}$ has not changed and the magnitude is less than the no-friction case. Therefore, less torque is required on link 2 to maintain the mechanism in equilibrium. We find that

$$M_{12} = (4 \text{ in})(95 \text{ lb}) = 380 \text{ in} \cdot \text{lb}. \qquad \textit{Ans.}$$

Note also that in Example 13.4 we have $F_{14}^{n} = 30 \cos 14 = 29.11$ lb, whereas it was 30.15 lb in Example 13.2 before friction was included. This indicates why the normal force without friction could *not* be multiplied by μ to obtain F_{14}^{t} and then added by superposition; friction changes both the normal and the tangential components of all forces.

Finally, note that in Example 13.4 the point of application of force F_{14} is located 3.76 in. to the left of point B. Fortunately, this is still within the width of block 4; therefore, the solution is valid. However, if point Q had been slightly higher, the point of application of force F_{14} might appear to fall outside of the width of block 4. In this case, block 4 would not be in equilibrium, but would be on the verge of tipping counterclockwise until prevented by additional contact at the top right corner. Block 4 would then become a four-force member with two unknown forces F_{14} at both the top right and the bottom left corners, both tipped by the friction angle ϕ, but tipped in opposite directions because one is upward and one is downward, and a new force polygon would be required for the solution for block 4 under those conditions.

EXAMPLE 13.5

Repeat the static force analysis of the cam–follower system analyzed in Example 13.3 (Fig. 13.10*a*), assuming that there is a coefficient of static friction of $\mu = 0.15$ between links 1 and 4 at both sliding bearings B and D. Friction in all other joints is considered negligible. Determine the minimum force necessary at A to hold the system in equilibrium.

SOLUTION

First note from Example 13.3, for the problem without friction. the sense of each of the normal force components, $\mathbf{F}_B^n$ and $\mathbf{F}_D^n$ were both found to act to the right (see Fig. 13.10*c*).

Next consider the problem statement carefully and decide the direction of the impending motion. As stated, the problem asks for the minimum force at A necessary for equilibrium; that is, the problem statement implies that if $\mathbf{F}_A$ were any smaller, the system would move downward. Thus, the impending motion is with both velocities $\mathbf{V}_{D4/1}$ and $\mathbf{V}_{B4/1}$ downward. Therefore, the two friction forces at B and D must both act upward from link 1 onto link 4.

Next, we redraw the free-body diagram of link 4 (Fig. 13.10*c*) to include the friction forces as illustrated in Fig. 13.13*a*. Here, because of static friction, the lines of action of $\mathbf{F}_B$ and $\mathbf{F}_D$ are both shown tipped through the angle ϕ, which is calculated from Eq. (13.11):

$$\phi = \tan^{-1} 0.15 = 8.5°.$$

In deciding the direction of tip of the angles ϕ, it is necessary to know the sense of each friction force (upward) and the sense of each normal force (toward the right) at B and D. This explains why the solution without friction must be done first.

Once the new lines of action of the forces $\mathbf{F}_B$ and $\mathbf{F}_D$ are known, the solution can proceed exactly as in Example 13.3. The graphic solution, with friction effects, is illustrated in Fig. 13.13*b*, where it is found that

$$F_B = 28.7 \text{ N}, \quad F_D = 6.57 \text{ N}, \quad F_A = 45.8 \text{ N}. \qquad \textit{Ans.}$$

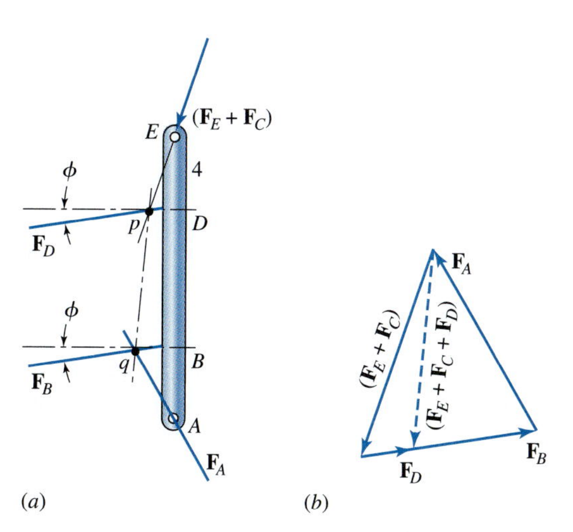

Figure 13.13 Graphic solution for Example 13.5: free-body diagram of link 4 with static friction.

We note from this example that the normal components of the two forces at B and D are now $F_B^n = 28.4$ N and $F_D^n = 6.50$ N and are different from the values determined without friction in Example 13.3. This should be a warning, whether working graphically or analytically, that it is incorrect to simply multiply the frictionless normal forces by the coefficient of friction to find the friction forces and then to add these to the frictionless solution. All forces may (and usually do) change magnitude when friction is included, and the problem must be completely reworked from the beginning with friction included. The effects of static or Coulomb friction *cannot* be added afterward by superposition.

We note also that if the problem statement had asked for the maximum force at A, the impending motion of link 4 would have been upward and the friction forces downward on link 4. This would have reversed the tilt of the two lines of action for both $\mathbf{F}_B$ and $\mathbf{F}_D$ and would have totally changed the final results. Now, in practice, if the actual value of the force $\mathbf{F}_A$ is between these minimum and maximum values, equilibrium will be maintained, and the values of other constraint forces will be between the two extreme values determined by this type of analysis. Also, if the value of $\mathbf{F}_A$ is slowly increased from the minimum toward the maximum value, the follower will remain in equilibrium until the maximum value is reached and then begin to move; this discontinuous action is sometimes referred to as "stiction."

13.11 SPUR- AND HELICAL-GEAR FORCE ANALYSIS

Figure 13.14*a* illustrates a pinion with center O_2 rotating clockwise at a speed of ω_2 and driving a gear with center at O_3 at a speed of ω_3. As discussed in Chapter 7, the reactions between the teeth occur along the pressure line AB, tipped by the pressure angle ϕ from the common tangent to the pitch circles. Free-body diagrams of the pinion and the gear are illustrated in Fig. 13.14*b*. The action of the pinion on the gear is indicated by the force $\mathbf{F}_{23}$ acting at the pitch point along the pressure line.* Since the gear is supported by its shaft, then from $\sum \mathbf{F} = \mathbf{0}$, an equal and opposite force $\mathbf{F}_{13}$ must act at the centerline of the shaft. A similar analysis of the pinion indicates that the same observations are true. In each case, the forces are equal in magnitude, opposite in direction, parallel, and in the same plane. On either gear, therefore, they form a couple.

Note that the free-body diagram of the pinion indicates the forces resolved into components. Here we employ the superscripts r and t to indicate the radial and tangential directions with respect to the pitch circle. It is expedient to use the same superscripts for the components of the force F_{12} that the shaft exerts on the pinion. The moment of the couple formed

* It is true that treating the force $\mathbf{F}_{23}$ in this manner ignores the possible effects of friction forces between the meshing gear teeth. Justification for this comes in four forms: (1) the actual point of contact is continually varying during the meshing cycle, but always remains near the pitch point, which is the instant center of velocity; thus, the relative motion between the teeth is close to true rolling motion with only a very small amount of slip. (2) The fact that the friction forces are continually changing in both magnitude and direction throughout the meshing cycle, and that the total force is often shared by more than one tooth in contact, makes a more exact analysis impractical. (3) The machined surfaces of the teeth and the fact that they are usually well lubricated produces a very small coefficient of friction. (4) Experimental data indicate that gear efficiencies are usually very high, often approaching 99%, demonstrating that any errors produced by ignoring friction are quite small.

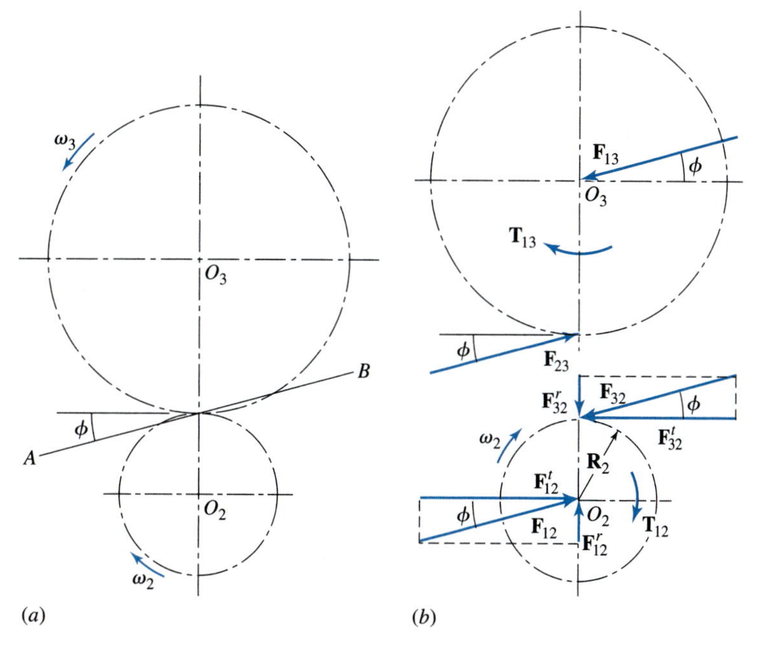

Figure 13.14 Forces on spur gears.

by $\mathbf{F}^t_{32}$ and $\mathbf{F}^t_{12}$ is in equilibrium with the torque $\mathbf{T}_{12}$ that must be applied by the shaft to drive the gearset. When the pitch radius of the pinion is designated R_2, then $\sum \mathbf{M}_{O_2} = \mathbf{0}$ indicates that the torque is

$$T_{12} = R_2 F^t_{32}. \tag{13.12}$$

Note that the radial force component $\mathbf{F}^r_{32}$ serves no purpose as far as the transmission of power is concerned. For this reason, $\mathbf{F}^t_{32}$ is frequently called the *transmitted* force.

In applications involving gears, the power transmitted and the shaft speeds are often specified. Remembering that power is the product of force times velocity or torque times angular velocity, we can find the relation between power and the transmitted force. Using the symbol P to denote power, we obtain

$$P = T_{12}\omega_2 = R_2 F^t_{32}\omega_2, \tag{13.13a}$$

which can be solved for the transmitted force F^t_{32} as

$$F^t_{32} = \frac{P}{R_2\omega_2}. \tag{13.13b}$$

In application of these formulae it is often necessary to remember that the U.S. customary units used for power are horsepower, abbreviated hp, where 1 hp = 33 000 ft · lb/min, and in SI units, power is measured in watts, abbreviated W, where 1 W = 1 N · m/s.

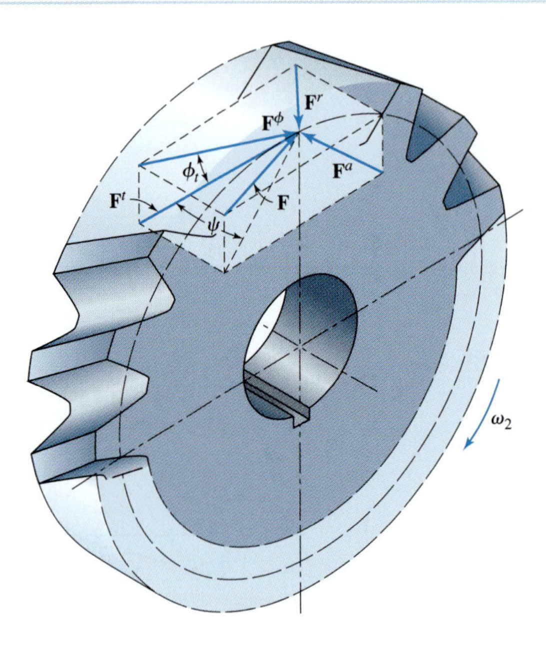

Figure 13.15 Force components on a helical gear at the tooth contact point.

Once the transmitted force is known, the following relations for spur gears are evident from Fig. 13.14*b*:

$$F_{32}^r = F_{32}^t \tan\phi \qquad \text{and} \qquad F_{32} = \frac{F_{32}^t}{\cos\phi}.$$

In the treatment of forces on helical gears it is convenient to determine the axial force, work with it independently, and treat the remaining force components the same as for straight spur gears. Figure 13.15 is a drawing of a helical gear with a portion of the face removed to illustrate the tooth contact force and its components acting at the pitch point. The gear is imagined to be driven clockwise under load. The driving gear has been removed and its effect replaced by the force shown acting on the teeth.

The resultant force $\mathbf{F}$ is shown divided into three components, $\mathbf{F}^a$, $\mathbf{F}^r$, and $\mathbf{F}^t$, which are the axial, radial, and tangential components, respectively. The tangential force component is the transmitted force and the force that is effective in transmitting torque. When the transverse pressure angle is designated ϕ_t and the helix angle ψ, the following relations are evident from Fig. 13.15:

$$\mathbf{F} = \mathbf{F}^a + \mathbf{F}^r + \mathbf{F}^t \tag{13.14}$$

$$F^a = F^t \tan\psi \tag{13.15}$$

$$F^r = F^t \tan\phi_t. \tag{13.16}$$

It is also expedient to make use of the resultant of $\mathbf{F}^r$ and $\mathbf{F}^t$. We shall designate this force $\mathbf{F}^\phi$, defined by the equation

$$\mathbf{F}^\phi = \mathbf{F}^r + \mathbf{F}^t. \tag{13.17}$$

EXAMPLE 13.6

A gear train is composed of three helical gears with shaft centers in line. The driver is a right-hand helical gear having a pitch radius of 2 in., a transverse pressure angle of 20°, and a helix angle of 30°. An idler gear in the train has the teeth cut left handed and has a pitch radius of 3.25 in. The idler transmits no power to its shaft. The driven gear in the train has the teeth cut right handed and has a pitch radius of 2.50 in. If the transmitted force is 600 lb, find the shaft forces acting on each gear. Gravitational forces can be neglected.

SOLUTION

First we consider only the axial forces, as previously suggested. For each mesh, the axial component of the reaction, from Eq. (13.15), is

$$F^a = F^t \tan\psi = (600 \text{ lb}) \tan 30° = 346 \text{ lb}.$$

Figure 13.16*a* is a top view of the three gears, looking down on the plane formed by the three axes of rotation. For each gear, rotation is considered about the *z* axis for this problem. In Fig. 13.16*b*, free-body diagrams of each of the three gears are drawn in projection and the three coordinate axes are shown. As indicated, the idler exerts a force $\mathbf{F}^a_{32}$ on the driver. This is resisted by the axial shaft force $\mathbf{F}^a_{12}$. The forces $\mathbf{F}^a_{12}$ and $\mathbf{F}^a_{32}$ form a couple that is resisted by the moment $\mathbf{T}^y_{12}$. Note that this moment is negative (clockwise) about the $+y$ axis. Consequently, it produces a bending moment in the shaft. The magnitude of this moment is

$$T^y_{12} = -R_2 F^a_{12} = -(2.00 \text{ in})(346 \text{ lb}) = -692 \text{ in} \cdot \text{lb}.$$

Turning our attention next to the idler, we see from Figs. 13.16*a* and 13.16*b* that the net axial force on the shaft of the idler is zero. The axial component of the force from the driver onto the idler is $\mathbf{F}^a_{23}$ and that from the driven gear onto the idler is $\mathbf{F}^a_{43}$. These two force components are equal and opposite and form a couple that is resisted by the bending moment in the shaft of the idler $\mathbf{T}^y_{13}$, of magnitude

$$T^y_{13} = -2R_3 F^a_{23} = -2(3.25 \text{ in})(346 \text{ lb}) = -2\,249 \text{ in} \cdot \text{lb}.$$

The driven gear has the axial force component $\mathbf{F}^a_{34}$ acting along its pitch line caused by the helix angle of the idler. This is resisted by the axial shaft reaction $\mathbf{F}^a_{14}$. As indicated, these forces form a couple that is resisted by the moment $\mathbf{T}^y_{14}$. The magnitude of this moment is

$$T^y_{14} = -R_4 F^a_{34} = -(2.50 \text{ in})(346 \text{ lb}) = -865 \text{ in} \cdot \text{lb}.$$

It is emphasized that the three resisting moments $\mathbf{T}^y_{12}$, $\mathbf{T}^y_{13}$, and $\mathbf{T}^y_{14}$ are caused solely by the axial components of the reactions between the gear teeth. These produce static bearing reactions and have no effect on the amount of power transmitted.

Now that all of the reactions due to the axial components have been determined, we turn our attention to the remaining force components and examine their effects as if they were operating independently of the axial forces. Free-body diagrams showing the force components in the plane of rotation for the driver, idler, and driven gears are illustrated in

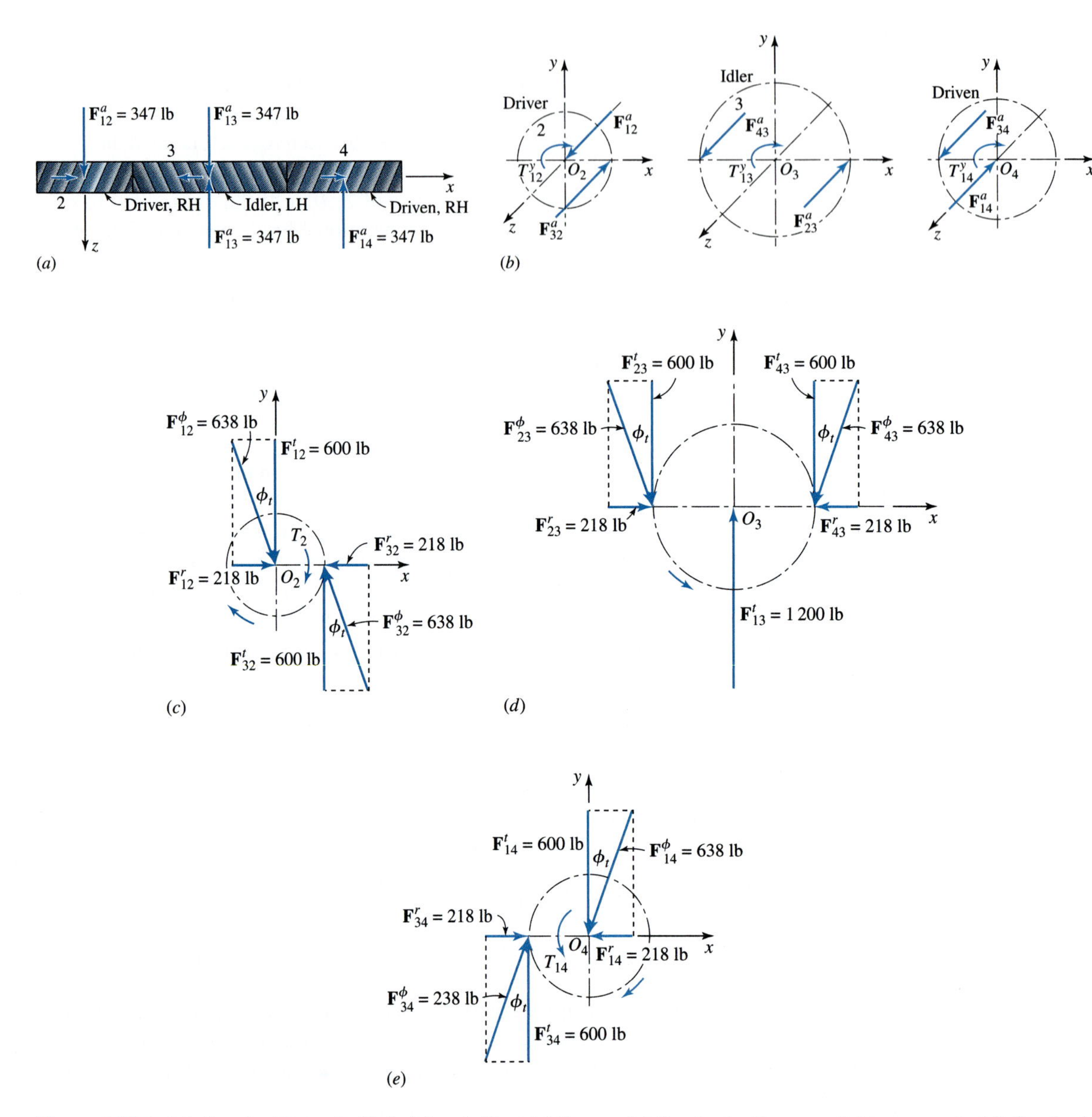

Figure 13.16 Solution for Example 13.6: (*a*) and (*b*) axial forces; (*c*) free-body diagram of driver gear 2; (*d*) free-body diagram of idler gear 3; and (*e*) free-body diagram of driven gear 4.

Figs. 13.16*c*, 13.16*d*, and 13.16*e*, respectively. These force components can be obtained graphically as shown or by applying Eqs. (13.12) and (13.16). It is not necessary to combine the components to determine the resultant forces, because the components are exactly those that are desired to proceed with machine design.

EXAMPLE 13.7

The planetary gear train illustrated in Fig. 13.17*a* has input shaft *a* that is driven by a torque of $\mathbf{T}_{a2} = -100\hat{\mathbf{k}}$ in · lb. Note that input shaft *a* is connected directly to gear 2 and that the planetary arm 3 is connected directly to the output shaft *b*. Shafts *a* and *b* rotate about the same axis but are not connected. Gear 6 is fixed to the stationary frame 1 (not shown). All gears have a diametral pitch of 10 teeth/in. and a pressure angle of 20°. Assuming that the forces act in a single plane and that gravitational forces and centrifugal forces on the planet gears can be neglected, perform a complete force analysis of the parts of the train and compute the magnitude and direction of the output torque delivered by shaft *b*.

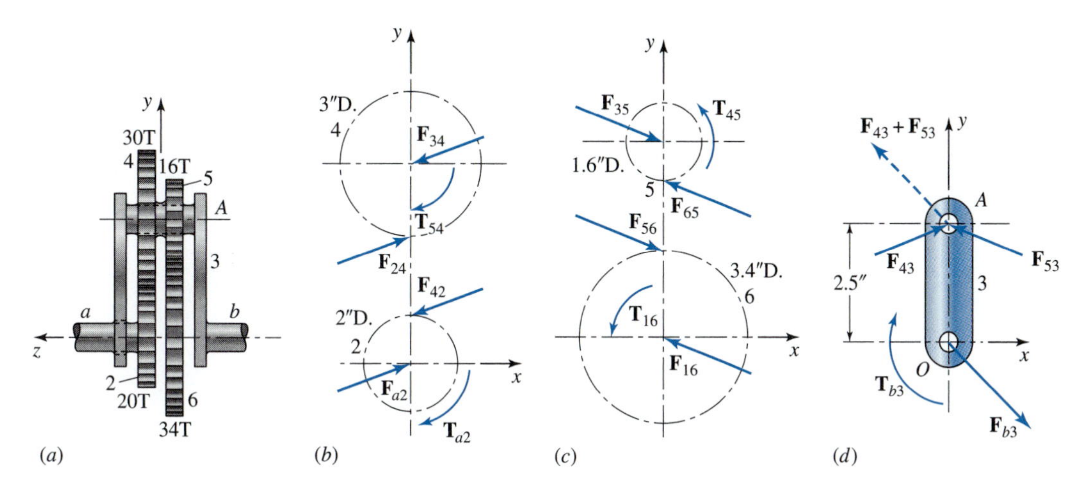

Figure 13.17 Solution for Example 13.7: (*a*) planetary gear train with tooth numbers; (*b*) free-body diagrams of gears 2 and 4; (*c*) free-body diagrams of gears 5 and 6; and (*d*) free-body diagram of planet carrier arm 3.

SOLUTION

The pitch radii of the gears are $R_2 = (20 \text{ teeth})/(2 \cdot 10 \text{ teeth/in}) = 1.00$ in and, similarly, $R_4 = 1.50$ in, $R_5 = 0.80$ in, and $R_6 = 1.70$ in. The distance between the centers of the meshing gear pairs is $(R_2 + R_4) = (1.00 \text{ in} + 1.50 \text{ in}) = 2.50 \text{ in} = (R_5 + R_6)$. Because the torque that the input shaft exerts on gear 2 is $T_{a2} = 100$ in · lb, the transmitted force is $F^t_{42} = T_{a2}/R_2 = (100 \text{ in} \cdot \text{lb})/(1 \text{ in}) = 100$ lb. Therefore, $F_{42} = F^t_{42}/\cos\phi = 100 \text{ lb}/\cos 20° = 106$ lb. The free-body diagram of gear 2 is illustrated in Fig. 13.17*b*. In vector form the results are

$$\mathbf{F}_{a2} = -\mathbf{F}_{42} = 106 \text{ lb}\angle 20°.$$

Figure 13.17*b* also illustrates the free-body diagram of gear 4. The forces are

$$\mathbf{F}_{24} = -\mathbf{F}_{34} = 106 \text{ lb}\angle 20°,$$

where $\mathbf{F}_{34}$ is the force of planet arm 3 against gear 4. Gears 4 and 5 are connected to each other, but turn freely on the planet arm shaft. Thus, $\mathbf{T}_{54}$ is the torque exerted by gear 5 onto gear 4. This torque is $T_{54} = (R_4)F^t_{24} = (1.50 \text{ in})(100 \text{ lb}) = 150 \text{ in} \cdot \text{lb cw}$.

Turning next to the free-body diagram of gear 5 in Fig. 13.17*c*, we first find $F^t_{65} = T_{45}/R_5 = (150 \text{ in} \cdot \text{lb})/(0.800 \text{ in}) = 188 \text{ lb}$. Therefore, $F_{65} = (188 \text{ lb})/\cos 20° = 200 \text{ lb}$. In vector form, the results for gear 5 are summarized as

$$\mathbf{F}_{65} = -\mathbf{F}_{35} = 200 \text{ lb}\angle 160°, \quad \mathbf{T}_{45} = 150\hat{\mathbf{k}} \text{ in} \cdot \text{lb}.$$

For gear 6, illustrated in Fig. 13.17*c*, we have

$$\mathbf{F}_{16} = -\mathbf{F}_{56} = 200 \text{ lb}\angle 160°$$

$$\mathbf{T}_{16} = R_6 F^t_{56}\hat{\mathbf{k}} = (1.70 \text{ in})(200 \text{ lb} \cos 20°)\hat{\mathbf{k}} = 319\hat{\mathbf{k}} \text{ in} \cdot \text{lb}.$$

Note that $\mathbf{F}_{16}$ and $\mathbf{T}_{16}$ are the force and torque, respectively, exerted by the frame on gear 6.

The free-body diagram of arm 3 is illustrated in Fig. 13.17*d*. As noted earlier, the forces are assumed to act in a single plane. The two forces $\mathbf{F}_{43}$ and $\mathbf{F}_{53}$ are

$$\mathbf{F}_{43} = -\mathbf{F}_{34} = 106 \text{ lb}\angle 20°, \qquad \mathbf{F}_{53} = -\mathbf{F}_{35} = 200 \text{ lb}\angle 160°$$

and can be summed to

$$\mathbf{F}_{43} + \mathbf{F}_{53} = 137 \text{ lb}\angle 130.2°.$$

We now find the output shaft reaction to be

$$\mathbf{F}_{b3} = -(\mathbf{F}_{43} + \mathbf{F}_{53}) = 137 \text{ lb}\angle -49.8°.$$

Using $\mathbf{R}_{AO} = 2.50\hat{\mathbf{j}}$ in and the equation

$$\sum \mathbf{M}_O = \mathbf{T}_{b3} + \mathbf{R}_{AO} \times (\mathbf{F}_{43} + \mathbf{F}_{53}) = \mathbf{0},$$

we find $\mathbf{T}_{b3} = -221\hat{\mathbf{k}}$ in $\cdot$ lb. Therefore, the output shaft torque is $\mathbf{T}_b = 221\hat{\mathbf{k}}$ in $\cdot$ lb.

13.12 STRAIGHT-TOOTH BEVEL-GEAR FORCE ANALYSIS

In determining the tooth forces on bevel gears, it is customary to use the forces that would occur at the mid-thickness of the tooth on the pitch cone. The resultant tangential force probably occurs somewhere between the midpoint and the large end of the tooth, but there will be only a small error in making this approximation. The tangential or transmitted force is then given by

$$F^t = \frac{T}{R}, \tag{13.18}$$

where R is the mid-radius of the pitch cone as illustrated in Fig. 13.18 and T is the shaft torque.

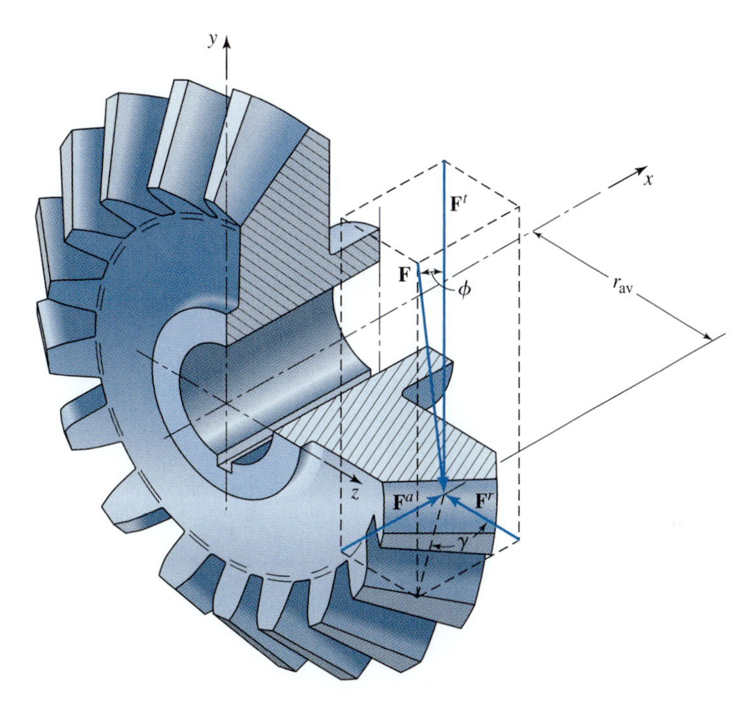

Figure 13.18 Force components on a straight-tooth bevel gear at the tooth contact point.

Figure 13.18 also illustrates all of the components of the resultant force acting at the midpoint of the tooth. The following relationships can be derived by inspection of Fig. 13.18:

$$\mathbf{F} = \mathbf{F}^a + \mathbf{F}^r + \mathbf{F}^t \tag{13.19}$$

$$F^r = F^t \tan\phi \cos\gamma \tag{13.20}$$

$$F^a = F^t \tan\phi \sin\gamma . \tag{13.21}$$

Note, as in the case of helical gears, that the axial force $\mathbf{F}^a$ results in a couple on the shaft that produces a bending moment.

EXAMPLE 13.8

The bevel pinion illustrated in Fig. 13.19 rotates at 600 rev/min in the direction indicated and transmits 5 hp to the gear. The mounting distances are shown, together with the locations of the bearings on each shaft. Bearings A and C are capable of taking both radial and axial loads, whereas bearings B and D are designed to receive only radial loads. The teeth of the gears have a 20° pressure angle. Find the components of the forces that the bearings exert on the shafts in the x, y, and z directions.

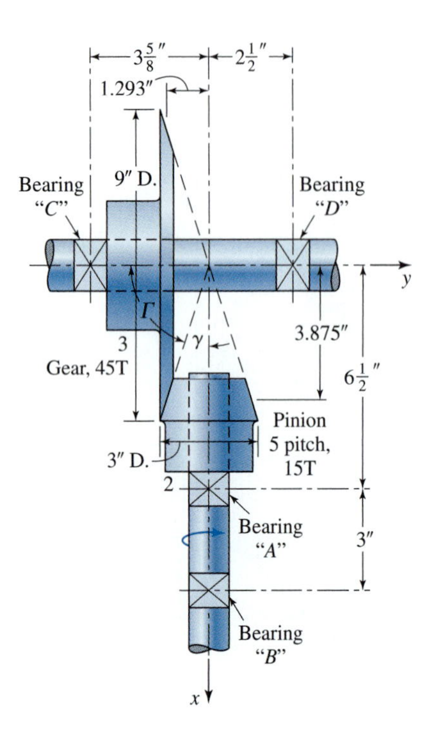

Figure 13.19 Bevel gearset and bearing locations for Example 13.8.

SOLUTION

The pitch angles for the pinion and gear are

$$\gamma = \tan^{-1}\left(\frac{1.5\text{ in}}{4.5\text{ in}}\right) = 18.4^\circ$$

$$\Gamma = \tan^{-1}\left(\frac{4.5\text{ in}}{1.5\text{ in}}\right) = 71.6^\circ.$$

The radii to the midpoints of the teeth are shown on the drawing and are $R_2 = 1.293$ in and $R_3 = 3.875$ in for the pinion and gear, respectively.

Let us first determine the forces acting on the pinion. Using Eq. (13.13*b*), we find the transmitted force to be

$$F_{32}^t = \frac{P}{R_2\omega_2} = \frac{(5\text{ hp})(33\,000\text{ ft}\cdot\text{lb/min/hp})(12\text{ in/ft})}{(1.293\text{ in})(600\text{ rev/min})(2\pi\text{ rad/rev})} = 406\text{ lb}.$$

This force acts in the negative z direction, that is, into the plane of Fig. 13.19. The radial and axial components of $\mathbf{F}_{32}$ are obtained from Eqs. (13.20) and (13.21),

$$F_{32}^r = 406\text{ lb}\tan 20^\circ\cos 18.4^\circ = 140\text{ lb}$$

$$F_{32}^a = 406\text{ lb}\tan 20^\circ\sin 18.4^\circ = 46.6\text{ lb}$$

where $\mathbf{F}_{32}^r$ acts in the positive y direction and $\mathbf{F}_{32}^a$ in the positive x direction.

These three forces are components of the total force $\mathbf{F}_{32}$. Thus,

$$\mathbf{F}_{32} = 46.6\hat{\mathbf{i}} + 140\hat{\mathbf{j}} - 406\hat{\mathbf{k}} \text{ lb}.$$

The torque applied to the pinion shaft is

$$\mathbf{T}_{12} = -\mathbf{R}_2 \times \mathbf{F}_{32}^t = -(-1.293\hat{\mathbf{j}} \text{ in}) \times (-406\hat{\mathbf{k}} \text{ lb}) = -525\hat{\mathbf{i}} \text{ in} \cdot \text{lb}.$$

A free-body diagram of the pinion and shaft is illustrated schematically in Fig. 13.20*a*. The dimensions, the torque $\mathbf{T}_{12}$, and the force $\mathbf{F}_{32}$ are known and the problem is to determine the bearing reactions $\mathbf{F}_A$ and $\mathbf{F}_B$. To determine $\mathbf{F}_B$ we sum moments about A, that is

$$\sum \mathbf{M}_A = \mathbf{T}_{12} + \mathbf{R}_{BA} \times \mathbf{F}_B + \mathbf{R}_{PA} \times \mathbf{F}_{32} = \mathbf{0}. \tag{1}$$

where the two position difference vectors are

$$\mathbf{R}_{BA} = 3.0\hat{\mathbf{i}} \text{ in} \quad \text{and} \quad \mathbf{R}_{PA} = -2.625\hat{\mathbf{i}} - 1.293\hat{\mathbf{j}} \text{ in}.$$

Therefore, the second and third terms of Eq. (1), respectively, are

$$\begin{aligned} \mathbf{R}_{BA} \times \mathbf{F}_B &= (3.0 \text{ in})\hat{\mathbf{i}} \times (F_B^y\hat{\mathbf{j}} + F_B^z\hat{\mathbf{k}}) \\ &= -(3.0 \text{ in})F_B^z\hat{\mathbf{j}} + (3.0 \text{ in})F_B^y\hat{\mathbf{k}} \end{aligned} \tag{2}$$

$$\begin{aligned} \mathbf{R}_{PA} \times \mathbf{F}_{32} &= (-2.625\hat{\mathbf{i}} - 1.293\hat{\mathbf{j}} \text{ in}) \times (46.6\hat{\mathbf{i}} + 140\hat{\mathbf{j}} - 406\hat{\mathbf{k}} \text{ lb}) \\ &= 525\hat{\mathbf{i}} - 1\,066\hat{\mathbf{j}} - 307\hat{\mathbf{k}} \text{ in} \cdot \text{lb}. \end{aligned} \tag{3}$$

Substituting the value of $\mathbf{T}_{12}$ and Eqs. (2) and (3) into Eq. (1) and solving, the bearing reaction at B is

$$\mathbf{F}_B = 102\hat{\mathbf{j}} - 355\hat{\mathbf{k}} \text{ lb}. \qquad \textit{Ans.}$$

The magnitude of the bearing reaction at B is 370 lb.

Next, to determine the bearing reaction at A, we write

$$\sum \mathbf{F} = \mathbf{F}_{32} + \mathbf{F}_A + \mathbf{F}_B = \mathbf{0}.$$

Then solving this equation, the bearing reaction at A is

$$\mathbf{F}_A = -46.6\hat{\mathbf{i}} - 242\hat{\mathbf{j}} + 761\hat{\mathbf{k}} \text{ lb}. \qquad \textit{Ans.}$$

The magnitude of the bearing reaction at A is 798 lb. The results are illustrated in Fig. 13.20*b*.

A similar procedure is used for the gear shaft. The results are displayed in Fig. 13.20*c*.

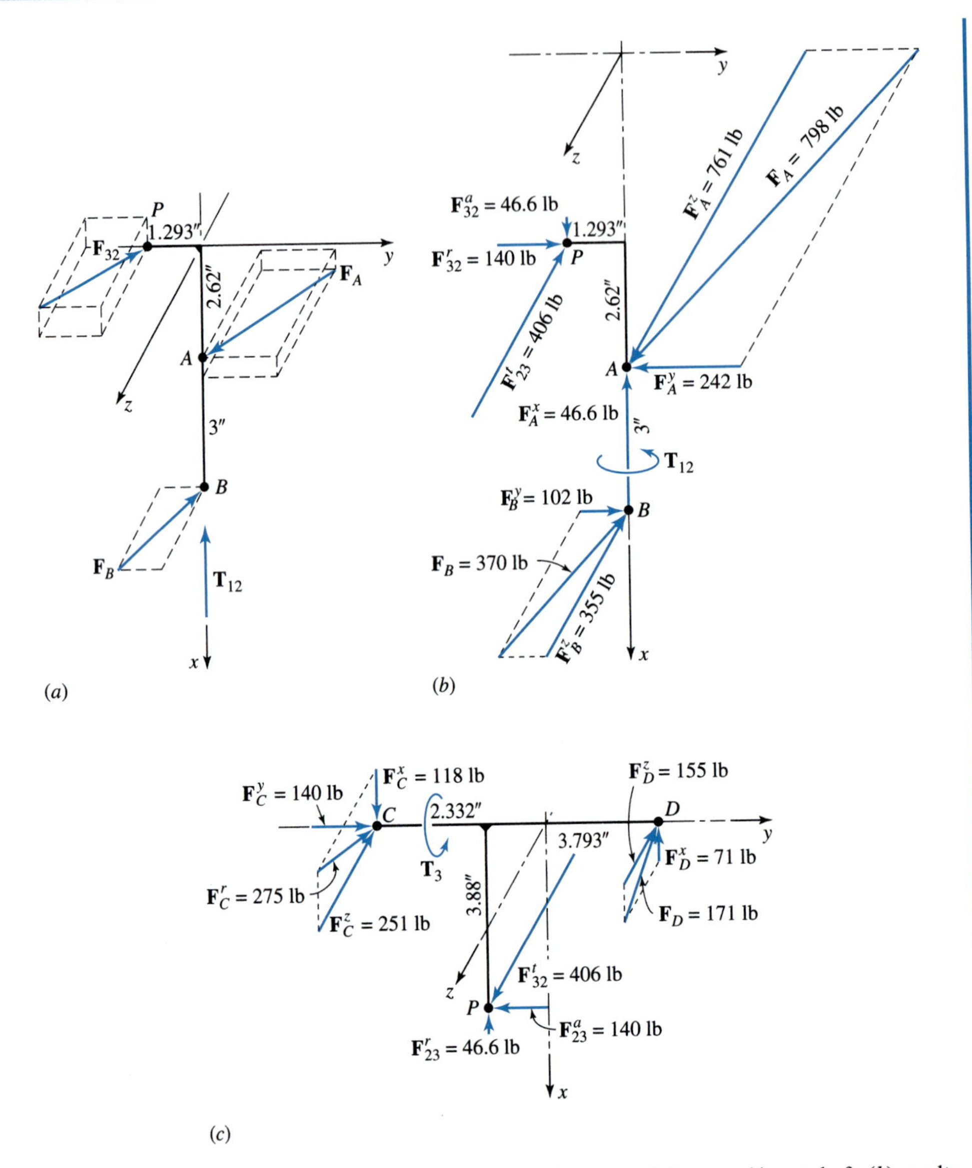

Figure 13.20 Solution for Example 13.8: (*a*) free-body diagrams of piston and input shaft; (*b*) results and component values on pinion and input shaft; and (*c*) free-body diagrams of gear and output shaft with results and component values.

13.13 THE METHOD OF VIRTUAL WORK

So far in Chapter 13 we have learned to analyze problems involving the equilibrium of mechanical systems by the application of Newton's laws. Another fundamentally different approach to force-analysis problems is based on the principle of virtual work, first proposed by the Swiss mathematician J. Bernoulli in the 18th century.

The method is based on an energy balance of the system that requires that the net change in internal energy during a small displacement must be equal to the difference between the work input to the system and the work output including the work done against friction, if any. Thus, for a system of rigid bodies in equilibrium under a system of applied forces, if given an arbitrary small displacement from equilibrium, the net change in the internal energy, denoted here by dU, is equal to the work dW done on the system:

$$dU = dW. \tag{13.22}$$

Work and change in internal energy are positive when work is done on the system, giving it increased internal energy and negative when energy is lost from the system, such as when it is dissipated through friction. If the system has no friction or other dissipation losses, energy is conserved and the net change in internal energy during a small displacement from equilibrium is zero.

Of course, such a method requires that we know how to calculate the work done by each force during the small *virtual displacement* chosen. If some force $\mathbf{F}$ acts at a point of application Q that undergoes a small displacement $d\mathbf{R}_Q$, then the work done by this force on the system is given by

$$dU = \mathbf{F} \cdot d\mathbf{R}_Q, \tag{13.23}$$

where dU is a scalar value, having units of work or energy, and is positive for work done onto the system and negative for work output from the system.

The displacement considered is called a *virtual* displacement because it need not be one that truly happens on the physical machine. It need only be a small displacement that is hypothetically possible and consistent with the constraints imposed on the system. The small displacement relationships of the system can be determined through the principles of kinematics covered in Part 1 of this book, and the input-to-output force relationships of the machine can therefore be determined. This will become clearer through the following example.

EXAMPLE 13.9

Repeat the static force analysis of the four-bar linkage analyzed in Example 13.1. Find the input crank torque $\mathbf{M}_{12}$ required for equilibrium. Friction effects and the weights of the links can be neglected.

SOLUTION

From Example 13.1 we recall that the position difference vector from O_4 to Q and the applied force at Q, respectively, are

$$\mathbf{R}_{QO_4} = 1.84\hat{\mathbf{i}} + 4.65\hat{\mathbf{j}} \text{ in.} \tag{1}$$

$$\mathbf{P} = 120 \text{ lb}\angle 220^\circ = -91.9\hat{\mathbf{i}} - 77.1\hat{\mathbf{j}} \text{ lb.} \tag{2}$$

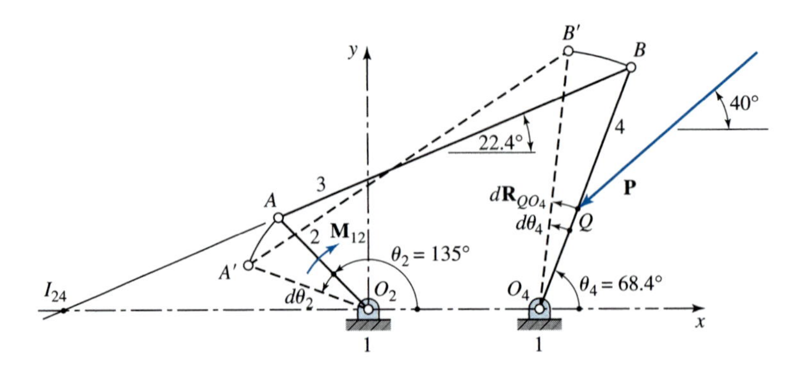

Figure 13.21 Virtual-work solution for Example 13.9.

Figure 13.21 presents a scale diagram of the system at the position in question. If a small virtual displacement $d\theta_2$ is given to the input crank 2, this results in a small displacement $d\theta_4$ of the output crank 4, along with a small displacement $d\mathbf{R}_{QO_4}$ of point Q as illustrated.

Assuming that these virtual displacements are small and happen in a small time increment dt, the velocity difference equation, Eq. (3.3), can be multiplied by dt to yield the relationship between $d\mathbf{R}_{QO_4}$ and $d\theta_4$ as follows:

$$\begin{aligned} d\mathbf{R}_{QO_4} &= d\theta_4\hat{\mathbf{k}} \times \mathbf{R}_{QO_4} = (d\theta_4\hat{\mathbf{k}} \text{ rad}) \times (1.84\hat{\mathbf{i}} + 4.65\hat{\mathbf{j}} \text{ in}) \\ &= (-4.65\hat{\mathbf{i}} + 1.84\hat{\mathbf{j}} \text{ in})d\theta_4. \end{aligned} \tag{3}$$

Because the input torque $\mathbf{M}_{12}$ and the applied force $\mathbf{P}$ are the only forces doing work during the chosen displacement, Eq. (13.22) can be written for this problem as

$$dU = M_{12}\hat{\mathbf{k}} \cdot d\theta_2\hat{\mathbf{k}} + \mathbf{P} \cdot d\mathbf{R}_{QO_4} = 0.$$

Substituting Eqs. (2) and (3) into this equation gives

$$M_{12}d\theta_2 + (-91.9\hat{\mathbf{i}} - 77.1\hat{\mathbf{j}} \text{ lb}) \cdot (-4.65\hat{\mathbf{i}} + 1.84\hat{\mathbf{j}} \text{ in})d\theta_4 = 0.$$

Then, dividing by $d\theta_2$, rearranging, and recognizing the first-order kinematic coefficient, we find

$$M_{12} = (-285.5 \text{ in} \cdot \text{lb})\frac{d\theta_4}{d\theta_2} = -285.5\theta_4' \text{ in} \cdot \text{lb}. \tag{4}$$

By dividing the numerator and denominator of this angular displacement ratio by dt, we recognize that this is equal to the angular-velocity ratio ω_4/ω_2, which can be determined from the angular velocity ratio theorem of Eq. (3.28).* Using the location of the instant

* This technique of treating small displacement ratios as velocity ratios or first-order kinematic coefficients can often be extremely helpful because, as we remember from Section 3.8, velocity relationships lead to linear equations rather than the nonlinear relations implied by position or displacement. Velocity polygons and instant-center methods are also helpful.

center I_{24} and measurements from Fig. 13.21, we find

$$\theta_4' = \frac{d\theta_4}{d\theta_2} = \frac{\omega_4}{\omega_2} = \frac{R_{I_{24}I_{12}}}{R_{I_{24}I_{14}}} = \frac{14.3 \text{ in}}{22.3 \text{ in}} = +0.641 \text{ rad/rad}.$$

Finally, substituting this result into Eq. (4) gives

$$\mathbf{M}_{12} = (-285.5\hat{\mathbf{k}} \text{ in} \cdot \text{lb})(+0.641 \text{ rad/rad}) = -183.1\hat{\mathbf{k}} \text{ in} \cdot \text{lb} \qquad \textit{Ans.}$$

One primary advantage of the method of virtual work for force analysis over the other methods demonstrated above comes in problems where the input-to-output force relationships are sought. Note that the constraint forces are not required in this solution technique because both their action and their reaction forces are internal to the system, both move through identical displacements, and thus their virtual-work contributions cancel each other. This would not be true for friction forces where the displacements would be different for the action and reaction force components, with this work difference representing the energy dissipated. Otherwise, internal constraint forces need not be considered because they cause no net virtual work.

13.14 EULER COLUMN FORMULA

As mentioned in Example 13.1, an important issue in machine design is the problem of buckling of a long, slender member subject to a compressive load. When a member buckles, it experiences bending and loses strength. The loss in strength can be very dramatic and for this reason a buckling failure in a structure can occur without warning and is often catastrophic. Figure 13.22 illustrates the buckling collapse of two support columns of the Harbor Expressway (Kobe, Japan). Here we can see that buckling is a very important

Figure 13.22 A catastrophic failure caused by buckling.

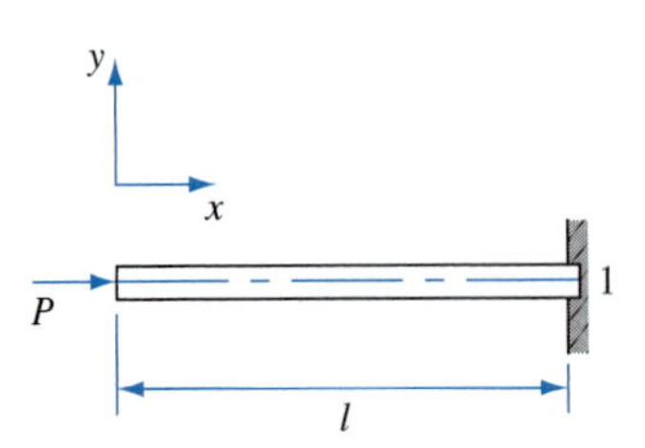

Figure 13.23 A long member subjected to a central compressive load P.

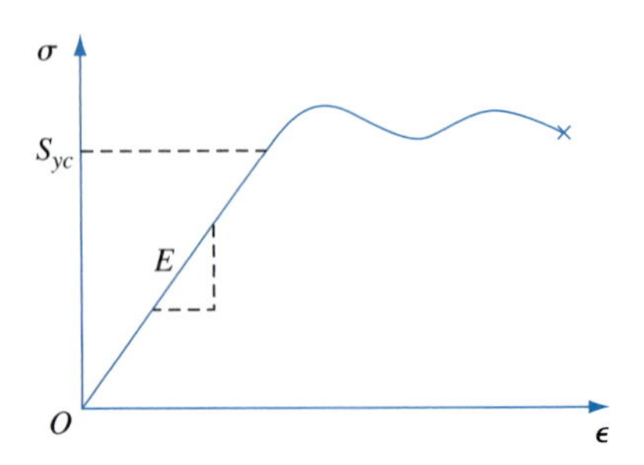

Figure 13.24 Stress versus strain for a mild steel bar in tension.

problem. The buckling of a member is also a fascinating topic in mechanical design and is the subject of this section and the following three sections.

Consider a long, slender member of length L with fixed-free end conditions, as illustrated in Fig. 13.23. Assume that the member is subject to the central compressive load P shown in the figure. The first problem is to derive an equation for the critical value of the load P, that is, the load which will cause the member to become unstable. This load is referred to as the critical load.

For small deflections and for homogeneous and isotropic materials, the curvature of the member caused by the compressive load P can be written as

$$\kappa = \frac{d^2y}{dx^2} = \frac{M}{EI}, \tag{13.24}$$

where d^2y/dx^2 is the second derivative of the transverse deflection, M is the bending moment at an arbitrary location x along the member, E is Young's modulus, and I is the second moment of area of the member cross-section at location x. The bending moment can be written as

$$M = -Py. \tag{13.25}$$

where y is the transverse deflection of the member at location x.

Young's modulus is the modulus of elasticity of the material of the member; it has units of stress (that is, lb/in^2 or Pa) and is a measure of the stiffness of the material. Consider a typical plot of stress σ versus strain ε for a mild steel bar in tension; see Fig. 13.24. Hooke's law states that, in the elastic region, the stress in the bar is proportional to the strain, with a proportionality constant of the modulus of elasticity. This can be written as

$$E = \sigma/\varepsilon.$$

A rule of thumb for the modulus of elasticity for: (i) a steel alloy is

$$E = 30 \text{ Mpsi} = 30\times10^6 \text{ lb/in}^2 \quad \text{or} \quad E = 207 \text{ GPa} = 207\times10^9 \text{ Pa}.$$

and (*ii*) a stainless steel is

$$E = 27.5 \text{ Mpsi} \quad \text{or} \quad E = 190 \text{ GPa}$$

Substituting Eq. (13.25) into Eq. (13.24), the second derivative of the transverse deflection of the member can be written as

$$\frac{d^2y}{dx^2} = -\frac{Py}{EI}.$$

Rearranging this equation gives the second-order differential equation for the member,

$$\frac{d^2y}{dx^2} + \left(\frac{P}{EI}\right)y = 0. \tag{13.26}$$

The solution to this differential equation, that is, the transverse deflection of the member, is

$$y = A\sin\left(\sqrt{\frac{P}{EI}}x\right) + B\cos\left(\sqrt{\frac{P}{EI}}x\right), \tag{13.27}$$

where A and B are the constants of integration and will be obtained from the support conditions at the ends of the member.

Pinned–pinned end conditions Consider the case where both ends of the member are pinned, as illustrated in Fig. 13.25 (also referred to in civil engineering as rounded–rounded ends).

The end conditions are: (*i*) at $x = 0$, the deflection $y = 0$, and (*ii*) at $x = L$, the deflection $y = 0$. Substituting $y = 0$ at $x = 0$ into Eq. (13.27) gives

$$0 = 0 + B\cos 0.$$

Therefore, the constant of integration is $B = 0$. Substituting this into Eq. (13.27) gives

$$y = A\sin\left(\sqrt{\frac{P}{EI}}x\right). \tag{13.28}$$

Then, substituting $y = 0$ at $x = L$ into this equation gives

$$0 = A\sin\left(\sqrt{\frac{P}{EI}}L\right).$$

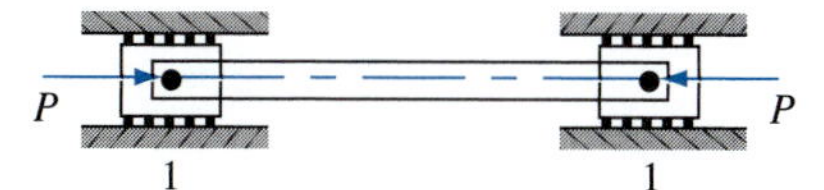

Figure 13.25 A member with pinned–pinned ends.

There are two solutions to this equation. Either the constant of integration $A = 0$, which is a trivial solution (it implies that the deflection is $y = 0$ for all values of x) and can be ignored, or

$$\sin\left(\sqrt{\frac{P}{EI}}L\right) = 0.$$

The solution to this equation is

$$\sqrt{\frac{P_{CR}}{EI}}L = n\pi \quad (n = 1, 2, 3, \ldots), \tag{13.29}$$

where P_{CR} is a compressive load that places the member in a condition of unstable equilibrium and is referred to as the *critical load*. The first critical load (that is, the lowest or minimum value of P_{CR}) is the most important from practical considerations. For $n = 1$, the critical load is

$$P_{CR} = \left(\frac{\pi^2}{L^2}\right)EI. \tag{13.30}$$

This equation is commonly referred to as *Euler's column formula* for a long, slender member with pinned–pinned ends. Note an interesting result that the strength of the material is not a factor; that is, the critical load depends only on: (*a*) the length L, (*b*) the second moment of area I, and (*c*) Young's modulus E. Therefore, if the member is steel, then using a stronger steel (with a higher compressive yield strength S_{yc}) will not help because all steel alloys have essentially the same modulus of elasticity E (approximately 30 Mpsi or 207 GPa).

Substituting Eq. (13.30) into Eq. (13.28), the transverse deflection of the member at the critical load can be written as

$$y = A\sin\left(\frac{\pi x}{L}\right), \tag{13.31}$$

which indicates that the deflection curve of the member (for pinned-pinned ends) is a half-sine wave, as illustrated in Fig. 13.26.

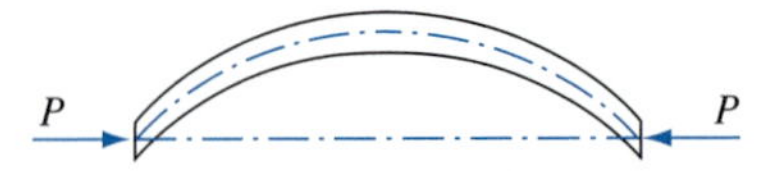

Figure 13.26 The deflection curve of a member with pinned-pinned ends is a half-sine wave.

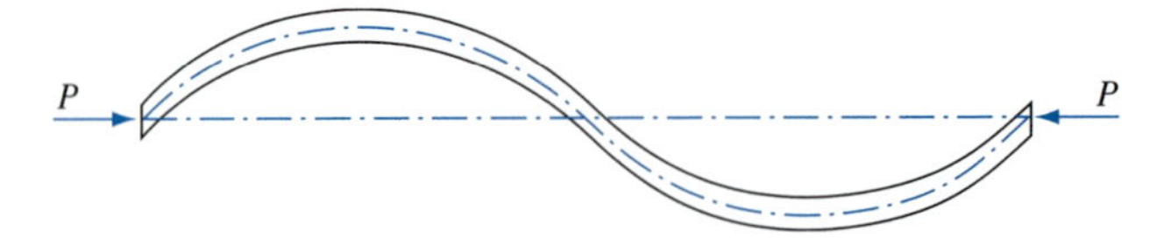

Figure 13.27 The deflection curve is a full sine wave.

Values of $n > 1$ in Eq. (13.29) result in deflection curves that cross the X-axis at points of inflection and are multiples of half-sine waves. For $n = 2$, it is a full sine wave, as illustrated in Fig. 13.27.

13.15 THE CRITICAL UNIT LOAD

Next we hope to obtain an expression for the critical load per unit cross-sectional area of the member, referred to as the *critical unit load*, which has the units of stress or strength. The second moment of area (also called the *area moment of inertia**) of the member cross-section in a plane perpendicular to the axis of the member can be written as

$$I = Ak^2,$$

where k is called the *radius of gyration*†. Rearranging this equation, the radius of gyration of the member cross-section can be written as

$$k = \sqrt{\frac{I}{A}}.$$

From this we define a nondimensional parameter called the *slenderness ratio*,

$$S_r = \frac{L}{k}. \tag{13.32}$$

The slenderness ratio, rather than the actual length L, is commonly used in classifying members in compression according to length categories.

Substituting this definition into Eq. (13.30), the first *critical load* can be written as

$$P_{\mathrm{CR}} = \frac{\pi^2 EA}{S_r^2}.$$

The *critical unit load* is defined by dividing this critical load by the cross-sectional area,

$$\frac{P_{\mathrm{CR}}}{A} = \frac{\pi^2 E}{S_r^2}. \tag{13.33}$$

The critical unit load represents the strength of a particular member under compression, rather than the strength of the material from which it is made, that is, the yield strength or the ultimate strength.

Including the pinned–pinned end conditions presented above, there are four other common pairs of end conditions for a member. These four are: (*i*) fixed–fixed ends, (*ii*) fixed–pinned ends, (*iii*) pinned–pinned ends, and (*iv*) fixed–free ends. The critical unit load

* Note that the *area* moment of inertia, shown here, is quite different from the *mass* moment of inertia of Section 14.3. Note the dimensions of length to the fourth power compared with mass length squared.

† Note that this is *not* the same as the term with similar name defined in Eq. (14.11).

for each of these four cases can be written in the form of Eq. (13.33) using an end-condition constant C; that is,

$$\frac{P_{\mathrm{CR}}}{A} = C\frac{\pi^2 E}{S_r^2}. \tag{13.34}$$

This equation is commonly referred to as the Euler column formula for different end conditions.

To determine the end-condition constant C, the right-hand side of this equation can be written as

$$C\frac{\pi^2 E}{S_r^2} = \frac{\pi^2 E}{(L_{\mathrm{EFF}}/k)^2},$$

where L_{EFF} denotes the effective length of the member. The effective length is the length of the half-sine wave for the equivalent member. Therefore, by rearranging this equation, the end condition constant can be expressed in terms of the effective length of the member as $C = (L/L_{\mathrm{EFF}})^2$. Also, using the American Institute of Steel Construction (AISC) recommended column end-condition effective length factor $\alpha = L_{\mathrm{EFF}}/L$, the end condition constant can be written as $C = 1/\alpha^2$.

Case (*i*). Fixed–fixed ends are illustrated in Fig. 13.28.

For the case of a member with fixed–fixed ends, the effective length of the member is $L_{\mathrm{EFF}} = 0.5L$. Substituting this value into the above equations, the end-condition constant is $C = 4$.

Case (*ii*). Fixed–pinned ends are illustrated in Fig. 13.29.

For the case of a member with fixed–pinned ends, the effective length of the member is $L_{\mathrm{EFF}} = 0.707L$. Using this value, the above equations give the end-condition constant $C = 2$.

Case (*iii*). Pinned–pinned ends, illustrated in Fig. 13.25.

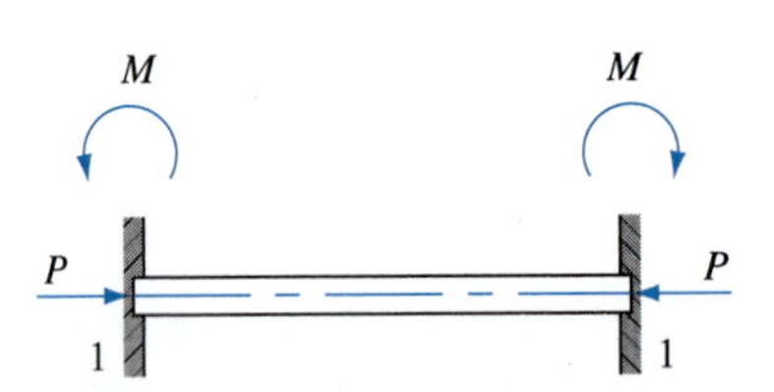

Figure 13.28 Fixed–fixed ends.

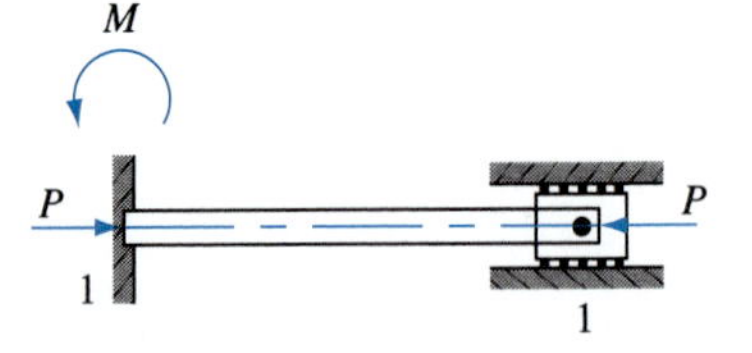

Figure 13.29 Fixed–pinned ends.

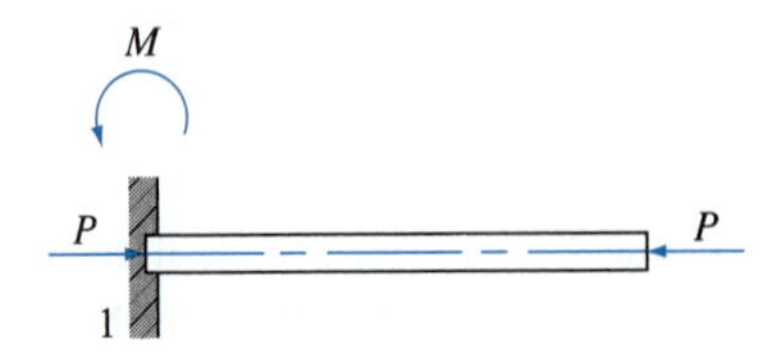

Figure 13.30 Fixed–free ends.

The case of a member with pinned–pinned ends is the case used for the above equations. The effective length of the member is $L_{\text{EFF}} = L$, and the end-condition constant is $C = 1$.

Case (*iv*). Fixed–free ends are illustrated in Fig. 13.30.

For the case of a member with fixed–free ends, the effective length of the member is $L_{\text{EFF}} = 2L$. With this value in the above equations, the end-condition constant is $C = 1/4$.

13.16 CRITICAL UNIT LOAD AND THE SLENDERNESS RATIO

A typical plot of the critical unit load versus the slenderness ratio of a long, slender member subjected to an axial compressive load P is illustrated in Fig. 13.31.

Note that there are two distinct curves illustrated in Fig. 13.31: (*i*) Euler's column formula; and (*ii*) *Johnson's parabolic equation**. The criterion for using Euler's column formula, Eq. (13.34), is

$$S_r > (S_r)_D, \tag{13.35}$$

where $(S_r)_D$ is the slenderness ratio at the point of tangency D between Euler's column formula and Johnson's parabolic equation. The critical unit load at the point of tangency is

$$\frac{P_{\text{CR}}}{\text{A}} = 0.5S_{yc}.$$

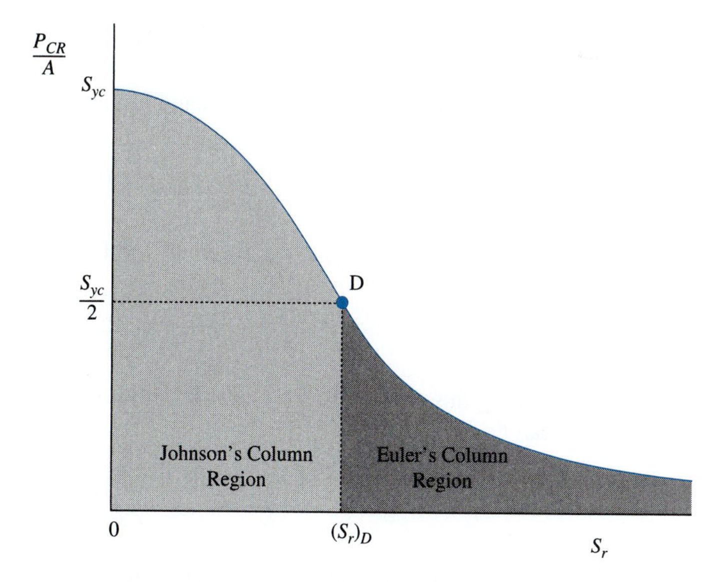

Figure 13.31 The critical unit load versus the slenderness ratio.

* Proposed by Prof. J. B. Johnson in 1898 and, independently, by Prof. A. Ostenfeld of Copenhagen.

Substituting this equation into Eq. (13.34) and rearranging, the slenderness ratio at the point of tangency, that is, when the slenderness ratio $S_r = (S_r)_D$, can be written as

$$(S_r)_D = \pi\sqrt{\frac{2CE}{S_{yc}}}. \tag{13.36}$$

If the slenderness ratio of the member is less than this value, $S_r < (S_r)_D$, then the criterion that is most commonly used for buckling is Johnson's parabolic equation.

13.17 JOHNSON'S PARABOLIC EQUATION

The equation of the parabola on the left side of Fig. 13.31 can be written as

$$\frac{P_{\text{CR}}}{\text{A}} = a + bS_r^2. \tag{13.37}$$

The focus here is to determine the constants a and b. There are two known data points in Fig. 13.31, namely, (*i*) for the slenderness ratio $S_r = 0$, the critical unit load is $P_{\text{CR}}/A = S_{yc}$, and (*ii*) for the slenderness ratio $S_r = (S_r)_D$, the critical load is $P_{\text{CR}}/A = 0.5S_{yc}$.

Substituting the first of these data points into Eq. (13.37), we find the coefficient

$$a = S_{yc}.$$

Substituting this result into Eq. (13.37), the critical unit load can be written as

$$\frac{P_{\text{CR}}}{\text{A}} = S_{yc} + bS_r^2. \tag{13.38}$$

Then substituting the second data point and Eq. (13.36) into Eq. (13.38) gives

$$0.5S_{yc} = S_{yc} + b\pi^2\frac{2CE}{S_{yc}}.$$

Finally, rearranging this equation, the coefficient b can be written as

$$b = -\frac{S_{yc}^2}{4CE\pi^2}.$$

Substituting this result into Eq. (13.38), Johnson's parabolic equation for the critical unit load can be written as

$$\frac{P_{\text{CR}}}{\text{A}} = S_{yc} - \frac{1}{CE}\left(\frac{S_{yc}S_r}{2\pi}\right)^2. \tag{13.39}$$

Therefore, the critical load for a member using Johnson's parabolic equation is

$$P_{\text{CR}} = A\left[S_{yc} - \frac{1}{CE}\left(\frac{S_{yc}S_r}{2\pi}\right)^2\right]. \tag{13.40}$$

EXAMPLE 13.10

For the four-bar linkage in the position illustrated in Fig. 13.32, a vertical load P is acting at point E on the horizontal coupler link 3 (i.e., the link is parallel to the X-axis). The coupler link is pinned to the vertical side links 2 and 4 at points A and B. The links are made of a steel alloy with a compressive yield strength $S_{yc} = 60$ kpsi and a modulus of elasticity $E = 30$ Mpsi. The side links have hollow circular cross-sections with 2-in. outside diameters and 0.25-in wall thicknesses. Assume that the value for the end-condition constant for both side links is $C = 1$. The known link lengths are $R_{AO} = 8$ ft, $R_{BA} = 6$ ft, $R_{BC} = 5$ ft, and $R_{EA} = 4$ ft. Determine: (*i*) the radii of gyration for links 2 and 4, (*ii*) the values of the slenderness ratios for links 2 and 4 and at the point of tangency between Euler's column formula and Johnson's parabolic equation, (*iii*) the critical loads for links 2 and 4, and (*iv*) the maximum value of load P if the factors of safety guarding against buckling for links 2 and 4 are both specified as $N = 2$.

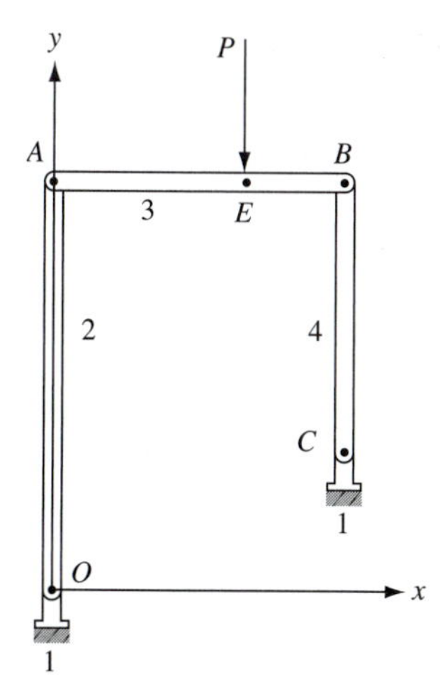

Figure 13.32 Example 13.10. A four-bar linkage subjected to a vertical load P.

SOLUTION

(*i*) The cross-sectional area of each side link is

$$A = \frac{\pi}{4}(D^2 - d^2) = \frac{\pi}{4}[(2 \text{ in})^2 - (1.5 \text{ in})^2] = 1.374 \text{ in}^2. \tag{1}$$

The area moment of inertia for the hollow circular cross-section of each side link is (see Appendix A, Table 4)

$$I = \frac{\pi}{64}(D^4 - d^4) = \frac{\pi}{64}[(2 \text{ in})^4 - (1.5 \text{ in})^4] = 0.537 \text{ in}^4. \tag{2}$$

The radius of gyration for each side link is

$$k = \sqrt{\frac{I}{A}} = \sqrt{\frac{0.537 \text{ in}^4}{1.374 \text{ in}^2}} = 0.625 \text{ in}. \qquad \textit{Ans.} \tag{3}$$

(*ii*) From the given end conditions for links 2 and 4, that is, pinned–pinned ends, the end condition constant is $C = 1.0 = (1/\alpha)^2$ and $\alpha = 1.0 = L/L_{\text{EFF}}$. This implies that the effective length L_{EFF} is equal to the actual length L for both link 2 and link 4. Therefore, the slenderness ratio of link 2 is

$$(S_r)_2 = \frac{(L_{\text{EFF}})_2}{k} = \frac{(8\text{ ft})\ 12\text{ in/ft}}{0.625\text{ in}} = 153.60 \qquad \textit{Ans.}\ (4a)$$

and the slenderness ratio of link 4 is

$$(S_r)_4 = \frac{(L_{\text{EFF}})_4}{k} = \frac{(5\text{ ft})\ 12\text{ in/ft}}{0.625\text{ in}} = 96.00. \qquad \textit{Ans.}\ (4b)$$

The slenderness ratio for either side link at the point of tangency is

$$(S_r)_D = \pi\sqrt{\frac{2CE}{S_{\text{yc}}}} = \pi\sqrt{\frac{2 \cdot 1.0(30 \times 10^6\text{ psi})}{60 \times 10^3\text{ psi}}} = 99.35. \qquad \textit{Ans.}\ (5)$$

(*iii*) The critical load is determined by first determining whether each link is considered an Euler column or a Johnson column. Note that the slenderness ratios given by Eqs. (4*a*), (4*b*), and (5) are

$$(S_r)_2 = 153.60, \quad (S_r)_4 = 96.00, \quad \text{and} \quad (S_r)_D = 99.35.$$

The criterion for using Euler's column formula is given in Eq. (13.35) as $S_r > (S_r)_D$. Therefore, the conclusion is that, because $153.60 > 99.35$, link 2 is regarded as an Euler column. The critical unit load for link 2, from Euler's column formula, Eq. (13.34), can be written as

$$\frac{(P_{\text{CR}})_2}{A} = \frac{C\pi^2 E}{(S_r)_2^2}. \qquad (6)$$

Substituting the known values into Eq. (6) gives

$$\frac{(P_{\text{CR}})_2}{1.374\text{ in}^2} = \frac{1.0\pi^2(30 \times 10^6\text{ psi})}{153.60^2},$$

Therefore, the critical load for link 2 is

$$(P_{\text{CR}})_2 = 17\,244\text{ lb}. \qquad \textit{Ans.}\ (7)$$

The criterion for using Johnson's parabolic equation is $S_r < (S_r)_D$. Therefore, because $96.00 < 99.35$, Johnson's parabolic equation must be used for link 4. The critical load for link 4 from Johnson's parabolic equation is given by Eq. (13.40), that is

$$(P_{\text{CR}})_4 = A\left[S_{yc} - \frac{1}{CE}\left(\frac{S_{yc}S_r}{2\pi}\right)^2\right].$$

Substituting the known values into this equation gives

$$(P_{CR})_4 = (1.374\ \text{in}^2)$$

$$\times \left[(60 \times 10^3\ \text{psi}) - \frac{1}{1.0(30 \times 10^6\ \text{psi})} \left(\frac{(60 \times 10^3\ \text{psi})96.00}{2\pi} \right)^2 \right].$$

Therefore, the critical load for link 4 is

$$(P_{CR})_4 = 43\,950\ \text{lb}. \qquad \textit{Ans.}\ (8)$$

(*iv*) For the given factor of safety guarding against buckling, an equation can be written for the applied load on either of the two side links as follows:

$$P_{APP} = \frac{P_{CR}}{N} = \frac{P_{CR}}{2}.$$

The following two cases must be compared.

Case (*a*): When the load acting at point A from link 3 onto link 2 is a maximum, the maximum applied load at point A can be written from Eq. (7) as

$$(P_{APP})_2 = \frac{(P_{CR})_2}{2} = \frac{17\,244\ \text{lb}}{2} = 8\,622\ \text{lb}.$$

To determine the load on link 2 from the applied load P, we take moments about point B. Therefore, the applied load P necessary to cause buckling of link 2 is

$$(P)_2 = \frac{R_{AB}(P_{APP})_2}{R_{EB}} = \frac{6\ \text{ft}\ (8\,622\ \text{lb})}{2\ \text{ft}} = 25\,866\ \text{lb}. \qquad (9a)$$

Case (*b*). When the load acting at point B from link 3 onto link 4 is a maximum, the maximum applied load at point B can be written from Eq. (8) as

$$(P_{APP})_4 = \frac{(P_{CR})_4}{2} = \frac{43\,950\ \text{lb}}{2} = 21\,975\ \text{lb}.$$

To determine the load on link 4 from the applied load P, we take moments about point A. Therefore, the applied load P necessary to cause buckling of link 4 is

$$(P)_4 = \frac{R_{BA}(P_{APP})_4}{R_{EA}} = \frac{6\ \text{ft}\ (21\,975\ \text{lb})}{4\ \text{ft}} = 32\,963\ \text{lb}. \qquad (9b)$$

To insure a factor of safety of at least 2 on both beams, we must take the minimum of the applied loads from Cases (*a*) and (*b*). That is, in comparing Eqs. (9*a*) and (9*b*), because $(P)_2 < (P)_4$ (that is, 25 866 lb < 32 963 lb), the maximum load P to guard against buckling of either link with a factor of safety of 2 is

$$P < 25\,866\ \text{lb}. \qquad \textit{Ans.}$$

EXAMPLE 13.11

A force P is applied vertically upward on the horizontal link 4 that is pinned to the ground at point B and pinned to the vertical link 2 at point A, as illustrated in Fig. 13.33. Link 2 is fixed to the ground at point D. Assume that links 2 and 4 are in static equilibrium. The distances are $R_{BA} = 1$ m and $R_{CA} = 3$ m and links 2 and 4 are made from a steel alloy with a compressive yield strength $S_{yc} = 415$ MPa and a modulus of elasticity $E = 207$ GPa. The length of the vertical link 2 is $R_{DA} = 2.75$ m and this link has a hollow circular cross-section with an outside diameter of 50 mm and a wall thickness of 6 mm. Using the theoretical value for the end-condition constant for link 2, determine: (*i*) the value of the slenderness ratio at the point of tangency between Euler's column formula and Johnson's parabolic formula; (*ii*) the critical load acting on link 2 and the critical unit load on link 2; and (*iii*) the force P for the factor of safety of link 2 to guard against buckling to be $N = 2$. (*iv*) If link 2 is replaced with a solid circular cross-section of the same material with diameter of 40 mm and the force $P = 70$ kN, then determine the maximum length of link 2 for the factor of safety to guard against buckling of the link to be $N = 2$.

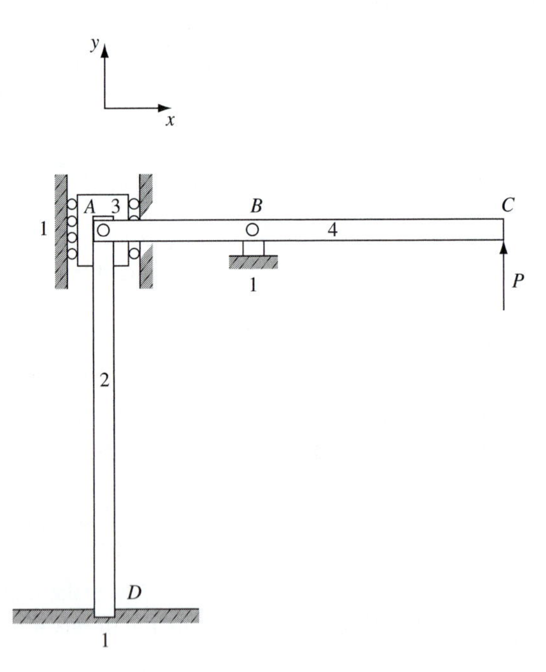

Figure 13.33 Example 13.11. Link 4 is pinned to the ground at point B and to links 2 and 3 at point A.

SOLUTION

(*i*) From the fixed–pinned end conditions of link 2, the end-condition constant for this link is $C = 2$. Therefore, the slenderness ratio for link 2 at the point of tangency is

$$(S_r)_D = \pi\sqrt{\frac{2CE}{S_{yc}}} = \pi\sqrt{\frac{2 \cdot 2(207 \times 10^9 \text{ Pa})}{415 \times 10^6 \text{ Pa}}} = 140.3. \qquad \textit{Ans.} \text{ (1)}$$

(*ii*) The cross-sectional area of link 2 can be written as

$$A = \frac{\pi}{4}(D^2 - d^2) = \frac{\pi}{4}[(0.050\ \text{m})^2 - (0.038\ \text{m})^2] = 0.000\ 829\ \text{m}^2.$$

The second moment of area of link 2 can be written as (see Appendix A, Table 4)

$$I = \frac{\pi}{64}(D^4 - d^4) = \frac{\pi}{64}[(0.050\ \text{m})^4 - (0.038\ \text{m})^4] = 0.2044 \times 10^{-6}\ \text{m}^4.$$

The radius of gyration of link 2 can be written as

$$k = \sqrt{\frac{I}{A}} = \sqrt{\frac{0.2044 \times 10^{-6}\ \text{m}^4}{0.000\ 829\ \text{m}^2}} = 0.0157\ \text{m}.$$

The slenderness ratio of link 2 is

$$S_r = \frac{L}{k} = \frac{2.75\ \text{m}}{0.0157\ \text{m}} = 175.2. \qquad (2)$$

To determine the critical load on link 2, we must first determine whether this link is an Euler column or a Johnson column. The criterion for using Johnson's parabolic equation is $S_r < (S_r)_D$. From Eqs. (1) and (2), we have $S_r > (S_r)_D$, that is, $175.2 > 140.3$. Therefore, link 2 is an Euler column.

The critical load on link 2, from the Euler column equation, is

$$P_{CR} = A\left[\frac{C\pi^2 E}{S_r^2}\right] = (0.000\ 829\ \text{m}^2)\left[\frac{2\pi^2(207 \times 10^9\ \text{Pa})}{175.2^2}\right] = 110.35\ \text{kN}.$$

Ans.

Therefore, the critical unit load on link 2 is

$$\frac{P_{CR}}{A} = \frac{110.35\ \text{kN}}{0.000\ 829\ \text{m}^2} = 133.1\ \text{MPa}. \qquad \textit{Ans.}$$

(*iii*) Links 2 and 4 are in static equilibrium. The free body diagram of link 4 is as illustrated in Fig. 13.34.

Taking moments about pin B gives

$$\mathbf{R}_{CB} \times \mathbf{P} = \mathbf{R}_{AB} \times \mathbf{F}_{24}^{y}.$$

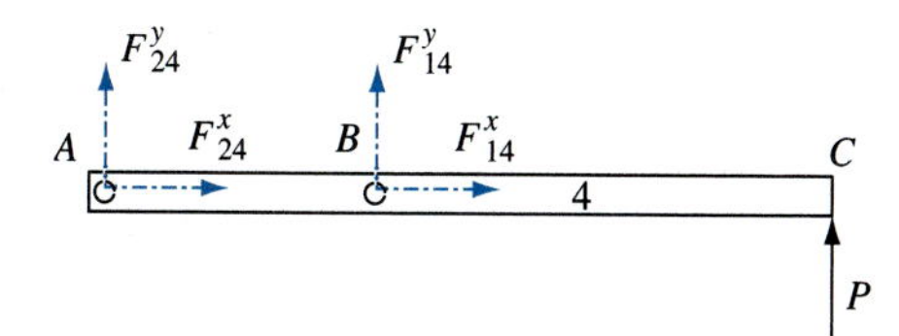

Figure 13.34 The free body diagram of link 4.

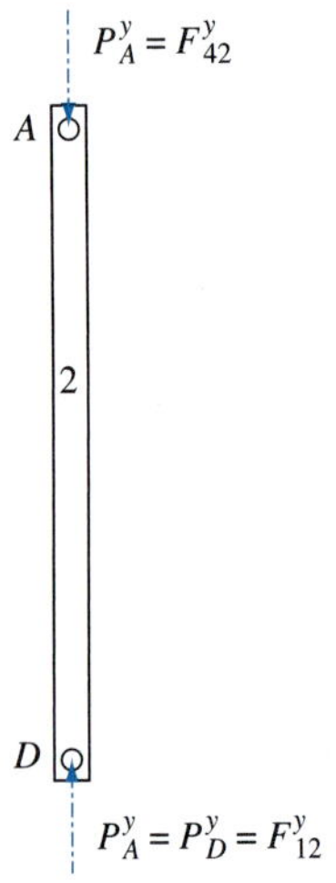

Figure 13.35 Link 2 is in compression.

Substituting the given data into this equation gives

$$P = \frac{(1\text{ m})F_{24}^y}{2\text{ m}} = \frac{F_{24}^y}{2}. \tag{3}$$

Link 2 is in compression, as illustrated in Fig. 13.35.

The reaction force $P_A^y = F_{42}^y$ is opposite to the reaction force F_{24}^y; therefore, the magnitude of P_A^y is equal to the magnitude of F_{24}^y. Equation (3) can be written as

$$P = \frac{P_A^y}{2}. \tag{4}$$

The factor of safety guarding against buckling for link 2 is defined as

$$N = \frac{P_{CR}}{P_A^y}, \tag{5}$$

where $P_A^y = F_{42}^y$ is the compressive load at point D on link 2, as illustrated in Fig. 13.35. Therefore, rearranging Eq. (5), the force P_A^y is

$$P_A^y = \frac{P_{CR}}{N} = \frac{110.35\text{ kN}}{2} = 55.175\text{ kN}. \tag{6}$$

Finally, substituting Eq. (6) into Eq. (4) gives

$$P = \frac{P_A^y}{2} = \frac{55.175\text{ kN}}{2} = 27.588\text{ kN}. \qquad \textit{Ans.}$$

(*iv*) Link 2 is a solid circular column with a factor of safety $N = 2$ and $P = 70$ kN.

Rearranging Eq. (4), the compressive force is

$$P_A^y = 2P = 2(70 \text{ kN}) = 140 \text{ kN}.$$

Rearranging Eq. (5), the critical load applied on link 2 is

$$P_{\text{CR}} = P_A^y \cdot N = 140 \text{ kN} \cdot 2 = 280 \text{ kN}.$$

First, assume the link is an Euler column. Then from Euler's column formula, see Eq. (13.34), the critical load applied on link 2 is

$$P_{\text{CR}} = A\left[\frac{C\pi^2 E}{S_r^2}\right]. \tag{7}$$

The cross-sectional area of link 2 is

$$A = \frac{\pi}{4}(D^2) = \frac{\pi}{4}(0.040 \text{ m})^2 = 0.001\,26 \text{ m}^2.$$

The second moment of area of the cross-section of link 2 is

$$I = \frac{\pi}{64}(D^4) = \frac{\pi}{64}(0.040 \text{ m})^4 = 0.126 \times 10^{-6} \text{ m}^4.$$

The radius of gyration of link 2 is

$$k = \sqrt{\frac{I}{A}} = \sqrt{\frac{0.126 \times 10^{-6} \text{ m}^4}{0.001\,26 \text{ m}^2}} = 0.010 \text{ m}.$$

Rearranging Eq. (7), the slenderness ratio of link 2 is

$$(S_r)_{\text{Euler}} = \sqrt{\frac{C\pi^2 EA}{P_{\text{CR}}}} = \sqrt{\frac{2\pi^2(207 \times 10^9 \text{ Pa})(0.001\,26 \text{ m}^2)}{280 \times 10^3 \text{ N}}} = 135.6.$$

Next, assume the link is a Johnson column. The critical unit load can be written from Johnson's parabolic equation, Eq. (13.39), as

$$\frac{P_{\text{CR}}}{A} = \left[S_{yc} - \frac{1}{CE}\left(\frac{S_{yc}S_r}{2\pi}\right)^2\right]. \tag{8}$$

Rearranging Eq. (8), the slenderness ratio can be written as

$$(S_r)_{\text{Johnson}} = \frac{2\pi}{S_{yc}}\sqrt{\left(S_{yc} - \frac{P_{\text{CR}}}{A}\right)CE}.$$

Substituting the known data into this equation gives

$$(S_r)_{\text{Johnson}} = \frac{2\pi}{(415 \times 10^6 \text{ Pa})}\sqrt{\left[(415 \times 10^6 \text{ Pa}) - \frac{280\,000 \text{ N}}{0.001\,26 \text{ m}^2}\right]2(207 \times 10^9 \text{ Pa})}$$

$$= 135.3.$$

The slenderness ratio at the point of tangency can be written as

$$(S_r)_D = \pi\sqrt{\frac{2CE}{S_{yc}}}.$$

Because the material properties are the same, this gives the same result as Eq. (1), that is, $(S_r)_D = 140.33$. Therefore, because $(S_r)_{\text{Johnson}} < (S_r)_{\text{Euler}} < (S_r)_D$, the link is a Johnson column. Using the slenderness ratio for the case of a Johnson column, that is, $S_r = (S_r)_{\text{Johnson}} = 135.3$, and rearranging Eq. (2), the maximum possible length of link 2 is

$$L = kS_r = (0.010 \text{ m})135.3 = 1.353 \text{ m}. \qquad \textit{Ans.}$$

EXAMPLE 13.12

A load P_C is acting on the horizontal link 2 at point C of the simple truss, as illustrated in Fig. 13.36. Link 2 is pinned to the ground at point O_2 and to the inclined link 3 at point B, whereas link 3 is pinned to the ground at point O_3. The lengths are $R_{CO_2} = 5$ ft, $R_{BO_2} = 4$ ft, and $R_{BO_3} = 5$ ft. The two links each have a solid circular cross-section with a diameter $D = 2$ in. Also, the two links are made from a steel alloy with a compressive yield

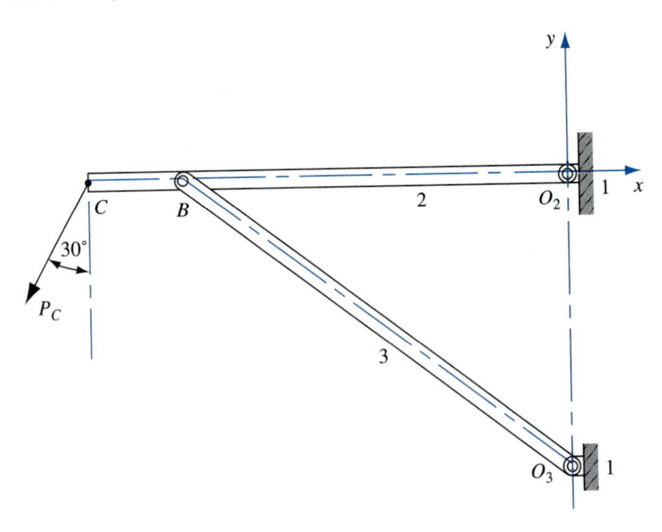

Figure 13.36 Example 13.12. A simple pinned truss.

strength $S_{yc} = 60$ kpsi, a tensile yield strength $S_{yt} = 50$ kpsi, and a modulus of elasticity $E = 30$ Mpsi.

Part 1: Using the theoretical value for the end-condition constants for the two links, determine (*i*) the value of the slenderness ratio at the point of tangency between Euler's column formula and Johnson's parabolic formula for link 3, (*ii*) the critical load and the critical unit load acting on link 3, and (*iii*) the load P_C for the factor of safety to guard against buckling of link 3 to be $N = 1$.
Part 2: If the load acting at point C is specified as $P_C = 24\,000$ lb, determine (*i*) the factor of safety for link 2, (*ii*) the factor of safety for link 3, and (*iii*) which link is most likely to fail first. Briefly explain your answer.

SOLUTION

Part 1.

(*i*) Links 2 and 3 are in static equilibrium. The free body diagram of link 2 is illustrated in Fig. 13.37. Note that the link is in tension because of the external load P_C.

Taking moments about the pin O_2 gives the y component of the load at pin B as

$$F_{32}^y = \frac{5\text{ ft}}{4\text{ ft}} P_C^y \quad \text{or} \quad F_{23}^y = -\frac{5}{4} P_C^y.$$

The free body diagram of link 3 is as illustrated in Fig. 13.38.

Taking moments about the pin O_3 gives

$$F_{23}^x = -\frac{4\text{ ft}}{3\text{ ft}} F_{23}^y = -\frac{4}{3}\left(-\frac{5}{4} P_C^y\right) = \frac{5}{3} P_C^y.$$

The magnitude of the force at pin B can be written as

$$F_{23} = \sqrt{\left(\frac{5}{3} P_C^y\right)^2 + \left(-\frac{5}{4} P_C^y\right)^2} = 5P_C^y\sqrt{\frac{1}{9} + \frac{1}{16}} = 5P_C^y\sqrt{\frac{16+9}{144}} = \frac{25}{12} P_C^y. \tag{1}$$

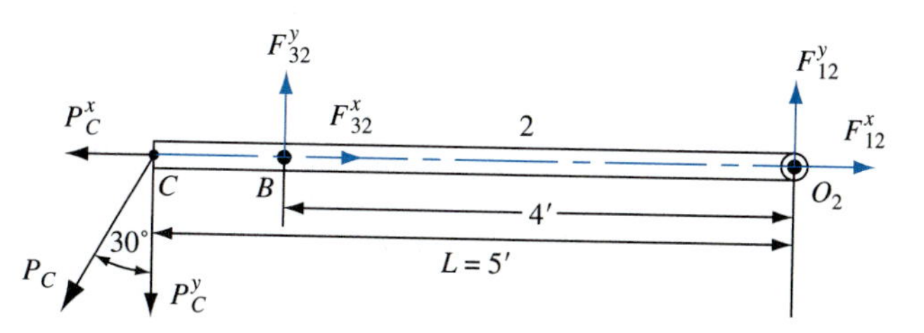

Figure 13.37 The free body diagram of link 2.

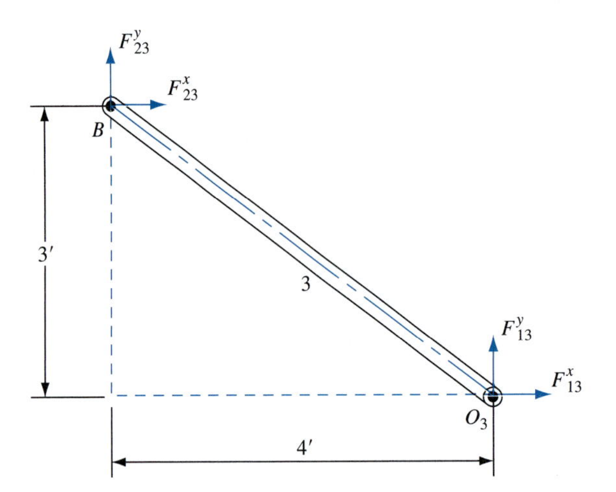

Figure 13.38 The free body diagram of link 3.

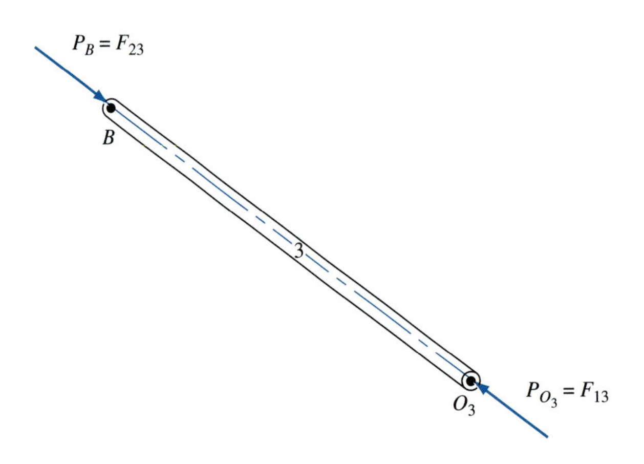

Figure 13.39 Link 3 is in compression.

The direction of this force at pin B, that is, the line of action, is

$$\angle F_{23} = \tan^{-1}\left(\frac{F_{23}^y}{F_{23}^x}\right) = \tan^{-1}\left(\frac{-5/4P_C^y}{5/3P_C^y}\right) = \tan^{-1}\left(\frac{-3}{4}\right) = -36.87°.$$

Because link 3 is a two-force member, the forces F_{23} and F_{13} must be equal, opposite, and collinear; that is, $F_{23} = F_{13}$, and link 3 is in compression because of the internal load P_B (see Fig. 13.39). Therefore, it is possible that link 3 could fail because of buckling.

We must check link 3 for the factor of safety to guard against buckling to be $N = 1$. The cross-sectional area of link 3 is

$$A = \frac{\pi}{4}D^2 = \frac{\pi}{4}(2 \text{ in})^2 = 3.1416 \text{ in}^2.$$

The second moment of area of link 3 can be written as

$$I = \frac{\pi}{64}D^4 = \frac{\pi}{64}(2\text{ in})^4 = 0.7854\text{ in}^4.$$

The radius of gyration of link 3 is

$$k = \sqrt{\frac{I}{A}} = \sqrt{\frac{0.785\,4\text{ in}^4}{3.141\,6\text{ in}^2}} = 0.500\text{ in}.$$

The slenderness ratio of link 3 is

$$S_r = \frac{L}{k} = \frac{5\text{ ft}\cdot 12\text{ in/ft}}{0.500\text{ in}} = 120. \tag{2}$$

From the pinned–pinned end conditions, the end-condition constant for link 3 is $C = 1.0$. Therefore, the slenderness ratio of link 3 at the point of tangency is

$$(S_r)_D = \pi\sqrt{\frac{2CE}{S_{yc}}} = \pi\sqrt{\frac{2\cdot 1.0(30\times 10^6\text{ psi})}{60\times 10^3\text{ psi}}} = 99.35. \qquad \textit{Ans.}\ (3)$$

(*ii*) To determine the critical load on link 3, we must first determine whether this link is an Euler column or a Johnson column. The criterion for using Johnson's parabolic equation is $S_r < (S_r)_D$. From Eqs. (2) and (3), we have $S_r > (S_r)_D$; that is, $120 > 99.35$. Therefore, link 3 is an Euler column. The critical load on link 3 is

$$P_{\text{CR}} = A\left[\frac{C\pi^2 E}{S_r^2}\right] = (3.141\,6\text{ in}^2)\left[\frac{1.0\pi^2(30\times 10^6\text{ psi})}{120^2}\right] = 64\,596\text{ lb}. \qquad \textit{Ans.}\ (4)$$

The critical unit load is

$$\frac{P_{\text{CR}}}{A} = \frac{64\,596\text{ lb}}{3.141\,6\text{ in}^2} = 20\,562\text{ lb/in}^2. \qquad \textit{Ans.}$$

(*iii*) If we identify $P_B = F_{23}$ as the compressive load at point B on link 3 (see Fig. 13.40), then, from the given factor of safety of $N = 1$ guarding against buckling for link 3, the applied load at pin B on link 3 can be written as

$$P_B = \frac{P_{\text{CR}}}{N} = \frac{P_{\text{CR}}}{1} = 64\,596\text{ lb}. \tag{5}$$

Therefore, the load $P_B (= F_{23})$ is equal to the critical load P_{CR}, as illustrated in Fig. 13.40. Note that a compressive load $P_B = 64\,596$ lb is the limiting load on link 3 to avoid buckling with a factor of safety of $N = 1$.

Rearranging Eq. (1), the vertical component of the load at point C is

$$P_C^y = \frac{12}{25}F_{23} = \frac{12}{25}P_B. \tag{6}$$

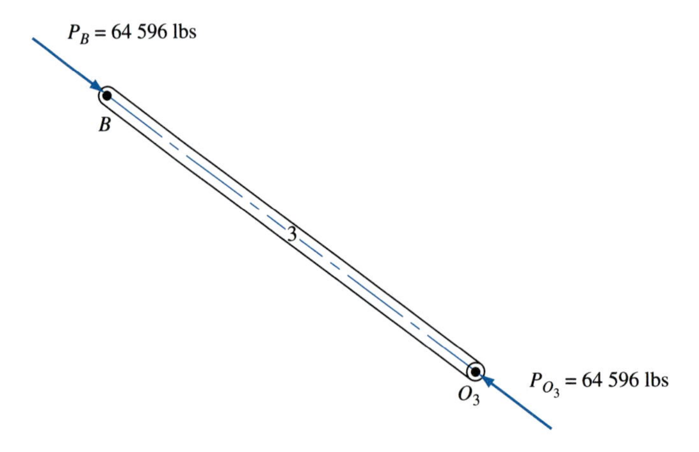

Figure 13.40 The compressive load acting on link 3.

Substituting Eq. (5) into Eq. (6), the vertical component of the load at point C is

$$P_C^Y = \frac{12}{25}(64\ 596\ \text{lb}) = 31\ 006\ \text{lb}.$$

Finally, the load acting at point C is

$$P_C = \frac{P_C^y}{\cos 30^\circ} = \frac{31\ 006\ \text{lb}}{0.866} = 35\ 803\ \text{lb}. \qquad \textit{Ans.}$$

Note that the load $P_C = 35\ 803$ lb is the load that ensures that the factor of safety to guard against buckling of link 3 is $N = 1$. A load $P_C > 35\ 803$ lb would cause link 3 to buckle; that is, the factor of safety would be $N < 1$.

Part 2.

(*i*) The x and y components of the load $P_C = 24\ 000$ lb are

$$P_C^x = P_C \sin 30^\circ = (24\ 000\ \text{lb})0.500 = 12\ 000\ \text{lb}$$

$$P_C^y = P_C \cos 30^\circ = (24\ 000\ \text{lb})0.866 = 20\ 785\ \text{lb}. \qquad (7)$$

Link 2 is subject to the tensile load P_C^x, which creates a tensile stress in the link. The factor of safety guarding against yielding of link 2 is defined as

$$N = \frac{S_{yt}}{\sigma},$$

where the normal stress caused by the tensile load can be written as

$$\sigma = \frac{P_C^x}{A} = \frac{12\ 000\ \text{lb}}{3.1416\ \text{in}^2} = 3\ 819.7\ \text{lb/in}^2.$$

Therefore, the factor of safety guarding against yielding of link 2 is

$$N = \frac{50\ 000 \text{ psi}}{3\ 819.7 \text{ psi}} = 13.1. \qquad \textit{Ans.}\ (8)$$

Note that this relatively high factor of safety indicates safety against tensile failure.

(*ii*) Substituting Eq. (7) into Eq. (6), the compressive force F_{23} exerted on link 3 is

$$F_{23} = \frac{25}{12} P_C^y = \frac{25}{12}(20\ 785 \text{ lb}) = 43\ 302 \text{ lb}. \qquad (9)$$

where force F_{23} is equal to force P_B. For link 3, the factor of safety to guard against buckling can be calculated by substituting Eqs. (4) and (9) into Eq. (5); that is,

$$N = \frac{P_{CR}}{P_B} = \frac{P_{CR}}{F_{23}} = \frac{64\ 596 \text{ lb}}{43\ 302 \text{ lb}} = 1.49. \qquad \textit{Ans.}\ (10)$$

(*iii*) Note that the factor of safety guarding against yielding of link 2 [the tensile member; see Eq. (8)] is much higher than the factor of safety guarding against buckling of link 3 [the compressive member; see Eq. (10)]. Therefore, the prediction is that link 3 will fail before link 2. *Ans.*

EXAMPLE 13.13

A vertically downward force $F = 15\ 000$ lb is applied at point B on member 2, which is pinned to the ground at point O_2 and to the vertical member 3 at point A, as illustrated in Fig. 13.41. The length of member 2, that is R_{BO_2}, is 60 in. and the ground pin O_3 is located at a distance b = 30 in. along the x-axis from the ground pin O_2. Members 2 and 3 are

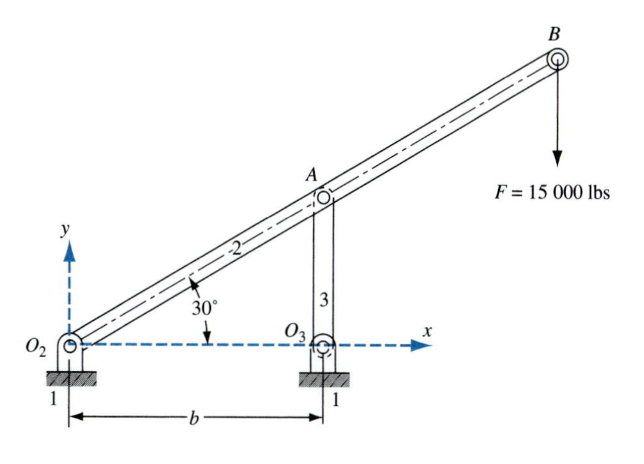

Figure 13.41 Example 13.13. The structure for a buckling analysis.

made from a steel alloy with a compressive yield strength $S_{yc} = 60$ kpsi and a modulus of elasticity $E = 30$ Mpsi. For buckling in the plane, member 3 behaves as a pinned–pinned member. Therefore, the theoretical end-condition constant is $C = 1$. For buckling out of the plane, member 3 behaves as a fixed–fixed member with a theoretical end-condition constant $C = 4$.

Part A: Assuming that the slenderness ratio of member 3 is such that Euler's column formula is valid, determine whether the following statements are true or false; give a brief explanation.

1. For members with a slenderness ratio greater than 10, it is safe to assume that the members designed for failure against yielding are automatically safe against buckling failure.
2. With all other design parameters unchanged, a member with fixed–free end conditions has a larger factor of safety guarding against buckling than the same member with pinned–pinned end conditions.
3. Two members with the same length, cross-section, and modulus of elasticity, but with different compressive yield strengths have the same critical buckling load.

Part B: Assume that member 3 has a hollow rectangular cross-section, as illustrated in Fig. 13.42*a*. Initially, it is not known whether member 3 will be attached to the ground at pin O_3 by either Case I or Case II illustrated in Fig. 13.42*b*. Using symbolic arguments (no numerical calculations are required), determine whether Case I or Case II leads to a higher factor of safety guarding against buckling. Assume the possibility of buckling failure in the plane of Fig. 13.42.

Part C: Assuming the worst case scenario, that is, the case with the lower factor of safety guarding against buckling, design the cross-section of member 3, that is, determine the dimension D, such that member 3 has a factor of safety guarding against buckling failure of $N = 2$. Note that designing the cross-section for the case with the lower factor of safety automatically ensures a safe design for the case with the higher factor of safety.

Part D: Determine the factor of safety of member 3 guarding against buckling failure. Use the dimensions of the rectangular cross-section obtained in Part C. Assume that member 3 is attached as in Case I and that there is the possibility of buckling out of the plane.

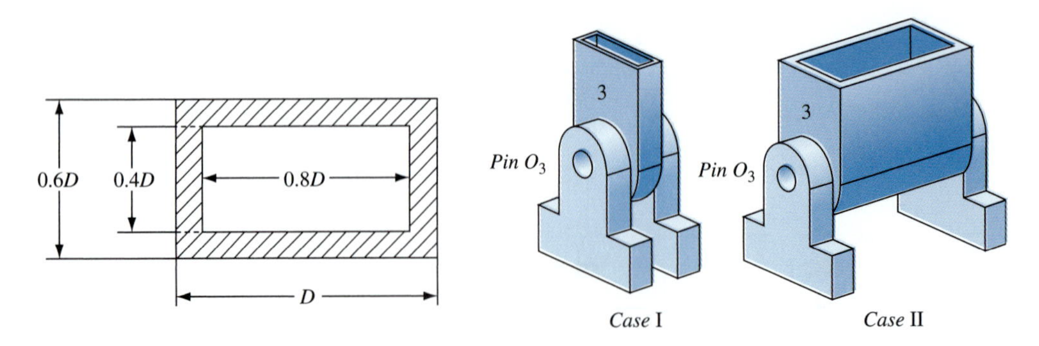

Figure 13.42 (*a*) The cross-section of member 3. (*b*) Case I and Case II.

Part E: Assume that member 3 is attached as shown in Case II and that there is the possibility of buckling in the plane. The problem is to determine a new value for b, that is, to change the location of pin O_3. The distance b cannot be less than 5 in and cannot be greater than 45 in. Using the dimensions of the rectangular cross-section obtained in Part C, determine an expression for the factor of safety guarding against buckling failure as a function of the distance b. Then use a computer program, such as MATLAB, to plot a graph of N versus b. Using this graph, determine the minimum value of b above which the member is safe against buckling failure.

SOLUTION

Part A:

1. For members with a slenderness ratio greater than 10 it is safe to assume that members designed for failure against yielding are automatically safe against buckling failure. FALSE. *Ans.*
2. With all other design parameters unchanged, a member with fixed–free end conditions has a larger factor of safety than the same member with pinned–pinned end conditions. FALSE. *Ans.*
3. Two members with the same length, cross-section, and modulus of elasticity but with different compressive yield strengths have the same value of critical buckling load. TRUE. *Ans.*

Part B: For Case I, the second moment of area for member 3 can be written as

$$I_{\text{Case I}} = \frac{0.6D(D)^3}{12} - \frac{0.4D(0.8D)^3}{12} = \frac{0.3952D^4}{12} = 0.032\,93D^4.$$

For Case II, the second moment of area for member 3 can be written as

$$I_{\text{Case II}} = \frac{D(0.6D)^3}{12} - \frac{0.8D(0.4D)^3}{12} = \frac{0.1648D^4}{12} = 0.013\,73D^4. \tag{1}$$

Therefore, the second moment of area of Case I is greater than the second moment of area of Case II, that is,

$$I_{\text{Case I}} > I_{\text{Case II}}. \tag{2}$$

Because the cross-sectional area is the same in both cases, Eq. (2) implies that the radius of gyration of Case I is greater than the radius of gyration of Case II, that is,

$$k_{\text{Case I}} > k_{\text{Case II}}. \tag{3}$$

Because the length of the member is the same in both cases, Eq. (3) implies that the slenderness ratio in Case I is less than the slenderness ratio in Case II, that is,

$$(S_r)_{\text{Case I}} < (S_r)_{\text{Case II}}.$$

Therefore, for either Euler's column formula or Johnson's parabolic equation, the critical load of Case I is greater than the critical load of Case II, that is,

$$(P_{\text{CR}})_{\text{Case I}} > (P_{\text{CR}})_{\text{Case II}}.$$

Because the applied load is the same in both cases, the factor of safety guarding against buckling for Case I is greater than the factor of safety guarding against buckling for Case II, that is,

$$(N)_{\text{Case I}} > (N)_{\text{Case II}}. \qquad \textit{Ans.}$$

This implies that, if the cross-sectional area of the member is designed for Case II, then the member will automatically be safe against buckling for Case I.

Part C: The cross-sectional area of member 3, that is, a hollow rectangular cross-section, can be written as

$$A = D(0.6D) - 0.8D(0.4D) = 0.28D^2.$$

From Eq. (1), the second moment of area for the member can be written as $I = 0.013\,73D^4$. The radius of gyration of the member can be written as

$$k = \sqrt{\frac{I}{A}} = \sqrt{\frac{0.013\,73D^4}{0.28D^2}} = 0.2215D. \tag{4}$$

Because $b = 30$ in, the length of member 3 is

$$L = b \tan 30^\circ = 17.321 \text{ in}.$$

Therefore, the slenderness ratio of the member is

$$S_{\text{r}} = \frac{L}{k} = \frac{17.321 \text{ in}}{0.2215D} = \frac{78.199 \text{ in}}{D}.$$

The slenderness ratio at the point of tangency can be written as

$$(S_r)_D = \pi\sqrt{\frac{2CE}{S_{yc}}}.$$

From the specified fixed–fixed end conditions, the end-condition constant is $C = 1$. Therefore, the slenderness ratio at the point of tangency is

$$(S_r)_D = \pi\sqrt{\frac{2 \cdot 1(30 \times 10^6 \text{ psi})}{60 \times 10^3 \text{ psi}}} = 99.3.$$

To determine the applied force at point A, we take moments about point O_2, which can be written as

$$(60 \text{ in}) \cos 30^\circ F = bP_{\text{APP}}.$$

Therefore, the applied force at point A is

$$P_{\text{APP}} = \frac{(60 \text{ in}) \cos 30°(15\,000 \text{ lb})}{30 \text{ in}} = 25\,981 \text{ lb}.$$

The factor of safety guarding against buckling is defined as

$$N = \frac{P_{\text{CR}}}{P_{\text{APP}}}.$$

Because $N = 2$, the critical load is

$$P_{\text{CR}} = P_{\text{APP}}N = (25\,981 \text{ lb})2 = 51\,962 \text{ lb}.$$

Note: At this point it is not known whether Euler's column formula or Johnson's parabolic equation is the proper equation to determine the dimensions of the cross-section of the member. If we assume that Johnson's parabolic equation is the proper equation, then the critical load is

$$P_{\text{CR}} = A\left[S_{\text{yc}} - \frac{1}{CE}\left(\frac{S_{\text{yc}}S_{\text{r}}}{2\pi}\right)^2\right].$$

Substituting the known values into this equation gives

$$51\,962 \text{ lb} = 0.28D^2\left[(60 \times 10^3 \text{ psi}) - \frac{1}{1(30 \times 10^6 \text{ psi})}\left(\frac{(60 \times 10^3 \text{ psi})78.199}{2\pi D}\right)^2\right].$$

Rearranging this equation gives

$$(16\,800 \text{ psi})D^2 = 51\,962 \text{ lb} + 5\,204 \text{ lb} = 57\,166 \text{ lb}.$$

Solving this equation, the dimension D is

$$D = 1.845 \text{ in}. \qquad \textit{Ans.}$$

As a check, using Eq. (4) the slenderness ratio is

$$S_{\text{r}} = \frac{L}{k} = \frac{17.321 \text{ in}}{0.2215(1.845 \text{ in})} = 42.384 < (S_r)_D = 99.3.$$

Therefore, Johnson's parabolic equation is proper for this case.
Part D: Because the dimensions of the rectangular cross-section are now known, the cross-sectional area of member 3 can be written as

$$A = 0.28D^2 = 0.9531 \text{ in}^2.$$

From Eq. (1), the moment of inertia of the hollow rectangular member can be written as

$$I = 0.013\,73D^4 = 0.1591 \text{ in}^4.$$

From Eq. (4), the radius of gyration for the member can be written as

$$k = 0.2215D = 0.4087 \text{ in.}$$

Because $b = 30$ in, the length of the member is

$$L = b \tan 30^\circ = 17.310 \text{ in.}$$

The slenderness ratio of the member is

$$S_r = \frac{L}{k} = \frac{17.310 \text{ in}}{0.4087 \text{ in}} = 42.354.$$

Because the member is attached as shown in Case I and there is the possibility of buckling out of the plane of Fig. 13.42, the end-condition constant is $C = 4$. Because $S_r < (S_r)_D$, Johnson's parabolic equation is appropriate and the critical load is

$$P_{CR} = A\left[S_{yc} - \frac{1}{CE}\left(\frac{S_{yc}S_r}{2\pi}\right)^2\right],$$

which gives

$$P_{CR} = 0.9531 \text{ in}^2\left[60\,000 \text{ psi} - \frac{1}{4(30 \times 10^6 \text{ psi})}\left(\frac{60\,000 \text{ psi} \cdot 42.353}{2\pi}\right)^2\right] = 55\,886 \text{ lb.}$$

Because the applied load $P_{APP} = 25\,981$ lb, the factor of safety is

$$N = \frac{55\,886 \text{ lb}}{25\,981 \text{ lb}} = 2.15. \qquad \textit{Ans.}$$

Part E: Let the distance b between pins O_2 and O_3 be varied between 5 and 45 in. The goal is to determine the minimum value of b such that the member does not fail in buckling. Assume that member 3 is located at an unknown distance b from pin O_2. Then force P_{APP} acting at point A can be obtained by taking moments about O_2, that is,

$$P_{APP} = \frac{(60 \text{ in} \cos 30^\circ)15{,}000 \text{ lb}}{b} = \frac{779\,420 \text{ in} \cdot \text{lb}}{b}. \tag{5}$$

As in Part D, the cross-sectional area of member 3 is $A = 0.9531$ in.^{2}, the moment of inertia is $I = 0.1591$ in.^{4}, and the radius of gyration of the member is $k = 0.4087$ in.

The length of the member is

$$L = b \tan 30^\circ = 0.5774b.$$

The slenderness ratio of the member is

$$S_r = \frac{L}{k} = \frac{0.5774b}{0.4087 \text{ in}} = (1.413 \text{ in}^{-1})b.$$

Note: Because b can vary between 5 and 45 in., the maximum value of the slenderness ratio of the member is $S_r = 63.58$, which is still less than $(S_r)_D = 99.3$. This implies that Johnson's parabolic equation is valid for all possible values of b. Therefore, the critical load can be written as

$$P_{CR} = A\left[S_{yc} - \frac{1}{CE}\left(\frac{S_{yc}S_r}{2\pi}\right)^2\right].$$

Substituting the known data into this equation, the critical load can be written as

$$P_{CR} = 0.9531\ \text{in}^2\left[(60 \times 10^3\ \text{psi}) - \frac{1}{1(30 \times 10^6\ \text{psi})}\left(\frac{(60 \times 10^3\ \text{psi})(1.413\ \text{in}^{-1})b}{2\pi}\right)^2\right]$$

or as

$$P_{CR} = 57\ 186\ \text{lb} - (5.784\ \text{psi})b^2. \tag{6}$$

The factor of safety guarding against buckling is defined as

$$N = \frac{P_{CR}}{P_{APP}}. \tag{7}$$

Substituting Eqs. (5) and (6) into Eq. (7), the factor of safety guarding against buckling can be written as

$$N = \frac{57\ 186\ \text{lb} - (5.784\ \text{psi})b^2}{779\ 420\ \text{in} \cdot \text{lb}/b} = \frac{(57\ 186\ \text{lb})b - (5.784\ \text{psi})b^3}{779\ 420\ \text{in} \cdot \text{lb}}$$

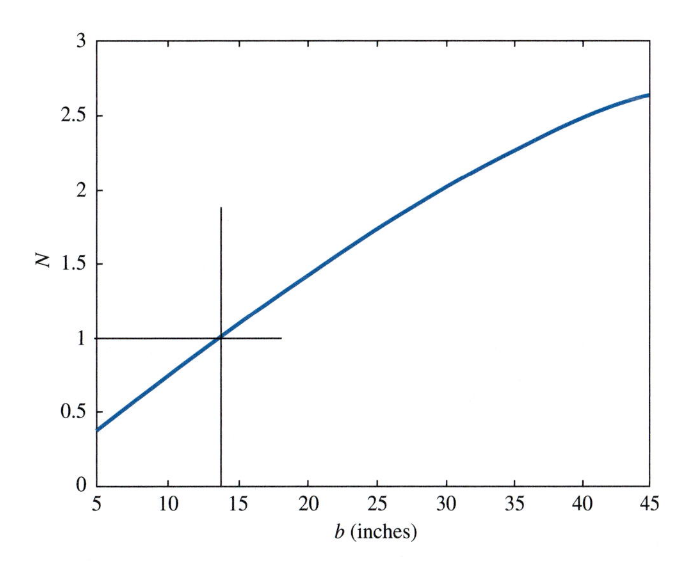

Figure 13.43 The plot of the factor of safety N versus dimension b.

or as

$$N = (0.07337 \text{ in}^{-1})b - (7.421 \times 10^{-6} \text{ in}^{-3})b^3.$$

The member will fail in buckling for all values of b for which the value of N is less than 1. A plot of N versus b is illustrated in Fig. 13.43. We observe from this plot that the factor of safety guarding against buckling is less than 1 for values of b less than 13.9 in. This implies failure in the buckling mode. *Ans.*

EXAMPLE 13.14

A solid aluminum member is subjected to the concentric axial load $P = 3$ kN, as illustrated in Fig. 13.44. The length of the member is $L = 180$ mm and the cross-section is rectangular with dimensions $b = 20$ mm and $h = 4$ mm. The aluminum has a modulus of elasticity $E = 75$ GPa and a compressive yield strength $S_{yc} = 150$ MPa. Use the AISC recommended end-condition constant for fixed–pinned end conditions.

(*i*) Is the member an Euler column or a Johnson column?
(*ii*) Determine the maximum load P_{CR} that can be applied before buckling will occur.
(*iii*) Determine the factor of safety against buckling of the member.

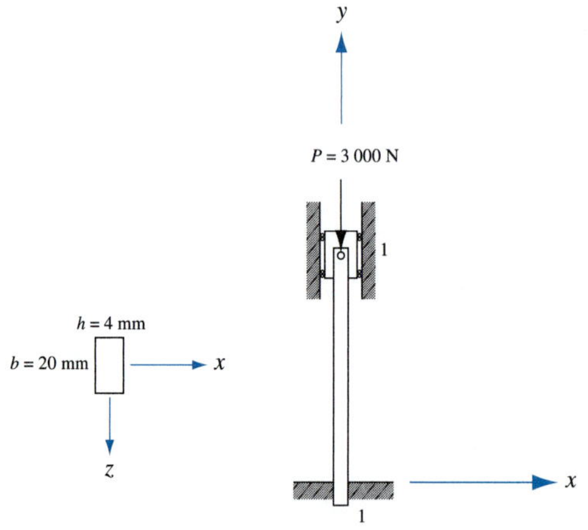

Figure 13.44 Example 13.14. A pinned–fixed member.

SOLUTION

The cross-sectional area of the member is

$$A = bh = (0.020 \text{ m})(0.004 \text{ m}) = 0.000\,080 \text{ m}^2.$$

The second moment of area (see Table 4 in Appendix A) is

$$I = I_{ZZ} = \frac{bh^3}{12} = \frac{(0.020 \text{ m})(0.004 \text{ m})^3}{12} = 1.067 \times 10^{-10} \text{m}^4.$$

Note here that the second moment of area has the smallest value allowable because the member will begin to buckle in the weakest plane. For this example, the member will begin to buckle in the plane of Fig. 13.44.

The radius of gyration of the member is

$$k = \sqrt{\frac{I}{A}} = \sqrt{\frac{1.067 \times 10^{-10}\ \text{m}^4}{0.000\ 080\ \text{m}^2}} = 1.155\ \text{mm}.$$

Using the AISC recommended beam end-condition effective length factor for fixed–pinned end conditions (that is, $\alpha = L_{\text{EFF}}/L = 0.707$; see Fig. 13.29), the end-condition constant is

$$C = \frac{1}{\alpha^2} = \frac{1}{0.707^2} = 2.$$

(*i*) The slenderness ratio of the member is

$$S_{\text{r}} = \frac{L}{k} = \frac{0.180\ \text{m}}{0.001\ 155\ \text{m}} = 155.84.$$

The slenderness ratio at the point of tangency is

$$(S_{\text{r}})_{\text{D}} = \pi\sqrt{\frac{2EC}{S_{\text{yc}}}} = \pi\sqrt{\frac{2(75 \times 10^9\ \text{Pa})2}{150 \times 10^6\ \text{Pa}}} = 140.5.$$

Because the slenderness ratio $S_{\text{r}} \geq (S_{\text{r}})_{\text{D}}$, the member is an Euler column. *Ans.*

(*ii*) Using Euler's column formula, the critical load is

$$P_{\text{CR}} = A\left[\frac{C\pi^2 E}{S_r^2}\right] = (0.000\ 080\ \text{m}^2)\left[\frac{2\pi^2(75 \times 10^9\ \text{Pa})}{155.84^2}\right] = 4\ 877\ \text{N}. \quad \textit{Ans.}$$

(*iii*) The factor of safety guarding against buckling is

$$N = \frac{P_{\text{CR}}}{P} = \frac{4\ 877\ \text{N}}{3\ 000\ \text{N}} 1.63. \qquad \textit{Ans.}$$

Therefore, the member should not fail in buckling.

EXAMPLE 13.15

For the simple truss illustrated in Fig. 13.45, a vertical load $P = 10$ kN is acting at point A in the negative y direction on the horizontal member 2 which is pinned to the vertical wall at point D. Assume that member 3 is pinned to member 2 at point B and also to the wall at point C. Use the theoretical value for the end-condition constant of member 3 and assume that this member is steel with a yield strength $S_{\text{yc}} = 370$ MPa and a modulus of elasticity $E = 207$ GPa.

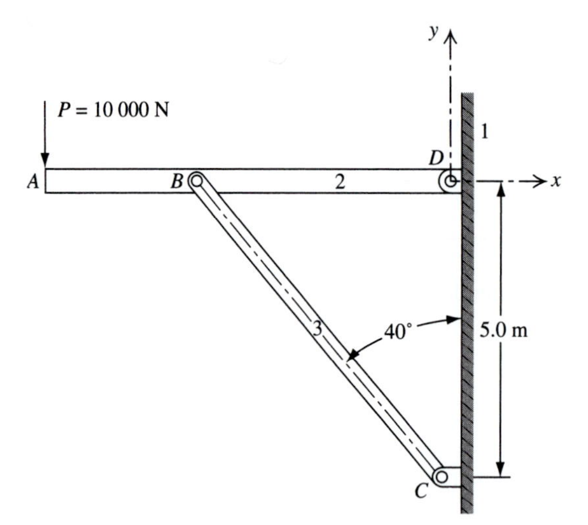

Figure 13.45 Example 13.15. A simple truss subjected to a vertical load P.

(*i*) Assume that member 3 has a solid circular cross-section with a constant diameter d. Determine the diameter d to ensure that the member has a factor of safety guarding against buckling of $N = 2$. Does Euler's column formula or Johnson's parabolic equation give the correct answer for the diameter?

(*ii*) Assume that member 3 has a square cross-section with a constant width t. Determine the width t to ensure that the member has a factor of safety guarding against buckling of $N = 2$. Does Euler's column formula or Johnson's parabolic equation give the correct answer for the width?

SOLUTION

First, determine the magnitude of the compressive load that is acting on member 3. The unknown lengths can be determined from the given geometry. The length of member 3 can be obtained from the trigonometric relation

$$L_3 = 5.0 \text{ m}/\cos 40° = 6.527 \text{ m}.$$

The distance from point B to point D is

$$R_{BD} = (5.0 \text{ m}) \tan 40° = 4.196 \text{ m}.$$

The force acting on member 3 can be obtained by summing moments about pin D, that is,

$$(7.0 \text{ m})(10\,000 \text{ N}) - (4.196 \text{ m})F_B^Y = 0.$$

Therefore, the vertical force acting on member 3 is

$$F_B^Y = 16\,683 \text{ N}.$$

The total compressive force P acting on member 3 can then be obtained from the geometry, that is,

$$P = \frac{16\ 683\ \text{N}}{\cos 40^\circ} = 21\ 778\ \text{N}.$$

(*i*) Member 3 has a circular cross-section with a constant diameter d. Therefore, from Table 4 of Appendix A, the cross-sectional area and second moment of area of the member are

$$A = \frac{\pi d^2}{4} \quad \text{and} \quad I = \frac{\pi d^4}{64}.$$

The radius of gyration of the member is

$$k = \sqrt{\frac{I}{A}} = \sqrt{\frac{\pi d^4/64}{\pi d^2/4}} = \frac{d}{4}\ \text{m}.$$

The end conditions of member 3 are pinned–pinned. Using the theoretical value for the end-condition constant (that is, $C = 1$ or $\alpha = 1$) gives $L_{\text{EFF}} = L$. Therefore, the slenderness ratio of member 3 is

$$S_r = \frac{L_{\text{EFF}}}{k} = \frac{6.527\ \text{m}}{d/4} = \frac{26.108\ \text{m}}{d}.$$

The factor of safety guarding against buckling is defined as

$$N = \frac{P_{\text{CR}}}{P}.$$

Rearranging this equation gives the critical load as

$$P_{\text{CR}} = NP = 2(21\ 778\ \text{N}) = 43\ 556\ \text{N}.$$

The slenderness ratio at the point of tangency is

$$(S_r)_D = \pi\sqrt{\frac{2E}{S_{yc}}} = \pi\sqrt{\frac{2(207 \times 10^9\ \text{Pa})}{370 \times 10^6\ \text{Pa}}} = 105.1.$$

The diameter d can be obtained from Euler's column formula; that is,

$$\frac{P_{\text{CR}}}{A} = \frac{\pi^2 E}{S_r^2}.$$

Substituting the known values into this equation gives

$$\frac{43\ 556\ \text{N}}{\pi d^2/4} = \frac{\pi^2(207 \times 10^9\ \text{Pa})}{(26.108\ \text{m}/d)^2}.$$

Solving for the diameter d gives

$$d = 0.066 \text{ m}. \tag{1}$$

Check: The diameter d can also be obtained from Johnson's parabolic equation, that is,

$$\frac{P_{\text{CR}}}{A} = S_{\text{yc}} - \frac{1}{E}\left(\frac{S_{\text{yc}}S_{\text{r}}}{2\pi}\right)^2. \tag{2}$$

Substituting the known values into Eq. (2) gives

$$\frac{43\,556 \text{ N}}{(\pi d^2/4)} = (370 \times 10^6 \text{ Pa}) - \frac{1}{207 \times 10^9 \text{ Pa}}\left(\frac{(370 \times 10^6 \text{ Pa})(26.108 \text{ m}/d)}{2\pi}\right)^2.$$

Solving this for the diameter d gives

$$d = 0.176 \text{ m}. \tag{3}$$

Now we must check which answer is valid, that is, Eq. (1) or (3). We recall that the slenderness ratio is

$$S_{\text{r}} = \frac{26.108 \text{ m}}{d}.$$

First, we check to determine whether Euler's column result, Eq. (1), is valid. The slenderness ratio is

$$(S_{\text{r}})_{\text{Euler}} = \frac{26.108 \text{ m}}{0.066 \text{ m}} = 395.6.$$

Because the slenderness ratio $(S_{\text{r}})_{\text{Euler}}$ is greater than $(S_{\text{r}})_D$, that is, because 395.6 is greater than 105.1, Euler's column formula gives a valid result. The diameter of the member is

$$d = 0.066 \text{ m}. \qquad \textit{Ans.}$$

Next we check to make sure that Johnson's parabolic equation result is not valid. The slenderness ratio from this result is

$$(S_{\text{r}})_{\text{Johnson}} = \frac{26.108 \text{ m}}{0.176 \text{ m}} = 148.3.$$

Because the slenderness ratio $(S_{\text{r}})_{\text{Johnson}}$ is greater than $(S_{\text{r}})_D$, that is, because 148.3 is greater than 105.1, Johnson's parabolic equation result is *not* valid.

(*ii*) Member 3 has a square cross-section with constant width t. Therefore, from Table 4, Appendix A, the cross-sectional area and second moment of area of the member are

$$A = t^2 \quad \text{and} \quad I = t^4/12.$$

The radius of gyration of the member is

$$k = \sqrt{\frac{I}{A}} = \sqrt{\frac{t^4/12}{t^2}} = 0.289t.$$

The slenderness ratio of member 3 is

$$S_r = \frac{L_{EFF}}{k} = \frac{6.527 \text{ m}}{0.289t} = \frac{22.6 \text{ m}}{t}.$$

First, t can be found using Euler's column formula, that is,

$$\frac{P_{CR}}{A} = \frac{\pi^2 E}{S_r^2}.$$

Substituting the known values into this equation gives

$$\frac{43\,556 \text{ N}}{t^2} = \frac{\pi^2 (207 \times 10^9 \text{ Pa})}{(22.6 \text{ m}/t)^2}.$$

Solving for the width t gives

$$t = 0.057 \text{ m}.$$

The width t can also be obtained from Johnson's parabolic equation, that is,

$$\frac{P_{CR}}{A} = S_{yc} - \frac{1}{E}\left(\frac{S_{yc} S_r}{2\pi}\right)^2.$$

Substituting the known values into this equation gives

$$\frac{43\,556 \text{ N}}{t^2} = (370 \times 10^6 \text{ Pa}) - \frac{1}{(207 \times 10^9 \text{ Pa})}\left(\frac{(370 \times 10^6 \text{ Pa})22.6 \text{ m}/t}{2\pi}\right)^2.$$

Therefore, the width t is

$$t = 0.152 \text{ m}.$$

To check which answer is valid, we recall that the slenderness ratio is

$$S_r = \frac{22.6 \text{ m}}{t}.$$

First, check to see whether the use of Euler's column formula is valid.

$$(S_r)_{\text{Euler}} = \frac{22.6 \text{ m}}{0.057 \text{ m}} = 396.5$$

Because the slenderness ratio $(S_r)_{\text{Euler}}$ is greater than $(S_r)_D$, that is, because 396.5 is greater than 105.1, Euler's column formula gives a valid result. The correct width of the member is

$$t = 0.057 \text{ m}. \qquad \textit{Ans.}$$

Next, we check to make sure that Johnson's parabolic equation result is not valid. The slenderness ratio from this result is

$$(S_r)_{\text{Johnson}} = \frac{22.6 \text{ m}}{0.152 \text{ m}} = 148.7.$$

Because the slenderness ratio $(S_r)_{\text{Johnson}}$ is greater than $(S_r)_D$, that is, because 148.7 is greater than 105.1, use of Johnson's parabolic equation is *not* valid.

13.18 REFERENCES

[1] J. E. Shigley and C. R. Mischke, *Mechanical Engineering Design*, 6th ed. New York: McGraw–Hill, 2001.
[2] Isaac Newton, *Principia*, 1687, translated by Andrew Motte, 1729.
[3] See, for example, M. J. Neale (ed.), *Tribology Handbook*. London: Butterworths, 1975, p. C8.
[4] See, for example, F. P. Beer and E. R. Johnston, *Vector Mechanics for Engineers*, 8th ed. New York: McGraw–Hill, 2007, pp. 440–6.

PROBLEMS

13.1 Figure P13.1 illustrates four mechanisms and the external forces and torques exerted on or by the mechanisms. Sketch the free-body diagram of each part of each mechanism. Do not attempt to show the magnitudes of the forces, except roughly, but do sketch them in their proper locations and orientations.

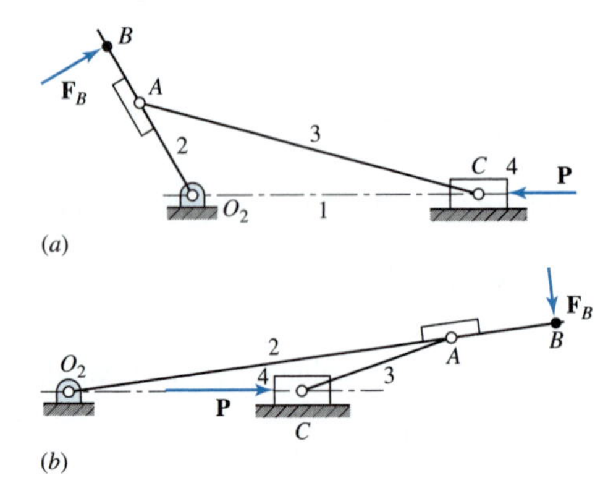

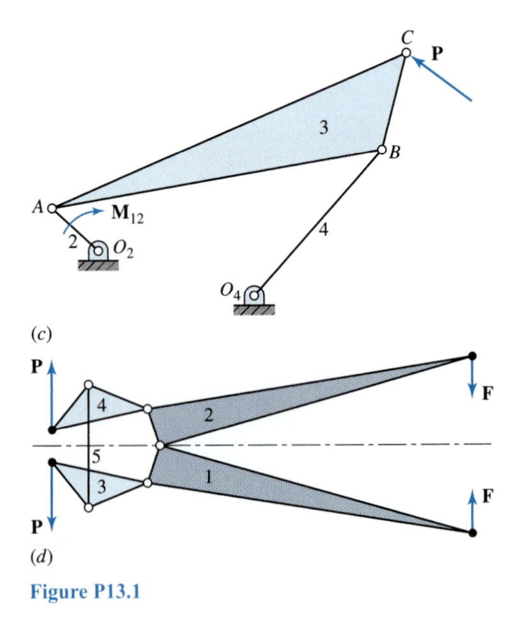

Figure P13.1

†13.2 What moment $\mathbf{M}_{12}$ must be applied to the crank of the mechanism illustrated in Fig. P13.2 if $P = 0.9$ kN?

* Unless otherwise stated, solve all problems without friction and without gravitational loads.

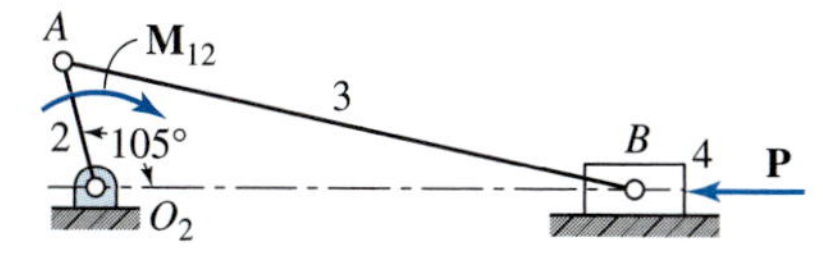

Figure P13.2 $R_{AO_2} = 75$ mm, $R_{BA} = 350$ mm.

†13.3 If $M_{12} = 100$ N·m for the mechanism illustrated in Fig. P13.2, what force **P** is required to maintain static equilibrium?

†13.4 Find the frame reactions and torque $\mathbf{M}_{12}$ necessary to maintain equilibrium of the four-bar linkage shown Fig. P13.4.

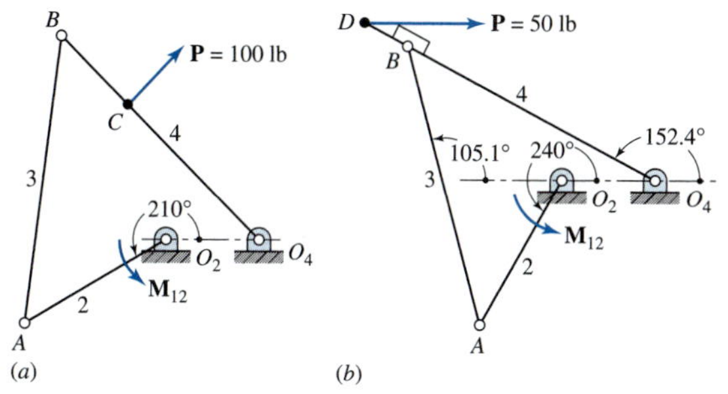

Figure P13.4 $R_{AO_2} = 3.5$ in; $R_{BA} = R_{BO_4} = 6$ in; $R_{CO_4} = 4$ in; $R_{O_2O_4} = 2$ in.

13.5 What torque must be applied to link 2 of the linkage illustrated in Fig. P13.4 to maintain static equilibrium?

13.6 Sketch a complete free-body diagram of each link of the linkage shown. What force **P** is necessary for equilibrium?

13.7 Determine the torque $\mathbf{M}_{12}$ required to drive slider 6 of Fig. P13.7 against a load of $P = 100$ lb at a crank angle of $\theta = 30°$ or as specified by your instructor.

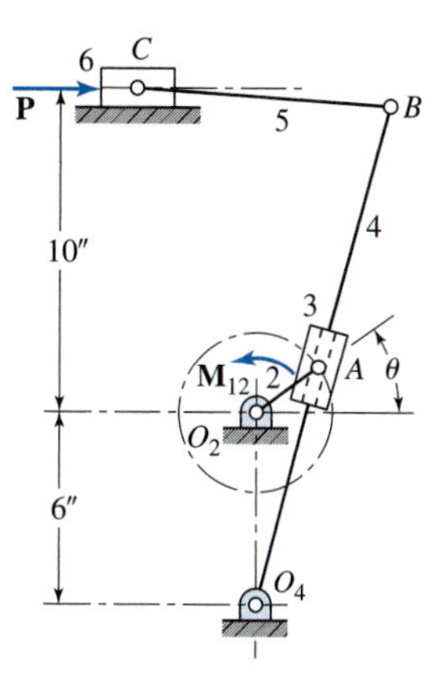

Figure P13.7 $R_{AO_2} = 2.5$ in; $R_{BO_4} = 16$ in; $R_{BC} = 8$ in.

†13.8 Sketch complete free-body diagrams for the illustrated four-bar linkage. What torque $\mathbf{M}_{12}$ must be applied to link 2 to maintain static equilibrium at the position shown?

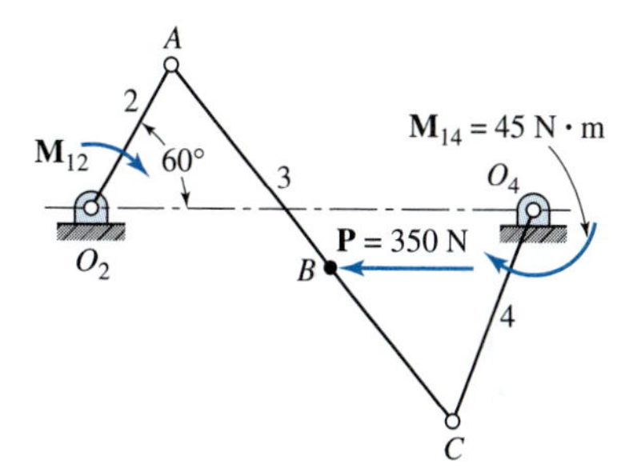

Figure P13.8 $R_{AO_2} = 200$ mm; $R_{BA} = 400$ mm; $R_{CA} = R_{O_4O_2} = 700$ mm; and $R_{CO_4} = 350$ mm.

13.9 Sketch free-body diagrams of each link and show in Fig. P13.9 all the forces acting. Find the magnitude and direction of the moment that must be applied to link 2 to drive the linkage against the forces shown.

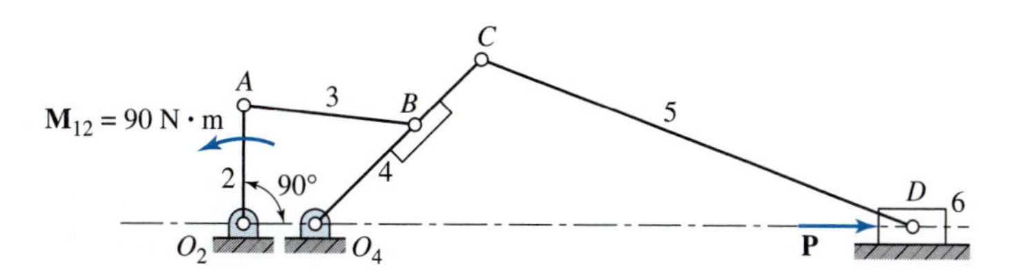

Figure P13.6 $R_{AO_2} = 100$ mm; $R_{BA} = 150$ mm; $R_{BO_4} = 125$ mm; $R_{CO_4} = 200$ mm; $R_{CD} = 400$ mm; and $R_{O_2O_4} = 60$ mm.

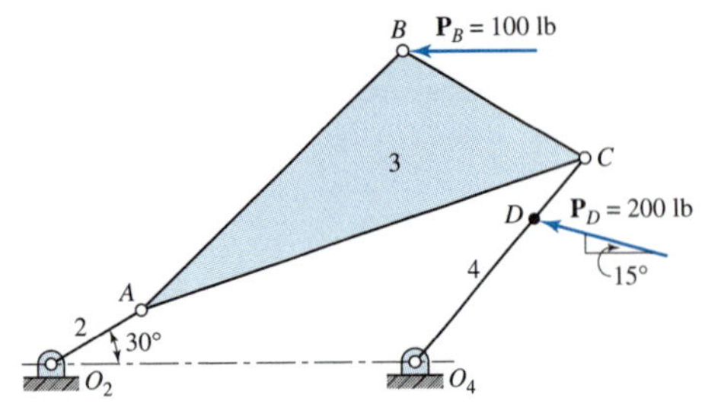

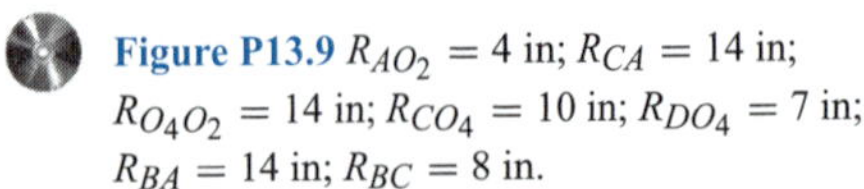
Figure P13.9 $R_{AO_2} = 4$ in; $R_{CA} = 14$ in; $R_{O_4O_2} = 14$ in; $R_{CO_4} = 10$ in; $R_{DO_4} = 7$ in; $R_{BA} = 14$ in; $R_{BC} = 8$ in.

13.10 Figure P13.10 illustrates a four-bar linkage with external forces applied at points B and C. Draw a free-body diagram of each link and show all the forces acting on each. Find the torque that must be applied to link 2 to maintain equilibrium.

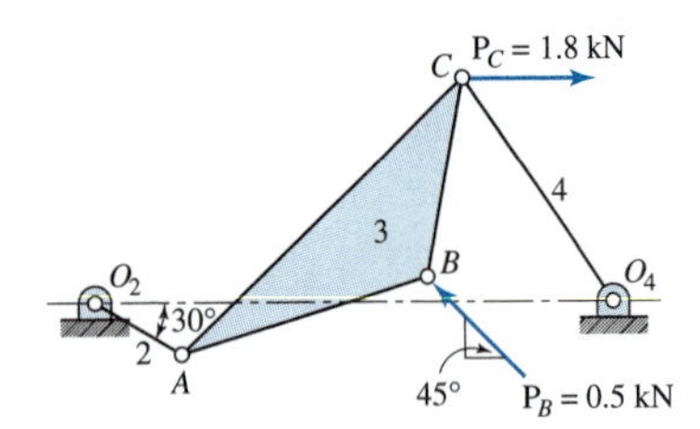

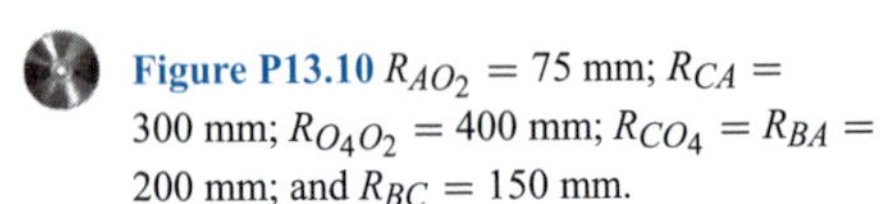
Figure P13.10 $R_{AO_2} = 75$ mm; $R_{CA} = 300$ mm; $R_{O_4O_2} = 400$ mm; $R_{CO_4} = R_{BA} = 200$ mm; and $R_{BC} = 150$ mm.

13.11 Draw a free-body diagram of each of the members of the mechanism illustrated in Fig. P13.11 and find the magnitudes and the directions of all the forces and moments. Compute the magnitude and direction of the torque that must be applied to link 2 to maintain static equilibrium.

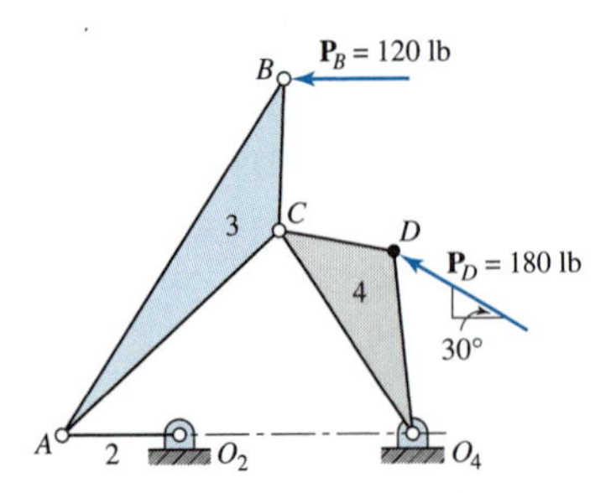

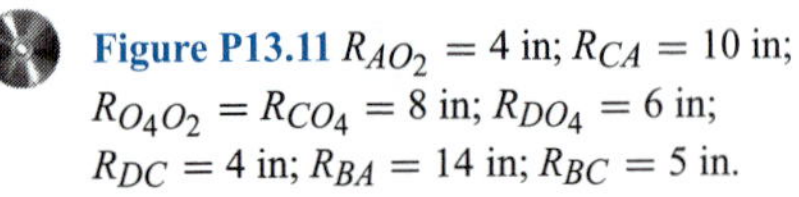
Figure P13.11 $R_{AO_2} = 4$ in; $R_{CA} = 10$ in; $R_{O_4O_2} = R_{CO_4} = 8$ in; $R_{DO_4} = 6$ in; $R_{DC} = 4$ in; $R_{BA} = 14$ in; $R_{BC} = 5$ in.

13.12 Determine the magnitude and direction of the torque that must be applied to link 2 to maintain static equilibrium.

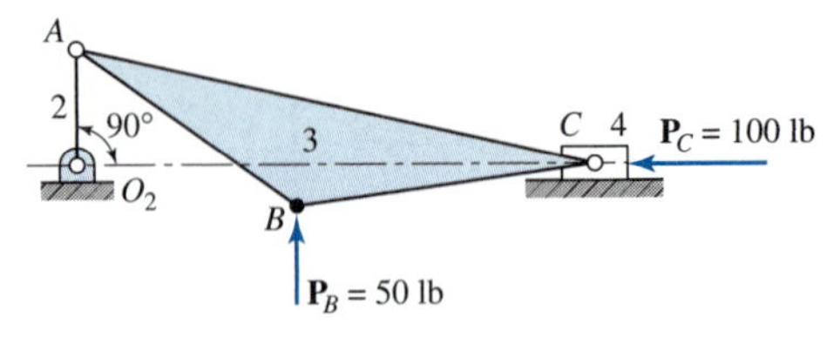

Figure P13.12 $R_{AO_2} = 3$ in; $R_{CA} = 14$ in; $R_{BA} = 7$ in; $R_{BC} = 8$ in.

13.13 Figure P13.13 shows a Figee floating crane with lemniscate boom configuration. Also shown is a schematic diagram of the crane. The lifting capacity is 16 T (where 1 T = 1 metric ton = 1,000 kg) including the grab, which is about 10 T. The maximum outreach is 30 m, which corresponds to the position $\theta_2 = 49°$. The minimum outreach is 10.5 m at $\theta_2 = 132°$. Other dimensions are given in the legend to Fig. P13.13. For the maximum outreach position and a grab load of 10 T (under standard gravity), find the bearing reactions at A, B, O_2, and O_4, as well as the moment $\mathbf{M}_{12}$ required. Note that the photograph shows a counterweight on link 2; neglect this weight and also the weights of the members.

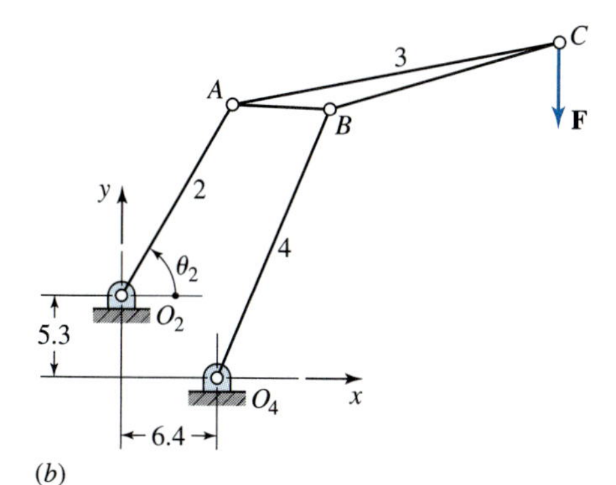

(b)

Figure P13.13 Figee floating crane with leminscate boom configuration. (Photograph by permission from B.V. Machinefabriek Figee, Haarlem, Holland). (*b*) Schematic diagram: $R_{AO_2} = 14.7$ m; $R_{BA} = 6.5$ m; $R_{BO_4} = 19.3$ m; $R_{CA} = 22.3$ m; $R_{CB} = 16$ m. (Dimensions by permission from B.V. Machinefabriek Figee, Haarlem, Holland).

13.14 Repeat Problem 13.13 for the minimum outreach position.

13.15 Repeat Problem 13.7 assuming coefficients of Coulomb friction $\mu_c = 0.20$ between links 1 and 6 and $\mu_c = 0.10$ between links 3 and 4. Determine the torque $\mathbf{M}_{12}$ necessary to drive the system, including friction, against the load **P**.

13.16 Repeat Problem 13.12 assuming a coefficient of static friction $\mu = 0.15$ between links 1 and 4. Determine the torque $\mathbf{M}_{12}$ necessary to overcome friction.

13.17 In each case shown, pinion 2 is the driver, gear 3 is an idler, and the gears have diametral pitch of 6 and 20° pressure angle. For each case, sketch the free-body diagram of gear 3 and show all forces acting. For (*a*), pinion 2 rotates at 600 rev/min and transmits 18 hp to the gearset. For (*b*) and (*c*), pinion 2 rotates at 900 rev/min and transmits 25 hp to the gearset.

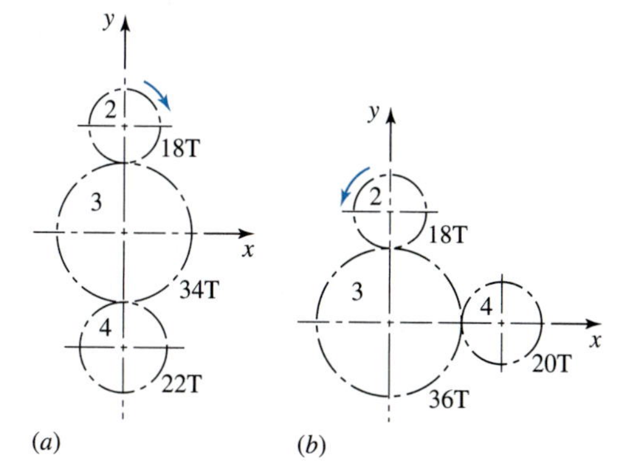

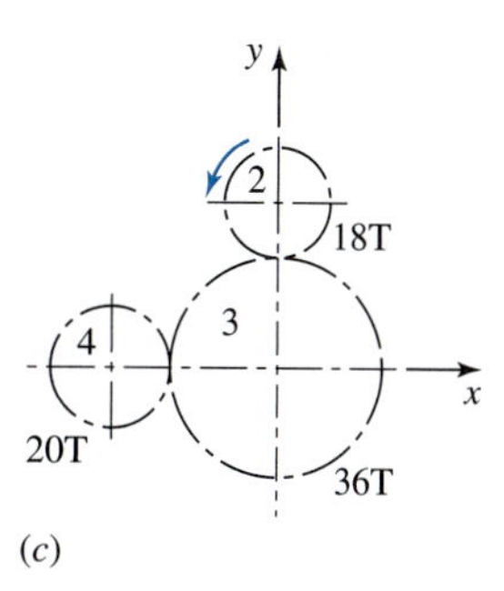

(*c*)

Figure P13.17

13.18 A 15-tooth spur pinion has a diametral pitch of 5 and 20° pressure angle, rotates at 600 rev/min, and drives a 60-tooth gear. The drive transmits 25 hp. Construct a free-body diagram of each gear showing upon it the tangential and radial components of the forces and their proper directions.

13.19 A 16-tooth pinion on shaft 2 rotates at 1 720 rev/min and transmits 5 hp to the double-reduction gear train. All gears have 20° pressure angle. The distances between centers of the bearings and gears for shaft 3 are shown in Fig. P13.19. Find the magnitude and direction of the radial force that each bearing exerts against the shaft.

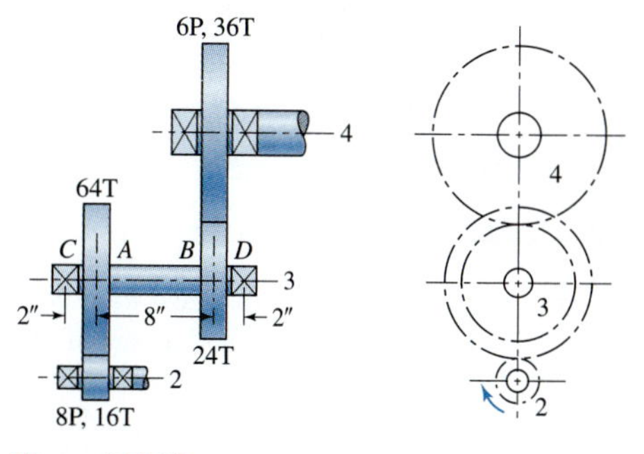

Figure P13.19

13.20 Solve Problem 13.17 if each pinion has right-hand helical teeth with a 30° helix angle and a 20° pressure angle. All gears in the train are helical and, of course, the normal diametral pitch is 6 teeth/in. for each case.

13.21 Analyze the gear shaft of Example 13.8 and find the bearing reactions $\mathbf{F}_C$ and $\mathbf{F}_D$.

13.22 In each of the bevel gear drives shown in Fig. P13.22, bearing A takes both thrust load and radial load, whereas bearing B takes only radial load. The teeth are cut with a 20° pressure angle. For (a) $\mathbf{T}_2 = -180\hat{\mathbf{i}}$ in · lb and for (b) $\mathbf{T}_2 = -240\hat{\mathbf{k}}$ in · lb. Compute the bearing loads for each case.

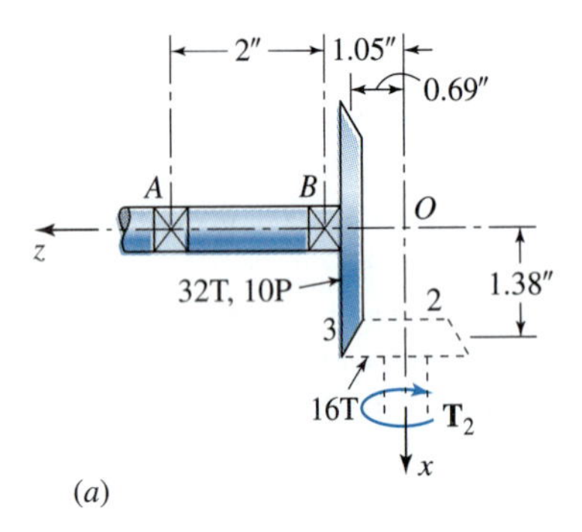

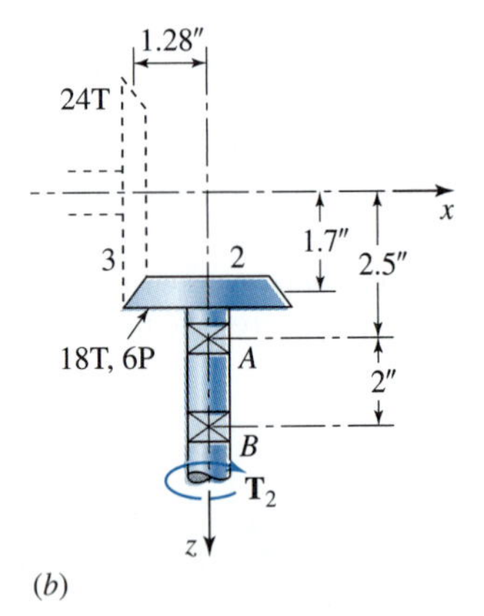

Figure P13.22

13.23 Figure P13.23 shows a gear train composed of a pair of helical gears and a pair of straight-tooth bevel gears. Shaft 4 is the output of the train and delivers 6 hp to the load at a speed of 370 rev/min. All gears have pressure angles of 20°. If bearing E is to take both thrust load and radial load, whereas bearing F is to take only radial load, determine the force that each bearing exerts against shaft 4.

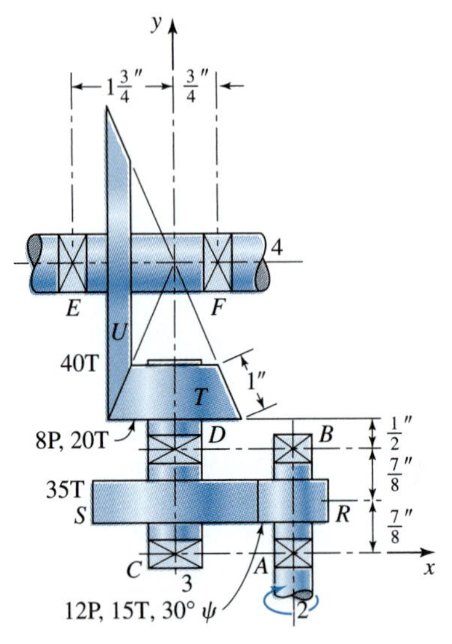

Figure P13.23

13.24 Using the data of Problem 13.23, find the forces exerted by bearings C and D onto shaft 3. Which of these bearings should take the thrust load if the shaft is to be loaded in compression?

13.25 Use the method of virtual work to solve the slider–crank mechanism of Problem 13.2.

13.26 Use the method of virtual work to solve the four-bar linkage of Problem 13.4.

13.27 Use the method of virtual work to analyze the crank–shaper linkage of Problem 13.7. Given that the load remains constant at $\mathbf{P} = 100\hat{\mathbf{i}}$ lb, find and plot a graph of the crank torque M_{12} for all positions in the cycle using increments of 30° for the input crank.

13.28 Use the method of virtual work to solve the four-bar linkage of Problem 13.10.

13.29 A car (link 2) that weighs 2,000 lb is slowly backing a 1,000 lb trailer (link 3) up a 30° inclined ramp, as illustrated in Fig. P13.29. The car wheels are of 13-in. radius, and the trailer wheels have 10-in. radius; the center of the hitch ball is also 13 in. above the roadway. The centers of mass of the car and trailer are located at G_2 and G_3, respectively, and gravity acts vertically downward in Fig. P13.29. The weights of the wheels and friction in the bearings are considered negligible. Assume that there are no brakes

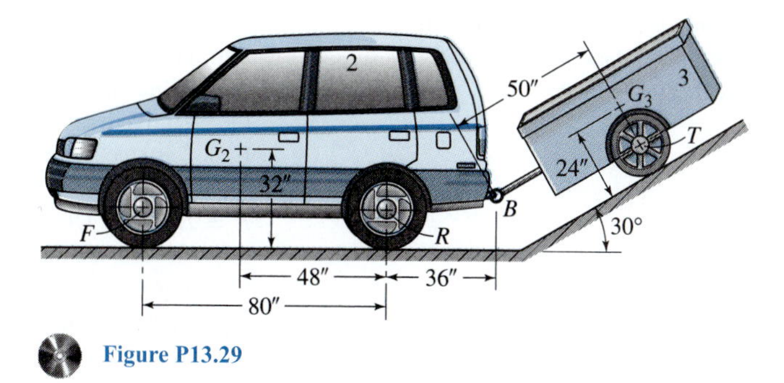

Figure P13.29

applied on the car or on the trailer and that the car has front-wheel drive. Determine the loads on each of the wheels and the minimum coefficient of static friction between the driving wheels and the road to avoid slipping.

13.30 Repeat Problem 13.29 assuming that the car has rear-wheel drive rather than front-wheel drive.

13.31 The low-speed disk cam with oscillating flat-face follower illustrated in Fig. P13.31 is driven at a constant shaft speed. The displacement curve for the cam has a full-rise cycloidal motion, defined by Eq. (6.13) with parameters $L = 30°$, $\beta = 30°$, and a prime circle radius $R_0 = 30$ mm; the instant pictured is at $\theta_2 = 112.5°$. A force of $F_C = 8$ N is applied at point C and continues at 45° from the face of the follower, as demonstrated. Use the virtual-work approach to determine the moment $\mathbf{M}_{12}$ required on the crankshaft at the instant shown to produce this motion.

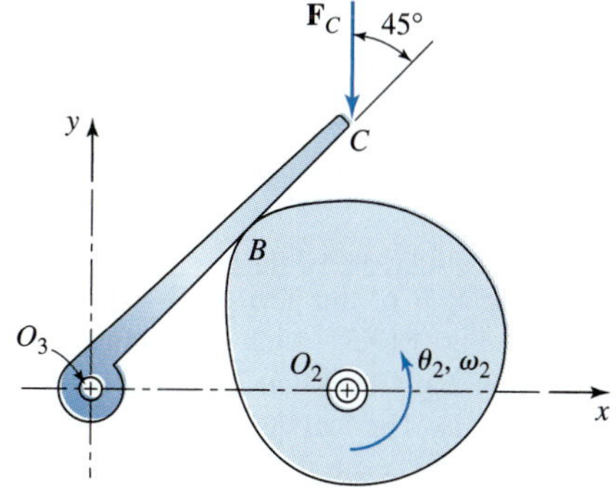

Figure P13.31 $R_{O_2O_3} = 50$ mm; $R_B = 42$ mm; $R_C = 150$ mm.

13.32 Repeat Problem 13.31 for the entire lift portion of the cycle, finding $\mathbf{M}_{12}$ as a function of θ_2.

13.33 A disk 3 of radius R is being slowly rolled under a pivoted bar 2 driven by an applied torque T as illustrated in Fig. P13.33. Assume a coefficient of static friction of μ between the disk and ground and that all other joints are frictionless. A force $\mathbf{F}$ is acting vertically downward on the bar at a distance d from the pivot O_2. Assume that the weights of the links are negligible in comparison to F. Find an equation for the torque $\mathbf{T}$ required as a function of the distance $x = R_{CO_2}$ and an equation for the final distance x that is reached when friction no longer allows further movement.

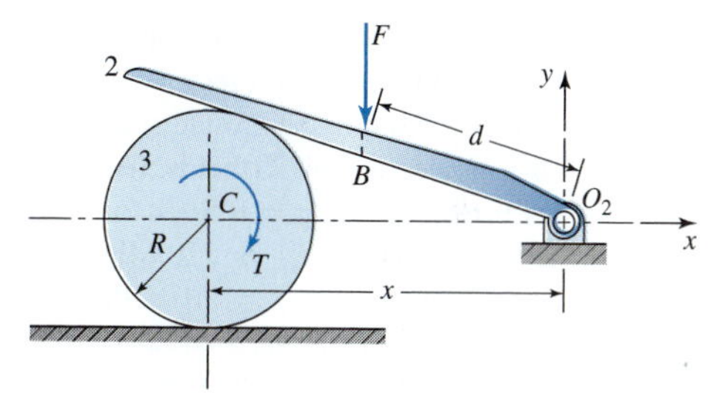

Figure P13.33

13.34 Links 2 and 3 are pinned together at B and a constant vertical load $P = 800$ kN is applied at B, as illustrated in Fig. P13.34. Link 2 is fixed in the ground at A and link 3 is pinned to the ground at C. The length of link 2 is 8 m and it has a 150-mm solid square cross-section. The length of link 3 is 0.5 m and it has a solid circular cross-section with diameter D. Both links are made from a steel with a modulus of elasticity $E = 207$ GPa and a compressive yield strength $S_{yc} = 200$ MPa. Using the theoretical values for the end condition constants of each link, determine: (i) the slenderness ratio, the critical load, and the factor of safety guarding against buckling of link 2 and (*ii*) the minimum diameter D_{min} of link 3 if the static factor of safety guarding against buckling is to be $N = 2$.

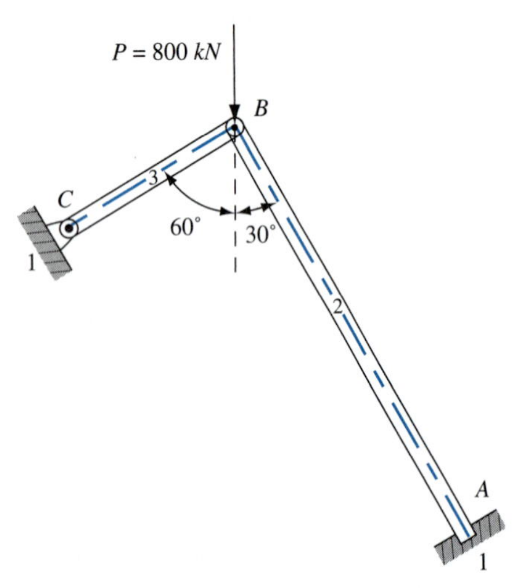

Figure P13.34 Two links *AB* and *BC* pinned at *B*.

13.35 The horizontal link 2 is subjected to the load $F = 150$ kN at *C*, as illustrated in Fig. P13.35. The link is supported by the solid circular aluminum link 3. The lengths of the links are $L_2 = R_{CA} = 5$ m, $R_{BA} = 3$ m, and $L_3 = R_{BD} = 3$ m. The end *D* of link 3 is fixed in the ground and the opposite end *B* is pinned to link 2 (that is, the effective length of the link is $L_{EFF} = 0.5L_2$). For aluminum, the yield strength is $S_{yc} = 370$ MPa and the modulus of elasticity is $E = 207$ GPa. Determine the diameter *d* of the solid circular cross-section of link 3 to ensure that the static factor of safety is $N = 2.5$.

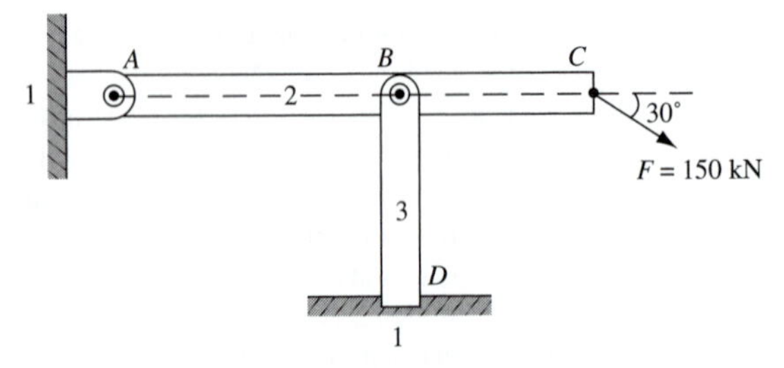

Figure P13.35 The link *BD* supporting the link *AC*.

13.36 The horizontal link 2 is subjected to the inclined load $F = 8\ 000$ N at *C*, as illustrated in Fig. P13.36. The link is supported by a solid circular cross-section link 3 whose length $L_3 = BD = 5$ m. The end *D* of the vertical link 3 is fixed in the ground link and the end *B* supports link 2 (that is, the effective length of the link is $L_{EFF} = 0.5L_3$). Link 3 is steel with a compressive yield strength $S_{yc} = 370$ MPa and a modulus of elasticity $E = 207$ GPa. Determine the diameter *d* of link 3 to ensure that the factor of safety guarding against buckling is $N = 2.5$. Also, answer the following statements true or false and briefly give your reasons. (*i*) The slenderness ratio at the point of tangency between Euler's column formula and Johnson's parabolic equation does not depend on the geometry of the column. (*ii*) Under the same loading conditions, a link with pinned–pinned ends will give a higher factor of safety against buckling than an identical link with fixed–fixed ends. (*iii*) If the slenderness ratio $S_r = (S_r)_D$ at the point of tangency, then the critical unit load does not depend on the yield strength of the column material.

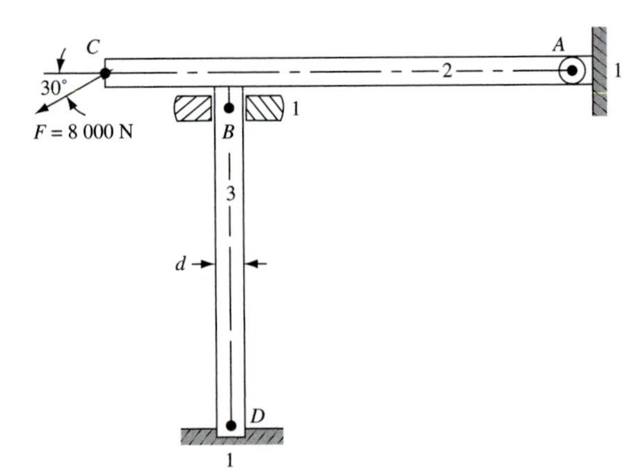

Figure P13.36 The link *AC* supported by the link *BD*.

13.37 A load $\mathbf{P}_A$ is acting at *A* and a load $\mathbf{P}_B$ is acting at *B* of the horizontal link 3, as illustrated in Fig. P13.37. Link 3 is pinned to the vertical link 2 at *O* and link 2 is fixed in the ground link 1 at *D*. The lengths are $AO = 4$ ft, $OB = 2$ ft, and $DO = 6$ ft. Both links have solid circular cross-sections with diameter $D = 2$ in and are made from a steel alloy with a compressive yield strength $S_{yc} = 85 \times 10^3$ psi, a tensile yield strength $S_{yt} = 75 \times 10^3$ psi, and a modulus of elasticity $E = 30 \times 10^6$ psi. Assuming that links 2

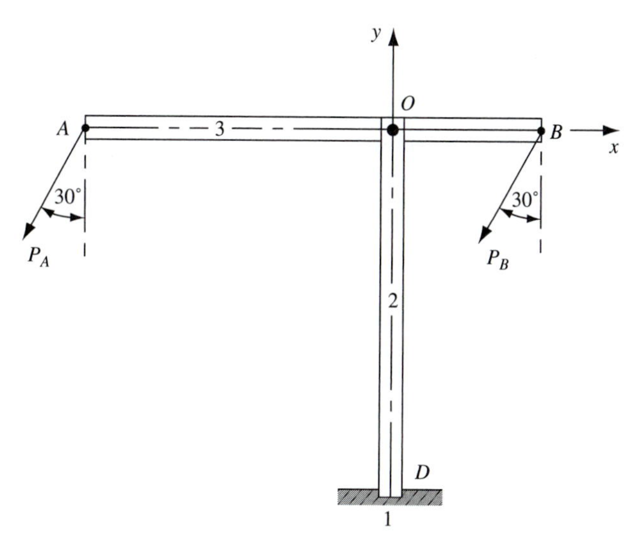

Figure P13.37 The link AB supporting two loads.

and 3 are in static equilibrium and using the theoretical value for the end-condition constant for link 2, determine: (i) the magnitude of the force $\mathbf{P}_B$ that is acting as shown at B if $P_A = 30\,000$ lb, (ii) the critical load, the critical unit load, and the factor of safety to guard against buckling for link 2; and (iii) the diameter of a solid circular cross-section for link 2 that will ensure the factor of safety guarding against buckling of the link is $N = 4$.

13.38 The horizontal link 2 is subjected to the load $P = 5\,000$ N as illustrated in Fig. P13.38. This link is supported by the vertical link 3, which has a constant circular cross-section. The lengths are $AC = 5$ m, $AB = 4$ m, and $DB = L_3 = 5$ m. For the vertical link 3, the end D is fixed in the ground link and the end B supports link 2 (that is, the effective length of link 3 is $L_{EFF} = 0.5L_3$). The yield strength and the modulus of elasticity for the aluminum link 3 are $S_y = 370$ MPa and $E = 207$ GPa, respectively. Determine the diameter d of link 3 to ensure that the static factor of safety guarding against buckling is $N = 2.5$.

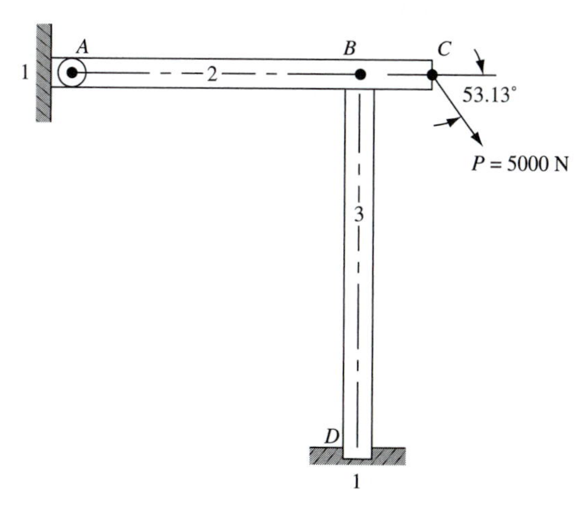

Figure P13.38 Link 3 supporting link AC.

13.39 The horizontal link 2 is pinned to the vertical wall at A and pinned to link 3 at B as illustrated in Fig. P13.39. The opposite end of link 3 is pinned to the wall at C. A vertical force $P = 25$ kN is acting on link 2 at B. Link 2 has a 20×30 mm solid rectangular cross-section and link 3 has a 40×40 mm solid square cross-section. The length of link 3 is $BC = 1.2$ m and the angle $\angle ABC = 30°$. The two links are made from a steel alloy with a tensile yield strength $S_{yt} = 190$ MPa, a compressive yield strength $S_{yc} = 205$ MPa, and a modulus of elasticity $E = 207$ GPa. Using the theoretical value for the end-condition constant for link 3, determine: (*i*) the value of the slenderness ratio at the point of tangency between Euler's column formula and Johnson's parabolic formula, (*ii*) the critical load and the factor of safety guarding against buckling of link 3, and (*iii*) the minimum width of the square cross-section of link 3 for the factor of safety to guard against buckling to be $N = 1$.

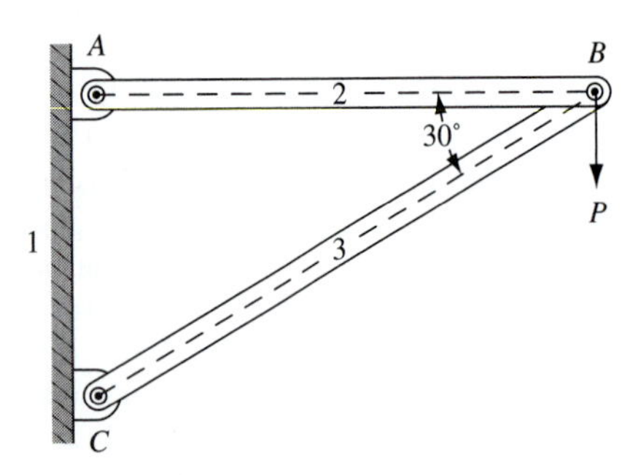

Figure P13.39 A simple truss.

13.40 The link $BC = 1.2$ m and 25 mm square cross-section is fixed in the vertical wall at C and pinned at B to a circular steel cable AB with diameter $d = 20$ mm as illustrated in Fig. P13.40. The distance $AC = 0.7$ m. The mass m of a container, suspended from pin B, produces a gravitational load at B, which results in the moment at point C in the wall $M_C = 8\ 000$ Nm ccw. The yield strength and modulus of elasticity of the steel cable AB and the steel link BC are $S_y = 370$ MPa and $E = 207$ GPa, respectively. Given that $m = 2\,000$ kg, determine: (*i*) the tension in the cable AB and the factor of safety guarding against tensile failure, (*ii*) the compressive load acting in link BC, and (*iii*) the factor of safety guarding against buckling of link BC. (Use the theoretical value of the end-condition constant assuming that the link has fixed–pinned ends.) If $M_C = 10\ 000$ Nm ccw, then determine the maximum mass of a container that can be suspended from pin B before buckling of link BC will begin (that is, the factor of safety guarding against buckling failure is $N = 1$).

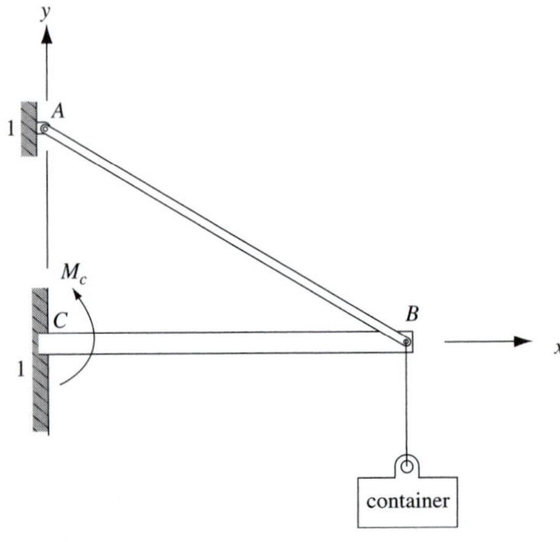

Figure P13.40 A structure supporting a load.

13.41 A vertically upward force F is applied at C in the horizontal link 4, as illustrated in Fig. P13.41. The link is pinned to the ground at B and pinned to the vertical link 2 at A. The lengths $AB = 2$ ft and $AC = 5$ ft and links 2 and 4 are made from a steel alloy with a compressive yield strength $S_{yc} = 60 \times 10^3$ psi and a modulus of elasticity $E = 30 \times 10^6$ psi. Link 2 has a hollow circular

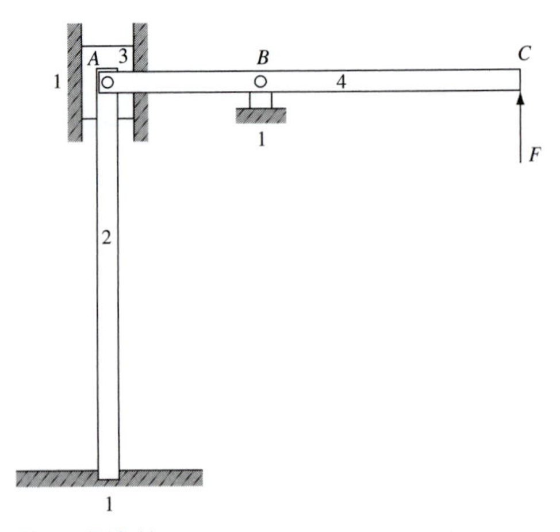

Figure P13.41 The link AC supporting a load F.

cross-section with an outside diameter $D = 2$ in, wall thickness $t = 0.25$ in, and length $L = 5$ ft. Using the theoretical value for the end-condition constant for link 2, determine: (*i*) the value of the slenderness ratio at the point of tangency between Euler's column formula and Johnson's parabolic formula, (*ii*) the critical load acting on link 2, and (*iii*) the force F for the factor of safety of link 2 to guard against buckling to be $N = 1$. (*iv*) If link 2 has a solid circular cross-section with diameter $D = 3$ in. and $F = 20\,000$ lbs, then determine the maximum length of link 2 for the factor of safety to guard against buckling to be $N = 2$.

13.42 A force F is acting at C perpendicular to link 2, as illustrated in Fig. P13.42. The end A is pinned to the ground and the supporting link 3 is pinned to link 2 at B and pinned to the ground at D. The lengths are $AC = 200$ mm and $AD = 150$ mm. Link 3 has a circular cross-section with diameter $D = 5$ mm. Both links are a steel alloy with a compressive yield strength $S_{yc} = 420$ MPa and a modulus of elasticity $E = 206$ GPa. Using the theoretical value for the end-condition constant for link 3, determine: (*i*) the slenderness ratio at the point of tangency between Euler's column formula and Johnson's parabolic formula, (*ii*) the critical load and the critical unit load acting on the link, and (*iii*) the force F for the factor of safety to guard against buckling to be $N = 1$. If the force $F = 3\,000$ N then for link 3 determine: (*a*) the critical load for the factor of safety to guard against buckling $N = 1$ and (*b*) the slenderness ratio.

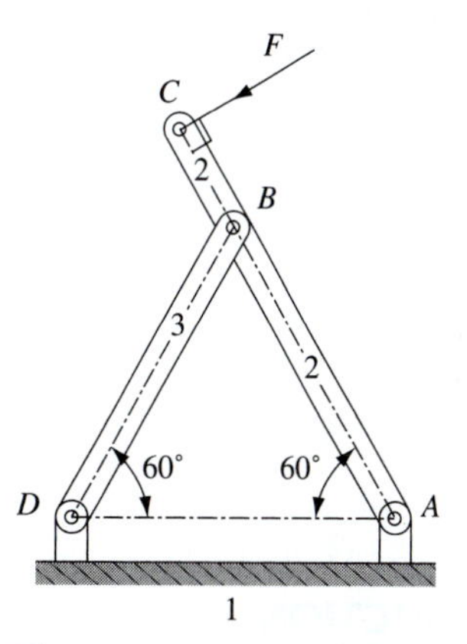

Figure P13.42 A simple truss.

14 DYNAMIC FORCE ANALYSIS

14.1 INTRODUCTION

In Chapter 13 we studied the forces in machine systems in which all forces on the bodies were in balance and therefore the systems were in either static or dynamic equilibrium. However, in real machines this is seldom, if ever, the case except when the machine is stopped. We learned in Chapter 4 that although the input crank of a machine may be driven at constant speed, this does not mean that all points of the input crank have constant velocity vectors or that other members of the machine will operate at constant speeds. In general, there will be accelerations and therefore machines with moving parts having mass will not be in equilibrium.

Of course, techniques for static-force analysis are important, not only because stationary structures must be designed to withstand their imposed loads, but also because they introduce concepts and approaches that can be built upon and extended to nonequilibrium situations. That introduces the purpose of Chapter 14: to learn how much acceleration will result from a system of unbalanced forces and also to learn how these *dynamic* forces can be assessed for systems that are not in equilibrium.

14.2 CENTROID AND CENTER OF MASS

We recall from Section 13.2 that Newton's laws set forth the relationships between the net unbalanced force on a *particle*, its mass, and its acceleration. For Chapter 13, because we were only studying systems in equilibrium, we made use of that relationship for entire rigid bodies, arguing that they are made up of collections of particles and that the action and the reaction forces between the particles cancel each other. In Chapter 14, we must be more careful. We must remember that each of these particles may have acceleration and that the

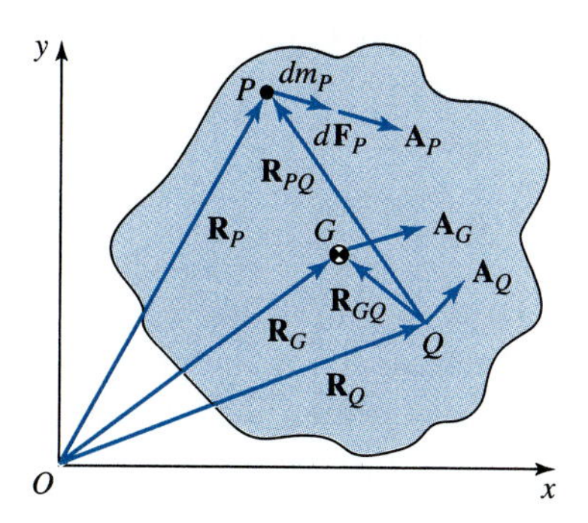

Figure 14.1 Particle of mass dm_P at location P on a rigid body.

accelerations of these particles may all be different from each other. Important questions include the following: Which of these many point accelerations are we to use? And why?

Referring to the rigid body illustrated in Fig. 14.1, we consider a particle with mass dm_P at some arbitrary point P in the rigid body. Equation (13.1) tells us that the net unbalanced force $d\mathbf{F}_P$ on this particle is proportional to its mass and its absolute acceleration $\mathbf{A}_P$, that is,

$$d\mathbf{F}_P = \mathbf{A}_P dm_P. \tag{a}$$

Our task now is to sum these effects, that is, to integrate over all particles of the body and to put the result in some usable form for rigid bodies other than single particles. As we did in Chapter 13, we can conclude that the action and the reaction forces between particles of the body balance each other and therefore cancel in the process of the summation. The only net remaining forces are the constraint forces, those whose reactions are on some other body than this one. Thus, integrating Eq. (*a*) over all particles of mass in our rigid body, we obtain

$$\Sigma\mathbf{F} = \int \mathbf{A}_P dm_P. \tag{b}$$

We find it difficult to perform this integration because each particle has a different acceleration. However, if we assume that we know (or can find) the acceleration of one particular point of the body—say, point Q—and also the angular velocity $\boldsymbol{\omega}$ and angular acceleration $\boldsymbol{\alpha}$ of the body, then we can express the various accelerations of all other particles of the body in terms of their acceleration differences from that of point Q. Thus, from Eqs. (4.5), we write

$$\mathbf{A}_P = \mathbf{A}_Q + \boldsymbol{\omega} \times (\boldsymbol{\omega} \times \mathbf{R}_{PQ}) + \boldsymbol{\alpha} \times \mathbf{R}_{PQ}. \tag{c}$$

Substituting this equation into Eq. (b), the sum of the forces can be written as

$$\Sigma\mathbf{F} = \mathbf{A}_Q \int dm_P + \boldsymbol{\omega} \times \left(\boldsymbol{\omega} \times \int \mathbf{R}_{PQ} dm_P\right) + \boldsymbol{\alpha} \times \int \mathbf{R}_{PQ} dm_P, \tag{d}$$

where we have factored out of the integrals all quantities that are the same for all particles of the body.

We easily recognize the first integral of Eq. (*d*); the summation of all particle masses gives the total mass of the body:

$$\int dm_P = m. \tag{14.1}$$

The second and third terms of Eq. (*d*) both contain another integral, which does not look as easy. However, because of its frequent appearance in the study of mechanics, a particular point G having a location given by the position vector $\mathbf{R}_G$ has been defined by the following integral equation:

$$\int \mathbf{R}_P dm_P = m\mathbf{R}_G. \tag{14.2}$$

This point G is called the *centroid* or the *center of mass* of the body. More will be said about this special point in the material below.

Substituting Eqs. (14.1) and (14.2) into Eq. (*d*) gives

$$\Sigma\mathbf{F} = m\mathbf{A}_Q + \boldsymbol{\omega} \times [\boldsymbol{\omega} \times (m\mathbf{R}_{GQ})] + \boldsymbol{\alpha} \times (m\mathbf{R}_{GQ}),$$

which can be written as

$$\Sigma\mathbf{F} = m[\mathbf{A}_Q + \boldsymbol{\omega} \times (\boldsymbol{\omega} \times \mathbf{R}_{GQ}) + \boldsymbol{\alpha} \times \mathbf{R}_{GQ}].$$

Finally, recognizing the sum of the three acceleration terms in the square bracket, we find

$$\Sigma\mathbf{F} = m\mathbf{A}_G. \tag{14.3}$$

This important equation is the integrated form of Newton's law for a particle, now extended to a rigid body. Note that it is the same equation as in Eq. (13.1). However, careful derivation of the acceleration term was not done there because we were treating static problems where accelerations were to be set to zero.

We have now answered the question raised earlier in this section. Recognizing that each particle of a rigid body may have a different acceleration, which one should be used? Equation (14.3) demonstrates clearly that the absolute acceleration of the center of mass of the body is the proper acceleration to be used in Newton's law. That particular point, and no other, is the proper point for which Newton's law for a rigid body has the same form as for a single particle.

In solving engineering problems, we frequently find that forces are distributed in some manner over a line, over an area, or over a volume. The resultant of these distributed forces is usually not too difficult to find. To have the equivalent effect, this resultant must act at the centroid of the system. Thus, *the centroid of a system is a point at which a system of distributed forces may be considered concentrated with exactly the same effect.*

Instead of a system of forces, we may have a distributed mass, as in the above derivation. Then, by *center of mass* we mean *the point at which the mass may be considered concentrated* and that the effect is the same.

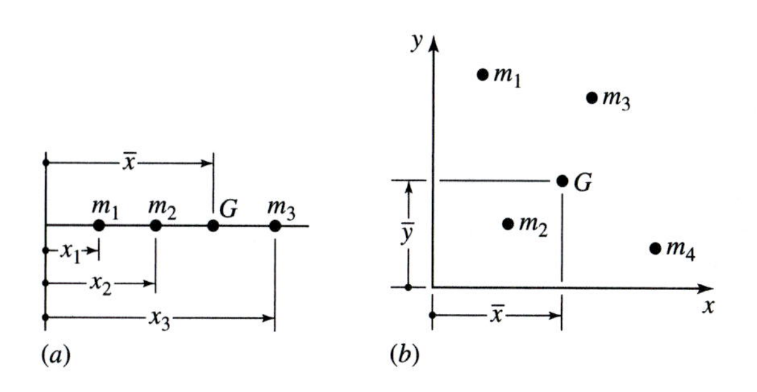

Figure 14.2 (*a*) Particles of mass distributed along a line. (*b*) Particles of mass distributed in a plane.

In Fig. 14.2*a*, a series of particles with masses are shown located at various positions along a line. The center of mass G is located at

$$\bar{x} = \frac{m_1x_1 + m_2x_2 + m_3x_3}{m_1 + m_2 + m_3} = \frac{\Sigma m_i x_i}{\Sigma m_i}. \tag{14.4}$$

In Fig. 14.2*b*, the mass particles m_i are located at various positions $\mathbf{R}_i$ in a plane. The location of the center of mass G is now given by

$$\mathbf{R}_G = \frac{m_1\mathbf{R}_1 + m_2\mathbf{R}_2 + m_3\mathbf{R}_3 + m_4\mathbf{R}_4}{m_1 + m_2 + m_3 + m_4} = \frac{\Sigma m_i \mathbf{R}_i}{\Sigma m_i}. \tag{14.5}$$

This procedure can also be extended to particles distributed in a volume simply using Eq. (14.5) and treating position vectors $\mathbf{R}_i$ as three dimensional rather than two dimensional.

When the mass is distributed continuously along a line or over a plane or volume, the concept of summation in Eq. (14.5) is replaced by integration over infinitesimal particles of mass. The result is

$$\mathbf{R}_G = \frac{1}{m}\int \mathbf{R}dm, \tag{14.6}$$

where m is the total mass of the body considered. It is from this definition that Eq. (14.2) was obtained.

When mass is evenly distributed over a plane area or a volume, the center of mass can often be found by symmetry. Figure 14.3 illustrates the locations of the centers of mass for a circular solid, a rectangular solid, and a triangular solid. Each is assumed to have constant thickness and uniform density distribution.

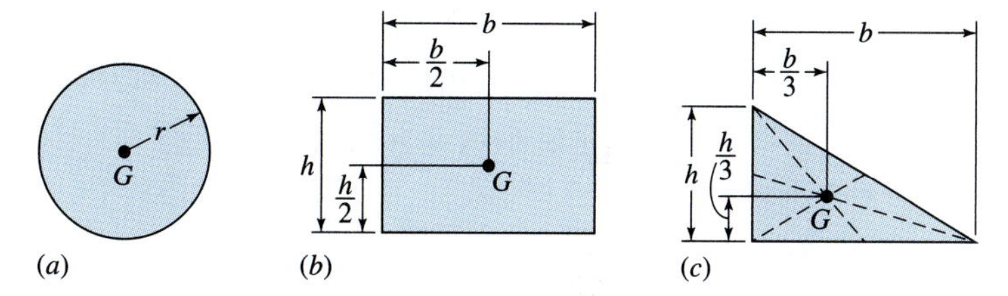

Figure 14.3 Center of mass location for (*a*) a right circular solid, (*b*) a rectangular solid, and (*c*) a triangular prism.

When a body is of a more irregular shape, the center of mass can often still be determined by considering it a combination of simpler subshapes, as illustrated in the following example.

EXAMPLE 14.1

The volume illustrated in Fig. 14.4 consists of a rectangular solid, less a circular through hole, plus a triangular plate of a different thickness. The dimensions are as shown in Fig. 14.4, and the entire part is made of cast iron. Find the center of mass of this composite shape.

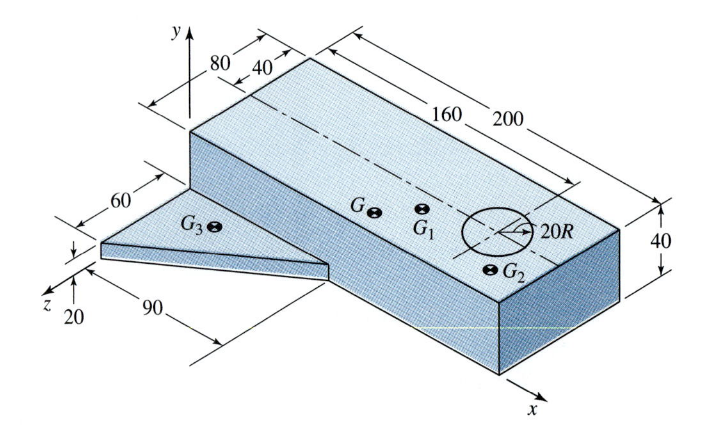

Figure 14.4 Composite shape for Example 14.1 with dimensions in millimeters.

SOLUTION

We can find the masses and the centers of mass of each of the three subshapes with the help of the formulae given in Fig. 14.3. For the rectangular solid subshape we obtain*

$$m_1 = (80\text{ mm})(200\text{ mm})(40\text{ mm})(\rho\text{ kg/mm}^3) = 640\,000\rho\text{ kg}$$

$$\mathbf{R}_{G_1} = 100\hat{\mathbf{i}} + 20\hat{\mathbf{j}} - 40\hat{\mathbf{k}}\text{ mm}$$

For the hole we treat the mass as negative, giving

$$m_2 = \pi(20\text{ mm})^2(40\text{ mm})(-\rho\text{ kg/mm}^3) = -50\,265\rho\text{ kg}$$

$$\mathbf{R}_{G_2} = 160\hat{\mathbf{i}} + 20\hat{\mathbf{j}} - 40\hat{\mathbf{k}}\text{ mm.}$$

* Note that this example is done in SI units; in U.S. customary units, pounds, slugs, or other units might have been used. However, this is not stressed here, because the density cancels in the use of Eq. (14.5) below.

Finally, for the triangular subshape we find

$$m_3 = 0.5(60\text{ mm})(90\text{ mm})(20\text{ mm})(\rho\text{ kg/mm}^3) = 54\,000\rho\text{ kg}$$

$$\mathbf{R}_{G_3} = 30\hat{\mathbf{i}} + 10\hat{\mathbf{j}} + 20\hat{\mathbf{k}}\text{ mm}.$$

Now, because we are completing a process of integration, we can combine the values of the subscripts; that is, from Eq. (14.5) we can write

$$\mathbf{R}_G = \frac{m_1\mathbf{R}_1 + m_2\mathbf{R}_2 + m_3\mathbf{R}_3}{m_1 + m_2 + m_3}.$$

This gives the following result:

$$\mathbf{R}_G = 89.4\hat{\mathbf{i}} + 19.2\hat{\mathbf{j}} - 35.0\hat{\mathbf{k}}\text{ mm}. \qquad \textit{Ans.}$$

14.3 MASS MOMENTS AND PRODUCTS OF INERTIA

Another problem that often arises when forces are distributed over an area or volume is that of calculating their moment about a specified point or axis of rotation. Sometimes the force intensity varies according to its distance from the point or axis of rotation. Although we will save a more thorough derivation of these equations until Section 14.11 and later, we will point out here that such problems always give rise to integrals of the form $\int$ (distance)$^2 dm$.

In three-dimensional problems, three such integrals are defined as follows*:

$$\begin{aligned}
I^{xx} &= \int (\hat{\mathbf{i}} \times \mathbf{R}) \cdot (\hat{\mathbf{i}} \times \mathbf{R})dm = \int [(R^y)^2 + (R^z)^2]dm \\
I^{yy} &= \int (\hat{\mathbf{j}} \times \mathbf{R}) \cdot (\hat{\mathbf{j}} \times \mathbf{R})dm = \int [(R^z)^2 + (R^x)^2]dm \\
I^{zz} &= \int (\hat{\mathbf{k}} \times \mathbf{R}) \cdot (\hat{\mathbf{k}} \times \mathbf{R})dm = \int [(R^x)^2 + (R^y)^2]dm.
\end{aligned} \tag{14.7}$$

These three integrals are called the *mass moments of inertia* of the body. Another three similar integrals are

$$\begin{aligned}
I^{xy} = I^{yx} &= \int (\hat{\mathbf{i}} \times \mathbf{R}) \cdot (\hat{\mathbf{j}} \times \mathbf{R})dm = -\int (R^x R^y)dm \\
I^{yz} = I^{zy} &= \int (\hat{\mathbf{j}} \times \mathbf{R}) \cdot (\hat{\mathbf{k}} \times \mathbf{R})dm = -\int (R^y R^z)dm \\
I^{zx} = I^{xz} &= \int (\hat{\mathbf{k}} \times \mathbf{R}) \cdot (\hat{\mathbf{i}} \times \mathbf{R})dm = -\int (R^z R^x)dm
\end{aligned} \tag{14.8}$$

* It should be carefully noted here that these integrals are not the same as those called *area moments of inertia* or *second moments of area*, which are integrals over dA, a differential area, rather than integrals over dm, a differential mass. These other integrals also arise in problems involving forces distributed over an area or volume, but are different. In two-dimensional problems of constant thickness, however, they are easily related because dA times the thickness times the mass density yields dm for the integral.

and these three integrals are called the *mass products of inertia* of the body. Sometimes it is convenient to arrange these mass moments of inertia and mass products of inertia into a symmetric square array or matrix format called the *inertia tensor* of the body:

$$\mathbf{I} = \begin{bmatrix} I^{xx} & -I^{xy} & -I^{xz} \\ -I^{yx} & I^{yy} & -I^{yz} \\ -I^{zx} & -I^{zy} & I^{zz} \end{bmatrix} \tag{14.9}$$

A careful look at the above integrals will indicate that they represent the mass distribution of the body with respect to the coordinate system about which they are determined, but that they change if evaluated in a different coordinate system. To keep their meaning direct and simple, we assume that the coordinate system chosen for each body is attached to that body in a convenient location and orientation. Therefore, for rigid bodies, the mass moments and products of inertia are constant properties of the body and its mass distribution and they do not change when the body moves; they do, however, depend on the coordinate system chosen.

An interesting property of these integrals is that it is always possible to choose the coordinate system so that its origin is located at the center of mass of the body and oriented such that all of the mass products of inertia become zero. Such a choice of the coordinate axes of the body is called its *principal axes*, and the corresponding values of Eqs. (14.7) are then called the *principal mass moments of inertia*. A variety of simple geometric solids, the orientations of their principal axes, and formulae for their principal mass moments of inertia are included in Appendix A (see Table 5).

If we note that mass moments of inertia have units of mass times distance squared, it seems natural to define a radius value for the body as

$$I_G = k^2 m \quad \text{or} \quad k = \sqrt{\frac{I_G}{m}}. \tag{14.10}$$

This distance k is called the *radius of gyration* of the body, and it is always calculated or measured from the center of mass of the part about one of the principal axes. For three-dimensional motions of parts there are three radii of gyration, k^x, k^y, and k^z, associated with the three principal axes, I^{xx}, I^{yy}, and I^{zz}.

It is often necessary to determine the moments and products of inertia of bodies, which are composed of several simpler subshapes for which formulae are known, such as those given in Table 5 in Appendix A. The easiest method of finding these is to compute the mass moments about the principal axes of each subshape, then to shift the origins of each to the mass center of the composite body, and then to sum the results. This requires that we develop methods of redefining mass moments and products of inertia when the axes are translated to a new position. The form of the *transfer*, or *parallel-axis theorem* for mass moment of inertia, is written

$$I = I_G + md^2, \tag{14.11}$$

where I_G is one of the principal mass moments of inertia about some known principal axis and I is the mass moment of inertia about a parallel axis at distance d from that principal axis. Equation (14.11) must only be used for *translation* of inertia axes starting from a principal axis. Also, the rotation of these axes results in the introduction of product of inertia terms. More will be said on general transformations of inertia axes in Section 14.11.

EXAMPLE 14.2

Figure 14.5 illustrates a connecting rod made of ductile iron with density of 0.260 lb/in^3. Find the mass moment of inertia about the z axis.

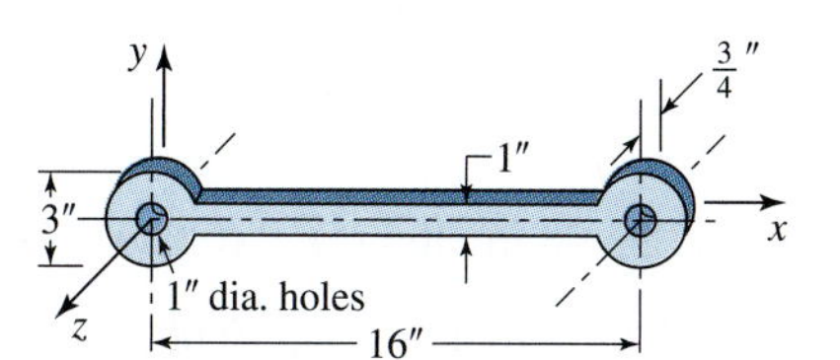

Figure 14.5 Connecting-rod shape for Example 14.2.

SOLUTION

We solve by finding the mass moment of inertia of each of the cylinders at the ends of the rod and of the central prismatic bar, all taken about their own mass centers. Then we use the parallel-axis theorem to transfer each to the z axis.

The mass of each hollow cylinder is

$$\begin{aligned} m_{\text{cyl}} &= \rho\pi(r_o^2 - r_i^2)l \\ &= \frac{(0.260\ \text{lb/in}^3)\pi[(1.5\ \text{in})^2 - (0.5\ \text{in})^2](0.75\ \text{in})}{386\ \text{in/s}^2} \\ &= 0.003\ 17\ \text{lb}\cdot\text{s}^2/\text{in} \end{aligned}$$

The mass of the central prismatic bar is

$$\begin{aligned} m_{\text{bar}} &= \rho whl = \frac{(0.260\ \text{lb/in}^3)(13\ \text{in})(1\ \text{in})(0.75\ \text{in})}{386\ \text{in/s}^2} \\ &= 0.006\ 57\ \text{lb}\cdot\text{s}^2/\text{in} \end{aligned}$$

From Table 5 in Appendix A, we find the mass moment of inertia of each element is

$$\begin{aligned} I_{\text{cyl}} &= \frac{m(r_o^2 + r_i^2)}{2} \\ &= \frac{(0.003\ 17\ \text{lb}\cdot\text{s}^2/\text{in})[(1.5\ \text{in})^2 + (0.5\ \text{in})^2]}{2} \\ &= 0.003\ 96\ \text{in}\cdot\text{lb}\cdot\text{s}^2 \end{aligned}$$

and

$$\begin{aligned} I_{\text{bar}} &= \frac{m(w^2 + h^2)}{12} \\ &= \frac{(0.006\ 57\ \text{lb} \cdot \text{s}^2/\text{in})[(1\ \text{in})^2 + (13\ \text{in})^2]}{12} \\ &= 0.093\ 08\ \text{in} \cdot \text{lb} \cdot \text{s}^2 \end{aligned}$$

Then, using Eq. (14.11) to transfer the axes, the mass moment of inertia about the z axis is

$$\begin{aligned} I^{zz} &= I_{\text{cyl}} + (I_{\text{bar}} + m_{\text{bar}} d_{\text{bar}}^2) + (I_{\text{cyl}} + m_{\text{cyl}} d_{\text{cyl}}^2) \\ &= (0.003\ 96\ \text{in} \cdot \text{lb} \cdot \text{s}^2) + [(0.093\ 08\ \text{in} \cdot \text{lb} \cdot \text{s}^2) + (0.006\ 57\ \text{lb} \cdot \text{s}^2/\text{in})(8\ \text{in})^2] \\ &\qquad + [(0.003\ 96\ \text{in} \cdot \text{lb} \cdot \text{s}^2) + (0.003\ 17\ \text{lb} \cdot \text{s}^2/\text{in})(16\ \text{in})^2] \\ &= 1.333\ \text{in} \cdot \text{lb} \cdot \text{s}^2 \qquad \textit{Ans.} \end{aligned}$$

It will be noted in this example that only one mass moment of inertia I^{zz} was requested or determined. This does not mean that I^{xx} and I^{yy} are zero, but rather that they will likely not be needed for further analysis. In problems with only planar motion, only I^{zz} is needed because I^{xx} and I^{yy} are used only with rotations out of the xy plane. These other mass moments and products of inertia are used in Sections 14.10 and later, where we treat problems with spatial motion, and they are determined in identical fashion.

14.4 INERTIA FORCES AND D'ALEMBERT'S PRINCIPLE

Next, let us consider a moving rigid body of mass m acted upon by any system of forces, say $\mathbf{F}_1$, $\mathbf{F}_2$, and $\mathbf{F}_3$, as illustrated in Fig. 14.6*a*. We designate the center of mass of the body

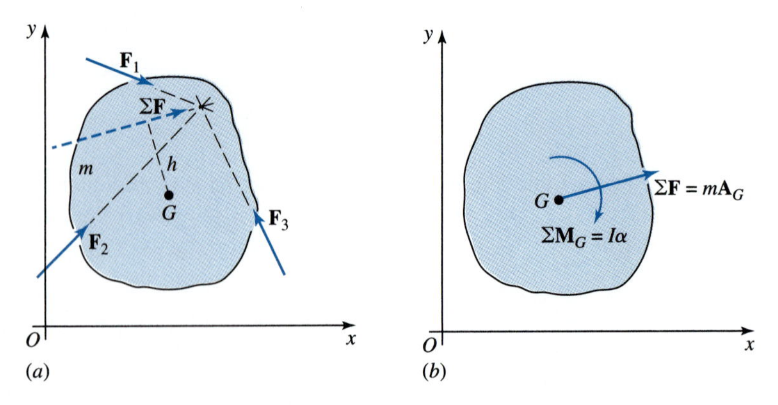

Figure 14.6 (*a*) An unbalanced set of forces on a rigid body. (*b*) The accelerations that result from the unbalanced forces.

as point G, and we find the resultant of the system of forces from the equation

$$\Sigma\mathbf{F} = \mathbf{F}_1 + \mathbf{F}_2 + \mathbf{F}_3.$$

In the general case, the line of action of this resultant will *not* be through the mass center but will be displaced by some distance, illustrated in Fig. 14.6*a* as distance h. We demonstrated in Eq. (14.3) that the effect of this unbalanced force system is to produce an acceleration of the center of mass of the body:

$$\sum \mathbf{F}_{ij} = m_j \mathbf{A}_{G_j}. \tag{14.12}$$

In a very similar way, it has been proven that the unbalanced moment effect of this resultant force about the center of mass causes angular acceleration of the body that obeys the equation:

$$\sum \mathbf{M}_G = I_{G_j} \boldsymbol{\alpha}_j. \tag{14.13}$$

However, this equation is restricted to use in taking moments about the center of mass G. It cannot be used for taking moments about an arbitrary point.

The quantity $\Sigma\mathbf{F}$ is the resultant of all external forces acting upon the body, and $\Sigma\mathbf{M}_G$ is the sum of all applied external moments and the moments of all externally applied forces about point G. The mass moment of inertia is designated as I_G, signifying that it must be taken with respect to the mass center G.

Equations (14.12) and (14.13) demonstrate that when an unbalanced system of forces acts upon a rigid body, the body experiences a rectilinear acceleration $\boldsymbol{A}_G$ of its mass center in the same direction as the resultant force $\Sigma\mathbf{F}$. The body also experiences an angular acceleration $\boldsymbol{\alpha}$ in the same direction as the resultant moment $\Sigma\mathbf{M}_G$, caused by the moments of the forces and the torques about the mass center. This situation is illustrated in Fig. 14.6*b*. If the forces and moments are known, Eqs. (14.12) and (14.13) may be used to determine the resulting acceleration pattern—that is, the resulting motion—of the body.

During engineering design, however, the motions of the machine members are often specified in advance by other machine requirements. The problem then is this: given the motions of the machine elements, what forces are required to produce these motions? The problem requires (1) a kinematic analysis to determine the translational and rotational accelerations of the various members and (2) definitions of the actual shapes, dimensions, and material specifications to determine the centroids and mass moments of inertia of the members. In the examples presented here, only the results of the kinematic analysis are included because methods of finding these have been presented in Chapter 4. The selection of the materials, shapes, and many of the dimensions of machine members form the subject of machine design and are not further discussed here.

In the dynamic analysis of machines, the acceleration vectors are usually known; therefore, an alternative form of Eqs. (14.12) and (14.13) is often convenient in determining the forces required to produce these known accelerations. Thus, we can write

$$\Sigma\mathbf{F} + (-m\mathbf{A}_G) = \mathbf{0} \tag{14.14}$$

and

$$\Sigma \mathbf{M}_G + (-I_G \boldsymbol{\alpha}) = \mathbf{0}. \tag{14.15}$$

Both of these are vector equations applying to the planar motion of a rigid body. Equation (14.14) states that the vector sum of all external forces acting upon the body plus the fictitious force $-m\mathbf{A}_G$ sum to zero. This new fictitious force $-m\mathbf{A}_G$ is called an *inertia force*. It has the same line of action as the absolute acceleration $\mathbf{A}_G$, but is opposite in sense. Equation (14.15) states that the sum of all external moments and the moments of all external forces acting upon the body about an axis through G and perpendicular to the plane of motion plus the fictitious torque $-I_G\boldsymbol{\alpha}$ sum to zero. This new fictitious torque $-I_G\boldsymbol{\alpha}$ is called an *inertia torque*. The inertia torque is opposite in sense to the angular acceleration vector $\boldsymbol{\alpha}$. We recall that Newton's first law states that a body perseveres in its state of uniform motion except when compelled to change by impressed forces; in other words, bodies resist any change in motion. In a sense, we can picture the fictitious inertia force and inertia torque vectors as resistances of the body to the change of motion required by the net unbalanced forces.

Equations (14.14) and (14.15) are known as *d'Alembert's principle*, because d'Alembert* was the first to call attention to the fact that addition of the inertia force and inertia torque to the real system of forces and torques enables a solution from the equations of static equilibrium. We note that the equations can also be written

$$\Sigma \mathbf{F} = \mathbf{0} \quad \text{and} \quad \Sigma \mathbf{M} = \mathbf{0}, \tag{14.16}$$

where it is understood that both the external and the inertia forces and torques are to be included in the summations. Equations (14.16) are useful because they permit us to take the summation of moments about any axis perpendicular to the plane of motion.

D'Alembert's principle is summarized as follows: The vector sum of all external forces and inertia forces acting upon a system of rigid bodies is zero. The vector sum of all external moments and inertia torques acting upon a system of rigid bodies is also separately zero.

When a graphical solution by a force polygon is desired, Eqs. (14.16) can be combined. In Fig. 14.7*a*, link 3 is acted upon by the external forces $\mathbf{F}_{23}$ and $\mathbf{F}_{43}$. The resultant $\mathbf{F}_{23}+\mathbf{F}_{43}$ produces an acceleration of the center of mass $\mathbf{A}_G$ and an angular acceleration of the link $\boldsymbol{\alpha}_3$ because the line of action of the resultant does not pass through the center of mass. Representing the inertia torque $-I_G\boldsymbol{\alpha}_3$ as a couple, as illustrated in Fig. 14.7*b*, we intentionally choose the two forces of this couple to be $\pm m_3\mathbf{A}_G$. For the moment of the couple to be of magnitude $-I_G\alpha_3$, the distance between the forces of the couple must be

$$h = \frac{I_G \alpha_3}{m_3 A_G}. \tag{14.17}$$

Because of this particular choice for the couple, one force of the couple exactly balances the inertia force itself and leaves only a single force, as illustrated in Fig. 14.7*c*. This force includes the combined effects of the inertia force and the inertia torque, yet appears as only a single inertia force offset by the distance h to give the effect of the inertia torque.

* Jean leRond d'Alembert (1717–1783).

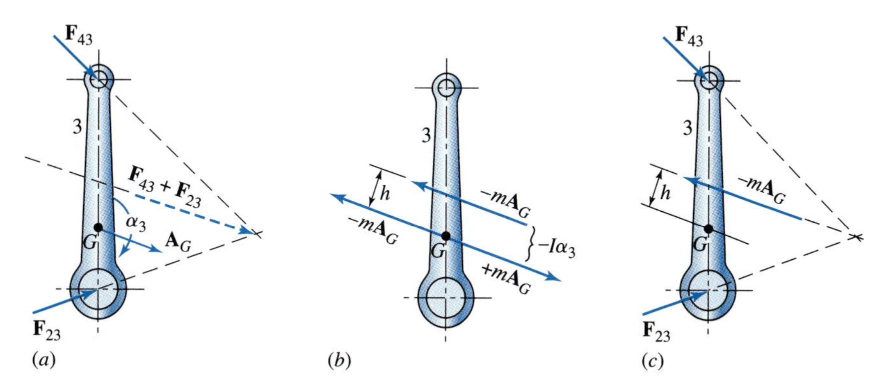

Figure 14.7 (*a*) Unbalanced forces and resulting accelerations. (*b*) Inertia force and inertia couple. (*c*) Inertia force offset from center of mass.

EXAMPLE 14.3

For the mechanism illustrated in Fig. 14.8*a*, point A has a constant velocity $V_A = 12.6$ ft/s. The mechanism moves in a horizontal plane with gravity normal to the plane of motion. The weight and principal mass moment of inertia of coupler link 3 are 2.20 lb and $I_{G_3} = 0.047\,9$ in · lb · s^2, respectively. Determine the force $\mathbf{F}_A$ required for dynamic equilibrium of the mechanism. Assume that the weights of links 2 and 4 and friction in the mechanism are negligible.

SOLUTION

A kinematic analysis provides the acceleration information illustrated in the polygon of Fig. 14.8*b*. The acceleration of the mass center of the coupler link is $A_G = 444$ ft/s^2 and

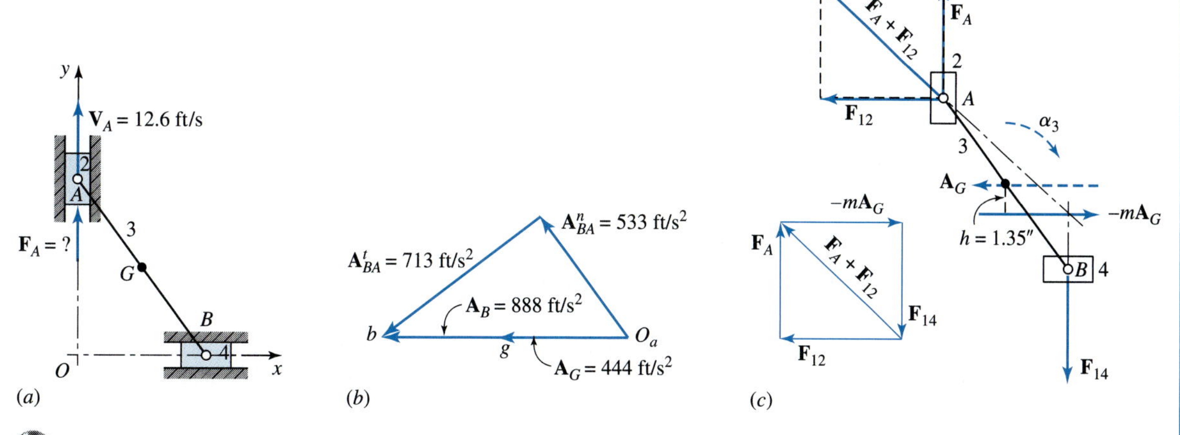

Figure 14.8 Solution for Example 14.3. (*a*) Scale drawing with $R_{BA} = 10$ in, $R_{GA} = 5$ in, $R_{AO} = 8$ in, and $R_{BO} = 6$ in. (*b*) Acceleration polygon. (*c*) Free-body diagram and force polygon.

the angular acceleration of this link is

$$\alpha_3 = \frac{A^t_{BA}}{R_{BA}} = \frac{(713 \text{ ft/s}^2)(12 \text{ in/ft})}{10 \text{ in}} = 856 \text{ rad/s}^2 \text{ cw}.$$

The mass of the coupler link is $m_3 = (2.20 \text{ lb})/(386 \text{ in/s}^2) = 0.005\,70 \text{ lb·s}^2/\text{in}$. Substituting this information into Eq. (14.17), the offset distance is

$$h = \frac{(0.047\,9 \text{ in} \cdot \text{lb} \cdot \text{s}^2)(856 \text{ rad/s}^2)}{(0.005\,70 \text{ lb} \cdot \text{s}^2/\text{in})(444 \text{ ft/s}^2)(12 \text{ in/ft})} = 1.35 \text{ in}$$

The free-body diagram of links 2, 3, and 4 and the resulting force polygon are illustrated in Fig. 14.8*c*. Note that the inertia force $-m\mathbf{A}_G$ is offset from G by the distance h so as to include a counterclockwise moment $-I_G\alpha_3$ about G and with the inertia force in the opposite sense to $\mathbf{A}_G$. The constraint reaction at B is $\mathbf{F}_{14}$ and, with no friction, is vertically downward. The forces at A are the constraint horizontal reaction $\mathbf{F}_{12}$ and the vertical actuating force $\mathbf{F}_A$. Recognizing this as a four-force member, as demonstrated in Section 13.8, we find the concurrency point at the intersection of the offset inertia force and $\mathbf{F}_{14}$, with the directions of both being known. The line of action of the total force $\mathbf{F}_{12} + \mathbf{F}_A$ at A must pass through this point of concurrency. This fact permits construction of the force polygon, where the unknown forces $\mathbf{F}_{12}$ and $\mathbf{F}_A$, having known directions, are found as components of $\mathbf{F}_{12} + \mathbf{F}_A$. The actuating force $\mathbf{F}_A$ is determined by measurement to be

$$\mathbf{F}_A = 27\hat{\mathbf{j}} \text{ lb.} \qquad \textit{Ans.}$$

EXAMPLE 14.4

As an example of dynamic force analysis using SI units, we use the four-bar linkage of Fig. 14.9. The required data, based on a complete kinematic analysis, are illustrated in Fig. 14.9 and in the legend. At the crank angle shown and assuming that gravity and friction effects are negligible, determine all the constraint forces and the driving torque required to produce the acceleration conditions specified.

SOLUTION

We start with the following kinematic information.

$$\mathbf{R}_{AO_2} = 60 \text{ mm}\angle 115^\circ = -25.4\hat{\mathbf{i}} + 54.4\hat{\mathbf{j}} \text{ mm},$$

$$\mathbf{R}_{G_3A} = 90 \text{ mm}\angle 48.7^\circ = 59.4\hat{\mathbf{i}} + 67.6\hat{\mathbf{j}} \text{ mm}$$

$$\mathbf{R}_{BA} = 220 \text{ mm}\angle 18.7^\circ = 208.0\hat{\mathbf{i}} + 70.5\hat{\mathbf{j}} \text{ mm},$$

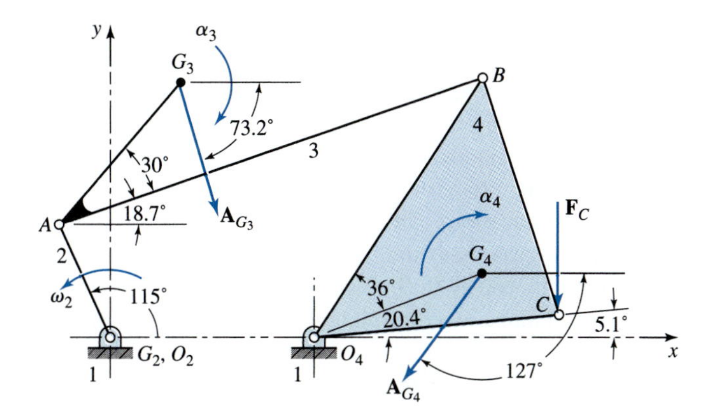

Figure 14.9 Example 14.4: $R_{AO_2} = 60$ mm, $R_{O_4O_2} = 100$ mm, $R_{BA} = 220$ mm, $R_{BO_4} = 150$ mm, $R_{CO_4} = R_{CB} = 120$ mm, $R_{G_3A} = 90$ mm, $R_{G_4O_4} = 90$ mm, $m_3 = 1.5$ kg, $m_4 = 5$ kg, $I_{G_2} = 0.025$ kg $\cdot$ m^2, $I_{G_3} = 0.012$ kg $\cdot$ m^2, $I_{G_4} = 0.054$ kg $\cdot$ m^2, $\boldsymbol{\alpha}_2 = \mathbf{0}$, $\boldsymbol{\alpha}_3 = 119\hat{\mathbf{k}}$ rad/s^2, $\boldsymbol{\alpha}_4 = -625\hat{\mathbf{k}}$ rad/s^2, $\mathbf{A}_{G_3} = 162$ m/s$^2\angle -73.2°$, $\mathbf{A}_{G_4} = 104$ m/s$^2\angle 233°$, $\mathbf{F}_C = -0.8\hat{\mathbf{j}}$ kN.

$$\mathbf{R}_{BO_4} = 150 \text{ mm}\angle 56.4° = 83.0\hat{\mathbf{i}} + 125.0\hat{\mathbf{j}} \text{ mm}$$

$$\mathbf{R}_{G_4O_4} = 90 \text{ mm}\angle 20.4° = 84.4\hat{\mathbf{i}} + 31.4\hat{\mathbf{j}} \text{ mm},$$

$$\mathbf{R}_{CO_4} = 120 \text{ mm}\angle 5.1° = 120.0\hat{\mathbf{i}} + 10.7\hat{\mathbf{j}} \text{ mm}$$

$$\boldsymbol{\alpha}_2 = \mathbf{0}, \quad \boldsymbol{\alpha}_3 = -119\hat{\mathbf{k}} \text{ rad/s}^2, \quad \boldsymbol{\alpha}_4 = -625\hat{\mathbf{k}} \text{ rad/s}^2$$

$$\mathbf{A}_{G_2} = \mathbf{0}, \quad \mathbf{A}_{G_3} = 46.8\hat{\mathbf{i}} - 155\hat{\mathbf{j}}\text{m/s}^2, \quad \mathbf{A}_{G_4} = -62.6\hat{\mathbf{i}} - 83.1\hat{\mathbf{j}} \text{ m/s}^2$$

Next we calculate the inertia forces and inertia torques. Because the solution is analytical, we do not need to calculate offset distances nor do we replace the inertia torques by couples. The six equations are

$$-m_2\mathbf{A}_{G_2} = \mathbf{0}$$

$$-m_3\mathbf{A}_{G_3} = -(1.5 \text{ kg})(46.8\hat{\mathbf{i}} - 155\hat{\mathbf{j}} \text{ m/s}^2) = -70.2\hat{\mathbf{i}} + 233\hat{\mathbf{j}} \text{ N}$$

$$-m_4\mathbf{A}_{G_4} = -(5.0 \text{ kg})(-62.6\hat{\mathbf{i}} - 83.1\hat{\mathbf{j}} \text{ m/s}^2) = 313\hat{\mathbf{i}} + 415\hat{\mathbf{j}} \text{ N}$$

$$-I_{G_2}\boldsymbol{\alpha}_2 = \mathbf{0}$$

$$-I_{G_3}\boldsymbol{\alpha}_3 = -(0.012 \text{ kg} \cdot \text{m}^2)(-119\hat{\mathbf{k}} \text{ rad/s}^2) = 1.43\hat{\mathbf{k}} \text{ N} \cdot \text{m}$$

and

$$-I_{G_4}\boldsymbol{\alpha}_4 = -(0.054 \text{ kg} \cdot \text{m}^2)(-625\hat{\mathbf{k}} \text{ rad/s}^2) = 33.8\hat{\mathbf{k}} \text{ N} \cdot \text{m}.$$

In considering our next step, we should at least make sketches of the free-body diagrams of the separate links. In so doing, we discover a problem. Each link has at least four unknown quantities, the magnitudes and directions of the reaction forces at each pivot. We know that because this analysis is planar, there are only three useful scalar equations for each body, the horizontal and vertical components of force and the in-plane moment equation. Worse yet, combining the bodies (say, links 3 and 4) into a single free-body diagram does not improve this. Can something be done or is this problem not solvable?

Fortunately, considering the two free-body diagrams for links 3 and 4, we note that there are a total of six unknown force components between them. There are also six total scalar equations available, two components of the forces and one for moments on each of the two free-body diagrams. The problem is solvable, but these six equations must be solved simultaneously to find the six unknowns. Therefore, considering the free-body diagram of link 4 alone, we formulate the summation of moments about point O_4:

$$\Sigma \mathbf{M}_{O_4} = \mathbf{R}_{G_4O_4} \times \left(-m_4\mathbf{A}_{G_4}\right) + (-I_{G_4}\boldsymbol{\alpha}_4) + \mathbf{R}_{CO_4} \times \mathbf{F}_C + \mathbf{F}_{BO_4} \times \mathbf{F}_{34} = \mathbf{0}. \quad (1)$$

Also, considering the free-body diagram of link 3 alone, we formulate the summation of moments about point A:

$$\Sigma \mathbf{M}_A = \mathbf{R}_{G_3A} \times (-m_3\mathbf{A}_{\mathbf{G_3}}) + (-I_{G_3}\boldsymbol{\alpha}_3) + \mathbf{R}_{BA} \times \mathbf{F}_{43} = \mathbf{0}. \quad (2)$$

Remembering that $\mathbf{F}_{43} = -\mathbf{F}_{34}$, the in-plane components of $\mathbf{F}_{34}$ are the only two unknowns and they are shared by these two equations.

The individual terms of the two equations are

$$\mathbf{R}_{G_4O_4} \times (-m_4\mathbf{A}_{G_4}) = (84.4\hat{\mathbf{i}} + 31.4\hat{\mathbf{j}} \text{ mm}) \times (313\hat{\mathbf{i}} + 415\hat{\mathbf{j}} \text{ N}) = 25.2\hat{\mathbf{k}} \text{ N} \cdot \text{m}$$

$$\mathbf{R}_{CO_4} \times \mathbf{F}_C = (120\hat{\mathbf{i}} + 10.7\hat{\mathbf{j}} \text{ mm}) \times (-800\hat{\mathbf{j}} \text{ N}) = -96.0\hat{\mathbf{k}} \text{ N} \cdot \text{m}$$

$$\mathbf{R}_{BO_4} \times \mathbf{F}_{34} = (83\hat{\mathbf{i}} + 125\hat{\mathbf{j}} \text{ mm}) \times (F_{34}^x\hat{\mathbf{i}} + F_{34}^y\hat{\mathbf{j}} \text{ kN})$$
$$= (-125F_{34}^x + 83F_{34}^y)\hat{\mathbf{k}} \text{ N} \cdot \text{m}$$

and

$$\mathbf{R}_{G_3A} \times (-m_3\mathbf{A}_{G_3}) = (59.4\hat{\mathbf{i}} + 67.6\hat{\mathbf{j}} \text{ mm}) \times (-70\hat{\mathbf{i}} + 233\hat{\mathbf{j}} \text{ N}) = 18.6\hat{\mathbf{k}} \text{ N} \cdot \text{m}$$

$$\mathbf{R}_{BA} \times \mathbf{F}_{43} = (208\hat{\mathbf{i}} + 70.5\hat{\mathbf{j}} \text{ mm}) \times (-F_{34}^x\hat{\mathbf{i}} - F_{34}^y\hat{\mathbf{j}} \text{ kN})$$
$$= (70.5F_{34}^x - 208F_{34}^y)\hat{\mathbf{k}} \text{ N} \cdot \text{m}$$

Then, substituting these values into Eqs. (1) and (2) above gives

$$\Sigma \mathbf{M}_{O_4} = 25.2\hat{\mathbf{k}} + 33.8\hat{\mathbf{k}} - 96.0\hat{\mathbf{k}} + (-125\text{F}_{34}^x + 83F_{34}^y)\hat{\mathbf{k}} = \mathbf{0}$$

and

$$\Sigma \mathbf{M}_A = 18.6\hat{\mathbf{k}} + 1.43\hat{\mathbf{k}} + (70.5\text{F}_{34}^x - 208F_{34}^y)\hat{\mathbf{k}} = \mathbf{0}.$$

Rearranging terms gives two equations in two unknowns, that is,

$$-125F_{34}^x + 83F_{34}^y = 37.0 \text{ N} \cdot \text{m}$$

and

$$70.5F_{34}^{x} - 208F_{34}^{y} = -20.0\ \text{N}\cdot\text{m}$$

Solving simultaneously, the two components of the force vector $\mathbf{F}_{34}$ are

$$F_{34}^{x} = -0.300\ \text{kN} = -300\ \text{N} \quad \text{and} \quad F_{34}^{y} = -0.005\ 39\ \text{kN} = -5.39\ \text{N}.$$

Therefore, the force vector is

$$F_{34} = -300\hat{\mathbf{i}} - 5.39\hat{\mathbf{j}} = 300\ \text{N}\angle - 179.0^\circ. \qquad \textit{Ans.}$$

Next, summing forces on link 4 yields the equation

$$\Sigma\mathbf{F}_{i4} = \mathbf{F}_{14} + \mathbf{F}_{34} + \mathbf{F}_{C} + (-m_4\mathbf{A}_{G_4}) = \mathbf{0}$$

and, upon solving, we find the force vector

$$\begin{aligned}\mathbf{F}_{14} &= -\mathbf{F}_{34} - \mathbf{F}_{C} - (-m_4\mathbf{A}_{G_4})\\ &= -(-300\hat{\mathbf{i}} - 5.39\hat{\mathbf{j}}\ \text{N}) - (-800\hat{\mathbf{j}}\ \text{N}) - (313\hat{\mathbf{i}} + 415\hat{\mathbf{j}}\ \text{N})\\ &= -13\hat{\mathbf{i}} + 390\hat{\mathbf{j}} = 390\ \text{N}\angle 91.9^\circ \end{aligned} \qquad \textit{Ans.}$$

Similarly, summing forces on link 3 yields

$$\Sigma\mathbf{F}_{13} = \mathbf{F}_{23} + \mathbf{F}_{43} + (-m_3\mathbf{A}_{G_3}) = \mathbf{0}$$

and, solving this, the force vector is

$$\begin{aligned}\mathbf{F}_{23} &= -(-\mathbf{F}_{34}) - (-m_3\mathbf{A}_{G_3})\\ &= -(300\hat{\mathbf{i}} + 5.39\hat{\mathbf{j}}\ \text{N}) - (-70.2\hat{\mathbf{i}} + 233\hat{\mathbf{j}}\ \text{N})\\ &= -230\hat{\mathbf{i}} - 238\hat{\mathbf{j}} = 331\ \text{N}\angle - 134^\circ \end{aligned} \qquad \textit{Ans.}$$

Now, for link 2, we have

$$\Sigma\mathbf{F}_{i2} = \mathbf{F}_{12} + \mathbf{F}_{32} + (-m_2\mathbf{A}_{G_2}) = \mathbf{0}$$

or

$$\mathbf{F}_{12} = -\mathbf{F}_{32} = \mathbf{F}_{23} = -230\hat{\mathbf{i}} - 238\hat{\mathbf{j}} = 331\ \text{N}\angle - 134.0^\circ.$$

Summing moments on link 2 about O_2, gives

$$\Sigma\mathbf{M}_{O_2} = \mathbf{R}_{AO_2} \times \mathbf{F}_{32} + \mathbf{M}_{12} + (-I_{G2}\boldsymbol{\alpha}_2) = \mathbf{0}$$

And solving for the driving crank torque: gives

$$\mathbf{M}_{12} = -\mathbf{R}_{AO_2} \times \mathbf{F}_{32} = -(-25.4\hat{\mathbf{i}} + 54.4\hat{\mathbf{j}}\ \text{mm}) \times (230\hat{\mathbf{i}} + 238\hat{\mathbf{j}}\ \text{N}) = 18.6\hat{\mathbf{k}}\ \text{N}\cdot\text{m}$$

Ans.

14.5 THE PRINCIPLE OF SUPERPOSITION

Linear systems are those in which effect is proportional to cause. This means that the response or output of a linear system is directly proportional to the drive or input to the system. An example of a linear system is a spring, where the deflection (output) is directly proportional to the force (input) exerted on the spring.

The *principle of superposition* may be used to solve problems involving linear systems by considering each of the inputs to the system separately. If the system is linear, the responses to each of these inputs can be summed or superposed on each other to determine the total response of the system. Thus, the principle of superposition states that *for a linear system the individual responses to several disturbances, or driving functions, can be superposed on each other to obtain the total response of the system.*

The principle of superposition does not apply to nonlinear systems. Some examples of nonlinear systems, where superposition may not be used, are systems with static or Coulomb friction, systems with clearances or backlash, or systems with springs that change stiffness as they are deflected.

We have now reviewed all of the principles necessary for making a complete dynamic-force analysis of a planar motion mechanism. The steps in using the principle of superposition for making such an analysis are summarized as follows:

1. Perform a kinematic analysis of the mechanism. Locate the center of mass of each link and find its acceleration; also find the angular acceleration of each link.
2. If the inertia forces are attached to all links simultaneously, along with other applied forces and moments, then there are often no two- or three-force members and it may become difficult to find the lines of action of unknown constraint forces. Instead of doing this, it is sometimes more convenient to ignore the masses and applied forces and moments on all but one or two links and to leave other links as two- or three-force members. By choosing in this manner, a solution may become possible for the constraint forces caused by the masses or applied forces and moments being considered, but without those caused by the masses and applied forces and moments being ignored.
3. Those masses and applied forces and moments considered in Step 2 can now be ignored while a solution is obtained for additional constraint force components caused by some of the previously ignored masses or applied forces and moments. This process can be continued until constraint force components caused by all masses and all applied forces and moments are found.
4. The results of Steps 2 and 3 can now be vectorially added to obtain the resultant forces and torques on each link caused by the combined effects of all masses and all applied forces and moments.

This process of superposition is demonstrated in the following example:

EXAMPLE 14.5

Using the principle of superposition, make a complete dynamic-force analysis of the four-bar linkage illustrated in Fig. 14.10. The known information is included in the figure legend.

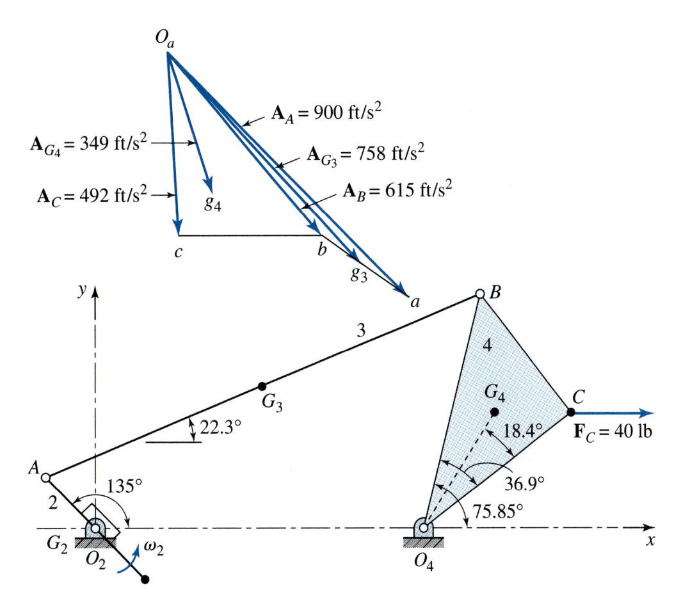

Figure 14.10 Example 14.5: $R_{AO_2} = 3$ in, $R_{O_4O_2} = 14$ in, $R_{BA} = 20$ in, $R_{BO_4} = 10$ in, $R_{CO_4} = 8$ in, $R_{CB} = 6$ in, $R_{G_3A} = 10$ in, $R_{G_4O_4} = 5.69$ in, $w_3 = 7.13$ lb, $w_4 = 3.42$ lb, $I_{G_2} = 0.25$ in · lb · s², $I_{G_3} = 0.625$ in · lb · s², $I_{G_4} = 0.037$ in · lb · s², $\omega_2 = 60$ rad/s ccw, and $\alpha_2 = 0$.

SOLUTION

STEP 1. All of the details of a complete kinematic analysis of the mechanism are not included here, but the resulting acceleration polygon is illustrated in Fig. 14.10. The numerical results are shown on the acceleration polygon in case you wish to verify them. From the methods of Chapter 4, the angular accelerations of link 3 and 4 are

$$\alpha_3 = 148 \text{ rad/s}^2 \text{ ccw} \quad \text{and} \quad \alpha_4 = 604 \text{ rad/s}^2 \text{ cw}$$

STEP 2. Because the center of mass of link 2 is located at ground pivot O_2, the major portion of the force analysis is concerned with links 3 and 4. Free-body diagrams of links 4 and 3 are illustrated separately in Figs. 14.11 and 14.12, respectively. Note that these diagrams are arranged in pseudo-equation form to emphasize the concept of superposition. Thus, in each figure, the forces in (*a*) plus those in (*b*) and (*c*) produce the results illustrated in (*d*). The action and reaction forces in Figs. 14.11 and 14.12 are also correlated; for example, $\mathbf{F}'_{34}$ in Fig. 14.11*a* is equal to $-\mathbf{F}'_{43}$ in Fig. 14.12*a* and so on. The following analysis is not difficult, but it is complex; it is important to read it slowly and examine the figures carefully, detail by detail.

We start by considering the effects of the mass of link 4 alone, while ignoring the masses of links 2 and 3. These effects are illustrated in Figs. 14.11*a* and 14.12*a*. Proceeding in accordance with our earlier investigations, we make the following calculations:

$$I_{G_4}\alpha_4 = (0.037 \text{ in} \cdot \text{lb} \cdot \text{s}^2)(604 \text{ rad/s}^2) = 22.3 \text{ in} \cdot \text{lb} \cdot \text{cw}$$

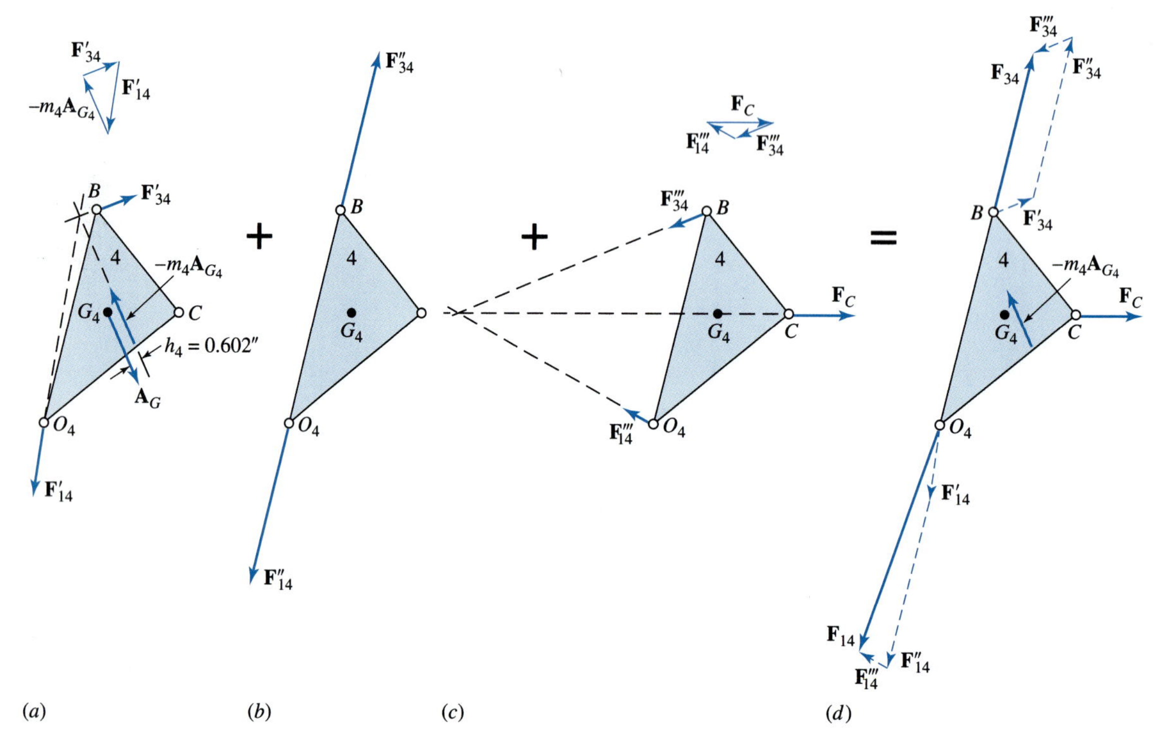

Figure 14.11 Free-body diagrams of link 4 of Example 14.5 showing superposition of forces: (*a*) $\mathbf{F}'_{34} = 24.3$ lb, $\mathbf{F}'_{14} = 44.3$ lb; (*b*) $\mathbf{F}''_{34} = -\mathbf{F}''_{14} = 94.8$ lb; (*c*) $\mathbf{F}'''_{34} = 25$ lb, $\mathbf{F}'''_{14} = 19.3$ lb; (*d*) $\mathbf{F}_{34} = 94.3$ lb, and $\mathbf{F}_{14} = 132$ lb.

$$m_4 A_{G_4} = \frac{3.42 \text{ lb}}{32.2 \text{ ft/s}^2}(349 \text{ ft/s}^2) = 37.1 \text{ lb}$$

$$h_4 = \frac{I_{G_4}\alpha_4}{m_4 A_{G_4}} = \frac{22.3 \text{ in} \cdot \text{lb}}{37.1 \text{ lb}} = 0.602 \text{ in}$$

Now the inertia force $-m_4 A_{G_4} = 37.1$ lb is positioned on the free-body diagram of link 4 opposite in direction to $\mathbf{A}_{G_4}$ and offset from G_4 by the distance h_4. The direction of the offset is to the right of G_4 so that the inertia force $-m_4\mathbf{A}_{G_4}$ produces a counterclockwise inertia torque $-I_{G_4}\alpha_4$ about G_4 opposite to the clockwise sense of $\boldsymbol{\alpha}_4$. The line of action of $\mathbf{F}'_{34}$ is taken along link 3 for part (*a*) of our superposition approach, because link 3 is a two-force member in this step. The intersection of the inertia force $-m_4\mathbf{A}_{G_4}$ and $\mathbf{F}'_{34}$ gives the concurrency point and establishes the line of action of $\mathbf{F}'_{14}$. The force polygon for part (*a*) for link 4 can now be constructed and the magnitudes of $\mathbf{F}'_{34}$ and $\mathbf{F}'_{14}$ can be determined. These values are illustrated in the legend to Fig. 14.11.

We proceed next to Fig. 14.12*a*. The forces $\mathbf{F}'_{43}$ and $\mathbf{F}'_{23}$ now become known from the preceding analysis.

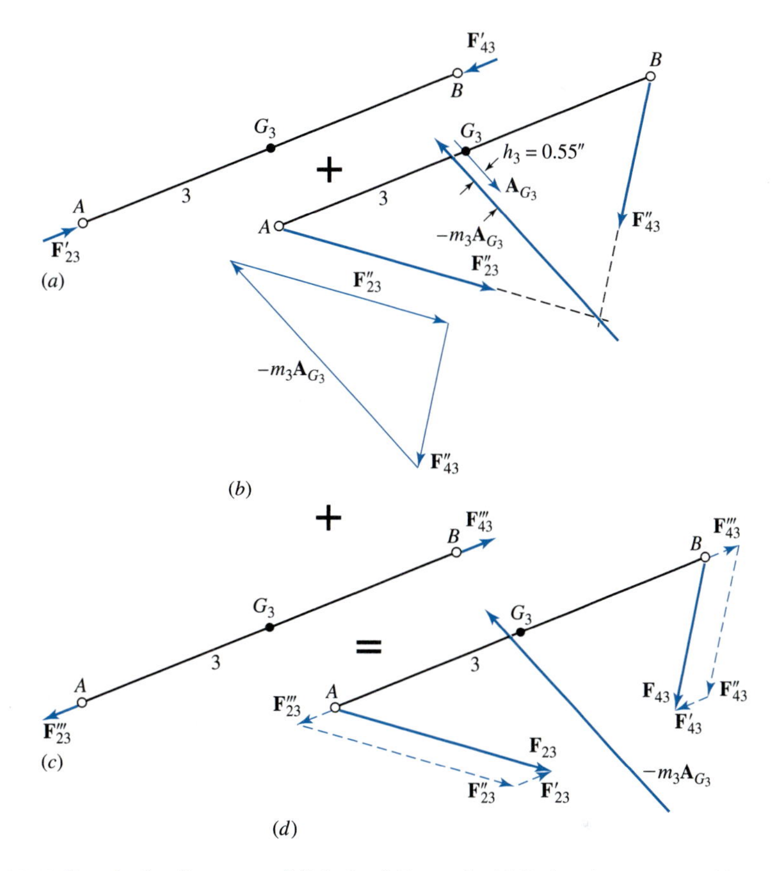

Figure 14.12 Free-body diagrams of link 3 of Example 14.5 showing superposition of forces: (*a*) $\mathbf{F}'_{23} = \mathbf{F}'_{43} = 24.3$ lb; (*b*) $\mathbf{F}''_{23} = 145$ lb, $\mathbf{F}''_{43} = 94.8$ lb; (*c*) $\mathbf{F}'''_{23} = \mathbf{F}'''_{43} = 25$ lb; (*d*) $\mathbf{F}_{23} = 145$ lb, $\mathbf{F}_{43} = 94.3$ lb.

STEP 3. Next we wish to consider the effects of the mass of link 3 alone. Going to Fig. 14.12*b*, we make the following calculations:

$$I_{G_3}\alpha_3 = (0.625 \text{ in} \cdot \text{lb} \cdot \text{s}^2)(148 \text{ rad/s}^2) = 92.5 \text{ in} \cdot \text{lb} \cdot \text{ccw}$$

$$m_3 A_{G_3} = \frac{7.13 \text{ lb}}{32.2 \text{ ft/s}^2}(758 \text{ ft/s}^2) = 168 \text{ lb}$$

$$h_3 = \frac{I_{G_3}\alpha_3}{m_3 A_{G_3}} = \frac{92.5 \text{ in} \cdot \text{lb}}{168 \text{ lb}} = 0.550 \text{ in}$$

Now we position the inertia force $-m_3 A_{G_3} = 168$ lb on the free-body diagram of link 3 opposite in direction to $\mathbf{A}_{G_3}$ and offset by a distance of $h_3 = 0.550$ in from G_3 so as to produce a clockwise torque about G_3, opposite to the counterclockwise sense of $\boldsymbol{\alpha}_3$. The line of action of $\mathbf{F}''_{43}$ is along link 4 for part (*b*) of our superposition approach, because the mass of link 4 was considered in Step 2 and link 4 is now a two-force member in this step.

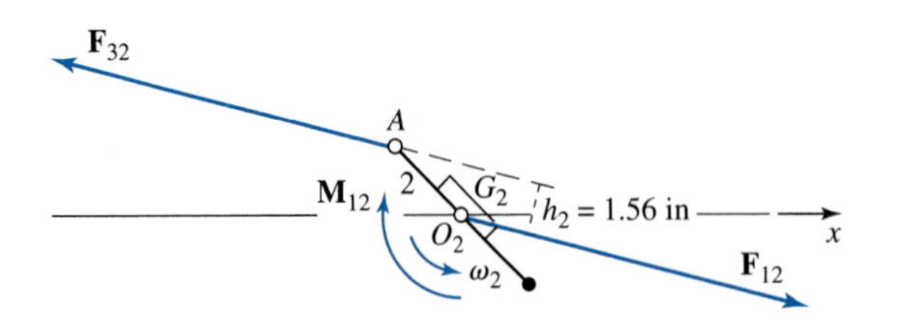

Figure 14.13 Free-body diagram of link 2 of Example 14.5: $F_{32} = F_{12} = 145$ lb, $M_{12} = 226$ in · lb.

The intersection of the line of action of $\mathbf{F}''_{43}$ and the inertia force $-m_3\mathbf{A}_{G_3}$ gives the point of concurrency. Thus, the line of action of $\mathbf{F}''_{43}$ becomes known and the force polygon for link 3 can be constructed. The resulting magnitudes of $\mathbf{F}''_{43}$ and $\mathbf{F}''_{23}$ are included in the legend to Fig. 14.12.

In Fig. 14.11*b* the forces $\mathbf{F}''_{34}$ and $\mathbf{F}''_{14}$ become known from the preceding Step 3 analysis.

STEP 4. Figure 14.11*c* illustrates the results of the static-force analysis with the applied force $F_c = 40$ lb as the given loading, but with no inertia forces. Recognizing that link 3 is again a two-force member, the force polygon in Fig. 14.11*c* determines the values of the forces acting on link 4. From these the magnitudes and directions of the forces, $\mathbf{F}'''_{43}$ and $\mathbf{F}'''_{23}$ acting on link 3 as illustrated in Fig. 14.12*c* are determined.

STEP 5. The next step is to perform the superposition of the results obtained in Steps 2, 3, and 4; this is done by the vector additions illustrated in part (*d*) of each figure. The analysis is completed by taking the resultant force $\mathbf{F}_{23}$ from Fig. 14.12*d* and applying the negative value, $\mathbf{F}_{32}$, to link 2. This is illustrated in Fig. 14.13. The offset distance h_2 is determined by measurement to be $h_2 = 1.56$ in. Therefore, the external torque to be applied to link 2 is

$$M_{12} = h_2 F_{32} = (1.56 \text{ in})(145 \text{ lb}) = 226 \text{ in} \cdot \text{lb cw}$$

Note that this torque is opposite in sense to the direction of rotation of link 2; this is not true for the entire cycle of operation, but it can occur at particular crank angles. This torque must sometimes be in the reverse direction to continue rotation with constant input velocity.

14.6 PLANAR ROTATION ABOUT A FIXED CENTER

The previous sections have dealt with the case of dynamic forces for a rigid body having general planar motion. It is important to emphasize that the equations and methods of analysis investigated in these sections are general and apply to *all* problems with planar motion. It will be interesting now to study the application of these methods to a special case, that of a rigid body rotating about a fixed center.

Let us consider a rigid body constrained as illustrated in Fig. 14.14*a* to rotate with an angular velocity $\boldsymbol{\omega}$ about some fixed center O not coincident with its center of mass G. A system of forces (not shown) is applied to the body, causing it to undergo an angular acceleration $\boldsymbol{\alpha}$. This motion of the body implies that the mass center G has normal and tangential components of acceleration $\mathbf{A}^n_{GO}$ and $\mathbf{A}^t_{GO}$ whose magnitudes are $R_{GO}\omega^2$ and $R_{GO}\alpha$, respectively, as shown in Fig. 14.14*b*. Thus, if we resolve the resultant external force into normal and tangential components, these components must have magnitudes

$$F^n = mR_{GO}\omega^2 \quad \text{and} \quad F^t = mR_{GO}\alpha, \tag{a}$$

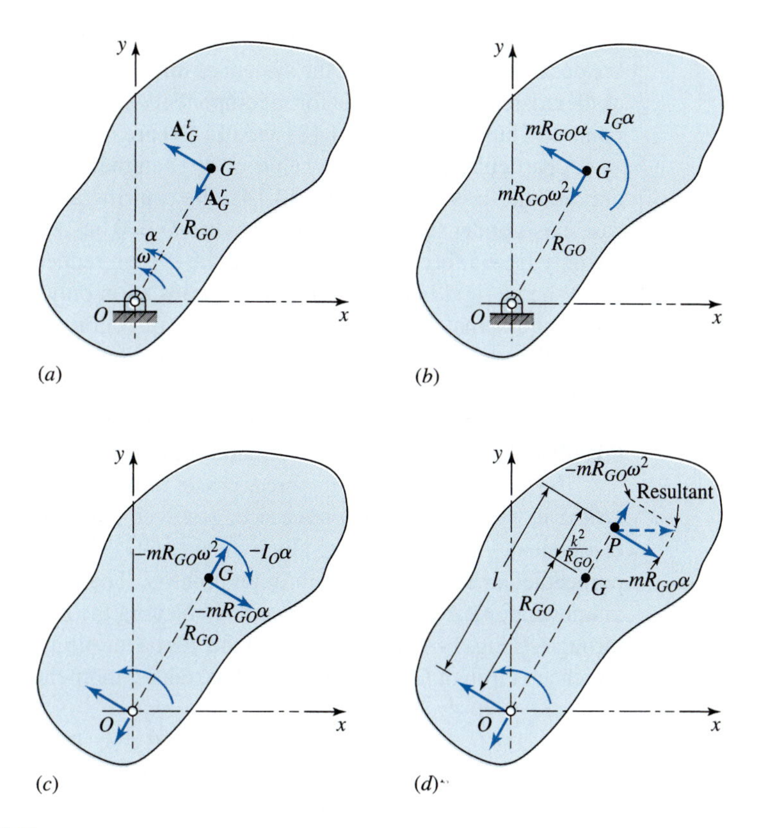

Figure 14.14

in accordance with Eq. (14.12). In addition, Eq. (14.13) states that an external torque must exist to create the angular acceleration and that the magnitude of this torque must be $T_G = I_G\alpha$. If we now sum the moments of these forces about O, we have

$$\Sigma M_O = I_G\alpha + R_{GO}(mR_{GO}\alpha) = (I_G + mR_{GO}^2)\alpha. \tag{b}$$

But the quantity in parentheses in Eq. (*b*) is identical in form to Eq. (14.11) and transfers the moment of inertia to another axis not coincident with the center of mass. Therefore, Eq. (*b*) can be written in the form

$$\Sigma M_O = I_O\alpha.$$

Equations (14.14) and (14.15) then become

$$\Sigma \mathbf{F} - m\mathbf{A}_G = \mathbf{0} \tag{14.18}$$

and

$$\Sigma \mathbf{M}_O - I_O\boldsymbol{\alpha} = \mathbf{0} \tag{14.19}$$

by including the inertia force $-m\mathbf{A}_G$ and inertia torque $-I_O\boldsymbol{\alpha}$ as illustrated in Fig. 14.14*c*. We observe particularly that the system of forces does *not* reduce to a single couple because of the existence of the inertia force component $-m\mathbf{R}_{GO}\omega^2$, which has no moment arm about point O. Thus, both Eqs. (14.18) and (14.19) are necessary.

A particular case arises when $\boldsymbol{\alpha} = \mathbf{0}$. Then the external moment $\Sigma\mathbf{M}_O$ is zero and the only inertia force is, from Fig. 14.14*c*, the centrifugal force $-m\mathbf{R}_{GO}\omega^2$. A second special case exists under starting conditions, when $\omega = 0$ but α is not zero. Under these conditions the only inertia force is $-mR_{GO}\alpha$, and the system reduces to a single couple.

When a rigid body has a motion of translation only, the resultant inertia force and the resultant external force share the same line of action, which passes through the center of mass of the body. When a rigid body has rotation and angular acceleration, the resultant inertia force and resultant external force have the same line of action; however, this line does *not* pass through the center of mass but is offset from it. Let us now locate a point on this offset line of action of the resultant of the inertia forces.

The resultant of the inertia forces passes through some point P of Fig. 14.14*d* on the line OG or its extension. This force can be resolved into two components, one of which is the component $-m\mathbf{R}_{GO}\omega^2$ acting along the line OG. The other component is $-mR_{GO}\alpha$ acting perpendicular to OG but not through point G. The distance, designated l, from O to the unknown point P can be determined by equating the moment of the component $-mR_{GO}\alpha$ through P (Fig. 14.14*d*) to the sum of the inertia torque and the moment of the inertia forces, which act through G (Fig. 14.14*c*). Thus, taking moments about O in each figure, we have

$$(-mR_{GO}\alpha)l = -I_G\alpha + (-mR_{GO}\alpha)R_{GO}$$

or

$$l = \frac{I_G}{mR_{GO}} + R_{GO}.$$

Remembering the definition of the radius of gyration and substituting the value of I_G from Eq. (14.10) gives

$$l = \frac{k^2}{R_{GO}} + R_{GO}. \tag{14.20}$$

The point P located by Eq. (14.20) and illustrated in Fig. 14.14*d* is called the *center of percussion*. As indicated, the resultant inertia force passes through P and, consequently, the inertia force has zero moment about the center of percussion. If an external force is applied at P, perpendicular to OG, an angular acceleration α will result, but the bearing reaction at O will be zero except for the component caused by the centrifugal inertia force $-mR_{GO}\omega^2$. It is the usual practice in shock testing machines to apply the force at the center of percussion to eliminate the tangential bearing reaction that would otherwise be caused by the externally applied shock load.

As another example, suppose we consider the impact of a baseball against a bat. If we crudely approximate the bat as a cylindrical rod of length L, then its center of mass is located at approximately $R_G = L/2$, and its radius of gyration is approximately $k = L/\sqrt{12}$. Substituting these values into Eq. (14.20) gives the location of the center of percussion as

$l = 2L/3$. This demonstrates why that location on the bat is referred to as the "sweet spot"; hitting the ball at that point produces no dynamic reaction force on the hands.

Equation (14.20) demonstrates that the location of the center of percussion is independent of the values of ω and α.

If the axis of rotation is coincident with the center of mass, $R_{GO} = 0$ and Eq. (14.20) indicates that $l = \infty$. Under these conditions there is no resultant inertia force, but rather a resultant inertia couple $-I_G\alpha$.

14.7 SHAKING FORCES AND MOMENTS

Of special interest to the designer are the forces transmitted to the frame or foundation of a machine owing to the inertia of the moving links. When these forces vary in magnitude or direction, they tend to shake or vibrate the machine (and the frame); consequently, such effects are called *shaking forces* and *shaking moments*.

If we consider some machine, say a four-bar linkage for example, with links 2, 3, and 4 as the moving members and link 1 as the frame, then taking the entire group of moving parts as a system, not including the frame, and draw a free-body diagram, we can immediately write

$$\Sigma\mathbf{F} = \mathbf{F}_{12} + \mathbf{F}_{14} + (-m_2\mathbf{A}_{G_2}) + (-m_3\mathbf{A}_{G_3}) + (-m_4\mathbf{A}_{G_4}) = \mathbf{0}.$$

Using $\mathbf{F}_S$ as a symbol for the resulting shaking force, we define this as equal to the resultant of all the reaction forces on the ground link 1,

$$\mathbf{F}_s = \mathbf{F}_{21} + \mathbf{F}_{41}.$$

Therefore, from the previous equation, we have

$$\mathbf{F}_s = (-m_2\mathbf{A}_{G_2}) + (-m_3\mathbf{A}_{G_3}) + (-m_4\mathbf{A}_{G_4}). \tag{14.21}$$

Generalizing from this example, we can write for any machine that the shaking force is

$$\mathbf{F}_s = \sum(-m_j\mathbf{A}_{G_j}). \tag{14.22}$$

This makes sense because if we consider a free-body diagram of the entire machine *including the frame,* all other applied and constraint forces have equal and opposite reaction forces and these cancel within the free-body system. Only the inertia forces, having no reactions, are ultimately external to the system and remain unbalanced.* These are not balanced by reaction forces and produce unbalanced shaking effects between the frame and whatever bench or other surface on which it is mounted. These are the forces that require that the machine be fastened down to prevent it from moving.

---

* The same arguments can be made about gravitational, magnetic, electric, or other noncontacting force fields. However, because these are usually static forces, it is not customary to include them in the definition of shaking force.

A similar derivation can be made for unbalanced moments. Using the symbol $\mathbf{M}_s$ for the shaking moment, we take the summation of moments about the coordinate system origin and find

$$\mathbf{M}_s = \sum \left[\mathbf{R}_{G_j} \times (-m_j \mathbf{A}_{G_j})\right] + \sum (-I_{G_j} \boldsymbol{\alpha}_j). \tag{14.23}$$

14.8 COMPLEX ALGEBRA APPROACH

In this section, we investigate the dynamic force analysis of a mechanism using the complex algebra approach. The general procedure that is adopted is as follows:

1. Perform a complete kinematic analysis.
2. Draw a free-body diagram of each link in the mechanism.
3. Write the dynamic force equations for each moving link.
4. Corresponding to each input angle compute items such as the following:
   a. The axial and the transverse forces exerted on each moving link;
   b. The force components exerted on the ground bearings (here referred to as the bearing loads);
   c. The inertia torque; and
   d. The shaking moment on the foundation.

To demonstrate the complex algebra approach, we will consider the four-bar linkage illustrated in Fig. 14.15.

For a specified angular velocity and acceleration of input link 2, the angular velocity and accelerations of coupler link 3 and output link 4 are determined following the methods of Chapters 3 and 4. The accelerations of the mass centers of links 2, 3, and 4 are now expressed in coordinates attached to and aligned along the respective links.

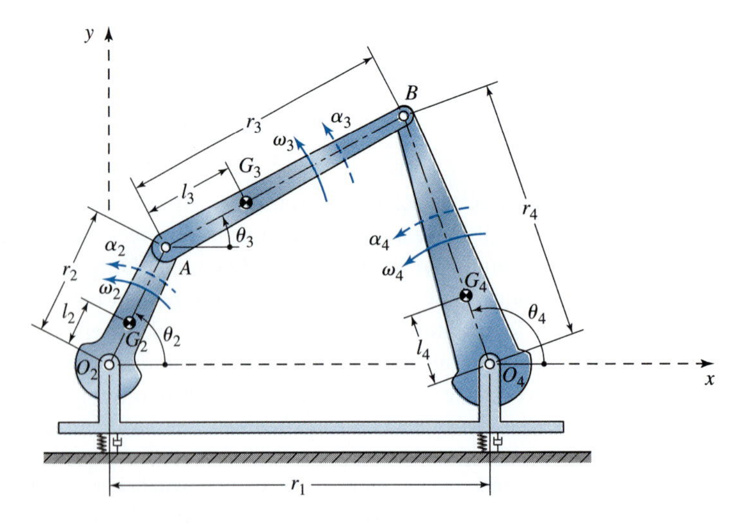

Figure 14.15 Four-bar linkage.

The acceleration of the mass center of input link 2 can be written as

$$\mathbf{A}_{G_2} = l_2(-\omega_2^2 + j\alpha_2)e^{j\theta_2} = (A_{G_2}^{x2} + jA_{G_2}^{y2})e^{j\theta_2}. \tag{14.24a}$$

Therefore, the x_2 and y_2 components of the acceleration, respectively, are

$$A_{G_2}^{x2} = -l_2\omega_2^2 \quad \text{and} \quad A_{G_2}^{y2} = -l_2\alpha_2. \tag{14.24b}$$

The acceleration of the mass center of coupler link 3 can be written as

$$\mathbf{A}_{G_3} = r_2(-\omega_2^2 + j\alpha_2)e^{j\theta_2} + l_3(-\omega_3^2 + j\alpha_3)e^{j\theta_3}$$

or as

$$\mathbf{A}_{G_3} = \left[r_2(-\omega_2^2 + j\alpha_2)e^{j\eta_{23}} + l_3(-\omega_3^2 + j\alpha_3)\right]e^{j\theta_3} \tag{14.25}$$

where $\eta_{23} = \theta_2 - \theta_3$. The acceleration of the mass center of the coupler link can also be written as

$$\mathbf{A}_{G_3} = (A_{G_3}^{x3} + jA_{G_3}^{y3})e^{j\theta_3}. \tag{14.26}$$

Equating Eqs. (14.25) and (14.26), the x_3 and y_3 components of the acceleration are

$$A_{G_3}^{x3} = r_2(-\omega_2^2 \cos\eta_{23} - \alpha_2 \sin\eta_{23}) - l_3\omega_3^2 \tag{14.27a}$$

$$A_{G_3}^{y3} = r_2(-\omega_2^2 \sin\eta_{23} + \alpha_2 \cos\eta_{23}) - l_3\alpha_3. \tag{14.27b}$$

The acceleration of the mass center of output link 4 can be written as

$$\mathbf{A}_{G_4} = r_2(-\omega_4^2 + j\alpha_4)e^{j\theta_4} = (A_{G_4}^{x4} + jA_{G_4}^{y4})e^{j\theta_4}. \tag{14.28a}$$

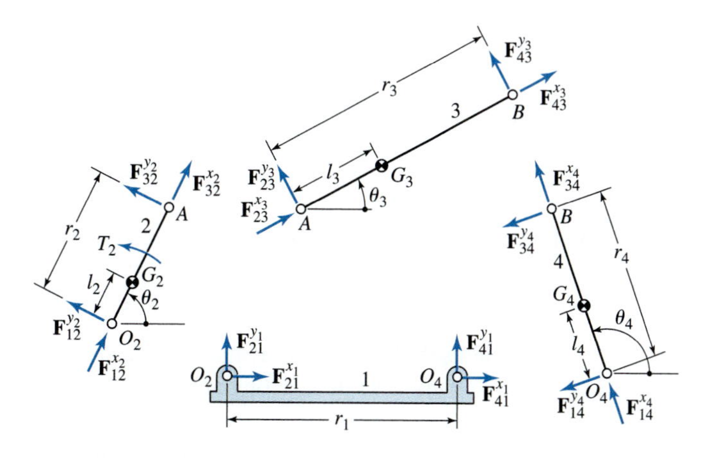

Figure 14.16 Free-body diagrams for the four-bar linkage.

Therefore, the x_4 and y_4 components of the acceleration, respectively, are

$$A_{G_4}^{x4} = -l_4\omega_4^2 \quad \text{and} \quad A_{G_4}^{y4} = -l_4\alpha_4. \tag{14.28b}$$

The free-body diagrams of the four links are presented in Fig. 14.16.

For the input link, the sum of the forces gives

$$\left[(F_{12}^{x2} + F_{32}^{x2}) + j(F_{12}^{y2} + F_{32}^{y2})\right] e^{j\theta_2} = m_2(A_{G_2}^{x2} + jA_{G_2}^{y2})e^{j\theta_2} \tag{14.29a}$$

and the sum of the moments about the bearing O_2 gives

$$r_2 F_{32}^{y2} + T_2 = (I_{G_2} + m_2 l_2^2)\alpha_2. \tag{14.29b}$$

For the coupler link, the sum of the forces gives

$$\left[(F_{23}^{x3} + F_{43}^{x3}) + j(F_{23}^{y3} + F_{43}^{y3})\right] e^{j\theta_3} = m_3(A_{G_3}^{x3} + jA_{G_3}^{y3})e^{j\theta_3} \tag{14.30a}$$

and the sum of the moments about the mass center G_3 gives

$$F_{43}^{y3}(r_3 - l_3) - F_{23}^{y3} l_3 = I_{G_3}\alpha_3. \tag{14.30b}$$

For the output link, the sum of the forces gives

$$\left[(F_{14}^{x4} + F_{34}^{x4}) + j(F_{14}^{y4} + F_{34}^{y4})\right] e^{j\theta_4} = m_4(A_{G_4}^{x4} + jA_{G_4}^{y4})e^{j\theta_4} \tag{14.31a}$$

and the sum of the moments about the bearing O_4 gives

$$r_4 F_{34}^{y4} = (I_{G_3} + m_4 l_4^2)\alpha_4. \tag{14.31b}$$

For the revolute joint A, the sum of the forces gives

$$(F_{32}^{x2} + jF_{32}^{y2})e^{j\theta_2} + (F_{23}^{x3} + jF_{23}^{y3})e^{j\theta_3} = 0 \tag{14.32a}$$

and for the revolute joint B, the sum of the forces gives

$$(F_{43}^{x3} + jF_{43}^{y3})e^{j\theta_3} + (F_{34}^{x4} + jF_{34}^{y4})e^{j\theta_4} = 0 \tag{14.32b}$$

For the bearing O_2, the sum of the forces gives

$$(F_{21}^{x1} + jF_{21}^{y1}) + (F_{12}^{x2} + jF_{12}^{y2})e^{j\theta_2} = 0 \tag{14.32c}$$

and for the bearing O_4, the sum of the forces gives

$$(F_{41}^{x1} + jF_{41}^{y1}) + (F_{14}^{x4} + jF_{14}^{y4})e^{j\theta_4} = 0. \tag{14.32d}$$

To obtain solutions for the dynamic force analysis, we proceed as follows. From Eq. (14.31b), the transverse force acting at point B on output link 4 is

$$F_{34}^{y4} = (I_{G_4} + m_4 l_4^2)\frac{\alpha_4}{r_4}. \tag{14.33}$$

Equating the real and imaginary parts of Eq. (14.29*a*), the x_2 and y_2 components of the forces, respectively, are

$$F_{12}^{x_2} + F_{32}^{x_2} = m_2 A_{G_2}^{x_2} \tag{14.34a}$$

$$F_{12}^{y_2} + F_{32}^{y_2} = m_2 A_{G_2}^{y_2}. \tag{14.34b}$$

Equating the real and imaginary parts of Eq. (14.30*a*), the x_3 and y_3 components of the forces, respectively, are

$$F_{23}^{x_3} + F_{43}^{x_3} = m_3 A_{G_3}^{x_3} \tag{14.35a}$$

$$F_{23}^{y_3} + F_{43}^{y_3} = m_3 A_{G_3}^{y_3}. \tag{14.35b}$$

Equating the real and imaginary parts of Eq. (14.31*a*), the x_4 and y_4 components of the forces, respectively, are

$$F_{14}^{x_4} + F_{34}^{x_4} = m_4 A_{G_4}^{x_4} \tag{14.36a}$$

$$F_{14}^{y_4} + F_{34}^{y_4} = m_4 A_{G_4}^{y_4}. \tag{14.36b}$$

Substituting Eq. (14.35*b*) into Eq. (14.30*b*), the transverse force acting at point B on the coupler link is

$$F_{43}^{y_3} = \frac{1}{r_3}(I_{G_e}\alpha_3 + m_3 l_3 A_{G_3}^{y_3}). \tag{14.37a}$$

Then, from Eq. (14.35*b*), the transverse force acting at point A on the coupler link is

$$F_{23}^{y_3} = m_3 A_{G_3}^{y_3} - F_{43}^{y_3}. \tag{14.37b}$$

Equation (14.32*b*) can be written as

$$(F_{43}^{x_3} + jF_{43}^{y_3})e^{j\eta_{34}} + (F_{34}^{x_4} + jF_{34}^{y_4}) = 0 \tag{14.38a}$$

or as

$$(F_{43}^{x_3} + jF_{43}^{y_3}) + (F_{34}^{x_4} + jF_{34}^{y_4})e^{-j\eta_{34}} = 0, \tag{14.38b}$$

where $\eta_{34} = \theta_3 - \theta_4$. Equating the imaginary parts of Eq. (14.38*a*) and equating the imaginary parts of Eq. (14.38*b*), respectively, gives

$$F_{43}^{x_3} \sin\eta_{34} + F_{43}^{y_3} \cos\eta_{34} + F_{34}^{y_4} = 0 \tag{14.39a}$$

$$F_{43}^{y_3} - (F_{34}^{x_4} \sin\eta_{34} - F_{34}^{y_4} \cos\eta_{34}) = 0. \tag{14.39b}$$

Then, from these two equations we have

$$F_{43}^{x_3} = -(F_{43}^{y_3} \cot\eta_{34} + F_{34}^{y_4} \csc\eta_{34}) \tag{14.40a}$$

$$F_{34}^{x_4} = F_{43}^{y_3} \csc\eta_{34} + F_{34}^{y_4} \cot\eta_{34}. \tag{14.40b}$$

Then, from Eqs. (14.35*a*) and (14.36*a*), the axial forces acting on links 3 and 4, respectively, are

$$F_{23}^{x3} = m_3 A_{G_3}^{x3} - F_{43}^{x3} \tag{14.41a}$$

$$F_{14}^{x4} = m_4 A_{G_4}^{x4} - F_{34}^{x4}. \tag{14.41b}$$

Rewriting Eq. (14.32*a*) as

$$F_{32}^{x2} + jF_{32}^{y2} = -(F_{23}^{x3} + jF_{23}^{y3})e^{-j\eta_{23}} \tag{14.42}$$

and equating the real and imaginary parts, respectively, gives

$$F_{32}^{x2} = -F_{23}^{x3} \cos \eta_{23} - F_{23}^{y3} \sin \eta_{23} \tag{14.43a}$$

$$F_{32}^{y2} = F_{23}^{x3} \sin \eta_{23} - F_{23}^{y3} \cos \eta_{23}. \tag{14.43b}$$

The input torque (sometimes referred to as the inertia torque) can now be obtained by writing Eq. (14.29*b*) as

$$T_2 = -r_2 F_{32}^{y2} + (I_{G_2} + m_2 l_2^2)\alpha_2. \tag{14.44}$$

If the four-bar linkage is in *steady-state operation*, that is, if the input angular velocity is constant, then substituting $\alpha_2 = 0$ into Eq. (14.44) gives

$$T_2 = -r_2 F_{32}^{y2}. \tag{14.29}$$

From Eqs. (14.34) and (14.36), we have

$$F_{12}^{x2} = m_2 A_{G_2}^{x2} - F_{32}^{x2} \quad \text{and} \quad F_{14}^{x4} = m_4 A_{G_4}^{x4} - F_{34}^{x4} \tag{14.46a}$$

$$F_{12}^{y2} = m_2 A_{G_2}^{y2} - F_{32}^{y2} \quad \text{and} \quad F_{14}^{y4} = m_4 A_{G_4}^{y4} - F_{34}^{y4}. \tag{14.46b}$$

If we write Eqs. (14.32*c*) and (14.32*d*), respectively, as

$$(F_{21}^{x1} + jF_{21}^{y1}) = -(F_{12}^{x2} + jF_{12}^{y2})e^{j\theta_2}$$

$$(F_{41}^{x1} + jF_{41}^{y1}) = -(F_{14}^{x4} + jF_{14}^{y4})e^{j\theta_4},$$

then the x and y components of the loads at the bearing O_2, respectively, are

$$F_{21}^{x1} = -F_{12}^{x2} \cos \theta_2 + F_{12}^{y2} \sin \theta_2 \quad \text{and} \quad F_{21}^{y1} = -F_{12}^{x2} \sin \theta_2 - F_{12}^{y2} \cos \theta_2 \tag{14.47a}$$

and the x and y components of the loads at the bearing O_4, respectively, are

$$F_{41}^{x1} = -F_{14}^{x4} \cos \theta_4 + F_{14}^{y4} \sin \theta_4 \quad \text{and} \quad F_{41}^{y1} = -F_{14}^{x4} \sin \theta_4 - F_{14}^{y4} \cos \theta_4. \tag{14.47b}$$

Finally, the dynamic shaking moment can be written as

$$M_s = r_1 F_{41}^{y1}.$$

Substituting Eq. (14.47*b*) into this equation gives

$$M_s = -r_1(F_{14}^{x4} \sin \theta_4 + F_{14}^{y4} \cos \theta_4). \tag{14.48}$$

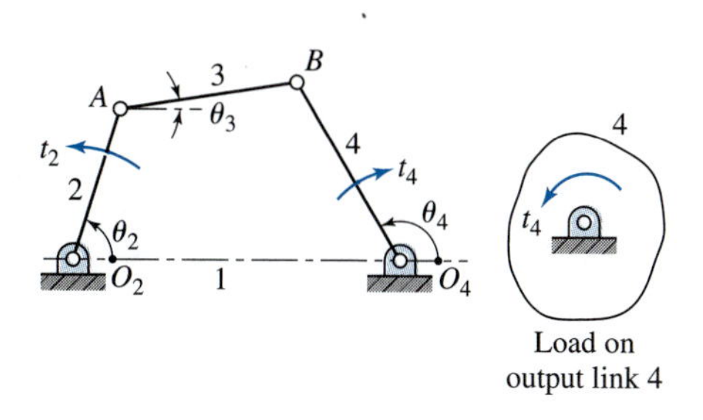

Figure 14.17 Output torque of the four-bar linkage.

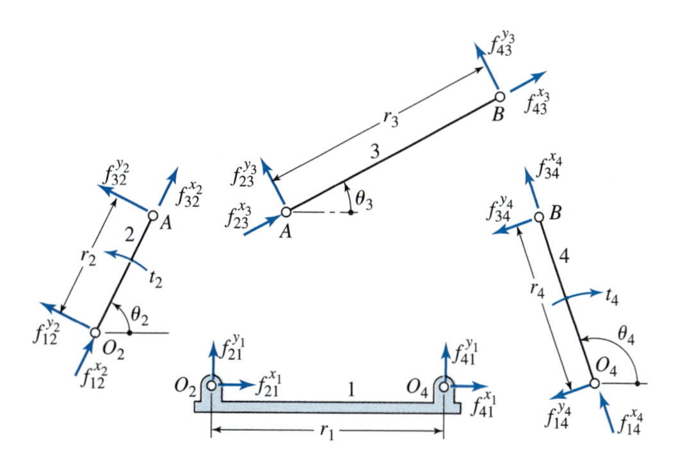

Figure 14.18 Free-body diagrams for the four-bar linkage.

Let us consider the special case where the four-bar linkage is in static equilibrium because of the output torque t_4 illustrated in Fig. 14.17, that is, the torque exerted on the load by the output axis through the revolute joint at O_4. The free-body diagrams are as illustrated in Fig. 14.18.

From the dynamic force analysis, we can now determine the following as functions of the input angle:

1. The axial and the transverse force components exerted on the three moving links.
2. The bearing loads.
3. The inertia torque.
4. The shaking moment.

From Eqs. (14.29), the sum of the forces on link 2 and the sum of the moments about the bearing O_2 are

$$\left[(f_{12}^{x_2} + f_{32}^{x_2}) + j(f_{12}^{y_2} + f_{32}^{y_2})\right] e^{j\theta_2} = 0 \tag{14.49a}$$

$$r_2 f_{32}^{y_2} + t_2 = 0. \tag{14.49b}$$

From Eqs. (14.30), the sum of the forces on link 3 and the sum of the moments about the mass center G_3 are

$$\left[(f_{23}^{x3}+f_{43}^{x3})+j(f_{23}^{y3}+f_{43}^{y3})\right]e^{j\theta_3}=0 \tag{14.50a}$$

$$r_3 f_{43}^{y3}=0. \tag{14.50b}$$

From Eq. (14.31), the sum of the forces on link 4 and the sum of the moments about the bearing O_4 are

$$\left[(f_{14}^{x4}+f_{34}^{x4})+j(f_{14}^{y4}+f_{34}^{y4})\right]e^{j\theta_4}=0 \tag{14.51a}$$

$$r_4 f_{34}^{y4}+t_4=0. \tag{14.51b}$$

For the revolute joint A, a sum of the forces gives

$$(f_{32}^{x2}+jf_{32}^{y2})e^{j\theta_2}+(f_{23}^{x3}+jf_{23}^{y3})e^{j\theta_3}=0 \tag{14.52a}$$

and for the revolute joint B, a sum of the forces gives

$$(f_{43}^{x3}+jf_{43}^{y3})e^{j\theta_3}+(f_{34}^{x4}+jf_{34}^{y4})e^{j\theta_4}=0. \tag{14.52b}$$

For the bearing O_2, a sum of the forces gives

$$(f_{21}^{x1}+jf_{21}^{y1})+(f_{12}^{x2}+jf_{12}^{y2})e^{i\theta_2}=0 \tag{14.53a}$$

$$(f_{41}^{x1}+jf_{41}^{y1})+(f_{14}^{x4}+jf_{14}^{y4})e^{i\theta_4}=0. \tag{14.53b}$$

From Eqs. (14.50a), the axial and transverse forces at point A on the coupler link are

$$f_{23}^{x3}=-f_{43}^{x3} \quad \text{and} \quad f_{23}^{y3}=-f_{43}^{y3}=0. \tag{14.54}$$

From Eq. (14.51b), the transverse force acting at point B on the output link 4 is

$$f_{34}^{y4}=-\frac{t_4}{r_4}. \tag{14.55}$$

Substituting Eq. (14.54) into Eq. (14.52a) and rearranging gives

$$f_{32}^{x2}+jf_{32}^{y2}=-f_{23}^{x3}e^{-j\eta_{23}}.$$

Equating the real and imaginary parts of this equation, respectively, gives

$$f_{32}^{x2}=f_{43}^{x3}\cos\eta_{23} \quad \text{and} \quad f_{32}^{y2}=-f_{43}^{x3}\sin\eta_{23}. \tag{14.56}$$

Equation (14.52b) can be written as

$$f_{43}^{x3}e^{j\eta_{34}}+(f_{34}^{x4}+jf_{34}^{y4})=0.$$

Equating the real and imaginary parts of this equation, respectively, gives

$$f_{43}^{x3}=-f_{23}^{x3}=-\frac{t_4}{r_4}\frac{1}{\sin\eta_{34}} \quad \text{and} \quad f_{43}^{x3}\cos\eta_{34}+f_{34}^{x4}=0. \tag{14.57}$$

Rearranging these two equations, the axial force on the output link can be written as

$$f_{34}^{x4} = \frac{t_4}{r_4} \cot \eta_{34}. \tag{14.58}$$

From Eqs. (14.49*a*) and (14.51*a*), we have

$$f_{12}^{x2} = -f_{32}^{x2} \quad \text{and} \quad f_{12}^{y2} = -f_{32}^{y2} \tag{14.59a}$$

$$f_{14}^{x4} = -f_{34}^{x4} \quad \text{and} \quad f_{14}^{y4} = -f_{34}^{y4}. \tag{14.59b}$$

Adding Eqs. (14.52*a*) and (14.53*a*) and with the aid of Eqs. (14.59*a*), we obtain

$$f_{21}^{x1} + jf_{21}^{y1} = f_{43}^{x3} e^{j\theta_3}.$$

Equating the real and imaginary parts of this equation, respectively, gives

$$f_{21}^{x1} = f_{43}^{x3} \cos \theta_3 \quad \text{and} \quad f_{21}^{y1} = f_{43}^{x3} \sin \theta_3. \tag{14.60}$$

Adding Eqs. (14.52*b*) and (14.53*b*) and with the aid of Eqs. (14.59*b*), we obtain

$$f_{41}^{x1} + jf_{41}^{y1} = f_{43}^{x3} e^{j\theta_3}.$$

Equating the real and imaginary parts of this equation, respectively, gives

$$f_{41}^{x1} = f_{43}^{x3} \cos \theta_3 \quad \text{and} \quad f_{41}^{y1} = f_{43}^{x3} \sin \theta_3. \tag{14.61}$$

Substituting Eq. (14.57) into Eq. (14.56), the transverse force acting at point *A* on the input link can be written as

$$f_{32}^{y2} = \frac{t_4}{r_4} \frac{\sin \eta_{23}}{\sin \eta_{34}}.$$

Then, substituting this equation into (14.49*b*) and rearranging, the input torque can be written as

$$t_2 = \frac{r_2}{r_4} \frac{\sin \eta_{23}}{\sin \eta_{34}} t_4. \tag{14.62}$$

Recall from Sections 1.10 and 3.20 that the mechanical advantage of a four-bar linkage is defined as the ratio of the output torque to the input torque. Therefore, rearranging Eq. (14.62), the mechanical advantage of the four-bar linkage can be written as

$$MA = \frac{t_4}{t_2} = -\frac{r_4}{r_2} \frac{\sin \eta_{34}}{\sin \eta_{23}}.$$

Note that this result is consistent with Eq. (3.40).

Finally, the shaking moment can be written with the aid of Eq. (14.61) as

$$m_s = r_1 f_{41}^{y1} = r_1 f_{43}^{x3} \sin \theta_3.$$

Substituting Eq. (14.57) into this equation, the shaking moment can be written as

$$m_s = \frac{r_1}{r_4}\frac{\sin\theta_3}{\sin\eta_{34}}t_4. \tag{14.63}$$

If friction in the mechanism is ignored, then the total load, which will be denoted here by an asterisk superscript, can be determined as the sum of the static load and the dynamic load; this is the principle of superposition. Therefore, the load torque is

$$T_4^* = t_4 + T_4. \tag{14.64}$$

The axial and transverse loads at bearing O_2, respectively, are

$$(F_{12}^{x2})^* = -f_{32}^{x2} + F_{12}^{x2}, \quad (F_{12}^{y2})^* = -f_{32}^{y2} + F_{12}^{y2}. \tag{14.65}$$

For the input link, the axial and transverse loads at joint A, respectively, are

$$(F_{32}^{x2})^* = f_{32}^{x2} + F_{32}^{x2}, \quad (F_{32}^{y2})^* = f_{32}^{y2} + F_{32}^{y2}. \tag{14.66a}$$

For the coupler link, the axial and transverse loads at joint A, respectively, are

$$(F_{23}^{x3})^* = -f_{43}^{x3} + F_{23}^{x3}, \quad (F_{23}^{y3})^* = F_{23}^{y3}. \tag{14.66b}$$

For the coupler link, the axial and transverse loads at joint B, respectively, are

$$(F_{43}^{x3})^* = -f_{43}^{x3} + F_{43}^{x3}, \quad (F_{43}^{y3})^* = F_{43}^{y3}. \tag{14.67a}$$

For the output link, the axial and transverse loads at joint B, respectively, are

$$(F_{34}^{x4})^* = -f_{34}^{x4} + F_{34}^{x4}, \quad (F_{34}^{y4})^* = f_{34}^{y4} + F_{34}^{y4}. \tag{14.67b}$$

The axial and transverse loads at bearing O_4, respectively, are

$$(F_{14}^{x4})^* = -f_{34}^{x4} + F_{14}^{x4}, \quad (F_{14}^{y4})^* = -f_{34}^{y4} + F_{14}^{y4}. \tag{14.68}$$

The x and y components of the load at bearing O_2, respectively, are

$$(F_{21}^{x1})^* = f_{21}^{x1} + F_{21}^{x1} \quad \text{and} \quad (F_{21}^{y1})^* = f_{21}^{y1} + F_{21}^{y1}. \tag{14.69a}$$

Therefore, the magnitude and the orientation, respectively, are

$$(F_{21})^* = \sqrt{(F_{21}^{x1})^{*2} + (F_{21}^{y1})^{*2}} \quad \text{and} \quad \psi_2 = \tan_2^{-1}\left(\frac{(F_{21}^{y1})^*}{(F_{21}^{x1})^*}\right). \tag{14.69b}$$

The x and y components of the load at bearing O_4, respectively, are

$$(F_{41}^{x1})^* = f_{41}^{x1} + F_{41}^{x1} \quad \text{and} \quad (F_{41}^{y1})^* = f_{41}^{y1} + F_{41}^{y1} \tag{14.70a}$$

Therefore, the magnitude and orientation, respectively, are

$$(F_{41})^* = \sqrt{(F_{41}^{x_1})^{*2} + (F_{41}^{y_1})^{*2}} \quad \text{and} \quad \psi_4 = \tan_2^{-1}\left(\frac{(F_{41}^{y_1})^*}{(F_{41}^{x_1})^*}\right). \tag{14.70b}$$

The total driving torque to input link 2 is

$$T_2^* = T_2 + t_2,$$

where the inertia input torque T_2 and the static input torque t_2 are given by Eqs. (14.44) and (14.62), respectively.

The total shaking moment is

$$M_s^* = M_s + m_s,$$

where the dynamic shaking M_s and the static shaking moment m_s are given by Eqs. (14.48) and (14.63), respectively.

For stable operation we could vary the link lengths l_1, l_2, and l_3 such that the power $P = T_2^* \omega_2$, discussed in the following section, is maximized and the shaking moment is minimized. For a better suspension system we should try to minimize the vertical components of the loads at the two ground bearings, that is, $(F_{21}^{y_1})^*$ and $(F_{41}^{y_1})^*$.

14.9 EQUATION OF MOTION FROM POWER EQUATION

This section presents a method of obtaining the equation of motion of a mechanism based on energy considerations. We will first write the power equation for the mechanism and then express this equation in terms of kinematic coefficients, as discussed in Chapters 3 and 4.

The work-energy equation for a mechanism can be written as

$$W = \Delta T + \Delta U + W_f, \tag{14.71}$$

where

W = the net work input to the mechanism, that is, the work input less the work output

ΔT = the change in the kinetic energy of the moving links

ΔU = the change in the potential energy stored in the mechanism

W_f = the energy dissipated through damping and friction.

Differentiating Eq. (14.71) with respect to time gives

$$P = \frac{dW}{dt} = \frac{dT}{dt} + \frac{dU}{dt} + \frac{dW_f}{dt}, \tag{14.72}$$

where P is the *net power input* to the mechanism. Equation (14.72) is commonly referred to as the *power equation*. The net power is a scalar quantity and can be written as

$$P = Q\dot{\psi}, \tag{14.73}$$

where Q is the generalized input force (that is, the force or the torque acting on the input) and $\dot{\psi}$ is the generalized input velocity (that is, either the rectilinear or angular velocity of the input degree of freedom). If the product of these two signed quantities is positive—that is, if the signed quantities are acting in the same direction—then power is going into the system. If the product is negative—that is, if the two signed quantities are acting in opposite directions—then power is being removed from the system.

The kinetic energy of a link in the mechanism, say link j, can be written as

$$T_j = \tfrac{1}{2}m_j V_{G_j}^2 + \tfrac{1}{2}I_{G_j}\omega_j^2. \tag{14.74}$$

Recall from Section 3.12 that the velocity of the mass center of link j can be written in terms of first-order kinematic coefficients as

$$\mathbf{V}_{G_j} = (x'_{G_j}\hat{\mathbf{i}} + y'_{G_j}\hat{\mathbf{j}})\dot{\psi} \tag{14.75a}$$

and the angular velocity of link j can be written as

$$\omega_j = \theta'_j\dot{\psi}. \tag{14.75b}$$

Substituting Eqs. (14.75) into Eq. (14.74), the kinetic energy of link j can be written as

$$T_j = \tfrac{1}{2}m_j(x_{G_j}'^2 + y_{G_j}'^2)\dot{\psi}^2 + \tfrac{1}{2}I_{G_j}\theta_j'^2\dot{\psi}^2. \tag{14.76}$$

Differentiating this equation with respect to time, the time rate of change in the kinetic energy of link j can be written as

$$\frac{dT_j}{dt} = A_j\dot{\psi}\ddot{\psi} + B_j\dot{\psi}^3, \tag{14.77}$$

where

$$A_j = m_j(x_{G_j}'^2 + y_{G_j}'^2) + I_{G_j}\theta_j'^2 \tag{14.78a}$$

$$B_j = m_j(x'_{G_j}x''_{G_j} + y'_{G_j}y''_{G_j}) + I_{G_j}\theta'_j\theta''_j. \tag{14.78b}$$

Comparing Eq. (14.78a) and Eq. (14.76), we see that the kinetic energy can be written as

$$T_j = \frac{1}{2}A_j\dot{\psi}^2. \tag{14.79}$$

From Eqs. (14.78a) and (14.78b), we also note the relationship between the coefficients, that is,

$$B_j = \frac{1}{2}\frac{dA_j}{d\psi}. \tag{14.80}$$

For a mechanism with n links and with link 1 fixed, the time rate of change in kinetic energy can be written, from Eq. (14.77), as

$$\frac{dT}{dt} = \sum_{j=2}^{n} A_j\dot{\psi}\ddot{\psi} + \sum_{j=2}^{n} B_j\dot{\psi}^3. \tag{14.81}$$

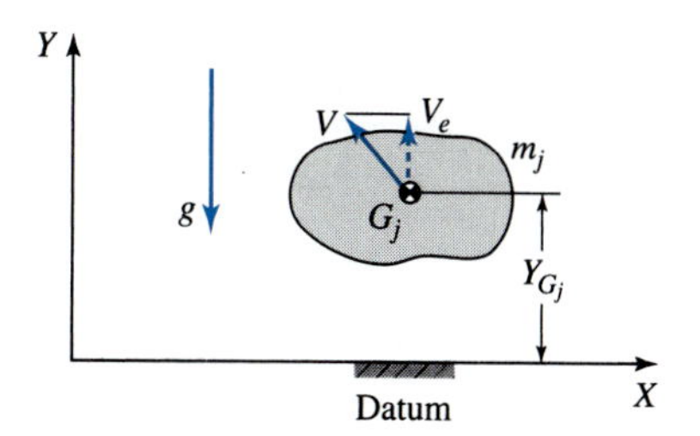

Figure 14.19 Potential energy caused by elevation.

Next we consider the potential energy stored in link j of the mechanism:

1. *Caused by gravity*. If we assume that the gravitational force acts in the negative Y direction, then the potential energy caused by gravity is

$$U_{gj} = m_j g Y_{G_j}, \tag{14.82}$$

where m_j is the mass of the link, g is the gravitational constant, and Y_{G_j} is the height of the mass center above an arbitrarily chosen datum at the origin, as illustrated in Fig. 14.19. Differentiating Eq. (14.82) with respect to time, the rate of change in the potential energy of link j can be written as

$$\frac{dU_{gj}}{dt} = m_j g y'_{G_j} \dot{\psi}, \tag{14.83a}$$

where the first-order kinematic coefficient is

$$y'_{G_j} = \frac{dY_{G_j}}{d\psi}. \tag{14.83b}$$

Therefore, the time rate of change in the potential energy stored in the mechanism is

$$\frac{dU_g}{dt} = \sum_{j=2}^{n} m_j g y'_{G_j} \dot{\psi}. \tag{14.84}$$

2. *Caused by a rectilinear spring*. If the mechanism contains a spring element, then the potential energy stored in the spring is

$$U_s = k(r_s - r_0)^2, \tag{14.85}$$

where k is the spring rate, r_0 is the free length of the spring, and r_s is the actual length of the spring, as illustrated in Fig. 14.20. Differentiating Eq. (14.85) with respect to time, the time rate of change in the potential energy stored in the spring can be written as

$$\frac{dU_s}{dt} = k(r_s - r_0) r'_s \dot{\psi}, \tag{14.86a}$$

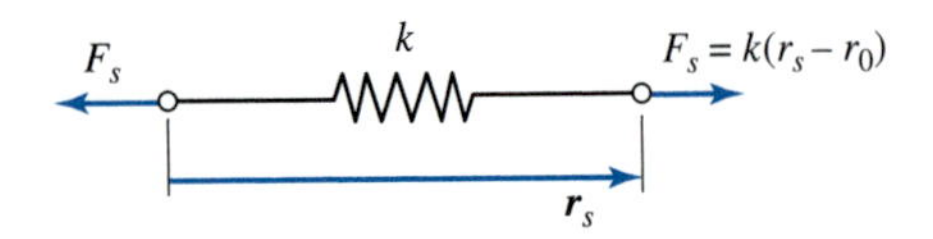

Figure 14.20 Vector across a rectilinear spring.

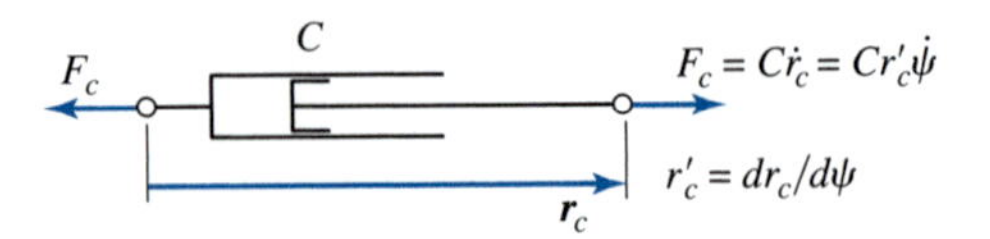

Figure 14.21 Vector across a damper.

where the first-order kinematic coefficient of the spring is

$$r'_s = \frac{dr_s}{d\psi}. \tag{14.86b}$$

Next, we consider the dissipative effects caused by the following:

1. *A viscous damper.* The work done in overcoming the damping effect is

$$W_f = C\dot{r}_c \Delta r_c, \tag{14.87}$$

where C is the damping coefficient and r_c is the length of the vector across the damper, as illustrated in Fig. 14.21. Differentiating Eq. (14.87) with respect to time, the time rate of change of the dissipative effect can be written as

$$\frac{dW_f}{dt} = Cr_c'^2 \dot{\psi}^2, \tag{14.88a}$$

where the first-order kinematic coefficient of the damper is

$$r'_c = \frac{dr_c}{d\psi}. \tag{14.88b}$$

2. *Coulomb friction.* The work done in overcoming Coulomb friction is

$$W_f = \mu N \Delta r_f, \tag{14.89}$$

where μ is the coefficient of friction, N is the normal force between the contacting surfaces on links 2 and 3, and $\mathbf{r}_f$ is the vector along the link to the point of contact illustrated as point C in Fig. 14.22. Differentiating Eq. (14.89) with respect to time, the time rate of change of the dissipative effect can be written as

$$\frac{dW_f}{dt} = \mu N |r'_j \dot{\psi}|, \tag{14.90a}$$

where the first-order kinematic coefficient is

$$r'_j = \frac{dr_f}{d\psi}. \tag{14.90b}$$

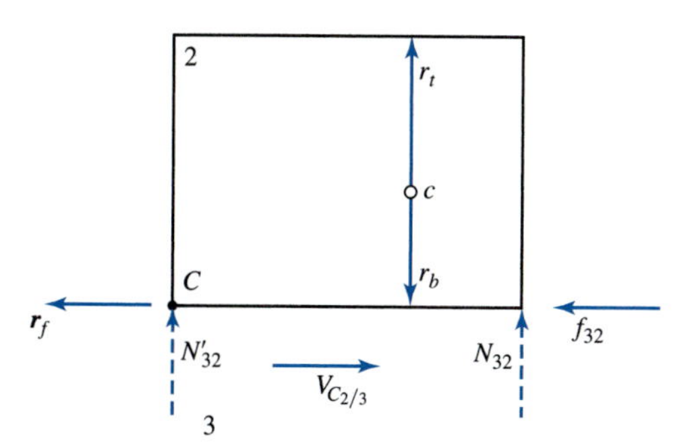

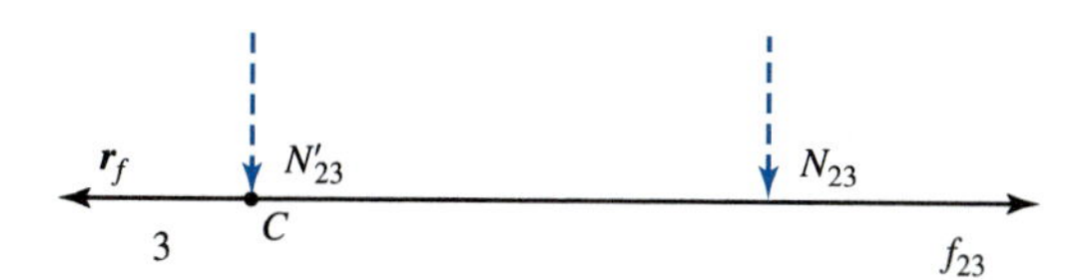

Figure 14.22 Vector to the point of contact.

Substituting Eqs. (14.73), (14.81), (14.84), (14.86*a*), (14.88*a*), and (14.90*a*) into Eq. (14.72), the net power can be written as

$$Q\dot{\psi} = \sum_{j=2}^{n} A_j \dot{\psi}\ddot{\psi} + \sum_{j=2}^{n} B_j \dot{\psi}^3 + \sum_{j=2}^{n} m_j g y'_{G_j} \dot{\psi} + k(r_s - r_0) r'_s \dot{\psi} + C r_c'^2 \dot{\psi}^2 + \mu N |r'_j \dot{\psi}| \tag{14.91}$$

We note that each term in this equation contains the generalized input velocity $\dot{\psi}$. If we divide through by this common factor, then we obtain the *equation of motion* for the mechanism, that is,

$$Q = \sum_{j=2}^{n} A_j \ddot{\psi} + \sum_{j=2}^{n} B_j \dot{\psi}^2 + \sum_{j=2}^{n} m_j g y'_{G_j} + k(r_s - r_0) r'_s + C r_c'^2 \dot{\psi} + \mu N |r'_j|. \tag{14.92}$$

This result is also valid when the generalized input velocity is zero, that is, when the input link is instantaneously stopped or when the mechanism is starting from rest. In this case, Eq. (14.92) reduces to

$$Q = \sum_{j=2}^{n} A_j \ddot{\psi} + \sum_{j=2}^{n} m_j g y'_{G_j} + k(r_s - r_0) r'_s + \mu N |r'_j| \tag{14.93}$$

and for this case there is no necessity to calculate $\sum_{j=2}^{n} B_j$ or the first-order kinematic coefficient of the viscous damper.

We will consider two cases for the input motion: (1) input rotation, that is, $\dot{\psi} = \dot{\theta}_i$, and (2) input translation, that is, $\dot{\psi} = \dot{r}_i$.

1. If the input motion is a rotation, then Eq. (14.92) can be written as

$$M_i = \sum_{j=2}^{n} A_j \ddot{\theta}_i + \sum_{j=2}^{n} B_j \dot{\theta}_i^2 + \sum_{j=2}^{n} m_j g y'_{G_j} + k(r_s - r_0)r'_s + Cr_c'^2 \dot{\theta}_i + \mu N |r'_j|, \tag{14.94}$$

where M_i is the torque acting at the input. Note that the coefficient $\sum_{j=2}^{n} A_j$ must have units of moment of inertia and is referred to as the *equivalent mass moment of inertia* of the system and denoted I_{EQ}. This moment of inertia, if thought of as concentrated on the input link alone, would have the same kinetic energy as the entire mechanism. From Eq. (14.79), the kinetic energy could then be written as

$$T = \frac{1}{2} I_{EQ} \dot{\theta}_i^2. \tag{14.95}$$

The equation of motion, that is, Eq. (14.94), could then be written as

$$M_i = I_{EQ} \ddot{\theta}_i + \frac{1}{2} \frac{dI_{EQ}}{d\theta_i} \dot{\theta}_i^2 + \sum_{j=2}^{n} m_j g y'_{G_j} + k(r_s - r_0)r'_s + Cr_c'^2 \dot{\theta}_i + \mu N |r'_f|. \tag{14.96}$$

2. If the input motion is translation only, then Eq. (14.92) can be written as

$$F_i = \sum_{j=2}^{n} A_j \ddot{r}_i + \sum_{j=2}^{n} B_j \dot{r}_i^2 + \sum_{j=2}^{n} m_j g y'_{G_j} + k(r_s - r_0)r'_s + Cr_c'^2 \dot{r}_i + \mu N |r'_f|, \tag{14.97}$$

where F_i is the force acting at the input. For this case, the coefficient $\sum_{j=2}^{n} A_j$ has units of mass and is referred to as the *equivalent mass* of the system and denoted m_{EQ}. This mass, if thought of as concentrated on the input link alone, would have the same kinetic energy as the entire mechanism. From Eq. (14.79), the kinetic energy could then be written as

$$T = \frac{1}{2} m_{EQ} \dot{r}_i^2. \tag{14.98}$$

The equation of motion, that is, Eq. (14.97), could then be written as

$$F_i = m_{EQ} \ddot{r}_i + \frac{1}{2} \frac{dm_{EQ}}{dr_i} \dot{r}_i^2 + \sum_{j=2}^{n} m_j g y'_{G_j} + k(r_s - r_0)r'_s + Cr_c'^2 \dot{r}_i + \mu N |r'_f|. \tag{14.99}$$

EXAMPLE 14.6

For the parallelogram four-bar mechanism illustrated in Fig. 14.23, determine the torque M_2 that must be applied to input link 2 to produce an angular velocity of $\omega_2 = 10$ rad/s ccw and an angular acceleration $\alpha_2 = 10$ rad/s^2 cw. In the position illustrated in Fig. 14.23, $R_{AD} = R_{BE} = 0.8$ m. The free length of the spring is $r_0 = 0.6$ m, the spring rate is $k = 500$ N/m, and the damping constant is $C = 10$ N · s/m. The masses of the links are $m_2 = 2.5$ kg, $m_3 = 5$ kg, and $m_4 = 2.5$ kg, and the center of mass of each link is coincident with its geometric center. The mass moments of inertia of the links are $I_{G_2} = I_{G_4} = 0.25$ kg · m^2 and $I_{G_3} = 4.5$ kg · m^2. Assume that gravity acts vertically downward and that friction can be ignored.

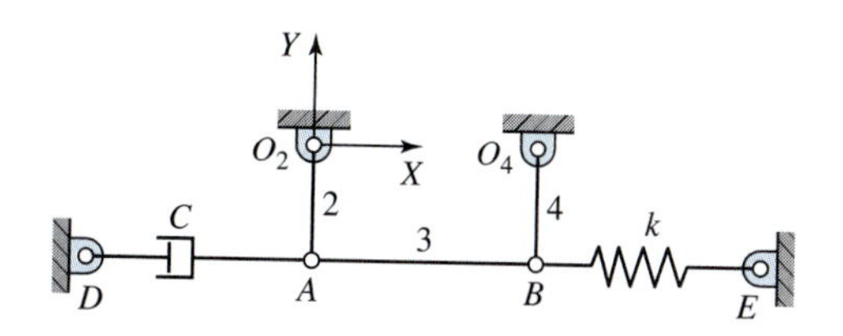

Figure 14.23 Example 14.6. The link dimensions are $r_1 = 0.8$ m, $r_2 = 0.4$ m, $r_3 = 0.8$ m, and $r_4 = 0.4$ m.

SOLUTION

From the kinematic analysis of the four-bar linkage presented in Sections 3.12 and 4.12, the first- and second-order kinematic coefficients of links 3 and 4 are

$$\theta_3' = 0, \quad \theta_4' = 1 \text{ rad/rad}, \quad \theta_3'' = 0, \quad \text{and} \quad \theta_4'' = 0. \tag{1}$$

Recall that the angular velocities and angular accelerations of links 3 and 4 can be written as

$$\omega_3 = \theta_3'\omega_2 \quad \text{and} \quad \omega_4 = \theta_4'\omega_2$$
$$\alpha_3 = \theta_3''\omega_2^2 + \theta_3'\alpha_2 \quad \text{and} \quad \alpha_4 = \theta_4''\omega_2^2 + \theta_4'\alpha_2. \tag{2}$$

Substituting Eqs. (1) and the input angular velocity and angular acceleration into Eqs. (2) gives

$$\omega_3 = 0, \quad \omega_4 = 10 \text{ rad/s ccw}, \quad \alpha_3 = 0, \quad \text{and} \quad \alpha_4 = 10 \text{ rad/s}^2 \text{ cw}. \tag{3}$$

The x and y components of the position of the mass center G_2 with respect to the ground pivot O_2 are

$$x_{G_2} = \frac{1}{2}r_2\cos\theta_2 \quad \text{and} \quad y_{G_2} = \frac{1}{2}r_2\sin\theta_2. \tag{4}$$

Differentiating these equations twice with respect to θ_2 and substituting $\theta_2 = 270°$, the first- and second-order kinematic coefficients of the mass center of link 2 are

$$x'_{G_2} = 0.2 \text{ m/rad}, \quad y'_{G_2} = 0, \quad x''_{G_2} = 0, \quad y''_{G_2} = 0.2 \text{ m/rad}^2. \tag{5}$$

The x and y components of the position of the mass center G_3 with respect to the ground pivot O_2 can be written as

$$x_{G_3} = r_2 \cos\theta_2 + \frac{1}{2} r_3 \cos\theta_3 \quad \text{and} \quad y_{G_3} = r_2 \sin\theta_2 + \frac{1}{2} r_3 \sin\theta_3. \tag{6}$$

Differentiating these equations twice with respect to θ_2 and substituting $\theta_2 = 270°$ and $\theta_3 = 0°$, the first- and second-order kinematic coefficients of the mass center of link 3 are

$$x'_{G_3} = 0.4 \text{ m/rad}, \quad y'_{G_3} = 0, \quad x''_{G_3} = 0, \quad y''_{G_3} = 0.4 \text{ m/rad}^2. \tag{7}$$

The first- and second-order kinematic coefficients of the mass center of link 4 are the same as for the mass center of link 2, that is,

$$x'_{G4} = x'_{G_2} = 0.2 \text{ m/rad}, \quad y'_{G4} = y'_{G_2} = 0, \quad x''_{G4} = x''_{G2} = 0, \quad y''_{G4} = y''_{G_2} = 0.2 \text{ m/rad}^2. \tag{8}$$

To obtain the first-order kinematic coefficient for the spring, the vector loop-closure equation, illustrated in Fig. 14.24, can be written as

$$\mathbf{r}_4 + \mathbf{r}_s - \mathbf{r}_6 = \mathbf{0}. \tag{9}$$

The two scalar equations are

$$r_4 \cos\theta_4 + r_s \cos\theta_s - r_6 \cos\theta_6 = 0$$
$$r_4 \sin\theta_4 + r_s \sin\theta_s - r_6 \sin\theta_6 = 0. \tag{10}$$

Differentiating Eqs. (10) with respect to the input θ_2 gives

$$-r_4 \sin\theta_4\theta'_4 - r_s \sin\theta_s\theta'_s + r'_s \cos\theta_s = 0 \tag{11a}$$
$$r_4 \cos\theta_4\theta'_4 + r_s \cos\theta_s\theta'_s + r'_s \sin\theta_s = 0. \tag{11b}$$

Substituting $\theta_s = 0°$, $\theta_4 = 270°$ and $\theta'_4 = 1$ rad/rad into Eq. (11a), the first-order kinematic coefficient for the spring is

$$r'_s = -r_4\theta'_4 = -0.4 \text{ m/rad}. \tag{12}$$

To obtain the first-order kinematic coefficient for the damper, the vector loop-closure equation, illustrated in Fig. 14.25, can be written as

$$\mathbf{r}_2 + \mathbf{r}_c - \mathbf{r}_8 = \mathbf{0}. \tag{13}$$

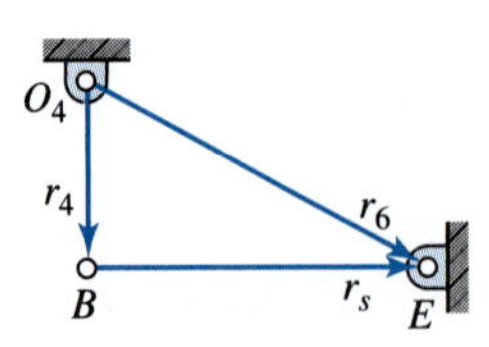

Figure 14.24 Vector loop for the spring.

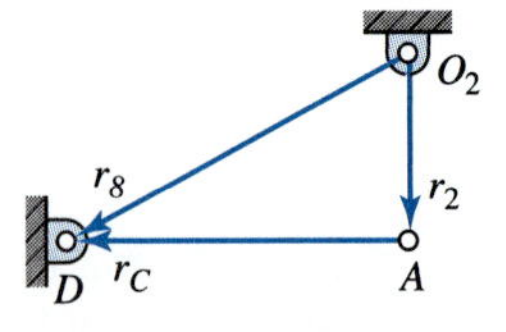

Figure 14.25 Vector loop for the viscous damper.

The two scalar equations are

$$r_2 \cos\theta_2 + r_c \cos\theta_c - r_8 \cos\theta_8 = 0 \tag{14a}$$

$$r_2 \sin\theta_2 + r_c \sin\theta_c - r_8 \sin\theta_8 = 0. \tag{14b}$$

Differentiating Eqs. (14) with respect to the input θ_2 gives

$$-r_2 \sin\theta_2 - r_c \sin\theta_c\theta_c' + r_c' \cos\theta_c = 0 \tag{15a}$$

$$r_2 \cos\theta_2 + r_c \cos\theta_c\theta_c' + r_c' \sin\theta_c = 0. \tag{15b}$$

Substituting $\theta_2 = 270°$ and $\theta_c = 180°$ into Eq. (15*a*), the first-order kinematic coefficient for the damper is

$$r_c' = r_2 = 0.4 \text{ m/rad}. \tag{16}$$

The equivalent moment of inertia for the four-bar linkage is

$$I_{EQ} = \sum_{j=2}^{4} A_j.$$

From Eq. (14.78*a*), the equivalent moment of inertia can be written as

$$I_{EQ} = m_2(x_{G_2}'^2 + y_{G_2}'^2) + I_{G_2} + m_3(x_{G_3}'^2 + y_{G_3}'^2) + I_{G_3}\theta_3'^2 + m_4(x_{G_4}'^2 + y_{G_4}'^2) + I_{G_4}\theta_4'^2.$$

Substituting the known data and the kinematic coefficients into this equation gives

$$I_{EQ} = 0.1 + 0.25 + 0.8 + 0.1 + 0.25 = 1.5 \text{ kg}\cdot\text{m}^2. \tag{17}$$

From Eq. (14.78*b*), we can write

$$\sum_{j=2}^{4} B_j = m_2(x_{G_2}'x_{G_2}'' + y_{G_2}'y_{G_2}'') + m_3(x_{G_3}'x_{G_3}'' + y_{G_3}'y_{G_3}'') + I_{G_3}\theta_3'\theta_3''$$

$$+m_4(x_{G_4}'x_{G_4}'' + y_{G_4}'y_{G_4}'') + I_{G_4}\theta_4'\theta_4''. \tag{18}$$

For the given position, the kinematic coefficients are given by Eqs. (1), (5), (7), and (8). Substituting these kinematic coefficients into Eq. (18) gives

$$\sum_{j=2}^{n} B_j = 0. \tag{19}$$

Then, substituting the known data, the kinematic coefficients, and Eqs. (17) and (19) into Eq. (14.96), the equation of motion can be written

$$M_2 = (1.5 \text{ kg}\cdot\text{m}^2)\ddot{\theta}_2 + k(r_s - r_0)r_s' + Cr_c'^2\dot{\theta}_2.$$

Finally, substituting the known data and Eqs. (12) and (16) into this equation, the input torque is

$$M_2 = -15\ \text{N}\cdot\text{m} + (500\ \text{N/m})(0.8\ \text{m} - 0.6\ \text{m})(-0.4\ \text{m/rad})$$
$$+ (10\ \text{N}\cdot\text{s/m})\cdot(0.4\ \text{m/rad})^2\cdot(10\ \text{rad/s})$$
$$M_2 = -39\ \text{N}\cdot\text{m} = 39\ \text{N}\cdot\text{m cw.} \qquad \textit{Ans.}$$

Because the answer is negative, the input torque acts in the direction opposite to the input angular velocity, that is, clockwise. Note that this is necessary to achieve the clockwise input angular acceleration specified. Also note that a clockwise input torque of $M_2 = 24\ \text{N}\cdot\text{m}$ would yield a zero input acceleration at the instant under study.

14.10 MEASURING MASS MOMENT OF INERTIA

Sometimes the shapes of machine parts are so complicated that it is extremely tedious and time consuming to calculate the mass moment(s) of inertia. Consider, for example, the problem of finding the mass moment of inertia of an automobile body about a vertical axis through its center of mass. For such problems it is usually possible to determine the mass moment of inertia of a body by observing its dynamic behavior caused by a known rotational disturbance.

Many bodies, connecting rods and cranks for example, are shaped so that their masses can be assumed to lie in a single plane. Once such bodies have been weighed and their mass centers located, they can be suspended like a pendulum and caused to oscillate. The mass moment of inertia of such a body can then be computed from an observation of its period or frequency of oscillation. For best experimental results, the part should be suspended with the pivot located close to, but not coincident with, the center of mass. It is not usually necessary to drill a hole to suspend the body; for example, a spoked wheel or a gear can be suspended on a knife edge at its rim.

When the body of Fig. 14.26*a* is displaced through an angle θ, a gravity force mg acts at G. Summing moments about the pivot O gives

$$\sum M_o = -mg(r_G \sin\theta) - I_o\ddot{\theta} = 0. \qquad (a)$$

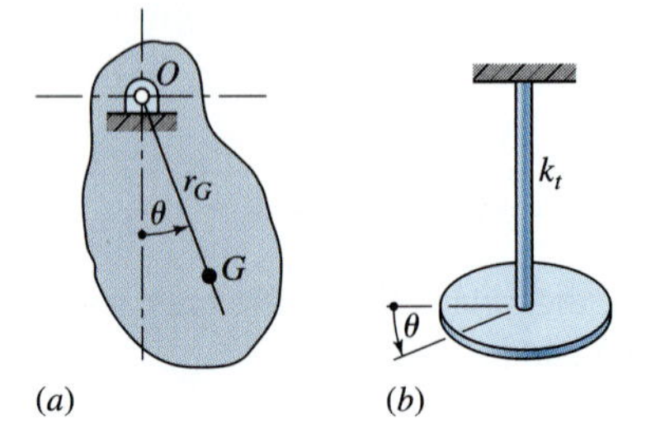

Figure 14.26 (*a*) Simple pendulum. (*b*) Torsional pendulum.

We intend that the pendulum be displaced only through a small angle, so that $\sin\theta$ can be approximated by θ. Equation (a) can then be written as

$$\ddot{\theta} + \frac{mgr_G}{I_o}\theta = 0. \tag{b}$$

This second-order, linear differential equation has the well-known solution

$$\theta = c_1 \sin\sqrt{\frac{mgr_G}{I_o}}t + c_2 \cos\sqrt{\frac{mgr_G}{I_o}}t, \tag{c}$$

where c_1 and c_2 are constants of integration.

We shall start the pendulum motion by displacing it through a small angle θ_0 and releasing it with no initial velocity from this position. Thus, at time $t = 0$, we obtain $\theta = \theta_0$ and $\dot{\theta} = 0$. Substituting these conditions into Eq. (c) and its first time derivative enables us to evaluate the two constants. They are $c_1 = 0$ and $c_2 = \theta_0$. Therefore,

$$\theta = \theta_0 \cos\sqrt{\frac{mgr_G}{I_o}}t. \tag{14.100}$$

Because a cosine function repeats itself every 360° or 2π radians, the period of the motion is

$$\tau = 2\pi\sqrt{\frac{I_o}{mgr_G}}. \tag{d}$$

Therefore, the mass moment of inertia of the body about the pivot O is

$$I_O = mgr_G\left(\frac{\tau}{2\pi}\right)^2. \tag{14.101}$$

This equation demonstrates that the body must be weighed to get mg, the distance r_G must be measured, and then the pendulum must be suspended and oscillated so that the period τ can be observed; Eq. (14.101) can then be solved to give I_O about O. If the moment of inertia about the mass center is desired, it can be obtained by using the parallel axis theorem, Eq. (14.11).

Figure 14.26b illustrates how the mass moment of inertia can be determined without actually weighing the body. The body is connected to a slender rod or wire at the mass center. The torsional stiffness k_t of the rod or wire is defined as the torque necessary to twist the rod through a unit angle. If the body of Fig. 14.26b is turned through a small angle θ and released, the equation of motion becomes

$$\ddot{\theta} + \frac{k_t}{I_G}\theta = 0 \tag{e}$$

This is similar to Eq. (b) and, with the same starting conditions, has the solution

$$\theta = \theta_0 \cos\sqrt{\frac{k_t}{I_G}}t. \tag{14.102}$$

The period of oscillation is then

$$\tau = 2\pi\sqrt{\frac{I_G}{k_t}} \tag{f}$$

and

$$I_G = k_t\left(\frac{\tau}{2\pi}\right)^2. \tag{14.103}$$

The torsional stiffness is often known or can be computed from a knowledge of the length and diameter of the rod or wire and its material. Then the oscillation of the body can be observed and Eq. (14.103) can be used to compute the mass moment of inertia I_G. Alternatively, when the torsional stiffness k_t is unknown, a body with known mass moment of inertia I_G can be mounted and Eq. (14.103) can be used to determine k_t.

A *trifilar pendulum,* also called a *three-string torsional pendulum*, illustrated in Fig. 14.27, can provide a very accurate method of measuring mass moment of inertia. Three strings of equal length support a lightweight platform and are equally spaced about its center. A round platform serves just as well as the triangular one shown. The part whose mass moment of inertia is to be determined is carefully placed on the platform so that the center of mass of the object coincides with the platform center. The platform is then made to oscillate, and the number of oscillations is counted over a specified period of time.[1]

The notation for the three-string torsional pendulum analysis is as follows.

m = mass of the part

m_p = mass of the platform

I_G = mass moment of inertia of the part

I_p = mass moment of inertia of the platform

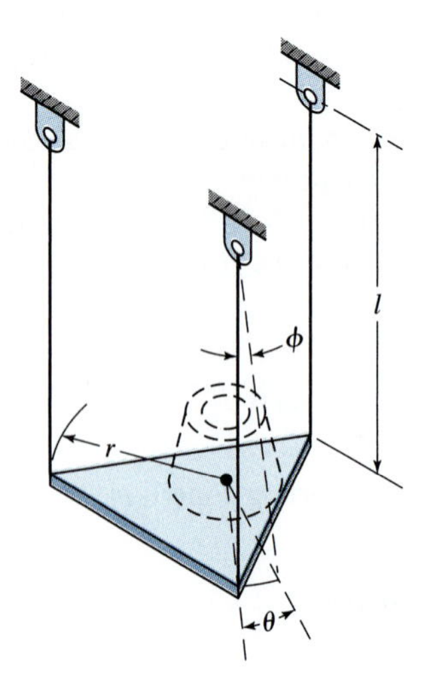

Figure 14.27 Trifilar pendulum.

$r =$ platform radius

$\theta =$ platform angular displacement

$l =$ string length

$\phi =$ string angular displacement

$z =$ vertical axis through the center of the platform

We begin by writing the summation of moments about the z axis. This gives

$$\sum M^z = -r(m + m_p)g \sin\phi - (I_G + I_p)\ddot{\theta} = 0. \tag{g}$$

Because we are assuming small displacements, the sine of an angle can be approximated by the angle itself. Therefore,

$$\phi \approx \frac{r}{l}\theta \tag{h}$$

and Eq. (g) becomes

$$\ddot{\theta} + \frac{(m + m_p)gr^2}{(I_G + I_p)l}\theta = 0. \tag{i}$$

This equation can be solved in the same manner as Eq. (b). The result is

$$I_G + I_p = \frac{(m + m_p)gr^2}{l}\left(\frac{\tau}{2\pi}\right). \tag{14.104}$$

This equation should be used first with an empty platform to find I_p. With m_p and I_p known, the equation can then be used to find I_G of the part being measured.

14.11 TRANSFORMATION OF INERTIA AXES

A brief review will demonstrate that all examples presented so far have been limited to planar motion. It is now time to extend our study to include spatial problems. The basic principles are not new, but the problems presented may seem more complex because of our difficulty in visualizing in three dimensions. In addition, our previous treatment of angular motion was not presented in detail enough to deal with three-dimensional rotations.

Up to this point, we have continually used what are called the *principal mass moments of inertia*. In a coordinate system with origin at the center of mass and with axes aligned in the principal axis directions, all *mass products of inertia* [Eqs. (14.8)] are zero. This choice of axes has been used to simplify the form of the equations from that required for other choices.

In Eq. (14.11), the parallel-axis formula, we demonstrated how translation of axes could be performed. However, up to now, we have said little of how inertia properties could be rotated to a coordinate system that is not parallel to the principal axes. This is the purpose of this section.

Let us assume that the principal axes are aligned along unit vector directions $\hat{\mathbf{i}}$, $\hat{\mathbf{j}}$, and $\hat{\mathbf{k}}$, and that the new rotated axes desired have directions denoted by $\hat{\mathbf{i}}'$, $\hat{\mathbf{j}}'$, and $\hat{\mathbf{k}}'$. Then, any point in the principal axis coordinate system is located by the position vector $\mathbf{R}$, and by $\mathbf{R}'$ in the rotated system. Because both are descriptions of the same point position, $\mathbf{R}' = \mathbf{R}$ and

$$R^{x'}\hat{\mathbf{i}}' + R^{y'}\hat{\mathbf{j}}' + R^{z'}\hat{\mathbf{k}}' = R^{x}\hat{\mathbf{i}} + R^{y}\hat{\mathbf{j}} + R^{z}\hat{\mathbf{k}}.$$

Now, by taking dot products of this equation with $\hat{\mathbf{i}}'$, $\hat{\mathbf{j}}'$, and $\hat{\mathbf{k}}'$, respectively, we find the transformation equations between the two sets of axes. They are

$$\begin{aligned} R^{x'} &= (\hat{\mathbf{i}}' \cdot \hat{\mathbf{i}})R^{x} + (\hat{\mathbf{i}}' \cdot \hat{\mathbf{j}})R^{y} + (\hat{\mathbf{i}}' \cdot \hat{\mathbf{k}})R^{z} \\ R^{y'} &= (\hat{\mathbf{j}}' \cdot \hat{\mathbf{i}})R^{x} + (\hat{\mathbf{j}}' \cdot \hat{\mathbf{j}})R^{y} + (\hat{\mathbf{j}}' \cdot \hat{\mathbf{k}})R^{z} \\ R^{z'} &= (\hat{\mathbf{k}}' \cdot \hat{\mathbf{i}})R^{x} + (\hat{\mathbf{k}}' \cdot \hat{\mathbf{j}})R^{y} + (\hat{\mathbf{k}}' \cdot \hat{\mathbf{k}})R^{z}. \end{aligned} \tag{a}$$

But we may remember that the dot product of two unit vectors is equal to the cosine of the angle between them. For example, if the angle between $\hat{\mathbf{i}}'$ and $\hat{\mathbf{j}}$ is denoted by $\theta_{i'j}$, then

$$\hat{\mathbf{i}}' \cdot \hat{\mathbf{j}} = \cos\theta_{i'j}.$$

Thus, Eqs. (a) become

$$\begin{aligned} R^{x'} &= \cos\theta_{i'i}R^{x} + \cos\theta_{i'j}R^{y} + \cos\theta_{i'k}R^{z} \\ R^{y'} &= \cos\theta_{j'i}R^{x} + \cos\theta_{j'j}R^{y} + \cos\theta_{j'k}R^{z} \\ R^{z'} &= \cos\theta_{k'i}R^{x} + \cos\theta_{k'j}R^{y} + \cos\theta_{k'k}R^{z}. \end{aligned} \tag{b}$$

By substituting these into the definitions of mass moments of inertia in Eqs. (14.7), expanding, and then performing the integration, we get

$$\begin{aligned} I^{x'x'} &= \cos^2\theta_{i'i}I^{xx} + \cos^2\theta_{i'j}I^{yy} + \cos^2\theta_{i'k}I^{zz} \\ &\quad + 2\cos\theta_{i'i}\cos\theta_{i'j}I^{xy} + 2\cos\theta_{i'i}\cos\theta_{i'k}I^{xz} + 2\cos\theta_{i'j}\cos\theta_{i'k}I^{yz} \\ I^{y'y'} &= \cos^2\theta_{j'i}I^{xx} + \cos^2\theta_{j'j}I^{yy} + \cos^2\theta_{j'k}I^{zz} \\ &\quad + 2\cos\theta_{j'i}\cos\theta_{j'j}I^{xy} + 2\cos\theta_{j'i}\cos\theta_{j'k}I^{xz} + 2\cos\theta_{j'j}\cos\theta_{j'k}I^{yz} \\ I^{z'z'} &= \cos^2\theta_{k'i}I^{xx} + \cos^2\theta_{k'j}I^{yy} + \cos^2\theta_{k'k}I^{zz} \\ &\quad + 2\cos\theta_{k'i}\cos\theta_{k'j}I^{xy} + 2\cos\theta_{k'i}\cos\theta_{k'k}I^{xz} + 2\cos\theta_{k'j}\cos\theta_{k'k}I^{yz}. \end{aligned} \tag{14.105}$$

Similarly, for the mass products of inertia, starting from Eqs. (14.8), we get

$$\begin{aligned} I^{x'y'} &= -\cos\theta_{i'i}\cos\theta_{j'i}I^{xx} - \cos\theta_{i'j}\cos\theta_{j'j}I^{yy} - \cos\theta_{i'k}\cos\theta_{j'k}I^{zz} \\ &\quad + (\cos\theta_{i'i}\cos\theta_{j'j} + \cos\theta_{i'j}\cos\theta_{j'i})I^{xy} \\ &\quad + (\cos\theta_{i'j}\cos\theta_{j'k} + \cos\theta_{i'k}\cos\theta_{j'j})I^{yz} \\ &\quad + (\cos\theta_{i'k}\cos\theta_{j'i} + \cos\theta_{i'i}\cos\theta_{j'k})I^{zx} \end{aligned} \tag{14.106a}$$

$$\begin{aligned} I^{y'z'} &= -\cos\theta_{j'i}\cos\theta_{k'i}I^{xx} - \cos\theta_{j'j}\cos\theta_{k'j}I^{yy} - \cos\theta_{j'k}\cos\theta_{k'k}I^{zz} \\ &\quad + (\cos\theta_{j'i}\cos\theta_{k'j} + \cos\theta_{j'j}\cos\theta_{k'i})I^{xy} \\ &\quad + (\cos\theta_{j'j}\cos\theta_{k'k} + \cos\theta_{j'k}\cos\theta_{k'j})I^{yz} \\ &\quad + (\cos\theta_{j'k}\cos\theta_{k'i} + \cos\theta_{j'i}\cos\theta_{k'k})I^{zx} \end{aligned} \tag{14.106b}$$

$$\begin{aligned} I^{z'x'} &= -\cos\theta_{k'i}\cos\theta_{i'i}I^{xx} - \cos\theta_{k'j}\cos\theta_{i'j}I^{yy} - \cos\theta_{k'k}\cos\theta_{i'k}I^{zz} \\ &\quad + (\cos\theta_{k'i}\cos\theta_{i'j} + \cos\theta_{k'j}\cos\theta_{i'i})I^{xy} \\ &\quad + (\cos\theta_{k'j}\cos\theta_{i'k} + \cos\theta_{k'k}\cos\theta_{i'j})I^{yz} \\ &\quad + (\cos\theta_{k'k}\cos\theta_{i'i} + \cos\theta_{k'i}\cos\theta_{i'k})I^{zx}. \end{aligned} \tag{14.106c}$$

Once we have rotated away from the principal axes or if we had started with other than the principal axes, then the situation is different. The parallel-axis formula of Eq. (14.11) is no longer sufficient for the translation of axes. Let us assume that the mass moments and products of inertia are known in one coordinate system and are desired in another coordinate system, which is translated from the first by the equations

$$\begin{aligned} R^{x''} &= R^{x'} + d^{x'} \\ R^{y''} &= R^{y'} + d^{y'} \\ R^{z''} &= R^{z'} + d^{z'}. \end{aligned} \tag{14.107}$$

Then, substituting into the definition of I^{xx} of Eq. (14.7), for example, and integrating,

$$\begin{aligned} I^{x''x''} &= \int [(R^{y''})^2 + (R^{z''})^2]dm \\ &= \int [(R^{y'} + d^{y'})^2 + (R^{z'} + d^{z'})^2]dm \\ &= \int [(R^{y'})^2 + (R^{z'})^2]dm + 2\left(\int R^{y'}dm\right)d^{y'} \\ &\quad + 2\left(\int R^{z'}dm\right)d^{z'} + (d^{y'})\int dm + (d^{z'})^2\int dm \\ &= I^{x'x'} + 2mR_G^{y'}d^{y'} + 2mR_G^{z'}d^{z'} + m(d^{y'})^2 + m(d^{z'})^2 \end{aligned}$$

Proceeding in this manner through each of Eqs. (14.7) and (14.8), we find

$$\begin{aligned}
I^{x''x''} &= I^{x'x'} = 2mR_G^{y'}d^{y'} + 2mR_G^{z'}d^{z'} + m(d^{y'})^2 + m(d^{z'})^2 \\
I^{y''y''} &= I^{y'y'} = 2mR_G^{z'}d^{z'} + 2mR_G^{x'}d^{x'} + m(d^{z'})^2 + m(d^{x'})^2 \\
I^{z''z''} &= I^{z'z'} = 2mR_G^{x'}d^{x'} + 2mR_G^{y'}d^{y'} + m(d^{x'})^2 + m(d^{y'})^2 \\
I^{x''y''} &= I^{x'y'} = mR_G^{x'}d^{y'} + mR_G^{y'}d^{x'} + md^{x'}d^{y'} \\
I^{y''z'} &= I^{y'z'} = mR_G^{y'}d^{z'} + mR_G^{z'}d^{y'} + md^{y'}d^{z'} \\
I^{z''x''} &= I^{z'x'} = mR_G^{z'}d^{x'} + mR_G^{x'}d^{z'} + md^{z'}d^{x'}.
\end{aligned} \tag{14.108}$$

EXAMPLE 14.7

A circular steel disk is fastened at its center at an angle of 30° to a shaft, as illustrated in Fig. 14.28. Using 0.282 lb/in^3 as the density of steel, find the mass moments and products of inertia about the center of mass and aligned along the axes of the shaft.

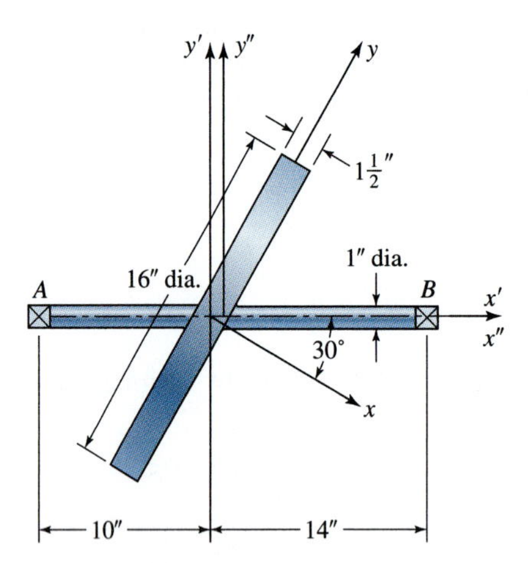

Figure 14.28 Example 14.7. Shaft with swashplate.

SOLUTION

First, we calculate the mass of the disk and the principal mass moments of inertia of the disk using the equations of Table 5 in Appendix A. These are

$$m = \frac{0.282 \text{ lb/in}^3}{386 \text{ in/s}^2}\pi(8 \text{ in})^2(1.5 \text{ in}) = 220 \text{ lb} \cdot \text{s}^2/\text{in}$$

$$I^{xx} = \frac{(0.220 \text{ lb} \cdot \text{s}^2/\text{in})(8 \text{ in})^2}{2} = 7.04 \text{ in} \cdot \text{lb} \cdot \text{s}^2$$

$$I^{yy} = I^{zz} = \frac{(0.220 \text{ lb} \cdot \text{s}^2/\text{in})(8 \text{ in})^2}{4} = 3.52 \text{ in} \cdot \text{lb} \cdot \text{s}^2.$$

The mass products of inertia are zero for the disk because the axes used are principal axes.

From Fig. 14.28 we now find the direction cosines between the coordinate systems. We find these to be

$$\begin{aligned} &\cos\theta_{i'i} = 0.866, && \cos\theta_{i'j} = 0.500, && \cos\theta_{i'k} = 0.000 \\ &\cos\theta_{j'i} = -0.500, && \cos\theta_{j'j} = 0.866, && \cos\theta_{j'k} = 0.000 \\ &\cos\theta_{k'i} = 0.000, && \cos\theta_{k'j} = 0.000, && \cos\theta_{k'k} = 1.000. \end{aligned}$$

Substituting these results into Eqs. (14.105) and (14.106) gives the mass moments and products of inertia aligned with the axes of the shaft. These are

$$\begin{aligned} I^{x'x'} &= (0.866)^2(7.04 \text{ in} \cdot \text{lb} \cdot \text{s}^2) + (0.500)^2(3.52 \text{ in} \cdot \text{lb} \cdot \text{s}^2) = 6.16 \text{ in} \cdot \text{lb} \cdot \text{s}^2 \\ I^{y'y'} &= (-0.500)^2(7.04 \text{ in} \cdot \text{lb} \cdot \text{s}^2) + (0.866)^2(3.52 \text{ in} \cdot \text{lb} \cdot \text{s}^2) = 4.40 \text{ in} \cdot \text{lb} \cdot \text{s}^2 \\ I^{z'z'} &= I^{zz} = 3.52 \text{ in} \cdot \text{lb} \cdot \text{s}^2 \\ I^{x'y'} &= -(0.866)(-0.500)(7.04 \text{ in} \cdot \text{lb} \cdot \text{s}^2) \\ &\quad - (0.500)(0.866)(3.52 \text{ in} \cdot \text{lb} \cdot \text{s}^2) = 1.52 \text{ in} \cdot \text{lb} \cdot \text{s}^2 \\ I^{y'z'} &= I^{z'x'} = 0. \end{aligned}$$

Next, we find the mass of the shaft and the principal mass moments of inertia of the shaft using the equations of Table 5 in Appendix A. These are

$$\begin{aligned} m &= \frac{0.282 \text{ lb/in}^3}{386 \text{ in/s}^2}\pi(0.5 \text{ in})^2(24 \text{ in}) = 0.013\,8 \text{ lb} \cdot \text{s}^2/\text{in} \\ I^{xx} &= \frac{(0.013\,8 \text{ lb} \cdot \text{s}^2/\text{in})(0.5 \text{ in})^2}{2} = 0.001\,72 \text{ in} \cdot \text{lb} \cdot \text{s}^2 \\ I^{yy} = I^{zz} &= \frac{(0.013\,8 \text{ lb} \cdot \text{s}^2/\text{in})[3(0.5 \text{ in})^2 + (24 \text{ in})^2]}{12} = 0.662 \text{ in} \cdot \text{lb} \cdot \text{s}^2. \end{aligned}$$

The mass products of inertia are zero because these are principal axes. Translating the values to the same axes as the disk, Eq. (14.7) gives

$$\begin{aligned} I^{x'x'} &= 0.001\,72 \text{ in} \cdot \text{lb} \cdot \text{s}^2 \\ I^{y'y'} &= I^{z'z'} = (0.662 \text{ in} \cdot \text{lb} \cdot \text{s}^2) + (0.113\,8 \text{ lb} \cdot \text{s}^2/\text{in})(2 \text{ in})^2 = 0.717 \text{ in} \cdot \text{lb} \cdot \text{s}^2. \end{aligned}$$

The values of the disk and the shaft are now both in the same coordinate system and are combined to give

$$
\begin{aligned}
I^{x'x'} &= (6.16 \text{ in} \cdot \text{lb} \cdot \text{s}^2) + (0.001\,72 \text{ in} \cdot \text{lb} \cdot \text{s}^2) = 6.16 \text{ in} \cdot \text{lb} \cdot \text{s}^2 \\
I^{y'y'} &= (4.40 \text{ in} \cdot \text{lb} \cdot \text{s}^2) + (0.717 \text{ in} \cdot \text{lb} \cdot \text{s}^2) = 5.12 \text{ in} \cdot \text{lb} \cdot \text{s}^2 \\
I^{z'z'} &= (3.52 \text{ in} \cdot \text{lb} \cdot \text{s}^2) + (0.717 \text{ in} \cdot \text{lb} \cdot \text{s}^2) = 4.24 \text{ in} \cdot \text{lb} \cdot \text{s}^2 \\
I^{x'y'} &= (1.52 \text{ in} \cdot \text{lb} \cdot \text{s}^2) + 0 = 1.52 \text{ in} \cdot \text{lb} \cdot \text{s}^2 \\
I^{y'z'} &= I^{z'x'} = 0.
\end{aligned}
$$

These are now translated to the center of mass using Eqs. (14.108):

$$
\begin{aligned}
m &= (0.220 \text{ lb} \cdot \text{s}^2/\text{in}) + (0.013\,8 \text{ lb} \cdot \text{s}^2/\text{in}) = 0.234 \text{ lb} \cdot \text{s}^2/\text{in}. \\
R_G^{x'} &= \frac{(0.220 \text{ lb} \cdot \text{s}^2/\text{in})(0) + (0.013\,8 \text{ lb} \cdot \text{s}^2/\text{in})(2.0 \text{ in})}{0.234 \text{ lb} \cdot \text{s}^2/\text{in}} = 0.118 \text{ in} \\
R_G^{y'} &= R_G^{z'} = 0 \\
d^{x'} &= -0.127 \text{ in} \\
d^{y'} &= d^{z'} = 0 \\
I^{x''x''} &= 6.16 \text{ in} \cdot \text{lb} \cdot \text{s}^2 && \textit{Ans.} \\
I^{y''y''} &= (5.12 \text{ in} \cdot \text{lb} \cdot \text{s}^2) + 2(0.235 \text{ lb} \cdot \text{s}^2/\text{in})(0.118)(-0.118) + (0.234 \text{ in})(-0.118 \text{ in})^2 \\
&= 5.12 \text{ in} \cdot \text{lb} \cdot \text{s}^2 && \textit{Ans.} \\
I^{z''z''} &= (4.24 \text{ in} \cdot \text{lb} \cdot \text{s}^2) + 2(0.234 \text{ lb} \cdot \text{s}^2/\text{in})(0.118 \text{ in})(-0.118 \text{ in}) \\
&\quad + (0.234 \text{ in} \cdot \text{lb} \cdot \text{s}^2)(-0.118 \text{ in})^2 \\
&= 4.24 \text{ in} \cdot \text{lb} \cdot \text{s}^2 && \textit{Ans.} \\
I^{x''y''} &= 1.52 \text{ in} \cdot \text{lb} \cdot \text{s}^2 && \textit{Ans.} \\
I^{y''z''} &= I^{z''x''} = 0. && \textit{Ans.}
\end{aligned}
$$

14.12 EULER'S EQUATIONS OF MOTION

In Section 14.2 we presented a detailed derivation that showed the form of Newton's law for a rigid body by integrating over all particles of the body. This demonstrated that the center of mass of the body was the single unique point of the body where Newton's law had the same form as that for a single particle. Repeating Eq. (14.12), this was

$$\sum \mathbf{F}_{ij} = m_j \mathbf{A}_{G_j}. \tag{14.109}$$

Taking much less care with derivation, we then gave the "equivalent" rotational form, Eq. (14.13), as

$$\sum \mathbf{M}_{Gij} = I_{G_j}\boldsymbol{\alpha}_j.$$

Although this equation is valid for all problems with only planar motion, the equation is oversimplified and is *not* always valid for problems with spatial motion. Our task now is to more carefully derive an appropriate equation that *is* valid for problems that include three-dimensional effects.

We start, as before, with Newton's law written for a single differential particle at location P on a rigid body. This can be written as

$$d\mathbf{F} = \mathbf{A}_P dm,$$

where $d\mathbf{F}$ is the net unbalanced force on the particle, $\mathbf{A}_P$ is the absolute acceleration of the particle; and dm is the mass of the particle. Next, we find the net unbalanced moment that this particle contributes to the body by taking the moment of its net unbalanced force about the center of mass of the body, point G:

$$\begin{aligned}\mathbf{R}_{PG} \times d\mathbf{F} &= \mathbf{R}_{PG} \times \mathbf{A}_P dm \\ &= \mathbf{R}_{PG} \times [\mathbf{A}_G + \boldsymbol{\omega} \times (\boldsymbol{\omega} \times \mathbf{R}_{PG}) + \boldsymbol{\alpha} \times \mathbf{R}_{PG}]dm.\end{aligned}$$

On the left-hand side of this equation is the net unbalanced moment contribution of this single particle. If we rearrange the terms on the right-hand side, the equation becomes

$$d\mathbf{M} = \mathbf{R}_{PG} \times \mathbf{A}_G dm + \mathbf{R}_{PG} \times [\boldsymbol{\omega} \times (\boldsymbol{\omega} \times \mathbf{R}_{PG})]dm + \mathbf{R}_{PG} \times (\boldsymbol{\alpha} \times \mathbf{R}_{PG})dm. \qquad (a)$$

We now recall the following vector identity for the triple cross-product of three arbitrary vectors, say $\mathbf{A}$, $\mathbf{B}$, and $\mathbf{C}$,

$$\mathbf{A} \times (\mathbf{B} \times \mathbf{C}) = (\mathbf{A} \cdot \mathbf{C})\mathbf{B} - (\mathbf{A} \cdot \mathbf{B})\mathbf{C}.$$

Using this identity on the second term on the right-hand side of Eq. (a), we have

$$\begin{aligned}d\mathbf{M} = \mathbf{R}_{PG} &\times \mathbf{A}_G dm + \mathbf{R}_{PG} \times (\boldsymbol{\omega} \cdot \mathbf{R}_{PG})\boldsymbol{\omega} dm \\ &- \mathbf{R}_{PG} \times (\boldsymbol{\omega} \cdot \boldsymbol{\omega})\mathbf{R}_{PG} dm + \mathbf{R}_{PG} \times (\boldsymbol{\alpha} \times \mathbf{R}_{PG})dm. \qquad (b)\end{aligned}$$

Now we wish to integrate this equation over all particles of the body. We will look at this term by term. The integral of the moment contributions of the individual particles gives the net externally applied unbalanced moment on the body:

$$\int d\mathbf{M} = \sum \mathbf{M}_{Gij}. \qquad (c)$$

Recognizing that $\mathbf{A}_G$ is common for all particles, we see that the first term on the right-hand side of Eq. (b) integrates to

$$\int \mathbf{R}_{PG} \times \mathbf{A}_G dm = \int \mathbf{R}_{PG} dm \times \mathbf{A}_G = \mathbf{0}, \qquad (d)$$

where we have taken advantage of the definition of the center of mass and noted that $\int \mathbf{R}_{PG} dm = m\mathbf{R}_{GG} = \mathbf{0}$.

The second term on the right-hand side of Eq. (b) can be expanded according to its components along the coordinate axes and integrated. The integrals will be recognized as the mass moments and products of inertia of the body. Thus,

$$\begin{aligned}\int \mathbf{R}_{PG} \times (\boldsymbol{\omega}\cdot\mathbf{R}_{PG})\boldsymbol{\omega} dm &= [-(I_G^{yy} - I_G^{zz})\omega^y\omega^z - I_G^{yz}(\omega^y\omega^y - \omega^z\omega^z) + (I_G^{xy}\omega^z - I_G^{xz}\omega^y)\omega^x]\hat{\mathbf{i}} \\ &= [-(I_G^{zz} - I_G^{xx})\omega^z\omega^x - I_G^{zx}(\omega^z\omega^z - \omega^x\omega^x) + (I_G^{yz}\omega^x - I_G^{yx}\omega^z)\omega^y]\hat{\mathbf{j}} \\ &= [-(I_G^{xx} - I_G^{yy})\omega^x\omega^y - I_G^{xy}(\omega^x\omega^x - \omega^y\omega^y) + (I_G^{zx}\omega^y - I_G^{zy}\omega^x)\omega^z]\hat{\mathbf{k}}.\end{aligned} \qquad (e)$$

Because $\mathbf{R}_{PG} \times \mathbf{R}_{PG} = \mathbf{0}$ for every particle, the third term on the right-hand side of Eq. (b) integrates to

$$(\boldsymbol{\omega}\cdot\boldsymbol{\omega})\int \mathbf{R}_{PG} \times \mathbf{R}_{PG} dm = \mathbf{0}. \qquad (f)$$

The final term on the right-hand side of Eq. (b) is also expanded according to its components along the coordinate axes and integrated. Thus,

$$\begin{aligned}\int \mathbf{R}_{PG} \times (\boldsymbol{\alpha} \times \mathbf{R}_{PG}) dm = &(I_G^{xx}\alpha^x - I_G^{xy}\alpha^y - I_G^{xz}\alpha^z)\hat{\mathbf{i}} \\ &+ (I_G^{yx}\alpha^x - I_G^{yy}\alpha^y - I_G^{yz}\alpha^z)\hat{\mathbf{j}} \\ &+ (I_G^{zx}\alpha^x - I_G^{zy}\alpha^y - I_G^{zz}\alpha^z)\hat{\mathbf{k}}.\end{aligned} \qquad (g)$$

When we now substitute Eqs. (c) through (g) into Eq. (b) and equate components in the $\hat{\mathbf{i}}$, $\hat{\mathbf{j}}$, and $\hat{\mathbf{k}}$ directions, we obtain the following three equations:

$$\begin{aligned}\sum M_{G_{ij}}^x = &I_G^{xx}\alpha^x - I_G^{xy}\alpha^y - I_G^{zx}\alpha^z \\ &+ (I_G^{xy}\omega^z - I_G^{zx}\omega^y)\omega^x - (I_G^{yy} - I_G^{zz})\omega^y\omega^z - I_G^{yz}(\omega^y\omega^y - \omega^z\omega^z) \\ \sum M_{G_{ij}}^y = &-I_G^{xy}\alpha^x + I_G^{yy}\alpha^y - I_G^{yz}\alpha^z \\ &+ (I_G^{yz}\omega^x - I_G^{xy}\omega^z)\omega^y - (I_G^{zz} - I_G^{xx})\omega^z\omega^x - I_G^{zx}(\omega^z\omega^z - \omega^x\omega^x) \\ \sum M_{G_{ij}}^z = &-I_G^{xz}\alpha^x - I_G^{yz}\alpha^y + I_G^{zz}\alpha^z \\ &+ (I_G^{zx}\omega^y - I_G^{yz}\omega^x)\omega^z - (I_G^{xx} - I_G^{yy})\omega^x\omega^y - I_G^{xy}(\omega^x\omega^x - \omega^y\omega^y).\end{aligned} \qquad (14.110)$$

This set of equations is the most general case of the moment equation. It allows for angular velocities and angular accelerations in three dimensions. Its only restriction is that the summation of moments and the mass moments and products of inertia must be taken about the mass center of the body.

If we further require that the mass moments of inertia be measured around the principal axes of the body, then the mass products of inertia are zero. The above equations then reduce to

$$\begin{aligned}
\sum M_{G_{ij}}^{x} &= I_G^{xx}\alpha^x - (I_G^{yy} - I_G^{zz})\omega^y\omega^z \\
\sum M_{G_{ij}}^{y} &= I_G^{yy}\alpha^y - (I_G^{zz} - I_G^{xx})\omega^z\omega^x \\
\sum M_{G_{ij}}^{z} &= I_G^{zz}\alpha^z - (I_G^{xx} - I_G^{yy})\omega^x\omega^y.
\end{aligned} \quad (14.111)$$

This important set of equations is called *Euler's equations of motion*. It should be emphasized that they govern any three-dimensional rotational motion of a rigid body, but that the principal axes of inertia must be used for the x-, y-, and z-component directions.*

EXAMPLE 14.8

The shaft with the angularly mounted disk that was analyzed in Example 14.7 is illustrated again in Fig. 14.29. At the instant shown, the shaft is rotating at a speed of 1500 rev/min cw, and this speed is being reduced at a rate of 100 rev/min/s. The shaft is supported by radial bearings at A and B. Calculate the bearing reactions.

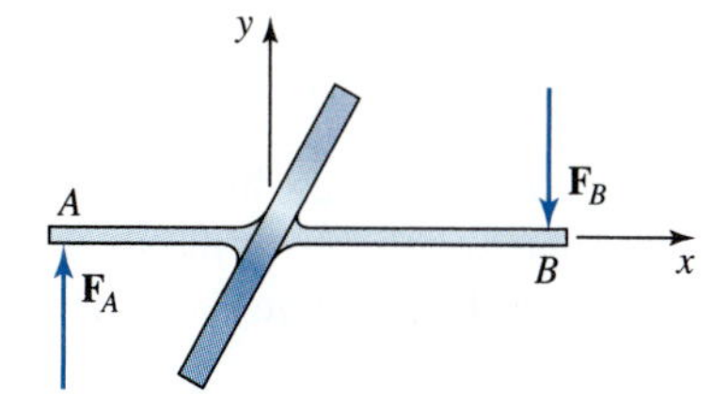

Figure 14.29 Example 14.8. Shaft with swashplate.

SOLUTION

From the choice of axes illustrated in Fig. 14.29 and the data specified in the problem statement, the angular velocity and acceleration, respectively, of the shaft are

$$\boldsymbol{\omega} = -(1500\ \text{rev/min})(2\pi\ \text{rad/rev})/(60\ \text{s/min})\hat{\mathbf{i}} = -157\hat{\mathbf{i}}\ \text{rad/s}$$

$$\boldsymbol{\alpha} = +(100\ \text{rev/min/s})(2\pi\ \text{rad/rev})/(60\ \text{s/min})\hat{\mathbf{i}} = +10.5\hat{\mathbf{i}}\ \text{rad/s}^2.$$

The mass moments and products of inertia were determined in Example 14.7.

Substituting these values into Eqs. (14.110) gives

$$\begin{aligned}
\sum M^x &= (6.16\ \text{in}\cdot\text{lb}\text{-s}^2)(10.5\ \text{rad/s}^2) = 64.7\ \text{in}\cdot\text{lb} \\
\sum M^y &= -(1.52\ \text{in}\cdot\text{lb}\text{-s}^2)(10.5\ \text{rad/s}^2) = -16\ \text{in}\cdot\text{lb} \\
\sum M^z &= -(1.52\ \text{in}\cdot\text{lb}\text{-s}^2)(-157\ \text{rad/s})^2 = -37\ 500\ \text{in}\cdot\text{lb}
\end{aligned}$$

* It should now be recognized that, when we restricted ourselves to motions in the xy plane, it was a special case of Eq. (14.111) that we were using in Eq. (14.13) and what followed.

or

$$\sum \mathbf{M} = 64.7\hat{\mathbf{i}} - 16.0\hat{\mathbf{j}} - 37\,500\hat{\mathbf{k}} \text{ in} \cdot \text{lb}.$$

The $\sum M^x$ component is the result of the shaft torque causing the deceleration. The other two components are provided by forces at the bearings A and B. From these we find, by taking moments about the mass center, that

$$\sum M^y = (10 \text{ in})(F_A^z) - (14 \text{ in})(F_B^z) = -16.0 \text{ in} \cdot \text{lb}$$

$$\sum M^z = -(10 \text{ in})(F_A^y) + (14 \text{ in})(F_B^y) = -37\,500 \text{ in} \cdot \text{lb}.$$

Remembering that $(F_A^y) + (F_B^y) = 0$ and $(F_A^a) + (F_B^z) = 0$ for equilibrium, then we find that

$$F_A^z = -F_B^z = (-16.0 \text{ in} \cdot \text{lb})/(24 \text{ in}) = -0.667 \text{ lb} \approx 0$$

$$F_A^y = -F_B^y = (37\,500 \text{ in} \cdot \text{lb})/(24 \text{ in}) = 1\,560 \text{ lb}.$$

Thus, the two bearing reaction forces on the shaft are

$$\mathbf{F}_A = 1\,560\hat{\mathbf{j}} \text{ lb} \qquad \textit{Ans.}$$

$$\mathbf{F}_B = -1\,560\hat{\mathbf{j}} \text{ lb}. \qquad \textit{Ans.}$$

Note that these bearing reactions are caused entirely by the z component of the inertia moment. This, in turn, is caused by the angular velocity of the shaft and not by its angular acceleration; the bearing reaction loads are still quite large even when the shaft is operating at constant speed.

Note also that this example exhibits only planar motion, but if it had been analyzed by the methods of Section 14.6, we would have found *no* net bearing reaction forces! This should be sufficient warning of the danger of ignoring the third dimension in dynamic-force analysis, based solely on the justification that the *motion* is planar.

14.13 IMPULSE AND MOMENTUM

If we consider that both force and acceleration may be functions of time, we can multiply both sides of Eq. (14.109) by dt and integrate between two chosen times, t_1 and t_2. This results in

$$\sum \int_{t_1}^{t_2} \mathbf{F}_{ij}(t)dt = m_j \int_{t_1}^{t_2} \mathbf{A}_{G_j}(t)dt = m_j\mathbf{V}_{G_j}(t_2) - m_j\mathbf{V}_{G_j}(t_1), \qquad (a)$$

where $\mathbf{V}_{G_j}(t_2)$ and $\mathbf{V}_{G_j}(t_1)$ are the velocities of the centers of mass at times t_2 and t_1, respectively.

The product of the mass and the velocity of the center of mass of a moving body is called its *momentum*. Momentum is a vector quantity and is given the symbol **L**. Thus, for body j, the momentum is

$$\mathbf{L}_j(t) = m_j\mathbf{V}_{G_j}(t). \qquad (14.112)$$

Using this definition, Eq. (a) becomes

$$\sum \int_{t_1}^{t_2} \mathbf{F}_{ij}(t)dt = \mathbf{L}_j(t_2) - \mathbf{L}_j(t_1) = \Delta\mathbf{L}_j. \tag{14.113}$$

The integral of a force over an interval of time is called the *impulse* of the force. Therefore, Eq. (14.113) expresses the *principle of impulse and momentum*, which is that *the total impulse on a rigid body is equal to its change in momentum during the same time interval.* Conversely, *any change in momentum of a rigid body over a time interval is caused by the total impulse of the external forces on that body.*

If we now make the time interval between t_1 and t_2 infinitesimal, divide Eq. (14.113) by this interval, and take the limit, it becomes

$$\sum \mathbf{F}_{ij} = \frac{d\mathbf{L}_j}{dt}. \tag{14.114}$$

Therefore, *the resultant external force acting upon a rigid body is equal to the time rate of change of its momentum.* If there is no net external force acting on a body, then there can be no change in its momentum. Thus, with no net external force,

$$\frac{d\mathbf{L}_j}{dt} = \mathbf{0} \quad \text{or} \quad \mathbf{L}_j = \text{constant} \tag{14.115}$$

is a statement of the law of *conservation of momentum.*

14.14 ANGULAR IMPULSE AND ANGULAR MOMENTUM

When a body translates, its motion is described in terms of its velocity and acceleration, and we are interested in the forces that produce this motion. When a body rotates, we are interested in the moments, and the motion is described by angular velocity and angular acceleration. Similarly, in dealing with rotational motion, we must consider terms such as *angular impulse* or *moment of impulse* and *angular momentum*, also called *moment of momentum*.

Angular momentum is a vector quantity and is usually given the symbol $\mathbf{H}$. For a particle P of mass dm at location $\mathbf{R}_P$ with momentum $d\mathbf{L}$, the moment of momentum $d\mathbf{H}$ of this single particle is defined, as implied by its name, as

$$d\mathbf{H} = \mathbf{R}_P \times d\mathbf{L} = \mathbf{R}_P \times \mathbf{V}_P dm. \tag{a}$$

Analogous to the development in Section 14.12, we can express this equation as

$$\begin{aligned} d\mathbf{H} &= \mathbf{R}_P \times (\mathbf{V}_G + \boldsymbol{\omega} \times \mathbf{R}_{PG})dm \\ &= \mathbf{R}_P \times \mathbf{V}_G dm + \mathbf{R}_P \times (\boldsymbol{\omega} \times \mathbf{R}_{PG})dm \\ &= \mathbf{R}_P \times \mathbf{V}_G dm + \mathbf{R}_G \times (\boldsymbol{\omega} \times \mathbf{R}_{PG})dm + \mathbf{R}_{PG} \times (\boldsymbol{\omega} \times \mathbf{R}_{PG})dm. \end{aligned}$$

This equation can now be integrated over all particles of a rigid body:

$$\int d\mathbf{H} = \int \mathbf{R}_P dm \times \mathbf{V}_G + \mathbf{R}_G \times \left(\boldsymbol{\omega} \times \int \mathbf{R}_{PG} dm\right) + \int \mathbf{R}_{PG} \times (\boldsymbol{\omega} \times \mathbf{R}_{PG})dm. \tag{b}$$

The left-hand side of this equation integrates to give the total angular momentum of the body about the origin of the coordinates:

$$\int d\mathbf{H} = \mathbf{H}_O. \tag{c}$$

Recognizing the definition of the center of mass, we see that the first two terms on the right-hand side of Eq. (*b*) become

$$\int \mathbf{R}_P dm \times \mathbf{V}_G = m\mathbf{R}_G \times \mathbf{V}_G = \mathbf{R}_G \times \mathbf{L} \tag{d}$$

$$\mathbf{R}_G \times \left(\boldsymbol{\omega} \times \int \mathbf{R}_{PG} dm\right) = \mathbf{R}_G \times (\boldsymbol{\omega} \times \mathbf{R}_{GG} m) = \mathbf{0}. \tag{e}$$

As in Section 14.12, the final term of Eq. (*b*) must be expanded according to its components along the coordinate axes and integrated. Thus,

$$\begin{aligned}\int \mathbf{R}_{PG} \times (\boldsymbol{\omega} \times \mathbf{R}_{PG}) dm = {} & (I_G^{xx}\omega^x - I_G^{xy}\omega^y - I_G^{xz}\omega^z)\hat{\mathbf{i}} \\ & + (-I_G^{yx}\omega^x - I_G^{yy}\omega^y - I_G^{yz}\omega^z)\hat{\mathbf{j}} \\ & + (-I_G^{zx}\omega^x - I_G^{zy}\omega^y - I_G^{zz}\omega^z)\hat{\mathbf{k}}.\end{aligned} \tag{f}$$

Upon substituting Eqs. (*c*) through (*f*) into Eq. (*b*), we obtain

$$\begin{aligned}\mathbf{H}_O = {} & \mathbf{R}_G \times \mathbf{L} \\ & + (+I_G^{xx}\omega^x - I_G^{xy}\omega^y - I_G^{xz}\omega^z)\hat{\mathbf{i}} \\ & + (-I_G^{yx}\omega^x - I_G^{yy}\omega^y - I_G^{yz}\omega^z)\hat{\mathbf{j}} \\ & + (-I_G^{zx}\omega^x - I_G^{zy}\omega^y - I_G^{zz}\omega^z)\hat{\mathbf{k}}.\end{aligned} \tag{g}$$

We can recognize that, if we had chosen a coordinate system with origin at the center of mass of the body, then $\mathbf{R}_G = \mathbf{0}$, and the above equation gives the angular momentum as

$$\mathbf{H}_G = H_G^x + H_G^y + H_G^z, \tag{14.116}$$

where

$$\begin{aligned}H_G^x &= +I_G^{xx}\omega^x - I_G^{xy}\omega^y - I_G^{xz}\omega^z \\ H_G^y &= -I_G^{yx}\omega^x + I_G^{yy}\omega^y - I_G^{yz}\omega^z \\ H_G^z &= -I_G^{zx}\omega^x - I_G^{zy}\omega^y + I_G^{zz}\omega^z.\end{aligned} \tag{14.117}$$

However, when moments are taken about a point other than the center of mass, then the moment of momentum about an arbitrary point O is

$$\mathbf{H}_O = \mathbf{H}_G + \mathbf{R}_{GO} \times \mathbf{L}. \tag{14.118}$$

Now, as we did in Eq. (a) above, we can also take the cross-product between the position vector of a particle and the terms of Newton's laws. Using an inertial coordinate system,

$$\sum \mathbf{R}_P \times d\mathbf{F} = \mathbf{R}_P \times \mathbf{A}_P dm \tag{h}$$

If we consider that both force and acceleration may be functions of time, we can multiply both sides of this equation by dt and integrate between t_1 and t_2. The result is

$$\sum \int_{t_1}^{t_2} \mathbf{R}_P \times d\mathbf{F}(t)dt = \mathbf{R}_P \times \mathbf{V}_P(t_2)dm - \mathbf{R}_P \times \mathbf{V}_P(t_1)dm.$$

After integrating over all particles of a rigid body, this yields

$$\sum \int_{t_1}^{t_2} \mathbf{M}_{ij}(t)dt = \mathbf{H}(t_2) - \mathbf{H}(t_1) = \Delta\mathbf{H}. \tag{14.119}$$

The integral on the left-hand side of this equation is the net result of all external moment impulses that occur on the body in the time interval t_1 to t_2, and is called the *angular impulse*. On the right is the change in the moment of angular momentum that occurs over the same time interval as a result of the angular impulse.

If we now make the interval between t_1 and t_2 an infinitesimal, divide Eq. (14.119) by this time interval, and take the limit, it becomes

$$\Sigma\mathbf{M}_{ij} = \frac{d\mathbf{H}}{dt}. \tag{14.120}$$

Therefore, *the resultant external moment acting upon a rigid body is equal to the time rate of change of its angular momentum*. If there is no net external moment acting on a body, then there can be no change in its angular momentum. Thus, with no net external moment,

$$\frac{d\mathbf{H}}{dt} = \mathbf{0} \quad \text{or} \quad \mathbf{H} = \text{constant} \tag{14.121}$$

is a statement of the *law of conservation of angular momentum*.

It would be wise now to review this section mentally and to take careful note of how the coordinate axes may be selected for a particular application. This review will show that, whether stated or not, all results that were derived from Newton's law depend on the use of an inertial coordinate system. For example, the angular velocity used in Eq. (14.117) must be taken with respect to an absolute coordinate system so that the derivative in Eq. (14.120) will contain the required absolute angular acceleration terms. Yet this seems contradictory, because the integration performed in finding the mass moments and products of inertia must often be done in a coordinate system attached to the body itself.

If the coordinate axes are chosen stationary, then the moments and products of inertia used in finding the angular momentum must be transformed to an inertial coordinate system, using the methods of Section 14.10, and will therefore become functions of time. This will complicate the use of Eqs. (14.119), (14.120), and (14.121). For this reason, it is usually preferable to choose the x, y, and z components to be directed along axes fixed in the moving body.

Equations (14.119), (14.120), and (14.121), being vector equations, can be expressed in any coordinate system, including body-fixed axes, and are still correct. This has the great advantage that the mass moments and products of inertia are constants for a rigid body. However, it must be kept in mind that the $\hat{\mathbf{i}}$, $\hat{\mathbf{j}}$, and $\hat{\mathbf{k}}$ unit vectors are moving. They are functions of time in Eq. (14.116), and they have derivatives when using Eqs. (14.120) or (14.121). Therefore, the components of Eqs. (14.121), for example, cannot be found directly by differentiating Eqs. (14.117), but will be

$$\sum M_G^x = I_G^{xx}\alpha^x - I_G^{xy}\alpha^y - I_G^{xz}\alpha^z + \omega^y H^z - \omega^z H^y$$

$$\sum M_G^y = -I_G^{yx}\alpha^x - I_G^{yy}\alpha^y - I_G^{yz}\alpha^z + \omega^z H^x - \omega^x H^z$$

$$\sum M_G^z = -I_G^{zx}\alpha^x - I_G^{zy}\alpha^y + I_G^{zz}\alpha^z + \omega^x H^y - \omega^y H^x. \tag{14.122}$$

Here, we must be careful to find the *components* of $\boldsymbol{\omega}$ and $\boldsymbol{\alpha}$ along the axes of the body, although they are absolute angular velocity and acceleration terms. In addition, we must remember when results are interpreted that these moment components are expressed along the body axes.

EXAMPLE 14.9

Figure 14.30 illustrates a hypothetical problem typical of the situations occurring in the design or analysis of machines in which gyroscopic forces must be considered. A round plate designated body 2 rotates about its central axis with a constant angular velocity $\boldsymbol{\omega}_2 = 5\hat{\mathbf{k}}$ rad/s. Mounted on this revolving round plate are two bearings A and B, which retain a shaft and a mass 3 rotating with an angular velocity $\boldsymbol{\omega}_{3/2} = 350\hat{\mathbf{i}}$ rad/s with respect to the

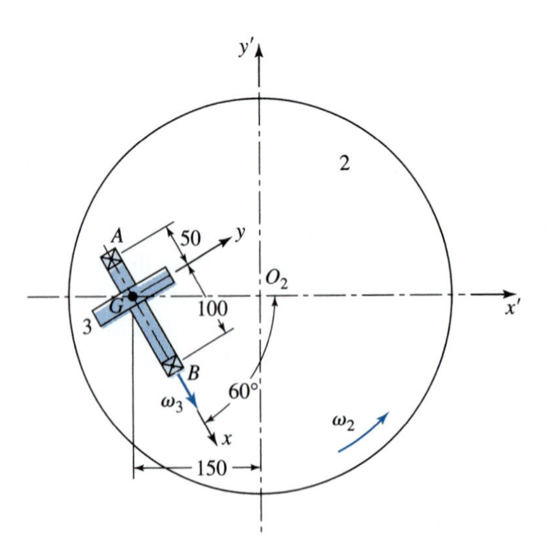

Figure 14.30 Example 14.9. Offset flywheel on a turntable, with dimensions in millimeters.

rotating plate. The rotating body 3 has a mass of 4.5 kg and a radius of gyration of 50 mm about its spin axis. Its center of mass is located at G. Assume that the weight of the shaft is negligible, that body 3 rotates at a constant angular velocity, and that bearing B can only support a radial load. Calculate the bearing reactions at A and B.

SOLUTION

After considering the above discussion, we choose the local body coordinate system illustrated by axes x and y in Fig. 14.30. Note that this coordinate system is fixed to body 2, not body 3, and it is not stationary. Yet because body 3 is assumed to be radially symmetric about this x axis, the mass properties remain constant. Note also that the origin is chosen to be at the center of mass and that the axes are principal axes of inertia.

From the data given and a simple kinematic analysis, we have

$$I_{G_3}^{xx} = m_3(k_3^x)^2 = (4.5\text{ kg})(0.050\text{ m})^2 = 0.011\,25\text{ kg}\cdot\text{m}^2$$

$$\boldsymbol{\omega}_2 = 5\hat{\mathbf{k}}\text{ rad/s}$$

$$\boldsymbol{\omega}_3 = 350\hat{\mathbf{i}} + 5\hat{\mathbf{k}}\text{ rad/s}$$

$$\boldsymbol{\alpha}_2 = \boldsymbol{\alpha}_3 = \mathbf{0}$$

$$\mathbf{R}_{\text{GO}_2} = -(150\cos 60°)\hat{\mathbf{i}} - (150\sin 60°)\hat{\mathbf{j}}\text{ mm}$$

$$= -0.075\hat{\mathbf{i}} - 0.130\hat{\mathbf{j}}\text{ m}$$

$$\mathbf{V}_G = \boldsymbol{\omega}_2 \times \mathbf{R}_{GO_2} = 0.650\hat{\mathbf{i}} - 0.375\hat{\mathbf{j}}\text{ m/s}$$

$$\mathbf{R}_{\text{AG}} = -0.050\hat{\mathbf{i}}\text{ m}$$

$$\mathbf{R}_{\text{BG}} = 0.100\hat{\mathbf{i}}\text{ m}$$

and the time rates of change of the unit vectors are

$$\frac{d\hat{\mathbf{i}}}{dt} = \boldsymbol{\omega}_2 \times \hat{\mathbf{i}} = 5\hat{\mathbf{j}}\text{ rad/s}$$

$$\frac{d\hat{\mathbf{j}}}{dt} = \boldsymbol{\omega}_2 \times \hat{\mathbf{j}} = -5\hat{\mathbf{i}}\text{ rad/s}$$

$$\frac{d\hat{\mathbf{k}}}{dt} = \boldsymbol{\omega}_2 \times \hat{\mathbf{k}} = \mathbf{0}.$$

From Eq. (14.112) we find the momentum, which is

$$\mathbf{L}_3 = m_3\mathbf{V}_G = (4.5\text{ kg})(0.650\hat{\mathbf{i}} - 0.375\hat{\mathbf{j}}\text{ m/s})$$

$$= 2.93\hat{\mathbf{i}} - 1.69\hat{\mathbf{j}}\text{ kg}\cdot\text{m/s}.$$

From Eq. (14.114), we find the dynamic force effects attributable to the motion of the center of mass. Remembering that $\hat{\mathbf{i}}$ and $\hat{\mathbf{j}}$ are moving and have derivatives, this equation gives

$$\sum \mathbf{F}_{i3} = \frac{d\mathbf{L}_3}{dt}$$

$$\mathbf{F}_{A_{23}} + \mathbf{F}_{B_{23}} = 2.93\frac{d\hat{\mathbf{i}}}{dt} - 1.69\frac{d\hat{\mathbf{j}}}{dt} = 8.45\hat{\mathbf{i}} + 14.65\hat{\mathbf{j}}\ \text{N}$$

$$F^x_{A_{23}} + F^x_{B_{23}} = 8.45N$$

$$F^y_{A_{23}} + F^y_{B_{23}} = 14.65N$$

$$F^z_{A_{23}} + F^z_{B_{23}} = 0. \qquad (1)$$

From Eqs. (14.117) and (14.116), we find the angular momentum, which is

$$H^x_{G_3} = (0.011\ 25\ \text{kg}\cdot\text{m}^2)(350\ \text{rad/s}) = 3.94\ \text{kg}\cdot\text{m}^2/\text{s}$$

$$H^y_{G_3} = 0$$

$$H^z_{G_3} = I^{zz}_G(5\ \text{rad/s}) = 5I^{zz}_G\ \text{kg}\cdot\text{m}^2/\text{s}$$

$$\mathbf{H}_{G_3} = 3.94\hat{\mathbf{i}} + 5I^{zz}_G\hat{\mathbf{k}}\ \text{kg}\cdot\text{m}^2/\text{s}.$$

We can find the dynamic moment effects from Eq. (14.120).

$$\sum \mathbf{M}_{G_3} = \frac{d\mathbf{H}_{G_3}}{dt}$$

$$\mathbf{R}_{AG} \times \mathbf{F}_{A_{23}} + \mathbf{R}_{BG} \times \mathbf{F}_{B_{23}} = 3.94\frac{d\hat{\mathbf{i}}}{dt} + 5I^{zz}_G\frac{d\hat{\mathbf{k}}}{dt}\ \text{N}\cdot\text{m}$$

$$(-0.050\hat{\mathbf{i}}\ \text{m}) \times \mathbf{F}_{A_{23}} + (0.100\hat{\mathbf{i}}\ \text{m}) \times \mathbf{F}_{B_{23}} = 3.94(5\hat{\mathbf{j}})\ \text{N}\cdot\text{m}$$

Equating the $\hat{\mathbf{j}}$ and $\hat{\mathbf{k}}$ components, respectively, gives

$$0.050F^z_{A_{23}} - 0.100F^z_{B_{23}} = 19.7\ \text{N}\cdot\text{m}$$

$$-0.050F^y_{A_{23}} + 0.100F^y_{B_{23}} = 0. \qquad (2)$$

Because bearing B can only support a radial load, $F^x_{B_{23}} = 0$. Therefore, solving Eqs. (1) and (2) simultaneously, we find

$$F^x_{A_{23}} = 8.45\ \text{N}$$

$$F^y_{A_{23}} = 9.77\ \text{N}$$

$$F^z_{A_{23}} = 131\ \text{N}$$

$$F^y_{B_{23}} = 4.88\ \text{N}$$

$$F^z_{B_{23}} = -131\ \text{N}.$$

Therefore, the total bearing reactions at A and B are

$$\mathbf{F}_{A_{23}} = 8.45\hat{\mathbf{i}} + 9.77\hat{\mathbf{j}} + 131\hat{\mathbf{k}} \text{ N} \qquad \textit{Ans.}$$

$$\mathbf{F}_{B_{23}} = 4.88\hat{\mathbf{j}} - 131\hat{\mathbf{k}} \text{ N}. \qquad \textit{Ans.}$$

We note that the effect of the gyroscopic couple is to lift the front bearing upward and to push the rear bearing against the plate. We can also note that there is no x component in $\sum \mathbf{M}_{23}$; this justifies our assumption that $\boldsymbol{\alpha}_3 = \mathbf{0}$ although there is no motor controlling that shaft.

The general moment equation is given by Eq. (14.120) and can be written as

$$\sum \mathbf{M}_O = \dot{\mathbf{H}}_O, \tag{14.123}$$

where O denotes either a fixed point or the center of mass of the system. Recall that, in the derivation, the time derivative of the angular momentum is taken with respect to an absolute coordinate system. In many problems it is helpful to express the time derivative of the angular momentum in terms of components measured relative to a moving coordinate system x, y, z, which has an angular velocity $\boldsymbol{\Omega}$. The relation between the time derivative of the angular momentum in the moving x, y, z system and the absolute time derivative can be written as

$$\dot{\mathbf{H}}_O = \left(\frac{d\mathbf{H}_O}{dt}\right)_{xyz} + \boldsymbol{\Omega} \times \mathbf{H}_O.$$

The first term on the right-hand side represents that part of $\dot{\mathbf{H}}_O$ caused by the change in the magnitude of $\mathbf{H}_O$ and the cross-product term represents that part caused by the change in the direction of $\mathbf{H}_O$. An application of this equation to a dynamic problem is demonstrated in the following example.

EXAMPLE 14.10

Figure 14.31 illustrates bevel gear A rolling around the fixed horizontal bevel gear B. Gear A is also free to rotate about the shaft OG with an angular speed of ω_1. The shaft OG is attached to a vertical shaft by a clevis pin at point O and is initially at rest when it is given a constant angular acceleration $\dot{\Omega}$. If the weight of gear A is 20 lb and the weight of shaft OG is negligible, determine the tangential force and the normal force between the two gears as a function of the angle θ.

SOLUTION

The coordinate system illustrated by the x and y axes in Fig. 14.31 is fixed to shaft OG and it is not stationary. Because gear A is assumed to be symmetric about the x axis, its mass properties remain constant. Note also that the origin of the coordinate system is chosen at point O and that the axes are principal axes of inertia.

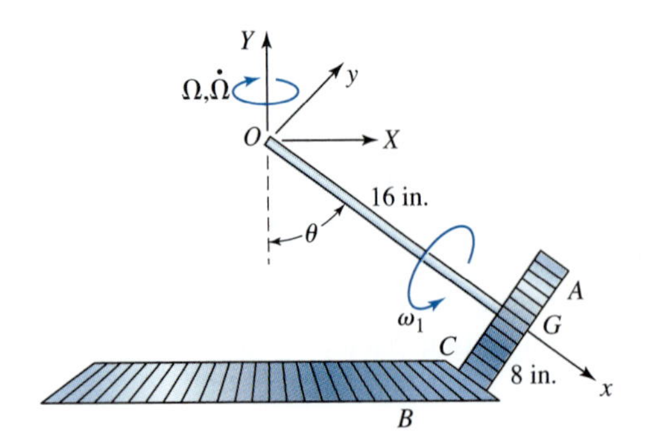

Figure 14.31 Example 14.10. Meshing bevel gears, with dimensions in inches.

The mass moments of inertia of gear A about point O are

$$I_O^{xx} = 1/2mR^2 = 1/2\left(\frac{20\text{ lb}}{386\text{ in}/\text{s}^2}\right)(8\text{ in})^2 = 1.658\text{ in}\cdot\text{lb}\cdot\text{s}^2 \tag{1}$$

$$I_O^{yy} = I_O^{zz} = I_G^{yy} + ml^2 = 1/4mR^2 + ml^2$$
$$= \left(\frac{20\text{ lb}}{386\text{ in}/\text{s}^2}\right)\left(\frac{(8\text{ in})^2}{4} + (16\text{ in})^2\right) = 14.093\text{ in}\cdot\text{lb}\cdot\text{s}^2 \tag{2}$$

and, because these are principal axes, the products of inertia are

$$I_O^{yy} = I_O^{yz} = I_G^{xz} = 0. \tag{3}$$

The angular velocity and angular acceleration of gear A, respectively, are

$$\boldsymbol{\omega} = (\omega_1 + \Omega\cos\theta)\hat{\mathbf{i}} - \Omega\sin\theta\hat{\mathbf{j}}\text{ rad/s} \tag{4}$$

$$\boldsymbol{\alpha} = (\dot{\omega}_1 + \dot{\Omega}\cos\theta)\hat{\mathbf{i}} - \dot{\Omega}\sin\theta\hat{\mathbf{j}} + \omega_1\Omega\sin\theta\hat{\mathbf{k}}\text{ rad/s}^2. \tag{5}$$

The velocity of the point of contact between the two gears, point C, can be written as

$$\mathbf{V}_C = \mathbf{V}_O + \boldsymbol{\omega}\times\mathbf{R}_{CO},$$

where $\mathbf{R}_{CO} = 16\hat{\mathbf{i}} - 8\hat{\mathbf{j}}$. Because the velocity of points O and C are zero, then this equation can be written, with the aid of Eq. (4), as

$$\boldsymbol{\omega}\times\mathbf{R}_{CO} = [-(8\text{ in})(\omega_1 + \Omega\cos\theta) + (16\text{ in})\Omega\sin\theta]\hat{\mathbf{k}} = \mathbf{0}.$$

Rearranging this equation, the angular velocity of gear A relative to the shaft OG is found to be

$$\omega_1 = \Omega(2\sin\theta - \cos\theta). \tag{6}$$

Differentiating this equation with respect to time gives

$$\dot{\omega}_1 = \dot{\Omega}(2\sin\theta - \cos\theta). \tag{7}$$

Substituting Eqs. (6) and (7) into Eqs. (4) and (5), respectively, gives

$$\boldsymbol{\omega} = \Omega \sin\theta(2\hat{\mathbf{i}} - \hat{\mathbf{j}}) \text{ rad/s} \tag{8}$$

$$\boldsymbol{\alpha} = \dot{\Omega} \sin\theta(2\hat{\mathbf{i}} - \hat{\mathbf{j}}) + \Omega^2 \sin\theta(2\sin\theta - \cos\theta)\hat{\mathbf{k}} \text{ rad/s.} \tag{9}$$

From Eq. (14.123), the time rate of change of the angular momentum of gear A can be written as

$$\dot{\mathbf{H}}_O = \frac{\mathbf{H}_O}{dt}\boldsymbol{\omega} \times \dot{\mathbf{H}}_O \tag{10}$$

From Eqs. (14.116) and (14.117), the angular momentum of gear A about the fixed point O can be written as

$$\mathbf{H}_O = I_O^{xx}\omega^x\hat{\mathbf{i}} + I_O^{yy}\omega^y\hat{\mathbf{j}} + I_O^{zz}\omega^z\hat{\mathbf{k}}. \tag{11}$$

Substituting Eqs. (1), (2), and (8) into this equation gives

$$\begin{aligned}\mathbf{H}_O &= 1.658(2\Omega\sin\theta)\hat{\mathbf{i}} + 14.093(-\Omega\sin\theta)\hat{\mathbf{j}} \\ &= \Omega\sin\theta(3.316\hat{\mathbf{i}} - 14.093\hat{\mathbf{j}}) \text{ in}\cdot\text{lb}\cdot\text{s.}\end{aligned} \tag{12}$$

Differentiating Eq. (11) with respect to time gives

$$\frac{d\mathbf{H}_O}{dt} = I_O^{xx}\alpha^x\hat{\mathbf{i}} + I_O^{yy}\alpha^y\hat{\mathbf{j}} + I_O^{zz}\alpha^z\hat{\mathbf{k}}.$$

Substituting Eqs. (1), (2), and (9) into this equation gives

$$\frac{d\mathbf{H}_O}{dt} = 1.658(2\dot{\Omega}\sin\theta)\hat{\mathbf{i}} + 14.093(-\dot{\Omega}\sin\theta)\hat{\mathbf{j}} + 14.093[\Omega^2\sin\theta(2\sin\theta - \cos\theta)]\hat{\mathbf{k}},$$

which can be written as

$$\frac{d\mathbf{H}_O}{dt} = \sin\theta[3.316\dot{\Omega}\hat{\mathbf{i}} - 14.093\dot{\Omega}\hat{\mathbf{j}} + 14.093\Omega^2(2\sin\theta - \cos\theta)\hat{\mathbf{k}}] \text{ in}\cdot\text{lb.} \tag{13}$$

Substituting Eqs. (8), (12), and (13) into Eq. (10) gives

$$\mathbf{H}_O = \sin\theta[3.316\dot{\Omega}\hat{\mathbf{i}} - 14.093\dot{\Omega}\hat{\mathbf{j}} + \Omega^2(3.316\sin\theta - 14.093\cos\theta)\hat{\mathbf{k}}] \text{ in}\cdot\text{lb.} \tag{14}$$

From the free-body diagram of the shaft and the gear illustrated in Fig. 14.32, the sum of the moments about the clevis pin O can be written as

$$\begin{aligned}\sum \mathbf{M}_O &= M_O^y\hat{\mathbf{j}} + \mathbf{R}_{GO} \times 20(\cos\theta\hat{\mathbf{i}} - \sin\theta\hat{\mathbf{j}}) + \mathbf{R}_{CO} \times (N\hat{\mathbf{j}} + f\hat{\mathbf{k}}) \\ &= -8f\hat{\mathbf{i}} + (M_O^y - 16f)\hat{\mathbf{j}} + (-320\sin\theta + 16N)\hat{\mathbf{k}} \text{ in}\cdot\text{lb,}\end{aligned} \tag{15}$$

where f is the tangential force between the two gears at the point of contact C.

Equating the $\hat{\mathbf{i}}$ components of Eqs. (14) and (15) gives

$$-8f = 3.316\sin\theta\dot{\Omega}.$$

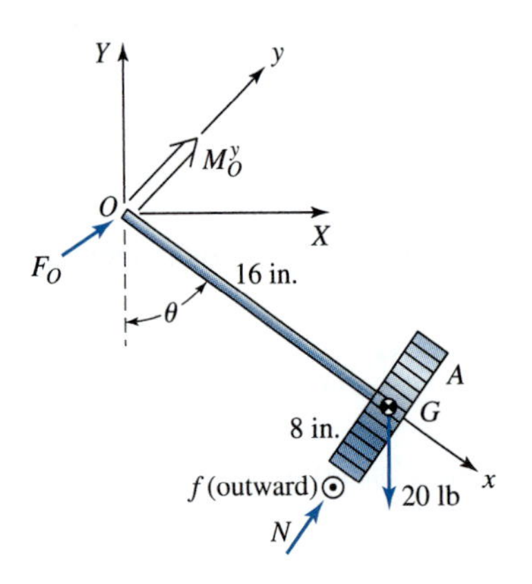

Figure 14.32 Free-body diagram of shaft OG and gear A.

Therefore, the tangential force between the two gears is

$$f = -0.415 \sin\theta\dot{\Omega} \text{ lb} \qquad \textit{Ans.}$$

Equating the $\hat{\mathbf{j}}$ components of Eqs. (14) and (15) gives

$$M_O^y - 16f = -14.093 \sin\theta\dot{\Omega}.$$

Substituting the tangential force into this equation, and rearranging, the moment about point O can be written as

$$M_O^y = -20.725 \sin\theta\dot{\Omega} \text{ in} \cdot \text{lb}.$$

Finally, equating the $\hat{\mathbf{k}}$ components of Eqs. (14) and (15) gives

$$-320 \sin\theta + 16N = \Omega^2 \sin\theta(3.316 \sin\theta - 14.093 \cos\theta).$$

Rearranging this equation, the normal reaction force can be written as

$$N = 20 \sin\theta + \left(\frac{\Omega}{4}\right)^2 \sin\theta(3.316 \sin\theta - 14.093 \cos\theta) \text{ lb}. \qquad \textit{Ans.}$$

Note that the first term on the right-hand side of this equation represents the static gravitational effect, and the second term represents two opposing inertial effects. If gear A did not mesh with the fixed gear B, then the angular velocity ω_1 would be zero and the centripetal acceleration of point G would decrease the normal reaction force N. Because of ω_1, however, a gyroscopic moment is created that has the effect of forcing gear A downward. Note that if $\theta = \tan^{-1}\left(\frac{14.093}{3.316}\right) = 76.76°$, then the two dynamic effects counterbalance each other and the normal reaction force reduces to its static value of $N = 20 \sin\theta$ lb.

14.15 REFERENCES

[1] Additional details can be found in F. E. Fisher and H. H. Alford, *Instrumentation for Mechanical Analysis*, The University of Michigan Summer Conferences, Ann Arbor, MI, 1977, p. 129. The analysis presented here is by permission of the authors.

PROBLEMS*

14.1 The steel bell crank illustrated Fig. P14.1 is used as an oscillating cam follower. Using 0.282 lb/in^3 for the density of steel, find the mass moment of inertia of the lever about an axis through O.

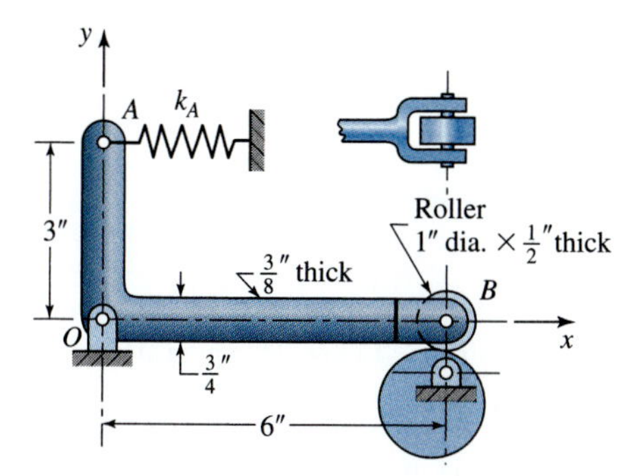

Figure P14.1

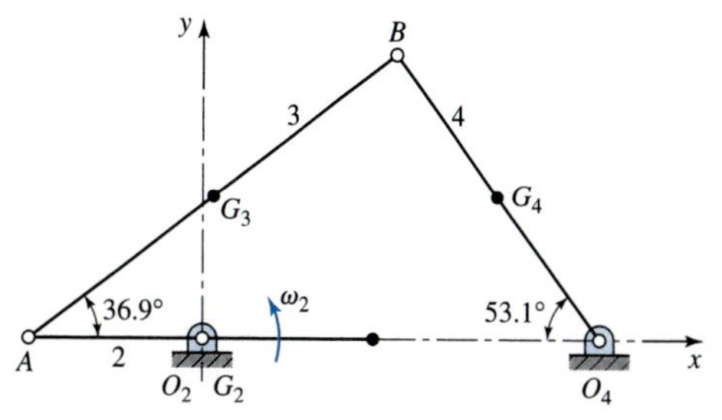

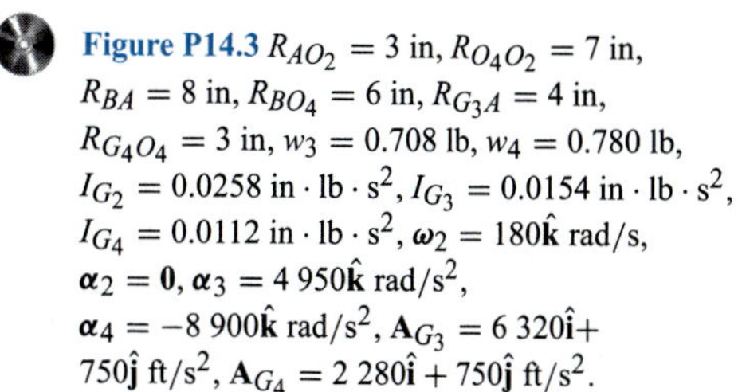

Figure P14.3 $R_{AO_2} = 3$ in, $R_{O_4O_2} = 7$ in, $R_{BA} = 8$ in, $R_{BO_4} = 6$ in, $R_{G_3A} = 4$ in, $R_{G_4O_4} = 3$ in, $w_3 = 0.708$ lb, $w_4 = 0.780$ lb, $I_{G_2} = 0.0258$ in · lb · s^2, $I_{G_3} = 0.0154$ in · lb · s^2, $I_{G_4} = 0.0112$ in · lb · s^2, $\boldsymbol{\omega}_2 = 180\hat{\mathbf{k}}$ rad/s, $\boldsymbol{\alpha}_2 = \mathbf{0}$, $\boldsymbol{\alpha}_3 = 4\,950\hat{\mathbf{k}}$ rad/s^2, $\boldsymbol{\alpha}_4 = -8\,900\hat{\mathbf{k}}$ rad/s^2, $\mathbf{A}_{G_3} = 6\,320\hat{\mathbf{i}} + 750\hat{\mathbf{j}}$ ft/s^2, $\mathbf{A}_{G_4} = 2\,280\hat{\mathbf{i}} + 750\hat{\mathbf{j}}$ ft/s^2.

14.2 A 5- by 50- by 300-mm steel bar has two round steel disks, each 50 mm in diameter and 20 mm long, welded to one end as shown. A small hole is drilled 25 mm from the other end. The density of steel is 7.80 Mg/m^3. Find the mass moment of inertia of this weldment about an axis through the hole.

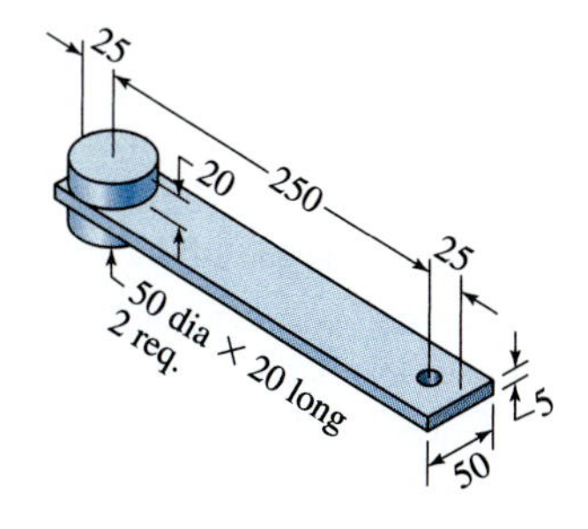

Figure P14.2 Dimensions are in millimeters.

14.3 Find the external torque that must be applied to link 2 of the four-bar linkage shown to drive it at the given velocity.

14.4 Crank 2 of the four-bar linkage illustrated in Fig. P14.4 is balanced. For the given angular velocity of link 2, find the forces acting at each joint and the external torque that must be applied to link 2.

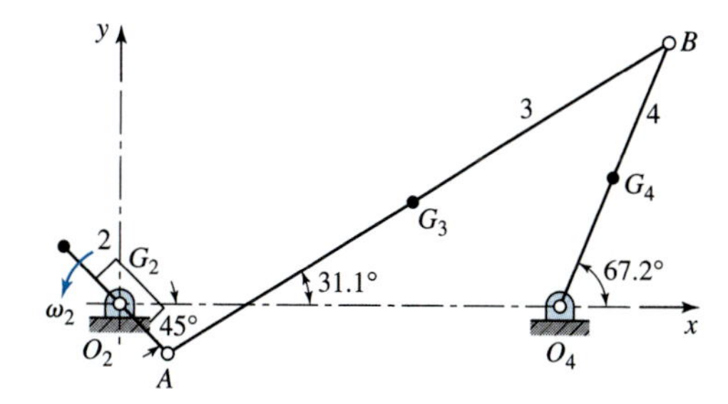

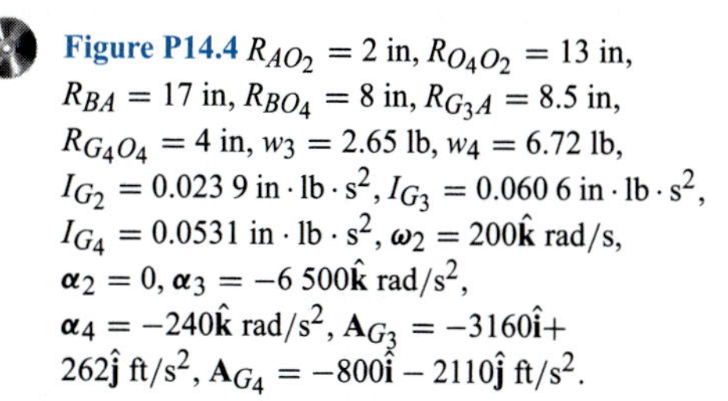

Figure P14.4 $R_{AO_2} = 2$ in, $R_{O_4O_2} = 13$ in, $R_{BA} = 17$ in, $R_{BO_4} = 8$ in, $R_{G_3A} = 8.5$ in, $R_{G_4O_4} = 4$ in, $w_3 = 2.65$ lb, $w_4 = 6.72$ lb, $I_{G_2} = 0.023\,9$ in · lb · s^2, $I_{G_3} = 0.060\,6$ in · lb · s^2, $I_{G_4} = 0.0531$ in · lb · s^2, $\boldsymbol{\omega}_2 = 200\hat{\mathbf{k}}$ rad/s, $\boldsymbol{\alpha}_2 = 0$, $\boldsymbol{\alpha}_3 = -6\,500\hat{\mathbf{k}}$ rad/s^2, $\boldsymbol{\alpha}_4 = -240\hat{\mathbf{k}}$ rad/s^2, $\mathbf{A}_{G_3} = -3160\hat{\mathbf{i}} + 262\hat{\mathbf{j}}$ ft/s^2, $\mathbf{A}_{G_4} = -800\hat{\mathbf{i}} - 2110\hat{\mathbf{j}}$ ft/s^2.

14.5 For the angular velocity of crank 2 given in Fig. P14.5, find the reactions at each joint and the external torque applied to the crank.

---

* Unless instructed otherwise, solve all problems without friction and without gravitational loads.

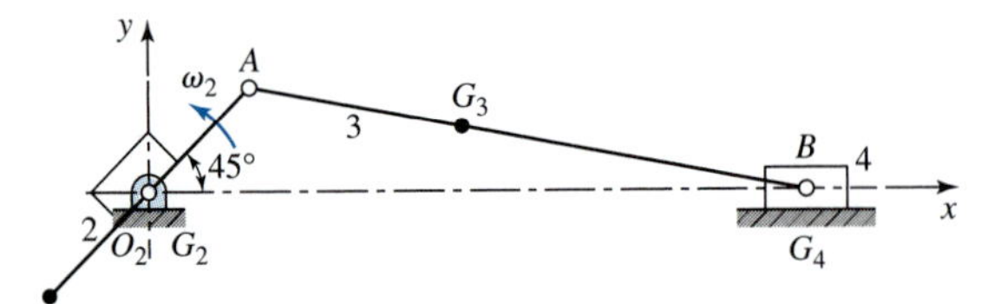

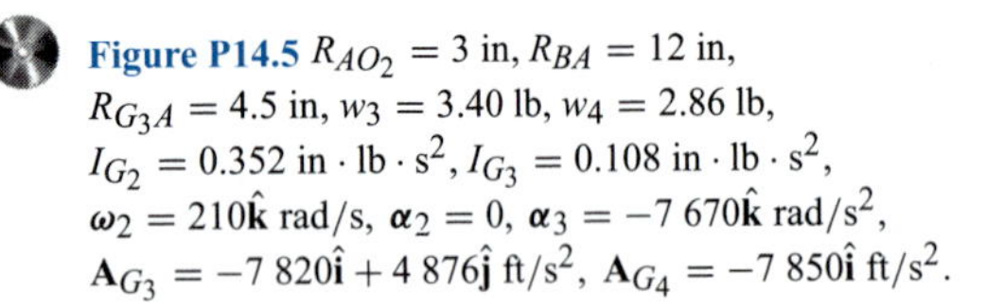

Figure P14.5 $R_{AO_2} = 3$ in, $R_{BA} = 12$ in, $R_{G_3A} = 4.5$ in, $w_3 = 3.40$ lb, $w_4 = 2.86$ lb, $I_{G_2} = 0.352$ in · lb · s^2, $I_{G_3} = 0.108$ in · lb · s^2, $\boldsymbol{\omega}_2 = 210\hat{\mathbf{k}}$ rad/s, $\boldsymbol{\alpha}_2 = 0$, $\boldsymbol{\alpha}_3 = -7\,670\hat{\mathbf{k}}$ rad/s^2, $\mathbf{A}_{G_3} = -7\,820\hat{\mathbf{i}} + 4\,876\hat{\mathbf{j}}$ ft/s^2, $\mathbf{A}_{G_4} = -7\,850\hat{\mathbf{i}}$ ft/s^2.

14.6 Figure P14.6 illustrates a slider–crank mechanism with an external force $\mathbf{F}_B$ applied to the piston. For the given crank velocity, find the reaction forces in the joints and the crank torque.

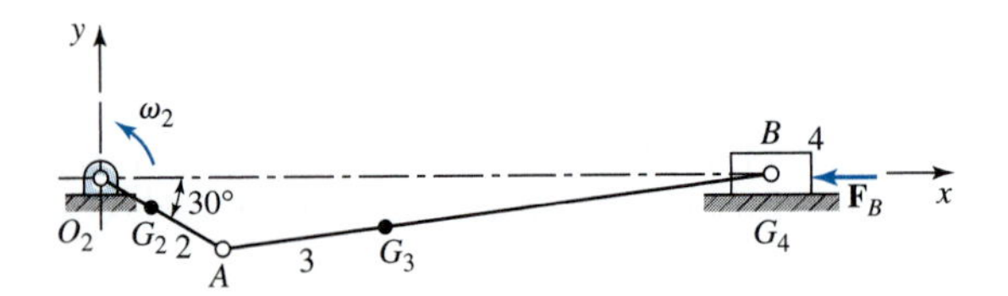

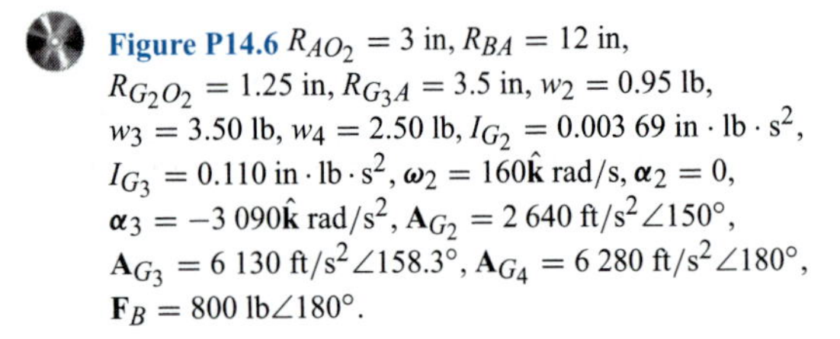

Figure P14.6 $R_{AO_2} = 3$ in, $R_{BA} = 12$ in, $R_{G_2O_2} = 1.25$ in, $R_{G_3A} = 3.5$ in, $w_2 = 0.95$ lb, $w_3 = 3.50$ lb, $w_4 = 2.50$ lb, $I_{G_2} = 0.003\,69$ in · lb · s^2, $I_{G_3} = 0.110$ in · lb · s^2, $\boldsymbol{\omega}_2 = 160\hat{\mathbf{k}}$ rad/s, $\boldsymbol{\alpha}_2 = 0$, $\boldsymbol{\alpha}_3 = -3\,090\hat{\mathbf{k}}$ rad/s^2, $\mathbf{A}_{G_2} = 2\,640$ ft/s$^2\angle 150°$, $\mathbf{A}_{G_3} = 6\,130$ ft/s$^2\angle 158.3°$, $\mathbf{A}_{G_4} = 6\,280$ ft/s$^2\angle 180°$, $\mathbf{F}_B = 800$ lb$\angle 180°$.

14.7 The following data apply to the four-bar linkage illustrated in Fig. P14.7: $R_{AO_2} = 0.3$ m, $R_{O_4O_2} = 0.9$ m, $R_{BA} = 1.5$ m, $R_{BO_4} = 0.8$ m, $R_{CA} = 0.85$ m, $\theta_C = 33°$, $R_{DO_2} = 0.4$ m, $\theta_D = 53°$, $R_{G_2O_2} = 0$, $R_{G_3A} = 0.65$ m, $\alpha = 16°$, $R_{G_4O_4} = 0.45$ m, $\beta = 17°$, $m_2 = 5.2$ kg, $m_3 = 65.8$ kg, $m_4 = 21.8$ kg, $I_{G_2} = 2.3$ kg·m^2, $I_{G_3} = 4.2$ kg·m^2, $I_{G_4} = 2.51$ kg · m^2. A kinematic analysis at $\theta_2 = 53°$, $\boldsymbol{\omega}_2 = 12\hat{\mathbf{k}}$ rad/s ccw, and $\boldsymbol{\alpha}_2 = 0$, gives $\theta_3 = 0.7°$, $\theta_4 = 20.4°$, $\alpha_3 = 85.6$ rad/s^2 cw, $\alpha_4 = 712$ rad/s^2 cw, $\mathbf{A}_{G_3} = 96.4$ m/s$^2\angle 259°$, and $\mathbf{A}_{G_4} = 97.8$ m/s$^2\angle 270°$. Find all pin reactions and the torque applied to link 2.

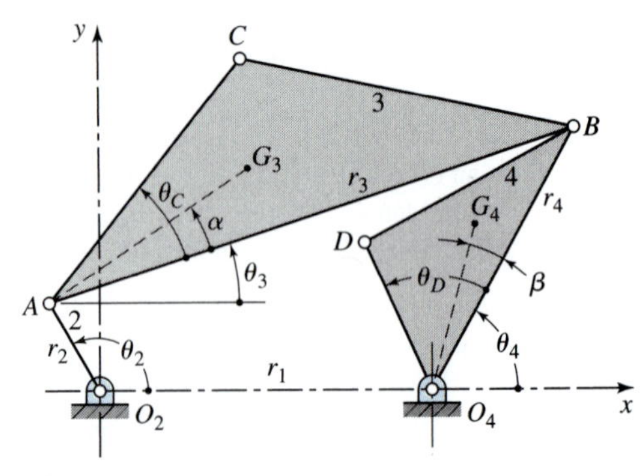

Figure P14.7

14.8 Solve Problem 14.7 with an additional external force $\mathbf{F}_D = 12$ kN$\angle 0°$ acting at point D.

14.9 Make a complete kinematic and dynamic analysis of the four-bar linkage of Problem 14.7 using the same data, but with $\theta_2 = 170°$, $\omega_2 = 12$ rad/s ccw, $\alpha_2 = 0$, and an external force $\mathbf{F}_D = 8.94$ kN$\angle 64.3°$ acting at point D.

14.10 Repeat Problem 14.9 using $\theta_2 = 200°$, $\omega_2 = 12$ rad/s ccw, $\alpha_2 = 0$, and an external force $\mathbf{F}_C = 8.49$ kN$\angle 45°$ acting at point C.

14.11 At $\theta_2 = 270°$, $\omega_2 = 18$ rad/s ccw, $\alpha_2 = 0$, a kinematic analysis of the linkage whose geometry is given in Problem 14.7 gives $\theta_3 = 46.6°$, $\theta_4 = 80.5°$, $\alpha_3 = 178$ rad/s^2 cw, $\alpha_4 = 256$ rad/s^2 cw, $\mathbf{A}_{G_3} = 112$ m/s$^2\angle 22.7°$, $\mathbf{A}_{G_4} = 119$ m/s$^2\angle 352.5°$. An external force $\mathbf{F}_D = 8.94$ kN$\angle 64.3°$ acts at point D. Make a complete dynamic analysis of the linkage.

14.12 The following data apply to a four-bar linkage similar to the one illustrated in Fig. P14.7: $R_{AO_2} = 120$ mm, $R_{O_4O_2} = 300$ mm, $R_{BA} = 320$ mm, $R_{BO_4} = 250$ mm, $R_{CA} = 360$ mm, $\theta_C = 15°$, $R_{DO_4} = 0$, $\theta_D = 0°$, $R_{G_2O_2} = 0$, $R_{G_3A} = 200$ mm, $\alpha = 8°$, $R_{G_4O_4} = 125$ mm, $\beta = 0°$, $m_2 = 0.5$ kg, $m_3 = 4$ kg, $m_4 = 1.5$ kg, $I_{G_2} = 0.005$ N·m·s^2, $I_{G_3} = 0.011$ N·m·s^2, $I_{G_4} = 0.002\,3$ N·m·s^2. A kinematic analysis at $\theta_2 = 90°$, and $\omega_2 = 32$ rad/s ccw with $\alpha_2 = 0$ gives $\theta_3 = 23.9°$, $\theta_4 = 91.7°$, $\alpha_3 = 221$ rad/s^2 ccw, $\alpha_4 = 122$ rad/s^2 ccw, $\mathbf{A}_{G_3} = 88.6$ m/s$^2\angle 255°$, and $\mathbf{A}_{G_4} = 32.6$ m/s$^2\angle 244°$. Using an external force $\mathbf{F}_C = 632$ kN$\angle 342°$ acting at point C, make a complete dynamic analysis of the linkage.

14.13 Repeat Problem 14.12 at $\theta_2 = 260°$. Analyze both the kinematics and the dynamics of the system at this position.

14.14 Repeat Problem 14.13 at $\theta_2 = 300°$.

14.15 Analyze the dynamics of the offset slider–crank mechanism illustrated in Fig. P14.15 using the following data: $a = 0.06$ m, $R_{AO_2} = 0.1$ m, $R_{BA} = 0.38$ m, $R_{CA} = 0.4$ m, $\theta_C = 32°$, $R_{G_3A} = 0.26$ m, $\alpha = 22°$, $m_2 = 2.5$ kg, $m_3 = 7.4$ kg, $m_4 = 2.5$ kg, $I_{G_2} = 0.005$ N · m · s^2, $I_{G_3} = 0.013$ N · m · s^2, $\theta_2 = 120°$, and $\omega_2 = 18$ rad/s cw with $\alpha_2 = 0$, $\mathbf{F}_B = -2\ 000\hat{\mathbf{i}}$ N, $\mathbf{F}_C = -1\ 000\hat{\mathbf{i}}$ N. Assume a balanced crank and no friction forces.

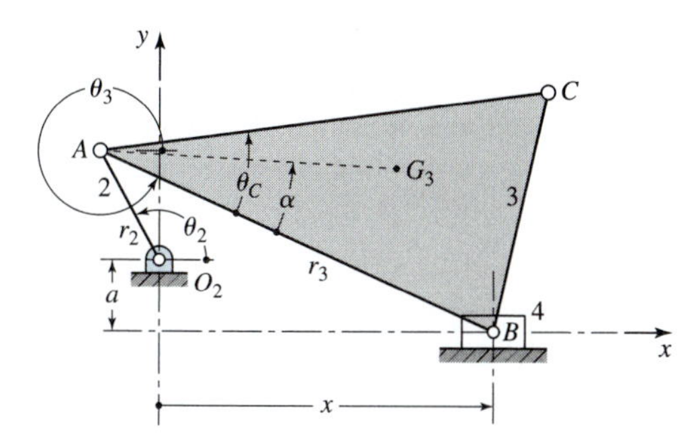

Figure P14.15

14.16 Analyze the system of Problem 14.15 for a complete rotation of the crank. Use $\mathbf{F}_C = \mathbf{0}$ and $\mathbf{F}_B = -1\ 000\hat{\mathbf{N}}$ when $\mathbf{V}_B$ is toward the right, but use $\mathbf{F}_B = \mathbf{0}$ when $\mathbf{V}_B$ is toward the left. Plot a graph of M_{12} versus θ_2.

14.17 A slider–crank mechanism similar to that of Problem 14.15 has $a = 0$, $R_{AO_2} = 0.1$ m, $R_{BA} = 0.45$ m, $R_{CB} = 0$, $\theta_C = 0$, $R_{G_3A} = 0.2$ m, $\alpha = 0$, $m_2 = 1.5$ kg, $m_3 = 3.5$ kg, $m_4 = 1.2$ kg, $I_{G_2} = 0.010$ N · m · s^2, $I_{G_3} = 0.060$ N · m · s^2, and $M_{12} = 60$ N · m. Corresponding to $\theta_2 = 120°$ and $\omega_2 = 24$ rad/s cw with $\alpha_2 = 0$, a kinematic analysis gives $\theta_3 = -9°$, $R_B = 0.374$ m, $\alpha_3 = 89.3$ rad/s^2 ccw, $\mathbf{A}_B = 40.6\hat{\mathbf{i}}$ m/s^2, and $\mathbf{A}_{G_3} = 40.6\hat{\mathbf{i}} - 22.6\hat{\mathbf{j}}$ m/s^2. Assume link 2 is balanced and find $\mathbf{F}_{14}$ and $\mathbf{F}_{23}$.

14.18 Repeat Problem 14.17 for $\theta_2 = 240°$. The results of a kinematic analysis are $\theta_3 = 11.1°$, $R_B = 0.392$ m, $\alpha_3 = 112$ rad/s^2 cw, $\mathbf{A}_B = 35.2\hat{\mathbf{i}}$ m/s^2, and $\mathbf{A}_{G_3} = 31.6\hat{\mathbf{i}} - 27.7\hat{\mathbf{j}}$ m/s^2.

14.19 A slider-crank mechanism, as in Problem 14.15, has $a = 0.008$ m, $R_{AO_2} = 0.25$ m, $R_{BA} = 1.25$ m, $R_{CA} = 1.0$ m, $\theta_C = -38°$, $R_{G_3A} = 0.75$ m, $\alpha = -18°$, $m_2 = 10$ kg, $m_3 = 140$ kg, $m_4 = 50$ kg, $I_{G_2} = 2.0$ N · m · s^2, and $I_{G_3} = 8.42$ N · m · s^2, and has a balanced crank. Make a complete kinematic and dynamic analysis of this system at $\theta_2 = 120°$ with $\omega_2 = 6$ rad/s ccw and $\alpha_2 = 0$, using $\mathbf{F}_B = 50$ kN$\angle 180°$ and $\mathbf{F}_C = 80$ kN$\angle -60°$.

14.20 Cranks 2 and 4 of the cross-linkage illustrated in Fig. P14.20 are balanced. The dimensions of the linkage are $R_{AO_2} = 6$ in, $R_{O_4O_2} = 18$ in, $R_{BA} = 18$ in, $R_{BO_4} = 6$ in, $R_{CA} = 24$ in, and $R_{G_3A} = 12$ in. Also, $w_3 = 4$ lb, $I_{G_2} = I_{G_4} = 0.063$ in · lb · s^2, and $I_{G_3} = 0.497$ in · lb · s^2. Corresponding to the position shown, and with $\omega_2 = 10$ rad/s ccw and $\alpha_2 = 0$, a kinematic analysis gives results of $\omega_3 = 1.43$ rad/s cw, $\omega_4 = 11.43$ rad/s cw, $\alpha_3 = \alpha_4 = 84.8$ rad/s^2 ccw, and $\mathbf{A}_{G_3} = 25.92\hat{\mathbf{i}} + 24.58\hat{\mathbf{j}}$ ft/s^2. Find the driving torque and the pin reactions with $\mathbf{F}_C = -30\hat{\mathbf{j}}$ lb.

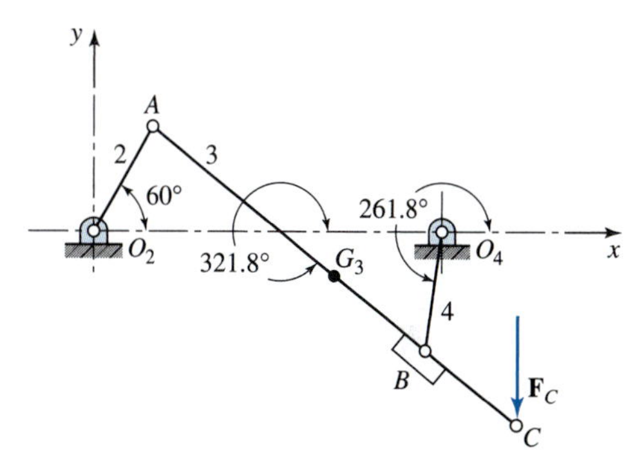

Figure P14.20

14.21 Find the driving torque and the pin reactions for the mechanism of Problem 14.20 under the same dynamic conditions, but with crank 4 as the driver.

14.22 A kinematic analysis of the mechanism of Problem 14.20 at $\theta_2 = 210°$ with $\omega_2 = 10$ rad/s ccw and $\alpha_2 = 0$ gives $\theta_3 = 14.7°$, $\theta_4 = 164.7°$, $\omega_3 = 4.73$ rad/s ccw, $\omega_4 = 5.27$ rad/s cw, $\alpha_3 = \alpha_4 = 10.39$ rad/s cw, and $\mathbf{A}_{G_3} = 26$ ft/s$^2\angle 20.85°$. Compute the crank torque and the pin reactions for this posture using the same force $\mathbf{F}_C$ as in Problem 14.20.

14.23 Figure P14.23 illustrates a linkage with an extended coupler having an external force of $\mathbf{F}_C$ acting during a portion of the cycle. The dimensions of the linkage are $R_{AO_2} = 16$ in, $R_{O_4O_2} = R_{BA} = 40$ in,

$R_{BO_4} = 56$ in, $R_{G_3A} = 32$ in, and $R_{G_4O_2} = 20$ in. Also, $w_3 = 222$ lb, $w_4 = 208$ lb, $I_{G_3} = 226$ in · lb · s^2, and $I_{G_4} = 264$ in · lb · s^2, and the crank is balanced. Make a kinematic and dynamic analysis for a complete rotation of the crank using $\omega_2 = 10$ rad/s ccw, $\mathbf{F}_C = -500\hat{\mathbf{i}} + 886\hat{\mathbf{j}}$ lb for $90° \leq \theta_2 \leq 300°$, $\mathbf{F}_C = \mathbf{0}$ otherwise.

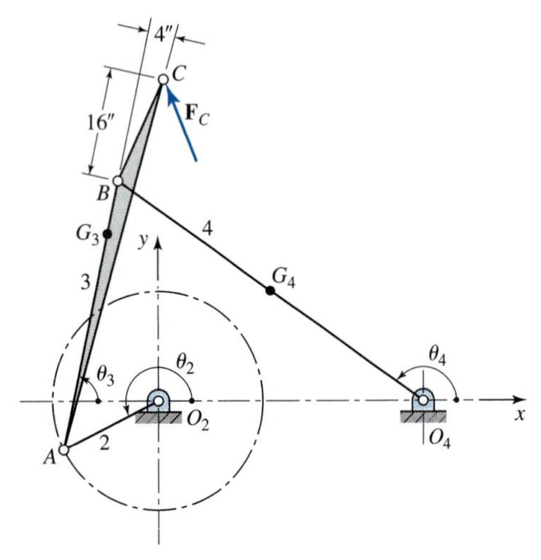

Figure P14.23

14.24 Figure P14.24 illustrates a motor geared to a shaft on which a flywheel is mounted. The mass moments of inertia of the parts are as follows: flywheel, $I = 2.73$ in · lb · s^2; flywheel shaft, $I = 0.0155$ in·lb·s^2; gear, $I = 0.172$ in·lb·s^2; pinion, $I = 0.00349$ in·lb·s^2; motor, $I = 0.0864$ in·lb·s^2. If the motor has a starting torque of 75 in · lb, what is the angular acceleration of the flywheel shaft at the instant the motor is turned on?

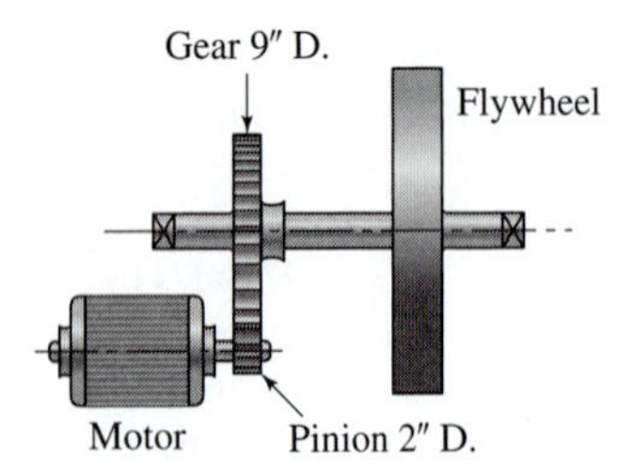

Figure P14.24

14.25 The disk cam of Problem 13.31 is driven at a constant input shaft speed of $\omega_2 = 20$ rad/s ccw. Both the cam and the follower have been balanced so that the centers of mass of each are located at their respective fixed pivots. The mass of the cam is 0.075 kg with radius of gyration of 30 mm and for the follower the mass is 0.030 kg with radius of gyration of 35 mm. Determine the moment $\mathbf{M}_{12}$ required on the camshaft at the instant illustrated in the figure to produce this motion.

14.26 Repeat Problem 14.25 with a shaft speed of $\omega_2 = 40$ rad/s ccw.

14.27 A rotating drum is pivoted at O_2 and is decelerated by the double-shoe brake mechanism illustrated in Fig. P14.27. The mass of the drum is 230 lb and its radius of gyration is 5.66 in. The brake is actuated by force $\mathbf{P} = -100\hat{\mathbf{j}}$ lb, and it is assumed that the contact between the two shoes and the drum act at points C and D, where the coefficients of Coulomb friction are $\mu = 0.300$. Determine the angular acceleration of the drum and the reaction force at the fixed pivot $\mathbf{F}_{12}$.

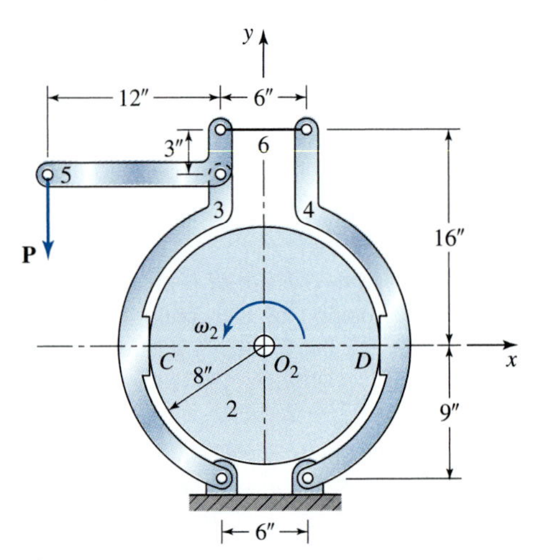

Figure P14.27

14.28 For the mechanism illustrated in Fig. P14.28, the dimensions are $R_{G_2O_4} = 0.15$ m, $R_{EG_2} = 0.20$ m, and the length of link 4 is 0.20 m, symmetric about O_4. The ground bearing is midway between E and G_2. There is a torque T_2 acting on input link 2 and a torque T_4 acting on link 4. Link 2 is in translation with a velocity of $\mathbf{V}_{G_2} = 0.114\ 8\hat{\mathbf{j}}$ m/s and an acceleration of $\mathbf{A}_{G_2} = -0.35\hat{\mathbf{j}}$ m/s and the line connecting mass centers G_2 and G_3 is horizontal. The kinematic coefficients are $\theta_3' = -11.5$ rad/m, $\theta_3'' = -380$ rad/m, $R_{43}' = +2$ m/m, and $R_{43}'' = +40$ m/m^2, (where $\mathbf{R}_{43}$ is the vector from G_3

to G_4). The acceleration of the mass center of link 3 is $\mathbf{A}_{G_3} = +1.053\hat{\mathbf{i}} + 0.432\hat{\mathbf{j}}$ m/s^2. The masses and second moments of mass of the moving links are $m_2 = m_4 = 0.5$ kg, $m_3 = 1$ kg, $I_{G_2} = I_{G_4} = 2$ kg · m^2, and $I_{G_3} = 5$ kg · m^2. Assume that gravity acts in the negative Z direction and that the effects of friction can be neglected. Determine the unknown internal reaction forces and the unknown torques T_2 and T_4.

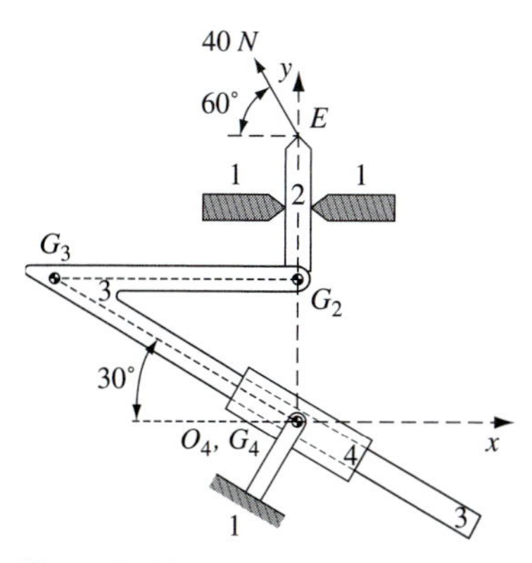

Figure P14.28 A planar mechanism.

14.29 For the mechanism illustrated in Fig. P14.29, the kinematic coefficients are $\theta_3' = -2$ rad/m, $\theta_3'' = -6.928$ rad/m^2, $R_4' = -1.732$ m/m, and $R_4'' = -8$ m/m^2. The velocity and the acceleration of the input link 2 are $\mathbf{V}_{A_2} = -5\hat{\mathbf{j}}$ m/s and $\mathbf{A}_{A_2} = -20\hat{\mathbf{j}}$ m/s^2 and the force acting on link 2 is $\mathbf{F} = -200\hat{\mathbf{j}}$ N. The length of link 3 is $R_{BA} = 1$ m and the distance $R_{G_3G_2} = 0.5$ m. A linear spring is attached between points O and A with a free length $L = 0.5$ m and a spring constant $K = 2\,500$ N/m. A viscous damper with a damping coefficient $C = 45$ N·s/m is connected between the ground and link 4. The masses and mass moments of inertia of the links are $m_2 = 0.75$ kg, $m_3 = 2.0$ kg, $m_4 = 1.5$ kg, $I_{G_2} = 0.25$ N · m · s^2, $I_{G_3} = 1.0$ N · m · s^2 and $I_{G_4} = 0.35$ N·m·s^2. Assume that gravity acts in the negative Y direction (as illustrated in Fig. P14.29) and the effects of friction in the mechanism can be neglected.

(i) Determine the first-order kinematic coefficients of the spring and the viscous damper.

(ii) Determine the equivalent mass of the mechanism.

(iii) Determine the magnitude and direction of the horizontal force P that is acting at point on link 4.

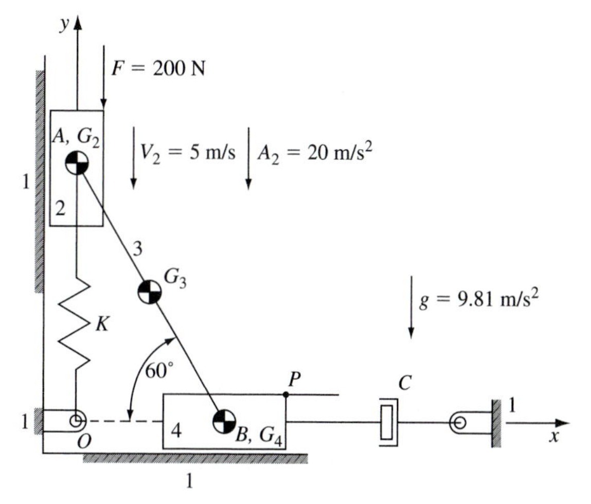

Figure P14.29 A double-slider mechanism.

14.30 Consider the four-bar linkage of Problem 14.20, modified as illustrated in Fig. P14.30. The linkage includes a spring and a viscous damper, as shown. The spring has a stiffness $k = 12$ lb/in and a free length $R_0 = 4.5$ in. The viscous damper has a damping coefficient $C = 0.25$ lb · s/in. The external force acting at point C of coupler link 3 is $F_C = 125$ lb vertically downward (that is, in the negative Y direction). The input crank is rotating with a constant angular velocity $\omega_2 = 10$ rad/s ccw and the angular acceleration of link 3 is $\alpha_3 = 84.8$ rad/s^2 ccw; the acceleration of the mass center of coupler link 3 is $\mathbf{A}_{G_3} = 310\hat{\mathbf{i}} + 295\hat{\mathbf{j}}$ in/s^2. The masses and the second moments of mass are as specified in Problem 14.20 with the exception that the weight of link 3 is $w_3 = 10$ lb. Assume that the locations of the centers of mass of links 2 and 4 are coincident with the ground pivots O_2 and O_4, respectively, and the center of mass of link 3 is as indicated by G_3 in Fig. P14.30. Also assume that gravity acts vertically downward (that is, in the negative Y direction) and the effects of friction in the mechanism can be neglected.

1. Write the equation of motion for the mechanism.
2. Determine the equivalent mass moment of inertia of the mechanism.
3. Determine the driving torque T_2 on the input crank 2 from the equation of motion.

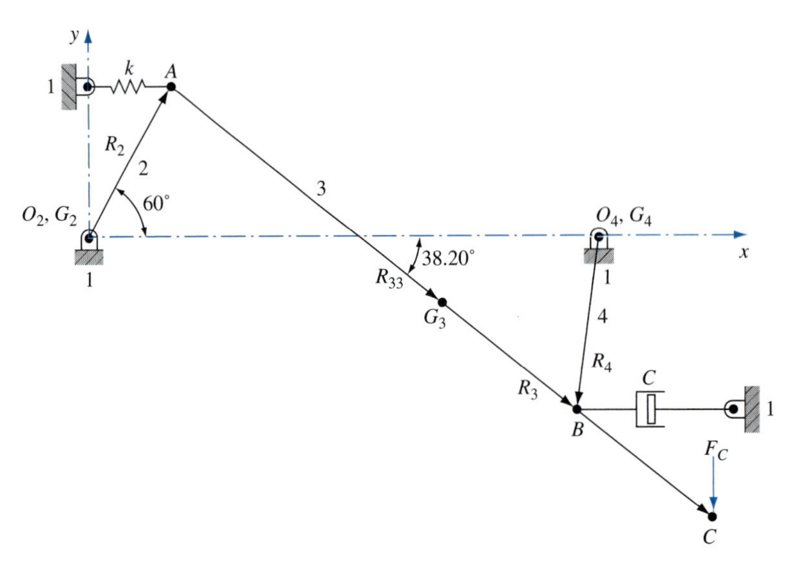

Figure P14.30 The mechanism modified from Problem 14.20.

14.31 For the scotch-yoke mechanism in the position illustrated in Fig. P14.31, an external force $\mathbf{P} = 125\hat{\mathbf{j}}$ N is acting on link 4, and an unknown torque T_2 is acting on the input link 2. The length $R_2 = 1$ m, the angle $\varphi = 30°$, and the angular velocity and acceleration of link 2 are $\boldsymbol{\omega}_2 = 15\,\hat{\mathbf{k}}$ rad/s and $\boldsymbol{\alpha}_2 = 2\,\hat{\mathbf{k}}$ rad/s^2, respectively. The accelerations of the centers of mass of the links are $\mathbf{A}_{G_2} = -5.4\hat{\mathbf{i}} + 11.3\hat{\mathbf{j}}$ m/s^2, $\mathbf{A}_{G_3} = -10.8\hat{\mathbf{i}} + 22.6\hat{\mathbf{j}}$ m/s^2, and $\mathbf{A}_{G_4} = 22.6\hat{\jmath}$ m/s^2. The centers of mass of links 2 and 3 are at the geometric centers of links 2 and 3, respectively. The masses and mass moments of inertia of the links are $m_2 = 5$ kg, $m_3 = 5$ kg, $m_4 = 15$ kg, $I_{G_2} = 0.02$ N · m · s^2, $I_{G_3} = 0.12$ N · m · s^2 and $I_{G_4} = 0.08$ N · m · s^2. Gravity is acting vertically downward (that is, $g = 9.81$ m/s^2 in the negative Y direction). Assume that friction in the mechanism can be neglected. (*i*) Draw free-body diagrams of all moving links of the mechanism. (*ii*) Write the governing equations for all moving links. List all unknown variables. (*iii*) Determine the magnitudes and directions of all internal reaction forces. (*iv*) Determine the magnitude and the direction of the torque T_2. (*v*) Indicate the point(s) of contact of link 4 with ground link 1.

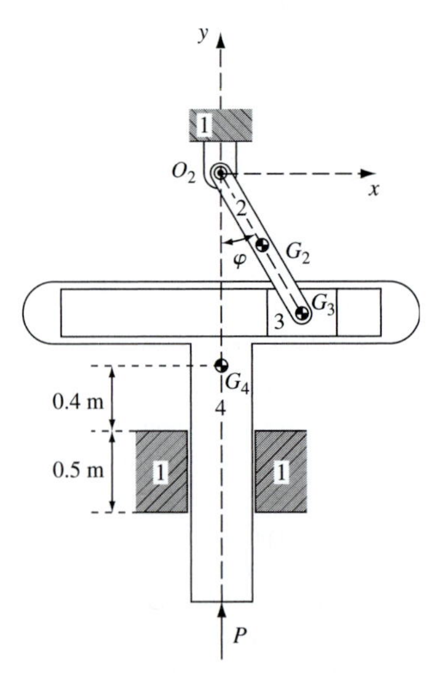

Figure P14.31 A Scotch-yoke mechanism.

14.32 For the parallelogram four-bar linkage in the position illustrated in Fig. P14.32, the angular velocity and acceleration of the input link 2 are $\omega_2 = 2$ rad/s ccw and $\alpha_2 = 1$ rad/s^2 ccw, respectively. The distances $R_{BO_2} = R_{AO_4} = 0.2$ m,

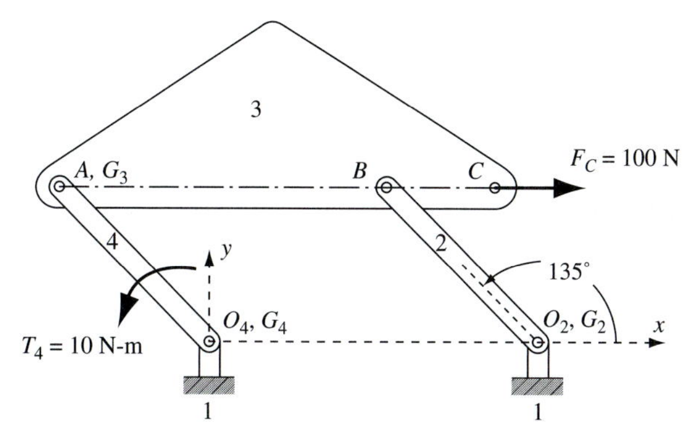

Figure P14.32 A parallelogram four-bar linkage.

$R_{BA} = R_{O_2O_4} = 0.3$ m, and $R_{CB} = 0.1$ m. The force $F_C = 100$ N acts at point C on link 3 in the X direction and a counterclockwise torque $T_4 = 10$ N · m acts on link 4. The masses and the second moments of mass are $m_2 = m_4 = 0.5$ kg, $I_{G_2} = I_{G_4} = 2$ kg · m^2, $m_3 = 1$ kg, and $I_{G_3} = 5$ kg · m^2. The mass centers of links 2 and 4 are coincident with pins O_2 and O_4 and the mass center of link 3 is coincident with pin A. Gravity acts into the page (in the negative Z direction) and friction can be neglected. The first- and second-order kinematic coefficients of the mass center of link 3 are $X'_{G_3} = 0.141$ m/rad, $Y'_{G_3} = -0.141$ m/rad, $X''_{G_3} = 0.141$ m/rad^2, and $Y''_{G_3} = -0.141$ m/rad^2. (*i*) Determine the acceleration of the mass center of link 3. (*ii*) Draw the free-body diagrams for links 2, 3, and 4. List all unknown variables. (*iii*) Determine the magnitudes and the directions of the internal reaction forces $\mathbf{F}_{23}$ and $\mathbf{F}_{43}$. (*iv*) Determine the magnitude and the direction of the input torque $\mathbf{T}_2$.

14.33 For the mechanism in the position shown, the velocity and acceleration of link 2 are $\mathbf{V}_2 = -10\hat{\mathbf{i}}$ m/s and $\mathbf{A}_2 = 10\hat{\mathbf{i}}$ m/s^2, respectively. The length of link 3 is $R_{CG_3} = 3.5$ m and the distance $R_{OG_3} = 2.5$ m. The first- and second-order kinematic coefficients of link 3 are $\theta'_3 = +0.200$ rad/m and $\theta''_3 = -0.1386$ rad/m^2. The force $F_C = 10$ N acts in the negative Y direction at point C on link 3 and the line of action of an unknown force P is acting on link 2 parallel to the X axis, as illustrated in Fig. P14.33. The masses and the second moments of mass of links 2 and 3 are $m_2 = 3$ kg, $m_3 = 5$ kg, $I_{G_2} = 1.5$ kg · m^2, and $I_{G_3} = 7.5$ kg · m^2. Gravity acts in the negative Y direction and there is

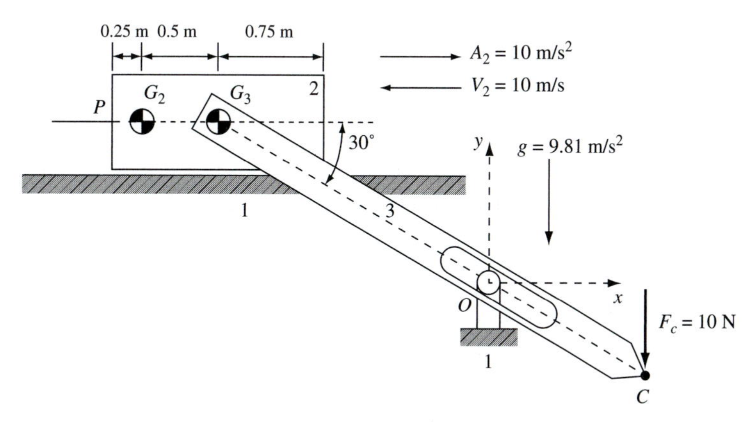

Figure P14.33 A planar mechanism.

no friction in the mechanism. (*i*) Draw the free-body diagrams of links 2 and 3. (*ii*) Write the governing equations for links 2 and 3. List all unknown variables. (*iii*) Determine the magnitudes and directions of all internal reaction forces. (*iv*) Determine the magnitude and sense of the force P acting on link 2. (*v*) Indicate the point(s) of contact of link 2 with ground link 1.

14.34 Consider the slider–crank mechanism of Problem 14.5. The designer proposes to modify this mechanism by including a linear spring and a viscous damper, as illustrated in Fig. P14.34. The spring, with a stiffness $K_S = 20$ lb/in and an unstretched length $r_0 = 3$ in, is attached from the ground to point E on input link 2. In the given position, the spring is parallel to the x axis. The damper with coefficient $C = 7$ lb · s/in is attached between the ground pivot O_2 and pin B on link 4. At the input position $\theta_2 = 45°$, the motor driving input link 2 is applying a torque $T_2 = 60$ lb · in ccw, causing the link to rotate with an angular velocity $\omega_2 = 100$ rad/s ccw and an angular acceleration $\alpha_2 = 10$ rad/s^2 ccw. An external horizontal force $\mathbf{F}_B$ is acting at pin B on link 4, as illustrated in Fig. P14.34. The variable positions, velocities, and accelerations of the mechanism have been determined and are provided in Table P14.34.

Assume: (*i*) Gravity acts in the negative y direction. (*ii*) The location of the center of mass of link 2 is coincident with the ground pivot O_2 and the center of mass of link 4 is coincident with pin B. The location of the center of mass of link 3 is as indicated in Fig. P14.5. (*iii*) The effects of friction in the mechanism can be neglected. (*iv*) The weight and mass moment of inertia of each link are as given in Fig. P14.5. Determine: (*i*) the first- and second-order kinematic coefficients of the mechanism that are necessary for the power equation, (*ii*) the equivalent mass moment of inertia of the mechanism, (*iii*) the equation of motion for the mechanism, and (*iv*) the magnitude and direction of the external force F_B acting on link 4 when the mechanism is in the given position.

14.35 For the mechanism in the position illustrated in Fig. P14.35, the velocity and acceleration of the input link 2 are $\mathbf{V}_{G_2} = 7\hat{\mathbf{i}}$ m/s and $\mathbf{A}_{G_2} = -2\hat{\mathbf{i}}$ m/s^2, respectively. The first- and second-order kinematic coefficients of links 3 and 4 are $\theta_3' = 0$, $\theta_4' = -1.0$ rad/m, $\theta_3'' = 1.0$ rad/m^2, and $\theta_4'' = 1.0$ rad/m^2. The radius of massless link 4, which is rolling on the ground link, is $R_4 = 1$ m, the length of link 3 is $R_{G_2A} = 6$ m, and the radius of the ground link is $R_1 = 2$ m. The free length and spring rate of the spring, the damping constant of the viscous damper, the masses, and the mass moments of inertia of links 2 and 3 (about their mass centers) are as shown in Table P14.35.

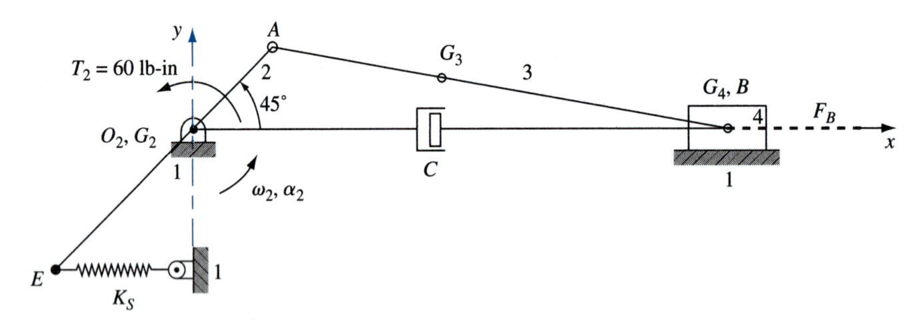

Figure P14.34 The mechanism modified from Problem P14.5.

TABLE P14.34

θ_2 deg	θ_3 deg	R_{AO_2} in	R_{BA} in	R_{EO_2} in	ω_3 rad/s	V_{B4} in/s	α_3 rad/s^2	A_{B4} in/s^2
45	−10.18	3	12	5	−17.96	−250.23	1 736.2	−21 361

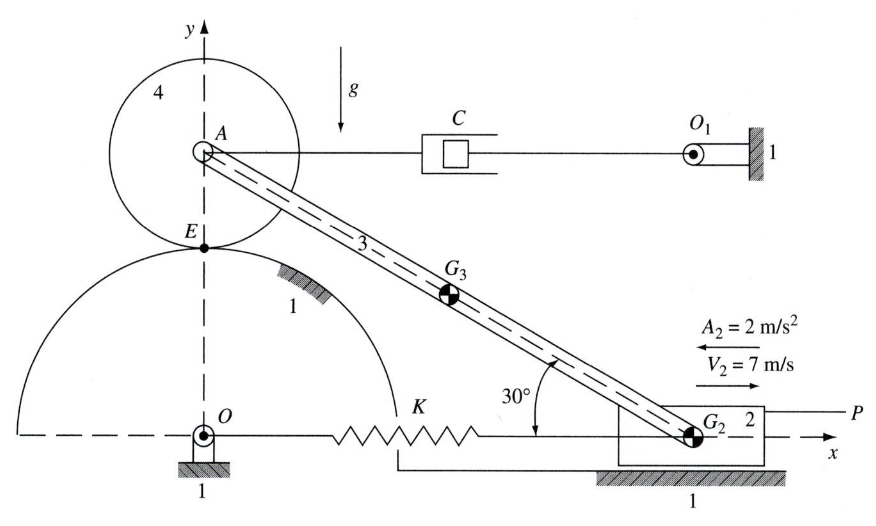

Figure P14.35 A planar mechanism.

TABLE P14.35

R_O m	k N/m	C N·s/m	m_2 kg	m_3 kg	I_{G_2} kg · m²	I_{G_3} kg · m²
3	25	15	1.20	0.80	0.25	0.10

Assume that gravity acts in the negative Y direction and the effects of friction can be neglected. Determine: (*i*) the kinematic coefficients r'_S, r'_C, x'_{G_3}, y'_{G_3}, x''_{G_3}, and y''_{G_3}, (*ii*) the equivalent mass of the mechanism, (*iii*) the equation of motion for the mechanism in symbolic form, and (*iv*) the magnitude and direction of the horizontal external force P acting on link 2.

14.36 For the mechanism in the position illustrated in Fig. P14.36, link 3 is horizontal. The constant angular velocity of the input link 2, which rolls without slipping on the inclined plane, is $\omega_2 = 20$ rad/s cw. The first- and second-order kinematic coefficients of links 3 and 4 are $\theta'_3 = -0.125$ rad/rad, $R'_4 = 1.299$ m/rad, $\theta''_3 = 0$, and $R''_4 = 0.094$ m/rad². The radius of link 2 is $R = 1.5$ m, the length of link 3 is $R_{BA} = R_{G_4G_2} = 6$ m, and $R_{AO_S} = 2.5$ m. The free length of the spring is 3 m, the spring rate is $k = 25$ N/m, and the damping constant of the viscous damper is $C = 15$ N · s/m. The masses and mass moments of inertia of links 2 and 4 are $m_2 = 7$ kg, $m_4 = 4$ kg, $I_{G_2} = 1.8$ kg·m², and $I_{G_4} = 22$ kg · m². Assume that the mass of link 3 is negligible compared with the masses of links 2 and 4, the effects of friction can be neglected, and gravity acts vertically downward, as illustrated in Fig. P14.36. Determine: (*i*) the first- and second-order kinematic coefficients of the mass centers of links 2 and 4, (*ii*) the equivalent mass moment of inertia of the mechanism, and (*iii*) the magnitude and direction of the input torque acting on link 2.

14.37 Figure P14.37 illustrates a two-throw opposed-crank crankshaft mounted in bearings at A and G. Each crank has an eccentric weight of 6 lb, which may be considered as located at a radius of 2 in. from the axis of rotation and at the center of each throw (points C and E). It is proposed to locate weights at B and F to reduce the bearing reactions, caused by the rotating eccentric cranks, to zero. If these weights are to be mounted 3 in. from the axis of rotation, how much must they weigh?

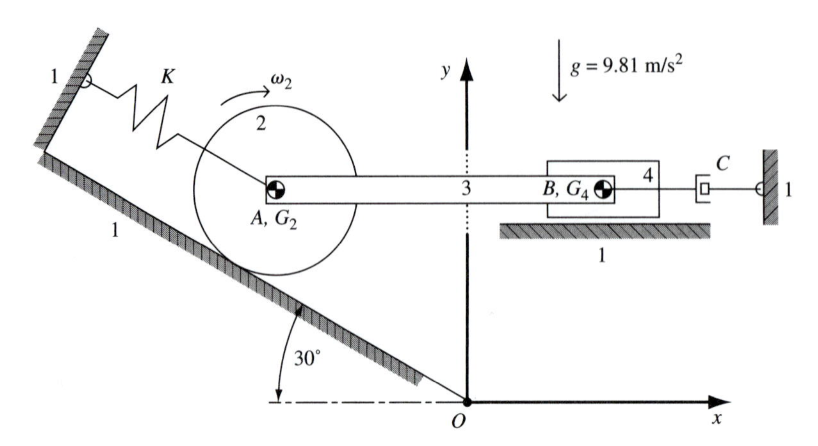

Figure P14.36 A planar mechanism.

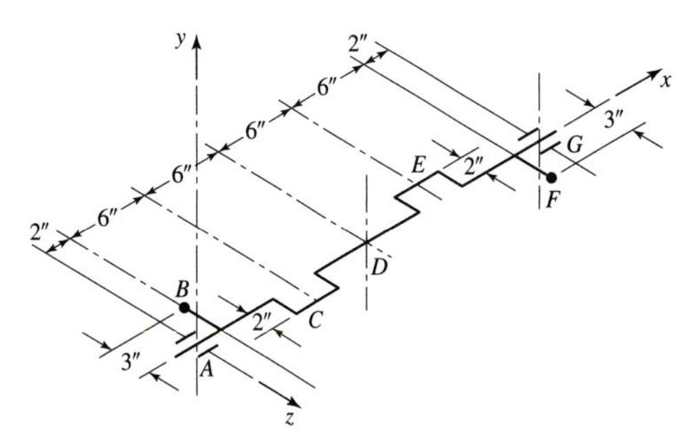

Figure P14.37

14.38 Figure P14.38 illustrates a two-throw crankshaft, mounted in bearings at A and F,with the cranks spaced 90° apart. Each crank may be considered to have an eccentric weight of 6 lb at the center of the throw and 2 in. from the axis of rotation. It is proposed to eliminate the rotating bearing reactions, which the crank would cause, by mounting additional correction weights on 3-in. arms at points B and E. Calculate the magnitudes and angular locations of these weights.

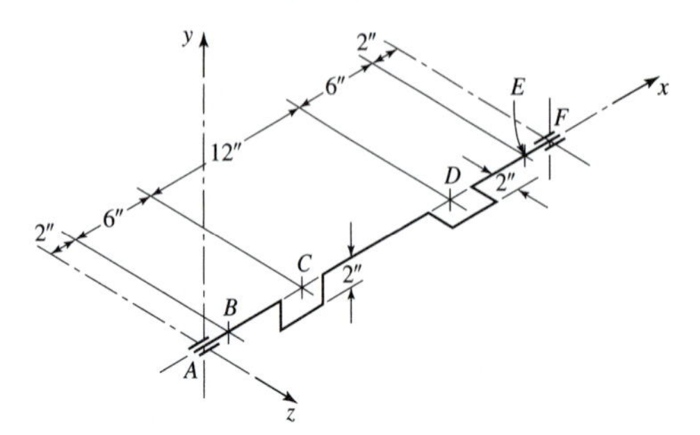

Figure P14.38

14.39 Solve Problem 14.38 with the angle between the two throws reduced from 90° to 0°.

14.40 The connecting rod illustrated in Fig. P14.40 weighs 7.90 lb and is pivoted on a knife edge and caused to oscillate as a pendulum. The rod is observed to complete 64.5 oscillations in 1 min. Determine the mass moment of inertia of the rod about its own center of mass.

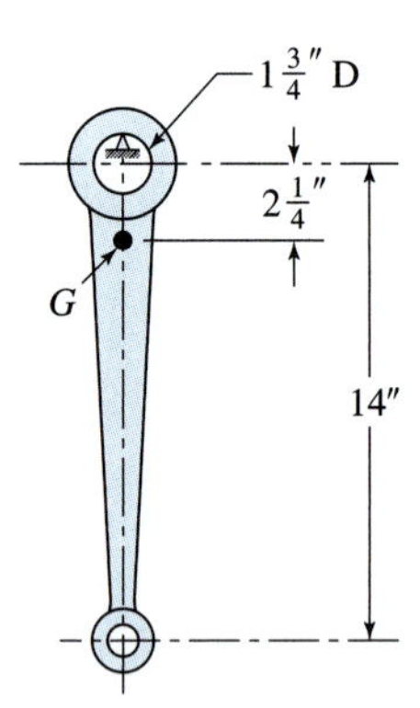

Figure P14.40

14.41 A gear is suspended on a knife edge at the rim as illustrated in Fig. P14.41 and caused to oscillate as a pendulum. Its period of oscillation is observed to be 1.08 s. Assume that the center of mass and the axis of rotation are coincident. If the weight of the gear is 41 lb, find the mass moment of inertia and the radius of gyration of the gear.

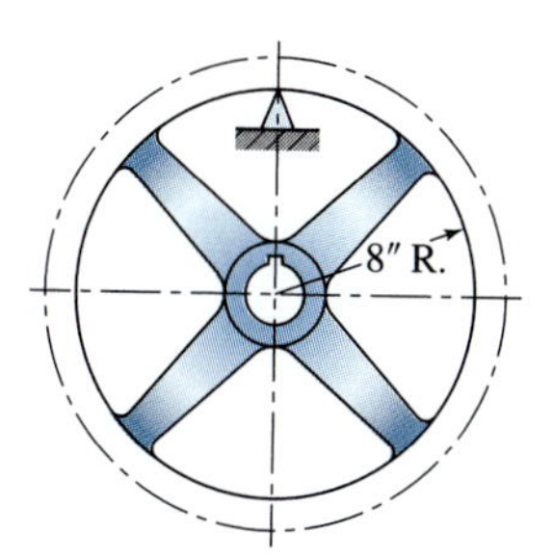

Figure P14.41

14.42 Figure P14.42 illustrates a wheel whose mass moment of inertia I is to be determined. The wheel is mounted on a shaft in bearings with very low frictional resistance to rotation. At one end of the shaft and on the outboard side of the bearings is connected a rod with a weight W_b secured to its end. It is possible to measure the mass moment of inertia of the wheel by displacing the weight W_b from equilibrium and permitting the assembly to oscillate. If the weight of the pendulum arm is neglected, show that the mass moment of inertia of the wheel can be obtained from the equation

$$I = W_b l \left(\frac{\tau^2}{4\pi^2} - \frac{l}{g} \right).$$

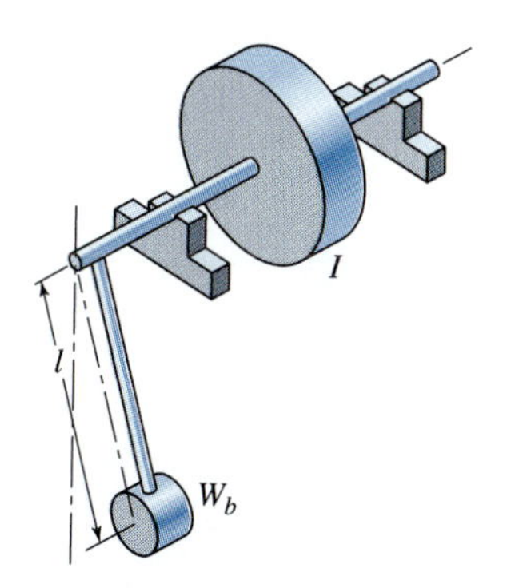

Figure P14.42

14.43 If the weight of the pendulum arm is not neglected in Problem 14.42, but is assumed to be uniformly distributed over the length l, show that the mass moment of inertia of the wheel can be obtained from the equation

$$I = l \left[\frac{\tau^2}{4\pi^2} \left(W_b + \frac{W_a}{2} \right) - \frac{l}{g} \left(W_b + \frac{W_a}{3} \right) \right],$$

where W_a is the weight of the arm.

14.44 Wheel 2 in Fig. P14.44 is a round disk that rotates about a vertical axis z through its center. The wheel carries a pin B at a distance R from the axis of rotation of the wheel, about which link 3 is free to rotate. Link 3 has its center of mass G located at a distance r from the vertical axis through B, and it has a weight W_3 and a mass moment of inertia I_G about its own mass center. The wheel rotates at an angular velocity ω_2 with link 3 fully extended. Develop an expression for the angular velocity ω_3 that link 3 would acquire if the wheel were suddenly stopped.

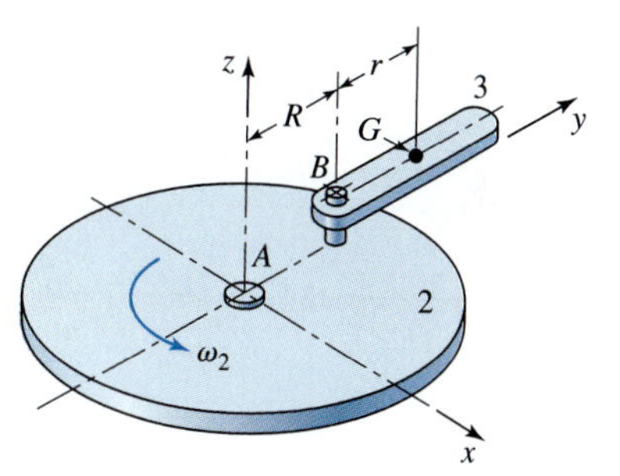

Figure P14.44

14.45 Repeat Problem 14.44, except assume that the wheel rotates with link 3 radially inward. Under these conditions, is there a value for the distance r for which the resulting angular velocity ω_3 is zero?

14.46 Figure P14.46 illustrates a planetary gear-reduction unit that utilizes 7-pitch spur gears cut on the 20° full-depth system. All parts are steel with density 0.282 lb/in^3. The arm is rectangular and is 4 in. wide by 14 in. long with a 4-in. diameter central hub and two 3-in. diameter planetary hubs. The segment separating the planet gears is a 0.5 × 4 in. diameter cylinder. The inertia of the gears can be obtained by treating them as cylinders equal in diameter to their respective pitch circles. The input to the reducer is driven with 25 hp at 600 rev/min. The mass moment of inertia of the resisting load is 5.83 in · lb · s^2. Calculate the bearing reactions on the input, output, and planetary shafts. As a designer, what forces would you use in designing the mounting bolts? Why?

14.47 It frequently happens in motor-driven machinery that the greatest torque is exerted when the motor is first turned on, because of the fact that some motors are capable of delivering more starting torque than running torque. Analyze the bearing reactions of

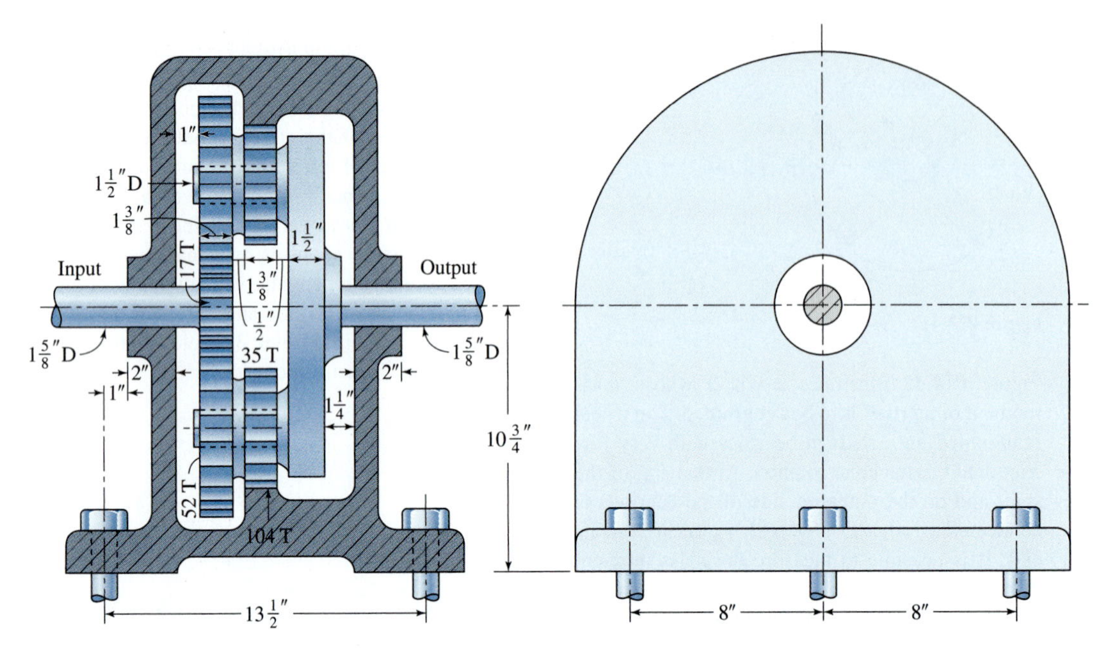

Figure P14.46

Problem 14.46 again, but this time use a starting torque equal to 250% of the full-load torque. Assume a normal-load torque and a speed of zero. How does this starting condition affect the forces on the mounting bolts?

14.48 The gear-reduction unit of Problem 14.46 is running at 600 rev/min when the motor is suddenly turned off, without changing the resisting-load torque. Solve Problem P14.46 for this condition.

14.49 The differential gear train illustrated in Fig. P14.49 has gear 1 fixed and is driven by rotating shaft 5 at 500 rev/min in the direction shown. Gear 2 has fixed bearings constraining it to rotate about the positive y axis, which remains vertical; this is the output shaft. Gears 3 and 4 have bearings connecting them to the ends of the carrier arm, which is integral with shaft 5. The pitch diameters of gears 1 and 5 are both 8.0 in., whereas the pitch diameters of gears 3 and 4 are both 6.0 in. All gears have 20° pressure angles and are each 0.75 in. thick, and all are made of steel with density 0.286 lb/in^3. The mass of shaft 5 and all gravitational loads are negligible. The output shaft torque loading is $\mathbf{T} = -100\hat{\mathbf{j}}$ ft·lb as shown. Note that the coordinate axes shown rotate with input shaft 5. Determine the driving torque required and the forces and moments in each of the bearings. (*Hint*: It is reasonable to assume through symmetry that $F^t_{13} = F^t_{14}$. It is also necessary to recognize that only compressive loads, not tension, can be transmitted between gear teeth.)

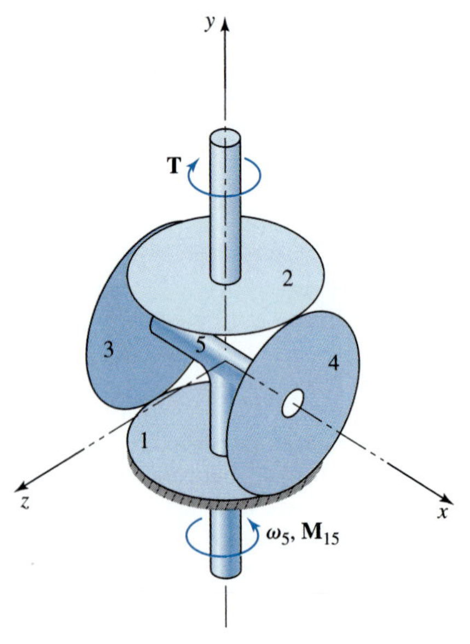

Figure P14.49

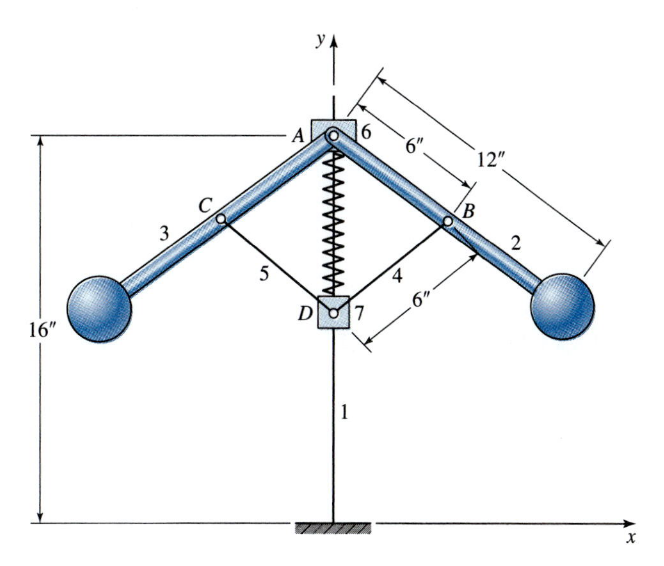

Figure P14.50

14.50 Figure P14.50 illustrates a flyball governor. Arms 2 and 3 are pivoted to block 6, which remains at the height shown but is free to rotate around the y axis. Block 7 also rotates about and is free to slide along the y axis. Links 4 and 5 are pivoted at both ends between the two arms and block 7. The two balls at the ends of links 2 and 3 weigh 3.5 lb each, and all other masses are negligible in comparison; gravity acts in the $-\hat{\mathbf{j}}$ direction. The spring between links 6 and 7 has a stiffness of 1.0 lb/in. and would be unloaded if block 7 were at a height of $\mathbf{R}_D = 11\hat{\mathbf{j}}$ in. All moving links rotate about the y axis with angular velocities of $\omega\hat{\mathbf{j}}$. Make a graph of the height R_D versus the rotational speed ω rev/min, assuming that changes in speed are slow.

15 Vibration Analysis

The existence of vibrating elements in any mechanical system produces unwanted noise, high stresses, wear, poor reliability, and, frequently, premature failure of one or more of the parts. The moving parts of all machines are inherently vibration producers, and for this reason engineers must expect vibrations to exist in the devices they design. But there is a great deal they can do during the design of the system to anticipate a vibration problem and to minimize its undesirable effects.

Sometimes it is necessary to build a vibratory system into a machine—a vibratory conveyor, for example. Under these conditions the engineer must understand the mechanics of vibration to obtain an optimal design.

15.1 DIFFERENTIAL EQUATIONS OF MOTION

Any motion that exactly repeats itself after a certain interval of time is a periodic motion and is called a *vibration*. Vibrations may be either free or forced. A mechanical element is said to have a *free vibration* if the periodic motion continues after the cause of the original disturbance is removed, but if a vibratory motion persists because of the continuing existence of a disturbing force, then it is called a *forced vibration*. Any free vibration of a mechanical system will eventually cease because of loss of energy. In vibration analysis we often take account of these energy losses using a single factor called the *damping factor*. Thus, a heavily damped system is one in which the vibration decays rapidly. The *period* of a vibration is the time for a single event or cycle; the *frequency* is the number of cycles or periods occurring in unit time. The *natural frequency* is the frequency of a free vibration. If the forcing frequency becomes equal to the natural frequency of a system, then *resonance* is said to occur.

We shall also use the term *steady-state vibration* to indicate that a motion is repeating itself exactly in each successive cycle and *transient vibration* to indicate a vibratory-type motion that is changing in character. If a periodic force operates on a mechanical system, the resulting motion will be transient in character when the force first begins to act, but after an interval of time the transient will decay, owing to damping, and the resulting motion is termed a *steady-state vibration*.

The word *response* is frequently used in discussing vibratory systems. The words *response*, *behavior*, and *performance* have roughly the same meaning when used in dynamic analysis. Thus, we can apply an external force having a sine-wave relationship with time to a vibrating system to determine how the system "responds," or "behaves," when the frequency of the force is varied. A plot using the vibration amplitude along the ordinate and the forcing frequency along the abscissa is then described as a *performance* or *response* curve for the system. Sometimes it is useful to apply arbitrary input disturbances or forces to a system. These may not resemble the force characteristics that a real system would receive in use at all; yet the response of the system to these arbitrary disturbances can provide much useful information about the system.

Vibration analysis is sometimes called *elastic-body analysis* or *deformable-body analysis*, because, as we shall see, a mechanical system must have elasticity to allow vibration. When a rotating shaft has a torsional vibration, this means that a mark on the circumference at one end of the shaft is successively ahead of and then behind a corresponding mark on the other end of the shaft. In order words, torsional vibration of a shaft is the alternate twisting and untwisting of the rotating material and requires elasticity for its existence. We shall begin our study of vibration by assuming that elastic parts have no mass and that heavy parts are absolutely rigid, that is, they have no elasticity. Of course these assumptions are never true, and so, in the course of our studies, we must also learn to correct for the effects of making these assumptions.

Figure 15.1 illustrates an idealized vibrating system having a mass m guided to move only in the x direction. The mass is connected to a fixed frame through the spring k and the dashpot c. The assumptions used are as follows:

1. The spring and the dashpot are massless.
2. The mass is absolutely rigid.
3. All damping is concentrated in the dashpot.

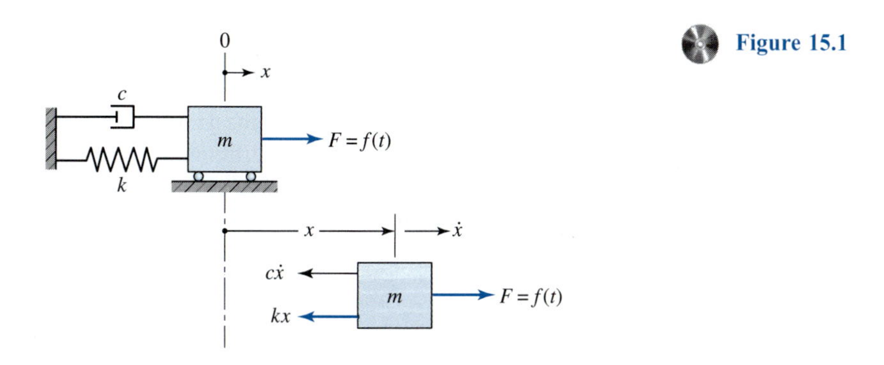

Figure 15.1

It turns out that a great many mechanical systems can be analyzed quite accurately using these assumptions.

The elasticity of the system of Fig. 15.1 is completely represented by the spring. The *stiffness* or *scale* is designated as k and defined by the equation

$$k = \frac{F}{x}, \tag{15.1}$$

where F is the force required to deflect the spring a distance x.

Similarly, the friction, or damping, is assumed to be entirely viscous damping (later we shall examine other kinds of friction) and is designated using a *coefficient of viscous damping* c. Thus,

$$c = \frac{F}{\dot{x}}, \tag{15.2}$$

where F is the force required to move the mass at a velocity of $\dot{x}$.

You should note that we are using the notation for a time derivative that is customary in vibration theory. Thus,

$$\dot{x} = \frac{dx}{dt} \quad \text{and} \quad \ddot{x} = \frac{d^2x}{dt^2}.$$

This is called *Newton's notation*, because Newton was the first to use it. It is used only when the derivatives are with respect to time.

The vibrating system of Fig. 15.1 has one degree of freedom, because the position of the mass can be completely defined by a single coordinate. An external force $F = f(t)$ is shown acting upon the mass. Thus, this system is classified as a forced, single-degree-of-freedom system with damping.

The equation of motion of the system is written by displacing the mass in the positive direction, giving it a positive velocity, and then summing all the forces, including the inertia force. As illustrated in Fig. 15.1, there are three forces acting: a spring force kx acting in the negative direction, a damping force $c\dot{x}$ also acting in the negative direction, and an external force $f(t)$ acting in the positive direction. Summing these forces together with the inertia force gives

$$\sum F = -kx - c\dot{x} + f(t) + (-m\ddot{x}) = 0$$

or

$$m\ddot{x} + c\dot{x} + kx = f(t). \tag{15.3}$$

Equation (15.3) is an important equation in dynamic analysis, and we shall eventually solve it for many specialized conditions. It is a linear differential equation of the second order.

Consider next the idealized torsional vibrating system of Fig. 15.2. Here a disk having a mass moment of inertia I is mounted upon the end of a weightless shaft having a torsional spring constant k, defined by

$$k = \frac{T}{\theta}, \tag{15.4}$$

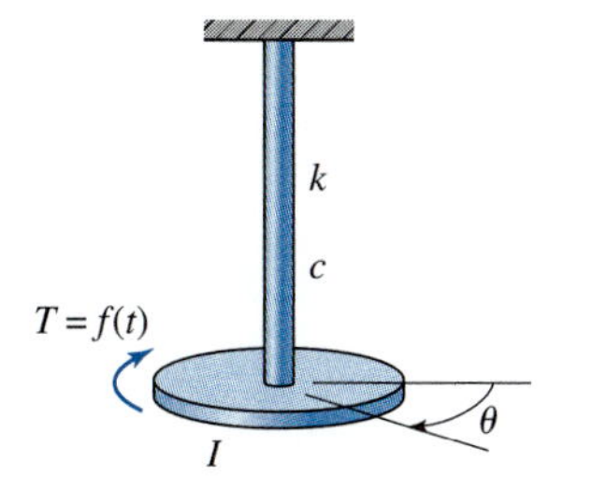

Figure 15.2

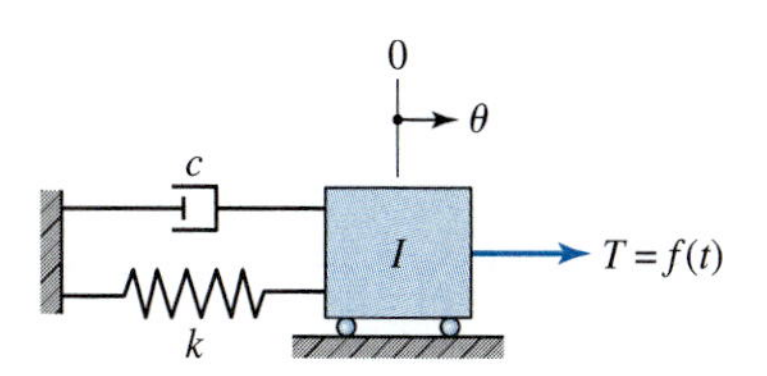

Figure 15.3

where T is the torque necessary to produce an angular deflection θ of the shaft. In a similar manner, the torsional viscous damping coefficient is defined by

$$c = \frac{T}{\dot{\theta}}. \tag{15.5}$$

Note that we are using the same symbols for denoting these torsional parameters as for rectilinear ones. The nature of the problem or of the differential equation will usually reveal whether the system is rectilinear or torsional, and so there will be no confusion in this usage.

Next, designating an external torque forcing function by $T = f(t)$, we find that the differential equation for the torsional system is

$$\sum T = -k\theta - c\dot{\theta} + f(t) + (-I\ddot{\theta}) = 0$$

or

$$I\ddot{\theta} + c\dot{\theta} + k\theta = f(t), \tag{15.6}$$

which is of the same form as Eq. (15.3). Thus, with appropriate substitutions, the solution of Eq. (15.6) will be the same as that of Eq. (15.3). Note, too, that this means we can *simulate* a torsional system, as in Fig. 15.3, merely by substituting torsional notation for the usual rectilinear notation.

It has been said that mathematics is a human invention and, therefore, that it can never perfectly describe nature.[1] But the linear differential equation frequently does such an excellent job of simulating the action of a mechanical system that one wonders whether differential equations are not an exception to this rule. Perhaps people just happened to discover them, as they have other curiosities of nature.

EXAMPLE 15.1

Figure 15.4*a* illustrates a vibrating system in which a time-dependent displacement $y = y(t)$ excites a spring-mass system through a viscous dashpot. Write the differential equation of this system.

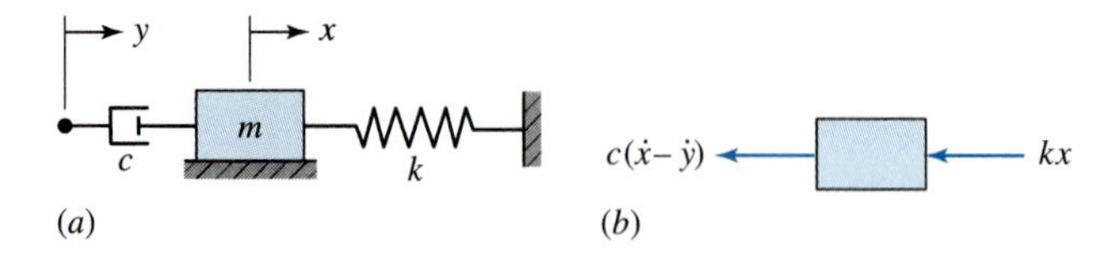

Figure 15.4 (*a*) The system. (*b*) Forces acting on the mass; note that these forces are acting in the negative direction because of the arbitrary assumptions that were made.

SOLUTION

To write the differential equation, we assume any arbitrary relationship between the coordinates and their first derivatives, say, $x > y$ and $\dot{x} > \dot{y}$. We also assume senses for x and $\dot{x}$, say, $x > 0$ and $\dot{x} > 0$. Regardless of the assumptions made, the result is the same; try it. With these assumptions the forces acting on the mass are as illustrated in Fig. 15.4*b*. Summing the forces in the x direction and including the inertia forces gives

$$\sum F = -kx - c\,(\dot{x} - \dot{y}) + (-m\ddot{x}) = 0. \tag{1}$$

Such equations are usually rearranged with the derivative terms on the left-hand side and the forcing term on the right, that is,

$$m\ddot{x} + c\dot{x} + kx = c\dot{y}. \qquad \textit{Ans.}$$

15.2 A VERTICAL MODEL

The system of Fig. 15.1 moves in the horizontal direction, and, as a result, gravity has no effect on its motion. Let us now turn the system to the vertical direction and, also incidentally, make the friction and the external forces zero. This yields the model of Fig. 15.5. We choose the origin of the coordinate system corresponding to the equilibrium position of the mass, and we choose the symbol x to denote the vertical displacement. Thus, when the mass is given a positive displacement, the spring force is made up of two components. One of these is $k\delta_{st}$ where $\delta_{st} = W/k$ is the distance the spring is deflected when the weight is suspended from it. The other component is kx. Summing the forces acting on the mass gives

$$\sum F = -k(x + \delta_{st}) + W + (-m\ddot{x}) = 0.$$

Rearranging this equation gives

$$m\ddot{x} + kx = W - k\delta_{st}$$

and, because of the definition $\delta_{st} = W/k$, this equation reduces to the homogeneous form

$$m\ddot{x} + kx = 0. \tag{15.7}$$

Note that Eq. (15.3) is identical to Eq. (15.7) if the damping force $c\dot{x}$ and the external force $f(t)$ are made zero.

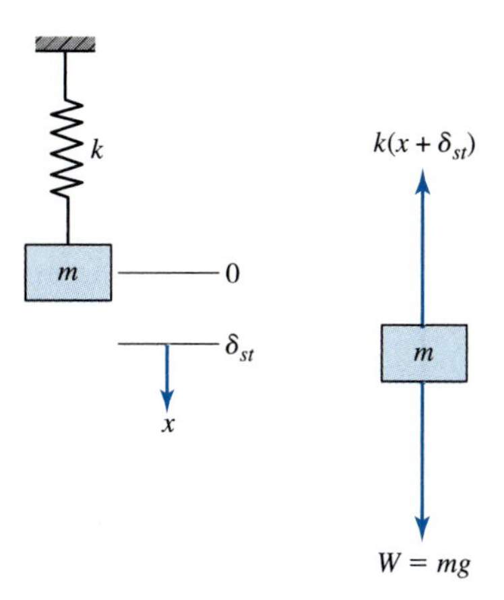

Figure 15.5

15.3 SOLUTION OF THE DIFFERENTIAL EQUATION

If Eq. (15.7) is arranged in the form

$$\ddot{x} = -\frac{k}{m}x, \tag{a}$$

then we can see that the function used for x must be of the same form as the negative of its second derivative. One such function is

$$x = A \sin bt \tag{b}$$

for which

$$\dot{x} = Ab \cos bt \tag{c}$$

and

$$\ddot{x} = -Ab^2 \sin bt. \tag{d}$$

Substituting Eqs. (*b*) and (*d*) into Eq. (*a*) produces

$$-Ab^2 \sin bt = -\frac{k}{m}A \sin bt. \tag{e}$$

Then dividing both sides by $(-A \sin bt)$ leaves

$$b^2 = \frac{k}{m}.$$

Thus, Eq. (*b*) is a solution, provided we set $b = \sqrt{k/m}$. By an exactly parallel development, it is clear that

$$x = B \cos bt \tag{f}$$

is also a solution to Eq. (a). A general solution is obtained by adding Eqs. (b) and (f), that is,

$$x = A \sin \sqrt{\frac{k}{m}} t + B \cos \sqrt{\frac{k}{m}} t. \tag{g}$$

Mathematically, the constants A and B are the constants of integration, because, theoretically, we could integrate Eq. (a) twice to get the solution. Unfortunately, most differential equations cannot be solved in this manner, and we must resort to cleverness to obtain the solution. Physically, the constants A and B represent the manner in which the vibration was started, or, to put it another way, the state of the motion at the instant $t = 0$.

We define the quantity

$$\omega_n = \sqrt{\frac{k}{m}}, \tag{15.8}$$

which is called the *natural frequency* of the system; it has units of radians per second.

The solution can now be written in the form

$$x = A \sin \omega_n t + B \cos \omega_n t. \tag{15.9}$$

As an illustration of the meaning of the constants A and B, suppose we move the mass by a distance x_0 and release it with zero velocity at the instant $t = 0$. Then the starting conditions are

$$\text{at } t = 0, \quad x = x_0, \quad \dot{x} = 0.$$

Substituting the first of these into Eq. (15.9) gives

$$x_0 = A(0) + B(1)$$

or

$$B = x_0.$$

Next, taking the first time derivative of Eq. (15.9) gives

$$\dot{x} = A\omega_n \cos \omega_n t - B\omega_n \sin \omega_n t \tag{h}$$

and substituting the other condition into this equation yields

$$0 = A\omega_n(1) - B\omega_n(0)$$

or

$$A = 0.$$

Then, substituting A and B back into Eq. (15.9) gives the result

$$x = x_o \cos \omega_n t. \tag{15.10}$$

If we start the motion with the initial conditions,

$$\text{at } t = 0, \quad x = 0, \ \dot{x} = v_0,$$

the result is

$$x = \frac{v_0}{\omega_n} \sin \omega_n t. \tag{15.11}$$

But if we start the motion with

$$\text{at } t = 0, \quad x = x_0, \ \dot{x} = v_0,$$

the result is

$$x = \frac{v_0}{\omega_n} \sin \omega_n t + x_0 \cos \omega_n t, \tag{15.12}$$

which is the most general form of the solution. We can substitute Eq. (15.12) together with its second derivative back into the differential equation and so demonstrate its validity.

The three solutions [Eqs. (15.10), (15.11), and (15.12)] are represented graphically in Fig. 15.6 using *phasors* to generate the trigonometric functions. Phasors are not vectors in the classic sense, because they can also be manipulated in ways that are not defined for vectors. They are complex numbers, however, and they can be added and subtracted just as vectors can.

The ordinate of the graph of Fig. 15.6 is the displacement x, and the abscissa can be considered as the time axis or as the angular displacement $\omega_n t$ of the phasors for a given time after the motion has commenced. The phasors x_0 and v_0/ω_n are shown in their initial positions, and as time passes, these rotate counterclockwise with an angular velocity of ω_n and generate the displacement curves shown. Figure 15.6 illustrates that the phasor x_0 starts from a maximum positive displacement and the phasor v_0/ω_n starts from a zero displacement. These, therefore, are very special, and the most general form is that given by Eq. (15.12), in which motion begins at some intermediate point.

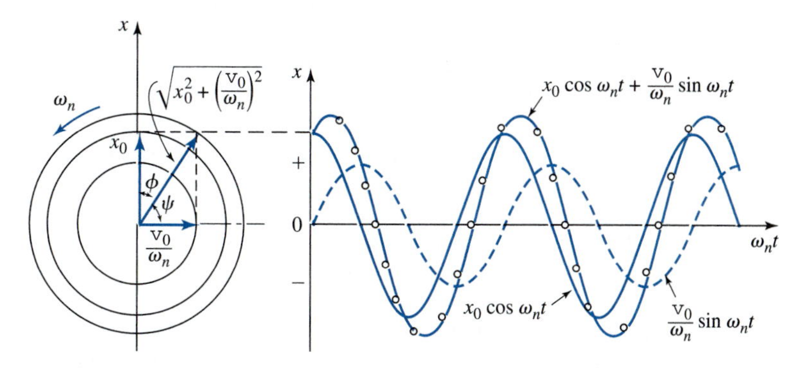

Figure 15.6 The use of phasors to generate time-displacement diagrams.

In Eq. (15.8), the quantity $\omega_n = \sqrt{k/m}$ was defined as the natural frequency of the system. For most systems, the natural frequency is a constant because the mass and the spring constant do not vary. Because one cycle is equal to 2π radians, the period of a free vibration is

$$\tau = \frac{2\pi}{\omega_n} = 2\pi\sqrt{\frac{m}{k}}, \tag{15.13}$$

where τ is usually in seconds per cycle. Although the natural units for ω_n are radians per second, frequency can also be defined as the reciprocal of the period and this gives

$$f = \frac{\omega_n}{2\pi} = \frac{1}{2\pi}\sqrt{\frac{k}{m}}, \tag{15.14}$$

where f is expressed in hertz (cycles/s or s^{-1}) and is abbreviated as Hz.

A study of Fig. 15.6 indicates that one should also be able to express the motion by the equation*

$$x = X_0 \cos(\omega_n t - \phi), \tag{15.15}$$

where X_0 and ϕ are the constants of integration whose values depend upon the initial conditions. These constants can be obtained directly from the trigonometry of Fig. 15.6 and are

$$X_0 = \sqrt{x_0^2 + \left(\frac{v_0}{\omega_n}\right)^2} \quad \text{and} \quad \phi = \tan^{-1}\frac{v_0}{\omega_n x_0}. \tag{15.16}$$

Equation (15.15) can now be written in the form

$$x = \sqrt{x_0^2 + \left(\frac{v_0}{\omega_n}\right)^2}\cos(\omega_n t - \phi). \tag{15.17}$$

This is a particularly convenient form of the equation because the coefficient is the *amplitude* of the vibration. The amplitude is the maximum displacement of the mass. The angle ϕ is called a *phase angle*, and it denotes the angular lag of the motion with respect to the cosine function.

The velocity and acceleration can be obtained by successively differentiating Eq. (15.15), that is,

$$\dot{x} = -X_0\omega_n \sin(\omega_n t - \phi) \quad \text{and} \quad \ddot{x} = -X_0\omega_n^2 \cos(\omega_n t - \phi),$$

* The motion can also be expressed in the form

$$x = X_0 \sin(\omega_n t + \psi),$$

where X_0 and ψ are the constants of integration. This is probably not a good way of expressing it, however, because in the study of forced vibration, it might imply that the output can lead the input, which, of course, is not possible.

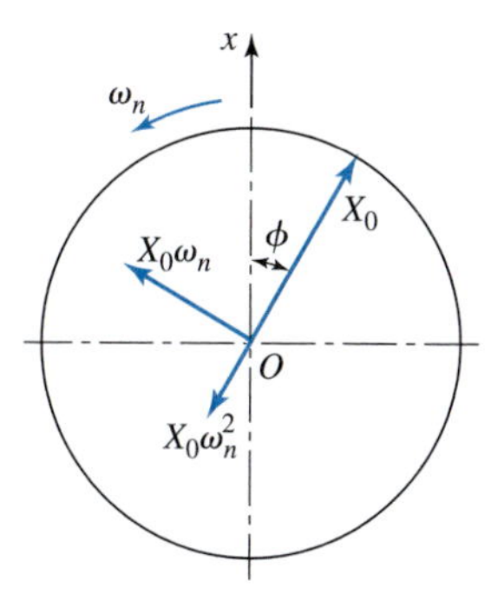

Figure 15.7 Phase relationship of displacement, velocity, and acceleration.

where the velocity amplitude is $X_0\omega_n$ and the amplitude of the acceleration is $X_0\omega_n^2$. The displacement, velocity, and acceleration phasors are illustrated in Fig. 15.7 to illustrate their phase relationships. All three phasors maintain fixed phase angles as they rotate together at the constant angular velocity ω_n. The velocity leads the displacement by 90° and the acceleration is 180° out of phase with the displacement.

15.4 STEP INPUT FORCING

Let us again consider the vibrating system of Fig. 15.1, this time adding to the system a force F applied to the mass which, in general, might be a function of time. As a start, however, let us assume that this force is constant and acting in the positive x direction. As before, we consider the damping to be zero. For this condition, Eq. (15.3) is written

$$\ddot{x} + \frac{k}{m}x = \frac{F}{m}. \tag{15.18}$$

The general solution to this equation is

$$x = A\cos\omega_n t + B\sin\omega_n t + \frac{F}{k}, \tag{15.19}$$

where A and B are the constants of integration and where the natural frequency is

$$\omega_n = \sqrt{\frac{k}{m}}, \tag{a}$$

as before. We note that the quantity F/k has units of length. The physical interpretation of this is that a force F applied to a spring of stiffness k would produce an elongation (or deformation) of the spring of magnitude F/k.

It is interesting to consider starting the motion when the system is at rest. Thus, with the system in equilibrium at $x = 0$, we apply a constant force F at the instant $t = 0$ and observe the behavior of the system. Because the system is motionless at the instant the force is applied, the starting conditions are $x = 0$ and $\dot{x} = 0$ when $t = 0$. Substituting these conditions into Eq. (15.19) produces the following values for the two constants of

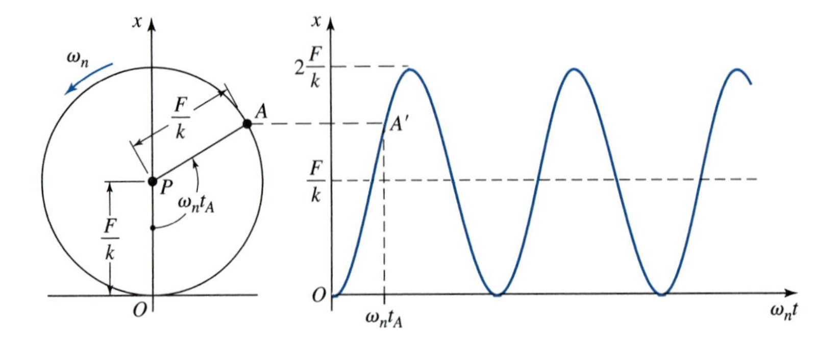

Figure 15.8 Response of an undamped vibrating system to a constant force.

integration:

$$A = -\frac{F}{k} \quad \text{and} \quad B = 0.$$

The equation of motion is obtained by substituting these back into Eq. (15.19). This gives

$$x = \frac{F}{k}(1 - \cos \omega_n t). \tag{15.20}$$

This equation is plotted in Fig. 15.8 to illustrate how the system behaves. Figure 15.8 illustrates that the application of a constant force F produces a vibration of amplitude F/k about a position of equilibrium displaced a distance F/k from the origin. This is evident from Eq. (15.20) because it contains a positive constant term,

$$x_1 = \frac{F}{k},$$

and a negative trigonometric term,

$$x_2 = -\frac{F}{k} \cos \omega_n t.$$

For $t = 0$ these two terms become

$$x_1 = \frac{F}{k} \quad \text{and} \quad x_2 = -\frac{F}{k}.$$

On the phase diagram (Fig. 15.8) the motion is represented by the phasor PA of length F/k rotating counterclockwise at ω_n rad/s. This phasor starts from the position PO when $t = 0$, rotates about P as a center, and generates the circle of radius F/k.

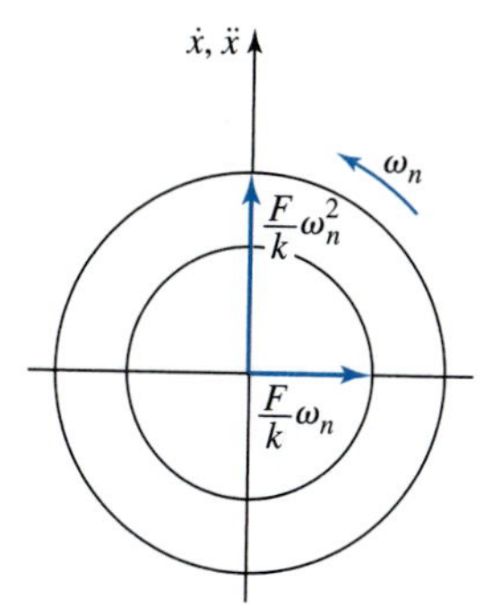

Figure 15.9 Starting positions of the velocity and acceleration phasors.

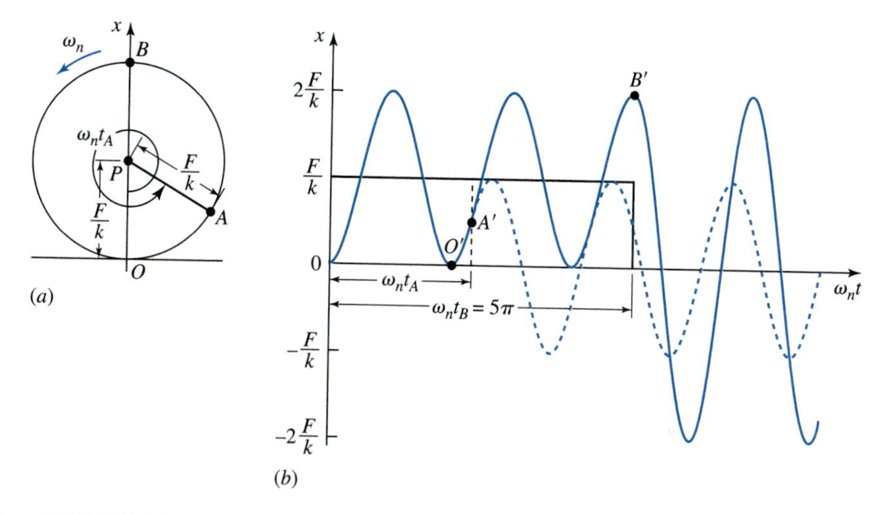

Figure 15.10 (*a*) Phase diagram. (*b*) Displacement diagram.

The velocity and acceleration are obtained by successive differentiation of Eq. (15.20) and are

$$\dot{x} = \frac{F}{k}\omega_n \sin \omega_n t \tag{15.21}$$

$$\ddot{x} = \frac{F}{k}\omega_n^2 \cos \omega_n t. \tag{15.22}$$

These can be represented as phasors too, as illustrated in Fig. 15.9. Their starting positions are determined by calculating their values for $\omega_n t = 0$. Velocity–time and acceleration–time graphs can also be plotted for these phasors employing the same methods used to obtain the displacement–time plot of Fig. 15.8.

Figure 15.10 illustrates the results when the force is just as suddenly removed. If, for example, the force is removed at time $t = 5\pi/\omega_n$ when the phasor occupies the position PB and the displacement diagram has progressed to B', the resulting motion has an amplitude about the zero axis of $2F/k$. The diagram shows an amplitude of F/k about the zero axis if the force is removed at time $t = 9\pi/4\omega_n$ at position A' on the displacement diagram. Finally, note that if the force is removed at $t = 2\pi/\omega_n$ at point O' on the displacement diagram, the resulting motion is zero.

15.5 PHASE-PLANE REPRESENTATION

The phase-plane method is a graphical means of solving transient vibration problems that is quite easy to understand and use. The method eliminates the necessity for solving differential equations, some of which can be very difficult, and even enables solutions to be obtained when the functions involved are not expressed in algebraic form. Engineers must concern themselves as much with transient disturbances and motions of machine parts as with steady-state motions. The phase-plane method presents the physics of the problem with so much clarity that it will serve as an excellent vehicle for the study of mechanical transients.

Before introducing the details of the phase-plane method, it will be of value to demonstrate how the displacement-time and the velocity-time relations are generated by a single rotating phasor. We have already observed that a free undamped vibrating system has an equation of motion, which can be expressed in the form

$$x = X_0 \cos(\omega_n t - \phi), \tag{15.23}$$

and that its velocity is

$$\dot{x} = -X_0\omega_n \sin(\omega_n t - \phi). \tag{15.24}$$

The displacement, as given by Eq. (15.23), can be represented by the projection on a vertical axis of a phasor of length X_0 rotating at ω_n rad/s in the counterclockwise direction (Fig. 15.11*a*). The angle $(\omega_n t - \phi)$, in this example, is measured from the vertical axis. Similarly, the velocity can be represented on the same vertical axis by the projection of another phasor of length $X_0\omega_n$ rotating at the same angular velocity but leading X_0 by a phase angle of 90°, as illustrated in Fig. 15.11*a*. Therefore, the angular location of the velocity phasor is measured from the horizontal axis. If we take the coordinate system containing the velocity phasor and rotate it backward (clockwise) through an angle of 90°, then the velocity

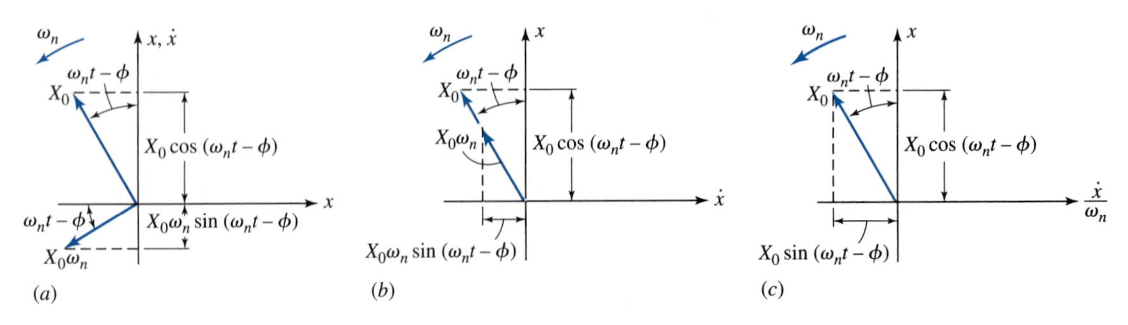

Figure 15.11

and displacement phasors will be colinear and their angular locations can be measured from the same vertical axis. This step has been taken in Fig. 15.11*b*, where the displacement is still measured along the same vertical axis. But, having rotated the coordinate system in which the velocity is measured, we see that the velocity is obtained by projecting the $X_0\omega_n$ phasor to the horizontal axis. Thus, we can now measure velocities on a separate axis from displacements. Note, too, that the direction of positive velocities is to the right.

Our final step is taken by noting that the velocity phasor differs in length from the displacement phasor by the constant factor ω_n. Thus, instead of plotting velocities, if we plot the quantity $\dot{x}/\omega_n$, then we will have a quantity that is proportional to velocity. This step has been taken in Fig. 15.11*c*, where the horizontal axis is designated the $\dot{x}/\omega_n$ axis. With this change the projection of the phasor X_0 on the x axis gives the displacement and its projection on the $\dot{x}/\omega_n$ axis gives a quantity that is directly proportional to the velocity.

The displacement-time and the velocity-time graphs of a free undamped vibration have been plotted on Fig. 15.12 to demonstrate how the vector diagram is related to them. A point A on the phase diagram corresponds with A' on the displacement diagram and with A'' on the velocity diagram. Note that quantities obtained from the velocity plot must be multiplied by ω_n to obtain the actual velocity.

It is possible to arrive at these same conclusions in a different manner. If Eq. (15.23) is squared, we obtain

$$x^2 = X_0^2 \cos^2(\omega_n t - \phi). \tag{a}$$

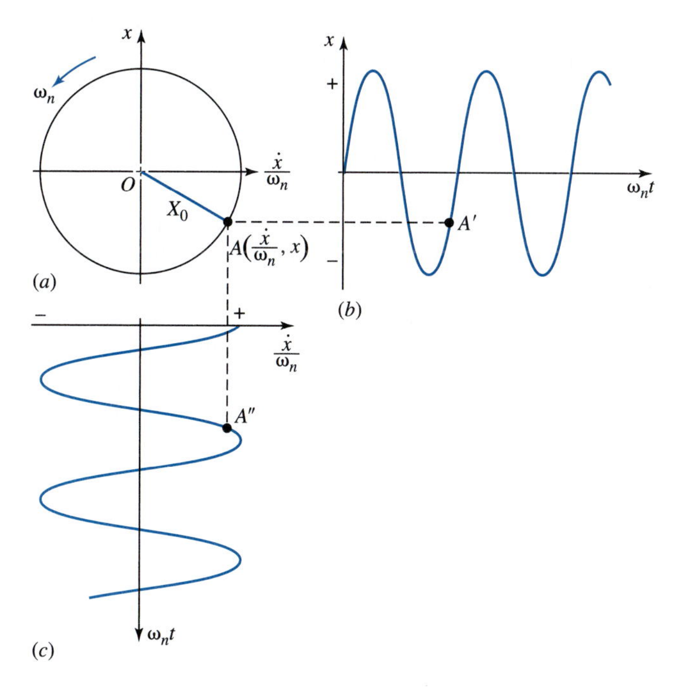

Figure 15.12 (*a*) Phase diagram. (*b*) Displacement diagram. (*c*) Velocity diagram.

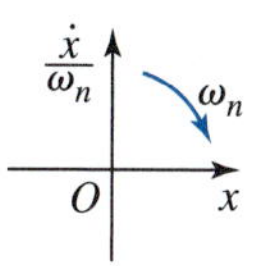

Figure 15.13

Next, dividing Eq. (15.24) by ω_n and squaring the result gives

$$\left(\frac{\dot{x}}{\omega_n}\right)^2 = X_0^2 \sin^2(\omega_n t - \phi). \tag{b}$$

Then, adding Eqs. (a) and (b) gives

$$x^2 + \left(\frac{\dot{x}}{\omega_n}\right)^2 = X_0^2. \tag{15.25}$$

This is the equation of a circle having the amplitude X_0 as its radius and with its center at the origin of a coordinate system having the axes x and $\dot{x}/\omega_n$. Thus, Eq. (15.25) describes the circle of Fig. 15.12a, where OA is the amplitude X_0 of the motion. The coordinate system $x, \dot{x}/\omega_n$ defines the position of points in a region called the *phase plane*.

The phase plane is widely used in the solution of nonlinear differential equations. Such equations occur frequently in the study of vibrations and feedback-control systems. When the phase plane is used to solve nonlinear differential equations, it is customary to arrange the axes as illustrated in Fig. 15.13, with ω_n considered positive in the clockwise direction instead of the counterclockwise direction as we are using it here. This arrangement of the axes seems more logical, because $\dot{x}/\omega_n$ is a function of x. However, for the analysis of transient disturbances to mechanical systems and for analyzing cam mechanisms, the arrangement of the axes as in Fig. 15.12 is more appropriate, and this is the one we shall employ throughout Chapter 15. The reason for this is that we shall employ the phase-plane method not as an end in itself, as is done when it is used for the solution of nonlinear problems, but as a tool to obtain the transient response.

15.6 PHASE-PLANE ANALYSIS

As a first example of the use of the phase-plane method, we shall consider the vibrating system of Fig. 15.14. This is a spring-mass system having a spring attached to a frame that can be positioned at will. To begin the motion we might consider the system as initially at rest and then, at $t = 0$, suddenly move the frame a distance x_1 to the right. The effect of this sudden motion is to compress the spring an amount x_1 and to shift the equilibrium position of the mass a distance x_1 to the right. For $t < 0$, the mass is in equilibrium at the position illustrated in Fig. 15.14. For $t > 0$, the equilibrium position of the mass is a distance x_1 to the right of the position shown. If the motion of the frame occurs instantaneously, the initial conditions are

$$x = -x_1 \quad \text{and} \quad \dot{x} = 0 \text{ at } t = 0, \tag{a}$$

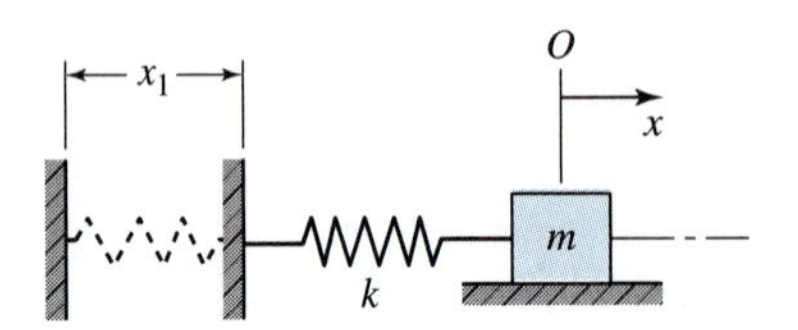

Figure 15.14

where the motion is now measured from the shifted equilibrium position. If these initial conditions are used to evaluate the constants of integration of Eq. (15.15), we obtain

$$X_0 = -x_1 \quad \text{and} \quad \phi = 0$$

so that

$$x = -x_1 \cos \omega_n t. \tag{b}$$

Next, note that the application of a constant force to an undamped spring-mass system gave as the equation of motion [Eq. (15.20)]

$$x = \frac{F}{k} - \frac{F}{k} \cos \omega_n t. \tag{c}$$

By rearranging, we obtain

$$x - \frac{F}{k} = -\frac{F}{k} \cos \omega_n t. \tag{d}$$

If we let $F/k = x_1$, then Eq. (d) becomes

$$x - x_1 = -x_1 \cos \omega_n t. \tag{e}$$

But if the origin of x is shifted a distance x_1, then Eq. (e) is the same as Eq. (b). Thus, shifting of the frame of a vibrating system through a distance x_1 is equivalent to adding a constant force $F = kx_1$ acting upon the mass. This problem is illustrated in Fig. 15.15, where the earlier origin is O and the new origin is O_1. These are separated, then, by the distance x_1. At the instant $t = 0$, the origin O is shifted to O_1, initiating the motion. A phasor O_1A, equal in magnitude to x_1, begins its rotation from the position O_1O. Rotating at ω_n rad/s, the projection of this phasor on the x axis describes the displacement of the mass. This phasor then continues to rotate until something else happens to the system. Arriving at point A, it has traversed an angle $\omega_n t_A$, and the corresponding point on the displacement-time diagram is A'.

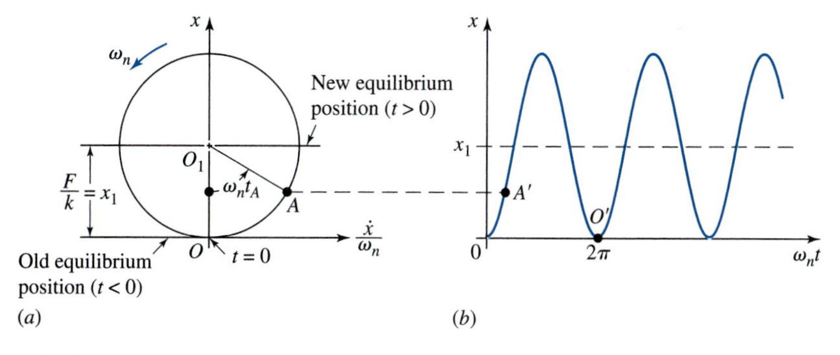

Figure 15.15 (a) Phase-plane diagram. (b) Displacement-time diagram.

LIVERPOOL JOHN MOORES UNIVERSITY

The phase-plane method permits us to move the origin in any manner we may choose and at any instant in time that we choose. In the example above, the origin was shifted a distance x_1, and we have seen that this is equivalent to the sudden application of a force $F = kx_1$ applied to the mass in the positive direction. Suppose that we permit the phasor to make one complete revolution of 360°. It then generates one complete cycle of displacement and returns to point O. If, at this instant, we shift the frame back to its original position, we might reasonably ask: What are the displacement and velocity at the instant of making this shift? The phase-plane diagram illustrates that both the displacement and the velocity are zero at the end of a cycle. The displacement and velocity are exactly the quantities we require to determine the motion of the system in the next era. Because these are both zero, there is no motion after returning the frame to its original position. It will be recalled that these are exactly the results that we obtained in analyzing square-ware forcing functions when the force endured for an integral number of cycles. Thus, the phasor generates the displacement diagram through the angle $\omega_n t = 2\pi$ in this case, and at this point we return the frame to its original position. The initial conditions are now $x = 0$, $\dot{x} = 0$; consequently, the mass stops its motion completely.

Let us now create a vibration by shifting the frame from $x = 0$ to $x = x_1$, waiting a period of time Δt, and then shifting the frame back to $x = 0$. This constitutes a square-wave forcing function, or a step disturbance to the system, and the phase-plane and displacement diagrams for such a motion are illustrated in Fig. 15.16. The motion can be described in three eras. During the first, from t_0 to t_1, everything is at rest. At $t = t_1$, the frame suddenly moves to the right a distance $x = x_1$, compressing the spring. The duration of the second era is from t_1 to $t_2(\Delta t)$, and at time t_2 the frame suddenly returns to its original position. The third era represents time when $t \geq t_2$. The position of the frame during these three eras is illustrated in the displacement diagram. The displacement diagram also describes the motion of the mass, and the phase-plane diagram explains why it moves as it does. Between t_0 and t_1 the mass is at rest and nothing happens. At time t_1, the frame shifts from O to O_1 on the phase plane diagram, which is the distance x_1. If we designate the original frame position as the origin of the $x, \dot{x}/\omega_n$ system, then the conditions at $t = t_1$ are $x = +x_1$,

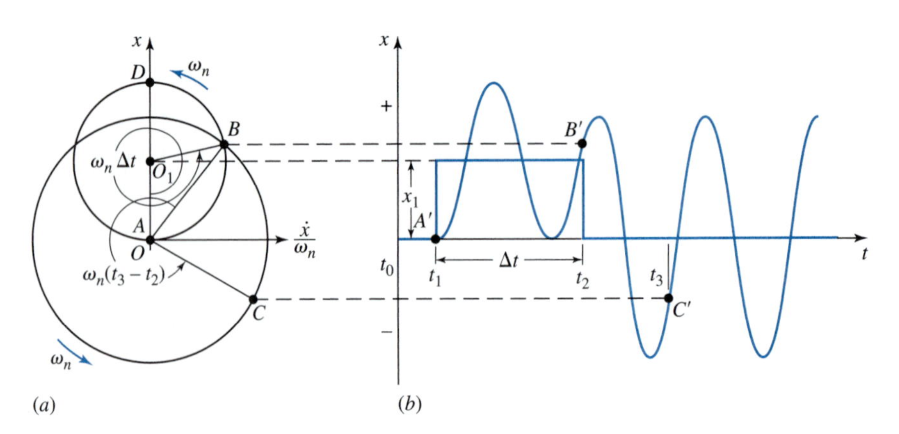

Figure 15.16 A step disturbance.

$\dot{x} = 0$. However, the motion of the mass about O is equal to its motion about O_1 plus the distance from O to O_1. Therefore,

$$x = -x_1 \cos \omega_n t' + x_1 = x_1(1 - \cos \omega_n t'). \tag{f}$$

Also,

$$\dot{x} = x_1 \omega_n \sin \omega_n t', \tag{g}$$

where t' is understood to begin at time t_1. As illustrated in Fig. 15.16*b*, the mass vibrates about the position $x = x_1$ during this era. The vibration begins at point A on the phase-plane diagram and continues as a free vibration for the time Δt. During this period, the line O_1A rotates counterclockwise with angular velocity ω_n rad/s until at time t_2 it occupies the position O_1B. At this instant, the third era begins when the frame suddenly returns through the distance x_1 to its original position. Thus, the starting conditions for the third era are the same as the ending conditions for the second and are

$$x_2 = x_1(1 - \cos \omega_n \Delta t) \quad \text{and} \quad \dot{x}_2 = x_1 \omega_n \sin \omega_n \Delta t. \tag{h}$$

Substituting these conditions into Eq. (15.12) and rearranging gives the equation of motion for the third era as

$$x = x_1(1 - \cos \omega_n \Delta t) \cos \omega_n t'' + x_1 \sin \omega_n \Delta t \sin \omega_n t'', \tag{i}$$

where t'' is the time measured from the start of the third era. Equation (i) can be transformed into an equivalent expression containing only a single trigonometric term and a phase angle as in Eq. (15.15), but we shall not do so here. The third era, we have seen, begins at point B on the phase-plane diagram (B' on the displacement diagram); the motion of point B as it moves about a circle with center at O in the counterclockwise direction describes the motion of the mass. Thus, at instant t_3 the line OB will have rotated through the angle $\omega_n(t_3 - t_2)$ and be located at C. The corresponding point on the displacement diagram is C'.

The extension of the phase-plane method to any number of steps, taken in either or both the positive or the negative x direction, should now be apparent. In each case, the starting conditions for the next era are taken equal to the ending conditions for the previous era. The equations of motion should be written for each era with time counted from the start of that era. You can now understand that if a third era of Fig. 15.16 begins at point D on the phase-plane diagram, then the resulting motion will have an amplitude twice as large as that in the second era.

15.7 TRANSIENT DISTURBANCES

Any action that destroys the static equilibrium of a vibrating system may be called a *disturbance* to that system. A *transient disturbance* is any action that endures for only a relatively short period of time. The analyses in the several preceding sections have dealt with transient disturbances having a stepwise relationship to time. Because all machine parts have elasticity and inertia, forces do not come into existence instantaneously in real life.* Consequently, we can usually expect to encounter forcing functions that vary smoothly with time. Although the step forcing function is not true to nature, it is our purpose in this section to demonstrate how the step function is used with the phase-plane method to obtain a good approximation of the vibration of a system excited by a "natural" disturbance.

The procedure is to plot the disturbance as a function of time, to divide this into steps, and then to use the steps successively to make a phase-plane plot. The resulting displacement and velocity diagrams can then be obtained by graphically projecting points from the phase-plane diagram, as previously explained. It turns out that very accurate results can frequently be obtained using only a small number of steps. Of course, as in any graphical solution, better results are obtained when a large number of steps are employed and when the work is plotted to a larger scale. It is difficult to set up general rules for selecting the size of the steps to be used. For slowly vibrating systems and for relatively smooth forcing functions, the step width can be quite large, but even a slow system will require narrow steps if the forcing function has numerous sharp peaks and valleys, that is, if it has a great deal of frequency content. For smooth forcing functions and slowly vibrating systems, a step width such that the phasor sweeps out an angle of 180° is probably about the largest that one should use. It is a good idea to check the step width during the construction of the phase-plane diagram. Too great a width will cause a discontinuity in the slope of two curves at the point of adjacency of the two. If this occurs, then the step can immediately be narrowed and the procedure resumed.

Figure 15.17 illustrates how to find the heights of the steps. The first step for the forcing function in Fig. 15.17 has been given a width of Δt_1 and a height of h_1. This height is obtained by constructing a horizontal line across the force curve such that the areas of the two shaded triangles are equal. For a step width of Δt_1, the phasor sweeps out an angle of $\omega_n \Delta t_1$. Thus, the angular rotation of the phasor is obtained simply by multiplying the step width in seconds by the natural frequency of the system.

Figure 15.18 illustrates a four-step forcing function that we may assume has been deduced from a natural function. The width and height of the steps have been plotted to

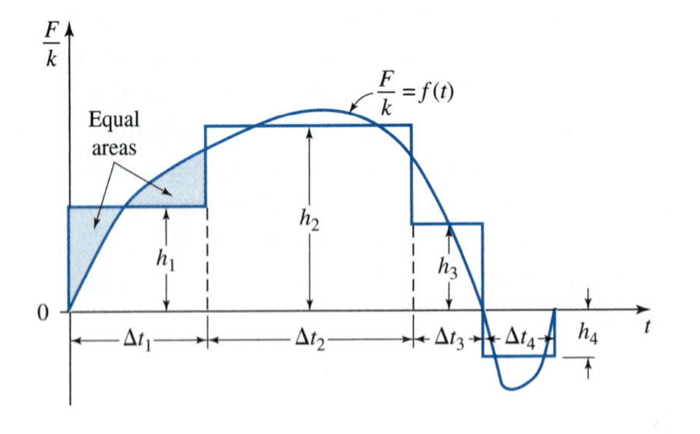

Figure 15.17 Finding the heights of the steps.

* If a force could truly be applied instantaneously, then, according to Newton, an acceleration would occur instantaneously. It would still take at least one instant to integrate this to become a velocity and another instant to integrate into a change in position. Finally, according to Hooke's law, our elastic body would display a force.

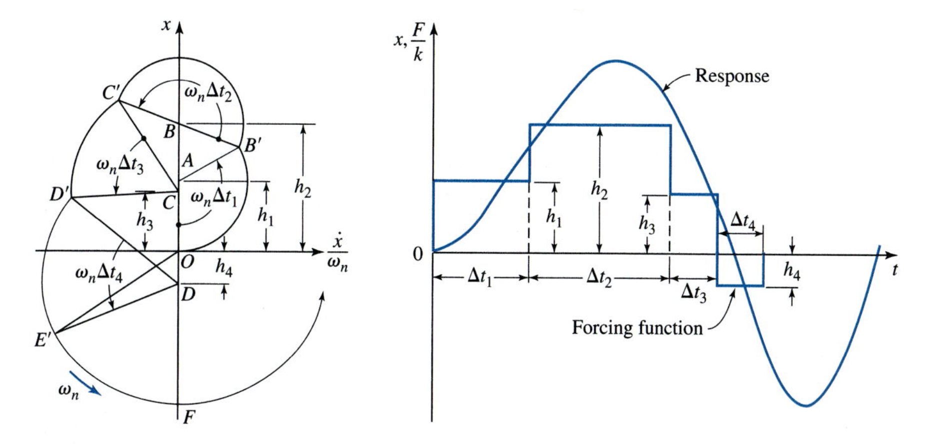

Figure 15.18 Construction of the phase-plane and displacement diagrams for a four-step forcing function.

scale and are Δt_1 and h_1 for the first step, Δt_2 and h_2 for the second step, and so on. Note that the fourth step has a negative h_4. The motion begins at $t = 0$ when a phasor of length h_1 starts rotating about A as a center from the initial position AO. This phasor rotates through an angle $\omega_n \Delta t_1$ and generates the portion of the response curve contained in the interval Δt_1. At the end of this period of time the second step begins with the center of rotation of the phasor shifting from A to B. The length of the phasor for the duration of the second step is the distance BB'. This phasor then rotates through the angle $\omega_n \Delta t_2$ about a center at B until it arrives at the position BC'. At this instant, the third step begins. The center of rotation shifts to C, and the phasor CC' rotates through the angle $\omega_n \Delta t_3$. At the end of the fourth step the phasor has arrived at E'. The center of rotation now shifts to the origin O, and the motion continues as a free vibration of amplitude OE' until something else (not shown) happens to the system.

EXAMPLE 15.2

Measurements on a mechanical vibrating system show that the mass has a weight of 16.90 lb and that the springs can be combined to give an equivalent spring rate of 30 lb/in. This system is observed to vibrate quite freely, and so damping is neglected. A transient force resembling the first half-cycle of a sine wave operates on the system. The maximum value of the force is 10 lb, and it is applied for 0.120 s. Determine the response.

SOLUTION

The natural frequency of the system is

$$\omega_n = \sqrt{\frac{k}{m}} = \sqrt{\frac{(30 \text{ lb/in})}{(16.90 \text{ lb})/(386 \text{ in/s}^2)}} = 26.2 \text{ rad/s}.$$

The period and frequency are

$$\tau = \frac{2\pi}{\omega_n} = \frac{2\pi \text{ rad/cycle}}{26.2 \text{ rad/s}} = 0.240 \text{ s/cycle} \quad \text{and} \quad f = \frac{1}{\tau} = \frac{1}{0.240 \text{ s/cycle}} = 4.17 \text{ Hz}.$$

We shall first determine the response by replacing the entire forcing function by a single step disturbance. The height of the step should be the time average of the forcing function for the duration of the step. The average ordinate of a half-cycle of the sine wave is

$$h_1 = \frac{1}{\pi}\int_0^{\pi}\left[\frac{F_{max}}{k}\sin bt\right]d\,(bt) = 0.637\frac{F_{max}}{k} = 0.637\frac{10 \text{ lb}}{30 \text{ lb/in}} = 0.212 \text{ in.} \qquad (1)$$

This is all we need to solve the problem. The graphical solution is illustrated in Fig. 15.19, together with the force and the step, all plotted to the same time scale. As shown, the amplitude of a vibration resulting from a single step is $X_0 = 0.424$ in.

To check the accuracy of a single step, we shall solve the example again using three steps. These steps need not be equal, but it will be convenient in this example to make them so. For the three equal steps Eq. (1) is written as

$$h_1 = h_3 = \frac{1}{\pi/3}\int_0^{\pi/3}\left[\frac{F_{max}}{k}\sin bt\right]d\,(bt) = \frac{3}{2\pi}\left(\frac{F_{max}}{k}\right) = \frac{3}{2\pi}\left(\frac{10 \text{ lb}}{30 \text{ lb/in}}\right) = 0.159 \text{ in.}$$

$$h_2 = \frac{1}{\pi/3}\int_{\pi/3}^{2\pi/3}\left[\frac{F_{max}}{k}\sin bt\right]d\,(bt) = \frac{3}{\pi}\left(\frac{F_{max}}{k}\right) = \frac{3}{\pi}\left(\frac{10 \text{ lb}}{30 \text{ lb/in}}\right) = 0.318 \text{ in.}$$

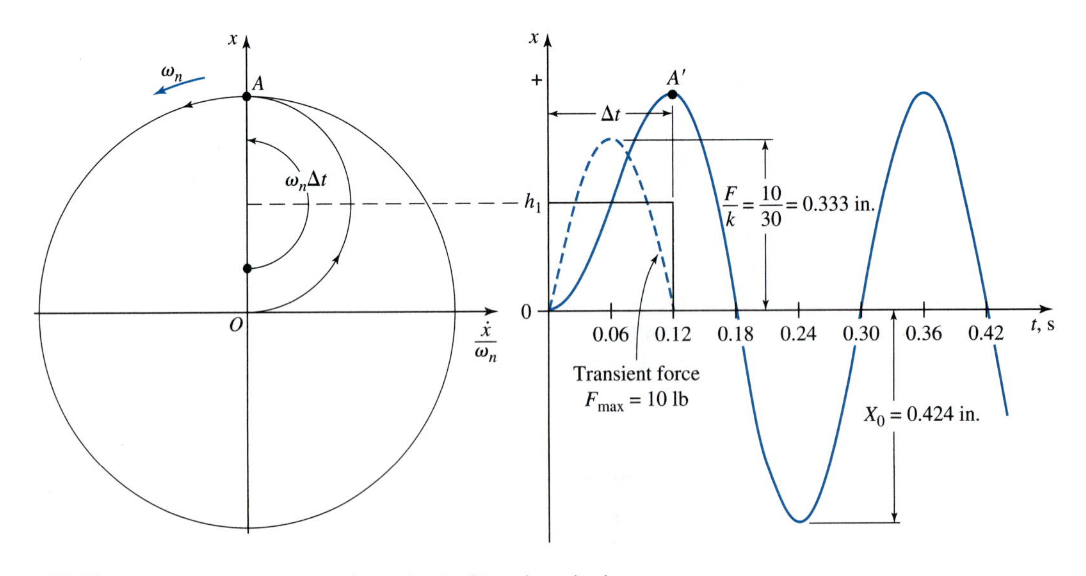

Figure 15.19 Example 15.2. Response determined with only a single step.

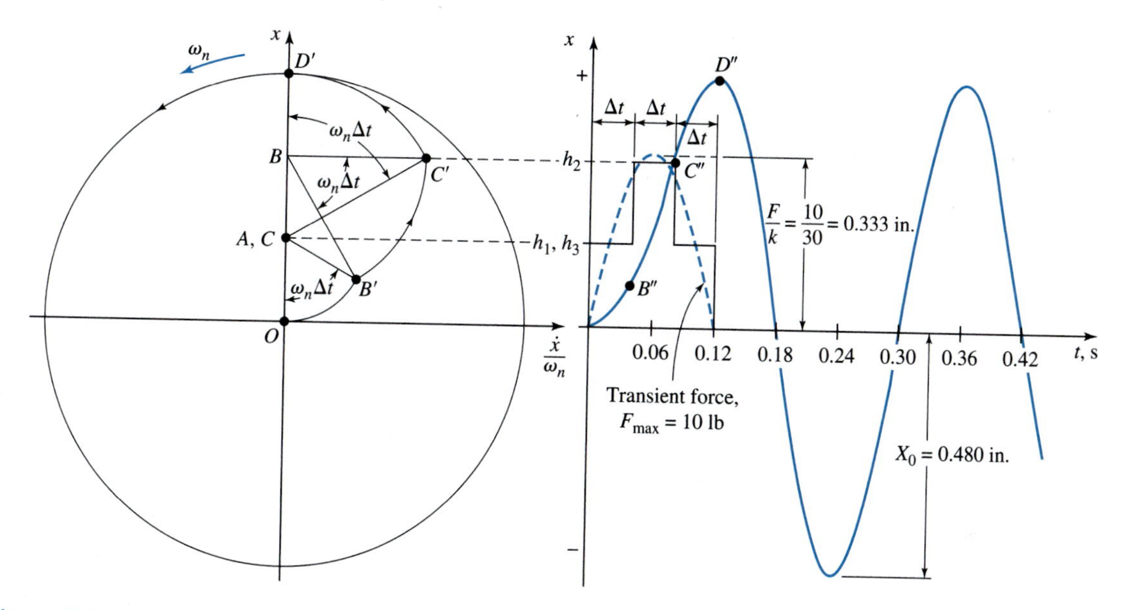

Figure 15.20 Example 15.2. Response determined in three steps. The first step has a height of h_1 and is taken from O *to* B' with a radius AO; the second has a height h_2 and is from B' *to* C' with a radius BB'; the third has a height h_3 and is from C' *to* D' with a radius CC'. The vibration is a free motion after point D'.

The construction and results are illustrated in Fig. 15.20. Note that the amplitude is slightly larger than that for a single step. In this case, it is very doubtful whether the extra labor of a five-step solution would be justified.

15.8 FREE VIBRATION WITH VISCOUS DAMPING

Lumping all the friction in a system and representing it as purely viscous damping is the method most widely used in the simulation of vibration in mechanical systems. This is primarily because it results in a linear differential equation that is convenient to solve, not necessarily because this truly represents the real system. The differential equation for an unforced, viscously damped system was derived previously in Eq. (15.3) and, with no externally applied force, this is

$$m\ddot{x} + c\dot{x} + kx = 0.$$

Dividing by m gives

$$\ddot{x} + \frac{c}{m}\dot{x} + \frac{k}{m}x = 0. \tag{15.26}$$

Inspection of this equation indicates that the time variation of x and its derivatives must be alike if the relation is to be satisfied. The exponential function satisfies this requirement;

so it is not unreasonable to guess that the solution might be in the form,

$$x = Ae^{st}, \tag{a}$$

where A and s are values still to be determined. The first and second time derivatives of Eq. (a) are

$$\dot{x} = Ase^{st} \quad \text{and} \quad \ddot{x} = As^2e^{st}.$$

Substituting these into Eq. (15.26) and dividing out the common factors gives

$$s^2 + \frac{c}{m}s + \frac{k}{m} = 0. \tag{b}$$

Thus, Eq. (a) is a solution of Eq. (15.26), provided that the quantity s is selected to satisfy Eq. (b). The roots of Eq. (b) are

$$s = -\frac{c}{2m} \pm \sqrt{\left(\frac{c}{2m}\right)^2 - \frac{k}{m}}, \tag{c}$$

and so Eq. (a) can be written

$$x = Ae^{s_1 t} + Be^{s_2 t}, \tag{d}$$

where s_1 and s_2 are the two roots of Eq. (c), and A and B are the constants of integration.

The value of damping that makes the radical of Eq. (c) zero has special significance; we shall call it the *critical-damping coefficient* and designate it by the symbol c_c. With critical damping the radical is zero, and so

$$\frac{c_c}{2m} = \sqrt{\frac{k}{m}} = \omega_n \quad \text{or} \quad c_c = 2m\omega_n. \tag{15.27}$$

It is also convenient to define a *damping ratio* ζ, which is the ratio of the actual to the critical damping. Thus,

$$\zeta = \frac{c}{c_c} = \frac{c}{2m\omega_n}. \tag{15.28}$$

After some algebraic manipulation Eq. (c) can be written

$$s = \left(-\zeta \pm \sqrt{\zeta^2 - 1}\right)\omega_n. \tag{e}$$

In the case in which the actual damping is larger than the critical, $\zeta > 1$ and the radical is real. This is analogous to movement of the mass in a very thick viscous fluid, such as molasses, and there is no vibration. Although this case does have application in certain mechanical systems, because of space limitations we shall pursue it no further here. It is not difficult to learn that if the mass is displaced and released in a system with more than critical damping, the mass will return slowly to its position of equilibrium without overshooting.

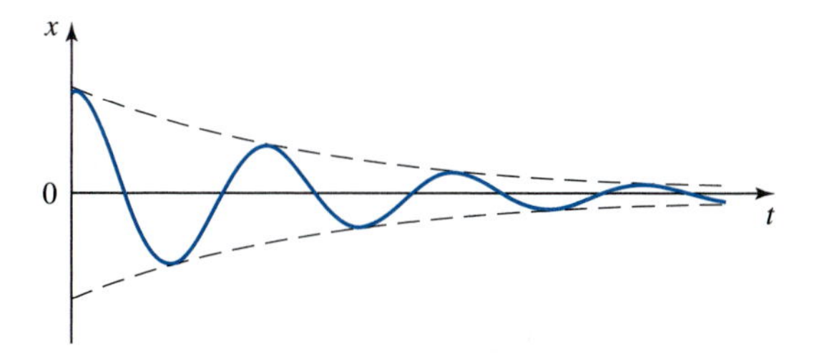

Figure 15.21

When the damping is less than critical, $\zeta < 1$ and the radical of Eq. (e) is imaginary. It is then written as

$$s = \left(-\zeta \pm j\sqrt{1-\zeta^2}\right)\omega_n, \tag{f}$$

where we have used the operator notation $j = \sqrt{-1}$. Substituting these roots into Eq. (d) gives

$$x = e^{-\zeta\omega_n t}\left(Ae^{j\sqrt{1-\zeta^2}\omega_n t} + Be^{-j\sqrt{1-\zeta^2}\omega_n t}\right). \tag{g}$$

Because A and B are complex conjugates, this equation can be transformed into

$$x = e^{-\zeta\omega_n t}[(A+B)\cos\omega_d t + j(A-B)\sin\omega_d t], \tag{h}$$

where ω_d is the natural frequency of the damped vibration, and

$$\omega_d = \omega_n\sqrt{1-\zeta^2}. \tag{15.29}$$

It is expedient to transform Eq. (h) into a form having only a single trigonometric term. Making this transformation gives

$$x = X_0 e^{-\zeta\omega_n t}\cos(\omega_d t - \phi), \tag{15.30}$$

where X_0 and ϕ are the new constants of integration. It is apparent that Eq. (15.30) reduces to Eq. (15.17) if the damping is made zero. The constants of integration can also be determined in exactly the same manner.

Equation (15.30) is the product of a trigonometric function and a decreasing exponential function. The resulting motion is therefore oscillatory with an exponentially decreasing amplitude, as illustrated in Fig. 15.21. The frequency does *not* depend upon amplitude and is less than that of an undamped system, as is indicated by the factor $\sqrt{1-\zeta^2}$ in Eq. (15.29).

15.9 DAMPING OBTAINED BY EXPERIMENT

The classical method of obtaining the damping coefficient is by an experiment in which the system is disturbed in some manner—say, by hitting it with a hammer—and then recording the decaying response by means of a strain gauge, rotary potentiometer, solar cell, or other transducer. If the friction is mostly viscous, the result should resemble Fig. 15.21. The rate of decay can then be measured, and, by means of the analysis to follow, the viscous-damping coefficient can be calculated.

The recording also provides a qualitative guide to the predominating friction. In many mechanical systems, recordings reveal that the first portion of the decay is curved, as in Fig. 15.21, but the smaller part decays at a linear rate instead. A linear rate of decay identifies Coulomb, or sliding, friction. Still another type of decay sometimes determined is curved at the beginning but then flattens out for small amplitudes and requires a much greater time to decay to zero. This kind of damping is proportional to the square of the velocity. Most mechanical systems have several kinds of friction present, and the investigator is usually interested only in the predominant kind. For this reason, he or she should analyze that portion of the decay record in which the amplitudes are closest to those actually experienced in normal operation of the system.

Perhaps the best method of obtaining an average damping coefficient is to use quite a number of cycles of decay, if they can be obtained, rather than a single cycle. If we take any response curve, such as that of Fig. 15.9, and measure the amplitude of the nth and also of the $(n+N)$th cycle, then these measurements are taken when the cosine term of Eq. (15.30) is approximately unity; so,

$$x_n = X_0 e^{-\zeta\omega_n t_n} \quad \text{and} \quad x_{n+N} = X_0 e^{-\zeta\omega_n(t_n+N\tau)},$$

where τ is the period of vibration and N is the number of cycles of motion between the amplitude measurements. The *logarithmic decrement* δ_N is defined as the natural logarithm of the ratio of these two amplitudes and is

$$\delta_N = \ln\frac{x_n}{x_{n+N}} = \ln\frac{X_0 e^{-\zeta\omega_n t_n}}{X_0 e^{-\zeta\omega_n(t_n+N\tau)}} = \ln e^{\zeta\omega_n N\tau} = \zeta\omega_n N\tau. \tag{15.31}$$

The period of the vibration is

$$\tau = \frac{2\pi}{\omega_d} = \frac{2\pi}{\omega_n\sqrt{1-\zeta^2}}. \tag{15.32}$$

Therefore, the logarithmic decrement can be written in the form

$$\delta = \frac{\delta_N}{N} = \frac{2\pi\zeta}{\sqrt{1-\zeta^2}}. \tag{15.33}$$

Measurements of many damping ratios indicate that a value of under 20% can be expected for most machine systems, with a value of 10% or less being the most probable. For this range of values the radical in Eq. (15.33) can be taken as approximately unity, giving

$$\delta \approx 2\pi\zeta \tag{15.34}$$

as an approximate formula.

EXAMPLE 15.3

Let the vibrating system of Example 15.2 have a dashpot attached that exerts a force of 0.25 lb on the mass when the mass has a velocity of 1 in/s. Find the critical damping constant, the logarithmic decrement, and the ratio of two consecutive maxima.

SOLUTION

The data from Example 15.2 are repeated here for convenience:

$$W = 16.90 \text{ lb}, \quad k = 30 \text{ lb/in}, \quad \omega_n = 26.2 \text{ rad/s}.$$

The critical damping constant is

$$c_c = 2m\omega_n = 2\left(\frac{16.90 \text{ lb}}{386 \text{ in/s}^2}\right)(26.2 \text{ rad/s}) = 2.29 \text{ lb} \cdot \text{s/in}. \qquad \textit{Ans.}$$

The viscous damping constant is $c = F/\dot{x} = 0.25$ lb · s/in. Therefore, the damping ratio is

$$\zeta = \frac{c}{c_c} = \frac{0.25 \text{ lb} \cdot \text{s/in}}{2.29 \text{ lb} \cdot \text{s/in}} = 0.109.$$

The logarithmic decrement is obtained from Eq. (15.33):

$$\delta = \frac{2\pi\zeta}{\sqrt{1-\zeta^2}} = \frac{2\pi\,(0.109)}{\sqrt{1-(0.109)^2}} = 0.689. \qquad \textit{Ans.}$$

The ratio of two consecutive maxima is

$$\frac{x_{n+1}}{x_n} = e^{-\delta} = e^{-0.689} = 0.502. \qquad \textit{Ans.}$$

Therefore, for this amount of damping, each amplitude is approximately 50% of that of the previous cycle.

15.10 PHASE-PLANE REPRESENTATION OF DAMPED VIBRATION

We have seen that when undamped mechanical systems are subjected to transient forces, the resulting motion (mathematically, at least) endures forever without decreasing in amplitude. We know, however, that energy losses always exist, although sometimes only in minute amounts, and that these losses eventually cause vibration to stop. In the case of vibrating systems known to be acted upon by transient forces, it is often desirable to introduce additional damping as one means of decreasing the number of cycles of vibration. For this reason, the phase-plane solution of a damped vibration is particularly important.

We have seen that the phase-plane diagram of an undamped free vibration is generated by a phasor of constant length rotating at a constant angular velocity. In the case of a damped free vibration, the phase-plane diagram is generated by a phasor whose length is decreasing exponentially. Thus, the trajectory for such a motion is a spiral instead of a circle.

It turns out that a simple spiral will give incorrect values for the velocity if plotted on the same $x, \dot{x}/\omega_n$ axes as used for undamped free vibration. The reason for this is that the

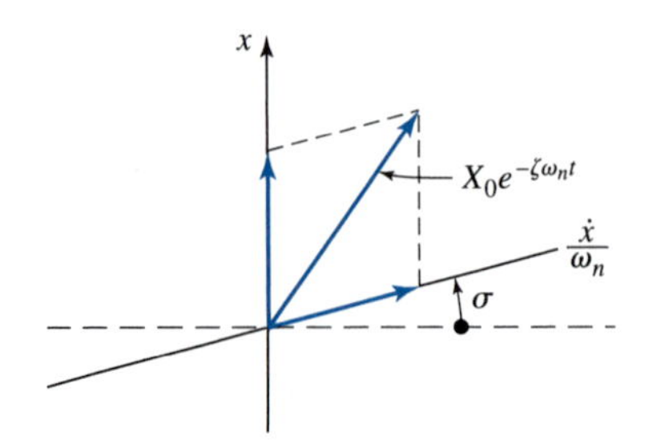

Figure 15.22 The angle σ is the damping phase angle and is applied only to the velocity term.

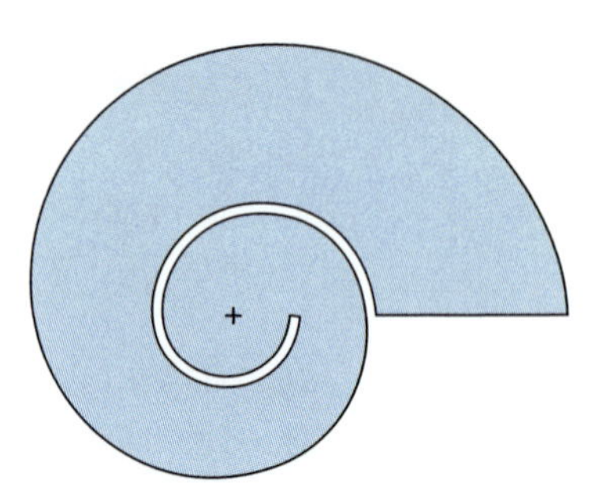

Figure 15.23 A spiral template for $\zeta = 0.15$.

phase relationship between the velocity and displacement is not 90° as it is for an undamped system. Instead, for a damped system, the phase angle depends upon the amount of damping present.

An easy way to obtain a phase-plane diagram for damped motion is to tilt or rotate the $\dot{x}/\omega_n$ axis through an angle σ, as illustrated in Fig. 15.22. The angle σ is called the *damping phase angle*, and the following analysis demonstrates how this angle comes about.

Taking the time derivative of Eq. (15.30) gives

$$\dot{x} = -X_0\zeta\omega_n e^{-\zeta\omega_n t}\cos\left(\sqrt{1-\zeta^2}\omega_n t - \phi\right) - X_0\sqrt{1-\zeta^2}\omega_n e^{-\zeta\omega_n t}\sin\left(\sqrt{1-\zeta^2}\omega_n t - \phi\right)$$

or

$$\frac{\dot{x}}{\omega_n} = -X_0 e^{-\zeta\omega_n t}\left[\zeta\cos\left(\sqrt{1-\zeta^2}\omega_n t - \phi\right) + \sqrt{1-\zeta^2}\sin\left(\sqrt{1-\zeta^2}\omega_n t - \phi\right)\right]. \qquad (a)$$

Designating $\cos\sigma = \sqrt{1-\zeta^2}$ and $\sin\sigma = \zeta$ and noting the trigonometric identity

$$\sin(\alpha+\beta) = \sin\alpha\cos\beta + \cos\alpha\sin\beta,$$

we see that Eq. (a) can be written as

$$\frac{\dot{x}}{\omega_n} = -X_0 e^{-\zeta\omega_n t}\sin\left(\sqrt{1-\zeta^2}\omega_n t - \phi + \sigma\right). \qquad (15.35)$$

If the damping is zero, note that $\sqrt{1-\zeta^2} = 1$, $e^{-\zeta\omega_n t} = 1$, and $\sigma = 0$, and Eqs. (15.30) and (15.35) reduce to the same set of equations that we employed to develop the phase-plane analysis of undamped motion. As illustrated in Fig. 15.22, the $\dot{x}/\omega_n$ axis should be rotated counterclockwise through the angle σ in plotting the phase-plane diagram. The velocity diagram is then obtained by projecting perpendicular to the $\dot{x}/\omega_n$ axis.

In analyzing a damped system that is acted upon by transient forces by the phase-plane method, it is very helpful to construct transparent spiral templates, as illustrated in Fig. 15.23. These are made from a sheet of clear plastic that is placed over a drawing of the spiral so that the spiral and its center can be scribed on the plastic using the point of a pair of dividers. The scribed line can then be trimmed with a pair of manicurist's scissors or a sharp razor blade. Because only one template is needed for each damping ratio, they can be stored for use in future problems. Do not forget to label each template with the value of the damping ratio for which it was constructed.

The spirals can be plotted directly from a table calculated using Eq. (15.30), but this is very tedious. A much more rapid method, which is accurate enough for graphical purposes, is to approximate each quadrant of the spiral with a circular arc. To do this it is necessary only to calculate the change in the length of the phasor in a quarter of a turn. From Eq. (15.30), the length of the phasor is $x = X_0 c^{-\zeta\omega_n t}$ and its angular velocity is $\omega_n\sqrt{1-\zeta^2}$. Therefore, for 90° rotation,

$$\omega_n\sqrt{1-\zeta^2}t = \frac{\pi}{2} \quad \text{or} \quad \omega_n t = \frac{\pi}{2\sqrt{1-\zeta^2}}. \tag{b}$$

Thus, the length, after rotation through 90°, is

$$x_{90^\circ} = X_0 e^{-\zeta\pi/2\sqrt{1-\zeta^2}}. \tag{c}$$

To demonstrate the construction of the spirals, let us employ a damping ratio $\zeta = 0.15$ and begin with a phasor 2.50 in. long. Then, after 90° of rotation, the length, from Eq. (c), is

$$x_{90^\circ} = (2.50)\, e^{-0.15\pi/2}\sqrt{1-(0.15)^2} = 1.97 \text{ in.}$$

The construction is illustrated in Fig. 15.24 and is explained as follows: construct the axes 1 and 2 at right angles to each other and lay off 2.50 in. to A and 1.97 in. to B on axes 1 and 2, respectively. Draw two lines through the origin at 45° to the axes. The perpendicular bisector of AB intersects one of these lines at P_1; using P_1 as a center, strike an arc from A to B. A line P_1B crosses another 45° line defining the center P_2 of arc BC. This process is continued in the same fashion with point P_3 defining the center of the arc CD, and so on.

The template is used by placing its center at the same point from which a circle arc would be constructed for undamped motion. It is then turned until the template spiral coincides with the point from which the spiral diagram is to be started.

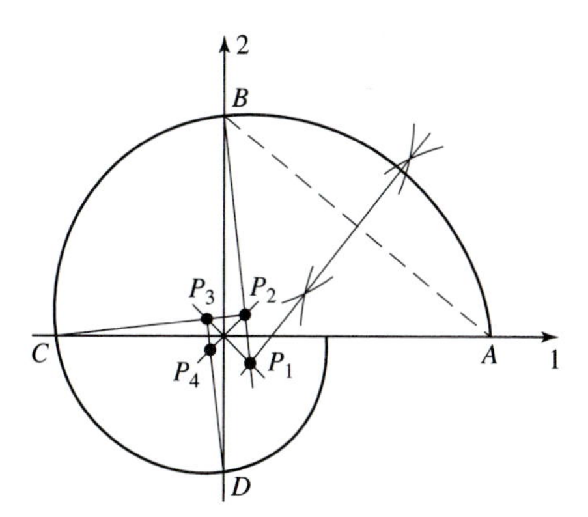

Figure 15.24

EXAMPLE 15.4

A potential vibrating system has an 80-lb mass mounted upon springs with an equivalent spring rate of 360 lb/in. A dashpot is included in the system and is estimated to produce 15% of the critical damping. A transient force on the system is assumed to act in three steps as follows: 720 lb for 0.0508 s, −270 lb for 0.0635 s, and 180 lb for 0.0381 s. The negative sign on the second step force simply means that its direction is opposite to the direction of the first and third steps. Determine the response of the system assuming no motion when the force function begins to act.

SOLUTION

The undamped natural frequency is

$$\omega_n = \sqrt{\frac{k}{m}} = \sqrt{\frac{(360 \text{ lb/in})}{(80 \text{ lb})/\left(386 \text{ in/s}^2\right)}} = 41.7 \text{ rad/s}$$

and so the damped natural frequency is

$$\omega_d = \omega_n\sqrt{1-\zeta^2} = (41.7 \text{ rad/s})\sqrt{1-(0.15)^2} = 41.2 \text{ rad/s}.$$

The period of the motion is $\tau = (2\pi \text{ rad/cycle})/\omega_d = 0.152$ s/cycle. The F/k values are

$$\frac{F_1}{k} = \frac{720 \text{ lb}}{360 \text{ lb/in}} = 2.00 \text{ in.}, \qquad \frac{F_2}{k} = \frac{270 \text{ lb}}{360 \text{ lb/in}} = 0.75 \text{ in.}, \qquad \frac{F_3}{k} = \frac{180 \text{ lb}}{360 \text{ lb/in}} = 0.50 \text{ in.}$$

These three steps are plotted in Fig. 15.25 to scale. The angular duration of each step for the phase-plane diagram is obtained by multiplying the damped natural frequency by the

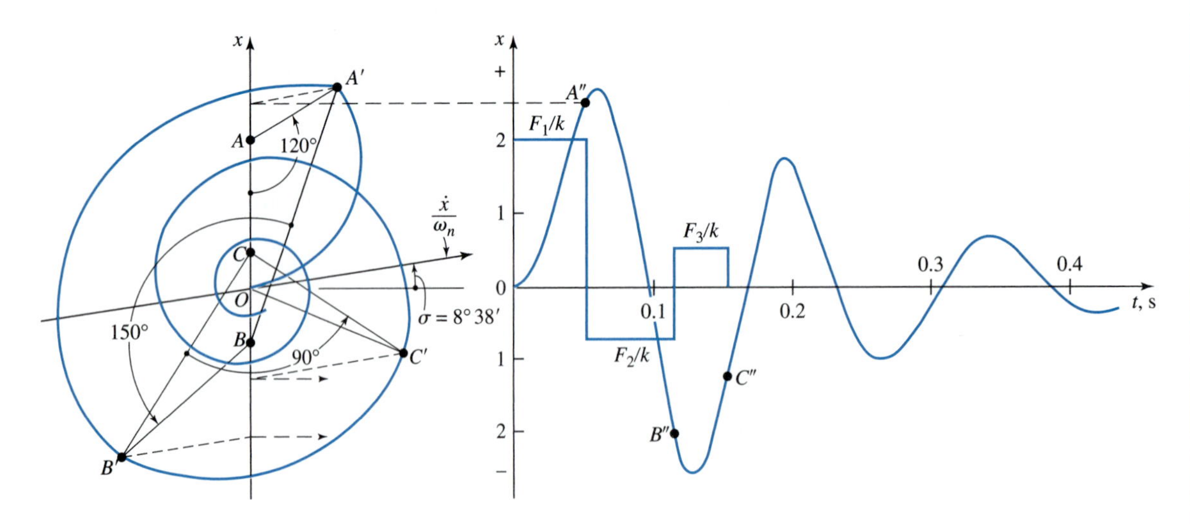

Figure 15.25 Example 15.4.

time of each step and converting to degrees. This gives

$$\text{Step1} \quad \omega_d \Delta t_1 \frac{180^\circ}{\pi \text{ rad}} = (41.2 \text{ rad/s})(0.0508 \text{ s}) \frac{180^\circ}{\pi \text{ rad}} = 120^\circ$$

$$\text{Step2} \quad \omega_d \Delta t_2 \frac{180^\circ}{\pi \text{ rad}} = (41.2 \text{ rad/s})(0.0635 \text{ s}) \frac{180^\circ}{\pi \text{ rad}} = 150^\circ$$

$$\text{Step3} \quad \omega_d \Delta t_3 \frac{180^\circ}{\pi \text{ rad}} = (41.2 \text{ rad/s})(0.0381 \text{ s}) \frac{180^\circ}{\pi \text{ rad}} = 90^\circ.$$

The $\dot{x}/\omega_n$ axis must be turned $\sigma = \sin^{-1}(0.15) = 8.63^\circ$ in the counterclockwise direction, as indicated in the phase-plane diagram.

The construction of Fig. 15.25 is explained as follows: project F_1/k to A on the x axis. A phasor AO, with center at A, begins rotating at the instant $t = 0$. This phasor rotates through an angle of 120° to AA' while exponentially decreasing its length. At A', the step function changes the origin of the phasor to B, and it rotates 150° about this point to B', also exponentially decreasing. At B' the origin again shifts, this time to C, and the phasor rotates 90° to C'. Now the origin shifts to O, and the phasor continues its rotation with O as the origin until the motion becomes too small to follow. Note that various points on the spiral diagram are projected parallel to the $\dot{x}/\omega_n$ axis, to the x axis, and then horizontally to the displacement diagram. In a similar manner, the velocity diagram, although not shown, would be projected perpendicular to the $\dot{x}/\omega_n$ axis. Points on the spiral would project parallel to the x axis until they intersect the $\dot{x}/\omega_n$ axis. The displacement diagram illustrates the resulting response. The maximum amplitude occurs on the first positive half-cycle and is 2.72 in.

15.11 RESPONSE TO PERIODIC FORCING

Differential equations of the form

$$m\ddot{x} + kx = ky(\omega t), \tag{15.36}$$

where $y(\omega t)$ is a periodic forcing function, occur often in mechanical systems. Let us define $\omega t = n\pi$, where n is an integer; we shall then be concerned in this section only with small values of n, say, n less than 8. An example of Eq. (15.36) is illustrated in Fig. 15.26 in which a roller riding in the groove of a face cam generates the motion from point O (Fig. 15.26*b*) of

$$y(\omega t) = y_0(1 - \cos \omega t). \tag{15.37}$$

This, then, is the motion applied to the left end of a spring k connected to a sliding mass m with friction assumed to be zero. Note, in Fig. 15.26*b*, that we choose Eq. (15.37) to represent the driving function because we have assumed that the roller is in its extreme left position when the cam begins its rotation.

Alternatively, we could have chosen point A in Fig. 15.26*b* to correspond to the start of the motion. For this choice the driving function would have been

$$y_A(\omega t) = y_0 \sin \omega t.$$

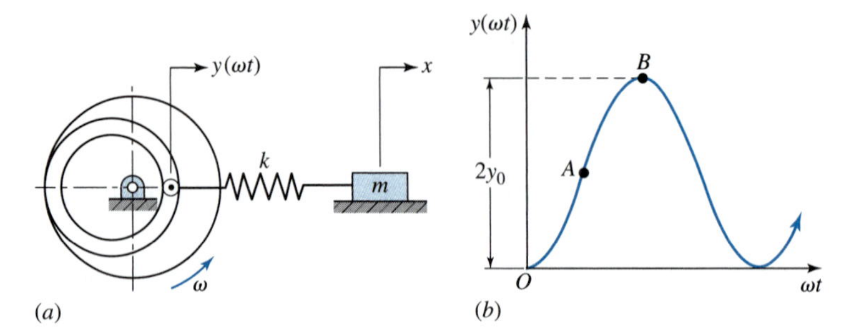

Figure 15.26

One of the reasons for introducing this problem is to illustrate an interesting relationship: Figure 15.26*a* represents a single-degree-of-freedom vibrating system; but, when Eq. (15.36) or a similar equation is solved, the solution often looks more like that of a two-degree-of-freedom system. Novices are sometimes quite surprised when they first view this solution and are unable to explain what is happening.

Substituting Eq. (15.37) into Eq. (15.36) gives the equation to be solved as

$$m\ddot{x} + kx = ky_0(1 - \cos\omega t). \tag{15.38}$$

The complementary solution is

$$x' = A\cos\omega_n t + B\sin\omega_n t$$

and the particular solution is

$$x'' = C + D\cos\omega t. \tag{a}$$

The second derivative of this equation is

$$\ddot{x}'' = -D\omega^2\cos\omega t. \tag{b}$$

By substituting Eqs. (*a*) and (*b*) into Eq. (15.38), we find that $C = y_0$ and

$$D = -\frac{ky_0}{k - m\omega^2} = -\frac{y_0}{1 - (\omega^2/\omega_n^2)}.$$

Therefore, the complete solution is

$$x = A\cos\omega_n t + B\sin\omega_n t + y_0\left[1 - \frac{\cos\omega t}{1 - (\omega^2/\omega_n^2)}\right]. \tag{15.39}$$

The velocity is

$$\dot{x} = -A\omega_n\sin\omega_n t + B\omega_n\cos\omega_n t + \frac{y_0\sin\omega t}{1 - (\omega^2/\omega_n^2)}. \tag{15.40}$$

Because we are starting from rest, where $x(0) = \dot{x}(0) = 0$, Eqs. (15.39) and (15.40) give

$$A = \frac{(\omega^2/\omega_n^2)y_0 \cos \omega_n t}{1 - (\omega^2/\omega_n^2)} \quad \text{and} \quad B = 0.$$

Equation (15.39) then becomes

$$x = \frac{(\omega^2/\omega_n^2)y_0 \cos \omega_n t}{1 - (\omega^2/\omega_n^2)} - \frac{y_0 \cos \omega t}{1 - (\omega^2/\omega_n^2)} + y_0. \tag{15.41}$$

The first term on the right-hand side of Eq. (15.41) is called the *starting transient.* Note that this is a vibration at the natural frequency ω_n, not at the forcing frequency ω. The usual physical system will contain a certain amount of friction, which, as we shall see in the sections to follow, will cause this term to die out after a certain period of time. The second and third terms on the right represent the *steady-state* solution and these contain another component of the vibration at the forcing frequency ω. Thus, the total motion is composed of a constant plus two vibrations of differing amplitudes and frequencies. If Fig. 15.26*a* were a cam-and-follower system, spring k would probably be quite stiff; consequently, ω_n would be quite large compared with ω. This would cause the amplitude of the starting transient to be relatively small, and a phase-diagram plot would resemble Fig. 15.27. Here we can see that the motion is formed by the sum of a fixed phasor of length y_0 and two phasors of different lengths and different rotational velocities.

A computer solution of a particular case in which $\omega_n = 3\omega$ is illustrated in Fig. 15.28. This is also of interest because of the nearly perfect dwells that occur at the extremes of the motion. One can easily prove the existence of these dwells by constructing a plot similar to Fig. 15.27 or by substituting $\omega_n = 3\omega$ in Eq. (15.41) and investigating the solution in the region for which ωt is zero or π.

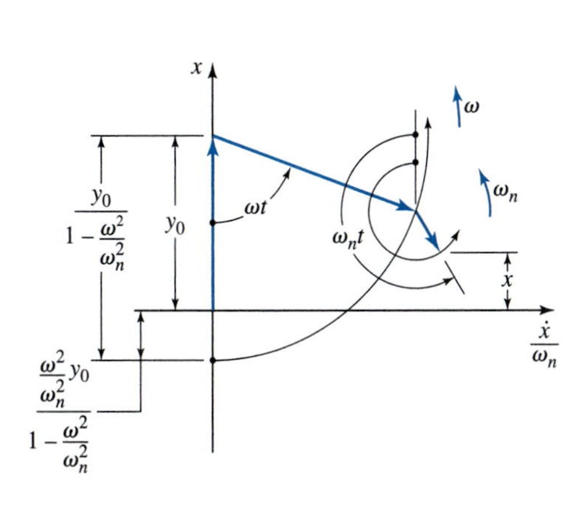

Figure 15.27

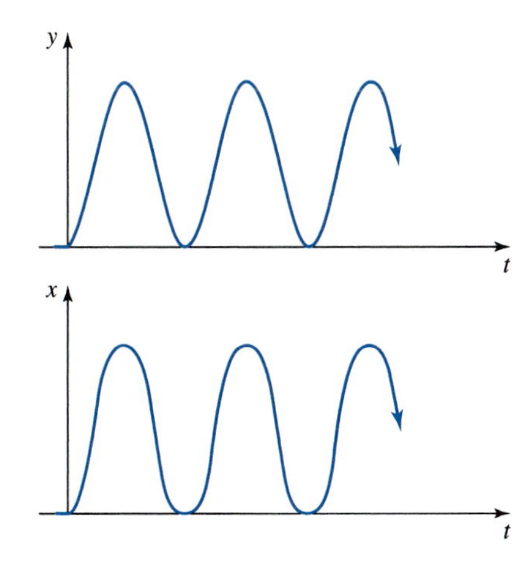

Figure 15.28 Computer solution of Eq. (15.38) for $\omega_n = 3\omega$; the amplitude scales for x and y are equal.

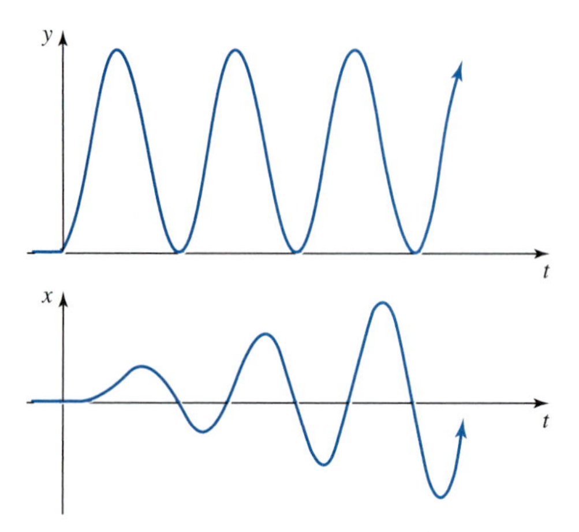

Figure 15.29 Response at resonance. The amplitude scale for x has been greatly increased to retain the plot within reasonable bounds.

Examination of Eq. (15.41) indicates that, for $\omega/\omega_n = 0$, the solution becomes

$$x = y_0 (1 - \cos \omega t),$$

which is the rigid-body solution. If we replace the spring with a rigid member, then k and, therefore, ω_n become very large, and so the mass exactly follows the cam motion.

Resonance is defined by the case in which $\omega = \omega_n$. Rearranging Eq. (15.41) to

$$x = y_0 \frac{(\omega^2/\omega_n^2) \cos \omega_n t - \cos \omega t + 1 - (\omega^2/\omega_n^2)}{1 - (\omega^2/\omega_n^2)},$$

we see that $x = 0/0$ for $\omega/\omega_n = 1$. However, L'Hôpital's rule enables us to find the solution; thus,

$$\lim_{\omega \to \omega_n} (x) = y_0 \left(1 - \cos \omega_n t + \frac{\omega_n t}{2} \sin \omega_n t \right).$$

A plot of this equation, obtained by solving the differential equation on the computer, is illustrated in Fig. 15.29. Note that the sine term grows and very quickly predominates in the response.

15.12 HARMONIC FORCING

We have seen that the action of any transient forcing function on a damped system is to create a vibration, but the vibration decays when the applied force is removed. A machine element that is connected to or is a part of any moving machinery is often subject to forces that vary periodically with time. Because all metal machine parts have both mass and elasticity, the opportunity for vibration always exists. Many machines do operate at fairly constant speeds and uniform output, and it is not difficult to see that vibratory forces may exist that have a fairly constant amplitude over a period of time. Of course, these varying forces do change in magnitude when the machine speed or output changes, but there is a rather broad class of vibration problems that can be analyzed and corrected using the assumption of a periodically varying force of constant amplitude. Sometimes these forces exhibit a time characteristic that is very similar to that of a sine wave. At other times they are quite complex and must be analyzed as a Fourier series. Such a series is the sum of a number of sine and cosine waves, and the resultant motion is the sum of the responses to the individual terms. In this text we shall study only the motion resulting from the application of a single sinusoidal force.

In Section 15.11 we examined an undamped system subjected to a periodic force and discovered that the solution contained components of the motion at two frequencies. One of these components contained the forcing frequency, whereas the other contained the natural frequency. Actual systems always have damping present, and this causes the component at the (damped) natural frequency to become insignificant after a period of time; the motion that remains contains only the driving frequency and is termed the *steady-state motion.*

To illustrate steady-state motion, we shall solve the equation

$$m\ddot{x} + c\dot{x} + kx = F_0 \cos \omega t. \tag{15.42}$$

The solution is in two parts, as before. The first, or complementary, part is the solution to the homogeneous equation

$$m\ddot{x} + c\dot{x} + kx = 0. \tag{15.43}$$

This we solved in Section 15.17, and we obtained [see Eq. (h)]

$$x' = e^{-\zeta\omega_n t}\left(C_1 \cos \omega_d t + C_2 \sin \omega_d t\right), \tag{15.44}$$

where C_1 *and* C_2 are the constants of integration. Because the forcing is harmonic, the particular part is obtained by assuming a solution in the form

$$x'' = A \cos \omega t + B \sin \omega t, \tag{15.45}$$

where ω is the frequency of the forcing function and where A and B are constants yet to be determined, but are not the constants of integration. Taking successive time derivatives of Eq. (15.45) produces

$$\dot{x}'' = -A\omega \sin \omega t + B \cos \omega t \tag{a}$$

$$\ddot{x}'' = -A\omega^2 \cos \omega t + B\omega^2 \sin \omega t \tag{b}$$

and substituting Eqs. (15.45), (a), and (b) back into the differential equation yields

$$(kA + c\omega B - m\omega^2 A) \cos \omega t + (kB + c\omega A - m\omega^2 B) \sin \omega t = F_0 \cos \omega t.$$

Next, separating the terms, we obtain

$$\begin{aligned} (k - m\omega^2)A + c\omega B &= F_0 \\ -c\omega A + (k - m\omega^2)B &= 0 \end{aligned}\ . \tag{c}$$

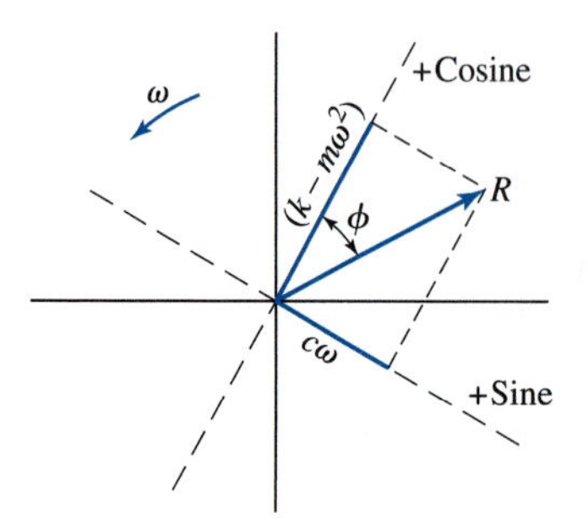

Figure 15.30

Solving Eqs. (c) for A and B and substituting the results back into Eq. (15.45), we obtain

$$x'' = \frac{F_0}{(k - m\omega^2)^2 + c^2\omega^2}[(k - m\omega^2)\cos\omega t + c\omega\sin\omega t]. \tag{d}$$

Now, when we create the phase diagram (Fig. 15.30) and plot the amplitudes of the two trigonometric terms, we find the resultant is

$$R = \sqrt{(k - m\omega^2)^2 + c^2\omega^2}, \tag{e}$$

lagging the direction of the positive cosine by a phase angle of

$$\phi = \tan^{-1}\left(\frac{c\omega}{k - m\omega^2}\right). \tag{15.46}$$

Substituting Eq. (e) into Eq. (d) and dividing out common factors gives

$$x'' = \frac{F_0\cos(\omega t - \phi)}{\sqrt{(k - m\omega^2)^2 + c^2\omega^2}} \tag{15.47}$$

and so the complete solution is

$$x = x' + x'' = e^{-\zeta\omega_n t}(C_1\cos\omega_d t + C_2\sin\omega_d t) + \frac{F_0\cos(\omega t - \phi)}{\sqrt{(k - m\omega^2)^2 + c^2\omega^2}}. \tag{15.48}$$

The first part of Eq. (15.48) is called the *transient* part because the exponential term will cause it to decay in a short period of time. When this happens, the second term will be all that remains and, because this has neither a beginning nor an end, it is called the *steady-state solution.*

When we are interested only in the steady-state term, we shall write the solution in the form of Eq. (15.47), omitting the prime marks on x.

Now use write Eq. (15.47) in the form

$$x = X\cos(\omega t - \phi) \tag{15.49}$$

and find the successive derivatives to be

$$\dot{x} = -\omega X \sin(\omega t - \phi) \tag{f}$$

$$\ddot{x} = -\omega^2 X \cos(\omega t - \phi). \tag{g}$$

These are then substituted into Eq. (15.42), and the forcing term is brought to the left-hand side of the equation. This gives

$$-m\omega^2 X \cos(\omega t - \phi) - c\omega X \sin(\omega t - \phi) + kX \cos(\omega t - \phi) - F_0 \cos \omega t = 0. \tag{h}$$

The amplitudes of the trigonometric terms are identified as follows:

$$\begin{aligned} m\omega^2 X &= \text{inertia force} \\ c\omega X &= \text{damping force} \\ kX &= \text{spring force} \\ F_0 &= \text{exciting force.} \end{aligned}$$

Our next step is to plot these phasors on the phase diagram. We are interested only in the steady-state solution; it has no beginning and no end and so, although the phasors have a definite phase relation with each other, they can exist anywhere on the diagram at the instant we examine them.

We begin Fig. 15.31*a* by selecting one of the directions, say, the direction of $\cos(\omega t - \phi)$. This automatically defines the direction of $\sin(\omega t - \phi)$. The direction of $\cos \omega t$ is ahead of $\cos(\omega t - \phi)$ by the angle ϕ. Having defined this set of directions, we lay off each phasor in the proper direction, being sure not to forget the signs.

Because Eq. (*h*) states that the sum of these four phasors is zero, we can also construct a phasor or vector polygon of them. This has been done in Fig. 15.31*b*; from the trigonometry,

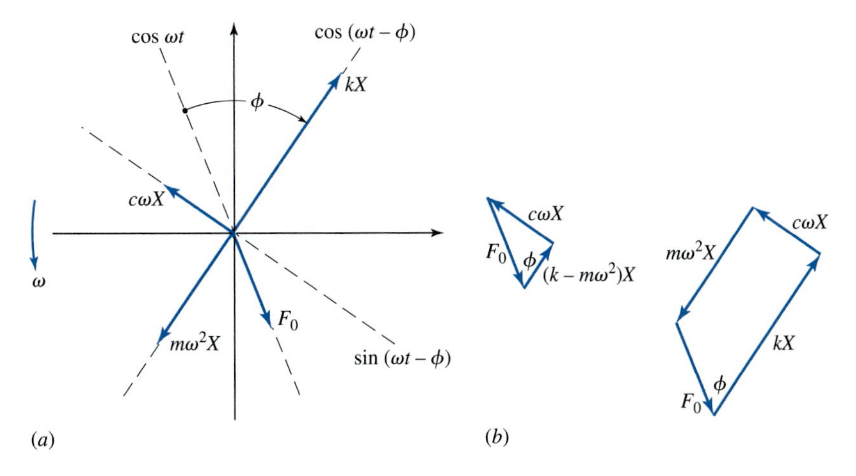

Figure 15.31

we see

$$\left[\left(k - m\omega^2\right)^2 + c^2\omega^2\right]X^2 = F_0^2$$

or

$$X = \frac{F_0}{\sqrt{\left(k - m\omega^2\right)^2 + c^2\omega^2}} \qquad (i)$$

and

$$\phi = \tan^{-1}\left(\frac{c\omega}{k - m\omega^2}\right). \qquad (15.50)$$

Consequently, Eq. (15.49) becomes

$$x = \frac{F_0 \cos(\omega t - \phi)}{\sqrt{\left(k - m\omega^2\right)^2 + c^2\omega^2}}. \qquad (15.51)$$

These equations can be simplified by introducing the expressions

$$\omega_n = \sqrt{\frac{k}{m}}, \qquad \zeta = \frac{c}{c_c}, \qquad c_c = 2m\omega_n.$$

They can then be written in the form

$$\frac{X}{F_0/k} = \frac{1}{\sqrt{(1 - \omega^2/\omega_n^2)^2 + (2\zeta\omega/\omega_n)^2}} \qquad (15.52)$$

and

$$\phi = \tan^{-1}\left(\frac{2\zeta\omega/\omega_n}{1 - \omega^2/\omega_n^2}\right), \qquad (15.53)$$

which is a dimensionless form that is convenient for analysis. It is interesting to note that the quantity F_0/k is the deflection that a spring of rate k would experience if acted upon by the force F_0.

If various values are selected for the frequency and damping ratios, curves can be plotted showing the effects upon the motion of varying these quantities. This has been done in Figs. 15.32 and 15.33. Note that the amplitude is very large near resonance (that is, where ω/ω_n is near unity) for small damping ratios. Note, too, that the peak amplitudes occur slightly before the resonance point is reached. Figure 15.33 also indicates that the phase angle changes very rapidly near the resonance condition.

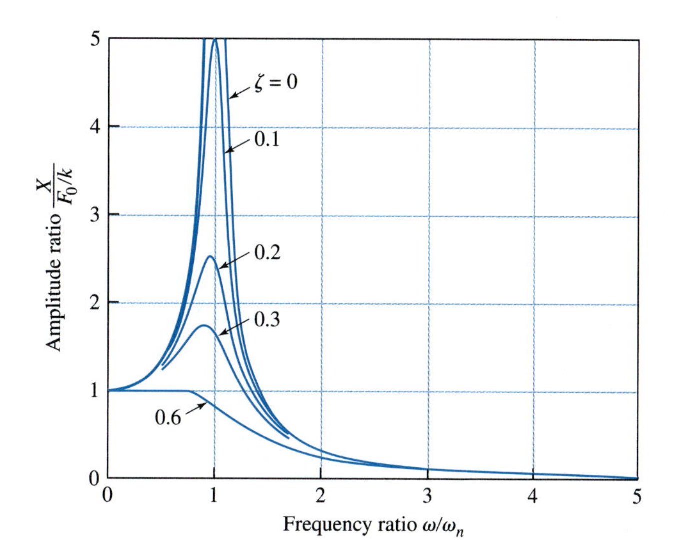

Figure 15.32 Relative displacement of a damped forced system as a function of the damping and frequency ratios.

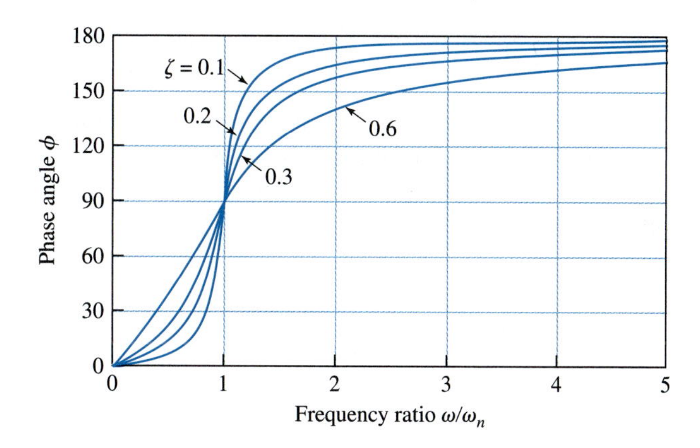

Figure 15.33 Relationship of the phase angle to the damping and frequency ratios.

15.13 FORCING CAUSED BY UNBALANCE

Quite frequently, the exciting force in a machine (see Chapter 17) is caused by the rapid rotation of a small unbalanced mass. In this case, the differential equation is written

$$m\ddot{x} + c\dot{x} + kx = m_u e\omega^2 \cos \omega t, \tag{15.54}$$

where m = vibrating mass
m_u = unbalanced mass.
e = eccentricity

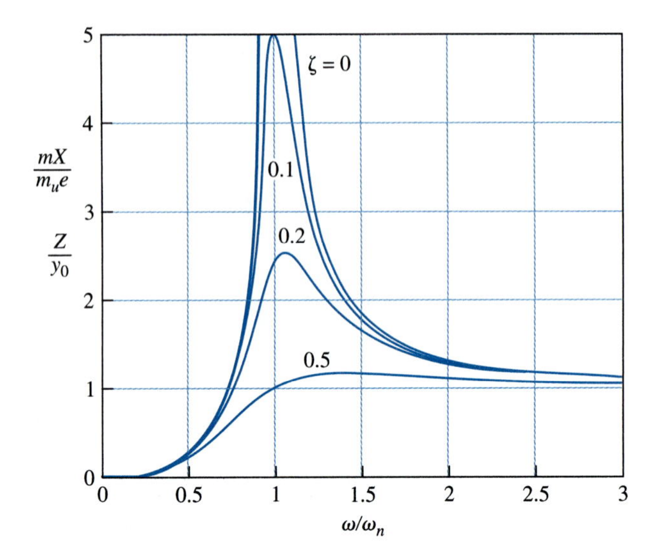

Figure 15.34 A plot of magnification factor versus frequency ratio for two systems. In one system the main mass is excited by a rotating unbalanced mass m_u. The chart gives the nondimensional absolute response of the main mass. In the other system the mass is excited by moving the far end of the spring and dashpot, which is connected to the mass, with harmonic motion. The chart gives the nondimensional response Z/y_0 relative to the far end of the spring–dashpot combination.

The solution to Eq. (15.54) is the same as that for Eq. (15.51) with F_0 replaced by $m_u e\omega^2$. Thus, the motion of the main mass is

$$x = \frac{m_u e\omega^2 \cos(\omega t - \phi)}{\sqrt{(k - m\omega^2)^2 + c^2\omega^2}}. \tag{15.55}$$

If we make the same transformation we used to obtain Eq. (15.52), we get

$$\frac{mX}{m_u e} = \frac{\omega^2/\omega_n^2}{\sqrt{[1 - (\omega^2/\omega_n^2)]^2 + (2\zeta\omega/\omega_n)^2}}. \tag{15.56}$$

This is plotted in Fig. 15.34. This graph illustrates that for high speeds where the frequency ratio is greater than unity, we can reduce the amplitude only by reducing the mass and eccentricity of the rotating unbalance.

15.14 RELATIVE MOTION

In many cases—in vibration-measuring instruments, for example—it is useful to know the response of a system relative to another moving system. Thus, in Fig. 15.35, we might wish to learn the relative response z instead of the absolute response x.

The differential equation for this system is

$$m\ddot{x} + c\dot{x} + kx = ky + c\dot{y}. \tag{15.57}$$

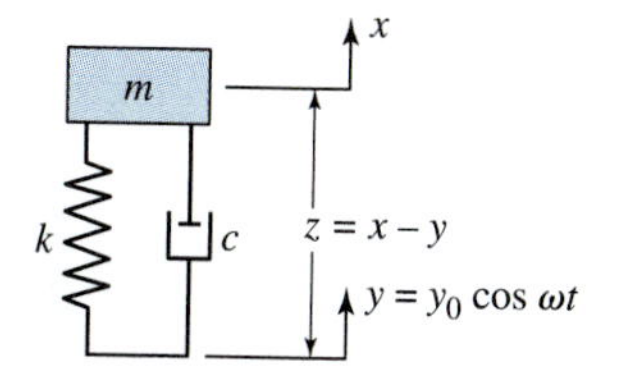

Figure 15.35

Note that because $\dot{z} = \dot{x} - \dot{y}$ and $\ddot{z} = \ddot{x} - \ddot{y}$, then written in terms of z and y, Eq. (15.57) becomes

$$m\ddot{z} + c\dot{z} + kz = -m\ddot{y}$$

or

$$m\ddot{z} + c\dot{z} + kz = -m\omega^2 y_0 \cos \omega t. \tag{15.58}$$

So the solution is the same as that of Eq. (15.54), and we can write

$$\frac{Z}{y_0} = \frac{\omega^2/\omega_n^2}{\sqrt{[1 - (\omega^2/\omega_n^2)]^2 + (2\zeta\omega/\omega_n)^2}}. \tag{15.59}$$

The response chart of Fig. 15.34 also applies to the relative response of such a system.

15.15 ISOLATION

The investigations so far enable the engineer to design mechanical equipment so as to minimize vibration and other dynamic problems, but because machines are inherently vibration generators, in many cases it is impractical to eliminate all such motion. In such cases, a more practical and economical vibration approach is that of reducing as much as possible the difficulties that the vibratory motion causes. This may take any of several directions, depending upon the nature of the problem. For example, everything economically feasible may have been done toward elimination of vibrations in a machine, but the residuals may still be strong and cause objectionable noise by transmitting these vibrations to the base structure. In another case, an item of equipment, such as a computer monitor, which is not of itself a vibration generator, may receive objectionable vibrations from another source. Both of these problems can be solved by isolating the equipment from the support. In analyzing these problems we may be interested, therefore, in the *isolation of forces* or in the *isolation of motions*, as illustrated in Fig. 15.36.

In Figs. 15.36*a* and 15.36*c* the forces acting against the support must be transmitted through the spring and dashpot because these are the only connections. The spring and damping forces are phasors that are always at right angles to each other, and so the force transmitted through these connections is

$$F_u = \sqrt{(kX^2) + (c\omega X)^2} = kX\sqrt{1 + \left(\frac{2\zeta\omega}{\omega_n}\right)^2}. \tag{15.60}$$

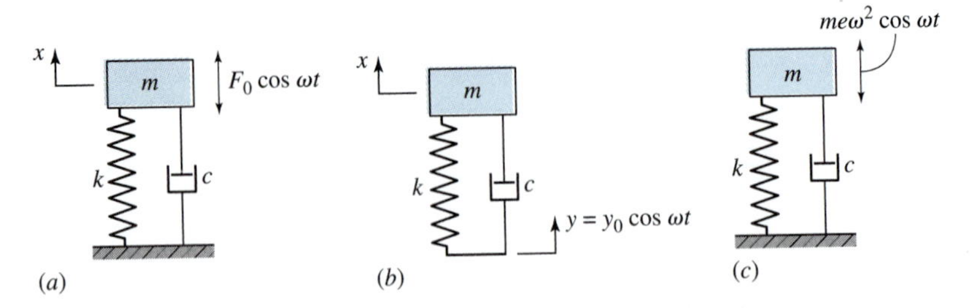

Figure 15.36 In (*a*) the exciting-force amplitude is constant, but in (*c*) it depends upon ω because it is caused by a rotating unbalance. In both (*a*) and (*c*) the support is to be isolated from the vibrating mass.

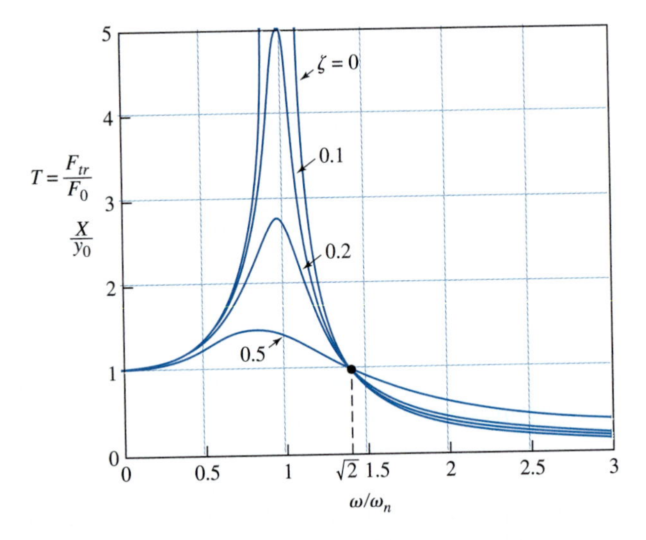

Figure 15.37 This is a plot of force transmissibility versus frequency ratio for a system in which a steady-state periodic forcing function is applied directly to the mass. The transmissibility is the percentage of the exciting force that is transmitted to the frame. This is also a plot of the amplitude-magnification factor for a system in which the far end of a spring and dashpot are driven by a sinusoidal displacement of amplitude y_0. The graph gives the nondimensional absolute response of the mass.

The *transmissibility* T is a nondimensional ratio that defines the percentage of the exciting force transmitted to the frame. For the system of Fig. 15.36*a*, X is given by Eq. (15.52), and so the transmissibility is

$$T = \frac{F_u}{F_0} = \frac{\sqrt{1 + (2\zeta\omega/\omega_n)^2}}{\sqrt{\left[1 - (\omega^2/\omega_n^2)\right]^2 + (2\zeta\omega/\omega_n)^2}}. \tag{15.61}$$

Figure 15.37 illustrates that the isolator must be designed so that ω/ω_n is large and the damping is small.

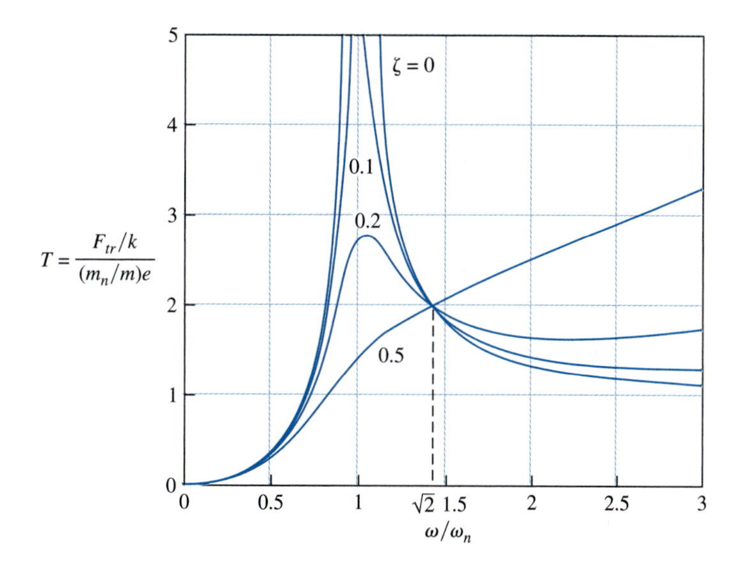

Figure 15.38 A plot of acceleration transmissibility versus frequency ratio. In a system in which the exciting force is produced by a rotating unbalanced mass, this plot gives the percentage of this force transmitted to the frame of the machine.

Using Eqs. (15.56) and (15.60), we find that the transmissibility for the system of Fig. 15.36*c* is

$$T = \frac{F_{tr}/k}{em_n/m} = \frac{\left(\omega^2/\omega_n^2\right)\sqrt{1 + (2\zeta\omega/\omega_n)^2}}{\sqrt{\left[1 - \left(\omega^2/\omega_n^2\right)\right]^2 + (2\zeta\omega/\omega_n)^2}}. \tag{15.62}$$

The plot of Fig. 15.38 is discouraging. For large values of ω/ω_n the damping actually serves to increase the transmissibility. The only practical solution, in general, is to balance the equipment as perfectly as possible to minimize the product em_n.

We shall choose the complex-operator method for the solution of the system of Fig. 15.36*b*. Note that this system is the same as that of Section 15.14, except that now we are interested in the absolute motion of the mass. We begin by defining the forcing function as

$$y = y_0 e^{j\omega t}.$$

Then,

$$\dot{y} = j\omega y_0 e^{j\omega t}.$$

Also, assuming a solution in the form

$$x = Xe^{j\omega t}$$

gives velocity and acceleration of the form

$$\dot{x} = j\omega X e^{j\omega t} \quad \text{and} \quad \ddot{x} = -\omega^2 X e^{j\omega t}.$$

Substituting all these into Eq. (15.57) and dividing through by the exponential produces,

$$-m\omega^2 X + jc\omega X + kX = ky_0 + jc\omega y_0,$$

and so

$$\frac{X}{y_0} = \frac{k + jc\omega}{(k - m\omega^2) + jc\omega}. \tag{a}$$

The product of two complex values is a new quantity whose magnitude is the product of the magnitudes of the original complex values and whose angle is the sum of the angles of the original complex values. Complex values can also be divided; in this case, we divide their magnitudes and subtract their angles.

In this case, we are not interested in the phase and so Eq. (a) becomes

$$\frac{X}{y_0} = \frac{\sqrt{k^2 + c^2\omega^2}}{\sqrt{\left(k - m\omega^2\right)^2 + c^2\omega^2}}.$$

Finally, dividing both numerator and denominator by m, we obtain

$$\frac{X}{y_0} = \frac{\sqrt{1 + (2\zeta\omega/\omega_n)^2}}{\sqrt{\left[1 - (\omega^2/\omega_n^2)\right]^2 + (2\zeta\omega/\omega_n)^2}}, \tag{15.63}$$

which matches Eq. (15.61) and is plotted in Fig. 15.37.

15.16 RAYLEIGH'S METHOD

For any single-degree-of-freedom system, the quantity W/k represents the static deflection y_0 of a spring of stiffness k caused by the weight W acting upon it. Rearranging the natural frequency equation by substituting $m = W/g$ gives

$$\omega_n = \sqrt{\frac{k}{m}} = \sqrt{\frac{g}{W/k}} = \sqrt{\frac{g}{y_0}}. \tag{15.64}$$

Equation (15.64) is, incidentally, very useful for determining natural frequencies of mechanical systems, because static deflections can usually be measured quite easily.

The static deflection y_0 of a cantilever beam caused by a concentrated weight W on the end is[2]

$$y_0 = \frac{Wl^3}{3EI},$$

where l is the length of the cantilever, E is its modulus of elasticity, and I is the second moment of area of its cross-section. Substituting this value into Eq. (15.64) produces

$$\omega_n = \sqrt{\frac{g}{y_0}} = \sqrt{\frac{3EIg}{Wl^3}}$$

so that the natural frequency in hertz is

$$f = \frac{1}{2\pi}\sqrt{\frac{3EIg}{Wl^3}}.$$

In our investigation of the mechanics of vibration we have been concerned with systems whose motions can be described using a single coordinate. In the cases of vibrating beams and rotating shafts, there may be many masses involved or the mass may be distributed. A coupled set of differential equations must be written, with one equation for each mass or each element of mass of the system, and these equations must be solved simultaneously if we are to obtain the equations of motion of a multimass system. Although a number of optional approaches are available, here we shall present an energy method, which is due to Lord Rayleigh[3], because of its importance in the study of vibration.

It is probable that Eq. (15.64) first suggested to Rayleigh the idea of employing the static deflection to find the natural frequency of a system. If we consider a freely vibrating system without damping then, during its motion, no energy is added to the system nor is any lost. Yet when the mass has velocity, kinetic energy exists, and when the spring is compressed or extended, potential energy exists. Because no energy is added or taken away, the maximum kinetic energy of a system must be the same as the maximum potential energy. This is the basis of Rayleigh's method; a mathematical statement of it is

$$T_{\max} = V_{\max}, \tag{15.65}$$

where T denotes the kinetic energy and V denotes the potential energy in the system.

In order to see how it works, let us apply the method to the simple cantilever beam. We begin by assuming that the motion is harmonic and of the form

$$y = y_0 \sin \omega_n t. \tag{a}$$

The potential energy is a maximum when the spring is fully extended or compressed and occurs when $\sin \omega_n t = 1$. It is

$$V_{\max} = \tfrac{1}{2} W y_0. \tag{b}$$

The velocity of the weight is given by

$$\dot{y} = y_0 \omega_n \cos \omega_n t. \tag{c}$$

The kinetic energy reaches a maximum when the velocity is a maximum, that is, when $\cos \omega_n t = 1$. The maximum kinetic energy is

$$T_{\max} = \tfrac{1}{2} m \dot{y}_{\max}^2 = \tfrac{1}{2}\frac{W}{g}(y_0 \omega_n)^2. \tag{d}$$

Applying Eq. (15.65),

$$\tfrac{1}{2}\frac{W}{g}(y_0\omega_n)^2 = \tfrac{1}{2}Wy_0 \quad \text{or} \quad \omega_n = \sqrt{\frac{g}{y_0}}, \tag{e}$$

which is identical with Eq. (15.64).

It is true in multimass systems that the dynamic deflection curves are not the same as the static deflection curves. The importance of the method, however, is that any *reasonable* deflection curve can be used in the process. The static deflection curve is a reasonable one and hence it gives a good approximation. Rayleigh also shows that the correct curve will always give the lowest value for the natural frequency (a lower bound), although we shall not demonstrate this fact here. It is sufficient to know that if many different deflection curves are assumed, the one giving the lowest natural frequency is the best.

Rayleigh's method is applied to a multimass system composed of weights W_1, W_2, W_3, and so on, by assuming, as before, a deflection of each mass according to the equation

$$y_i = y_{0i} \sin \omega_n t. \tag{f}$$

The maximum deflections are therefore $y_{0_1}, y_{0_2}, y_{0_3}$, and so on, and the maximum velocities are $y_{0_1}\omega_n, y_{0_2}\omega_n, y_{0_3}\omega_n$, and so on. The maximum potential energy for the system is

$$V_{\text{max}} = \tfrac{1}{2}W_1 y_{01} + \tfrac{1}{2}W_2 y_{02} + \tfrac{1}{2}W_2 y_{03} + \cdots = \tfrac{1}{2}\sum W_i y_{0i}. \tag{g}$$

The maximum kinetic energy for the system is

$$T_{\text{max}} = \frac{1}{2g}W_1 y_{01}^2\omega_n^2 + \frac{1}{2g}W_2 y_{02}^2\omega_n^2 + \frac{1}{2g}W_3 y_{03}^2\omega_n^2 + \cdots = \frac{\omega_n^2}{2g}\sum W_i y_{0i}^2. \tag{h}$$

Applying Rayleigh's principle and solving for the frequency yields

$$\omega_n = \sqrt{\frac{g\sum W_i y_{0i}}{\sum W_i y_{0i}^2}} \quad \text{or} \quad f = \frac{1}{2\pi}\sqrt{\frac{g\sum W_i y_{0i}}{\sum W_i y_{0i}^2}}. \tag{15.66}$$

Equation (15.66), commonly called the Rayleigh–Ritz equation, can be applied to a beam consisting of several masses or to a rotating shaft having masses in the form of gears, pulleys, flywheels, and the like mounted upon it. If the speed of the rotating shaft should become equal to the natural frequency given by Eq. (15.66), then violent vibrations will occur. For this reason it is common practice to designate this frequency the *critical speed.*

15.17 FIRST AND SECOND CRITICAL SPEEDS OF A SHAFT

This section presents an approach to obtain exact solutions to the first and second critical speeds of a rotating shaft using the method of *influence coefficients*.

The deflection at location i of a shaft caused by a load of unit magnitude at location j is denoted a_{ij} and is referred to as an *influence coefficient*. Note the reciprocal relationship between the definition of an influence coefficient and the definition of spring stiffness, that is,

$$a_{ij} = \frac{1}{k_{ij}}. \tag{15.67}$$

So, for a shaft of negligible mass supporting two disks with mass (for example, gears, pulleys, or flywheels), the influence coefficients are as follows:

a_{11} = the deflection at location 1 caused by a load of unit magnitude at location 1

a_{12} = the deflection at location 1 caused by a load of unit magnitude at location 2

a_{21} = the deflection at location 2 caused by a load of unit magnitude at location 1

a_{22} = the deflection at location 2 caused by a load of unit magnitude at location 2.

Maxwell's Reciprocity Theorem This theorem states that the deflection at location i caused by a unit load at location j is equal to the deflection at location j caused by a unit load at location i.

$$a_{ij} = a_{ji} \tag{15.68}$$

The total deflection at location 1 caused by both disks can be written, by superposition, as

$$y_1 = a_{11}W_1 + a_{12}W_2. \tag{15.69}$$

Similarly, the total deflection at location 2 caused by both disks can be written as

$$y_2 = a_{21}W_1 + a_{22}W_2. \tag{15.70}$$

Therefore, the total deflections at locations 1 and 2 can be written from Eqs. (15.69) and (15.70) as

$$y_1 = a_{11}(m_1y_1\omega^2) + a_{12}(m_2y_2\omega^2)$$

and

$$y_2 = a_{21}(m_1y_1\omega^2) + a_{22}(m_2y_2\omega^2).$$

Rearranging these two equations and writing them in matrix form gives

$$\begin{bmatrix} a_{11}m_1\omega^2 - 1 & a_{12}m_2\omega^2 \\ a_{21}m_1\omega^2 & a_{22}m_2\omega^2 - 1 \end{bmatrix} \begin{bmatrix} y_1 \\ y_2 \end{bmatrix} = \begin{bmatrix} 0 \\ 0 \end{bmatrix}. \tag{15.71}$$

A nontrivial solution requires that the determinant of the (2×2) coefficient matrix must be zero, that is,

$$(a_{11}m_1\omega^2 - 1)(a_{22}m_2\omega^2 - 1) - (a_{12}m_2\omega^2)(a_{21}m_1\omega^2) = 0,$$

which can be written as

$$m_1 m_2(a_{11}a_{22})\omega^4 - (a_{11}m_1 + a_{22}m_2\omega^2) + 1 - m_1 m_2(a_{12}a_{21})\omega^4 = 0.$$

Rearranging this equation and dividing by ω^4, we have

$$1/\omega^4 - (a_{11}m_1 + a_{22}m_2)/\omega^2 + m_1 m_2(a_{11}a_{22} - a_{12}a_{21}) = 0. \tag{15.72}$$

Equation (15.72) is commonly referred to as the *frequency equation*. If we substitute $X = 1/\omega^2$, then we have a quadratic equation in X, that is,

$$X^2 - (a_{11}m_1 + a_{22}m_2)X + m_1 m_2(a_{11}a_{22} - a_{12}a_{21}) = 0. \tag{15.73}$$

Then the two solutions for X can be written as

$$X_1, X_2 = \left(\frac{a_{11}m_1 + a_{22}m_2}{2}\right) \pm \sqrt{\left(\frac{a_{11}m_1 + a_{22}m_2}{2}\right)^2 - m_1 m_2(a_{11}a_{22} - a_{12}a_{21})}. \tag{15.74}$$

This equation is commonly referred to as the *exact equation* for the first two critical speeds of the system.

This procedure can easily be extended to a rotating shaft supporting n disks. In this case, Eq. (15.71) is an $(n \times n)$ matrix and Eq. (15.73) is an nth order polynomial in $X = 1/\omega^2$. The n critical speeds of the system can then be determined from this polynomial.

Dunkerley's Method The Dunkerley equation is an approximation for the first critical speed of a rotating shaft supporting several mass disks. Note that the coefficient of X in Eq. (15.73) is the sum of the roots of the polynomial; therefore, for two mass disks, that is,

$$\frac{1}{\omega_1^2} + \frac{1}{\omega_2^2} = a_{11}m_1 + a_{22}m_2. \tag{15.75}$$

This is a single equation in two unknowns, ω_1 and ω_2. To obtain an approximation to the first critical speed, assume that

$$\frac{1}{\omega_1^2} \gg \frac{1}{\omega_2^2}. \tag{15.76}$$

Therefore, Eq. (15.75) can be written as

$$\frac{1}{\omega_1^2} = a_{11}m_1 + a_{22}m_2. \tag{15.77}$$

From Eq. (15.77), the Dunkerley approximation for the first critical speed can be written as

$$\omega_1 = \frac{1}{\sqrt{a_{11}m_1 + a_{22}m_2}}. \tag{15.78}$$

Note that the Dunkerley approximation will always underestimate the first critical speed (a lower bound). However, in general, the Rayleigh–Ritz equation [see Eq. (15.66)] will overestimate the first critical speed (an upper bound).

EXAMPLE 15.5

A steel shaft that is 4 ft long and has 0.5-in diameter is simply supported in two bearings as illustrated in Fig. 15.39. Two 80-lb gears are attached to the shaft. One gear is 12 in to the right of the left bearing and the other is 12 in to the left of the right bearing. The mass of the shaft is negligible compared with the mass of the gears. The influence coefficients are known to be $a_{11} = a_{22} = 5.504 \times 10^{-5}$ in/lb and $a_{12} = a_{21} = 4.278 \times 10^{-5}$ in/lb. Determine the first critical speed of the system using (1) the Rayleigh equation, (2) the exact equation, and (3) the Dunkerley equation.

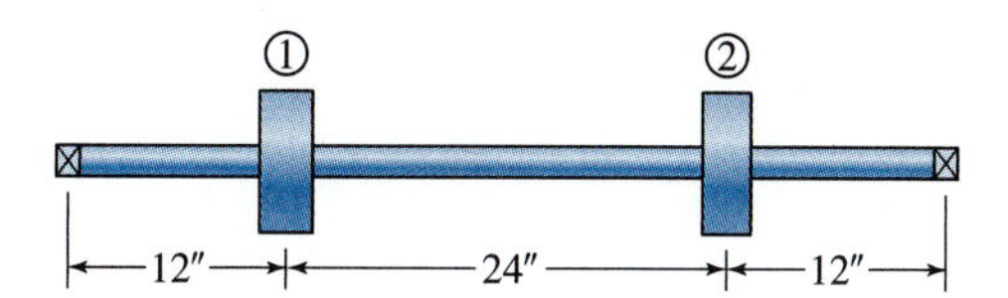

Figure 15.39 Example 15.5.

SOLUTION

(1) Rayleigh's equation [see Eq. (15.66)] can be written as

$$\omega_1^2 = \frac{g(W_1 y_1 + W_2 y_2)}{W_1 y_1^2 + W_2 y_2^2}. \tag{1}$$

Because the two gears have the same weight, $W_1 = W_2$, and Eq. (1) can be written

$$\omega_1^2 = \frac{g(y_1 + y_2)}{y_1^2 + y_2^2}. \tag{2}$$

The total deflections at locations 1 and 2 from Eqs. (15.69) and (15.70) are

$$y_1 = a_{11} W_1 + a_{12} W_2 \tag{3}$$

$$y_2 = a_{21} W_1 + a_{22} W_2. \tag{4}$$

Note from symmetry that the deflection is $y_1 = y_2 = y$; therefore, Eq. (2) can be written

$$\omega_1^2 = \frac{g}{y}. \tag{5}$$

Substituting the given data into Eq. (3) or (4) gives

$$y = y_1 = y_2 = 80\ \text{lb}(a_{11} + a_{12}). \tag{6a}$$

Therefore, the deflection is

$$y = 80\ \text{lb}(5.504 + 4.278)10^{-5}\ \text{in/lb} = 782.56 \cdot 10^{-5}\ \text{in.} \tag{6b}$$

Substituting Eq. (6*b*) into Eq. (5) gives

$$\omega_1^2 = \frac{386\ \text{in/s}^2 \times 10^5}{782.56\ \text{in}} = 49\,325\ \text{rad}^2/\text{s}^2$$

Therefore, the first critical speed by Rayleigh's method is

$$\omega_1 = 222\ \text{rad/s.} \qquad \textit{Ans.}$$

(2) From Eq. (15.74), the exact equation for the first two critical speeds of the system is

$$X_1, X_2 = \left(\frac{a_{11}m_1 + a_{22}m_2}{2}\right) \pm \sqrt{\left(\frac{a_{11}m_1 + a_{22}m_2}{2}\right)^2 - m_1 m_2(a_{11}a_{22} - a_{12}a_{21})}. \tag{7}$$

Substituting the given data into this equation gives

$$X_1, X_2 = 1.141 \cdot 10^{-5}\ \text{s}^2 \pm \sqrt{(1.141^2 - 0.516) \cdot 10^{-10}}\ \text{s}^2, \tag{8a}$$

which can be written as

$$\frac{1}{\omega_1^2}, \frac{1}{\omega_2^2} = (1.141 \pm 0.886) \cdot 10^{-5}\ \text{s}^2 \tag{8b}$$

or as

$$\omega_1^2 = 49\,334\ \text{rad}^2/\text{s}^2 \quad \text{and} \quad \omega_2^2 = 392\,157\ \text{rad}^2/\text{s}^2. \tag{8c}$$

Therefore, the first and second critical speeds from the exact equation are

$$\omega_1 = 222\ \text{rad/s} \quad \text{and} \quad \omega_2 = 626\ \text{rad/s}. \qquad \textit{Ans.}$$

Note that for this example the answer for the first critical speed from the Rayleigh equation is the same as the answer from the exact equation.

(3) The Dunkerley equation for the rotating shaft supporting the two mass discs is

$$\frac{1}{\omega_1^2} = a_{11}m_1 + a_{22}m_2. \tag{9}$$

Substituting the given data into this equation gives

$$\frac{1}{\omega_1^2} = 2\left(\frac{80\ \text{lb}}{386\ \text{in/s}^2}\right) 5.500 \cdot 10^{-5}\ \text{in/lb} = 2.280 \cdot 10^{-5}\ \text{s}^2. \tag{10}$$

Therefore, the first critical speed for the system is

$$\omega_1 = 209\,\text{rad/s}. \qquad \textit{Ans.}$$

Note that this answer is less than the first critical speed obtained from the exact equation.

EXAMPLE 15.6

The rotating shaft illustrated in Fig. 15.40 supports two flywheels, each of mass m. The first critical speed of the system has been determined experimentally to be 300 rad/s and the first critical speed of the shaft alone is 600 rad/s. If one of the flywheels is removed from the shaft, then determine the first critical speed of the system.

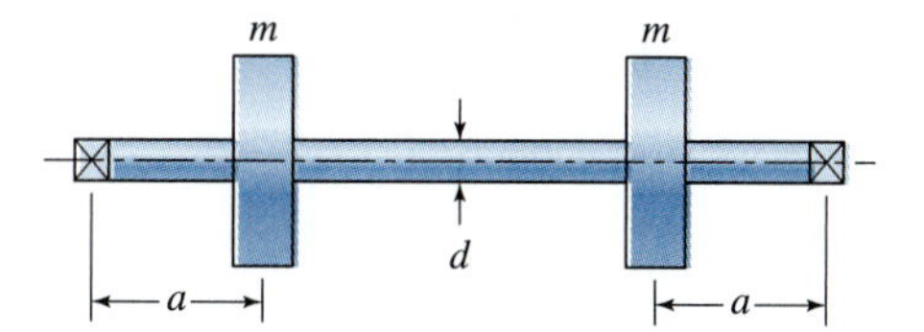

Figure 15.40 Example 15.6.

SOLUTION

The first natural frequency of the shaft carrying the two flywheels can be written as

$$\frac{1}{\omega_1^2} = \frac{1}{\omega_{11}^2} + \frac{1}{\omega_{22}^2} + \frac{1}{\omega_{ss}^2} \tag{1a}$$

or as

$$\frac{1}{\omega_1^2} = ma_{11} + ma_{22} + \frac{1}{\omega_{ss}^2}, \tag{1b}$$

where ω_{ss} is the critical speed of the shaft alone.

From symmetry we note that the influence coefficient $a_{11} = a_{22}$; therefore, Eq. (1b) can be written as

$$2ma_{11} = \frac{1}{\omega_1^2} - \frac{1}{\omega_{ss}^2}. \tag{2}$$

Substituting the given critical speeds into Eq. (2) gives

$$2ma_{11} = \frac{1}{(300\,\text{rad/s})^2} - \frac{1}{(600\,\text{rad/s})^2} = \frac{1}{120\,000}\,\text{s}^2. \tag{3}$$

For the massless shaft with only one flywheel, the first natural frequency can be written from Eq. ($1a$) as

$$\frac{1}{\omega_1^2} = \frac{1}{\omega_{11}^2} + \frac{1}{\omega_{ss}^2} \tag{4}$$

or as

$$\frac{1}{\omega_1^2} = ma_{11} + \frac{1}{\omega_{ss}^2}. \tag{5}$$

Substituting Eq. (3) into Eq. (5) gives

$$\frac{1}{\omega_1^2} = \frac{1}{2}\left[\frac{1}{120\,000}\,\mathrm{s}^2\right] + \frac{1}{\omega_{ss}^2}. \tag{6}$$

Finally, substituting the critical speed of the shaft $\omega_{ss} = 600$ rad/s into Eq. (6) gives

$$\frac{1}{\omega_1^2} = \frac{1}{2}\left[\frac{1}{120\,000}\,\mathrm{s}^2\right] + \frac{1}{(600\,\mathrm{rad/s})^2} = 6.944 \cdot 10^{-6}\,\mathrm{s}^2. \tag{7}$$

Therefore, the first critical speed of the system is

$$\omega_1 = 379.5\,\mathrm{rad/s}. \qquad \textit{Ans.}$$

In the investigation of rectilinear vibrations to follow, we shall see that there are as many natural frequencies in a multimass vibrating system as there are degrees of freedom. Similarly, a three-mass torsional system will have three degrees of freedom and, consequently, three natural frequencies. Equation (15.66) gives only the first or lowest of these frequencies.

It is probable that the critical speed can be determined by Eq. (15.66), for most cases, within about 5%. This is because of the assumption that the static and dynamic deflection curves are identical. Bearings, couplings, belts, and so on all have an effect upon the spring constant and the damping. Shaft vibration usually occurs over an appreciable range of shaft speed, and for this reason Eq. (15.66) is accurate enough for many engineering purposes.

As a general rule, shafts have many discontinuities—such as shoulders, keyways, holes, and grooves—to locate and secure various gears, pulleys, and other shaft-mounted masses. These diametral changes usually require deflection analysis using numerical integration. For such problems, Simpson's rule is easy to apply using a computer or calculator or by hand calculations.[4]

15.18 TORSIONAL SYSTEMS

Figure 15.41 illustrates a shaft supported on bearings at A and B with two masses connected at the ends. The masses represent any rotating machine parts—an engine and its flywheel, for example. We wish to study the possibility of free vibration of the system when it rotates at constant angular velocity. To investigate the motion of each mass, it is necessary to picture

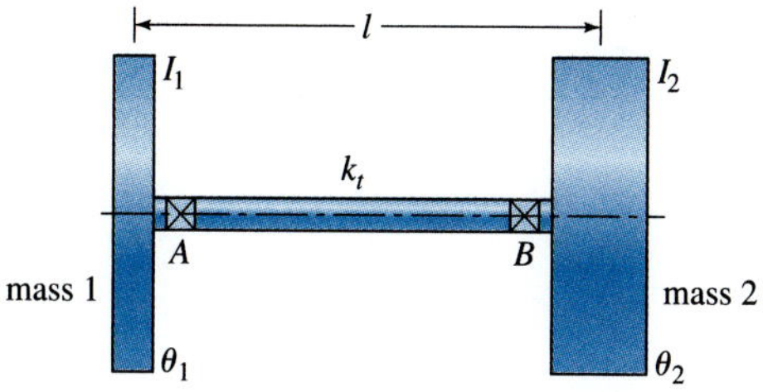

Figure 15.41

a reference system fixed to the shaft and rotating with the shaft at the same angular velocity. Then we can measure the angular displacement of either mass by finding the instantaneous angular location of a mark on the mass relative to one of the rotating axes. Thus, we define θ_1 and θ_2 as the angular displacements of mass 1 and mass 2, respectively, with respect to the rotating axes.

Now, assuming no damping, Eq. (15.6) is written for each mass,

$$\begin{aligned} I_1\ddot{\theta}_1 + k_t(\theta_1 - \theta_2) &= 0 \\ I_2\ddot{\theta}_2 + k_t(\theta_2 - \theta_1) &= 0 \end{aligned}, \tag{a}$$

where the angle $(\theta_1 - \theta_2)$ represents the total twist of the shaft. A solution is assumed in the form

$$\theta_i = \gamma_i \sin \omega_n t.$$

Then, the angular acceleration is

$$\ddot{\theta}_i = -\gamma_i \omega_n^2 \sin \omega_n t.$$

If these equations are substituted into Eqs. (*a*), we obtain

$$\begin{aligned} (k_t - I_1\omega_n^2)\gamma_1 - k_t\gamma_2 &= 0 \\ -k_t\gamma_1 + (k_t - I_2\omega_n^2)\gamma_2 &= 0 \end{aligned}, \tag{b}$$

which can be written in matrix form as

$$\begin{bmatrix} k_t - I_1\omega_n^2 & -k_t \\ -k_t & k_t - I_2\omega_n^2 \end{bmatrix} \begin{bmatrix} \gamma_1 \\ \gamma_2 \end{bmatrix} = \begin{bmatrix} 0 \\ 0 \end{bmatrix}.$$

For a nontrivial solution, the determinant of the coefficient matrix must be zero; that is,

$$(k_t - I_1\omega_n^2)(k_t - I_2\omega_n^2) - k_t^2 = 0,$$

so that

$$\omega_n^2 \left(\omega_n^2 - k_t \frac{I_1 + I_2}{I_1 I_2} \right) = 0$$

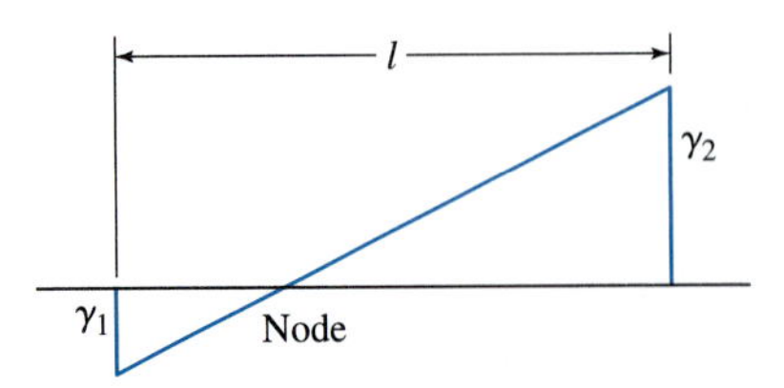

Figure 15.42

and

$$\omega_{n1} = 0 \quad \text{and} \quad \omega_{n2} = \sqrt{k_t \frac{I_1 + I_2}{I_1 I_2}}. \tag{15.79}$$

Substituting $\omega_{n1} = 0$ into Eqs. (b), we find that $\gamma_1 = \gamma_2$. Therefore, the masses rotate together without any relative displacement and there is no vibration.

Substituting the value of ω_{n2} into Eqs. (b), it is determined that

$$\gamma_1 = -\frac{I_2}{I_1}\gamma_2,$$

indicating that the motion of the second mass is opposite to that of the first and proportional to I_1/I_2. Figure 15.42 is a graphical representation of the vibration. Here, l is the distance between the two masses and γ_1 and γ_2 are the instantaneous angular displacements plotted to scale. If a line is drawn connecting the ends of the angular displacements, this line crosses the axis at a *node*, which is a location on the shaft having zero angular displacement. It is convenient to designate the configuration of the vibrating system as a *mode* of vibration. This system has two modes of vibration, corresponding to the two frequencies, although we have seen that one of them is degenerate. A system of n masses would have n modes corresponding to n different frequencies.

For multimass torsional systems, the *Holzer tabulation method* can be used to find all the natural frequencies. It is easy to use and avoids solving the simultaneous differential equations.[5]

15.19 REFERENCES

[1] P. W. Bridgman, *The Logic of Modern Physics*. New York: Macmillan, 1946, p. 62.

[2] J. E. Shigley and C. R. Mischke, *Mechanical Engineering Design*, 6th ed. New York: McGraw–Hill, 2001, Table E-9, pp. 1179–86.

[3] J. W. Strutt (Baron Rayleigh), *Theory of Sound.* Republished New York: Dover, 1945. This book is in two volumes and is the classic treatise on the theory of vibrations. It was originally published in 1877–1878.

[4] See J. E. Shigley and C. R. Mischke, *ibid*, pp. 142–3, for methods and examples of such calculations.

[5] A detailed description of this method with a worked-out numerical example can be found in T. S. Sankar and R. B. Bhat, "Viration and Control of Vibration," in J. E. Shigley and C. R. Mischke (eds.), *Standard Handbook of Machine Design*. New York: McGraw–Hill, 1986, pp. 38.22–7.

PROBLEMS

†15.1 Derive the differential equation of motion for each of the systems illustrated in Fig. P15.1 and write the formula for the natural frequency ω_n for each system.

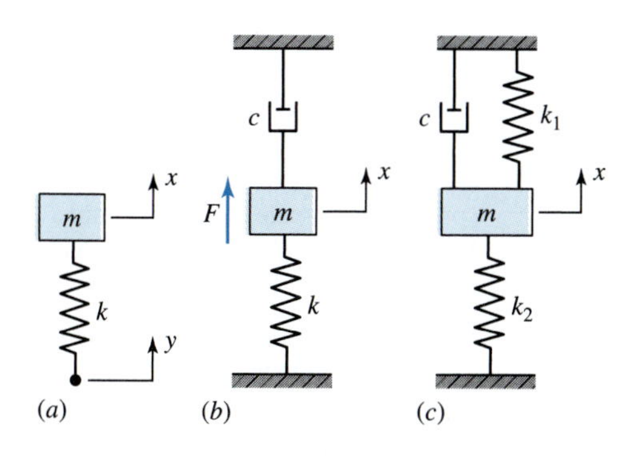

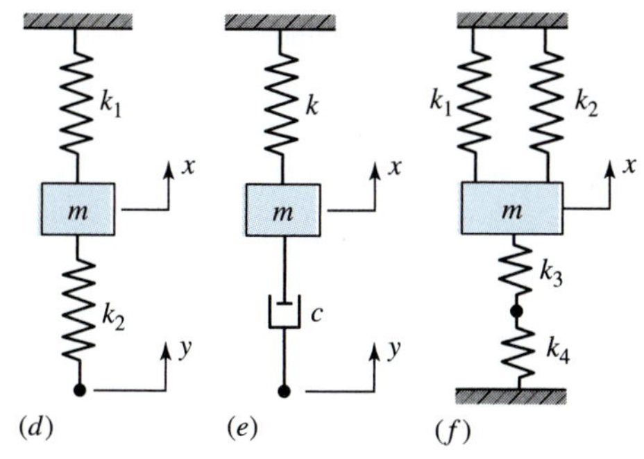

Figure P15.1

†15.2 Evaluate the constants of integration of the solution to the differential equation for an undamped free system, using the following sets of starting conditions:

(a) $x = x_0, \dot{x} = 0$

(b) $x = 0, \dot{x} = v_0$

(c) $x = x_0, \ddot{x} = a_0$

(d) $x = x_0, \dddot{x} = b_0$.

For each case, transform the solution to a form containing a single trigonometric term.

†15.3 A system like Fig. 15.5 has $m = 1$ kg and an equation of motion $x = 20 \cos(8\pi t - \pi/4)$ mm. Determine the following:

(a) The spring constant k

(b) The static deflection δ_{st}

(c) The period

(d) The natural frequency in hertz

(e) The velocity and acceleration at the instant $t = 0.20$ s

(f) The spring force at $t = 0.20$ s.

Plot a phase diagram to scale showing the displacement, velocity, acceleration, and spring-force phasors at the instant $t = 0.20$ s.

†15.4 The weight W_1 in Fig. P15.4 drops through the distance h and collides with W_2 with plastic impact (a coefficient of restitution of zero). Derive the differential equation of motion of the system and determine the amplitude of the resulting motion of W_2.

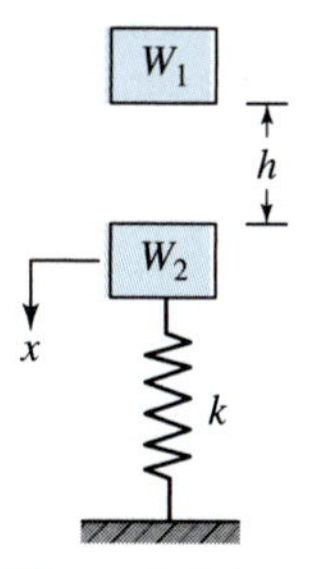

Figure P15.4

15.5 The vibrating system illustrated in Fig. P15.5 has $k_1 = k_3 = 875$ N/m, $k_2 = 1\,750$ N/m, and $W = 40$ N. What is the natural frequency in hertz?

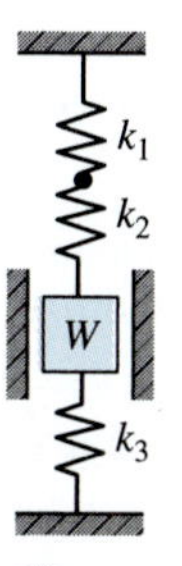

Figure P15.5

15.6 Figure P15.6 illustrates a weight $W = 15$ lb connected to a pivoted rod, which is assumed to be weightless and very rigid. A spring having a rate of $k = 60$ lb/in. is connected to the center of the rod and holds the system in static equilibrium at the position shown. Assuming that the rod can vibrate with a small amplitude, determine the period of the motion.

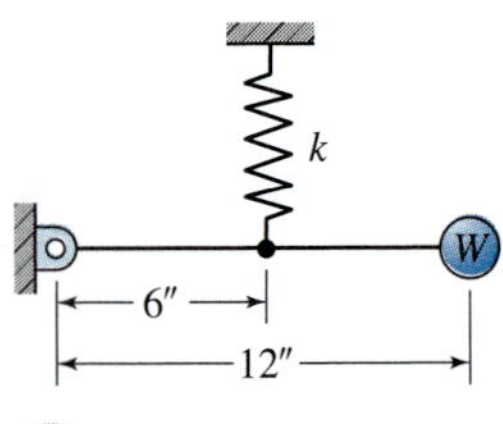

Figure P15.6

15.7 Figure P15.7 illustrates an upside-down pendulum of length l retained by two springs connected a distance a from the pivot. The springs have been positioned such that the pendulum is in static equilibrium when it is in the vertical position.

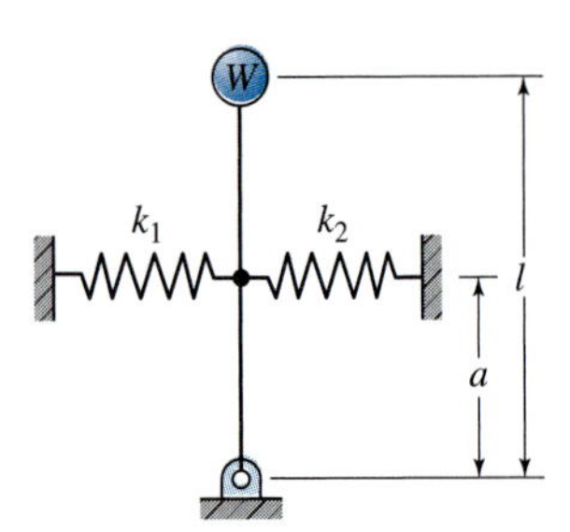

Figure P15.7

(a) For small amplitudes, find the natural frequency of this system.

(b) Find the ratio l/a at which the system becomes unstable.

15.8 (a) Write the differential equation for the system illustrated in Fig. P15.8 and find the natural frequency.

(b) Find the response x if y is a step input of height y_0.

(c) Find the relative response $z = x - y$ to the step input of part (b).

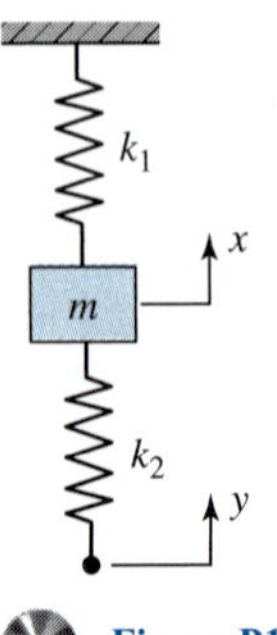

Figure P15.8

15.9 An undamped vibrating system consists of a spring whose scale is 35 kN/m and a mass of 1.2 kg. A step force $F = 50$ N is exerted on the mass for 0.040 s.

(a) Write the equations of motion of the system for the era in which the force acts and for the era that follows.

(b) What are the amplitudes in each era?

(c) Sketch a time plot of the displacement.

15.10 Figure P15.10 illustrates a round shaft whose torsional spring constant is k_t in · lb/rad connecting two wheels having mass moments of inertia I_1 *and* I_2. Show that the system is likely to vibrate torsionally with a frequency of

$$\omega_n = \sqrt{\frac{k_t\,(I_1 + I_2)}{I_1 I_2}}.$$

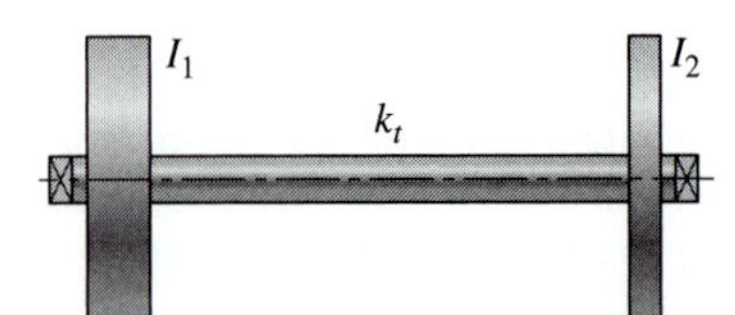

Figure P15.10

15.11 A motor is connected to a flywheel by a 5/8-in diameter steel shaft 36 in long, as shown. Using the methods of Chapter 15, it can be demonstrated that the torsional spring constant of the shaft is 4 700 in · lb/rad. The mass moments of inertia of the motor and flywheel are 24.0 and 56.0 in · lb·s^2, respectively. The motor is turned on for 2 s, and during this period it exerts a constant torque of 200 *in* · *lb* on the shaft.

(a) What speed in revolutions per minute does the shaft attain?

(b) What is the natural frequency of vibration of the system?

(c) Assuming no damping, what is the amplitude of the vibration of the system in degrees during the first era? During the second era?

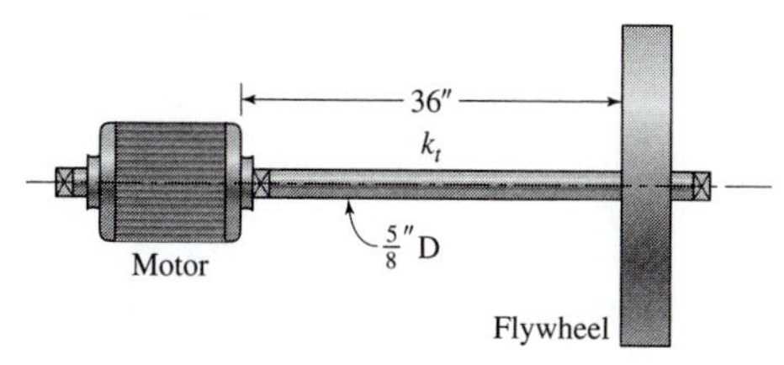

Figure P15.11

15.12 The weight of the mass of a vibrating system is 10 lb, and it has a natural frequency of 1 Hz. Using the phase-plane method, plot the response of the system to the force function illustrated in Fig. P15.12. What is the final amplitude of the motion?

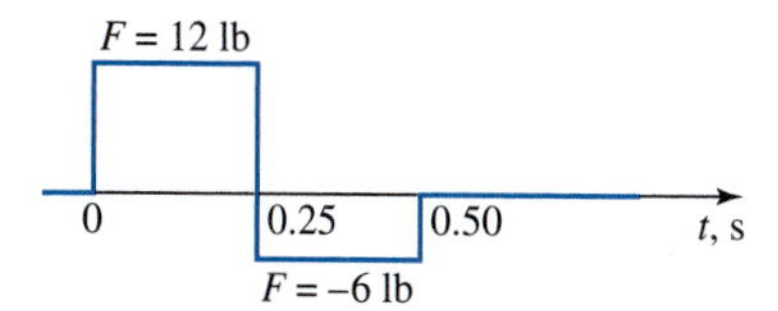

Figure P15.12

15.13 An undamped vibrating system has a spring scale of 200 lb/in and a weight of 50 lb. Find the response and the final amplitude of vibration of the system if it is acted upon by the forcing function illustrated in Fig. P15.13. Use the phase-plane method.

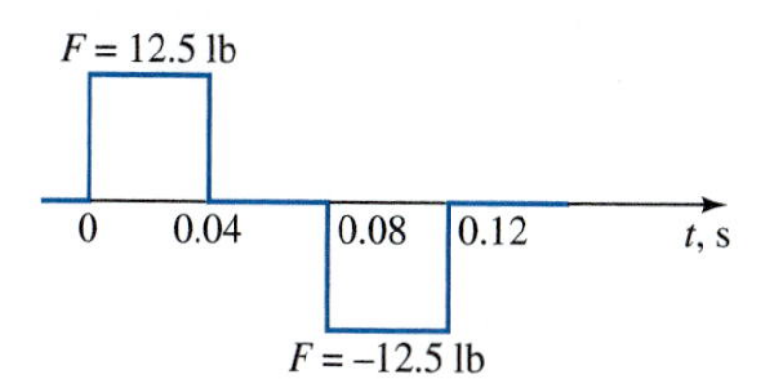

Figure P15.13

15.14 A vibrating system has a spring $k = 400$ lb/in and a weight of $W = 80$ lb. Plot the response of this system to the forcing function illustrated in Fig. P15.14:

(a) Using three steps

(b) Using six steps.

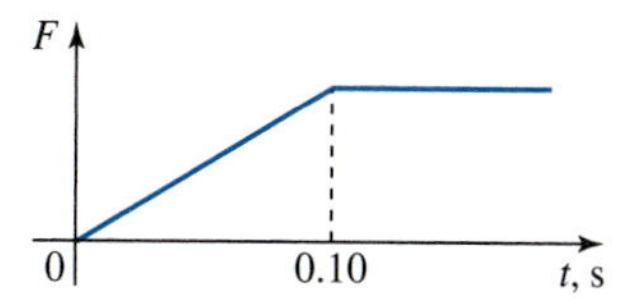

Figure P15.14 $F_{max} = 200$ lb.

15.15 (a) What is the value of the coefficient of critical damping for a spring-mass-damper system in which $k = 56$ kN/m and $m = 40$ kg?

(b) If the actual damping is 20% of critical, what is the natural frequency of the system?

(c) What is the period of the damped system?

(d) What is the value of the logarithmic decrement?

15.16 A vibrating system has a spring $k = 3.5$ kN/m and a mass $m = 15$ kg. When disturbed, it was observed that the amplitude decayed to one fourth of its original value in 4.80 s. Find the damping coefficient and the damping factor.

15.17 A vibrating system has $k = 300$ lb/in, $W = 90$ lb, and damping equal to 20% of critical.

(a) What is the damped natural frequency ω_d of the system?

(b) What are the period and the logarithmic decrement?

15.18 Solve Problem 15.14 using damping equal to 15% of critical.

15.19 A damped vibrating system has an undamped natural frequency of 10 Hz and a weight of 800 lb. The damping ratio is 0.15. Using the phase-plane method, determine the response of the system to the forcing function illustrated in Fig. P15.19.

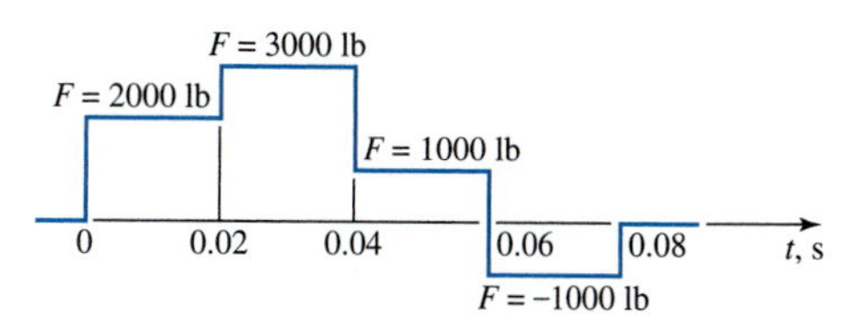

Figure P15.19

15.20 A vibrating system has a spring rate of 3 000 lb/in, a damping factor of 55 lb · s/in, and a weight of 800 lb. It is excited by a harmonically varying force $F_0 = 100$ lb at a frequency of 435 cycles per minute.

(a) Calculate the amplitude of the forced vibration and the phase angle between the vibration and the force.

(b) Plot several cycles of the displacement-time and force-time diagrams.

15.21 A spring-mounted mass has $k = 525$ kN/m, $c = 9640\ N \cdot s/m$, and $m = 360$ kg. This system is excited by a force having an amplitude of 450 N at a frequency of 4.80 Hz. Find the amplitude and phase angle of the resulting vibration and plot several cycles of the force-time and displacement-time diagrams.

15.22 When a 6,000-lb press is mounted upon structural-steel floor beams, it causes them to deflect 0.75 in. If the press has a reciprocating unbalance of 420 lb and it operates at a speed of 80 rev/min, how much of the force will be transmitted from the floor beams to other parts of the building? Assume no damping. Can this mounting be improved?

15.23 Four vibration mounts are used to support a 450-kg machine that has a rotating unbalance of $0.35\ kg \cdot m$ and runs at 300 rev/min. The vibration mounts have damping equal to 30% of critical. What must the spring constant of the mounting be if 20% of the exciting force is transmitted to the foundation? What is the resulting amplitude of motion of the machine?

15.24 A 600-mm-long steel shaft is simply supported by two bearings at A and C, as illustrated in Fig. P15.24. Flywheels 1 and 2 are attached to the shaft at locations *B* and *D*, respectively. Flywheel 1 at location *B* weighs 50 N, flywheel 2 at location D weighs 20 N, and the weight of the shaft can be neglected. The known stiffness coefficients are $k_{11} = 25\ 000$ N/m, $k_{12} = 50\ 000$ N/m, and $k_{22} = 40\ 000$ N/m. Determine (*i*) the first and second critical speeds of the shaft using the exact solution and the first critical speed using (*ii*) the Dunkerley and (*iii*) Rayleigh–Ritz approximations. (*iv*) If flywheel 2 is then placed at location *B* and flywheel 1 is placed at location *D*, determine the first critical speed of the new system using the Dunkerley approximation.

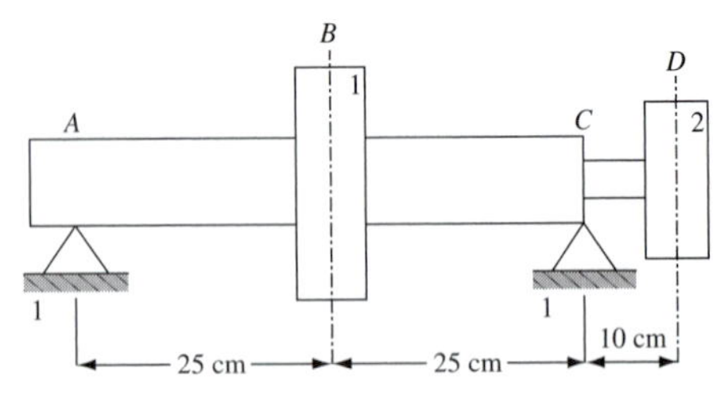

Figure P15.24 A steel shaft simply supported by two rolling element bearings.

15.25 The first critical speeds of a rotating shaft with two mass disks, obtained from three different mathematical techniques, are 110, 112, and 100 rad/s, respectively. (*i*) Which values correspond to the first critical speed of the shaft from the exact solution, the Dunkerley approximation, and the Rayleigh–Ritz approximation? (*ii*) If the influence coefficients are $a_{11} = a_{22} = 10^{-4}$ m/N and the masses of the two disks are the same, that is, $m_1 = m_2 = m$, then use the Dunkerley approximation to calculate the mass *m*. (*iii*) If the influence coefficients are $a_{11} = a_{22} = 10^{-4}$ m/N and the masses of the two disks are specified as $m_1 = m_2 = m = 0.5$ kg, use the Rayleigh–Ritz approximation to calculate the influence coefficient a_{12}.

15.26 A steel shaft is simply supported by two rolling element bearings at *A* and *B*, as illustrated in Fig. P15.26. The length of the shaft is 1.45 m and two flywheels with weight 300 N are attached to the shaft at the locations shown. One flywheel is 0.35 m to the right of the left bearing at *A* and the other flywheel is 0.35 m to the left of the right bearing at *B*. The weight of the shaft can be neglected. The influence coefficients are specified as $a_{11} = 126 \times 10^{-5}$ mm/N and $a_{21} = 92.5 \times 10^{-5}$ mm/N. (*i*) Determine the first and second critical speeds of the shaft using the exact solution. Determine the first critical speed of the shaft using: (*ii*) the Dunkerley approximation and (*iii*) the Rayleigh–Ritz equation.

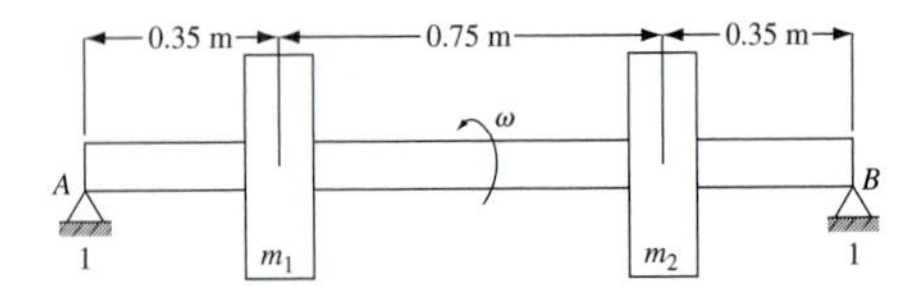

Figure P15.26 A steel shaft simply supported by two rolling element bearings.

15.27 A steel shaft is simply supported by two rolling element bearings at A and C, as illustrated in Fig. P15.26. The length of the shaft is 0.6 m and two flywheels are attached to the shaft at the locations B and D as shown. The flywheel at location B weighs 200 N and the flywheel at location D weighs 90 N. The weight of the shaft can be neglected. It was determined that with flywheel 1 alone the first critical speed of the shaft is 800 rad/s and with flywheel 2 alone the first critical speed of the shaft is 1 200 rad/s. (i) Determine the first critical speed for the two-mass system. (ii) If the two flywheels are interchanged (that is, flywheel 2 is placed at location B and flywheel 1 is placed at location D), determine the first critical speed of the new system using the Dunkerley approximation.

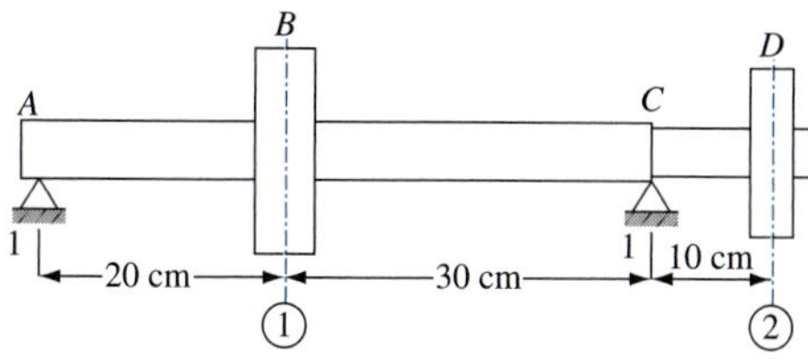

Figure P15.27 A steel shaft simply supported by two rolling element bearings.

15.28 A steel shaft, which is 50 in. in length, is simply supported by two bearings at B and D, as illustrated in Fig. P15.28. Flywheels 1 and 2 are attached to the shaft at A and C, respectively. The flywheel at location A weighs 15 lbs, the flywheel at location C weighs 30 lbs, and the weight of the shaft can be neglected. The stiffness coefficients are specified as $k_{11} = 2.5 \cdot 10^4$ lb/in. and $k_{22} = 4.0 \cdot 10^4$ lb/in. (i) Determine the first critical speed for the two-mass system using the Dunkerley approximation. (ii) If the two flywheels are interchanged (that is, flywheel 2 is placed at location A and flywheel 1 is placed at location C), determine the first critical speed of the new system.

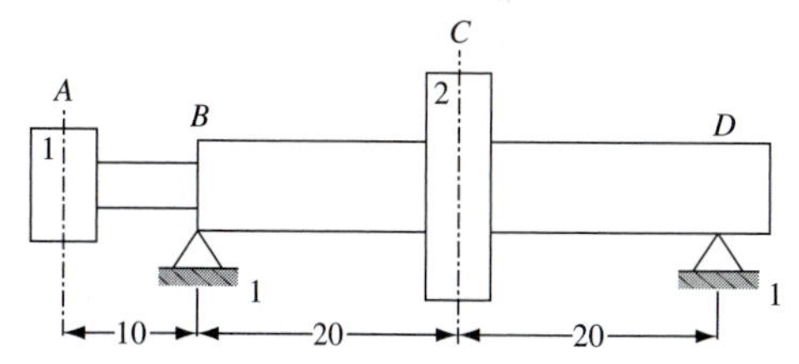

Figure P15.28 A steel shaft simply supported by two rolling element bearings.

16 Dynamics of Reciprocating Engines

The purpose of Chapter 16 is to apply fundamentals—kinematic and dynamic analysis—in a complete investigation of a particular category of machines. The reciprocating engine has been selected for this purpose because it has reached a high state of development and is of more general interest than most other machines. For our purposes, however, another machine or type of machines involving interesting dynamic situations would serve just as well. The primary objective of Chapter 16 is to demonstrate methods of applying fundamentals to the dynamic analysis of machines.

16.1 ENGINE TYPES

The descriptions and characteristics of all the engine types that have been conceived and constructed over the years would fill many books. Here, our purpose is to describe very briefly a few of the engine types that are currently in general use. The exposition is not intended to be complete. Furthermore, because you are anticipated to be mechanically inclined and generally familiar with internal combustion engines, the primary purpose of this section is merely to record things that you already know and to furnish a nomenclature for the balance of Chapter 16.

In Chapter 16 we classify engines according to their intended application, the combustion cycle used, and the number and arrangement of cylinders. Thus, we refer to aircraft engines, automotive engines, marine engines, and stationary engines, for example, all so named because of the application for which they are designed. Similarly, one might have in mind an engine designed on the basis of the *Otto cycle*, in which the fuel and air are mixed before compression and in which combustion takes place with no excess air, or on the basis of the *Diesel cycle*, in which the fuel is injected near the end of compression and combustion takes place with a good deal of excess air. The Otto-cycle engine uses quite volatile fuels and ignition is by spark, but the Diesel-cycle engine operates on fuels of lower volatility and ignition occurs because of compression.

Both Diesel- and Otto-cycle engines may use either a *two-stroke cycle* or a *four-stroke cycle*, depending upon the number of piston strokes comprising the complete combustion cycle. In the four-stroke cycle the exhaust ports open near the end of each power stroke to permit exhaust gases to flow out during the exhaust stroke. Next, the exhaust ports close and inlet ports open and permit entry of a fuel–air mixture during the suction stroke. Then, all ports are closed and the piston compresses the fuel–air mixture. Finally, the fuel–air mixture is ignited, causing a expansion or power stroke, and the cycle begins again. Note that each four-stroke cycle requires two revolutions of the crank. In contrast, many outboard or lawn mower engines use a two-stroke cycle or are simply two-stroke engines. Note that a two-stroke engine has only an expansion stroke and a compression stroke and that both occur during each revolution of the crank.

The four-stroke engine uses four piston strokes in a single combustion cycle corresponding to two revolutions of the crank. The events corresponding to the four strokes are (1) suction, or the input stroke; (2) the compression stroke; (3) expansion, or the power stroke; and (4) the exhaust stroke.

Multicylinder engines are also broadly classified according to how the cylinders are arranged with respect to each other and the crankshaft. Thus, an *in-line* engine is one in which all cylinder axes form a single plane containing the crankshaft and, in that plane, the pistons are all on the same side of the crankshaft. Figure 16.1 is a schematic drawing of a three-cylinder in-line engine with the three cranks spaced at 120° of crank rotation. A firing-order diagram for four-stroke operation is included for interest.

Figure 16.2 illustrates a cutaway view of a five-cylinder in-line engine for a recent passenger car. We can see here that the plane formed by the centerlines of the cylinders is not mounted vertically to keep the height of the engine smaller to fit low-profile styling requirements. Figure 16.2 also includes a good view of the relative location of the camshaft and the overhead valves.

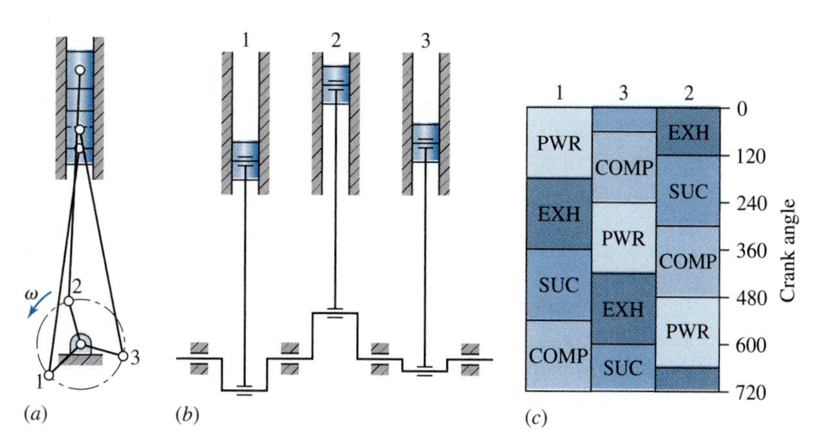

Figure 16.1 A three-cylinder in-line engine: (*a*) front view; (*b*) side view; (*c*) firing order.

Figure 16.2 A five-cylinder in-line engine from an Audi Coupe Quattro. (Courtesy of Audi of America, Inc., Troy, MI.)

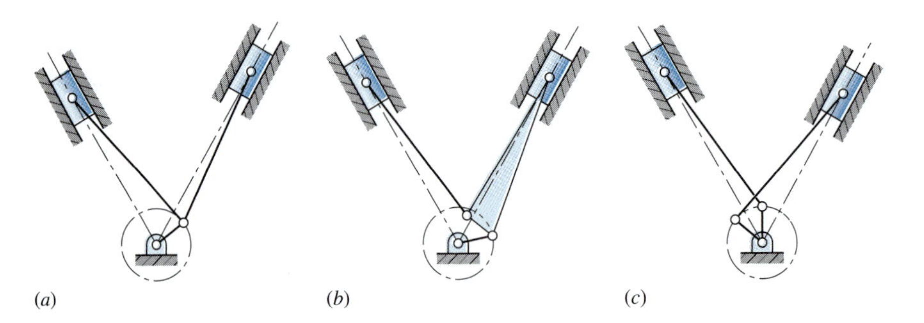

Figure 16.3 Crank arrangements of different V engines: (*a*) single crank throw per pair of cylinders—connecting rods interlock with each other and are of *fork-and-blade design*; (*b*) single crank throw per pair of cylinders—the master connecting rod carries a bearing for the *articulated rod*; (*c*) separate crank throws connect to staggered rods and pistons.

A V-type engine uses two banks of one or more in-line cylinders, all connected to a single crankshaft. Figure 16.3 illustrates some common crank arrangements. The pistons in the right and left banks of Figs. 16.3*a* and 16.3*b* may be in the same plane, but those in Fig. 16.3*c* are in different planes.

If the V angle is increased to 180°, the result is called an *opposed-piston engine*. An opposed engine may have the two piston axes coincident or offset, and the rods may connect to the same crank throw or to separate crank throws 180° apart.

A *radial* engine is one having multiple cylinder axes arranged radially about the crank center. Radial engines use a master connecting rod for one cylinder and the remaining pistons are connected to the master rod by articulated rods, somewhat the same as for the V engine of Fig. 16.3*b*.

Figures 16.4 to 16.6 illustrate, respectively, the piston-connecting rod assembly, the crankshaft, and the block of a V6 truck engine. These are included as typical of engine design to demonstrate the form of important parts of an engine. The following specifications also give a general idea of the performance and design characteristics of typical engines, together with the sizes of parts used in them.

The reader will note that all quantities throughout Chapter 16 are quoted in SI rather than U.S. customary units. Because of the international nature of the automotive and related industries, engines today are all designed in SI units, even in the United States. Although the units of horsepower and gallons are still seen in advertisements, the power of an engine is measured in kilowatts (kW) and engine displacement is quoted in liters (L)* by knowledgeable engineers throughout the industry.

GMC Truck and Coach Division, General Motors Corporation, Pontiac, Michigan One of the GMC V6 truck engines is illustrated in Fig. 16.7. These engines are manufactured in four displacements, and they include one model, a V12 (11.5 L), which is described as a "twin six" because many of the V6 parts are interchangeable with it. Data included here

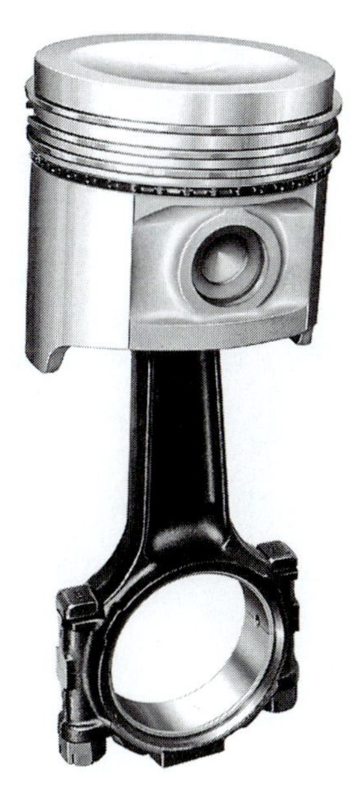

Figure 16.4 A piston–connecting rod assembly for a 5.75-L V6 truck engine (Courtesy of GMC Truck and Coach Division, General Motors Corporation, Pontiac, MI.)

* The unit for volume of an engine or cylinder is the liter. The international symbol for liter is lower case "l," which can easily be confused with the numeral "1." Therefore the symbol "L" is used throughout this text.

Figure 16.5 A cast crankshaft for a 5-L V6 truck engine (Courtesy of GMC Truck and Coach Division, General Motors Corporation, Pontiac, MI.)

Figure 16.6 Block for a 5-L V6 truck engine. The same casting is used for a 5.75-L engine by boring for larger pistons (Courtesy of GMC Truck and Coach Division, General Motors Corporation, Pontiac, MI.)

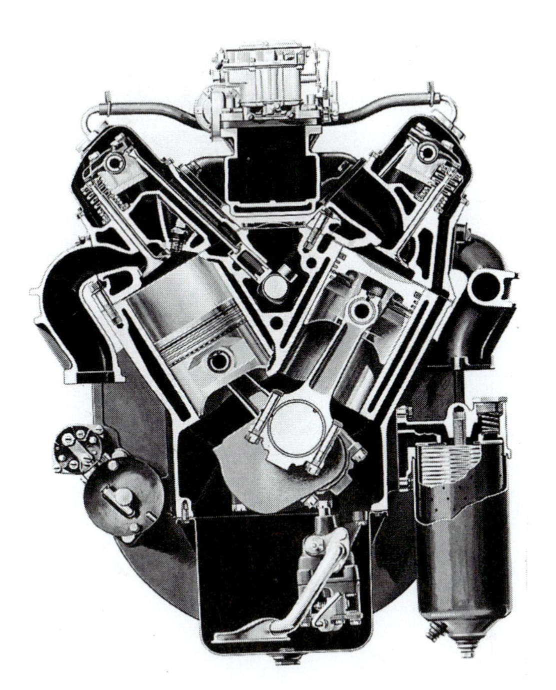

Figure 16.7 Cross-sectional view of a 6.57-L V6 truck engine (Courtesy of GMC Truck and Coach Division, General Motors Corporation, Pontiac, MI.)

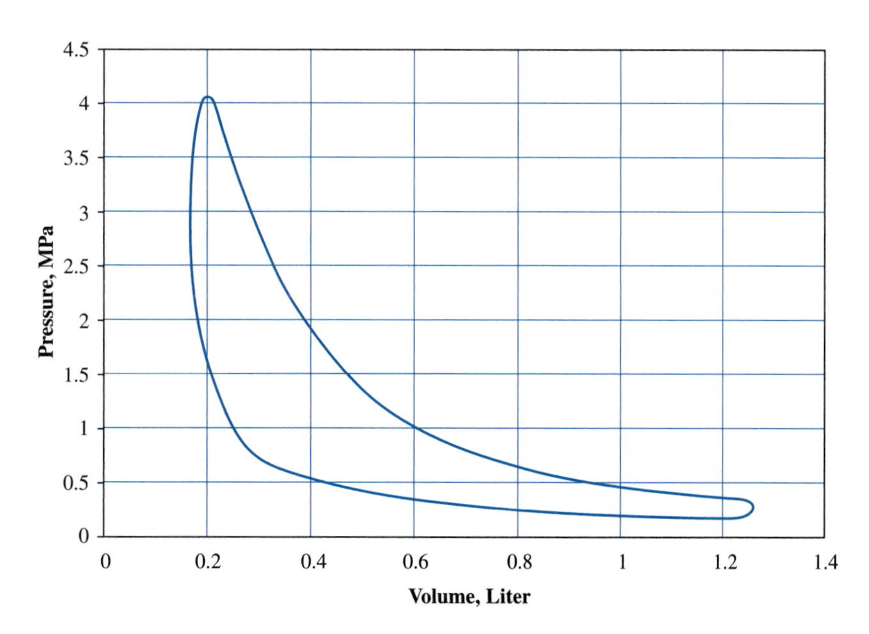

Figure 16.8 Typical indicator diagram for a 6.57-L V6 truck engine; conditions unknown. (Courtesy of GMC Truck and Coach Division, General Motors Corporation, Pontiac, MI.)

are for the 6.57-L engine. Typical performance curves are exhibited in Figs. 16.8, 16.9, and 16.10. The specifications are as follows: 60° V design; bore = 123.8 mm; stroke = 90.4 mm; connecting rod length = 182.6 mm; compression ratio = 7.50:1; cylinders are numbered 1,

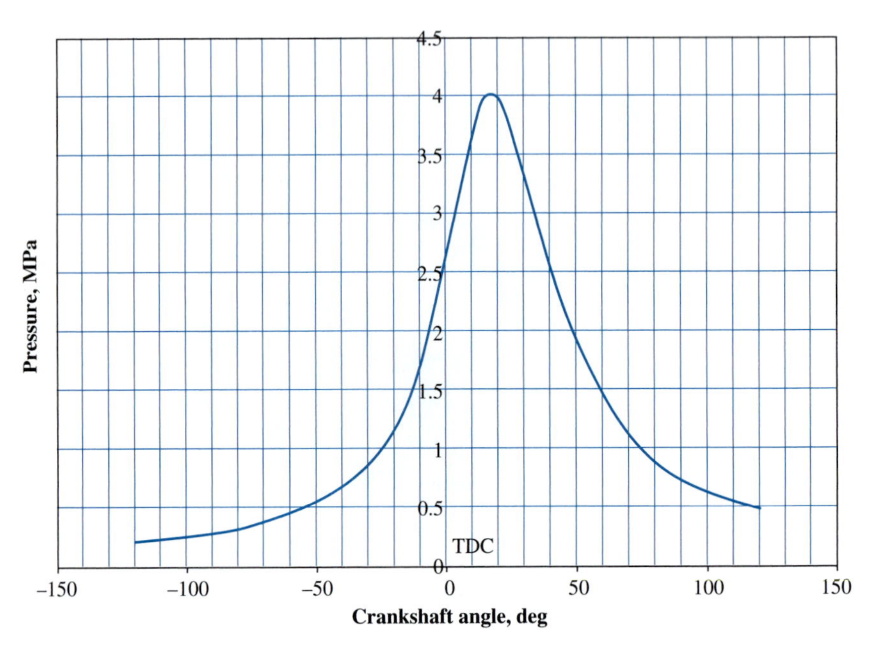

Figure 16.9 A pressure–crank angle curve for the 6.57-L V6 truck engine; these data were taken from a running engine. (Courtesy of GMC Truck and Coach Division, General Motors Corporation, Pontiac, MI.)

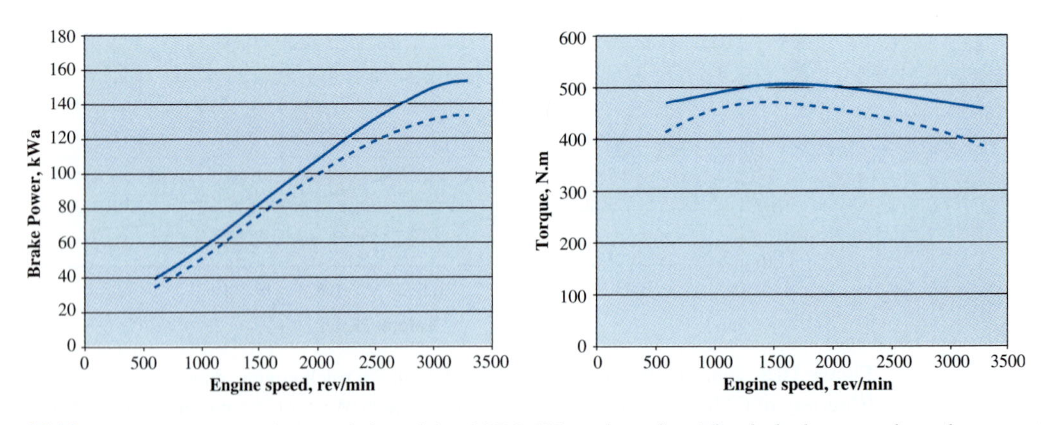

Figure 16.10 Power and torque characteristics of the 6.57-L V6 truck engine. The dashed curves show the net output as installed; the solid curves show the maximum output without accessories. Note that maximum torque occurs at very low engine speed. (Courtesy of GMC Truck and Coach Division, General Motors Corporation, Pontiac, MI.)

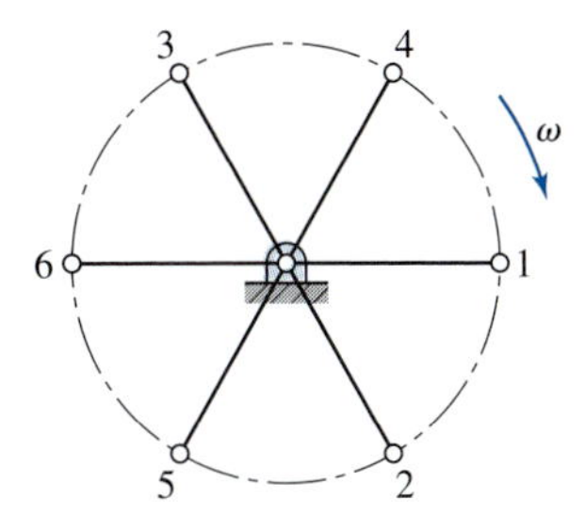

Figure 16.11 Front view of a V6 engine showing crank arrangement and direction of rotation.

3, and 5 from front to rear on the left bank and 2, 4, and 6 on the right bank; firing order is 1, 6, 5, 4, 3, 2; the crank arrangement is illustrated in Fig. 16.11.

16.2 INDICATOR DIAGRAMS

Experimentally, an instrument called an *engine indicator* is used to measure the variation in pressure within a cylinder. This instrument constructs a graph, during operation of the engine, which is known as an *indicator diagram*. Known constraints of the indicator make it possible to study the diagram and determine the relationship between the gas pressure and the crank angle for the particular running conditions in existence at the time the diagram is recorded.

When an engine is in the design stage, it is necessary to estimate a diagram from theoretical considerations. From such an approximation, a pilot model of the proposed engine is designed and built, and the actual indicator diagram is taken and compared with the theoretically estimated one. This provides much useful information for the design of the production model.

An indicator diagram for the ideal air-standard cycle is illustrated in Fig. 16.12 for a four-stroke engine. During compression, the cylinder volume changes from v_1 to v_2 and the cylinder pressure changes from p_1 to p_2. The relationship, at any point of this stroke, is given by the polytropic gas law as

$$pv^k = p_1 v_1^k = \text{constant}. \tag{16.1}$$

In an actual indicator card, the corners at points 2 and 3 are rounded and the line joining these points is curved. This is explained by the fact that combustion is not instantaneous and ignition occurs before the end of the compression stroke. An actual card is also rounded at points 4 and 1 because the valves do not operate simultaneously.

The polytropic exponent k in Eq. (16.1) is often taken to be about 1.30 for both compression and expansion, although differences probably exist.

The relationship between the power developed and the dimensions of the engine is given by

$$P_b = \frac{p_b lan\,(1\,000\,\text{N/m}^2/\text{kPa})}{(1\,000\,\text{N}\cdot\text{m/s/kW})(1\,000\,\text{mm/m})\,(60\,\text{s / min})} = \frac{p_b lan}{60\,000}, \tag{16.2}$$

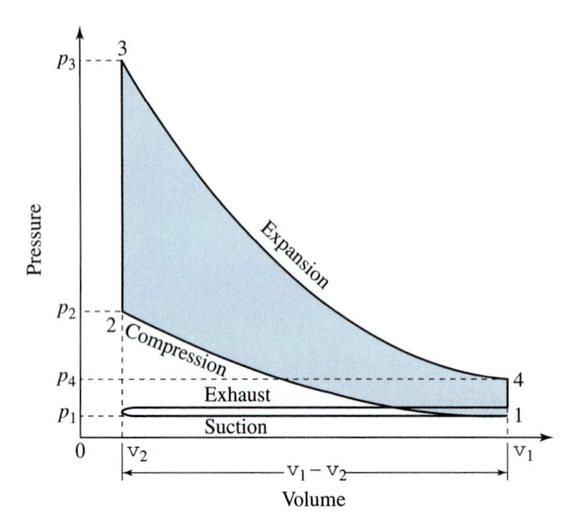

Figure 16.12 An ideal indicator diagram for a four-stroke engine.

where

P_b = brake power per cylinder (kW)
p_b = brake mean effective pressure (kPa)
l = length of stroke (mm)
a = piston area (m^2)
n = number of working strokes per minute.

The amount of power that can be obtained from 1 L of piston displacement varies considerably, depending upon the engine type. For typical automotive engines it ranges from about 25 to 45 kW/L, with a current average of perhaps 32. On the other hand, many marine diesel engines have ratios varying from 4.5 to 9 kW/L. About the best that can be done in designing a new engine is to use standard references to discover what others have done with the same types of engines and then choose a value that seems to be reasonably attainable.

For many engines the bore-to-stroke ratio varies from about 0.75 to 1.00. The tendency in automotive-engine design seems to be toward shorter-stroke engines to reduce engine height. Decisions on bore–stroke ratio and power-per-unit displacement volume are helpful in solving Eq. (16.2) to obtain suitable dimensions when the power, speed, and number of cylinders have been decided upon.

The ratio of the brake mean effective pressure p_b to the indicated mean effective pressure p_i, obtained experimentally from an indicator card, is the mechanical efficiency e_m:

$$e_m = \frac{p_b}{p_i}. \tag{16.3}$$

Differences between a theoretical and an experimentally determined indicator diagram can be accounted for by applying a correction called a *card factor*. The card factor is defined by

$$f_c = \frac{p_i}{p_i'}, \tag{16.4}$$

where p_i' is the theoretical indicated mean effective pressure and the card factor is usually about 0.90 to 0.95.

Because the *compression ratio* (Fig. 16.12) is defined as

$$R = \frac{v_1}{v_2}, \tag{16.5}$$

the work done during compression is

$$U_c = \int_{v_2}^{v_1} p dv = p_1 v_1^k \int_{v_2}^{v_1} \frac{dv}{v^k} = \frac{p_1 v_1}{k-1}(R^{k-1} - 1). \tag{a}$$

The displacement volume can be written as

$$v_1 - v_2 = v_1 - \frac{v_1}{R} = \frac{v_1(R-1)}{R}. \tag{b}$$

Substituting v_1 from Eq. (*b*) into Eq. (*a*), the work done during compression can be written as

$$U_c = \frac{p_1(v_1 - v_2)}{k-1}\frac{R^k - R}{R-1}. \tag{c}$$

The work done during expansion is the area under the curve between points 3 and 4 of Fig. 16.12. This is determined in the same manner; the result is

$$U_e = \frac{p_4(v_1 - v_2)}{k-1}\frac{R^k - R}{R-1}. \tag{d}$$

The net amount of work accomplished in a cycle is the difference in Eqs. (*c*) and (*d*), and it must be equal to the product of the theoretical indicated mean effective pressure and the displacement volume. That is,

$$U = U_e - U_c = \frac{p_4(v_1 - v_2)}{k-1}\frac{R^k - R}{R-1} - \frac{p_1(v_1 - v_2)}{k-1}\frac{R^k - R}{R-1} = p_i'(v_1 - v_2). \tag{16.6}$$

If the exponent is the same for expansion as for compression, Eq. (16.6) can be solved to give

$$p_4 = p_1'(k-1)\frac{R-1}{R^k - R} + p_1. \tag{e}$$

Substituting the theoretical indicated mean effective pressure p_i' [see Eq. (16.4)] into Eq. (*e*) gives

$$p_4 = (k-1)\frac{R-1}{R^k - R}\frac{p_i}{f_c} + p_1. \tag{16.7}$$

Equations (16.1) and (16.7) can be used to create the theoretical indicator diagram. The corners are then rounded off so that the pressure at point 3 is made about 75% of that given by Eq. (16.1). As a check, the area of the diagram can be measured and divided by the displacement volume; the result should equal the indicated mean effective pressure.

16.3 DYNAMIC ANALYSIS—GENERAL

The balance of Chapter 16 is devoted to an analysis of the dynamics of a single-cylinder engine. To simplify this work, it is assumed that the engine is running at a constant crankshaft speed and that gravitational forces and friction forces can be neglected in comparison with dynamic-force effects. The gas forces and inertia forces are determined separately; then these forces are combined, using the principle of superposition (see Section 14.5), to obtain the total bearing forces and crankshaft torque.

The subject of engine balancing is treated in Chapter 17.

16.4 GAS FORCES

In this section we temporarily assume that the moving parts of our engine are massless so that gravity and inertia forces and torques are zero and there is no friction. These assumptions allow us to trace the force effects of the gas pressure from the piston to the crankshaft without the complicating effects of other forces. These other effects, except friction, will be added later, using superposition.

In Chapter 13, both graphic and analytic methods for finding the static forces in a mechanism were presented. Either of those approaches can be used to solve this gas-force problem. The advantage of the analytic methods is that they can be programmed for automatic computation throughout a cycle. In contrast, the graphic technique must be repeated for each crank position until a complete cycle of operation (720° for a four-stroke engine) is completed. Because we prefer not to duplicate the studies of Chapter 13, we present here an algebraic approach.

In Fig. 16.13 we designate the crank angle as ωt, taken positive in the counterclockwise direction, and the connecting-rod angle as ϕ, taken positive when the crank angle is in the first or second quadrant ($R^y_{AO_2} > 0$), as shown. A relation between these two angles is seen from Fig. 16.13,

$$l \sin\phi = r \sin\omega t \quad \text{or} \quad \sin\phi = \frac{r}{l}\sin\omega t, \tag{a}$$

where r and l designate the lengths of the crank and the connecting rod, respectively. Designating the piston position by the coordinate $x = R_{BO_2}$, we find

$$x = r\cos\omega t + l\cos\phi = r\cos\omega t + l\sqrt{1 - \left(\frac{r}{l}\sin\omega t\right)^2}. \tag{16.8}$$

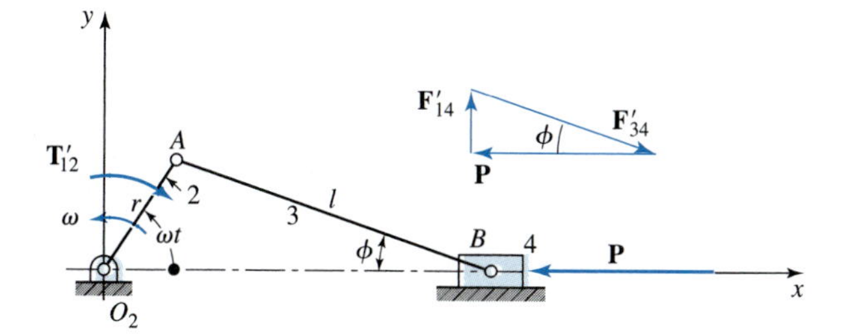

Figure 16.13

For most engines the ratio $r/l \leq 1/4$, and so the maximum value of the second term under the radical is about $1/16$ or less. Expanding the radical using the binomial expansion gives

$$\sqrt{1-\left(\frac{r}{l}\sin\omega t\right)^2} \approx 1-\frac{r^2}{2l^2}\sin^2\omega t, \tag{b}$$

where only the first two terms have been retained. Then using the trigonometric identity,

$$\sin^2\omega t = \frac{1-\cos 2\omega t}{2}, \tag{c}$$

Eq. (16.8) can be written as

$$x = l - \frac{r^2}{4l} + r\left(\cos\omega t + \frac{r}{4l}\cos 2\omega t\right). \tag{16.9}$$

Differentiating this equation twice with respect to time, the velocity and acceleration of the piston are

$$\dot{x} = -r\omega\left(\sin\omega t + \frac{r}{2l}\sin 2\omega t\right) \tag{16.10}$$

$$\ddot{x} = -r\alpha\left(\sin\omega t + \frac{r}{2l}\sin 2\omega t\right) - r\omega^2\left(\cos\omega t + \frac{r}{l}\cos 2\omega t\right). \tag{16.11}$$

Referring again to Fig. 16.13, we designate a gas-force vector **P** as defined or obtained using the methods of Section 16.2. Reactions caused by this force are designated using a single prime. Thus, $\mathbf{F}'_{14}$ is the force from the cylinder wall acting against the piston and $\mathbf{F}'_{34}$ is the force of the connecting rod acting onto the piston at the piston pin. The force polygon in Fig. 16.13 illustrates the relation among **P**, $\mathbf{F}'_{14}$, and $\mathbf{F}'_{34}$. Thus, we have

$$\mathbf{F}'_{14} = P\tan\phi\hat{\mathbf{j}}. \tag{16.12}$$

The quantity $\tan\phi$ appears frequently in expressions throughout Chapter 16. It is therefore convenient to develop an expression in terms of the crank angle ωt. Thus,

$$\tan\phi = \frac{(r/l)\sin\omega t}{\cos\phi} = \frac{(r/l)\sin\omega t}{\sqrt{1-[(r/l)\sin\omega t]^2}}. \tag{d}$$

Now, using the binomial expansion again, we find that

$$\frac{1}{\sqrt{1-[(r/l)\sin\omega t]^2}} \approx 1 + \frac{r^2}{2l^2}\sin^2\omega t, \tag{e}$$

where only the first two terms have been retained. Then substituting Eq. (*e*) into Eq. (*d*) we have

$$\tan\phi = \frac{r}{l}\sin\omega t\left(1 + \frac{r^2}{2l^2}\sin^2\omega t\right). \tag{16.13}$$

The trigonometry of Fig. 16.13 indicates that the *wrist-pin* (piston-pin) bearing force has a magnitude of

$$F'_{34} = \frac{P}{\cos\phi} = \frac{P}{\sqrt{1-[(r/l)\sin\omega t]^2}} = P\left(1 + \frac{r^2}{2l^2}\sin^2\omega t\right) \tag{f}$$

or, in vector notation,

$$\mathbf{F}'_{34} = P\hat{\mathbf{i}} - F'_{14}\hat{\mathbf{j}} = P\hat{\mathbf{i}} - P\tan\phi\hat{\mathbf{j}}. \tag{16.14}$$

By taking moments about the crank center, we find the torque $\mathbf{T}'_{21}$ delivered by the crank to the shaft is the product of the force $\mathbf{F}'_{14}$ and the piston coordinate x, that is,

$$\mathbf{T}'_{21} = F'_{14}x\hat{\mathbf{k}}.$$

Substituting Eqs. (16.9) and (16.12) into this equation and using Eq. (16.13), the torque is

$$\mathbf{T}'_{21} = P\left(\frac{r}{l}\sin\omega t\right)\left(1 + \frac{r^2}{2l^2}\sin^2\omega t\right)\left[l - \frac{r^2}{4l} + r\left(\cos\omega t + \frac{r}{4l}\cos 2\omega t\right)\right]\hat{\mathbf{k}}. \tag{g}$$

When the terms of Eq. (*g*) are multiplied, with only a very small error we can neglect those containing second or higher powers of r/l. Therefore, Eq. (*g*) can be written as

$$\mathbf{T}'_{21} = Pr\sin\omega t\left(1 + \frac{r}{l}\cos\omega t\right)\hat{\mathbf{k}}. \tag{16.15}$$

This is the torque delivered to the crankshaft by the gas force; the counterclockwise direction is positive.

16.5 EQUIVALENT MASSES

Problems P14.5 and P14.6, at the end of Chapter 14, focused on the slider–crank mechanism, which is the basis for most engine mechanisms. The dynamics of this mechanism were analyzed using the methods presented in Chapter 14. In Chapter 16 we are concerned with the same problem. However, here we will show certain simplifications that are customarily used to reduce the complexity of the algebraic solution process. These simplifications are approximations and do introduce certain errors. In Chapter 16 we will demonstrate the simplifications and comment on the errors that they introduce.

In analyzing the inertia forces caused by the connecting rod of an engine, it is often convenient to picture a portion of the mass of the rod as concentrated at the crankpin A with the remaining portion lumped at the wrist pin B (see Fig. 16.14). The reason for this is that the crankpin moves on a circle and the wrist pin on a straight line. Both of these motions are quite easy to analyze. However, the center of mass G_3 of the connecting rod is somewhere between the crankpin and the wrist pin, and its motion is more complex and consequently more difficult to determine accurately in algebraic form.

The mass of the connecting rod m_3 is truly located at the centroid G_3. However, we can mentally divide this mass into two mass particles; one particle, m_{3B}, is concentrated at the wrist pin B, whereas the other particle, m_{3P}, is concentrated at the center of percussion P (see Section 14.6) for oscillation of the rod about point B.The division of these masses is dynamically equivalent to the original rod if (1) the total mass is the same, (2) the position of the centroid G_3 is unchanged, and (3) the mass moment of inertia is the same. Writing these three conditions in equation form gives

$$m_3 = m_{3B} + m_{3P} \tag{a}$$

$$m_{3B} l_B = m_{3P} l_P \tag{b}$$

and

$$I_G = m_{3B} l_B^2 + m_{3P} l_P^2. \tag{c}$$

These three equations can be solved to give the mass particles m_{3B} and m_{3P}, and the location of the center of percussion, l_P. The procedure is as follows. Solving Eqs. (*a*) and (*b*) simultaneously gives the mass particles at the crankpin A and the wrist pin B:

$$m_{3B} = m_3 \frac{l_P}{l_B + l_P} \quad \text{and} \quad m_{3P} = m_3 \frac{l_B}{l_B + l_P}. \tag{16.16}$$

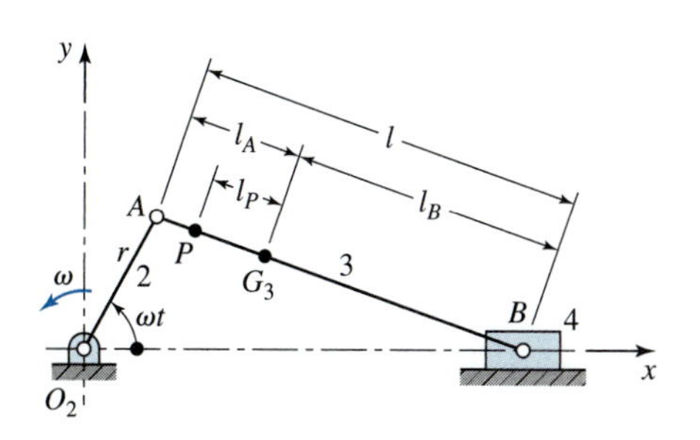

Figure 16.14

Then substituting Eq. (16.16) into Eq. (c) gives

$$I_G = m_3 \frac{l_P}{l_B + l_P} l_B^2 + m_3 \frac{l_B}{l_B + l_P} l_P^2 = m_3 l_P l_B \tag{d}$$

or

$$l_P l_B = \frac{I_G}{m_3}. \tag{e}$$

This equation demonstrates that the two distances l_P and l_B are dependent on each other. Therefore, if l_B is known, then l_P becomes set by Eq. (e). The result can be shown to be purely geometrical by substituting $I_G = m_3 k_G^2$ into the equation and rearranging to give

$$l_P = \frac{k_G^2}{l_B}. \tag{16.17}$$

Note that this answer is in complete agreement with Eq. (14.20).

In the usual connecting rod, the center of percussion P is close to crankpin A and it is usual to assume that they are coincident. Thus, if we let $l_P = l_A$, Eqs. (16.16) reduce to

$$m_{3B} = \frac{m_3 l_A}{l} \quad \text{and} \quad m_{3A} = \frac{m_3 l_B}{l}. \tag{16.18}$$

We note again that the equivalent masses, obtained by Eqs. (16.18), are not exact because of the assumption made, but they are close enough for ordinary connecting rods. This approximation, however, would not be valid for the master connecting rod of a radial engine, because the crankpin end has bearings for all of the other connecting rods as well as its own bearing.

For estimating and checking purposes, about two thirds of the mass of the usual connecting rod is concentrated at crankpin A and the remaining one third is concentrated at wrist pin B.

Figure 16.15 illustrates an engine linkage in which the mass of the crank m_2 is not balanced, as evidenced by the fact that the center of gravity G_2 is located outward along the crank a distance r_G from the axis of rotation. In the inertia force analysis, another simplification is made by locating an equivalent mass m_{2A} at the crankpin. For equivalence, this requires

$$m_2 r_G = m_{2A} r \quad \text{or} \quad m_{2A} = m_2 \frac{r_G}{r}. \tag{16.19}$$

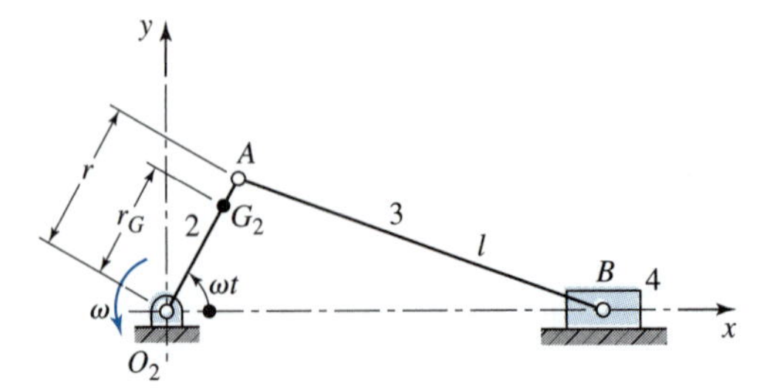

Figure 16.15

16.6 INERTIA FORCES

Using the methods of the preceding section, we begin by locating equivalent masses at the crankpins and at the wrist pin. Thus,

$$m_A = m_{2A} + m_{3A} \tag{16.20}$$

$$m_B = m_{3B} + m_4 \tag{16.21}$$

Equation (16.20) states that the mass m_A located at the crankpin is made up of the equivalent masses m_{2A} of the crank and m_{3A} of part of the connecting rod. Of course, if the crank is balanced (Chapter 17), including balancing the m_{3A} portion of the connecting rod, all of its mass is located at the axis of rotation and m_A is then zero. Equation (16.21) indicates that the reciprocating mass m_B located at the wrist pin is composed of the equivalent mass m_{3B} of the other part of the connecting rod and the mass m_4 of the piston assembly.

Figure 16.16 illustrates the (unbalanced) slider–crank mechanism with masses m_A and m_B located at points A and B, respectively. If we designate the angular velocity of the crank as ω and the angular acceleration as α, the position vector of the crankpin relative to the origin O_2 is

$$\mathbf{R}_A = r\cos\omega t\hat{\mathbf{i}} + r\sin\omega t\hat{\mathbf{j}}. \tag{a}$$

Differentiating this equation twice with respect to time, the acceleration of point A is

$$\mathbf{A}_A = (-r\alpha\sin\omega t - r\omega^2\cos\omega t)\hat{\mathbf{i}} + (r\alpha\cos\omega t - r\omega^2\sin\omega t)\hat{\mathbf{j}}. \tag{16.22}$$

The inertia force of the rotating parts is then

$$-m_A\mathbf{A}_A = m_A r(\alpha\sin\omega t + \omega^2\cos\omega t)\hat{\mathbf{i}} + m_A r(-\alpha\cos\omega t + \omega^2\sin\omega t)\hat{\mathbf{j}}. \tag{16.23}$$

Because the analysis is usually made at constant angular velocity ($\alpha = 0$), Eq. (16.23) reduces to

$$-m_A\mathbf{A}_A = m_A r\omega^2(\cos\omega t\hat{\mathbf{i}} + \sin\omega t\hat{\mathbf{j}}). \tag{16.24}$$

The acceleration of the piston has already been determined [Eq. (16.11)] and is repeated here, for convenience, as

$$\mathbf{A}_B = \left[-r\alpha\left(\sin\omega t + \frac{r}{2l}\sin 2\omega t\right) - r\omega^2\left(\cos\omega t + \frac{r}{l}\cos 2\omega t\right)\right]\hat{\mathbf{i}}. \tag{16.25}$$

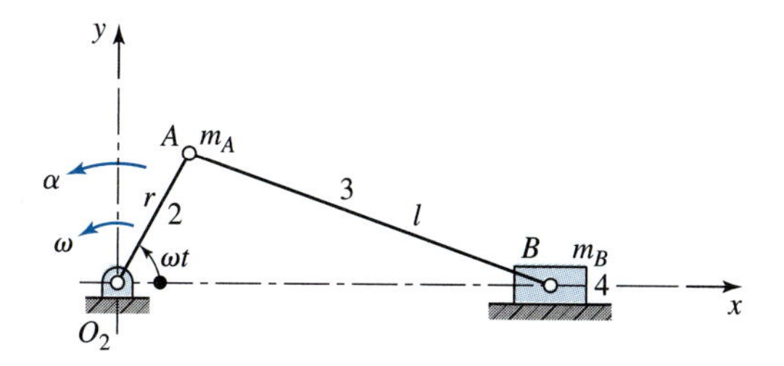

Figure 16.16

Therefore, the inertia force of the reciprocating parts is

$$-m_B\mathbf{A}_B = \left[m_B r\alpha\left(\sin\omega t + \frac{r}{2l}\sin 2\omega t\right) + m_B r\omega^2\left(\cos\omega t + \frac{r}{l}\cos 2\omega t\right)\right]\hat{\mathbf{i}}. \quad (16.26)$$

For constant angular velocity of the crank, this is

$$-m_B\mathbf{A}_B = m_B r\omega^2\left(\cos\omega t + \frac{r}{l}\cos 2\omega t\right)\hat{\mathbf{i}}. \quad (16.27)$$

Adding Eqs. (16.24) and (16.27) gives the total inertia force for all of the moving parts (for constant angular velocity). The components in the x and y directions are

$$F^x = (m_A + m_B)r\omega^2\cos\omega t + \left(m_B\frac{r}{l}\right)r\omega^2\cos 2\omega t \quad (16.28)$$

and

$$F^y = m_A r\omega^2\sin\omega t. \quad (16.29)$$

It is customary to refer to the portion of the force occurring at the frequency ω rad/s as the *primary inertia force* and the portion occurring at 2ω rad/s as the *secondary inertia force*. We note that the x component, which is in the direction of the cylinder axis, has a primary part varying directly with the crankshaft speed and a secondary part varying at twice the crankshaft speed. On the other hand, the y component, perpendicular to the cylinder axis, has only a primary part and it therefore varies directly with the crankshaft speed.

We proceed now to a determination of the inertia torque. As illustrated in Fig. 16.17, the inertia force caused by the mass at the crankpin A has no moment about O_2 and, therefore, produces no inertia torque. Consequently, we need consider only the inertia force given by Eq. (16.27) caused by the reciprocating mass m_B.

From the force polygon of Fig. 16.17, the inertia torque exerted by the engine on the crankshaft is

$$\mathbf{T}''_{21} = -(-m_B\ddot{x}\tan\phi)x\hat{\mathbf{k}}. \quad (b)$$

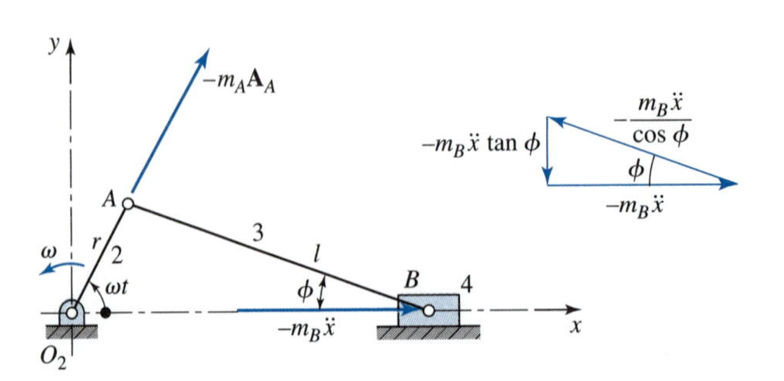

Figure 16.17

Expressions for x, $\ddot{x}$, and $\tan\phi$ appear in Section 16.8 [see Eqs. (16.9), (16.11) and (*d*)]. Making appropriate substitutions for these yields the following equation for the torque:

$$\mathbf{T}''_{21} = -m_B r\omega^2 \left(\cos\omega t + \frac{r}{l}\cos 2\omega t\right)\left[l - \frac{r^2}{4l} + r\left(\cos\omega t + \frac{r}{4l}\cos 2\omega t\right)\right]$$
$$\cdot \frac{r}{l}\sin\omega t\left(1 + \frac{r^2}{2l^2}\sin^2\omega t\right)\hat{\mathbf{k}}. \tag{c}$$

Terms that are proportional to the second and higher powers of r/l are now neglected in performing the indicated multiplications. Equation (*c*) can therefore be written as

$$\mathbf{T}''_{21} = -m_B r^2\omega^2 \sin\omega t\left(\frac{r}{2l} + \cos\omega t + \frac{3r}{2l}\cos 2\omega t\right)\hat{\mathbf{k}}. \tag{d}$$

Then, using the identities

$$2\sin\omega t\cos 2\omega t = \sin 3\omega t - \sin\omega t \tag{e}$$

and

$$2\sin\omega t\cos\omega t = \sin 2\omega t \tag{f}$$

results in an equation having only sine terms, and Eq. (*d*) finally becomes

$$\mathbf{T}''_{21} = \frac{m_B}{2}r^2\omega^2\left(\frac{r}{2l}\sin\omega t - \sin 2\omega t - \frac{3r}{2l}\sin 3\omega t\right)\hat{\mathbf{k}}. \tag{16.30}$$

This is the inertia torque exerted by the engine on the crankshaft in the positive direction. A clockwise, or negative, inertia torque of the same magnitude is, of course, exerted on the frame of the engine.

The assumed distribution of the connecting-rod mass results in a moment of inertia that is greater than the true value. Consequently, the torque given by Eq. (16.30) is not the exact value. In addition, terms proportional to the second- and higher-order powers of r/l were dropped in simplifying Eq. (*c*). These two approximations are of about the same magnitude and are quite small for ordinary connecting rods having $r/l \le 1/4$.

16.7 BEARING LOADS IN A SINGLE-CYLINDER ENGINE

The designer of a reciprocating engine must know the values of the forces acting upon the bearings and how these forces vary during a cycle of operation. These are necessary to proportion and select the bearings properly and are also needed for the design of other engine parts. This section is an investigation of the force exerted by the piston against the cylinder wall and the forces acting against the piston pin and against the crankpin. The main bearing forces are investigated in a later section because they depend upon the action within the cylinders of the engine.

The resultant total bearing loads are made up of the following components:

1. Gas-force components, designated by a single prime;
2. Inertia force caused by the mass m_4 of the piston assembly, designated by a double prime;
3. Inertia force of that part m_{3B} of the connecting rod assigned to the wrist-pin end, designated by a triple prime; and
4. Connecting-rod inertia force of that part m_{3A} at the crankpin end, designated by a quadruple prime.

Equations for the gas-force components have been determined in Section 16.4, and references will be made to them in finding the total bearing loads.

Figure 16.18 is a graphical analysis of the forces in the engine mechanism with zero gas force and subjected only to an inertia force resulting from the mass m_4 of the piston assembly. Figure 16.18*a* illustrates the position of the mechanism selected for analysis, and the inertia force $-m_4\mathbf{A}_B$ is shown acting upon the piston. In Fig. 16.18*b* the free-body diagram of the piston is illustrated together with the polygon from which the forces are obtained. Figures 16.18*c* through 16.18*e*, respectively, illustrate the free-body diagrams showing forces acting upon the connecting rod, crank, and frame.

In Fig. 16.18*e* we note that the torque $\mathbf{T}''_{21}$ balances the couple formed by the forces $\mathbf{F}''_{41}$ and $\mathbf{F}''^{y}_{21}$. But the force $\mathbf{F}''^{x}_{21}$ at the crank center remains unopposed by any other force. This is a very important observation that we shall reserve for discussion in a separate section.

The following forces are of interest to us:

1. The force $\mathbf{F}''_{41}$ of the piston against the cylinder wall;

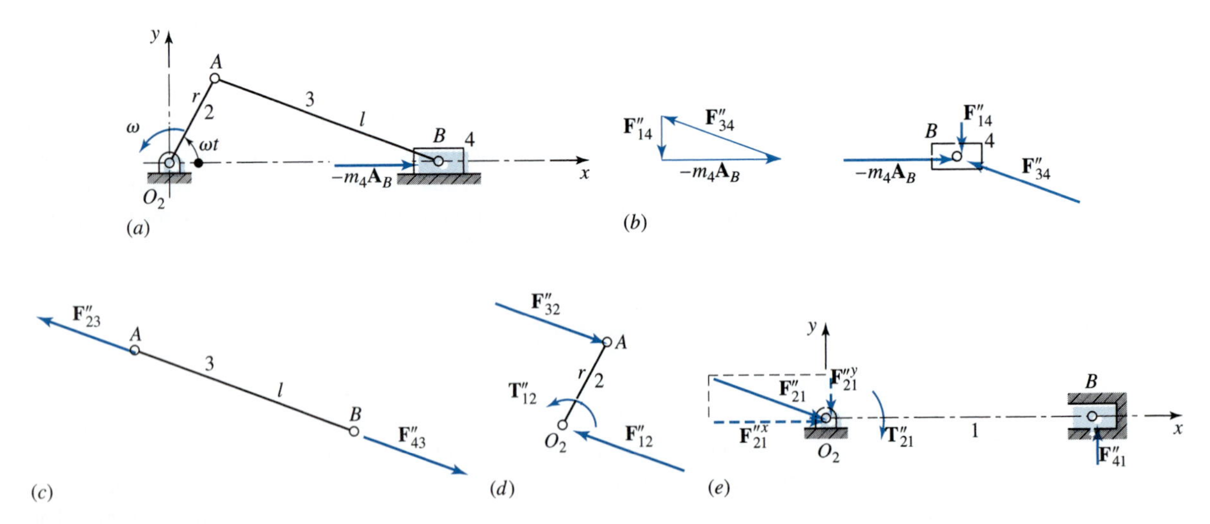

Figure 16.18 Graphical analysis of the forces in the engine mechanism with only the inertia force caused by the mass m_4 of the piston assembly considered.

2. The force $\mathbf{F}''_{34}$ of the connecting rod against the wrist pin;
3. The force $\mathbf{F}''_{32}$ of the connecting rod against the crankpin;
4. The force $\mathbf{F}''_{12}$ of the engine block against the crank.

By methods similar to those used earlier in Chapter 16, the analytical expressions are determined to be

$$\mathbf{F}''_{41} = -m_4\ddot{x}\tan\phi\hat{\mathbf{j}} \tag{16.31}$$

$$\mathbf{F}''_{34} = m_4\ddot{x}\hat{\mathbf{i}} - m_4\ddot{x}\tan\phi\hat{\mathbf{j}} \tag{16.32}$$

$$\mathbf{F}''_{32} = -\mathbf{F}''_{34} \tag{16.33}$$

$$\mathbf{F}''_{12} = -\mathbf{F}''_{32} = \mathbf{F}''_{34}, \tag{16.34}$$

where $\ddot{x}$ is the acceleration of the piston as given by Eq. (16.11) and m_4 is the mass of the piston assembly. The quantity $\tan\phi$ can be evaluated in terms of the crank angle by use of Eq. (16.13).

Figure 16.19 illustrates only those forces that result because of that part m_{3B} of the mass of the connecting rod that is assumed to be located at the wrist-pin center. Thus, Fig. 16.19*b* is a free-body diagram of the connecting rod showing the inertia force $-m_{3B}\mathbf{A}_B$ acting at the wrist-pin end.

We now note that it is incorrect when finding the bearing loads to add m_{3B} and m_4 together and then to compute a resultant inertia force, although such a procedure would seem to be simpler. The reason for this is that m_4 is the mass of the piston assembly and the corresponding inertia force acts on the piston side of the wrist pin. But m_{3B} is part of the connecting-rod mass, and hence its inertia force acts on the connecting-rod side of the wrist

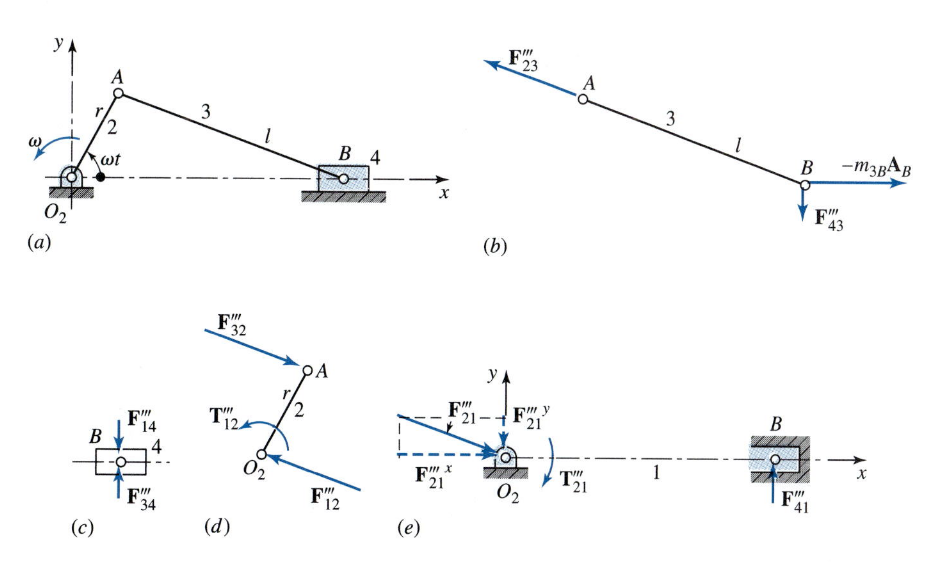

Figure 16.19 Graphical analysis of the forces resulting solely from the mass m_{3B} of the connecting rod, assumed to be concentrated at the wrist-pin end.

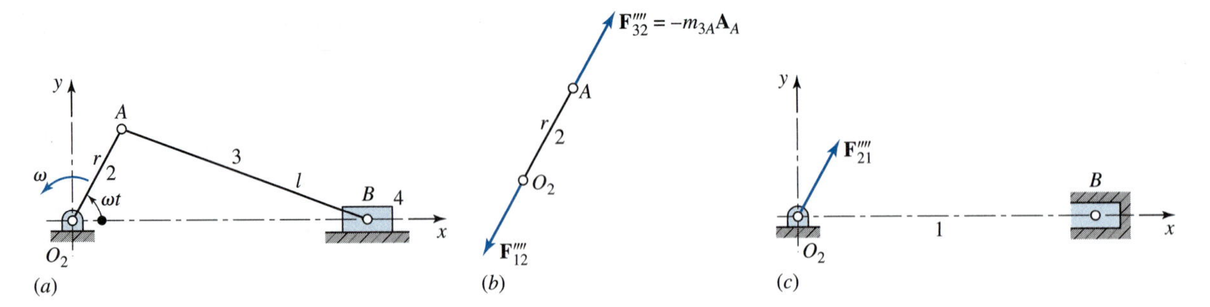

Figure 16.20 Graphical analysis of the forces resulting solely from the mass m_{3A} of the connecting rod, assumed to be concentrated at the crankpin end.

pin. Thus, adding the two would yield correct results for the crankpin load and the force of the piston against the cylinder wall but would give *incorrect* results for the wrist-pin load.

The forces on the piston pin, the crank, and the frame resulting from the inertia of mass m_{3B} are illustrated in Fig. 16.19*c*, Figures 16.19*d*, and 16.19*e*, respectively. The equations for these forces for a crank having uniform angular velocity are determined to be

$$\mathbf{F}_{41}''' = -m_{3B}\ddot{x}\tan\phi\hat{\mathbf{j}} \tag{16.35}$$

$$\mathbf{F}_{34}''' = \mathbf{F}_{41}''' \tag{16.36}$$

$$\mathbf{F}_{32}''' = -m_{3B}\ddot{x}\hat{\mathbf{i}} + m_{3B}\ddot{x}\tan\phi\hat{\mathbf{j}} \tag{16.37}$$

$$\mathbf{F}_{12}''' = -\mathbf{F}_{32}'''. \tag{16.38}$$

Figure 16.20 illustrates the forces that result from that part m_{3A} of the connecting-rod mass that is assumed concentrated at the crankpin end. Whereas a counterweight attached to the crank balances the reaction at O_2, it cannot make $F_{32}'''\,'$ zero. Thus, the crankpin force exists regardless of whether the rotating mass of the connecting rod is balanced. This force is

$$\mathbf{F}_{32}'''\,' = m_{3A}r\omega^2\left(\cos\omega t\hat{\mathbf{i}} + \sin\omega t\hat{\mathbf{j}}\right). \tag{16.39}$$

The last step is to sum the above expressions to obtain the resultant bearing loads. The total force of the piston against the cylinder wall, for example, is determined by summing Eqs. (16.12), (16.31), and (16.35), with due regard for subscripts and signs. When simplified, the result is

$$\mathbf{F}_{41} = \mathbf{F}_{41}' + \mathbf{F}_{41}'' + \mathbf{F}_{41}''' = -\left[(m_{3B} + m_4)\ddot{x} + P\right]\tan\phi\hat{\mathbf{j}}. \tag{16.40}$$

The forces on the wrist pin, the crankpin, and the crankshaft are determined in a similar manner and are

$$\mathbf{F}_{34} = (m_4\ddot{x} + P)\hat{\mathbf{i}} - \left[(m_{3B} + m_4)\ddot{x} + P\right]\tan\phi\hat{\mathbf{j}} \tag{16.41}$$

$$\mathbf{F}_{32} = [m_{3A} r\omega^2 \cos\omega t - (m_{3B} + m_4)\ddot{x} - P]\hat{\mathbf{i}} + \{m_{3A} r\omega^2 \sin\omega t + [(m_{3B} + m_4)\ddot{x} + P]\tan\phi\}\hat{\mathbf{j}} \tag{16.42}$$

$$\mathbf{F}_{21} = \mathbf{F}_{32}. \tag{16.43}$$

16.8 CRANKSHAFT TORQUE

The torque delivered by the crankshaft to the load is called the *crankshaft torque*, and it is the negative of the moment of the couple formed by the forces $\mathbf{F}_{41}$ and $\mathbf{F}^y_{21}$. Therefore, it is obtained from the equation

$$\mathbf{T}_{21} = -F_{41} x\hat{\mathbf{k}} = [(m_{3B} + m_4)\ddot{x} + P]\,x\tan\phi\hat{\mathbf{k}}. \tag{16.44}$$

16.9 SHAKING FORCES OF ENGINES

The inertia force caused by the reciprocating masses is illustrated acting in the positive direction in Fig. 16.21*a*. In Fig. 16.21*b* the forces acting upon the engine block caused by these inertia forces are illustrated. The resultant forces are $\mathbf{F}_{21}$, the force exerted by the crankshaft on the main bearings, and a positive couple formed by the forces $\mathbf{F}_{41}$ and $\mathbf{F}^y_{21}$. The force $\mathbf{F}^x_{21} = -m_B\mathbf{A}_B$ is termed the *shaking force,* and the couple $T = xF_{41}$ is called the *shaking moment*. As indicated by Eqs. (16.27) and (16.30), the magnitude and the direction of this force and couple change with ωt; consequently, the shaking force induces rectilinear vibration of the block in the x direction, and the shaking couple induces a torsional vibration of the block about the crank center.

A graphical representation of the inertia force is possible if Eq. (16.27) is rearranged as

$$F = m_B r\omega^2 \cos\omega t + m_B r\omega^2 \frac{r}{l}\cos 2\omega t, \tag{16.45}$$

where $F = F^x_{21}$ for simplicity of notation. The first term of Eq. (16.45) is represented by the x projection of a vector of length $m_B r\omega^2$ rotating at ω rad/s. This is the primary part of the inertia force. The second term is similarly represented by the $x$projection of a vector of length $m_B r\omega^2(r/l)$ rotating at 2ω rad/s; this is the secondary part. Such a diagram is illustrated in Fig. 16.22 for $r/l = \frac{1}{4}$. The total inertia or shaking force is the algebraic sum of the horizontal projections of these two vectors.

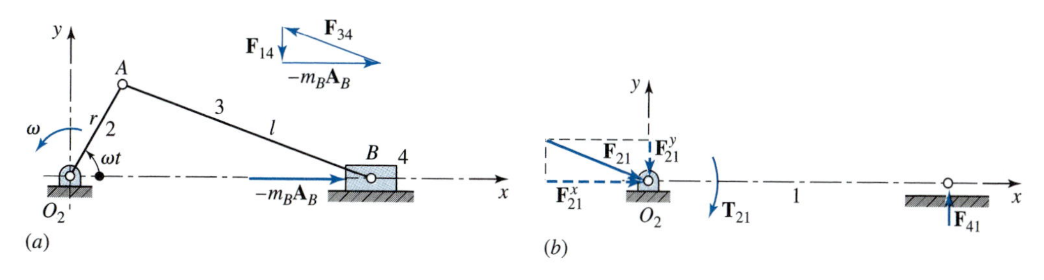

Figure 16.21 Dynamic forces caused by the reciprocating masses; the primes have been omitted for clarity.

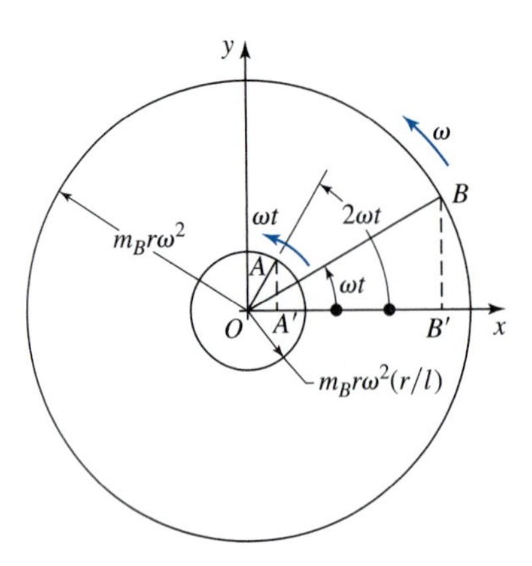

Figure 16.22 Circle diagram illustrating inertia forces. The total shaking force is $OA' + OB'$.

16.10 COMPUTATION HINTS

This section contains suggestions for using computers or programmable calculators in solving the dynamics of engine mechanisms. Many of the ideas, however, are also useful for readers using nonprogrammable calculators, as well as for checking purposes.

Indicator Diagrams It would be very convenient if a subprogram for computing the gas forces could be devised and the results used directly in a main program to compute all the resultant bearing forces and crankshaft torques. Unfortunately, the theoretical indicator diagram must be manipulated by hand to obtain a reasonable approximation to the experimental data. This manipulation can be done graphically or with a computer having a graphic display. The procedure is illustrated by the following example.

EXAMPLE 16.1

Determine the pressure-versus-piston displacement relation for a six-cylinder engine having a displacement of 2.30 L, a compression ratio of 8, and a brake power of 42.5 kW at 2,400 rev/min. Use a mechanical efficiency of 75%, a card factor of 0.85, a suction pressure of 100 kPa, and a polytropic exponent of 1.30.

SOLUTION

Rearranging Eq. (16.2), we find the brake mean effective pressure as follows:

$$p_b = \frac{(42.5\ \text{kW})\ (1000(\text{N}\cdot\text{m/s})/\text{kW})\ (60\ \text{s/min})}{\left(0.002\ 30\ \text{m}^3\right)\ (2\ 400\ \text{rev/min})\ (0.5\ \text{stroke/rev})\ \left(1\ 000(N/\text{m}^2)/\text{kPa}\right)} = 923.9\ \text{kPa}.$$

Then, from Eq. (16.3), the indicated mean effective pressure is

$$p_i = \frac{p_b}{e_m} = \frac{923.9\ \text{kPa}}{0.75} = 1\ 232\ \text{kPa}.$$

Next, we determine the pressure p_4 on the theoretical diagram of Fig. 16.12. Substituting the information into Eq. (16.7) gives

$$p_4 = (1.3 - 1)\left(\frac{8-1}{8^{1.3}-8}\right)\frac{1\,232\text{ kPa}}{0.85} + 100\text{ kPa} = 355\text{ kPa}.$$

The volume difference $v_1 - v_2$ in Fig. 16.12 is the volume swept out by the piston. Therefore,

$$v_1 - v_2 = la = \frac{2.30\text{ L}}{6\text{ cyl}} = 0.383\text{ L/cyl}.$$

Substituting this displacement volume and $R = 8$ into Eq. (b) of Section 16.2 gives

$$v_1 = \frac{R(v_1 - v_2)}{R-1} = \frac{8(0.383\text{ L})}{8-1} = 0.438\text{ L}.$$

Therefore,

$$v_2 = 0.438\text{ L} - 0.383\text{ L} = 0.055\text{ L}.$$

Then the percentage clearance C is

$$C = \frac{v_2}{v_1 - v_2}(100)\% = \frac{0.055\text{ L}}{0.383\text{ L}}(100) = 14.4\%.$$

Expressing volumes as percentages of the displacement volume enables us to write Eq. (16.1) in the form

$$p_x(X + C)^k = p_1(100 + C)^k,$$

where X is the percentage of piston travel measured from the head end of the stroke. Thus, the formula

$$p_x = p_1\left(\frac{100 + C}{X + C}\right)^k = 1\,232\text{ kPa}\left(\frac{114.4}{X + 14.4}\right)^{1.3} \tag{1}$$

is used to compute the pressure during the compression stroke for any piston position between $X = 0$ and $X = 100\%$. For the expansion stroke, Eq. (16.1) becomes

$$p_x = p_4\left(\frac{100 + C}{X + C}\right)^k = 355\text{ kPa}\left(\frac{114.4}{X + 14.4}\right)^{1.3}. \tag{2}$$

Equations (1) and (2) are easy to program for machine computation. The results can be displayed and recorded, or printed, for graphical use. Alternatively, the results can be displayed for hand calculation.

Figure 16.23 illustrates the plotted results of the computation using $\Delta X = 5\%$ increments. Note particularly how the results are rounded to obtain a smooth indicator diagram. This rounding, of course, produces results that are not exactly duplicated in subsequent trials. The greatest differences occur in the vicinity of point B.

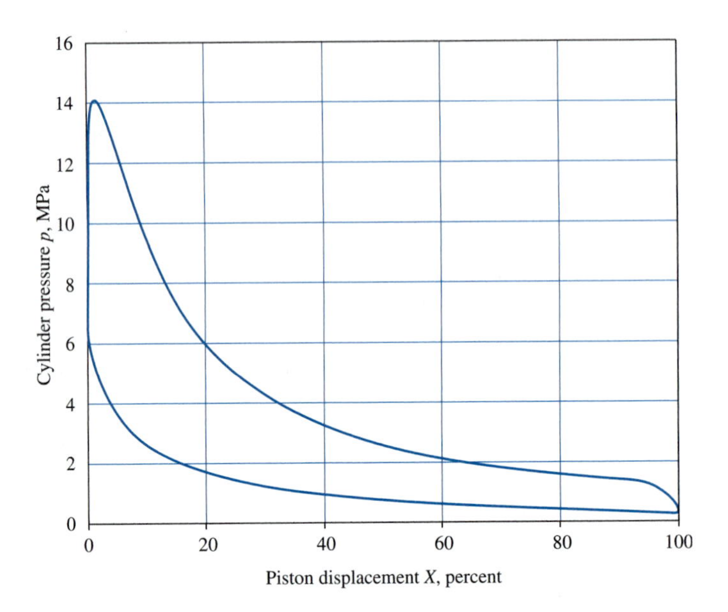

Figure 16.23 The diagram was rounded by hand. The peak cylinder pressure is about 75% of the maximum computed pressure at the beginning of the expansion stroke.

Force Analysis In a computer analysis the values of pressure are read from a diagram like Fig. 16.23. Most analysts prefer to tabulate these data; a table is usually constructed with the first column containing values of the crank angle ωt. For a four-stroke engine, values of this angle are entered from 0° to 720°.

Values of x corresponding to each ωt are obtained from Eq. (16.9). Then the corresponding piston displacement X in percentage is obtained from the equation

$$X = \frac{r + l - x}{2r}(100\%). \tag{16.46}$$

Some care must be taken in tabulating X and the corresponding pressures. Then the gas forces corresponding to each value of ωt are computed using the piston area.

The balance of the analysis is perfectly straightforward; we use Eqs. (16.11), (16.13), and (16.40) through (16.44), in that order.

PROBLEMS

16.1 A one-cylinder, four-stroke engine has a compression ratio of 7.6 and develops brake power of 2.25 kW at 3 000 rev/min. The crank length is 22 mm with a 60-mm bore. Develop and plot a rounded indicator diagram using a card factor of 0.90, a mechanical efficiency of 72%, a suction pressure of 100 kPa, and a polytropic exponent of 1.30.

16.2 Construct a rounded indicator diagram for a four-cylinder, four-stroke gasoline engine having a 85-mm bore, a 90-mm stroke, and a compression ratio of 6.25. The operating conditions to be used are 22.4 kW at 1,900 rev/min. Use a mechanical efficiency of 72%, a card factor of 0.90, a suction pressure of 100 kPa, and a polytropic exponent of 1.30.

16.3 Construct an indicator diagram for a V6 four-stroke gasoline engine having a 100-mm bore, a 90-mm stroke, and a compression ratio of 8.40. The engine develops 150 kW at 4,400 rev/min. Use a mechanical efficiency of 72%, a card factor of 0.88, a suction pressure of 100 kPa, and a polytropic exponent of 1.30.

16.4 A single-cylinder, two-stroke gasoline engine develops 30 kW at 4,500 rev/min. The engine has an 80-mm bore, a stroke of 70 mm, and a compression ratio of 7.0. Develop a rounded indicator diagram for this engine using a card factor of 0.990, a mechanical efficiency of 65%, a suction pressure of 100 kPa, and a polytropic exponent of 1.30.

16.5 The engine of Problem 16.1 has a connecting rod 80 mm long and a mass of 0.100 kg, with the mass center 10 mm from the crankpin end. Piston mass is 0.180 kg. Find the bearing reactions and the crankshaft torque during the expansion stroke corresponding to a piston displacement of $X = 30\%$ ($\omega t = 60°$). To find p_e, see the answer to Problem 16.1 in Appendix B.

16.6 Repeat Problem 16.5, but do the computations for the compression stroke ($\omega t = 660°$).

16.7 Make a complete force analysis of the engine of Problem 16.5. Plot a graph of the crankshaft torque versus crank angle for 720° of crank rotation.

16.8 The engine of Problem 16.3 uses a connecting rod 300 mm long. The masses are $m_{3A} = 0.80$ kg, $m_{3B} = 0.38$ kg, and $m_4 = 1.64$ kg. Find all the bearing reactions and the crankshaft torque for one cylinder of the engine during the expansion stroke at a piston displacement of $X = 30\%$ ($\omega t = 63.2°$). The pressure should be obtained from the indicator diagram, Fig. AP16.3 in Appendix B.

16.9 Repeat Problem 16.8, but do the computations for the same position in the compression stroke ($\omega t = 656.8°$).

16.10 Additional data for the engine of Problem 16.4 are $l_3 = 110$ mm, $R_{G_3A} = 15$ mm, $m_4 = 0.24$ kg, and $m_3 = 0.13$ kg. Make a complete force analysis of the engine and plot a graph of the crankshaft torque versus crank angle for 360° of crank rotation.

16.11 The four-stroke engine of Problem 16.1 has a stroke of 66 mm and a connecting rod length of 183 mm. The mass of the rod is 0.386 kg, and the center of mass center is 42 mm from the crankpin. The piston assembly has mass of 0.576 kg. Make a complete force analysis for one cylinder of this engine for 720° of crank rotation. Use 110 kPa for the exhaust pressure and 70 kPa for the suction pressure. Plot a graph to show the variation of the crankshaft torque with the crank angle. Use Fig. 16.23 for the pressures.

17 Balancing

Balancing is the technique of correcting or eliminating unwanted inertia forces and moments in rotating machinery. In previous chapters we have seen that shaking forces on the frame can vary significantly during a cycle of operation. Such forces can cause vibrations that at times may reach dangerous amplitudes. Even if they are not dangerous, vibrations increase component stresses and subject bearings to repeated loads that may cause parts to fail prematurely by fatigue. Thus, in the design of machinery it is not sufficient merely to avoid operation near the critical speeds; we must eliminate, or at least reduce, the dynamic forces that produce these vibrations in the first place.

Tolerances used in the manufacture of machinery are typically set as close as possible without prohibitively increasing the cost of manufacture. In general, however, it is more economical to produce parts that are not quite true and then to subject them to a balancing procedure than it is to produce such perfect parts that no correction is needed. Because of this, each part produced is an individual case, in that no two parts can normally be expected to require the same corrective measures. Thus, determining the unbalance and the required corrections is the principal topic in the study of balancing.

17.1 STATIC UNBALANCE

The arrangement illustrated in Fig. 17.1*a* consists of a disk-and-shaft combination resting on rigid horizontal rails so that the shaft, which is assumed to be perfectly straight, can roll without friction. A reference system *xyz* is attached to the disk and moves with it. Simple experiments to determine whether the disk is statically unbalanced can be conducted as follows. We can roll the disk gently by hand and permit it to coast until it comes to rest. Then we can mark with chalk the lowest point of the periphery of the disk. After we repeat this four or five times, if the chalk marks are scattered at different places around

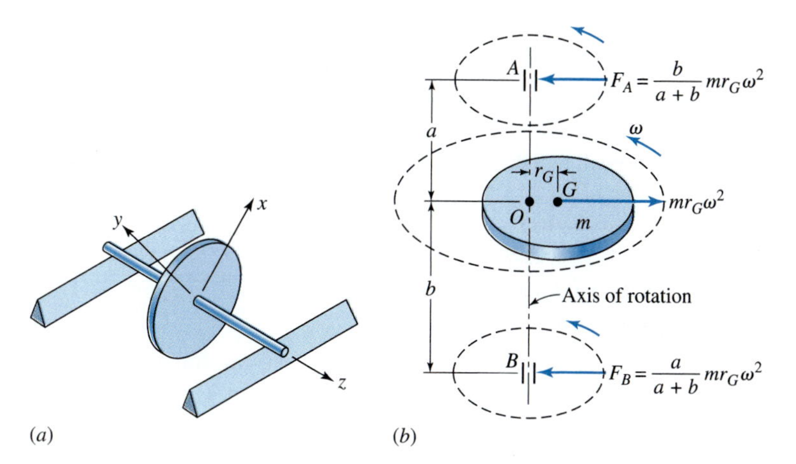

Figure 17.1

the periphery, the disk is in static balance. If all of the chalk marks are clustered, the disk is statically unbalanced, which means that the axis of the shaft and the center of mass of the disk are not coincident. The position of the chalk marks with respect to the *xy* system indicates the angular location of the unbalance but *not* the amount.

It is unlikely that any of the marks will be located 180° from the remaining ones, although it is theoretically possible to obtain static equilibrium with the unbalance above the shaft axis.

If static unbalance is found to exist, it can be corrected by drilling out material at the chalk mark cluster or by adding mass to the periphery 180° from the cluster. Because the amount of unbalance is unknown, these corrections must be made by trial and error.

17.2 EQUATIONS OF MOTION

If an unbalanced disk and shaft of mass m is mounted in bearings and caused to rotate at an angular speed of ω then a centrifugal force of magnitude $mr_G\omega^2$ exists (where r_G is the distance of the mass center of the disk from the shaft axis, that is, eccentricity), as illustrated in Fig. 17.1*b*. This force acting on the shaft produces the rotating bearing reactions illustrated in Fig. 17.1*b*.

To determine the equation of motion of the system, we specify m as the total mass and m_u as the unbalanced mass. We also let k signify the shaft stiffness, a value that describes the magnitude of a force necessary to bend the shaft a unit distance when applied at O. Thus, k has units of pounds per inch or Newtons per meter. Let c be the coefficient of viscous damping as defined in Eq. (15.2). Selecting any x coordinate normal to the shaft axis, we can now write

$$\sum F_O = -kx - c\dot{x} - m\ddot{x} + m_u r_G \omega^2 \cos \omega t = 0$$

or, upon rearranging,

$$m\ddot{x} + c\dot{x} + kx = m_u r_G \omega^2 \cos \omega t. \tag{a}$$

Recall that the solution to this differential equation was studied in Sections 15.12 and 15.13. From Eq. (15.55), the deflection of the shaft can be written as

$$x = \frac{m_u r_G \omega^2 \cos(\omega t - \phi)}{\sqrt{(k - m\omega^2)^2 + c^2\omega^2}}, \tag{b}$$

where ϕ is the phase angle between the force $m_u r_G \omega^2$ and the amplitude X of the shaft vibration. The *phase angle* given by Eq. (15.50) and is repeated here for the convenience of the reader:

$$\phi = \tan^{-1}\left(\frac{c\omega}{k - m\omega^2}\right). \tag{c}$$

Certain simplifications can be made with Eq. (b) to clarify its meaning.

First, consider the term $k - m\omega^2$ in the denominator. If this term were zero, the amplitude of x would be very large because it would be limited only by the damping constant c, which is usually very small. The value of ω that makes the term $k - m\omega^2$ equal to zero is denoted ω_n and called the *natural angular velocity*, *critical speed*, or *natural frequency*. Therefore,

$$\omega_n = \sqrt{\frac{k}{m}}, \tag{17.1}$$

which agrees with the definition in Section 15.2 [see Eq. (15.8)].

In the study of free or unforced vibrations (Section 15.8), it is found that a certain value of the viscous damping factor c will result in no vibration at all. This special value is called the *critical coefficient of viscous damping* [see Eq. (15.27)], that is,

$$c_c = 2m\omega_n \tag{17.2}$$

The *damping ratio* ζ is the ratio of the actual to the critical damping, see Eq. (15.28), that is

$$\zeta = \frac{c}{c_c} = \frac{c}{2m\omega_n}. \tag{17.3}$$

For most machine systems in which damping is not deliberately introduced, ζ is generally in the range $0.015 \leq \zeta \leq 0.120$.

Next, note that Eq. (b) can be expressed in the form given by Eq. (15.15) or Eq. (15.23), that is,

$$x = X \cos(\omega t - \phi). \tag{d}$$

If we now divide the numerator and denominator of the amplitude X of Eq. (b) by k, designate the eccentricity as $e = r_G$, and introduce Eqs. (17.1) and (17.3), we obtain the ratio given by Eq. (15.56), that is,

$$\frac{mX}{m_u e} = \frac{(\omega/\omega_n)^2}{\sqrt{(1-\omega^2/\omega_n^2)^2 + (2\zeta\omega/\omega_n)^2}}. \tag{17.4}$$

This is the equation for the amplitude ratio of the vibration of a rotating disk-and-shaft combination. If we neglect damping, that is, set $\zeta = 0$, let $m = m_u$, and replace e with r_G (where r_G is the eccentricity of the unbalance), then the amplitude of the vibration corresponding to any frequency ratio ω/ω_n can be written from Eq. (17.4) as

$$X = r_G \frac{(\omega/\omega_n)^2}{1-(\omega/\omega_n)^2}. \tag{17.5}$$

Now if, in Fig. 17.1*b*, we designate O the intersection of the center of the shaft and the central plane of the disk and G the mass center of the disk, we can draw some interesting conclusions by plotting Eq. (17.5). This is done in Fig. 17.2, where the amplitude is plotted on the vertical axis and the frequency ratio along the abscissa. The natural frequency is ω_n, which corresponds to the critical speed, whereas ω is the actual speed of rotation of the shaft.

When rotation is just beginning, ω is much less than ω_n and the graph indicates that the amplitude of the vibration is very small. As the shaft speed increases, the amplitude also increases and theoretically becomes infinite at the critical speed. As the shaft goes through the critical speed, the amplitude changes to a negative value and decreases as the shaft speed increases further. The graph indicates that the amplitude never returns to zero, no matter how much the shaft speed is increased, but reaches a limiting value of $-r_G$. Note that in this high-speed range the disk is rotating about its own center of mass, which has then become coincident with the centerline of the bearings.

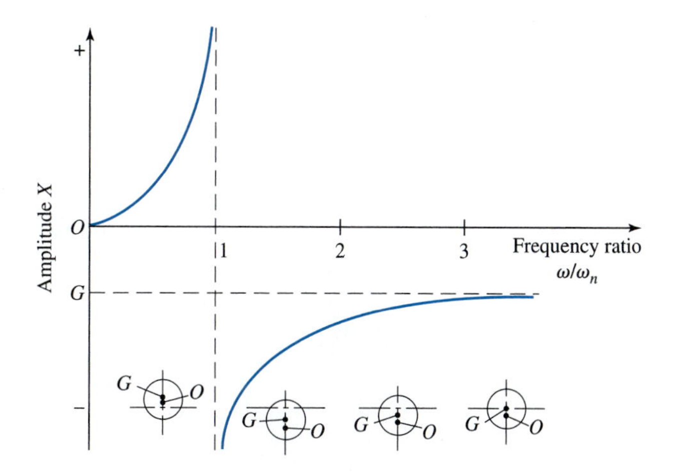

Figure 17.2 The small illustrations below the graph indicate the relative position of three points for various frequency ratios. The mass center of the disk is at G; the center of the shaft is at O; and the axis of rotation is at the intersection of the centerlines. Thus, both amplitude and phase relationships are illustrated.

The preceding discussion demonstrates that statically unbalanced rotating systems produce undesirable vibrations and rotating bearing reactions. Using static balancing equipment, the eccentricity r_G can be reduced, but it is impossible in practice to make it exactly zero. Therefore, no matter how small r_G is made (even if it was zero), trouble can always be expected when $\omega = \omega_n$. When the operating speed must be higher than the natural frequency, the machine should be designed so that, on startup, it passes through the natural frequency quickly to avoid dangerous vibrations from building.

17.3 STATIC BALANCING MACHINES

The purpose of a balancing machine is first to indicate whether a mechanical part is in balance. If it is out of balance, the machine should measure the unbalance by indicating its *magnitude* and *location*.

Static balancing machines are used only for parts whose axial dimensions are small, such as gears, fans, and impellers, and the machines are often called *single-plane balancers* because the mass must lie practically in a single plane. In the sections to follow, we discuss balancing in several planes, but it is important to note here that if several disks are to be mounted upon a shaft that is to rotate, the parts should be individually statically balanced before mounting. Although it is possible, instead, to balance the assembly in two planes after the parts are mounted, additional bending moments inevitably come into existence when this is done.

Static balancing is essentially a weighing process in which the part is acted upon by either a gravity force or a centrifugal force. We have seen that the disk and shaft of the preceding section could be balanced by placing it on parallel rails, rocking it, and permitting it to seek equilibrium. In this case, the location of the unbalance is determined through the aid of the force of gravity. Another method of balancing the disk is to rotate it at a predetermined speed. Then the bearing reactions can be measured and their magnitudes used to indicate

Figure 17.3 A helicopter-rotor assembly balancer. (Courtesy of Micro-Poise Measurement Systems, LLC, Akron, OH.)

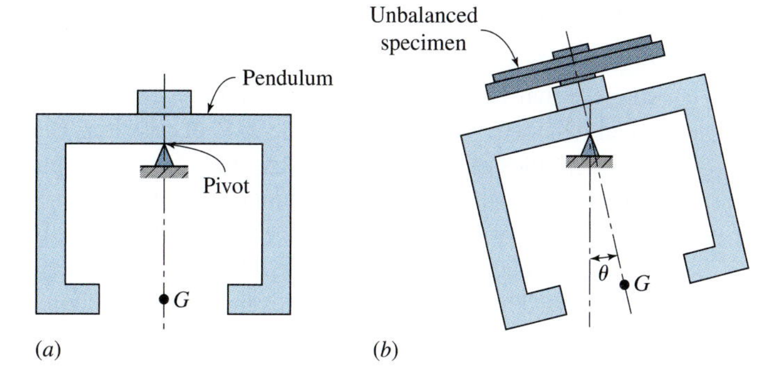

Figure 17.4 Operation of a static balancing machine.

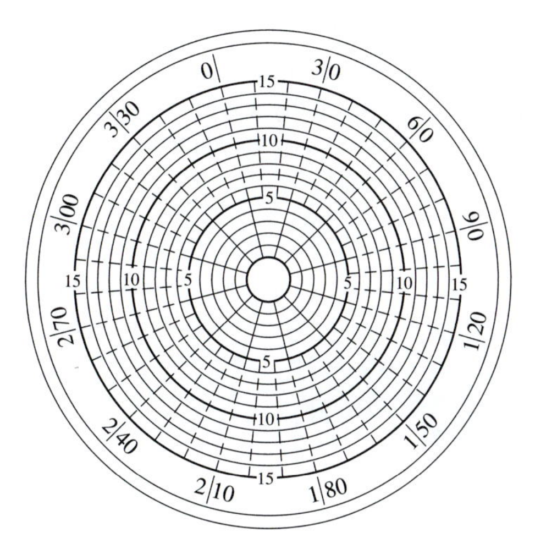

Figure 17.5 Drawing of the universal level used in the Micro-Poise balancer. The numbers on the periphery are degrees; the radial distances are calibrated in units proportional to inch-ounces. The position of the bubble indicates both the direction and the magnitude of the unbalance. (Micro-Poise Measurement Systems, LLC, Akron, OH.)

the amount of unbalance. Because the part is rotating while the measurements are taken, a stroboscope is used to indicate the location of the required correction.

When machine parts are manufactured in large quantities, a balancer is required that will measure both the amount and the location of the unbalance and give the correction directly and quickly. Also, time can be saved if it is not necessary to make the part rotate. Such a balancing machine is illustrated in Fig. 17.3. This machine is essentially a pendulum that can tilt in any direction, as illustrated by the schematic drawing of Fig. 17.4*a*.

When an unbalanced specimen is mounted on the platform of the balancing machine, the pendulum tilts. The direction of the tilt gives the location of the unbalance, whereas the tilt angle θ (Fig. 17.4*b*) indicates the magnitude of the unbalance. Some damping is employed to reduce oscillations of the pendulum. Figure 17.5 illustrates a universal level that is mounted on the platform of the balancer. A bubble, at the center when in balance, moves and indicates both the location and the magnitude of the correction.

17.4 DYNAMIC UNBALANCE

Figure 17.6 illustrates a long rotor that is to be mounted in bearings at A and B. We might suppose that two equal masses m_1 and m_2 are placed at opposite ends of the rotor and at equal distances r_1 and r_2 from the axis of rotation. Because the masses are equal and on opposite sides of the rotational axis, the rotor can be placed on rails as described earlier to demonstrate that it is statically balanced in all angular positions.

If the rotor of Fig. 17.6 is placed in bearings and caused to rotate at an angular velocity of ω, then centrifugal forces $m_1 r_1 \omega^2$ and $m_2 r_2 \omega^2$ act, respectively, at m_1 and m_2 on the rotor ends. These centrifugal forces produce the unequal bearing reactions $\mathbf{F}_A$ and $\mathbf{F}_B$, and the entire system of forces rotates with the rotor at the angular velocity ω. Thus, a part may be statically balanced and at the same time dynamically unbalanced (Fig. 17.7).

In the general case, accurate distribution of the mass along the axis of a part depends upon the configuration of the part, but inaccuracy occurs in machining and also in casting and in forging. Other errors or unbalances may be caused by improper boring, keys, and assembly. It is the designer's responsibility to design a rotating part so that a line joining all mass centers will be a straight line coinciding with the axis of rotation. However, perfect parts and perfect assembly are seldom attained; consequently, a line from one end of the part to the other, joining all mass centers, is usually a spatial curve that may only occasionally cross or coincide with the axis of rotation. An unbalanced part, therefore, will usually be

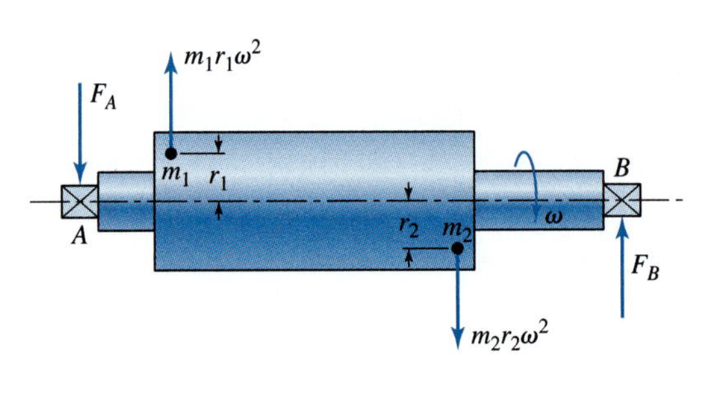

Figure 17.6 If $m_1 = m_2$ and $r_1 = r_2$, the rotor is statically balanced but dynamically unbalanced.

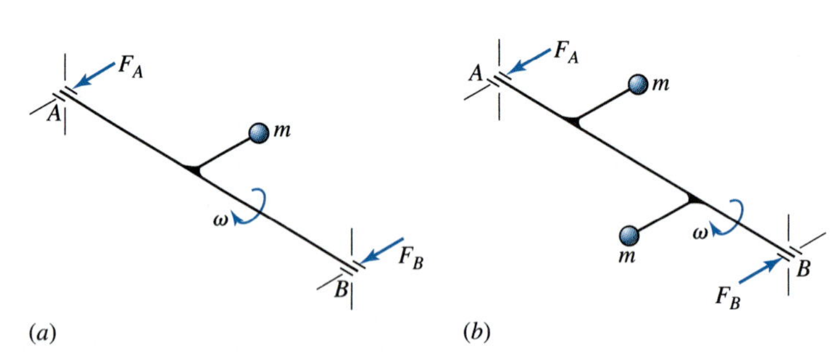

Figure 17.7 (*a*) Static unbalance; when the shaft rotates, both bearing reactions are in the same plane and in the same direction. (*b*) Dynamic unbalance; when the shaft rotates, the unbalance creates a couple tending to turn the shaft end over end. The shaft is in equilibrium because of the opposite couple formed by the bearing reactions. Note that the bearing reactions are still in the same plane but in opposite directions.

out of balance, both statically and dynamically. This is the most general kind of unbalance, and when the part is supported by two bearings, we can then expect the magnitudes as well as the directions of the rotating bearing reactions to be different.

17.5 ANALYSIS OF UNBALANCE

In this section we will see how to analyze any unbalanced rotating system and determine the proper corrections using graphical methods, vector methods, and computer or calculator methods.

Graphic Analysis The two equations

$$\sum \mathbf{F} = \mathbf{0} \quad \text{and} \quad \sum \mathbf{M} = \mathbf{0} \tag{a}$$

are used to determine the amount and orientation of the corrections. We begin by noting that each centrifugal force $m\mathbf{R}\omega^2$ of an eccentric rotating mass is proportional to its product, $m\mathbf{R}$. Thus, for Fig. 17.8*a*, vector quantities proportional to the centrifugal force of each of the three masses, $m_1\mathbf{R}_1$, $m_2\mathbf{R}_2$, and $m_3\mathbf{R}_3$, will act in radial directions, as indicated. The first of Eqs. (*a*) is applied by constructing a force polygon (Fig. 17.8*b*). Because this polygon requires another vector, $m_c\mathbf{R}_c$, for closure, the magnitude of the correction is $m_c\mathbf{R}_c$ and its orientation is parallel to $\mathbf{R}_c$. The three masses of Fig. 17.8 are assumed to rotate in a single plane and so this is the correction for a case of static unbalance.

When the rotating masses are in different planes, both of Eqs. (*a*) must be used. Figure 17.9*a* is an end view of a shaft having mounted upon it the three masses m_1, m_2, and m_3 at the radial distances R_1, R_2, and R_3, respectively. Figure 17.9*b* is a side view of the same shaft showing left and right correction planes and the axial distances to the three masses. We want to find the magnitudes and angular orientations of the corrections to be added in each correction plane.

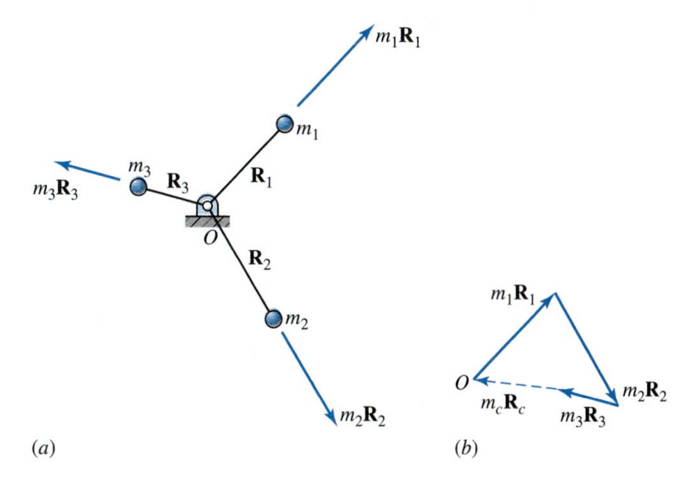

Figure 17.8 (*a*) A three-mass system rotating in a single plane. (*b*) Centrifugal force polygon gives $m_c\mathbf{R}_c$ as the required correction.

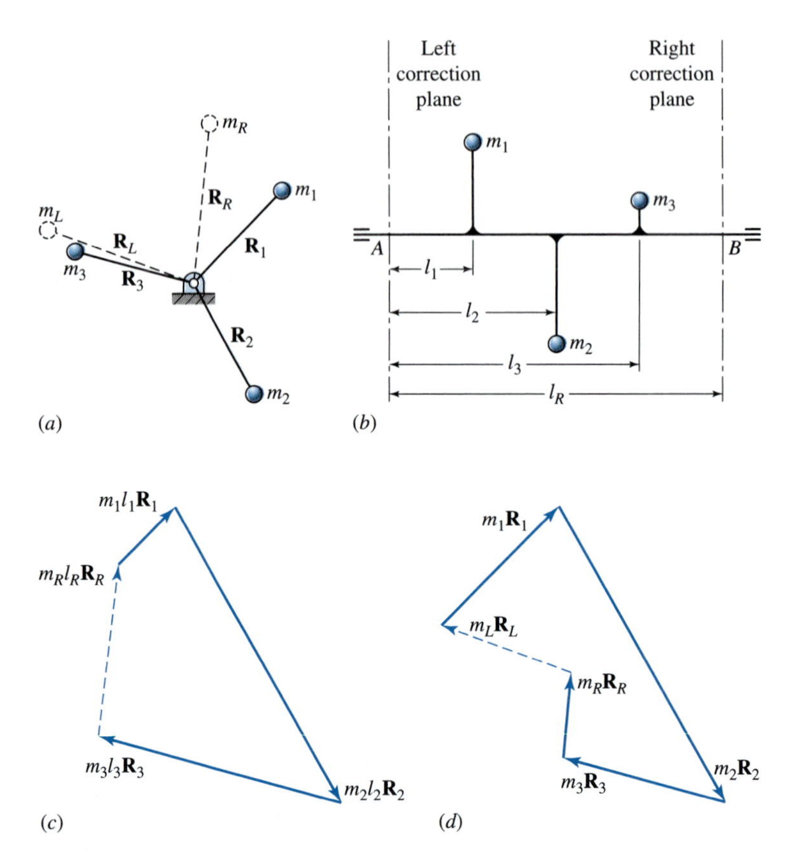

Figure 17.9 Graphical analysis of unbalance.

The first step in the solution is to take the sum of the moments of the centrifugal forces, including the corrections, about some point. We choose to take this sum about point A in the left correction plane to eliminate the moment of the left correction mass. Thus, applying the second of Eqs. (a) gives

$$\sum \mathbf{M}_A = m_1 l_1 \mathbf{R}_1 + m_2 l_2 \mathbf{R}_2 + m_3 l_3 \mathbf{R}_3 + m_R l_R \mathbf{R}_R = \mathbf{0}. \tag{b}$$

This is a vector equation in which the directions of the vectors are parallel, respectively, to the vectors $\mathbf{R}_j$ in Fig. 17.9a. Consequently, the moment polygon in Fig. 17.9c can be constructed. The closing vector $m_R l_R \mathbf{R}_R$ gives the magnitude and orientation of the correction required for the right-hand correction plane. Next, the equation

$$\sum \mathbf{F} = m_1 \mathbf{R}_1 + m_2 \mathbf{R}_2 + m_3 \mathbf{R}_3 + m_R \mathbf{R}_R + m_L \mathbf{R}_L = \mathbf{0} \tag{c}$$

can be written. By constructing the force polygon of Fig. 17.9d, this equation is solved for the left-hand correction $m_L \mathbf{R}_L$.

Although Fig. 17.9*c* has been called a moment polygon, it is worth noting that the vectors making up this polygon consist of quantities proportional to the moment magnitudes and the position vector directions. A true moment polygon would require rotating this polygon 90° cw, because each moment vector is really equal to $l\hat{\mathbf{k}} \times m\mathbf{R}\omega^2$.

After solution, the separate quantities m_R, $\mathbf{R}_R$, and m_L, $\mathbf{R}_L$ can be determined because, in general, the magnitudes of $\mathbf{R}_R$ and $\mathbf{R}_L$ are specified in the problem statement.

Vector Analysis The following two examples illustrate the vector algebra approach.

EXAMPLE 17.1

Figure 17.10 represents a rotating system that has been idealized for illustrative purposes. A weightless shaft that is supported in bearings at A and B rotates with an angular velocity $\boldsymbol{\omega} = 100\,\hat{\mathbf{i}}$ rad/s. Three mass particles, with weights $w_1 = 2$ oz, $w_2 = 1$ oz, and $w_3 = 1.5$ oz, are connected to the shaft and rotate with it, causing an unbalance. Determine the bearing reactions at A and B for the orientation shown.

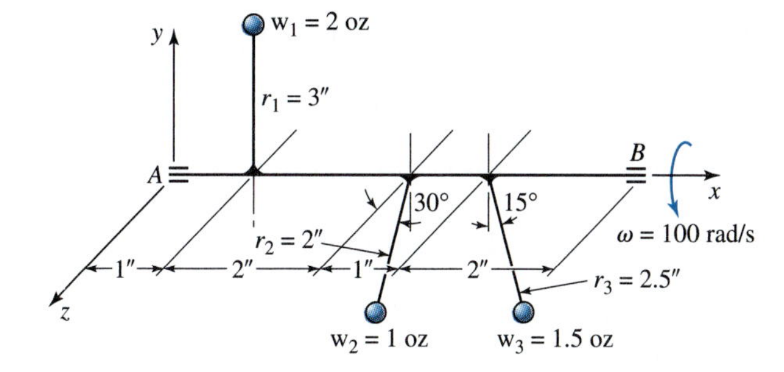

Figure 17.10 Example 17.1.

SOLUTION

The centrifugal force caused by each rotating mass particle is

$$m_1 r_1 \omega^2 = \frac{(2\text{ oz})(3\text{ in})(100\text{ rad/s})^2}{(386\text{ in/s}^2)(16\text{ oz/lb})} = 9.72\text{ lb}$$

$$m_2 r_2 \omega^2 = \frac{(1\text{ oz})(2\text{ in})(100\text{ rad/s})^2}{(386\text{ in/s}^2)(16\text{ oz/lb})} = 3.24\text{ lb}$$

and

$$m_3 r_3 \omega^2 = \frac{(1.5\text{ oz})(2.5\text{ in})(100\text{ rad/s})^2}{(386\text{ in/s}^2)(16\text{ oz/lb})} = 6.07\text{ lb}.$$

These three forces are parallel to the yz plane and can be written in vector form as

$$\mathbf{F}_1 = m_1 r_1 \omega^2 \angle\theta_1 = 9.72\text{ lb}\angle 0° = 9.72\hat{\mathbf{j}}\text{ lb}$$

$$\mathbf{F}_2 = m_2 r_2 \omega^2 \angle\theta_2 = 3.24\text{ lb}\angle 120° = -1.62\hat{\mathbf{j}} + 2.81\hat{\mathbf{k}}\text{ lb}$$

and

$$\mathbf{F}_3 = m_3 r_3 \omega^2 \angle \theta_3 = 6.07 \text{ lb}\angle 195^\circ = -5.86\hat{\mathbf{j}} - 1.57\hat{\mathbf{k}} \text{ lb}.$$

where the angles θ_1, θ_2, and θ_3, are measured counterclockwise from the y axis when viewed from the positive end of the x axis. The moments of these forces taken about the bearing at A must be balanced by the moment of the bearing reaction at B. Therefore,

$$\sum \mathbf{M}_A = 1\hat{\mathbf{i}} \times (9.72\hat{\mathbf{j}}) + 3\hat{\mathbf{i}} \times (-1.62\hat{\mathbf{j}} + 2.81\hat{\mathbf{k}})$$
$$+ 4\hat{\mathbf{i}} \times (-5.86\hat{\mathbf{j}} - 1.57\hat{\mathbf{k}}) + 6\hat{\mathbf{i}} \times (\mathbf{F}_B) = \mathbf{0}.$$

Solving this equation, the bearing reaction at B is

$$\mathbf{F}_B = 3.10\hat{\mathbf{j}} - 0.36\hat{\mathbf{k}} \text{ lb} = 3.12 \text{ lb}\angle -6.6^\circ. \qquad \textit{Ans.}$$

To find the bearing reaction at A we repeat the analysis. Taking moments about B gives

$$\sum \mathbf{M}_B = -2\hat{\mathbf{i}} \times (-5.86\hat{\mathbf{j}} - 1.57\hat{\mathbf{k}}) + (-3\hat{\mathbf{i}}) \times (-1.62\hat{\mathbf{j}} + 2.81\hat{\mathbf{k}})$$
$$+ (-5\hat{\mathbf{i}}) \times (9.72\hat{\mathbf{j}}) + (-6\hat{\mathbf{i}}) \times (\mathbf{F}_A) = \mathbf{0}$$

Then solving this equation, the bearing reaction at A is

$$\mathbf{F}_A = -5.34\hat{\mathbf{j}} - 0.88\hat{\mathbf{k}} \text{ lb} = 5.41 \text{ lb}\angle 189.4^\circ. \qquad \textit{Ans.}$$

Note that these are rotating reactions and that static, or stationary, components caused by gravity are not included.

EXAMPLE 17.2

The angular speed of the system illustrated in Fig. 17.11 is 750 rev/min. Determine (*a*) the bearing reactions at A and B and (*b*) the magnitude and location of a correcting (or balancing) mass to be placed at a radius of 0.25 m.

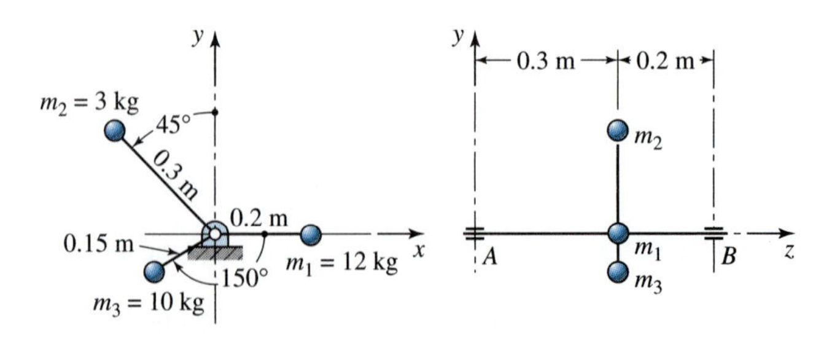

 **Figure 17.11** Example 17.2.

SOLUTION

(a) The angular velocity of this system is

$$\omega = (750 \text{ rev/min})(2\pi \text{ rad/rev})/(60 \text{ s/min}) = 78.54 \text{ rad/s}.$$

The centrifugal forces caused by the masses are

$$F_1 = m_1 r_1 \omega^2 = (12 \text{ kg})(0.2 \text{ m})(78.5 \text{ rad/s})^2(10^{-3} \text{ kN/N}) = 14.80 \text{ kN}$$

$$F_2 = m_2 r_2 \omega^2 = (3 \text{ kg})(0.3 \text{ m})(78.5 \text{ rad/s})^2(10^{-3} \text{ kN/N}) = 5.55 \text{ kN}$$

and

$$F_3 = m_3 r_3 \omega^2 = (10 \text{ kg})(0.15 \text{ m})(78.5 \text{ rad/s})^2(10^{-3} \text{ kN/N}) = 9.25 \text{ kN}.$$

These forces can be written in vector form as

$$\begin{aligned}
\mathbf{F}_1 &= 14.80 \text{ kN}\angle 0^\circ = 14.80\hat{\mathbf{i}} \text{ kN} \\
\mathbf{F}_2 &= 5.55 \text{ kN}\angle 135^\circ = -3.93\hat{\mathbf{i}} + 3.93\hat{\mathbf{j}} \text{ kN} \\
\mathbf{F}_3 &= 9.25 \text{ kN}\angle -150^\circ = -8.01\hat{\mathbf{i}} - 4.63\hat{\mathbf{j}} \text{ kN}.
\end{aligned}$$

To find the bearing reaction at B we take moments about the bearing A. This equation is written as

$$\begin{aligned}
\sum \mathbf{M}_A = 0.3\hat{\mathbf{k}} \text{ m} \times [(14.80\hat{\mathbf{i}} \text{ kN}) + (-3.93\hat{\mathbf{i}} + 3.93\hat{\mathbf{j}} \text{ kN}) \\
+ (-8.00\hat{\mathbf{i}} - 4.63\hat{\mathbf{j}} \text{ kN})] + 0.5\hat{\mathbf{k}} \text{ m} \times \mathbf{F}_B = \mathbf{0}.
\end{aligned}$$

Taking the cross products and rearranging gives

$$0.5\hat{\mathbf{k}} \text{ m} \times \mathbf{F}_B = -0.21\hat{\mathbf{i}} - 0.86\hat{\mathbf{j}} \text{ kN} \cdot \text{m}.$$

Solving this equation for F_B gives

$$\mathbf{F}_B = -1.73\hat{\mathbf{i}} + 0.42\hat{\mathbf{j}} \text{ kN} = 1.78 \text{ kN}\angle 166.35^\circ. \qquad \textit{Ans.}$$

Summing forces gives

$$\mathbf{F}_A + \mathbf{F}_1 + \mathbf{F}_2 + \mathbf{F}_3 + \mathbf{F}_B = \mathbf{0}$$

Therefore, the reaction at A is

$$\begin{aligned}
\mathbf{F}_A = -14.80\hat{\mathbf{i}} \text{ kN} - (-3.93\hat{\mathbf{i}} + 3.93\hat{\mathbf{j}} \text{ kN}) - (-8.01\hat{\mathbf{i}} - 4.63\hat{\mathbf{j}} \text{ kN}) \\
- (1.73\hat{\mathbf{i}} + 0.42\hat{\mathbf{j}} \text{ kN})
\end{aligned}$$

or

$$\mathbf{F}_A = -1.14\hat{\mathbf{i}} + 0.28\hat{\mathbf{j}}\ \text{kN} = 1.18\ \text{kN}\angle 166.32^\circ. \qquad \textit{Ans.}$$

(b) Let $\mathbf{F}_C$ be the correcting force. Then for zero bearing reactions

$$\sum \mathbf{F} = \mathbf{F}_1 + \mathbf{F}_2 + \mathbf{F}_3 + \mathbf{F}_C = \mathbf{0}.$$

Thus,

$$\begin{aligned}\mathbf{F}_C &= -14.80\hat{\mathbf{i}}\ \text{kN} - (-3.93\hat{\mathbf{i}} + 3.93\hat{\mathbf{j}}\ \text{kN}) - (-8.00\hat{\mathbf{i}} - 4.63\hat{\mathbf{j}}\ \text{kN}) \\ &= -2.87\hat{\mathbf{i}} + 0.70\hat{\mathbf{j}}\ \text{kN} = 2.95\ \text{kN}\angle 166.29^\circ.\end{aligned}$$

Therefore, the correcting mass should be placed at 166.29°. *Ans.*
The correcting mass at a radius of 0.25 m is

$$m_C = \frac{F_C}{r_C\omega^2} = \frac{2.95(10)^3\ \text{N}}{0.25\ \text{m}(78.54\ \text{rad/s})^2} = 1.91\ \text{kg}. \qquad \textit{Ans.}$$

Scalar Equations For a rotating system with n discrete mass particles and two correcting planes ($j = 1, 2$) we can write the following four scalar equations:

$$\sum_{i=1}^{n} m_i r_i \cos\phi_i + \sum_{j=1}^{2} m_{cj} r_{cj} \cos\phi_{cj} = 0 \tag{17.6a}$$

$$\sum_{i=1}^{n} m_i r_i \sin\phi_i + \sum_{j=1}^{2} m_{cj} r_{cj} \sin\phi_{cj} = 0 \tag{17.6b}$$

$$\sum_{i=1}^{n} z_i m_i r_i \cos\phi_i + \sum_{j=1}^{2} z_{cj} m_{cj} r_{cj} \cos\phi_{cj} = 0 \tag{17.7a}$$

$$\sum_{i=1}^{n} z_i m_i r_i \sin\phi_i + \sum_{j=1}^{2} z_{cj} m_{cj} r_{cj} \sin\phi_{cj} = 0 \tag{17.7b}$$

Note that the equations do not depend on the angular velocity of the rotating system; that is, if the system is balanced at one speed then it is balanced for all speeds.

For a distributed mass system and two correcting planes we can write the following four scalar equations:

$$F^x_{A_{21}} + F^x_{B_{21}} + \omega^2 \sum_{j=1}^{2} m_{cj} r_{cj} \cos\phi_{cj} = 0 \tag{17.8a}$$

$$F^y_{A_{21}} + F^y_{B_{21}} + \omega^2 \sum_{j=1}^{2} m_{cj} r_{cj} \sin\phi_{cj} = 0 \tag{17.8b}$$

$$z_A F^x_{A_{21}} + z_B F^x_{B_{21}} + \omega^2 \sum_{j=1}^{2} z_j m_{cj} r_{cj} \cos\phi_{cj} = 0 \tag{17.9a}$$

$$z_A F^y_{A_{21}} + z_B F^y_{B_{21}} + \omega^2 \sum_{j=1}^{2} z_j m_{cj} r_{cj} \sin\phi_{cj} = 0 \tag{17.9b}$$

A Direct Method of Dynamic Balancing To avoid solving four equations in four unknowns, that is Eqs. (17.6) and (17.7) or Eqs. (17.8) and (17.9), we can take moments about one correction plane and solve for two unknowns. Then we can take moments about the other correction plane and solve for the remaining two unknowns. Recall that if a rotating system is dynamically balanced, then it is automatically statically balanced. However, if the system is statically balanced there is no guarantee that it is dynamically balanced (see Section 17.4).

EXAMPLE 17.3

The distributed mass system illustrated in Fig. 17.12 has been tested for unbalance by rotating the rotor at an angular velocity of 100 rad/s. The bearing reactions on the frame at the two bearings A and B are as illustrated in Fig. 17.12. Determine the magnitudes and locations of the correcting masses to be removed in the specified planes 1 and 2 to achieve dynamic balance. The correcting masses are to be located at the radii $r_1 = r_2 = r = 4$ in.

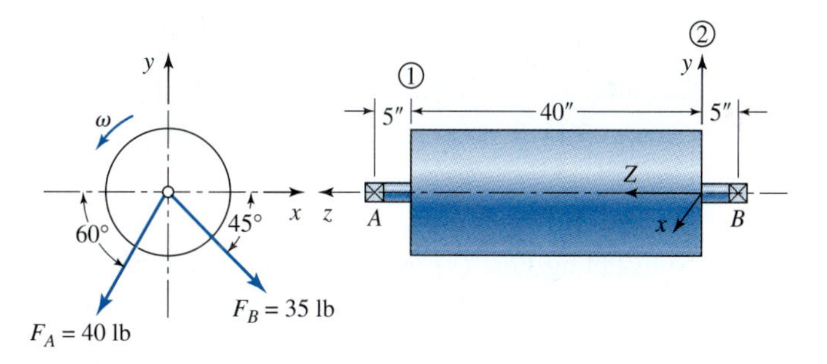

Figure 17.12 Example 17.3.

SOLUTION

The x and y components of the forces acting on the frame at the bearings A and B are

$$F^x_{A_{21}} = 40\cos 240^\circ \text{ lb} = -20 \text{ lb} \tag{1a}$$

$$F^y_{A_{21}} = 40\sin 240^\circ \text{ lb} = -20\sqrt{3} \text{ lb} \tag{1b}$$

$$F^x_{B_{21}} = 35\cos(-45^\circ) \text{ lb} = 17.5\sqrt{2} \text{ lb} \tag{1c}$$

$$F^y_{B_{21}} = 35\sin(-45^\circ) \text{ lb} = -17.5\sqrt{2} \text{ lb.} \tag{1d}$$

Equations (17.8) and (17.9) can be written as

$$m_1 r\omega^2 \cos\phi_1 + m_2 r\omega^2 \cos\phi_2 = -F^x_{A21} - F^x_{B21} \tag{2a}$$

$$m_1 r\omega^2 \sin\phi_1 + m_2 r\omega^2 \sin\phi_2 = -F^y_{A21} - F^y_{B21} \tag{2b}$$

$$z_1 m_1 r\omega^2 \cos\phi_1 + z_2 m_2 r\omega^2 \cos\phi_2 = -z_A F^x_{A21} - z_B F^x_{B21} \tag{3a}$$

$$z_1 m_1 r\omega^2 \sin\phi_1 + z_2 m_2 r\omega^2 \sin\phi_2 = -z_A F^y_{A21} - z_B F^y_{B21}. \tag{3b}$$

To simplify the analysis, we take moments about one of the two balancing planes. Here we take moments about plane 2 and note that the distances are

$$z_1 = +40 \text{ in.}, \quad z_2 = 0, \quad z_A = +45 \text{ in.}, \quad \text{and } z_B = -5 \text{ in.} \tag{4}$$

Substituting Eqs. (4) and the known data into Eqs. (2) and (3) gives

$$(4 \text{ in})(100 \text{ rad/s})^2(m_1 \cos\phi_1 + m_2 \cos\phi_2) = -4.749 \text{ lb} \tag{5a}$$

$$(4 \text{ in})(100 \text{ rad/s})^2(m_1 \sin\phi_1 + m_2 \sin\phi_2) = 59.390 \text{ lb} \tag{5b}$$

$$(40 \text{ in})(4 \text{ in})(100 \text{ rad/s})^2 m_1 \cos\phi_1 = 1\,023.744 \text{ in}\cdot\text{lb} \tag{6a}$$

$$(40 \text{ in})(4 \text{ in})(100 \text{ rad/s})^2 m_1 \sin\phi_1 = 1\,435.102 \text{ in}\cdot\text{lb}. \tag{6b}$$

Dividing Eq. (6*b*) by Eq. (6*a*), the location of the first correcting mass is

$$\phi_1 = \tan^{-1}\left(\frac{1\,435.102 \text{ in}\cdot\text{lb}}{1\,023.744 \text{ in}\cdot\text{lb}}\right) = 54.5°. \tag{7a}$$

Substituting Eq. (7*a*) into Eq. (6*a*) or (6*b*), the first correcting mass is

$$m_1 = \frac{(1\,435.102 \text{ in}\cdot\text{lb})(386 \text{ in/s}^2)}{(1\,600\,000 \text{ in}^2/\text{s}^2)\sin 54.5°} = 0.425 \text{ lb}. \tag{7b}$$

Substituting Eqs. (7) into Eqs. (5), rearranging, and dividing Eq. (5*b*) by Eq. (5*a*), the location of the second correcting mass is

$$\phi_2 = \tan^{-1}\left(\frac{23.535 \text{ lb}}{-30.324 \text{ lb}}\right) = 142.2°. \tag{8a}$$

Substituting Eq. (8*a*) into Eq. (5*a*) or (5*b*), the second correcting mass is

$$m_2 = \frac{(-30.342 \text{ lb})(386 \text{ in/s}^2)}{(40\,000 \text{ in/s}^2)\cos 142.2°} = 0.370 \text{ lb}. \qquad \textit{Ans.} \ (8b)$$

Note that the answers given by Eq. (7*a*) and (8*a*) are for adding mass to the system. To balance the rotating system by removing mass we must do the following:

At Plane 1:

Remove mass $m_1 = 0.425$ lb at the angle $54.5° + 180° = 234.5° = -125.5°$ *Ans.*

At Plane 2:

Remove mass $m_2 = 0.370$ lb at the angle $142.2° + 180° = 322.2° = -37.8°$ *Ans.*

Computer Solution For a computer analysis it is convenient to choose the xy plane as the plane of rotation with z as the axis of rotation, as illustrated in Fig. 17.13. In this manner, the unbalance vectors $m_i\mathbf{R}_i$ and the two correction vectors, $m_L\mathbf{R}_L$ in the left plane and $m_R\mathbf{R}_R$ in the right plane, can be expressed in the two-dimensional polar notation $m\mathbf{R} = mR\angle\theta$. This makes it easy to use the polar-rectangular conversion feature and its inverse, found on many programmable calculators.

Note that Fig. 17.13 has $m_1, m_2, \ldots, m_N$ unbalances. By solving Eqs. (b) and (c) in the section above entitled Graphic Analysis for the corrections, we have

$$m_L\mathbf{R}_L = -\sum_{i=1}^{N} \frac{m_i l_i}{l}\mathbf{R}_i \tag{17.10}$$

$$m_R\mathbf{R}_R = -m_L\mathbf{R}_L - \sum_{i=1}^{N} m_i\mathbf{R}_i. \tag{17.11}$$

These two equations can easily be programmed for computer solution. If a programmable calculator is used, it is suggested that the summation key be employed with each term of the summation entered using a user-defined key.

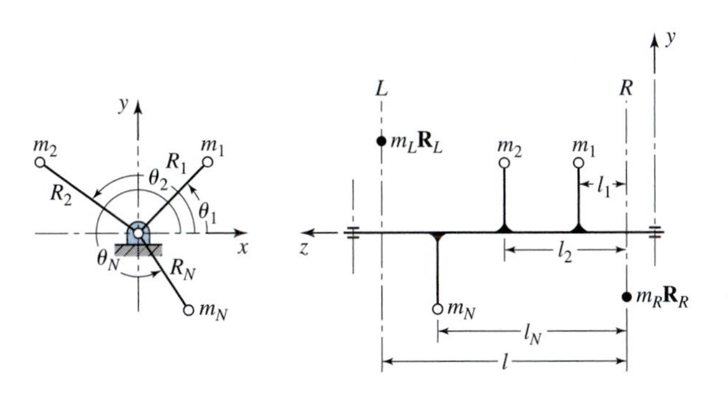

Figure 17.13 Notation for numerical solution; corrections are not shown in the end view.

17.6 DYNAMIC BALANCING

The units in which each unbalance is measured have customarily been the ounce · inch (oz · in.), the gram · centimeter (g · cm), or the inconsistent unit of gram · inch (g · in.). If correct practice is followed in the use of SI units, the most appropriate unite of unbalance is the milligram · meter (mg · m) because prefixes in multiples of 1,000 are preferred in SI; thus, the prefix centi- is not recommended. Furthermore, not more than one prefix should be used in a compound unit; preferably, the first-named quantity should be prefixed. Thus, neither the gram · centimeter nor the kilogram · millimeter, both acceptable in size, should be used. In this text we use the ounce · inch and the milligram · meter for units of unbalance.

We have seen that static balancing is sufficient for rotating disks, wheels, gears, and the like, where the mass can be assumed to exist in a single rotating plane. In the case of longer-axis machine elements, such as turbine rotors or motor armatures, the unbalanced centrifugal forces result in a couple whose effect tends to cause the rotor to turn end over end. The purpose of balancing is to measure this unbalanced couple and to add a new couple in the opposite direction of the same magnitude. The new couple is introduced by the addition of masses in two preselected correction planes or by subtracting (drilling out) masses from these two planes. To require balancing, a rotor usually has both static and dynamic unbalance; consequently, the correction masses, their radial orientations, or both are not the same for the two correction masses; also their radial orientations need not be the same for the two correction planes. This means that the angular separation of the correction masses on the two planes is usually not 180°. Thus, to balance a rotor, we must measure the magnitude and the angular orientation of the correction mass for each of the two correction planes.

Three methods of measuring the corrections for two planes are in general use, the *pivoted-cradle*, the *nodal-point*, and the *mechanical-compensation* methods.

Figure 17.14 illustrates a pivoted-cradle balancing machine with a specimen to be balanced, mounted on half-bearings or rollers attached to a cradle. The right end of the specimen is connected to a drive motor through a universal joint. The cradle can be rocked about either of two points that are adjusted to coincide with the two correction planes on the specimen to be balanced. In Fig. 17.14, the left pivot is shown in the released position and the cradle and specimen are free to rock or oscillate about the right pivot, which is illustrated in the locked position. Springs and dashpots are secured at each end of the cradle to provide a single-degree-of-freedom vibrating system. Often, these are made adjustable so that the natural frequency can be tuned to the motor speed. Also illustrated are amplitude indicators at each end of the cradle. These transducers are differential transformers or they may consist of a permanent magnet mounted on the cradle that moves relative to a stationary coil to generate a voltage proportional to the unbalance.

With the pivots located in the two correction planes, the operator can lock either pivot and take readings of the amount and orientation angle of the correction. The readings

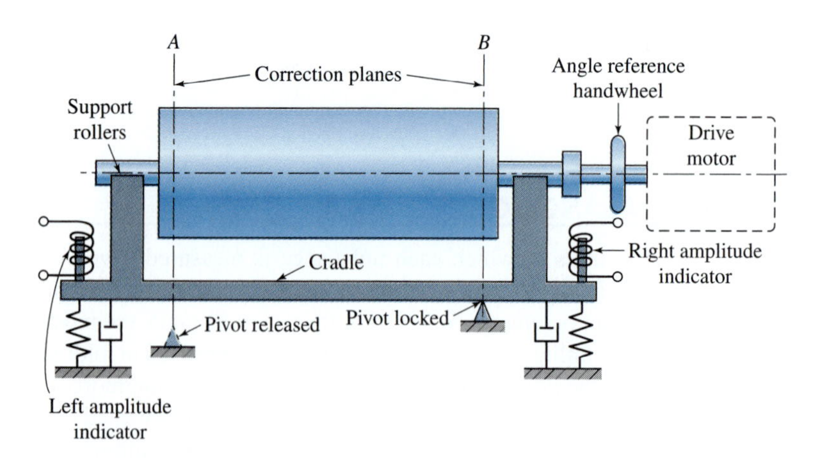

Figure 17.14 Schematic drawing of a pivoted-cradle balancing machine.

obtained are completely independent of the measurements taken in the other correction plane because an unbalance in the plane of the locked pivot has no moment about that pivot. With the right-hand pivot locked, an unbalance correctable in the left correction plane causes vibration whose amplitude is measured by the left amplitude indicator. Once this correction is measured, the right-hand pivot is released, the left pivot is locked, and another set of measurements is made for the right-hand correction plane using the right-hand amplitude indicator.

The relation between the amount of unbalance and the measured amplitude is given by Eq. (17.4). Rearranging that equation and substituting r for e, the amplitude of the motion can be written as

$$X = \frac{m_u r(\omega/\omega_n)^2}{m\sqrt{(1-\omega^2/\omega_n^2)^2 + (2\zeta\omega/\omega_n)^2}}, \tag{17.12}$$

where $m_u r$ is the unbalance and m is the mass of the cradle and the specimen. This equation demonstrates that the amplitude X is directly proportional to the unbalance $m_u r$. Figure 17.15*a* illustrates a plot of this equation for a particular damping ratio ζ and demonstrates that the machine is most sensitive near resonance ($\omega = \omega_n$), because in this region the greatest amplitude is recorded for a given unbalance. Damping is deliberately introduced in balancing machines to filter noise and other vibrations that might affect the results. Damping also helps to maintain calibration against effects of temperature and other environmental conditions.

Not shown in the balancing machine of Fig. 17.14 is a sine-wave signal generator that is attached to the drive shaft. If the resulting sine-wave signal generator is compared on a dual-beam oscilloscope with the wave generated by one of the amplitude indicators, a phase difference is found. The angular phase difference is the angular orientation of the unbalance. In a balancing machine, an electronic phasemeter measures the phase angle and gives the result on another meter calibrated in degrees. To locate the correction on the specimen (Fig. 17.14), the angular reference handwheel is turned by hand until the indicated angle is in line with a reference pointer. This places the heavy side of the specimen in a preselected position and permits the correction to be made.

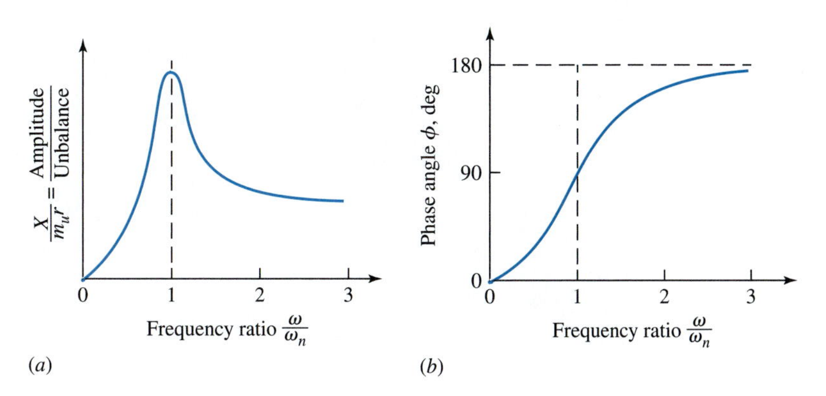

Figure 17.15

Substituting Eqs. (17.1) and (17.3) into Eq. (*c*) of Section 17.2 and rearranging, the phase angle can be written in parametric form as

$$\phi = \tan^{-1} \frac{2\zeta\omega/\omega_n}{1 - \omega^2/\omega_n^2}. \tag{17.13}$$

A plot of this equation for a single damping ratio and for various frequency ratios is illustrated in Fig. 17.15*b*. This curve indicates that, at resonance, when the speed ω of the shaft matches the natural frequency ω_n of the system, the displacement lags the unbalance by the angle $\phi = 90°$. If the top of the specimen is turning away from the operator, the unbalance is horizontal and directly in front of the operator when the displacement is maximum downward. Figure 17.15*b* also illustrates that the angular orientation approaches 180° as the shaft speed ω is increased above resonance.

17.7 BALANCING MACHINES

A pivoted-cradle balancing machine for high-speed production is illustrated in Fig. 17.16. The shaft-mounted signal generator can be seen at the extreme left.

Figure 17.16 A Tinius Olsen static-dynamic pivoted-cradle balancing machine with specimen mounted for balancing. (Tinius Olsen Testing Machine Company, Willow Grove, PA.)

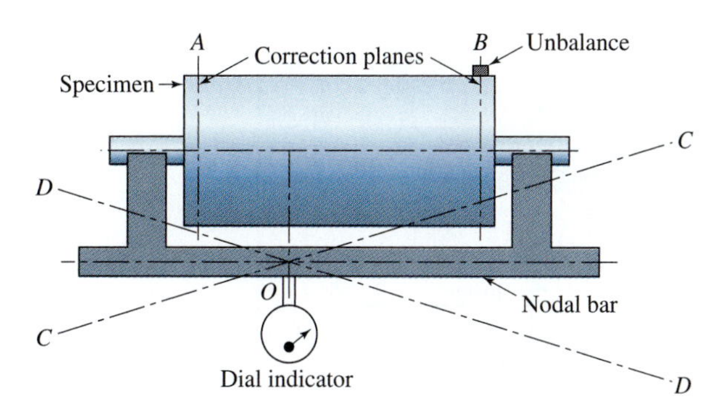

Figure 17.17 Plane separation by the nodal-point method. The nodal bar experiences the same vibration as the specimen.

Nodal-Point Balancing Plane separation using a point of zero or minimum vibration is called the *nodal-point method of balancing.* To see how this method is used, examine Fig. 17.17. Here, the specimen to be balanced is shown mounted on bearings that are fastened to a nodal bar. We assume that the specimen is already balanced in the left-hand correction plane and that an unbalance still exists in the right-hand plane, as shown. Because of this unbalance, a vibration of the entire assembly takes place, causing the nodal bar to oscillate about some point O, occupying first position CC and then DD. Point O is easily located by sliding a dial indicator along the nodal bar; a point of zero motion or minimum motion is readily determined. This is the null or nodal point. Its location is the center of oscillation for a center of percussion (Section 14.6) in the right-hand correction plane.

We assumed, at the beginning of this discussion, that no unbalance existed in the left-hand correction plane. However, if unbalance is present, its magnitude is given by the dial indicator located at the nodal point just determined. Thus, by locating the dial indicator at this nodal point, we measure the unbalance in the left-hand plane without any interference from that in the right-hand plane. In a similar manner, another nodal point can be determined that will measure only the unbalance in the right-hand correction plane without any interference from that in the left-hand plane.

In commercial balancing machines employing the nodal-point principle, the plane separation is accomplished in electrical networks. Typical of these is the Micro Dynamic Balancer; a schematic is illustrated in Fig. 17.18. On this machine a switching knob selects either correction plane and displays the unbalance on a voltmeter, which is calibrated in appropriate unbalance units.

The computer of Fig. 17.18 contains a filter that eliminates bearing noise and other frequencies not related to the unbalance. A multiplying network is used to give the sensitivity desired and to cause the meter to read in preselected balancing units. The strobe light is driven by an oscillator that is synchronized to the rotor speed.

The rotor is driven at a speed that is much greater than the natural frequency of the system and, because the damping is quite small, Fig. 17.15*b* illustrates that the phase angle is approximately 180°. Marked on the right-hand end of the rotor are degrees or numbers that are readable and stationary under the strobe light during rotation of the rotor. Thus, it is only necessary to observe the particular station number of the degree marking under the strobe light to locate the heavy spot. When the switch is shifted to the other correction plane,

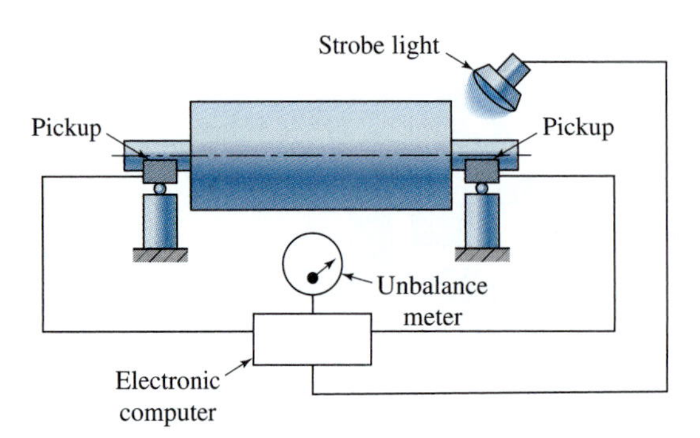

Figure 17.18 Diagram of the electrical circuit in a Micro Dynamic Balancer. (Schenck RoTec Corp., Orion, MI.)

the meter again reads the amount and the strobe light illuminates the station. Sometimes as few as five station numbers distributed uniformly around the periphery are adequate for balancing.

The direction of the vibration is horizontal, and the phase angle is nearly 180°. Thus, rotation such that the top of the rotor moves away from the operator will cause the heavy spot to be in a horizontal plane and on the near side of the axis when illuminated by the strobe lamp. A pointer is usually placed here to indicate its location. If, during production balancing, it is determined that the phase angle is less than 180°, the pointer can be shifted slightly to indicate the proper position to observe.

Mechanical Compensation

An unbalanced rotating rotor mounted in a balancing machine develops a vibration. One can introduce in the balancing machine counterforces in each correction plane that exactly balance the forces causing the vibration. The result of introducing these forces is a smooth-running system. Upon stopping, the location and amount of each counterforce is measured to give the exact correction required. This is called *mechanical compensation*.

When mechanical compensation is used, the speed of the rotor during balancing is not important because the equipment is in calibration for all speeds. The rotor may be driven by a belt, through a universal joint, or it may be self-driven if, for example, it is a gasoline engine. The electronic equipment is simple, no built-in damping is necessary, and the machine is easy to operate because the unbalances in both correction planes are measured simultaneously and the magnitudes and orientations are read directly.

We can understand how mechanical compensation is applied by examining Fig. 17.19a. Looking at the end of the rotor, we see one of the correction planes with the unbalance to be corrected (represented by wr). Two compensator weights are also illustrated in Fig. 17.19a. All three of these weights are to rotate with the same angular velocity ω, but the position of the compensator weights relative to one another and their position relative to the unbalanced weight can be varied by two controls. One of these controls changes the angle α, that is, the angle between the compensator weights. The other control changes the angular position of the compensator weights relative to the unbalance, that is, the angle β. The knob that changes the angle β is the *orientation control* and, when the rotor is compensated (balanced) in this plane, a pointer on the knob indicates the exact angular orientation of the unbalance. The knob that changes the angle α is the

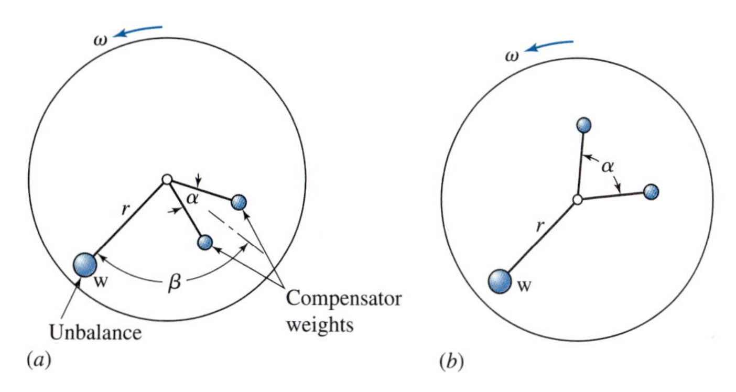

Figure 17.19 Correction plane viewed along the axis of rotation to illustrate the unbalance and the compensator weights: (*a*) position of compensator weights increases the vibration; (*b*) compensated.

amount control, and it also gives a direct reading when the rotor unbalance is compensated. The magnitude of the vibration is measured electrically and displayed on a voltmeter. Thus, compensation is found when the controls are manipulated to make the voltmeter read zero.

17.8 FIELD BALANCING WITH A PROGRAMMABLE CALCULATOR*

Field balancing is necessary for very large rotors for which balancing machines are impractical. Despite the fact that high-speed rotors are balanced in the shop during manufacture, it is frequently necessary to rebalance them in the field because of slight deformations brought on by shipping, creep, or high operating temperatures.

It is possible to balance a machine in the field by balancing a single plane at a time. But cross-coupling effects and correction-plane interference often require balancing each end of a rotor two or three times to obtain satisfactory results. Some machines may require as much as an hour to bring them up to full speed, resulting in even more delays in the balancing procedure.

Both Rathbone and Thearle[1] have developed methods of two-plane field balancing that can be expressed in complex-number notation and solved using a programmable calculator. The time saved using a programmable calculator is several hours when compared with graphical methods or analysis with complex numbers using an ordinary scientific calculator.

In the analysis that follows, boldface letters are used to represent complex numbers; for example,

$$\mathbf{R} = R\angle\theta = Re^{j\theta} = x + jy.$$

In Fig. 17.20, unknown unbalances $\mathbf{M}_L$ and $\mathbf{M}_R$ are assumed to exist in the left- and right-hand correction planes, respectively. The magnitudes of these unbalances are M_L and

* The authors are grateful to Prof. W. B. Fagerstrom, University of Delaware, for contributing some of the ideas of this section.

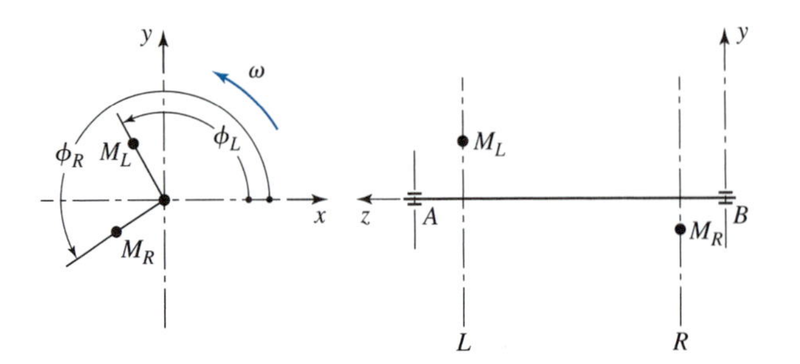

Figure 17.20 Notation for two-plane field balancing. The xy system is the rotating reference.

M_R, and they are located at angles ϕ_L and ϕ_R from the rotating reference. When these unbalances have been found, their negatives are located in the left and right planes to achieve balance.

The rotating unbalances $\mathbf{M}_L$ and $\mathbf{M}_R$ produce disturbances at bearings A and B. Using commercial field-balancing equipment, it is possible to measure the amplitudes and the angular locations of these disturbances. The notation $\mathbf{X} = X\angle\phi$, with appropriate subscripts, is used to designate these quantities.

In field balancing, three runs or tests are made, as follows:

- *First run:* Measure vector $\mathbf{X}_A = X_A\angle\phi_A$, at bearing A and vector $\mathbf{X}_B = X_B\angle\phi_B$, at bearing B due only to the original unbalances $\mathbf{M}_L = M_L\angle\phi_L$ and $\mathbf{M}_R = M_R\angle\phi_R$.
- *Second run:* Add trial mass $\mathbf{m}_L = m_L\angle\theta_L$ to the left correction plane and measure the amplitudes $\mathbf{X}_{AL} = X_{AL}\angle\phi_{AL}$, and $\mathbf{X}_{BL} = X_{BL}\angle\phi_{BL}$, at the left and right bearings (A and B), respectively.
- *Third run:* Remove the trial mass $\mathbf{m}_L = m_L\angle\theta_L$. Add trial mass $\mathbf{m}_R = m_R\angle\theta_R$ to the right-hand correction plane and again measure the bearing amplitudes. These results are designated $\mathbf{X}_{AR} = X_{AR}\angle\phi_{AR}$ for bearing A, and $\mathbf{X}_{BR} = X_{BR}\angle\phi_{BR}$ for bearing B.

Note in the above runs that the term "trial mass" means the same as a trial unbalance and a unit distance from the axis of rotation is used.

To develop the equations for the unbalance being sought, we first define *complex stiffness* as the amplitude that would result at either bearing caused by a unit unbalance located at the intersection of the rotating reference mark and one of the correction planes. Thus, we must find the complex stiffnesses $\mathbf{A}_L$ and $\mathbf{B}_L$ caused by a unit unbalance located at the intersection of the rotating reference mark and plane L. In addition, we require the complex stiffnesses $\mathbf{A}_R$ and $\mathbf{B}_R$ caused by a unit unbalance located at the intersection of the rotating reference mark and plane R.

If these stiffnesses were available, we could write the following sets of complex equations:

$$\mathbf{X}_{AL} = \mathbf{X}_A + \mathbf{A}_L\mathbf{M}_L, \quad \mathbf{X}_{BL} = \mathbf{X}_B + \mathbf{B}_L\mathbf{m}_L \tag{a}$$

$$\mathbf{X}_{AR} = \mathbf{X}_A + \mathbf{A}_R\mathbf{M}_R, \quad \mathbf{X}_{BR} = \mathbf{X}_B + \mathbf{B}_R\mathbf{m}_R. \tag{b}$$

After the three runs are made, the stiffnesses are the only unknowns in these equations. Therefore,

$$\mathbf{A}_L = \frac{\mathbf{X}_{AL} - \mathbf{X}_A}{\mathbf{m}_L}, \quad \mathbf{B}_L = \frac{\mathbf{X}_{BL} - \mathbf{X}_B}{\mathbf{m}_L}$$

$$\mathbf{A}_R = \frac{\mathbf{X}_{AR} - X_A}{\mathbf{m}_R}, \quad \mathbf{B}_R = \frac{\mathbf{X}_{BR} - \mathbf{X}_B}{\mathbf{m}_R}. \tag{17.14}$$

Then, from the definition of these stiffnesses, we have from the first run

$$\mathbf{X}_A = \mathbf{A}_L\mathbf{M}_L + \mathbf{A}_R\mathbf{M}_R, \quad \mathbf{X}_B = \mathbf{B}_L\mathbf{M}_L + \mathbf{B}_R\mathbf{M}_R. \tag{c}$$

Solving this pair of equations simultaneously gives

$$\mathbf{M}_L = \frac{\mathbf{X}_A\mathbf{B}_R - \mathbf{A}_R\mathbf{X}_B}{\mathbf{A}_L\mathbf{B}_R - \mathbf{A}_R\mathbf{B}_L}, \quad \mathbf{M}_R = \frac{\mathbf{A}_L\mathbf{X}_B - \mathbf{X}_A\mathbf{B}_L}{\mathbf{A}_L\mathbf{B}_R - \mathbf{A}_R\mathbf{B}_L}. \tag{17.15}$$

These equations can be programmed, either in complex polar form or in complex rectangular form. The suggestions that follow were formed assuming a complex rectangular form for the solution.

Because the original data are formulated in polar coordinates, a subprogram should be written to transform the data into rectangular coordinates before storage.

The equations reveal that complex subtraction, division, and multiplication are used often. These operations can be set up as subprograms to be called from the main program. If $\mathbf{A} = a + jb$ and $\mathbf{B} = c + jd$, the formula for complex subtraction is

$$\mathbf{A} - \mathbf{B} = (a - c) + j(b - d). \tag{17.16}$$

For complex multiplication, the formula is

$$\mathbf{AB} = (ac - bd) + j(bc + ad) \tag{17.17}$$

and for complex division the formula is

$$\frac{\mathbf{A}}{\mathbf{B}} = \frac{(ac + bd) + j(bc - ad)}{c^2 + d^2}. \tag{17.18}$$

With these subprograms it is an easy matter to program Eqs. (17.14) and (17.15).

As a check on your programming, you may wish to use the following data: $\mathbf{X}_A = 8.6\angle 63^\circ$, $\mathbf{X}_B = 6.5\angle 206^\circ$, $\mathbf{M}_L = 10\angle 270^\circ$, $\mathbf{M}_R = 12\angle 180^\circ$, $\mathbf{X}_{AL} = 5.9\angle 123^\circ$, $\mathbf{X}_{BL} = 4.5\angle 228^\circ$, $\mathbf{X}_{AR} = 6.2\angle 36^\circ$, and $\mathbf{X}_{BR} = 10.4\angle 162^\circ$. The correct results are $\mathbf{M}_L = 10.76\angle 146.6^\circ$ and $\mathbf{M}_R = 6.20\angle 245.4^\circ$.

According to Fagerstrom, the vibration angles used can be expressed in two different systems. The first is the rotating-protractor-stationary-mark system (RPSM). This is the system used in the preceding analysis and is the one a theoretician would prefer. In actual practice, however, it is usually easier to have the protractor stationary and use a rotating mark like a key or keyway. This is called the rotating-mark-stationary-protractor system (RMSP). The only difference between the two systems is in the sign of the vibration angle, but there is no sign change on the trial or correction masses.

17.9 BALANCING A SINGLE-CYLINDER ENGINE

The rotating masses in a single-cylinder engine can be balanced using the methods already discussed in Chapter 17. The reciprocating masses, however, cannot be completely balanced, and so our studies in this section are really concerned with minimizing the unbalance.

Although the reciprocating masses cannot be totally balanced using a simple counterweight, it is possible to modify the shaking forces (see Section 16.9) by unbalancing the rotating masses. As an example of this let us add a counterweight opposite the crankpin whose mass exceeds the rotating mass by one half of the reciprocating mass (from one-half to two-thirds of the reciprocating mass is usually added to the counterweight to alter the balance characteristics in single-cylinder engines). We designate the mass of the counterweight m_C, substitute this mass into Eq. (16.24), and use a negative sign because the counterweight is opposite the crankpin; then the inertia force caused by this counterweight is

$$\mathbf{F}_C = -m_C r\omega^2 \cos\omega t\hat{\mathbf{i}} - m_C r\omega^2 \sin\omega t\hat{\mathbf{j}}. \tag{a}$$

Note that the balancing mass and the crankpin both have the same radius. Designating by m_A and m_B the masses of the rotating and the reciprocating parts, respectively, as in Chapter 16, then, according to the supposition above, we have

$$m_C = m_A + \frac{1}{2}m_B. \tag{b}$$

Equation (*a*) can now be written

$$\mathbf{F}_C = -(m_A + \frac{1}{2}m_B)r\omega^2 \cos\omega t\hat{\mathbf{i}} - (m_A + \frac{1}{2}m_B)r\omega^2 \sin\omega t\hat{\mathbf{j}}. \tag{c}$$

The inertia force caused by the rotating and reciprocating masses is, from Eqs. (16.28) and (16.29),

$$\mathbf{F}_{A,B} = F^x\hat{\mathbf{i}} + F^y\hat{\mathbf{j}} = \left[(m_A + m_B)r\omega^2 \cos\omega t + m_B r\omega^2 \frac{r}{l}\cos 2\omega t\right]\hat{\mathbf{i}} + m_A r\omega^2 \sin\omega t\hat{\mathbf{j}}. \tag{d}$$

Adding Eqs. (*c*) and (*d*), the resultant inertia force can be written as

$$\mathbf{F} = \left(\frac{1}{2}m_B r\omega^2 \cos\omega t + m_B r\omega^2 \frac{r}{l}\cos 2\omega t\right)\hat{\mathbf{i}} - \frac{1}{2}m_B r\omega^2 \sin\omega t\hat{\mathbf{j}}$$

or as

$$\mathbf{F} = \left(\frac{1}{2}m_B r\omega^2 \cos\omega t + m_B r\omega^2 \frac{r}{l}\cos 2\omega t\right)\hat{\mathbf{i}} - \frac{1}{2}m_B r\omega^2 \sin\omega t\hat{\mathbf{j}}. \tag{17.19}$$

The first term is called the *primary component*. This component has a magnitude $\frac{1}{2}m_B r\omega^2$ and can be represented as a backward (clockwise)-rotating vector with angular velocity ω. The second term is called the *secondary component*. This component is the x projection of a vector of length $m_B r\omega^2(r/l)$ rotating forward (counterclockwise) with an angular velocity of 2ω.

The maximum inertia force occurs when $\omega t = 0$. Substituting this condition into Eq. (17.19) gives

$$F_{\max} = m_B r \omega^2 \left(\frac{r}{l} + \frac{1}{2} \right) \tag{e}$$

because $\cos \omega t = \cos 2\omega t = 1$ and $\sin \omega t = 0$ when $\omega t = 0$. Before the extra counterweight is added, the maximum inertia force is

$$F_{\max} = m_B r \omega^2 \left(\frac{r}{l} + 1 \right). \tag{f}$$

Therefore, in this instance, the effect of the additional counterweight is to reduce the maximum shaking force by 50% of the primary component and to add axial inertia forces where formerly none existed. Equation (17.19) is plotted as a polar diagram in Fig. 17.21 for an r/l value of 1/4. Here, the vector OA rotates counterclockwise at angular velocity of 2ω. The horizontal projection of this vector OA' is the secondary component. The vector OB, the primary component, rotates clockwise at an angular velocity ω. The total shaking force F is shown for the 30° position and is the sum of the vectors OB and $BB' = OA'$.

Imaginary-Mass Approach

Stevensen has refined and extended a method of engine balancing that is here called the *imaginary-mass approach*.[2] It is possible that the method is known in some circles as the *virtual-rotor approach* because it uses what might be called a virtual rotor that counterrotates to accommodate part of the piston effect in a reciprocating engine.

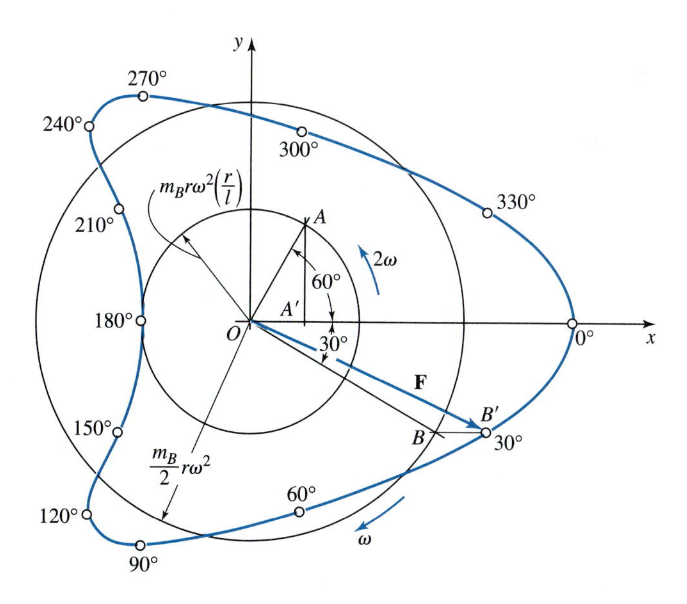

Figure 17.21 Polar diagram of inertia forces in a single-cylinder engine for $r/l = 1/4$. The counterweight includes one half of the reciprocating mass.

Before going into details, it is necessary to explain a change in the method of viewing the crank circle of an engine. In developing the imaginary-mass approach in this section and the section to follow, we use the coordinate system of Fig. 17.22*a*. This might appear to be a left-hand system because the y axis is located clockwise from the x axis and because positive rotation is shown as clockwise. We adopt this notation because it has been used for so long by the automotive industry.* If you prefer, you can think of this system as a right-hand, three-dimensional system viewed from the negative z axis.

The imaginary-mass approach uses two fictitious masses, each equal to half the equivalent reciprocating mass at the particular harmonic frequency studied. The purpose of these fictitious masses is to replace the effects of the reciprocating mass. These imaginary masses rotate about the crank center in opposite directions and with equal velocities. They are arranged so that they come together at both the top dead center (TDC) and the bottom dead center (BDC), as illustrated in Fig. 17.22*a*. The mass $+\frac{1}{2}m_B$ rotates *with* the crank motion; the other mass rotates opposite to the crank motion. The mass rotating with the crank is designated in Fig. 17.22*a* by a plus sign, and the mass rotating in the opposite direction is designated by a minus sign. The center of mass of these two rotating masses always lies on the cylinder axis.

The imaginary-mass approach was conceived because the piston motion and the resulting inertia force can always be represented by a Fourier series. Such a series has an infinite number of terms, each term representing a simple harmonic motion having a known frequency and amplitude. It turns out that the higher-frequency amplitudes are so small that they can be neglected and hence only a small number of the lower-frequency amplitudes are needed. Also, the odd harmonics (third, fifth, etc.) are not present because of the symmetry of the piston motion.

Each harmonic (the first, second, fourth, and so on) is represented by a pair of imaginary masses. The angular velocities of these masses are $\pm\omega$ for the first harmonic, $\pm 2\omega$ for the

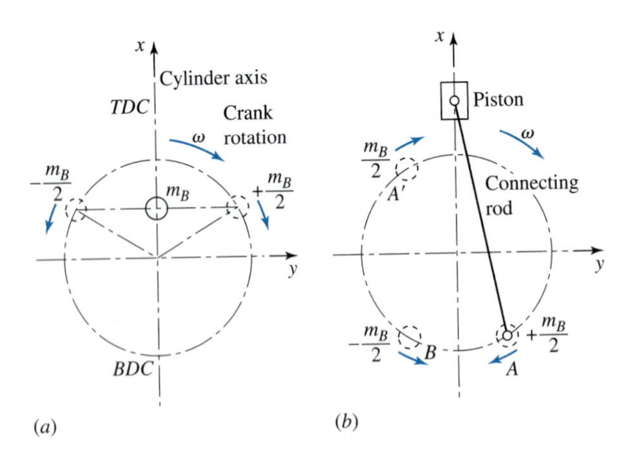

Figure 17.22 Note that the axes are right-handed but the view of the crank circle is from the negative z axis.

* Readers who are antique automobile buffs will understand this convention because it is the direction in which an antique engine is hand cranked.

second, $\pm 4\omega$ for the fourth, and so on. It is rarely necessary to consider the sixth or higher harmonics.

Stevensen gives the following rule for placing the imaginary masses:

> For any given position of the cranks the positions of the imaginary masses are found, first, by determining the angles of travel of each crank from its top dead center and, second, by moving their imaginary masses, one clockwise and the other counterclockwise, by angles equal to the crank angle times the number of the harmonic.

All of these angles must be measured from the same dead-center crank position.

Let us apply this approach to the single-cylinder engine, considering only the first harmonic. In Fig. 17.22*b* the mass $+\frac{1}{2}m_B$ at A rotates at the angular velocity ω with the crank, whereas mass $-\frac{1}{2}m_B$ at B rotates at the angular velocity $-\omega$ opposite to crank rotation. The imaginary mass at A can be balanced by adding an equal mass at A' to rotate with the crankshaft. However, the mass at B can be balanced neither by the addition nor by the subtraction of mass from any part of the crankshaft because it is rotating in the opposite direction. When half the mass of the reciprocating parts is balanced in this manner—that is, by adding the mass at A'—the unbalanced part of the first harmonic, caused by the mass at B, causes the engine to vibrate in the plane of rotation equally in all directions like a true unbalanced rotating mass.

It is interesting to know that in one-cylinder motorcycle engines a fore-and-aft unbalance is less objectionable than an up-and-down unbalance. For this reason, such engines are overbalanced by using a counterweight whose mass is more than half the reciprocating mass.

It is impossible to balance the second and higher harmonics with masses rotating at crankshaft speeds, because the frequency of the unbalance is higher than that of the crankshaft rotation. Balancing of second harmonics has been accomplished using shafts geared to run at twice the engine crankshaft speed, as in the case of the 1976 Plymouth Arrow engine, but at the cost of complication. It is not usually done.

For ready reference, we now summarize the inertia forces in the single-cylinder engine, with balancing masses, in Table 17.1. The expressions shown have been obtained from Eqs. (16.24) and (16.27), rewritten so that the effect of the second harmonic is presented as a mass equal to $\frac{1}{2}m_Br/l$ reciprocating at a speed of 2ω. Note that the subscript C is used to designate counterweights (balancing masses) and their radii. Because the centrifugal balancing masses will be selected and placed to counterbalance the centrifugal forces, the

TABLE 17.1 One-cylinder engine inertia forces

Type	Equivalent mass	Radius	Along cylinder axis (x)	Across cylinder axis (y)
Centrifugal	m_A	r	$m_Ar\omega^2\cos\omega t$	$m_Ar\omega^2\sin\omega t$
	m_{AC}	r_C	$m_{AC}r_C\omega^2\cos(\omega t+\pi)$	$m_{AC}r_C\omega^2\sin(\omega t+\pi)$
Reciprocating	m_B	r	$m_Br\omega^2\cos\omega t$	0
First harmonic	m_{BC}	r_C	$m_{BC}r_C\omega^2\cos(\omega t+\pi)$	$m_{BC}r_C\omega^2\sin(\omega t+\pi)$
Second harmonic	$\dfrac{m_Br}{4l}$	r	$\dfrac{m_Br}{4l}(r)(2\omega)^2\cos 2\omega t$	0

only unbalance that results along the cylinder axis will be the sum of the last three entries. Similarly, the only unbalance across the cylinder axis will be the value in the fourth row. The maximum values of the unbalance in these two directions can be predetermined in any desired ratio to each other, as indicated previously, and a solution can be obtained for m_{BC} at radius r_C.

If this approach is used to include the effect of the fourth harmonic, there will result an additional mass of $m_B r^3/16l^3$ reciprocating at the speed of 2ω and a mass of $-m_B r^3/64l^3$ reciprocating at a speed of 4ω, illustrating the decreasing significance of the higher harmonics.

17.10 BALANCING MULTICYLINDER ENGINES

To obtain a basic understanding of the balancing problem in multicylinder engines, let us consider a two-cylinder in-line engine having cranks 180° apart and rotating parts already balanced by counterweights. Such an engine is illustrated in Fig. 17.23. Applying the imaginary-mass approach for the first harmonic results in the diagram of Fig. 17.23*a*. Figure 17.23*a* illustrates that masses +1 and +2, rotating clockwise, balance each other, as do masses −1 and −2, rotating counterclockwise. Thus, the first harmonic forces are inherently balanced for this crank arrangement. Figure 17.23*b* illustrates, however, that these forces are not in the same plane. For this reason, unbalanced couples will be set up that tend to rotate the engine about the y axis. The values of these couples can be determined using the force expressions in Table 17.1 together with the coupling distance because the equations can be applied to each cylinder separately. It is possible to balance the couple because of the real rotating masses as well as the imaginary half masses that rotate with the engine; however, the couple resulting from the half mass of the first harmonic that is counterrotating cannot be balanced.

Figure 17.24*a* illustrates the location of the imaginary masses for the second harmonic using Stevensen's rule. This diagram illustrates that the second-harmonic forces are not

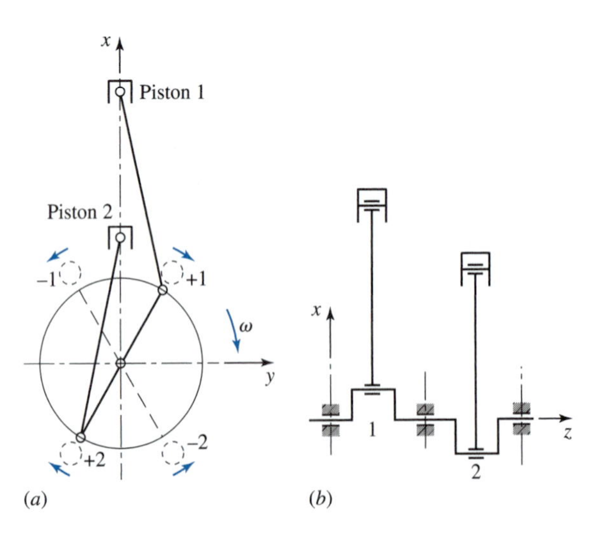

Figure 17.23 (*a*) First harmonics; (*b*) a two-throw crankshaft with three main bearings.

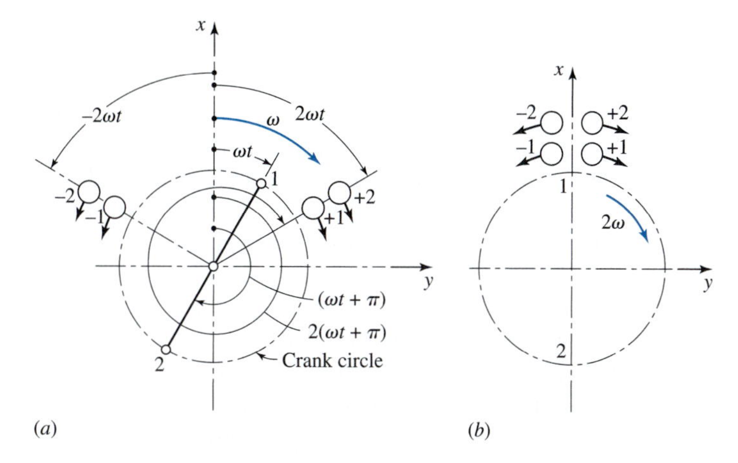

Figure 17.24 Second-harmonic positions of the imaginary masses: (*a*) positions for same crank angle in Fig. 17.22*a*; (*b*) extreme, or dead-center, positions.

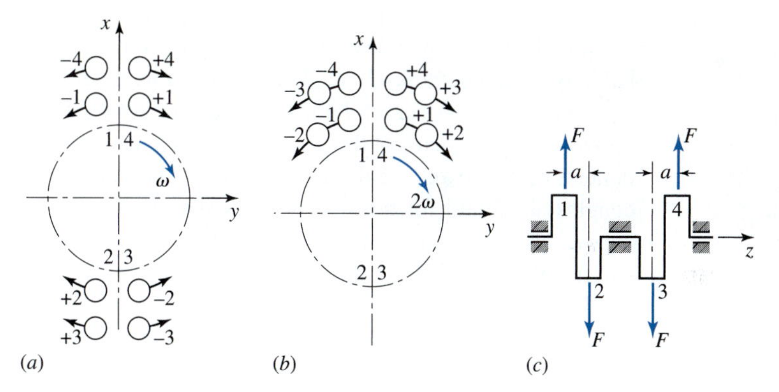

Figure 17.25 Four-cycle engine: (*a*) first-harmonic positions; (*b*) second-harmonic positions; (*c*) crankshaft with first-harmonic couples added.

balanced. Because the greatest unbalance occurs at the dead-center positions, the diagrams are usually drawn for this extreme position, with crank 1 at TDC as in Fig. 17.24*b*. This unbalance causes a vibration in the *xz* plane having the frequency 2ω.

The diagram for the fourth harmonics, not shown, is the same as in Fig. 17.24*b* but, of course, the speed is 4ω.

Four-Cylinder Engine

A four-cylinder in-line engine with cranks spaced 180° apart is illustrated in Fig. 17.25*c*. This engine can be treated as two two-cylinder engines placed back to back. Thus, the first harmonic forces still balance and, in addition, from Fig. 17.25*a* and Fig. 17.25*c*, the first-harmonic couples also balance. These couples tend, however, to deflect the center bearing of a three-bearing crankshaft up and down and to bend the center of a two-bearing shaft in the same manner.

Figure 17.25*b* illustrates that when cranks 1 and 4 are at TDC, all the masses representing the second harmonic traveling in both directions accumulate at TDC, giving an unbalanced force. The center of mass of all the masses is always on the x axis, and so the unbalanced second harmonics cause a vertical vibration with a frequency of twice the engine speed. This characteristic is typical of all four-cylinder engines with this crank arrangement. Because the masses and forces all act in the same direction, there is no coupling action.

A diagram of the fourth harmonics is identical to that of Fig. 17.25*b*, and the effects are the same, but they do have a higher frequency and exert less force.

Three-Cylinder Engine A three-cylinder in-line engine with cranks spaced 120° apart is illustrated in Fig. 17.26. Note that the cylinders are numbered according to the order in which they arrive at TDC. Figure 17.27 illustrates that the first, second, and fourth harmonic forces are completely balanced, and only the sixth harmonic forces are completely unbalanced. These unbalanced forces tend to create a vibration in the plane of the centerlines of the cylinders, but the magnitude of the forces is very small and can be neglected as far as vibration is concerned.

An analysis of the couples of the first harmonic forces indicates that when crank 1 is at TDC (Fig. 17.26), there is a vertical component of the forces on cranks 2 and 3 equal in magnitude to half of the force on crank 1. The resultant of these two downward components is equivalent to a force downward, equal in magnitude to the force on crank 1 and located halfway between cranks 2 and 3. Thus, a couple is set up with an arm equal to the distance between the center of crank 1 and the centerline between cranks 2 and 3. At the same time, the horizontal components of the $+2$ and -2 forces cancel each other, as do the horizontal components of the $+3$ and -3 forces (Fig. 17.27). Therefore, no horizontal couple exists. Similar couples are found for both the second and the fourth harmonics. Thus, a three-cylinder engine, although inherently balanced for forces in the first, second, and fourth harmonics, still is not free from vibration caused by the presence of couples at these harmonics.

Six-Cylinder Engine If a six-cylinder in-line engine is conceived as a combination of two three-cylinder engines placed back to back with parallel cylinders, it has the same inherent balance of the first, second, and fourth harmonics. By virtue of symmetry, the couples of each three-cylinder engine act in opposite directions and balance the other. These couples,

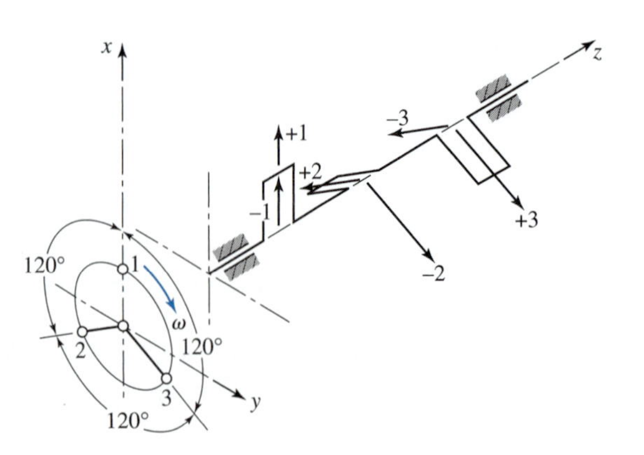

Figure 17.26 Crank arrangement of three-cylinder engine; first-harmonic forces shown.

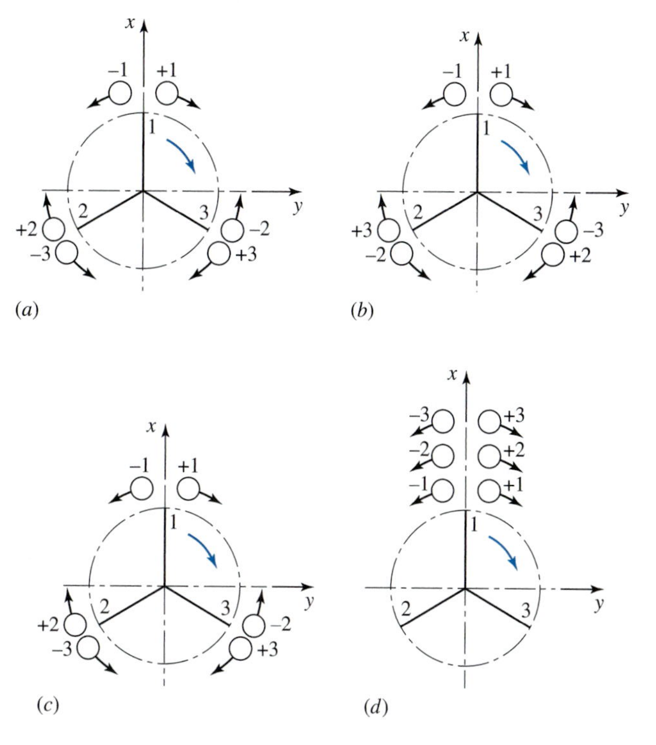

Figure 17.27 Positions of the imaginary masses for the three-cylinder engine: (*a*) first harmonics; (*b*) second harmonics; (*c*) fourth harmonics; and (*d*) sixth harmonics.

although perfectly balanced, tend to bend the crankshaft and crankcase and necessitate the use of rigid construction for high-speed operation. As before, the sixth harmonic forces are completely unbalanced and tend to create a vibration in the vertical plane with a frequency of 6ω. The magnitude of these forces, however, is very small and practically negligible as a source of vibration.

Other Engines Taking into consideration the cylinder arrangement and crank spacing permits a great many configurations. For any combination, the balancing situation can be investigated to any harmonic desired by the methods outlined in this section. Particular attention must be paid when analyzing the part of Stevensen's rule that calls for determining the angle of travel from the TDC of the cylinder under consideration and moving the imaginary masses through the appropriate angles from that same TDC. This is especially important when radial and opposed-piston engines are investigated.

As practice problems you may wish to use these methods to confirm the following facts:

1. In a three-cylinder radial engine with one crank and three connecting rods having the same crankpin, the negative masses are inherently balanced for the first harmonic forces, whereas the positive masses are always located at the crankpin.

These two findings are inherently true for all radial engines. Also, because the radial engine has its cylinders in a single plane, unbalanced couples do not occur. The three-cylinder engine has unbalanced forces in the second and higher harmonics.

2. A two-cylinder opposed-piston engine with a crank spacing of 180° is balanced for forces in the first, second, and fourth harmonics but unbalanced for couples.
3. A four-cylinder in-line engine with cranks at 90° is balanced for forces in the first harmonic but unbalanced for couples. In the second harmonic it is balanced for both forces and couples.
4. An eight-cylinder in-line engine with the cranks at 90° is inherently balanced for both forces and couples in the first and second harmonics but unbalanced in the fourth harmonic.
5. An eight-cylinder V-engine with cranks at 90° is inherently balanced for forces in the first and second harmonics and for couples in the second harmonic. The unbalanced couples in the first harmonic can be balanced by counterweights that introduce an equal and opposite couple. Such an engine is unbalanced for forces in the fourth harmonic.

17.11 ANALYTICAL TECHNIQUE FOR BALANCING MULTICYLINDER RECIPROCATING ENGINES

First Harmonic Forces The x and y components of the resultant of the first harmonic forces for any multicylinder reciprocating engine can be written in the form

$$\begin{aligned} F_P^x &= A\cos\theta + B\sin\theta \\ F_P^y &= C\cos\theta + D\sin\theta \end{aligned} \tag{17.20}$$

where

$$A = \sum_{i=1}^{n} P_i \cos(\psi_i - \phi_i)\cos\psi_i \tag{17.21a}$$

$$B = \sum_{i=1}^{n} P_i \sin(\psi_i - \phi_i)\cos\psi_i \tag{17.21b}$$

$$C = \sum_{i=1}^{n} P_i \cos(\psi_i - \phi_i)\sin\psi_i \tag{17.21c}$$

and

$$D = \sum_{i=1}^{n} P_i \sin(\psi_i - \phi_i)\sin\psi_i. \tag{17.21d}$$

where n is the number of cylinders, $P_i = m_i r_i \omega^2$ is the inertia force of piston i, m_i is the mass of piston i, r_i is the length of crank i, ω is the constant angular velocity of the crankshaft, ψ_i is the angle from the X axis to the line of sliding of piston i, and ϕ_i is the angle from the reference line on crank 1 to crank i (by definition $\phi_1 = 0$). Note that if the

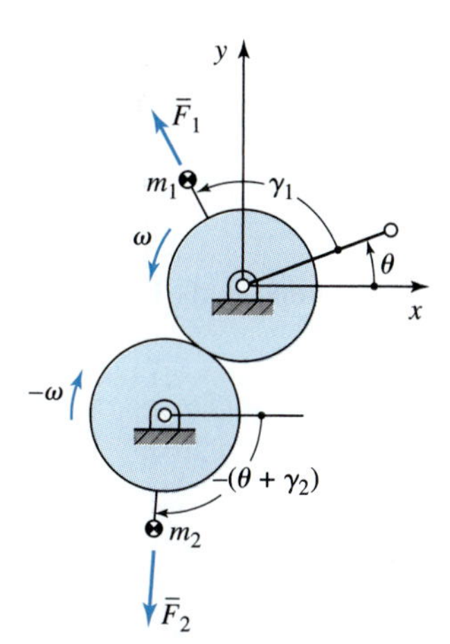

Figure 17.28 General arrangement of counterrotating masses for balancing the primary forces.

coefficients A, B, C, and D are all zero, then the x and y components of the resultant of the first harmonic forces are zero; that is, there is no primary shaking force. This would be the case, for example, with a six-cylinder in-line engine (see the previous section).

If the coefficients A, B, C, and D are not all zero, then the first harmonic forces can always be balanced by a pair of rotating masses, or in some special cases by a single mass, as illustrated in Fig. 17.28. The correcting masses are rotating with the same speed as the crankshaft. The correcting mass m_1 creates a correcting inertia force F_1 at an angle of $(\theta+\gamma_1)$ from the x axis and the correcting mass m_2 creates a correcting inertia force F_2 at an angle of $-(\theta+\gamma_2)$ from the x axis. For balance, these two correcting forces plus the resultant of the first harmonic forces (the primary shaking force) must be equal to zero; that is,

$$\begin{aligned} A\cos\theta + B\sin\theta + F_1\cos(\theta+\gamma_1) + F_2\cos[-(\theta+\gamma_2)] &= 0 \\ C\cos\theta + D\sin\theta + F_1\sin(\theta+\gamma_1) + F_2\sin[-(\theta+\gamma_2)] &= 0 \end{aligned}. \tag{17.22}$$

Expanding these two equations, in terms of functions of θ, γ_1, and γ_2, and rearranging gives

$$\begin{aligned} (F_1\cos\gamma_1 + F_2\cos\gamma_2)\cos\theta - (F_1\sin\gamma_1 + F_2\sin\gamma_2)\sin\theta &= -A\cos\theta - B\sin\theta \\ (F_1\sin\gamma_1 - F_2\sin\gamma_2)\cos\theta + (F_1\cos\gamma_1 - F_2\cos\gamma_2)\sin\theta &= -C\cos\theta - D\sin\theta \end{aligned}. \tag{17.23}$$

To satisfy Eqs. (17.23), for all values of the crank angle θ, the necessary conditions are

$$\begin{aligned} F_1\cos\gamma_1 + F_2\cos\gamma_2 &= -A \\ F_1\sin\gamma_1 + F_2\sin\gamma_2 &= B \\ F_1\sin\gamma_1 - F_2\sin\gamma_2 &= -C \\ F_1\cos\gamma_1 - F_2\cos\gamma_2 &= -D \end{aligned}. \tag{17.24}$$

Solving Eqs. (17.24), the forces are

$$\begin{aligned} F_1 &= \frac{1}{2}\sqrt{(A+D)^2 + (B-C)^2} \\ F_2 &= \frac{1}{2}\sqrt{(A-D)^2 + (B+C)^2} \end{aligned} \tag{17.25}$$

and the orientation angles are given by

$$\begin{aligned} \tan\gamma_1 &= (B - C)/-(D + A) \\ \tan\gamma_2 &= (B + C)/(D - A) \end{aligned} . \tag{17.26}$$

Second Harmonic Forces The x and y components of the resultant of the second harmonic forces can be written in the form

$$\begin{aligned} F_S^x &= A' \cos 2\theta + B' \sin 2\theta \\ F_S^y &= C' \cos 2\theta + D' \sin 2\theta\text{'} \end{aligned} \tag{17.27}$$

where

$$A' = \sum_{i=1}^{n} Q_i \cos 2(\psi_i - \phi_i) \cos \psi_i$$

$$B' = \sum_{i=1}^{n} Q_i \sin 2(\psi_i - \phi_i) \cos \psi_i$$

$$C' = \sum_{i=1}^{n} Q_i \cos 2(\psi_i - \phi_i) \sin \psi_i$$

and

$$D' = \sum_{i=1}^{n} Q_i \sin 2(\psi_i - \phi_i) \sin \psi_i,$$

where the inertia force $Q_i = m_i(r_i/l_i)r_i\omega_i^2 = (r_i/l_i)P_i$. We have stated previously that, in general, $r_i/l_i < 1/4$; therefore, $Q_i < P_i$, which indicates that the magnitudes of the second harmonic forces are much smaller than the magnitudes of the first harmonic forces.

The second harmonic forces can always be balanced by a pair of gears rotating at twice the crankshaft speed, as illustrated in Fig. 17.29. The mass m_1' creates a force F_1' at an

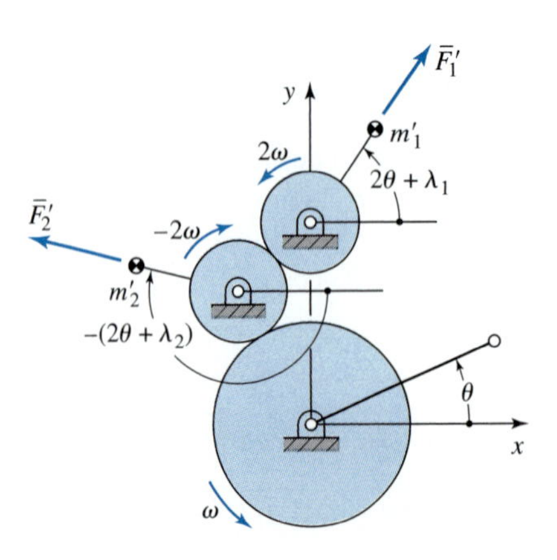

Figure 17.29 General arrangement of counterrotating masses for balancing the secondary forces.

orientation angle of $(2\theta + \lambda_1)$ from the x axis, and the mass m'_2 creates a force F'_2 at an orientation angle of $-(2\theta + \lambda_2)$ from the x axis. For balance, these two forces plus the resultant of the second harmonic forces must be equal to zero.

Note that the analysis for the second harmonic forces is exactly the same as for the first harmonic forces, with A, B, C, and D replaced by A', B', C', and D', θ replaced by 2θ, and the orientation angles γ_1 and γ_2 replaced by the orientation angles λ_1 and λ_2, respectively. The two correcting forces can be written as

$$F'_1 = \frac{1}{2}\sqrt{(A' + D')^2 + (B' - C')^2}$$

and

$$F'_2 = \frac{1}{2}\sqrt{(A' - D')^2 + (B' + C')^2}.$$

The two orientation angles can be written as

$$\tan \lambda_1 = \frac{B' - C'}{-(A' + D')}$$

and

$$\tan \lambda_2 = \frac{B' + C'}{-(A' - D')}.$$

EXAMPLE 17.4

Determine the magnitudes and orientations of the forces created by the correcting masses needed to balance the first and second harmonic shaking forces of the three-cylinder engine illustrated in Fig. 17.30.

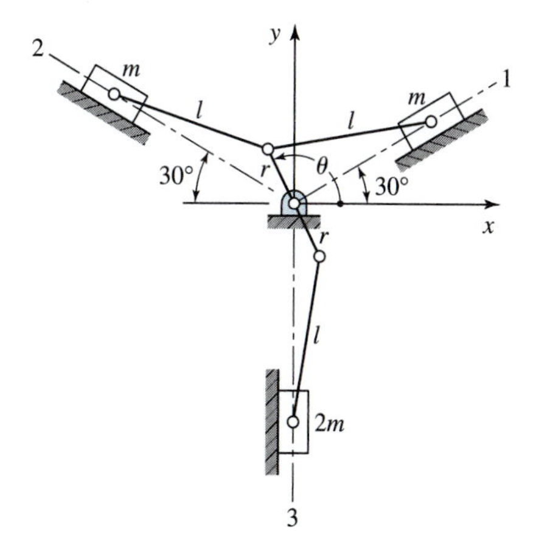

Figure 17.30 Example 17.4: Three-cylinder arrangement.

SOLUTION

The resultant first harmonic force vector is

$$\mathbf{F}_P = 1.5P\cos\theta\hat{\mathbf{i}} - 1.5P\sin\theta\hat{\mathbf{j}}, \tag{1}$$

where

$$P = mr\omega^2.$$

Comparing Eq. (1) with Eqs. (17.20), the coefficients are

$$A = 1.5P, \quad B = 0, \quad C = 0, \quad D = -1.5P.$$

Therefore, from Eqs. (17.25), the forces are

$$F_1 = \frac{1}{2}\sqrt{(A+D)^2 + (B-C)^2} = 0$$

$$F_2 = \frac{1}{2}\sqrt{(A-D)^2 + (B+C)^2} = 1.5P.$$

Because the first force is zero, there is only one correcting mass. The orientation angle for this mass, from the second of Eqs. (17.26), is

$$\tan\gamma_2 = (B+C)/(D-A) = 0/-3P = 0.$$

Therefore, because the denominator (the cosine of the angle) is negative, the result is

$$\gamma_2 = 180°. \qquad \textit{Ans.}$$

The conclusion is that we need one mass rotating opposite to the crankshaft (at crankshaft speed) and oriented at an angle $-(\theta + 180°)$. Figure 17.31 illustrates the orientation of the correcting mass for $\theta = 0°$. If the correcting mass m_2 is to be placed at a

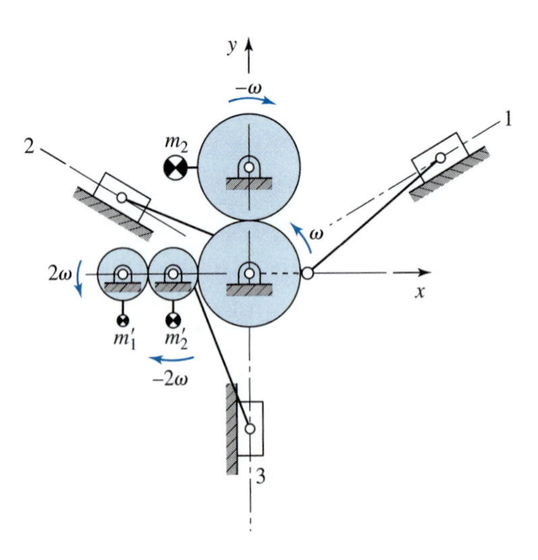

Figure 17.31 Example 17.4: Balancing arrangement (shown at $\theta = 0°$).

radius r_2, then the inertia force is

$$F_2 = m_2 r_2 \omega^2 = 1.5 m r \omega^2.$$

Therefore, the product is

$$m_2 r_2 = 1.5 m r. \qquad \textit{Ans.}$$

The second harmonic force vector is

$$\mathbf{F}_s = 1.5 Q \sin 2\theta \hat{\mathbf{i}} + 2.5 Q \cos 2\theta \hat{\mathbf{j}}, \qquad (2)$$

where

$$Q = \frac{m r^2 \omega^2}{l}.$$

Comparing Eq. (2) with Eqs. (17.23), the coefficients are

$$A' = 0, \quad B' = 1.5Q, \quad C' = 2.5Q, \quad D' = 0.$$

Therefore, the inertia forces, from Eqs. (17.25), are

$$F_1' = \frac{1}{2}\sqrt{(A' + D')^2 + (B' - C')^2} = 0.5Q$$

$$F_2' = \frac{1}{2}\sqrt{(A' - D')^2 + (B' + C')^2} = 2.0Q.$$

The forces can be expressed as

$$F_1' = m_1' r_1' (2\omega)^2 = 0.5 m \frac{r^2}{l} \omega^2$$

$$F_2' = m_2' r_2' (2\omega)^2 = 2 m \frac{r^2}{l} \omega^2.$$

Therefore, the correcting mass m_1' can be placed at a radius r_1' such that

$$m_1' r_1' = 0.125 m r^2 / l \qquad \textit{Ans.}$$

and the correcting mass m_2' can be placed at a radius r_2' such that

$$m_2' r_2' = 0.5 m r^2 / l. \qquad \textit{Ans.}$$

From Eq. (17.26*a*), the first orientation angle is

$$\tan \lambda_1 = (B' - C')/-(D' + A') = -1.0Q/0 = \infty.$$

Therefore, the answer is $\lambda_1 = 270°$, because the numerator (the sine of the angle) is negative. From Eq. (17.25*b*), the second orientation angle is

$$\tan \lambda_2 = (B' + C')/(D' - A') = +4.0Q/0 = \infty.$$

The answer is $\lambda_2 = 90°$, because the numerator (or the sine of the angle) is positive.

The conclusion is that we need the arrangement illustrated in Fig. 17.31. The mass m'_1 is located on a shaft turning in the same direction as the crankshaft and at twice the crankshaft speed. The orientation angle is $2\theta + 270°$. The mass m'_2 is located on a shaft turning opposite to the crankshaft and at twice the crankshaft speed. The orientation angle is $-(2\theta + 90°)$. *Ans.*

17.12 BALANCING LINKAGES[3]

The two problems that arise in balancing linkages are balancing the shaking force and balancing the shaking moment. Lowen and Berkof note that very few studies have been reported on the problem of balancing the shaking moment. This problem is discussed further in Section 17.13.

In force balancing a linkage, we must concern ourselves with the position of the total center of mass. If a way can be found to cause this total center of mass to remain stationary, the vector sum of all the frame forces will always be zero. Berkof and Lowen[4] have listed five methods of force balancing:

1. The method of static balancing, in which concentrated link masses are replaced by systems of masses that are statically equivalent;
2. The method of principal vectors, in which an analytical expression is obtained for the center of mass and then manipulated to learn how its trajectory can be influenced;
3. The method of linearly independent vectors, in which the center of mass of a mechanism is made stationary, causing the coefficients of the time-dependent terms of the equation describing the trajectory of the total center of mass to vanish;
4. The use of cam-driven masses to keep the total center of mass stationary; and
5. The addition of an axially symmetric duplicate mechanism by which the new combined total center of mass is made stationary.

Here we present only the Berkof–Lowen method,[5] which employs the method of linearly independent vectors. The method is developed completely for the four-bar linkage, but the final results alone are given for a typical six-bar linkage. The procedure is as follows. First, find the equation that describes the trajectory of the total center of mass of the linkage. This equation will contain certain terms whose coefficients are time dependent. Then the total center of mass is made stationary by changing the positions of the individual link masses so that the coefficients of all time-dependent terms vanish. To accomplish this, it is necessary to write the equation in such a form that the time-dependent unit vectors contained in the equation are independent.

In Fig. 17.32 a general four-bar linkage is shown having link mass m_2 located at G_2, m_3 located at G_3, and m_4 located at G_4. The total mass M of the four-bar linkage is

$$M = m_2 + m_3 + m_4 \tag{a}$$

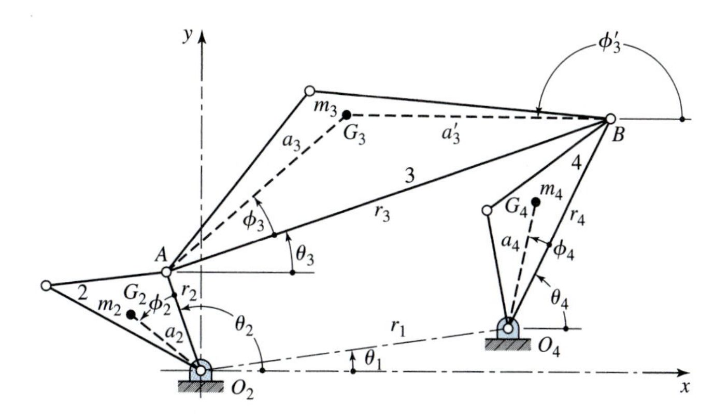

Figure 17.32 Four-bar linkage showing arbitrary positions of the link masses.

We begin by defining the position of the total center of mass of the linkage by the vector $\mathbf{r}_s$,

$$\mathbf{r}_s = \frac{1}{M}(m_2\mathbf{r}_{s_2} + m_3\mathbf{r}_{s_3} + m_4\mathbf{r}_{s_4}), \tag{b}$$

where $\mathbf{r}_{s_2}$, $\mathbf{r}_{s_3}$, and $\mathbf{r}_{s_4}$ are the vectors that describe the positions of m_2, m_3, and m_4, respectively, in the global xy coordinate system. The coordinates a_i, ϕ_i describe the positions of the mass centers within each link. Thus, from Fig. 17.32,

$$\begin{aligned} \mathbf{r}_{s_2} &= a_2 e^{j(\theta_2+\phi_2)} \\ \mathbf{r}_{s_3} &= r_2 e^{j\theta_2} + a_3 e^{j(\theta_3+\phi_3)} \\ \mathbf{r}_{s_4} &= r_1 e^{j\theta_1} + a_4 e^{j(\theta_4+\phi_4)} \end{aligned}. \tag{c}$$

Substituting Eqs. (c) into (b) gives

$$M\mathbf{r}_s = (m_2 a_2 e^{j\phi_2} + m_3 r_2)e^{j\theta_2} + (m_3 a_3 e^{j\phi_3})e^{j\theta_3} + (m_4 a_4 e^{j\phi_4})e^{j\theta_4} + m_4 r_1 e^{j\theta_1}, \tag{d}$$

where we have used the identity $e^{j\alpha}e^{j\beta} = e^{j(\alpha+\beta)}$.

For a four-bar linkage the vector loop-closure equation can be written as

$$r_2 e^{j\theta_2} + r_3 e^{j\theta_3} - r_4 e^{j\theta_4} - r_1 e^{j\theta_1} = 0. \tag{e}$$

Thus, the time-dependent terms $e^{j\theta_2}$, $e^{j\theta_3}$, and $e^{j\theta_4}$ in Eq. (d) are not independent. To make them so, we solve Eq. (e) for one of the unit vectors, say $e^{j\theta_3}$, and substitute the result into Eq. (d). Thus,

$$e^{j\theta_3} = \frac{1}{r_3}(r_1 e^{j\theta_1} - r_2 e^{j\theta_2} + r_4 e^{j\theta_4}) \tag{f}$$

and Eq. (d) now becomes

$$\begin{aligned}M\mathbf{r}_s = &\left(m_2a_2e^{j\phi_2} + m_3r_2 - m_3a_3\frac{r_2}{r_3}e^{j\phi_3}\right)e^{j\theta_2} + \left(m_4a_4e^{j\phi_4} + m_3a_3\frac{r_4}{r_3}e^{j\phi_3}\right)e^{j\theta_4}\\ &+ \left(m_4r_1 + m_3a_3\frac{r_1}{r_3}e^{j\phi_3}\right)e^{j\theta_1}\end{aligned}. \quad (g)$$

Equation (g) shows that the center of mass will be stationary at the position

$$\mathbf{r}_s = \frac{r_1}{r_3M}(m_4r_3 + m_3a_3e^{j\phi_3})e^{j\theta_1} \quad (17.28)$$

if we make the following coefficients of the time-dependent terms vanish:

$$m_2a_2e^{j\phi_2} + m_3r_2 - m_3a_3\frac{r_2}{r_3}e^{j\phi_3} = 0 \quad (h)$$

$$m_4a_4e^{j\phi_4} + m_3a_3\frac{r_4}{r_3}e^{j\phi_3} = 0. \quad (i)$$

But Eq. (h) can be simplified by locating G_3 from point B instead of point A (Fig. 17.32). Thus,

$$a_3e^{j\phi_3} = r_3 + a_3'e^{j\phi_3'}.$$

With this substitution Eq. (h) becomes

$$m_2a_2e^{j\phi_2} - m_3a_3'\frac{r_2}{r_3}e^{j\phi_3'} = 0. \quad (j)$$

Equations (i) and (j) must be satisfied to obtain total force balance. These equations yield the two sets of conditions:

$$\begin{aligned}m_2a_2 = m_3a_3'\frac{r_2}{r_3} \quad &\text{and} \quad \phi_2 = \phi_3'\\ m_4a_4 = m_3a_3\frac{r_4}{r_3} \quad &\text{and} \quad \phi_4 = \phi_3 + \pi\end{aligned}. \quad (17.29)$$

A study of these conditions indicates that the mass and its location can be specified in advance for any single link; then full balance can be obtained by rearranging the mass of the other two links.

The usual problem in balancing a four-bar linkage is that the link lengths r_i are specified in advance because of the functional requirements. For this situation, counterweights can be added to the input and output links to redistribute their masses, while the geometry of the coupler link is undisturbed.

When adding counterweights, the following relations must be satisfied,

$$m_ia_i\angle\phi_i = m_i^\circ a_i^\circ\angle\phi_i^\circ + m_i^*a_i^*\angle\phi_i^*, \quad (17.30)$$

where $m_i^\circ, a_i^\circ, \phi_i^\circ$ are the parameters of the unbalanced linkage, m_i^*, a_i^*, ϕ_i^* are the parameters of the counterweights, and m_i, a_i, ϕ_i are the parameters that result from Eqs. (17.29). A second condition that must generally be satisfied is

$$m_i = m_i^\circ + m_i^*. \tag{17.31}$$

If the solution to a balancing problem can remain as the mass distance product $m_i^* a_i^*$, Eq. (17.31) need not be used and Eq. (17.30) can be solved to yield

$$m_i^* a_i^* = \sqrt{(m_i a_i)^2 + (m_i^o a_i^o)^2 - 2(m_i a_i)(m_i^o a_i^o)\cos(\phi_i - \phi_i^o)} \tag{17.32}$$

$$\phi_i^* = \tan^{-1}\left(\frac{m_i a_i \sin\phi_i - m_i^o a_i^o \sin\phi_i^o}{m_i a_i \cos\phi_i - m_i^o a_i^o \cos\phi_i^o}\right) \tag{17.33}$$

Figure 17.33 illustrates a typical six-bar linkage and the notation. For this, the Berkof–Lowen conditions for total balance are

$$\begin{aligned} m_2\frac{a_2}{r_2}e^{j\phi_2} &= m_5\frac{a_5'}{r_5}\frac{b_2}{r_2}e^{j(\phi_1+\alpha_2)} + m_3\frac{a_3'}{r_3}e^{j\phi_1} \\ m_4\frac{a_4}{r_4}e^{j\phi_4} &= m_6\frac{a_6'}{r_6}\frac{b_4}{r_4}e^{j\phi_6'} - m_3\frac{a_3}{r_3}e^{j(\phi_3+\alpha_4)} \\ m_5\frac{a_5}{r_5}e^{j\phi_5} &= -m_6\frac{a_6}{r_6}e^{j\phi_6} \end{aligned}. \tag{17.34}$$

Similar relations can be devised for other six-bar linkages. For total balance, Eqs. (17.34) demonstrate that a certain mass-geometry relation between links 5 and 6 must be satisfied, after which the masses of any two links and their locations can be specified. Balance is then achieved by a redistribution of the masses of the remaining three movable links.

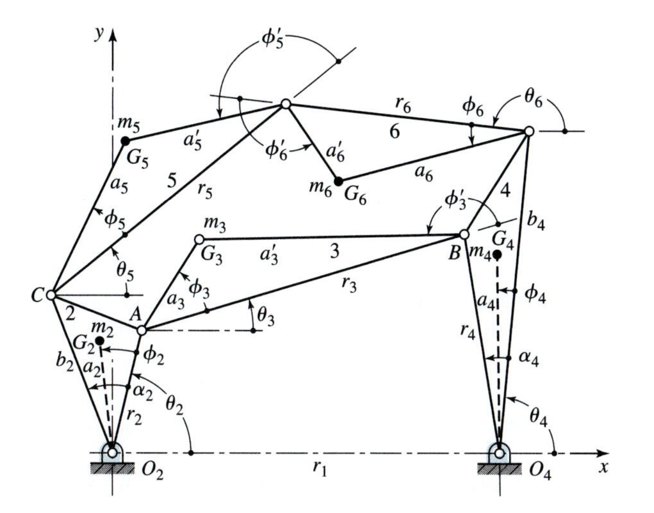

Figure 17.33 Notation for a six-bar linkage.

It is important to note that the addition of counterweights to balance the shaking forces will probably increase the internal bearing forces as well as the shaking moment. Thus, only a partial balance may represent the best compromise among these three effects.

EXAMPLE 17.5

Table 17.2 is a tabulation of the dimensions, masses, and locations of the mass centers of a four-bar mechanism having link 2 as the input and link 4 as the output. Complete force balancing is to be achieved by adding counterweights to the input and the output links. Find the mass-distance value and the angular orientation of each counterweight.

TABLE 17.2 Parameters of an unbalanced four-bar linkage

Link i	1	2	3	4
r_i, mm	140	50	150	75
a_i^o, mm	—	25	80	40
ϕ_i^o	—	0°	15°	0°
a'^o_i, mm	—	—	75.6	—
ϕ'^o_i	—	—	164.1°	—
m_i^o, kg	—	0.046	0.125	0.054

SOLUTION

From Eqs. (17.29) we first find

$$m_2 a_2 = m_3^o a'^o_3 \frac{r_2}{r_3} = (0.125\ \text{kg})(75.6\ \text{mm})\frac{50\ \text{mm}}{150\ \text{mm}} = 3.150\ \text{g}\cdot\text{m}$$

$$\phi_2 = \phi_3' = 164.1°$$

$$m_4 a_4 = m_4^o a_3^o \frac{r_4}{r_3} = (0.125\ \text{kg})(80\ \text{mm})\frac{75\ \text{mm}}{150\ \text{mm}} = 5.000\ \text{g}\cdot\text{m}$$

$$\phi_4 = \phi_4^o + 180° = 15° + 180° = 195°.$$

Note that m_2a_2 and m_4a_4 are the mass-distance values *after* the counterweights have been added. Also, note that the link 3 parameters will not be altered. We next compute

$$m_2^o a_2^o = (0.046\ \text{kg})(25\ \text{mm}) = 1.150\ \text{g}\cdot\text{m} \quad \text{and}$$

$$m_4^o a_4^o = (0.054\ \text{kg})(40\ \text{mm}) = 2.160\ \text{g}\cdot\text{m}.$$ *Ans.*

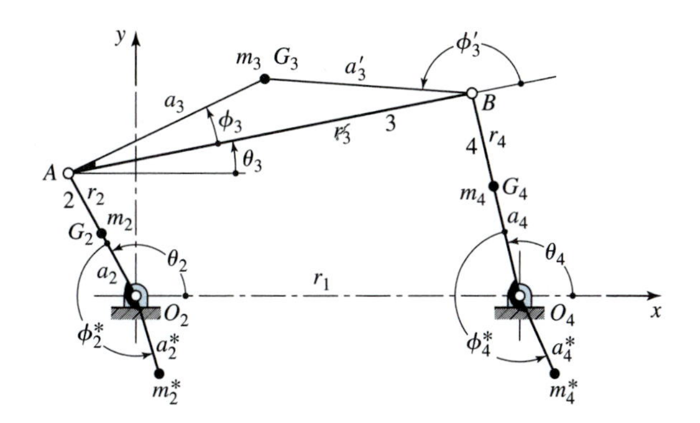

Figure 17.34 Four-bar crank-and-rocker linkage showing counterweights added to input and output links to achieve complete force balance.

Using Eq. (17.32), we compute the mass-distance value for the link 2 counterweight as

$$m_2^* a_2^* = \sqrt{(m_2 a_2)^2 + (m_2^o a_2^o)^2 - 2(m_2 a_2)(m_2^o a_2^o)\cos(\phi_2 - \phi_2^o)}$$

$$= \sqrt{(3.150\ \text{g}\cdot\text{m})^2 + (1.150\ \text{g}\cdot\text{m})^2 - 2(3.150\ \text{g}\cdot\text{m})(1.150\ \text{g}\cdot\text{m})\cos(164.1^\circ - 0^\circ)}$$

$$= 4.268\ \text{g}\cdot\text{m}. \qquad \textit{Ans.}$$

From Eq. (17.33), we find the orientation of this counterweight to be

$$\phi_2^* = \tan^{-1}\left(\frac{m_2 a_2 \sin\phi_2 - m_2^o a_2^o \sin\phi_2^o}{m_2 a_2 \cos\phi_2 - m_2^o a_2^o \cos\phi_2^o}\right)$$

$$= \tan^{-1}\left(\frac{(3.150\ \text{g}\cdot\text{m})\sin 164.1^\circ - (1.150\ \text{g}\cdot\text{m})\sin 0^\circ}{(3.150\ \text{g}\cdot\text{m})\cos 164.1^\circ - (1.150\ \text{g}\cdot\text{m})\cos 0^\circ}\right) = 168.3^\circ. \qquad \textit{Ans.}$$

Using the same procedure for link 4 yields

$$m_4^* a_4^* = 7.11\ \text{g}\cdot\text{m} \quad \text{at} \quad \phi_4^* = 190.5^\circ. \qquad \textit{Ans.}$$

Figure 17.34 is a scale drawing of the complete linkage with the two counterweights added.

17.13 BALANCING OF MACHINES[6]

In the Section 17.12 we learned how to force balance a simple linkage using two or more counterweights, depending upon the number of links composing the linkage. Unfortunately, this does not balance the shaking moment and, in fact, may even make it worse because of the addition of the counterweights. If a machine is imagined to be composed of several mechanisms, one might consider balancing the machine by balancing each mechanism separately. But this may not result in the best balance for the machine, because the addition of a large number of counterweights may cause the inertia torque to be completely unacceptable. Furthermore, unbalance of one mechanism may counteract the unbalance in another, eliminating the need for some of the counterweights in the first place.

Stevensen demonstrates that any single harmonic of unbalanced forces, moments of forces, and torques in a machine can be balanced by the addition of six counterweights. They are arranged on three shafts, two per shaft, driven at the constant speed of the harmonic, and have axes parallel, respectively, to each of the three mutually perpendicular axes through the center of mass of the machine. The method is too complex to be included here, but it is worthwhile to look at the overall approach.

Using the methods of this book together with a computer, the rectilinear and angular accelerations of each of the moving mass centers of a machine are computed for points throughout a cycle of motion. The masses and mass moments of inertia of the machine must also be computed or determined experimentally. Then the inertia forces, the inertia torques, and the moments of the forces are computed with reference to the three mutually perpendicular coordinate axes through the center of mass of the machine. When these are summed for each posture in the cycle, six functions of time are the result, three for the forces and three for the moments. With the aid of a computer it is then possible to use numerical harmonic analysis to define the component harmonics of the unbalanced forces parallel to the three axes and of the unbalanced moments about the three axes.

To balance a single harmonic, each component of the unbalance of the machine is represented in the form $A \cos \omega t + B \sin \omega t$ with appropriate subscripts. Six equations of equilibrium are then written, which include the unbalances as well as the effects of the six unknown counterweights. These equations are arranged so that each of the $\sin \omega t$ and $\cos \omega t$ terms is multiplied by parenthetical coefficients. Balance is then achieved by setting the parenthetical terms in each equation equal to zero, much in the same manner as in the preceding section. This results in 12 equations, all linear, in 12 unknowns. With the locations for the balancing counterweights on the three shafts specified, the 12 equations can be solved for the six mr products and the six phase angles needed for the six balancing weights. Stevensen goes on to demonstrate that when fewer than the necessary three shafts are available, it becomes necessary to optimize some effect of the unbalance, such as the motion of a point on the machine.

17.14 REFERENES

[1] T. C. Rathbone, "Turbine Vibration and Balancing," *Trans. ASME* 1929: p. 267; E. L. Thearle, "Dynamic Balancing in the Field," *Trans. ASME* 1934, p. 745.

[2] The presentation here is from Prof. E. N. Stevensen, University of Hartford, class notes, with his permission. Although some changes have been made to conform to the notation of this book, the material is all Stevensen's. He refers to Maleev and Lichty [V. L. Maleev, *Internal Combustion Engines*, New York: McGraw–Hill, 1993; and L. C. Lichty, *Internal Combustion Engines*, 5th ed., New York: McGraw–Hill, 1939] and states that the method first came to his attention in both the Maleev and the Lichty books.

[3] Those who wish to investigate this topic in detail should begin with the following reference, in which an entire issue is devoted to the subject of linkage balancing: G. G. Lowen and R. S. Berkof. 1968. Survey of Investigations into the Balancing of Linkages. *J. Mech.* 3:221. This issue contains 11 translations on the subject from the German and Russian literature.

[4] R. S. Berkof and G. G. Lowen, "A New Method for Completely Force Balancing Simple Linkages," *J. Eng. Ind. Trans. ASME B*, Vol. 91, pp. 21–26, 1969.

[5] *Ibid.*
[6] The material for this section is from E. N. Stevensen, Jr. "Balancing of Machines," *J. Eng. Ind. Trans. ASME B*, Vol. 95, pp. 650–656, 1973. It is included with the advice and consent of Professor Stevensen.

PROBLEMS

17.1 Determine the bearing reactions at A and B for the system illustrated in Fig. P17.1 if the speed is 300 rev/min. Determine the magnitude and the angular orientation of the balancing mass if it is located at a radius of 50 mm.

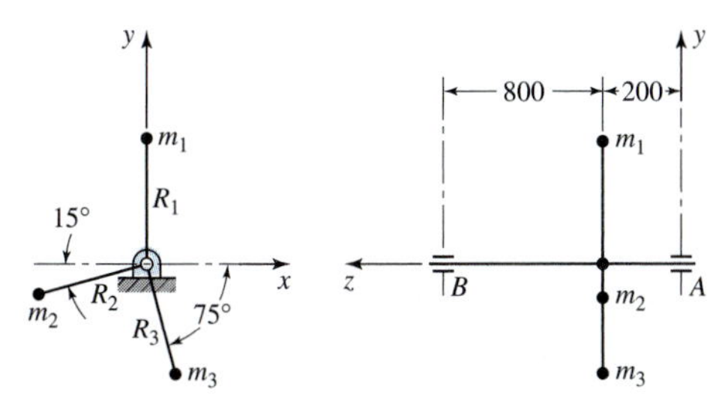

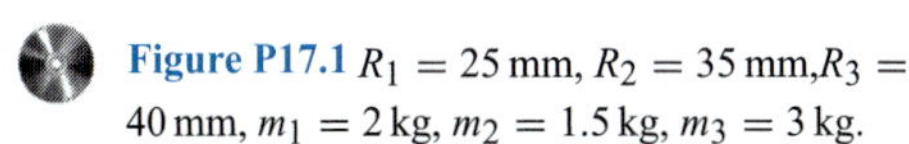
Figure P17.1 $R_1 = 25$ mm, $R_2 = 35$ mm, $R_3 = 40$ mm, $m_1 = 2$ kg, $m_2 = 1.5$ kg, $m_3 = 3$ kg.

17.2 Figure P17.2 illustrates three weights connected to a shaft that rotates in bearings at A and B. Determine the magnitude of the bearing reactions if the shaft speed is 300 rev/min. A counterweight is to be located at a radius of 10 in. Find the value of the weight and its angular orientation.

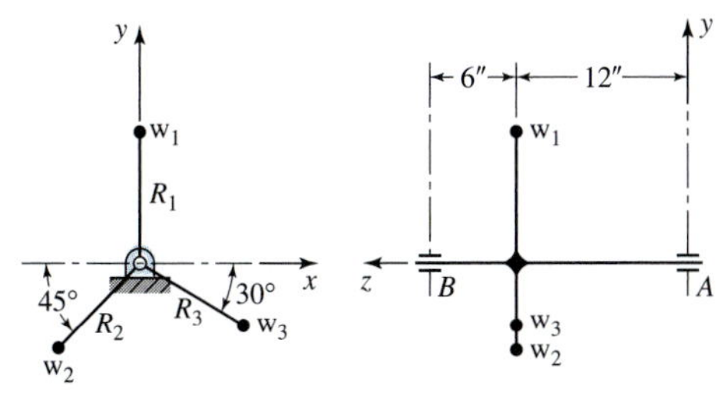

Figure P17.2 $R_1 = 8$ in, $R_2 = 12$ in, $R_3 = 6$ in, $w_1 = 2$ oz, $w_2 = 1.5$ oz, $w_3 = 3$ oz.

†17.3 Figure P17.3 illustrates two weights connected to a rotating shaft and mounted outboard of bearings A and B. If the shaft rotates at 120 rev/min, what are the magnitudes of the bearing reactions at A and B? Suppose the system is to be balanced by reducing a weight at a radius of 5 in. Determine the amount and the angular orientation of the weight to be removed.

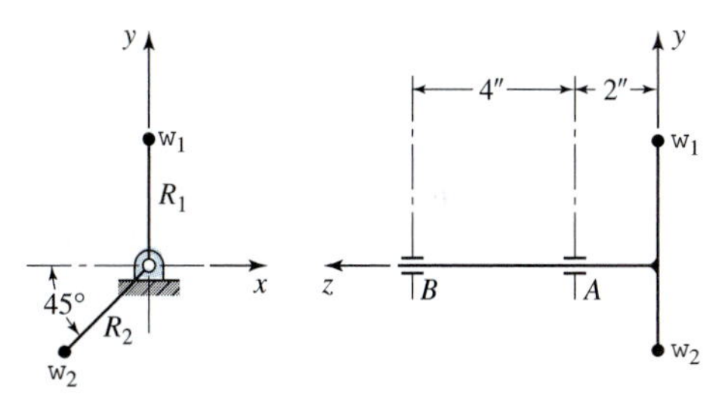

Figure P17.3 $R_1 = 4$ in, $R_2 = 6$ in, $w_1 = 4$ lb, $w_2 = 3$ lb.

17.4 For a speed of 220 rev/min, calculate the magnitudes and relative angular orientations of the bearing reactions at A and B for the two-mass system shown in Fig. P17.4.

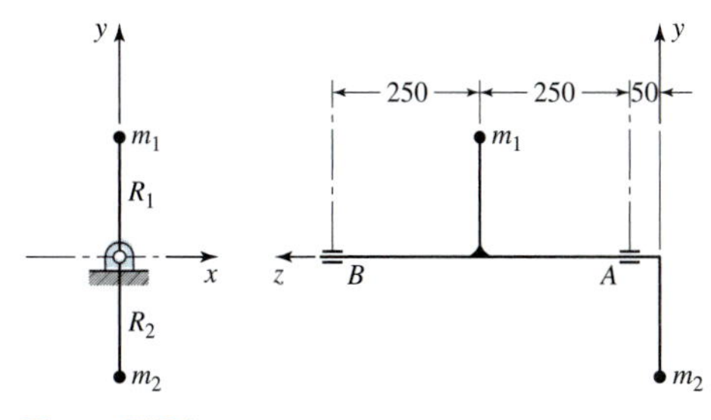

Figure P17.4 $R_1 = 60$ mm, $R_2 = 40$ mm, $m_1 = 2$ kg, $m_2 = 1.5$ kg.

17.5 The rotating system illustrated in Fig. P17.5 has $R_1 = R_2 = 60$ mm, $a = c = 300$ mm, $b = 600$ mm, $m_1 = 1$ kg, and $m_2 = 3$ kg. Find the bearing reactions at A and B and their angular orientations measured from a rotating reference mark if the shaft speed is 100 rev/min.

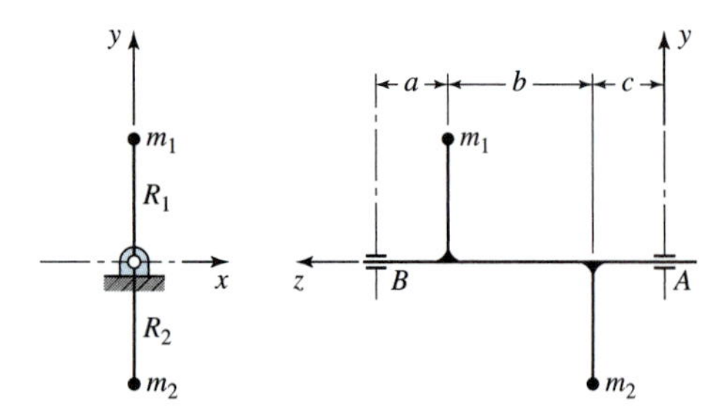

Figure P17.5

17.6 The rotating shaft illustrated in Fig. P17.5 supports two masses, m_1 and m_2, whose weights are 4 lb and 5 lb, respectively. The dimensions are $R_1 = 4$ in, $R_2 = 3$ in, $a = 2$ in, $b = 8$ in, and $c = 3$ in. Find the magnitudes of the rotating bearing reactions at A and B and their angular orientations measured from a rotating reference mark if the shaft speed is 360 rev/min.

17.7 The shaft illustrated in Fig. P17.7 is to be balanced by placing masses in the correction planes L and R. The weights of the three masses m_1, m_2, and m_3 are 4 oz, 3 oz, and 4 oz, respectively. The dimensions are $R_1 = 5$ in, $R_2 = 4$ in, $R_3 = 5$ in, $a = 1$ in, $b = e = 8$ in, $c = 10$ in, and $d = 9$ in. Calculate the magnitudes of the corrections in ounce-inches and their angular orientations.

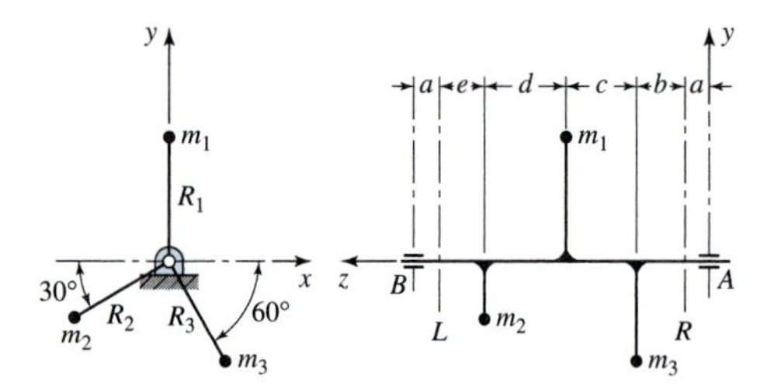

Figure P17.7

17.8 The shaft of Problem 17.7 is to be balanced by removing weight from the two correction planes. Determine the corrections to be subtracted in ounce-inches and their angular orientations.

†17.9 The shaft illustrated in Fig. P17.7 is to be balanced by subtracting masses in the two correction planes, L and R. The three masses are $m_1 = 6$ g, $m_2 = 7$ g, and $m_3 = 5$ g. The dimensions are $R_1 = 125$ mm, $R_2 = 150$ mm, $R_3 = 100$ mm, $a = 25$ mm, $b = 300$ mm, $c = 600$ mm, $d = 150$ mm, and $e = 75$ mm. Calculate the magnitudes and angular orientations of the corrections.

17.10 Repeat Problem 17.9 if masses are to be added in the two correction planes.

17.11 Solve the two-plane balancing problem as stated in Section 17.8.

17.12 A rotor to be balanced in the field yielded an amplitude of 5 at an angle of 142° at the left-hand bearing and an amplitude of 3 at an angle of −22° at the right-hand bearing because of unbalance. To correct this, a trial mass of 12 was added to the left-hand correction plane at an angle of 210° from the rotating reference. A second run then gave left-hand and right-hand responses of $8\angle 160°$ and $4\angle 260°$, respectively. The first trial mass was then removed and a second mass of 6 added to the right-hand correction plane at an angle of −70°. The responses to this were $2\angle 74°$ and $4.5\angle -80°$ for the left- and right-hand bearings, respectively. Determine the original unbalances.

18 CAM DYNAMICS

18.1 RIGID- AND ELASTIC-BODY CAM SYSTEMS

Figure 18.1*a* is a cross-sectional view illustrating the overhead valve arrangement in an automotive engine. In analyzing the dynamics of this or any other cam system, we would expect to determine the contact force at the cam surface, the spring force, and the cam-shaft torque, all for a complete rotation of the cam. In one method of analysis all parts of the cam-follower train, consisting of the push rod, the rocker arm, and the valve stem, together with the cam shaft, are assumed to be rigid. If this is an accurate assumption and if the speed of the cam-follower train is moderate, then such an analysis will usually produce quite satisfactory results. In any event, such a rigid-body analysis should always be attempted as the first iteration.

Sometimes the speeds are so high or the members are so elastic (perhaps because of extreme lengths or slenderness) that an elastic-body analysis must be used. This fact is usually discovered when troubles are encountered with the system. Such troubles are usually evidenced by noise, chatter, unusual wear, poor product performance, or perhaps fatigue failure of some of the parts. In other cases, laboratory investigation of the operation of a prototype system may reveal substantial differences between the theoretical and the observed performance.

Figure 18.1*b* is a mathematical model of an elastic-body cam system. Here m_3 represents the mass of the cam and a portion of the cam shaft. The motion machined into the cam profile is the coordinate $y(\theta)$, a function of the cam-shaft angle θ. The masses m_1 and m_2 and the stiffnesses k_2 and k_3 are lumped characteristics of the follower train. The stiffness of the follower retaining spring is k_1 and the bending stiffness of the cam shaft is k_4. The dashpots $c_i (i = 1, 2, 3, 4, \text{ and } 5)$ are inserted to represent the effects of friction, which, in the analysis, may indicate either viscous damping or sliding friction or any combination

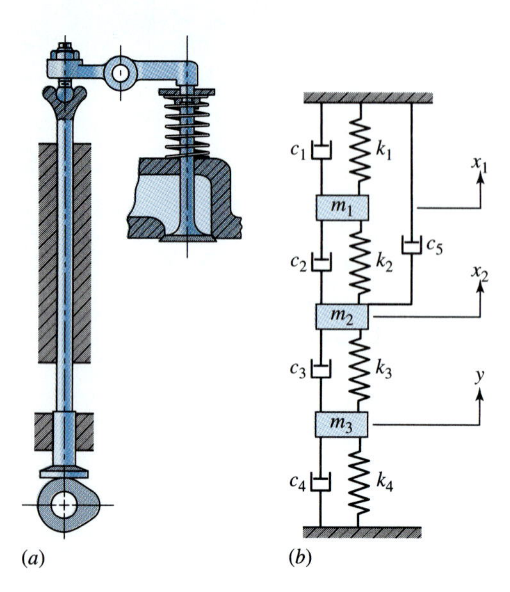

Figure 18.1 An overhead valve arrangement for an automotive engine.

of the two. The system of Fig. 18.1*b* is a rather sophisticated one requiring the solution of three simultaneous differential equations. This will not be presented in Chapter 18; instead, we will focus our attention on simpler systems.

18.2 ANALYSIS OF AN ECCENTRIC CAM

An eccentric is the name given to a circular plate cam with the cam-shaft mounted off-center. The distance e between the center of the disk and the center of the shaft is called the eccentricity. Figure 18.2*a* illustrates a simple reciprocating follower eccentric-cam system. It consists of an eccentric plate cam, a flat-face follower mass, and a retaining spring of stiffness k. The coordinate y designates the motion of the follower as long as the cam remains in contact with the follower. We arbitrarily select $y = 0$ at the bottom of the stroke. Then the kinematic quantities of interest are

$$y = e - e\cos\omega t, \qquad \dot{y} = e\omega\sin\omega t, \qquad \ddot{y} = e\omega^2\cos\omega t, \tag{18.1}$$

where ωt is the cam angle θ.

To make a rigid-body analysis, we assume no friction and construct a free-body diagram of the follower. In Fig. 18.2b, F_{23} is the cam contact force and F_S is the spring force. In general, F_{23} and F_S do not have the same line of action, and so a pair of frame forces, $F_{13,A}$ and $F_{13,B}$, is exerted at bearings A and B.

Before writing the equation of motion, let us investigate the spring force in more detail. Recall that the *spring stiffness k*, also called the *spring rate*, is defined as the amount of force necessary to deform the spring a unit length. Thus, the units of k are usually expressed in Newtons per meter or pounds per inch. The purpose of the spring is to keep or retain the follower in contact with the cam. Thus, the spring should exert some force even at the bottom of the stroke, where it is extended the most. This force, called the *preload P*, is the

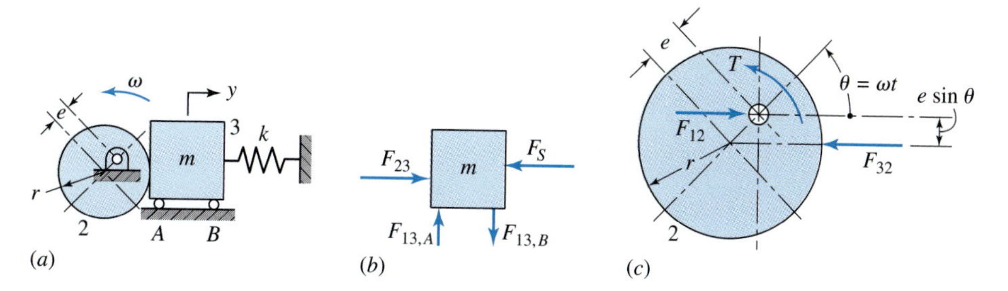

Figure 18.2 (*a*) Eccentric plate cam and flat-face follower; (*b*) free-body diagram of the follower; (*c*) free-body diagram of the cam.

force exerted by the spring when $y = 0$. Thus, $P = k\delta$, where δ is the deformation through which the spring must be compressed to assemble it.

Summing forces on the follower mass in the y direction gives

$$\sum F^y = F_{23} - k(y + \delta) = m\ddot{y}. \tag{a}$$

Note that the contact force F_{23} can have only a positive value. Rearranging this equation gives

$$m\ddot{y} + ky = F_{23} - k\delta \quad \text{or} \quad m\ddot{y} + ky = F_{23} - P. \tag{b}$$

Then, substituting the first and third equations of Eqs. (18.1) into this equation and rearranging, the contact force can be written as

$$F_{23} = (m\omega^2 - k)e\cos\omega t + (ke + P). \tag{18.2}$$

This equation and Fig. 18.3*a* demonstrate that the contact force F_{23} consists of a constant term $ke + P$ with a cosine wave superimposed on it. The maximum occurs at $\theta = 0°$ and the minimum occurs at $\theta = 180°$. The cosine, or variable, component has an amplitude that depends upon the square of the cam-shaft speed. Thus, as the speed increases, this term increases at a greater rate. At a certain speed, the contact force could become zero at or near $\theta = 180°$. When this happens, there is usually some impact between the cam and the follower, resulting in clicking, rattling, or very noisy operation. In effect, the sluggishness, or inertia, of the follower prevents it from following the cam. The result is often called *jump* or *float*. The noise occurs when contact is reestablished. Of course, the purpose of the retaining spring is to prevent this. Because the contact force consists of a cosine wave superimposed on a constant term, all we must do to prevent jump is to move, or elevate, the cosine wave away from the zero position. To do this we can increase the constant term $ke + P$ by increasing the preload P, the spring rate k, or both.

Having learned that jump begins at $\cos\omega t = -1$ with zero contact force (that is, $F_{23} = 0$), we can solve Eq. (18.2) for the jump speed. The result is

$$\omega = \sqrt{\frac{2ke + P}{me}}. \tag{18.3}$$

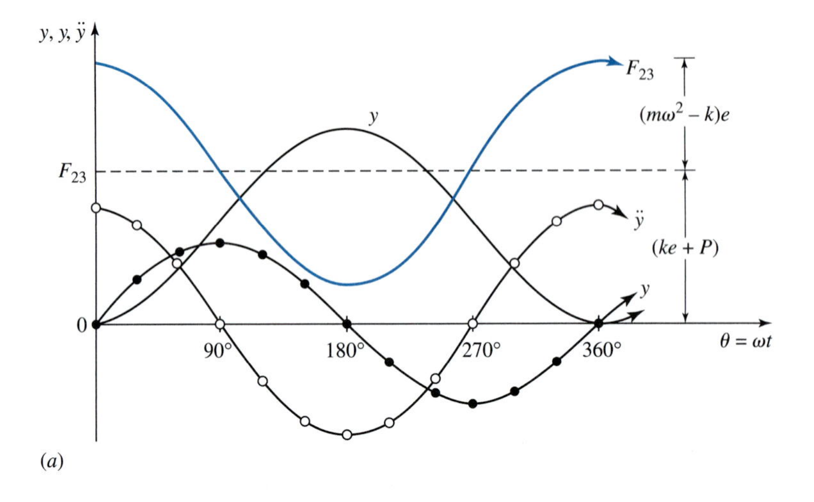

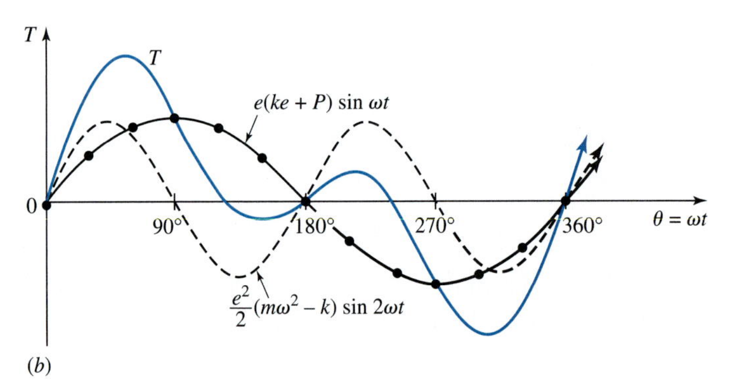

Figure 18.3 (*a*) Plot of displacement, velocity, acceleration, and contact force for an eccentric cam system; (*b*) graph of torque components and total cam-shaft torque.

Using the same procedure, we find that jump will not occur for the range of preload,

$$P > e(m\omega^2 - 2k). \tag{18.4}$$

Figure 18.2*c* is a free-body diagram of the cam. The torque T, applied by the shaft onto the cam, is

$$T = F_{23} e \sin \omega t.$$

Then substituting Eqs. (18.2) into this equation gives

$$T = [(m\omega^2 - k)e \cos \omega t + (ke + P)]e \sin \omega t,$$

which can be written, through a trigonometric identity, as

$$T = e(ke + P)\sin\omega t + {}^1\!/\!_2 e^2(m\omega^2 - k)\sin 2\omega t. \tag{18.5}$$

A plot of this equation is presented in Fig. 18.3*b*. Note that the torque consists of a double-frequency component, whose amplitude is a function of the cam velocity squared, superimposed on a single-frequency component, whose amplitude is independent of velocity. In this example, the area of the torque-displacement diagram in the positive T direction is the same as in the negative T direction. This means that the energy required to drive the follower in the forward direction is recovered when the follower returns. A flywheel, or inertia, on the cam shaft can be used to store and release this fluctuating energy requirement. Of course, if an external load is connected in some manner to the follower system, the energy required to drive this load will raise the torque curve in the positive direction and increase the area in the positive loop of the T curve.

EXAMPLE 18.1

A cam-and-follower mechanism similar to Fig. 18.2*a* has the cam machined so that it will move the follower to the right through a distance of 40 mm with parabolic motion in 120° of cam rotation, dwell for 30°, then return with parabolic motion to the starting position in the remaining cam angle. The spring rate is 5 kN/m, and the mechanism is assembled with a 35-N preload. The follower mass is 18 kg. Assume no friction. (*a*) Without computing numeric values, sketch approximate graphs of the displacement, the acceleration, and the cam-contact force, all versus cam angle for the full cycle of events from $\theta = 0°$ to $\theta = 360°$ of cam rotation. On this graph, indicate where jump or liftoff is most likely to begin. (*b*) At what speed would jump begin? Show all calculations.

SOLUTION

a. The cam-contact force can be written from Eq. (*a*) as

$$F = ky + P + m\ddot{y}, \tag{1}$$

which is composed of the term $m\ddot{y}$ that varies with the acceleration, the term ky that varies with the displacement, and the constant term P. Figure 18.4 illustrates the displacement diagram of the cam motion described, along with the acceleration $\ddot{y}$, and the contact force F. Note that jump will first occur at $\theta = \omega t = 60°$ because this is the first position where F approaches zero.

b. Liftoff will occur at the half-point of the rise where (for $\beta = 120°$) the cam angle is $\theta = \beta/2 = 60°$ when the acceleration becomes negative. The acceleration at this position [see Eq. (6.6*c*)] is

$$\ddot{y} = -\frac{4L\omega^2}{\beta^2} = -\frac{4(0.040\text{ m})\omega^2}{[120°\ (\pi\text{ rad}/180°)]^2} = (-0.0365\text{ m/rad}^2)\omega^2.$$

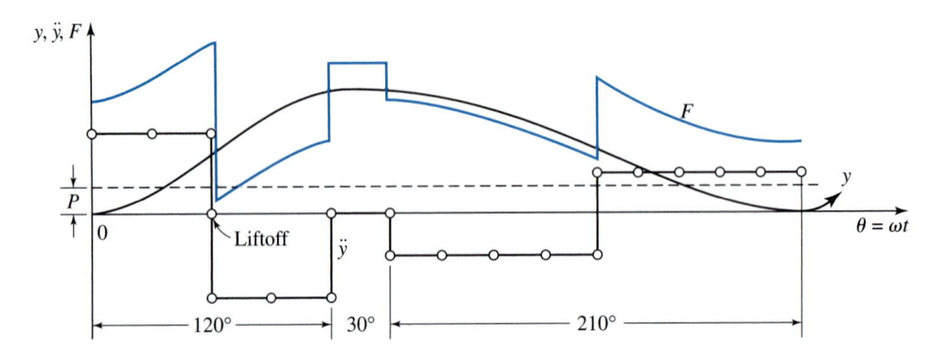

Figure 18.4 Example 18.1.

Substituting this value, $P = 35$ N, and $ky = (5 \text{ kN/m})(20 \text{ mm}) = 100$ N into Eq. (1) with $F = 0$ gives

$$0 = 100 \text{ N} + 35 \text{ N} + (18 \text{ kg})(-0.0365 \text{ m/s}^2)\omega^2.$$

Then rearranging this equation gives

$$\omega = \sqrt{\frac{100 \text{ N} + 35 \text{ N}}{(18 \text{ kg})(0.0365 \text{ m})}} = 14.3 \text{ rad/s} = 137 \text{ rev/min}. \qquad \textit{Ans.}$$

18.3 EFFECT OF SLIDING FRICTION

Let F_μ be the force of sliding (Coulomb) friction as defined by Eq. (13.10). Because the friction force is always opposite in direction to the velocity, let us define a sign function as follows:

$$\operatorname{sgn} \dot{y} = \begin{cases} +1 & \text{for } \dot{y} \geq 0 \\ -1 & \text{for } \dot{y} < 0 \end{cases}. \tag{18.6}$$

With this notation, Eq. (*a*) of Section 18.2 can be written as

$$\sum F^y = F_{23} - F_\mu \operatorname{sgn} \dot{y} - k(y + \delta) - m\ddot{y} = 0$$

or

$$F_{23} = F_\mu \operatorname{sgn} \dot{y} + k(y + \delta) + m\ddot{y}. \tag{18.7}$$

This equation is plotted for simple harmonic motion with no dwells in Fig. 18.5. By studying both parts of this diagram we note that F_μ is positive when $\dot{y}$ is positive and we see how F_{23} is obtained by graphically summing the four component curves.

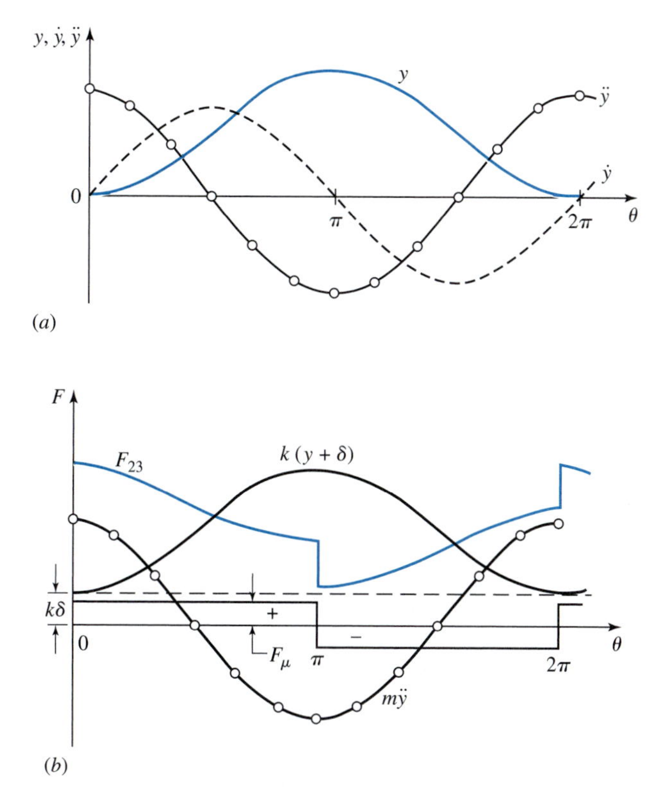

Figure 18.5 Effect of sliding friction on a cam system with harmonic motion: (*a*) graph of displacement, velocity, and acceleration for one motion cycle; (*b*) graph of force components F_μ, $k\delta$, $k(y+\delta)$, $m\ddot{y}$, and the resultant contact force F_{23}.

18.4 ANALYSIS OF DISK CAM WITH RECIPROCATING ROLLER FOLLOWER

In Chapter 13 we analyzed a cam system incorporating a reciprocating roller follower. In this section we present an analytical approach to a similar problem in which sliding friction is also included. The geometry of such a system is illustrated in Fig. 18.6*a*. In the analysis to follow, the effect of follower weight on bearings *B* and *C* is neglected.

Figure 18.6*b* is a free-body diagram of the follower and roller. If $y(\theta)$ is some motion machined into the cam and $\theta = \omega t$ is the cam angle, at $y = 0$ the follower is at the bottom of its stroke and so $O_2A = R + r$. Therefore,

$$a = R + r + y. \tag{18.8}$$

In Fig. 18.6*b*, the roller contact force forms an angle ϕ, the pressure angle, with the y axis. Because the direction of the force $\mathbf{F}_{23}$ is the same as the normal to the contacting surfaces, the intersection of this line with the x axis is the common instant center of the cam and follower. This means that the velocity of this point is the same no matter whether it is

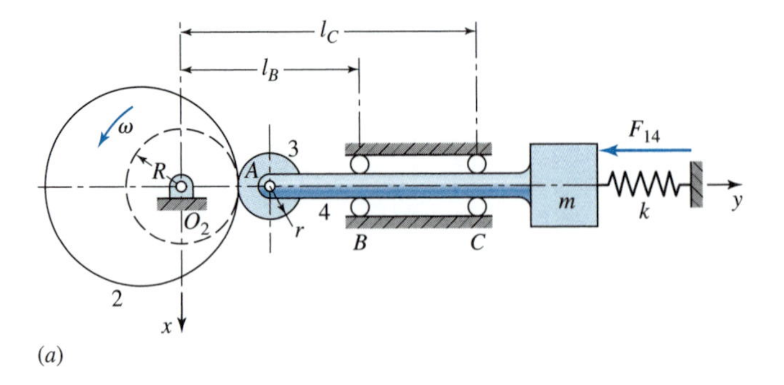

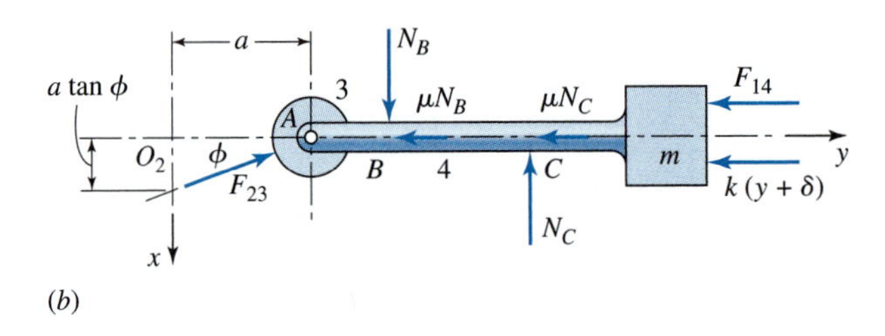

Figure 18.6 (*a*) Plate cam driving a reciprocating roller–follower system. (*b*) Free-body diagram of the follower system.

considered a point on the follower or a point on the cam. Therefore,

$$\dot{y} = a\omega \tan \phi$$

and so

$$\tan \phi = \frac{\dot{y}}{a\omega} = \frac{y'}{a}, \tag{18.9}$$

where y' is the first-order kinematic coefficient of the cam motion.

In the analysis to follow, the two bearing reactions are N_B and N_C, the coefficient of sliding friction is μ, and δ is the precompression of the retaining spring. Summing forces in the x and y directions gives

$$\sum F^x_{i3,4} = -F^x_{23} + N_B - N_C = 0 \tag{a}$$

and

$$\sum F^y_{i3,4} = F^y_{23} - \mu \operatorname{sgn} \dot{y}(N_B + N_C) - F_{14} - k(y+\delta) - m\ddot{y} = 0. \tag{b}$$

A third equation is obtained by taking moments about A:

$$\sum M_A = -N_B(l_B - a) + N_C(l_c - a) = 0. \tag{c}$$

With the help of Eq. (18.9), these three equations can be solved for the unknowns F_{23}, N_B, and N_C.

First, we solve Eq. (c) for N_C. This gives

$$N_C = N_B \frac{l_B - a}{l_C - a}. \tag{d}$$

Now we substitute Eq. (d) into Eq. (a) and solve for the bearing reaction N_B. The result is

$$N_B = \frac{l_C - a}{l_C - l_B} F_{23}^x. \tag{e}$$

Substituting $F_{23}^x = F_{23}^y \tan\phi$, into this equation and using Eq. (18.9) gives

$$N_B = \frac{F_{23}^y (l_C - a)\tan\phi}{l_C - l_B} = \frac{(l_C - a)y'}{(l_C - l_B)a} F_{23}^y. \tag{18.10}$$

Next, we substitute Eqs. (d) and (18.10) into the friction term of Eq. (b):

$$\mu \operatorname{sgn} \dot{y}(N_B + N_C) = y'\mu \operatorname{sgn} \dot{y} \left[\frac{l_C + l_B - 2a}{(l_C - l_B)a}\right] F_{23}^y. \tag{f}$$

Substituting this result back into Eq. (b) and solving for F_{23}^y gives

$$F_{23}^y = \frac{F_{14} + k(y + \delta) + m\ddot{y}}{1 - y'\mu \operatorname{sgn} \dot{y}\left[\frac{l_C + l_B - 2a}{(l_C - l_B)a}\right]}. \tag{18.11}$$

For computer or calculator solution, a simple computation for the sgn function is

$$\operatorname{sgn} \dot{y} = \frac{\dot{y}}{|\dot{y}|}. \tag{18.12}$$

Finally, the cam-shaft torque is

$$T = -a \tan\phi F_{23}^y = -y' F_{23}^y. \tag{18.13}$$

The equations of this section require the kinematic expressions for the appropriate rise and return motions, developed in Chapter 6.

18.5 ANALYSIS OF ELASTIC CAM SYSTEMS

Figure 18.7 illustrates the effect of follower elasticity upon the actual measured displacement and velocity of a follower system driven by a cycloidal cam. To see what has happened, we must compare these diagrams with the theoretical ones in Chapter 6. Although the effect of elasticity is more pronounced for the velocity, it is usually the variation of the displacement, especially at the top of rise, that causes the most trouble in practical situations. These troubles are usually evidenced by poor or unreliable product quality when elastic systems are used in manufacturing or assembly lines, and they result in noise, unusual wear, and fatigue failure.

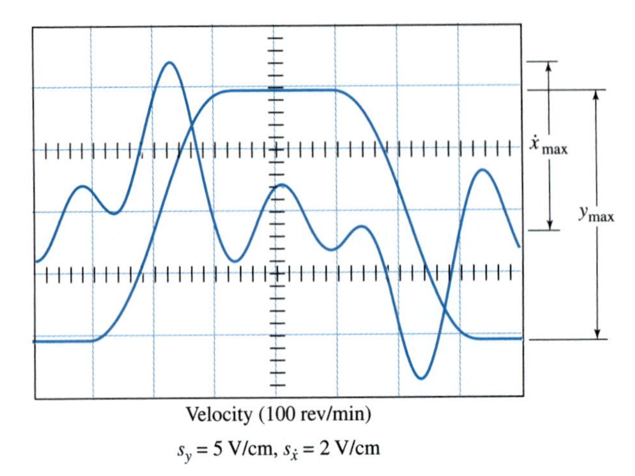

Figure 18.7 Photograph of the oscilloscope traces of the measured displacement and velocity of a dwell–rise–dwell–return cam-and-follower system machined for cycloidal motion. The zero axis of the displacement diagram has been translated downward to obtain a larger diagram in the space available.

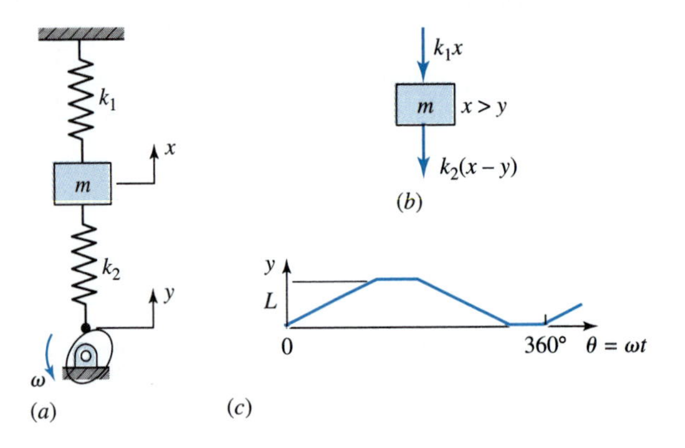

Figure 18.8 (*a*) Undamped model of a cam-and-follower system. (*b*) Free-body diagram of the follower mass. (*c*) Displacement diagram.

A complete analysis of elastic cam systems requires a good background in vibration analysis. To avoid the necessity for this background while still developing a basic understanding, we will use an extremely simplified cam system with a linear-motion cam. It must be observed, however, that such a cam system would never be used for real high-speed applications.

In Fig. 18.8*a*, k_1 is the stiffness of the retaining spring, m is the lumped mass of the follower, and k_2 represents the stiffness of the follower. Because the follower is usually a rod or a lever, k_2 is many times greater than k_1.

Spring k_1 is assembled with a preload. The coordinate x of the follower motion is chosen at the equilibrium position of the mass after spring k_1 is assembled. Thus, k_1 and k_2 exert equal and opposite preload forces on the mass. Assuming no friction, the free-body diagram of the mass is as illustrated in Fig. 18.8*b*. To determine the directions of the forces the coordinate x, representing the motion of the follower, has been assumed to be larger than the coordinate y, representing the motion machined into the cam. However, the same result is obtained if y is assumed to be larger than x.

Using Fig. 18.8*b*, we find the equation of motion to be

$$\sum F = -k_1 x - k_2(x - y) - m\ddot{x} = 0 \tag{a}$$

or

$$\ddot{x} + \frac{k_1 + k_2}{m} x = \frac{k_2}{m} y. \tag{18.14}$$

This is the differential equation for the motion of the follower. It can be solved as indicated in Chapter 15 when the function y is specified. This equation can be solved piecewise for each cam segment; that is, the ending conditions for one segment of motion must be used as the beginning or starting conditions for the next segment.

Let us analyze the first segment of motion using uniform motion, as illustrated in Fig. 18.8*c*. First we use the notation

$$\omega_n = \sqrt{\frac{k_1 + k_2}{m}}. \tag{18.15}$$

We should not confuse ω_n with the angular velocity of the cam ω. The quantity ω_n is called the *undamped natural frequency*. The units of ω_n and ω are both radians per second.

Equation (18.14) can now be written as

$$\ddot{x} + \omega_n^2 x = \frac{k_2 y}{m}. \tag{18.16}$$

The solution to this equation is

$$x = A \cos \omega_n t + B \sin \omega_n t + \frac{k_2 y}{m\omega_n^2}, \tag{b}$$

where, for the linear rise segment of duration β, the cam motion is

$$y = \frac{L}{\beta}\theta = \frac{L\omega t}{\beta}. \tag{18.17}$$

Of course Eq. (18.17) is valid only during the rise segment. We can verify Eq. (*b*) as the solution by substituting it and its second derivative into Eq. (18.16).

The first derivative of Eq. (*b*) is

$$\dot{x} = -A\omega_n \sin \omega_n t + B\omega_n \cos \omega_n t + \frac{k_2 \dot{y}}{m\omega_n^2}. \tag{c}$$

For $t = 0$ at the beginning of rise with $x = \dot{x} = 0$, we find from Eqs. (*b*) and (*c*) that

$$A = 0 \text{ and } B = -\frac{k_2 \dot{y}}{m\omega_n^3}.$$

Thus, Eq. (*b*) becomes

$$x = \frac{k_2}{m\omega_n^2}\left(y - \frac{\dot{y}}{\omega_n} \sin \omega_n t\right). \tag{18.18}$$

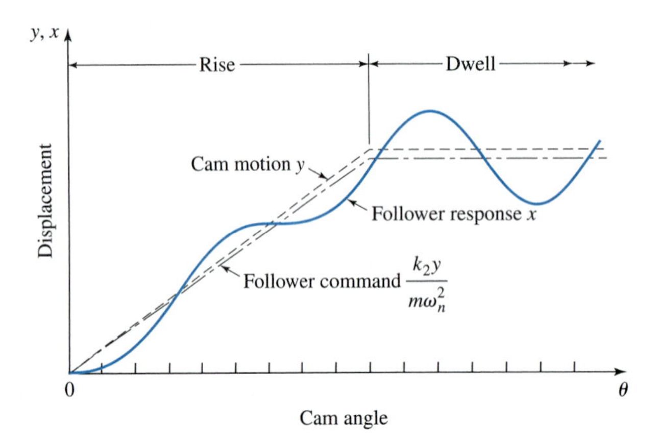

Figure 18.9 Displacement diagram of a uniform-motion cam mechanism showing the follower response.

This equation is plotted in Fig. 18.9. Note that the motion consists of a negative sine term superimposed upon a ramp representing the uniform rise. Because of the additional compression of spring k_2 during the rise, the ramp term $k_2 y / m\omega_n^2$, called the follower command in Fig. 18.9, becomes less than the intended cam rise motion y.

After the end of rise, Eqs. (18.16) through (18.18) are no longer valid, and a second segment of motion, a dwell, begins. The follower response for this segment is illustrated in Fig. 18.9, but we will not solve for it.

Equation (18.18) demonstrates that the vibration amplitude $\dot{y}/\omega_n$ can be reduced by making ω_n large, and Eq. (18.15) demonstrates that this can be done by increasing k_2, which means that a very rigid follower should be used.

18.6 UNBALANCE, SPRING SURGE, AND WINDUP

As illustrated in Fig. 18.10*a*, a disk cam produces unbalance because its mass is not symmetric about the axis of rotation. This means that two sets of vibratory forces exist, one caused by the eccentric cam mass and the other caused by the reaction of the follower against the cam. By keeping these effects in mind during design, the engineer can do much to guard against difficulties during operation.

Figures 18.10*b* and 18.10*c* illustrate that face and cylindrical cams have good balance characteristics. For this reason, these are good choices when high-speed operation is involved.

Spring Surge Texts on spring design demonstrate that helical springs may themselves vibrate because of the mass of the spring coils. This vibration within the retaining spring, called *spring surge*, has been photographed with high-speed cameras and the results have been exhibited in slow motion. When serious vibrations exist, a clear wave motion can be seen traveling up and down the spring. For example, poorly designed automotive valve springs operating near their critical frequency permit the valve to open for short intervals during the period the valve is supposed to be closed. Such conditions result in poor operation of the engine and rapid fatigue failure of the springs themselves.

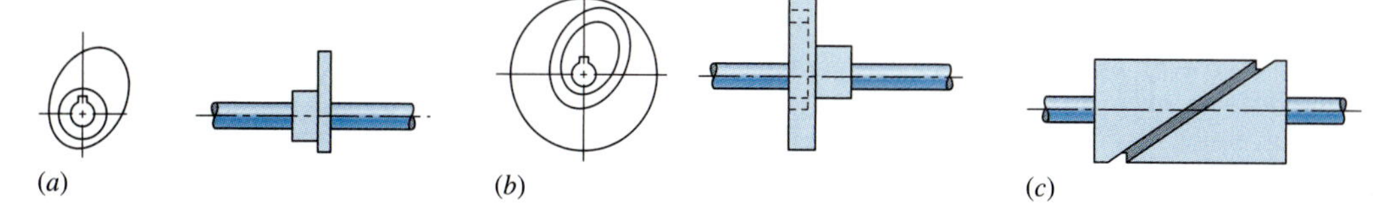

Figure 18.10 (*a*) A disk cam is inherently unbalanced. (*b*) A face cam is usually well balanced. (*c*) A cylindrical cam has good balance.

Windup Figure 18.3*b* is a plot of cam-shaft torque illustrating that the shaft exerts torque on the cam during a portion of the cycle and that the cam exerts torque on the shaft during another portion of the cycle. This varying torque requirement may cause the shaft to twist, or wind up, as the torque increases during follower rise. Also, during this period the angular cam velocity is slowed and so is the follower velocity. Near the end of rise, the energy stored in the shaft by the windup is released, causing both the follower velocity and the acceleration to rise above normal values. The resulting kick may produce follower jump or impact. This effect is most pronounced when heavy loads are being moved by the follower, when the follower moves at high speed, and when the shaft is flexible.

In most cases a flywheel must be employed in a cam system to provide for varying torque requirements. Cam-shaft windup can be prevented to a large extent by mounting the flywheel as close as possible to the cam. Mounting it a long distance from the cam may actually make matters worse.

PROBLEMS

18.1 In Fig. P18.1*a*, the mass m is constrained to move only in the vertical direction. The circular cam has an eccentricity of 2 in, a speed of 20 rad/s, and a weight of 8 lb. Neglecting friction, find the angle $\theta = \omega t$ at the instant the cam jumps.

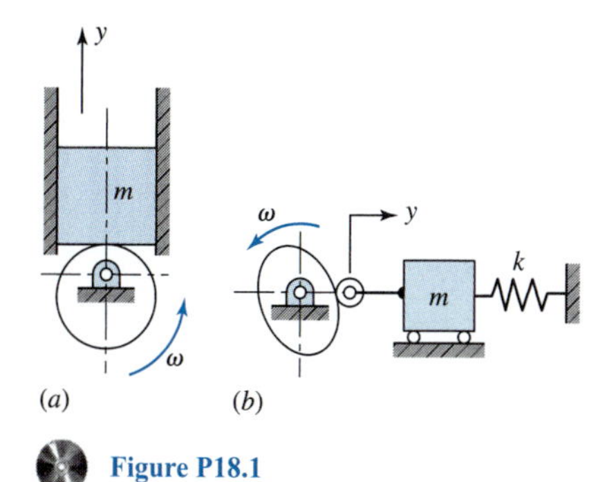

Figure P18.1

18.2 In Fig. P18.1*a*, the mass m is driven up and down by the eccentric cam and it has a weight of 10 lb. The cam eccentricity is 1 in. Assume no friction.

(a) Derive the equation for the contact force.

(b) Find the cam velocity ω corresponding to the beginning of the cam jump.

18.3 In Fig. P18.1*a*, the slider has a mass of 2.5 kg. The cam is a simple eccentric and causes the slider to rise 25 mm with no friction. At what cam speed in revolutions per minute will the slider first lose contact with the cam? Sketch a graph of the contact force at this speed for 360° of cam rotation.

18.4 The cam-and-follower system illustrated in Fig. P18.1b has $k = 1$ kN/m, $m = 0.90$ kg, $y = 15 - 15\cos\omega t$ mm, and $\omega = 60$ rad/s. The retaining spring is assembled with a preload of 2.5 N.

(a) Compute the maximum and minimum values of the contact force.

(b) If the follower is found to jump off the cam, compute the angle $\theta = \omega t$ corresponding to the very beginning of jump.

18.5 Figure P18.1b illustrates the model of a cam-and-follower system. The motion machined into the cam is to move the mass to the right through a distance of 2 in with parabolic motion in 150° of cam rotation, dwell for 30°, return to the starting position with simple harmonic motion, and dwell for the remaining 30° of cam angle. There is no friction or damping. The spring rate is 40 lb/in and the spring preload is 6 lb, corresponding to the $y = 0$ position. The mass is 36 lb.

(a) Sketch a displacement diagram showing the follower motion for the entire 360° of cam rotation. Without computing numerical values, superimpose graphs of the acceleration and cam contact force onto the same axes. Show where jump is most likely to begin.

(b) At what speed in revolutions per minute would jump begin?

18.6 A cam-and-follower mechanism is illustrated in abstract form in Fig. P18.1b. The cam is cut so that it causes the mass to move to the right a distance of 25 mm with harmonic motion in 150° of cam rotation, dwell for 30°, and then return to the starting position in the remaining 180° of cam rotation, also with harmonic motion. The spring is assembled with a 22 N preload and it has a rate of 4.4 kN/m. The follower mass is 17.5 kg. Compute the cam speed in revolutions per minute at which jump would begin.

18.7 Figure P18.7 illustrates a lever OAB driven by a cam cut to give the roller a rise of 1 in with parabolic motion and a parabolic return with no dwells. The lever and roller are to be assumed weightless, and there is no friction. Calculate the jump speed if $l = 5$ in and the mass weighs 5 lb.

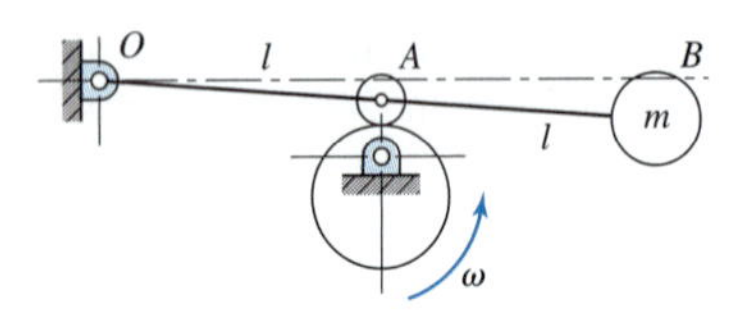

Figure P18.7

18.8 A cam-and-follower system similar to the one in Fig. 18.6 uses a plate cam driven at a speed of 600 rev/min and employs simple harmonic rise and parabolic return motions. The events are rise in 150°, dwell for 30°, and return in 180°. The retaining spring has a rate $k = 14$ kN/m with a precompression of 12.5 mm. The follower has a mass of 1.6 kg. The external load is related to the follower motion y by the equation $F = 0.325 - 10.75y$, where y is in meters and F is in kilonewtons. Dimensions corresponding to Fig. 18.6 are $R = 20$ mm, $r = 5$ mm, $l_B = 60$ mm, and $l_C = 90$ mm. Using a rise of $L = 20$ mm and assuming no friction, plot the displacement, cam-shaft torque, and radial component of the cam force for one complete revolution of the cam.

18.9 Repeat Problem 18.8 with a speed of 900 rev/min and $F = 0.110 + 10.75y$ kN, where y is in meters and the coefficient of sliding friction is $\mu = 0.025$.

18.10 A plate cam drives a reciprocating roller follower through the distance $L = 1.25$ in with parabolic motion in 120° of cam rotation, dwells for 30°, and returns with cycloidal motion in 120°, followed by dwells for the remaining cam angle. The external load on the follower is $F_{14} = 36$ lb during the rise and zero during the dwells and the return. In the notation of Fig. 18.6, $R = 3$ in, $r = 1$ in, $l_B = 6$ in, $l_C = 8$ in, and $k = 150$ lb/in. The spring is assembled with a preload of 37.5 lb when the follower is at the bottom of its stroke. The weight of the follower is 1.8 lb and the cam velocity is 140 rad/s. Assuming no friction, plot the displacement, the torque exerted on the cam by the shaft, and the radial component of the contact force exerted by the roller against the cam surface for one complete cycle of motion.

18.11 Repeat Problem 18.10 if friction exists with $\mu = 0.04$ and the cycloidal return takes place in 180°.

19 Flywheels, Governors, and Gyroscopes

19.1 DYNAMIC THEORY OF FLYWHEELS

A flywheel is an energy storage device. It absorbs mechanical energy by increasing its angular velocity and delivers energy by decreasing its angular velocity. Commonly, a flywheel is used to smooth the flow of energy between a power source and its load. If the load happens to be a punch press, for example, the actual punching operation requires energy for only a small fraction of its motion cycle. As another example, if the power source happens to be a two-cylinder, four-stroke engine, the engine delivers energy during only about half of its motion cycle. Other applications involve using a flywheel to absorb braking energy and deliver acceleration energy for an automobile and to act as energy-smoothing devices for electric utilities as well as solar- and wind-power-generating facilities. Electric railways have long used regenerative braking by feeding braking energy back into power lines, but newly introduced and stronger materials now make the flywheel more feasible for such purposes.

Figure 19.1 is a mathematical representation of a flywheel. The flywheel, whose motion is measured by the angular coordinate θ, has a mass moment of inertia I. An input torque T_i, corresponding to a coordinate θ_i, will cause the flywheel speed to increase. A load or output torque T_o, with corresponding coordinate θ_o, will absorb energy from the flywheel and cause it to slow down. If the work going into the system is considered positive and work output is negative, the equation of motion of the flywheel is

$$\sum M = T_i(\theta_i, \dot{\theta}_i) - T_o(\theta_o, \dot{\theta}_o) - I\ddot{\theta} = 0,$$

which can be written as

$$I\alpha = T_i(\theta_i, \omega_i) - T_o(\theta_o, \omega_o). \qquad (a)$$

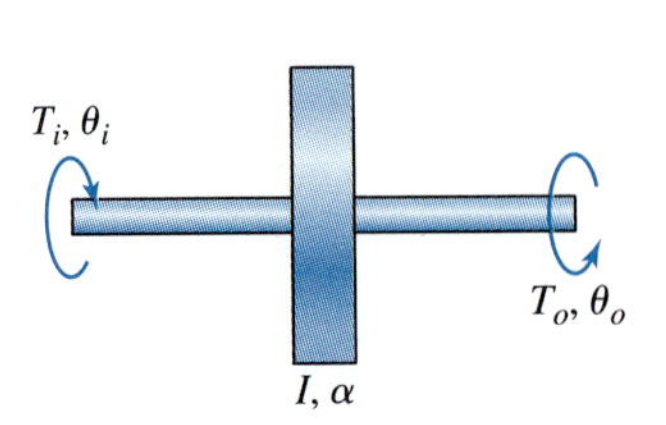

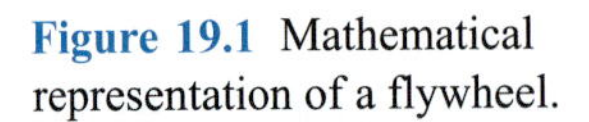
Figure 19.1 Mathematical representation of a flywheel.

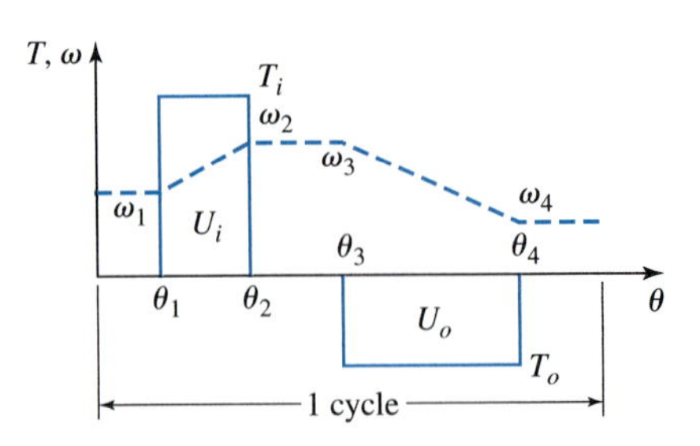

Figure 19.2

Note that both T_i and T_o may depend for their values on their angular positions θ_i and θ_o, as well as their angular velocities ω_i and ω_o. Typically, the torque characteristic depends upon only one of these, either position or velocity. Thus, the torque delivered by an induction motor depends on the speed of the motor. In fact, it is common practice for the manufacturers of electric motors to publish charts detailing the torque-speed characteristics of their motors.

When the input and output torque functions are given, Eq. (*a*) can be solved for the motion of the flywheel using well-known techniques for solving linear and nonlinear differential equations. The resulting equations can easily be solved using the methods of Chapter 15. In Chapter 19 we assume rigid shafting, giving $\theta_i = \theta = \theta_o$. Thus, Eq. (*a*) becomes

$$I\alpha = T_i(\theta, \omega) - T_o(\theta, \omega). \tag{b}$$

When the two torque functions are known and the starting values of the displacement θ and velocity ω are given, Eq. (*b*) can be solved for θ, ω, and α as functions of time. However, typically, we are not really interested in the instantaneous values of the kinematic parameters. Primarily, we want to know the overall performance of the flywheel. The important questions include what is the mass moment of inertia; how do we match the power source to the load to get an optimum size of motor or engine; and what are the resulting performance characteristics of the system?

To gain insight into the problem, consider the hypothetical situation that is diagrammed in Fig. 19.2. An input power source subjects a flywheel to a constant torque T_i, while the output shaft rotates from θ_1 to θ_2. This is a positive torque and is plotted upward. Equation (*b*) indicates that a positive acceleration α will be the result, and so the shaft angular velocity increases from ω_1 to ω_2. As shown, the shaft now rotates from θ_2 to θ_3 with zero torque and, hence, from Eq. (*b*), with zero angular acceleration. Therefore, $\omega_3 = \omega_2$. From θ_3 to θ_4, a load, or output torque, of constant magnitude is applied, causing the shaft to slow down from ω_3 to ω_4. Note that the output torque is plotted in the negative direction in accordance with Eq. (*b*).

The work input to the flywheel is the area of the rectangle between θ_1 and θ_2, or

$$U_i = T_i(\theta_2 - \theta_1). \tag{c}$$

The work output of the flywheel is the area of the rectangle from θ_3 to θ_4, or

$$U_o = T_o(\theta_4 - \theta_3). \tag{d}$$

If U_o is greater than U_i, the load uses more energy than has been delivered to the flywheel and so ω_4 will be less than ω_1. If U_o is equal to U_i, then ω_4 will be equal to ω_1 because the gain and loss are equal; we are assuming no friction losses. Finally, if U_i is greater than U_o, then ω_4 will be greater than ω_1.

We can also write these relations in terms of kinetic energy. At $\theta = \theta_1$, the flywheel has an angular velocity of ω_1 rad/s, and so its kinetic energy is

$$U_1 = \tfrac{1}{2} I \omega_1^2. \tag{e}$$

At $\theta = \theta_2$, the angular velocity is ω_2, and so its kinetic energy is

$$U_2 = \tfrac{1}{2} I \omega_2^2. \tag{f}$$

Therefore, the change in the kinetic energy is

$$U_2 - U_1 = \tfrac{1}{2} I (\omega_2^2 - \omega_1^2). \tag{19.1}$$

19.2 INTEGRATION TECHNIQUE

Many of the torque-displacement functions encountered in practical engineering situations are so complicated that they must be integrated by approximate methods. Figure 19.3, for example, is a plot of the engine torque for one cycle of motion of a single-cylinder engine. Because a part of the torque curve is negative, the flywheel must return part of the energy back to the engine. Approximate integration of this curve for a cycle of 4π rad yields a mean torque T_m available to drive a load.

A simple integration routine is Simpson's rule (Section 15.17); this approximation can be handled on any computer and is simple enough for even the crudest programmable

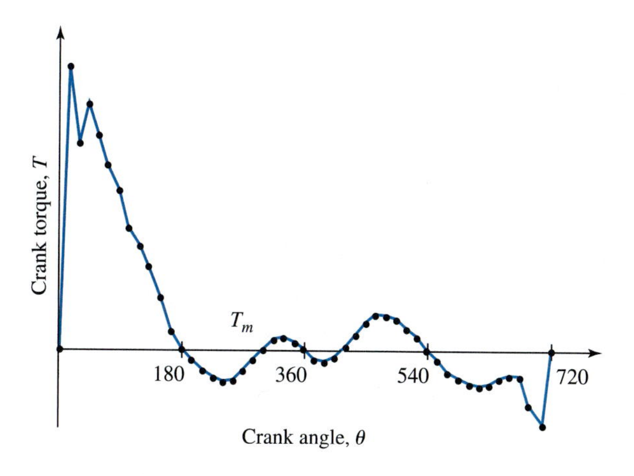

Figure 19.3 Relation between torque and crank angle for a one-cylinder, four-stroke internal combustion engine.

calculators. In fact, this routine is often found as a part of the library for many calculators and personal computers. The equation used is

$$\int_{x_0}^{x_n} f(x)dx = \frac{h}{3}(f_0 + 4f_1 + 2f_2 + 4f_3 + 2f_4 + \cdots + 2f_{n-2} + 4f_{n-1} + f_n), \tag{19.2}$$

where

$$h = (x_n - x_0)/n \quad \text{with} \quad x_n > x_0$$

and n is the number of subintervals used, $2, 4, 6, \ldots$, which must be even. If memory is limited, Eq. (19.2) can be solved in two or more steps, say from 0 to $n/2$ and then from $n/2$ to n.

For flywheel applications it is convenient to define a *coefficient of speed fluctuation* as

$$C_s = \frac{\omega_2 - \omega_1}{\omega}, \tag{19.3}$$

where ω is the nominal or average angular velocity, given by

$$\omega = \tfrac{1}{2}(\omega_2 + \omega_1). \tag{19.4}$$

Equation (19.1) can be factored to give the change in the kinetic energy as

$$U_2 - U_1 = \tfrac{1}{2}I(\omega_2 - \omega_1)(\omega_2 + \omega_1).$$

Substituting Eqs. (19.3) and (19.4) into this equation gives

$$U_2 - U_1 = C_s I \omega^2, \tag{19.5}$$

which can be used to obtain an appropriate flywheel inertia corresponding to the change in the kinetic energy.

EXAMPLE 19.1

Table 19.1 lists values of the torque plotted in Fig. 19.3. The nominal speed of the engine is to be 250 rad/s.

(a) Integrate the torque-displacement function for one cycle and find the energy that can be delivered to the load during the cycle.
(b) Determine the mean torque T_m (see Fig. 19.3).
(c) The greatest energy fluctuation will occur approximately between $\theta = 15°$ and $\theta = 150°$ on the $T_i - T_o$ diagram (see Fig. 19.3) and note that $T_o = -T_m$. Using a coefficient of speed fluctuation of $C_s = 0.1$, find a suitable value for the flywheel inertia.
(d) Find ω_2 and ω_1.

TABLE 19.1 Example 19.1: Torque data for Figure 19.3

θ_i deg	T_i in · lb	θ_i deg	T_i in · lb	θ_i deg	T_i in · lb	θ_i deg	T_i in · lb	θ_i deg	T_i in · lb
0	0	150	532	300	−8	450	242	600	−355
15	2800	165	184	315	89	465	310	615	−371
30	2090	180	0	330	125	480	323	630	−362
45	2430	195	−107	345	85	495	280	645	−312
60	2160	210	−206	360	0	510	206	660	−272
75	1840	225	−280	375	−85	525	107	675	−274
90	1590	240	−323	390	−125	540	0	690	−548
105	1210	255	−310	405	−89	555	−107	705	−760
120	1066	270	−242	420	8	570	−206		
135	803	285	−126	435	126	585	−292		

SOLUTION

(a) Using $n = 48$ and $h = 4\pi/48$, we enter the data of Table 19.1 into our calculator and compute the integral by Simpson's rule as defined in Eq. (19.2); the amount of energy that can be delivered to the load during the cycle is

$$U = 3490 \text{ in} \cdot \text{lb}. \qquad \textit{Ans.}$$

(b) The mean torque is

$$T_m = 3490 \text{ in.} \cdot \text{lb}/4\pi = 278 \text{ in} \cdot \text{lb}. \qquad \textit{Ans.}$$

The largest positive loop in the torque-displacement diagram occurs between $\theta = 0°$ and $\theta = 180°$. We select this loop as yielding the largest speed change. Subtracting 278 in.·lb from the values in Table 19.1 for this loop gives, respectively,

$$-278, 2\,522, 1\,812, 2\,152, 1\,882, 1\,562, 1\,312, 932, 788, 535, 254, -94,$$
$$\text{and } -278 \text{in} \cdot \text{lb}.$$

Entering the data into Simpson's rule again, using $n = 12$ and $h = \pi/12$, the change in kinetic energy is

$$U_2 - U_1 = 3663 \text{ in.} \cdot \text{lb}.$$

Rearranging Eq. (19.5), the mass moment of inertia of the flywheel can be written as

$$I = \frac{U_2 - U_1}{C_s \omega^2}.$$

Substituting the known information into this equation, a suitable value for the flywheel inertia is

$$I = \frac{3663 \text{ in} \cdot \text{lb}}{0.1(250 \text{ rad/s})^2} = 0.586 \text{ in} \cdot \text{lb} \cdot \text{s}^2. \qquad \textit{Ans.}$$

(c) Equations (19.3) and (19.4) can now be solved simultaneously for ω_2 and ω_1. Substituting appropriate values into these two equations yields

$$\omega_2 = \tfrac{1}{2}(2 + C_s)\omega = \tfrac{1}{2}(2 + 0.1)250 \text{ rad/s} = 262.5 \text{ rad/s} \qquad \textit{Ans.}$$

$$\omega_1 = 2\omega - \omega_2 = 2(250 \text{ rad/s}) - 262.5 \text{ rad/s} = 237.5 \text{ rad/s}. \qquad \textit{Ans.}$$

These two speeds occur at $\theta = 180°$ and $\theta = 0$, respectively.

19.3 MULTICYLINDER ENGINE TORQUE SUMMATION

After the torque-displacement relation has been defined for a single cylinder, it is easy to assume these for multicylinder engines. If, for example, we wish to find the torque function for a three-cylinder engine, then we examine the firing order illustrated in Fig. 16.1 and write

$$T_{total} = T_\theta + T_{\theta+240} + T_{\theta+480} \qquad (a)$$

because the torque events are spaced 240° apart. Here θ is the crank angle of the first cylinder.

Applying this same procedure for a four-cylinder engine, along with using the torque values in Table 19.1, gives the results in Table 19.2. Figure 19.4 illustrates how this torque

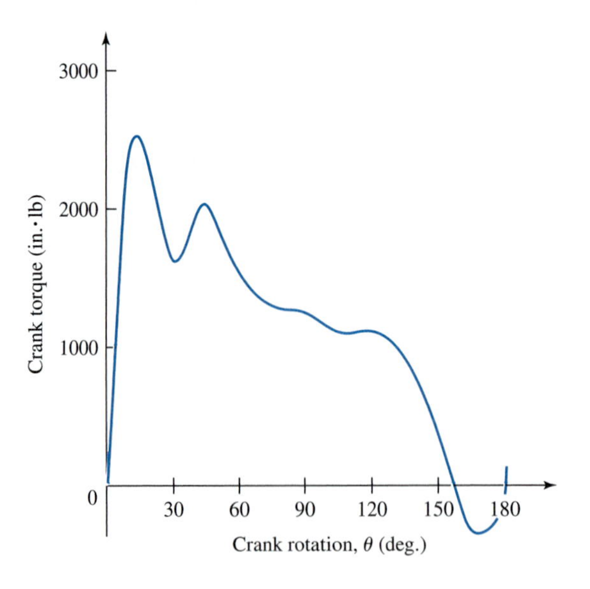

Figure 19.4 Relation between torque and crank angle for a four-cylinder, four-stroke internal combustion engine.

varies during each 180° of crank rotation.

TABLE 19.2 Torque data for a four-cylinder, four-cycle internal combustion engine

θ_i deg	T_θ in · lb	$T_{\theta+180}$ in · lb	$T_{\theta+360}$ in · lb	$T_{\theta+540}$ in · lb	T_{total} in · lb
0	0	0	0	0	0
15	2800	−107	−85	−107	2501
30	2090	−206	−125	−206	1553
45	2430	−280	−89	−292	1769
60	2160	−323	8	−355	1490
75	1840	−310	126	−371	1285
90	1590	−242	242	−362	1228
105	1210	−126	310	−312	1082
120	1066	−8	323	−272	1109
135	803	89	280	−274	898
150	532	125	206	−548	315
165	184	85	107	−760	−384

19.4 CLASSIFICATION OF GOVERNORS

So far in Chapter 19 we have learned that flywheels are used to regulate speed over short intervals of time, such as a single revolution or the duration of an engine cycle. *Governors*, too, are devices which are used to regulate speed. In contrast to flywheels, however, governors are used to regulate speed over much longer intervals of time; in fact, they are intended to maintain a balance between the energy supplied to a moving system and the external load or resistance applied to that system.

When the speed of a machine must, over its lifetime, always be kept at the same level, or approximately so, then a shaft-mounted mechanical device may be an appropriate speed regulator. Such governors may be classified as follows:

- Centrifugal governors or
- Inertia governors.

As the name indicates, centrifugal force plays the dominant role in a centrifugal governor. In inertia governors it is the angular acceleration, or change in speed, that dominates the regulating action.

The availability today of a wide variety of low-priced solid-state electronic devices and transducers makes it possible to regulate mechanical systems to a finer degree and at less cost than with the older all-mechanical governors. The electronic governor also has the advantage that the speed to be regulated can be changed quite easily and at will.

19.5 CENTRIFUGAL GOVERNORS

Figure 19.5 illustrates a simple spring-controlled centrifugal shaft governor. Masses attached to the bell-crank levers are driven outward by centrifugal force against springs. The motion of the shaft-mounted sleeve is dependent on the motions of the masses and the ratio of the bell-crank lengths.

Selecting the nomenclature of Fig. 19.6, we let

k = spring rate,

P = spring force,

W = sleeve weight acting along the vertical shaft,

r = instantaneous radial position of mass center, and

a = position of mass center at zero spring force.

Then the spring force at any position r is

$$P = k(r - a). \tag{a}$$

Taking moments about pivot O gives

$$\sum M_O = Pl_2 \cos\alpha + \frac{W}{2} l_1 \cos\alpha + (-m\ddot{r}) l_2 \cos\alpha = 0. \tag{b}$$

Now, with $\ddot{r} = r\omega^2$ and $m = W/g$, this gives

$$\frac{W}{g} r\omega^2 l_2 = Pl_2 + \frac{W}{2} l_1. \tag{c}$$

The controlling force can be written as

$$F = \frac{W}{g} r\omega^2. \tag{d}$$

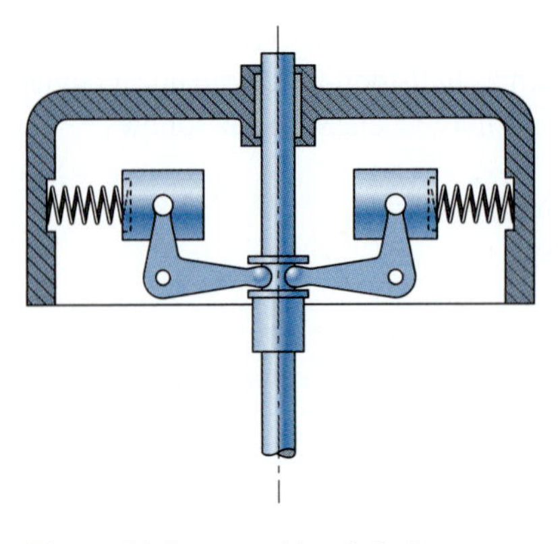

Figure 19.5 A centrifugal shaft governor.

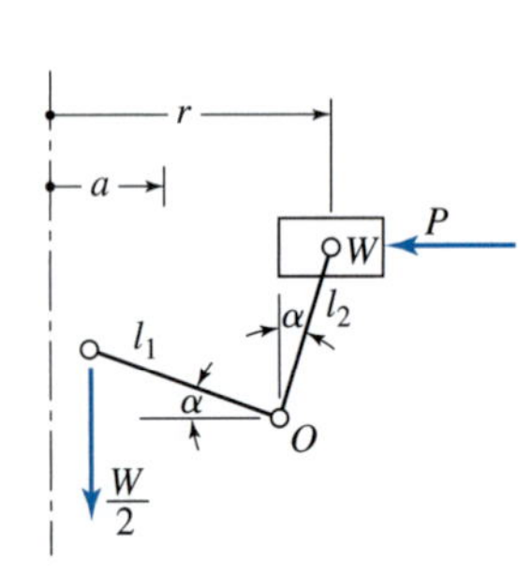

Figure 19.6

Then substituting this expression into Eq. (c), and using Eq. (a), the controlling force is

$$F = P + \frac{W}{2}\frac{l_1}{l_2} = k(r - a) + \frac{W}{2}\frac{l_1}{l_2}. \tag{19.6}$$

Equating Eqs. (d) and (19.6) gives

$$\frac{W}{g}r\omega^2 = k(r - a) + \frac{W}{2}\frac{l_1}{l_2}.$$

Then, rearranging this equation, the corresponding angular velocity can be written as

$$\omega^2 = \frac{kg}{Wr}\left[r - \left(a - \frac{W}{2k}\frac{l_1}{l_2}\right)\right]. \tag{19.7}$$

19.6 INERTIA GOVERNORS

In a centrifugal governor an increase in speed causes the rotating masses to move radially outward, as we have seen. Thus, it is the radial (normal) component of the acceleration that is primarily responsible for creating the controlling force. In an inertia shaft governor, as illustrated in Fig. 19.7, the mass pivot is located at point A very close to the shaft center at O. Thus, the radial component of acceleration is smaller and much less effective. But a sudden change of speed, producing an angular acceleration, will cause the masses to lag and produce tension in the spring. This spring tension produces a transverse (tangential) acceleration force. Consequently, it is the angular acceleration that determines the position of the masses.

Compared with a centrifugal governor, the inertia type is more sensitive because it acts at the very beginning of a speed change.

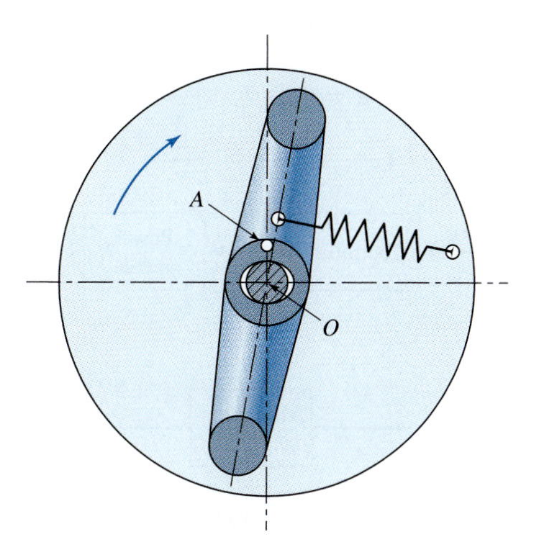

Figure 19.7 An inertia governor.

19.7 MECHANICAL CONTROL SYSTEMS

Many mechanical control systems are represented schematically, as illustrated in Fig. 19.8. Here, θ_i and θ_o represent any set of input and output functions. In the case of a governor, θ_i represents the desired speed and θ_o the actual output speed. For control systems in general, the input and output functions could represent force, torque, or displacement, either rectilinear or angular.

The system illustrated in Fig. 19.8 is called a *closed-loop* or *feedback control system* because the output value θ_o is fed back to the detector at the input so as to measure error ε which is the difference between the input and the output. The purpose of the controller is to cause this error to become as close to zero as possible. The mechanical characteristics of the system, that is, the mechanical clearances, friction, inertias, and stiffnesses, sometimes cause the output to differ somewhat from the input, and so it is the designer's responsibility to examine these mechanical effects in an effort to minimize the error for all operating conditions.

The differential equation for a control system is always written as an expression of the dynamic equilibrium of the elements of the system. Thus, for the system of Fig. 19.8, we express mathematically that the inertia torque is equal to the sum of all other torques acting upon the system. If the load has both inertia and viscous damping, then the expression is

$$I\ddot{\theta}_o = -c\dot{\theta}_o + k'f(\varepsilon). \tag{a}$$

The function $f(\varepsilon)$ depends upon the characteristics of the controller and the power source. A very simple system would result if the components had characteristics such that the relation between $f(\varepsilon)$ and ε is linear. For this condition we can make the substitution

$$k'f(\varepsilon) = k\varepsilon. \tag{b}$$

Substituting Eq. (*b*) into Eq. (*a*) and rearranging gives

$$I\ddot{\theta}_o + c\dot{\theta}_o = k\varepsilon. \tag{c}$$

By definition, we have

$$\varepsilon = \theta_i - \theta_o. \tag{d}$$

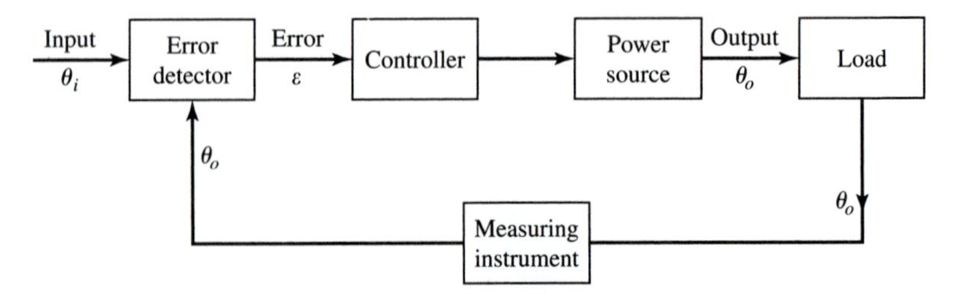

Figure 19.8 Block diagram of a closed-loop control system.

Therefore, Eq. (c) can be written as

$$I\ddot{\theta}_o + c\dot{\theta}_o + k\theta_o = k\theta_i, \tag{19.8}$$

This second-order differential equation can be simplified by the following substitutions:

$$\omega_n = \sqrt{\frac{k}{I}} \tag{19.9}$$

$$2\zeta\omega_n = \frac{c}{I} \tag{19.10}$$

and

$$\zeta = \frac{c}{2\sqrt{kI}}, \tag{19.11}$$

where

I = mass moment of inertia,

c = torsional damping coefficient,

k = torsional stiffness,

ω_n = natural frequency, and

ζ = damping ratio, c/c_{cr}.

After a review of this notation for linear systems in Chapter 15, Eq. (19.8) can be written in the form

$$\ddot{\theta}_o + 2\zeta\omega_n\dot{\theta}_o + \omega_n^2\theta_o = \omega_n^2\theta_i. \tag{19.12}$$

19.8 STANDARD INPUT FUNCTIONS

There is a great deal of useful information that can be gained from a mathematical analysis as well as a laboratory analysis of feedback control systems when standard input functions are used to study their performance. Standard input functions result in differential equations that are much easier to analyze mathematically than they would be if the actual operating conditions were used as input. Furthermore, the use of standard inputs makes it possible to compare the performance of different control systems.

One of the most useful standard input functions is the *unit-step function* illustrated in Fig. 19.9a. This function is not continuous and, consequently, we cannot define initial conditions at $t = 0$. It is customary to specify the conditions at $t = 0+$ and $t = 0-$, where the signs indicate the conditions for values of time slightly greater than or slightly less than zero. Thus, for the unit-step function we have

$$\theta_i = 0, \quad \dot{\theta}_i = 0 \quad \text{when } t = 0-$$

$$\theta_i = 1, \quad \dot{\theta}_i = 0 \quad \text{when } t = 0+.$$

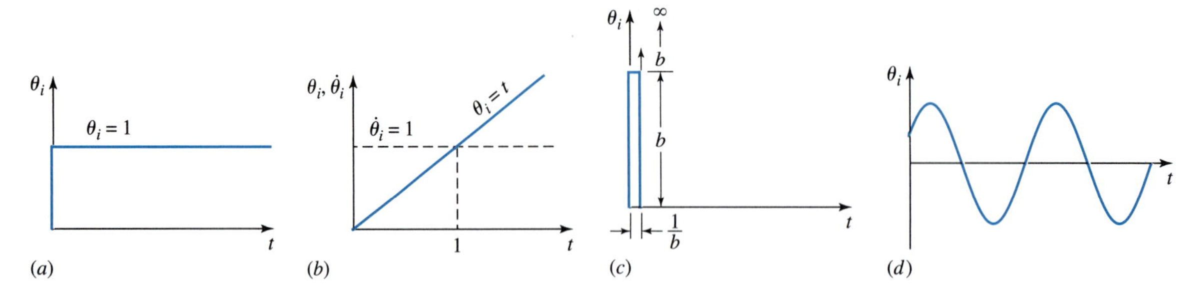

Figure 19.9 Standard input functions. (*a*) Unit-step function. (*b*) Unit-step velocity function. (*c*) Unit-impulse function. (*d*) Harmonic input function.

The performance resulting from the application of these input conditions is called the *unit-step response*.

Another standard input function is the *unit-step velocity function*, illustrated in Fig. 19.9*b*. This function is sometimes called the *unit-slope ramp function* because of the form of its displacement diagram. This function is defined as follows:

$$\theta_i = 0, \quad \dot{\theta}_{\mathrm{i}} = 0 \text{ when } t = 0-$$
$$\theta_i = t, \quad \dot{\theta}_{\mathrm{i}} = 1 \text{ when } t = 0+.$$

Also illustrated in Figs. 19.9*c* and 19.9*d* are the unit-impulse function and the harmonic input function, both of which are useful in studying a system's response.

It sometimes happens in control systems that the input function is relatively constant but the output is subjected to suddenly varying applied forces or torques. For example, an automatically controlled machine tool may run into a deeper cut, causing a greater load torque to be exerted on the machine, thus tending to slow it down. The differential equation for this condition is written as

$$I\ddot{\theta}_o + c\dot{\theta}_o + k\theta_o = k\theta_i - T. \tag{a}$$

When this equation is simplified by substituting the values from Eqs. (19.9), (19.10), and (19.11) and rearranged, it becomes

$$\ddot{\theta}_o + 2\zeta\omega_n\dot{\theta}_o + \omega_n^2\theta_o = \omega_n^2\theta_i - \frac{T}{I}. \tag{19.13}$$

When the solution to Eq. (19.12) is obtained for a given function θ_i, the solution to Eq. (19.13) can be obtained by superposition.

19.9 SOLUTION OF LINEAR DIFFERENTIAL EQUATIONS

In the analysis of feedback control systems, nth-order linear differential equations are frequently encountered. The general form of these equations is

$$a_n \frac{d^n\theta}{dt^n} + a_{n-1} \frac{d^{n-1}\theta}{dt^{n-1}} + \cdots + a_1 \frac{d\theta}{dt} + a_0\theta = f(t). \tag{19.14}$$

The function $f(t)$ is the driving or forcing function, and so t and θ are the independent and dependent variables, respectively. The coefficients $a_0, a_1, \ldots, a_n$ are constants and are independent of t and θ.

The solution to equations of the form of Eq. (19.14) is composed of two parts (Chapter 15). The first part is called the *complimentary solution*, and it is the solution to the homogeneous equation

$$a_n \frac{d^n\theta}{dt^n} + a_{n-1} \frac{d^{n-1}\theta}{dt^{n-1}} + \cdots + a_1 \frac{d\theta}{dt} + a_0\theta = 0. \tag{19.15}$$

The complimentary solution is also called the *transient solution* in the literature of automatic controls because, for a damped stable system, this part of the solution dies away. If the control system should happen to be unstable, then we shall find that the controlled quantity increases without limit. Note that the transient solution is obtained by making the forcing function zero.

The other part of the solution is called the *particular solution* by mathematicians and the *steady-state solution* by control engineers. It is any particular solution of Eq. (19.14).

The complete solution is the sum of the transient and the steady-state solutions. The transient part will contain n arbitrary constants that are evaluated from the initial conditions. The amplitude of the transient solution depends upon both the forcing function and the initial conditions, but all other characteristics of the transient solution are completely independent of the forcing function. We shall find, for linear systems, that neither the forcing function nor the amplitude has any effect on the system stability, this being dependent only on the transient portion of the solution.

The Transient Solution The following steps are used to obtain the transient solution:

1. Set the forcing function equal to zero and arrange the equation in the form of Eq. (19.15).
2. Assume a solution of the form

$$\theta = Ae^{st}. \tag{19.16}$$

3. Substitute Eq. (19.16) and its derivatives into the differential equation, and simplify to obtain the characteristic equation.
4. Solve the characteristic equation for its roots.
5. Obtain the transient solution by substituting the roots back into the assumed solution.

As an example of this procedure, let us solve Eq. (19.12) for the transient solution. Following Step 1, we write

$$\ddot{\theta}_o + 2\zeta\omega_n\dot{\theta}_o + \omega_n^2\theta_o = 0 \tag{a}$$

and for Step 2 we have

$$\theta_o = Ae^{st}, \quad \dot{\theta}_o = Ase^{st}, \quad \ddot{\theta}_o = As^2e^{st}. \tag{b}$$

For Step 3, we substitute Eqs. (b) into Eq. (a). This gives

$$As^2e^{st} + 2\zeta\omega_n Ase^{st} + \omega_n^2 Ae^{st} = 0. \tag{d}$$

The characteristic equation is then obtained by dividing out common terms, that is,

$$s^2 + 2\zeta\omega_n s + \omega_n^2 = 0. \tag{d}$$

Step 4 is to solve this equation and obtain the roots.

$$\begin{aligned} s_1 &= -\zeta\omega_n + \omega_n\sqrt{\zeta^2 - 1} \\ s_2 &= -\zeta\omega_n - \omega_n\sqrt{\zeta^2 - 1} \end{aligned} \tag{e}$$

We shall not consider the situation in which $\zeta > 1$. Consequently, for $\zeta < 1$, the radical in Eq. (e) has an imaginary value, and we prefer to write the roots in the form

$$\begin{aligned} s_1 &= -\zeta\omega_n + j\omega_n\sqrt{1 - \zeta^2} \\ s_2 &= -\zeta\omega_n - j\omega_n\sqrt{1 - \zeta^2} \end{aligned}, \tag{f}$$

where $j = \sqrt{-1}$. Finally, for Step 5, we substitute the roots back into the assumed solution. Then, after factoring, the transient solution becomes

$$\theta_{o,t} = e^{-\zeta\omega_n t}\left(Ae^{j\omega_n\sqrt{1-\zeta^2}t} + Be^{-j\omega_n\sqrt{1-\zeta^2}t}\right), \tag{g}$$

where A and B are the constants of integration and are two in number, the same as the order of the highest derivative in the differential equation. These cannot be evaluated until the complete solution is found. At that time, they will be determined from the initial conditions. We can, however, transform Eq. (g) into trigonometric form using DeMoivre's theorem. The result of this transformation is

$$\theta_{o,t} = e^{-\zeta\omega_n t}\left(A\cos\omega_n\sqrt{1 - \zeta^2}t + B\sin\omega_n\sqrt{1 - \zeta^2}t\right). \tag{19.17}$$

The coefficient $e^{-\zeta\omega_n t}$ of Eq. (19.17) indicates its transient nature, because this term becomes approximately zero for large values of t.

It can sometimes happen that two or more roots of the characteristic equation are identical. For the case of two identical roots, $s_1 = s_2$, the transient solution is

$$\theta_{o,t} = Ae^{s_1 t} + Bte^{s_1 t}. \tag{h}$$

For other special cases the reader is advised to refer to a text that focuses on the solution of linear differential equations.

The Steady-State Solution Here we shall solve Eq. (19.12) for various input conditions. For the unit-step function, $\theta_i = 1$. Substituting this function into Eq. (19.12) gives

$$\ddot{\theta}_o + 2\zeta\omega_n\dot{\theta}_o + \omega_n^2\theta_o = \omega_n^2. \tag{i}$$

The solution is

$$\theta_{o,s} = 1, \tag{19.18}$$

which can be verified by substituting Eq. (19.18) and its derivatives into Eq. (i).

For the unit-slope ramp function, $\theta_i = t$, substituting this function into Eq. (19.12) gives

$$\ddot{\theta}_o + 2\zeta\omega_n\dot{\theta}_o + \omega_n^2\theta_o = \omega_n^2 t. \tag{j}$$

We assume a solution of the form

$$\theta_{o,s} = At + B. \tag{k}$$

The successive derivatives are

$$\dot{\theta}_{o,s} = A, \quad \ddot{\theta}_{o,s} = 0.$$

Substituting the assumed solution and its derivatives into Eq. (j) gives

$$2\zeta\omega_n A + \omega_n^2(At + B) = \omega_n^2 t.$$

We now arrange the equation into the form

$$(2\zeta\omega_n A + \omega_n^2 B) + (\omega_n^2 A)t = \omega_n^2 t$$

and solve for A and B. The constants of integration are

$$A = 1, \quad B = -\frac{2\zeta}{\omega_n}.$$

Substituting these into Eq. (k), the steady-state solution for a unit step velocity function is

$$\theta_{o,s} = t - \frac{2\zeta}{\omega_n}. \tag{19.19}$$

The Complete Solution The complete solution is the sum of the transient and the steady-state solution terms. Thus,

$$\theta_o = \theta_{o,t} + \theta_{o,s}.$$

Therefore, adding Eqs. (19.17) and (19.18) gives

$$\theta_o = e^{-\zeta\omega_n t}\left(C_1 \cos\omega_n\sqrt{1-\zeta^2}t + C_2 \sin\omega_n\sqrt{1-\zeta^2}t\right) + 1. \tag{19.20}$$

The initial conditions for $t = 0+$ can now be applied to evaluate the constants C_1 and C_2. Imposing the condition that $\theta_o = 0$ at the beginning of the step gives

$$0 = 1(C_1 \cos 0 + C_2 \sin 0) + 1$$

or

$$C_1 = -1.$$

The second condition to be imposed is that $\dot{\theta}_o = 0$ at $t = 0+$. To apply this condition it is necessary to take the derivative of Eq. (19.20). This is

$$\dot{\theta}_o = -\zeta\omega_n e^{-\zeta\omega_n t}\left(C_1 \cos \omega_n\sqrt{1-\zeta^2}t + C_2 \sin \omega_n\sqrt{1-\zeta^2}t\right)$$
$$+ e^{-\zeta\omega_n t}\left(-C_1\omega_n\sqrt{1-\zeta^2}\sin \omega_n\sqrt{1-\zeta^2}t + C_2\omega_n\sqrt{1-\zeta^2}\cos \omega_n\sqrt{1-\zeta^2}t\right).$$

Substituting $\dot{\theta}_o = 0$ and $t = 0$ gives

$$0 = -\zeta\omega_n(C_1 \cos 0 + C_2 \sin 0) + 1\left(-C_1\omega_n\sqrt{1-\zeta^2}\sin 0 + C_2\omega_n\sqrt{1-\zeta^2}\cos 0\right).$$

Substituting the value of C_1 and solving yields

$$C_2 = -\frac{\zeta}{\sqrt{1-\zeta^2}}.$$

Substituting the values of C_1 and C_2 into Eq. (19.20) and rearranging, the complete solution can be written as

$$\theta_o = 1 - e^{-\zeta\omega_n t}\left(\cos \omega_n\sqrt{1-\zeta^2}t + \sin \omega_n\sqrt{1-\zeta^2}t\right). \tag{19.21}$$

If transformed to a single trigonometric term and a phase angle as demonstrated in Chapter 15, the complete solution is

$$\theta_o = 1 - \frac{e^{-\zeta\omega_n t}}{\sqrt{1-\zeta^2}}\cos\left(\omega_n\sqrt{1-\zeta^2}t - \phi\right), \tag{19.22}$$

where

$$\phi = \tan^{-1}\left(\frac{\zeta}{\sqrt{1-\zeta^2}}\right).$$

The complete solution for the unit-step velocity function input is obtained in a similar manner. The initial conditions to be applied to evaluate the constants of integration are

$$\theta_o = 0 \quad \text{and} \quad \dot{\theta}_o = 0 \text{ when } t = 0-.$$

The equation to be solved for the integration constants is obtained by adding Eqs. (19.17) and (19.19); it is

$$\theta_o = e^{-\zeta\omega_n t}\left(C_1 \cos\omega_n\sqrt{1-\zeta^2}t + C_2 \sin\omega_n\sqrt{1-\zeta^2}t\right) + t - \frac{2\zeta}{\omega_n}. \tag{19.23}$$

The two constants are found to be

$$C_1 = \frac{2\zeta}{\omega_n} \quad \text{and} \quad C_2 = -\frac{1-2\zeta^2}{\omega_n\sqrt{1-\zeta^2}}$$

and the complete solution is

$$\theta_o = t - \frac{2\zeta}{\omega_n} + e^{-\zeta\omega_n t}\left(\frac{2\zeta}{\omega_n}\cos\omega_n\sqrt{1-\zeta^2}t - \frac{1-2\zeta^2}{\omega_n\sqrt{1-\zeta^2}}\sin\omega_n\sqrt{1-\zeta^2}t\right) \tag{19.24}$$

or

$$\theta_o = t - \frac{2\zeta}{\omega_n} + \frac{e^{-\zeta\omega_n t}}{\omega_n\sqrt{1-\zeta^2}}\cos\left(\omega_n\sqrt{1-\zeta^2}t - \phi\right), \tag{19.25}$$

where the phase angle is

$$\phi = \tan^{-1}\left(\frac{2\zeta^2-1}{2\zeta\sqrt{1-\zeta^2}}\right).$$

19.10 ANALYSIS OF PROPORTIONAL-ERROR FEEDBACK SYSTEMS

As the name implies, a proportional-error feedback control system is one that operates by applying a correction that is directly proportional to any error that might exist. Such a system must be analyzed to determine the following:

1. The response time, or the time required to reach steady-state operation after the application of a disturbance to the system. Because a system with viscous damping never reaches a steady-state condition, this is usually defined as the time required for the transient to decay to 5% of its initial value.
2. The natural frequency.
3. The steady-state error, that is, the difference between the output and input during steady-state operation.
4. The maximum overshoot, or the maximum deviation between output and input during transient conditions.

Output response of the system to a unit-step input is given by Eq. (19.22) and has been plotted in Fig. 19.10 for various values of the damping ratio. The abscissa is measured in dimensionless time $\omega_n t/2\pi$, and so the curves can be applied to any physical system. Thus, if the undamped natural frequency ω_n is known for a given system, then the response time can be calculated simply by multiplying dimensionless time by the quantity by $2\pi/\omega_n$. The

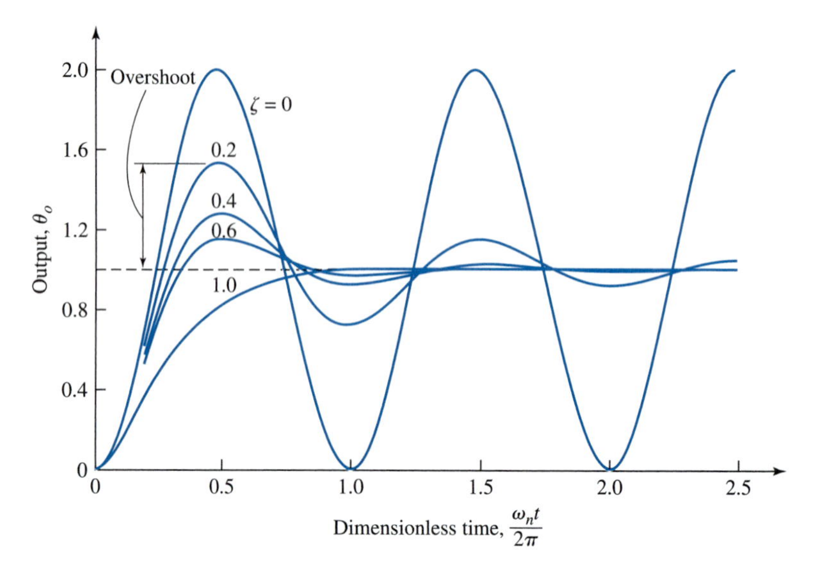

Figure 19.10 Response to a unit-step input.

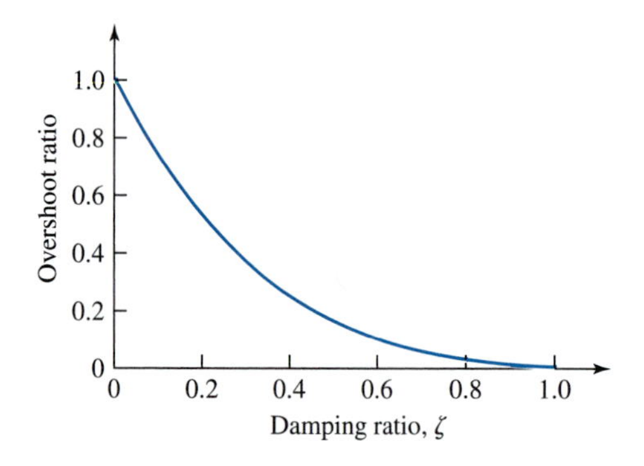

Figure 19.11

response for zero damping is included for its academic interest; an automatic control system with no damping at all would give completely unsatisfactory performance.

The response for critical damping $\zeta = 1$ shows no overshoot, and the steady-state condition is reached at about $\omega_n t/2\pi = 1.0$.

It is convenient to define an *overshoot ratio* as the ratio of overshoot with damping to the overshoot that would exist if no damping were present. Figure 19.11 illustrates a plot of this ratio versus the damping ratio. For $\zeta = 0.40$, the overshoot ratio is approximately 0.25; this quantity can also be read from Fig. 19.10. Note that the steady-state condition is not reached for $\zeta = 0.40$ until about $\omega_n t/2\pi = 2.0$. The response time, therefore, is twice that for critical damping.

Because the response time is proportional to $\omega_n t/2\pi$, we can make this time short simply by making the undamped natural frequency, ω_n, large.

In practice, ζ is usually between 0.4 and 0.7 in a physical system. If a certain deviation from steady-state is permitted, say 5%, then a system with $\zeta = 0.7$ will reach steady state before one having $\zeta = 1.0$ because the early portion of its response curve is steeper.

Figure 19.10 demonstrates that the steady-state value of the output is unity. Consequently, there is no steady-state error for step-input functions when applied to proportional-error feedback systems.

Local Disturbance Another look at the behavior of a system can be obtained by considering that the input signal is constant or zero and that the load is subjected to a disturbance. Selecting the zero input condition, for simplicity, Eq. (19.13) then becomes

$$\ddot{\theta}_o + 2\zeta\omega_n\dot{\theta}_o + \omega_n^2\theta_o = -\frac{T}{I}, \tag{19.26}$$

where T represents a constant torque suddenly applied to the input. The steady-state component of the solution is

$$\theta_{o,s} = -\frac{T}{I\omega_n^2}, \tag{a}$$

so that the complete solution must be

$$\theta_o = e^{-\zeta\omega_n t}\left(C_1 \cos\omega_n\sqrt{1-\zeta^2}t + C_2 \sin\omega_n\sqrt{1-\zeta^2}t\right) - \frac{T}{I\omega_n^2}. \tag{b}$$

The initial conditions are $\theta_o = 0$ and $\dot{\theta}_o = 0$ at $t = 0$. Using these conditions, we can solve Eq. (b) for the two constants

$$C_1 = \frac{T}{I\omega_n^2} \quad \text{and} \quad C_2 = \frac{\zeta T}{I\omega_n^2\sqrt{1-\zeta^2}}.$$

Substituting these into Eq. (b) and transforming to a single trigonometric term, the result is

$$\theta_o = \frac{T}{I\omega_n^2}\left[\frac{e^{-\zeta\omega_n t}}{\sqrt{1-\zeta^2}}\cos\left(\omega_n\sqrt{1-\zeta^2}t - \phi\right) - 1\right], \tag{19.27}$$

where

$$\phi = \tan^{-1}\left(\frac{\zeta}{\sqrt{1-\zeta^2}}\right).$$

Equation (19.27) is plotted in Fig. 19.12 for three values of the damping ratio to illustrate what happens. As shown, the disturbance decays at a rate that is dependent on the amount of damping. It finally reaches a steady-state condition that is not zero, but is of an amount

$$\varepsilon_s = \frac{T}{I\omega_n^2}. \tag{19.28}$$

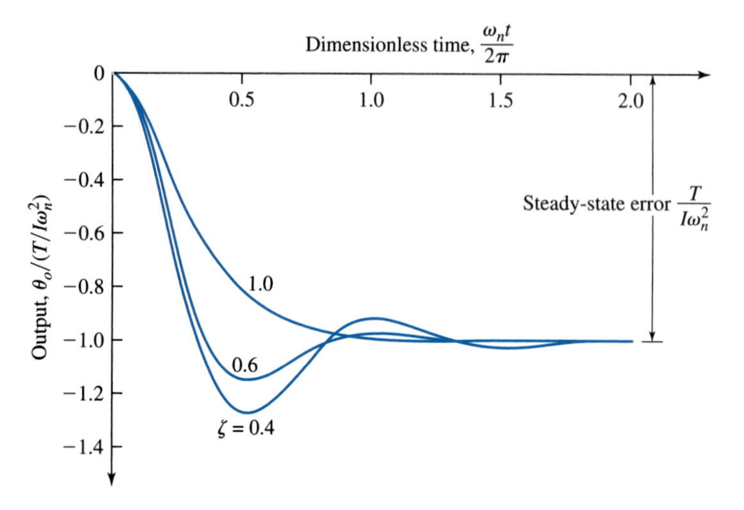

Figure 19.12 Response of proportional-error feedback control to a load disturbance.

This is the steady-state error. If we substitute $\omega_n^2 = k/I$, the equation becomes

$$\varepsilon_s = \frac{T}{k}. \tag{19.29}$$

Hence, the only method of reducing the magnitude of this error is to increase the gain k of the system.

We have seen that the control system is defined by the parameters I, c, and k. Of these three, the gain is usually the easiest to change. Some variation in damping is usually possible, but the employment of dashpots or friction dampers is not often a good solution. Variation in the inertia I is the most difficult change to make because this is fixed by the design of the driven element and, furthermore, improvement always requires a decrease in inertia.

Unit-Step Velocity Input For the unit-step velocity input we have $\theta_i = t$, and from Eq. (19.19)

$$\theta_{o,s} = t - \frac{2\zeta}{\omega_n}$$

after the transient decays. Therefore, the steady-state error is

$$\varepsilon_s = \frac{2\zeta}{\omega_n} = \frac{c}{k}, \tag{19.30}$$

which is obtained by substitution of the value of ζ from Eq. (19.11). This error persists only as long as the input function signals for a constant velocity. Again we see that the magnitude can be reduced by increasing the gain k.

The widely used automotive cruise-control system is an excellent example of an electromechanical governor. A transducer is attached to the speedometer cable, and the electrical output of this transducer is the signal θ_o fed to the error detector of Fig. 19.8. In some cases, magnets are mounted on the driveshaft of the car to activate the transducer. In the cruise-control system, the error detector is an electronic regulator, usually mounted under the dashboard. The regulator is turned on by an engagement switch under, or near, the steering wheel. A power unit is connected to the carburetor throttle linkage; the power unit is controlled by the regulator and gets its power from a vacuum port on the engine. Such systems have one or two brake-release switches, as well as the engagement switch. The accelerator pedal can also be used to override the system.

19.11 INTRODUCTION TO GYROSCOPES

A gyroscope may be defined as a rigid body capable of three-dimensional rotation with high angular velocity about any axis that passes through a fixed point called the center, which may or may not be its center of mass. A child's toy top fits this definition and is a form of gyroscope; its fixed point is the point of contact of the top with the floor or table on which it spins.

The usual form of a gyroscope is a mechanical device in which the essential part is a rotor having a heavy rim spinning at high speed about a fixed point on the rotor axis. The rotor is then mounted so as to turn freely about its center of mass by means of double gimbals called a Cardan suspension; an example is pictured in Fig. 19.13.

The gyroscope has fascinated students of mechanics and applied mathematics for many years. In fact, once the rotor is set spinning, a gyroscope appears to act like a device possessing intelligence. If we attempt to move some of its parts, it seems not only to resist

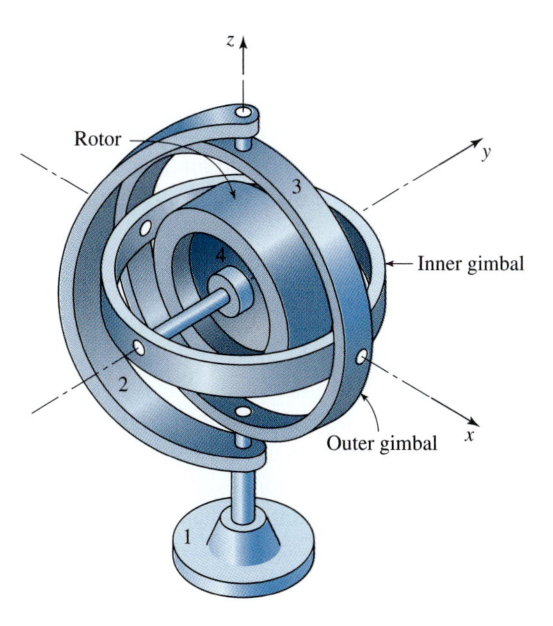

Figure 19.13 A laboratory gyroscope.

this motion but also even to evade it. We shall see that it apparently fails to conform to the laws of static equilibrium and gravitation.

The early history of the gyroscope is rather obscure. Probably the earliest gyroscope of the type now in use was constructed by Bohnenburger in Germany in 1817.[1] In 1852, Leon Foucault, of Paris, constructed a very refined version to demonstrate the rotation of the earth; it was Foucault who named the instrument the gyroscope from the Greek words *gyros*, circle or ring, and *skopien*, to view.[2] The mathematical foundations of gyroscopic theory were laid by Leonhard Euler in 1765 in his work on the dynamics of rigid bodies.[3] Gyroscopes were not put to practical and industrial use until the beginning of the 20th century in the United States, at which time the gyroscopic compass, the ship stabilizer, and the monorail car were all invented. Subsequent uses of the gyroscope as a turn-and-bank indicator, artificial horizon, and automatic pilot in aircraft and missiles are well known.

We may also become concerned with gyroscopic effects in the design of machines, although not always intentionally. Such effects are present in the riding of a motorcycle or bicycle; they are always present, owing to the rotating masses, when an airplane or automobile is making a turn. Sometimes these gyroscopic effects are desirable, but more often they are undesirable and designers must account for them in their selection of bearings and rotating parts. It is certainly true that, as machine speeds increase to higher and higher values and as factors of safety decrease, we must stop neglecting gyroscopic forces in our design calculations because their values will become more significant.

19.12 THE MOTION OF A GYROSCOPE

Although we have noted above that a gyroscope seems to have intelligence and appears to avoid compliance with the fundamental laws of mechanics, this is not truly the case. In fact, we have thoroughly covered the basic theory involved in Chapter 14, where we studied dynamic forces with spatial motion. Still, because gyroscopic forces are of increasing importance in high-speed machines, we will look at them again and try to explain the apparent paradoxes they seem to raise.

To provide a vehicle for the explanation of the simpler motions of a gyroscope, we will consider a series of experiments to be performed on the one pictured in Fig. 19.13. In the following discussion we assume that the rotor is already spinning rapidly and that any pivot friction is negligible.

As a first experiment, we transport the entire gyroscope about the table or around the room. We find that although we travel along a curved path, the orientation of the rotor's axis of rotation does not change as we move. This is a consequence of the law of conservation of angular momentum. If the orientation of the axis of rotation is to change, then the orientation of the angular momentum vector must also change. However, this requires an externally applied moment that, with the three sets of frictionless bearings, have not been supplied for this experiment. Therefore, the orientation of the rotation axis does not change.

For a second experiment, while the rotor is still spinning, we lift the inner gimbal out of its bearings and move it about. We again find that it can be translated anywhere but that we meet with definite resistance if we attempt to rotate the axis of spin. In other words, the rotor persists in maintaining its plane of rotation.

For our third experiment, we replace the inner gimbal back into its bearings with the axis of *spin* horizontal, as illustrated in Fig. 19.13. If we now apply a steady downward force to the inner gimbal at one end of the spin axis, say by pushing on it with a pencil, we find that the end of the spin axis does not move downward as we might expect. Not only do we meet with resistance to the force of the pencil, but also the outer gimbal begins to rotate about the vertical axis, causing the rotor to skew around in the horizontal plane, and it continues this rotation until we remove the force of the pencil. This skewing motion of the spin axis is called *precession*.

Although it may seem strange and unexpected, this precession motion is in strict obedience to the laws of dynamic equilibrium as expressed by the Euler equations of motion in Eqs. (14.110). Yet it is easier to understand and explain this phenomena if we again think in terms of the angular momentum vector. The downward force applied by the pencil to the inner gimbal produces a net external moment **M** on the rotor shaft through a pair of equal and opposite forces at the bearings and results in a time rate of change of the angular momentum vector **H**. As we demonstrated in Eq. (14.120), the time rate of change of the angular momentum can be written as

$$\frac{d\mathbf{H}}{dt} = \sum \mathbf{M}. \tag{19.31}$$

Because this applied external moment cannot change the rate of spin of the rotor about its own axis, it changes the angular momentum by making the rotor rotate about the vertical axis as well, thus causing the precession.

As our next set of experiments we repeat the previous experiment, watching carefully the directions involved. We first cause the rotor to spin in the positive direction, that is, with its angular velocity vector in the positive y direction. If we next apply a positive torque (moment vector in the positive x direction) using downward force of our pencil on the negative y end of the rotor axis, the precession (rotation) of the outer gimbal is found to be in the negative z direction. Further experiments indicate that either a negative spin velocity or a negative moment caused by the pencil results in a positive direction for the precession.

For a final experiment we apply a moment to the outer gimbal in an attempt to cause the rotor to rotate about the z axis. Such an attempt meets with definite resistance and causes the inner gimbal and the spin axis to rotate. Note again in this case that the angular momentum vector is changing as the result of the application of external torque. If the spin axis starts in the vertical position (aligned along z), however, then the gyroscope is in stable equilibrium and the outer and inner gimbals can be turned together quite freely.

19.13 STEADY OR REGULAR PRECESSION

Let us now suppose that a heavy rotor is spinning with a constant angular velocity of $\boldsymbol{\omega}_s$ about a spin axis that is tipped at a constant angle θ from the vertical. At the same time, let us assume that the spin axis has a constant angular velocity of $\boldsymbol{\omega}_p$, as illustrated in Fig. 19.14. The state of motion just described is called *steady* or *regular* precession.

We should take careful note of several things in Fig. 19.14. First, we have shown the rotor at a position for which the inner gimbal is rotated to quite a different angle than in

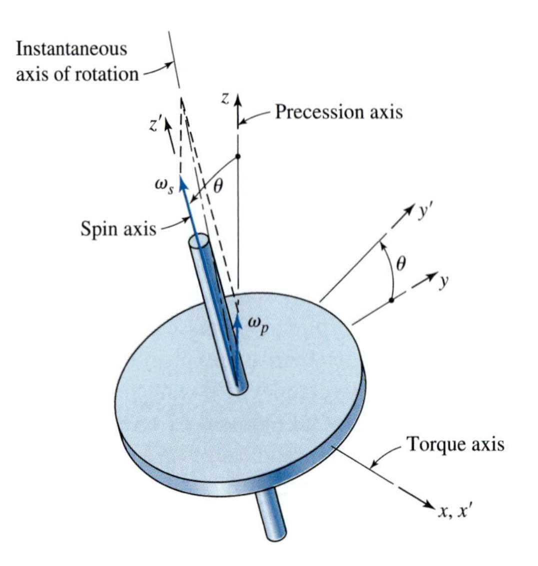

Figure 19.14 Angular velocity components of a gyroscope rotor.

Fig. 19.13, yet the xyz coordinate system is still oriented with x along the pivot axis between the inner and outer gimbals, and z is still along the true axis of precession. Second, we have designated a new coordinate system $x'y'z'$ with the x' axis coincident with x and the z' axis aligned along the axis of the rotor, the spin axis. Third, we should note that neither of these two coordinate systems is stationary and, similarly, that $x'y'z'$ is not attached to the rotor and does not experience the angular velocity $\boldsymbol{\omega}_s$. The $x'y'z'$ coordinate system remains fixed to the inner gimbal, whereas xyz remains fixed to the outer gimbal.

Despite this, we can see that, for a disk-shape rotor mounted as described, the $x'y'z'$ axes are the principal axes of inertia of the rotor. We will designate the corresponding principal mass moments of inertia $I^{z'z'} = I^s$ and $I^{x'x'} = I^{y'y'} = I$.

Recognize that the constant angular spin velocity $\boldsymbol{\omega}_s$ is not truly an absolute angular velocity; it is the "relative" or *apparent* angular velocity of the rotor with respect to the inner gimbal. If we designate the rotor as body 4, with the inner and outer gimbals being links 3 and 2, respectively, then

$$\boldsymbol{\omega}_{4/3} = \boldsymbol{\omega}_5 = \omega_5 \hat{\mathbf{k}}'. \tag{a}$$

Similarly, the angular velocity of precession is really the angular velocity of the outer gimbal.

$$\boldsymbol{\omega}_2 = \boldsymbol{\omega}_p = \omega_p \hat{\mathbf{k}} = (\omega_p \sin\theta)\hat{\mathbf{j}}' + (\omega_p \cos\theta)\hat{\mathbf{k}}' \tag{b}$$

Because we are interested in steady precession here, the angle θ has been assumed constant; thus, there is no rotation between the inner and outer gimbals.

$$\boldsymbol{\omega}_{3/2} = \dot{\theta}\hat{\mathbf{i}}' = 0. \tag{c}$$

We can now use Eqs. (a) through (c) to find the absolute angular velocity of the rotor.*

$$\begin{aligned}\boldsymbol{\omega}_4 &= \boldsymbol{\omega}_2 + \boldsymbol{\omega}_{3/2} + \boldsymbol{\omega}_{4/3} = \boldsymbol{\omega}_p + \boldsymbol{\omega}_s \\ &= (\omega_p \sin\theta)\hat{\mathbf{j}}' + (\omega_s + \omega_p \cos\theta)\hat{\mathbf{k}}'\end{aligned} \tag{19.32}$$

If we assume that the masses of the gimbals are negligible in comparison with the rotor, then the angular momentum of the system is

$$\mathbf{H} = I(\omega_p \sin\theta)\hat{\mathbf{j}}' + I^s(\omega_s + \omega_p \cos\theta)\hat{\mathbf{k}}'. \tag{19.33}$$

To find the time rate of change of the angular momentum, we must recognize that $\hat{\mathbf{j}}'$ and $\hat{\mathbf{k}}'$ are not constant; they rotate with the angular velocity of the $x'y'z'$ coordinate system. From Eqs. (b) and (c), this is

$$\begin{aligned}\boldsymbol{\omega}_3 &= \boldsymbol{\omega}_2 + \boldsymbol{\omega}_{3/2} = \boldsymbol{\omega}_p \\ &= (\omega_p \sin\theta)\hat{\mathbf{j}}' + (\omega_p \cos\theta)\hat{\mathbf{k}}'\end{aligned} \quad . \tag{d}$$

Thus, the time rate of change of $\hat{\mathbf{j}}'$ and $\hat{\mathbf{k}}'$ are

$$\frac{d\hat{\mathbf{j}}'}{dt} = \boldsymbol{\omega}_p \times \hat{\mathbf{j}}' = -(\omega_p \cos\theta)\hat{\mathbf{i}}' \tag{e}$$

$$\frac{d\hat{\mathbf{k}}'}{dt} = \boldsymbol{\omega}_p \times \hat{\mathbf{k}}' = (\omega_p \sin\theta)\hat{\mathbf{i}}'. \tag{f}$$

Finally, using these to differentiate Eq. (19.33), the net external moment that must be applied to the rotor to sustain the steady precession motion can be written as

$$\begin{aligned}\sum \mathbf{M} &= \frac{d\mathbf{H}}{dt} = \boldsymbol{\omega}_p \times \mathbf{H} \\ &= -I(\omega_p \sin\theta)(\omega_p \cos\theta)\hat{\mathbf{i}}' + I^s(\omega_s + \omega_p \cos\theta)(\omega_p \sin\theta)\hat{\mathbf{i}}'\end{aligned} \tag{19.34}$$

or as

$$\sum \mathbf{M} = [I^s\omega_s + (I^s - I)\omega_p \cos\theta]\omega_p \sin\theta\hat{\mathbf{i}}' \tag{19.35}$$

or as

$$\sum \mathbf{M} = \left[I^s + (I^s - I)\frac{\omega_p}{\omega_s}\cos\theta\right](\boldsymbol{\omega}_p \times \boldsymbol{\omega}_s). \tag{19.36}$$

---

* We should note here that the true instantaneous axis of rotation of the rotor is not the spin axis, but accounts for the precession rotation also; this axis is illustrated in Fig. 19.14. It may be our tendency to picture the spin axis as the true axis of rotation of the rotor that leads us to the intuitive feeling that a gyroscope does not follow the laws of mechanics.

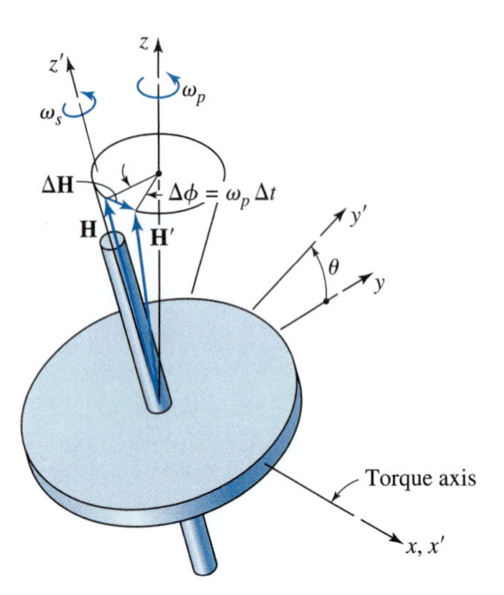

Figure 19.15 Angular momentum vector of a gyroscope rotor during steady precession.

We note that when the angular velocity of spin ω_s is much larger than that of precession ω_p, which is usually the case, then the second term in the square brackets of Eqs. (19.35) and (19.36) is negligible with respect to the first. Therefore, the two equations reduce to

$$\sum \mathbf{M} = I^s \omega_p \omega_s \sin\theta \hat{\mathbf{i}}' = I^s (\boldsymbol{\omega}_p \times \boldsymbol{\omega}_s). \tag{19.37}$$

Figure 19.15 illustrates the same rotor in the same orientation as Fig. 19.14, but this time the angular momentum vector **H** rather than the angular velocities is shown. We have already noted that the angular momentum vector includes the effects of both the spin and the precession angular velocities. We can also see that it continually precesses, sweeping out a cone with apex angle θ about the precession axis.

In Fig. 19.15 we see the angular momentum vector **H** at some instant t and also its changed orientation $\mathbf{H}'$ after a short time interval Δt, and we note that it has not changed magnitude, only orientation. Spanning the tips of these two vectors we see the vector change in angular momentum $\Delta\mathbf{H}$ over this short time interval. We note that in the limit as Δt approaches zero, the direction of the $\Delta\mathbf{H}$ vector approaches the direction of the positive x and x' axes. This is totally consistent with Eq. (19.35), which indicates that $\Delta\mathbf{H}/\Delta t$ is a vector in the $\hat{\mathbf{i}}'$ direction.

To maintain a steady precession, Eq. (19.36) demonstrates that an external moment must be continually applied to the rotor; if this moment is not applied, regular precession will not continue. Note that the axis of the applied moment must be along the x' axis, which is continually changing during the precession. Note also that the sense of the moment is the same as that which would seem required to increase angle θ. Thus, we see that there really is no paradox at all; the externally applied moment $\sum \mathbf{M}$ about the x axis causes a change in angular momentum in exactly the same direction as the moment is applied.

19.14 FORCED PRECESSION

We have already observed that as speeds are increased in modern machinery, the engineer must be mindful of the increased importance of gyroscopic torques. Common mechanical equipment, which, in the past, was not thought of as exhibiting gyroscopic effects, do in fact often experience such torques, and these will become more significant as speeds increase. The purpose of this section is to present a few examples that, although they may not look like the standard gyroscope, do present gyroscopic torques that should be considered during the design of the equipment.

Much was presented in the earlier sections on the phenomena of precession. This type of motion is almost certainly accompanied by gyroscopic torques. Yet, lest we think that precession is only an unintended wobbling motion of a toy top, we will look at examples where this precession is recognized and even designed into the operation of some mechanical devices.

Let us first consider a vehicle such as a train, automobile, or racing car, moving at high speed on a straight road. The gyroscopic effect of the spinning wheels is to keep the vehicle moving straight ahead and to resist changes in direction. But when an external moment is applied that forces the wheel to change its direction, gyroscopic reaction forces immediately come into play.

To study this gyroscopic reaction in the case of vehicles, let us consider a pair of wheels connected by an axle, rounding a curve as illustrated in Fig. 19.16. This wheel ensemble may be considered a gyroscope. Such rounding of the curve is a forced precession of the wheel-axle assembly around a vertical axis through the center of *curvature* of the track.

EXAMPLE 19.2

A pair of wheels of radius r and combined mass moment of inertia I^s about their axis of rotation are connected by a straight axle. This assembly is rounding a curve of constant radius R with the center of the axle traveling at a velocity $\mathbf{V}$, as illustrated in Fig. 19.16. For simplicity it is assumed that the roadbed is not banked. Find the gyroscopic torque exerted on the wheel-axle assembly.

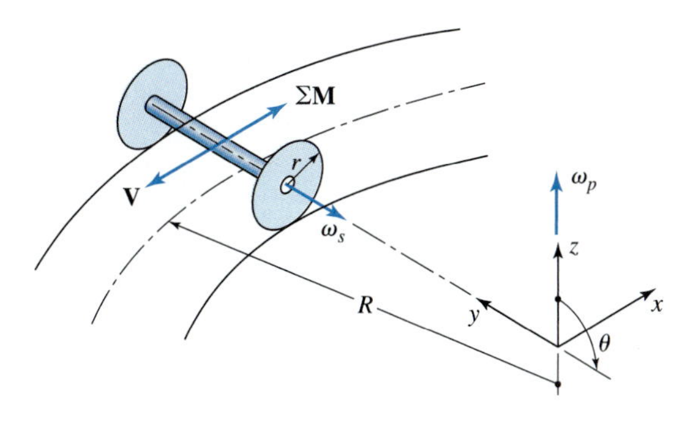

Figure 19.16 Example 19.2. Automobile axle and tires following a curved path.

SOLUTION

From the radius of the turn and the velocity, the angular velocity of precession of the assembly is

$$\omega_p = V/R \tag{1}$$

in the direction illustrated in Fig. 19.16. Similarly, from the radius of the wheels and the velocity, the average angular velocity of spin for the two wheels is

$$\omega_s = V/r, \tag{2}$$

which is also illustrated in Fig. 19.16.

From Fig. 19.16 we see that the angle of the precession cone, the angle between $\boldsymbol{\omega}_p$ and $\boldsymbol{\omega}_s$, is 90°. Substituting this value into Eq. (19.36) gives

$$\sum \mathbf{M} = I^s \boldsymbol{\omega}_p \times \boldsymbol{\omega}_s.$$

Then, substituting Eqs. (1) and (2) into this result gives

$$\sum \mathbf{M} = \frac{I^s V^2}{Rr}\hat{\mathbf{i}}. \qquad Ans.(4)$$

This is the additional external moment applied on the wheel-axle assembly caused by the gyroscopic action of the wheels while rounding the turn (forced precession) over and above the static and other steady-state dynamic loads. We note that the direction of this additional moment is such as to increase the tendency of the vehicle to roll over during the turn. This moment is applied by the tires by increasing the upward force on the outside tire and decreasing the upward force on the inside tire.

To make a quantitative comparison of the gyroscopic and centrifugal moments on the vehicle, let us consider the case of a racing car. The weight of the car including the driver is $W = 1,800$ lb, and the height of the center of mass above the road is $h = 20$ in. We estimate that the radius of each wheel is $r = 18$ in, the radius of gyration is $k = 15$ in, and the weight of each wheel is $w = 100$ lb. From these data we can determine the centrifugal moment **T** tending to cause rollover of the vehicle. Taking moments about the contact point of the outer tire, the answer is

$$\mathbf{T} = \frac{hWV^2}{gR}\hat{\mathbf{i}} = \frac{(20\text{ in})(1\,800\text{ lb})V^2}{gR}\hat{\mathbf{i}} = 36\,000\frac{V^2}{gR}\hat{\mathbf{i}}\text{ in}\cdot\text{lb}. \tag{5}$$

Because the vehicle has two axles and each has two wheels, the combined mass moment of inertia is

$$I^s = 2mk^2 = \frac{2(100\text{ lb})(15\text{ in})^2}{g} = \frac{45\,000\text{ in}^2\cdot\text{lb}}{g}. \tag{6}$$

Then, using Eq. (4) for two axles, the additional external moment applied on the wheel-axle assembly caused by the gyroscopic action of the wheels while rounding the turn is

$$\sum \mathbf{M} = 2\frac{I^s V^2}{Rr}\hat{\mathbf{i}} = 2\frac{(45\,000\text{ in}^2\cdot\text{lb})V^2}{gR(18\text{ in})}\hat{\mathbf{i}} = (5\,000\text{ in}\cdot\text{lb})\frac{V^2}{gR}\hat{\mathbf{i}} = 0.138\,9\mathbf{T}. \tag{7}$$

Thus, we see that the gyroscopic effects of the tires add almost 14% to the tendency of the vehicle to roll over in a turn, and we see that this is independent of both the radius of the turn and the speed of the vehicle.

We note from this example that the problem itself looks nothing like a gyroscope, yet it has gyroscopic effects and these may be skipped in an oversimplified analysis. We note also that the precession motion is not just an unexpected result of the motion characteristics of the system. It is knowingly caused by the driver who steers the car; it is a forced precession.

Another gyroscopic effect in automobiles is that caused by the flywheel. Because the rotation of the flywheel of a rear-wheel-drive vehicle is along the longitudinal axis and because the flywheel rotates counterclockwise as viewed from the rear, the spin vector ω_s points toward the rear of the vehicle. When the vehicle makes a turn, the axis of the flywheel is forced to precess about a vertical axis, as in the previous example. This forced precession brings into existence an applied gyroscopic moment about a horizontal axis through the center of the turn. The effect of this gyroscopic moment is to produce a bending moment on the driveshaft, tending to bend it in a vertical plane. The size of this gyroscopic moment causing bending in the driveshaft is usually of minor importance compared with the torsion loading because of the relatively small mass of the flywheel.

In case we might think that gyroscopic forces are always a disadvantage with which we must cope, we shall now look at a problem in which the gyroscopic effect is put to good use.

EXAMPLE 19.3

The edge mill illustrated schematically in Fig. 19.17 is a gyroscopic grinder used for crushing ore, seeds, grain, and so on. It consists of a large steel pan in which one or more heavy conical rollers, called *mullers*, roll without slipping on the bottom of the pan and at the same time revolve about a vertical shaft passing through the central axis of the pan. The mullers rotate about either horizontal or inclined axles that are attached to the vertical shaft, which is rotated under power. The radius of a muller is $r = 18$ in and the length of its axle is $l = 30$ in; its weight is $W = 2,200$ lb and its mass moment of inertia can be approximated by that for a cylindrical disk. Assuming that the vertical shaft is driven at a constant angular velocity of $\omega_p = 40$ rev/min and that the muller rolls without slipping, find the optimum inclination angle θ for the muller axle, which maximizes the crushing force between the muller and the pan.

SOLUTION

First, we must use the assumption of rolling without slip to find the angular velocity of the muller. From Fig. 19.17 we see that the angular velocity of the muller axis, link 3, is

$$\boldsymbol{\omega}_p = \boldsymbol{\omega}_3 = 40\hat{\mathbf{j}} \text{ rev/min} = 4.19\hat{\mathbf{j}} \text{ rad/s}, \tag{1}$$

whereas the angular velocity of spin of the muller about its axle is

$$\boldsymbol{\omega}_s = \boldsymbol{\omega}_{4/3} = \omega_s(-\sin\theta\hat{\mathbf{i}} + \cos\theta\hat{\mathbf{j}}). \tag{2}$$

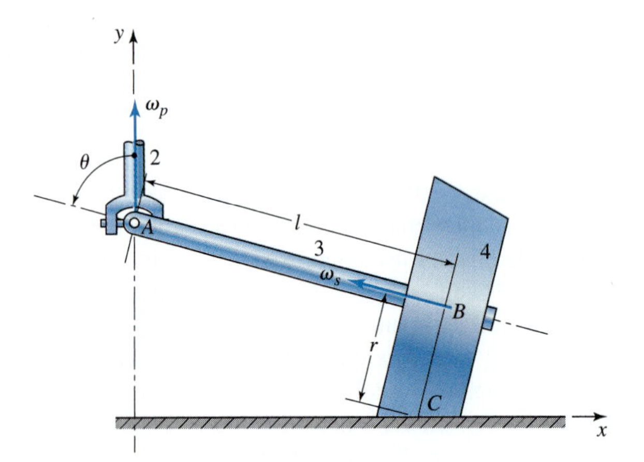

Figure 19.17 Example 19.3. Schematic diagram of an edge mill.

Because the muller undergoes precession in addition to the spin about its axle, the absolute angular velocity of the muller is

$$\boldsymbol{\omega}_4 = \boldsymbol{\omega}_3 + \boldsymbol{\omega}_{4/3} = (-\omega_s \sin\theta)\hat{\mathbf{i}} + (4.19 + \omega_s \cos\theta)\hat{\mathbf{j}}. \tag{3}$$

Because point A is fixed, the velocity of point B of link 3 is

$$\begin{aligned} \mathbf{V}_B &= \boldsymbol{\omega}_3 \times \mathbf{R}_{BA} \\ &= (4.19\hat{\mathbf{j}} \text{ rad/s}) \times (30\sin\theta\hat{\mathbf{i}} - 30\cos\theta\hat{\mathbf{j}} \text{ in}). \\ &= -125.7\sin\theta\hat{\mathbf{k}} \text{ in/s} \end{aligned} \tag{4}$$

Also, because there is no slip at point C, the velocity of point B of link 4 is

$$\begin{aligned} \mathbf{V}_B &= \boldsymbol{\omega}_4 \times \mathbf{R}_{BC} \\ &= [(-\omega_s \sin\theta)\hat{\mathbf{i}} + (4.19 + \omega_s \cos\theta)\hat{\mathbf{j}}] \times (18\cos\theta\hat{\mathbf{i}} + 18\sin\theta\hat{\mathbf{j}} \text{ in}) \ . \\ &= -18(\omega_s + 4.19\cos\theta)\hat{\mathbf{k}} \end{aligned} \tag{5}$$

Equating Eqs. (4) and (5), the angular velocity of spin of the muller about its axle is

$$\omega_s = 6.98\sin\theta - 4.19\cos\theta \text{ rad/s}. \tag{6}$$

From Eqs. (1) and (2), the cross-product is

$$\begin{aligned} (\boldsymbol{\omega}_p \times \boldsymbol{\omega}_s) &= (4.19\hat{\mathbf{j}} \text{ rad/s}) \times \boldsymbol{\omega}_s(-\sin\theta\hat{\mathbf{i}} + \cos\theta\hat{\mathbf{j}}) \\ &= 4.19\omega_s \sin\theta\hat{\mathbf{k}} \end{aligned} .$$

Then, substituting Eq. (6) into this result gives

$$(\boldsymbol{\omega}_p \times \boldsymbol{\omega}_s) = 4.19(6.98\sin\theta - 4.19\cos\theta)\sin\theta\hat{\mathbf{k}} \text{ rad/s}^2. \tag{7}$$

Using the formula for a round disk from Table 5 in Appendix A, the mass moment of inertia of the muller is

$$I^s = \frac{mr^2}{2} = \frac{(2\,200 \text{ lb})(18 \text{ in})^2}{(386 \text{ in/s}^2)2} = 923.3 \text{ in} \cdot \text{lb} \cdot \text{s}^2$$

$$I = \frac{mr^2}{4} + ml^2 = 5\,590 \text{ in} \cdot \text{lb} \cdot \text{s}^2. \tag{8}$$

Then substituting Eqs. (7) and (8) into Eq. (19.36), the net externally applied moment on the muller is

$$\sum \mathbf{M} = \left[923.3 + \frac{-4\,667(4.19)\cos\theta}{6.98\sin\theta - 4.19\cos\theta}\right](6.98\sin\theta - 4.19\cos\theta)4.19\sin\theta\hat{\mathbf{k}}.$$

Rearranging this equation, the moment applied to the muller can be written as

$$\sum \mathbf{M} = (27\,001\sin^2\theta - 98\,082\sin\theta\cos\theta)\hat{\mathbf{k}} \text{ in} \cdot \text{lb}. \tag{9}$$

This moment must be externally applied to the muller to sustain the forced precession. It must come from the net effect of the weight of the muller and the crushing force between the muller and the pan. Formulating these effects for the other side of the equation, we get

$$\begin{aligned}\sum \mathbf{M} &= \mathbf{R}_{BA} \times \mathbf{W}_4 + \mathbf{R}_{CA} \times \mathbf{F}_c \\ &= (30\sin\theta\hat{\mathbf{i}} - 30\cos\theta\hat{\mathbf{j}} \text{ in}) \times (-2\,200\hat{\mathbf{j}} \text{ lb}) \\ &\quad + [(30\sin\theta - 18\cos\theta)\hat{\mathbf{i}} + (18\sin\theta - 30\cos\theta)\hat{\mathbf{j}} \text{ in}] \times (F_c\hat{\mathbf{j}}) \\ &= (-66\,000\sin\theta + 30F_c\sin\theta - 18F_c\cos\theta)\hat{\mathbf{k}} \text{ in} \cdot \text{lb}\end{aligned} \tag{10}$$

Equating Eqs. (9) and (10), the crushing force becomes

$$F_c = \frac{27\,001\sin^2\theta - 98\,082\sin\theta\cos\theta + 66\,000\sin\theta}{30\sin\theta - 18\cos\theta}. \tag{11}$$

We note that the crushing force is a function of the angle θ of inclination of the muller axis; we now hope to choose this angle θ to maximize the crushing force. To do this, we differentiate Eq. (11) with respect to θ and set the result equal to zero; that is,

$$\begin{aligned}&(54\,002\sin\theta\cos\theta - 98\,082\cos^2\theta + 98\,082\sin^2\theta + 66\,000\cos\theta)(30\sin\theta - 18\cos\theta) \\ &\quad -(30\cos\theta + 18\sin\theta)(27\,001\sin^2\theta - 98\,082\sin\theta\cos\theta + 66\,000\sin\theta) = 0\end{aligned}$$

This equation simplifies to

$$\begin{aligned}&2\,456\,442\sin^3\theta + 810\,030\sin^2\theta\cos\theta \\ &\quad -972\,036\sin\theta\cos^2\theta + 1\,765\,476\cos^3\theta - 1\,188\,000 = 0\end{aligned}$$

and can be solved numerically (the equation has multiple roots). When this is done, the root that maximizes the crushing force F_c of Eq. (11) is found to be

$$\theta = 115.9^\circ. \qquad \textit{Ans.}$$

This is the optimum angle of inclination of the muller axle. Substituting this value into Eq. (11), the crushing force of the mill is

$$F_c = 3\,437 \text{ lb}. \qquad \textit{Ans.}$$

We should note that this crushing force is more than one and one half times the weight of the muller; the additional force is attributable to choosing the inclination angle θ for which both the gyroscopic and the centrifugal force effects contribute as much as possible to the crushing action of the mill. There are other designs for crushing machines in which the pan rotates under a muller that has a fixed axis. In such a design there is no gyroscopic action and the crushing action is caused solely by the weight of the muller.

An airplane propeller spinning at high speed is another example of a mechanical system exhibiting gyroscopic torque effects. In the case of a single propeller airplane, a turn in compass heading, for example, is a forced precession of the propeller's spin axis about a vertical axis. If the propeller is rotating clockwise as viewed from the rear, this forced precession will induce a gyroscopic moment, causing the nose of the plane to move upward or downward as the heading is changed to the left or right, respectively. Turning the nose upward rather suddenly will cause the plane to turn to the right, whereas a sudden turn downward will cause a turn to the left.

Note that it is not the propeller forces that cause this effect, but the spinning mass and its gyroscopic effect during a change in direction of the plane (forced precession). The very substantial spinning mass of a radial engine (see Section 1.8, Fig. 1.22*b*) for early aircraft markedly exaggerated this gyroscopic effect and was, at least in part, responsible for the disappearance of the use of rotary engines on airplanes. Because the effect comes from the precession of the spinning mass and not the propeller, does the same danger not exist from the spinning mass of the rotor of a turbojet engine? When an airplane is equipped with two propellers rotating at equal speeds in opposite directions (or with counterrotating turbines), however, the gyroscopic torques of the two can annul each other and leave no appreciable net effect on the plane as a whole.

19.15 REFERENCES

[1] J. G. F. Bohnenburger, "Beschreibung einer Maschine zur Erläuterung der Geseze der Umdrehung der Erde um ihre Axe, um der Veränderung der Lage der letzteren," *Tübinger Blätter für Naturwissenschaften und Arzneikunde*, 3, 1817, 72–83.
[2] J. B. Scarborough, *The Gyroscope, Theory and Applications*. New York: Interscience, 1958.
[3] L. Euler, "Theoria motus corporum solidorum seu rigidorum," *Commentarii Academiae Scientiarum Imperialis Petropolitanae*, 1765.

PROBLEMS

19.1 Table P19.1 lists the output torque for a one cylinder engine running at 4 600 rev/min.

(a) Find the mean output torque.

(b) Determine the mass moment of inertia of an appropriate flywheel using $C_s = 0.025$.

Table P19.1 Torque data for Problem 19.1

θ_i deg	T_i N · m	θ_i deg	T_i N · m	θ_i deg	T_i N · m	θ_i deg	T_i N · m
0	0	180	0	360	0	540	0
10	17	190	−344	370	−145	550	−344
20	812	200	−540	380	−150	560	−540
30	963	210	−576	390	7	570	−577
40	1016	220	−570	400	164	580	−572
50	937	230	−638	410	235	590	−643
60	774	240	−785	420	203	600	−793
70	641	250	−879	430	490	610	−893
80	697	260	−814	440	424	620	−836
90	849	270	−571	450	571	630	−605
100	1031	280	−324	460	814	640	−379
110	1027	290	−190	470	879	650	−264
120	902	300	−203	480	785	660	−300
130	712	310	−235	490	638	670	−368
140	607	320	−164	500	570	680	−334
150	594	330	−7	510	576	690	−198
160	544	340	150	520	540	700	−56
170	345	350	145	530	344	710	−2

19.2 Using the data of Table 19.2, determine the moment of inertia for a flywheel for a two-cylinder 90°V engine having a single crank. Use $C_s = 0.0125$ and a nominal speed of 4 600 rev/min. If a cylindrical or disk-type flywheel is to be used, what should be the thickness if it is made of steel and has an outside diameter of 400 mm? Use $\rho = 7.8\ \text{Mg/m}^3$ as the density of steel.

19.3 Using the data of Table 19.1, find the mean output torque and the flywheel inertia required for a three-cylinder in-line engine corresponding to a nominal speed of 2 400 rev/min. Use $C_s = 0.03$.

19.4 The load torque required by a 200-ton punch press is displayed in Table P19.4 for one revolution of the flywheel. The flywheel is to have a nominal angular velocity of 2 400 rev/min and to be designed for a coefficient of speed fluctuation of 0.075.

(a) Determine the mean motor torque required at the flywheel shaft and the motor horsepower needed, assuming a constant torque-speed characteristic for the motor.

(b) Find the moment of inertia needed for the flywheel.

Table P19.4 Torque data for Problem 19.4

θ_i deg	T_i in · lb	θ_i deg	T_i in · lb	θ_i deg	T_i in · lb	θ_i deg	T_i in · lb
0	857	90	7888	180	1801	270	857
10	857	100	8317	190	1629	280	857
20	857	110	8488	200	1458	290	857
30	857	120	8574	210	1372	300	857
40	857	130	8403	220	1115	310	857
50	1287	140	7717	230	1029	320	857
60	2572	150	3515	240	943	330	857
70	5144	160	2144	250	857	340	857
80	6859	170	1972	260	857	350	857

19.5 Find T_m for the four-cylinder engine whose torque displacement is that of Fig. 19.4.

19.6 A pendulum mill is illustrated schematically in Fig. P19.6. In such a mill, grinding is done by a conical muller that is free to spin about a pendulous axle that, in turn, is connected to a powered vertical shaft by a Hooke universal joint. The muller presses against the inner wall of a heavy steel pan, and it rolls around the inside of the pan without slipping. The weight of the muller is $W = 980$ lb; its principal mass moments of inertia are $I^s = 121$ in · lb · s^2 and $I = 88$ in · lb · s^2. The length of the muller axle is $l = R_{GA} = 40$ in and the radius of the muller at its center of mass is $R_{GB} = 10$ in. Assuming that the vertical shaft is to be inclined at $\theta = 30°$ and will be driven at a constant angular velocity of

$\omega_p = 240$ rev/min, find the crushing force between the muller and the pan. Also determine the minimum angular velocity ω_p required to ensure contact between the muller and the pan.

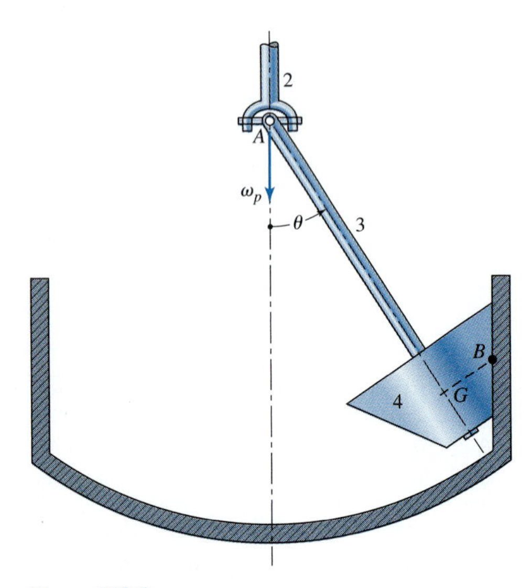

Figure P19.6

19.7 Use the gyroscopic formulae of Chapter 19 to solve again the problem presented in Example 14.9 of Chapter 14.

19.8 The oscillating fan illustrated in Fig. P19.8 precesses sinusoidally according to the equation $\theta_p = \beta \sin 1.5t$, where $\beta = 30°$; the fan blade spins at $\boldsymbol{\omega}_s = 1\,800\hat{\mathbf{i}}$ rev/min. The weight of the fan and motor armature is 5.25 lb, and other masses can be assumed negligible; gravity acts in the $-\hat{\mathbf{j}}$ direction. The principal mass moments of inertia are $I^s = 0.065$ in · lb · s^2 and $I = 0.025$ in · lb · s^2; the center of mass is located at $R_{GC} = 4$ in to the front of the precession axis. Determine the maximum moment M^z that must be accounted for in the clamped tilting pivot at C.

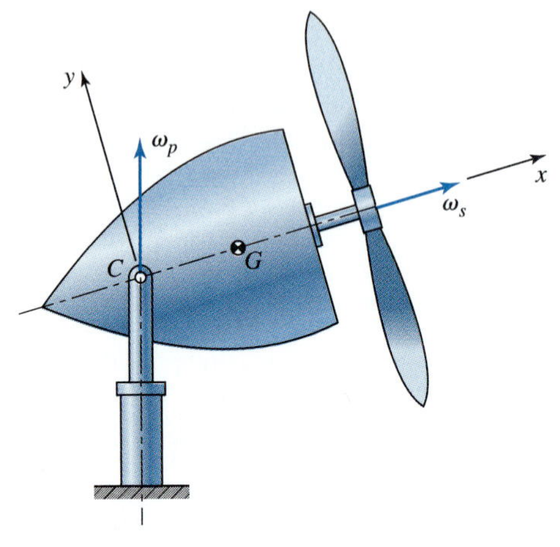

Figure P19.8

19.9 The propeller of an outboard motorboat is spinning at high speed and is caused to precess by steering to the right or left. Do the gyroscopic effects tend to raise or lower the rear of the boat? What is the effect and is it of notable size?

19.10 A large and very high-speed turbine is to operate at an angular velocity of $\omega = 18\,000$ rev/min and will have a rotor with a principal mass moment of inertia of $I^s = 225$ in · lb · s^2. It has been suggested that because this turbine will be installed at the North Pole with its axis horizontal, perhaps the rotation of the earth will cause gyroscopic loads on it bearings. Estimate the size of these additional loads.

Appendix A: Tables

TABLE 1 Standard SI prefixes*,†

Name	Symbol	Factor
exa	E	1 000 000 000 000 000 000 = 10^{18}
peta	P	1 000 000 000 000 000 = 10^{15}
tera	T	1 000 000 000 000 = 10^{12}
giga	G	1 000 000 000 = 10^{9}
mega	M	1 000 000 = 10^{6}
kilo	k	1 000 = 10^{3}
hecto‡	h	100 = 10^{2}
deka‡	da	10 = 10^{1}
deci‡	d	0.1 = 10^{-1}
centi‡	c	0.01 = 10^{-2}
milli	m	0.001 = 10^{-3}
micro	μ	0.000 001 = 10^{-6}
nano	n	0.000 000 001 = 10^{-9}
pico	p	0.000 000 000 001 = 10^{-12}
femto	f	0.000 000 000 000 001 = 10^{-15}
atto	a	0.000 000 000 000 000 001 = 10^{-18}

*If possible, multiple and submultiple prefixes are used in steps of 1 000. For example, lengths are specified in millimeters, meters, or kilometers. In a combination unit, prefixes are only used in the numerator; for example, meganewton per square meter (MN/m^2) is used, but not Newton per square millimeter (N/mm^2).

†Spaces are used in SI instead of commas to group numbers to avoid confusion with the practice in some countries of using commas for decimal points.

‡Not recommended but sometimes encountered.

TABLE 2 Conversion from U.S. customary units to SI units

To convert from	To	Multiply by	
		Accurate*	Common
Foot (ft)	Meter (m)	0.304 800 0*	0.305
Horsepower (hp)	Watt (W)	745.699 9	746
Inch (in)	Meter (m)	0.025 400 0*	0.025 4
Mile, U.S. statute (mi)	Meter (m)	1 609.344*	1 610
Pound force (lb)	Newton (N)	4.448 222	4.45
Pound mass (lb)	Kilogram (kg)	0.453 592 4	0.454
Pound · foot (lb · ft)	Newton · meter (N · m)	1.355 818	1.36
	Joule (J)	1.355 818	1.36
Pound · foot/second (lb · ft/s)	Watt (W)	1.355 818	1.36
Pound · inch (lb · in)	Newton · meter (N · m)	0.112 984 8	0.113
	Joule (J)	0.112 984 8	0.113
Pound · inch/second (lb · in/s)	Watt (W)	0.112 984 8	0.113
Pound/ft^2 (lb/ft^2)	Pascal (Pa)	47.880 26	47.9
Pound/in^2 (lb/in^2), (psi)	Pascal (Pa)	6 894.757	6 890
Revolutions/min (rev/min)	Radian/second (rad/s)	0.104 719 8	0.105
Ton, short (2 000 lb)	Kilogram (kg)	907.184 7	907

*An asterisk indicates that the coversion is exact.

TABLE 3 Conversion from SI units to U.S. customary units

To convert from	To	Multiply by	
		Accurate	Common
Joule (J)	Pound · foot (lb · ft)	0.737 562 1	0.738
	Pound · inch (lb · in)	8.850 746	8.85
Kilogram (kg)	Pound mass (lb)	2.204 623	2.20
	Ton, short (2 000 lb)	0.001 102 311	0.001 10
Meter (m)	Foot (ft)	3.280 840	3.28
	Inch (in)	39.370 08	39.4
	Mile (mi)	0.000 621 371	0.000 621
Newton (N)	Pound (lb)	0.224 808 9	0.225
Newton · meter (N · m)	Pound · foot (lb · ft)	0.737 562 1	0.738
	Pound · inch (lb · in)	8.850 746	8.85
Newton · meter/second (N · m/s)	Horsepower (hp)	0.001 341 022	0.001 34
Pascal (Pa)	Pound/foot^2 (lb/ft^2)	0.020 885 43	0.020 9
	Pound/inch^2 (lb/in^2), (psi)	0.000 145 037 7	0.000 145
Radian/second (rad/s)	Revolutions/minute (rev/min)	9.549 297	9.55
Watt (W)	Horsepower (hp)	0.001 341 022	0.001 34
	Pound · foot/second (lb · ft/s)	0.737 562 1	0.738
	Pound · inch/second (lb · in/s)	8.850 744	8.85

TABLE 4 Properties of areas

A = area
I = area moment of inertia
J = polar area moment of inertia
k = centroidal radius of gyration
$\bar{y}$ = centroidal distance from base

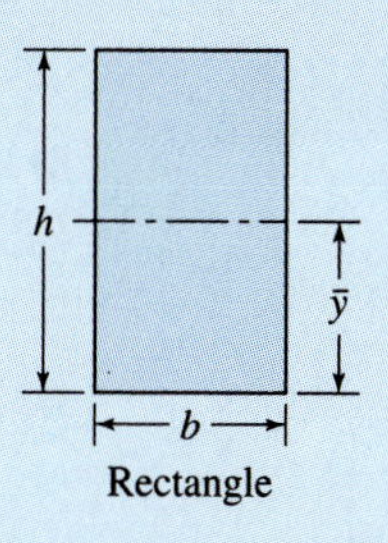

Rectangle

$A = bh$ $\quad k = 0.289h$
$I = bh^3/12$ $\quad \bar{y} = h/2$

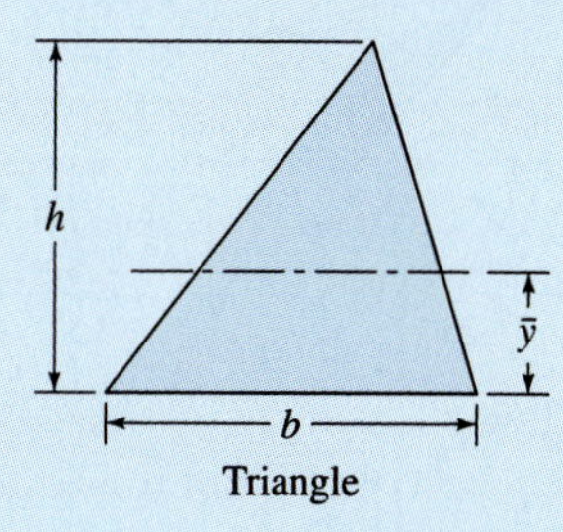

Triangle

$A = bh/2$ $\quad k = 0.236h$
$I = bh^3/36$ $\quad \bar{y} = h/3$

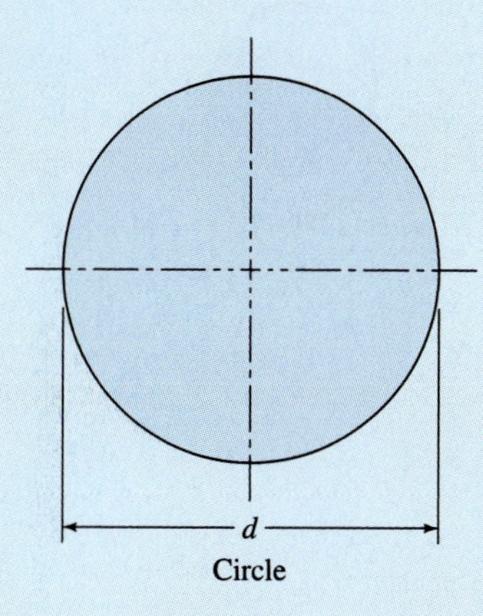

Circle

$A = \pi d^2/4$ $\quad k = d/4$
$I = \pi d^4/64$ $\quad \bar{y} = d/2$
$J = \pi d^4/32$

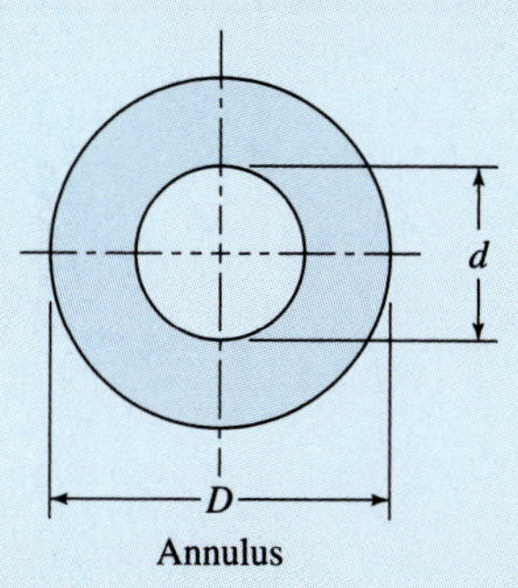

Annulus

$A = \pi(D^2 - d^2)/4$ $\quad k = \sqrt{D^2 + d^2}/4$
$I = \pi(D^4 - d^4)/64$ $\quad \bar{y} = D/2$
$J = \pi(D^4 - d^4)/32$

TABLE 5 Mass moments of inertia

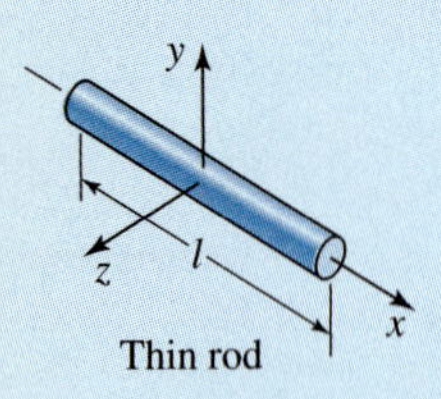

Thin rod

$m = \rho\pi r^2 l$
$I^{yy} = I^{zz} = ml^2/12$

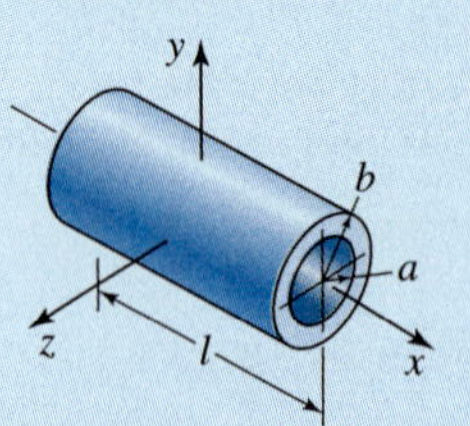

Annular cylinder

$m = \rho\pi(b^2 - a^2)l$
$I^{xx} = m(a^2 + b^2)/2$
$I^{yy} = I^{zz} = m(3a^2 + 3b^2 + l^2)/12$

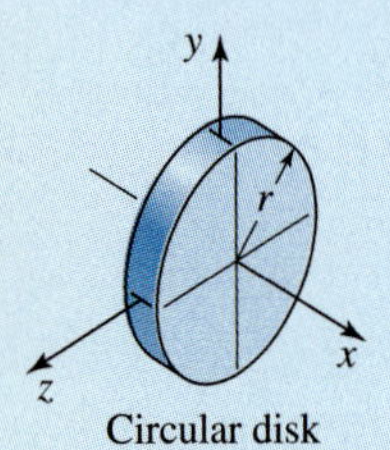

Circular disk

$m = \rho\pi r^2 t$
$I^{xx} = mr^2/2$
$I^{yy} = I^{zz} = mr^2/4$

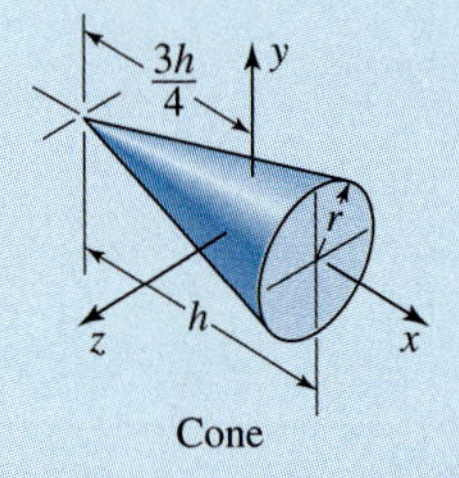

Cone

$m = \rho\pi r^2 h/3$
$I^{xx} = 3mr^2/10$
$I^{yy} = I^{zz} = m(12r^2 + 3h^2)/80$

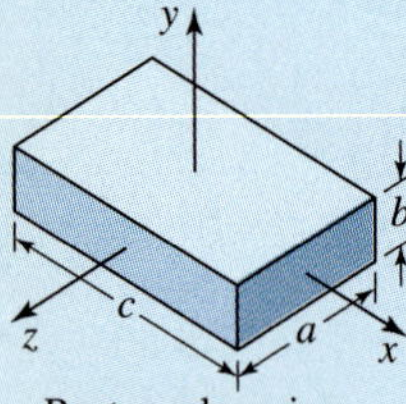

Rectangular prism

$m = \rho abc$
$I^{xx} = m(a^2 + b^2)/12$
$I^{yy} = m(a^2 + c^2)/12$
$I^{zz} = m(b^2 + c^2)/12$

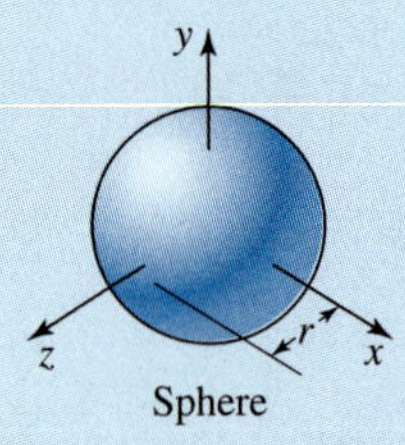

Sphere

$m = \rho 4\pi r^3/3$
$I^{xx} = I^{yy} = I^{zz} = 2mr^2/5$

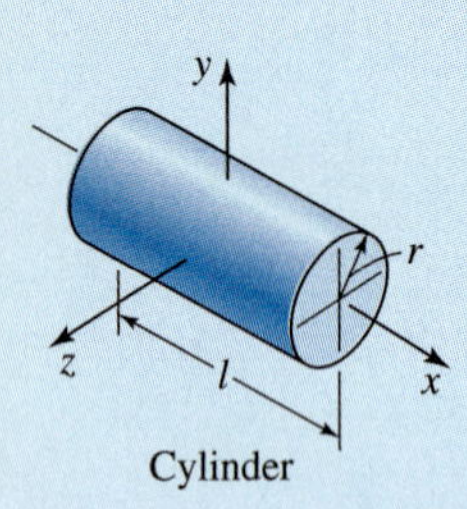

Cylinder

$m = \rho\pi r^2 l$
$I^{xx} = mr^2/2$
$I^{yy} = I^{zz} = m(3r^2 + l^2)/12$

TABLE 6 Involute function

φ (deg)	Inv (φ)	Inv ($\varphi + 0.1°$)	Inv ($\varphi + 0.2°$)	Inv ($\varphi + 0.3°$)	Inv ($\varphi + 0.4°$)
0.0	0.000000	0.000000	0.000000	0.000000	0.000000
0.5	0.000000	0.000000	0.000000	0.000000	0.000001
1.0	0.000002	0.000002	0.000003	0.000004	0.000005
1.5	0.000006	0.000007	0.000009	0.000010	0.000012
2.0	0.000014	0.000016	0.000019	0.000022	0.000025
2.5	0.000028	0.000031	0.000035	0.000039	0.000043
3.0	0.000048	0.000053	0.000058	0.000064	0.000070
3.5	0.000076	0.000083	0.000090	0.000097	0.000105
4.0	0.000114	0.000122	0.000132	0.000141	0.000151
4.5	0.000162	0.000173	0.000184	0.000197	0.000209
5.0	0.000222	0.000236	0.000250	0.000265	0.000280
5.5	0.000296	0.000312	0.000329	0.000347	0.000366
6.0	0.000384	0.000404	0.000424	0.000445	0.000467
6.5	0.000489	0.000512	0.000536	0.000560	0.000586
7.0	0.000612	0.000638	0.000666	0.000694	0.000723
7.5	0.000753	0.000783	0.000815	0.000847	0.000880
8.0	0.000914	0.000949	0.000985	0.001022	0.001059
8.5	0.001098	0.001137	0.001178	0.001219	0.001283
9.0	0.001305	0.001349	0.001394	0.001440	0.001488
9.5	0.001536	0.001586	0.001636	0.001688	0.001740
10.0	0.001794	0.001849	0.001905	0.001962	0.002020
10.5	0.002079	0.002140	0.002202	0.002265	0.002329
11.0	0.002394	0.002461	0.002528	0.002598	0.002668
11.5	0.002739	0.002812	0.002894	0.002962	0.003039
12.0	0.003117	0.003197	0.003277	0.003360	0.003443
12.5	0.003529	0.003615	0.003712	0.003792	0.003882
13.0	0.003975	0.004069	0.004164	0.004261	0.004359
13.5	0.004459	0.004561	0.004664	0.004768	0.004874
14.0	0.004982	0.005091	0.005202	0.005315	0.005429
14.5	0.005545	0.005662	0.005782	0.005903	0.006025
15.0	0.006150	0.006276	0.006404	0.006534	0.006665
15.5	0.006799	0.006934	0.007071	0.007209	0.007350
16.0	0.007493	0.007637	0.007784	0.007932	0.008082
16.5	0.008234	0.008388	0.008544	0.008702	0.008863
17.0	0.009025	0.009189	0.009355	0.009523	0.009694
17.5	0.009866	0.010041	0.010217	0.010396	0.010577
18.0	0.010760	0.010946	0.011133	0.011323	0.011515
18.5	0.011709	0.011906	0.012105	0.012306	0.012509
19.0	0.012715	0.012923	0.013134	0.013346	0.013562
19.5	0.013779	0.013999	0.014222	0.014447	0.014674
20.0	0.014904	0.015137	0.015372	0.015609	0.015850
20.5	0.016092	0.016337	0.016585	0.016836	0.017089
21.0	0.017345	0.017603	0.017865	0.018129	0.018395
21.5	0.018665	0.018937	0.019212	0.019490	0.019770
22.0	0.020054	0.020340	0.020630	0.020921	0.021216
22.5	0.021514	0.021815	0.022119	0.022426	0.022736

TABLE 6 *(continued)*

φ (deg)	Inv (φ)	Inv ($\varphi + 0.1°$)	Inv ($\varphi + 0.2°$)	Inv ($\varphi + 0.3°$)	Inv ($\varphi + 0.4°$)
23.0	0.023049	0.023365	0.023684	0.024006	0.024332
23.5	0.024660	0.024992	0.025326	0.025664	0.026005
24.0	0.026350	0.026697	0.027048	0.027402	0.027760
24.5	0.028121	0.028485	0.028852	0.029223	0.029598
25.0	0.029975	0.030357	0.030741	0.031130	0.031521
25.5	0.031917	0.032315	0.032718	0.033124	0.033534
26.0	0.033947	0.034364	0.034785	0.035209	0.035637
26.5	0.036069	0.036505	0.036945	0.037388	0.037835
27.0	0.038287	0.038696	0.039201	0.039664	0.040131
27.5	0.040602	0.041076	0.041556	0.042039	0.042526
28.0	0.043017	0.043513	0.044012	0.044516	0.045024
28.5	0.045537	0.046054	0.046575	0.047100	0.047630
29.0	0.048164	0.048702	0.049245	0.049792	0.050344
29.5	0.050901	0.051462	0.052027	0.052597	0.053172
30.0	0.053751	0.054336	0.054924	0.055519	0.056116
30.5	0.056720	0.057267	0.057940	0.058558	0.059181
31.0	0.059809	0.060441	0.061779	0.061721	0.062369
31.5	0.063022	0.063680	0.064343	0.065012	0.065685
32.0	0.066364	0.067048	0.067738	0.068432	0.069133
32.5	0.069838	0.070549	0.071266	0.071988	0.072716
33.0	0.073449	0.074188	0.074932	0.075683	0.076439
33.5	0.077200	0.077968	0.078741	0.079520	0.080305
34.0	0.081097	0.081974	0.082697	0.083506	0.084321
34.5	0.085142	0.085970	0.086804	0.087644	0.088490
35.0	0.089342	0.090201	0.091066	0.091938	0.092816
35.5	0.093701	0.094592	0.095490	0.096395	0.097306
36.0	0.098224	0.099149	0.100080	0.101019	0.101964
36.5	0.102916	0.103875	0.104841	0.105814	0.106795
37.0	0.107782	0.108777	0.109779	0.110788	0.111805
37.5	0.112828	0.113860	0.114899	0.115945	0.116999
38.0	0.118060	0.119130	0.120207	0.121291	0.122384
38.5	0.123484	0.124592	0.125709	0.126833	0.127965
39.0	0.129106	0.130254	0.131411	0.132576	0.133749
39.5	0.134931	0.136122	0.137320	0.138528	0.139743
40.0	0.140968	0.142201	0.143443	0.144694	0.145954
40.5	0.147222	0.148500	0.149787	0.151082	0.152387
41.0	0.153702	0.155025	0.156358	0.157700	0.159052
41.5	0.160414	0.161785	0.163165	0.164556	0.165956
42.0	0.167366	0.168786	0.170216	0.171656	0.173106
42.5	0.174566	0.176037	0.177518	0.179009	0.180511
43.0	0.182023	0.183546	0.185080	0.186625	0.188180
43.5	0.189746	0.191324	0.192912	0.194511	0.196122
44.0	0.197744	0.199377	0.201022	0.202678	0.204346
44.5	0.206026	0.207717	0.209420	0.211135	0.212863
45.0	0.214602				

Appendix B

Answers to Selected Problems

1.3 $\gamma_{\min} = 53.1°$; $\gamma_{\max} = 98.1°$; at $\theta = 40.1°$, $\gamma = 59.1°$; at $\theta = 228.6°$, $\gamma = 90.9°$
1.5 (*a*) $m = 1$; (*b*) $m = 1$; (*c*) $m = 0$; (*d*) $m = 1$
1.7 Seven variations
1.9 $m = 1$
1.11 $m = 1$
1.13 $j_1 = 5, j_2 = 1$; no slipping at A
1.15 $Q = 1.10$
1.17 0.069 44 in. to left
1.19 One solution is $r_1 = 0.823$ m, $r_2 = 0.250$ m, $r_3 = 0.700$ m, $r_4 = 0500$ m

2.1 Spiral
2.3 $\mathbf{R}_{QP} = -7\hat{\mathbf{i}} - 14\hat{\mathbf{j}}$
2.5 $\mathbf{R}_A = -4.5a\hat{\mathbf{i}}$
2.7 Clockwise; $\mathbf{R} = 4\angle 0°$; $t = 20$; $\mathbf{R} = 404\angle 0°$; $\Delta\mathbf{R} = 400\angle 0°$
2.9 $\Delta\mathbf{R}_{P_3} = -2.121\hat{\mathbf{i}}_1 + 3.879\hat{\mathbf{j}}_1$; $\Delta\mathbf{R}_{P_3/2} = 3\hat{\mathbf{i}}_2$
2.11 $\Delta\mathbf{R}_Q = 1.902\hat{\mathbf{i}} + 1.098\hat{\mathbf{j}}$ in $= 2.196$ in $\angle 30°$
2.13 $R_C = 2.5\cos\theta_2 + \sqrt{48 - 5\sin\theta_2 - 6.25\sin^2\theta_2}$
2.15 Two loops; one constraint
2.17 Two loops; two constraints
2.19 Two loops; one constraints
2.21 Two loops; two constraints
2.25 $\theta_2 = \pm(2k+1)\pi/2 = \pm 90°, \pm 270°, \ldots$
2.27 (*a*) $\theta_2 = 29°$, $\theta_4 = 58°$; $\theta_2' = 248°$, $\theta_4' = 136°$; (*b*) $\Delta\theta_4 = 78°$; (*c*) $\gamma = 29°$, $\gamma' = 68°$.
2.29 $\Delta\theta_{drive} = 180° + \sin^{-1}\left(\dfrac{e}{r_3 - r_2}\right) - \sin^{-1}\left(\dfrac{e}{r_3 + r_2}\right)$;
$\Delta\theta_{return} = 180° + \sin^{-1}\left(\dfrac{e}{r_3 + r_2}\right) - \sin^{-1}\left(\dfrac{e}{r_3 - r_2}\right)$; assuming driving with B sliding to the right, the crank should rotate clockwise.

3.1 $\dot{\mathbf{R}} = 314$ in/s $\angle 162°$.
3.3 $\mathbf{V}_{BA} = \mathbf{V}_{B_3/2} = 82.6$ mi/h N 25° E
3.5 (*a*) $d = 1\,400$ mm; (*b*) $\mathbf{V}_{AB} = 60\hat{\mathbf{j}}$ m/s; $\mathbf{V}_{BA} = -60\hat{\mathbf{j}}$ m/s; $\omega_2 = 200$ rad/s cw
3.7 (*a*) Straight line at N 48° E; (*b*) no change

3.9 $\boldsymbol{\omega}_3 = 1.43$ rad/s ccw; $\boldsymbol{\omega}_4 = 15.40$ rad/s ccw
3.11 $\mathbf{V}_C = 23.7$ ft/s $\angle 284°$; $\boldsymbol{\omega}_3 = 0.33$ rad/s ccw
3.13 $\mathbf{V}_C = 0.402$ m/s $\angle 151°$; $\mathbf{V}_D = 0.290$ m/s $\angle 249°$
3.15 $\mathbf{V}_B = 4.79$ m/s $\angle 96°$; $\boldsymbol{\omega}_3 = \boldsymbol{\omega}_4 = 22$ rad/s ccw
3.17 $\boldsymbol{\omega}_6 = 4$ rad/s ccw; $\mathbf{V}_B = 0.964$ ft/s $\angle 180°$; $\mathbf{V}_C = 2.02$ ft/s $\angle 208°$; $\mathbf{V}_D = 2.02$ ft/s $\angle 206°$
3.19 $\boldsymbol{\omega}_3 = 3.23$ rad/s ccw; $\mathbf{V}_B = 16.9$ ft/s $\angle -56°$
3.21 $\mathbf{V}_C = 9.03$ m/s $\angle 138°$
3.23 $\mathbf{V}_B = 35.5$ ft/s $\angle 240°$; $\mathbf{V}_C = 40.6$ ft/s $\angle 267°$; $\mathbf{V}_D = 31.6$ ft/s $\angle -60°$
3.25 $\mathbf{V}_B = 1.04$ ft/s $\angle -23°$
3.27 $\boldsymbol{\omega}_3 = \boldsymbol{\omega}_4 = 14.4$ rad/s ccw; $\boldsymbol{\omega}_5 = \boldsymbol{\omega}_6 = 9.67$ rad/s cw; $\mathbf{V}_E = 77.4$ in/s $\angle -99°$
3.29 $\boldsymbol{\omega}_4 = 4.35$ rad/s ccw
3.31 $\mathbf{V}_A = 1.30$ in/s $\angle 180°$; $\mathbf{V}_B = 5.78$ in/s $\angle 180°$
3.32 $\boldsymbol{\omega}_3 = 30.0$ rad/s cw
3.41 $\theta_4' = -0.0577$ rad/in, $r_4' = -1.1547$ in/in; $\omega_3 = 0.520$ rad/s ccw, $\omega_4 = 0.173$ rad/s cw
3.43 $\theta_3' = 0.0577$ rad/in, $r_4' = 1.1547$ in/in; $\omega_3 = 0.289$ rad/s cw, $V_{3/4} = -5.774$ in/s
3.45 $r_3' = -1.1547$ m/m, $\theta_3' = 14.43$ rad/m, $\theta_4' = -43.30$ rad/m; $\omega_3 = 2.165$ rad/s ccw, $\omega_4 = 6.495$ rad/s cw, $V_E = 0.150$ m/s $\angle 150°$, $V_{B_4/3} = 0.173$ m/s $\angle -60°$
3.47 $\theta_3' = -0.25$ rad/in, $\theta_4' = 0$; $\omega_3 = 3.75$ rad/s cw, $\omega_4 = 0$; $\Delta = 0$ when $\theta_3 = \theta_4 = 68.907°$
3.49 $\theta_3' = -0.200$ rad/rad, $\theta_4' = 0$, $\theta_5' = -0.500$ rad/rad; $\omega_3 = 1.00$ rad/s cw, $\omega_4 = 0$, $\omega_5 = 2.50$ rad/s cw

4.1 $\ddot{\mathbf{R}} = -4\hat{\mathbf{i}}$ in/s^2
4.3 $\hat{\mathbf{u}}^t = 0.300\,71\hat{\mathbf{i}} - 0.953\,72\hat{\mathbf{j}}$; $A^n = -43.678$ mm/s^2; $A^t = 12.581$ mm/s^2; $\rho = -405.4$ mm
4.5 $\mathbf{A}_A = -7\,200\hat{\mathbf{i}} + 2\,400\hat{\mathbf{j}}$ m/s^2
4.7 $\mathbf{V}_B = 12.0$ ft/s $\angle 270°$; $\mathbf{V}_C = 8.367$ ft/s $\angle 12°$; $\mathbf{A}_B = 395.1$ ft/s^2 $\angle 165°$; $\mathbf{A}_C = 210.2$ ft/s^2 $\angle 240°$
4.9 $\boldsymbol{\omega}_2 = 38.64$ rad/s cw; $\boldsymbol{\alpha}_2 = 5\,571.3$ rad/s^2 cw
4.11 $\boldsymbol{\alpha}_3 = 563.3$ rad/s^2 ccw; $\boldsymbol{\alpha}_4 = 123.7$ rad/s^2 ccw
4.13 $\mathbf{A}_C = 3\,104.4$ ft/s^2 $\angle 114.4°$; $\boldsymbol{\alpha}_3 = 1\,741.6$ rad/s^2 ccw; $\boldsymbol{\alpha}_4 = 3\,055.8$ rad/s^2 ccw
4.15 $\mathbf{A}_C = 2\,604$ ft/s^2 $\angle -69°$; $\boldsymbol{\alpha}_4 = 1\,495$ rad/s^2 ccw
4.17 $\mathbf{A}_B = 16.7$ ft/s^2 $\angle 0°$; $\boldsymbol{\alpha}_3 = 17.49$ rad/s^2 ccw; $\boldsymbol{\alpha}_6 = 10.81$ rad/s^2 cw
4.19 $\boldsymbol{\alpha}_2 = 4\,181$ rad/s^2 ccw
4.21 $\mathbf{A}_C = 450.6$ m/s^2 $\angle 255.6°$; $\boldsymbol{\alpha}_3 = 74.08$ rad/s^2 cw
4.23 $\mathbf{A}_B = 2\,440.6$ ft/s^2 $\angle 240°$; $\mathbf{A}_D = 4\,031$ ft/s^2 $\angle 120°$
4.25 $\theta_3 = 171°$; $\theta_4 = 196°$; $\boldsymbol{\omega}_3 = 70.5$ rad/s ccw; $\boldsymbol{\omega}_4 = 47.6$ rad/s ccw; $\boldsymbol{\alpha}_3 = 3\,200$ rad/s^2 ccw; $\boldsymbol{\alpha}_4 = 3\,331$ rad/s^2 ccw
4.27 $\theta_3 = 28.3°$; $\theta_4 = 55.9°$; $\boldsymbol{\omega}_3 = 0.633$ rad/s cw; $\boldsymbol{\omega}_4 = 2.16$ rad/s cw; $\boldsymbol{\alpha}_3 = 7.82$ rad/s^2 ccw; $\boldsymbol{\alpha}_4 = 6.70$ rad/s^2 ccw
4.29 $\theta_3 = 38.4°$; $\theta_4 = 155.6°$; $\boldsymbol{\omega}_3 = 6.86$ rad/s cw; $\boldsymbol{\omega}_4 = 1.23$ rad/s cw; $\boldsymbol{\alpha}_3 = 62.5$ rad/s^2 ccw; $\boldsymbol{\alpha}_4 = 96.5$ rad/s^2 cw

4.31 $\mathbf{V}_B = 184$ in/s $\angle -19.1°$; $\mathbf{A}_B = 2\,703$ in/s^2 $\angle -172.5°$; $\boldsymbol{\omega}_4 = 6.57$ rad/s cw; $\boldsymbol{\alpha}_4 = 86.4$ rad/s^2 ccw

4.33 $\mathbf{A}_E = 602.7$ ft/s^2 $\angle -107.2°$

4.35 $\mathbf{A}_B = 1.69$ ft/s^2 $\angle -102.1°$; $\alpha_3 = 1.247$ rad/s^2 cw.

4.37 $\mathbf{A}_{C_4} = 50.0$ ft/s^2 $\angle 91°$; $\alpha_3 = 6.0$ rad/s^2 ccw

4.39 $\mathbf{A}_G = 29.3$ ft/s^2 $\angle -64.5$; $\boldsymbol{\alpha}_5 = 0$; $\boldsymbol{\alpha}_6 = 110$ rad/s^2 cw

4.41 $\theta_3'' = 0.240$ rad/rad^2, $\theta_4'' = 0.150$ rad/rad^2, $\theta_5'' = 0$; $\alpha_3 = 6.0$ rad/s^2 ccw, $\alpha_4 = 3.75$ rad/s^2 ccw; $\alpha_5 = 0$

4.47 $\rho_C = 43.77$ m; $v = 56.584$ m/s; $A_I = 2\,070.97$ m/s^2; $V_A = 48.971$ m/s; $V_C = 44.103$ m/s; $A_A = 757.89$ m/s^2; $A_C = 378.95$ m/s^2.

4.49 $\rho_B = 3$ in; $\rho_E = 4.24$ in; $V_B = 10.0$ in/s; $V_E = 14.1$ in/s; $v = 13.3$ in/s; $A_B = 56.05$ in/s^2; $A_E = 165.58$ in/s^2; $A_I = 133.3$ in/s^2.

5.1 $\mathbf{V}_{P_4} = 2.355$ m/s $\angle 15.6°$, $\mathbf{A}_{P_4} = 125.7$ m/s^2 $\angle -66.6°$

5.3 $\omega_3 = 41.64$ rad/s cw; $\omega_5 = 47.20$ rad/s ccw

5.5 $\mathbf{V}_{P_3} = 47.17$ in/s $\angle -175.1°$; $\mathbf{A}_{P_3} = 447.21$ in/s^2$\angle -78.7°$

5.7 $\theta_4 = 101.39°$; $\theta_{5/4} = 61.45°$

5.9 $\omega_4 =$ rad/s cw; $\omega_{5/4} = 17.18$ rad/s ccw

5.11 $\alpha_4 = 91.94$ rad/s^2 ccw; $\alpha_{5/4} = 37.82$ rad/s^2 ccw;

6.3 Face $= 195$ mm from pivot

6.5 $\dot{y}\left(\frac{\beta}{2}\right) = \frac{\pi L}{2\beta}\omega$; $\dddot{y}\left(\frac{\beta}{2}\right) = -\frac{\pi^3 L}{2\beta^3}\omega^3$; $\ddot{y}\,(0) = \frac{\pi^2 L}{2\beta^2}\omega^2$; $\ddot{y}\,(\beta) = -\frac{\pi^2 L}{2\beta^2}\omega^2$

6.7 AB: dwell, $L_1 = 0$, $\beta_1 = 60°$; BC: full-rise eighth-order polynomial motion, Eq. (6.14), $L_2 = 2.5$ in., $\beta_2 = 62.442°$; CD: half-harmonic return motion, Eq. (6.20), $L_3 = 0.082\,4$ in., $\beta_3 = 7.750°$; DE: uniform motion, $L_4 = 1.0$ in., $\beta_4 = 60°$; EA: half-cycloidal return motion, Eq. (6.25), $L_5 = 1.417\,6$ in., $\beta_5 = 169.808°$

6.9 $t_{AB} = 0.025$ s; $\dot{y}_{\max} = 506$ in/s; $\dot{y}_{\min} = -40$ in/s; $\ddot{y}_{\max} = 19\,457$ in/s^2; $\ddot{y}_{\min} = -19\,457$ in/s^2

6.11 $\dot{y}_{\max} = 41.9$ rad/s; $\ddot{y}_{\max} = 7\,900$ rad/s^2

6.13 Face width $= 2.200$ in.; $\rho_{\min} = 3.000$ in.

6.15 $R_0 > 9.09$ in.; face width > 13.34 in.

6.17 $\phi_{\max} = 12°$; $R_r < 14.5$ in.

6.19 $R_0 > 56$ mm; $\ddot{y}_{\max} = 37$ m/s^2

6.21 $R_0 > 71$ mm; $\ddot{y}_{\max} = 39.5$ m/s^2

6.23 $u = (R_0 + R_C + y)\sin\theta + y'\cos\theta$; $v = (R_0 + R_C + y)\cos\theta - y'\sin\theta$

6.27 (*i*) $y = 6.25$ mm, $y' = 21.65$ mm/rad, $y'' = 25.0$ mm/rad^2, $y''' = -86.60$ mm/rad^3; (*ii*) $u_{cam} = 36.56$ mm, $v_{cam} = 8.76$ mm; (*iii*) $\rho = -87.02$ mm; (*iv*) $\hat{\mathbf{u}}^t = 0.746\hat{\mathbf{i}} - 0.666\hat{\mathbf{j}}$, $\hat{\mathbf{u}}^n = 0.666\hat{\mathbf{i}} + 0.746\hat{\mathbf{j}}$; (*v*) $\phi = 8.2°$

7.1 $P = 16$ teeth/in.

7.3 $m = 2$ mm/tooth

7.5 $P = 0.897\,6$ teeth/in; $D = 44.563$ in.
7.7 $m = 12.732$ mm/tooth; $D = 458.4$ mm
7.9 $D = 9.191$ in.
7.11 $N_2 = 17$ teeth; $N_3 = 51$ teeth
7.13 $a = 0.250$ in; $d = 0.312\,5$ in; $c = 0.062\,5$ in; $p_c = 0.785\,4$ in/tooth; $p_b = 0.738\,0$ in/tooth; $t = 0.392\,7$ in; $R_2 = 3.000$ in, $R_3 = 4.500$ in; $r_2 = 2.819$ in; $r_3 = 4.229$ in; $CP = 0.625$ in; $PD = 0.591$ in; $m_c = 1.647$ teeth
7.15 $\alpha_2 = 18.6°$; $\alpha_3 = 6.3°$; $\beta_2 = 16.0°$; $\beta_3 = 5.5°$; $m_c = 1.63$ teeth
7.17 $CP = 18.930$ mm; $PD = 16.243$ mm; $m_c = 1.54$ teeth
7.19 (*a*) $CP = 1.462$ in; $PD = 1.196$ in; $m_c = 1.80$ teeth
(*b*) $CP = 1.096$ in; $PD = 1.196$ in; $m_c = 1.55$ teeth; no change in pressure angle
7.25 $t_b = 17.142$ mm; $t_a = 6.737$ mm; $\varphi_a = 32.8°$
7.27 $t_b = 1.146$ in.
7.29 $t_b = 0.159\,4$ in $\varphi_a = 35.3$; $t_a = 0.041\,5$ in.
7.31 (*a*) 0.168 2 in; (*b*) 9.827 0 in.
7.33 $m_c = 1.66$
7.35 $m_c = 1.77$
7.37 $\phi = 26.24°$
7.39 $a_3 = 1.344$ in.
7.41 $a'_3 = 0.712\,3$ in, $d'_3 = 1.637\,7$ in, $a'_2 = 1.287\,7$ in, $d'_2 = 1.062\,3$ in, $e_2 = -0.062\,3$ in, $e_3 = -0.637\,7$ in, $m'_c = 1.49$ teeth

8.1 $p_t = 0.524$ in/tooth, $p_n = 0.370$ in/tooth, $P_n = 8.49$ teeth/in, $R_2 = 1.25$ in, $R_3 = 2.0$ in, $N_{e2} = 42.43$ teeth, $N_{e3} = 67.88$ teeth
8.3 $\psi_b = 33.34°$, $m_x = 1.67$ teeth
8.5 $N_3 = 72$ teeth, $R_2 = 2.0$ in, $R_3 = 9.0$ in, $p_n = 0.723$ in/tooth, $P_n = 4.345$ teeth/in, $F \approx 3.700$ in.
8.7 $m_n = 1.79$ teeth, $m_{\text{total}} = 2.87$ teeth
8.9 $\psi_3 = 15°$ RH, $N_3 = 39$ teeth, $R_2 = 0.966$ in, $R_3 = 1.682$ in.
8.11 $\gamma_2 = 18.4°$, $\gamma_3 = 71.6°$
8.13 $\gamma_2 = 27.0°$, $\gamma_3 = 93.0°$
8.15 $R_2 = 1.063$ in, $R_3 = 1.750$ in, $\gamma_2 = 34.8°$, $\gamma_3 = 70.2°$, $a_2 = 0.168\,1$ in, $a_3 = 0.081\,9$ in, $d_2 = 0.105\,4$ in, $d_3 = 0.191\,6$ in, $F = 0.558$ in, $N_{e2} = 20.71$, $N_{e3} = 82.54$
8.17 $N_2 = 1$ tooth, $R_2 = 1.725$ in, $N_3 = 60$ teeth, $R_3 = 4.775$ in.
8.19 $N_3 = 123.6$ teeth, $R_3 = 7.871$ in, $\psi = 20.0°$, $R_2 = 0.525$ in, $R_2 + R_3 = 8.396$ in.

9.1 $\omega_8 = 68.18$ rev/min ccw; $\theta'_{82} = 5/88$
9.3 $\omega_9 = 11.84$ rev/min cw
9.5 One solution: $N_3 = 30T, N_4 = 25T, N_5 = 30T, N_6 = 20T, N_7 = 25T, N_8 = 35T, N_{10} = 40T$
9.7 $\omega_7 = 222.1$ rev/min ccw
9.9 $\omega_2 = 644.8$ rev/min cw
9.11 77.3%, $\omega_3 = (-5/22)\omega_2$ or opposite in direction; replace gears 4 and 5 with a single 44-tooth gear

9.13 (*a*) $N_5 = 84T$, $R_A = 156$ mm; (*b*) $\omega_A = 8.08$ rev/min ccw
9.15 (*a*) $\omega_R = 651.3$ rev/min, $\omega_L = 693.3$ rev/min; (*b*) $\omega_A = 672.3$ rev/min
9.17 $\omega_A = (9/34)\omega_2$; Lévai type-F
9.19 59.1%

10.1 For six points, 0.170, 1.464, 3.706, 6.294, 8.536, and 9.830
10.3 Typical solution: $r_2 = 7.4$ in., $r_3 = 20.9$ in., $e = 8$ in
10.5 Typical solution: $r_1 = 7.63$ ft, $r_2 = 3.22$ ft, $r_3 = 8.48$ ft
10.7 Typical solution: O_2 at $x = -1\,790$ mm, $y = 320$ mm, $r_2 = 360$ mm, $r_3 = 1\,990$ mm
10.9 Typical solution: $O_2A = AB = O_4B = O_2O_4$ and spring AO_4 with chosen free length
10.13 and 10.23 $r_2/r_1 = 1.834$, $r_3/r_1 = 2.238$, $r_4/r_1 = -0.693$
10.15 and 10.25 $r_2/r_1 = -3.499$, $r_3/r_1 = 0.878$, $r_4/r_1 = 3.399$
10.17 and 10.27 $r_2/r_1 = 0.625$, $r_3/r_1 = 1.309$, $r_4/r_1 = -0.401$
10.19 and 10.29 $r_2/r_1 = -1.801$, $r_3/r_1 = 0.908$, $r_4/r_1 = 1.274$
10.21 and 10.31 $r_2/r_1 = -0.610$, $r_3/r_1 = 0.565$, $r_4/r_1 = 0.380$

11.1 $m = 2$ including one idle freedom. Path of B is the intersection of cylinder of radius BA about y axis and sphere of radius BO_3 about O_3.
11.3 $\boldsymbol{\omega}_2 = -2.58\hat{\mathbf{j}}$ rad/s; $\boldsymbol{\omega}_3 = 1.16\hat{\mathbf{i}} - 0.09\hat{\mathbf{j}} + 0.64\hat{\mathbf{k}}$ rad/s; $\mathbf{V}_B = -0.096\hat{\mathbf{i}} - 0.050\hat{\mathbf{j}} + 0.167\hat{\mathbf{k}}$ m/s
11.5 and 11.7 $\boldsymbol{\omega}_3 = \boldsymbol{\omega}_4 = -25.7\hat{\mathbf{i}}$ rad/s; $\mathbf{V}_A = 180\hat{\mathbf{j}}$ in/s; $\mathbf{V}_B = -231\hat{\mathbf{k}}$ in/s; $\boldsymbol{\alpha}_3 = 1\,543\,\hat{\mathbf{j}}$ rad/s^2; $\boldsymbol{\alpha}_4 = -343\hat{\mathbf{i}}$ rad/s^2; $\mathbf{A}_A = 10\,800\hat{\mathbf{i}}$ in/s^2; $\mathbf{A}_B = -5\,950\hat{\mathbf{j}} - 3\,087\hat{\mathbf{k}}$ in/s^2
11.9 $\Delta\theta_4 = 46.5°$; $Q = 1.01$
11.11 and 11.13 $\boldsymbol{\omega}_3 = 4.08\hat{\mathbf{i}} - 7.07\hat{\mathbf{j}} + 3.34\hat{\mathbf{k}}$ rad/s; $\boldsymbol{\omega}_4 = 10.80\hat{\mathbf{i}}$ rad/s; $\mathbf{V}_A = -1.15\hat{\mathbf{i}} - 0.65\hat{\mathbf{j}}$ m/s; $\mathbf{V}_B = 1.21\hat{\mathbf{j}} + 2.56\hat{\mathbf{k}}$ m/s; $\boldsymbol{\alpha}_3 = 273\hat{\mathbf{i}} + 115\hat{\mathbf{j}} - 148\hat{\mathbf{k}}$ rad/s^2; $\boldsymbol{\alpha}_4 = 130\hat{\mathbf{i}}$ rad/s^2; $\mathbf{A}_A = 23.3\hat{\mathbf{i}} - 41.4\hat{\mathbf{j}}$ m/s^2; $\mathbf{A}_B = -13.04\hat{\mathbf{j}} + 43.91\hat{\mathbf{k}}$m/s^2
11.15 $\boldsymbol{\omega}_2 = 20.8\hat{\mathbf{i}} - 12.0\hat{\mathbf{j}}$ rad/s; $\boldsymbol{\omega}_3 = 2.31\hat{\mathbf{i}} + 6.66\hat{\mathbf{j}} - 3.23\hat{\mathbf{k}}$ rad/s; $\mathbf{V}_A = 12.0\hat{\mathbf{i}} + 20.8\hat{\mathbf{j}} + 41.6\hat{\mathbf{k}}$ in/s; $\mathbf{V}_B = 13.8\hat{\mathbf{i}}$ in/s
11.17 and 11.19 $\boldsymbol{\omega}_2 = 12.0\hat{\mathbf{i}} - 20.8\hat{\mathbf{j}}$ rad/s; $\boldsymbol{\omega}_3 = 1.33\hat{\mathbf{i}} + 6.91\hat{\mathbf{j}} - 1.82\hat{\mathbf{k}}$ rad/s; $\mathbf{V}_A = 20.8\hat{\mathbf{i}} + 12.0\hat{\mathbf{j}} + 41.6\hat{\mathbf{k}}$ in/s; $\mathbf{V}_B = 26.1\hat{\mathbf{i}}$ in/s
11.21 and 11.23 $\mathbf{V}_B = 112$ in/s; $\omega_3 = -4.19\hat{\mathbf{i}} + 19.4\hat{\mathbf{j}} - 5.47\hat{\mathbf{k}}$rad/s; $\omega_4 = 19.4\hat{\mathbf{j}}$rad/s

12.1 $T_{15} = \begin{bmatrix} 0.866 & -0.500 & 0 & 17.32 \text{ in} \\ -0.500 & -0.866 & 0 & 0 \\ 0 & 0 & -1 & 8.00 \text{ in} \\ 0 & 0 & 0 & 1 \end{bmatrix}$; $\mathbf{R} = 17.32\hat{\mathbf{i}}_1 + 6.50\hat{\mathbf{k}}_1$ in.

12.3 $T_{15} = \begin{bmatrix} 0 & -1 & 0 & 180 \text{ mm} \\ 0 & 0 & -1 & -100 \text{ mm} \\ 1 & 0 & 0 & 450 \text{ mm} \\ 0 & 0 & 0 & 1 \end{bmatrix}$; $\mathbf{R} = 180\hat{\mathbf{i}}_1 - 145\hat{\mathbf{j}}_1 + 450\hat{\mathbf{k}}_1$ mm

12.5 $\mathbf{V} = -1.750\hat{\mathbf{i}}_1 + 0.433\hat{\mathbf{j}}_1$ in/s; $\mathbf{A} = -0.541\hat{\mathbf{i}}_1 - 0.088\hat{\mathbf{j}}_1$ in/s^2

12.7 $\phi_2 = \cos^{-1}\left(0.020t^2 + 0.050t - 0.875\right)$ with $-180° \le \phi_2 \le 0$

$\phi_1 = \tan^{-1}\left[\dfrac{3\,(1+\cos\phi_2) - 4\sin\phi_2}{4\,(1+\cos\phi_2) + 3\sin\phi_2}\right]$; $\phi_3 = 8$ in; $\phi_4 = \phi_1 + \phi_2 - 36.87°$

12.9 $\tau_1 = 25$ in · lb; $\tau_2 = 100\sin\phi_1 + 25$ in · lb; $\tau_3 = -5$ lb;
$\tau_4 = -100\sin\phi_1 - 100\sin(\phi_1+\phi_2) - 25$ in · lb

13.3 $P = 1\,460$ N

13.5 $\mathbf{M}_{12} = 90.6\hat{\mathbf{k}}$ in · lb

13.7 $\mathbf{M}_{12} = 383\hat{\mathbf{k}}$ in · lb

13.9 $\mathbf{M}_{12} = 252\hat{\mathbf{k}}$ in · lb

13.11 $\mathbf{M}_{12} = -756\hat{\mathbf{k}}$ in · lb, $\mathbf{F}_{12} = \mathbf{F}_{23} = 226$ lb $\angle 56.7°$, $\mathbf{F}_{14} = 318$ lb $\angle -61.4°$,
$\mathbf{F}_{34} = 189$ lb $\angle 88.8°$

13.13 $\mathbf{F}_{12} = \mathbf{F}_{23} = 307$ kN $\angle 230.4°$; $\mathbf{F}_{14} = \mathbf{F}_{43} = 387$ kN $\angle 59.7°$;
$\mathbf{M}_{12} = -113\hat{\mathbf{k}}$ kN · m

13.15 $\mathbf{M}_{12} = 426\hat{\mathbf{k}}$ in · lb

13.17 (*a*) $\mathbf{F}_{13} = 2\,520$ lb $\angle 0°$; (*b*) $\mathbf{F}_{13} = 1\,050$ lb $\angle 225°$; (*c*) $\mathbf{F}_{13} = 2\,250$ lb $\angle 315°$

13.19 $\mathbf{F}_C = 216$ lb $\angle 189°$; $\mathbf{F}_D = 350$ lb $\angle 163.4°$

13.21 $\mathbf{F}_C = 118\hat{\mathbf{i}} + 140\hat{\mathbf{j}} - 251\hat{\mathbf{k}}$ lb; $\mathbf{F}_D = -71\hat{\mathbf{i}} + 155\hat{\mathbf{k}}$ lb

13.23 $\mathbf{F}_E = 162\hat{\mathbf{i}} - 191\hat{\mathbf{j}} + 354\hat{\mathbf{k}}$ lb; $\mathbf{F}_F = 110\hat{\mathbf{j}} + 144\hat{\mathbf{k}}$ lb

13.25 $\mathbf{M}_{12} = -61.5\hat{\mathbf{k}}$ N · m

13.29 $\mathbf{F}_F = 378\hat{\mathbf{i}} + 1\,106\hat{\mathbf{j}}$ lb; $\mathbf{F}_R = 1\,239\hat{\mathbf{j}}$ lb; $\mathbf{F}_T = -378\hat{\mathbf{i}} + 655\hat{\mathbf{j}}$ lb; $\mu \ge 0.34$

13.31 $\mathbf{M}_{12} = 0.170\hat{\mathbf{k}}$ N · m

13.33 $T = F_B d\,(R/X)^2$; $X \ge R\sqrt{1 + 1/\mu^2}$

13.35 $d = 51.5$ mm

13.37 (*i*) $P_B = 60\,000$ lb; (*ii*) $P_{cr} = 89\,717$ lb, $P_{cr}/A = 28\,558$ lb/in^2, $N = 1.15$;
(*iii*) $D = 2.77$ in.

13.39 (*i*) $(S_r)_D = 141.18$; (*ii*) $P_{cr} = 239.14$ kN, $N = 4.78$; (*iii*) $b = 25.5$ mm

13.41 (*i*) $(S_r)_D = 140.50$; (*ii*) $P_{cr} = 63\,213$ lb; (*iii*) $F = 42\,142$ lb; (*iv*) $L = 16.5$ ft

14.1 $I_O = 0.030\,8$ in · lb · s^2

14.3 $\mathbf{M}_{12} = -192\hat{\mathbf{k}}$ in · lb

14.5 $\mathbf{F}_{23} = \mathbf{F}_{12} = 1\,525$ lb $\angle 183°$, $\mathbf{F}_{14} = 444$ lb $\angle -90°$, $\mathbf{F}_{34} = 826$ lb $\angle 147.5°$,
$\mathbf{M}_{12} = 3\,080\hat{\mathbf{k}}$ in · lb

14.7 $\mathbf{F}_{23} = \mathbf{F}_{12} = 9.827$ kN $\angle -20.5°$, $\mathbf{F}_{14} = 11.521$ kN $\angle -154.7°$,
$\mathbf{F}_{34} = 11.171$ kN $\angle -14.5°$, $\mathbf{M}_{12} = -2\,907\hat{\mathbf{k}}$ N · m

14.9 $\boldsymbol{\omega}_3 = 3.02\hat{\mathbf{k}}$ rad/s, $\boldsymbol{\omega}_4 = 3.61\hat{\mathbf{k}}$ rad/s, $\boldsymbol{\alpha}_3 = -0.39\hat{\mathbf{k}}$ rad/s^2, $\boldsymbol{\alpha}_4 = -40.82\hat{\mathbf{k}}$ rad/s^2,
$\mathbf{A}_{G_3} = 40.37$ m/s$^2 \angle -17.16°$, $\mathbf{A}_{G_4} = 19.27$ m/s$^2 \angle -8.48°$,
$\mathbf{F}_{23} = \mathbf{F}_{12} = 2.56$ kN $\angle -129.7°$, $\mathbf{F}_{14} = 6.97$ kN $\angle -84.1°$,
$\mathbf{F}_{34} = 4.34$ kN $\angle 196°$, $\mathbf{M}_{12} = 668\hat{\mathbf{k}}$ N · m

14.11 $\mathbf{F}_{23} = \mathbf{F}_{12} = 9.43$ kN $\angle 10.7°$, $\mathbf{F}_{14} = 9.26$ kN $\angle -98.8°$, $\mathbf{F}_{34} = 2.70$ kN $\angle 156°$,
$\mathbf{M}_{12} = 2\,780\hat{\mathbf{k}}$ N · m

14.13 $\boldsymbol{\omega}_3 = 10.6\hat{\mathbf{k}}$ rad/s, $\boldsymbol{\omega}_4 = -4.3\hat{\mathbf{k}}$ rad/s, $\boldsymbol{\alpha}_3 = -200\hat{\mathbf{k}}$ rad/s^2, $\boldsymbol{\alpha}_4 = -352\hat{\mathbf{k}}$ rad/s^2, $\mathbf{A}_{G_3} = 102.4$ m/s$^2\angle 59.00°$, $\mathbf{A}_{G_4} = 43.94$ m/s$^2\angle 43.92°$, $\mathbf{F}_{23} = \mathbf{F}_{12} = 169$ N $\angle 47.0°$, $\mathbf{F}_{34} = 659$ N $\angle -39.9°$, $\mathbf{F}_{14} = 655$ N $\angle 134.4°$, $\mathbf{M}_{12} = 11.0\hat{\mathbf{k}}$ N·m

14.15 $\boldsymbol{\omega}_3 = -2.57\hat{\mathbf{k}}$ rad/s, $\mathbf{V}_B = 1.18\hat{\mathbf{i}}$ m/s, $\boldsymbol{\alpha}_3 = 77.19\hat{\mathbf{k}}$ rad/s^2, $\mathbf{A}_B = 25.16\hat{\mathbf{i}}$ m/s^2, $\mathbf{A}_{G_3} = 16.52$ m/s$^2\angle -28.85°$, $\mathbf{F}_{23} = \mathbf{F}_{12} = 3.25$ kN $\angle -12.42°$, $\mathbf{F}_{34} = 2.16$ kN $\angle -17.21°$, $\mathbf{F}_{14} = 0.639\hat{\mathbf{j}}$ kN, $\mathbf{M}_{12} = -241\hat{\mathbf{k}}$ N·m

14.17 $\mathbf{F}_{23} = 191\hat{\mathbf{i}} - 78\hat{\mathbf{j}}$ N, $\mathbf{F}_{14} = -1\hat{\mathbf{j}}$ N

14.19 $\boldsymbol{\omega}_3 = 0.611\hat{\mathbf{k}}$ rad/s, $\mathbf{V}_B = -1.16\hat{\mathbf{i}}$ m/s, $\boldsymbol{\alpha}_3 = 6.28\hat{\mathbf{k}}$ rad/s^2, $\mathbf{A}_B = 5.46\hat{\mathbf{i}}$ m/s^2, $\mathbf{F}_{12} = \mathbf{F}_{23} = 55.6$ kN $\angle 78.41°$, $\mathbf{F}_{14} = 14.3\hat{\mathbf{j}}$ kN, $\mathbf{M}_{12} = -9.240\hat{\mathbf{k}}$ kN·m

14.21 $\mathbf{F}_{12} = \mathbf{F}_{23} = 8.83$ lb $\angle -120°$, $\mathbf{F}_{14} = \mathbf{F}_{43} = 41.41$ lb $\angle 79.4°$, $\mathbf{M}_{14} = 15.77\hat{\mathbf{k}}$ in·lb

14.23 At $\theta_2 = 0°$, $\theta_3 = 120°$, $\theta_4 = 141.8°$, $\omega_3 = 6.67$ rad/s cw, $\omega_4 = 6.67$ rad/s cw, $\alpha_3 = 141$ rad/s^2 cw, $\alpha_4 = 64.1$ rad/s^2 cw, $\mathbf{F}_{21} = 6\,734$ lb $\angle -56°$, $\mathbf{F}_{21} = 7\,883$ lb $\angle 142.8°$, $\mathbf{M}_{12} = 7\,468\hat{\mathbf{k}}$ ft·lb

14.25 $\mathbf{M}_{12} = 0.168\hat{\mathbf{k}}$ N·m

14.27 $\boldsymbol{\alpha}_2 = -262\hat{\mathbf{k}}$ rad/s^2, $\mathbf{F}_{12} = 600\hat{\mathbf{i}} + 180\hat{\mathbf{j}}$ lb

14.29 (*i*) $R'_S = 1$ m/m, $R'_C = 1.732$ m/m, (*ii*) $m_{EQ} = 11.25$ kg, (*iii*) $P = -705.66$ N

14.31 (*iii*) $\mathbf{F}_{12} = -81\hat{\mathbf{i}} + 628\hat{\mathbf{j}}$ N, $\mathbf{F}_{23} = -54\hat{\mathbf{i}} + 523\hat{\mathbf{j}}$ N, $\mathbf{F}_{34} = 361\hat{\mathbf{j}}$ N, $\mathbf{F}_{14} = \pm 361\hat{\mathbf{i}}$ N (ccw couple). (*iv*) $\mathbf{T}_2 = 230\hat{\mathbf{k}}$ N·m. (*v*) Contact at top right and bottom left.

14.33 (*iii*) $\mathbf{F}_{12} = 119.2\hat{\mathbf{j}}$ N, $\mathbf{F}_{23} = 67.7\hat{\mathbf{i}} + 89.8\hat{\mathbf{j}}$ N, $\mathbf{F}_{13} = 35.5$ N $\angle -120°$. (*iv*) $\mathbf{P} = 97.7\hat{\mathbf{i}}$ N. (*v*) Contact at 0.125 m to right of G_3.

14.35 (*i*) $r'_S = 1$ m/m, $r'_C = -1$ m/m, $x'_{G_3} = 1$ m/m, $y'_{G_3} = 0$, $x''_{G_3} = -1.5$ m/m^2, $y''_{G_3} = -2.598$ m/m^2. (*ii*) $m_{EQ} = 2$ kg. (*iv*) $\mathbf{P} = 97.1\hat{\mathbf{i}}$ N.

14.37 $w_B = w_F = 1.33$ lb

14.39 $w_B = w_E = 2.67$ lb, $\theta_B = \theta_E = 180°$

14.41 $I_G = 2.89$ in·lb·s^2, $k_G = 5.22$ in.

14.45 $\omega_3 = \left(1 - 3R/4r\right)\omega_2$; $\omega_3 = 0$ for $r = 3R/4$

14.47 $\alpha_{in} = 512$ rad/s^2, $F_{2A} = 338.1$ lb, $F_{11} = 221.4$ lb; unbalanced moment on housing of 4 190 in.·lb must be absorbed by mounting bolts during startup.

14.49 $\mathbf{F}_{12} = -87.4\hat{\mathbf{j}}$ lb, $\mathbf{M}_{12} = -1\,200\hat{\mathbf{j}}$ in·lb, $\mathbf{F}_{35} = -238\hat{\mathbf{i}} - 300\hat{\mathbf{k}}$ lb, $\mathbf{M}_{35} = 258\hat{\mathbf{k}}$ in·lb, $\mathbf{F}_{45} = 238\hat{\mathbf{i}} + 300\hat{\mathbf{k}}$ lb, $\mathbf{M}_{45} = -258\hat{\mathbf{k}}$ in·lb, $\mathbf{F}_{15} = \mathbf{0}$, $\mathbf{M}_{15} = 2\,400\hat{\mathbf{j}}$ in·lb

15.3 (*a*) 632 N/m; (*b*) 15.5 mm; (*c*) 0.25 s/rev; (*d*) 4 Hz; (*e*) $\dot{x} = 0.448$ m/s; $\ddot{x} = 5.74$ m/s^2; (*f*) -5.74 N

15.5 3.010 Hz

15.7 (*a*) $\omega_n = \sqrt{\dfrac{g}{\ell}\left[\dfrac{(k_1 + k_2)\,a^2}{W\ell} - 1\right]}$, (*b*) $\dfrac{\ell}{a} \geq (k_1 + k_2)\,\dfrac{a}{W}$

15.9 (*a*) First era: $0 \leq t \leq 0.040$ s, $x = (1.429 \text{ mm})\,(1 - \cos 171t)$; second era: $t > 0.040$ s, $x = (0.773 \text{ mm})\cos(171t - 74.3°)$ (*b*) First era: $X = 1.429$ mm; second era: $X = 0.773$ mm

15.11 (*a*) 47.7 rev/min; (*b*) 16.7 rad/s; (*c*) 1.71°, 3.18°

15.13 0.177 in.

15.15 (*a*) $c_c = 2\,993$ N · s/m; (*b*) $\omega_n = 37.4$ rad/s; (*c*) $\omega_d = 36.7$ rad/s; (*d*) $\delta = 1.28$
15.17 (*a*) $\omega_d = 35.15$ rad/s, (*b*) $\tau = 0.179$ s/cycle, $\delta = 1.28$
15.21 $X = 1.280$ mm, $\phi = 55.8°$
15.23 $k = 2\,637$ kN/m, $X = 0.151$ mm
15.25 (*i*) Upper limit ($\omega_1 = 112$ rad/s) is Rayleigh–Ritz, lower limit ($\omega_1 = 100$ rad/s) is Dunkerley. Exact is $\omega_1 = 110$ rad/s. (*ii*) $m = 0.5$ kg. (*iii*) $a_{12} = 5.94 \times 10^{-5}$ m/N
15.27 (*i*) $\omega_1 = 665.6$ rad/s (ii) $\omega_1 = 667.2$ rad/s

16.1 At $X = 30\%$, $p_e = 1\,512$ kPa, $p_c = 338$ kPa.
16.3 See figure.

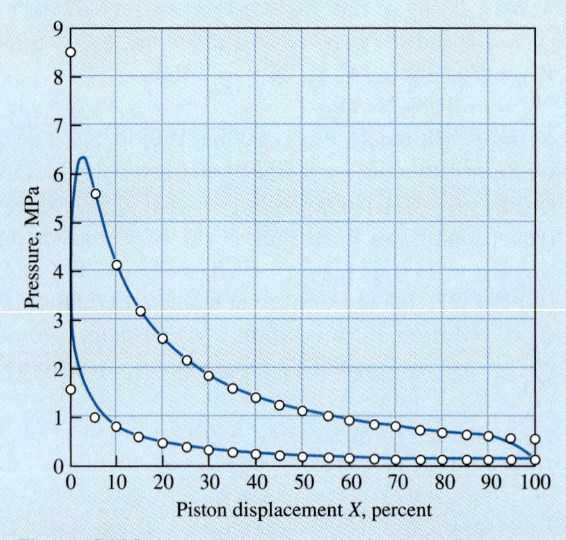

Figure AP16.3

16.5 $\mathbf{F}_{41} = -1010\hat{\mathbf{j}}$ N, $\mathbf{F}_{34} = 4\,254$ N $\angle - 13.7°$, $\mathbf{F}_{32} = 4\,299$ N $\angle 166.4°$, $\mathbf{T}_{21} = 89.59\hat{\mathbf{k}}$ N · m
16.9 $\mathbf{F}_{41} = -574\hat{\mathbf{j}}$ N, $\mathbf{F}_{34} = 2\,992$ N $\angle 191°$, $\mathbf{F}_{32} = 9\,479$ N $\angle - 35.7°$, $\mathbf{T}_{21} = 182\hat{\mathbf{k}}$ N · m
16.11 At $\omega t = 60°$, $X = 40.9\%$, $P = 4\,445$ N, $\ddot{x} = -1\,335$ m/s^2, $F_{14} = 561$ N, $T_{21} = 110.6$ N · m

17.1 $\mathbf{F}_A = 64.7 \angle 76.1°$ N, $\mathbf{F}_B = 16.2 \angle 76.1°$ N, $m_c = 1.64$ kg, $\theta_C = 76.1°$
17.3 $\mathbf{F}_A = 8.06 \angle - 14.4°$ lb, $\mathbf{F}_B = 2.69 \angle 165.5°$ lb, $w_C = -2.63$ lb at $\theta_C = -14.4°$
17.5 $\mathbf{F}_A = 13.16 \angle 90°$ N, $\mathbf{F}_B = 0$
17.7 $m_L\mathbf{R}_L = 5.98$ oz · in $\angle - 16.5°$, $m_R\mathbf{R}_R = 7.33$ oz · in $\angle 136.8°$
17.9 Remove $m_L\mathbf{R}_L = 782.1$ g · mm $\angle 180.4°$ and $m_R\mathbf{R}_R = 236.8$ g · mm $\angle 301.2°$
17.11 See Section 17.8 for answers.

18.1 119°
18.3 189 rev/min
18.5 Jump begins at $\theta = 75°$; $\omega = 196.3$ rev/min
18.7 21.8 rad/s
18.9 At $\theta = 120°$, $F_{32}^{y} = -572$ N, $T = -4.04$ N · m; at $\theta = 225°$, $F_{32}^{y} = -608$ N, $T = 3.87$ N · m
18.11 At $\theta = 60°$, $F_{32}^{y} = -271$ lb, $T = 324$ in · lb; at $\theta = 255°$, $F_{32}^{y} = -156$ lb, $T = -27$ in · lb

19.1 $T_m = 70.9$ N · m, $I = 0.154$ N · m · s^2
19.3 $T_m = 278$ in · lb, $I = 1.842$ in · lb · s^2
19.5 $T_m = 1\,111$ in · lb
19.6 $F = 3\,319\,000$ lb; $\omega_p \geq 33.9$ rev/min
19.8 $M^z = 2.2$ in · lb
19.10 $\sum M = -4.9\hat{\mathbf{k}}$ in · lb (negligible)

INDEX